AF252062

Trends in
Chemistry of Materials

Selected Research Papers of C N R Rao

IISc Centenary Lecture Series

Trends in
Chemistry of Materials

Selected Research Papers of C N R Rao

C N R Rao

Indian Institute of Science
and
Jawaharlal Nehru Centre for Advanced Scientific Research
Bangalore, India.

NEW JERSEY · LONDON · SINGAPORE · BEIJING · SHANGHAI · HONG KONG · TAIPEI · CHENNAI

Published by

World Scientific Publishing Co. Pte. Ltd.

5 Toh Tuck Link, Singapore 596224

USA office: 27 Warren Street, Suite 401-402, Hackensack, NJ 07601

UK office: 57 Shelton Street, Covent Garden, London WC2H 9HE

British Library Cataloguing-in-Publication Data
A catalogue record for this book is available from the British Library.

WSPC-IISc Centenary Lecture Series – Vol. 1
TRENDS IN CHEMISTRY OF MATERIALS
Selected Research Papers of C N R Rao

ISBN-13 978-981-283-383-9
ISBN-10 981-283-383-8

Printed in Singapore by World Scientific Printers

Professor C.N.R. Rao

FOREWORD

The Centenary Lectures at the Indian Institute of Science were initiated to mark the beginning of the 100th year of the institution. This lecture series aims to bring scientists of the highest accomplishment to the Institute, in order to inspire and inform a new generation of researchers. In beginning this lecture series, we did not have to look far for our inaugural speaker. Professor C.N.R. Rao is not only one of the world's leading solid state chemists and a pioneer in many areas of materials science, he is also a part of the Institute's folklore. He led the institution with unmatched distinction as the Director during the period 1984-1994 and was also responsible for building one of the most vigorous centres in the world in the area of solid state and materials chemistry.

The publication of the Centenary Lectures as a reprint volume, will provide a valuable resource, in the form of a uniquely personal account of the growth of an important field of science. Professor C.N.R. Rao's trail of research, over a period of half a century, moves from studies of metal oxides, at a time when the area was far from the mainstream, to the remarkably exciting period of the past twenty years which have seen the field of materials chemistry move to centre stage. This volume should convey to readers some of the sense of excitement and challenge that has pervaded the field of materials chemistry in recent times.

Professor P. Balaram
Director
Indian Institute of Science

PREFACE

When I started working on the chemistry of solid materials nearly 50 years ago, there were very few practitioners in the subject. Much of the preoccupation of those who worked in this area was on point defects in solids and nonstoichiometry. J.S. Anderson (Oxford), who was a source of inspiration to many of us, was seriously interested in the relation between nonstoichiometry and defects. Nonstoichiometry was a problem that plagued chemists for a long time since people were dismayed by compounds such as Ti_4O_7 and Pr_6O_{11} with unusual cation:anion ratios. In the late 1960's, it was established that these arose because of certain novel structural manifestations, typical of them being crystallographic shear planes and defect complexes. The subject of solid state chemistry grew slowly over the years, with people working on the structure and properties of materials. In particular, research on transition metal oxides was pursued by a number of solid state chemists, and the effort in organic solid state was marginal.

Transition metal oxides are fascinating because of the very wide range of properties and structures exhibited by them. I myself got into this area in the early 1960s. Two major researchers around that time were, my dear friends John Goodenough and Paul Hagenmuller. Several important discoveries were made in this area during that period. For example, metallic oxides such as ReO_3 were reported. ReO_3 looks like copper and conducts like copper. Oxides undergoing transitions from the insulating state to the metallic state were discovered, the most prominent amongst them being V_2O_3, showing a 10-million fold jump in resistivity around 150 K. My interest in such transitions was aroused by my association with J.M. Honig. V_2O_3 continues to be a problem of interest even today. Research on other types of materials such as chalcogenides and pnictides was also pursued by some workers, but oxides of different structures, specially those belonging to the perovskite family, received greater attention. It was really enjoyable to work in this area and I specially recall my association with Nevil Mott whom I consider to be the father of modern of solid state and materials science.

The biggest explosion in materials chemistry and physics occurred in late 1986 when high-temperature superconductivity was discovered in a lanthanum cuprate, a material which was a ceramic and on which a few chemists had worked earlier. As stated in a report of the US National Academy of Sciences, this discovery changed the role of chemistry in the study of materials, and materials chemistry became a more significant part of materials science. It is around this time that even chemists started to consider solid state chemistry as an integral and important part of main-stream chemistry.

In the last thirty years or so, the subject of solid state chemistry has got transformed into materials chemistry by absorbing various features of modern chemical science. The materials investigated by chemists are no longer limited to inorganic materials but include a variety of organic materials. Synthesis has become a major aspect of materials chemistry. Materials chemists employ a variety of chemical strategies, soft chemical approaches, in particular, for synthesis. Studies of structure, properties and phenomena and relating structure to properties are important aspects of materials chemistry.

Computation and theory are used extensively. Supramolecular strategies are employed to design materials. Materials chemistry is now truly interdisciplinary, not only within chemical science itself but also by having interfaces with biology and other subjects. It has immediate connections with physics, engineering and technology.

In this collection, I have included a set of my papers which represent some of the highlights of materials chemistry. There is a section on oxidic materials, which includes high-temperature superconductivity, colossal magnetoresistance, electronic phase separation and multiferroics. I could have included other areas as well as materials such as nitrides but could not do so because of limitation of space. We have, in fact, employed novel methods for making gallium nitride, boron nitride and such materials, by using precursors and the urea decomposition route. There is a section dealing with open-frame work and hybrid materials of which the latter has a great future since one can make use of the rigidity of inorganic structures and the functionality and flexibility of the organic residues to design materials with novel properties. I have had the pleasure of carrying out some collaborative work on hybrid frameworks with A.K. Cheetham. Nanomaterials, the new fascination, constitute a large section of the book. I have not been able to include many papers on organic materials since I have not worked in this area extensively.

The book was intended to be a selection of my research papers which in turn also presents the flavor of modern materials chemistry. I trust that I have succeeded, at least partly, in doing so. I realize that it is difficult to be up-to-date in a field like this since new materials, methods and phenomena are discovered constantly. Even as I write this preface, new results of vital importance have been obtained in my laboratory which cannot be included in this volume. I earnestly hope that the reader will find the papers in this volume interesting and informative, and representative of the subject. It has been a pleasure to see the subject grow in the last half a century, and to grow with it. I have no doubt that materials chemistry will continue to develop and unravel new horizons in the years to come.

I am most thankful to the Indian Institute of Science, specially its Director, Professor P. Balaram, for asking me to deliver the inaugural centenary lecture of the Institute. I owe much to the Institute where I have done most of my research work, and to the fine students and coworkers I have had over the years. It is to be noted that the centenary of the Indian Institute of Science coincides with my 75th birth anniversary and 50 years of research as an independent faculty member.

C.N.R. Rao
Bangalore
2008

BRIEF BIOGRAPHY OF PROFESSOR C.N.R. RAO

Prof. C.N.R. Rao (born on 30 June 1934, Bangalore, India) is the National Research Professor as well as Honorary President and Linus Pauling Research Professor at the Jawaharlal Nehru Centre for Advanced Scientific Research. He is also an Honorary Professor at the Indian Institute of Science. His main research interests are in solid state and materials chemistry. He is an author of over 1400 research papers and 41 books. He received the M.Sc. degree from Banaras, Ph.D. from Purdue, D.Sc. from Mysore universities and has received honoris causa doctorate degrees from 46 universities including Purdue, Bordeaux, Banaras, Delhi, Mysore, IIT Bombay, IIT Kharagpur, Northwestern, Notre Dame, Novosibirsk, Oxford, Stellenbosch, Uppsala, Wales, Wroclaw, Caen, Khartoum, Calcutta, Sri Venkateswara University and Desikottama from Visva-Bharati.

Prof. Rao is a member of all the major science academies in the world including the Royal Society, London, the National Academy of Sciences, U.S.A., the Russian Academy of Sciences, French Academy of Sciences, Japan Academy as well as the Polish, Czechoslovakian, Serbian, Slovenian, Brazil, Spanish, Korean and African Academies and the American Philosophical Society. He is a Member of the Pontifical Academy of Sciences, Foreign Member of Academia Europaea and Foreign Fellow of the Royal Society of Canada. He is on the editorial boards of several leading professional journals.

Among the various medals, honours and awards received by him, mention must be made of the Marlow Medal of the Faraday Society (1967), Bhatnagar Prize (1968), Jawaharlal Nehru Fellowship (1973), Padma Shri (1974), Centennial Foreign Fellowship of the American Chemical Society (1976), Royal Society of Chemistry (London) Medal (1981), Padma Vibhushan (1985), Honorary Fellowship of the Royal Society of Chemistry, London (1989), Hevrovsky Gold Medal of the Czechoslovak Academy (1989), Blackett Lectureship of the Royal Society (1991), Einstein Gold Medal of UNESCO (1996), Linnett Professorship of the University of Cambridge (1998), Centenary Medal of the Royal Society of Chemistry, London (2000), the Hughes Medal of the Royal Society, London, for original discovery in physical sciences (2000), Karnataka Ratna (2001) by the Karnataka Government, the Order of Scientific Merit (Grand-Cross) from the President of Brazil (2002), Gauss Professorship of Germany (2003) and the Somiya Award of the International Union of Materials Research (2004). He is the first recipient of the India Science Award by the Government of India (2005) and received the Dan David Prize for science in the future dimension for his research in Materials Science. He was named as Chemical Pioneer by the American Institute of Chemists (2005), "Chevalier de la Légion d'Honneur" by the President of the French Republic (2005) and received the Honorary Fellowship of the Institute of Physics,

London (2006) and Honorary Fellowship of St. Catherine's College, Oxford (2007). The Nikkei Prize for science was conferred on him recently (2008).

Prof. Rao is Chairman, Scientific Advisory Council to the Prime Minister, past President of The Academy of Sciences for the Developing World (TWAS), Member of the Atomic Energy Commission of India and Chairman, Indo-Japan Science Council. He is Founder-President of both the Chemical Research Society of India and of the Materials Research Society of India. Prof. Rao was President of the Indian National Science Academy (1985-86), the Indian Academy of Sciences (1989-91), the International Union of Pure and Applied Chemistry (1985-97), the Indian Science Congress Association (1987-88), and Chairman, Advisory Board of the Council of Scientific and Industrial Research (India). He was the Director of the Indian Institute of Science (1984-94), Chairman of the Science Advisory Council to Prime Minister Rajiv Gandhi (1985-89) and Chairman, Scientific Advisory Committee to the Union Cabinet (1997-98) and Albert Einstein Research Professor (1995-99).

ACKNOWLEDGEMENTS

I express deep sense of gratitude to the various publishers for giving copyright permission to reproduce the following articles in the book:

American Chemical Society: Articles: 4, 8, 13, 14, 21, 22, 26, 27, 29, 30, 31, 35, 37, 38, 42, 44, 45, 50 and 51.

American Institute of Physics: Article: 43.

American Physical Society: Article: 46.

Council of Scientific & Industrial Research, New Delhi: Article: 3.

Elsevier: Articles: 2, 6, 11, 17, 48, 49, 52 and 53.

Institute of Physics, London: Article (s): 25, 57, 58 and 59.

Narosa, Delhi: Article: 7

Royal Society, London: Articles: 16, 18 and 23.

Royal Society of Chemistry, London: Articles: 1, 5, 9, 10, 15, 19, 24, 28, 32, 33, 34, 39, 40, 41, 47, 54, 55, 56 and 60.

Taylor & Fransis: Article: 12.

Wiley-VCH Verlag GmbH & Co. KGaA: Articles: 20 and 36.

I thank my students, colleagues and coworkers in the Indian Institute of Science, Jawaharlal Nehru Centre for Advanced Scientific Research and elsewhere for their support and collaboration. My secretaries Ms. Shashi and Ms. Sudha have provided invaluable support. I am thankful to Kaniska Biswas for assistance in preparing the manuscript.

I thank Dr. K.K. Phua and others of the World Scientific Publishing Company for enthusiastically producing the book.

Lastly, I thank my dear wife, Indu, for her support and encouragement through many decades.

C.N.R. Rao

CONTENTS

I. SOME HIGHLIGHTS OF MATERIALS CHEMISTRY

Commentary by C.N.R. Rao

MATERIALS CHEMISTRY as we understand it today is relatively of recent origin. A few decades ago, the subject generally included cement, steel, and a few other topics of an applied nature. In the last 30 years of so, the subject has emerged to be recognized as an important new direction in modern chemistry, having incorporated all the salient aspects of solid state chemistry as well.[1,2] While solid state chemistry may be considered to represent the chemical counterpart of solid state physics, materials chemistry deals with structure, response, and function and has the ultimate purpose of developing novel materials and of understanding structure-property relations as well as phenomena related to a wide range of materials. The materials can be organic, inorganic, or biological and can be in any condensed state of matter. With this description, it becomes difficult to classify materials as in the earlier years when it was common to classify them as ceramics, metals, organics and so on. We now have organic metals, superconductors, and nonlinear materials. It is probably more convenient to classify materials based on properties or phenomena. For example, porous solids, superconductors and ferroics cover all types of chemical constituents. It is, however, common to distinguish molecular solids from extended solids, as they represent two limiting descriptions.

Materials chemistry contains all the elements of modern chemistry. These include synthesis, structure, dynamics, and properties. In synthesis, one employs all possible methods and conditions from high-temperature and high-pressure techniques to mild solution methods (chimie douce or soft chemistry).[3] Chemical methods generally tend to be more delicate, often yielding novel, metastable products. Kinetic rather than thermodynamic control of reactions favors the formation of such structures. Supramolecular organization provides new ways of designing materials.[4]

All available methods of diffraction, microscopy, and spectroscopy are used for structure elucidation in present-day materials chemistry.[1,2] For detailed structure determination, even powders suffice for the most part because of the advances in diffraction profile analysis. These advances in structural tools enable more meaningful correlations of structure with properties and phenomena. Catalysis is becoming more of a science partly because of our ability to unravel the structures and surfaces of catalysts. Phase transitions of all varieties[5] are being investigated more and more by chemists.

In this section, I have included a few of my papers dealing with synthesis, defects and certain properties of oxidic materials and fullerenes, besides a general article dealing with important directions in materials chemistry. There is also an article dealing with experimental charge densities in organic molecular crystals. These articles should indicate the diversity and breadth of coverage in materials chemistry.

References

1. C.N.R. Rao and J. Gopalakrishnan, *New Directions in Solid State Chemistry*, Cambridge University Press, Second Edition, 1997.
2. C.N.R. Rao (Ed.), *Chemistry of Advanced Materials*, IUPAC 21st Century Monograph, Blackwell, Oxford, 1993.
3. C.N.R. Rao, *Chemical Approaches to the Synthesis of Inorganic Materials*, John Wiley, New York, 1994.
4. W. Jones and C.N.R. Rao (Eds.), *Supramolecular Organization and Materials Design*, Cambridge University Press, 2002.
5. C.N.R. Rao and K.J. Rao, *Phase Transitions in Solids*, McGraw Hill, New York, 1978.

Novel materials, materials design and synthetic strategies: recent advances and new directions†

C. N. R. Rao‡

Solid State and Structural Chemistry Unit, Indian Institute of Science, Bangalore 560 012, India and Chemistry & Physics of Materials Unit, Jawaharlal Nehru Centre for Advanced Scientific Research, Jakkur P.O., Bangalore 560 064, India

Received 5th May 1998, Accepted 12th June 1998

There have been major advances in solid state and materials chemistry in the last two decades and the subject is growing rapidly. In this account, a few of the important aspects of materials chemistry of interest to the author are presented. Accordingly, transition metal oxides, which constitute the most fascinating class of inorganic materials, receive greater attention. Metal–insulator transitions in oxides, high temperature superconductivity in cuprates and colossal magnetoresistance in manganates are discussed at some length and the outstanding problems indicated. We then discuss certain other important classes of materials which include molecular materials, biomolecular materials and porous solids. Recent developments in synthetic strategies for inorganic materials are reviewed. Some results on metal nanoparticles and nanotubes are briefly presented. The overview, which is essentially intended to provide a flavour of the subject and show how it works, lists references to many crucial reviews in the recent literature.

I have been working in solid state and materials chemistry for over four decades. When I first got interested in the subject, it was not an integral part of main-stream chemistry. In the first book on solid state chemistry[1a] that I edited in 1970, it was stated in the preface, that 'Solid state chemistry has not become part of the formal training programmes in chemistry. Being one of the frontiers of chemistry, it has a tremendous future and undoubtedly demands the active involvement of many more chemists'. In 1986, in the first edition of the book *New Directions in Solid State Chemistry*,[1b] the preface states, 'At the present time, solid state chemistry is mainly concerned with the development of new methods of synthesis, new ways of identifying and characterizing materials and of describing their structure, and above all, with new strategies of tailor-making materials with desired and controllable properties, be they electronic, magnetic, dielectric, optical, absorptive or catalytic. It is heartening that the subject is increasingly coming to be recognized as an emerging area of chemical science'. In the last few years, solid state chemistry has given room to the broader area of materials chemistry. The subject has come of age and there are many reviews and books dealing with it.[2] I am delighted to have this opportunity to present an overview of the subject in this first materials chemistry discussion meeting. In so doing, I will try to present the highlights of some of the areas and refer to important reviews in the literature, but I cannot help dealing with those aspects which are of personal interest to me in somewhat greater detail. It is indeed impractical to cover every aspect of this vast subject in

Table 1 A description of materials chemistry

Constituent units	State[b]	Function	Advanced technology
atoms molecules[a] ions[a]	crystalline (molecular, ionic, polymeric, metallic *etc.*) non-crystalline (glasses) clusters and nanomaterials liquid crystalline	miniaturization selectivity and recognition transformation transduction transport energy storage	nanolithography microelectronics magnets sensors and transducers photonic devices energy devices porous solids and membranes micromachines

[a]Inorganic or organic. [b]In pure (monophasic) form or in the form of aggregates or composites.

an article. Furthermore, all research in solid state chemistry is not necessarily related to materials development. Materials chemistry, as distinct from solid state chemistry, deals with structure, response and function, and has an ultimate technological objective. I illustrate this aspect of materials chemistry in Table 1.

If one were to list the most important discoveries in solid state and materials science in the last decade, it would include high-temperature cuprate superconductors (1986), fullerenes and related materials (1990), mesoporous silica (1992) and colossal magnetoresistance (CMR) in manganates (1993). Three of these deal with metal oxides. I have been involved in research on metal oxides for many years and it has been exciting to witness the increasing importance gained by these materials. I shall deal with certain aspects of the chemistry of transition metal oxides at some length. This discussion is presented as a personal account and would also serve as a case-study. I will indicate the developments in molecular materials, porous solids and biomaterials and highlight out the role of chemical synthesis in inorganic systems. I will end the article with a discussion of some of my recent interests in nanotubes and metal nanoparticles.

Transition metal oxides

Transition metal oxides constitute the most fascinating class of materials, exhibiting a variety of structures and properties.[3] The metal–oxygen bond can vary anywhere between highly ionic to covalent or metallic. The unusual properties of transition metal oxides are clearly due to the unique nature of the outer d-electrons. The phenomenal range of electronic and magnetic properties exhibited by transition metal oxides is noteworthy. Thus, the electrical resistivity in oxide materials spans the extraordinary range of 10^{-10} to 10^{20} ohm cm. We have oxides with metallic properties (*e.g.* RuO_2, ReO_3, $LaNiO_3$) at one end of the range and oxides with highly insulating behaviour (*e.g.* $BaTiO_3$) at the other. There are also

†Basis of the presentation given at Materials Chemistry Discussion No. 1, 24–26 September 1998, ICMCB, University of Bordeaux, France.
‡Also at the Materials Research Laboratory, University of California, Santa Barbara, CA 93106-5050, USA.

oxides that traverse both these regimes with changes in temperature, pressure, or composition (*e.g.* V_2O_3, $La_{1-x}Sr_xVO_3$). Interesting electronic properties also arise from charge density waves (*e.g.* $K_{0.3}MoO_3$), charge-ordering (*e.g.* Fe_3O_4) and defect ordering (*e.g.* $Ca_2Mn_2O_5$, $Ca_2Fe_2O_5$). Oxides with diverse magnetic properties anywhere from ferromagnetism (*e.g.* CrO_2, $La_{0.5}Sr_{0.5}MnO_3$) to antiferromagnetism (*e.g.* NiO, $LaCrO_3$) are known. Many oxides possess switchable orientation states as in ferroelectric (*e.g.* $BaTiO_3$, $KNbO_3$) and ferroelastic [*e.g.* $Gd_2(MoO_4)_3$] materials. Then, there are a variety of oxide bronzes showing a gamut of properties.[4] Superconductivity in transition metal oxides has been known for some time, but the highest T_c reached was around 13 K; we now have oxides with T_cs in the region of 160 K. The discovery of high T_c superconductors[5] focused worldwide scientific attention on the chemistry of metal oxides and at the same time revealed the inadequacy of our understanding of these materials.

The unusual properties of transition metal oxides that distinguish them from the metallic elements and alloys, covalent semiconductors and ionic insulators are due to several factors. (a) Oxides of d-block transition elements have narrow electronic bands, because of the small overlap between the metal d and the oxygen p orbitals. The bandwidths are typically of the order of 1–2 eV (rather than 5–15 eV as in most metals). (b) Electron correlation effects play an important role, as expected because of the narrow electronic bands. The local electronic structure can be described in terms of atomic-like states [*e.g.* Cu^+ (d^{10}), Cu^{2+} (d^9) and Cu^{3+} (d^8) for Cu in CuO] as in the Heitler–London limit. (c) The polarizability of oxygen is also of importance. The divalent oxide ion, O^{2-}, does not exactly describe the state of oxygen and configurations such as O^- have to be included, especially in the solid state. This gives rise to polaronic and bipolaronic effects. Species such as O^-, which are oxygen holes with a p^5 configuration instead of the filled p^6 configuration of O^{2-}, can be mobile and correlated. (d) Many transition metal oxides are not truly three-dimensional, but have low-dimensional features. For example, La_2CuO_4 and La_2NiO_4 with the K_2NiF_4 structure are two-dimensional compared to $LaCuO_3$ and $LaNiO_3$, which are three-dimensional perovskites. Because of the varied features of individual oxides, it has not been possible to establish satisfactory theoretical models for complex transition metal oxides. I started working on transition metal oxides in the late 1950s and my early preoccupation was with the structures, defects and phase transitions of oxides such as TiO_2, Pr_6O_{11} and Bi_2O_3. Soon I became interested in the electronic and magnetic properties of oxides, especially those with the perovskite structure. The perovskite structure dominates all classes of materials, whether they are dielectrics or superconductors.

Transition metal oxides that transform from the metallic state to an insulating or semiconducting state are of great interest. It has always amazed me how a simple oxide like V_2O_3 conducts like copper at ordinary temperatures and becomes like wood on cooling it to 150 K. Interest in these transitions has blossomed over the years and the subject has itself become a frontier area of investigation.[6,7] Typical transitions of this type in oxide materials are: (a) pressure-induced transitions, as in NiO, in which the pressure increases the wavefunction overlap between neighbors to induce a change from localized to itinerant behavior of electrons; (b) transitions as in Fe_3O_4 involving charge-ordering; (c) transitions in $LaCoO_3$ that are initially induced because of the different spin configurations of the transition metal ion; electron transfer between the two spin states initiates a process that eventually renders the oxide metallic around 1200 K; (d) transitions as in EuO arising from the disappearance of spin polarization band-splitting effects when the ferromagnetic Curie temperature is reached; (e) compositionally induced transitions, as in $La_{1-x}Sr_xCoO_3$ and $LaNi_{1-x}Mn_xO_3$, in which changes of band

structure in the vicinity of the Fermi level are brought about by a change in composition or are due to disorder-induced localization; (f) transitions in two-dimensional systems, such as La_2NiO_4, in which Ni—O—Ni interactions can only occur in the *ab* plane (unlike in the three-dimensional analogue in $LaNiO_3$) and (g) temperature-induced transitions in a large class of oxides such as Ti_2O_3, VO_2 and V_2O_3. Although I have only mentioned metal–insulator transitions in oxides, it should be noted that these transitions are found in a variety of other systems such as doped semiconductors, metal-ammonia solutions and expanded metals. Global aspects of metal–insulator transitions are truly fascinating.[6,7] In spite of considerable effort, we do not yet understand many aspects of these transitions, specially the 10-million fold jump in resistivity at the transition in V_2O_3.

The perovskite structure is ideal for 180° cation–anion-cation interactions and there are no cation–cation interactions. One can vary the covalency (or the interaction between the metal and oxygen orbitals) in this structure by varying the central cation and/or the transition metal ion and obtain a wide variation in properties. Interestingly, a high proportion of oxide materials of technological importance (*e.g.* ferroelectrics, superconductors, CMR materials) possess the perovskite structure. If we examine the electronic and magnetic properties of the family of perovskites of the type $LnMO_3$ (Ln, rare earth; M, first-row transition metal), we find a great variety. The titanates and nickelates possessing low-spin trivalent transition metal ions are metallic or nearly metallic. On the other hand, $LnCrO_3$, $LnMnO_3$ and $LnFeO_3$ are antiferromagnetic insulators. $LaCoO_3$ and other cobaltates are interesting, in that Co^{3+} can be in the low-spin (t_{2g}^6) or the high spin ($t_{2g}^4e_g^2$) state. While at low temperatures, $LaCoO_3$ is a diamagnetic insulator, with increasing temperature, the high-spin population increases eventually resulting in a phase transition due to the ordering of the two spin states. At high temperatures (*ca.* 1000 K), the material becomes metallic, resulting from the electron transfer between different spin/oxidation states of cobalt. Substituting for Ln in $LnCoO_3$ with a divalent ion such as Sr^{2+} or Ba^{2+}, as in $Ln_{1-x}Sr_xCoO_3$, progressively renders it metallic and ferromagnetic due to the fast electron transfer between the Co^{3+} and Co^{4+} states. We have investigated these cobaltates by a variety of methods including ^{57}Co Mössbauer spectroscopy, ferromagnetic resonance, photoemission spectroscopy and EXAFS. Electron transport properties of $LnCoO_3$ doped lightly with Sr^{2+} and other similar oxide systems can be understood in terms of variable range hopping as in disordered systems. The metal–nonmetal transition in $Ln_{1-x}Sr_xCoO_3$ [Fig. 1(a)] is a situation similar to impurity doping.[8] Another system showing metal-nonmetal transitions is $LaNi_{1-x}M_xO_3$ (M = Mn, Co, *etc.*), where $LaNiO_3$ ($x = 0.0$) is metallic [see Fig. 1(b) for the Mn system]. On increasing x, the material becomes an insulator.[8] Interestingly, in all such systems, the coefficient of resistivity changes sign across a

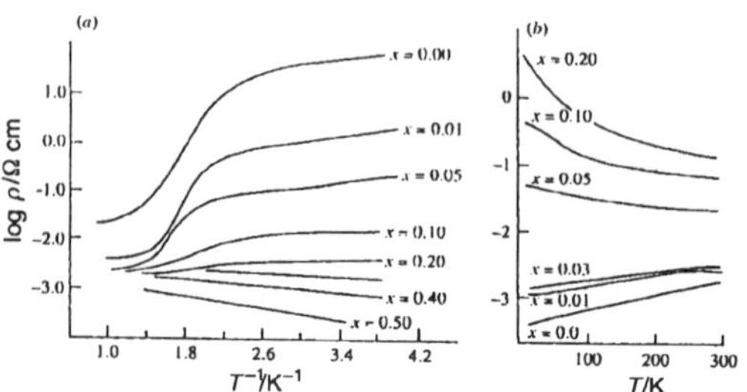

Fig. 1 Compositionally controlled metal–insulator transitions in (a) $La_{1-x}Sr_xCoO_3$ and (b) $LaNi_{1-x}Mn_xO_3$ [from Rao (reproduced with permission from ref. 8)].

universal value of resistivity, corresponding to Mott's minimum metallic conductivity.

Studies of the properties of quasi two-dimensional oxides with the K_2NiF_4 structure were initiated in this laboratory as early as 1970. It was found that Cu in Ln_2CuO_4 had no magnetic moment and that La_2CuO_4 was antiferromagnetic. Magnetic and electrical properties of La_2NiO_4 and its oxygen-excess nonstoichiometry were investigated. Solid solutions of La_2CuO_4 (T) and Nd_2CuO_4 (T′) showed an interesting phase diagram. When we wrote an overview article[9] on the structure and properties of two-dimensional oxides with the K_2NiF_4 structure in 1984, little did we realize then that La_2CuO_4 was going to become famous because of its high-temperature superconductivity. The effect of dimensionality on the electronic and magnetic properties of layered nickel oxides (e.g. $LaNiO_3$, $La_4Ni_3O_{10}$, $La_3Ni_2O_7$, La_2NiO_4 with decreasing dimensionality from 3 to 2) was examined some time ago. Materials of this type have the general formula $(LaO)(LaMO_3)_n$ where the (LaO) is the rock salt layer and the $LaMO_3$ is the perovskite layer.[10] When M = Ni, the $n = \infty$ member ($LaNiO_3$) is metallic, but the metallicity decreases as n decreases. The $n = 1$ material has the K_2NiF_4 (two-dimensional) structure. One of the more fascinating families of such materials prepared by us was of the type $(SrO)(La_{1-x}Sr_xMnO_3)_n$. These materials have now become familiar because of the colossal magnetoresistance recently discovered in them.[11] In $La_{2-x}Sr_xNiO_4$, the electronic and magnetic properties depend crucially on x and on the oxygen stoichiometry.

Of the many other oxide systems we have investigated, I would like to briefly discuss oxides forming intergrowth structures. Several metal oxide systems exhibit chemically well defined recurrent intergrowth structures with large periodicities, rather than random solid solutions with variable composition. The ordered intergrowth structures themselves frequently show the presence of wrong sequences. High resolution electron microscopy (HREM) enables direct examination of the extent to which a particular ordered arrangement repeats itself, and the presence of different sequences of intergrowths, often of unit cell dimensions. Selected area electron diffraction, which forms an essential part of HREM, provides useful information (not generally provided by X-ray diffraction) regarding the presence of supercells due to the formation of intergrowth structures. Several systems forming ordered intergrowth structures have been discovered and they generally exhibit homology.[12]

Aurivillius described a family of oxides of the general formula $Bi_2A_{n-1}B_nO_{3n+3}$ where the perovskite slabs, $(A_{n-1}B_nO_{3n+1})^{2-}$, n octahedra thick, are interleaved by $(Bi_2O_2)^{2+}$ layers. Typical members of this family are Bi_2WO_6 ($n = 1$), $Bi_3Ti_{1.5}W_{0.5}O_9$ ($n = 2$), $Bi_4Ti_3CrO_{12}$ ($n = 3$) and $Bi_5Ti_3CrO_{15}$ ($n = 4$). These oxides form intergrowth structure of the general formula $Bi_4A_{m+n-2}B_{m+n}O_{3(m+n)+6}$ involving alternate stacking of two Aurivillius oxides with different n values (Fig. 2). The method of preparation simply involves heating a mixture of the component metal oxides around 1000 K. Ordered intergrowth structures with (m, n) values of (1,2), (2,3) and (3,4) have been fully characterized by X-ray diffraction and HREM. What is amazing is that such intergrowth structures with long-range order are indeed formed while either member (m and n) can exist as a stable entity. These materials seem to be truly representative of recurrent intergrowth. The periodicity found in recurrent intergrowth solids formed by the Aurivillius family of oxides is indeed impressive.[12] WO_3 forms tetragonal, hexagonal or perovskite-type bronzes by interaction with alkali and other metals. The family of intergrowth tungsten bronzes (ITB) involving the intergrowth of nWO_3 slabs and one to three strips of the hexagonal tungsten bronze (HTB), first described by Kihlborg, is of relevance to the discussion here. In the intergrowth

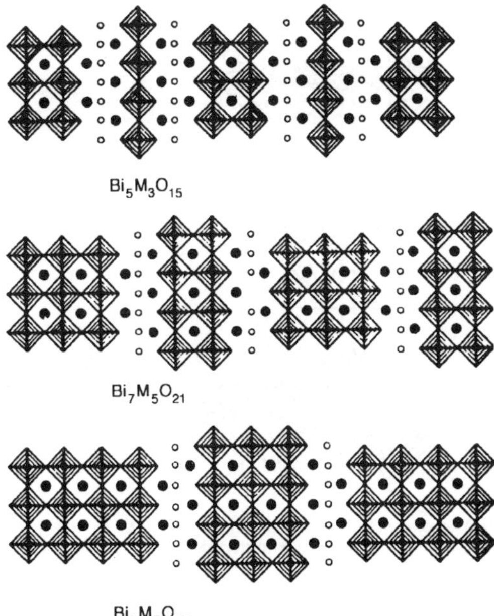

Fig. 2 Different types of intergrowth structures formed by the Aurivillius family of bismuth oxides.

tungsten bronzes of the general formula M_xWO_3, x is generally 0.1 or less and depends on whether the HTB strip is one or two tunnels wide. In the ITB phases of bismuth characterized by us some time ago, the HTB strips were always one tunnel wide. Displacement of adjacent tunnel rows due to the tilting of WO_3 octahedra often results in doubling of the long-period axis of the ITB. Evidence for the ordering of the intercalating bismuth atoms in the tunnels is found in terms of satellites around the superlattice spots in the electron diffraction patterns. The occurrence of recurrent intergrowth structures continues to fascinate chemists and crystallographers alike.

Superconducting cuprates

Late in 1986 the solid state community received the big news of the discovery of a high temperature oxide superconductor.[5] The material that was found to be superconducting with a T_c of ca. 30 K was based on the cuprate La_2CuO_4 (Fig. 3). In February 1987, we decided to look for an yttrium cuprate in the Y–Ba–Cu–O system as a candidate for high T_c superconductivity. Wu et al.[13] announced an yttrium cuprate with superconductivity above liquid nitrogen temperature around March 1987. Our strategy to prepare such a material was different. Wu et al. obtained a mixture of black and green compounds in their effort to make an yttrium cuprate doped with Ba. The black compound was superconducting ($T_c \approx 90$ K), but they did not know its composition or structure. Since we knew that Y_2CuO_4 could not be made, we started examining the $Y_{3-x}Ba_{3+x}Cu_6O_{14}$ system. We obtained 90 K superconductivity in the $x = 1$ composition corresponding to $YBa_2Cu_3O_7$. When we independently discovered 90 K superconductivity in March 1987, we knew the composition and the approximate structure.[14]

We worked on a variety of cuprates but unfortunately the area was so competitive and overcrowded that it became difficult to get due credit for much of the work. In spite of these limitations, we made several contributions to this area. These include, besides the independent synthesis and charac-

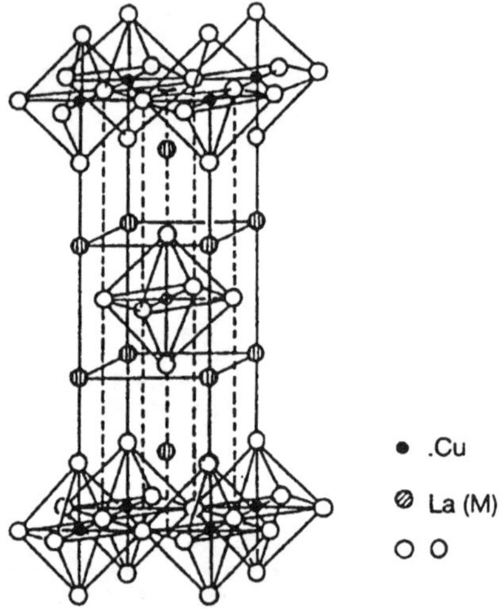

Fig. 3 Structure of La_2CuO_4 and $La_{2-x}A_xCuO_4$ (A = Sr or Ba). The CuO_6 octahedra are Jahn–Teller distorted.

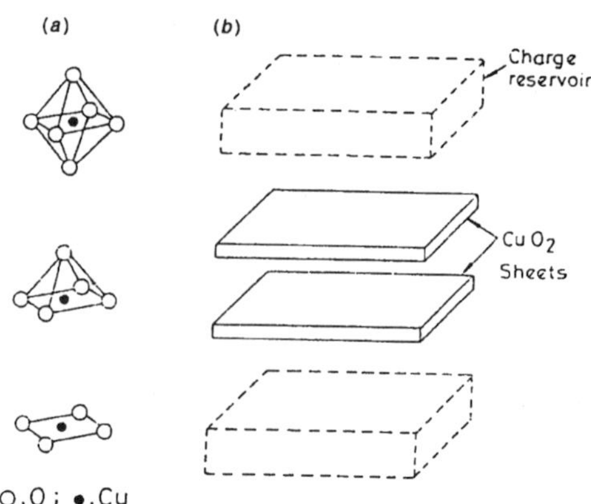

Fig. 4 (a) Cu–O polyhedra found in superconducting cuprates. (b) Schematic representation of the cuprates.

terization of $YBa_2Cu_3O_7$, the first observation of non-resonant microwave absorption in high T_c cuprates, metastability of $YBa_2Cu_3O_{7-\delta}$ (in the 60 K regime, $\delta = 0.3$–0.4) and its transformation to a novel type of $YBa_2Cu_4O_8$, the importance of Cu–O charge transfer energy in relation to the superconductivity of cuprates and the nature of charge carriers, optimization of hole concentration in cuprates and the relation between T_c and hole concentration. On the synthetic front, we made some contributions which we can consider to be Bangalore brand. For example, we could make a series of superconductors in the Tl–Ca–Sr–Cu–O system by substitution of Tl by Pb or Ca by a rare earth metal; a new family of superconductors of the type $TlSr_{n+1}Cu_nO_{2n+3}$ ($n = 1$ and $n = 2$) wherein Sr is partly substituted by a rare earth or Tl by Pb was discovered. Another noteworthy contribution was the synthesis of bismuth cuprate superconductors without superlattice modulation. This was significant since all the Bi cuprates showed such modulation. The modulation could be removed by suitable cation substitution (Pb/Bi) and the adjustment of oxygen stoichiometry. An interesting aspect was the synthesis and characterization of oxyanion derivatives of cuprate superconductors containing anions such as CO_3^{2-}, SO_4^{2-}, PO_4^{3-} and BO_3^{3-}, as integral parts of the oxide framework. Besides the mercury cuprates with the highest T_c to date, many interesting cuprates have been synthesized in the last 4–5 years.[15,16] These include copper oxycarbonates, copper oxyhalides, $Sr_2CuO_2F_{2+\delta}$ and ladder cuprates. Here, I must place on record the enormous contributions of Raveau and coworkers[17] to the synthesis and characterization of cuprate superconductors.

The several families of cuprate superconductors with holes as carriers all have the general features shown in Fig. 4. These features are common to the mercury cuprate showing the highest T_c as well. Electron superconductors of the type $Nd_{2-x}M_xCuO_4$ (M = Ce, Th) with the T′ structure possess square-planar CuO_4 units instead of the elongated octahedra in the T structure. The hole superconductors generally have the square pyramidal units. Many of the cuprate families have an antiferromagnetic insulator member at one end of the composition. What is more interesting is that all the cuprates are at a metal–insulator boundary. Some of them undergo a metal–nonmetal transition as a function of composition (e.g. $Bi_2Ca_{1-x}Y_xSr_2Cu_2O_8$ and $TlY_{1-x}Ca_xSr_2Cu_2O_7$ with change in x). The essential feature of the cuprates is the presence of

CuO_2 sheets with or without apical oxygens. The mobile charge carriers in the cuprates are in the CuO_2 sheets. All the cuprates have charge reservoirs as exemplified by the Cu–O chains in the 123 and 124 cuprates and the TIO, BiO and HgO layers in the other cuprates. That the CuO_2 sheets are the seat of high-temperature superconductivity is demonstrated by the fact that intercalation of iodine between BiO layers in the bismuth cuprates does not affect the superconducting transition temperature while introduction of fluorite layers between the CuO_2 sheets adversely affects superconductivity. In the different series of cuprates with varying number of CuO_2 sheets studied hitherto, the T_c reaches a maximum when $n = 3$ except in single thallium layer cuprates where the maximum is at $n = 4$. The infinite layered cuprates, where the CuO_2 sheets are separated by alkaline earth and other cations, show T_c values in the 40–110 K range. Superconductivity in these materials appears to be due to the presence of Sr–O defect layers corresponding to the insertion of $Sr_3O_{2\pm x}$ blocks. We shall briefly present the common features of cuprate superconductors[18] and examine some of the outstanding problems.

The chemistry of the majority of the cuprate superconductors is governed by oxygen holes. (Rouxel[19] has carried out fine work with sulfur holes in metal sulfides). The excess positive charge in cuprates can be represented in terms of the formal valence of copper, which in the absence of holes will be $+2$ in the CuO_2 sheets. In hole superconducting cuprates, it is generally around $+2.2$. In electron superconductors, it would be less than $+2$ as expected. The actual concentration of holes, n_h, in the CuO_2 sheets in $La_{2-x}A_xCuO_4$, $YBa_2Cu_3O_7$ and Bi cuprates is readily determined by redox titrations. In the 123 cuprates, the concentration of mobile holes in the CuO_2 sheets can be delineated from that in the Cu–O chains. Determination of n_h in thallium cuprates poses some problems, but in single Tl–O layer cuprates, chemical methods have been developed to obtain reasonable estimates. Generally, T_c in a given family of cuprates reaches a maximum value at an optimal value of n_h as shown in Fig. 5; the maximum is around $n_h \approx 0.2$ in most cuprates. Notice that the points in the underdoped region in Fig. 5 fall close to a straight line, but deviations occur in the overdoped region. Single layer thallium cuprates also show this behaviour. In $Tl_{1-y}Pb_yY_{1-x}Ca_xSr_2Cu_2O_7$ where the substitution of Tl^{3+} by Pb^{4+} has an effect opposite to that due to the substitution of Y^{3+} by Ca^{2+}, the T_c becomes a maximum at an optimal value of $(x-y)$, which is a measure of the hole concentration. By suitably manipulating x and y, the T_c of this system can be increased from 85–90 K to 110 K.

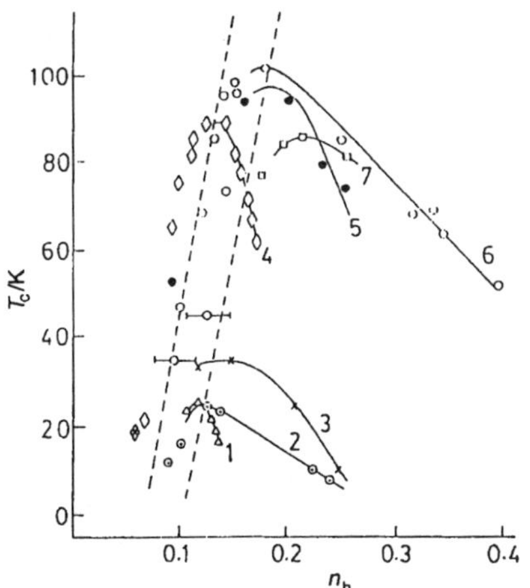

Fig. 5 Variation of T_c with the hole concentration, n_h: 1,2,6 and 7, Bi cuprates; 5, 123 cuprates; 3, $La_{2-x}Sr_xCuO_4$; 4, Tl cuprates (from Rao and Ganguli[18a]).

It is possible to obtain useful correlations between T_c and the Cu–O distances, both in-plane and apical. The Cu and oxygen Madelung energies also provide good correlations.[18a] The T_cs vary sensibly with the Cu–O charge-transfer energy as determined by core-level intensities. In spite of such relations with structural parameters, we are far from understanding the high T_c cuprates, particularly in the normal state.[18] The doped and undoped cuprates are both strongly correlated and we do not fully understand the strong correlation in these materials. Many puzzles remain. (i) For example, we do not fully understand the global composition–temperature phase diagram and the apparent lack of symmetry between hole and electron superconductors. (ii) The cuprates are unusual metals in-plane and insulators perpendicular to it; on cooling, they become three-dimensional superconductors. The anomalous temperature variation of resistivity in cuprates is noteworthy (Fig. 6), particularly the extraordinary linear variation of the ab plane metallic resistivity. (iii) The ac optical conductivity data do not conform the behavior expected of metals. (iv) The absence of splitting in the energy of states coupled by bilayer tunneling (as found from ARPES) is surprising. All the above observations cannot be reconciled with the picture of an anisotropic metal. (v) The linear resistivity (ab plane) with respect to temperature, with zero-intercept and nearly constant slope (independent of T_c) cannot be explained easily. The anisotropy is large and temperature dependent. The Hall resistivity varies inversely with temperature. (vi) Photoemission spectra suggest that the quasiparticles are not well defined in the normal state. (vii) Above all, there appears to be a characteristic low energy scale in cuprates. These pseudogap effects are indeed unique to cuprates. (viii) The superconducting state also has some unusual features, such as anisotropic order parameter, and well defined quasiparticles. Even a doped ladder cuprate has been rendered superconducting. Clearly, the status of our theoretical understanding of cuprate superconductors is far from satisfactory. It appears that cuprates constitute an experimentally overdetermined system. In the meantime, if one discovers a room-temperature superconductor, it will be an experimentalist's joy, but a theoretician's nightmare. It is noteworthy that the high T_c cuprates are finding applications in electronic devices, although their use in fabricating generators and magnets may take time.

Colossal magnetoresistance, charge ordering and related properties of rare earth manganates

Giant magnetoresistance was known to occur in bilayer and granular metallic materials. Perovskite manganates of the general formula $Ln_{1-x}A_xMnO_3$ (Ln = rare earth, A = alkaline earth) have created wide interest because they exhibit colossal magnetoresistance (CMR). These oxides become ferromagnetic at an optimal value of x (or Mn^{4+} content) and undergo an insulator–metal (I–M) transition around the ferromagnetic T_c. These properties are attributed to double exchange associated with electron hopping from Mn^{3+} to Mn^{4+}. The double-exchange which favours itinerant electron behavior is opposed by the Jahn–Teller distortion due to the presence of Mn^{3+}. The manganates show charge ordering especially when the average size of the A-site cations is small. Charge ordering also competes with double exchange and favours insulating behavior. Thus, the manganates exhibit a variety of properties associated with spin, charge as well as orbital ordering. In what follows, we discuss some important aspects of colossal magnetoresistance, charge ordering and related properties of the manganates.[8,11,20]

$LaMnO_3$ is an insulator with an orthorhombically distorted perovskite structure ($b > a > c\sqrt{2}$) due to Jahn–Teller distortion. Typically, as prepared samples of $LaMnO_3$ contain some Mn^{4+}. $LaMnO_3$ with a small proportion of Mn^{4+} ($<5\%$) becomes antiferromagnetically ordered at low temperatures ($T_N \approx 150$ K). When the La^{3+} in $LaMnO_3$ is progressively substituted by a divalent cation as in $La_{1-x}A_xMnO_3$ (A = Ca, Sr or Ba), it becomes ferromagnetic with a well defined Curie temperature, T_c, and metallic below T_c. The saturation moment is typically *ca.* 3.8 μ_B which is close to the theoretical estimate based on localized spin-only moments, suggesting that the carriers are spin-polarized. Below T_c, the manganates exhibit metal-like conductivity. Fig. 7 illustrates how the ferromagnetism and the I–M transition occur around T_c in a typical $La_{1-x}Ca_xMnO_3$ composition. The simultaneous observation of itinerant electron behavior and ferromagnetism in the manganates is explained by Zener's double-exchange mechanism. The basic process in this mechanism is the hopping of a d hole from Mn^{4+} (d^3, t_{2g}^3, $S = 3/2$) to Mn^{3+} (d^4, $t_{2g}^3 e_g^1$, $S = 2$) *via* the oxygen, so that the Mn^{4+} and Mn^{3+} ions change places. The t_{2g}^3 electrons of the Mn^{3+} ion are localized on the Mn site giving rise to a local spin of 3/2, but the e_g state, which is hybridized with the oxygen 2p state, can be localized or itinerant. There is a strong Hund's rule interaction between the e_g and the t_{2g}^3 electrons. Goodenough pointed out that the ferromagnetism is governed not only by double exchange,

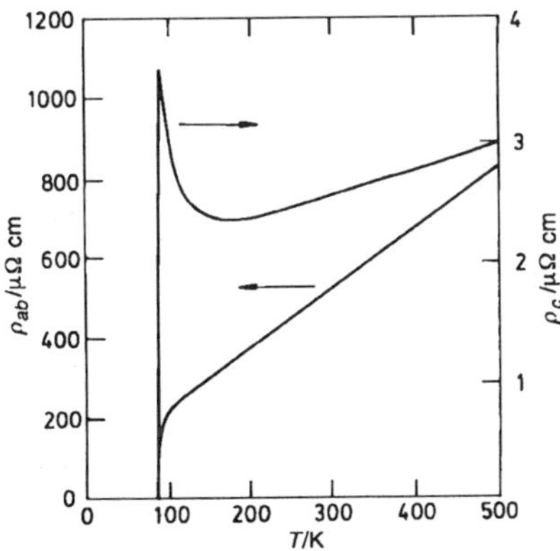

Fig. 6 Temperature variation of resistivity of $Bi_2Sr_2CaCu_2O_8$ single crystal along the ab plane and c-axis [data from Ong *et al.* as quoted by Ramakrishnan[18c] reproduced with permission from ref. 18(*c*)].

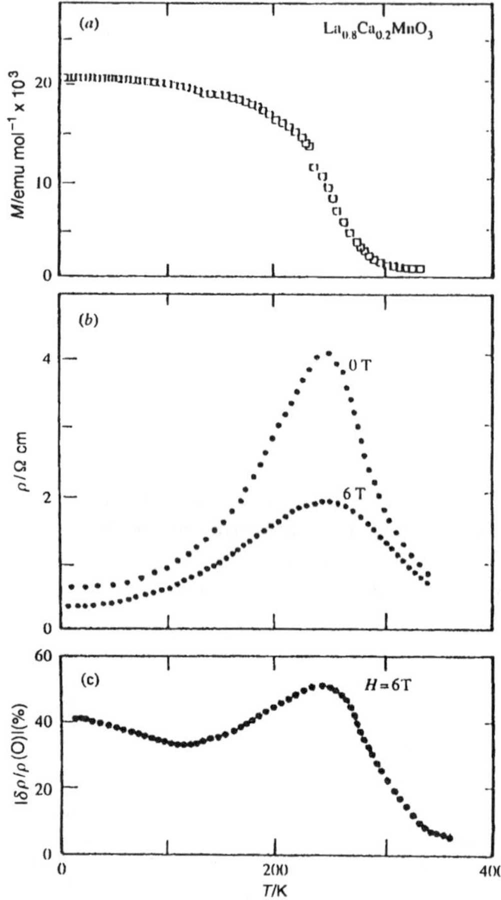

Fig. 7 Temperature variation of (*a*) the magnetization, (*b*) the resistivity (at 0 and 6 T) and (*c*) the magnetoresistance in $La_{0.8}Ca_{0.2}MnO_3$ [from Rao (reproduced with permission from ref. 8)].

but also by the nature of the superexchange interactions. It should be noted that the $Mn^{3+}-O-Mn^{4+}$ superexchange interaction is ferromagnetic while both the $Mn^{3+}-O-Mn^{3+}$ and $Mn^{4+}-O-Mn^{4+}$ interactions are antiferromagnetic.

The orthorhombic distortion of $LaMnO_3$ decreases as La is progressively substituted by a divalent ion and until it becomes rhombohedral or pseudocubic. The Mn–O–Mn angle approaches 180° as the distortion decreases. When the size of the A-site cations is small, orthorhombic distortion is favoured. The key factor responsible for the distortion in $LaMnO_3$ is the Jahn–Teller distortion due to Mn^{3+} ions. Accordingly, $LaMnO_3$ has three Mn–O distances (1.91, 1.96 and 2.19 Å). Across the I–M transition, the Jahn–Teller distortion decreases in $Ln_{1-x}A_xMnO_3$, the distortion being greater in the insulating regime. The structural parameters, specifically the oxygen thermal parameters, show a significant change across the M–I transition. Clearly, Mn^{4+} ions decrease the Jahn–Teller distortion, increase the itinerancy of electrons through double exchange and increase the tolerance factor of the perovskite structure. Temperature–composition phase diagrams for $La_{1-x}A_xMnO_3$ have been worked out with increasing x or Mn^{4+} content. A composition of $x \approx 0.3$ is ideal to observe a good I–M transition and negative CMR. The magnitude of CMR is maximum at the I–M transition (Fig. 7) and scales

with the field-induced magnetization. CMR is found in cation-deficient $LaMnO_3$ containing Mn^{4+} as well, and the features of the magnetic and metal–insulator transitions are similar to those in Fig. 7.

CMR is found over a wide range of compositions of $La_{1-x}Ca_xMnO_3$ ($x<0.5$). However, this material shows significant lattice effects in particular for low x. EXAFS studies reveal significant changes in the local structure in the $x=0.3$ composition in the 80–300 K range which are attributed to the formation of small polarons due to the Jahn–Teller distortion when $T>T_c$. Lattice polaron formation and local structural distortion have important implications on the electron transport (high resistivity) in these materials. $La_{1-x}Sr_xMnO_3$ is somewhat different from $La_{1-x}Ca_xMnO_3$, the A-site ion size being larger. The ferromagnetic metallic regime in $La_{1-x}Sr_xMnO_3$ extends over larger values of x and the Curie temperatures are higher. Magnetization in the manganates decreases as T^2 and the resistivity increases as T^2. In $La_{1-x}Sr_xMnO_3$ ($x=0.17$), the local spin moments and charge carriers couple strongly to changes in the structure and the structure can be switched by the application of a magnetic field depending on the temperature. The magnetic field which causes this crossover from the orthorhombic to the rhombohedral structure is rather low (*ca.* 2 T), showing thereby that a very small energy difference exists between these two structures.

Hydrostatic pressure stabilizes the ferromagnetic metallic state in $La_{1-x}A_xMnO_3$ (*i.e.* increases T_c). A similar effect can be brought about by increasing the radius of the A-site cation. There is a direct relationship between T_c and the average radius of the A-site cation, $<r_A>$, the T_c increasing with $<r_A>$. A small $<r_A>$ gives rise to a distortion of the MnO_6 octahedra by bending the Mn–O–Mn bond and in turn causing the narrowing of the e_g bandwidth to a greater extent. Accordingly, yttrium substitution in $La_{1-x}A_xMnO_3$ decreases T_c, and enhances the magnitudes of resistivity as well as magnetoresistance. By plotting the T_c against $<r_A>$, one obtains a phase diagram separating the ferromagnetic metal and the paramagnetic insulator regimes. In Fig. 8, we show a phase diagram obtained in this manner. In the lower left-hand corner, we see the ferromagnetic insulator regime. CMR generally decreases with increase in $<r_A>$, just as the peak resistivity at the I–M transition.

The effect of particle size on the electron transport and magnetic properties of polycrystalline $La_{0.7}Ca_{0.3}MnO_3$ has been investigated. Although T_c decreases with decreasing particle size, the magnetoresistance (MR) is insensitive to particle size. The I–M transition becomes broader when the particle size is small, but the %MR is nearly constant over a

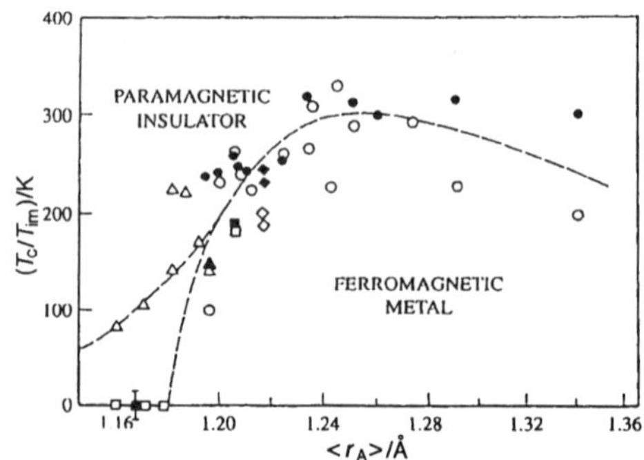

Fig. 8 Variation of the ferromagnetic T_c or the metal–insulator transition temperature, T_{im}, in $Ln_{1-x}A_xMnO_3$ with weighted average radius of the A site cations, $<r_A>$ [from Rao (reproduced with permission from ref. 8)].

wide temperature range. The low-temperature MR is significantly dependent on the grain size. For larger grain sizes, the MR at the lowest temperature decreases eventually becoming zero as in good single crystals. Samples of manganates with different particle sizes prepared with different heat treatments often show large differences between the ferromagnetic T_c and the temperature of the I–M transition. Grain boundaries appear to contribute substantially to magnetoresistance in polycrystalline materials, in particular at $T \ll T_c$.

The effects of dimensionality on CMR and related properties have been studied by examining the $(SrO)(La_{1-x}Sr_xMnO_3)_n$ family. The $n=1$ member which is two dimensional with the K_2NiF_4 structure, is an insulator, while the $n=\infty$ member is the three-dimensional perovskite showing CMR and other properties. The $n=2$ member exhibits the sharpest I–M transition and high MR. In $La_{1.4}Sr_{1.6}Mn_2O_7$, the current perpendicular to the MnO_2 planes is large, particularly close to T_c, and the barrier to transport provided by the intervening insulating layers is removed by the magnetic field as shown by Tokura *et al.* Interlayer transport seems to involve tunneling. We see some comparable features between the layered cuprate superconductors and the layered manganates. The big difference is that electrons are spin polarized in the latter.

Studies of the electrical, magnetic and other properties of $Ln_{1-x}A_xMnO_3$ systems have revealed certain unusual features with respect to the charge carriers in these oxides. The manganates exhibit high resistivities even in the metallic state, particularly at low temperatures. The values of resistivities are considerably higher than Mott's value of maximum metallic resistivity in many of these materials, the resistivity reaching constant values as high as 10^3–10^4 ohm cm at low temperatures. A clear picture of the nature of carriers responsible for the high resistivity is yet to emerge. The cause of the observed temperature dependence of resistivity at low temperatures is also not entirely clear. Another unusual feature of the manganates is related to the occurrence of large MR in a small field at relatively high temperatures. The scale of Zeeman energy associated with a field of 1 T in these spin systems is ≈ 5 K. This is much too small compared to the scale of thermal energy (200–300 K) where one sees such a large MR.

That double exchange is necessary for CMR in the manganates is shown by a study of the $LnMn_{1-x}Cr_xO_3$ system where Cr^{3+} replaces Mn^{4+}. In fact, double exchange probably suffices to explain most properties of $La_{1-x}Sr_xMnO_3$. The basic question that remains is why these manganates are insulating when $T>T_c$. The anomalous resistance has been explained by considering electron–phonon interactions arising from the Jahn–Teller splitting of the Mn d-levels. Jahn–Teller distortion localizes conduction electrons as polarons, especially in narrow band materials like $La_{1-x}Ca_xMnO_3$. As the temperature decreases, the transfer integral t_{ij} increases and the ratio of the Jahn–Teller energy to t_{ij} decreases. The coupling varies with temperature and composition, and increases across the metal–insulator transition. The large effects of local lattice distortion and crystal structure symmetry on electron transport and magnetism show that one cannot explain the unusual behavior without taking the lattice into account, especially in systems such as $La_{1-x}Ca_xMnO_3$. The observed T_c (*ca.* 250–400 K) in most manganates is considerably lower than that estimated from the measured spin-wave stiffness constant implying that the ferromagnetic exchange arising from the double exchange mechanism is not the only factor that affects T_c. The electrical resistivity of the rare earth manganates across the I–M transition is indeed difficult to understand especially at low temperatures based on double exchange or electron–phonon interaction alone. We also do not understand why CMR actually occurs in these oxides.[20] There is however every expectation that CMR will be exploited in the recording and sensor industries. We need to explore pure as well as composite materials showing CMR at ambient temperatures

at low magnetic fields. The exploration should not be limited to rare earth manganates since many other inorganic materials, including $Fe_{1-x}Cu_xCr_2S_4$ and EuO_x, exhibit large MR effects.

Marginal metallicity in oxide materials

We had earlier looked briefly at the metal–insulator transitions in oxides of the type $La_{1-x}Sr_xCoO_3$ or $LaNi_{1-x}Mn_xO_3$ where the temperature coefficient of resistance changes sign across a universal value of conductivity, *viz.* Mott's minimum metallic conductivity. These oxides not only obey the minimum metallic conductivity criterion arising from consideration of disorder, but also the $n_c^{1/3}a \approx 0.25$ criterion (n_c = critical carrier concentration at the transition) arising from electron interactions. The same situation holds for the cuprate superconductors which follow both these criteria. The rare earth manganates, $Ln_{1-x}A_xMnO_3$, exhibiting colossal magnetoresistance, on the other hand, are quite different in that many of them exhibit high residual resistivities far above Mott's value. The nature of carriers in the manganates would be much more difficult to describe than those in cuprate superconductors, which are themselves far from being understood. What is to be noted is that three of the most fascinating phenomena in materials, namely insulator–metal transitions, high temperature superconductivity and CMR are exhibited by oxide materials which are marginally metallic. Marginal metallicity appears to be a universal feature,[21] which requires further probing. It is interesting to speculate whether a universal minimum electron diffusivity due to localized states is present in these materials, besides the extended states.

Charge ordering in the manganates

Charge ordering is not new to transition metal oxides. Fe_3O_4 undergoes the Verwey transition with a resistivity anomaly due to charge ordering at 120 K and ferrimagnetic ordering at 860 K. Charge ordering of the Mn^{3+} and Mn^{4+} ions occurs in the perovskite manganates. $Ln_{1-x}A_xMnO_3$ is more interesting and is associated with dramatic changes in properties.[8,11,20] Charge ordering in the manganates is strongly dependent on the average radius of the A-site cations, $<r_A>$. The ferromagnetic T_c increases with increase in $<r_A>$ while charge ordering is favoured by small $<r_A>$. Charge ordering and double exchange are thus competing interactions. We shall examine charge ordering in $Ln_{0.5}A_{0.5}MnO_3$ with equal proportions of Mn^{3+} and Mn^{4+}.

When $<r_A>$ is sufficiently large, as in $La_{0.5}Sr_{0.5}MnO_3$ ($<r_A> = 1.26$ Å), the manganates become ferromagnetic and undergo an insulator–metal transition around T_c as described earlier. $Nd_{0.5}Sr_{0.5}MnO_3$ with a slightly smaller $<r_A>$ (1.236 Å), is ferromagnetic and metallic at 250 K (T_c), but transforms to a charge-ordered (CO), antiferromagnetic state at 150 K as shown by Tokura.[20] The charge-ordering transition temperature, T_{CO}, is the same as T_N. There is a significant specific heat anomaly at T_{CO} (T_N) and the resistivity increases markedly due to the metal–insulator (FM–CO) transition. This situation is to be contrasted with that of $Nd_{0.5}Ca_{0.5}MnO_3$ ($<r_A> = 1.17$ Å) which exhibits a T_{CO} of 250 K in the paramagnetic state exhibiting a small heat capacity anomaly. $Nd_{0.5}Ca_{0.5}MnO_3$ is an insulator at all temperatures. $Y_{0.5}Ca_{0.5}MnO_3$ with a $<r_A>$ of 1.13 Å also gets charge ordered in the paramagnetic state (T_{CO}, 240 K) and becomes antiferromagnetic at 140 K (T_N). In Fig. 9 we show a simple diagram where we delineate the different types of behavior of the manganates depending on the $<r_A>$. The charge ordered state of $Nd_{0.5}Sr_{0.5}MnO_3$ is melted by the application of a magnetic field, rendering the material metallic. On the other hand, a magnetic field of 6 T has no effect on the charge-ordered insulating state of $Y_{0.5}Ca_{0.5}MnO_3$. The charge ordered states in $Nd_{0.5}Sr_{0.5}MnO_3$ and $Nd_{0.5}(Y_{0.5})Ca_{0.5}MnO_3$ are

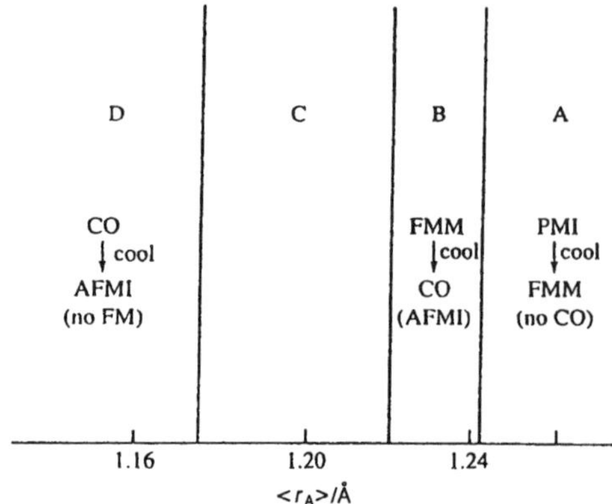

Fig. 9 Schematic diagram showing the different types of behavior in $La_{0.5}A_{0.5}MnO_3$. Region A corresponds to manganates showing CMR while region B corresponds to manganates of the type $Nd_{0.5}Sr_{0.5}MnO_3$ transforming from the ferromagnetic metallic (FMM) state to a charge-ordered (CO) state on cooling. Region D corresponds to manganates where the CO state is generally found in the paramagnetic insulating (PMI) state (*e.g.* $Nd_{0.5}Ca_{0.5}MnO_3$ and $Y_{0.5}Ca_{0.5}MnO_3$). Region C is expected to show complex behavior. AFMI stands for the antiferromagnetic insulating state [from Rao (reproduced with permission from ref. 8)].

clearly of different kinds, the difference being caused by the difference in $<r_A>$.

The occurrence of two types of CO states can be understood qualitatively in terms of the variation of the exchange couplings, J_{FM} and J_{AFM}, and the single-ion Jahn–Teller energy (E_{JT}) with $<r_A>$. While J_{FM} and J_{AFM} are expected to decrease with a decrease in $<r_A>$, albeit with different slopes, E_{JT} would be insensitive to $<r_A>$. In the small $<r_A>$ regime, a cooperative Jahn–Teller effect involving long-range elastic strain would dominate charge ordering, while at moderate values of $<r_A>$ (when $J_{AFM}>E_{JT}$, J_{FM}), the e_g electrons which are localized magnetically lower the configuration energy by charge ordering. Such a CO state, as exemplified by $Nd_{0.5}Sr_{0.5}MnO_3$, would be sensitive to magnetic fields unlike the manganates such as $Y_{0.5}Ca_{0.5}MnO_3$ in the small $<r_A>$ regime. In $Nd_{0.5}Sr_{0.5}MnO_3$, the gain in Zeeman energy resulting from the application of a magnetic field stabilizes the ferromagnetic metallic state over the antiferromagnetic insulating state. The general phase diagram in the T–$<r_A>$ plane has been constructed by Kumar and Rao based on these considerations.[8] Although we understand the gross features of charge-ordering and associated properties in manganates, we do not have a proper theory or model to explain all the varied aspects of these fascinating materials.

The high sensitivity of charge ordering in manganates to $<r_A>$ cannot be understood on the basis of the variation of the Mn–O–Mn angle and the Mn–O distance. This is because these geometrical changes do not give rise to sufficiently large changes in bandwidth as are necessary to explain the significant changes in the charge ordering temperature and the Curie temperature with $<r_A>$. The study of the complex regime (region C in Fig. 9), has yielded fascinating results. Thus, $Nd_{0.25}La_{0.25}Ca_{0.5}MnO_3$ ($<r_A> = 1.19$ Å) undergoes a reentrant transition from an incipient charge-ordered state to a FMM state on cooling, driven by a first order phase transition.[22] The charge-ordering gap collapses at the transition. Charge ordering in the manganates needs to be pursued further, especially regarding the marked effects of electric fields and chemical substitution. The role of the A-site ion mismatch on CO and FM states also needs attention.

Porous solids

Porous inorganic materials have many applications in catalytic and separation technologies. The subject has grown explosively with 8000 or more literature citations. The synthesis of microporous solids with connectivities and pore chemistry different from zeolitic materials has attracted considerable attention. A variety of such open framework structures, in particular Al and Ga phosphates as well as many other metal phosphates prepared in the presence of structure directing agents, have been characterized. The work of Cheetham, Ferey, Flanigen, Stucky and Thomas in this direction is noteworthy.[23a] Zeolites and zeolite-like materials[23,24] with micropores in the 5–20 Å range have contributed much to the advances in catalysis. There have been many break-throughs in the design and synthesis of these molecular seives with well defined crystalline architecture and there is a continuing quest for extra-large pore molecular sieves.[24] Several types of new materials including tunnel and layer structures of porous manganese oxides have been reported.[25] The progress has resulted mainly from our understanding of the role of structure-directing agents. The discovery of mesoporous silica (pore diameter 20–100 Å) by Mobil chemists added a new dimension to the study of porous solids. The synthesis of mesoporous materials also makes use of structure-directing surfactants[24,26] (cationic, anionic, and neutral) and a variety of mesoporous oxidic materials (*e.g.* ZrO_2, TiO_2, $AlPO_4$, aluminoborates), have been prepared and characterized.[26,27] In Fig. 10, the X-ray diffraction patterns of the three forms of $CaAlBO_4$ are shown.

An appropriate integration of hydrogen bond interactions at the inorganic–organic interface and the use of sol–gel and emulsion chemistry has enabled the synthesis of a large class of porous materials.[28a] Today, we have a better understanding of the structure, topology and phase transitions of mesoporous solids. Thus, block copolymers have been used to prepare mesoporous materials with large pore sizes[28a] (>30 nm). There is also some understanding of the lamellar–hexagonal-cubic phase transitions in mesoporous oxides[28b] (Fig. 11). Derivatized mesoporous materials have been explored for potential applications in catalysis and other areas.[29a] It has been found recently that transition metal complexes encapsu-

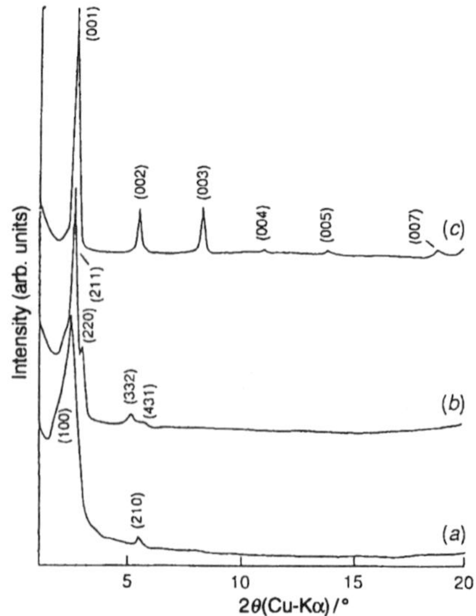

Fig. 10 X-Ray diffraction patterns of (*a*) hexagonal, (*b*) cubic and (*c*) lamellar forms of mesoporous calcium aluminoborate (from Ayyappan and Rao[27]).

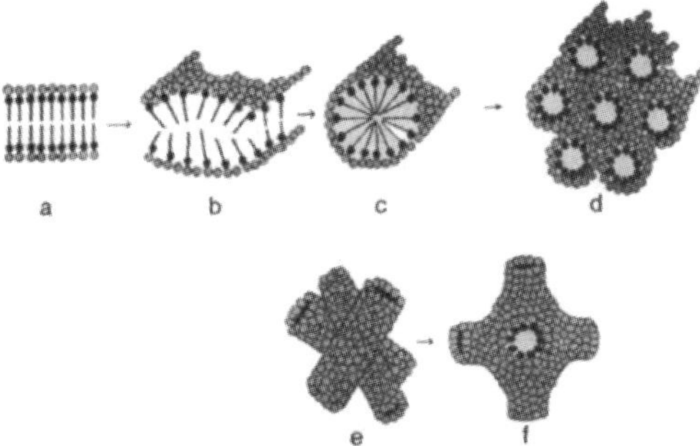

Fig. 11 Phase transitions in mesoporous solids: a–d, lamellar–hexagonal; e–f, hexagonal–cubic. The circular objects around the surfactant assemblies are the metal-oxo species (from Neeraj and Rao[28b]).

lated in cubic mesoporous phases show high catalytic activity in oxidation reactions.[29b]

Organic inclusion compounds and clathrates have been known for a long time. While these compounds are still being investigated, there have been efforts to synthesise novel organic structures by supramolecular means.[30] A recent example in this direction is the noncovalent synthesis of a novel channel structure formed by trithiocyanuric acid and bipyridine.[30b] The channels can accommodate benzene, xylenes and other molecules (Fig. 12) and the process is reversible.

Synthesis of inorganic materials

Although rational design and synthesis have remained primary objectives of materials chemistry, we are far from achieving this goal universally. There are many recent instances where rational synthesis has worked (*e.g.*, porous solids), but by and large the success has been limited. Thus, one may obtain the right structure and composition, but not the properties. There are very few instances where both the structure and properties of the material obtained are as desired. The synthesis of $ZrP_{2-x}V_xO_7$ solid solutions showing zero or negative thermal expansion is one such example.[31] The synthesis of $HgBa_2Ca_2Cu_3O_8$ with the highest superconducting T_c (*ca.* 160 K under pressure) is another example.[15,32] The preparation of modulation free bismuth cuprate superconductors by the substitution of Bi by Pb is also a good example.[33] One may

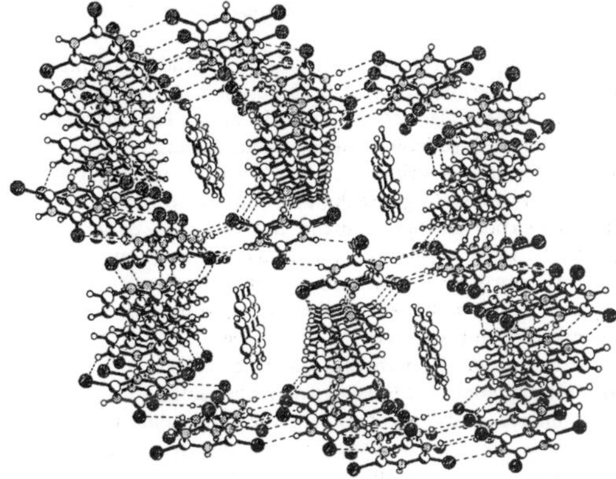

Fig. 12 Noncovalent synthesis of a layered network with large cavity based on hydrogen bonding between trithiocyanuric acid and bipyridine. Three-dimensional channels with benzene molecules can be seen [from Pedireddi *et al.* (reproduced with permission from ref. 30(*b*))].

also cite examples of the synthesis of ionic conductors and other materials. Most inorganic materials are still discovered accidentally, the cases of high T_c cuprates and CMR manganates being well known. What is noteworthy is the development of new strategies for synthesis, particularly those involving soft chemical routes. The subject has been reviewed adequately in the recent literature.[16,34,35]

The methods of soft chemistry include sol–gel, electrochemical, hydrothermal, intercalation and ion-exchange processes. Many of these methods are employed routinely for the synthesis of ceramic materials.[36,37] There have been recent reviews of the electrochemical methods,[38] intercalation reactions,[39] and the sol–gel technique.[40] The sol–gel method has been particularly effective with wide-ranging applications in ceramics, catalysts, porous solids and composites[41] and has given rise to fine precursor chemistry. Hydrothermal synthesis has been employed for the synthesis of oxidic materials under mild conditions[42] and most of the porous solids and open-framework structures using organic templates are prepared hydrothermally.[43] The advent of supramolecular chemistry has started to make an impact on synthesis,[16] mesoporous solids being well known examples.[43]

Many of the traditional methods continue to be exploited to synthesise novel materials. In the area of cuprates, the synthesis of a superconducting ladder cuprate and of carbonato- and halocuprates is noteworthy. High pressure methods have been particularly useful in the synthesis of certain cuprates and other materials.[36,44] The combustion method has come of age for the synthesis of oxidic and other materials.[45] Microwave synthesis is becoming popular[46] while sonochemistry has begun to be exploited.[47]

In the oxide literature, there are several reviews specific to the different families, such as layered transition metal oxides[48] and metal phosphonates.[49] Synthesis of metal nitrides has been discussed by Di Salvo.[50] Precursor synthesis of oxides, chalcogenides and other materials is being pursued with vigour.[16,36] Thus, single-molecule precursors of chalcogenide containing compound semiconductors have been developed.[51] Kanatzidis[52] has reviewed the application of molten polychalcophosphate fluxes for the synthesis of complex metal thiophosphates and selenophosphates.

In our own work over the years,[53] we have found the use of precursor carbonates to be effective in the synthesis of many oxide materials. Electrochemical techniques have been useful in oxidation reactions (*e.g.*, preparation of $SrCoO_3$, ferromagnetic $LaMnO_3$ containing *ca.* 30% Mn^{4+}, $La_2NiO_{4.25}$, $La_2CuO_{4+\delta}$). Intercalation and deintercalation of amines can be used to produce novel phases of oxides such as WO_3 and MoO_3. Topochemical oxidation, reduction and dehydration reactions also give rise to novel oxide phases, which cannot be prepared otherwise. Thus, $La_2Ni_2O_5$ can only be made by the reduction of $LaNiO_3$. Special mention must be made of the simple chemical technique of nebulized spray pyrolysis of solutions of organometallic precursors, to obtain epitaxial films of complex oxide materials; this technique can also be used to prepare nanoscale oxide powders.[54] Epitaxial films of PZT, $LaMnO_3$, $LaNiO_3$ and other oxide materials have been prepared by this method. Good films of copper and other metallic systems have also been obtained.

Molecular materials

Molecular materials, especially organic ones, have gained prominence in the last few years, with practical applications as optical and electronic materials already becoming possible. This rich area of research has benefitted from attempts to carry out rational design based on crystal engineering, including supramolecular chemistry[55,56] and the interest to investigate the structure and chemistry of the organic solid state.[57] The conventional areas of polymer research have become more

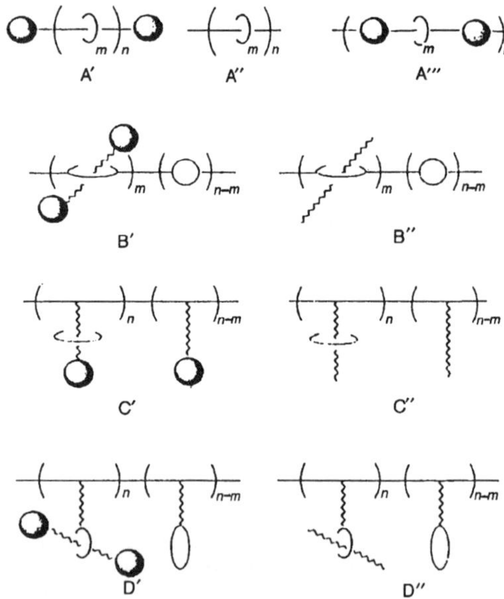

Fig. 13 Various types of main chain and side chain polyrotaxanes: blocking group (shaded circle); hollow circle (cyclic component); ellipses (cyclics threaded by linear species) [from Gong and Gibson (reproduced with permission from ref. 58)].

focussed on developing new synthetic methods and controlling polymerization. Since architecture and functionality at a molecular level control many of the properties of polymers, there are efforts to synthesise polymers with well defined topologies. Polymers involving rotaxanes, catenanes, rods, dendrimers and hyperbranched systems[58,59] are some recent examples of such efforts. Recognizing that hydrogen bonding is the driving force for the formation of crown ether based polyrotaxanes, mechanically linked polymers have been prepared, with the possibility of molecular shuttles at polymeric level (Fig. 13). Standard synthetic routes for dendrimers (Fig. 14) and hyperbranched polymers suggest many new possibilities.

Molecular magnetic materials constitute an interesting area of research, although practical applications may not be feasible in the short term. Synthesis and design of molecular ferromagnets, ferrimagnets and weak ferromagnets, providing permanent magnetization with fairly high T_c values are the main objectives of this effort. While purely organic ferromagnets have only reached modest T_cs below 2 K, materials incorporating transition metals have shown promise [$e.g.$ V(TCNE)$_x$]. Molecular magnetic materials generally make use of free radicals such as nitroxides, high-spin metal complex magnetic clusters or a combination of radicals and metal complexes.[60,61] There has been a renaissance in spin-crossover materials with emphasis on active elements for memory devices and light-induced spin changes.[62] The study of diradicals and radical pairs has been strengthened by our understanding of how to stabilize the triplet state.[61] C$_{60}$–TDAE ($T_c \approx 16$ K) and related compounds have attracted some attention (TDEA = tetrakisdimethylaminoethylene). BEDT-TTF salts with magnetic anions have been discovered recently.[63] Organometallic materials based on neutral nitroxide radicals and charge-transfer (CT) complexes derived from the radicals have been examined in some detail.[64]

Semiconducting, metallic and superconducting molecular materials have been investigated by several workers in the last two decades. New types of TTF type molecules, transition metal complexes with elongated π ligands, neutral radicals and Te-containing π donors have been synthesized and an organic superconductor with an organic anion, β''-(ET)$_2$SF$_5$CH$_2$CF$_2$SO$_3$, has been discovered.[65] Cation radical salts of BEDT-TTF and similar donors as well as BEDT-TTF salts are still a fertile ground for the production of new materials.[63,65] Synthesis of linearly extended TTF analogues[66] as well as of new TCNQ and DCNQ acceptors[67] have been reported. The use of organic semiconductors as active layers in thin films holds promise, α-sexithiophene being an example.[68,69]

High performance photonic devices have been fabricated from conjugated polymers such as poly(p-phenylenevinylene), polyaniline and polythiophene.[70] The devices include diodes, light-emitting diodes, photodiodes, field effect transistors, polymer grid triodes, light emitting electrochemical cells, optocouplers and lasers. The performance levels of many of these organic devices have reached those of the inorganic counterparts. The high photoluminescence efficiency and large cross section for stimulated emission of semiconducting polymers persist upto high concentrations (unlike dyes). By combination with InGaN, hybrid lasers can now be fabricated. It is gratifying that plastic electronics is moving rapidly from fundamental research to industrial applications.

The area of molecular nonlinear optics has been rejuvenated by efforts to investigate three-dimensional multipolar systems, functionalized polymers as optoelectronic materials, near infrared optical parametric oscillators and related aspects.[71] There have been some advances in chromophore design for second-order nonlinear optical materials;[72] these include one-dimensional CT molecules, octopolar compounds and organometallics. Some of the polydiacetylenes and poly(p-phenylenevinylene)s appear to possess the required properties for use as third-order nonlinear optical materials for photonic switching.[73]

Increasing interest in molecular materials has made an impact on the direction of organic chemistry as well, giving rise to large-scale activity in this area. The discovery of fullerenes and fullerene-based materials has contributed to some interesting developments.[74–76] Organic–inorganic hybrids offer many interesting possibilities, with the composites exhibiting properties and phenomena not found in either component.[77a] Two types of materials have been classified, organic–inorganic units held together by weak bonds and organometallic polymers.[77b] Silicon based hybrids and related materials have been reviewed by Livage.[40] In Fig. 15, we show some examples of polyhedral oligomeric silsesquioxanes used as building blocks for the sol–gel synthesis of hybrid materials. It is indeed gratifying that organic molecular devices have become a reality. The concepts and directions suggested by Whitesides[78a] have been useful in designing devices. Special mention must be made of the devices developed by Willner[78b] making use of self-organized thiols and other molecular systems. In this connection, the device-related studies of functional polymers by Wegner and others are also relevant.

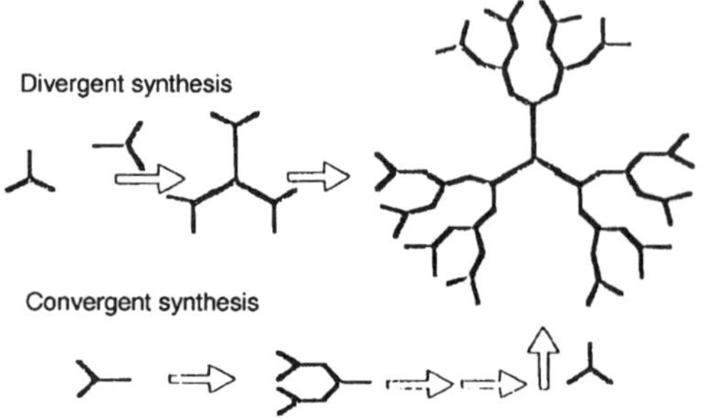

Fig. 14 Synthetic routes to dendrimers [from Hobson and Harrison (reproduced with permission from ref. 59)].

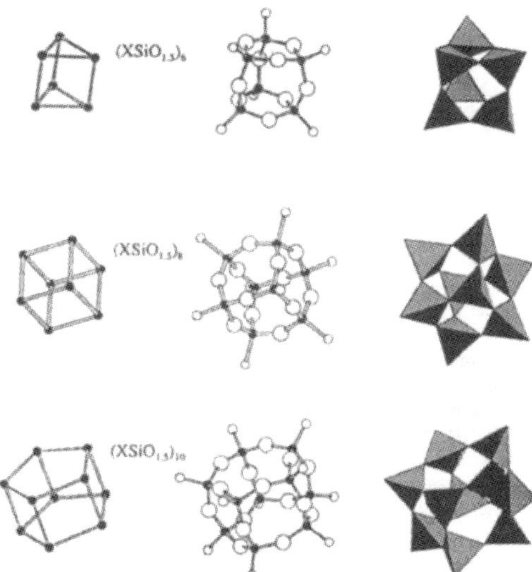

Fig. 15 Examples of silsesquioxanes used as building blocks for the sol–gel synthesis of hybrid materials [from Livage (reproduced with permission from ref. 60)].

Biomolecular materials

Biomolecular materials constitute a class of materials at the interface between biology and materials science. They not only provide an insight to Nature's ways, but also suggest new methods of synthesis and design of materials. Thus we can learn much from a study of the principles of structure–function relations in mineralized biological materials, the various forms of calcium carbonate forming the basis for many things that happen.[79,80] Some of the materials such as shells can, in principle, be considered to be part of traditional ceramic materials. The synthesis of mesoporous silica and related materials also owes its origins to biomineralization. Biomineralization has inspired the synthesis of nanocomposites, nanoparticles and nanoparticulate films.[81,82] Collagen fibrillar structure in mineralized and nonmineralized tissue, in particular the ultrastructure orientation of mineral crystals in bone relative to that of the collagen fibril, is an important problem receiving attention.[83]

Some of the recent developments in biomolecular materials include designed drug delivery systems consisting of self-assemblies of lipid and biocompatible lipids,[84] protein-based biological motors,[85] formation of new types of liposomes (*e.g.* spherical liposomes) and emulsions, tissue engineering, biogels, higher-order self-assemblies in biomaterials, and new methods of biopolymer synthesis.[86] Assembly of soft biofunctional and biocompatible interfaces of solids by the deposition of ultrathin soft polymer films, supported lipid–protein bilayers or membranes separated from solids by soft polymer cushions are other areas of vital interest.[87] The ability of polymer molecules attached at one end to a surface to prevent or enhance protein adsorption is being examined.[88]

Glasses and ceramics can bond to living tissues if there is bioactive layer. The development of a bioactive hydroxyapatite layer *in vivo* at body temperature is therefore an important problem.[89] Materials with the highest level of bioactivity develop a silica layer that enhances the formation of such a layer. Such sol–gel processes are used to produce bioactive coatings, powders and substrates which allow molecular control over the incorporation and behavior of proteins and cells with applications as sensors and implants. Sol–gel encapsulation of biomolecules within silica matrices has encompassed enzymes, metalloproteins, photoactive biomolecules and even whole cells.[78] Recently, interfacing electronic materials with lipids and proteins has shown promise in several fields, including biosensors.[90]

Nanomaterials

The synthesis, characterization and properties of nanomaterials have become very active areas of research in the last few years. In particular, nanostructured materials assembled by means of supramolecular organization offer many exciting possibilities. These include self-assembled monolayers and multilayers with different functionalities, intercalation in preassembled layered hosts and inorganic three-dimensional networks. The reader is referred to the special issue of *Chemistry of Materials*[91] for an overview of present day interests. There are many recent reviews on the varied aspects of nanomaterials. The work of Alivisatos[92] on the structural transitions, elec-

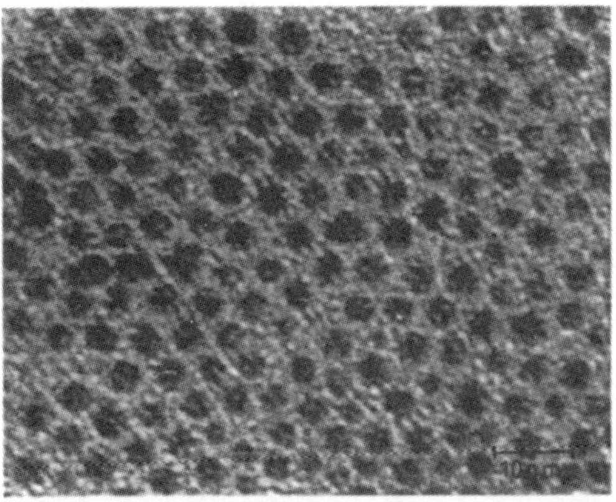

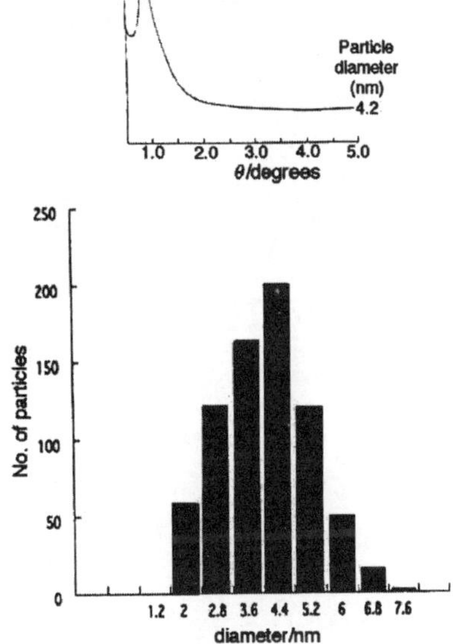

Fig. 16 TEM image of a nanocrystalline array of thiol-derivatized Au particles [from Sarathy *et al.* (reproduced with permission from ref. 96(*b*))].

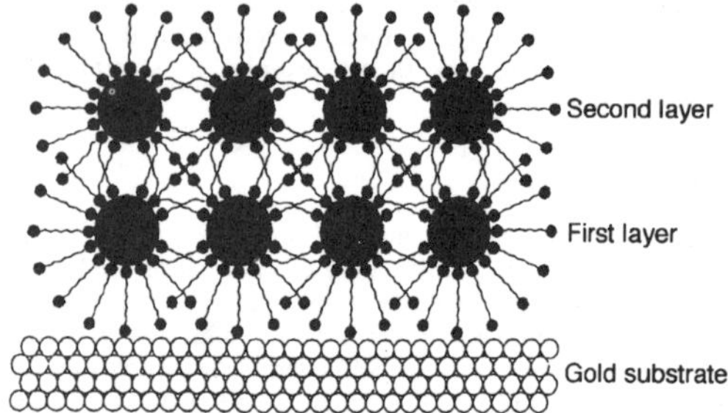

Fig. 17 Schematic representation of successive deposition of layers of metal (Pt or Au) nanoparticles and dithiols.

tronic properties and related aspects of nanoparticles of CdS, InAs and such materials is particularly noteworthy.

We have been interested in investigating the size-dependent electronic structure and reactivity of metal clusters deposited on solid substrates. Thus, we have shown that when the cluster size is small ($\lesssim 1$ nm), an energy gap opens up.[93a] Bimetallic clusters show additive effects due to alloying and cluster size[93b] in their electronic properties. Small metal clusters of Cu, Ni and Pd show enhanced chemical reactivity with respect to CO and other molecules.[94] Metal clusters and colloids, especially those with protective ligands, have been reviewed in relation to nanomaterials.[95] We have recently developed methods of preparing nanoparticles of various metals as well as nanocrystalline arrays of thiolized nanoparticles of Au, Ag and Pt.[96] In Fig. 16, we show the TEM image of thiol-derivatized Au

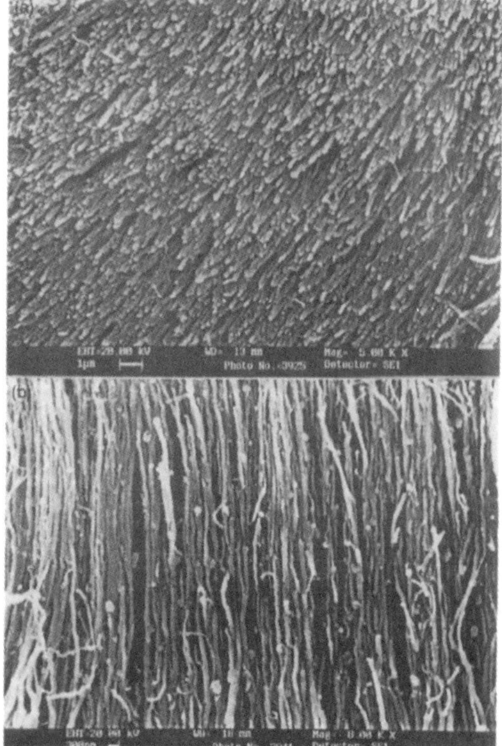

Fig. 18 SEM images of aligned-nanotube bundles obtained by the pyrolysis of ferrocene (from Rao *et al.*[99b]).

nanoparticles forming a nanocrystalline array. More interestingly, by using dithiols, it has been possible to accomplish layer-by-layer deposition of dithiol–metal nanoparticle films (Fig. 17). This is somewhat similar to the layer-by-layer self assembly of polyelectrolyte–inorganic nanoparticle sandwich films.[97] Such superlattices involving vertical organization of arrays of metal quantum dots may have novel properties.

Unprecedented interest has developed in carbon science ever since the discovery of fullerenes and nanotubes. Solid state properties of C_{60} and C_{70} have been of interest to the author since 1991. Some of the aspects investigated in this laboratory include orientational phase transitions, amorphization and polymerization under pressure, molecular magnetism of the TDAE derivative *etc.*[74] We have been working on the synthesis and characterization of carbon nanotubes for some time.[74] Opening, filling, closing and functionalizing carbon nanotubes have been accomplished.[98] Since metal particles are essential as catalysts to prepare nanotubes by the pyrolysis of hydrocarbons, we have employed organometallic precursors to generate nanotubes.[99a] Single-wall nanotubes have been obtained by the pyrolysis of metallocene or $Fe(CO)_5$–hydrocarbon mixtures under controlled conditions. It has also been possible to obtain copious quantities of aligned-nanotube bundles by the pyrolysis of ferrocene (Fig. 18) precursors. C–N and B–C–N nanotubes have been prepared by precursor pyrolysis as well.[99a] By using acid-treated carbon nanotubes as templates, ceramic oxide nanotubes have been prepared.[100] The procedure involves coating the acid-treated nanotubes by metal alkoxides and such precursors, and burning off the carbon.

Concluding remarks

In a brief overview of this type, it has not been possible to do justice to the work of many leading practitioners of materials chemistry or cover all the known classes of materials and phenomena. Thus I have not covered chalcogenides, nitrides and the like. An area that has not been discussed is noncrystalline materials, especially glasses, which have been of vital interest to materials chemists for many years. The subject is growing rapidly because of the increasing number of applications. Amorphous carbon, amorphous silicon, fluoride glasses, superionic glasses and metallic glasses are some of the amorphous materials of great interest.[101] The glass transition continues to arouse interest.[102] It is an unusual phase transition indeed. The subject of phase transitions itself is of great importance in materials chemistry and a proper understanding of the subject is essential to appreciate many of the materials properties.[103] Ionic conductors, energy storage materials and dielectrics constitute important areas of vital technological relevance. Recent developments in solid state electrochemistry are nicely covered in the book edited by Bruce.[104] An aspect of materials chemistry that is being increasingly exploited is computer simulation and modelling of structures, surfaces, processes and mechanisms.[105] Simulation and modelling techniques are particularly useful in understanding phenomena and structures in situations where experimentation is difficult (*e.g.* molecular processes in zeolites). Another development that deserves notice is the atomic layer-by-layer synthesis of inorganic materials,[106] which suggests the possibility of making use of oxides and other materials in integrated circuits. In spite of the brevity, I trust that the article has succeeded in communicating the nature of present-day materials chemistry. I believe that the references to many of the recent reviews will be useful to students, teachers and practitioners of the subject.

References

1 (*a*) *Modern Aspects of Solid State Chemistry*, ed. C. N. R. Rao, Plenum Press, New York, 1970; (*b*) C. N. R. Rao and

J. Gopalakrishnan, *New Directions in Solid State Chemistry*, first edition, Cambridge University Press, 1986 (second edition 1997).

2 (*a*) A. R. West, *Solid State Chemistry and Applications*, Wiley, Chichester, 1984; (*b*) *Solid State Chemistry*, ed. A. K. Cheetham and P. Day, Clarendon Press, Oxford 1987, 1992; (*c*) *Chemistry of Advanced Materials*, ed. C. N. R. Rao, Blackwell, Oxford, 1994; (*d*) L. V. Interrante and M. J. Hampden-Smith, *Chemistry of Advanced Materials*, Wiley-VCH, 1998.

3 (*a*) J. B. Goodenough, *Prog. Solid State Chem.*, 1971, **5**, 149; (*b*) C. N. R. Rao and B. Raveau, *Transition Metal Oxides*, second edition, Wiley-VCH, New York, 1998.

4 M. Greenblatt, *Chem. Rev.*, 1988, **88**, 31.

5 J. G. Bednorz and K. A. Müller, *Z. Phys. B*, 1986, **64**, 189.

6 (*a*) *Metal-insulator Transitions*, ed. P. P. Edwards and C. N. R. Rao, Taylor and Francis, London, 1985; (*b*) *Metal-insulator Transition Revisited*, ed. P. P. Edwards and C. N. R. Rao, Taylor and Francis, London, 1995.

7 (*a*) J. M. Honig and L. L. Van Zandt, *Annu. Rev. Mater. Sci.*, 1975, **5**, 225; (*b*) J. M. Honig and J. Spalek, *Proc. Ind. Nat. Sci. Acad. A*, 1986, **52**, 232.

8 C. N. R. Rao, *Philos. Trans. R. Soc. London*, 1998, **356**, 23.

9 P. Ganguly and C. N. R. Rao, *J. Solid State Chem.*, 1984, **53**, 193.

10 C. N. R. Rao, P. Ganguly, K. K. Singh and R. A. Mohanram, *J. Solid State Chem.*, 1988, **72**, 14.

11 C. N. R. Rao and A. K. Cheetham, *Adv. Mater.*, 1997, **9**, 1009 and references therein.

12 C. N. R. Rao and J. M. Thomas, *Acc. Chem. Res.*, 1985, **18**, 113.

13 M. K. Wu *et al.*, *Phys. Rev. Lett.*, 1987, **58**, 908.

14 C. N. R. Rao *et al.*, *Nature*, 1987, **326**, 856.

15 M. Hervieu, *Curr. Opin. Solid State Mater. Sci.*, 1996, **1**, 29.

16 J. Gopalakrishnan, N. S. P. Bhuvanesh and K. K. Rangan, *Curr. Opin. Solid State Mater. Sci.*, 1996, **1**, 285.

17 B. Raveau, C. Michel, M. Hervieu and D. Groult, *Crystal Chemistry of High T_c Superconducting Copper Oxides*, Springer, Berlin, 1991.

18 (*a*) C. N. R. Rao and A. K. Ganguli, *Chem. Soc. Rev.*, 1995, **24**, 1; (*b*) C. N. R. Rao, *Philos. Trans. R. Soc. London*, 1991, **336**, 595; (*c*) T. V. Ramakrishnan, in *Critical Problems in Physics*, Princeton University Press, 1996.

19 J. Rouxel, *Curr. Sci. (India)*, 1997, **73**, 31.

20 (*a*) *Colossal Magnetoresistance, Charge-ordering and Related Aspects of Manganese Oxides*, ed. C. N. R. Rao and B. Raveau, World Scientific, Singapore, 1998; (*b*) Y. Tokura in ref. 20(a).

21 C. N. R. Rao, *Chem. Commun.*, 1996, 2217.

22 A. Arulraj, A. Biswas, A. K. Raychaudhuri, C. N. R. Rao, P. M. Woodward, T. Vogt, D. E. Cox and A. K. Cheetham, *Phys. Rev. B*, 1998, **57**, R8115.

23 (*a*) S. Natarajan, M. Eswaramoorthy, A. K. Cheetham and C. N. R. Rao, *Chem. Commun.*, 1998, 1561 and references therein; (*b*) S. I. Zones and M. E. Davis, *Curr. Opin. Solid State Mater. Sci.*, 1996, **1**, 107; (*c*) J. M. Thomas, *Angew. Chem., Int. Ed. Engl.*, 1994, **33**, 913; (*d*) G. Ozin, *Adv. Mater.*, 1992, **4**, 612.

24 (*a*) C. J. Brinker, *Curr. Opin. Solid State Mater. Sci.*, 1996, **1**, 795; (*b*) M. E. Davis, *Chem. Eur. J.*, 1997, **3**, 1745.

25 S. L. Suib, *Curr. Opin. Solid State Mater Sci.*, 1998, **3**, 63.

26 J. S. Beck and J. C. Vartuli, *Curr. Opin. Solid State Mater. Sci.*, 1996, **1**, 76.

27 S. Ayyappan and C. N. R. Rao, *Chem. Commun.*, 1997, 575 and references therein.

28 (*a*) D. Zhao, P. Yang, Q. Huo, B. F. Chmelka and G. D. Stucky, *Curr. Opin. Solid State Mater. Sci.*, 1998, **3**, 111; (*b*) Neeraj and C. N. R. Rao, *J. Mater. Chem.*, 1998, **8**, 1631.

29 (*a*) T. Maschmeyer, *Curr. Opin. Solid State Mater. Sci.*, 1998, **3**, 71; (*b*) M. Eswaramoorthy, Neeraj and C. N. R. Rao, *Chem. Commun.*, 1998, 615.

30 (*a*) O. M. Yagi, G. Li and H. Li, *Nature*, 1995, **378**, 703; (*b*) V. R. Pedireddi, S. Chatterji, A. Ranganathan and C. N. R. Rao, *J. Am. Chem. Soc.*, 1997, **119**, 10867.

31 A. W. Sleight, *Endeavour*, 1995, **19**, 64.

32 E. V. Antipov, S. M. Loureno, C. Chaillout, J. J. Capponi, P. Bordet, J. L. Tholence, S. N. Putlin and M. Marezio, *Physica C*, 1993, **215**, 1.

33 V. Manivannan, J. Gopalakrishnan and C. N. R. Rao, *Phys. Rev. B*, 1991, **43**, 8686.

34 J. Gopalakrishnan, *Chem. Mater.*, 1995, **7**, 1265.

35 *Soft Chemistry Routes to New Materials—Chemie Douce*, ed. J. Rouxel, M. Tournoux and R. Brec, Trans. Tech. Publications, Aedermannsdorf, Switzerland, 1994.

36 (*a*) *Preparative Solid State Chemistry*, ed. P. Hagenmuller, Academic Press, New York, 1997, p. 2; (*b*) C. N. R. Rao, *Chemical Approaches to the Synthesis of Inorganic Materials*, John Wiley, Chichester, 1994; (*c*) J. D. Corbett, in *Solid State Chemistry—Techniques*, ed. A. K. Cheetham and P. Day, Clarendon Press, Oxford, 1987.

37 D. Segal, *J. Mater. Chem.*, 1997, **7**, 1297.

38 J. C. Grenier, M. Pouchard and A. Wattiaux, *Curr. Opin. Solid State Mater. Sci.*, 1996, **1**, 233.

39 G. Ouvrad and D. Guyomard, *Curr. Opin. Solid State Mater. Sci.*, 1996, **1**, 260.

40 J. Livage, *Curr. Opin. Solid State Mater. Sci.*, 1997, **2**, 132.

41 See special issue of *Chem. Mater.*, 1997, **9**, 2247–2670.

42 M. S. Whittingham, *Curr. Opin. Solid State Mater. Sci.*, 1996, **1**, 227.

43 M. E. Davis and I. E. Maxwell, *Curr. Opin. Solid State Mater. Sci.*, 1996, **1**, 55.

44 M. Takano and A. Onodera, *Curr. Opin. Solid State Mater. Sci.*, 1997, **2**, 166.

45 K. C. Patil, S. T. Aruna and S. Ekambaram, *Curr. Opin. Solid State Mater. Sci.*, 1997, **2**, 158.

46 (*a*) K. J. Rao and P. D. Ramesh, *Bull Mater. Sci.*, 1995, **18**, 447; (*b*) D. M. P. Mingos, *Chem. Soc. Rev.*, 1998, **27**, 213.

47 D. Peters, *J. Mater. Chem.*, 1996, **6**, 1605.

48 P. A. Salvador, T. O. Mason, M. E. Hagerman and K. R. Poeppelmeier, in *Chemistry of Advanced Materials: An Overview*, ed L. V. Interrante and M. J. Hampden-Smith, Wiley—VCH, New York, 1998.

49 A. Clearfield, *Curr. Opin. Solid State Mater. Sci.*, 1996, **1**, 268.

50 F. J. Di Salvo, *Curr. Opin. Solid State Mater. Sci.*, 1996, **1**, 241.

51 P. O'Brien and R. Nomura, *J. Mater. Chem.*, 1995, **5**, 1761.

52 M. G. Kanatzidis, *Curr. Opin. Solid State Mater. Sci.*, 1997, **2**, 139.

53 C. N. R. Rao, *Pure Appl. Chem.*, 1994, **66**, 1765; 1997, **69**, 199.

54 (*a*) H. N. Aiyer, A. R. Raju, G. N. Subbanna and C. N. R. Rao, *Chem. Mater.*, 1997, **9**, 755; (*b*) P. Murugavel, M. Kalaiselvam, A. R. Raju and C. N. R. Rao, *J. Mater. Chem.*, 1997, **7**, 1433.

55 A. Gavezzotti, *Curr. Opin. Solid State Mater. Sci.*, 1996, **1**, 501.

56 G. R. Desiraju, *Curr. Opin. Solid State Mater. Sci.*, 1997, **2**, 451.

57 See special issue of *Chem. Mater.* (dedicated to M.C. Etter), 1994, **6**, 1087–1461.

58 C. Gong and H. W. Gibson, *Curr. Opin. Solid State Mater. Sci.*, 1997, **2**, 647.

59 L. C. Hobson and R. M. Harrison, *Curr. Opin. Solid State Mater. Sci.*, 1997, **2**, 683.

60 D. Gatteschi, *Curr. Opin. Solid State Mater. Sci.*, 1996, **1**, 192.

61 K. Matsuda and H. Iwamura, *Curr. Opin. Solid State Mater. Sci.*, 1997, **2**, 446.

62 O. Kahn, *Curr. Opin. Solid State Mater. Sci.*, 1996, **1**, 547.

63 P. Day and M. Kurmod, *J. Mater. Chem.*, 1997, **7**, 1291.

64 S. Nakatsuji and H. Auzai, *J. Mater. Chem.*, 1997, **7**, 2161.

65 H. Kobayasbi, *Curr. Opin. Solid State Mater. Sci.*, 1997, **2**, 440.

66 J. Roncali, *J. Mater. Chem.*, 1997, **7**, 2307.

67 N. Martin, J. Segura and C. Sevane, *J. Mater. Chem.*, 1997, **7**, 1661.

68 F. Garnier, *Curr. Opin. Solid State Mater. Sci.*, 1997, **2**, 455.

69 H. E. Katz, *J. Mater. Chem.*, 1997, **7**, 369.

70 A. J. Heeger and M. A. Diaz-Garcia, *Curr. Opin. Solid State Mater. Sci.*, 1998, **3**, 16.

71 J. Zyss and J-F, Nicoud, *Curr. Opin. Solid State Mater. Sci.*, 1996, **1**, 533.

72 T. Verbiest, S. Houbrechts, M. Kauranen, K. Clays and A. Persons, *J. Mater. Chem.*, 1997, **7**, 2175.

73 B. Luther-Davies and M. Samic, *Curr. Opin. Solid State Mater. Sci.*, 1997, **2**, 213.

74 C. N. R. Rao, R. Seshadri, R. Sen and A. Govindaraj, *Mater. Sci. Eng. Rep.*, 1995, **R15**, 209; also see *Curr. Opin. Solid State Mater. Sci.*, 1996, **1**, 279.

75 K. Prassides, *Curr. Opin. Solid State Mater. Sci.*, 1997, **2**, 433.

76 M. Prato, *J. Mater. Chem.*, 1997, **7**, 1097.

77 (*a*) Y. Chujo, *Curr. Opin. Solid State Mater. Sci.*, 1996, **1**, 806; (*b*) P. Judeinstein and C. Sanchez, *J. Mater. Chem.*, 1996, **6**, 511.

78 (*a*) See for example G. M. Whitesides, *Acc. Chem. Res.*, 1995, **28**, 37, 219; (*b*) I. Willner, *Acc. Chem. Res.*, 1997, **30**, 347.

79 *Biomimetic Materials Chemistry*, ed. S. Mann, VCH Publishers, Weinheim, 1996.

80 S. Weiner and L. Addadi, *J. Mater. Chem*, 1997, **7**, 689.

81 S. Manne and L. A. Aksay, *Curr. Opin. Solid State Mater. Sci.*, 1997, **2**, 358.

82 J.H. Fendler, *Curr. Opin. Solid State Mater. Sci.*, 1997, **2**, 365.

83 A. Veis, *Curr. Opin. Solid State Mater. Sci.*, 1997, **2**, 370.

84 D. D. Lasic and D. Papahadzopouls, *Curr. Opin. Solid State Mater. Sci.*, 1996, **1**, 392.

85 F. Gittes and C. F. Schmidt, *Curr. Opin. Solid State Mater. Sci.*, 1996, **1**, 412.

86 C.R. Safinya and L. Addadi, *Curr. Opin. Solid State Mater. Sci.*, 1996, **1**, 387.
87 J. Rädler and E. Sackmann, *Curr. Opin. Solid State Mater. Sci.*, 1997, **2**, 330.
88 I. Szleifer, *Curr. Opin. Solid State Mater. Sci.*, 1997, **2**, 337.
89 L. L. Hench, *Curr. Opin. Solid State Mater. Sci.*, 1997, **2**, 604.
90 J. A. Zasadzinski, *Curr. Opin. Solid State Mater. Sci.*, 1997, **2**, 345.
91 See special issue of *Chem. Mater.*, 1996, **8**, 1569–2193.
92 P. Alivisatos, *J. Phys. Chem.*, 1996, **100**, 13226.
93 (a) C. P. Vinod, G. U. Kulkarni and C. N. R. Rao, *Chem. Phys. Lett.*, 1998, **289**, 329. (b) K. R. Harikumar, S. Ghosh and C. N. R. Rao, *J. Phys. Chem.*, 1997, **101**, 536.
94 A. K. Santra, S. Ghosh and C. N. R. Rao, *Langmuir*, 1994, **10**, 3937.
95 G. Schmid and G. L. Hornyak, *Curr. Opin. Solid State Mater. Sci.*, 1997, **2**, 204.
96 (a) S. Ayyappan, R. S. Gopalan, G. N. Subbanna and C. N. R. Rao, *J. Mater. Res.*, 1997, **12**, 398; (b) K. V. Sarathy, G. Raina, R. T. Yadav, G. U. Kulkarni and C. N. R. Rao, *J. Phys. Chem.*, 1997, **101**, 9876.
97 J. H. Fendler, *Chem. Mater.*, 1996, **8**, 1616.
98 B. C. Satishkumar, A. Govindaraj and C. N. R. Rao, *J. Phys. B*, 1996, **29**, 4925.
99 (a) R. Sen, A. Govindaraj and C. N. R. Rao, *Chem. Mater.*, 1997, **9**, 2078; *Chem. Phys. Lett.*, 1998, **287**, 671; (b) C. N. R. Rao, R. Sen, B. C. Satishkumar and A. Govindaraj, *Chem. Commun.*, 1998, 1525.
100 C. N. R. Rao, B. C. Satishkumar and A. Govindaraj, *Chem. Commun.*, 1997, 1581.
101 S. R. Elliott and R. Street, *Curr. Opin. Solid State Mater. Sci.*, 1996, **1**, 555; 1997, **2**, 397.
102 C. A. Angell, *Curr. Opin. Solid State Mater. Sci.*, 1996, **1**, 578.
103 C. N. R. Rao, *Acc. Chem. Res.*, 1984, **17**, 83; see also *J. Mol. Struct.*, 1993, **292**, 229.
104 *Solid State Electrochemistry*, ed. P. G. Bruce, Cambridge University Press, 1995.
105 M. Stoneham and M. L. Klein, *Curr. Opin. Solid State Mater. Sci.*, 1996, **1**, 817; M. Stoneham and S. Panteldes, *Curr. Opin. Solid State Mater. Sci.*, 1997, **2**, 6.
106 A. Gupta, *Curr. Opin. Solid State Mater. Sci.*, 1997, **2**, 23.

Paper 8/04467H

Materials Science and Engineering, B18 (1993) 1-21

Critical Review

Chemical synthesis of solid inorganic materials*

C. N. R. Rao

CSIR Centre of Excellence in Chemistry and Solid State and Structural Chemistry Unit, Indian Institute of Science, Bangalore 560 012 (India)

(Received August 19, 1992; accepted in revised form September 1, 1992)

Abstract

Chemical methods of synthesis of materials play a crucial role in designing and discovering new materials and also in providing better and less cumbersome methods for preparing known materials. In this article, we shall discuss the chemical synthesis of inorganic solids, in particular oxidic materials. We shall first briefly examine the different classes of chemical reactions generally employed for synthesis and then discuss the various methods used along with several case studies and examples. In addition to the traditional ceramic method, the topics discussed include the combustion method (self-propagating high-temperature synthesis), the precursor method, topochemical routes, intercalation compounds, the ion-exchange method, the sol-gel process, the alkali-flux method, electrochemical methods, the pyrosol process and high pressure methods. The last topic includes hydrothermal synthesis of zeolitic materials. Intergrowth structures and superconducting cuprates are discussed in separate sections. It is hoped that the article will provide a useful survey of chemical methods of synthesis of inorganic materials and will serve as a ready reference to practitioners of the subject.

1. Introduction

There is much chemical ingenuity in the synthesis of solid materials [1-6] and this aspect of materials science is being increasingly recognized as a vital component of the subject. While tailor-making materials of the desired structure and properties remains the main goal of solid state chemistry and materials science, it is not always possible. One can evolve a rational approach to the synthesis of solids [7], but there is always an element of surprise which is encountered not uncommonly. A well known example of an oxide discovered serendipitously is $NaMo_4O_6$ containing condensed Mo_6 octahedral metal clusters [8]. This was discovered during attempts to prepare the lithium analogue of $NaZn_2Mo_3O_8$. Another serendipitous discovery was that of the phosphorus–tungsten bronze $Rb_xP_8W_{32}O_{112}$ formed by the reaction of phosphorus present in the silica of the ampoule during the preparation of rubidium–tungsten bronze [9]. Since the material could not be prepared in a platinum crucible, it was suspected that a constituent of the silica ampoule must have been incorporated. This discovery led to the

synthesis of several phosphorus–tungsten bronzes of the type $A_xP_4O_8(WO_3)_{2m}$. Chevrel phases of the type $A_xMo_6S_8$ ($A \equiv Cu$, Pb, La etc.) were also discovered accidentally [10].

Rational synthesis of materials requires an understanding of the principles of crystal chemistry, and of thermodynamics, phase equilibria and reaction kinetics. There are many examples of rational synthesis. A good example is SIALON [11] where aluminium and oxygen were partly substituted for silicon and nitrogen in Si_3N_4. The fast Na^+ ion conductor NASICON, $Na_3Zr_2PSi_2O_{12}$, was synthesized with a clear understanding of the coordination preferences of cations and the nature of oxide networks formed by them [12]. The zero-expansion ceramic $Ca_{0.5}Ti_2P_3O_{12}$ possessing the NASICON framework was later synthesized based on the idea that the property of zero-expansion would be exhibited by two or three coordination polyhedra linked in such a manner as to leave substantial empty space in the network [7]. Another example of rational synthesis is that of silicate-based porous materials, making use of organic templates to predetermine the pore or cage geometries [13]. A microporous phosphate of the formula $(Me_4N)_{1.3}(H_3O)_{0.7}Mo_4O_8(PO_4)_2 \cdot 2H_2O$ where the tetramethylammonium ions fill the voids in the three-

---

*Contribution No. 870 from Solid State & Structural Chemistry Unit.

0921-5107/93/$6.00

dimensional structure made up of Mo_4O_8 cubes and PO_4 tetrahedra, has been prepared in this manner [14].

A large variety of inorganic solid materials has been prepared in recent years by the traditional ceramic method, which involves mixing and grinding powders of the constituent oxides, carbonates and such compounds and heating them at high temperatures with intermediate grinding when necessary. A wide range of conditions, often bordering on the extreme, such as very high temperatures or pressures, very low oxygen fugacities and rapid quenching have been employed in materials synthesis. The low-temperature chemical routes, however, are of greater interest. The trend nowadays is to avoid brute-force methods in order to have better control of the structure, stoichiometry and phase purity. Noteworthy chemical methods of synthesis include the precursor method, coprecipitation and soft-chemistry routes, the combustion method, the sol-gel method, topochemical methods and high-pressure methods. In this article, we shall discuss the synthesis of inorganic solids by chemical methods with several examples, especially oxide materials including superconducting cuprates synthesized by these means. For the purpose of brevity we shall cite only some of the key references which guide the reader to the original literature.

2. Common reactions encountered in the synthesis of inorganic solids

Various types of chemical reactions have been used for the synthesis of solid materials. Corbett [3] has written an excellent article on the subject. Some of the common reactions employed for the synthesis of inorganic solids are listed below:

(1) decomposition

$$A(s) \longrightarrow B(s) + C(g)$$

(2) combination

$$A(s) + B(g) \longrightarrow C(s)$$

(3) metathetic (combining (1) and (2) above)

$$A(s) + B(g) \longrightarrow C(s) + D(g)$$

(4) addition

$$A(s) + B(s) \longrightarrow C(s)$$

$$A(s) + B(l) \longrightarrow C(s)$$

$$A(g) + B(g) \longrightarrow C(s)$$

(5) exchange

$$AX(s) + BY(s) \longrightarrow AY(s) + BX(s)$$

$$AX(s) + BY(g) \longrightarrow AY(s) + BX(g)$$

Typical examples of the above simple reactions include the following:

(1) $CaCO_3(s) \longrightarrow CaO(s) + CO_2(g)$

$$M_mO_n(s) \longrightarrow M_mO_{n-\delta}(s) + \frac{\delta}{2}O_2(g)$$
$$(M = metal)$$

(2) $YBa_2Cu_3O_6(s) + O_2(g) \longrightarrow YBa_2Cu_3O_7(s)$

(3) $Pr_6O_{11}(s) + 2H_2(g) \longrightarrow 3Pr_2O_3(s) + 2H_2O(g)$

(4) $ZnO(s) + Fe_2O_3(s) \longrightarrow ZnFe_2O_4(s)$

$$BaO(s) + TiO_2(s) \longrightarrow BaTiO_3(s)$$

$$3SiCl_4(g) + 4NH_3(g) \longrightarrow Si_3N_4(s) + 12HCl(g)$$

$$GaMe_3(g) + 4NH_3(g) \longrightarrow GaAs(s) + 3CH_4(g)$$

(5) $ZnS(s) + CdO(s) \longrightarrow CdS(s) + ZnO(s)$

$$MnCl_2(s) + 2HBr \longrightarrow MnBr_2(s) + 2HCl$$

More complex reactions involving more than one type of reaction are also commonly employed in solid state synthesis. For example, in the preparation of complex oxides it is common to carry out thermal decomposition of a compound followed by oxidation (in air or O_2) essentially in one step:

$$2Ca_{0.5}Mn_{0.5}CO_3(s) + \tfrac{1}{2}O_2(g) \longrightarrow CaMnO_3(s) + 2CO_2(g)$$

Vapour phase reactions also yield solid products in many instances. In chemical vapour transport reactions, a gaseous reagent acts as a carrier to transport a solid by transforming it into the vapour state:

$$MgO(s) + Cr_2O_3(s) \xrightarrow{O_2} MgCr_2O_4(s)$$

Typical transport reaction equilibria include the following:

$$ZnS(s) + I_2 \rightleftharpoons ZnI_2 + \tfrac{1}{2}S_2$$

$$TaOCl_2(s) + TaCl_5 \rightleftharpoons TaOCl_3 + TaCl_4$$

$$Nb_2O_5(s) + 3NbCl_5 \rightleftharpoons 5NbOCl_3$$

$$GaAs(s) + HCl \rightleftharpoons GaCl + \tfrac{1}{2}H_2 + As$$

In the last reaction, the starting reactants would be $AsCl_3$, gallium and H_2. Table 1 lists several examples of chemical transport systems.

Specific reagents and reaction conditions are employed to carry out various processes such as reduction, oxidation and halogenation in the synthesis of solids. For example, reduction of oxides is carried out in an atmosphere of (flowing) pure or dilute hydrogen (*e.g.* N_2–H_2 mixtures) or sometimes in an atmosphere of CO or CO–CO_2 mixtures. Reduction of oxides for the purpose of lowering the oxygen content is also achieved by heating oxides in argon or nitrogen; appli-

C. N. R. Rao / *Synthesis of solid inorganic materials*   3

TABLE 1. Examples of chemical transport

Solid	Transporting agent	Solid	Transporting agent
Nb_2O_5	Cl_2, $NbCl_5$	CrOCl	Cl_2
TiO_2	$I_2 + S_2$	$FeWO_4$	Cl_2
IrO_2	O_2	$MgFe_2O_4$	HCl
WO_3	H_2O	$CaNb_2O_6$	Cl_2, HCl
NbS_2	S	ZrOS	I_2
TaS_3	S	$LaTe_2$	I_2
$MnGeO_3$	HCl	V_nO_{2n-1}	$TeCl_4$
$MgTiO_3$	Cl_2	NbS_2Cl_2	$NbCl_5$

cation of vacuum at an appropriate temperature (vacuum annealing or decomposition at low pressures) is also used. The obvious means of reducing solid compounds by hydrogen is employed not only for reducing oxides, but also halides and other compounds. Thermal decomposition of metal halides also yields lower halides:

$$M_2O_3(s) + H_2(g) \longrightarrow 2MO(s) + H_2O(g)$$

$$(e.g. M \equiv Fe)$$

$$ABO_3(s) + H_2(g) \longrightarrow ABO_{2.5}(s) + \tfrac{1}{2}H_2O(g)$$

$$(e.g.\ LaCoO_3)$$

$$MCl_3(s) + H_2(g) \longrightarrow MCl_2(s) + HCl(g)$$

$$(e.g.\ M \equiv Fe)$$

$$MX_3 \xrightarrow{\text{heat}} MX_2 + \tfrac{1}{2}X_2 \qquad (e.g.\ M \equiv Cr)$$

Reduction of oxides can be accomplished by reacting with elemental carbon or with a metal. Reduction of halides is also carried out by metals:

$$2MCl_3 + M \longrightarrow 3MCl_2 \qquad (e.g.\ M \equiv Nd, Fe)$$

$$3MCl_4 + M'(s) \longrightarrow 3MCl_3(s) + M'Cl_3(g)$$

$$(e.g.\ M \equiv Hf, M' \equiv Al)$$

$$M_2O_5 + 3M \longrightarrow 5MO \qquad (e.g.\ M \equiv Nb)$$

Metal oxychlorides are obtained by heating oxides with Cl_2 (LaOCl from La_2O_3). Fluorination is generally carried out using elemental fluorine. There are examples where oxides are reacted with a fluoride such as BaF_2 to attain partial fluorination.

3. Ceramic method

The most common method of preparing solid materials is by reaction of the component materials in the solid state at elevated temperatures. Several oxides, sulphides, phosphides, etc., have been prepared by this method. Knowledge of the phase diagram is generally helpful in fixing the desired composition and conditions for synthesis. Some caution is necessary in choosing the container; platinum, silica and alumina containers are generally used for the synthesis of metal oxides, while graphite containers are employed for sulphides and other chalcogenides as well as pnictides. If one of the constituents is volatile or sensitive to the atmosphere, the reaction is carried out in sealed evacuated capsules. Most ceramic preparations require relatively high temperatures which are generally attained by resistance heating. Electric arc and skull techniques give temperatures up to 3300 K while high-power CO_2 lasers give temperatures up to 4300 K.

The ceramic method suffers from several disadvantages. When no melt is formed during the reaction, the entire reaction has to occur in the solid state, initially by a phase boundary reaction at the points of contact between the components and later by diffusion of the constituents through the product phase. As the reaction progresses, diffusion paths become increasingly longer and the reaction rate slower. The product interface between the reacting particles acts as a barrier. The reaction can be speeded up to some extent by intermittent grinding between heating cycles. There is no simple way of monitoring the progress of the reaction in the ceramic method. It is only by trial and error (by carrying out X-ray diffraction and other measurements periodically) that one decides on appropriate conditions that lead to completion of the reaction. Because of this difficulty, one frequently ends up with mixtures of reactants and products. Separation of the desired product from these mixtures is generally difficult, if not impossible. It is sometimes difficult to obtain a compositionally homogeneous product by the ceramic technique, even when the reaction proceeds almost to completion.

In spite of such limitations, ceramic techniques have been widely used for the synthesis of solid materials. Mention must be made, among others, of the use of this technique for the synthesis of rare earth monochalcogenides such as SmS and SmSe. The method involves heating the elements, first at lower temperatures (870–1170 K) in evacuated silica tubes; the contents are then homogenized, sealed in tantalum tubes and heated to around 2300 K by passing a high current through the tube [15].

Various modifications of the ceramic technique have been employed to overcome some of the limitations. One of these relates to decreasing the diffusion path lengths. In a polycrystalline mixture of reactants, the individual particles are approximately 10 μm in size, representing diffusion distances of roughly 10 000 unit cells. By using freeze-drying, spray-drying, coprecipitation, and sol-gel and other techniques, it is possible to reduce the particle size to a few hundred ångströms and thus effect a more intimate mixing of the reactants. In spray-drying, suitable constituents dissolved in a solvent are sprayed in the form of fine droplets into a hot chamber. The solvent evaporates instantaneously, leaving behind an intimate mixture of reactants, which on heating at elevated temperatures gives the product. In freeze-drying, reactants in a common solvent are frozen by immersing in liquid nitrogen and the solvent is removed at low pressures. In coprecipitation, the required metal cations taken as soluble salts (*e.g.* nitrates) are coprecipitated from a common medium, usually as hydroxides, carbonates, oxalates, formates or citrates. In actual practice, one takes oxides or carbonates of the relevant metals, digests them with an acid (usually HNO_3) and then the precipitating reagent is added to the solution. The precipitate obtained after drying is heated to the required temperature in a desired atmosphere to produce the final product. For example, tetraethylammonium oxalate has been used to prepare superconducting $YBa_2Cu_4O_8$. The decomposition temperatures of the precipitates are generally lower than the temperatures employed in the ceramic method.

4. Combustion method

Combustion synthesis or self-propagating high-temperature synthesis is a versatile method used for the synthesis of a variety of solids. The method makes use of a highly exothermic reaction between the reactants to produce a flame due to spontaneous combustion which then yields the desired product or its precursor in finely divided form. Borides, carbides, oxides, chalcogenides and other metal derivatives have been prepared by this method and the topic has been reviewed recently by Merzhanov [16]. In order for combustion to occur, one has to ensure that the initial mixture of reactants is highly dispersed and has high chemical energy. For example, one may add a fuel and an oxidizer when preparing oxides by the combustion method, both these additives being removed during combustion to yield only the product or its precursor. Thus, one can take a mixture of nitrates (oxidizer) of the desired metals along with a fuel (*e.g.* hydrazine, glycine or urea) in solution, evaporate the solution to dryness and heat the resulting solid to around 423 K to obtain spontaneous combustion, yielding an oxidic product in fine particulate form. Even if the desired product is not formed immediately after combustion, the fine particulate nature of the product facilitates its formation on further heating.

In order to carry out combustion synthesis, the powdered mixture of reactants (0.1–100 μm particle size) is generally placed in an appropriate gas medium which favours an exothermic reaction on ignition. The combustion temperature is anywhere between 1500 and 3500 K, depending on the reaction. Reaction times are very short since the desired product results soon after the combustion. A gas medium is not always necessary. This is so in the synthesis of borides, silicides and carbides where the elements are quite stable at high temperatures (*e.g.* $Ti + 2B \rightarrow TiB_2$). Combustion in a nitrogen atmosphere yields nitrides. Nitrides of various metals have been prepared in this manner. Azides have been used as sources of nitrogen.

The following are some typical combustion reactions:

$$MoO_3 + 2SiO_2 + 7Mg \rightarrow MoSi_2 + 7MgO$$

$$WO_3 + C + 2Al \rightarrow WC + Al_2O_3$$

$$TiO_2 + B_2O_3 + 5Mg \rightarrow TiB_2 + 5MgO$$

$$Ta \xrightarrow{N_2} Ta_2N \xrightarrow[\text{after burning}]{N_2} TaN$$

Recently, MoS_2 and other refractories have been prepared starting from halides [17].

Use of the combustion method in an atmosphere of air or oxygen to prepare complex metal oxides seems obvious. In the last three to four years, a variety of oxides have been prepared using nitrate mixtures with a fuel such as glycine or urea. It seems that almost any ternary or quaternary oxide can be prepared by this method. All the superconducting cuprates have been prepared by this method, although the resulting products in fine particulate form have to be heated to an appropriate high temperature in a desired atmosphere to obtain the final cuprate [18]. Table 2 lists typical materials prepared by the combustion method.

5. Precursor method

As mentioned earlier, diffusion distances for the reacting cations are rather large in the ceramic method. Diffusion distances are markedly reduced by incorporating the cations in the same solid precursor. Synthesis

TABLE 2. Typical materials prepared by the combustion method

Oxides	$BaTiO_3$, $LiNbO_3$, $PbMoO_4$, $Bi_4Ti_3O_{12}$, $BaFe_{12}O_{19}$, $YBa_2Cu_3O_7$
Carbides	TiC, Mo_2C, NbC
Borides	TiB_2, CrB_2, MoB_2, FeB
Silicides	$MoSi_2$, $TiSi_2$, $ZrSi_2$
Phosphides	NbP, MnP, TiP
Chalcogenides	WS_2, MoS_2, $MoSe_2$, TaS_2, $LaTaS_3$
Hydrides	TiH_2, NdH_2

of complex oxides by the decomposition of compound precursors has been known for some time. For example, thermal decomposition of $LaCo(CN)_6 \cdot 5H_2O$ and $LaFe(CN)_6 \cdot 6H_2O$ in air readily yields $LaCoO_3$ and $LaFeO_3$ respectively. $BaTiO_3$ can be prepared by the thermal decomposition of $Ba[TiO(C_2O_4)_2]$, while $LiCrO_2$ can be prepared from $Li[Cr(C_2O_4)_2(H_2O)_2]$. Ferrite spinels of the general formula MFe_2O_4 ($M \equiv Mg$, Mn, Ni, Co) are prepared by the thermal decomposition of acetate precursors of the type $M_3Fe_6(CH_3COO)_{17}O_3OH \cdot 12C_5H_5N$. Chromites of type MCr_2O_4 are obtained by the decomposition of $(NH_4)_2M(CrO_4)_2 \cdot 6H_2O$.

Carbonates of metals such as calcium, magnesium, manganese, iron, cobalt, zinc and cadmium are all isostructural, possessing the calcite structure. We can therefore prepare a large number of carbonate solid solutions containing two or more cations in different proportions [19, 20]. The rhombohedral a_R of the carbonate solid solutions varies systematically with the weighted mean cation radius (Fig. 1). Carbonate solid solutions are ideal precursors for the synthesis of monoxide solid solutions of rock-salt structure. For example, the carbonates are decomposed in vacuum or in flowing dry nitrogen, to obtain monoxides of the type $Mn_{1-x}M_xO$ ($M \equiv Mg$, Ca, Co or Cd) of rock-salt structure. Oxide solid solutions of $M \equiv Mg$, Ca and Co would require temperatures of 770–970 K for formation, while those containing cadmium are formed at still lower temperatures. The facile formation of rock-salt oxides by the decomposition of carbonates of calcite structure is due to the close (possible topotactic) relationship between the structures of calcite and rock-salt. The monoxide solid solutions can be used as precursors for preparing spinels and other complex oxides.

Besides monoxide solid solutions, a number of ternary and quaternary oxides of novel structures can be prepared by decomposing carbonate precursors containing the different cations in the required proportions. Thus, one can prepare $Ca_2Fe_2O_5$ and $CaFe_2O_4$ by heating the corresponding carbonate solid solutions in air at 1070 and 1270 K respectively for about 1 h. $CaFe_2O_5$ is a defect perovskite with ordered oxide ion vacancies and has the well known brown-millerite structure (Fig. 2) with the Fe^{3+} ions in alternate octahedral (O) and tetrahedral (T) sites. Two new oxides of similar compositions, $Ca_2Co_2O_5$ and $CaCo_2O_4$, have been prepared by decomposing the appropriate carbonate precursors in oxygen atmosphere around 940 K. Unlike $Ca_2Fe_2O_5$, in $Ca_2Mn_2O_5$, anion-vacancy ordering in the perovskite structure gives a square-pyramidal coordination (SP) around the transition metal ion (Fig. 2). One can also synthesize quaternary oxides, Ca_2FeCoO_5, $Ca_2Fe_{1.6}Mn_{0.4}O_5$, $Ca_3Fe_2MnO_8$ etc. belonging to the $A_nB_nO_{3n-1}$ family, by the carbonate precursor route. In the Ca–Fe–O system, there are several other oxides such as $CaFe_4O_7$, $CaFe_{12}O_{19}$ and $CaFe_2O_4(FeO)_n$ ($n = 1$, 2, 3) which can, in principle, be synthesized by starting from the appropriate carbonate solid solutions and decomposing them in a proper atmosphere.

A good example of multistep solid state synthesis achieved starting from carbonate solid solution precursors is provided by the $Ca_2Fe_{2-x}Mn_xO_5$ series of oxides. The structures of both the end members, $Ca_2Fe_2O_5$ and $Ca_2Mn_2O_5$, are derived from that of the perovskite (Fig. 2). Solid solutions between the two oxides would be expected to show oxygen vacancy

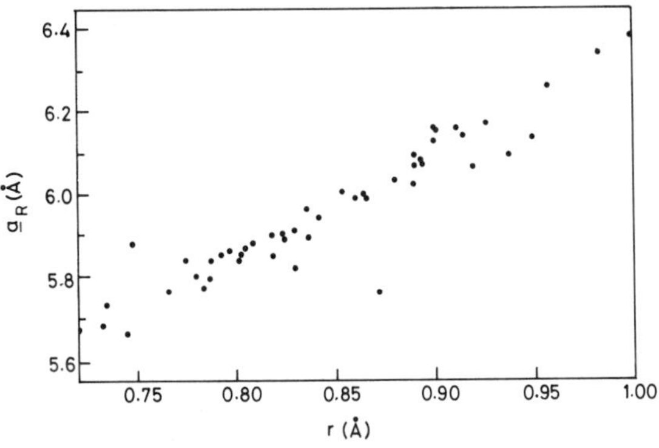

Fig. 1. Plot of the rhombohedral lattice parameters a_R of a variety of binary and ternary carbonates of calcite structure (e.g. Ca–M, Ca–M–M, Mg–M, M–M where M,M $\equiv$ Mn, Fe, Co, Cd, etc.) against the mean cation radius.

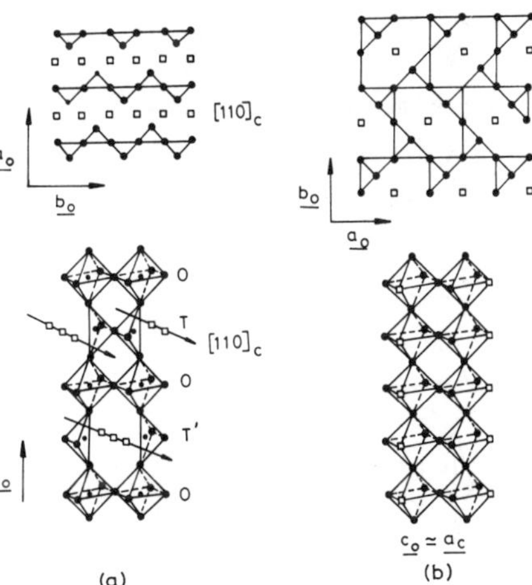

Fig.2. Structures of (a) $Ca_2Fe_2O_5$ (brown-millerite) and (b) $Ca_2Mn_2O_5$. Oxygen vacancy ordering in the a–b plane is also shown.

ordered superstructures with Fe^{3+} in octahedral (O) and tetrahedral (T) coordinations and Mn^{3+} in square-pyramidal (SP) coordination, but they cannot be prepared by the ceramic method. These solid solutions have indeed been prepared starting from the carbonate solid solutions, $Ca_2Fe_{2-x}Mn_x(CO_3)_4$. The carbonates decompose in air at around 1200–1350 K to give perovskite-like oxides, $Ca_2Fe_{2-x}Mn_xO_{6-y}$ ($y<1$). The compositions of the perovskites obtained with $x=2/3$ and 1 are $Ca_3Fe_2MnO_8$ and $Ca_3Fe_{1.5}Mn_{1.5}O_{8.1}$. X-ray and electron diffraction patterns show that they are members of the $A_nB_nO_{3n-1}$ homologous series with anion-vacancy ordered superstructures with $n=3$ ($A_3B_3O_{8+x}$). Careful reduction of $Ca_3Fe_2MnO_8$ in dilute hydrogen gives $Ca_3Fe_{4/3}Mn_{2/3}O_5 = Ca_3Fe_2MnO_{7.5}$ (Fig. 3). During this step, only Mn^{4+} in the parent oxides is topochemically reduced to Mn^{3+}, and Fe^{3+} remains unreduced. The most probable superstructure of $Ca_3Fe_2MnO_{7.5}$ involves SP, O and T polyhedra along the b-direction. On heating in vacuum at 1140 K, however, it transforms to the more stable brown-millerite structure with only O and T coordinations. In Fig. 4 we show typical oxides prepared from precursor solid solutions to illustrate the usefulness of the method.

A number of ternary and quaternary metal oxides of perovskite and related structures can be prepared by employing hydroxide, nitrate and cyanide solid solution precursors as well [20]. For example, hydroxide solid solutions of the general formula $Ln_{1-x}M_x(OH)_3$ where $Ln \equiv La$ or Nd and $M \equiv Al$, Cr, Fe, Co or Ni) and $La_{1-x-y}M'_xM''_y(OH)_3$ where (where $M' \equiv Ni$ and

$M'' \equiv Co$ or Cu) crystallizing in the rare earth trihydroxide structure are decomposed at relatively low temperatures (around 870 K) to yield $LaNiO_3$, $NdNiO_3$, $LaNi_{1-x}Co_xO_3$, $LaNi_{1-x}Cu_xO_3$ etc.

Making use of the fact that anhydrous alkaline earth nitrates $A(NO_3)_2$ ($A \equiv Ca$, Sr, Ba) and $Pb(NO_3)_2$ are isostructural, nitrate solid solutions of the formula $A_{1-x}Pb_x(NO_3)_2$ have been used as precursors for the preparation of ternary oxides such as $BaPbO_3$, Ba_2PbO_4, and Sr_2PbO_4. Quaternary oxides of type $LaFe_{0.5}Co_{0.5}O_3$ and $La_{0.5}Nd_{0.5}CoO_3$ which cannot be readily prepared by the ceramic method have been obtained by the decomposition of $LaFe_{0.5}Co_{0.5}(CN)_6 \cdot 5H_2O$ and $La_{0.5}Nd_{0.5}Co(CN)_6 \cdot 5H_2O$ respectively [20]. A hyponitrite precursor has been used to prepare superconducting $YBa_3Cu_3O_7$ free from $BaCO_3$ impurity [21].

Chevrel compounds of general formula $A_xMo_6S_8$ with $A \equiv Cu$, Pb, La etc. (Fig. 5) are generally prepared by the ceramic method. A novel precursor compound has been employed [22] to obtain these compounds by a one-step reduction as given by the reaction:

$$2A_x(NH_4)_yMo_3S_9 + 10H_2$$
$$\longrightarrow A_{2x}Mo_6S_8 + 10H_2S + 2yNH_3 + yH_2$$

Ammonium thiomolybdate, $(NH_4)_2MoS_4$, was reacted with the metal chloride (AX_n) to obtain the precursor compound.

Precursor solid solutions or compounds can be used to prepare metal alloys. Thus, Mo–W alloys have been prepared [23] by the hydrogen reduction of $(NH_4)_6[Mo_{7-x}W_xO_{24}]$. Metal alloys can be used as precursors to obtain the desired oxides on treatment with oxygen under appropriate conditions. For example, an Eu–Ba–Cu alloy has been oxidized at 1170 K to obtain superconducting $EuBa_2Cu_3O_7$ [24].

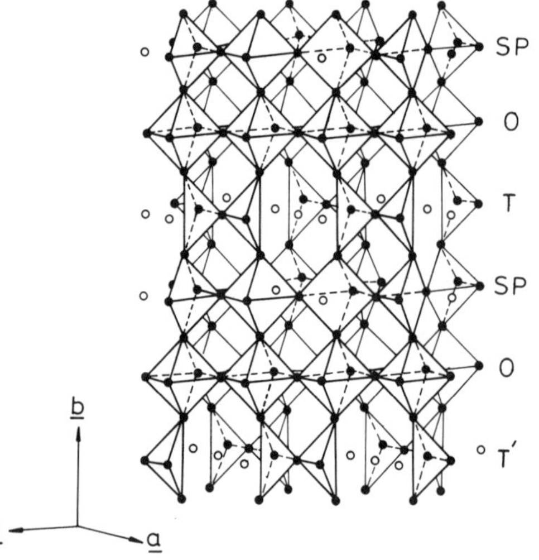

Fig. 3. $Ca_3Fe_2MnO_{7.5}$ obtained by the topotactic reduction of $Ca_3Fe_2MnO_8$. The latter is prepared by decomposition of the precursor carbonate, $Ca_2Fe_{4/3}Mn_{2/3}(CO_3)_4$.

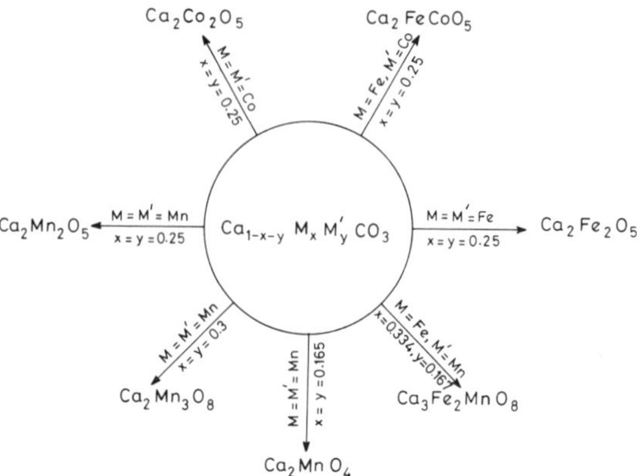

Fig. 4. Some of the complex oxides prepared by the decomposition of carbonate precursors.

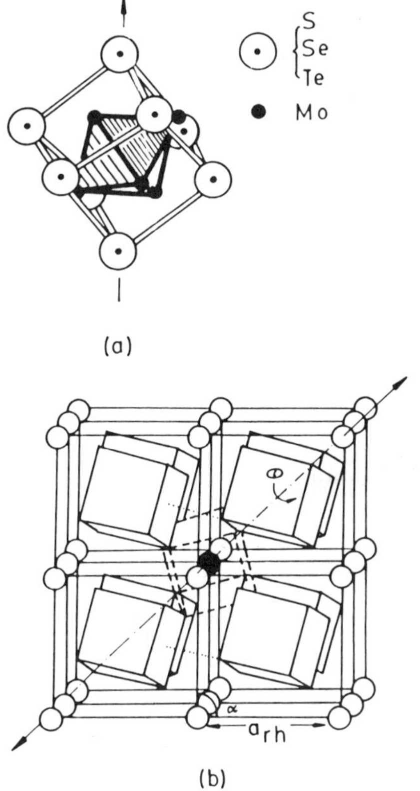

Fig. 5. Structure of Chevrel compounds, $A_xMo_6Ch_8$ (Ch $\equiv$ S, Se, Te): (a) Mo_6Ch_8 building block, (b) structure of Chevrel compounds where the A cation is shown by the closed circle.

Organometallic precursors are used in the synthesis of a variety of semiconducting materials and superconducting cuprates. Organoaluminium silicate precursors have been used for the synthesis of aluminosilicates [25], while polymeric methylsilylamines have been used to obtain SiC–Si_3N_4 fibres [26]. Silicon nitride can also be made using organometallic precursors.

6. Topochemical reactions

A solid state reaction is said to be topochemically controlled when the reactivity is controlled by the crystal structure rather than by the chemical nature of the constituents. The products obtained in many solid state decompositions are determined by topochemical factors, especially when the reaction occurs within the solid without the separation of a new phase [27, 28]. In topotactic solid state reactions, the atomic arrangement in the reactant crystal remains largely unaffected during the course of the reaction, except for changes in dimension in one or more directions. Dehydration of $WO_3·H_2O$ or $MoO_3·H_2O$ to give WO_3 or MoO_3 is one such reaction [29]. Dehydration of many other hydrates such as $VOPO_4·2H_2O$ and $HMoO_2PO_4·H_2O$ is also found to be topotactic. γ-FeOOH transforms to γ-Fe_2O_3 by treatment with an organic base. Intercalation reactions are generally topotactic in nature. Decomposition of WO_3, MoO_3 or TiO_2 to give lower mixed valent oxides (Magneli phases) is accommodated by the collapse of the structure in specific crystallographic directions. Decomposition of V_2O_5 to form V_6O_{13} is a similar reaction. The reduction of NiO to nickel metal proceeds topochemically [30].

6.1. Dehydration of $Mo_{1-x}W_xO_3·H_2O$

Many of the modern developments in solid state chemistry owe much to the investigations carried out on MoO_3 and WO_3, the crystallographic shear planes being a major discovery. WO_3 crystallizes in an ReO_3-like structure, but MoO_3 possesses a layered structure (Fig. 6). MoO_3 can be stabilized in the WO_3 structure by partly substituting tungsten for molybdenum. $Mo_{1-x}W_xO_3$ solid solutions can be prepared by the ceramic method (by heating MoO_3 and WO_3 in sealed tubes at around 870 K) or by the thermal decomposition of mixed ammonium metallates. These methods, however, do not always yield monophasic products owing to the difference in volatilities of MoO_3 and WO_3. We therefore sought to prepare $Mo_{1-x}W_xO_3$ by topochemical dehydration of the hydrates [31], the process being very gentle. $MoO_3·H_2O$ and $WO_3·H_2O$ are isostructural and the solid solutions between the two hydrates are prepared readily by adding a solution of MoO_3 and WO_3 in ammonia to hot 6 M HNO_3. The hydrates $Mo_{1-x}W_xO_3·H_2O$ crystallize in the same structure as $MoO_3·H_2O$ and $WO_3·H_2O$ with a monoclinic unit cell. The hydrate solid solutions undergo dehydration under mild conditions (around 500 K) yielding $Mo_{1-x}W_xO_3$ which crystallizes in the ReO_3-related structure of WO_3. The dehydration of these hydrates has been studied by *in situ* electron diffraction where the decomposition occurs owing to beam heating [31]. Electron diffraction patterns clearly show how $WO_3·H_2O$ transforms to WO_3 topotactically with the required orientational relationships. The mixed hydrates, $Mo_{1-x}W_xO_3·H_2O$, also form under topotactic dehydration with similar orientational relations. What is more interesting is that the dehydration of $MoO_3·H_2O$ under electron beam heating gives MoO_3 in the ReO_3 structure, instead of the expected layered structure [31]. The ReO_3 structure of MoO_3 is metastable and is produced only by topotactic dehydration under mild conditions. We believe that the preparation

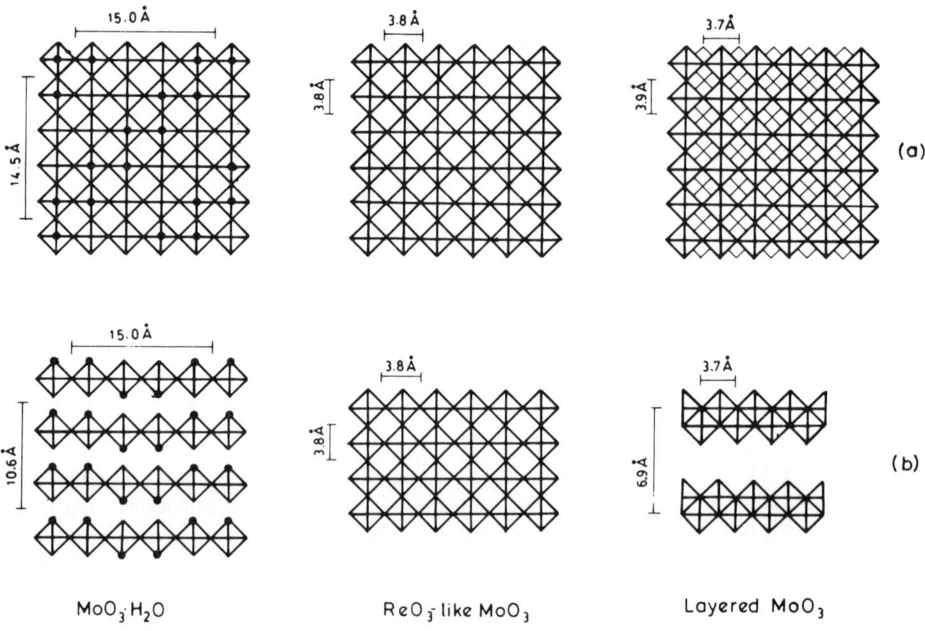

Fig. 6. Schematic representation of $MoO_3 \cdot H_2O$ (or $WO_3 \cdot H_2O$), ReO_3-like MoO_3 (or WO_3) and the layered structure of MoO_3.

of ReO_3-like MoO_3 by mild chemical processing is significant. Bulk quantities of MoO_3 in the ReO_3 structure have been prepared by mild dehydration of the hydrate [32].

6.2. Reduction of perovskite oxides

Reduction of ABO_3 perovskites to give $A_2B_2O_5$ and such defect oxides is found to be topochemical in many instances (*e.g.* $CaMnO_3$). We shall examine the reduction of $LaNiO_3$ and $LaCoO_3$ which crystallize in the rhombohedral perovskite structure. The occurrence of the $La_nNi_nO_{3n-1}$ homologous series was proposed some time ago [33] on the basis of a thermogravimetric study of the decomposition of $LaNiO_3$. However, it was not known whether a similar series exists in the case of $LaCoO_3$. Controlled reduction of $LaNiO_3$ and $LaCoO_3$ in hydrogen shows the formation of $La_2Ni_2O_5$ and $La_2Co_2O_5$ representing the $n=2$ members of the homologous series LaB_nO_{3n-1} ($B \equiv Co$ or Ni) [34]. $La_2Ni_2O_5$ can be prepared by the reduction of $LaNiO_3$ at 600 K in pure or dilute hydrogen, but $La_2Co_2O_5$ can only be prepared by the reduction of $LaCoO_3$ in dilute hydrogen at 670 K. Both the oxides can be oxidized back to the parent perovskites at low temperatures. Neither $La_2Ni_2O_5$ nor $La_2Co_2O_5$ can be prepared by the solid state reaction of La_2O_3 and the transition metal oxide. X-ray and electron diffraction data reveal that $La_2Co_2O_5$ adopts the brown-millerite structure. The X-ray diffraction pattern of $La_2Ni_2O_5$ is different

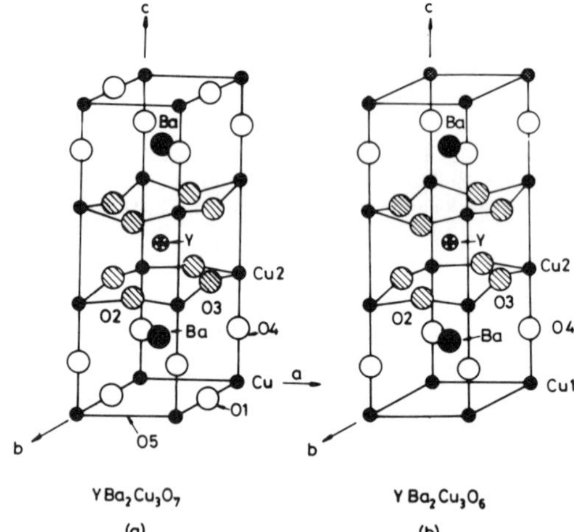

Fig. 7. Structures of $YBa_2Cu_3O_7$ and $YBa_2Cu_3O_6$.

from that of $La_2Co_2O_5$, but could be indexed on a tetragonal cell related to cubic perovskite. The formation of $La_2Co_2O_5$ and $La_2Ni_2O_5$ by the reduction of $LaCoO_3$ and $LaNiO_3$ respectively is certainly topochemical.

It should be noted that reduction of the high-temperature superconductor $YBa_2Cu_3O_7$ to $YBa_2Cu_3O_6$ (Fig. 7) is a topochemical process.

7. Intercalation compounds

Intercalation reactions of solids involve the insertion of a guest species (ion or molecule) into a solid host lattice without any major rearrangement of the solid structure:

$$x(\text{guest}) + \square_x[\text{host}] \rightleftharpoons (\text{guest})_x[\text{host}]$$

where $\square$ stands for a vacant lattice site. Redox intercalation reactions (*e.g.* Li_xTiS_2 where the lithium metal reduces the TiS_2 layers) can be written as

$$x(\text{guest})_x^+ + xe^- + \square_x[\text{host}] \rightleftharpoons (\text{guest})_x[\text{host}]$$

A variety of layered structures act as hosts. The general feature of these structures is that the interlayer interactions are weak while the intralayer bonding is strong. Intercalation compounds show interesting phase relations, staging (Fig. 8) being an important feature in some of them. Higher stages correspond to lower guest concentrations. Intercalation reactions have been reviewed by many authors [2, 35, 36] and we will present the essential features with a few typical examples.

Alkali metal intercalation involving a redox reaction is readily carried out electrochemically by using the host (MCh_2 dichalcogenide) as the cathode, the alkali metal as the anode and the non-aqueous solution of the alkali metal salt as the electrolyte:

$$\text{Na/NaI–propylenecarbonate/MCh}_2 \qquad (\text{Ch} \equiv \text{S, Se, Te})$$

$$\text{Li/LiClO}_4\text{–dioxolane/MCh}_2$$

The reaction is spontaneous if a reverse potential is applied or the cell is short circuited. Low alkali metal concentrations are obtained by using solutions of salts such as sodium or potassium naphthalide in THF or *n*-butyllithium in hexane:

$$x\text{C}_4\text{H}_9\text{Li} + \text{TiS}_2 \longrightarrow \text{Li}_x\text{TiS}_2 + \frac{x}{2}\,\text{C}_8\text{H}_{18}$$

Alkali metal intercalation in dichalcogenides is also achieved by direct reaction of the elements at around 1070 K (*e.g.* A_xMCh_2 where $\text{M} \equiv \text{V, Nb or Ta}$) in sealed tubes. Alkali metal intercalation compounds with dichalcogenides form hydrated phases, $\text{A}_x(\text{H}_2\text{O})_y\text{-MCh}_2$, just like some of the layered oxides (*e.g.*

$\text{VOPO}_4\cdot 2\text{H}_2\text{O}$, $\text{MoO}_3\cdot 2\text{H}_2\text{O}$). NH_3 is intercalated to dichalcogenides by direct reaction (by distilling liquid NH_3 into the dichalcogenide). Intercalation of organic compounds in dichalcogenides is carried out by thermal reaction at temperatures up to 470 K.

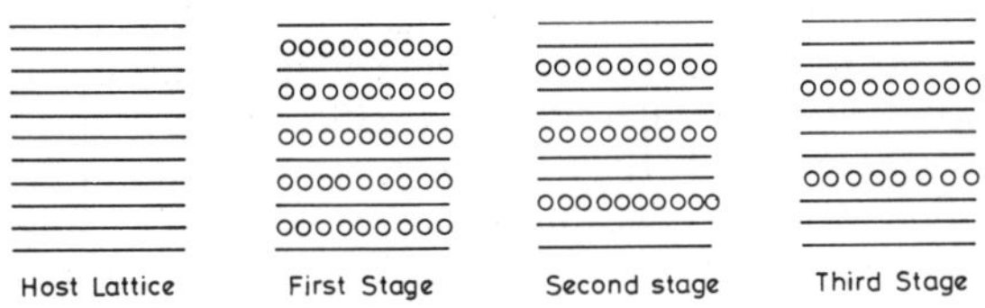

Metal phosphorus trisulphides undergo redox intercalation reactions just like the dichalcogenides and also ion exchange reactions, giving $(\text{guest})_x^+\text{-}[\text{M}_{1-x/2}\square_{x/2}\text{PS}_3]$. Metal oxyhalides (*e.g.* FeOCl) show intercalation reactions similar to the dichalcogenides. In addition, they undergo irreversible substitution reactions where the halogens of the top layer are substituted by other groups such as NHCH_3 and CH_3. Layered transition metal oxides such as MoO_3, V_2O_5, MOPO_4 and MoAsO_4 ($\text{M} \equiv \text{V, Nb, Ta}$) show reduction reactions similar to the dichalcogenides. Layered oxides of type AMO_2 and HTiNbO_5 and $\text{H}_2\text{Ti}_4\text{O}_9$ undergo ion-exchange and oxidative deintercalation reactions.

Ready deintercalation of lithium from LiMO_2 enables these materials to be used as cathodes in lithium cells. Delithiation occurs not only by electrochemical methods, but also by reaction with I_2 or Br_2 in solution phase. Lithium insertion in close-packed oxides such as TiO_2, Fe_2O_3 and Mn_3O_4 results in interesting structural changes. Delithiation gives rise to oxides in unusual metastable structures (*e.g.* VO_2 obtained from delithiation of LiVO_2). Delithiation of LiVS_2 gives VS_2 which cannot otherwise be prepared. Table 3 lists the important hosts and guests in intercalation compounds. In Table 4, we list lithium intercalated compounds to show the variety in this system.

Intercalation of sodium and potassium differs from that of lithium. In layered A_xMX_n, lithium is always octahedrally coordinated, while sodium and potassium occupy octahedral or trigonal prismatic sites; octahedral coordination is favoured by large values of x

Host Lattice   First Stage   Second stage   Third Stage

Fig. 8. Schematic diagram of staging in intercalation compounds. Guest molecules are represented by circles in between the layers (shown by lines).

TABLE 3. Examples of hosts and guests in intercalation compounds

Hosts	Guests
Neutral layers	
Graphite	FeCl_3, K, Br_2
MCh_2 ($\text{M} \equiv \text{Ti, Zr, Nb, Ta, etc.}$ $\text{Ch} \equiv \text{S, Se, Te}$)	Li, Na, NH_3, organic amines, CoCP_2
MPCh_3 ($\text{M} \equiv \text{Mg, V, Fe, Zn, etc.}$)	Li, CoCp_2
MoO_3, V_2O_5	H, alkali metal
MOPO_4 and MOAsO_4 ($\text{M} \equiv \text{V, Nb, Ta}$)	H_2O, pyridine, Li
MOCl and MOBr ($\text{M} \equiv \text{V, Fe, etc.}$)	Li, FeCp_2
Negatively charged layers	
$(\text{A})\text{MX}_2$ ($\text{M} \equiv \text{Ti, V, Cr, Fe, X} \equiv \text{O, S}$)	A=Group IA (Li, Na...
Layered silicates and clays	Organic compounds
$\text{M(HPO}_4)_2$ ($\text{M} \equiv \text{Ti, Zr...}$)	
$\text{K}_2\text{Ti}_4\text{O}_9$	

TABLE 4. Intercalation compounds of lithium

Host	Description	Reference
TiS_2	Li_xTiS_2, $0 < x \leqslant 1$	37
VS_2	Li_xVS_2, $0 < x \leqslant 1$. Phases obtained by deintercalation of Li from $LiVS_2$ using I_2–CH_3CN. Three different phase regions: $0.25 \leqslant x \leqslant 0.33$, $0.48 \leqslant x \leqslant 0.62$ and $0.85 \leqslant x \leqslant 1$ apart from VS_2	38
$NbS_2(3R)$	$Li_{0.5}NbS_2$ and $Li_{0.70}NbS_2$	37
MoS_3	Li_xMoS_3, $0 < x \leqslant 4$	39
MO_2	Li_xMO_2, $x \geqslant 1$ (M ≡ Mo, Ru, Os or Ir), MO_2 of rutile structure	40
TiO_2 (anatase)	Li_xTiO_2, $0 < x \leqslant 0.7$. $Li_{0.5}TiO_2$ transforms irreversibly to $LiTi_2O_4$ spinel at 770 K	41
CoO_2	Li_xCoO_2, $0 < x < 1$. Phases obtained by electrochemical delithiation of $LiCoO_2$	42
VO_2	(a) Li_xVO_2, $0 < x < 1$. Phases obtained by delithiation of $LiVO_2$ using Br_2–$CHCl_3$	43
	(b) Li_xVO_2, $0 < x < 2/3$. Lithiation using *n*-butyl lithium	44
Fe_2O_3	$Li_xFe_2O_3$, $0 < x < 2$. Anion array transforms from h.c.p. to c.c.p. on lithiation	45
Fe_3O_4	$Li_xFe_3O_4$, $0 < x < 2$. Fe_2O_4 subarray of the spinel structure remains intact	45
Mn_3O_4	$Li_xMn_3O_4$, $0 < x < 1.2$. Li insertion suppresses tetragonal distortion of Mn_3O_4	45
MoO_3	Li_xMoO_3, $0 < x < 1.55$	46
V_2O_5	$Li_xV_2O_5$, $0 < x < 1.1$. Intercalation of Li using LiI	47
ReO_3	Li_xReO_3, $0 < x < 2$. Three phases $0 < x \leqslant 0.35$, $x = 1$ and $1.8 \leqslant x \leqslant 2$	48

and low formal oxidation states of M. For smaller x and higher oxidation states of M, the coordination of sodium and potassium is trigonal prismatic. Intercalated caesium in MX_n is always trigonal prismatic. Intercalation of sodium and potassium in layered MX_2 oxides and sulphides results in structural transformations involving a change in the sequence of anion layer stacking.

Tungsten and molybdenum bronzes, A_xWO_3 and A_xMoO_3 (A ≡ K, Rb, Cs) are generally prepared by reaction of the alkali metals with the host oxide. Electrochemical methods are also employed for these preparations. A novel reaction that has been employed to prepare bronzes which are otherwise difficult to obtain involves the reaction of oxide host with anhydrous alkali iodides [49]:

$$Mo_{1-x}W_xO_3 + y(AI) \longrightarrow A_yMo_{1-x}W_xO_3 + \frac{y}{2}I_2$$

Atomic hydrogen has been inserted into many binary and ternary oxides. Recently, iodine has been intercalated into the superconducting cuprate, $Bi_2CaSr_2Cu_2O_8$, causing an expansion of the c-parameter of the unit cell, without destroying the superconductivity [50].

Chevrel phase compounds, $A_xMo_6Ch_8$ (Ch ≡ S, Se or Te), may also be considered as intercalation compounds. Mo_6S_8 can be prepared by acid-leaching copper from $Cu_xMo_6S_8$.

8. Ion exchange method

Ion-exchange in fast-ion conductors such as β-alumina is well known. The exchange can be carried out both in aqueous and molten salt conditions. Thus, sodium β-alumina has been exchanged with H_3O^+, NH_4^+ and other monovalent and divalent cations, giving rise to different β-aluminas [51]. Ion-exchange in inorganic solids is a general phenomenon, not restricted to fast ion conductors alone. The kinetic and thermodynamic aspects of ion-exchange in inorganic solids were examined by England *et al.* [52]. Their results reveal that ion-exchange is a phenomenon that occurs even when the diffusion coefficients are as small as approximately 10^{-12} cm^2 s^{-1}, at temperatures far below the sintering temperatures of solids. Ion-exchange occurs at a considerable rate in stoichiometric solids as well. Mobile ion vacancies introduced by non-stoichiometry or doping seem to be unnecessary for exchange to occur. Since the exchange occurs topochemically, it enables the preparation of metastable phases that are inaccessible by high-temperature reactions.

England *et al.* [52] have shown that a variety of metal oxides having layered, tunnel or close-packed structures can be ion exchanged in aqueous solutions or molten salt media to produce new phases. Typical examples are

$$\alpha\text{-NaCrO}_2 \xrightarrow[570\,\text{K, 24 h}]{\text{LiNO}_3} \alpha\text{-LiCrO}_2$$

$$\text{KAlO}_2 \xrightarrow{\text{AgNO}_3(\text{l})} \beta\text{-AgAlO}_2$$

$$\alpha\text{-LiFeO}_2 \xrightarrow{\text{CuCl}(\text{l})} \text{CuFeO}_2$$

The structure of the framework is largely retained during ion-exchange except for minor changes to accommodate the structural preferences of the incoming ion. Thus, when α-LiFeO$_2$ is converted to CuFeO$_2$ by exchange with molten CuCl, the structure changes from that of α-NaCrO$_2$ to that of delafossite to provide a linear anion coordination for Cu$^+$. Similarly, when KAlO$_2$ is converted to β-AgAlO$_2$ by ion-exchange, there is a change in structure from cristobalite type to ordered wurtzite type. The change probably occurs to provide a tetrahedral coordination for Ag$^+$.

An interesting ion-exchange reaction is the conversion of LiNbO$_3$ and LiTaO$_3$ to HNbO$_3$ and HTaO$_3$ respectively, by treatment with hot aqueous acid [53]. The exchange of Li$^+$ by protons is accompanied by a topotactic transformation of the rhombohedral LiNbO$_3$ structure to the cubic perovskite structure of HNbO$_3$. The mechanism suggested for the transformation is the reverse of the transformation of cubic ReO$_3$ to rhombohedral LiReO$_3$ and Li$_2$ReO$_3$ [50], involving a twisting of the octahedra along the [111] cubic direction so as to convert the 12-coordinated perovskite tunnel sites to two 6-coordinated sites in the rhombohedral structure. An interesting structural change accompanying ion-exchange is found in Na$_{0.7}$CoO$_2$ [54] where the anion layer sequence is ABBAA; cobalt ions occur in alternate interlayer octahedral sites and sodium ions in trigonal prismatic coordination in between the CoO$_2$ units. When this material is ion exchanged with LiCl, a metastable form of LiCoO$_2$ with the layer sequence ABCBA is obtained. The phase transforms irreversibly to the stable LiCoO$_2$ (ABCABC) at around 520 K.

A variety of inorganic solids has been exchanged with protons to give new phases, some of which exhibit high protonic conduction [2]. Typical of these are HTaWO$_6\cdot$H$_2$O, HMO$_3\cdot x$H$_2$O (M $\equiv$ Sb, Nb, Ta), pyrochlores and HTiNbO$_5$ [2]. Ion-exchange has also been reported in metal sulphides. For example, KFeS$_2$ undergoes topochemical exchange of potassium in aqueous solutions with alkaline earth metal cations to give new phases in which the [FeS$_{4/2}$] tetrahedral chain is preserved [55].

9. Sol-gel method

The sol-gel method has emerged to become an important means of preparing inorganic oxides in recent years. It is a wet chemical method and a multistep process involving both chemical and physical processes such as hydrolysis, polymerization, drying and densification. The name "sol-gel" is given to the process because of the distinctive increase in viscosity which occurs at a particular point in the sequence of steps. A sudden increase in viscosity is the common feature in sol-gel processing, indicating the onset of gel formation. In the sol-gel process, synthesis of inorganic oxides is achieved from inorganic or organometallic precursors (generally metal alkoxides). Most of the sol-gel literature deals with synthesis from alkoxides. The important features of the sol-gel techniques are better homogeneity compared with the traditional ceramic method, high purity, lower processing temperature, more uniform phase distribution in multicomponent systems, better size and morphological control, the possibility of preparing new crystalline and non-crystalline materials, and lastly easy preparation of thin films and coatings. The sol-gel method is widely used in ceramic technology and the subject has been adequately reviewed [56, 57]. We shall deal with the topic briefly for the purpose of completeness. The important steps in sol-gel synthesis are as follows.

Hydrolysis. The process of hydrolysis may start with a mixture of a metal alkoxide and water in a solvent (usually alcohol) at the ambient or a slightly elevated temperature. Acid or base catalysts are added to speed up the reaction.

Polymerization. This step involves condensation of adjacent molecules wherein H$_2$O and ROH are eliminated and metal oxide linkages are formed. Polymeric networks grow to colloidal dimensions in the liquid (sol) state.

Gelation. In this step, the polymeric networks link up to form a three-dimensional network throughout the liquid. The system becomes somewhat rigid, characteristic of a gel. The solvent as well as water and alcohol remain inside the pores of the gel. Aggregation of smaller polymeric units to the main network continues progressively on aging the gel.

Drying. Here, water and alcohol are removed at a moderate temperature (less than 470 K), leaving a hydroxylated metal oxide with residual organic content. If the objective is to prepare a high surface area of aerogel powder with low bulk density, the solvent is removed supercritically.

Dehydration. This step is carried out between 670 and 1070 K to drive off the organic residues and chemically bound water, yielding a glassy metal oxide with up to 20%–30% microporosity.

Densification. Temperatures in excess of 1270 K are used to form the dense oxide product.

The various steps in the sol-gel technique described above may or may not be strictly followed in practice. Thus, many complex metal oxides are prepared by a modified sol-gel route without actually preparing metal alkoxides. For example, a transition metal salt solution is converted into a gel by the addition of an appropriate organic reagent. In the case of cuprate superconductors, an equimolar proportion of citric acid is added to the solution of metal nitrates, followed by ethylenediamine until the solution attains a pH of 6–6.5. The blue sol is concentrated to obtain the gel. The xerogel is obtained by heating at approximately 420 K. The xerogel is decomposed at an appropriate temperature to obtain the cuprate.

The sol-gel technique has been used to prepare submicrometre metal oxide powders [58] with a narrow particle size distribution and unique particle shapes (*e.g.* Al_2O_3, TiO_2, ZrO_2, Fe_2O_3). Uniform SiO_2 spheres have been grown from aqueous solutions of colloidal SiO_2 [59]. Small metal clusters (*e.g.* nickel, copper, gold) have been prepared by *in situ* chemical reduction of metal salts [60]. Metal–ceramic composites (*e.g.* $Ni–Al_2O_3$, $Pt–ZrO_2$) can also be prepared in this manner. By employing several variants of the basic sol-gel technique, a number of multicomponent oxide systems have been prepared. Typical of these are, $SiO_2–B_2O_3$, $SiO_2–TiO_2$, $SiO–ZrO_2$, $SiO_2–Al_2O_3$, $ThO_2–UO_2$ [56, 57]. A variety of ternary and still more complex oxides such as $PbTiO_3$, $PbTi_{1-x}Zr_xO_3$ and NASICON have been prepared by this technique [56, 57]. In the last five years, different types of cuprate superconductors have been prepared by this method [61]. These include $YBa_2Cu_3O_7$, $YBa_2Cu_4O_8$, $Bi_2CaSr_2Cu_2O_8$ and $Pb_2Sr_2Ca_{1-x}Y_xCu_3O_8$.

10. Alkali flux method

The use of strong alkaline media, either in the form of solid fluxes or molten (or aqueous) solutions, has enabled the synthesis of novel oxides. The alkali flux method stabilizes higher oxidation states of the metal by providing an oxidizing atmosphere. Alkali car-

bonate fluxes have traditionally been used to prepare transition metal oxides such as $LaNiO_3$. A good example of an oxide synthesized in a strongly alkaline medium is the pyrochlore, $Pb_2(Ru_{2-x}Pb_x)O_{7-y}$ where Pb is in the 4 + state [62]; this oxide is a bifunctional electrocatalyst. The procedure for preparation involves bubbling oxygen through a solution of lead and rubidium salts in strong KOH at 320 K. The so-called alkaline hypochlorite method is used in many instances. For example, La_4NiO_{10} was prepared by bubbling Cl_2 gas through a NaOH solution of lanthanum and nickel nitrates.

Superconducting $La_2CuO_{4+\delta}$ has been prepared by reacting a mixture of La_2O_3 and CuO in molten KOH–NaOH around 520 K [63]. $YBa_2Cu_4O_8$ has been prepared by using a $Na_2CO_3–K_2CO_3$ flux in a flowing oxygen atmosphere [64]. KOH melt has been used to prepare superconducting $Ba_{1-x}K_xBiO_3$ [65].

11. Electrochemical method

Electrochemical methods have been employed to advantage for the synthesis of many solid materials [2, 3, 66, 67]. Typical of the materials prepared in this manner are metal borides, carbides, silicides, oxides and sulphides. Vanadate spinels of formula MV_2O_4 as well as tungsten bronzes A_xWO_3, have been prepared by the electrochemical route. Tungsten bronzes are obtained at the cathode when current is passed through two inert electrodes immersed in a molten solution of the alkali metal tungstate, A_2WO_4 and WO_3; oxygen is liberated at the anode [68]. Blue molybdenum bronzes have been prepared by fused salt electrolysis [69]. Monosulphides of uranium, gadolinium, thorium and other metals are obtained from a solution of the normal valent metal sulphide and chloride in an NaCl–KCl eutectic. LaB_6 is prepared by taking La_2O_3 and B_2O_3 in an $LiBO_2–LiF$ melt and using gold electrodes. Crystalline transition metal phosphides are prepared from solutions of oxides with alkali metal phosphates and halides.

As mentioned earlier, intercalation of alkali metals in host solids in readily accomplished electrochemically. It is easy to see how both intercalation (reduction of the host) and deintercalation (oxidation of the host) are processes suited for this method. Thus, lithium intercalation is carried out using a lithium anode and a lithium salt in a non-aqueous solvent

$$MS_2(s) + x\,Li^+ + xe^- \rightleftharpoons Li_xMS_2(s)$$

Superconducting $Ba_{1-x}K_xBiO_3$ has been prepared electrochemically.

Although the electrochemical method is old, the processes involved in the synthesis of various solids are not entirely understood. Generally one uses solvents whose decomposition potentials are high (*e.g.* alkali metal phosphates, borates, fluorides, *etc.*). Changes in melt composition could cause limitations in certain instances. There is considerable scope for investigating the chemistry in the electrochemical synthesis of solids.

12. High pressure methods

The use of high pressures for solid state synthesis has become increasingly common in recent years. With the development of high-pressure technology, commercial equipment permitting simultaneous use of both high-pressure and high-temperature conditions has become available. Reviews of various experimental aspects in this area have been published in the literature [2, 3, 70–72]. For the 1–10 kbar pressure range, the hydrothermal method is often employed. In this method, the reaction is carried out either in an open or a closed system. In the open system, the solid is in direct contact with the reacting gas (F_2, O_2 or N_2) which also serves as a pressure intensifier. A gold container is generally used in this type of synthesis. This method has been used for the synthesis of transition metal compounds such as RhO_2, PtO_2 and Na_2NiF_6 where the transition metal is in a high oxidation state. Hydrothermal high pressure synthesis under closed system conditions has been employed for the preparation of higher-valence metal oxides. An internal oxidant such as $KClO_3$ is added to the reactants, which on decomposition under reaction conditions provides the necessary oxygen pressure. For example, pyrochlores of palladium(IV) and platinum(IV), $Ln_2M_2O_7$, (Ln = rare earth) have been prepared by this method (970 K, 3 kbar). $(H_3O)Zr_2(PO_4)_3$ and a family of zero thermal expansion ceramics (*e.g.* $Ca_{0.5}Ti_2P_3O_{12}$) have also been prepared hydrothermally [73, 74]. Another good example is the synthesis of borates of aluminium, yttrium and such metals wherein the sesquioxides are reacted with boric acid [72]. Oxyfluorides have been prepared in HF medium [75].

Zeolites are generally prepared under hydrothermal conditions in the presence of alkali [13, 76, 77]. The alkali, the silica component and the source of aluminium are mixed in appropriate proportions and heated. The reactant mixture forms a hydrous gel which is then allowed to crystallize under pressure for several hours to several weeks between 330 and 470 K. In a typical synthesis, $Al_2O_3 \cdot 3H_2O$ dissolved in concentrated NaOH solution (20 N) is mixed with a 1 N solution of $Na_2SiO_3 \cdot 9H_2O$ to obtain a gel (of composition $2.1Na_2O \cdot Al_2O_3 \cdot 2.1SiO_2 \cdot 60H_2O$) which is then crystallized to give zeolite A. The Na_2O–SiO_2–Al_2O_3–H_2O system yields a large number of materials with the zeol-

itic framework. Under alkaline conditions, aluminium is present as $Al(OH)_4$ anions. The OH^- ions act as a mineralizing catalyst while the cations present in the reactant mixture determine the kinds of zeolite formed. Besides water, some inorganic salts are also encapsulated in some zeolites. Several zeolite structures are found in the K_2O–SiO_2–Al_2O_3–H_2O system as well. Li_2O, however, does not give rise to many microporous materials. Group IIA cations yield several zeolitic products.

Zeolitization in the presence of organic bases is useful for synthesizing silica-rich zeolites. Silicalite with a tetrahedral framework enclosing a three-dimensional system of channels (defined by 10 rings wide enough to absorb molecules up to 0.6 nm in diameter) has been synthesized by the reaction of tetrapropylammonium (TPA) hydroxide and a reactive form of silica between 370 and 470 K. The precursor crystals have the composition $(TPA)_2 0.48SiO_2 \cdot H_2O$ and the organic cation is removed by chemical reaction or thermal decomposition to yield microporous silicalite which may be considered to be a new polymorph of SiO_2 [77]. The clathrasil (silica analogue of a gas hydrate), dodecasil-1H, is prepared from an aqueous solution of tetramethoxysilane and $N(CH_3)_4OH$; after the addition of aminoadamantane, the solution is treated hydrothermally under nitrogen for four days at 470 K [78].

The use of template cations has enabled the synthesis of a variety of zeolite materials. Cations such as $(NMe_4)^+$ fit snugly into the cages (*e.g.* sodalite cages of sodalite and SAPO or gmelinite cages of zeolites omega). Neutral organic amines have also been used (*e.g.* in the synthesis of ZSM-5). Many new microporous materials, including those based on $AlPO_4$ (analogue of SiO_2), gallosilicates and aluminogerminates (analogues of aluminosilicates), have been prepared. $AlPO_4$-based materials are prepared by the crystallization of gels formed by adding an organic template to a mixture of active alumina, H_3PO_4 and water at a pH of 5–8 around 470 K.

Pressures in the range 10–150 kbar are commonly used for solid-state synthesis. In the piston-cylinder apparatus consisting of a tungsten carbide chamber and a piston assembly, the sample is contained in a suitable metal capsule surrounded by a pressure-transducer (pyrophyllite). Pressure is generated by moving the piston through the blind hole in the cylinder. A microfurnace made of graphite or molybdenum is incorporated in the design. Pressures up to 50 kbar and temperatures up to 1800 K are readily reached in a volume of 0.1 cm^3 using this design. In the anvil apparatus, first designed by Bridgman, the sample is subjected to pressure by simply squeezing it between two opposed anvils. Although pressures of around 200 kbar and temperatures up to 1300 K are reached in

                    C. N. R. Rao / Synthesis of solid inorganic materials

this technique, it is not popular for solid-state synthesis since only milligram quantities can be handled. An extension of the opposed anvil principle is the tetrahedral anvil design, where four massively supported anvils disposed tetrahedrally ram towards the centre where the sample is located in a pyrophyllite medium together with a heating arrangement. The multi-anvil design has been extended to cubic geometry, where six anvils act on the faces of a pyrophyllite cube located at the centre.

The belt apparatus provides the best high-pressure–high-temperature combination for solid-state synthesis. This apparatus, which was used for the synthesis of diamonds some years ago is a combination of the piston-cylinder and the opposed anvil designs. The apparatus consists of two conical pistons made of tungsten carbide, which ram through a specially shaped chamber from opposite directions. The chamber and pistons are laterally supported by several steel rings making it possible routinely to reach fairly high pressures (around 150 kbar) and high temperatures (approximately 2300 K). In the belt apparatus, the sample is contained in a noble metal capsule (a BN or MgO container is used for chalcogenides) and surrounded by pyrophyllite and a graphite sleeve, the latter serving as an internal heater. In a typical high-pressure run, the sample is loaded, the pressure raised to the desired value and then the temperature increased. After holding the pressure for about 30 min, the sample is quenched (400 K s^{-1}) while still under pressure. The pressure is released after the sample has cooled at room temperature.

High-pressure methods have been used for the synthesis of several materials that cannot possibly be made otherwise. In general, the formation of a new compound from its components requires that the new composition have a lower free energy than the sum of the free energies of the components. Pressure can aid in the lowering of free energy in different ways [70]. (a) Pressure delocalizes outer d electrons in transition-metal compounds by increasing the magnitude of coupling between the d electrons on neighbouring cations, thereby lowering the free energy. A typical example is the synthesis of $ACrO_3$ (A≡ Ca, Sr, Pb) perovskites and CrO_2. (b) Pressure stabilizes higher-valence states of transition metals, thus promoting the formation of a new phase. For example, in the Ca–Fe–O system only $CaFeO_{2.5}$ (brown-millerite) is stable under ambient pressures. Under high oxygen pressures, iron is oxidized to the 4^+ state and hence $CaFeO_3$ with the perovskite structure is formed. (c) Pressure can suppress the ferroelectric displacement of cations, thereby adding the synthesis of new phases. The synthesis of A_xMoO_3 bronzes, for example, requires populating the empty d orbitals centred on

molybdenum; at ambient pressures, MoO_3 is stabilized by a ferroelectric distortion of MoO_6 octahedra up to the melting point. (d) Pressure alters site-preference energies of cations, and facilitates the formation of new phases. For example, it is not possible to synthesize $A^{2+}Mn^{4+}O_3$ (A≡ Mg, Co, Zn) ilmenites because of the strong tetrahedral site preference of the divalent cations. One therefore obtains a mixture of $A[AMn]O_4$(spinel) + MnO_2(rutile) under atmospheric pressure instead of monophasic $AMnO_3$. However, the latter is formed at high pressures with a corundum-type structure in which both the A and Mn ions are in octahedral coordination. (e) Pressure can suppress the $6s^2$ core polarization in oxides containing isoelectronic Tl^+, Pb^{2+}, Bi^{3+} cations. For example, perovskite-type $PbSnO_3$ cannot be made at atmospheric pressure because the mixture of $PbO + SnO_2$ is more stable than the perovskite.

Stabilization of unusual oxidation states and spin states of transition metals is of considerable interest (*e.g.* $La_2Pd_2O_7$). Such stabilization can be rationalized by making use of correlations of structural factors with the electronic configuration. Six-coordinated high-spin iron(IV) has been stabilized in $La_{1.5}Sr_{0.5}Li_{0.5}Fe_{0.5}O_4$ which has the K_2NiF_4 structure [79]. The elongated FeO_6 octahedra and the presence of ionic Li–O bonds resulting from the K_2NiF_4 structure favour the high-spin iron(IV) state. The lithium and iron ions in this oxide are ordered in the a–b-plane, as evidenced by the supercell spots in the electron diffraction pattern. Such an oxide is prepared under oxidizing conditions. $CaFeO_3$ and $SrFeO_3$ prepared under oxygen pressure also contain octahedral iron(IV). While iron(IV) in $SrFeO_3$ is in the high-spin state with the e_g electron in the narrow σ^* band down to 4 K, iron(IV) in $CaFeO_3$ disproportionates to iron(III) and iron(V)below 290 K [80]. La_2LiFeO_6 prepared under high oxygen pressure has the perovskite structure with the iron in the pentavalent state [79].

Nickel in the $3+$ state is present in the perovskite $LaNiO_3$ which can be prepared at atmospheric pressure; other rare earth nickelates have been prepared at high oxygen pressures. Recently $NdNiO_3$ has been prepared by the sol-gel and other chemical routes [20, 81]. $MNiO_{3-x}$ (M ≡ Ba or Sr) prepared under high pressure contains nickel(IV) [82]. In $La_2Li_{0.5}Co_{0.5}O_4$, there is evidence for the transformation of the low-spin cobalt(III) to the intermediate-spin as well as high-spin states. The lithium and cobalt ions are ordered in the a–b-plane of this oxide of K_2NiF_4 structure; the highly elongated CoO_6 octahedra seem to stabilize the intermediate-spin state. Oxides in perovskite and K_2NiF_4 structures with trivalent copper have been prepared under high oxygen pressure [79]. High F_2 pressure has been employed to prepare Cs_2NiF_6 and other fluorides

[83]. Monel autoclaves have been used in such reactions of F_2.

As noted earlier, solid-state reactions are generally slow under ordinary pressures even when the product is thermodynamically stable. Pressure has a marked effect on the kinetics of the reaction, reducing the reaction times considerably, and at the same time giving more homogeneous and crystalline products. For instance, $LnFeO_3$, $LnRhO_3$ and $LnNiO_3$ ($Ln = $ rare earth) are prepared in a matter of hours under high-pressure–high-temperatue conditions, whereas at ambient pressure the reactions require several days ($LnFeO_3$ and $LnRhO_3$) or they do not occur at all ($LnNiO_3$). Thus $LnFeO_3$ is formed in 30 min at 50 kbar [84]. In several $(AX)(ABX_3)_n$ series of compounds, the end members ABX_3 and A_2BX_4, having the perovskite and K_2NiF_4 structures respectively, are formed at atmospheric pressures, but not the intermediate phases such as $A_3B_2X_7$ and $A_4B_3X_{10}$. Pressure facilitates the synthesis of such solids. $Sr_3Ru_2O_7$ and $Sr_4Ru_3O_{10}$ are formed in 15 min at 20 kbar and 1300 K. TaS_3, $NbSe_3$ and such solids are prepared in 30 min at 2 GPa and 970 K.

High-pressure methods have been employed in the synthesis of novel superconducting cuprates. A rudimentary example is the preparation of oxygen-excess La_2CuO_4 under high oxygen pressure. A more interesting example is the synthesis of the next homologue with two CuO_2 layers. $La_2Ca_{1-x}Sr_xCu_2O_6$ which had earlier been found to be an insulator was rendered superconducting by heating it under oxygen pressure [85]. $YBa_2Cu_4O_8$ was first prepared under high oxygen pressure, but this was soon found unnecessary [86]. However, superconducting cuprates with infinite CuO_2 layers of the type $Ca_{1-x}Sr_xCuO_2$ or $Sr_{1-x}Nd_xCuO_2$ can only be prepared under high hydrostatic pressures which help to give materials with shorter Cu–O bonds [87, 88]. It should be noted that $Ca_{1-x}Sr_xCuO_2$ prepared at ambient pressure is insulating [89].

13. The pyrosol process: a novel chemical method for depositing films

Pyrolysis of sprays is a well known method for depositing films. Thus, one can obtain films of oxidic materials such as CoO, ZnO and $YBa_2Cu_3O_7$ by the spray pyrolysis of solutions containing salts (*e.g.* nitrates) of the cations. A novel improvement in this technique is the pyrosol process involving the transport and subsequent pyrolysis of a spray generated by an ultrasonic atomizer [90]. When a high frequency (100 kHz–10 MHz range) ultrasonic beam is directed at a gas–liquid interface, a geyser is formed and the height of the geyser is proportional to the acoustic intensity.

Its formation is accompanied by the generation of a spray, resulting from the vibrations at the liquid surface and cavitation at the liquid–gas interface. The quantity of spray is a function of the intensity. Ultrasonic atomization is accomplished using an appropriate transducer made of PZT located at the bottom of the liquid container. A 500–1000 kHz transducer is generally adequate. The atomized spray which goes up in a column fixed to the liquid container is deposited onto a suitable solid substrate and then heat treated to obtain the film of the material concerned. The flow rate of the spray is controlled by the flow rate of air or any other gas. The liquid is heated to some extent, but its vaporization should be avoided.

The source liquid contains the relevant cations in the form of salts dissolved in an organic solvent. Organometallic compounds are often used (e.g. acetates, alkoxides, acetylacetonates, etc.). Proper gas flow is crucial to obtain satisfactory conditions for a good liquid spray. The pyrosol process is in some way between chemical vapour deposition and spray pyrolysis, but the choice of source compounds for the pyrosol process is much larger than that available for chemical vapour deposition. Films of a variety of materials have been obtained by the pyrosol method. The thickness of the films can be anywhere between a few hundred ångströms to a few micrometres. Table 5 lists typical materials prepared by this method [90]. Films of superconducting cuprates such as $YBa_2Cu_3O_7$ have also been prepared by the pyrosol process. Epitaxy has been observed in films deposited onto single-crystal substrates.

14. Intergrowth substrates

Several metal oxide systems exhibit chemically well defined recurrent intergrowth structures with large periodicities, rather than random solid solutions with variable composition. However, the ordered intergrowth structures themselves frequently show the presence of wrong sequences. The presence of wrong sequences or lamellae is best revealed by a technique that is more suited to the study of local structure. High resolution electron microscopy (HREM) enables direct examination of the extent to which a particular ordered arrangement repeats itself, the presence of different sequences of intergrowths, often of unit cell dimensions. Selected area electron diffraction, which forms an essential part of HREM, provides useful information (not generally provided by X-ray diffraction) regarding the presence of supercells due to the formation of intergrowth structures. Many systems forming ordered intergrowth structures have been discovered recently [91, 92]. These systems generally exhibit

TABLE 5. Typical films prepared by the pyrosol process

Material	Compound used	Solvent	Gas	Substrate[a]
Pt	Pt acetylacetonate	Acetylacetone	Air	Glass, Al_2O_3, Si (670 K)
In_2O_3	In acetylacetonate	Acetylacetone	Air	Glass, Al_2O_3, Si (770 K)
SnO_2	$SnCl_4$	Methanol	Air, N_2	Glass, Al_2O_3 (670 K)
$CdIn_2O_4$	In acetylacetonate + Cd acetate	Acetylacetone, methanol	Air	Glass, Al_2O_3 (710 K)
TiO_2	Butyl-orthotitanate	Butanol, acetylacetone	Air, N_2	Glass, steel (670 K)
γ-Fe_2O_3	Fe acetylacetonate	Butanol	Argon	Glass (760 K)
$(Ni,Zn)Fe_2O_4$	Ni, Zn, Fe acetylacetonates	Butanol	Air	Glass (770 K)
Al_2O_3	Al isopropoxide	Butanol	Air	Glass (920 K)

[a]Approximate substrate temperature is shown.

TABLE 6. Recurrent ordered intergrowth structures forming homologous series

(1) Barium ferrites
    (i) M_pY_q where $M = BaFe_{12}O_{19}$ and $Y = Ba_2Me_2Fe_{12}O_{22}$ ($Me \equiv Zn$, Ni etc.); $Ba_{2n+p}Me_{2n}Fe_{12(n+p)}O_{22n+19p}$ with $n = 1$–47 and $p = 2$–12
    (ii) MS_n where $S = Me_2Fe_4O_8$ with $n = 1, 2, 3, 4$
(2) Perovskites
    (i) $Bi_4A_{m+n-2}B_{m+n}O_{3(m+n)+6}$ formed by Aurivillius oxides of the type $Bi_2A_{n-1}B_nO_{3n+3}$
    (ii) $A_nB_nO_{3n+2}$ such as $(Na,Ca)_nNb_nO_{3n+2}$ with $n = 4$–4.5
    (iii) $A_{n+1}B_nO_{3n+1}$ as exemplified by Sr–Ti–O and La–Ni–O systems
(3) Tungsten bronzes
    (i) A_xWO_3 (ITB) with A = alkali metal, Bi etc. $(0.0 < x < 0.1)$
    (ii) $A_xM_xW_{1-x}O_3$ (bronzoids) with $M \equiv V$, Nb etc. $(0.0 < x < 0.1)$
    (iii) $A_xP_4O_8(WO_3)_{2m}$ with $m = 4$–10
    (iv) $K_xP_2O_4(WO_3)_{2m}$ with $m = 2$
    (v) $P_4O_8(WO_3)_{4m}$
    (vi) $ACu_3M_7O_{21}$ with $M \equiv Ta$, Nb
(4) Siliconiobates
    (i) $(A_3M_6Si_4O_{26})_n(A_3Nb_{8-x}M_xO_{21})$ with $A \equiv Ba$, Sr; $M \equiv Ta$, Nb
    (ii) $(Ba_3Nb_6Si_4O_{26})_nA_3Nb_xM_xO_{21}$ with A $\equiv$ K or Ba; $M \equiv Ti$, Ni, Zn, etc.
(5) Others
    (i) $(ATi_6O_{13})_n(ATi_4O_9)_m$ with $A \equiv Na$, $A \equiv Ba$
    (ii) La_2O_3–ThO_2 system

homology. In Table 6, various known intergrowth structures are listed.

If the ABO_3 perovskite structure is cut parallel to the (110) plane, slabs of the composition $A_{n-1}B_nO_{3n+2}$ are obtained; if these slabs are stacked, an extra sheet of A is introduced, giving rise to the family of oxides of the general formula $A_nB_nO_{3n+2}$. Typical members of this family are $Ca_2Nb_2O_7$ $(n = 4)$, $NaCa_4Nb_5O_{17}$ $(n = 5)$ and $Na_2Ca_4Nb_6O_{20}$ $(n = 6)$. HREM and X-ray studies show that an ordered intergrowth structure with $n = 4.5$ with the composition $NaCa_8Nb_9O_{31}$ corresponds to alternate stacking of $n = 4$ and $n = 5$ lamellae. What is curious is that $NaCa_8Nb_9O_{31}$ is prepared by the standard procedure of heating the mixture of component oxides and yet shows such extraordinary periodicity. Between $n = 4$ and 4.5, a large number of ordered solids are found with the b parameter of the unit cell ranging anywhere from 58.6 Å in the $n = 4.5$ compound to a few thousand ångströms in longer period structures. These solids seem to belong to a class of infinitely adaptive structures [2].

Aurivillius described the family of oxides of the general formula $Bi_2A_{n-1}B_nO_{3n+3}$ where the perovskite slabs, $(A_{n-1}B_nO_{3n+1})^{2-}$, n octahedra thick, are interleaved by $(Bi_2O_2)^{2+}$ layers. Typical members of this family are Bi_2WO_6 $(n = 1)$, $Bi_3Ti_{1.5}W_{0.5}O_9$ $(n = 2)$, $Bi_4Ti_3CrO_{12}$ $(n = 3)$ and $Bi_5Ti_3CrO_{15}$ $(n = 4)$. These oxides form intergrowth structure of the general formula $Bi_4A_{m+n-2}B_{m+n}O_{3(m+n)+6}$ involving alternate stacking of two Aurivillius oxides with different n values (Fig. 9). The method of preparation simply involves heating a mixture of the component metal oxides at around 1000 K. Ordered intergrowth structures with (m,n) values of $(1,2)$, $(2,3)$ and $(3,4)$ have been fully characterized by X-ray diffraction and HREM (see Fig. 10). What is amazing is that such intergrowth structures with long-range order are indeed formed while either member $(m$ and $n)$ can exist as a stable entity. These materials seem to be truly representative of recurrent intergrowth. The periodicity found in recurrent intergrowth solids formed by the Aurivillius family of oxides is indeed impressive.

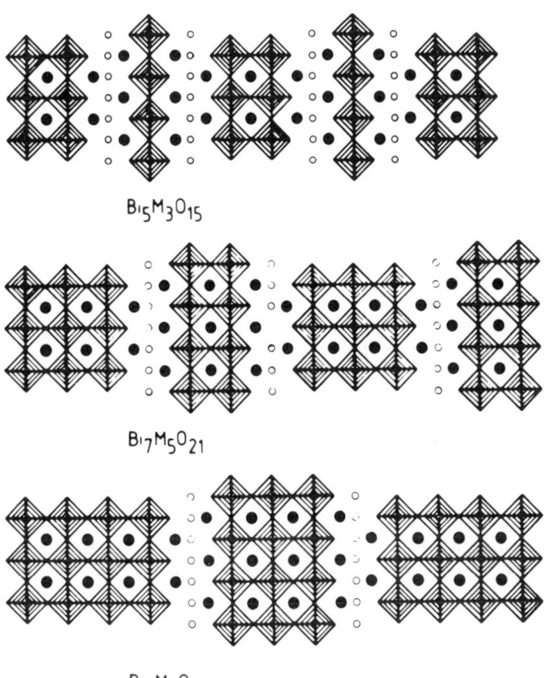

Fig. 9. Different types of intergrowth structures formed by the Aurivillius family of bismuth oxides.

WO_3 forms tetragonal, hexagonal or perovskite-type bronzes by interaction with alkali and other metals. The family of intergrowth tungsten bronzes (ITB) involving the intergrowth of nWO_3 slabs and one to three strips of the hexagonal tungsten bronze (HTB) is of relevance to our discussion here. In these intergrowth tungsten bronzes of the general formula M_xWO_3, x is generally 0.1 or less and depending on whether the HTB strip is one or two-tunnel wide, ITBs are classified as belonging to $(0, n)$ and $(1, n)$ series (Fig. 11). HTB strips of two-tunnel width seem to be most stable in ITBs and many ordered sequences of the $(0, n)$ and the $(1, n)$ series have been identified. ITB phases of bismuth were characterized some time ago in this laboratory. In this system, the HTB strips are always one-tunnel wide (Fig. 12). Displacement of adjacent tunnel rows due to the tilting of WO_3 octahedra often results in doubling of the long-period axis of the ITB. Evidence for the ordering of the intercalating bismuth atoms in the tunnels has been found in terms of satellites around the superlattice spots in the electron diffraction patterns. Among the other systems exhibiting ordered intergrowth, mention must be made of hexagonal barium ferrites M_pY_q ($M = BaFe_{12}O_{19}$ and $Y = Ba_2Me_{12}O_{22}$, where Me is Zn, Ni, Mg, *etc.*). A large number of intergrowth structures of this family have been identified.

15. Superconducting cuprates

Bednorz and Muller [93] discovered high T_c superconductivity (approximately 30 K) in $La_{2-x}Ba_xCuO_4$.

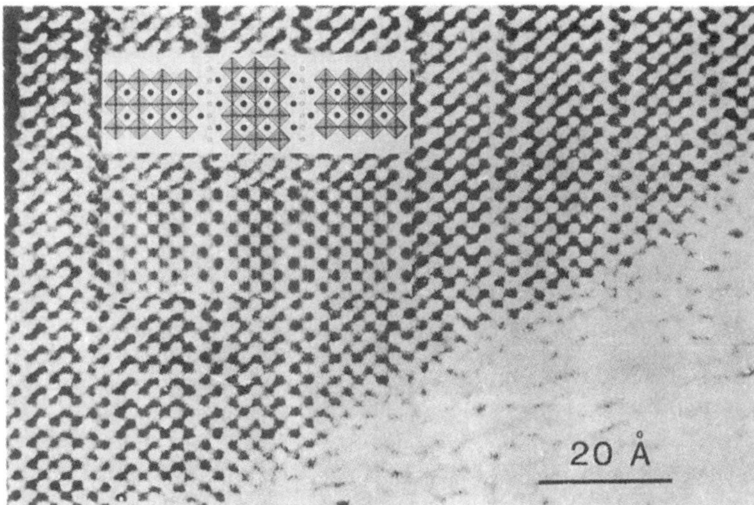

Fig. 10. High-resolution electron micrograph of a (3,4) intergrowth structure involving the Aurivillius phases $Bi_4Ti_3O_{12}$ ($n = 3$) and $Bi_5Ti_3CrO_{15}$ ($n = 4$).

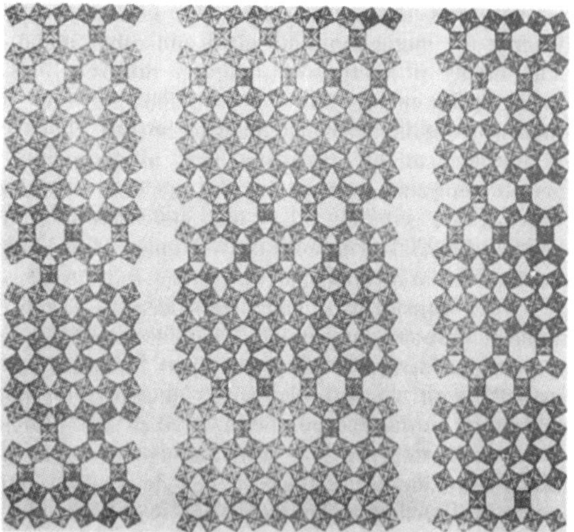

Fig. 11. Schematic drawing of (1,4), (1,5) and (1,6) intergrowth tungsten bronzes. Hexagonal tunnels of HTB strips separate the WO_3 slabs shown in polyhedral form (after Kihlborg [92]).

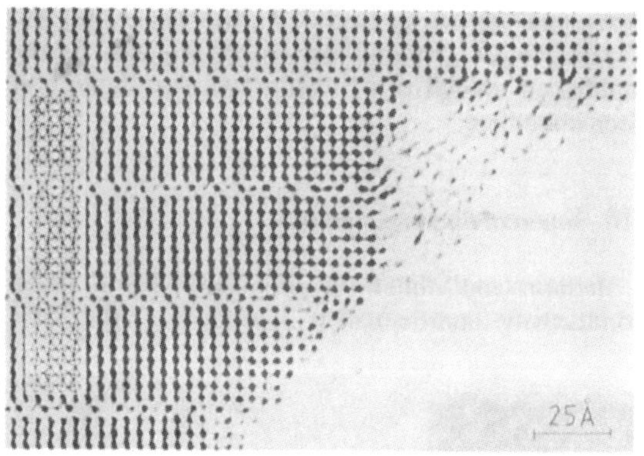

Fig. 12. High-resolution electron micrograph of Bi_xWO_3 intergrowth bronze. The dark circles between the WO_3 slabs represent bismuth atoms.

The discovery of a superconducting cuprate with T_c above 77 K created a sensation in early 1987. Wu *et al.* [94] who announced this discovery first made measurements on a mixture of oxides containing yttrium, barium and copper obtained in their efforts to obtain the yttrium-analogue of $La_{2-x}Ba_xCuO_4$. In this laboratory, we worked independently on the Y–Ba–Cu–O system on the basis of solid state chemistry [95]. We knew that Y_2CuO_4 could not be made and that substituting yttrium by barium in this cuprate was not the way to proceed. We therefore tried to make $Y_3Ba_3Cu_6O_{14}$ by analogy with the known $La_3Ba_3Cu_6O_{14}$ and varied the yttrium barium ratio as in $Y_{3-x}Ba_{3+x}Cu_6O_{14}$. By putting $x=1$, we obtained

$YBa_2Cu_3O_7$ ($T_c \approx 90$ K). We knew the structure had to be that of a defect perovskite from the beginning, because of the route we adopted for the synthesis.

We shall briefly examine some preparative aspects of the various types of cuprate superconductors [61, 96, 97]. The cuprates are ordinarily made by the traditional ceramic method (mix, grind and heat), which involves thoroughly mixing the various oxides and/or carbonates (or any other salt) in the desired proportion and heating the mixture (preferably in pellet form) at a high temperature. The mixture is ground again after some time and reheated until the desired product is formed, as indicated by X-ray diffraction. This method may not always yield the product with the desired structure, purity or oxygen stoichiometry. Variants of this method are often employed. For example, decomposing a mixture of nitrates has been found to yield a better product in the case of the 123 compounds by some workers; others prefer to use BaO_2 in place of $BaCO_3$ for the synthesis.

One of the problems with the bismuth cuprates is the difficulty of obtaining phasic purity (minimizing intergrowth of the different layered phases). The glass or the melt route has been employed to obtain better samples. This method involves preparing a glass by quenching the melt; the glass is then crystallized by heating it above the crystallization temperature. Thallium cuprates are best prepared in sealed tubes (gold or silver). Heating Tl_2O_3 with a matrix of the other oxides (already heated to 1100–1200 K) in a sealed tube is preferred by some workers. It is important that thallium cuprates are not prepared in open furnaces since Tl_2O_3 (which readily sublimes) is highly toxic. In order to obtain superconducting compositions corresponding to a particular copper content (number of CuO_2 sheets) by the ceramic method, one often has to start with various arbitrary compositions, especially in the case of the thallium cuprates. The real composition of a bismuth or a thallium cuprate superconductor is not likely to be anywhere near the starting composition. The actual composition has to be determined by analytical electron microscopy and other methods.

Heating oxidic materials under high oxygen pressures or in flowing oxygen often becomes necessary to attain the desired oxygen stoichiometry. Thus, La_2CuO_4 and $La_2Ca_{1-x}Sr_xCu_2O_6$ heated under high oxygen pressures become superconducting with T_c values of 40 and 60 K respectively. In the case of the 123 compounds, one of the problems is that they lose oxygen easily. Note that superconducting $LnBa_2Cu_3O_7$ (Ln $\equiv$ Y, rare earth) is orthorhombic while insulating $LnBa_2Cu_3O_{6.4}$ is tetragonal. It therefore becomes necessary to heat the material in an oxygen atmosphere below the orthorhombic–tetragonal transition temperature. Oxygen stoichiometry is, however, not a

C. N. R. Rao / Synthesis of solid inorganic materials

TABLE 7. Superconducting cuprates

Cuprate	Approximate T_c (K)	Methods of synthesis[a]
$L_{2-x}Sr_x(Ba_x)CuO_4$	35	Ceramic*, sol-gel, combustion
$La_2Ca_{1-x}Sr_xCu_2O_6$	60	Ceramic (high O_2 pressure)*
$La_2CuO_{4+\delta}$	35	Ceramic (high O_2 pressure)*, alkali-flux
$YBa_2Cu_3O_7$[b]	90	Ceramic (flowing O_2)*, sol-gel*, coprecipitation*, combustion
$YBa_2Cu_4O_8$[b]	80	Ceramic (high O_2 pressure), ceramic (with Na_2O_2)*, sol-gel*, coprecipitation*
$Bi_2CaSr_2Cu_2O_8$	90	Ceramic (air-quench)*, sol-gel*, combustion, melt (glass) route*
$Bi_2Ca_2Sr_2Cu_3O_{10}$	110	Ceramic*, sol-gel, melt route
$TlCaBa_2Cu_2O_{6+\delta}$	90	Ceramic (sealed Ag/Au tube)*
$TlCa_2Ba_2Cu_3O_{8+\delta}$	115	Ceramic (sealed Ag/Au tube)*
$Tl_2CaBa_2Cu_2O_8$	110	Ceramic (sealed Ag/Au tube)*
$Tl_2Ca_2Ba_2Cu_3O_{10}$	125	Ceramic (sealed Ag/Au tube)*
$Tl_{0.5}Pb_{0.5}CaSr_2Cu_2O_{6+\delta}$	110	Ceramic (sealed Ag/Au tube)*
$Pb_2Sr_2Ca_{1-x}Y_xCu_3O_8$	70	Ceramic (low O_2 partial pressure), sol-gel (low O_2 partial pressure)*
$Nd_{2-x}Ce_xCuO_4$	30	Ceramic (low O_2 partial pressure), coprecipitation (low O_2 partial pressure)*
$Ca_{1-x}Sr_xCuO_2$	40–110	Ceramic (high pressure)*
$Sr_{1-x}Nd_xCuO_2$	40–110	Ceramic (high pressure)*

[a]Recommended methods are indicated by asterisks.
[b]Other rare earth compounds of this type are also prepared by similar methods.

problem with the bismuth cuprates. The 124 superconductors were first prepared under high oxygen pressures, but it was later found that heating the oxide or nitrate mixture in the presence of Na_2O_2 in flowing oxygen is sufficient to obtain 124 compounds. Superconducting lead cuprates, however, can only be prepared in the presence of very little oxygen (N_2 with a small percentage of O_2). In the case of the electron superconductor $Nd_{2-x}Ce_xCuO_4$, it is necessary to heat the material in an oxygen-deficient atmosphere; otherwise, the electron given by cerium will merely result in an oxygen-excess material. It may be best to prepare $Nd_{2-x}Ce_xCuO_4$ by a suitable method (say decomposition of mixed oxalates or nitrates) and then reduce it with hydrogen.

The sol-gel method has been conveniently employed for the synthesis of 123 compounds such as $YBa_2Cu_3O_7$ and the bismuth cuprates. Materials prepared by such low-temperature methods have to be annealed or heated under suitable conditions to obtain the desired oxygen stoichiometry as well as the characteristic high T_c value. 124 cuprates, lead cuprates and even thallium cuprates have been made by the sol-gel method; the first two are particularly difficult to make by the ceramic method. Coprecipitation of all the cations in the form of a sparingly soluble salt such as carbonate in a proper medium (*e.g.* using tetraethylammonium oxalate), followed by thermal decomposition of the dried precipitate has been employed by many workers to prepare cuprates.

Several other novel strategies have been employed for the synthesis of superconducting cuprates; some of them were mentioned earlier while discussing the various methods. Especially noteworthy are the use of the combustion method and the alkali-flux method for cuprate synthesis. Superconducting infinite-layered cuprates seem to be possible only when prepared under high pressures because of bonding (structural) considerations [87, 88]. In Table 7 we list the various cuprate superconductors along with their properties and the preferred methods of synthesis.

Strategies where structure and bonding considerations are involved in the synthesis are generally more interesting. One such example is the synthesis of modulation-free superconducting bismuth cuprates [98]. Superconducting bismuth cuprates such as $Bi_2CaSr_2Cu_2O_8$ exhibit superlattice modulation. Since such modulation had something to do with the oxygen content in the Bi–O layers and lattice mismatch, Bi^{3+} was partly substituted by Pb^{2+} to eliminate the modulation, without losing the superconductivity.

16. Concluding remarks

Chemical synthesis of materials has become an extensive area with newer types of materials being prepared frequently by novel methods. We have reviewed some of the important methods and materials here and there are many more. For example, we have not discussed metal cluster or metal-rich compounds and pnictides; there are several interesting halides and chalcogenides as well. We have only touched on the chemistry involved with chemical transport. The ingenuity with which properties of oxides and other materials are modified drastically by appropriate substitutions or by modification of the structure also forms a part of synthetic strategies, just as the preparation of materials in different forms (such as fine powders) involves chemical inputs. While we have not brought in all these aspects of materials synthesis in this article, it is hoped that what has been covered gives the flavour of this vital area and helps to underscore the important role of chemistry in the synthesis of materials.

Acknowledgment

The author thanks the Indo-French Centre for Pure and Applied Research and CSIR (India) for support.

References

1 P. Hagenmuller (ed.), *Preparative Methods in Solid State Chemistry*, Academic Press, New York, 1972.

2 C. N. R. Rao and J. Gopalakrishnan, *New directions in Solid State Chemistry*, Cambridge University Press, Cambridge, 1986.

3 J. D. Corbett, in A. K. Cheetham and P. Day (eds.), *Solid State Chemistry — Techniques*, Clarendon Press, Oxford, 1987, p. 1.

4 F. J. Di Salvo, *Science, 247* (1990) 647.

5 A. K. Cheetham and P. Day (eds.), *Solid State Chemistry — Compounds*, Clarendon Press, Oxford, 1992.

6 C. N. R. Rao (ed.), *Chemistry of Advanced Materials*, IUPAC 21st Century Monograph Series, Blackwell, Oxford, 1992.

7 R. Roy, *Solid State Ionics, 32–33* (1989) 3.

8 C. Torardi and R. E. McCarley, *J. Am. Chem. Soc., 101* (1979) 3963.

9 J. P. Giroult, M. Goreaud, P. H. Labbe and B. Raveau, *Acta Crystallogr. B, 36* (1980) 2570.

10 K. Yvon, in E. Kaldis (ed.), *Current Topics in Materials Science, Vol. 3*, North-Holland, Amsterdam, 1979.

11 K. H. Jack, *Mater. Res. Bull., 13* (1978) 1327.

12 J. B. Goodenough, H. Y. P. Hong and J. A. Kafalas, *Mater. Res. Bull., 11* (1976) 203.

13 J. M. Newsam, in A. K. Cheetham and P. Day (eds.), *Solid State Chemistry — Compounds*, Clarendon Press, Oxford, 1992, p. 234.

14 R. C. Haushalter, K. G. Strohmaeu and F. W. Lai, *Science, 246* (1989) 1289.

15 A. Jayaraman, P. D. Dernier and L. D. Longinolti, *High Temp. High Pressures, 7* (1975) 1.

16 A. G. Merzhanov, in C. N. R. Rao (ed.), *Chemistry of Advanced Materials*, IUPAC 21st Century Monograph Series, Blackwell, Oxford, 1992, p. 19.

17 P. R. Bonneau, R. F. Jarvis, Jr., and R. B. Kaner, *Nature, 349* (1991) 510.

18 R. Mahesh, V. A. Pavate, Om Parkash and C. N. R. Rao, *Supercond. Sci. Technol., 5* (1992) 174.

19 K. Vidyasagar, J. Gopalakrishnan and C. N. R. Rao, *Inorg. Chem., 23* (1984) 1206.

20 C. N. R. Rao and J. Gopalakrishnan, *Acc. Chem. Res., 20* (1987) 20.

21 H. S. Horowitz, S. J. McLain and A. W. Sleight, *Science, 243* (1989) 66.

22 K. S. Nanjundaswamy, N. Y. Vasantacharya, J. Gopalakrishnan and C. N. R. Rao, *Inorg. Chem., 26* (1987) 4286.

23 A. K. Cheetham, *Nature, 228* (1980) 469.

24 K. Matsuzaki, A. Inoue, H. Kimura, K. Aoki and T. Masumoto, *Jpn. J. Appl. Phys., 26* (1987) L1310.

25 L. V. Interrante and A. G. Williams, *Polym. Prep. (Am. Chem. Soc. Div. Polym. Chem.), 25* (1984) 13.

26 D. Seyferth and G. H. Wiseman, *J. Ceram. Soc., 67* (1984) C132.

27 J. M. Thomas, *Philos. Trans. R. Soc. London, Ser. A, 277* (1974) 251.

28 G. W. Brindley, *Prog. Ceram. Sci., 3* (1963).

29 J. R. Gunter, *J. Solid State Chem., 5* (1972) 354.

30 A. Revcolevachi and G. Dhalenne, *Nature, 316* (1985) 335.

31 L. Ganapathi, A. Ramanan, J. Gopalakrishnan and C. N. R. Rao, *J. Chem. Soc. Chem. Commun.*, (1986) 62.

32 E. M. McCarron, *J. Chem. Soc. Chem. Commun.*, (1986) 336.

33 P. L. Gai and C. N. R. Rao, *Z. Naturfosch. A, 30* (1975) 1092.

34 K. Vidyasagar, A. Reller, J. Gopalakrishnan and C. N. R. Rao, *J. Chem. Soc. Chem. Commun.*, (1985) 7.

35 M. S. Whittingham and A. J. Jacobson (eds.), *Intercalation Chemistry*, Academic Press, New York, 1982.

36 A. J. Jacobson, in A. K. Cheetham and P. Day (eds.), *Solid State Chemistry — Compounds*, Clarendon Press, Oxford, 1992 and references listed therein, p. 182.

37 M. S. Whittingham, *Prog. Solid State Chem., 12* (1978) 41.

38 D. W. Murphy, C. Cros, F. J. Di Salvo and J. V. Waszezak, *Inorg. Chem., 16* (1977) 3027.

39 A. J. Jacobson, R. R. Chianelli, S. M. Rich and M. S. Whittingham, *Mater. Res. Bull., 14* (1979) 1437.

40 D. W. Murphy, F. J. Di Salvo, J. N. Carides and J. V. Waszezak, *Mater. Res. Bull., 13* (1978) 1395.

41 D. W. Murphy, M. Greenblatt, S. M. Zahurak, R. J. Cava, J. V. Waszezak and R. S. Hutton, *Rev. Chim. Miner., 19* (1982) 441.

42 K. Mizushima, P. C. Jones, P. J. Wiseman and J. B. Goodenough, *Mater. Res. Bull., 15* (1980) 783.

43 K. Vidyasagar and J. Gopalakrishnan, *J. Solid State Chem., 42* (1982) 217.

44 D. W. Murphy and P. A. Christian, *Science, 205* (1979) 651.

45 M. M. Thackeray, W. I. F. David and J. B. Goodenough, *Mater. Res. Bull., 17* (1982) 785; *18* (1983) 461.

46 D. G. Dickens and M. F. Pye, in M. S. Whittingham and A. J. Jacobson (eds.), *Intercalation Chemistry*, Academic Press, New York, 1982, p. 539.

47 P. G. Dickens, S. J. French, A. T. Hight and M. F. Pye, *Mater. Res. Bull., 14* (1979) 1295.

48 R. J. Cava, A. Santoro, D. W. Murphy, S. Zahurak and R. S. Roth, *J. Solid State Chem., 42* (1982) 251.

49 A. K. Ganguli, J. Gopalakrishnan and C. N. R. Rao, *J. Solid State Chem., 74* (1988) 228.

50 X. D. Xiang, S. McKernan and W. A. Vareka, *Nature, 348* (1990) 145.

51 B. C. Tofield, in M. S. Whittingham and A. J. Jacobson (eds.), *Intercalation Chemistry*, Academic Press, New York, 1982, p. 181.

52 W. A. England, J. B. Goodenough and P. J. Wiseman, *J. Solid State Chem., 49* (1983) 289.

53 C. E. Rice and J. L. Jackel, *J. Solid State Chem., 41* (1982) 308.

54 C. Delmas, J. J. Braconnier and P. Hagenmuller, *Mater. Res. Bull., 17*(1982) 117.

55 H. Boller, *Monatsh. Chem., 109*(1978) 975.

56 D. R. Uhlmann, B. J. Zelinski and G. E. Wnek, in C. J. Brinker, D. E. Clark and D. R. Ulrich (eds.), *Better Ceramics through Chemistry*, North-Holland, New York, 1984, p. 59.

57 B. J. Zelinski, C. J. Brinker, D. E. Clark and D. R. Ulrich (eds.), *Better Ceramics through Chemistry IV*, MRS symposium, vol. 180, Materials Research Society, Pittsburgh, PA, 1990, and earlier volumes.

58 E. Matijevic, in L. L. Hench and D. Ulrich (eds.), *Ultrastructure Processing of Ceramics, Glasses and Composites*, Wiley, New York, 1984, p. 123.

59 R. K. Iler, *The Chemistry of Silica*, Wiley, New York, 1979.

60 E. Matijevic, *Discuss. Faraday Soc., 92*(1991) 229.

61 C. N. R. Rao, R. Nagarajan and R. Vijayaraghawam, *Supercond. Sci. Technol.,* (1992) in press.

62 H. S. Horowitz, J. M. Longo and J. T. Lewandowski, *Mater. Res. Bull., 16*(1981) 489.

63 W. K. Ham, G. F. Holland and A. M. Stacy, *J. Am. Chem. Soc., 110*(1988) 5214.

64 R. J. Cava, J. J. Krajewski and W. F. Peck, *Nature, 338*(1989) 328.

65 L. F. Schneemeyer, J. K. Thomas and T. Siegrist, *Nature, 335* (1988) 421.

66 A. Wold and D. Bellawance, in P. Hagenmuller (ed.), *Preparative Methods in Solid State Chemistry*, Academic Press, new York, 1972, p. 279.

67 R. S. Feigelson, *Adv. Chem. Sci., 186*(1980) 243.

68 M. S. Whittingham and R. A. Huggins, in R. S. Roth and S. J. Schneider, Jr., (eds.), *Solid State Chemistry*, National Bureau of Standards, US Department of Commerce, Washington, DC, 1972, p. 133.

69 E. Banks and A. Wold, in C. N. R. Rao (ed.), *Solid State Chemistry*, Marcel Dekker, New York, 1974.

70 J. B. Goodenough, J. A. Kafalas and J. M. Longo, in P. Hagenmuller (ed.), *Preparative Methods in Solid State Chemistry*, Academic Press, New York, 1972, p. 2.

71 C. W. F. T. Pistorius, *Prog. Solid State Chem., 11*(1976) 1.

72 J. C. Joubert and J. Chenavas, in N. B. Hannay (ed.), *Treatise on Solid State Chemistry*, Vol. 5, Plenum, New York, 1975, p. 463.

73 R. Roy, D. K. Agarwal, J. Alamo and R. A. Roy, *Mater. Res. Bull., 19*(1984) 471.

74 M. A. Subramanian, B. D. Roberts and A. Clearfield, *Mater. Res. Bull., 19*(1984) 1471.

75 A. W. Sleight, *Inorg. Chem., 8*(1969) 1764.

76 R. M. Barrer, *Zeolites, 1*(1981) 130.

77 E. M. Flanigen, J. M. Bennett, R. W. Grose, J. P. Cohen, R. L. Patton, R. M. Kirchner and J. V. Smith, *Nature, 271* (1978) 512.

78 E. J. J. Groenen, N. C. M. Alma, A. G. T. M. Bastein, G. R. Hays, R. Huis and A. G. T. G. Kortbeck, *J. Chem. Soc. Chem. Commun.,* (1983) 1360.

79 G. Demazeau, B. Buffat, M. Pouchard and P. Hagenmuller, *J. Solid State Chem., 45*(1982) 881; *Z. Inorg. Allg. Chem., 491* (1982) 60.

80 M. Takano and Y. Takeda, *Bull. Inst. Chem. Res., (Kyoto Univ., Jpn.), 61* (1983) 406.

81 J. K. Vassilou, M. Hornbostel, R. Ziebarth and F. J. Di Salvo, *J. Solid State Chem., 81* (1989) 208.

82 Y. Takeda, F. Kanamaru, M. Shimada and M. Koizumi, *Acta Crystallogr. B, 32*(1976) 2464.

83 P. Hagenmuller (ed.), *Solid Inorganic Fluorides*, Academic Press, New York, 1985.

84 F. Sugawara, Y. Syono and S. Akimoto, *Mater. Res. Bull., 3* (1968) 529.

85 R. J. Cava, B. Battlogg, L. F. Schneemeyer *et al., Nature, 345* (1990) 602.

86 C. N. R. Rao, G. N. Subbanna, R. Nagarajan *et al., J. Solid State Chem., 88*(1990) 163.

87 M. Azuma, M. Tanako *et al., Nature, 356*(1992) 775.

88 M. Takano, Z. Hiroi, M. Azuma and Y. Takeda, in C. N. R. Rao (ed.), *Chemistry of High-Temperature Superconductors*, World Scientific, Singapore, 1992, p. 243.

89 T. Siegerist, S. M. Zahurak, D. W. Murphy and R. S. Roth, *Nature, 334*(1988) 155.

90 M. Langlett and J. C. Jobert, in C. N. R. Rao (ed.), *Chemistry of Advanced Materials*, IUPAC 21st Century Monograph Series, Blackwell, Oxford, 1992, p. 55.

91 C. N. R. Rao, *Bull. Mater. Sci., 7* (1985) 155; C. N. R. Rao and J. M. Thomas, *Acc. Chem. Res., 18*(1985) 113.

92 L. Kihlborg (ed.), Direct imaging of atoms in crystals and molecules, *Nobel Symposium, 47*(1979) 101.

93 J. G. Bednorz and K. A. Muller, *Z. Phys. B, 64*(1986) 189.

94 M. K. Wu, J. R. Ashburn, C. J. Torng, P. H. Hor, R.-L. Meng, L. L. Gao, Z. H. Huang, Y. Q. Wang and C. W. Chu, *Phys. Rev. Lett., 58*(1987) 908.

95 C. N. R. Rao, P. Ganguly, R. A. Mohan Ram, A. K. Raychaudhuri and K. Sreedhar, *Nature, 326*(1987) 856.

96 C. N. R. Rao, *Philos. Trans. R. Soc. London, 336*(1991) 596.

97 C. N. R. Rao (ed.), *Chemistry of High-Temperature Superconductors*, World Scientific, Singapore, 1992.

98 V. Manivannan, J. Gopalakrishnan and C. N. R. Rao, *Phys. Rev. B, 43*(1991) 8686.

Indian Journal of Chemistry
Vol. 23A, April 1984, pp. 265-284

Advances in Contemporary Research

Superstructures, Ordered Defects & Nonstoichiometry in Metal Oxides of Perovskite & Related Structures†

C N R RAO*‡

Jawaharlal Nehru Visiting Professor, Department of Physical Chemistry, University of Cambridge, Cambridge CB2 1EP, UK

and

J GOPALAKRISHNAN & K VIDYASAGAR

Solid State and Structural Chemistry Unit, Indian Institute of Science, Bangalore 560012

Received 14 *February* 1984

1. Introduction

Nonstoichiometry has been a problem of vital interest from the early beginnings of chemistry[1]. Nonstoichiometry was until recently explained in terms of equilibria involving point defects (vacancies and interstitials)[2,3], often invoking interaction of defects[4]. As more and more grossly nonstoichiometric compounds belonging to different families and possessing a variety of novel structural features were discovered, a new look at this old problem became necessary[5]. On the basis of careful structural studies, it was soon realized that the concept of point defects was most inadequate to explain nonstoichiometry in many of the metal oxides and other materials. Point defects may explain properties of solids with very small deviations in stoichiometry, but for most compounds where deviations in the cation or the anion site occupancy is sizeable (a small fraction of a percent or more), alternative models based on experimental findings have become possible. The models involve ordering of point defects or new structural features which eliminate point defects[6-9]. Ordering of point defects is found in simple solids such as TiO (where 15 % of both the cations and the anions are absent) and FeO where there is a relatively small cation deficiency. The formation of crystallographic shear planes[6,7] provides an important means of eliminating point defects and is found in oxides based on ReO_3 and TiO_2 structures, giving rise to homologous series of oxides such as Ti_nO_{2n-1}, W_nO_{3n-1} and W_nO_{3n-2}. It is interesting that the oxide of titanium with the composition $TiO_{1.9995}$ shows evidence of crystallographic shear planes rather than point defects. The subject of nonstoichiometry and defect structures has been reviewed widely in the last few years[8-11]. The main conclusion is that isolated point defects are rarely

of relevance to the chemistry of a large variety of nonstoichiometric inorganic compounds. Instead, the new structural principles of solid state inorganic chemistry enable us to rationalize the structures of nonstoichiometric compounds giving rise to new families of inorganic compounds and thereby transforming a subject of curiosity to a select bunch of chemists into one belonging to main stream chemistry.

In this article, we concern ourselves with the defect structures of metal oxides of the general formula ABO_3 possessing the perovskite structure, the topic being of great significance not only because of the wide variety of oxides crystallizing in structures related to the perovskite, but also because of the interesting electronic, magnetic, dielectric, catalytic and related properties exhibited by these metal oxides[12,13]. A number of perovskite oxides exhibit nonstoichiometry. Defects in these oxides can arise from cation deficiency (in A or B site), oxygen deficiency or oxygen excess. Such vacancies often give rise to ordered superstructures which are readily studied by electron diffraction and high resolution electron microscopy (HREM). X-ray diffraction is inadequate to examine vacancy ordering and aspects related to local structure such as stacking sequences and disorder. Electron diffraction is eminently suited to the study of superstructures formed by vacancy ordering. Electron diffraction forms an integral part of HREM which yields images of lattices in direct space and these images enable us to unravel the intricate details of local structure at the atomic or the unit cell level. We shall be discussing the results of HREM of defect perovskites throughout this article.

2. Metal oxides with the perovskite structure

The ideal perovskite (ABO_3) structure which is a simple cubic structure with the space group Pm3m, provides the basis for the structures of a large variety of inorganic solids. The perovskits structure is conventionally described as consisting of a BO_3 array

---

†Contribution No. 246 from the Solid State and Structural Chemistry Unit.

‡Permanent address: Indian Institute of Science, Bangalore 560012.

INDIAN J. CHEM., VOL. 23A, APRIL 1984

formed by corner-sharing BO_6 octahedra. The large A-cations occupy twelve-coordinated body-centre sites in the BO_3 network (Fig. 1a). The structure can be thought of as consisting of alternating BO_2 and AO layers stacked one over the other in the [001] direction (Fig. 1b). The geometrical (size) requirement for the perovskite structure is expressed in terms of Goldschmidt's tolerance factor,

$$t = (r_A + r_O)/\sqrt{2}(r_B + r_O)$$

where r_A, r_B and r_O are the ionic radii of A, B and oxide ions respectively. For the ideal structure, $t = 1$, but the perovskite structure occurs over a range of t with a lower limit of ~ 0.7. The perovskite structure also requires that A and B ions be stable in 12 and 6 (octahedral) coordinations respectively. In oxides, this limits the radii of A and B ions to $r_A > 0.9$ and $r_B > 0.51$ Å. For lower values of t, the cubic structure is distorted to optimize $A - O$ bond lengths. When $0.75 < t < 0.90$, the BO_6 octahedra are tilted cooperatively to give an enlarged orthorhombic unit cell (Fig. 1c). For t slightly less than unity, a rhombohedral distortion occurs as in $LaAlO_3$. In general, distortions from the ideal structure can be traced to cation displacement or tilting of the BO_6 octahedral or to a combination of both. Different kinds of octahedral tiltings have been classified and discussed[14,15].

Katz and Ward[16] have given an alternative description of the perovskite structure in terms of close packing of A and O ions. This description is more general and helps to understand the occurrence of the hexagonal polytypes of perovskites. An ABO_3 perovskite structure in Katz and Ward model is thought to consist of close-packed AO_3 layers (Fig. 2a) stacked one over the other; the B-cations occupy octahedral holes that are exclusively surrounded by oxygens. When the stacking of the AO_3 layers is all-cubic (ccc....), the B-cation octahedra share all their corners, giving rise to the perovskite structure (3C) (Fig. 2b). When the stacking of AO_3 is all-hexagonal

(hhh....), BO_6 octahedra share opposite faces forming infinite chains parallel to the hexagonal c-axis (Fig. 2c). Mixed stacking sequences of cubic and hexagonal stacking arrangements give rise to polytypes with larger periodicities. In Fig. 2d-f, we have shown the arrangements, cchcch (6H), chch (4H) and (chh)$_3$ (9H). The hexagonal stacking of AO_3 layers is favoured when the A-cation is large ($t > 1.0$). Hexagonal stacking results in considerable loss of the Madelung energy as a consequence of B-cation repulsion between face-shared octahedra. Several factors such as $B - B$ bonding, $B - O$ covalency and B-site vacancy which compensate for the loss in the Madelung energy, stabilize hexagonal stacking in preference to cubic stacking.

3. Cation-deficient perovskites

A-site vacancies: Since the BO_3 array in the perovskite structure forms a stable network, the large A-cations at twelve-coordinated sites can be missing either partially or wholly. The tungsten bronzes, A_xWO_3, which constitute an important family of nonstoichiometric oxides, may be regarded as A-site deficient perovskites[17]. Of the three different structures known for tungsten bronzes, hexagonal, tetragonal and cubic (Fig. 3), only the cubic one is related to the perovskite structure consisting of A-site vacancies. The cubic bronzes are referred to as perovskite tungsten bronzes (PTB) and two types of PTBs can be distinguished: those with small values of $x(x \ll 1.0)$ and others with large values of $x \leqslant 1.0$. PTBs with a large x-value show true bronze characteristics, viz. metallic behaviour and chemical inertness. An important question with regard to the non-stoichiometry of these phases is whether the A-site atoms and vacancies are ordered or not. An early neutron diffraction study[18] of $Na_{0.75}WO_3$ showed that sodium was ordered in six out of eight A-sites in a doubled perovskite cell. But the problem of ordering of A-side vacancies in metallic PTBs is not fully resolved.

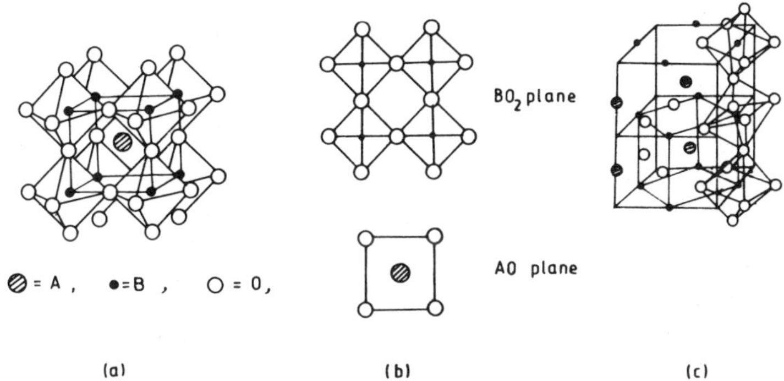

Fig. 1—(a) Ideal perovskite structure; (b) layer sequence in the perovskite structure parallel to (001); and (c) $GdFeO_3$ structure.

RAO *et al.*: METAL OXIDES OF PEROVSKITE & RELATED STRUCTURES

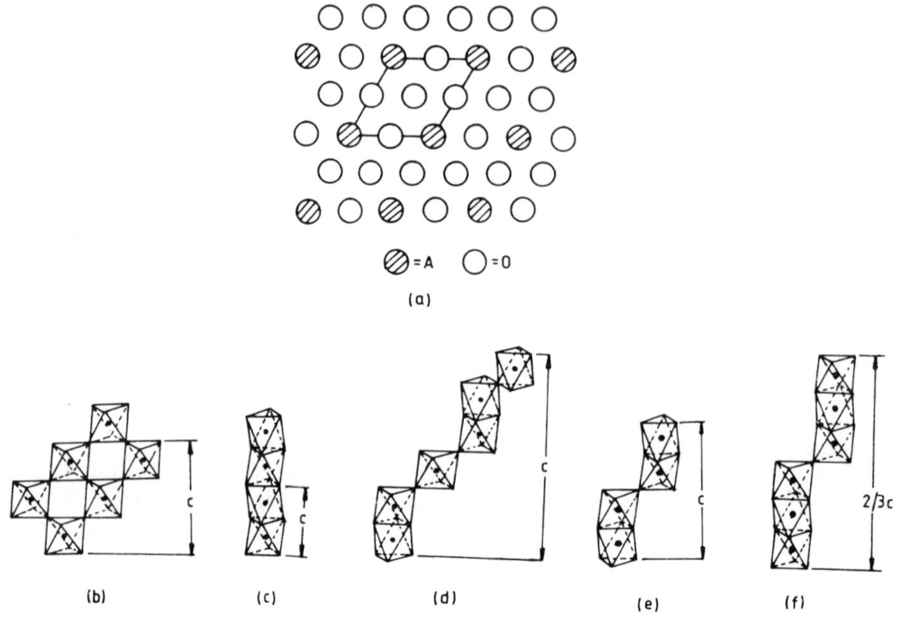

Fig. 2—(a) Close-packed AO_3 layer in perovskites; (b)-(f) BO_6 octahedra in different perovskite polytypes: (b) 3C; (c) 2H; (d) 6H; (e) 4H; and (f) 9R.

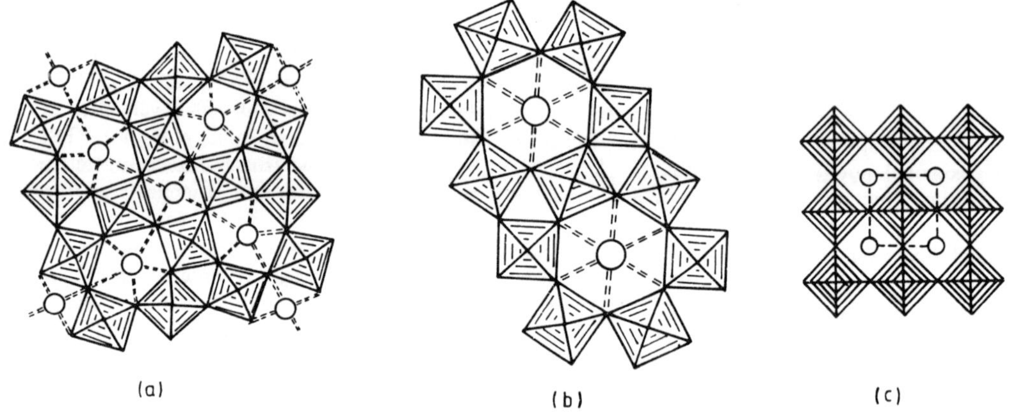

Fig. 3   Tungsten bronze structures: (a) tetragonal tungsten bronze (TTB); (b) hexagonal tungsten bronze (HTB); and (c) perovskite tungsten bronze (PTB).

In this context, it is relevant to note the difference between essentially ionic and essentially metallic solids as regards their nonstoichiometric behaviour[19]. In metallic systems, point defects would essentially be screened from one another by the conduction electrons; long-range defect interactions would be small and it is therefore possible that isolated random point defects or weakly interacting point defect clusters are present. This is probably the reason why in metallic PTBs no significant long-range ordering of vacancies is seen. But this is not the case in nonmetallic PTBs where one would expect A-site vacancy ordering.

Besides PTBs, A-site defective perovskite oxides are known to be formed when B = Ti, Nb, Ta and so on[7,13]. Such compounds exhibit metallic properties and perovskite structures when the B atom occurs in a low oxidation state. Compositions such as $A_{0.5}NbO_3$ (A = Ba, Pb etc.) where niobium is in the highest oxidation state adopt non-perovskite network structures. An interesting example[20,21] of a A-site defective perovskite is $Cu_{0.5}TaO_3$ which crystallizes in a pseudocubic perovskite structure. The unit cell is orthorhombic with $a = 7.523$, $b = 7.525$ and $c = 7.520$ Å and eight formula units per cell. Tantalum atoms form

INDIAN J. CHEM., VOL. 23A, APRIL 1984

a TaO_3 framework as in a cubic perovskite, while the copper atoms are ordered at the A-sites. Three of them are located in the middle of the cube edges, while the fourth is statistically distributed over the three face centres. Twelve oxygens around copper are arranged as three mutually perpendicular squares of different size: $Cu-O$ bond lengths are 2.763 and 3.923 Å.

Besides the perovskite, tetragonal and hexagonal tungsten bronzes, a new variety of tungsten bronzes formed by the intergrowth of WO_3 slabs (n-octahedra wide) with strips of hexagonal tungsten bronze (HTB), one to three tunnels wide, is noteworthy[22]. These intergrowth tungsten bronzes (ITB) can be fairly ordered as shown in Plate 1 or disordered as shown in Plate 2. The width of the HTB strips in Plate 1 is two tunnels wide and this is in fact the most stable configuration of alkali metal intergrowth tungsten bronzes. Recently we have synthesized[23] intergrowth tungsten bronzes of bismuth where the HTB strip is always one tunnel wide (Plate 3). What is interesting in the Bi_xWO_3 system is that when $x \leq 0.02$, a PTB phase seems to be formed, the ITBs being found only when $x > 0.02$. Furthermore, the ITB phase does not give place to the HTB at large x; the maximum value of x in this system is ~ 0.1.

WO_3 and ReO_3 may be regarded as the limiting cases of A-site vacancy perovskites. Both the oxides possess corner-linked framework of the octahedra, but unlike ReO_3, WO_3 is never cubic. It shows several polymorphic transitions starting from the low temperature triclinic structure to more symmetric forms with increasing temperature. The transitions arise from temperature-dependent displacements of the tungsten atom from the centre of the WO_6 octahedron[24].

B-site vacancies: B-site vacancies in perovskite oxides are energetically not favoured because of the large formal charge and the small size of B-site cations.

If vacancies are to occur at the B-sites, there must be other compensating factors such as $B-O$ covalency and $B-B$ interaction. While the covalency of the $B-O$ bond would increase with increasing charge and decreasing size of the B-cation, the $B-B$ interaction would be favoured by hexagonal rather than cubic stacking of AO_3 layers. Thus one would expect B-site vacancies to occur more frequently in perovskite oxides of highly charged B-cations possessing hexagonal polytypic structures. B-site vacancy perovskites are indeed far more common with hexagonal or mixed hexagonal-cubic AO_3 stackings than with exclusive cubic stacking of AO_3 layers. Rauser and Kemmler-Sack[25] have listed cubic perovskites exhibiting B-site vacancy. Some of the examples of B-site vacancy ordered perovskites are: $Ba_2Sm_{2/3}UO_6$ (5), $Ba_2Ce^{4+}_{3/4}Sb^{5+}O_6$ (7), $Ba_2Ce^{4+}Sb^{5+}_{4/5}O_6$ (9) and $Ba_2Sm^{3+}U^{6+}_{5/6}O_6$ (11), where the numbers in parentheses represent the number of B-cations per vacancy. Ordering of these vacancies is revealed by the formation of superstructures.

In Table 1, are listed some of the well-characterized B-site vacancy hexagonal perovskites. Normally, B-site vacancies are ordered between h-h layers where the BO_6 octahedra share faces. This is consistent with Pauling's rules for the sharing of coordination polyhedra, according to which corner-sharing is more favourable than face-sharing of polyhedra in essentially ionic structures[37]. Accordingly, $Ba_5Ta_4O_{15}$ adopts a five-layer (hhccc) sequence[30] wherein the octahedral site between h-h layers is vacant (Fig. 4a). The arrangement gives rise to isolated clusters of four octahedra that share each two opposite corners, $[Ta_4O_{15}]^{10-}$. The structure of $Ba_3Re_2O_9$ is similar[26] consisting of the (hhc)$_3$ layer sequence wherein the ReO_6 octahedra share corners, and the vacancy occurs at octahedral sites between h-h layers (Fig. 4b). Somewhat exceptional to this general trend[27,28,31] are

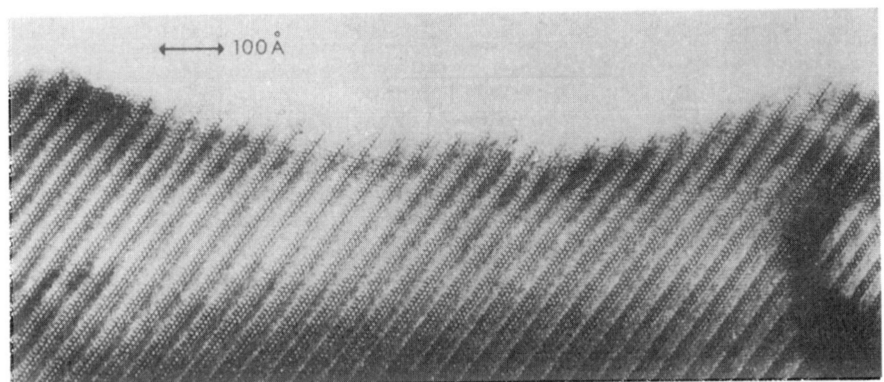

Plate 1—Lattice image of a highly ordered Cs_xWO_3 ITB (from Ref. 22). The narrow HTB strip can be readily distinguished from the WO_3 slabs. The ITB phase corresponds to the (1,7) system containing two tunnel-wide HTB strips separated by 7 WO_6 octahedra.

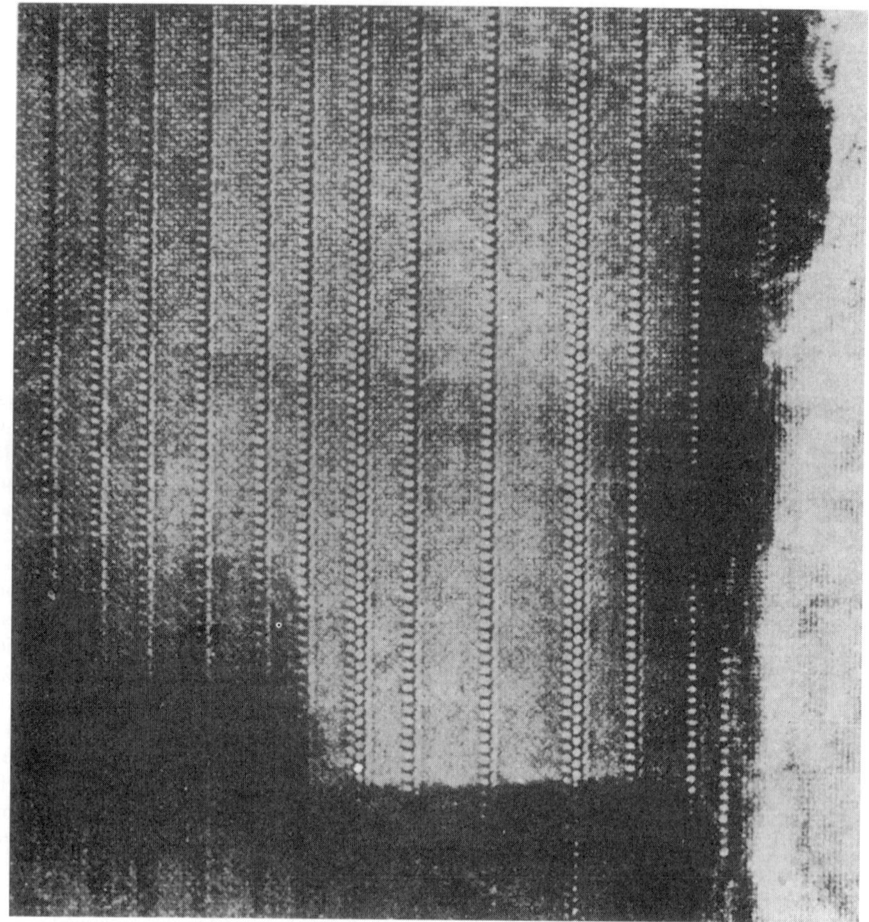

Plate 2—Lattice image of a disordered Cs_xWO_3 ITB showing different WO_3 slab widths (from ref. 22).

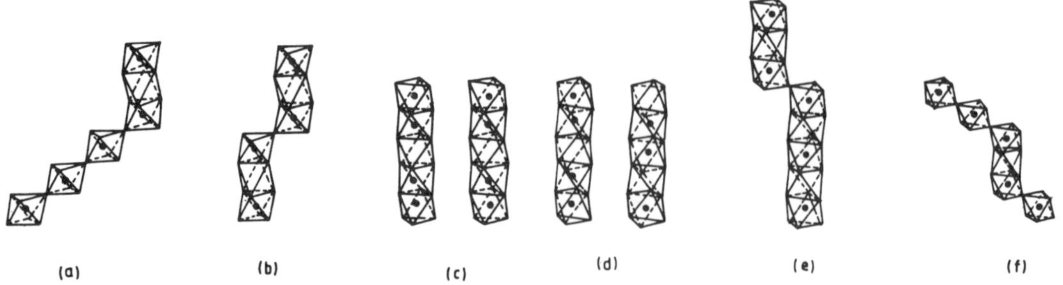

Fig. 4—Schematic illustration of linkage of BO_6 octahedra in hexagonal perovskites with B-site vacancy ordering: (a) $Ba_5Ta_4O_{15}$; (b) $Ba_3Re_2O_9$; (c) $Ba_3Te_2O_9$; (d) $Ba_3W_2O_9$; (e) $Ba_8Re_2W_3O_{24}$; and (f) $Ba_9Nb_6WO_{27}$.

INDIAN J. CHEM., VOL. 23A, APRIL 1984

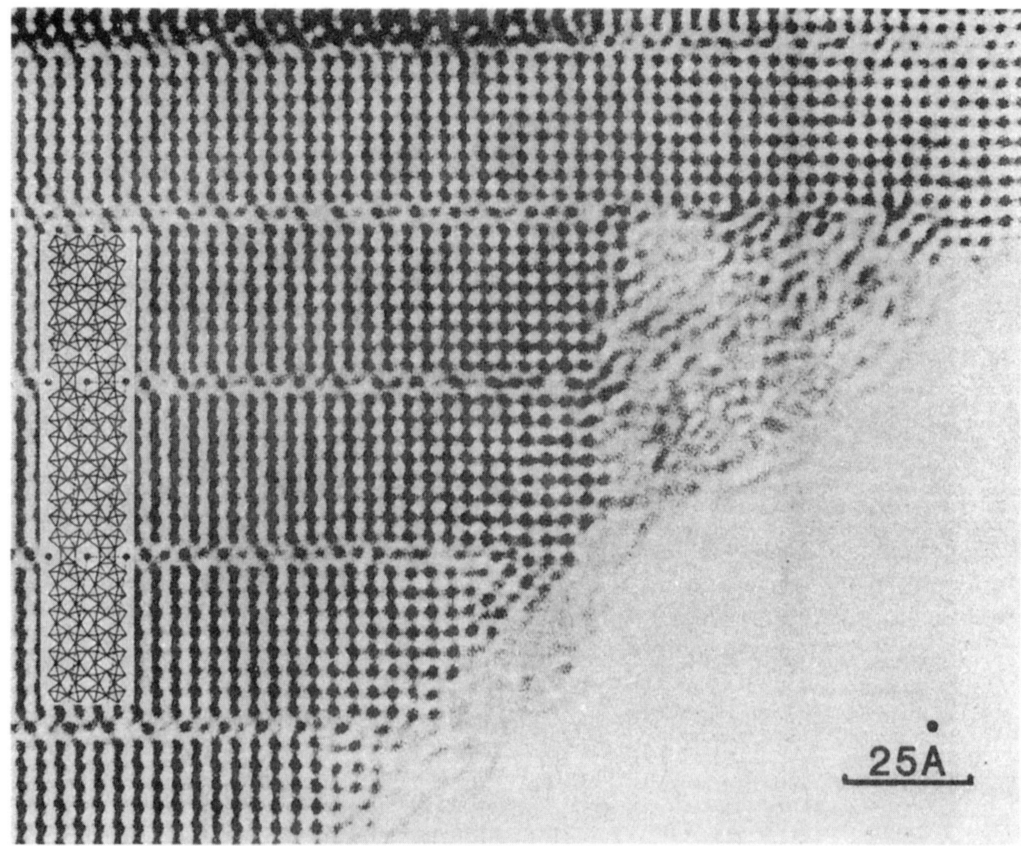

Plate 3—HREM (500kV) image of the ordered ITB phase of $Bi_{0.1}WO_3$ (nominal composition) showing Bi atoms in the tunnels (from ref. 23a).

$Ba_3W_2O_9$ and $Ba_3Te_2O_9$. Both these compounds adopt an all-hexagonal BaO_3 layer sequence. In $Ba_3W_2O_9$, two-thirds of the octahedral sites in every layer are filled by tungsten, while in $Ba_3Te_2O_9$, octahedral sites in two adjacent layers are fully occupied, every third layer being completely vacant. Both the orderings give rise to $[B_2O_9]^{6-}$ (B = W, Te) groups comprising two BO_6 octahedra sharing a common face (confacial bioctahedra) as shown in Figs. 4c and 4d. In both the structures, B-atoms are displaced considerably towards the vacancies, resulting in a minimization of B−B repulsion.

Oxide perovskites consisting of ordered B-site vacancies exhibit novel luminescence properties. $Ba_3W_2O_9$, for instance, shows an efficient blue (460 nm) photoluminescence[38] below 150K, unlike $BaWO_4$ and Ba_3WO_6 which consist of isolated WO_4 tetrahedra and WO_6 octahedra respectively. It has been suggested that clustering of WO_6 octahedra in $Ba_3W_2O_9$ is responsible for the efficient luminescence. More interestingly, B-site vacancy ordered perovskites

show different colour emissions, when they are doped with different rare-earth activators. For instance, when $Ba_{3-x}Sr_xLaScW_2O_{12}$, 12-layered (hhcc)$_3$ perovskite, is doped with Eu^{3+} and Tb^{3+} simultaneously, a green terbium emission ($Tb^{3+}:{}^5D_4 \rightarrow {}^7F_5$) at 547 nm and a red europium emission ($Eu^{3+}:{}^5D_0 \rightarrow {}^7F_2$) at 615 nm are seen. The former shows up when the phosphor is excited in the WO_6 charge-transfer band at 300 nm and the latter emission is seen when excitation occurs at 340 nm corresponding to charge-transfer from oxygen to Eu^{3+}.Similarly $Sr_3La_2W_2O_{12}$ doped with Eu^{3+} and Tm^{3+} shows red and blue emissions when excited suitably, and the same host doped with Er^{3+} and Tm^{3+} shows green and blue emissions[39,40].

The family of bismuth oxides of the general formula $Bi_mB_{m-1}O_{3m}$ (B = Ti, Nb and W), first described by Aurivillius[41], may be regarded as B-cation deficient perovskites. Typical members of this family are Bi_2WO_6, Bi_3TiNbO_9 and $Bi_4Ti_3O_{12}$. These phases adopt layer structures consisting of PbO-like $(Bi_2O_2)^{2+}$ layers which alternate in the orthorhombic

Table 1—Hexagonal Perovskite Oxides Exhibiting B-Site Vacancy-ordering

Compound	Structural details	Reference
$Ba_3Re_2O_9$	$(hhc)_3$: 9H. ReO_6 octahedra are corner-shared. Vacancy is found between h-h layers.	26
$Ba_3W_2O_9$	$(h)_6$: 6H. Isostructural with $Cs_3Tl_2Cl_9$. Clusters of (W_2O_9) formed by two WO_6 octahedra sharing a common face. Two-thirds of octahedral sites in every layer are filled.	27
$Ba_3Te_2O_9$	$(h)_6$: 6H. Isostructural with $Cs_3Fe_2F_9$. Clusters of (Te_2O_9) similar to (W_2O_9) clusters in $Ba_3W_2O_9$. Octahedral sites in two adjacent layers are completely filled by Te followed by an empty layer of octahedral sites.	28
$Ba_5M_4O_{15}$ ($M = Nb, Ta$)	(hhccc): 5H. B-site vacancy is ordered between two h layers.	29,30
$Ba_3LaM_3O_{12}$ $Ba_4M_2WO_{12}$ ($M = Nb, Ta$)	$(hhcc)_3$: 12H. B-site vacancy occurs at the centre of three face-connected octahedra.	31-33
$Ba_6B_2^{III}W_3O_{18}$ ($B^{III} = Gd-Lu, Y$)	$(hhcccc)_3$: 18H. Structure consists of three octahedra sharing faces which alternate with three octahedra sharing corners. B-site vacancy is ordered at the central octahedral site of the face-sharing groups of octahedra.	34
$Ba_8Re_2W_3O_{24}$	$(hhhhchhc)_3$: 24H. Structure consists of groups of three and five face-shared octahedra which are linked through corners (Fig. 4e). B-site vacancy is ordered in alternate face-shared octahedral sites.	35
$Ba_9Nb_6WO_{27}$	$(hhccchhcc)_3$: 27H. Structure consists of groups of three face-shared octahedra which are linked to one another through one and two corner-sharing octahedra alternatively (Fig. 4f). Central B-site in the group of three face-sharing octahedra is vacant.	36

c-direction with perovskite-like slabs of the formula $(A_{n-1}B_nO_{3n-1})^{2-}$ where $n = m - 1$. Bi_2WO_6 may also be considered to be the first member of the homologous series $Bi_2W_nO_{3n+3}$; the $n = 2$ member of this series, $Bi_2W_2O_9$, is known but the $n = 3$ member only occurs as isolated strips of intergrowth[42] in $Bi_2W_2O_9$.

4. Anion-deficient perovskites and vacancy-ordered structures

Anion vacancy nonstoichiometry in perovskite oxides is more common than that involving vacancies

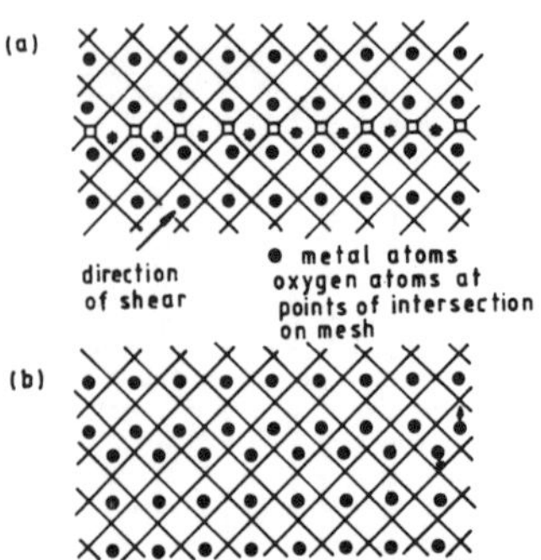

Fig. 5—Schematic representation of a shear plane (CS) formation: (a) idealized WO_3 structure showing anion vacancies aligned on a plane; and (b) sheared structure after vacancies are eliminated (following ref. 43).

in the cation sublattice. Anion vacancy nonstoichiometry is known with both ABO_3 and BO_3 frameworks. Nonstoichiometric tungsten trioxide, WO_{3-x}, is a typical example of the latter. Anion deficiency in WO_3 is accommodated by crystallographic shear (CS). In Fig. 5 is shown the formation of CS in WO_{3-x} schematically. When WO_3 becomes oxygen-deficient, oxygen vacancies are accumulated on specific planes (Fig. 5a) and this is accompanied by shearing of the lower part of the crystal relative to the upper part in the direction shown in Fig. 5a. This results in the oxygen atoms, marked by asterisk, being pushed to the position of oxygen vacancies. Vacancies are thus eliminated. Inspection of the sheared structure (Fig. 5b) reveals that cation coordination is restorted to six but the octahedra are now linked through edges instead of corners in the shear plane. This remarkable process of CS involving the elimination of anion vacancies results in the formation of the homologous series of phases, W_nO_{3n-2}, in the WO_{3-x} system, when the shear plane on $\{103\}$ recurs periodically.

An isolated CS plane is referred to as a Wadsley defect and a random array of CS planes is considered to constitute planar (extended) defects which are entirely different from point defects. It is obvious that when CS planes occur at regular intervals, the composition of the crystal is stoichiometric, whereas a random array of CS planes results in nonstoichiometric compositions. While we have invoked anion vacancies which are later annihilated in our description of CS plane formation, we must point out that vacancies are not essential precursors for the formation of CS planes. Accommodating anion-deficient nonstoichiometry through CS mechanism is a special feature restricted to d^0 metal oxides such as WO_3, Nb_2O_5 and TiO_2 which exhibit soft phonon modes. Soft phonon modes in metal oxides arise from soft metal-orxygen potentials which permit large cation relaxation. The latter

INDIAN J. CHEM., VOL. 23A, APRIL 1984

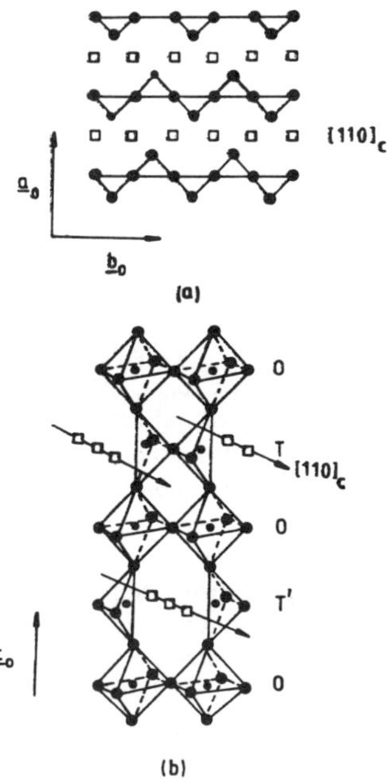

Fig. 6—Structure of brownmillerite: (a) vacancy ordering in the *a-b* plane; and (b) alternating sequence of octahedral and tetrahedral layers in the *c*-direction.

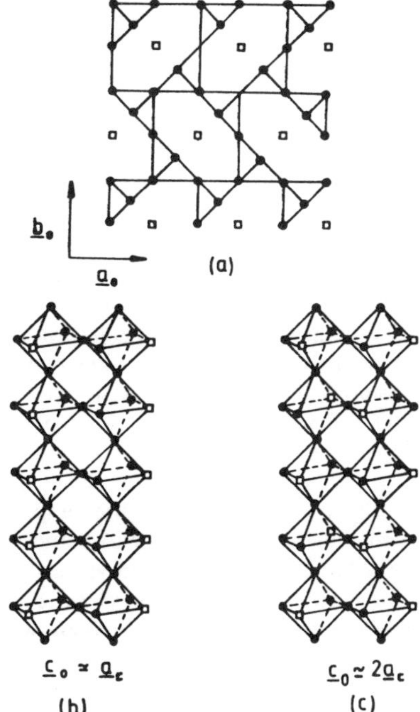

Fig. 7—Structures of $Ca_2Mn_2O_5$ and $Ca_2Co_2O_5$: (a) vacancy ordering in the *a-b* plane; vacancy ordering parallel to the *c*-direction in $Ca_2Mn_2O_5$; and $Ca_2Co_2O_5$ is shown in (b) and (c).

appears to be an essential prerequisite for CS plane formation[43]. CS planes are not formed for instance in SnO_{2-x} and TeO_{3-x} although both SnO_2 and TeO_3 crystallize in rutile and WO_3 related structures respectively.

Anion-deficient nonstoichiometry in ABO_{3-x} perovskites is not accommodated by the CS mechanism. The reason probably is that the constant A/B ratio required by the composition of perovskites prevents formation of CS planes. Defect-ordering in ABO_{3-x} oxides involves a conservative mechanism in the sense that the vacancies are assimilated into the structure resulting in large supercells of the basic perovskite structure. The type of superstructure formed depends however on the identity of the B-cation.

One of the best-characterized perovskite oxides with ordering of anion vacancies[44,45] is the brownmillerite structure exhibited by $Ca_2Fe_2O_5$ and Ca_2FeAlO_5. The compositions could be considered as anion-deficient perovskites with one-sixths of anion sites being vacant. The orthorhombic unit cell of brownmillerite structure ($a = 5.425$, $b = 5.598$ and $c = 14.768$ Å for $Ca_2Fe_2O_5$) arises because of vacancy ordering and is related to the cubic perovskite as $a \simeq \sqrt{2}a_c, b \simeq \sqrt{2}a_c, c \simeq 4a_c$. Oxygen vacancies are ordered in alternate (001) BO_2 planes of the cubic perovskite structure such that alternate [110] rows of oxygens are missing (Fig. 6a). The layer

sequence along the *c*-axis is therefore $CaO - FeO_2 - CaO - FeO...$ A slight shift of the iron atoms in the FeO layers leads to a tetrahedral coordination of iron in this layer resulting in alternate sheets of FeO_6 octahedra and FeO_4 tetrahedra (OTOT'....) in the *c*-direction (Fig. 6b). We have recently prepared a new oxide, Ca_2FeCoO_5, possessing the brownmillerite structure[46] wherein Fe^{3+} and Co^{3+} occupy tetrahedral and octahedral sites respectively.

The compounds $Ca_2Mn_2O_5$ and $Ca_2Co_2O_5$ which have compositions similar to brownmillerite have recently been reported[46-48], but their structures involve different modes of vacancy ordering. $Ca_2Mn_2O_5$ adopts an orthorhombic structure with $a = 5.432$, $b = 10.242$ and $c = 3.742$ Å, the relationship with the cubic perovskite structure being $a \simeq \sqrt{2}a_c, b \simeq \sqrt{2}a_c$ and $c \simeq a_c$. Oxygen vacancies in $Ca_2Mn_2O_5$ are ordered in every (001) BO_2 plane of the cubic perovskite such that one-half of the oxygen atoms in alternate [110] rows are removed (Fig. 7). The composition of the plane is now $MnO_{1.5}$. If these are stacked in the sequence $CaO - MnO_{1.5} - CaO - MnO_{1.5}...$, the structure of $Ca_2Mn_2O_5$ is obtained. An important feature of the structure which distinguishes it from brownmillerite is that manganese is square-pyramidally coordinated. It has been justifiably argued[47] that the $3d^4$-configuration of Mn^{3+} favours square-pyramidal coordination rather than tetrahedral coordination, thereby determining the anion vacancy ordering in $Ca_2Mn_2O_5$. $Ca_2Co_2O_5$ also crystallizes in an orthorhombic structure with $a = 11.12$, $b = 10.74$ and $c = 7.48$ Å. Comparison of these parameters with those of $Ca_2Mn_2O_5$ reveals a doubling of the a and c axes in $Ca_2Co_2O_5$. We propose that vacancy-ordering in $Ca_2Co_2O_5$ is similar to that in $Ca_2Mn_2O_5$, but the doubling of the c axis is likely to be due to the ordering scheme shown in Fig. 7. $Sr_2Co_2O_5$, on the other hand, crystallizes in an anion-deficient 2H perovskite structure at low temperatures and transforms to the brownmillerite structure[49,50] above 900°C. $SrCoO_{3-x}$ $(0 < x < 0.5)$ prepared under high oxygen pressures (50-2600 bars) adopts a cubic perovskite structure. The oxygen deficiency influences the unit cell parameter and the Curie temperature[51].

The compounds $SrTiO_{2.5}$, $SrVO_{2.5}$ and $KTiO_{2.5}$ are a few other anion vacancy perovskites[7] with the composition $ABO_{2.5}$. $KTiO_{2.5}$ shows little similarity to the perovskites, each titanium having a trigonal bipyramid oxygen coordination. X-ray diffraction studies of $SrTiO_{2.5}$ and $SrVO_{2.5}$ show cubic perovskite unit cells with $a \simeq 3.9$ Å, implying that anion vacancies are random. An electron diffraction study[52] of $SrTiO_{2.5}$ has revealed a six-fold superlattice ordering in the [111] direction. Subsequent work by Tofield[53] however did not substantiate the results. It is noteworthy that both $SrTiO_{3-x}$ and $SrVO_{3-x}$ exhibit metallic properties indicating that d-electrons associated with B-cations are itinerant. This accounts for the difference in the type of vacancy-ordering in these two oxides from the ordering in $CaMO_{2.5}$ (M = Mn, Fe, Co) where the d-electrons are localized.

In many of the ABO_{3-x} perovskites, non-stoichiometry spans the composition range $0 < x \le 0.5$ and we have so far discussed only the structures of the end members with $x = 0.5$. It is of interest to see the type of structures prevailing at intermediate compositions. In $SrFeO_{3-x}$, MacChesney *et al.*[54] reported a cubic structure when $0.0 < x < 0.12$, a tetragonally distorted perovskite structure when $0.15 < x < 0.28$ and a mixture of the perovskite and brownmillerite phases when $0.28 < x < 0.5$. $SrFeO_{2.75}$ was investigated by Tofield *et al.*[55] who found a supercell related to the perovskite as $a \simeq 2\sqrt{2}a_c$, $b \simeq 2a_c$ and $c \simeq 2\sqrt{2}a_c$ in the electron diffraction patterns. The vacancy-ordering suggested for this phase is related to that of brownmillerite; every other oxygen atom is lost in alternate $[110]_c$ strings so as to give five-coordination to half the iron atoms, the other half retaining octahedral coordination as in perovskite. In $Sr_xNd_{1-x}FeO_{3-y}$, superstructures involving random doubling of one of the perovskite cell axes (in a, b or c

directions) has been found[56]. This has been considered to be due to the existence of a microdomain texture wherein the domains intergrow randomly in three dimensions.

Reller *et al.*[57,58] have investigated in great detail the anion-deficient $CaMnO_{3-x}$ system over the range $0 < x \le 0.5$. Five distinct compositions, $CaMnO_{2.5}$, $CaMnO_{2.556}$, $CaMnO_{2.667}$, $CaMnO_{2.75}$ and $CaMnO_{2.80}$, with ordered oxygen vacancies have been identified and in all these structural features of the parent perovskite are preserved. The structure of $CaMnO_{2.5}$ possessing square-pyramidal coordination of Mn^{3+} was discussed earlier. In the other compositions, the proportion of square-pyramids relative to octahedra decreases as we go from $CaMnO_{2.5}$ to $CaMnO_{3.0}$. These structures may be pictured as superlattice repeats of the parent undistorted perovskite, but with the superlattice mesh rotated by an angle R in the (001) plane. In Fig. 8, is presented a comparison of the schematic arrangement of MnO_6 octahedra in an idealized cubic perovskite $CaMnO_3$ with the observed arrangement of the slightly distorted MnO_6 octahedra in $CaMnO_3$. In Fig. 9, the arrangements of MnO_6 octahedra and MnO_5 square-pyramids in $CaMnO_{2.8}$ are shown. The superlattice in $CaMnO_{2.8}$ is $\sqrt{5} \times \sqrt{5}$, but rotated by 26.5° and this structure represented in Fig. 10 gives tetragonal unit cell with $a = b = 8.34$ Å and $c = 7.46$ Å. In order to check this structure, lattice images have been computed and compared with the HREM image (Plate 4); the observed agreement tends to support the structural model. The structure of $CaMnO_{2.75}$ can be represented as $\sqrt{2} \times 2\sqrt{2}\, R\, 45°$ or

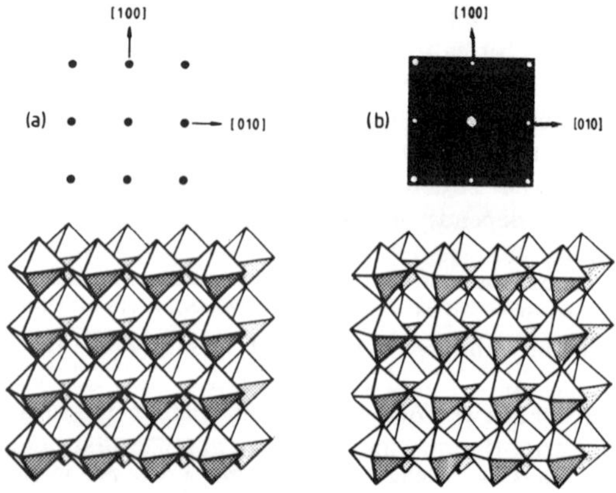

Fig. 8—(a) [001] zone axis electron diffraction pattern and schematic arrangement of MnO_6 octahedra in the idealized cubic perovskite, $CaMnO_3$ $(a = b = c = 3.73$ Å) (b) observed [001] zone axis electron diffraction and schematic arrangement of the slightly distorted MnO_6 octahedra of $CaMnO_3$ with $a = b = c = 7.46$ Å (after ref. 57).

INDIAN J. CHEM., VOL. 23A, APRIL 1984

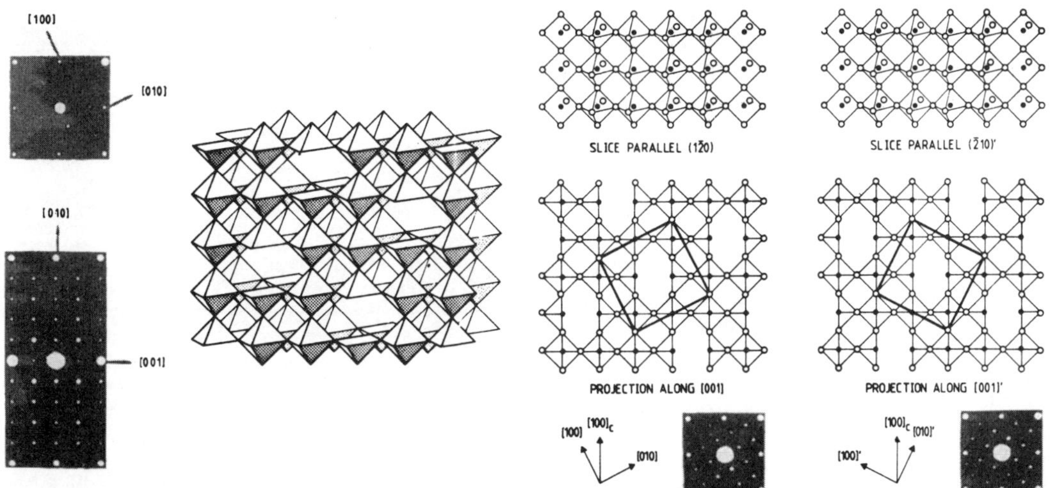

Fig. 9—Observed [001] and [100] zone axes electron diffraction patterns and the schematic arrangement of MnO_6 octahedra and MnO_5 square pyramids of $CaMnO_{2.8}$ with $a=b=8.34$ Å and $c=7.46$ Å (after ref. 57).

Fig. 10—Schematic representation of the structure and [001] zone axis electron diffraction patterns of $CaMnO_{2.8}$: Full circles, oxygen; closed circles, Mn; dotted broken circles, oxygen vacancies, Ca atoms are not shown for simplicity (after ref. 57).

$\sqrt{2}\times4\sqrt{2}\,R\,45°$, while that of $CaMnO_{2.667}$ is $\sqrt{2}\times3\sqrt{2}\,R\,45°$. Four ordered structures of $CaMnO_{2.5}$ have been observed, but the most common one is the $\sqrt{2}\times2\sqrt{2}\,R\,45°$ phase described by Poeppelmeier et al[47]. The relation between the defect structures of the members of the $CaMnO_{3-x}$ system and their catalytic activity has been commented upon by Reller et al[58].

Vacancy ordering in anion deficient $LaNiO_3$ was investigated by Gai and Rao[59]. Thermogravimetry in air and oxygen provided evidence for the formation of a series of phases pf the general formula $La_nNi_nO_{3n-1}$ with $n=7$, 9, 13 and 30 (Fig. 11). An electron diffraction study of the anion deficient samples

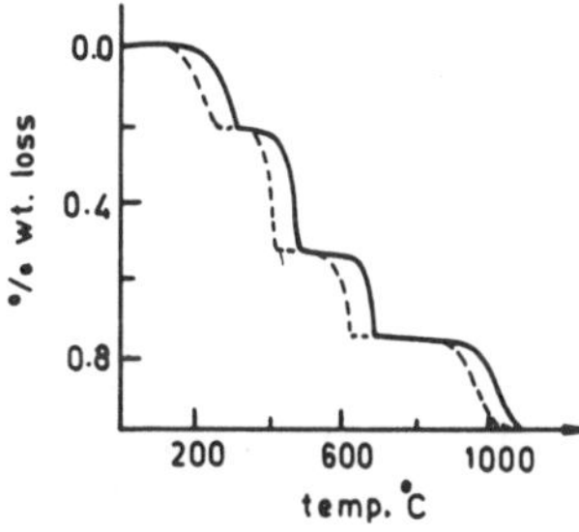

Fig. 11—Thermogram of $LaNiO_3$ showing stages and the relative ease with which oxygen is lost (after ref. 59).

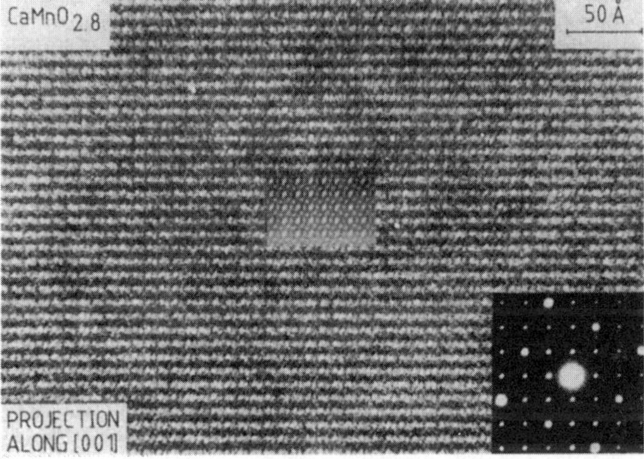

Plate 4—HREM image and calculated image (inset) of $CaMnO_{2.8}$ (from ref. 57).

suggested the formation of superstructures due to vacancy ordering. An orthorhombic or body-centred tetragonal cell with $a \simeq c \simeq 2\sqrt{2}a_c$ and $b \simeq 2a_c$ has been identified by electron diffraction. Crespin *et al.*[60] have recently reported that $La_2Ni_2O_5$ and $LaNiO_2$ are obtained by hydrogen reduction of $LaNiO_3$ around 550 and 670K respectively. The X-ray diffraction pattern of $La_2Ni_2O_5$, which corresponds to the $n = 2$ member of the $La_nNi_nO_{3n-1}$ series proposed by Gai and Rao[59], has been indexed on a monoclinic system with $a = 11.068, b = 11.168, c = 7.824$ Å and $\beta = 92.21°$ and a brownmillerite-type ordering of anion vacancies has been suggested. Since tetrahedral coordination of Ni^{2+} is unlikely to be present in oxides, we feel that anion vacancy ordering in $La_2Ni_2O_5$ may involve octahedral and square-planar coordination around nickel. Regarding $LaNiO_2$, there is much more doubt; it is rather strange that an oxide with nickel in the $1+$ state is stable in air.

Stabilities of $LaBO_3$ (B = V, Cr, Mn, Fe, Co and Ni) perovskites at 1270K in reducing atmosphere have been investigated by Nakamura *et al*[61] by thermogravimetry. The results reveal that the stabilities depend on the identity of the B atom, varying in the order, $LaVO_3 \sim LaCrO_3 > LaFeO_3 > LaMnO_3 > LaCoO_3 > LaNiO_3$. Under the experimental conditions employed in this study, $LaBO_{3-x}$ were however not obtained.

Grenier *et al.*[45,62] have investigated $Ca_xLa_{1-x}FeO_{3-y}$ (compositions intermediate between $Ca_2Fe_2O_5$ and $LaFeO_3$) and $CaTi_{1-x}Fe_xO_{3-y}$ (compositions intermediate between $Ca_2Fe_2O_5$ and $CaTiO_3$) systems in an attempt to characterize anion vacancy ordering at compositions intermediate between brownmillerite and perovskite. Their work has revealed the existence of a homologous series of phases of the general formula $A_nB_nO_{3n-1}$ (A = La, Ca; B = Ti, Fe). The structure of these phases consists of a succession of $(n-1)$ perovskite-like BO_6 sheets (O layers) and a sheet of MO_4 tetrahedra (T layers) in the c-direction. When $n = 2$ we have the brownmillerite phase where the sequence of octahedral and tetrahedral layers is ....OTOT'OTOT'.... where T and T' stand for two different orientations of tetrahedra in the structure (Fig. 6). When $n = \infty$, there will be no tetrahedral layers and the sequence is ....OOOO.... corresponding to the perovskite structure. Other members of this series are $Ca_2LaFe_3O_8$, $Ca_3Fe_2TiO_8$ ($n = 3$) and $Ca_4Fe_2Ti_2O_{11}$ ($n = 4$). The layer sequences in the $n = 3$ and $n = 4$ members are ....OOT OOT.... and ....OOOT OOOT.... respectively. An interesting member of this series is $Ca_4YFe_5O_{13}$ which corresponds to $n = 2.5$ in the series. The structure of this oxide, determined by high resolution (1 MeV) electron

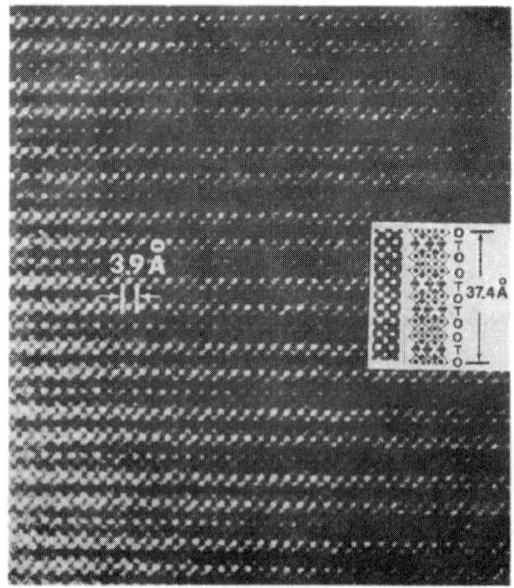

Plate 5—HREM image of $Ca_4YFe_5O_{13}$ showing ordered intergrowth of $n = 2$ and $n = 3$ units of the $A_nB_nO_{3n-1}$ series (from ref. 63).

microscopy[63], consists of a regular intergrowth of $n = 2$ and $n = 3$ units of the $A_nB_nO_{3n-1}$ series, the polyhedral layer sequence being ....OTOOTOTOOT.... (Plate 5). Other ordered intergrowth phases reported are $Ca_7Fe_6TiO_{18}$ ($n = 2.33$) and $Ca_5Fe_4TiO_{13}$ ($n = 2.5$).

Recent studies[64–66] have shown the non-stoichiometric behaviour in such Fe containing systems to be highly complex, the type of long-range ordering (or the absence of it) being dependent on sample preparation conditions. Intermediate compositions in the $Ca_xLa_{1-x}FeO_{3-y}$ ($2/3 < x < 1$) system exhibit two different types of behaviour. When the preparations are carried out in a non-oxidizing atmosphere at relatively low temperatures (1100°C), the material exhibits a disordered intergrowth structure comprising fragments of $Ca_2LaFe_3O_8$ ($n = 3$) and $Ca_2Fe_2O_5$ (Plate 6). When the samples are prepared at 1400°C in air, they become oxygen-rich and exhibit a totally different structural behaviour. Powder X-ray diffraction patterns of the latter samples show deceptively simple cubic perovskite structures with $a \simeq 3.85$ Å. Electron diffraction and HREM reveal the structures to be more complex consisting of microdomains of brownmillerite or $Ca_2LaFe_3O_8$ which are intergrown three-dimensionally so as to give an arrangement that appears cubic perovskite-like on the average. The dimension of individual domains is of the order $\sim 10^6$ Å^3 and decreases with increasing oxygen content. It is significant that even at the highest

INDIAN J. CHEM., VOL. 23A, APRIL 1984

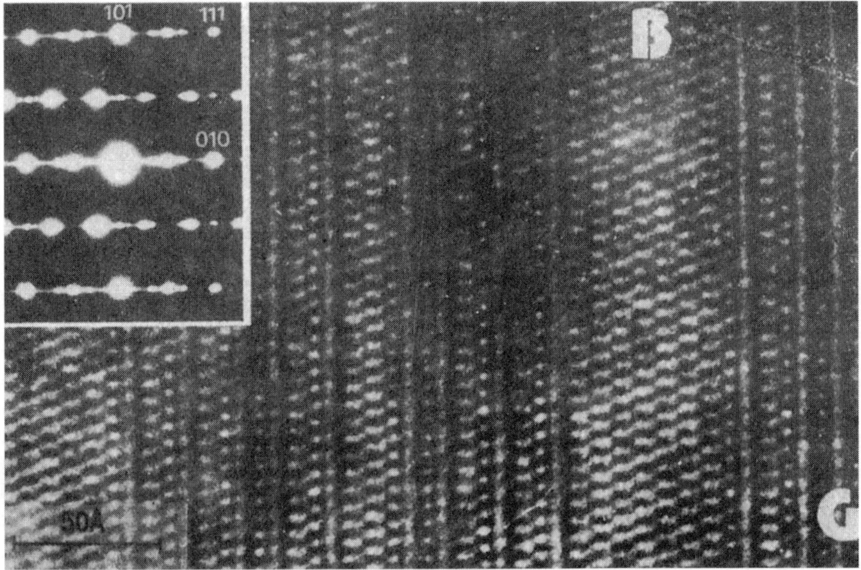

Plate 6—Lattice image of $Ca_xLa_{1-x}FeO_{3-y}$ showing disordered intergrowth. B stands for brownmillerite structure and G stands for $Ca_2LaFe_3O_8$ structure (from ref. 65).

temperatures of preparation the system does not exhibit classical nonstoichiometric behaviour consisting of randomly distributed point defects; apparent point defects are always structurally ordered giving microdomain texture of the material.

Anion-deficient hexagonal perovskites exhibit interesting structural behaviour involving polytypism. $BaMO_{3-x}$ (M = Mn, Fe, Co etc.) are prototype examples of this type of nonstoichiometric system. The nonstoichiometry is accommodated by these systems by adjusting the relative proportion of cubic and hexagonal stacking of AO_{3-x} layers. HREM provides direct verification of the stacking arrangement of the BaO_3 layers in $BaMO_3$ polytypes[67]; a typical image of a 4H polytype of $BaCrO_3$ is shown in Plate 7. 6H-$BaCrO_3$, 9R-$BaRuO_3$ and 4H-$BaRuO_3$ have also been identified by HREM. By employing HREM, the structure of $BaIrO_3$ has been solved[68]; this material is found to possess a 9H structure (rather than the 9R structure of $BaRuO_3$) admixed with 4H sequences.

The results of Negas and Roth[69] on the $BaMnO_{3-x}$ system are presented in Table 2 to illustrate the structural complexity of this system. We see that with decreasing oxygen content, cubic stacking increases in the structure. The structural changes accompanying oxygen nonstoichiometry are however not true polytypic transformations. The $BaFeO_{3-x}$ system[70,71] is equally complex and shows a different structural behaviour: $BaFeO_{2.5}$ (triclinic, brownmillerite type), $BaFeO_{2.62-2.64}$ (rhombohedral), $BaFeO_{2.67}$ (tri-

Table 2—Phase Relations of $BaMnO_{3-x}$ System in Air*

$BaMnO_{3-x}$	Temp °C	Layer sequence	% Cubic stacking
$x = 0$	1150	2H	0
$0 < x < 0.02$	1150-1300	15H	20
$0.03 < x < 0.05$	1300-1350	8H	25
$0.10 < x < 0.15$	1350-1475	6H	$33\frac{1}{3}$
$0.175 < x < 0.20$	1475-1550	10H	40
$x \sim 0.25$	1550	4H	50

*After ref. 69.

clinic), $BaFeO_{2.75-2.81}$ (tetragonal) and $BaFeO_{2.69-2.95}$ (6H hexagonal). A 12H phase has also been described but its composition is not known definitely. A structural investigation[72] of 6H $BaFeO_{2.79}$ shows that oxygen vacancies are unevenly distributed between the cubic and hexagonal layers which are arranged in the sequence $(cch)_2$. The hexagonal layers have the composition $BaO_{2.5}$ and the cubic layers $BaO_{2.835}$. The oxygen vacancy ordering in $BaO_{2.5}$ layer appears to be similar to the ordering in $SrO_{2.5}$ layers of $SrFeO_{2.5}$ where one [110] string of oxygen in every four is removed converting part of face-shared FeO_6 octahedra into corner-shared FeO_4 tetrahedra. The structure of 6H $BaFeO_{2.79}$ is to be contrasted with that of the manganese system. In 4H $Ba_{0.5}Sr_{0.5}MnO_{2.85}$, for instance, all the vacancies are found in the

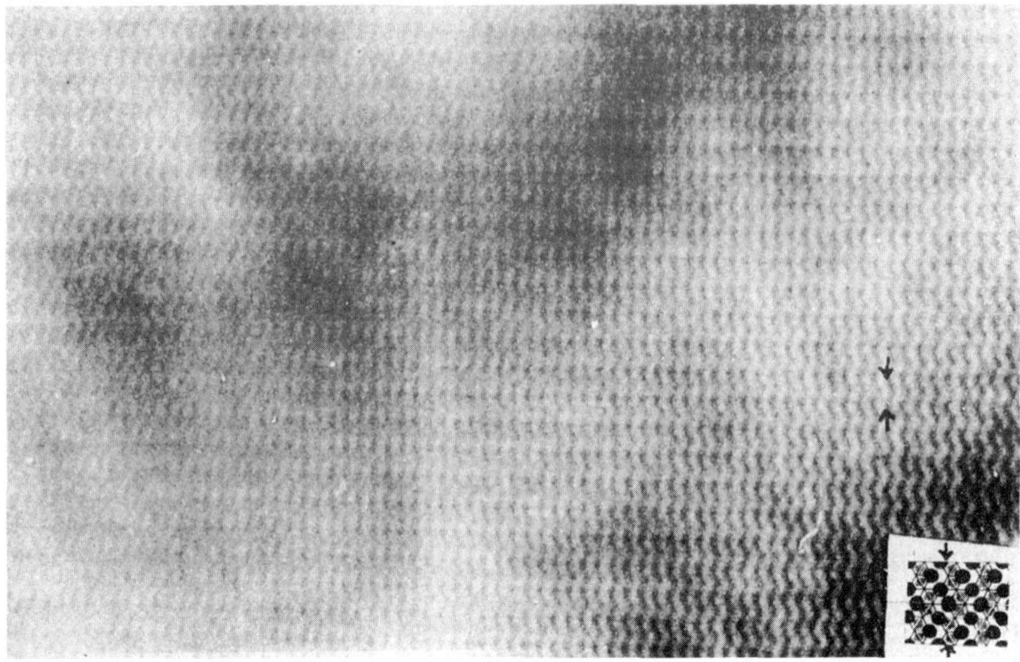

Plate 7—Lattice image of 4H-BaCrO$_3$ showing corrugative pattern of BaO$_3$ due to chch stacking (from ref. 67). A projection drawing is shown in the inset.

hexagonal layers, converting pairs of face-sharing octahedra in the parent 4H structure into edge-sharing trigonal bipyramides[73,74]. The difference in structural behaviour between the iron and manganese systems has been attributed to the difference in coordination tendencies of Fe^{3+} and Mn^{3+} high-spin Mn^{3+} (d^4) prefers trigonal bipyramidal or square-pyramidal environment while high-spin Fe^{3+} (d^5) accepts both octahedral and tetrahedral coordination.

Structures of the members of the BaCoO$_{3-x}$ system are interesting. Zanne *et al.*[75] found the following phases in this system at 1178 K: 2H BaCoO$_{3.0-2.85}$, 7H BaCoO$_{2.575-2.520}$, 12H BaCoO$_{2.49-2.43}$, 15H BaCoO$_{2.23-2.10}$ and an orthorhombic BaCoO$_{2.07}$. Among the various phases reported, a 12H phase is well-established in this system. The structure of 12H BaCoO$_{2.6}$ has been determined by HREM and powder neutron diffraction[76,77]. The layer sequence of this phase is (ccchh)$_2$. Oxygen vacancies are non-random in this structure being accommodated by replacing one-sixths of BaO$_3$ layers by BaO$_2$ layers (Fig. 12). As a consequence, one-third of the cobalt atoms are tetrahedrally coordinated by oxygen. The CoO$_4$ tetrahedra are linked through corners to strings of four face-shared octahedra that contain the remaining cobalt. Assuming that tetrahedral-site cobalts are 4 + and octahedral-site cobalts are 3 +, one arrives at the

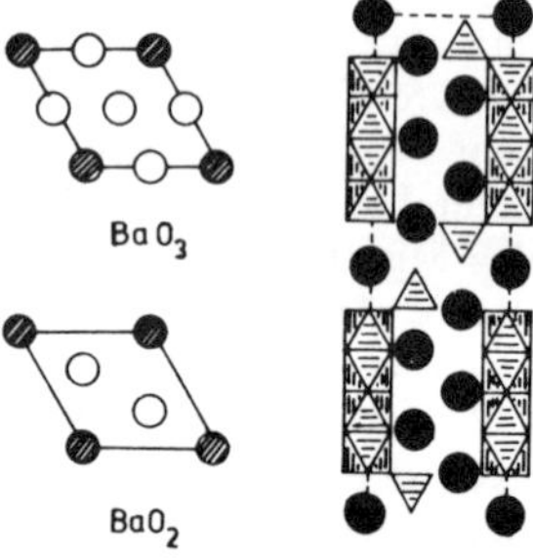

Fig. 12—Structure of BaCoO$_{2.61}$ (a) BaO$_3$ and BaO$_2$ layers and (b) (11.0) projection of the structure showing linkage of CoO$_4$ tetrahedra and CoO$_6$ octahedra (after ref. 77).

ideal composition of the 12H phase as BaCoO$_{2.67}$. Replacement of BaO$_3$ by BaO$_2$ layers leading to tetrahedral coordination of B-metal atoms is a novel way of accommodating oxygen deficiency in ABO$_3$ perovskites, which may be prevalent in other systems as well. BaVO$_{3-x}$ which crystallizes in a structure closely related to Ba$_3$V$_2$O$_8$ may belong to this category[78].

An interesting case[79] of a perovskite with high oxygen-deficiency ($x > 0.5$) where the B-metal is present in three different coordination environments is La$_3$Ba$_3$Cu$_6$O$_{18-y}$. Phases with varying y adopt a tetragonal cell related to the perovskite structure: $a_T \simeq \sqrt{2}a_c$ and $C_T \simeq 3a_c$ where a_c is the cubic perovskite parameter. The detailed structure of $La_3Ba_3Cu_6O_{14.1}$ has been determined. Oxygen vacancies are ordered in the AO layers of the perovskite structure and the layer sequence in the ordered structure along [001] is $Cu_2O_4 - A_2O - Cu_2O_4 - A_{2\,\,2} - Cu_2O_4 - A_2O$ .... This type of ordering of anion vacancies gives rise to three different coordination environments for copper: square-planar, square-pyramidal and distorted octahedral (Fig. 13). It has been suggested that Cu^{3+} ions occupy octahedral sites in this structure. It is clear from the foregoing examples that vacancy-ordering in anion-deficient $BaMO_{3-x}$ or $Ba_{1-y}Sr_yMO_{3-x}$ systems where M = Mn, Fe, Co or Cu is dictated by the coordination preference of the transition metal ion.

5. Anion-excess nonstoichiometry

Anion-excess nonstoichiometry in perovskite oxides is not as common as anion-deficient nonstoichiometry probably because introduction of interstitial oxygen in perovskite structure is energetically unfavourable. There are a few systems which show apparent oxygen-excess nonstoichiometry: $LaMnO_{3+x}$, $Ba_{1-x}La_xTiO_{3+x/2}$ and $EuTiO_{3+x}$. Tofield and Scott[80] investigated oxidative nonstoichiometry in $LaMnO_{3+x}$ by neutron diffraction. These workers suggest three possible models to accommodate oxygen excess in the perovskite structure: (i) interstitial oxygen at $(\frac{1}{2}00)$ face or $(\frac{111}{444})$ position of the cubic perovskite structure, (ii) cation vacancies at A and/or B sites leaving a perfect oxygen sublattice and (iii) formation of new oxidized phases. In $LaMnO_{3.12}$, diffraction results show that oxygen excess is accommodated by vacancies at A and B sites with partial elimination of La (as La_2O_3), the composition of the perovskite being $La_{0.94\,\,0.06}Mn_{0.98\,\,0.02}O_3$. Similar behaviour[81] is exhibited by $La_2TiCoO_{6.35}$, the composition being a mixture of a perovskite phase $La_{1.77}Ti^{4+}Co_{0.7}^{3+}Co_{0.3}^{2+}O_6$ and La_2O_3.

Where there is large oxygen excess, new phases possessing perovskite-related structures are formed. One of the ways of accommodating oxygen excess while retaining the features of the perovskite structure is to slice the cubic perovskite parallel to (110) face to give slabs of composition $(A_{n-1}B_nO_{3n+2})_\alpha$ and stack them one over the other by adding a layer of A-atoms in between. The process gives rise to a homologous series of general formula[11] $A_nB_nO_{3n+2}$ (Fig. 14). Important examples of this family are $Ca_2Nb_2O_7$ (ref. 82) and $La_2Ti_2O_7$ (ref. 83) which are $n=4$ members in the series. They may be regarded as oxidation products of $CaNbO_3$ and $LaTiO_3$ perovskites. Members of this family with n-values ranging from 1 to 6 have been characterized[84].

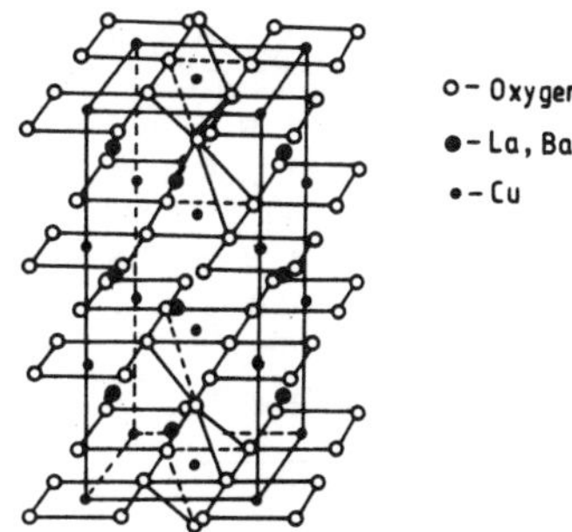

Fig. 13—Structure of $La_3Ba_3Cu_6O_{14.1}$ showing vacancy ordering and different coordination polyhedra around copper (after ref. 79).

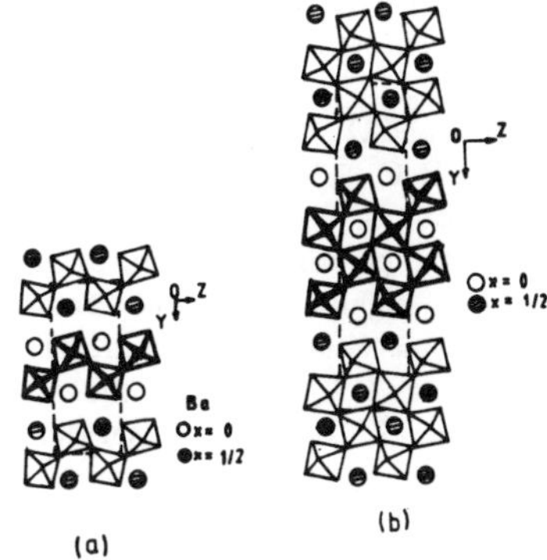

Fig. 14—Structure of $A_nB_nO_{3n+2}$ layered perovskites: (a) $n=2$ member and (b) $n=4$ member (after ref. 84).

6. Nonstoichiometry in oxides of structures related to perovskites

The first family of oxides that we consider here will be that related to the K_2NiF_4 structure. These oxides of the general formula A_2BO_4 contain alternating ABO_3 perovskite layers and AO rock-salt layers. Structural chemistry of these oxides has been reviewed recently[85]. Oxygen-excess nonstoichiometry is found in La_2NiO_4 where intergrowth phases containing $La_3Ni_2O_7$, $La_4Ni_3O_{10}$ etc. are found when the deviation is large[86]. Oxygen-deficient nonstoichiometry is found in Ca_2MnO_4 which can be reduced topotactically to $Ca_2MnO_{3.5}$ (Ref. 47). The unit cell dimensions are $a = 5.30$, $b = 10.05$ and $c = 12.24$Å, the relationship with the tetragonal K_2NiF_4 structure being $a \simeq \sqrt{2}a_T$, $b \simeq 2\sqrt{2}a_T$ and $c \simeq c_T$. The anion-vacancies are ordered along alternate $[110]_c$ rows as in $CaMnO_{2.5}$ to give sheets of (MnO_5) square-pyramids. Vidyasagar et al.[46] have recently characterized an oxide of the formula $Ca_2FeO_{3.5}$ with $a = 14.79$, $b = 13.71$ and $c = 12.19$Å. The structure of this oxide is different from that of $Ca_2MnO_{3.5}$. It is likely that the ordering of anion vacancies in $Ca_2FeO_{3.5}$ is similar to that in the brownmillerite structure consisting of

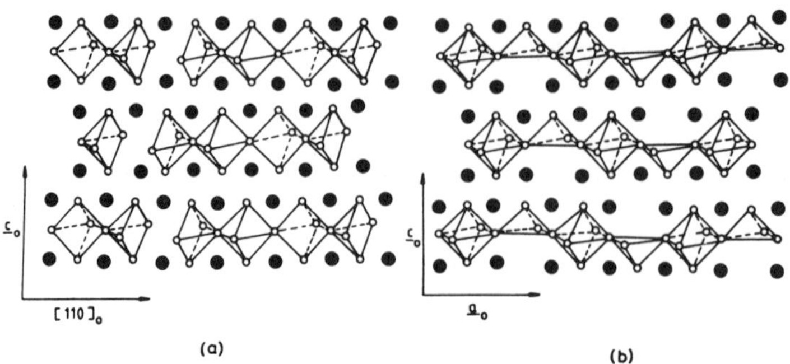

Fig. 15—Vacancy ordering in (a) $Ca_2MnO_{3.5}$ and (b) $Ca_2FeO_{3.5}$.

alternating rows of tetrahedra and octahedra in the perovskite-like layers of the K_2NiF_4 structure (Fig. 15).

Oxide pyrochlores, $A_2B_2O_6O'$, can tolerate vacancies at the A and O′ sites giving phases $A_2B_2O_6 \square$ (ABO_3) and $A \square B_2O_6 \square$ (AB_2O_6)[87]. A small number of $A_2B_2O_6$ compounds adopt the pyrochlore structure in preference to the perovskite structure when the A and B ions are highly polarizable but not too electropositive. Typical of these oxides are $Tl_2B_2O_6$ (B = Nb, Ta or U) and $A_2Sb_2O_6$. In the $A-Sb-O$ system[88] (A = Na, K or Ag), a series of non-stoichiometric compounds ASb_yO_z, with the maximum ratio of A : Sb depending on A, is known. Some of these compounds are superionic conductors. The defect pyrochlores of the composition AB_2O_6 will not be discussed here since they are not of direct relevance to this article.

7. Coherent intergrowth phases

In the last few years, the phenomenon of coherent intergrowth, wherein two or more phases possessing a common plane of nearly identical features intergrow randomly or in an ordered fashion, has emerged to become an important aspect of solid state inorganic chemistry. A variety of oxide systems belonging to the perovskite family show disordered intergrowth; disordered intergrowth of two or more phases necessarily involves nonstoichiometry if the parent phases have different compositions. Recurrent ordered intergrowth is found in some systems, $Ca_4YFe_5O_{13}$ being a typical example (Plate 5).

The $A_nB_nO_{3n+2}$ family shows an extraordinarily large number of ordered intergrowth phases when n is between 4 and 4.5 in the $Na-Ca-Nb-O$ system[84]; the oxide with $n = 4.5$ itself is a result of regular intergrowth between $n = 4$ and $n = 5$ compounds. When the n varies between 4.0 and 4.5 in these intergrowth phases, x in ABO_{3+x} varies between 3.444 and 3.500, but the large number of ordered phases found in this narrow range is truly remarkable. A typical ordered intergrowth phase (with a medium periodicity in this family) has the sequence 444544445 44454444445 (where 4 and 5 represent compositions corresponding to $n = 4$ and $n = 5$. In Plate 8 the HREM image of a member of this family is shown to illustrate the ordered recurrent intergrowth. Unlike the $A_nB_nO_{3n+2}$ family, the $A_{n+1}B_{3n+1}$ family obtained by cutting the perovskite structure into slabs along (100) planes does not form ordered intergrowth phases; instead disordered intergrowth is commonly observed as tÿified by the $Sr-Ti-O$ system[89]. In Plate 9 HREM image of $Sr_3Ti_2O_7$ containing lamellae of several other members of this series is shown. The La $-Ni-O$ system also shows disordered intergrowth of different members of the $A_{n+1}B_nO_{3n+1}$ family. Aurivillius phases on the other hand form intergrowth phases of the general formula $(Bi_2O_2)^{2+}$ $(A_{n-1}B_nO_{3n+1})^{2-}$ $(Bi_2O_2)^{2+}$ (A

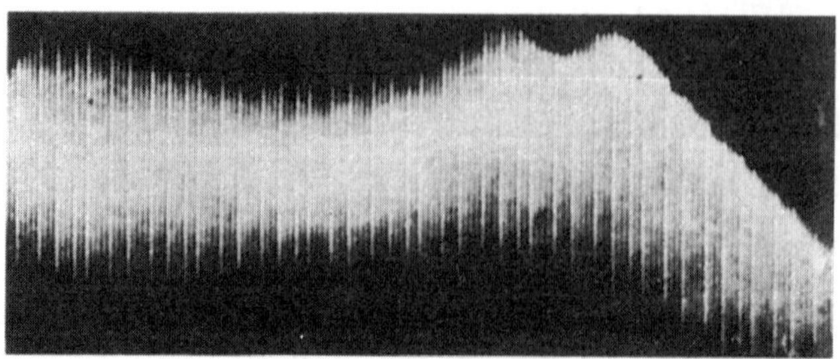

Plate 8—Lattice image of $A_nB_nO_{3n+2}$ oxides in the $Na-Ca-Nb-O$ system showing ordered recurrent intergrowth (from ref. 84).

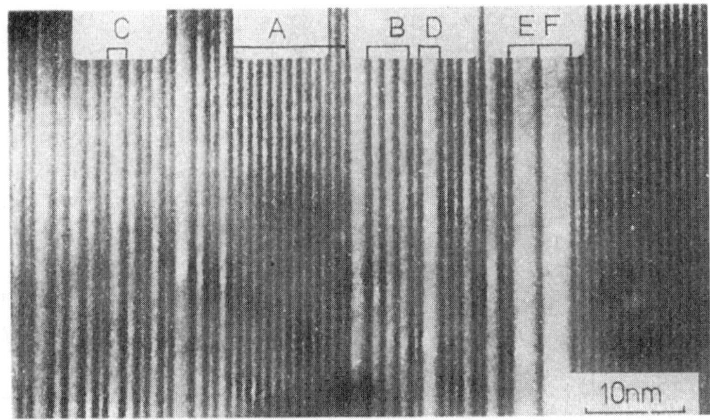

Plate 9—Lattice image of an oxide in the $Sr-Ti-O$ system showing disordered intergrowth among different members of the $Sr_{n+1}Ti_nO_{3n+1}$ family. A, $n=2$; B, $n=3$; C, $n=4$; D, $n=5$; E, $n=7$; and F, $n=8$ (from ref. 89).

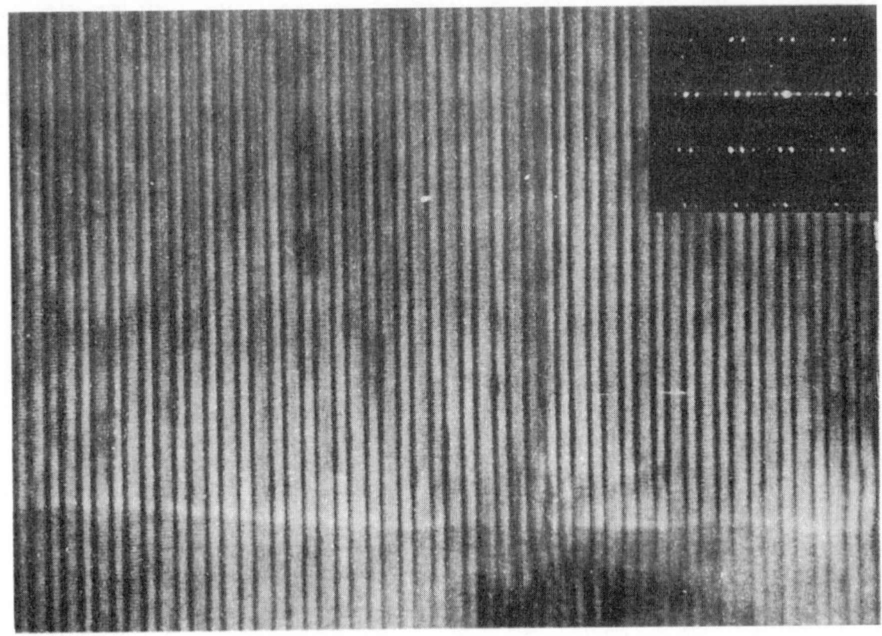

Plate 10—One-dimensional lattice image of $Bi_9Ti_6CrO_2$- showing ordered intergrowth between $n=3$ and $n=4$ members of the Aurivillius family.

$_nB_{n-1}O_{3n-4})^{2-}$. For example. $Bi_4Ti_3O_{12}$ ($n=3$) forms a regular ordered intergrowth phase with $n=4$ member $Bi_5Ti_3CrO_{15}$: in Plate 10 is shown a one-dimensional image of this compound to indicate the remarkable order in the intergrowth phase[90]. The image of a similar layered intergrowth phase between $Bi_4Ti_3O_{12}$ and $BaBi_4Ti_4O_{15}$ is shown in Plate 11. Long period Aurivillius phases show disordered intergrowth[91] as illustrated in Plate 12.

Barium ferrites of the general formula M_pY_q formed between $BaFe_{12}O_{19}$ (M. phase) and $Ba_2Me_2Fe_{12}O_{22}$ with $Me=Zn$, Co, Ni etc. (Y-phase) may also be considered here since the structures of these oxides consist of layers of barium and oxygen in a ratio of $1:3$ just as in the $\{111\}$ sheet of a cubic barium perovskite. This family of intergrowth phases starting from M_2Y to M_8Y_{12} with c-dimension periodicities going upto several hundred angstroms is found within the narrow

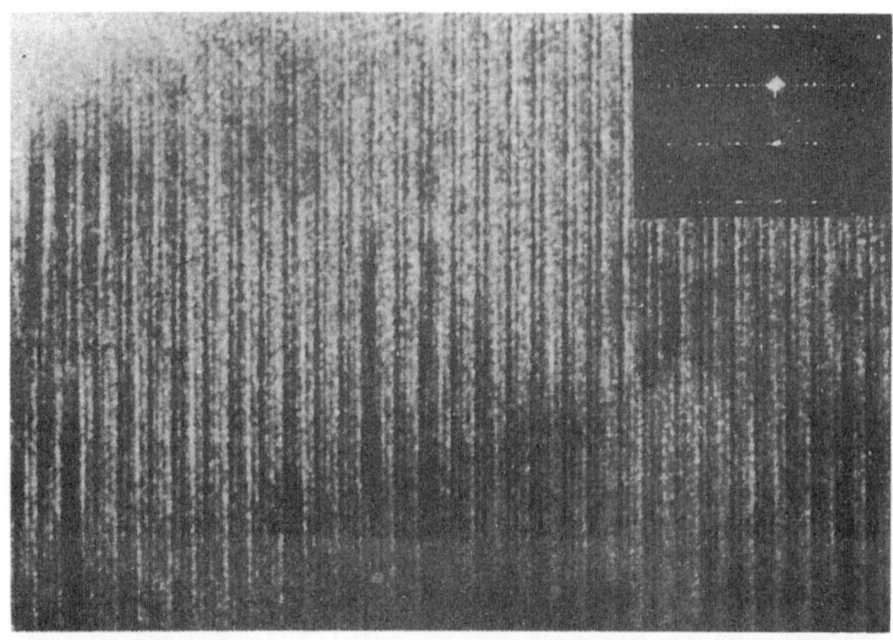

Plate 11—One-dimensional lattice image of $BaBi_8Ti-O_2$- showing ordered intergrowth of $n = 3$ and $n = 4$ members of the Aurivillius family.

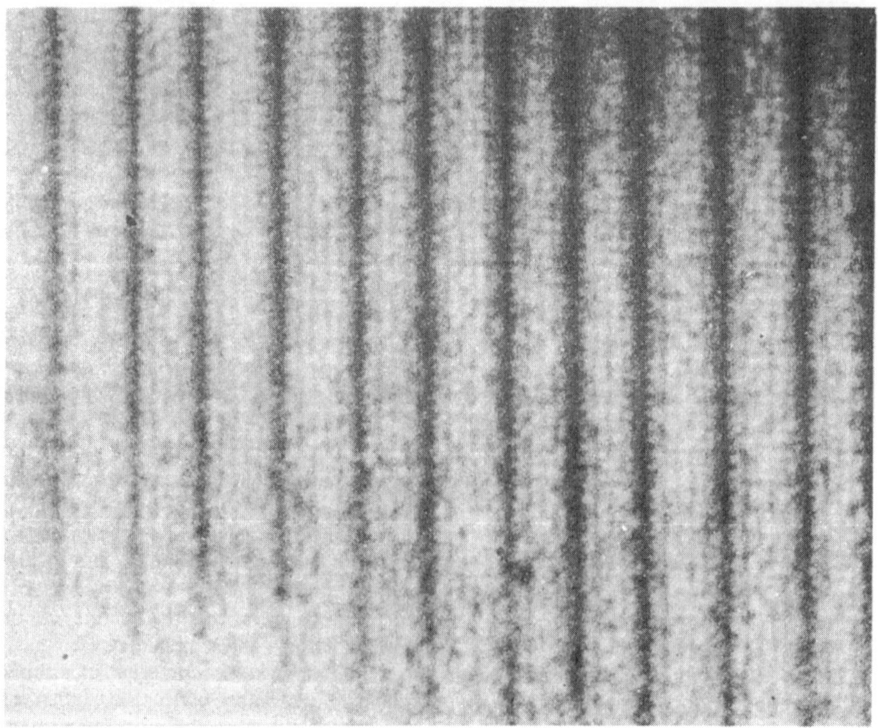

Plate 12—Disordered intergrowth of Aurivillius phases as illustrated in the case of $Bi_9Ti_3Fe_5O_2$- (from ref. 91).

INDIAN J. CHEM., VOL. 23A, APRIL 1984

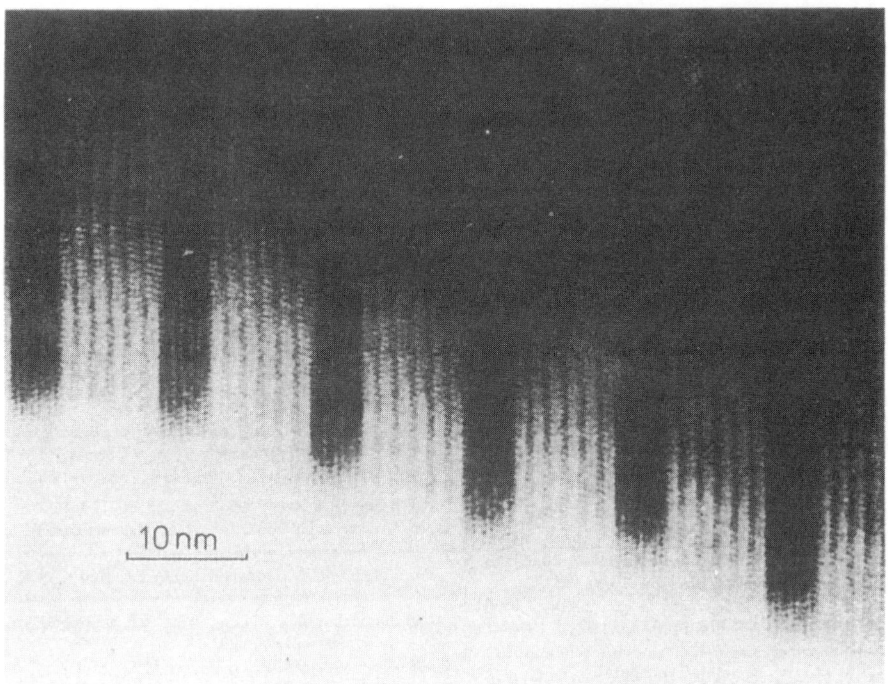

Plate 13—Lattice image of $MYMY_6$ intergrowth phase in the barium hexaferrite system (from ref. 92).

oxygen to metal ratio of 1.39179 and 1.42857. The large period intergrowth phases show considerable disorder. In Plate 13 we show the lattice image of a regular ordered intergrowth compound[92] belonging to this family.

It is truly remarkable that in none of the ordered or disordered intergrowth phases do we see any evidence of vacancies or other point defects. Nonstoichiometry is however present in all systems showing disordered intergrowth of two or more phases.

8. Concluding remarks

Defects in perovskite oxides can be due to cation vacancies (A or B site), anion vacancies or anion excess. Cation-deficient oxides such as A_xWO_3 give rise to oxide bronze structures, WO_3 itself representing the limiting case of the A-site deficient oxide; A-site vacancies are seldom ordered in these metallic systems. B-site vacancies are favoured in hexagonal perovskites and ordering of these vacancies gives rise to superstructures in some of the oxides.

Oxygen bacancies are commonly encountered in oxide perovskites. In ABO_{3-x}, unlike in WO_{3-x} and TiO_{2-x}, crystallographic shear planes are not found. Instead, a variety of superstructures are seen due to the ordering of vacancies. The brownmillerite phase of $Ca_2Fe_2O_5$ and oxides of the $CaMnO_{3-x}$ family are good examples of such vacancy-ordered structures. Complex intergrowth phases (ordered as well as disordered) involving brownmillerite and other related phases are commonly found in some of the anion-deficient oxides. Some of them also show polytypism due to different modes of stacking of the hexagonal and cubic layers.

Anion excess is found in some perovskites, though not commonly. $LaMnO_{3+x}$ is an example of an anion-excess perovskite with cation vacancies, while $LaTiO_{3.5}$ is a case of anion-excess perovskite with a layered structure. Anion excess also results in the formation of new structures, an example being the $A_nB_nO_{3n+2}$ system of oxides.

Ordering of point defects such as vacancies does not appear to be favoured in oxides with itinerant electrons (e.g. $SrTiO_{3-x}$). It is not clear that this is universal since oxides such as TiO exhibit vacancy ordering.

In spite of several investigations of defect perovskite systems in the last few years, there is much scope for study in this interesting aspect of solid state inorganic chemistry. The nature of defects as well as the ordering of defects in many systems are yet to be studied. For example, not much is known about the defect structures of $LaNiO_3$ and $LaCoO_3$, both of which are interesting materials. Relations between defect structures and properties, especially catalytic propoerties, of perovskite oxides can only be established based on systematic investigations of both structure and reactivity.

Acknowledgement

The authors thank the UGC, New Delhi and the Department of Science and Technology, Government of India, for supporting this work.

References

1 Rao C N R, *Chemica Scripta*, **19** (1982) 124.

2 Kroger F A, *The chemistry of imperfect crystals* (North-Holland, Amsterdam) 1964.

3 Libowitz G G, *Prog Solid State Chem*, Vol. 2, edited by H Reiss (Pergamon Press, Oxford) 1965.

4 Anderson J S, *Proc Roy Soc London*, **185** (1946) 67.

5 Anderson J S, *Modern aspects of solid state chemistry*, edited by C N R Rao (Plenum Press, New York) 1970.

6 Magneli A, *Arkiv Kemi*, **2** (1950) 513; see also *Acta Cryst*, **6** (1953) 495.

7 Wadsley A D, *Non-stoichiometric compounds*, edited by L Mandelcorn (Academic Press, New York) 1964.

8 Anderson J S, *Defects and transport in oxides*, edited by M S Seltzer & R I Jaffe (Plenum Press, New York) 1974.

9 *The chemistry of extended defects in nonmetallic solids*, edited by L Eyring & M O'Keeffe (North-Holland, Amsterdam) 1970.

10 *Nonstoichiometric oxides*, edited by O T Sorensen (Academic Press, New York) 1981.

11 Tilley R J D, *Chemical physics of solids and their surfaces*, Vol. 8, edited by M W Roberts & J M Thomas (Royal Society of Chemistry, London) 1980.

12 Rao C N R & Subba Rao G V, *Physica Status Solid*, **A1** (1970) 597.

13 Goodenough J B & Longo J M, *Landolt-Bornstein Tabellen*, New Series, Group III/Vol. 4a (Springer Verlag, Berlin) 1970.

14 Glazer A M & Megaw H D, *Phil Mag*, **25** (1972) 1119.

15 Glazer A M, *Acta Cryst*, **B28** (1972) 3384.

16 Katz L & Ward R, *Inorg Chem*, **3** (1964) 205.

17 Ekstrom T & Tilley R J D, *Chemica Scripta*, **16** (1980) 1.

18 Atoji M & Rundle R E, *J chem Phys*, **32** (1960) 627.

19 Anderson J S, *Surface and defect properties of solids*, Vol. 1, edited by M W Roberts & J M Thomas (The Chemical Society, London) 1972.

20 Longo J M & Sleight A W, *Materials Res Bull*, **10** (1975) 1273.

21 Vincent H, Bochu B, Aubert J J, Joubert C C & Marezio M, *J Solid State Chem*, **24** (1978) 245.

22 Kihlborg L, *Chemica Scripta*, **14** (1978) 187.

23 (a) Jefferson D A, Uppal M K, Smith D J, Gopalakrishnan J, Ramanan A & Rao C N R, *Materials Res Bull*, (1984) in print.

(b) Ramanan A, Gopalakrishnan J, Uppal M K, Jefferson D A & Rao C N R, *Proc Roy Soc London*, under publication.

24 Sundberg M, *Chem Commun*, (Univ Stockholm), **5** (1981).

25 Rauser G & Kemmler-Sack S, *J Solid State Chem*, **33** (1980) 135.

26 Calvo C, Ng N N & Chamberland B L, *Inorg Chem*, **17** (1978) 699.

27 Poeppelmeier K R, Jacobson A J & Longo J M, *Materials Res Bull*, **15** (1980) 339.

28 Jacobson A J, Scanlon J C, Poeppelmeier K R, Longo J M & Cox D E, *Materials Res Bull*, **16** (1981) 359.

29 Galasso F & Katz L, *Acta Cryst*, **14** (1961) 647.

30 Shannon J & Katz L, *Acta Cryst*, **B26** (1970) 102.

31 Kemmler-Sack S, *Z Anorg Allgem Chem*, **454** (1979) 63.

32 Kemmler-Sack S, *Z Anorg Allgem Chem*, **457** (1979) 157.

33 Kemmler-Sack S, *Z Anorg Allgem Chem*, **461** (1980) 151.

34 Schittenheim H J, Fadini A & Kemmler-Sack S, *Z Anorg Allgem Chem*, **457** (1979) 149.

35 Kemmler-Sack S & Treiber U, *Z Anorg Allgem Chem*, **451** (1979) 129.

36 Kemmler-Sack S & Treiber U, *Z Anorg Allgem Chem*, **462** (1980) 166.

37 Pauling L, *The nature of the chemical bond* (Cornell Univ. Press, New York) (1960) 559.

38 Blasse G & Dirksen G J, *J Solid State Chem*, **36** (1981) 124.

39 Brown R & Kemmler-Sack S, *Naturwissenchaften*, **70** (1983) 463.

40 Kemmler-Sack S & Ehmann A, *Naturwissenchaften*, **70** (1983) 250.

41 Aurivillius B, *Ark Kemi*, **1** (1949) 463, 499; Ark. Kemi, **2** (1950) 519.

42 Jefferson D A, Gopalakrishnan J & Ramanan A, *Materials Res Bull*, **17** (1982) 269.

43 Catlow C R A & James R, *Chemical physics of solids and their surfaces*, Vol. 8, edited by M W Roberts & J M Thomas (The Royal Society of Chemistry, London) 1980, 108.

44 Berggren J, *Acta chem Scand*, **25** (1971) 3616.

45 Grenier J C, Pouchard M & Hagenmuller P, *Structure and bonding*, **47** (1981) 1.

46 Vidyasagar K, Gopalakrishnan J & Rao C N R, *Inorg Chem*, (1984) in press.

47 Poeppelmeier K R, Leonowicz M E & Longo J M, *J Solid State Chem*, **44** (1982) 89.

48 Poeppelmeier K R, Leonowicz M E, Scanlon J C, Longo J M & Yelon W B, *J Solid State Chem*, **45** (1982) 71.

49 Takeda T & Watanabe H, *J Phys Soc Japan*, **33** (1972) 973.

50 Grenier J C, Ghodbane S, Demazeau G, Pouchard M & Hagenmuller P, *Materials Res Bull*, **14** (1979) 831.

51 Taguchi H, Shimada M & Koizumi M, *J Solid State Chem*, **29** (1979) 221.

52 Alario-Franco M A & Regi M V, *Nature*, **270** (1977) 706.

53 Tofield B C, *Nature*, **272** (1978) 713.

54 MacChesney J B, Sherwood R C & Potter J F, *J chem Phys*, **43** (1965) 1907.

55 Tofield B C, Greaves C & Fender B E F, *Materials Res Bull*, **10** (1975) 737.

56 Alario-Franco M A, Joubert J C & Levy J P, *Materials Res Bull*, **17** (1982) 733.

57 Reller A, Jefferson D A, Thomas J M & Uppal M K, *J phys Chem*, **87** (1983) 913.

58 Reller A, Thomas J M, Jefferson D A & Uppal M K, *Proc Roy Soc*, (1984) in print.

59 Gai P L & Rao C N R, *Z Naturforsch*, **30A** (1975) 1092.

60 Crespin M, Levitz P & Gatineau L, *J chem Soc Faraday* II, **79** (1983) 1181.

61 Nakamura T, Petzow G & Gauckler L J, *Materials Res Bull*, **14** (1979) 649.

62 Grenier J C, Darriet J, Pouchard M & Hagenmuller P, *Materials Res Bull*, **11** (1976) 1219.

63 Bando Y, Sekikawa Y, Nakamara H & Matsui Y, *Acta Cryst*, **A37** (1981) 723.

64 Alario-Franco M A, Henche M J R, Regi M V, Calbet J M G, Grenier J C, Wattiaux A & Hagenmuller P, *J Solid State Chem*, **46** (1983) 23.

65 Alario-Franco M A, Calbet J M G, Regi M V & Grenier J C, *J Solid State Chem*, **49** (1983) 219.

66 Calbet J M G, Regi M V, Alario-Franco M A & Grenier J C, *Materials Res Bull*, **18** (1983) 285.

INDIAN J. CHEM., VOL. 23A, APRIL 1984

67 Gai P L & Rao C N R, *Pramana*, **5** (1975) 274.

68 Gai P L, Jacobson A J & Rao C N R, *Inorg Chem*, **15** (1976) 480.

69 Negas T & Roth R S, *J Solid State Chem*, **3** (1971) 323.

70 Mori S, *J phys Soc Japan*, **28** (1970) 44.

71 Zanne M & Gleitzer C, *Bull Soc Chim Fr*, (1971) 1567.

72 Jacobson A J, *Acta Cryst*, **B32** (1976) 1087.

73 Jacobson A J & Horrox A J W, *Acta Cryst*, **B32** (1976) 1003.

74 Hutchison J L & Jacobson A J, *J Solid State Chem*, **20** (1977) 417.

75 Zanne M, Courtois A & Gleitzer C, *Bull Soc Chim Fr*, (1972) 4470.

76 Jacobson A J & Hutchison J L, *Chem Commun*, (1976) 116.

77 Jacobson A J & Hutchison J L, *J Solid State Chem*, **35** (1980) 334.

78 Palanisamy T, Gopalakrishnan J & Sastri M V C, *Z Anorg Allgem Chem*, **415** (1975) 275.

79 Er-Rahko L, Michel C, Provost J & Raveau B, *J Solid State Chem*, **37** (1981) 151.

80 Tofield B C & Scott W R, *J Solid State Chem*, **10** (1974) 183.

81 Ramadass N, Gopalakrishnan J & Sastri M V C, *J inorg nucl Chem*, **40** (1978) 1453.

82 Hervieu M, Studer F & Raveau B, *J Solid State Chem*, **22** (1977) 273.

83 Scheunemann K & Muller-Buschbaum H K, *J inorg nucl Chem*, **37** (1975) 1879, 2261.

84 Portier R, Carpy A, Fayard M & Galy J, *Phys Stat Solidi*, **A30** (1975) 683.

85 Ganguly P & Rao C N R, *J Solid State Chem*, (1984) in print.

86 Drennan J, Travares C P & Steele B C H, *Materials Res Bull*, **17** (1982) 621.

87 Subramanian M A, Aravamudan G & Subba Rao G V, *Prog Solid State Chem*, **15** (1983) 55.

88 Piffard Y & Tournox M, *Acta Cryst*, **35B** (1979) 1450 and references therein.

89 Tilley R J D, *J Solid State Chem*, **21** (1977) 293.

90 Ramanan A, Gopalakrishnan J, Smith D J, Jefferson D A & Rao C N R, unpublished results.

91 Hutchison J L, Anderson J S & Rao C N R, *Proc Roy Soc London*, **355** (1977) 301.

92 Anderson J S & Hutchison J L, *Contemp Phys*, **16** (1975) 443.

Intergrowth Structures: The Chemistry of Solid–Solid Interfaces

C. N. R. RAO*[†] and JOHN M. THOMAS*

Department of Physical Chemistry, University of Cambridge, Lensfield Road, Cambridge CB2 1EP, U.K.

Received July 26, 1984 (Revised Manuscript Received December 13, 1984)

Most chemists are familiar with two kinds of intergrowths: *epitaxy*, which involves the oriented overgrowth of one crystalline solid upon another, and *polytypism*, which arises when individual sheets in a layered material are stacked in different sequences. Epitaxy is currently of technological interest since it dictates the laying down of thin, single-crystal films of semiconductors on to a substratum suitable for the production of integrated microelectronic circuits. It also occupies a central role in the phenomena of biomineralization and urinary calculi where an inorganic crystal such as $CaCO_3$ or SiO_2 grows in registry with crystalline slivers of polysaccharide, protein, or purine.[1] Polytypism is less important technologically. Nonetheless, two extreme polytypic forms of ZnSe—one with hexagonal (ABAB), the other with cubic (ABCABC) packing—have[2] significantly different electronic band gaps (2.863 and 2.810 eV) and hence different luminscent and other photophysical properties. When polytypic regions are separated spatially on a nanometer scale in semiconducting materials such as GaAs and CdTe, so-called "*quantum wells*" may form. There are many other types of solid–solid interfaces of interest to the chemist. Although few of these are of immediate commercial relevance, they are all of considerable fundamental significance. Our discussions of such seemingly unrelated phenomena as *chiral turnover* in molecular crystals, of *infinitely adaptive inorganic structures*, of *coincidence boundaries* in zeolites, the origins of *gross nonstoichiometry*, and the occurrence of *modulated structures* all entail an understanding of crystalline intergrowths.

Almost all of the facts we discuss in relation to the phenomenology of intergrowths—in particular their structural characteristics—has been obtained by use of high-resolution electron microscopy (HREM). This technique has unique advantages (described elsewhere[3]) for the direct, real-space study of interfaces. Traditionally an interface within a crystalline solid has been pictured as a structural fault or planar defect. The most familiar is a *twin* plane, where the structures on either side of the plane may be mirror related. Another type of planar fault involves the rotation of one part of a crystal, on a specific plane, with respect to another. There are, in general, a number of lattice points at the interface which are common to the flanking structures that constitute that interface. An example is shown[4] in Figure 1. The presence of both the twin and the coincidence boundary involves an increase in the free energy of the crystal, so that such intergrowths are examples of nonequilibrium defects.[5] Provided their intrinsic energy is not excessively large, such faults, which are often introduced during crystal growth, can be kinetically stabilized and are eliminated only by prolonged annealing. Their occurrence does not give rise to local changes in the stoichiometry of the material within which they occur. New types of structures can be formed[6] by recurrent twinning. For example, a tunnel zeolite with overall hexagonal symmetry is produced by repeated twinning of zeolite Y, which has a cage-type structure with cubic symmetry.[6] In principle, III–V semiconductors crystallizing in diamond-like structures could, by recurrent twinning on the (111) plane, be converted to solids with novel electronic properties.

The same compositional invariance is true of polytypism also, but here other subtleties arise, and these are best appreciated after recalling what is meant in the broadest sense by *stacking* faults. It is well-known[7] that

[†] Jawaharlal Nehru Visiting Professor, University of Cambridge. Present address: Solid State Structural Chemistry Unit, Indian Institute of Science, Bangalore-560012, India.

(1) See R. J. P. Williams et al. in the special issue of *Philos. Trans. R. Soc. London, Ser. B*, **304**, 409–588 (1984).

(2) A. D. Yoffe, K. Howlett, and P. M. Williams, *Philos. Mag.* **25**, 247 (1971).

(3) J. M. Thomas and D. A. Jefferson, *Endeavour New Ser.*, **2**, 127 (1978); J. M. Thomas, *Ultramicroscopy*, **8**, 13 (1982); C. N. R. Rao, Sir C. V. Raman Lecture, Indian Institute of Science, 1983; J. M. Thomas, "Inorganic Chemistry Towards 21st Century", M. H. Chisholm, Ed., American Chemical Society, Washington, DC, 1983, ACS Symp. Ser. No. 211, p 445; G. R. Millward and J. M. Thomas, Proceedings of a NATO Advanced Study Institute on Surface Properties and Catalysis by Non-Metals and Oxides, J. P. Bonelle, et al., Ed., Dordrecht, 1983, p 19; C. N. R. Rao in "Solid-State Chemistry: A Perspective Report", Indian National Science Academy, Delhi, 1984; C. N. R. Rao, *Chem. Scr.*, **19**, 124 (1982).

(4) O. Terasaki, J. M. Thomas, and S. Ramdas, *J. Chem. Soc., Chem. Commun.*, 216 (1984); see also O. Terasaki, G. R. Millward, and J. M. Thomas, *Proc. R. Soc. London, Ser. A*, **395**, 153 (1984).

(5) J. M. Thomas, *Adv. Catal.*, **19**, 293 (1969); J. M. Thomas, *Chem. Br.*, **6**, 60 (1970); J. M. Thomas, *Endeavour*, **29**, 149 (1970).

(6) J. M. Thomas, M. Audier, and J. Klinowski, *J. Chem. Soc., Chem. Commun.*, 1221 (1981); M. Audier, J. M. Thomas, J. Klinowski, D. A. Jefferson, and L. A. Bursill, *J. Phys. Chem.*, **86**, 581 (1982).

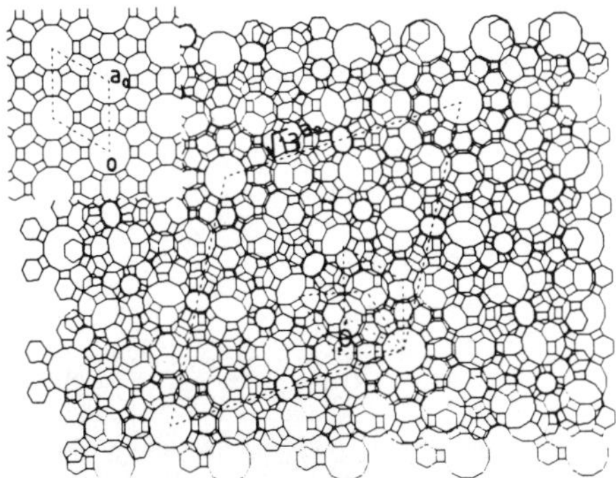

Figure 1. A coincidence boundary on (001) in zeolite L (idealized formula $K_6Na_3Al_9Si_{27}O_{72} \cdot 21H_2O$). The top part of the crystal is rotated[4] by 32.2° with respect to the bottom part, thereby generating the coincidence boundary, the repeat mesh of which bears a $13^{1/2} \cdot 13^{1/2}$ relationship to the parent mesh.

when layers of closely packed units, as in the metals, are stacked in a regular fashion, two simple alternatives are possible: ABABAB... (equivalently designed *hhh...*) and ABCABC... (or *cc...*) where the symbols *h* and *c* refer respectively to hexagonal and cubic stacking. By introducing a stacking fault, there are infractions to the regular sequencing. Thus a sequence *chcc*, which may be viewed as a regular cubic structure into which one fault has been inserted, gives rise to an intergrowth of cubic and hexagonal regions. Were this stacking fault inserted regularly, the resulting structure could be symbolized ...*chccchccchcc*... and this constitute one particular polytype. Silicon carbide, SiC, is known to exist in over 50 polytpic forms, the unit cell repeat distance in the stacking direction being very large (>- 1500 Å). Examples of this kind abound. YSeF, ZnS, CdI_2, the transition-metal chalcogenides, certain intermetallics, the chlorites, the spinelloids, and the micas all exhibit such behavior.

If the stacking fault does not extend across the entire plane on which it occurs but is restricted to narrow ribbons, the resulting solid may then be regarded as a parent structure laced with strips of a related daughter structure. Rhombohedral graphite, for example, which has ABC rather than ABAB stacking, as in the hexagonal polymorph, is of this kind: it does not tend to occur as a phase-pure rhombohedral polymorph. The fractional rhombohedral character of a given graphite specimen is increased (up to a maximum of about 30 %) by mechanical deformation by the introduction of dislocations which delineate the strips of stacking faults.[8]

Since stacking faults frequently do extend across entire planes, it is convenient to broaden this idea and thereby interrelate several different structures in terms of regular stacking sequences brought about by the recurrent insertion of a specific replacement vector parallel to the layer planes. We note that the notion of the "fault" now gives way to an intrinsic structural characteristic, a point to which we shall return fre-

(7) A. R. Verma and G. C. Trigunayat in "Solid State Chemistry", C. N. R. Rao, Ed., Marcel Dekker, New York, 1978; C. N. R. Rao, *Acc. Chem. Res.*, **17**, 83 (1984).

(8) S. Amelinckx, P. Delavignette, and M. Heerschap in "Chemistry and Physics of Carbon", P. Walker, Jr., Ed., Marcel Dekker, New York, 1965, Vol. 1, p 1.

Table I
Repeat Distances and Stacking Distances of Zeolites[a]

zeolite	formula (zeolitic water omitted)	repeat distance along stacking direction, Å	stacking sequence
cancrinite	$Na_6Al_6Si_6O_{24}$	5.1	AB
sodalite	$Na_6Al_6Si_6O_{24}$	7.7	ABC
offretite	$(Na_{27}Ca_7)_2Al_4Si_{14}O_{36}$	7.6	AAB
losod	$Na_{12}Al_{12}Si_{12}O_{48}$	10.5	ABAC
gmelinite	$(Na_2,Ca)_4Al_8Si_{16}O_{48}$	10.0	AABB
chabazite	$Ca_6Al_{12}Si_{24}O_{72}$	15.1	AABBCC
erionite	$(Na_2,Ca)_{4.5}Al_9Si_{27}O_{72}$	15.1	AABAAC
afghanite	$(Na_2,Ca,K_2)_{12}Al_{24}Si_{24}O_{96}$	21.4	ABABACAC
levyne	$Ca_9Al_{18}Si_{36}O_{108}$	23.0	AABCCABBC

[a] The zeolites named in the left-hand column (idealized) given for the framework and exchangeable cation may be pictured as built-up from the "sheet" shown in Figure 2a with the stacking sequence given in the right-hand column.

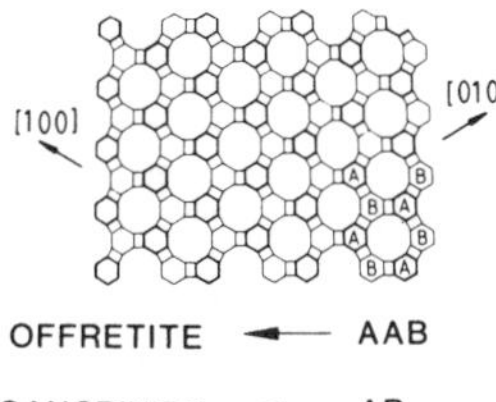

Figure 2. The structures of the zeolites listed in Table I may be regarded as having been derived from various, regular stacking sequences of the single sheet as shown. Thus an AAB sequence yields offretite, AB cancrinite, and so on. Each vertex is a tetrahedral site (T = Si^{4+} or Al^{3+} surrounded by four oxygens).[10]

quently. In particular, consider the family of zeolites,[9] known as the ABC-6 or chabazite group consisting of offretite, erionite, levyne, and several others in Table I. At first sight, judging from their stoichiometry (even when differences in exchangeable cation and in replaceable water are ignored), these appear to have little in common structurally. Yet on further scrutiny we notice that the tetrahedral (T) sites can be occupied either by Al^{3+} or Si^{4+} at the center pf TO_4 tetrahedra (all corner shared). Furthermore, each one of these zeolites is built of sheets such as that shown in Figure 2. For clarity, the positions of the cations and zeolitic water in these structures have been omitted; the vertices represent T sites and the straight lines joining them bridging oxygens. Viewed in this way, we see that the structures of these seemingly unrelated zeolites do have a familial pattern.[10,11]

The conceptual convenience of picturing structures to be made up of regular sequences of layers is not limited solely to the realm of the qualitative. It is already apparent that we can gain quantitative insights into the structural stability of known or hypothetical sequences of individual component sheets. This emerges from computational approaches[12] to the de-

(9) D. W. Breck, "Zeolite Molecular Sieves", Wiley Interscience, New York, 1965; R. M. Barrer, "Zeolites and Clay Minerals", Academic Press, New York, 1978; R. M. Barrer, "Hydrothermal Chemistry of Zeolites", Academic Press, New York, 1982.

(10) G. R. Millward and J. M. Thomas, *J. Chem. Soc., Chem. Commun.*, 77 (1984).

(11) W. M. Meier and D. H. Olson, "Atlas of Zeolite Structure Types", International Zeolite Association, Zurich, 1978.

(12) C. R. A. Catlow, J. M. Thomas, S. C. Parker, and D. A. Jefferson, *Nature (London)*, **295**, 658 (1982).

termination of lattice energies using parameterized values for A, ρ, and C in the potential $V(r)$ between two atoms separated by the distance r:

$$V(r) = A \exp(-r/\rho) - Cr^{-6} \qquad (1)$$

The pyroxenoid family[13] of silicates ($MSiO_3$, where M can be Mg^{2+}, Ca^{2+}, Fe^{2+}, Mn^{2+}, or mixtures thereof) illustrate this point. Good progress has also been made theoretically[14] in using the Ising model or the anisotropic nearest-neighbor Ising (ANNI)[15] model to interpret polytypism.

To summarize, we note that isolated interfaces within or between solids are regarded as structural faults, but that, in view of both experimental and computational evidence, it is often convenient to regard some structures as made up of regular sequences of "stacking faults".

Nonrecurrent Intergrowths

It is helpful to subdivide the materials that possess internal interfaces of a nonregular nature into those that, on the one hand, remain *compositionally invariant* and into those that, on the other, entail change of composition in crossing the interface. Since almost all crystals, subjected to the right stimulus during growth or therafter,[16] can be induced to undergo twinning, it is not profitable to pursue the various kinds of twinning situations that may arise. There are, however, related phenomena which have striking chemical consequences, especially in the organic solid state.

One such example is the occurrence of *enantiomeric intergrowths* in hexahelicenes. Hexahelicene has a chiral space group $P2_12_12_1$; yet, under certain circumstance, crystals of this compound grown from solution are not chiral, as expected, but essentially racemic, possessing slight enantiomeric excess. This is because the individual crystals are composed of intergrowths in which layers of pure (+) and (−) alternate, thereby generating an essentially racemic "composition". Detailed computations,[17] using atom–atom pairwise evaluation procedures,[18] demonstrate that preferential intergrowths should, and do indeed, occur on (100) planes. They also demonstrate that the extra energy required by the crystal to permit this *chiral turnover* on (100) is a relatively small fraction of the quantity needed for normal growth. In other organic molecular crystals like 1,8-dichloro-9-methylanthracene,[19] evidence[20] from microscopic examination and the nature of the products generated by UV irradiation confirm the occurrence of specific types of internal interfaces.

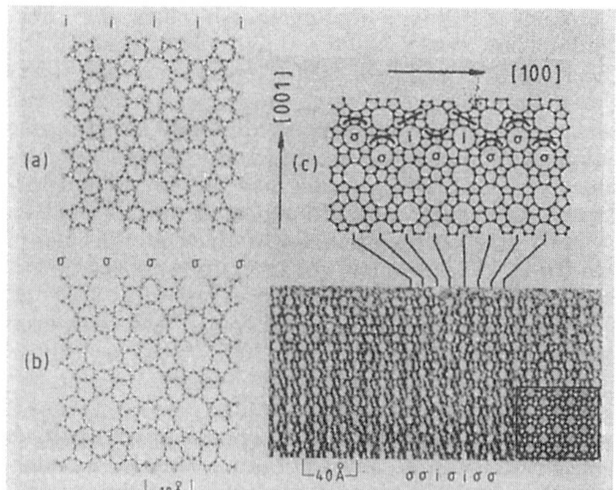

Figure 3. The distinction between (a) ZSM-5 and (b) ZSM-11. An example of a nonrecurrent intergrowth of ZSM-11 in ZSM-5 is shown in (c). Inset [bottom right in high resolution image] shows computer-simulated image.

Molecules situated on either side of such interfaces, which generally function as exciton traps, can photodimerize to yield products that cannot be generated at unfaulted regions of the crystal.

In Figure 1 we saw how zeolite L tends to exhibit internal coincidence boundaries across which the composition remains invariant. A synthetic porotectosilicate, ZSM-5, with much higher Si/Al ratios than naturally occurring zeolites, is a versatile shape-selective catalyst. It tends to form internal interfaces without change of composition.[21-23] To appreciate the subtlety of this process, it is helpful to summarize the structural features[24] of both ZSM-5 and ZSM-11, another shape-selective synthetic zeolite of identical composition: $M_nAl_nSi_{96-n}O_{192} \cdot 162H_2O$, where M is a monovalent cation. Each consists of systems of intersecting channels of diameter ca. 5.5 Å, and this is chiefly responsible for their shape selectivity. Figure 3 summarizes the distinguishing features of ZSM-5 and ZSM-11, which may be regarded as end members of an almost infinite family of regular intergrowths structures e.g., $\sigma\sigma i\sigma\sigma i$, $\sigma\sigma\sigma i\sigma i$, $iii\sigma i\sigma i$, $\sigma\sigma\sigma\sigma ii$, $iii\sigma\sigma$, etc., all of which would be polytypes possessing 60-Å repeats[22] in the [100] direction. Some evidence exists[21,22,25] for the occurrence of regular intergrowths; but to date, HREM has uncovered a greater propensity for nonrecurrent growths (Figure 3).

Other synthetic and naturally occurring silicates and aluminosilicates display similar tendencies to form nonrecurrent intergrowths but with more or less bias, depending upon the structural type and sample prehistory, toward the recurrently intergrown state. The sheet silicates mica, chloritoid, and stilphomelane fall into this category,[26] as do the naturally occurring py-

(13) M. Alario Franco, D. A. Jefferson, N. J. Pugh, J. M. Thomas, and A. C. Bishop, *Mater. Res. Bull.*, **15**, 73 (1980); J. M. Thomas, *New Sci.*, 580 (1980).

(14) S. Ramasesha and C. N. R. Rao, *Philos. Mag.*, **34**, 827 (1977); M. K. Uppal, S. Ramasesha, and C. N. R. Rao, *Acta Crystallogr., Sect. A*, **A36**, 356 (1980).

(15) J. Smith and V. Heine, *J. Phys. C*, in press; S. Ramasesha *Pramana*, **23**, 745 (1984); M. E. Fisher and W. Selke, *Phys. Rev. Lett.*, **44**, 1502 (1980).

(16) R. H. Martin and M. Debleckar, *Tetrahedron Lett.*, 3597 (1969); H. Wynberg, *Acc. Chem. Res.*, **4**, 65 (1971).

(17) S. Ramdas, J. M. Thomas, M. E. Jordan, and C. J. Eckhardt, *J. Phys. Chem.* **85**, 2421 (1981).

(18) A. I. Kitaigorodskii, *Chem. Soc. Rev.*, **7**, 133 (1978); S. Ramdas and J. M. Thomas, *Chem. Phys. Solids Their Surf.*, **7**, 31 (1978).

(19) S. Ramdas, J. M. Thomas, and M. J. Goringe, *J. Chem. Soc., Faraday Trans. 2*, **73**, 551 (1977).

(20) J. P. Desvergne, J. M. Thomas, J. O. Williams, and H. Bouas-Laurent, *J. Chem. Soc., Perkin Trans. 2*, 363 (1974).

(21) J. M. Thomas and G. R. Millard, *J. Chem. Soc., Chem. Commun.*, 1380 (1982).

(22) G. R. Millward, S. Ramdas, J. M. Thomas, and M. T. Barlow, *J. Chem. Soc., Faraday Trans. 2*, **79**, 1075 (1983).

(23) J. M. Thomas, G. R. Millward, S. Ramdas, and M. Audier in "Intrazeolite Chemistry" F. G. Dwyer and G. D. Stucky, Ed., American Chemical Society, Washington, DC, 1982, ACS Symp. Ser. No. 218, p 181.

(24) P. B. Weisz, *Pure Appl. Chem.*, **52**, 2091 (1980); G. T. Kokotailo and W. M. Meier, *Chem. Soc. Spec. Publ.*, **33**, 133 (1980).

(25) J. M. Thomas *Int. Conf. Catal., 8th, 1984*, 1, 31 (1984); *Pure Appl. Chem.*, in press.

roxenoids (encompassing pyroxene, wollastonite, rhodonite, and pyroxmangite). An instructive example is the synthetic mineral[12] $Mn_3MgSi_4O_{12}$, which, after short-duration annealing following crystallization from the melt, exhibits nonrecurrent intergrowths of pyroxenic and pyroxmanganitic structures. After longer annealing times, this material takes up an ordered (pyroxamangite) structure with repeat distances of 17.4 Å.

Intergrowths of amphiboles (double-chain silicates of width 9.5 Å) with analogous structures consisting of triple, quadrupole, or wider strips (typically 13.5 and 18.0 Å) of connected, pyroxene chains[7] have also been intensively studied, chiefly by HREM. These intergrowths are predominantly nonrecurrent; but examples of more ordered structures, entailing repeat distances of ca. 60 Å, are known.[28] In these examples, however, compositional differences arise in traversing the intergrowth; the idealized stoichiometry of a triple-chain feature accommodated within an amphibole such as nephrite jade is $Ca_2(Mg,Fe)_8Si_{12}O_{32}(OH)$ whereas the idealized stoichiometry of the host nephrite (actinolite end member) is $Ca_2(Mg,Fe)_5Si_8O_{22}(OH)_2$.

Solid–solid interfaces involving changes in composition are legion—all examples of epitaxy, whether they consist of catalytically active metals[29] or metal halide[30] supported on graphite, or of semiconducting elements (e.g., Ge) on optically transparent supports (BaF_2),[31] or of wheddelite on uric acid[32] in urine, fall into this category. Of greater relevance in the present context are the isolated planar faults generated as a result of *crystallographic shear* (CS).[33] The key points here are (i) that local coordination is invariant—thus for WO_3, six oxygens remain coordinated to each W, but at the CS fault itself which is introduced on reduction to WO_{3-x} ($x \leqslant 1.0$) corner sharing of octahedra is replaced by edge sharing—and (ii) that the displacement vector has a component perpendicular to the plane of the fault. If progressively more oxygen is removed from the parent oxide, more CS planes are introduced. These will not initially, take up a strictly recurrent pattern; but, just as with the pyroxenoids, quoted above, prolonged annealing generates recurrent intergrowths, and the resulting repeat distances are governed by the degree of gross nonstoichiometry that prevails. Typically for a composition $WO_{2.8}$ ($x = 0.2$) the separation distance is close to 33 Å. For nonstoichiometric oxide $M_nO_{3n-(m-1)}$ it can be shown that where the CS plane lies upon a (10 m) plane, the equilibrium CS plan spacing is $d_{CS} = n - [(m - {}^1/_2) a]/(m^2 + 1)^{1/2}$, where a is the length of the diagonal of the MO_6 octahedron (i.e., about 3.8 Å).

The oxides of Mo and W are able to form so-called Magneli phases[33,34] of general formula M_nO_{3n-1}, M_nO_{3n-2}, M_nO_{3n-3}, etc., with n typically ranging from 12 to 28. The $(3n - 1)$ homologous family results when CS planes

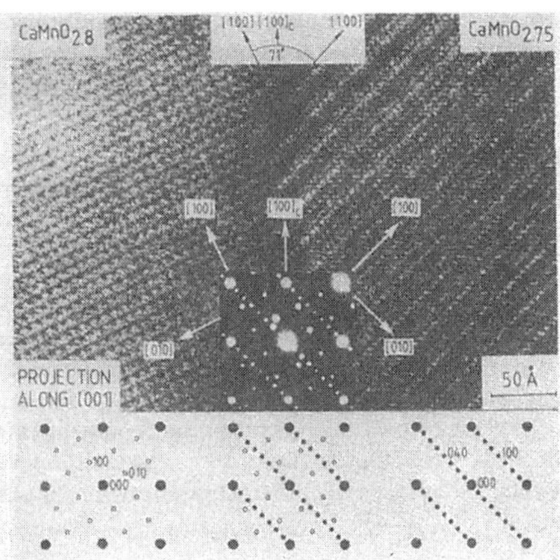

Figure 4. The HREM image and the selected area electron diffraction patterns of a typical example[42] of autoepitaxy, where one substoichiometric perovskite $CaMnO_{2.75}$ grows in coherent contact with its less fully reduced parent, $CaMnO_{2.8}$.

are on {102} of the parent oxide; the $(3n - 2)$ family when the CS planes are on {103}, and so on.

It is instructive to note that when TiO_2 and CrO_2 (both with the rutile structure) are rendered grossly nonstoichiometric, at the CS planes (where 6-fold coordination is preserved and edge sharing of MO_6 octahedra is replaced by face sharing); and at the (CS) itself the local structure is, in effect, a two-dimensional sliver of the corundum structure (local stoichiometry M_2O_3). Recognizing this fact, we now revert to the point made earlier; vis., that a single intergrowth (in this case a CS plane) is best regarded as a structural fault. But ordered, recurrent CS planes are best pictured as an integral feature of the structure of the resulting solid. This concept is encountered again below.

Noncurrent intergrowth is common not only in silicates but also in a variety of other inorganic solids, for example, β''-alumina; the same is true of β- and β''-gallia.[35] In Magneli phases of vanadium (general formula V_nO_{2n-1}), intergrowths of neighboring members has been observed.[36] In perovskite-related oxides of the formula $A_{n+1}B_nO_{3n+1}$, HREM studies have shown the coexistence of lamellae with different n values;[37] thus, the lattice image of $Sr_8Re_2O_7$ shows the presence of lamellae of oxides with $n = 2, 4, 5, 7,$ and 8. Similar random intergrowth of different members of a family is also found[38] in oxides of the general formula $(Bi_2O_2)^{2+}(A_{m-1}B_mO_{3m+1})^{2-}$, first described by Aurivillius.[39] Lattice images of $Ca_xLa_{1-x}FeO_{3-y}$ show disordered intergrowths of $Ca_2Fe_2O_5$ (brownmillerite structure) and $Ca_2LaFe_3O_8$.[40] Interesting nonrecurrent intergrowths have been recently discovered in the $CaMnO_{3-x}$ system.[41] Typical of these intergrowths are $CaMnO_{2.8}$–

(26) J. M. Thomas, D. A. Jefferson, L. G. Mallinson, D. J. Smith, and E. S. Crawford, *Chem. Scr.*, **14**, 167 (1978-79).

(27) J. B. Thompson, *Am. Mineral.*, **63**, 239 (1978).

(28) L. G. Mallinson, D. A. Jefferson, J. M. Thomas, and J. L. Hutchinson, *Philos. Trans. R. Soc.*, London, *Ser. A*, **295**, 537 (1980); D. R. Veblen, P. R. Buseck, and C. W. Burnham, *Science*, **198**, 359 (1977).

(29) E. L. Evans, O. P. Bahl, and J. M. Thomas, *Carbon*, **5**, 587 (1967).

(30) J. M. Thomas, G. R. Millward, R. Schlögl, and H. P. Boehm, *Mater. Res. Bull.*, **15**, 671 (1980).

(31) J. M. Gibson and J. M. Phillips, *Appl. Phys. Lett.*, **43**, 828 (1983).

(32) K. Lonsdale, *Nature* (London), **217**, 56 (1968).

(33) A. D. Wadsley, *Rev. Pure Appl. Chem.*, **5**, 165 (1955).

(34) A. Magneli, *Ark. Kemi*, **1**, 513 (1950).

(35) L. Ganapathi, G. N. Subbanna, J. Gopalakrishnan, and C. N. R. Rao, *J. Mater. Sci.*, in press.

(36) Y. Hirotsu, Y. Tsunashima, S. Nagakura, H. Kuwamoto, and H. Sato, *J. Solid State Chem.*, **43**, 33 (1982).

(37) R. J. D. Tilley, *J. Solid State Chem.*, **21**, 293 (1977).

(38) J. L. Hutchison, J. S. Anderson, and C. N. R. Rao, *Proc. R. Soc.*, London, *Ser. A*, **355**, 301 (1977).

(39) B. Aurivillius, *Ark. Kemi*, **2**, 519 (1950).

(40) M. Alario Franco, J. M. G. Calbet, M. V. Regi, and J. C. Grenier, *J. Solid State Chem.*, **49**, 285 (1983).

(41) A. Reller, D. A. Jefferson, J. M. Thomas, and M. K. Uppal, *J. Phys. Chem.*, **87**, 913 (1983); A. Reller, J. M. Thomas, D. A. Jefferson, and M. K. Uppal, *Proc. R. Soc.*, London, *Ser. A*, **394**, 223 (1984).

Table II
Typical Homologous Series Resulting from Recurrent
Intergrowth

$Ba_{2n+p}Me_{2n}Fe_{12(n+p)}O_{22n+19p}$ with Me = Zn, Ni, etc., n = 1–47, and p = 2–12

$BaFe_{12}O_{19}(Me_2Fe_4O_8)_n$ with n = 1–4

$Bi_4A_{2n-1}B_{2n+1}O_{6n+9}$ with A = Ba or Bi, B = Ti, Nb, Cr, etc., and n = 1–3

A_xWO_3 with A = alkali or alkaline earth metal, Bi, etc. (0.0 < x ≤ 0.1)

$A_xP_4O_8(WO_3)_{2m}$ with m = 4–10

$(Na,Ca)_nNb_nO_{3n+2}$ with n = 4–4.5

$(A_3Nb_6Si_4O_{26})_n(A_3Nb_{8-x}M_xO_{21})$ with A = Ba or Sr, M = Ti, Ni, Zn, etc., and n = 1–15

$CaMnO_3$, $CaMnO_{2.8}$–$CaMnO_{2.75}$ which may, in effect, be regarded as examples of autoepitaxy (Figure 4).

When there is a profusion of nonrecurrent intergrowths between minute regions of distinct structure and stoichiometry, the situation resembles that found in so-called *infinitely adaptive structures*. Such structures occur when, within certain compositional structural extremes, a given compositon has associated with it its own unique structure.[42] Infinitesimal changes in composition are accompanied by changes in the structure (e.g., changes in separation of CS planes or in the dimensional "blocks" formed when CS planes intersect). Good examples of infinitely adaptive structures are the solid solutions Ta_2O_5–WO_3 and $Ba_p(Fe_2S_4)_q$.

Recurrent Intergrowths

Many nonrecurrent intergrowths show a certain degree of order and it is often a matter of definition to stipulate how many repeats of a particular sequence constitute recurrent intergrowth. There are, however, several systems where recurrent intergrowth occurs over fairly large distances[44] and generates homologous series of structures with large unit cells (Table II).

If an intergrowth boundary is pictured as a perturbation, then recurrent intergrowths are sensibly classified, along with other, so-called *modulated structures*. This latter term is used to describe any periodic or partially periodic perturbation of a crystal structure within a repeat distance appreciably greater than the basic unit cell dimension—well-known examples in metallurgy and mineralogy are respectively Au–Mg alloys[43] and antigorite.[26]

An example of recurrent intergrowth is provided by the family of hexagonal barium ferrites, M_pY_q, formed by the components $BaFe_{12}O_{19}(M)$ and $Ba_2Me_2Fe_{12}O_{22}$ with Me = Zn, Ni, or Mg (Y).[45,46] We show in Figure 5 an electron micrograph of a typical ferrite. Barium ferrites exhibit a wide range of unit cell dimensions: for example, in MY, c = 26 Å and in M_8Y_{27}, c = 1455 Å. The complexity of these intergrowths is illustrated by $M_{12}Y_{47}$ which has the composition $Ba_{106}Ni_{94}Fe_{1268}$ with 12 M units in the primitive cell. The M_pY_q intergrowths are not formed when Fe^{3+} is replaced by Al^{3+} ions, suggesting the importance of magnetic interactions in stabilizing these large-repeat-distance structures.

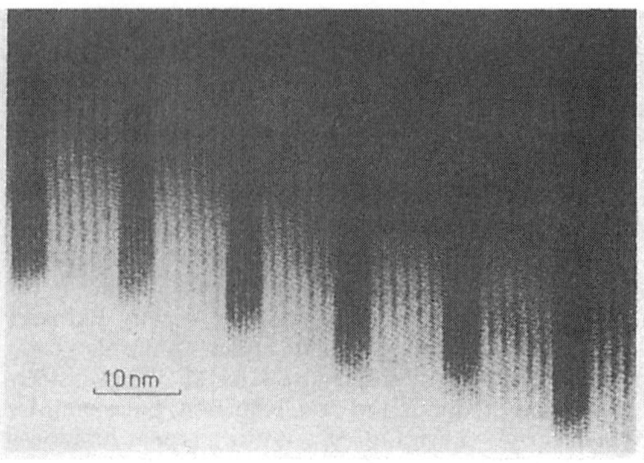

Figure 5. Lattice image of $MYMY_6$ intergrowth in the barium ferrite system.[46]

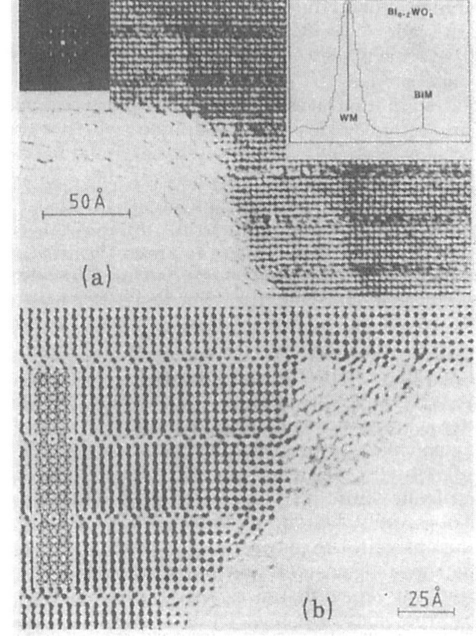

Figure 6. (a) HREM image of a crystal of nominal composition $Bi_{0.2}WO_3$ (obtained with a 200-kV microscope), selected area electron diffraction, and X-ray emission spectrum. The image corresponds to the [001] projection. (b) HREM image (500 kV) of a crystal of nominal composition $Bi_{0.1}WO_3$ again in the [001] projection. The structural model is also shown; dark circles in the model are Bi atoms, which are clearly visible in image.

An interesting class of recurrent intergrowths is formed when slabs of WO_3 cohere with strips of an hexagonal *tungsten* oxide *bronze* (HTB). These intergrowths, referred to as intergrowth tungsten oxide bronzes (ITB), show complex ordered sequences with large periodicities spanning fairly large distances.[47] Figure 6a shows the image of an ITB formed by bismuth.[48] Here WO_3 slabs intergrow with an HTB strip, one tunnel wide; and an electron-induced X-ray emission spectrum (given in the figure) establishes the

(42) J. S. Anderson, *J. Chem. Soc., Dalton Trans.*, 1107 (1973).
(43) O. Terasaki and D. Watanabe, *AIP Conf. Proc.*, **53**, 253 (1979).
(44) C. N. R. Rao and K. J. Rao, "Phase Transitions in Solids", McGraw-Hill, New York, 1978.
(45) J. A. Kohn, D. W. Eckart, and C. F. Cook, *Science*, **172**, 519 (1971).
(46) J. S. Anderson and J. L. Hutchison, *Contemp. Phys.*, **16**, 443 (1975).
(47) L. Kihlborg, *Chem. Scr.*, **14**, 187 (1978).
(48) D. A. Jefferson, M. K. Uppal, D. J. Smith, A. Ramanan, J. Gopalakrishnan, and C. N. R. Rao, *Mater. Res. Bull.*, **19**, 535 (1984); A. Ramanan, J. Gopalakrishnan, M. K. Uppal, D. A. Jefferson, and C. N. R. Rao, *Pr. R. Soc., London, Ser. A*, **395**, 127 (1984).

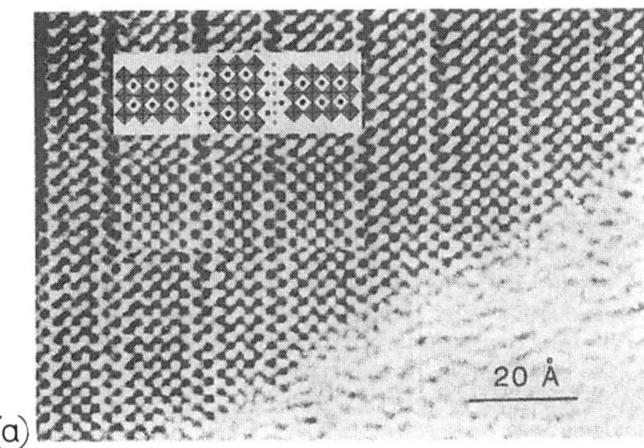

(a)

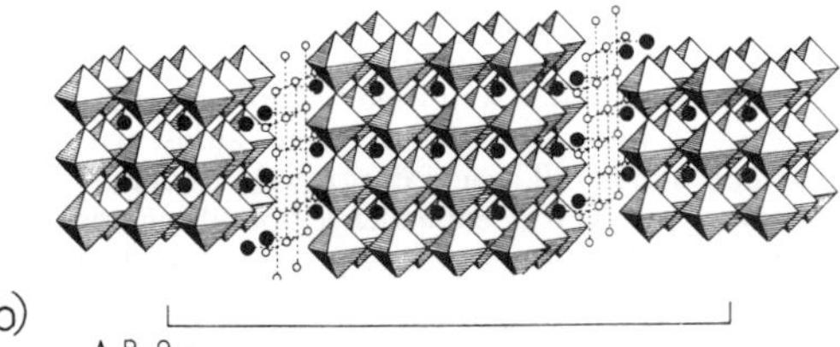

(b)

$A_9B_7O_{27}$

Figure 7. (a) High-resolution image of $Bi_9Ti_6CrO_{27}$ obtained with a 500-kV microscope. The computer-simulated image and structural model are shown as insets and in (b), where the general formula, $A_9B_7O_{27}$, for the $m = 3$, $m = 4$ is given. The dark circles separating the perovskite blocks are the Bi atoms at the interface. The computed image appears below the drawing of the structure in (a).

composition to be $Bi_{0.07}WO_3$, a composition which is also corroborated by X-ray photoelectron spectroscopic analysis. The high-resolution image of the ITB phase in Figure 6b clearly shows tunnels in the HTB strip occupied by bismuth atoms. Other "bronze-like" intergrowths have been reported.[49] The formation of these intergrowth bronzes is likely to be a consequence of the growth conditions and may well represent a case where the impurity-rejection model holds.[50] Long-period intergrowths of barium siliconiobates[51] and $(Na,Ca)_nNb_nO_{3n+2}$[52] seem to resemble the continuous series of ordered structures or the infinitely adaptive structures described by Anderson.[42]

Recurrent intergrowths formed by the *Aurivillius* group of oxides of the formula $(Bi_2O_2)^{2+}(A_{m-1}B_mO_{3m+1})^{2-}$ have been investigated in Cambridge and Bangalore. This system shows remarkable order even in the absence of magnetic or charge-ordering interactions and forms the homologous series $Bi_4A_{2n-1}B_{2n+1}O_{6n+9}$. Figure 7 shows the high-resolution image of $Bi_9Ti_6CrO_{27}$ formed by the intergrowth of $Bi_4Ti_3O_{12}$ ($m = 3$) and $Bi_5Ti_3CrO_{15}$ ($m = 4$), along with the computer-simulated image and the structural model, to illustrate the ordered arrangement of the $m = 3$ and $m = 4$ lamallae.[53] These

(49) M. Hervieu and B. Raveau, *J. Solid State Chem.*, **43**, 299 (1982); A. Benmoussa, D. Groult, F. Studer, and B. Raveau, *J. Solid State Chem.*, **42**, 221 (1982); J. P. Groult, M. Goreaud, Ph. Labbe, and B. Raveau, *J. Solid State Chem.*, **44**, 407 (1982).
(50) J. S. Anderson, J. M. Browne, and J. L. Hutchison, *Nature (London)*, **237**, 5351 (1972); R. J. D. Tilley, *Chem. Phys. Solids Their Surf.*, **8**, 166 (1980).
(51) F. Studer and B. Raveau, *Phys. Status Solids A*, **48**, 301 (1978).
(52) R. Portier, A. Capry, M. Fayard, and J. Galy, *Phys. Status Solidi A*, **30**, 683 (1975).
(53) J. Gopalakrishnan, A. Ramanan, C. N. R. Rao, D. A. Jefferson, and D. J. Smith, *J. Solid State Chem.*, **55**, 101 (1984).

systems represent cases where elastic forces alone seem to be responsible for the recurrent intergrowth.

Long-range repulsive interactions arising from elastic forces appear to be of great importance in the formation of recurrent intergrowths. Elastic interactions have indeed been considered to play a crucial role in governing polytypism, epitaxy, and the formation of infinitely adaptive structures. The average elastic energy per unit area per plane resulting from the uniform strain[54] induced by the coherent intergrowth of planes of two units A and B (giving rise to the composition AB) is given by

$$U = \frac{Yd}{4}(\epsilon_A{}^2 + \epsilon_B{}^2) \tag{2}$$

where ϵ_A and ϵ_B are the strains, Y is the elastic modulus, and d is the interplanar spacing. If we assume the planes to be nearly of the same lattice constant, a, U will be proportional to $(\Delta a/a)^2$ since $\epsilon_A = -\epsilon_B$ and $a(\epsilon_A + \epsilon_B) = \Delta a$. While this treatment explains how such a repulsive force could be responsible for long period intergrowths, an operational criterion can be derived by expressing the elastic strain energy in terms of the volume change accompanying the formation on the intergrowths. For the recurrent intergrowths formed by the oxides $(Bi_2O_2)^{2+}(A_{m-1}B_mO_{3m+1})^{2-}$, the volume change, $\Delta V'$, turns out to be

$$\Delta V' = 2V_B{}'\left(1 - \frac{a^2}{a'_B{}^2}\right)^2 + mk_mV_p{}'\left(1 - \frac{a^2}{a'_p{}^2}\right)^2 +$$
$$(m + 1)K_{m+1}V_p{}''\left(1 - \frac{a^2}{a''_p}\right)^2 \tag{3}$$

where $V_p{}'$ and $V_p{}''$ are the volumes of unconstrained perovskite slabs of m and $m + 1$ dimensions, k is related to the bulk modulus, and $V_B{}'$ represents the volumes of the unconstrained Bi_2O_2 layers. We find that $\Delta V'$ calculated for the various recurrent intergrowths of this family is smaller than the sum of the ΔV values for the component m and $m + 1$ oxides.[55] The $\Delta V'$ for $Ba-Bi_{12}Ti_{10}O_{39}$ which corresponds to the intergrowth of two $m = 3$ one $m = 4$ layers is close to the sum of ΔV values of the component (two 3 and one 4) oxides. In fact, we have not been able to obtain recurrent intergrowths with 3,3,4 sequences repeating over large distances.

Structure of the Interface

It is important to understand the nature of the elastic strain at the solid–solid interface in recurrent intergrowths. By a careful comparison of experimentally determined and computer-simulated high-resolution images, it has indeed been possible to do so.[56] In the recurrent intergrowth of $Bi_9Ti_6CrO_{27}$ (Figure 7) and related oxides, two layers of Bi_2O_2 constitute the interface. An examination of the micrographs indicates that the two Bi-atom layers at the interface are much closer to each other than expected, and such compression could arise from elastic strain. To investigate the nature of this strain from high-resolution images, we calculated[57] images for the following four models: (i)

(54) C. Kittel, *J. Solid State Comm.*, **25**, 519 (1978).
(55) K. Kikuchi, *Mater. Res. Bull.*, **14**, 1561 (1979).
(56) D. A. Jefferson, M. K. Uppal, D. J. Smith, and C. N. R. Rao, *Mater. Res. Bull.*, **19**, 1403 (1984).

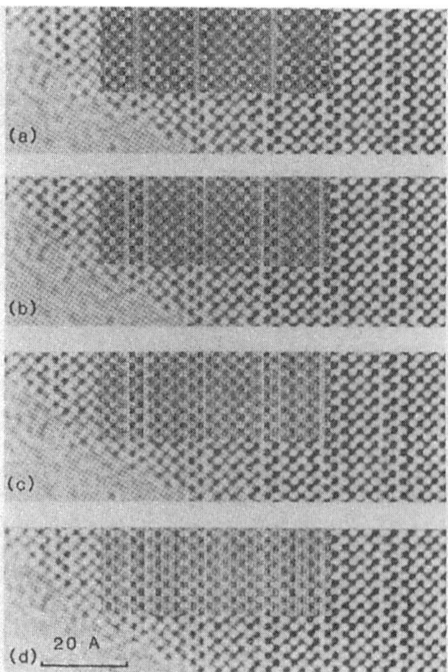

Figure 8. Image simulations of $BaBi_8Ti_7O_{27}$ compared with experimental image. $BaBi_8Ti_7O_{27}$ results from the recurrent intergrowths of $Bi_4Ti_3O_{12}$ ($m = 3$) and $BaBi_4Ti_4O_{15}$ ($m = 4$) of the $(Bi_2O_2)^{2+}$ $(A_{m-1}B_mO_{3m+1})^{2-}$ family. (a) Model i, (b) model ii, (c) model iii, and (d) model iv (see text). Model i predicts a light fringe between the two Bi layers at the interface while the observed image shows a zig-zag arrangement of Bi atoms resulting from a compression.

$m = 3$ and $m = 4$ lamellae intergrown without any movement of atoms from their ideal positions, (ii) Bi atoms at the edges of the lamellae shifted toward each other to the same extent, all other atoms retaining their positions in the idealized solid, (iii) movement of Bi atoms at the edges of the lamellae as in (ii) but coupled with 50 % expansion of the cation lattice within the perovskite lamellae in a direction perpendicular to the lamellae, and (iv) similar to (iii) but with 100 % expansion of the lattice within the lamellae. Comparison

(57) D. A. Jefferson, G. R. Millward, and J. M. Thomas, *Acta Crystallogr., Sect. A*, **A32**, 823 (1976); J. M. Thomas, D. A. Jefferson, and G. R. Millward, *J. Microsc. Spectrosc. Electron.*, **7**, 315 (1982).

of the calculated and observed images of $BaBi_8Ti_6O_{27}$ (Figure 8) clearly shows that model i is inappropriate. There is obviously compression of the Bi layers at the interface as in model ii; but the best match is found with model iii, where there is, in addition, some expansion of the cation rearrangement within the perovskite lamellae. Model iv with full expansion of the lamellae does not show the interface distinctly from the perovskite lamellae. While this is not appropriate for $BaBi_8Ti_7O_{27}$, some of the other intergrowths of this homologous series give high-resolution images which are best interpreted in terms of model iv. This study of the solid–solid interface in intergrowths constitutes a unique example of partial structure refinement using HREM.

Concluding Remarks

A wide diversity of solid-state features and phenomena can be viewed in a unified fashion in terms of the concept of intergrowths that occur either within or between solids. These intergrowths include chiral turnover, polytypism, coincidence boundaries, epitaxy, quantum wells, Magneli phases, gross nonstoichiometry, infinitely adaptive structures, and oxide bronzes. The structure of such solid–solid interfaces can often be directly imaged, at near-atomic resolution, by electron microscopy or deduced by computation with the aid of appropriate atom–atom potentials. There are merits in conceptually subdividing modulated structures, which are those that exhibit repeat distances much in excess of the unit cell dimension, into recurrent intergrowths of component substructures and of regarding polytypes as being derived from regularly arranged stacking faults. The importance of elastic forces in stabilizing intergrowth structures is quantitatively illustrated. Apart from connecting several, hitherto seemingly unrelated, topics in the chemistry of solids, this account emphasizes the value of high-resolution electron microscopy in gaining new insights into the structure and energetics of solids. It is clear that recurrent intergrowths may provide means of tailoring composition and therby of preparing solids possessing desired, finely tuned electronic, optical, dielectric, and magnetic properties. To achieve these ends, however, the nature of the long-range elastic forces responsible for recurrent intergrowths needs to be more fully understood.

Virtues of marginally metallic oxides

C. N. R. Rao

CSIR Centre of Excellence in Chemistry & Solid State and Structural Chemistry Unit, Indian Institute of Science, Bangalore 560 012, India and Materials Research Laboratory, University of California, Santa Barbara, CA 93106, USA

Transition-metal oxides at the metal–insulator boundary, especially those belonging to the perovskite family, exhibit fascinating phenomena such as insulator–metal transitions controlled by composition, high-temperature superconductivity and giant magnetoresistance (GMR). Interestingly, many of these marginally metallic oxides obey the established criteria for metallicity and have a finite density of states at the Fermi level. The perovskite manganates exhibiting GMR, on the other hand, are unusual in that they possess very high resistivities in the 'metallic' state and show no significant density of states at the Fermi level. Marginal metallicity in oxide systems is a problem of great complexity and contemporary interest and its understanding is of crucial significance to the diverse phenomena exhibited by these materials.

Properties of conventional metals such as gold and copper are familiar to most of us. One of these properties is the low electrical resistivity of metals. Some transition-metal compounds, in particular the oxides, exhibit low electrical resistivities just like conventional metals. The resistivity behaviour of the oxides such as ReO_3 and RuO_2 is indeed comparable to that of copper (Fig. 1). Metal oxides, however, exhibit a very wide range of electrical resistivity, anywhere between 10^{20} to 10^{-10} ohm cm, probably representing the widest range in any physical property known to date.[1] Besides the oxides with metallic resistivities and those with insulating properties, there are oxides which exhibit transitions from the insulating state to the metallic state. The insulator-to-metal (I–M) transition in V_2O_3 is well known for the unparalleled 10 or 100 million-fold jump in resistivity at 150 K; it is as if wood becomes copper at the transition temperature. The I–M transition in oxides can be brought about by changing the temperature, pressure or composition.[1,2] The perovskite oxides which exhibit compositionally controlled I–M transitions are especially noteworthy.[3] $La_{1-x}Sr_xCoO_3$ is an example of an oxide system which becomes metallic at a particular Sr concentration, the parent $LaCoO_3$ ($x = 0.0$) being an insulator at ordinary temperatures.

$LaNi_{1-x}Mn_xO_3$ is an example of an oxide system which becomes insulating as x is increased, $LaNiO_3$ ($x = 0.0$) being metallic. I shall discuss briefly the nature of such metal oxides to point out how the I–M transition occurs across a state which can be considered as barely metallic and yet obey the known criteria for metallicity.†

The cuprates which exhibit high-temperature superconductivity[4,5] are marginally metallic in the normal (non-superconducting) state and exhibit certain anomalous properties. What is even more interesting is that perovskite manganates of the general formula $La_{1-x}A_xMnO_3$ (Ln = rare earth, A = divalent ion such as Ca, Sr, Ba or Pb) exhibit high resistivities in what is considered to be the metallic state and do not seem to obey the criteria for metallicity. Marginal metallicity seems to be generally associated with all the oxides exhibiting giant magnetoresistance.[6,7] Thus, marginal metallicity of metal oxides is closely linked to three of the most important aspects of present-day solid-state science, namely insulator–metal transitions, superconductivity and giant magnetoresistance. In Fig. 2 we compare the temperature variation

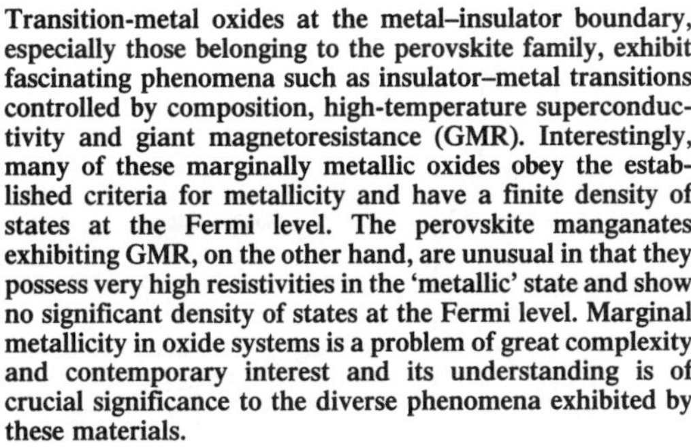

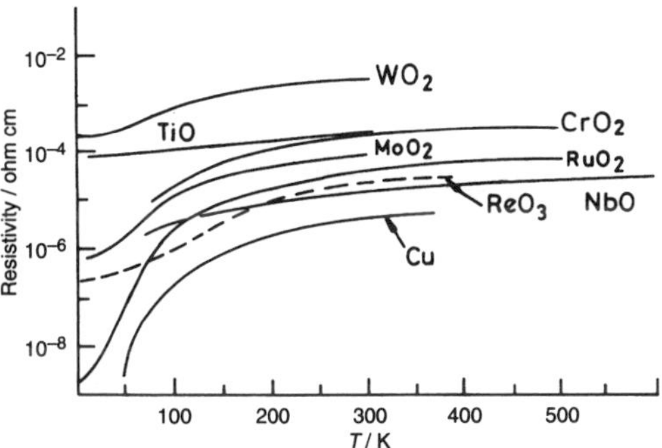

Fig 1 Resistivities of highly conducting (metallic) oxides

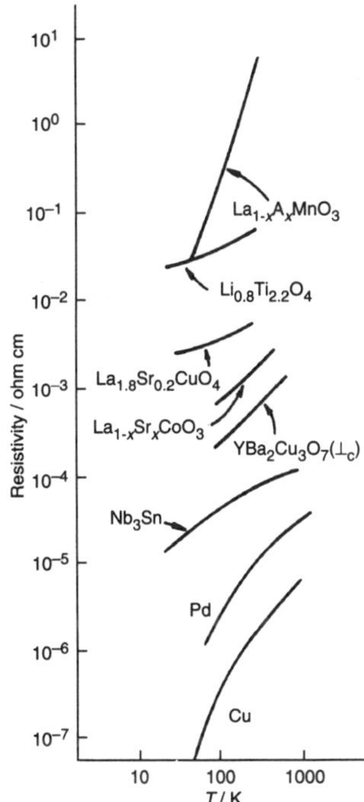

Fig 2 Resistivities of some of the oxides exhibiting marginal metallicity. Data on metallic Nb_3Sn, Pd and Cu are shown for comparison.

of resistivity of cuprates and other oxides to give an indication of the resistivity ranges encountered in marginally metallic oxides. In this article, I examine the nature of marginal metallicity encountered in transition-metal oxide systems after a brief discussion of the criteria for metallicity.

Operational criteria for metallicity

An important criterion for metallicity is the existence of the Fermi surface or the presence of a finite density of states at the Fermi level.† Two of the successful criteria for metallicity derived from theories of electron transport in solids are due to Mott.[8] Several years ago, Mott[9] pointed out how a material would be metallic at high charge carrier (electron) densities or small interparticle distances; in this state, the material would have a finite electrical conductivity at $T = 0$ K. On the other hand, at low carrier densities or large interparticle distances, the material would be an insulator with zero conductivity at 0 K. A discontinuous transition from the metallic state to the insulating state (Mott transition) can occur at a critical distance or a critical carrier concentration. At the transition, $n_c^{1/3} a_H \approx 0.25$, where n_c is the critical carrier concentration and a_H is the Bohr orbit radius of an isolated centre. A large number of systems, including doped semiconductors and metal–ammonia solutions, obey this relation.[2] The metallicity criterion as defined by this expression is essentially a manifestation of electron interactions or Coulomb correlation resulting in an energy gap at low densities.

Another criterion for metallicity arises from electron localization induced by disorder proposed by Anderson.[10] Electrons diffuse when the disorder is small, but at a critical disorder they do not diffuse (giving rise to zero conductivity). A transition from the metallic to the insulating state occurs as the disorder increases. Mott[11] proposed that there exists a minimum metallic conductivity, σ_{min}, or maximum metallic resistivity, ρ_{max}, for which the material may still be viewed as being metallic, prior to the localization of electrons due to disorder. The σ_{min} is given by $\rho 2/h\pi^3 l$, where l is the mean free path of the electrons. Mott's σ_{min} criterion assumes that the disorder-driven metal–insulator transition is discontinuous, but the scaling theory[12] predicts it to be continuous. It is, however, found that σ_{min} or ρ_{max} is a useful experimental criterion and represents the value of conductivity where the temperature coefficient of resistance changes sign from metal-like (+) to insulator-like (−) behaviour.[2,3] The value of Mott's ρ_{max} is around 1–10 mohm cm in oxide systems.

A useful way of describing and classifying transition-metal oxides in terms of their electronic properties is that due to Zaanen, Sawatzky and Allen (ZSA)[13] who consider the intra-atomic Coulomb strength, U, the ligand–metal charge-transfer energy, Δ, and the metal (d)–oxygen(p) hybridization strength, t_{pd}, to provide a electronic phase diagram (with U/t_{pd} plotted against Δ/t_{pd}). This diagram has been modified by Sarma to account for magnetic properties and doping effects. Such a modified phase diagram[14] is shown in Fig. 3. In this phase diagram, region D corresponds to d-band metals with $U < \Delta$, and $U < W$ where W is the d-band width and region C represents mixed-valent or p-type metals with $U > \Delta$. Mott insulators with $\Delta > U > W$ wherein U essentially determines the band gap are in region B. Charge-transfer insulators with $U > \Delta$ wherein the band gap is mainly controlled by Δ are in region A. Covalent insulators in region E are those where t_{pd} plays a crucial role; this region had not been identified in the ZSA phase diagram. Amongst the binary oxides, NiO is a charge-transfer insulator (region A) while ReO_3 is a d-band metal (D). Amongst the ternary oxides, $LaTiO_3$ is best described as a Mott insulator (B) while $LaNiO_3$ is a metal in the C region. Both $LaCoO_3$ and $LaMnO_3$ have mixed-valent ground states, but $LaCoO_3$ is closer to A than B, probably on the borderline, while $LaMnO_3$ is closer to B than A. La_2CuO_4, which is the parent cuprate for the numerous superconductor families, is described as a charge-transfer insulator with $\Delta \ll t_{pd}$.

Compositionally controlled metal–insulator transitions in perovskite oxides

In Fig. 4 we show typical compositionally controlled I–M transitions in two perovskite oxide systems. In $La_{1-x}Sr_xCoO_3$, the material becomes metallic as x is increased beyond a critical composition. A similar behaviour is seen in $La_{1-x}Sr_xVO_3$. Metallicity arises in these oxides because of fast hopping of

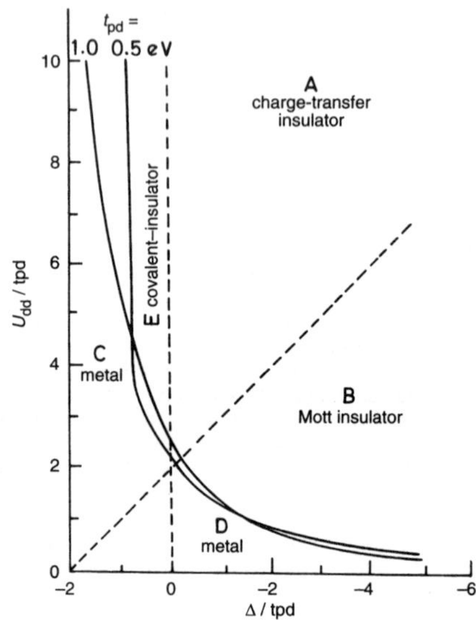

Fig. 3 ZSA phase diagram as modified by Sarma.[14] The solid lines separate the metallic and insulating regimes.

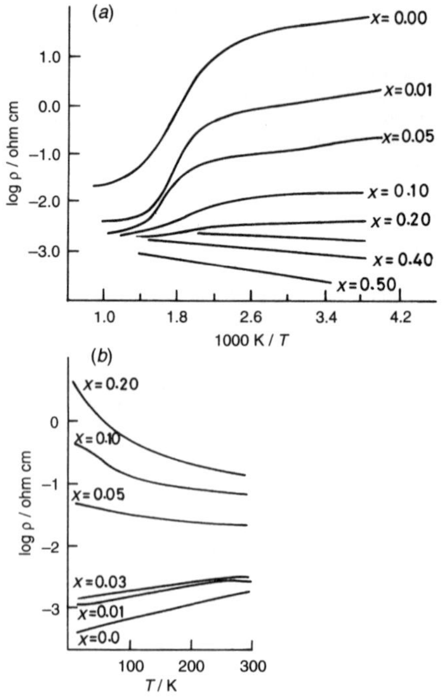

Fig. 4 Compositionally controlled insulator metal transitions in (a) $La_{1-x}Sr_xCoO_3$ and (b) $LaNi_{1-x}Mn_xO_3$ (from the author's laboratory)

electrons between the 3+ and 4+ states of the transition metal. It is noteworthy that the temperature coefficient of resistance changes sign around a resistivity value close to Mott's maximum resistivity, ρ_{max}. In $LaNi_{1-x}Mn_xO_3$ where the material becomes insulating with increasing x, the change from the metallic to the insulating state occurs again around Mott's ρ_{max}. The value of ρ_{max} or σ_{min} is known to scale with the critical carrier concentration,[15] n_c, as shown in Fig. 5 where log ρ_{max} is plotted against log n_c for a variety of systems including doped semiconductors. The points corresponding to the metal oxides exhibiting compositionally controlled metal–insulator transitions fall in line with the other electronic materials at the threshold of metallicity. It is noteworthy that these oxides obey the ρ_{max} or the σ_{min} criterion as well as the $n_c^{1/3}a_H \approx 0.25$ criterion. Although the oxides obey the σ_{min} criterion, it is possible that some of them may show a departure from this behaviour and become insulating at very low temperatures (say < 50 K), with the resistivity increasing with decreasing temperature[16] (*i.e.* conductivity is zero at 0 K). If so, σ_{min} serves as the high-temperature limit for the conductivity of these marginal metals.

Photoelectron spectroscopic studies[14] of $La_{1-x}Sr_xCoO_3$ in the valence band region show the presence of a finite density of states at the Fermi level in these marginally metallic oxides (Fig. 6). The oxygen K-absorption spectra of $La_{1-x}Sr_xCoO_3$ show a feature increasing in intensity and width with increasing x, suggesting a progressive creation of doped hole states with substantial oxygen p-character as expected of a charge-transfer insulator. These states overlap the top of the valence band above a critical concentration. A finite density of states at the Fermi level is seen in $LaNi_{1-x}Mn_xO_3$ as well, even slightly beyond the critical x value above which the material is insulating.

Cuprate superconductors

Most of the cuprates which exhibit high-temperature super-conductivity exhibit metallic resistivity (with the resistivity decreasing with temperature) before they become super-conducting (Fig. 7). The values of resistivities in the metallic regime before they become superconducting is in the range 2–5 mohm cm, not far from Mott's ρ_{max}. Interestingly, the cuprates conform to the $n_c^{1/3}a_H \approx 0.25$ relation (Fig. 8) and also follow the linear log ρ_{max}–log n_c plot (Fig. 5). Although these observations would suggest that the cuprates are similar to $La_{1-x}Sr_xVO_3$ and such oxides showing I–M transitions, they exhibit certain anomalous properties[18] in the metallic state, one of them being the linear variation of resistivity with temperature in the metallic state (see Fig. 7). High-energy spectroscopic studies[19] of the cuprates show that the oxygen 2p contribution in the hole doped states near the Fermi level is significant. The cuprates have strong correlation effects and there is as yet no unanimous view as to the best way of describing the marginally metallic state.

A novel feature of the cuprates is that they exhibit compositionally controlled metal–insulator transitions in the normal state ($T > T_c$) as can be seen in Fig. 9(*a*), (*b*). It is indeed curious that the insulating state transforms directly to the superconducting state in some compositions [Fig. 9(*b*)]. Such an insulator–metal transition is found in $La_{2-x}Sr_xCuO_4$ as well. The question that arises is whether there can be insulating and superconducting ground states without a metallic ground state in between. A careful comparison[20] shows that the Hall coefficients and other properties of such cuprate compositions are different from those of oxide systems such as $La_{1-x}Sr_xTiO_3$ which exhibit insulator–metal transitions. Clearly, the super-conducting cuprates are associated with a unique kind of a marginal metallic state.

Perovskite manganates exhibiting GMR

Of all the oxide systems considered here, the manganates of the general formula $Ln_{1-x}A_xMnO_3$ (Ln = La, Pr, Nd; A = Ca, Sr, Ba, Pb) possessing the perovskite structure are most unusual in

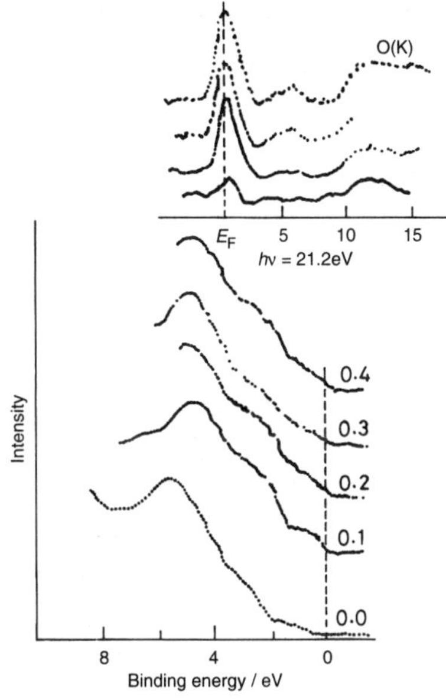

Fig. 6 Photoelectron spectra of $La_{1-x}Sr_xCoO_3$ in the valence region. The inset shows the oxygen K-absorption spectra (from ref. 14).

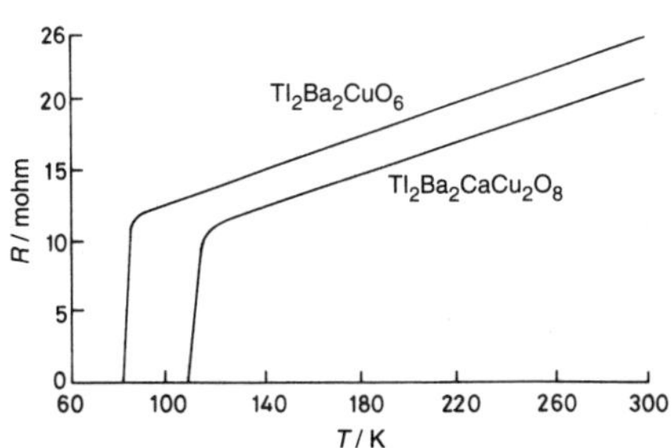

Fig. 7 Resistivity behaviour of cuprates showing linear temperature variation of resistivity in the normal (metallic) state. The linearity can be extended down to 0 K.

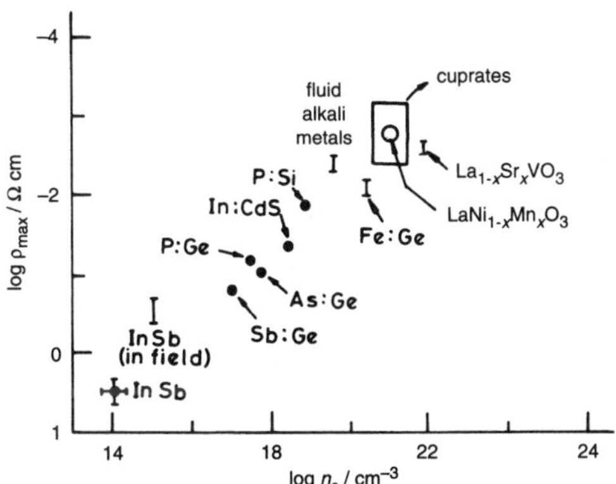

Fig. 5 A log–log plot of the maximum metallic resistivity, ρ_{max}, against the critical carrier density, n_c, at the insulator–metal transition

that they possess high resistivities in the metallic state which cannot be accounted for or explained on the basis of the criteria discussed earlier. Let us briefly examine the nature of the metallic state in these materials. In $La_{1-x}A_xMnO_3$, the Mn^{4+} content increases with x (when $x = 0.0$, Mn will be nominally in the 3+ state).‡ The Mn^{3+}–O–Mn^{4+} interaction is ferromagnetic, unlike the Mn^{3+}–O–Mn^{3+} and Mn^{4+}–O–Mn^{4+} interactions which are antiferromagnetic. The manganates become ferromagnetic above a critical concentration of Mn^{4+} (or value of x) and simultaneously exhibit metallic conductivity.§ The primary process involved is the hopping of a charge carrier as described by,

$$Mn^{3+}(i)\text{–}O\text{–}Mn^{4+}(j) \rightarrow Mn^{4+}(i)\text{–}O\text{–}Mn^{3+}(j)$$

where i and j are nearest neighbours. Zener[22] suggested that a paramagnetic to ferromagnetic transition should occur with a transition temperature T_c according to the relation $k_B T_c \approx x_h t Z$, where x_h is the Mn^{4+}(hole) concentration, t is the hopping amplitude and Z is the number of nearest neighbours. These materials exhibit a transition from an insulating to a metallic state as the temperature is lowered, the transition occurring close to the ferromagnetic Curie temperature, T_c. The material is insulating at $T > T_c$ because thermal fluctuations of the magnetic moments impede the motion of holes. The I–M transition in an $La_{1-x}Ca_xMnO_3$ composition is shown in Fig. 10. The resistivity peak in this figure arises from the I–M transition, with the temperature variation of resistivity being metal-like at $T < T_c$ (T_{IM}) and insulator-like when $T > T_c$ (T_{IM}).

When a magnetic field is applied to such a material the resistivity decreases enormously[23] (Fig. 10). This phenomenon, called giant magnetoresistance (GMR), is of great importance in magnetic recording and other technological applications. GMR-related aspects of the manganates are not within the scope of this article¶ and I shall restrict myself to the metallic state of these materials.

The resistivity of the manganates at the I–M transition is generally high, anywhere from 10 mohm cm to several ohm cm (see Fig. 2). No metal can have such resistivities, the observed values being much higher than Mott's ρ_{max}. Furthermore, the resistivity in the metallic regime $(T < T_c)$ in some of the manganates is comparable to or even higher than that in the insulating state $(T > T_c)$.‖ Residual resistivities at low temperatures (ca. 4 K) are also very large (several mohm cm), larger than in disordered metals. While the effects of the lattice and of magnetic polarons may contribute to such high resistivities, we note that electron correlation would play a major role in a Mott insulator such as $LaMnO_3$. What is curious is that photoelectron spectra in the valence-band region[14] of $La_{1-x}Sr_xMnO_3$ show the absence of any significant density of states in the 'metallic' compositions at ordinary temperatures as can be seen from Fig. 11. The oxygen K-absorption spectra show a progressive formation of hole states with the increase in x ca. 1 eV above the Fermi level (Fig. 11).

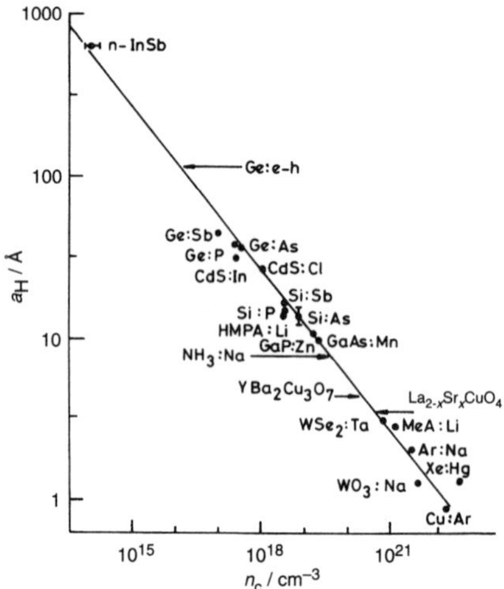

Fig. 8 Relation between the critical carrier concentration, n_c, and the Bohr radius, a_H. The solid line represents $n_c^{1/3} a_H = 0.26$ (from ref. 17).

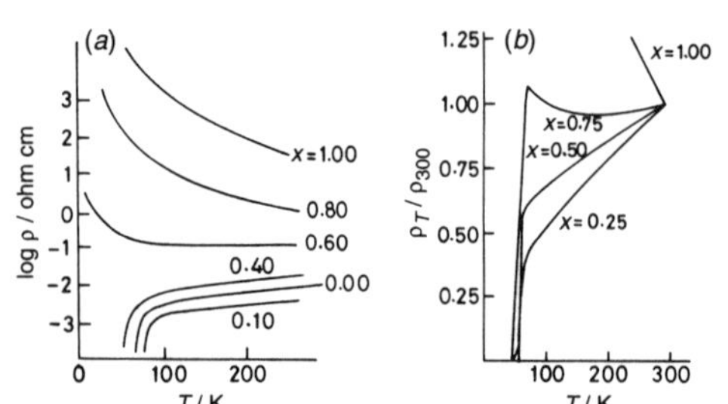

Fig. 9 Temperature variation of the resistivity of (a) superconducting $Bi_2Ca_{1-x}Y_xSr_2Cu_2O_{8+\delta}$ and (b) $TlCa_{1-x}Nd_xSr_2CuO_6$ (from the author's laboratory)

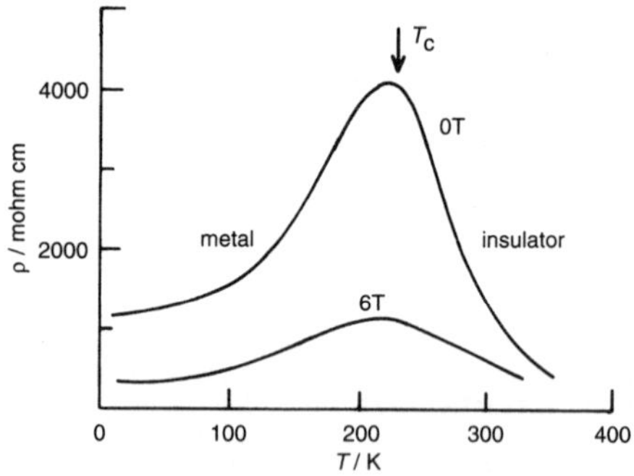

Fig. 10 Resistivity behaviour of $La_{0.9}Ca_{0.1}MnO_3$ at zero field (0 T) and in a magnetic field (6 T) (from ref. 23)

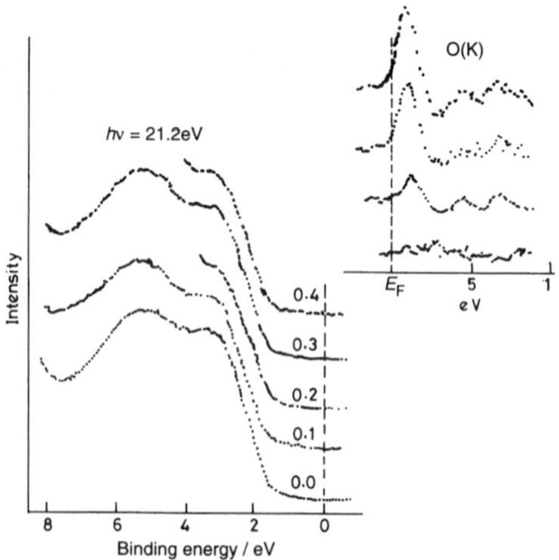

Fig. 11 Photoelectron spectra of $La_{1-x}Sr_xMnO_3$ in the valence region. Inset shows the oxygen K-absorption spectra (from ref. 14).

Concluding remarks

I have discussed three types of complex metal oxides which are marginally metallic. The first category of oxides exhibiting compositionally controlled insulator–metal transitions have a finite density of states at the Fermi level in the metallic state and obey the known criteria for metallicity. The second category involves the superconducting cuprates. They are marginally metallic in the normal state, obey the criteria for metallicity and undergo unusual insulator–metal and insulator–superconductor transitions, besides exhibiting certain anomalous properties in the metallic state. The first two categories of oxides seem to obey both the criteria of Mott arising from considerations of electron interactions (correlation) and of disorder. It is obvious that both these factors should be considered in describing these oxides. The importance of disorder in these oxides is also indicated by the fact that the phenomena are dependent on a critical concentration of one of the component species. Correlation is certainly important in all the systems considering that the parent oxides, $LaCoO_3$, $LaMnO_3$ and La_2CuO_4, all have mixed-valent ground states and fall in the category of Mott or charge-transfer insulators (Fig. 3). Yet, those who work on electron correlation models of oxides ignore disorder and *vice versa*. Unlike the two categories of oxides mentioned above, the manganates exhibiting giant magnetoresistance have high resistivities beyond the range of all known metallic oxides and show no evidence for any significant density of states at the Fermi level in the so-called 'metallic' state. The metallicity of the manganates is clearly of an entirely different category, not encountered hitherto in any other oxide system.

Clearly, there is need for models which include both electron interactions and disorder to describe marginally metallic oxides. This is not an easy task and requires new ideas. The problem does not end here. Besides electron interactions and disorder, other factors such as electron–lattice interactions, magnetic effects and the effect of finite temperature would have to be taken into account depending on the situation as shown schematically in Fig. 12. The theoretical complexity of the marginally metallic state is truly formidable, a proper description of either one of the factors alone (*e.g.* pure correlation or pure disorder) being fraught with many difficulties. Some workers have considered simple models which essentially assume the coexistence of localized and itinerant electrons to describe insulator–metal transitions[25,26] and superconductivity[26] in oxides. For example, Burdett[26] makes use of the interaction or crossing of two diabatic potential-energy curves for the insulating and metallic states to describe the two phenomena. Although educative, such models do not throw light on the nature of marginal metallicity. An effort to combine both electron interactions and disorder has been made recently by Logan *et al.*[27] Whether the metallic state has a universal

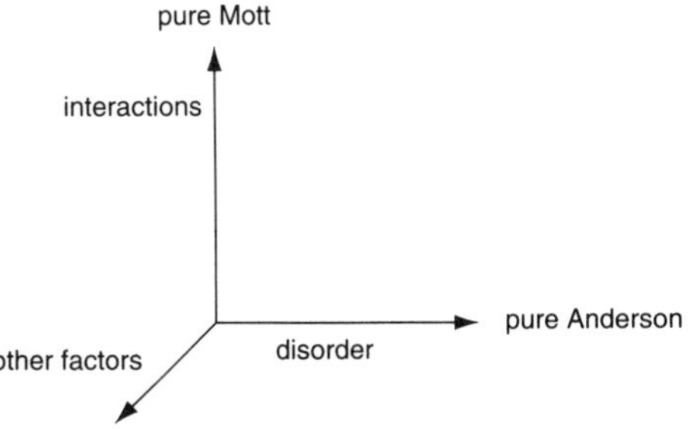

Fig. 12 Complexity of the problem of marginal metallicity (adapted from ref. 27). The oxides discussed in this article fall somewhere in the three-dimensional space indicated here. The 'other factors' include electron–lattice interaction, magnetic polaron and finite temperature effects.

minimum (electron) diffusity and hence minimum conductivity[28] due to localized states (in addition to extended electron states) is another question that needs to be explored.

Acknowledgements

The author thanks the Science Office of the European Union and the Department of Science and Technology, Government of India, for support of this research.

C. N. R. Rao is Albert Einstein Research Professor at the Indian Institute of Science and President of the Jawaharlal Nehru Centre for Advanced Scientific Research, Bangalore. He was born in Bangalore and educated in Mysore, Banaras and Purdue Universities. He was Visiting Professor at the Oxford and Cambridge Universities and is Honorary Professor at the University of Wales, Cardiff. He is a Fellow of the Royal Society, Foreign Associate of the US National Academy of Sciences and Member of the Pontifical Academy of Sciences. He is the recipient of the Marlow Medal and the Solid State Chemistry Medal of the Royal Society of Chemistry, of which he is an honorary fellow.

Footnotes

† An important experimental criterion for metallicity is the presence of a finite density of states at the Fermi level, which is readily established by photoelectron spectroscopy in the valence region and other high-energy spectroscopic techniques. Two of the most useful operational criteria for metallicity are those due to Mott discussed in the next section.

‡ $LaMnO_3$ as prepared in the laboratory by the solid-state reaction of the oxides and carbonates of the component metals at high temperatures, generally contains around 10% Mn^{4+}. The origin of Mn^{4+} is the presence of cation vacancies in both the A(La) and B(Mn) sites in roughly equal proportions. It cannot be due to oxygen excess since the perovskite structure cannot accommodate excess oxygen.[21]

§For small x, the material is an antiferromagnetic insulator; the same is true for large x ($x \geqslant 0.5$).

¶ The ferromagnetic T_c in $Ln_{1-x}A_xMnO_3$ increases markedly with the increase in the average radius of the A-site cations, $<r_A>$; accordingly, T_{IM} also increases with $<r_A>$. The magnitude of GMR as well as the peak resistivity at the I–M transition decrease with the increase in $<r_A>$. Increasing $<r_A>$ in these perovskites is analogous to increasing pressure.[24] It should also be noted that GMR is not confined to perovskites or manganates.[7]

‖ The magnitude of GMR itself increases with an increase in the peak resistivity (at T_{IM}). In other words, one needs a bad metal to observe good GMR. Furthermore, even formally antiferromagnetic compositions ($x >$ 0.5) show GMR because of the presence of ferromagnetic clusters.

References

1 C. N. R. Rao and B. Raveau, *Transition Metal Oxides*, VCH, Cambridge, 1995; C. N. R. Rao, *Annu. Rev. Phys. Chem.*, 1989, **40**, 291.
2 P. P. Edwards, T. V. Ramakrishnan and C. N. R. Rao, *J. Phys. Chem.*, 1995, **99**, 5228.
3 C. N. R. Rao and P. Ganguly, in *The metallic and the nonmetallic states of matter*, ed. P. P. Edwards and C. N. R. Rao, Taylor and Francis, London, 1985.
4 C. N. R. Rao, *Philos. Trans. R. Soc. London, A*, 1991, **336**, 595.
5 C. N. R. Rao and A. K. Ganguli, *Chem. Soc. Rev.*, 1995, **24**, 1; *Acta Crystallogr., Sect. B*, 1995, **51**, 604.
6 C. N. R. Rao and A. K. Cheetham, *Science.*, 1996, **272**, 369.
7 G. Briceno, H. Chang, X. Sun, P. G. Schultz and X. D. Xiang, *Science*, 1995, **270**, 273; T. Shimakawa, Y. Kubo and T. Manako, *Nature (London)*, 1996, **379**, 53.
8 N. F. Mott, *Metal–insulator transitions*, 2nd edn., Taylor and Francis, London, 1990.
9 N. F. Mott, *Philos. Mag.*, 1961, **6**, 287.
10 P. W. Anderson, *Phys. Rev.*, 1958, **109**, 1492.
11 N. F. Mott, *Philos. Mag.*, 1972, **26**, 1015.
12 E. Abrahams, P. W. Anderson, D. C. Liccardello and T. V. Ramakrishnan, *Phys. Rev. Lett.*, 1979, **42**, 693.
13 J. Zaanen, G. A. Sawatzky and J. W. Allen, *Phys. Rev. Lett.*, 1985, **55**, 418.

14 D. D. Sarma, in *Metal–insulator transitions revisited*, ed. P. P. Edwards and C. N. R. Rao, Taylor and Francis, London, 1995.
15 N. F. Mott, *Proc. R. Soc. London, A*, 1982, **382**, 1.
16 A. K. Raychaudhuri, *Phys. Rev. B*, 1991, **44**, 8572.
17 G. A. Thomas, *J. Phys. Chem.*, 1983, **88**, 3749; P. P. Edwards and M. J. Sienko, *Phys. Rev. B*, 1978, **17**, 2575.
18 T. V. Ramakrishnan and C. N. R. Rao, *J. Phys. Chem.*, 1989, **93**, 4414.
19 D. D. Sarma, in *Chemistry of high temperature superconductors*, ed. C. N. R. Rao, World Scientific, Singapore, 1991.
20 Y. Iye, in *Metal–insulator transitions revisited*, ed. P. P. Edwards and C. N. R. Rao, Taylor and Francis, London, 1995.
21 M. Hervieu, R. Mahesh, N. Rangavittal and C. N. R. Rao, *Eur. J. Solid State Inorg. Chem.*, 1995, **32**, 79.
22 C. Zener, *Phys. Rev.*, 1951, **82**, 403.
23 It is possible to increase the Mn^{4+} content in the parent $LaMnO_3$ and render it ferromagnetic. $LaMnO_3$ with $\geqslant 20\%$ Mn^{4+} shows the I–M transition and GMR: R. Mahesh, K. R. Kannan and C. N. R. Rao, *J. Solid State Chem.*, 1995, **114**, 294; R. Mahendiran, R. Mahesh, S. K. Tewari, N. Rangavittal, A. K. Raychaudhuri, T. V. Ramakrishnan and C. N. R. Rao, *Phys. Rev. B*, 1995, **53**, 3348.
24 R. Mahesh, R. Mahendiran, A. K. Raychaudhuri and C. N. R. Rao, *J. Solid State Chem.*, 1995, **120**, 204.
25 J. M. Honig and J. Spalek, in *Advances in Solid State Chemistry*, ed. C. N. R. Rao, Indian National Science Academy, New Delhi, 1986.
26 J. K. Burdett, *Acc. Chem. Res.*, 1995, **28**, 227.
27 D. E. Logan, Y. H. Szczechand and M. A. Tusch, in *Metal–insulator transitions revisited*, ed. P. P. Edwards and C. N. R. Rao, Taylor and Francis, London, 1995.
28 N. Kumar and A. M. Jayannavar, in *Metal–insulator transitions revisited*, ed. P. P. Edwards and C. N. R. Rao, Taylor and Francis, London, 1995.

Received, 21st March 1996; 6/01957I

ELSEVIER

Journal of Molecular Structure (Theochem) 500 (2000) 339–362

THEO
CHEM

www.elsevier.nl/locate/theochem

Experimental and theoretical electronic charge densities in molecular crystals

G.U. Kulkarni, R.S. Gopalan, C.N.R. Rao*

Chemistry and Physics of Materials Unit, Jawaharlal Nehru Centre for Advanced Scientific Research, Jakkur P.O., Bangalore 560 064, India

Abstract

Electronic charge density distribution in molecular systems has been described in terms of the topological properties. After briefly reviewing methods of obtaining charge densities from X-ray diffraction and theory, typical case studies are discussed. These studies include rings and cage systems, hydrogen bonded solids, polymorphic solids and molecular NLO materials. It is shown how combined experimental and theoretical investigations of charge densities in molecular crystals can provide useful insights into electronic structure and reactivity. © 2000 Elsevier Science B.V. All rights reserved.

Keywords: Electronic charge densities; Molecular crystals; Topological properties

1. Introduction

The description of charge distribution in crystalline lattices has come a long way since the first quantum model of the atom. It was known from early days that a quantitative account of the chemical bonds in molecules and crystals would require the calculation of the probability density of the electron cloud between atoms. The experimental possibility itself was considered soon after the discovery of X-ray diffraction. As early as 1915, Debye [1] stated "that experimental study of scattered radiation, in particular from light atoms, should get more attention, since along this way it should be possible to determine the arrangement of electrons in the atoms". Calculation of charge densities in molecules has been a preoccupation of theoretical chemists for sometime [2] and several charge density investigations of crystalline solids by both experiment and theory have been reported in the literature [3–11]. These comprise molecular crystals including non-linear materials, metallo-organic complexes and inorganic compounds. Koritsanszky [12] has provided a summary of charge density studies with reference to the topological properties and the electrostatic potential. The more recent literature has been surveyed by Spackman [13,14] and the charge density analysis in relation to metal–ligand and intermolecular interactions, has been discussed by Coppens [15].

In this article, we discuss electronic charge density in molecular systems as obtained from both experiment and theory. Besides introducing the topological analysis of charge density, we examine the experimental methods based on X-ray crystallography. In addition to dealing with the multipolar formalism for treating the experimental data, the program packages for orbital calculations in free molecules and crystals are mentioned. In particular, we discuss ring and cage systems, intermolecular hydrogen bonds, polymorphism and non-linear optical crystals.

* Corresponding author. Tel.: + 91-80-846-2762; fax: + 91-80-846-2766.
E-mail address: cnrrao@jncasr.ac.in (C.N.R. Rao).

 G.U. Kulkarni et al. / Journal of Molecular Structure (Theochem) 500 (2000) 339–362

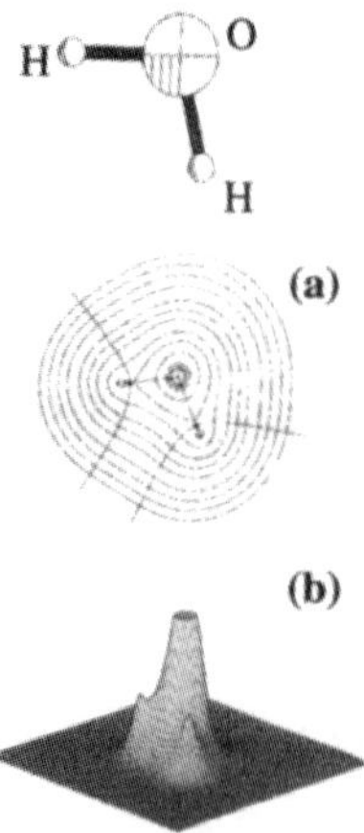

Fig. 1. Water: charge density in the molecular plane, (a) the contour map. The outermost contour has the value 0.0067 $e\text{Å}^{-3}$. The density increases almost exponentially for inner contours. The bond paths, the interatomic surfaces and the bond critical points are also indicated. (b) The relief map where the atom-cores are seen as peaks (reproduced with permission from Bader [2]).

2. Description of electronic charge density

The electronic charge density in an N-electron system is the probability per unit volume of finding any of the electrons in the phase space τ,

$$\rho(r) = N \int \psi^* \psi \, d\tau \qquad (1)$$

where ψ is the stationary state function; τ denotes the spin coordinates of all the electrons and the Cartesian coordinates of all N electrons but one [2]. It is expressed in $e\text{Å}^{-3}$ or au (1 au = 6.7483 $e\text{Å}^{-3}$). The description of electronic structure of a molecule in real space therefore relates to the charge density distribution around the constituent atoms. The density

Table 1
Critical points in molecular systems

Function	CP (rank, signature)	Chemical entity
$\rho(r)$	(3, −3)	Atom
$\rho(r)$, $V(r)$	(3, −1)	Bond
$\rho(r)$	(3, +1)	Ring
$\rho(r)$	(3, +3)	Cage
$\nabla^2\rho(r)$	(3, −3)	Lone-pairs
$V(r)$	(3, +3)	Lone-pairs

in a molecule can be conveniently modeled by partitioning into core, spherical valence and deformation valence around each atom [16],

$$\rho_{\text{atom}}(r) = \rho_{\text{core}}(r) + \rho_{\text{valence}}(r) + \rho_{\text{deformation}}(r, \theta, \phi) \qquad (2)$$

The topology of a charge distribution has many rich features—maxima, minima, saddles and nodes which help characterize intuitional elements such as atom cores, bonds and lone-pair electrons. As an example, the charge density distribution in water molecule [2] is depicted in Fig. 1 in the form of contour as well as relief maps. The density is maximum at the oxygen core position and decreases steeply towards the mid-region between oxygen and hydrogen reaching the minimum value at the 'critical point' ($\nabla\rho = 0$). This point carries maximum densities from the other two perpendicular directions. A quantitative description of charge density thus boils down to examining the number and the nature of such critical points in and around a molecule. A critical point (CP) is characterized not only by its density and location but also by the curvatures and the associated signs. The curvature of charge density at a point ($\nabla^2\rho$) well known as the Laplacian, is a measure of the charge concentration ($\nabla^2\rho < 0$) or depletion ($\nabla^2\rho > 0$) at that point. It is obtained as the sum of eigenvalues—λ_1, λ_2 and λ_3 of the Hessian matrix diagonalized against principal axes of curvature with λ_3 set along the internuclear vector. The rank of the matrix (3 for a stable molecular system) and the signature (sum of the signs of eigenvalues) imply the nature of the critical point. At a (3, − 3) CP, for example, all the curvatures are negative and ρ is locally the maximum. All the atom-cores exhibit (3, − 3) CPs in ρ (see Table 1). The relative magnitudes of the curvatures perpendicular to the bond direction (λ_3) determine the ellipticity [17] associated with a bond, $\epsilon = (\lambda_1/\lambda_2) - 1$.

An estimate of bond polarization [17] can be obtained from the location of the bond CP with respect to the internuclear vector,

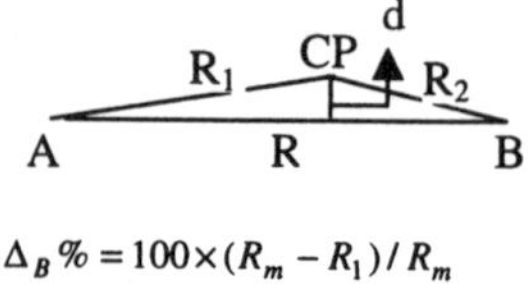

$$\Delta_B \% = 100 \times (R_m - R_1)/R_m$$

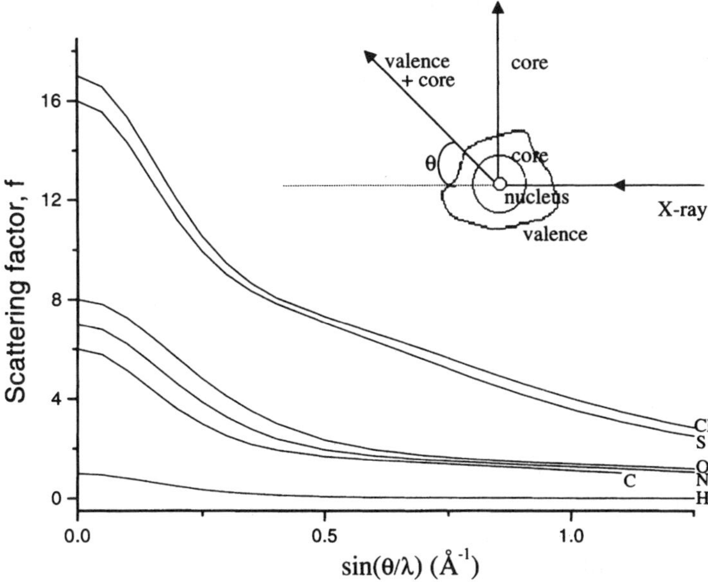

$$\sin(\theta/\lambda) \ (\text{Å}^{-1})$$

Fig. 2. Variations of the scattering factor with $\sin \theta/\lambda$ for different atoms. Above 0.5 Å^{-1}, the scattering due to valence electrons decreases gradually and the atom-core becomes visible. This is shown schematically in the inset. The scattering factors were obtained from the International Tables for Crystallography, Vol. IV, (1974) p. 71.

$$\Delta_B\% = 100 \times (R_m - R_1)/R_m \qquad (3)$$

where $R_m = (R_1 + R_2)/2$. The Δ value is used to describe relative electronegativities of the atoms involved. The strain involved in a bond can be estimated [17] in terms of the vertical displacement, d of the bond path from the internuclear vector

$$d = 2 \times \sqrt{s \times (s - R_1) \times (s - R_2) \times (s - R)}/R \qquad (4)$$

where $s = (R_1 + R_2 + R)/2$.

The electrostatic potential [18,19] generated by a molecule containing nuclear charges, z_i, placed at R_i, with a charge distribution $\rho(r)$, is given by

$$V(r) = \sum_i \frac{z_i}{|r - R_i|} - \int \frac{\rho(r')}{|r' - r|} d^3 r' \qquad (5)$$

It is more useful in describing attractive and repulsive interactions and also in determining the electrophilic and nucleophilic sites in molecules. The other quantity of interest is the kinetic energy density at the

critical point, G_{CP} [20] which has been obtained by

$$G_{CP} = \frac{3}{10}(3\pi)^{2/3} \rho_{CP}^{5/3} + \frac{\nabla^2 \rho_{CP}}{6} \qquad (6)$$

This quantity has been used in some cases for the calculation of hydrogen bond energies [21] and their classification [22].

3. Charge density from X-ray diffraction

Experimental determination of charge density relies mostly on X-ray diffraction although other techniques have been applied in some instances. X-ray diffraction arises from scattering by electrons and therefore carries information on the distribution of electronic charge in real space [10]. The intensity of a Bragg reflection, $I(\mathbf{h})$, at a given temperature, is proportional to the square of its structure factor,

$$I(\mathbf{h}) \propto |F(\mathbf{h})|^2 \propto \left| \sum_i f_i(\mathbf{h}) \, e^{2\pi i \mathbf{h} \cdot \mathbf{r}} \right|^2 \qquad (7)$$

where $f_i(\mathbf{h})$ is the scattering factor of the ith atom in the unit cell of volume, V. The charge density is obtained by the Fourier summation of the experimentally measured reflections

$$\rho(r) = \frac{1}{V} \sum_{\mathbf{h}} F(\mathbf{h})\, e^{-2\pi i \mathbf{h} \mathbf{r}} \qquad (8)$$

In conventional structure determination, $f_i(\mathbf{h})$ is approximated to the scattering factor from spherical electronic density, while for a complete description of bonding, accurate modeling of $f_i(\mathbf{h})$ becomes necessary. In parallel to $\rho(r)$ (see Eq. (2)),

$$f(\mathbf{h}) = f_{\text{core}}(\mathbf{h}) + f_{\text{valence}}(\mathbf{h}) + f_{\text{deformation}}(\mathbf{h}) \qquad (9)$$

Such a partitioning of $f(\mathbf{h})$ is justifiable in X-ray diffraction since one can select regions of reciprocal space where core scattering is predominant. In Fig. 2, we show variation of $f(\mathbf{h})$ with scattering angle, θ, for various elements. Each curve is composed of two regions. At low angles or Bragg vector ($\mathbf{h} = 2\sin\theta/\lambda$), $f(\mathbf{h})$ decreases steeply and above $\sim 0.5\ \text{Å}^{-1}$, the fall is gradual. The first part has contributions from both the atom-core and the valence density while the second arises mainly due to the core. Thus, X-ray diffraction facilitates extraction of the bonding or the deformation density,

$$\rho_{\text{deformation}} = \rho_{\text{total}} - \rho_{\text{promolecule}} \qquad (10)$$

where promolecule is obtained by the superposition of atoms without any interaction between them. This is called the (X–X) method. In the (X–N) method, the core positions along with the thermal parameters are obtained from a neutron diffraction experiment. The latter is particularly useful while dealing with hydrogen atom positions though it requires two data sets, which can be expensive besides having to grow bigger crystals. In recent years, the (X–X) method has become more popular.

Eq. (8) above necessitates data collection covering a wide range of the reciprocal space. For small unit cells with dimensions $\sim 30\ \text{Å}$, data collection up to moderately high resolution $(1.25\ \text{Å}^{-1})$ can be achieved using short wavelength radiations such as MoK_{α} $(0.71\ \text{Å})$. Moreover, data collection strategy critically depends on the type of the diffractometer. In the past, point detectors mounted on a four circle diffractometers were used for charge density measurements with the data collection extending in some

cases to a period of few weeks. Of late, area detectors or image plates are preferred over the conventional ones, as the experiments can be carried out faster with greater redundancy [23]. Area detectors in combination with synchrotron radiation are becoming increasingly popular [24,25]. Koritsanszky et al. [25] demonstrated that the charge density data can be collected within a day.

Thermal smearing of the charge density caused by atomic vibrations can hamper the extraction of subtle features of bonding,

$$f(T)|_{(h,k,l)} = f(0)\exp\left(-(b_{11}h^2 + b_{12}hk + b_{13}hl\right.$$
$$\left. + b_{22}k^2 + b_{23}kl + b_{33}l^2)\right) \qquad (11)$$

where

$$2\pi^2 a^{*2} U_{11} = b_{11} \qquad 2 \times 2\pi^2 a^* b^* U_{12} = b_{12} \qquad \text{etc.}$$

where U_{ij}s are the anisotropic displacement parameters and a^*, b^* and c^* are the reciprocal lattice vectors. Low temperature experiments at ~ 100 K, are carried out by allowing a stream of liquid nitrogen to fall on the crystal. In some cases however, much lower temperature (~ 20 K) has been achieved using one or two stage He-closed-cycle cryostats [9]. It is also necessary to choose a good quality, least mosaic crystal for charge density work.

4. Data refinement and computer codes

A preliminary knowledge of the crystal structure is important prior to a detailed charge density analysis. Direct methods are commonly used to solve structures in the spherical atom approximation. The most popular code is the SHELX from Sheldrick [26] which provides excellent graphical tools for visualization. The refinement of the atom positional parameters and anisotropic temperature factors are carried out by applying the full-matrix least-squares method on a data corrected if found necessary, for absorption and diffuse scattering. Hydrogen atoms are either fixed at idealized positions or located using the difference Fourier technique.

In the absence of inputs from neutron diffraction, a higher-order refinement of X-ray data $(>0.6\ \text{Å}^{-1})$ becomes essential to obtain accurate core positions and the associated thermal parameters (the X–X method). In this case, the hydrogen atom positions are often adjusted to the average neutron diffraction values [27] and are held there during the refinement. Often, the rigid bond test [28] is carried out and the parameters are corrected for translation–libration motions of the molecule [29]. The aspherical atom electron density is obtained in a local coordination system using the Hansen–Coppens formalism [16],

$$\rho(\mathbf{r}) = \rho_{\mathrm{c}}(r) + P_{\mathrm{v}}\rho_{\mathrm{v}}(\kappa r) + \sum_{l} R_{l}(\kappa' r) \sum_{m=-l}^{l} P_{lm} y_{lm}\left(\frac{\mathbf{r}}{r}\right)$$

$$(12)$$

Here, ρ_{c} and ρ_{v} are the spherically averaged Hartree–Fock core and valence densities, respectively, with ρ_{v} normalized to one electron. The Slater type radial functions $R_1 = N_1 r^n \exp(-\kappa' \xi r)$, modulated by the multipolar spherical harmonic angular functions y_{lm} define the deformation density. The population parameters, P_{v} and P_{lm}, are floated along with κ, κ' during the refinement. The kappa parameters control the expansion or the contraction of the radial part of the electron cloud with respect to the free atom. The multipoles on the first row atoms are generally refined up to octapole moments, while for the heavier ones, moments up to hexadecapole are used. Hydrogen atoms are restricted to dipole, although occasionally quadrupole moments are included in the refinement. The above formalism is well adopted in the recently developed user-friendly program package, XD [30]. Older codes such as MOLLY [16], VALRAY [31], LSEXP [32], POP [33] are also still in use. The quality of a refined model can be monitored based on the residuals and the goodness-of-fit apart from closely inspecting the deformation density maps.

The valence population coefficients P^i can be used to estimate the pseudo-atomic charges on the different atoms according to the equation,

$$q_i = n_i - P^i_{\mathrm{core}} - P^i_{\mathrm{valence}} \qquad (13)$$

where n_i is the total number of electrons of atom i. The molecular dipole moment is given by

$$\mathbf{p}_i = \sum_i z_i \mathbf{R}_i + \int_v \mathbf{r}\rho_i(\mathbf{r}_i)\,\mathrm{d}\mathbf{r} \qquad (14)$$

5. Theoretical methods

Electronic charge density distribution in a molecule or a crystal may be obtained by Hartree–Fock calculations [34]. It involves calculation of antisymmetrized many-electron wavefunction and minimizing the energy with respect to the coefficients of the one-electron wave function. When the energy is minimized, the wavefunction is said to have achieved self-consistency (SCF). Slater type atomic orbitals were used in the past which posed lot of difficulty in analytically integrating the polynomial functions. Use of gaussian functions in the radial part of the wavefunction has made HF method more applicable. Pople [35] has described the current status of ab initio quantum chemical models in his recent Nobel lecture. The accuracy that can be reached with these models critically depends on how many gaussians, polarization and diffuse functions make the basis set. In principle, using a full configuration interaction (FCI) with a large number of gaussians on each orbital should give the best results. For example, calculations performed using 6-311G**++ basis sets at FCI level of theory is common with small molecules. Computer program packages like GAUSSIAN [36] and GAMESS [37] are available for ab initio calculations on molecules. MOPAC [38] is used at a semiempirical level. Periodic Hartree–Fock calculations suitable for crystalline substances has been incorporated in CRYSTAL algorithm by Dovesi et al. [39]. It uses one-particle basis function made up of Bloch functions,

$$\chi_i(r, k) = (1/\sqrt{N}) \sum_t \chi_a^t \, \mathrm{e}^{ikt} \qquad (15)$$

Here χ_a^t refers to the ath atomic orbital in the unit cell of the crystal described by the lattice translation vector, t. The CRYSTAL code is also capable of calculating charge density in a solid using the density functional theory (DFT) at local density approximation (LDA) or at generalized gradient approximation (GGA).

The charge density obtained using the theoretical procedures can be usefully compared with that from X-ray diffraction by several means. Charge density maps either in total or in deformation provide the obvious tools to evaluate how well the two models agree. CRYSTAL95 [39] offers routines to calculate

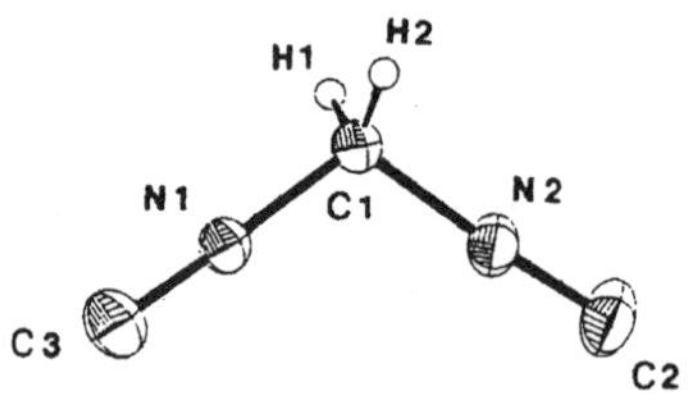

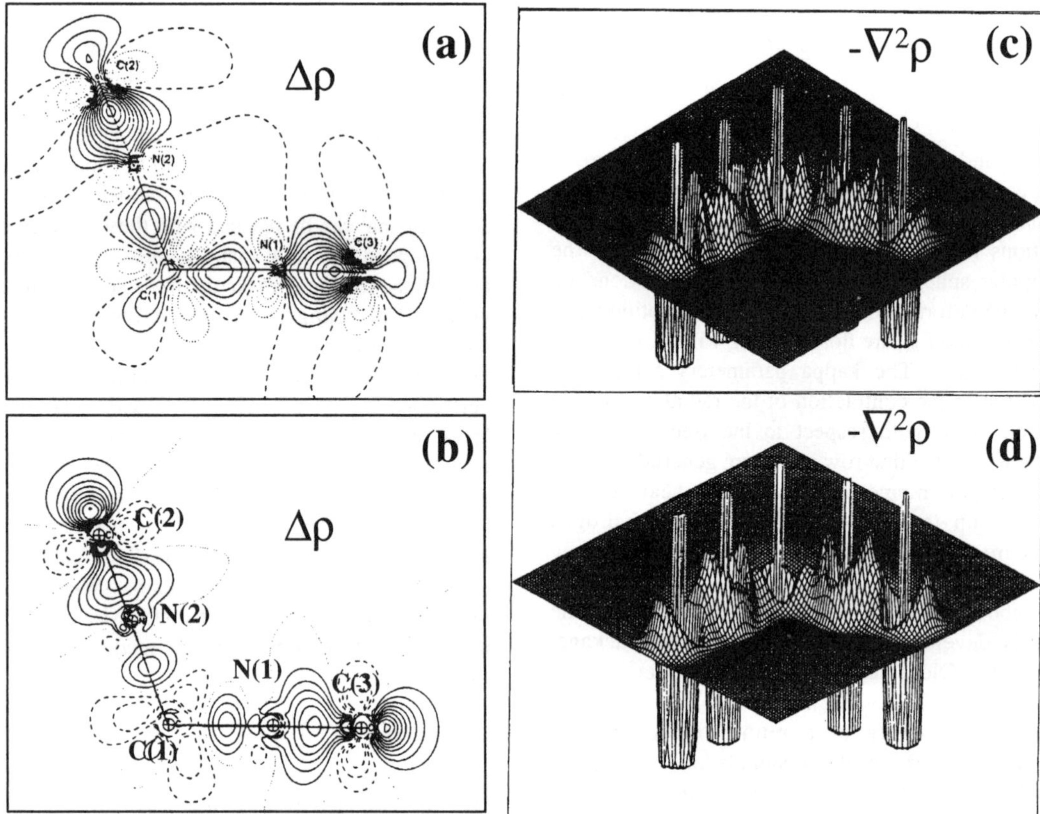

Fig. 3. Charge density in diisocyanomethane: deformation density maps in the molecular plane (a) experimental (b) theoretical (contours at $0.1\,e\text{Å}^{-3}$). The non-bonded regions of C(2) and C(3) are more depleted in (a) with the density migrating to the inside of the molecule. The corresponding Laplacians (range -20 to $250\,e\text{Å}^{-5}$) are shown in (c) and (d), respectively (reproduced with permission from Koritsanszky et al. [40]).

X-ray structure factors from the theoretical density. Using the theoretical structure factors, one can carry out multipolar refinement in parallel to experimental X-ray data and make in depth comparison of the topological properties from the two sets.

6. Charge density in bonds, rings and cages

Covalent bonds are associated with high charge densities $(1.5-3\,e\text{Å}^{-3})$ and negative Laplacians, while ionic bonds are characterized by small densities and positive Laplacians. Hydrogen bonds are

G.U. Kulkarni et al. / Journal of Molecular Structure (Theochem) 500 (2000) 339–362

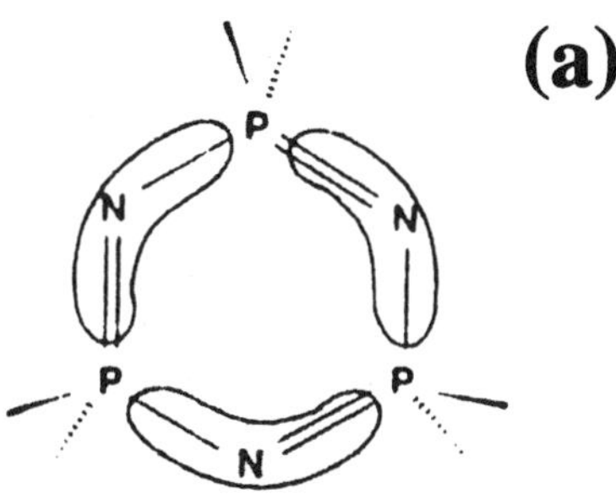

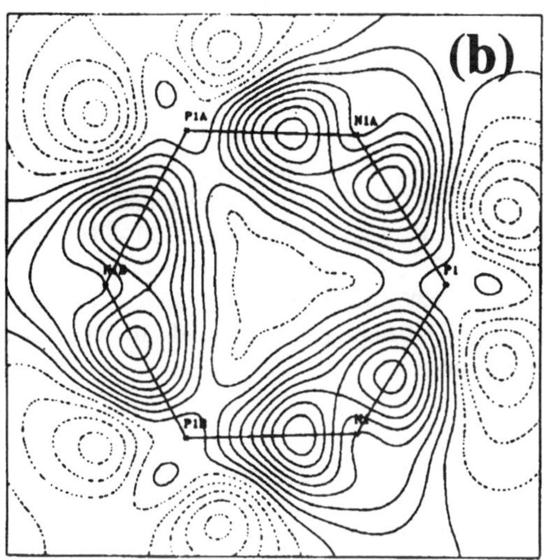

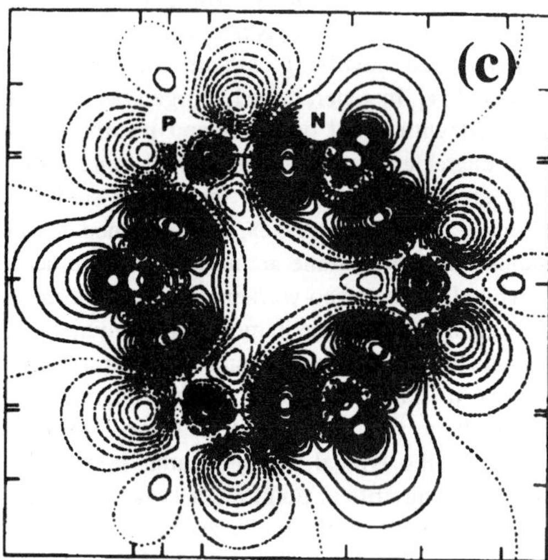

Fig. 4. The phosphazene ring: (a) island delocalization model predicting nodes in π-density at the phosphorus atoms (b) dynamic deformation density (at 0.1 $e\text{Å}^{-3}$) in the plane of the ring (c) theoretical deformation density (at 0.05 $e\text{Å}^{-3}$) of cyclic phosphazene was used as a model (reproduced with permission from Cameron et al. [43]).

associated with even smaller densities and Laplacians. Following Cramer and Kraka [17], the charge density at critical point, ρ_{CP}, is a measure of the bond strength in covalent bonds. Thus, a typical C–C bond carries a density of $\sim 1.7\ e\text{Å}^{-3}$ at the CP while a C=C bond exhibits much higher density $\sim 2.5\ e\text{Å}^{-3}$. Similarly, the density associated with a C≡C bond is $\sim 2.8\ e\text{Å}^{-3}$. The ellipticity of a bond, ϵ, is a measure of its extent of double bond character. For cylindrically symmetric bonds, the ellipticity is therefore zero, while in the case of ideal carbon double bonds, the theoretical estimate gives $\epsilon \sim 0.74$. These quantities, in combination with bond polarity (Δ), pseudo-atomic charges and the bent bond character (d) describe a bond quantitatively.

Let us examine the case of diisocyanomethane as a typical example. Isocyanides possess a formally divalent carbon atom with a coordination number of only one. Though they are divalent, their bond lengths are only slightly longer than ideal C≡N bond and therefore a resonance is expected. Koritsanszky et al. [40] analyzed topographs of electron density of diisocyanomethane (Fig. 3) derived from both experiment and theory. Refinements of the experimental data were carried out with different levels of constraints. The most restricted model contained C_{2v} symmetry on the tetrahedral carbon (C1), rotational symmetry on isocyano groups and thermal motion correction on all non-hydrogen atoms. This model gave the best convergence for the data. They also carried out ab initio calculations at the Hartree–Fock and MP2 levels, optimizing the molecule with 6-311++G(3d,3p) basis sets starting from X-ray structural data. Fig. 3 shows the deformation density and the Laplacians in the mean molecular plane from both theory and experiment. There is a striking difference between the two deformation density maps in that the interatomic regions of the experimental map are richer in electron density at the expense of charge in the non-bonded regions of the terminal carbon atoms (compare Fig. 3a and b). The isocyano bonds are also more polarized. The same is reflected in the Laplacians shown in Fig. 3c and d, where bonded charge concentration appears as a sharp peak in the experimental map. This has been taken to indicate a greater swing of the NC bond resonance towards the $N{\equiv}C$ bond.

Intramolecular bonding in benzene, triazine, phosphazene and such cyclic systems are of interest because of the varying degree of superposition of the s, p and d orbitals as the case may be. The Hückel rule predicts benzene to be aromatic with the π-electrons delocalized over the ring. According to the island delocalization model [41] for the phosphazene ring, the overlap of the d orbitals on the phosphorus atoms and the p orbitals on the nitrogen atoms in the ring would produce a π-system above and below the plane of the ring with nodes at each phosphorous atom (Fig. 4a). An extension of the π-system within and in the plane of the ring is also expected based on the π/π' model due to Craig and Paddock [42]. Cameron et al. [43] applied charge density methods to study small cyclic systems. For this purpose, they carried out high resolution X-ray diffraction at 200 K with MoK_α radiation on hexaazirdinylcyclo-triphosphazene crystallized from benzene. This system is particularly interesting in that, the solvent benzene gets securely trapped between two phosphazene molecules and offers one to use it as an internal standard. They observed that the benzene ring in plane density was symmetric with respect to the carbon–carbon bonds while the density in a plane perpendicular to the ring across the center of a bond showed elongation in the direction of the π-system as expected. In contrast, the in-plane density in phosphazene was highly polarized as shown in Fig. 4b. Nodes in density at phosphorus atoms can be clearly seen from the figure. The electron density appears to spread from one P–N bond through the nitrogen to the second N–P bond as predicted by the island delocalization model. Further, there is a considerable spread of electron density inside the ring validating the π'-bonding model. Theoretical electron densities determined from ab initio calculations using GAMESS [37] with 6-31G* basis set depict similar features in the deformation density (see Fig. 4c).

Charge density distribution in cage rings has been investigated in few cases. An adduct of C_{60} was studied by Irngartinger et al. [44] who discussed the degree of aromaticity. These workers have also carried out charge density measurements on a cubane derivative, methyl 3,4-difluorocubane-1-carboxylate [45]. This compound with the fluorine substituents fixed in a *cis*-like orientation to the rigid cubane cage (see Fig. 5) was expected to serve as a model for the *cis* isomer of 1,2-difluoroethylene. The latter along with its cousins such as *gauche*-1,2-difluoroethane is known to be energetically favorable compared to the *trans* isomer (*anti* conformer). Wilberg and co-workers [46] explained this '*cis/gauche* stability effect' and predicted the C–C bond to be bent due the strongly electronegative property of the fluorine substituents. Difference density maps in various diagonal planes of the cubane indeed show bending of the C–C bonds (Fig. 5). The CF–CF bond and the CH–CF bonds were found to be more bent $(d \sim 0.16\,\text{Å})$ than the CH–CH bonds $(d_{\text{average}} \sim 0.12\,\text{Å})$. This study clearly provides an evidence that strong electronegative substituents increase the bending of the bonds to which they are attached.

7. Charge density of hydrogen bonds

Hydrogen bonds can be classified on the basis of charge density. Alkorta and co-workers studied various types of hydrogen bonds including dihydrogen bonds [47], bifurcated hydrogen bonds [48], $H{\cdots}\pi$ interactions [49], inverse H bonds [50] and hydrogen bonds involving carbenes and silylenes as acceptors [51]. In general, the hydrogen bond CPs are associated with small densities and positive Laplacians characteristic of closed-shell interactions. Zhang et al. [52] carried out theoretical studies on hydrogen bonded complexes, with strained organic systems like tetrahedrane acting as pseudo-π-acceptors while Larsen and co-workers [53,54] have studied the strong hydrogen bond in methylammonium hydrogen maleate and benzoylacetone. Such strong bonds arise due to shared interactions exhibiting relatively large density $(\sim 1\,e\text{Å}^{-3})$ and negative Laplacians $(-7\,e\text{Å}^{-5})$ like typical intramolecular bonds. They also find a ring critical point (3, +1) near the center of the keto–enol dimer in benzoylacetone. Experimental charge density study of *cis*-$HMn(CO)_4PPh_3$

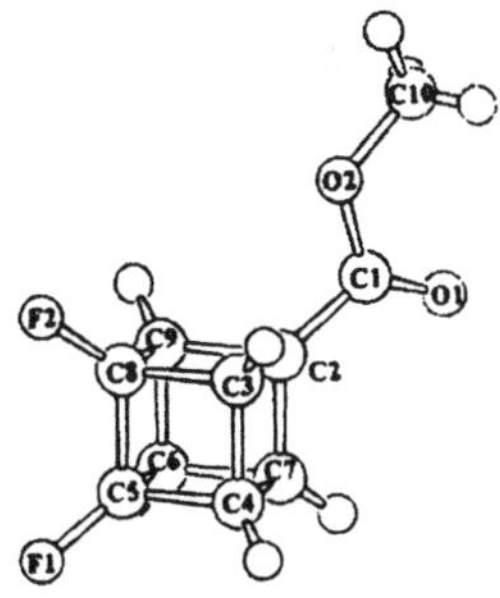

Methyl 3,4-difluorocubane-1-carboxylate

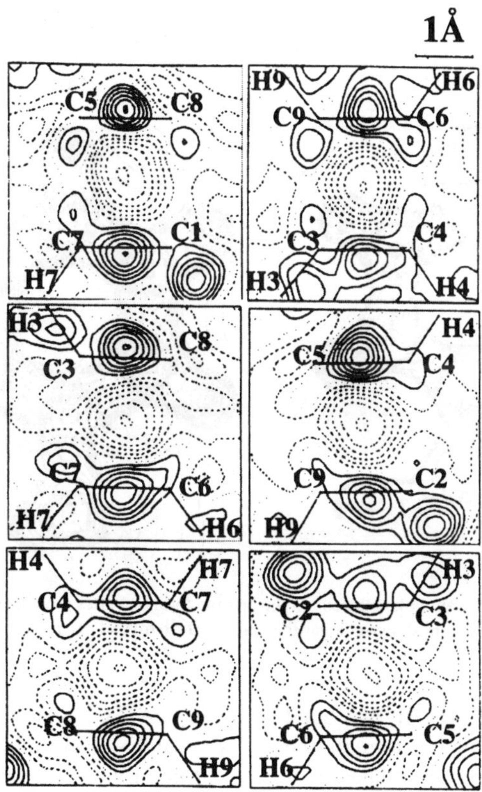

Fig. 5. Methyl 3,4-difluorocubane-1-carboxylate: difference density maps in various diagonal planes of the cubane cage. The maxima of the bond densities lie outside, implying that the cage bonds are bent (reproduced with permission from Irngartinger et al. [45]).

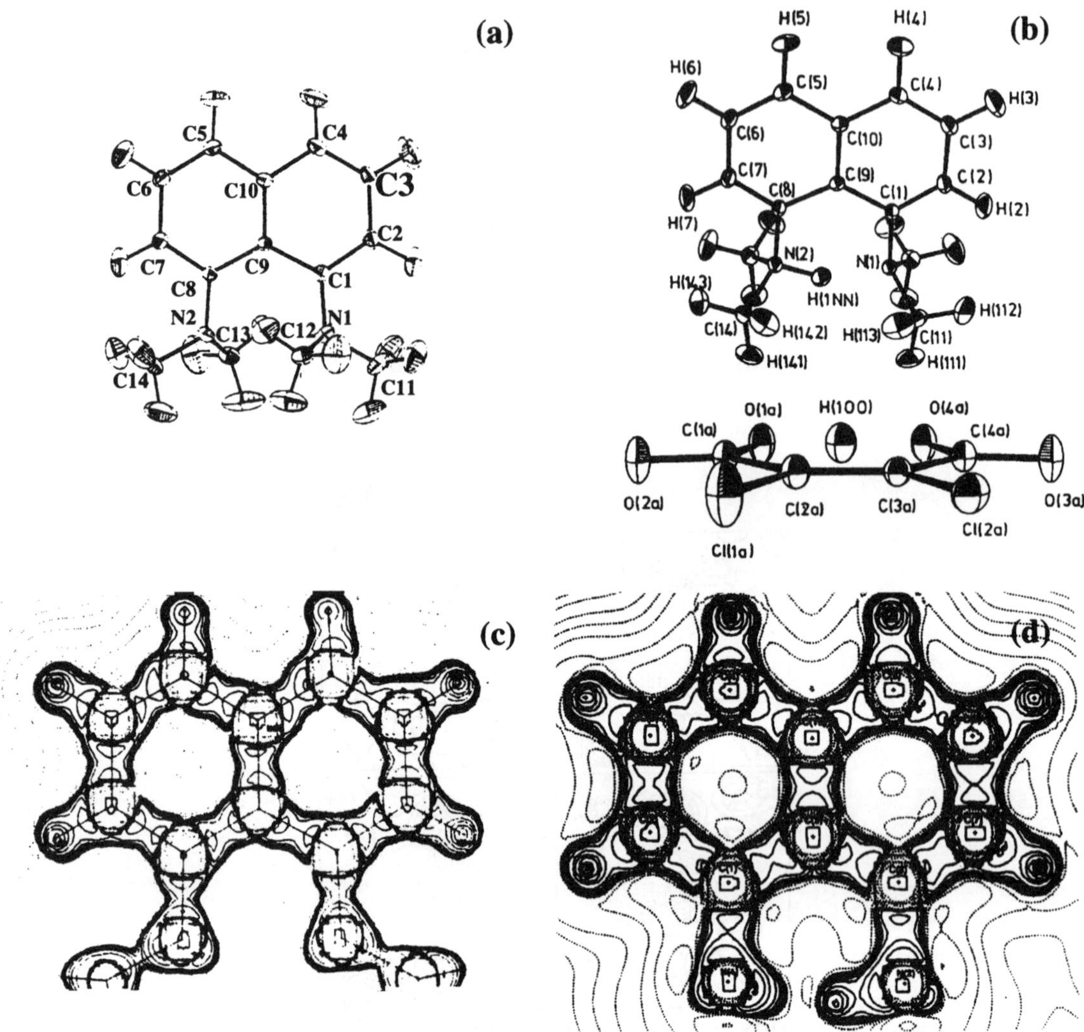

Fig. 6. A proton-sponge: (a) 1,8-Bis(dimethylamino)napthalene in the pristine form; (b) after protonation using 1,2-dichloro maleic acid. The contour maps of the corresponding Laplacians are shown at logarithmic intervals in (c) and (d), respectively (reproduced with permission from Mallinson et al. [56,57]).

has provided evidence for the C–H$\cdots$H–Mn bond [55]. They found that the hydrogen atom in the Mn–H bond is nucleophilic carrying a charge of $-0.4e$ while that in the C–H is electrophilic $(0.3e)$. The electrostatic part of the H$\cdots$H interaction energy was estimated to be 5.7 kcal/mol, which is in the range of H-bond interactions. It is characterized by a critical point carrying a small density $(0.066\,e\text{Å}^{-3})$ and a positive Laplacian $(0.79\,e\text{Å}^{-5})$, somewhat higher compared to a typical C–H$\cdots$O interaction.

Mallinson et al. [56,57] have carried out experimental and theoretical charge density determinations on a proton sponge compound, bis(dimethyamino)-napthalene (DMAN), in both pristine and protonated forms (Fig. 6). They used positional and thermal parameters of hydrogen atoms from an independent neutron diffraction experiment and treated the thermal

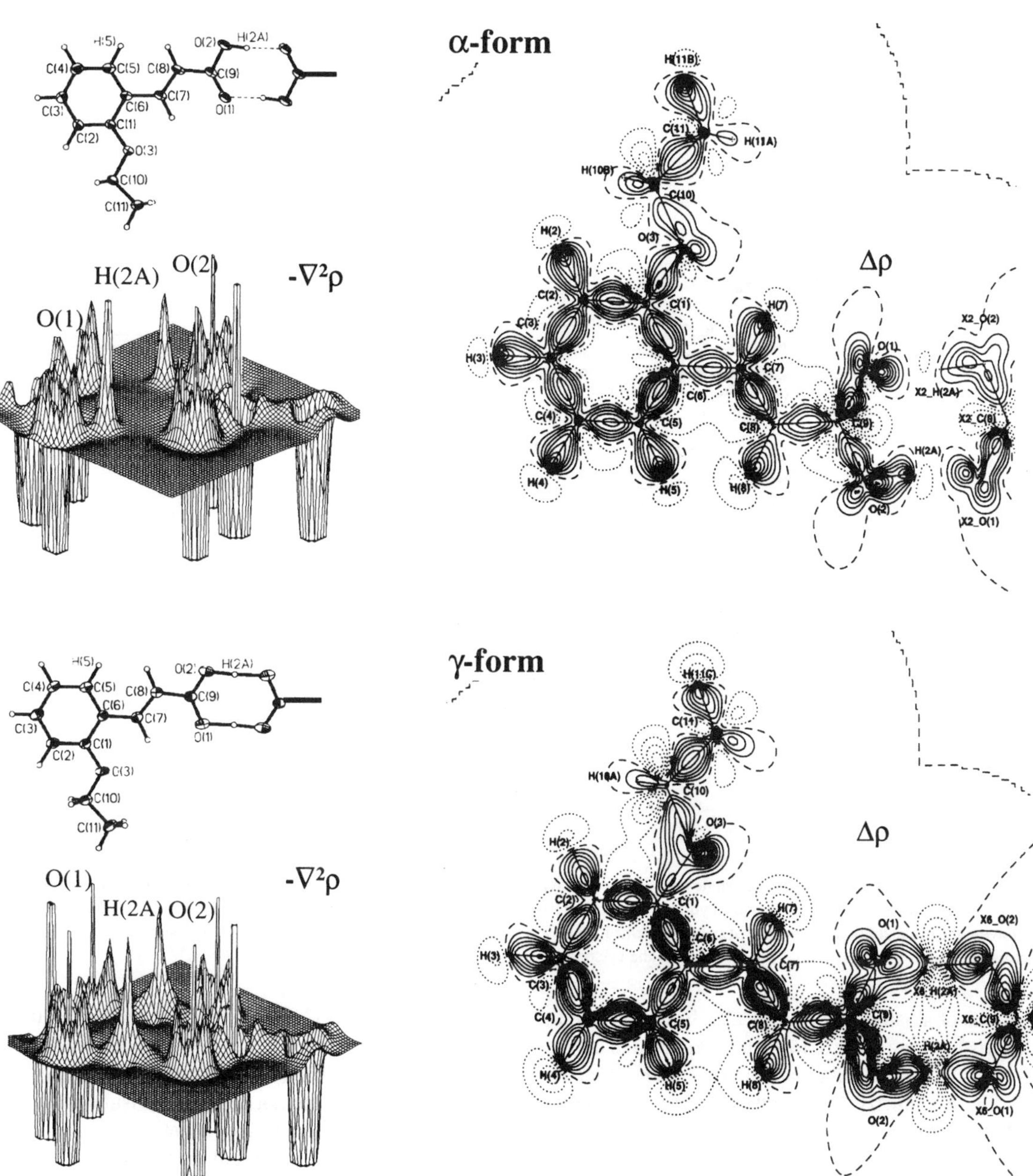

Fig. 7. Polymorphic forms of *o*-ethoxy cinnamic acid: molecular diagrams and deformation density maps close to the mean plane of the molecules in the α- and the γ-forms (contours at $0.12\,e\text{Å}^{-3}$). Subtle differences in the cinnamoyl bond and the hydrogen bond region are noticeable. The Laplacians of the intermolecular hydrogen bonds in the acid dimer are shown in the relief maps along side (range −250 to $250\,e\text{Å}^{-5}$).

G.U. Kulkarni et al. / Journal of Molecular Structure (Theochem) 500 (2000) 339–362

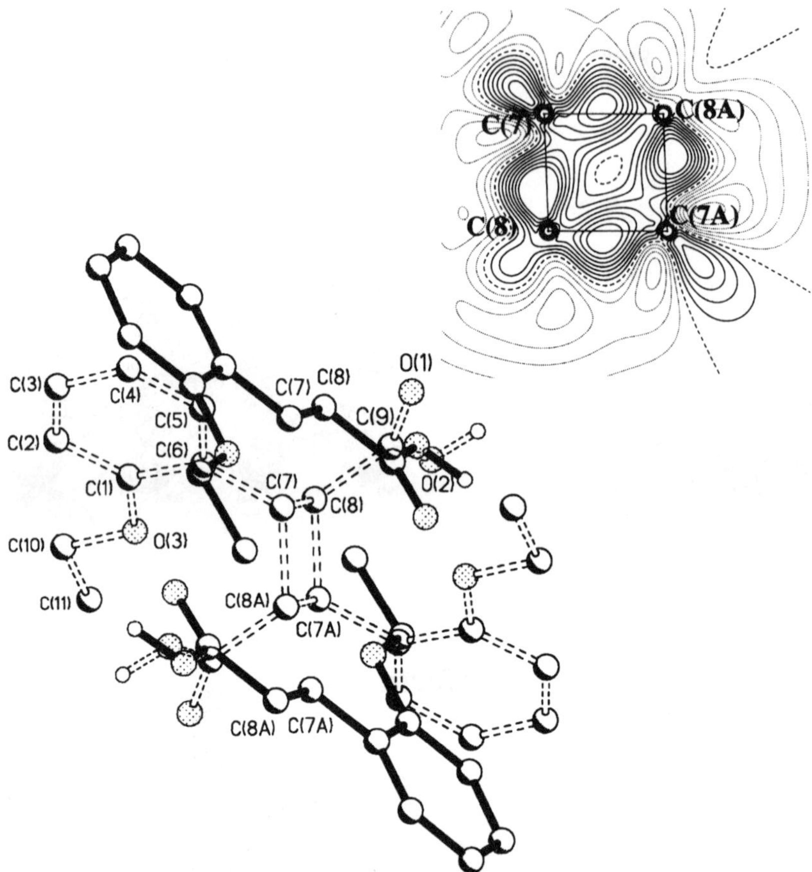

Fig. 8. Photodimerization in *o*-ethoxy cinnamic acid: Two centrosymmetric molecules of the α-form before (full lines) and after (dashed lines) the reaction. Hydrogen atoms other than the hydroxyl are omitted for the sake of clarity. The cinnamoyl double bonds are spaced at $\sim$4.3 Å and following the cycloaddition, a new pair of bonds is formed which are slightly longer ($\sim$1.57 Å) than typical C–C bonds. The inset shows the deformation density (at 0.075 $e\text{Å}^{-3}$) in the plane of the cyclobutane ring in the α-dimer.

motions of all atoms anisotropically. The molecule in the pristine form (Fig. 6a) which is expected to have a two-fold symmetry like naphthalene is unsymmetrical, the asymmetry being reflected only as a small difference in the Laplacian between the two rings. From Fig. 6c, we see that the shape of the contours of the corresponding regions of the two halves of the molecule are somewhat different. Accordingly, the Laplacian values of the N2–C8 and N1–C1 bonds are -15.9 and $-13.3\,e\text{Å}^{-5}$, respectively, while those of C6–C5 and C4–C3 are -21.2 and $-18.7\,e\text{Å}^{-5}$, respectively. Upon protonation (Fig. 6b), noticeable changes occur in bond lengths in the

molecule. The C–C bonds are shortened by $\sim$0.01 Å, while the C–N and C_{aromatic}–H bonds are lengthened by 0.04 and 0.02 Å, respectively. The $C_{\text{aliphatic}}$–H bonds however, do not change considerably. The atomic charges which are negative on the outer carbons decrease as a result of the migration of charge towards proton. These authors also carried out a ^{13}C NMR study in the solid state to show that the outer carbons are deshielded in the complexed proton sponge compared to the uncomplexed one and obtained useful correlations among ρ, $\nabla^2\rho$ and bond lengths. Another interesting aspect of this study is the curved interaction bond path joining two stacked

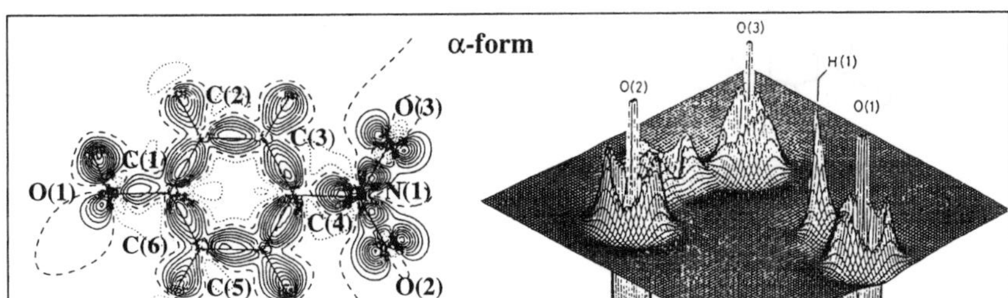

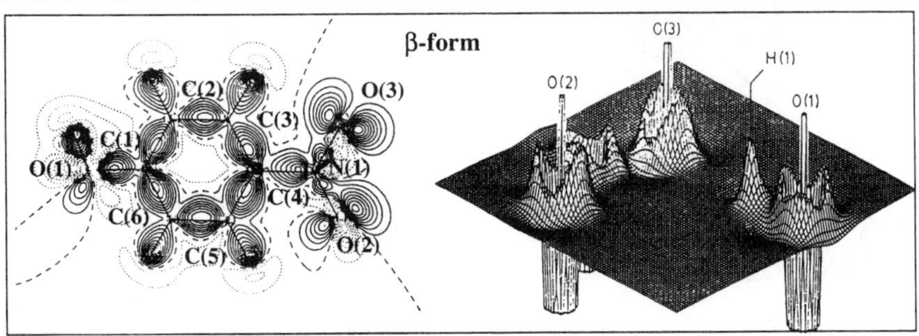

Fig. 9. Polymorphism in *p*-nitrophenol: static deformation density in the plane of the phenyl rings for the α- and the β-forms (contours at $0.1\ e\text{Å}^{-3}$). Intramolecular and lone-pair regions exhibit many differences. Relief maps of the Laplacians in the intermolecular hydrogen bond region are also shown (range -250 to $250\ e\text{Å}^{-5}$). In the α-form, H(1) bonds not only with O(3) but also with O(2) and N(1) of the neighboring nitro group (reproduced with permission from Kulkarni et al. [61]).

DMAN molecules attributed to the C–H···π interaction.

Intermolecular hydrogen bonds play a major role in deciding the properties of molecules in the solid state. Cinnamic acids for example, crystallize in two or three polymorphic forms [58], some being photoreactive forming cycloadditives while one polymorph may be photostable. We have carried out a charge density study using a CCD detector on the α- and γ-forms of *o*-ethoxy cinnamic acid (Fig. 7) obtained, respectively, from an ethyl acetate solution and slow cooling of an aqueous ethanol solution [59]. In the reactive α-form, the molecule is found to be quite planar, while in the photostable γ-form, the ethoxy and the cinnamoyl groups make angles of 6.5 and 3.5°, respectively, with the phenyl ring. Interestingly, the latter exhibits near-symmetric hydrogen bonds in the intermolecular region. The hydrogen H(2A) was located midway between the oxygens with the O–H and H···O distances of 1.23 and 1.39 Å, respectively. Accordingly, the ρ_{CP} values associated with the two bonds are comparable, ~1.6 and $0.8\ e\text{Å}^{-3}$,

respectively. Moreover, the intermolecular bond carries a negative Laplacian ($\sim -12.4\ e\text{Å}^{-5}$) like a hydroxy bond which is indicative of a highly shared interaction. In the α-form, on the other hand, the O–H and H···O distances are usual (0.96 and 1.68 Å, respectively) with ρ_{CP} of 2.2 and $0.32\ e\text{Å}^{-3}$, respectively. The H(2A)···O(1) hydrogen bond shows a small positive Laplacian of $4.81\ e\text{Å}^{-5}$ as is generally expected for a closed shell interaction. The other interesting aspect of this study is the inference on delocalization of the π-density in the α-molecule, from the cinnamoyl double bonds to the neighboring single bonds and across the phenyl group. This is also in compliance with the molecule being planar in this polymorph. A complete absence of such an effect in the γ-form was interpreted as due to the ionic nature induced by the symmetric hydrogen bonds. These factors influence the molecular geometry and the packing which in turn decide reactivity of a polymorphic form. The cinnamoyl bond undergoes (2 + 2) cycloaddition in the α-polymorph. Structural analysis indicated favorable approach of the

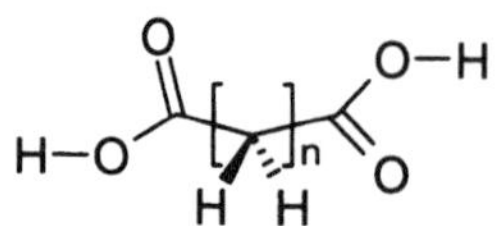

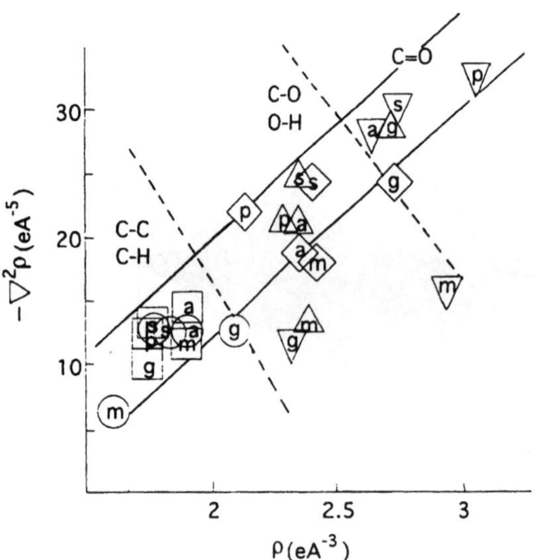

Fig. 10. Alkanedioic acids: variation of Laplacian with charge density at the critical point for various bonds, C–C, circle; C–H, square; C–O, up-triangle; C=O, down-triangle and O–H, rhombus. The letters m, s, g, a and p marked inside the data symbols represent bonds belonging to malonic, succinic, glutaric, adipic and pimelic acids, respectively. The different bond regions are delineated (reproduced with permission from Gopalan et al. [63]).

centrosymmetrically related cinnamoyl bonds in the photoactive α-form (Fig. 8). The deformation density of the cyclobutyl ring resulting from the photodimerization of α-form is shown in the inset of Fig. 8. It exhibits a (3, +1) CP at the inversion center of the ring associated with small density ($0.62\ e\text{Å}^{-3}$) and Laplacian ($6.6\ e\text{Å}^{-5}$). The densities of the ring bonds are small ($\sim 1.51\ e\text{Å}^{-3}$) implying that the bonds are weak. They are also associated with high ellipticity (~ 0.2) and polarization (6%). The contours of these bonds lie outside the interatomic vector, the vertical displacement being ~ 0.04 Å, characteristic of bent bonds in a strained ring [17,45].

p-Nitrophenol is known to show interesting photochemical activity only in one of its polymorphic forms. The α-form which crystallizes from benzene undergoes a topochemical transformation up on irradiation, changing its color from yellow to red, though structural changes associated with the transformation have been found to be insignificant [60]. On the other hand, the β-form obtained from aqueous solution, is light-stable. This has been the subject of a charge density study [61] (Fig. 9). It is found that the phenyl ring bonds in the α-form exhibit less density ($\sim 2.05\ e\text{Å}^{-3}$) compared to those in the β-form ($\sim 2.21\ e\text{Å}^{-3}$) as though charge had migrated outwardly in the former. Accordingly, the nitro and the hydroxyl bonds in the latter were found to carry relatively higher densities. The authors found many differences in the hydrogen bonding as well. In the β-form, there are four hydrogen bonds compared to six in the α-form. The striking difference between the two polymorphs is that in the α-form, the entire nitro group participates in hydrogen bonding with the neighboring hydroxyl hydrogen while in the β-form only the nitro oxygens involve in the hydrogen bonding. Interestingly upon photoirradiation of the α-form, charge density redistribution seems to occur with the reacted product assuming an intramolecular density similar to that found in the light stable β-form [62]. The authors have estimated the molecular dipole moments to be 18.0 and 21.5 Debye for the α- and β-forms, respectively and a somewhat lower moment in the case of the irradiated α-form, ~ 9.9 Debye.

The lattice cohesion of the first few members of the homologous series of aliphatic dicarboxylic acids has been investigated in terms of charge density [63]. These acids form an interesting class of organic compounds in that the structure and properties in the solid state exhibit undulatory behavior with the number of methylene groups being odd or even [64]. For example, the melting point alternates in the series malonic, succinic, glutaric, adipic, pimelic and so on. The study yielded interesting systematics in the charge densities and the Laplacians along the series. The ρ_{CP} values of C–C, C–O and O–H bonds increase from malonic (1.61, 2.27 and $2.45\ e\text{Å}^{-3}$, respectively) to glutaric (2.18, 2.72 and $2.7\ e\text{Å}^{-3}$, respectively) and decrease thereafter. The C=O and C–H bonds exhibit the opposite trend with glutaric acid carrying the minimum charge density ~ 2.33 and $1.66\ e\text{Å}^{-3}$, respectively. An overall

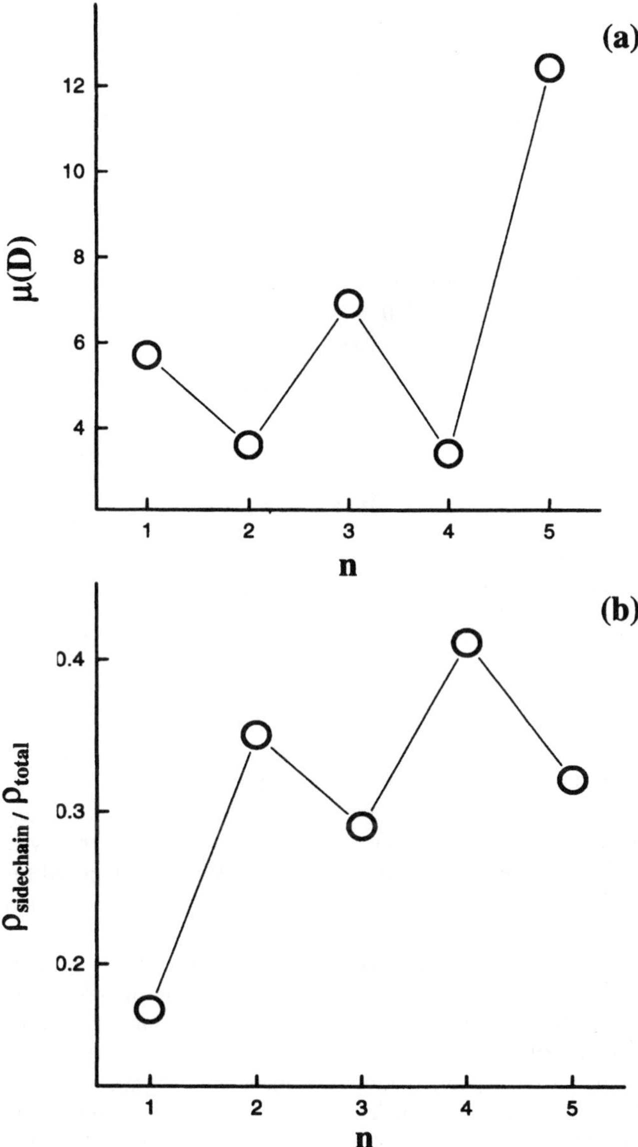

Fig. 11. (a) Molecular dipole moments of the dicarboxylic acids. Values obtained for the asymmetric units are shown since net dipole moment for an even acid vanishes due to center of symmetry. Here, n denotes the number of methylene groups in the acid. (b) A plot of the sum of ρ_{CP} obtained for the side-chain interactions normalized with the total ρ_{CP} due to intermolecular interactions, against the number of methylene groups, n, in the acid (reproduced with permission from Gopalan et al. [63]).

assessment of the charge distribution among various bonds can be made by plotting the Laplacian against the density for various bonds as shown in Fig. 10. We see that most bonds lie in a region where the Laplacian is roughly proportional to the bond density, as one would normally expect. Thus, the C–C and the C–H bonds fall in the first region of the plot while the C–O and the O–H bonds group fall in the second

 G.U. Kulkarni et al. / Journal of Molecular Structure (Theochem) 500 (2000) 339–362

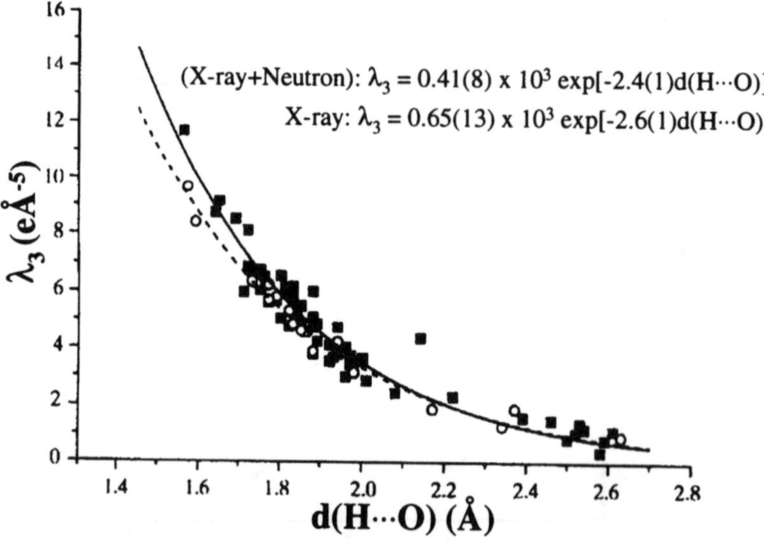

Fig. 12. Variation of the positive curvature, λ_3 at bond CP, with hydrogen contact distance, $d(H\cdots O)$. The data points obtained from a number of charge density investigations reported in the literature and were fitted using exponential equations. Circles and dashed line; joint X-ray and neutron, filled squares and solid line; X-ray data only. Among the three eigenvalues, λ_3 was found to be giving the best representation for hydrogen bonds having closed-shell interactions (reproduced with permission from Espinosa et al. [65]).

region. The C=O bonds are found in the third region. The C–O and C=O bonds of the malonic and the glutaric acids are somewhat unusual. In the case of the malonic acid, the Laplacians of the two bonds are noticeably lower. In glutaric acid, on the other hand, the Laplacian of the C=O bond is lower than that of the C–O bond, a trend which is reflected in the charge densities as well.

An important outcome of the study of the dicarboxylic acids is the analysis of the dipole moments of the asymmetric units in relation to intermolecular interactions in the crystal lattice (Fig. 11a). It is interesting that the dipole moments showed alternation along the series with the odd acids having higher moments than the even neighbors with pimelic acid exhibiting the highest moment, 12.7 Debye. The strength of molecular packing in the lattice was guided by the value of the sum of the ρ_{CP}s associated with side chain C–H$\cdots$O interactions expressed as a fraction of the total intermolecular density due to C–H$\cdots$O and O–H$\cdots$O interactions (Fig. 11b). Interestingly, there is an alternation in this value along the series with the even acids exhibiting higher values compared to their odd neighbors. Thus, it appears

that increased side-chain interactions in the even acids lead to distributed bond dipoles thereby decreasing the net dipole moment in the asymmetric unit. This may also be the reason for the relatively higher melting points in the even acids, since side-chain interactions as compared to the dimeric bonds play a decisive role in the cohesion of acid molecules in the solid state.

Hydrogen bonds have been classified by many workers in terms of varying degrees of shared and closed-shell interactions. Correlations between $\nabla^2\rho$ and ρ, as shown above in Fig. 10 find good agreement only within a group of like-hydrogen bonds. A more general approach has been suggested by Espinosa et al. [65] who used the variation of positive curvature along the interaction axis (λ_3) at CP with the hydrogen bond distance and obtained a well-defined behavior as shown in Fig. 12. The value of λ_3 was found to increase exponentially with decreasing hydrogen bond distances. It represents the overlap between the electron cloud of both H and O atoms at the critical point and is proportional to the kinetic energy density, G_{CP}.

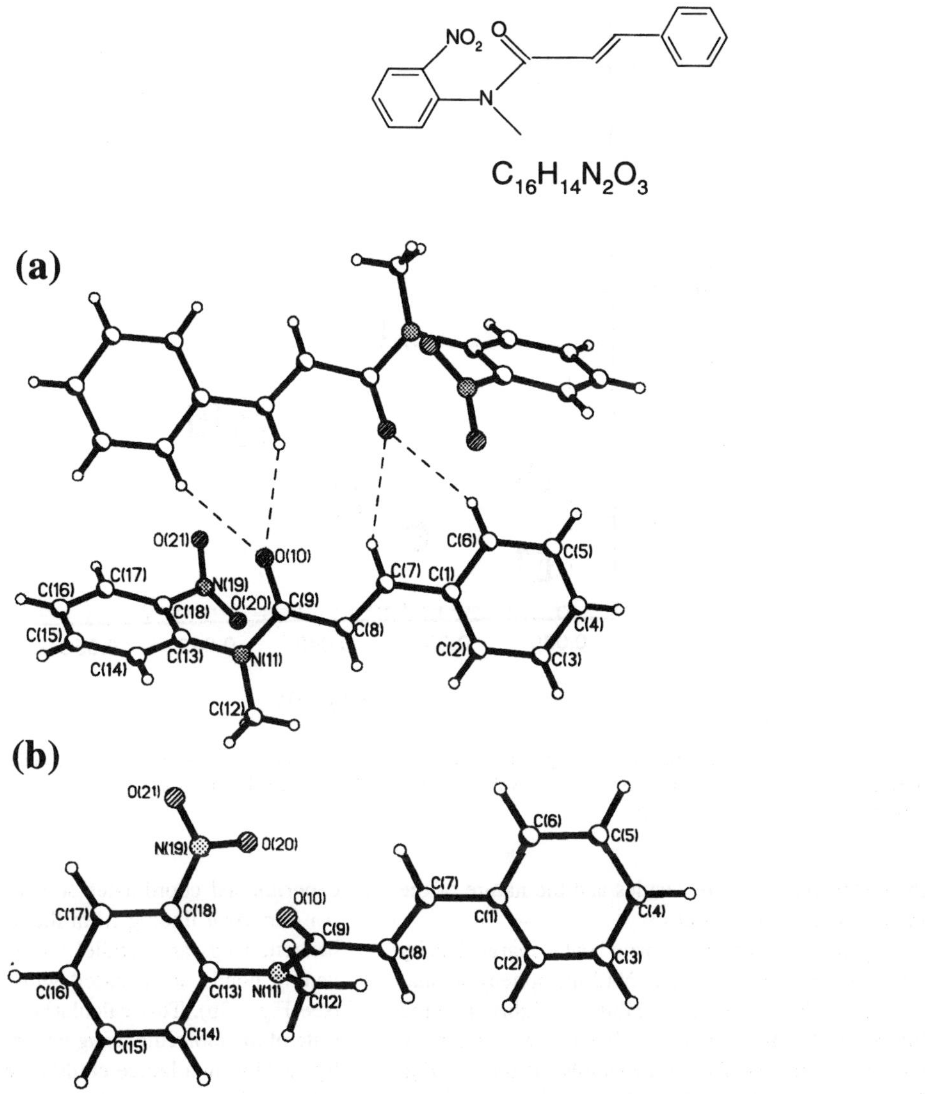

Fig. 13. Molecular structure of *N*-methyl-*N*-(2-nitrophenyl)cinnamanilide from (a) X-ray crystallography (bifurcated intermolecular hydrogen bonds are also shown) (b) AM1 calculation. Formula diagram is given in the inset (reproduced with permission from Gopalan et al. [68]).

8. Molecular packing in crystals

The geometry of a molecule in the solid state may deviate significantly from that in the free state, an effect which is mainly due to the constraints of packing in a lattice. For instance, the nitrobenzene molecule which is known to be planar in the free state is twisted by about 2° in the solid state across the nitro–benzene link. On the contrary, biphenyl [66] is non-planar in both gaseous and liquid states while it is planar [67] in the crystal at room temperature, exhibiting D_{2h} symmetry compared to D_2 in the former. Polymorphism in molecular crystals discussed above, is another subject of interest in this context. It is considered to be important to understand the symmetries prevailing in a molecular solid in terms of

356   G.U. Kulkarni et al. / Journal of Molecular Structure (Theochem) 500 (2000) 339–362

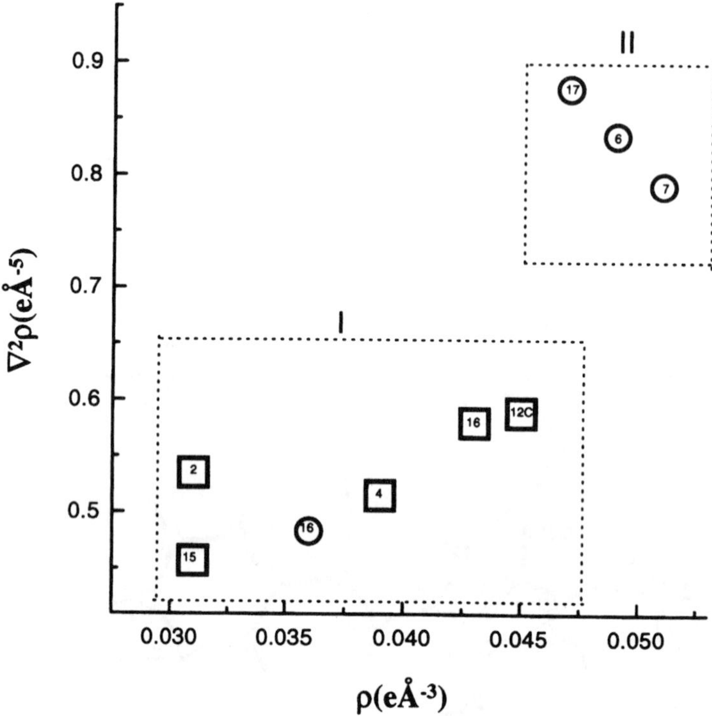

Fig. 14. Variation of the Laplacian with density at the critical points for various hydrogen bonds, $O_{amido}\cdots H$, circles; $O_{nitro}\cdots H$, squares. The numbers inside the symbols refer to the hydrogens involved in bonding. Regions I and II emphasize different trends in the plot (reproduced with permission from Gopalan et al. [68]).

the distortions of the molecules and the nature of the intermolecular interactions.

Gopalan et al. [68] have carried out a charge density investigation on N-methyl-N-(2-nitrophenyl)cinnamanilide. The molecule exhibits a highly favored bifurcated C–H$\cdots$O hydrogen bond ring structure in the lattice and is also considerably distorted (Fig. 13a). It is twisted at the amide-phenyl link with the cinnamide portion being nearly planar (torsion angle, C(2)–C(1)–C(7)–C(8), $-6.24°$). The nitrobenzene ring is highly non-planar, the nitro group being rotated with respect to the phenyl ring by $\sim$43.5°. The nitrobenzene group as a whole is twisted away from the mean cinnamide plane making an angle of 63.3°. The molecular structure in the lattice as discussed above was compared with that of a free molecule shown in Fig. 13b. The latter was obtained using AM1-PRECISE calculation of MOPAC [38] after optimizing the bond lengths, angles and torsion angles. The experimental coordinates served as the initial input. An important finding from the calculation is that the benzene rings are parallel within 8° while the intervening bonds are buckled with high torsion angles (see Fig. 13b). This calculation provides a reference state of the molecule using which constraints imposed by packing in a lattice could be examined.

Several intermolecular hydrogen bond contacts were examined. Four C–H$\cdots$O contacts were found to originate from the amidic oxygen (two involved in bifurcated bond) while six from the nitro-oxygens. The results of the charge density analysis is shown in Fig. 14. The $\nabla^2\rho$–ρ plot consists of two regions of interactions as shown—one region, where the hydrogen bonds exhibit low values of both the density and the Laplacian and the other, where both quantities are disproportionately higher. What is interesting is that all the bonds involving the nitro-oxygens belong to the first region and those from the amidic oxygen

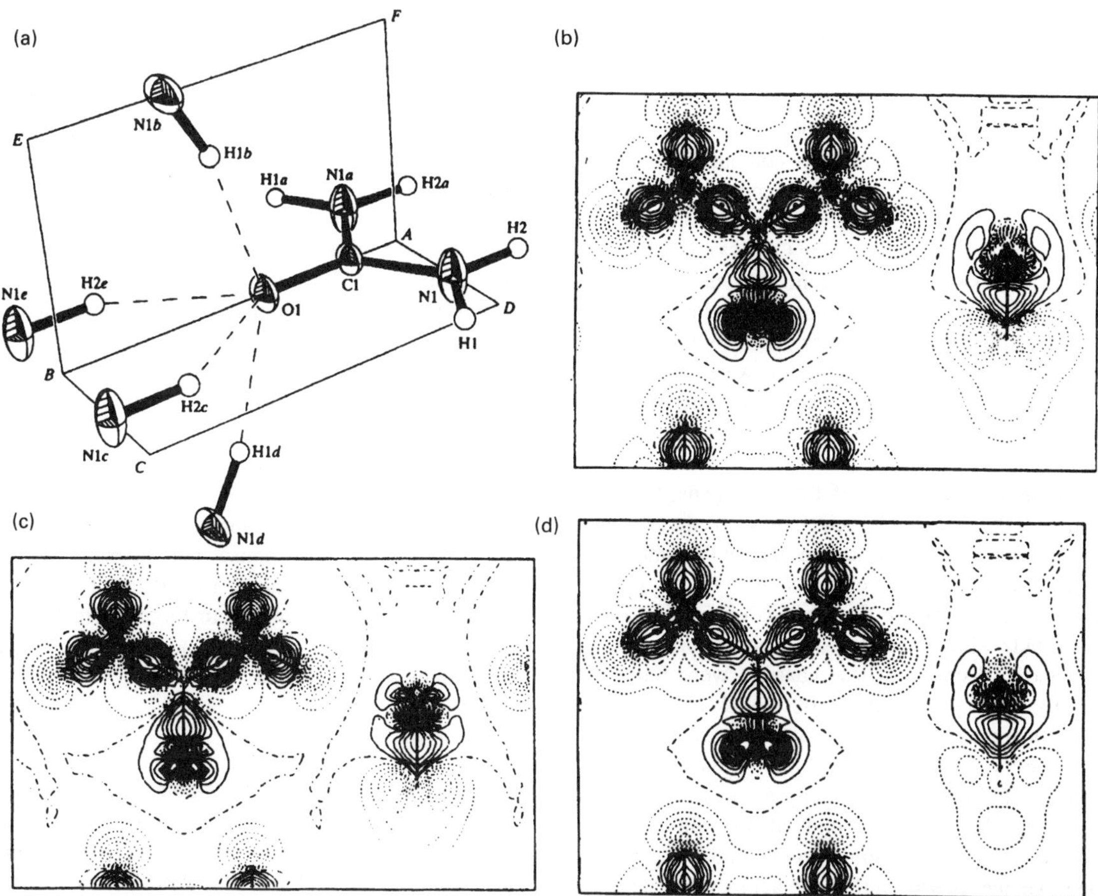

Fig. 15. (a) Intermolecular hydrogen bonds in urea crystal with displacement ellipsoids at 50% probability. (b) Static deformation density obtained from the multipolar analysis of the experimental data corrected for the thermal diffuse scattering. Theoretical deformation density obtained using (c) the Hartree–Fock method; (d) the DFT method by generalized gradient approximation (contours at 0.0675 $e\text{Å}^{-3}$) (reproduced with permission from Zavodnik et al. [69]).

fill the second region. Clearly, the amidic oxygen forms stronger hydrogen bonds compared to the more ionic nitro-oxygens. It is as though to favor bifurcated hydrogen bonding from the amidic oxygen, the molecule is highly twisted at the amide–nitrobenzene link.

Urea, despite being simple and highly symmetric (C_2), is known to crystallize in a non-centric crystal system ($P\text{-}42_1m$). The molecules are linked to each other through hydrogen bonds forming infinite tapes, adjacent tapes being held by orthogonal hydrogen bonds. This is perhaps the only example containing

a carbonyl group involved in four hydrogen bonds (Fig. 15a). Feil and co-workers [69] performed X-ray diffraction at 148 K and carried out charge density analysis on the data corrected for thermal diffuse scattering. They have also done orbital calculations in lattice using CRYSTAL95 [39] by both Hartee–Fock and DFT methods, the latter in both local density and generalized gradient approximations. The authors compared deformation densities and structure factors obtained from the experimental charge density analysis with those obtained from theoretical procedures. The deformation density maps from experiment and theory (compare Fig. 15b–d) were found to be in good agreement. The density functional theory seemed to yield slightly better results compared to the Hartree–Fock calculations.

Gatti et al. [70] have described the unusual hydrogen bonding in urea based on computations. They considered three cases of the urea molecule. A molecule as in the bulk structure with inputs from neutron diffraction, was computed using CRYSTAL92 [39] with 6-31 G** basis sets. In the second case, the free molecule was held at the crystal geometry and in the third, at more optimized geometry with only the C_{2v} constraint. Geometry optimization were done using GAUSSIAN92 with spherical harmonic gaussian functions. Compared to the free molecule, they found that the heavy atoms of the molecule in the bulk gain electrons at the expense of the hydrogens, the flux being $\sim 0.06e$. The fields created by such charge transfer polarizes the molecule in the bulk enhancing dipole moments. They observed that the ellipticity and ρ increase for the C–N bonds on passing from gas to bulk while for all the other bonds, both ellipticity and ρ decrease. These changes were found to comply with strengthening of the C–N bonds in terms of increased covalency and π-character. Other bonds in the molecule become more ionic on passing from gas to bulk. The ellipticity associated with the intermolecular hydrogen bonds along infinite planar tapes was found to be similar to that in the π-plane of the molecule. Based on this, the authors have argued that the π-conjugation propagates through these hydrogen bonds.

The above study has also shown that the carbonyl oxygen actually forms an active center for hydrogen bonding. The oxygen lone-pairs being electron-rich regions in the base, are characterized by high negative Laplacians in the free molecule (145.82 $e\text{Å}^{-5}$) while those associated with nitrogen carry much smaller Laplacians (54.85 $e\text{Å}^{-5}$). In the bulk, when oxygens get involved in hydrogen bonding, the lone-pair Laplacians were found to decrease to $\sim 136.62\ e\text{Å}^{-5}$. The two saddle points in between the lone-pairs which lie above and below the molecular plane are the second most electron-rich regions (89.26 $e\text{Å}^{-5}$) which are seen as maxima by hydrogens approaching perpendicularly. Accordingly, the Laplacian associated with the saddle point increases to 94.3 $e\text{Å}^{-5}$ in the bulk. The valence shell charge

concentration of oxygen therefore changes in such a way so as to form a torus of nearly uniform charge concentration in the non-bonded region. The authors conclude that the lengthening of the C=O bond in bulk urea by 0.13 Å accompanied by decreasing ellipticity and associated changes in the oxygen non-bonded regions is a fundamental step for the creation of a three-dimensional network of hydrogen bonds. This may be a more general mechanism in the formation of hydrogen bonded molecular crystals as was shown in the case of the cinnamanilide molecule.

9. Molecular NLO materials

In the last few years, several workers have analyzed charge density distribution in molecular crystals with non-linear optical (NLO) properties [71–74]. The NLO response can, in principle, be explained by an anharmonic distortion of the electron density distribution due to the electric field of an applied optical pulse. The polarization P induced in a molecule is

$$P = \mu + \alpha E + \beta E^2 + \cdots \tag{16}$$

where μ is the ground state dipole moment and α and β are the linear and the quadratic polarizabilities of the molecules, respectively. However, determination of polarizability from X-ray diffraction requires careful handling of data. Many of the studies on NLO crystals using charge density concentrate merely on the accurate extraction of phases associated with the structure factors [75–78],

$$\Delta\rho(\mathbf{r}) = V^{-1} \sum_{\mathbf{h}} [|F_\text{m}(\mathbf{h})|\exp i\varphi_\text{m}(\mathbf{h})$$

$$- |F_\text{s}(\mathbf{h})| \exp i\varphi_\text{s}(\mathbf{h})] \exp(-2\pi i\mathbf{h}\mathbf{r}) \tag{17}$$

where the subscripts m designate the atom-centered multipolar density model, and s, the spherically averaged free-atom superposition model. The deformation density therefore, can be written as the sum of an amplitude deformation density and a phase deformation density,

$$\Delta\rho = \Delta\rho(\Delta|F|) + \Delta\rho(\Delta\varphi) \tag{18}$$

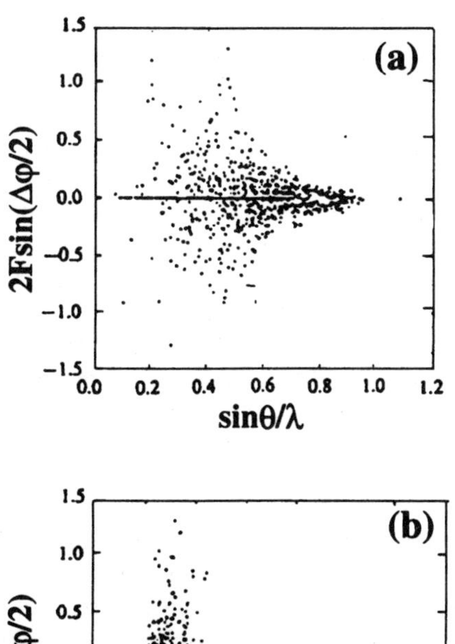

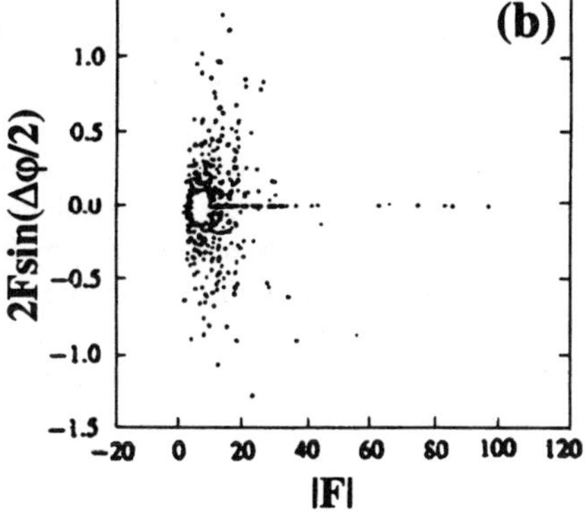

Fig. 16. Distribution of $2F \sin(\Delta\varphi/2)$ against (a) $(\sin\theta/\lambda$; and (b) $|F|$ (reproduced with permission from Hamazaoui et al. [76]).

where

$$\Delta\rho(\Delta|F|) = V^{-1} \sum (|F_m| - |F_s|) \exp(i\varphi_m)$$

$$\times \exp(-2\pi i \mathbf{hr}) \qquad (19)$$

and

$$\Delta\rho(\Delta\varphi) = V^{-1} \sum 2|F_s| \sin(\Delta\varphi/2) \exp[i(\varphi_s + \varphi_m$$

$$+ \pi)/2] \exp(-2\pi i \mathbf{hr}) \qquad (20)$$

For a given data set, it is instructive to examine the magnitude of $2F \sin(\Delta\varphi/2)$ in order to understand the effect of phase correction on the final solution. Fig. 16 shows the distribution of $2F \sin(\Delta\varphi/2)$ as a function of $\sin\theta/\lambda$ and $|F|$,

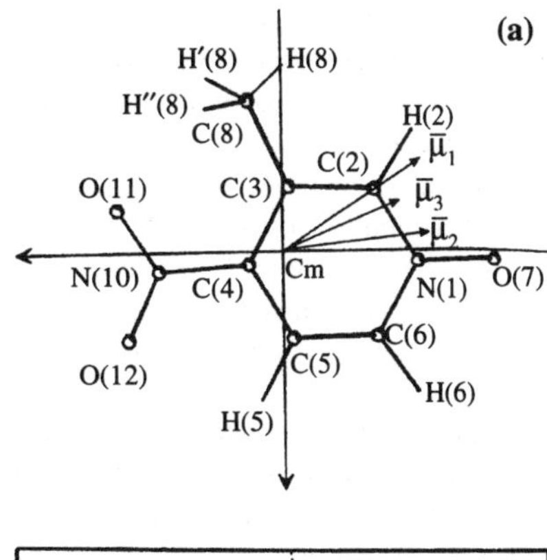

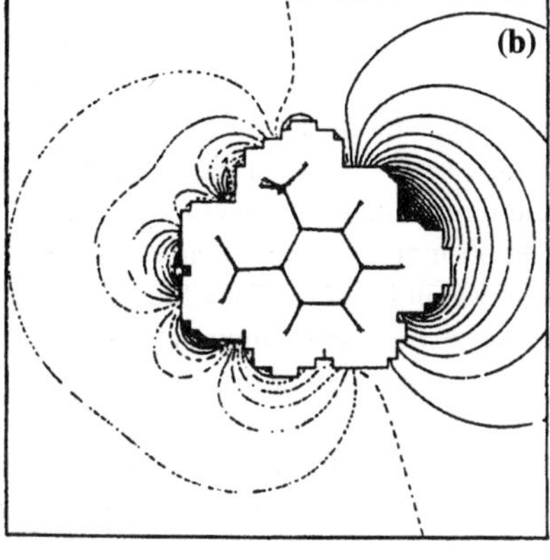

Fig. 17. (a) Orientation of molecular dipole moments in 3-methyl-4-nitropyridine *N*-oxide: μ_1, molecular dipole moment from direct integration methods; μ_2, from multipolar model; and μ_3 from semi-empirical calculation. (b) Electrostatic potential around the molecule in the plane of ring atoms. Contours at 0.2 kcal/mol (reproduced with permission from Hamazaoui et al. [79]).

from a data set due to Hamazaoui et al. [76]. It is apparent that the magnitude of phase correction is relatively higher for the low-order reflections (<0.6 Å^{-1}) as well as for weak- and medium-intensity reflections. Consequently, the measurement and the processing of the low angle weak reflections deserve special care. The authors have also shown that the overall charge density increases by $\sim$25% after phase correction. Besides phase correction, the authors carried out thermal analysis involving rigid bond tests. Such corrections are highly recommendable while determining charge density in non-centric crystals.

There has been some attempt to calculate of

hyperpolarizability from higher moments, but a more acceptable quantitative connection is yet to be established. Fkyerat et al. using octupole moments of the charge distribution in N-(4-nitrophenyl)-L-prolinol [73,74] estimated β to be 42.9 $e\text{Å}^{-3}$ and compared with the experimental result $\sim$39.14 $e\text{Å}^{-3}$. They concluded that NPP is a one-dimensional non-linear compound with β oriented almost along the main axis of the molecule making only 19.8° with the dipole moment vector.

Other studies on NLO crystals report topological analysis of deformation density and electrostatic potential and also on the calculation of molecular dipole moment. Hamazaoui et al. have carried out an experimental charge density study on 3-methyl-4-nitro-pyridine-N-oxide [79] and have computed the molecular dipole moments by three different procedures namely semi-empirical calculation, multipolar model and direct integration method. In all the methods, the dipole moment points nearly towards N-oxide group as shown in Fig. 17a. They found a close agreement between the semi-empirical and multipolar models while that from direct integration differed considerably in the direction of the dipole moment. The authors attributed this difference to the deconvolutability of thermal parameters from electron density in the case of multipolar refinement which is not possible in direct integration. The dipole moment of this molecule (1 Debye) was compared with that of nitropyridine N-oxide, a closely related molecule in which the methyl group in the *meta*-position is absent. The latter was found to have a much smaller dipole moment (0.4 Debye) [80]. On this basis, the authors concluded that the introduction of a methyl group in the *meta*-position favors intramolecular charge transfer. They have also determined the electrostatic potential in the plane of the ring which showed positive electrostatic potential around the N-oxide group and a negative electrostatic potential around the nitro group (Fig. 17b) thereby confirming the nature of charge transfer as found by the orientation of the molecular dipole moment.

10. Conclusions

In this article, we have dealt with the salient features of the electronic charge density distribution in molecular solids obtained by both theory and experiment. The importance of comparative experimental and theoretical studies of electron density is pointed out. The topological analysis of the density is illustrated taking representative examples. The chemical aspects of the charge distribution is described in terms of the deformation density, the Laplacian of the total density as well as the electrostatic potential. The polarization of the electron density in cyano bonds of diisocyanomethane, π/π' delocalization in the phosphazene ring and the bent cage bonds in cubane are some of the examples discussed. In addition, cases revealing the interplay between the intermolecular hydrogen bonding and properties are examined. These include proton sponges, polymorphic forms of cinnamic acid and *p*-nitrophenol where one of the forms exhibits photochemical reactivity. The role of intermolecular hydrogen bonding in lattice cohesion is discussed for a series of dicarboxylic acids. Molecular distortions in urea and cinnamanilide molecules are explained on the basis of the strength of the surrounding hydrogen bonds. Molecules exhibiting NLO properties in the solid state such as N-(4-nitrophenyl)-L-prolinol have been discussed, focusing on the intramolecular charge transfer leading to enhanced dipole moments.

References

[1] P. Debye, Dispersion of Roentgen rays, Ann. Phys. 46 (1915) 809.

[2] R.F.W. Bader, Atoms in Molecules—A Quantum Theory, Clanderon Press, Oxford, 1990.

[3] P. Coppens, Annu. Rev. Phys. Chem. 43 (1992) 663.

[4] D. Feil, J. Mol. Struct. 255 (1992) 221.

[5] F.L. Hirshfeld, in: A. Domenicano, I. Hargittai (Eds.), Accurate Molecular Structures. Their Determination and Importance, IUCr/Oxford University Press, Oxford, 1992, p. 237.

[6] F.L. Hirshfeld, Crystallogr. Rev. 2 (1991) 169.

[7] M.A. Spackman, Chem. Rev. 92 (1992) 1769.

[8] V.G. Tsirelson, R.P. Ozerov, J. Mol. Struct. 255 (1992) 335.

[9] G.A. Jeffrey, J.F. Piniella (Eds.), The Application of Charge Density Research to Chemistry and Drug Design Plenum Press, New York, 1991.

[10] P. Coppens, X-ray Charge Densities and Chemical Bonding, Oxford University Press, Oxford, 1997.

[11] R. Blessing (Ed.), Studies of electron distributions in molecules and crystals, Trans. Am. Crystallogr. Assoc., 26 (1990).

[12] T. Koritsanszky, in: W. Gans, A. Amann, J.C.A. Boeyens (Eds.), Fundamental Principles of Molecular Modelling, Plenum Press, New York, 1996, p. 143.

[13] M.A. Spackman, A.S. Brown, Annu. Rep. Prog. Chem. Sect. C: Phys. Chem. 91 (1994) 175.

[14] M.A. Spackman, Annu. Rep. Prog. Chem. Sect. C: Phys. Chem. 94 (1998) 177.

[15] P. Coppens, Acta Crystallogr. A54 (1998) 779.

[16] N.K. Hansen, P. Coppens, Acta Crystallogr. A34 (1978) 909.

[17] D. Cremer, E. Kraka, Croat. Chem. Acta 57 (1984) 1259.

[18] P. Politzer, D.G. Truhlar (Eds.), Chemical Applications of Atomic and Molecular Electrostatic Potentials Plenum Press, New York, 1981.

[19] J.S. Murray, K.D. Sen (Eds.), Molecular Electrostatic Potentials: Concepts and Applications Elsevier, Amsterdam, 1996.

[20] Y. Abramov, Acta Crystallogr. A53 (1997) 264.

[21] P. Coppens, Y. Abramov, M. Carducci, B. Korjov, I. Novozhilova, C. Alhambra, M.R. Pressprich, J. Am. Chem. Soc. 121 (1999) 2585.

[22] E. Espinosa, E. Mollins, C. Lecomte, Chem. Phys. Lett. 285 (1998) 170.

[23] P. Macchi, D.M. Prosperpio, A. Sironi, R. Soave, R. Destro, J. Appl. Crystallogr. 31 (1998) 583.

[24] A. Volkov, G. Wu, P. Coppens, J. Synchroton. Rad. 6 (1997) 1007.

[25] T. Koritsansky, R. Flaig, D. Zobel, H.-G. Crane, W. Morgenroth, P. Luger, Science 279 (1998) 356.

[26] G.M. Sheldrick, SHELX-76, Program for crystal structure determination, University of Göttingen, Germany.

[27] F.H. Allen, O. Kennard, D.G. Watson, L. Brammer, A.G. Orpen, R. Taylor, J. Chem. Soc. Perkin Trans. II (1987) S1.

[28] F.L. Hirshfeld, Acta Crystallogr. A32 (1976) 239.

[29] V. Schoemaker, K.N. Trueblood, Acta Cryst. B54 (1998) 507.

[30] T. Koritsansky, S.T. Howard, T. Richter, P.R. Mallinson, Z. Su, N.K. Hansen, XD, A computer program package for multipole refinement and analysis of charge densities from diffraction data, Cardiff, Glasgow, Buffalo, Nancy, Berlin.

[31] R.F. Stewart, M.A. Spackman, VALRAY User's Manual, Carnegie Mellon University, Pittsburgh, PA.

[32] H.L. Hirshfeld, Acta Crystallogr. A32 (1976) 239.

[33] B.M. Craven, H.P. Weber, X. He, Technical Report TR-87-2, Department of Crystallography, University of Pittsburgh, PA, 15260, 1987.

[34] W.J. Hehre, L. Radom, P.v.R. Schleyer, J.A. Pople, Ab initio Molecular Orbital Theory, Wiley, New York, 1986.

[35] J.A. Pople, Angew. Chem. Int. Ed. Engl. 38 (1999) 1894.

[36] M.J. Frisch, G.W. Trucks, H.B. Schlegel, P.M.W. Wong, J.B. Foresman, M.A. Robb, M. Head-Gordon, E.S. Replogle, R. Gomperts, J.L. Andres, K. Raghavachari, J.S. Binkley, C. Gonzalez, R.L. Martin, D.J. Fox, D.J. Defrees, J. Baker, J.J.P. Stewart, J.A. Pople, GAUSSIAN 92/DFT, Revision G.4, Gaussian Inc., Pittsburgh, PA, 1993.

[37] M.W. Schmidt, K.K. Baldridge, J.A. Boatz, S.T. Elbert, M.S. Gordon, J.H. Jensen, S. Koseki, N. Matsunaga, K.A. Nguyen, S.J. Su, T.L. Windus, M. Dupuis, J.A. Montgomery, Gamess, J. Comput. Chem. 14 (1993) 1347.

[38] J.J.P. Stewart, J. Comput.-Aided Mol. Des. 4 (1990) 1.

[39] R. Dovesi, V.R. Saunders, C. Roetti, M. Causa, N.M. Harrison, R. Orlando, R. Apra, CRYSTAL95 User's Manual, University of Turin, Turin, Italy, 1996.

[40] T. Koritsanszky, J. Buschmann, D. Lentz, P. Luger, G. Peretuo, M. Rottger, Chem. Eur. J. 5 (1999) 3413.

[41] M.J.S. Dewar, E.A.C. Lucken, M.A. Whitehead, J. Chem. Soc. (1960) 243.

[42] D.P. Craig, N.L. Paddock, J. Chem. Soc. (1962) 4118.

[43] T.S. Cameron, B. Boreka, W. Kwiatkowski, J. Am. Chem. Soc. 116 (1994) 1211.

[44] H. Irngartinger, A. Wesler, T. Oeser, Angew. Chem. Int. Ed. Engl. 38 (1999) 1279.

[45] H. Irngartinger, S. Strack, J. Am. Chem. Soc. 120 (1998) 5818.

[46] K.B. Wilberg, C.M. Hadad, C.M. Breneman, K.E. Laidig, M.A. Murcko, T.J. LePage, Science 252 (1991) 1266.

[47] I. Alkorta, J. Elguero, C. Foces-Foces, Chem. Commun. (1996) 1633.

[48] I. Rozas, I. Alkorta, J. Elguero, J. Phys. Chem. A 102 (1998) 9925.

[49] I. Rozas, I. Alkorta, J. Elguero, J. Phys. Chem. A 101 (1997) 9457.

[50] I. Rozas, I. Alkorta, J. Elguero, J. Phys. Chem. A 101 (1997) 4236.

[51] I. Alkorta, J. Elguero, J. Phys. Chem. (1996) 100.

[52] Y.H. Zhang, J.-K. Hao, X. Wang, W. Zhou, T.-H. Tang, J. Mol. Struct. (Theochem) 455 (1998) 85.

[53] D. Madsen, C. Flensburg, S. Larsen, J. Phys. Chem. A 102 (1998) 2177.

[54] G.K.H. Madsen, B.B. Iversen, F.K. Larsen, M. Kapon, G.M. Reisner, F.H. Herbstein, J. Am. Chem. Soc 120 (1998) 10040.

[55] Y.A. Abramov, L. Brammer, W.T. Klooster, R.M. Bullock, Inorg. Chem. 37 (1998) 6317.

[56] P.R. Mallinson, K. Wozniak, T. Garry, K.L. McCormak, J. Am. Chem. Soc. 119 (1997) 11502.

[57] P.R. Mallinson, K. Wozniak, C.C. Wilson, K.L. McCormak, D.M. Yufit, J. Am. Chem. Soc. 121 (1999) 4640.

[58] M.D. Cohen, G.M.J. Schmidt, F.I. Sonntag, J. Chem. Soc. (1964) 2000.

[59] R.S. Gopalan, G.U. Kulkarni, C.N.R. Rao, Acta Cryst. B (2000) in press.

[60] P. Coppens, G.M.J. Schmidt, Acta Crystallogr. 17 (1964) 222.

[61] G.U. Kulkarni, P. Kumaradhas, C.N.R. Rao, Chem. Mater. 10 (1998) 3498.

[62] P. Kumaradhas, R.S. Gopalan, G.U. Kulkarni, Proc. Indian Acad. Sci. (Chem. Sci.) 111 (1999) 569.

[63] R.S. Gopalan, P. Kumaradhas, G.U. Kulkarni, C.N.R. Rao, J. Mol. Struct. (2000) (in press).

[64] R.T. Morrison, R.N. Boyd, Organic Chemistry, 5th ed., Prentice-Hall, London, 1987 (p. 846).

[65] E. Espinosa, M. Souhassou, H. Lachekar, C. Lecomte, Acta Crystallogr. B55 (1999) 563.

[66] O. Bastiansen, Acta Chem. Scand. 3 (1949) 408.

[67] A. Chakrabarti, S. Yashonath, C.N.R. Rao, Mol. Phys. 84 (1995) 49.

[68] R.S. Gopalan, G.U. Kulkarni, E. Subramanian, S. Renganayaki, J. Mol. Struct. (2000) (in press).

[69] V. Zavodnik, A. Stash, V. Tsirelson, R. De Vries, D. Feil, Acta Crystallogr. B55 (1999) 45.

362   *G.U. Kulkarni et al. / Journal of Molecular Structure (Theochem) 500 (2000) 339–362*

[70] C. Gatti, V.R. Saunders, C. Roetti, J. Chem. Phys. 101 (1994) 10686.

[71] S.T. Howard, M.B. Hursthouse, C.W. Lehmann, P.R. Mallinson, C.S. Frampton, J. Chem. Phys. 97 (1992) 5616.

[72] E. Espinosa, C. Lecomte, E. Molins, S. Veintemillas, A. Cousson, W. Paulus, Acta Crystallogr. B52 (1996) 519.

[73] A. Fkyerat, A. Guelzim, F. Baert, W. Paulus, G. Heger, J. Zyss, A. Perigaud, Acta Crystallogr. B51 (1995) 197.

[74] A. Fkyerat, A. Guelzim, F. Baert, J. Zyss, A. Perigaud, Phys. Rev. B 53 (1996) 16236.

[75] M. Souhassou, C. Lecomte, R.H. Blessing, A. Aubry, M.-M. Rohmer, R. Wiest, M. Benard, M. Merraud, Acta Crystallogr. B47 (1991) 253.

[76] F. Hamazaoui, F. Baert, C. Wojcik, Acta Crystallogr. B52 (1996) 159.

[77] A. El-Haouzi, N.K. Hansen, C. Le Henaff, J. Protas, Acta Crystallogr. A52 (1996) 291.

[78] M.A. Spackman, P.G. Byrom, Acta Crystallogr. B53 (1997) 553.

[79] F. Hamazaoui, F. Baert, J. Zyss, J. Mater. Chem. 6 (1996) 1123.

[80] P. Coppens, Phys. Rev. Lett. 34 (1975) 98.

SOLID STATE PROPERTIES OF FULLERENES
AND RELATED MATERIALS

Ram Seshadri and C. N. R. Rao
Solid State and Structural Chemistry Unit
and
CSIR Centre of Excellence in Chemistry
Indian Institute of Science, Bangalore 560 012, India

1. Introduction

The discovery of fullerenes [1] was clearly a great landmark in carbon chemistry and materials science. The availability of ready methods of preparation of C_{60} and C_{70} by arc evaporation of graphite [2] has made it possible to investigate various aspects of these fascinating molecules. Of particular interest to us are the properties of C_{60} and C_{70} in the solid state. These include the phase transitions of C_{60} and C_{70} related to orientational ordering, superconductivity in alkali fullerides and ferromagnetism in the radical ion salt C_{60}.TDAE (TDAE = *tetrakis*-dimethylamino ethylene). In this article, we shall examine some of the important results obtained to-date on these solid state properties of fullerenes. In addition, we shall briefly describe some aspects of carbon nanotubes [3] and onions [4] which are new carbon nanostructures attracting considerable attention in recent months.

2. Preparation, purification and characterization of fullerenes and fullerene tubules

The fullerenes, C_{60} and C_{70}, are produced in the laboratory by the contact arc-evaporation of 6 mm graphite rods (e.g. Johnson Matthey, spectroscopic grade) in ~ 100 torr of helium in a water-cooled stainless steel chamber described previously [5]. The soluble material in the soot produced from the arc-evaporation is extracted with toluene using a Soxhlet apparatus. The pure fullerenes are obtained by chromatography on neutral alumina columns using hexanes as the eluant, or by the use of a simple filtration technique using charcoal-silica as the stationary phase and toluene as the eluant [5]. The fullerenes so prepared are characterized by UV/Vis spectroscopy and other techniques. FT-IR spectra of vacuum deposited fullerene films on KBr crystals also provide a means of characterization, just as do Raman spectra of films deposited on a silicon crystal. Ultraviolet and X-ray photoelectron spectra of fullerene films on gold foil provide information on the electron states [6]. Fig. 1 shows the ultraviolet valence band spectra of C_{60} and C_{70}. Considerable structure in the fullerene HOMO can be seen. In the case of C_{60}, the first two bands correspond to h_u and $h_g + g_g$.

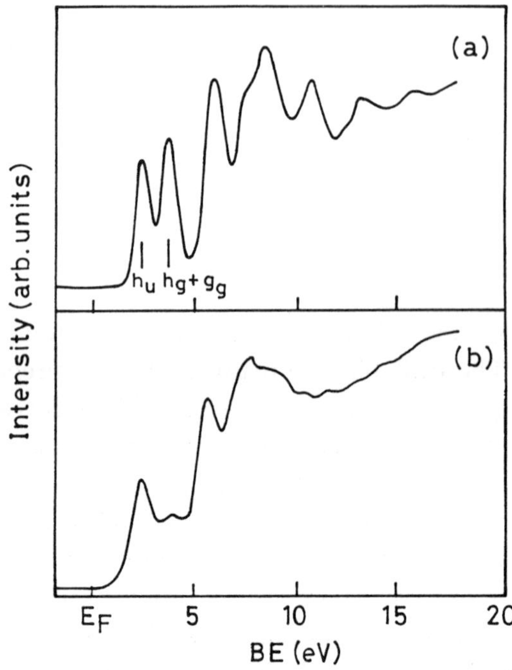

Fig. 1. UV photoelectron spectra of (a) C_{60} and (b) C_{70} (from Santra et al. [6]).

Carbon nanotubes (fullerene tubules) are prepared using the same contact-arc generator, except that dc currents are employed. At low powers, soot generation is minimal and the material deposited on the cathode is found to contain mostly carbon nanotubes and onions along with some nanocrystalline graphite. Heating the tubes in air or oxygen results in the nanocrystalline graphite burning away to give tubules with clean surfaces. This burning away of graphite particles also allows onions to be clearly seen.

3. Phase transitions in C_{60} and C_{70}

Symmetrical molecules generally show some degree of orientational disorder in the solid state. Thus most spherical tetrahedral molecules (e.g. CCl_4, neopentane, camphor) and even certain simple molecules such as H_2 exhibit transition from a phase where there is orientational disorder or free rotation, to an ordered phase. We would therefore expect C_{60} and C_{70} which are van der Waals solids comprising highly symmetrical molecules to show phase transitions associated with orientational ordering. Early single crystal X-ray diffraction studies on solid C_{60} at room temperature showed that large thermal parameters were necessary to obtain reasonable fits to the diffraction data [7]. These authors assumed that merohedral disorder of the C_{60} molecules would reduce the symmetry, thereby allowing the icosahedral point group of the molecule to be compatible with the cubic space group of the solid. Based on the narrow NMR linewidths of solid C_{60} at room temperature (Fig. 2) due to short orientational correlation times [8, 9], solid C_{60} was considered to be orientationally disordered at room temperature, the disorder being dynamic rather than static. X-ray diffraction and DSC measurements of Heiney et al. [10] showed that C_{60} undergoes a phase transition around 250 K (Fig. 3) on cooling from an orientationally ordered fcc phase to a sc phase where molecular rotation persists, but only along preferred axes. Temperature-dependent NMR studies show a bi-exponential behaviour of the orientational correlation time with two energies of activation above and below the transition (Fig. 4). Below the transition temperature, the molecules jump between preferred orientations, the barrier height being around 3000 K [8]. Molecular Dynamics (MD) simulation [11] as well as mean field theories [12] reproduce the experimental behaviour quite well. Though a simple L-J potential is sufficient for the phase transition to be modelled, charge-transfer effects are also important. Coulombic terms are introduced in modified L-J potentials by assigning a radius and potential to bond centres also. Such charge transfer effects manifest in the low temperature neutron diffraction

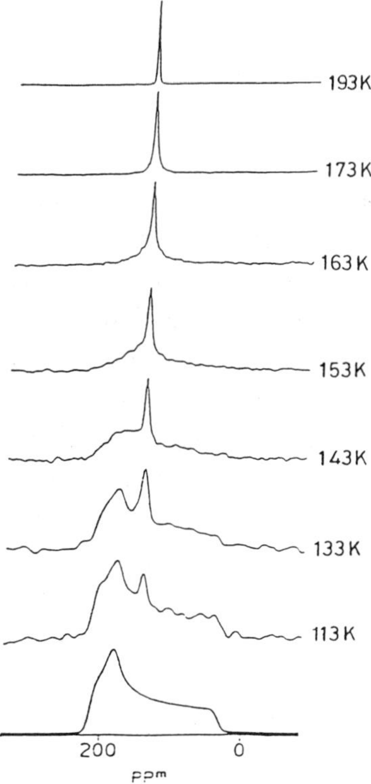

Fig. 2. NMR powder spectrum of C_{60} as a function of temperature. The bottommost spectrum is predicted for rigid molecules (from Tycko et al. [9]).

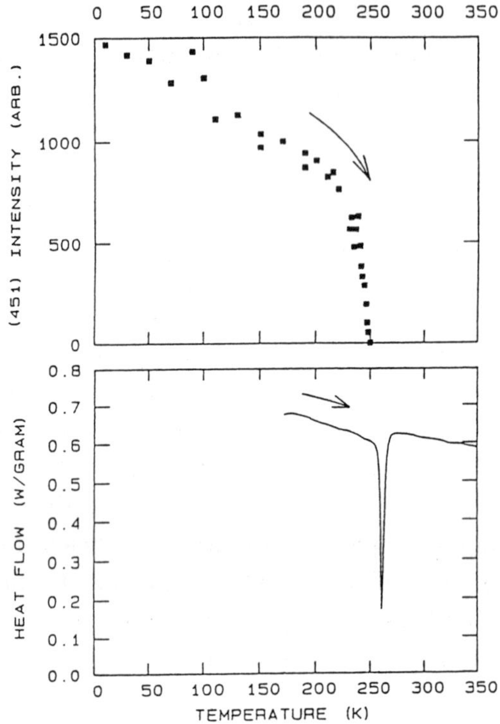

Fig. 3. (a) Intensity of the (451) reflection in the X-ray diffraction pattern of the low-temperature phase of C_{60} as a function of temperature; and (b) DSC trace (heating) of C_{60} (from Heiney et al. [10]).

experiments, which show that the electron-rich bonds of one molecule are proximal to the electron-poor five-membered rings of the adjacent molecule [13].

Raman spectroscopy of C_{60} under pressure [14] and other studies indicate the occurrence of an orientationally glassy state. Monte Carlo studies [15] show that the instantaneous cooling of the plastically-crystalline phase of C_{60} leads to an orientationally glassy phase (Fig. 5). The calculations predict an orientational glass transition temperature of ~ 80 K which is close to the experimental value of T_g reported in the literature. A glassy state in C_{60} has been experimentally observed using dilatometry [16]. High resolution neutron diffraction data show a continuous change in the lattice parameter around 80 K, indicating a glass transition [17].

C_{70} has lower symmetry than C_{60} and would therefore be expected to display a richer phase behavior. Vaughan et al. [18] reported two phase transitions of C_{70} at 337 K and 276 K. However, the structures of the phases are yet to be established unequivocally. At room temperature and above, both the fcc and the hcp phases seem to coexist, being energetically similar. Single crystals of as-grown C_{70} are usually twinned and unsuitable for X-ray diffraction studies. Based on diffraction as well as computer simulation studies, Verheijen et al. [19] report the occurrence of fcc, rhombohedral, hcp1, hcp2 and distorted hcp/monoclinic phases on going from high

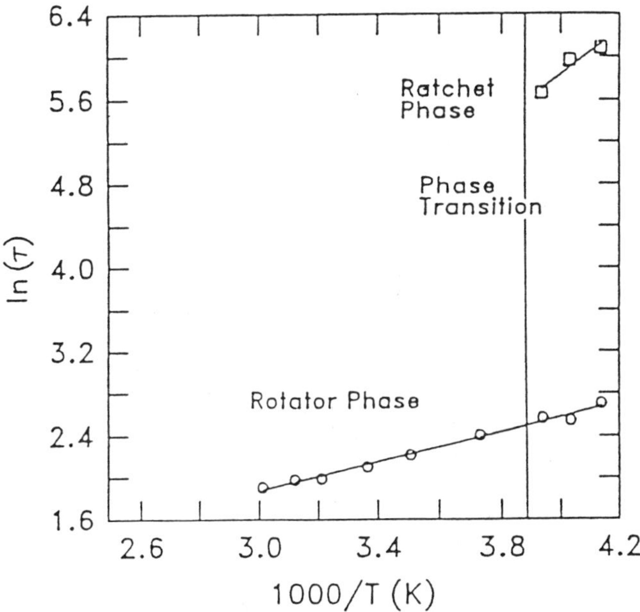

Fig. 4. Activation energy plot for the reorientation of C_{60} in the solid state showing biexponential behaviour (from Tycko et al. [9]).

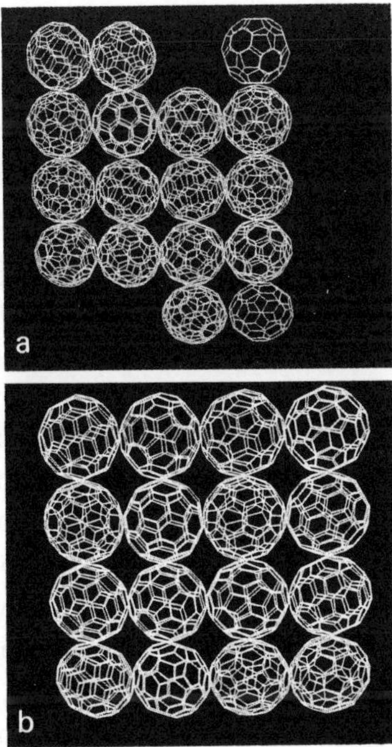

Fig. 5. Snapshots of C_{60} in (a) the high temperature disordered state and (b) the orientational glassy state (from Chakrabarti et al. [15]).

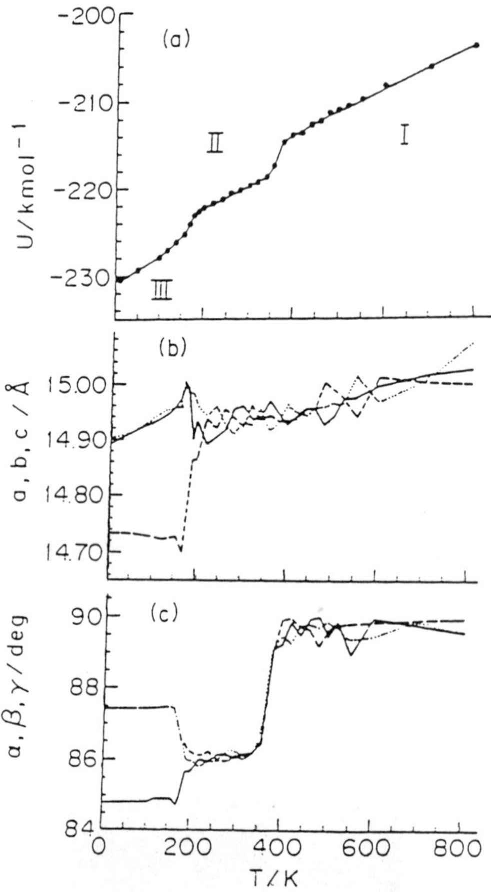

Fig. 6. Unit cell parameters of C_{70} from MD simulation (from Sprik et al. [20]).

to low temperatures. Molecular dynamics simulations [20] show that on cooling the high temperature rotator phase, a transition to a phase with trigonal symmetry occurs, followed by a transition to a monoclinic phase (Fig. 6).

Powder X-ray diffraction studies on C_{70} carried out by us [21] as a function of temperature show that the high temperature phase (> 350 K) is clearly fcc and sluggishly transforms to a hcp phase (possibly a mixture of two hcp phases) on cooling (~ 310 K). Below ~ 280 K, there is a transition to a more ordered monoclinic phase. Our X-ray diffraction studies could only establish the occurrence of two phase transitions at ~ 350 K and ~ 280 K. Investigations of the variable temperature infrared [22] and Raman [23] spectra show the occurrence of changes in linewidths across the orientational phase transitions around 340 K and 280 K as shown in Fig. 7. The Raman phonon frequencies harden on cooling across these transitions, possibly due to decreasing unit cell volume. The linewidths decrease on cooling due to the hindering of molecular motion resulting in a decrease in the rotational density of states. The changes in Raman frequency and linewidths across the phase transitions are

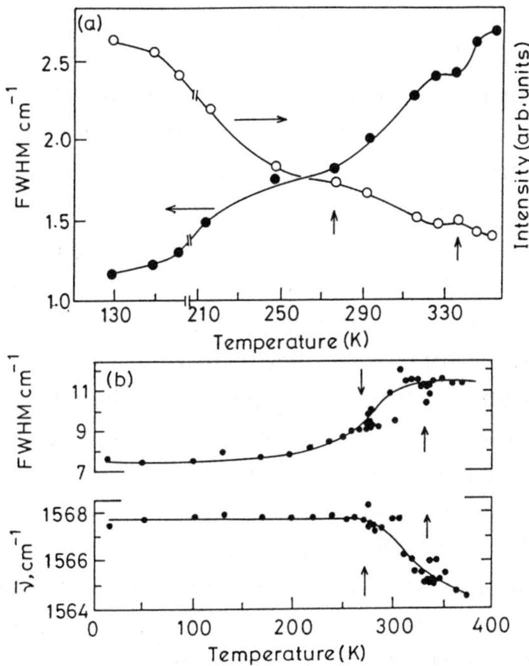

Fig. 7. (a) Temperature variation of the intensity and FWHM of the 643 cm^{-1} IR band of solid C$_{70}$ showing two transitions (from Varma et al. .[22]) and (b) Temperature variation of the FWHM and phonon frequency of the 1566 cm^{-1} Raman band of C$_{70}$ (from Chandrabhas et al. [23]).

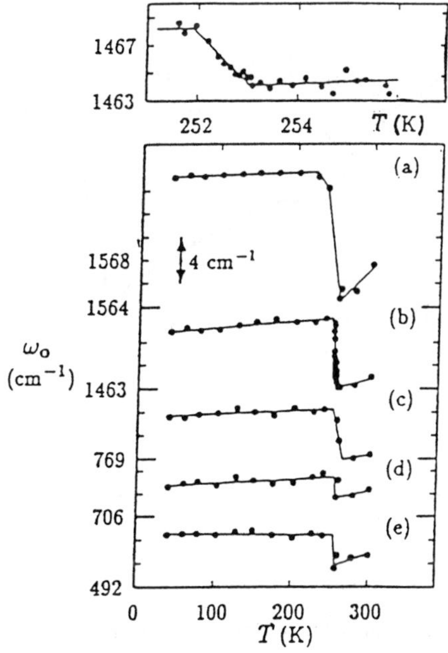

Fig. 8. Temperature dependence of the Raman phonon modes of C$_{60}$ (from van Loosdrecht et al. [24]).

continuous rather than abrupt. This appears to be true for C_{60} as well, contrary to the earlier findings of van Loosdrecht et al. [24] shown in Fig. 8. Infrared studies do show some evidence for the coexistence of phases in the 210–270 K region. Both IR and Raman spectra suggest the freezing out of the orientational order below 150 K.

The effect of pressure on the phase transitions in C_{60} and C_{70} is interesting. In the case of C_{60}, the temperature of the transition increases at a rate of $\sim$ 10 K kbar^{-1} [21, 25]. The DSC curve of Samara et al. [25] shows a shoulder beyond 6 kbar, indicating the presence of two states of nearly equal stability near ambient pressures. Recent measurements of the variation of resistance of C_{60} at different pressures [21] clearly show the occurrence of two distinct transitions (Fig. 9). Raman investigations on C_{60} single crystals under pressure [14] show that the pentagonal pinch mode undergoes considerable softening around 3.5 kbar. At higher pressures, the linewidth increases, till at around 130 Kbar, the line shape has almost merged into the background (Fig. 10), indicating an orientationally glassy state as found at low temperatures.

Under pressure, C_{70} shows three phase transitions. The highest transition temperature increases with pressure at a rate of 6.8 K kbar^{-1}, the lowest at 8.4 K kbar^{-1} and the intermediate one at 5.3 K kbar^{-1} [21] (Fig. 11). A recent DSC study [26], also shows indications of a transition at 330 K besides the previously reported ones at 280 K and 337 K. It appears that application of pressure delineates like phases of similar energies in both C_{60} and C_{70} giving rise to two or three orientational phase transitions respectively.

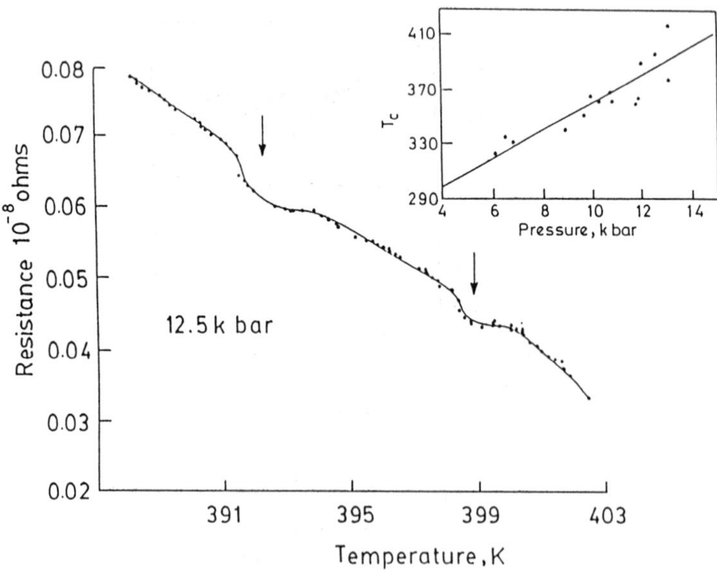

Fig. 9. Variation of resistance of C_{60} with temperature at 12.5 kbar showing two transitions. Inset shows the pressure variation of the average transition temperature (from Ramasesha et al. [21]).

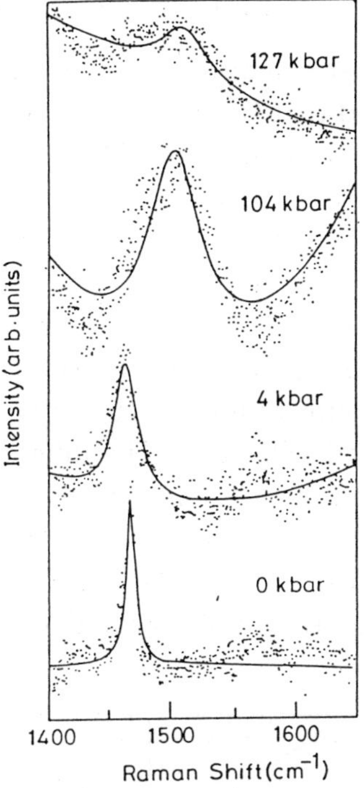

Fig. 10. Pentagonal pinch mode of a C_{60} crystal at different pressures suggesting the formation of an orientational glass (from Chandrabhas et al. [14]).

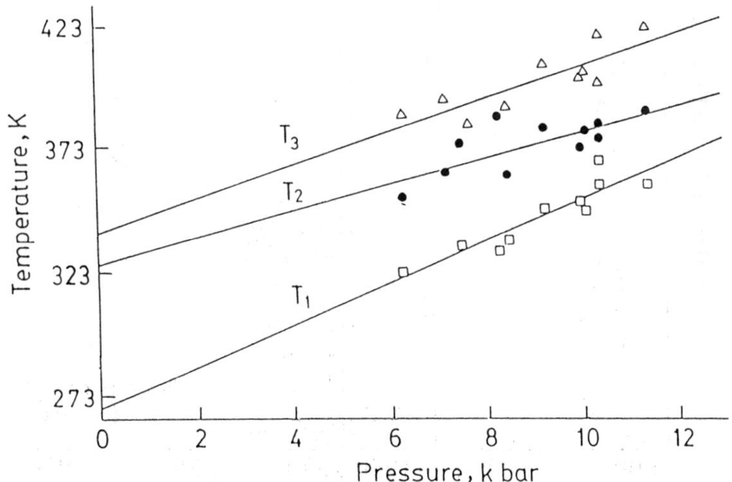

Fig. 11. Pressure dependence of the phase transition temperatures of C_{70} (from Ramasesha et al. [29]).

4. Amorphization of fullerenes under high pressures

There has been much interest in high pressure modifications of C_{60} and C_{70}. Early X-ray diffraction studies showed that C_{60} transforms to a lower symmetry structure at ~ 20 GPa under non-hydrostatic compression [27]. Raman and other studies [28, 29] have shown that C_{60} goes to an amorphous carbon phase at pressures higher than 22 GPa. The amorphous phase shows evidence for sp^3 carbons and are hence considered to result from a chemical reaction. Raman and photoluminescence studies [30] of C_{70} show that at around 12 GPa, only a single broad sp^2 carbon feature is seen, possibly due to the cage being distorted. However, this transformation is wholly reversible and release of pressure results in the original ambient pressure PL and Raman spectrum being restored (Fig. 12). The different behaviour of C_{60} and C_{70} with respect to pressure induced distortions is likely to be related to the lower symmetry of the latter. The elongated shape of C_{70} allows it to be easily deformed by pinching around the central 'waist' of the molecule. The exact nature of this distortion requires detailed structural studies as well as simulations.

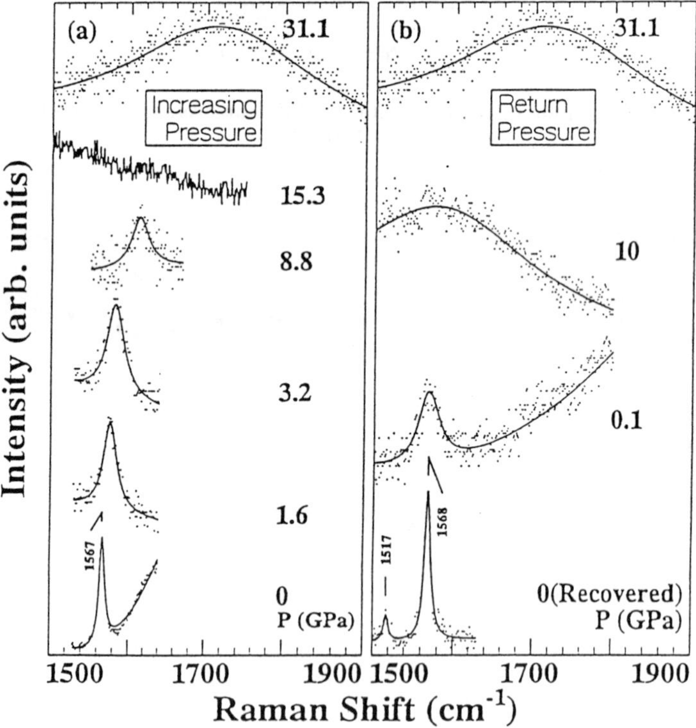

Fig. 12. Raman spectra of a C_{70} crystal at different pressures showing reversible amorphization (from Chandrabhas et al. [30]).

5. Photoluminescence of C_{60} under pressure

High pressure studies on C_{60} single crystals performed in this laboratory [31] show that with increasing pressure the photoluminescence band, initially centred around 1.6 eV is gradually red-shifted till, at around 3.2 GPa, the band vanishes into the background (Fig. 13). The crystal can be observed between the diamond anvils, to turn from deep red to black at around the same pressure. Since the C_{60} molecule shows very little structural distortion at these pressures, the closing of the PL gap can be interpreted as arising from the broadening and overlap of the LUMO and HOMO of the molecule or the valence and conduction band of the solid. Such a broadening is to be expected because decreased interball distance would increase the interball hopping term. The role of symmetry distortions allowing electronic excitations which are otherwise forbidden as a mechanism for the closing of the observed PL band gap, are ruled out by the observation that activation energy for conduction decreases with increasing pressure. Such studies have definite implications for the strengths of electron-phonon coupling and therefore, for superconductivity and other low temperature ordering phenomena in doped fullerene phases.

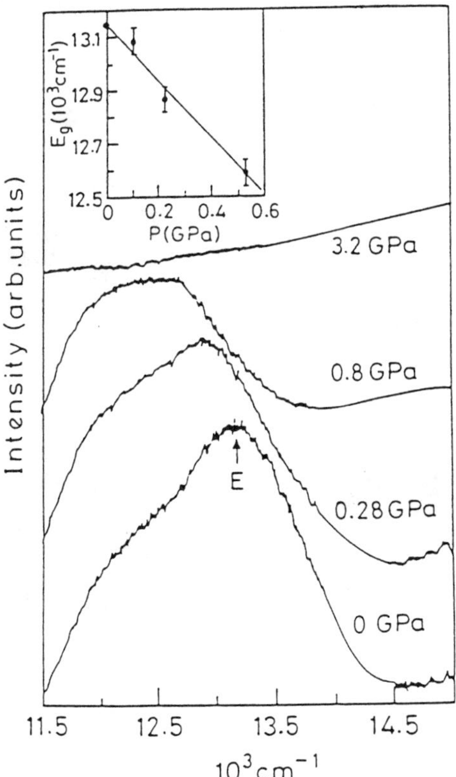

Fig. 13. PL band of C_{60} as a function of pressure: Inset shows the band gap as a function of pressure (from Sood et al. [31]).

6. Superconductivity in alkali and alkaline earth doped C_{60}

The presence of five-membered rings in the C_{60} cage makes the molecule a good electron acceptor and up to six electrons can be doped into the t_{1u} LUMO and six more in the t_{1g}. Fcc C_{60} can accommodate alkali metal ions in its tetrahedral and octahedral voids. Early doping studies on films of C_{60} showed K_xC_{60} to be metallic [32]. Soon, the discovery of superconductivity in K_xC_{60} followed [33]. In Fig. 14 we show the structures of the different alkali metal derivative of C_{60}. magnetic susceptibility studies showed that A_3C_{60} is a line phase with the maximum Meissner fraction [34]. A_6C_{60} (bcc) A_4C_{60} (bct) phases are semiconducting unlike the A_3C_{60} (fcc) phases which are metallic and become superconducting [35, 36]. Rb_2CsC_{60} shows a T_C of 31 K and this is highest T_c known to date in an organic superconductor [37]. X-ray and ultra-violet photoelectron spectroscopy (Fig. 15) have helped to understand the nature of electron doping into the C_{60} LUMO [38]. The LUMO is seen to gradually fill up with increasing exposure of C_{60} to potassium. However, continued exposure results in the LUMO band being shifted well below E_F giving rise to an insulator corresponding to K_6C_{60}. Photoemission results are, however, complicated by the fact the K_3C_{60} is a line phase so that low amounts of K exposure actually correspond to simple mixtures of K_3C_{60} and C_{60}. With transition metals, such LUMO level filling is not observed.

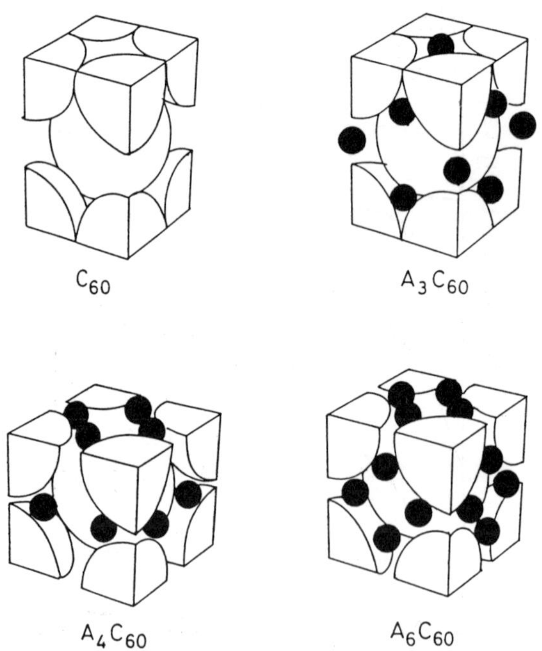

Fig. 14. Structures of (a) fcc-C_{60}, (b) fcc-A_3C_{60}, (c) bct-A_4C_{60} and (d) bcc-A_6C_{60} in the bct representation (from Murphy et al. [39]).

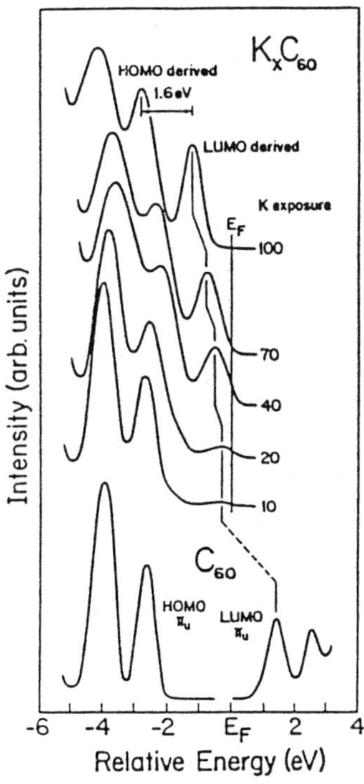

Fig. 15. Photoemission spectra for solid C$_{60}$ as a function of K exposure (arbitrary). The bottom-most curve shows both the occupied and unoccupied density of states (from Benning et al. [38]).

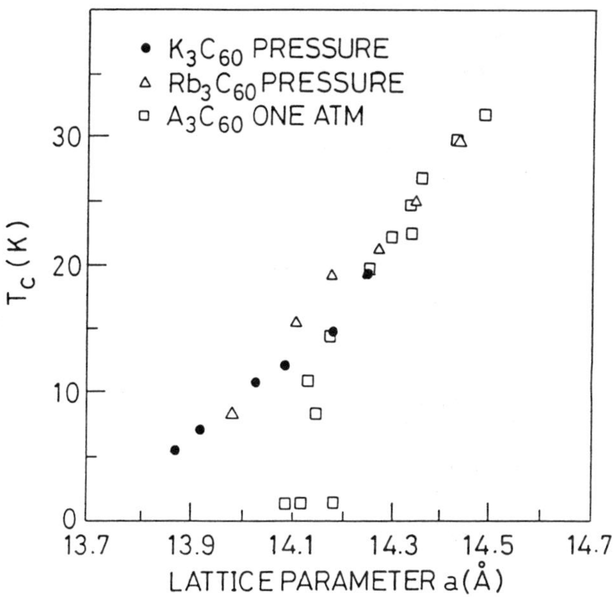

Fig. 16. Plot of T$_c$ vs. fcc-a parameter for a series of A$_3$C$_{60}$ phases including both high pressure as well as alloying data (from Murphy et al. [39]).

On the contrary, the metal d-states grow near E_F and some degree of partial charge-transfer from the metal clusters to the fullerene is observed [6]. The superconducting T_c of the A_3C_{60} phases increases with the increasing unit cell a-parameter (Fig. 16). Results from high-pressure studies as well as from alloying in the A site coincide, indicating that T_c is only a function of the interball separation. Increasing the interball separation tends to sharpen the density of states near E_F [39]. It appears that intramolecular phonon modes are responsible for the superconducting ground state in these materials. Varma et al. [40] have treated superconductivity of A_3C_{60} as a case of simple BCS theory with intermediate electron-phonon coupling, but involving a strong intramolecular H_g phonon mode whose strength is ~ 1000 K. Alkaline earth doped phases, Ca_3C_{60} [41] and surprisingly, Ba_6C_{60} [42], are also superconducting. Both these phases have partially filled C_{60} t_{1g} orbitals which in the Ba_6C_{60} case implies that charge-transfer from the metal to C_{60} is not complete. Superconductivity in the doped phases of the higher fullerenes has not been reported. A possible reason is that few, if any, of the higher fullerenes are as isotropic as C_{60} and also, none of them has a triply degenerate LUMO.

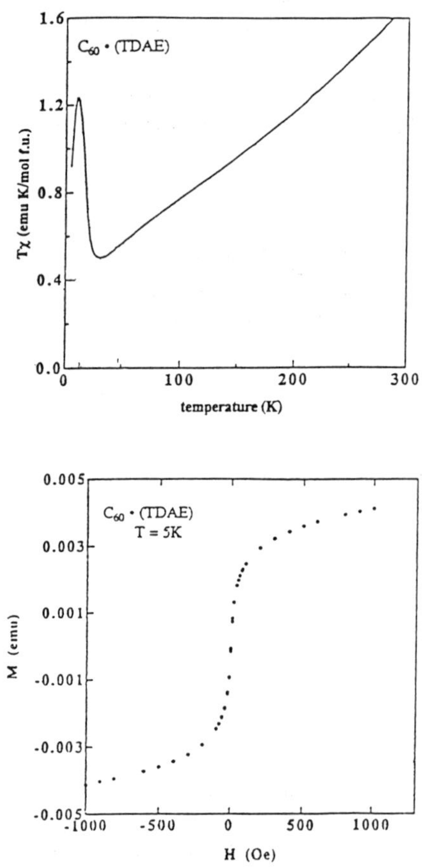

Fig. 17. (a) χT vs. T plot for C_{60}.TDAE showing transition at 16 K; and (b) Magnetization of C_{60}.TDAE at 5 K (from Allemand et al. [44]).

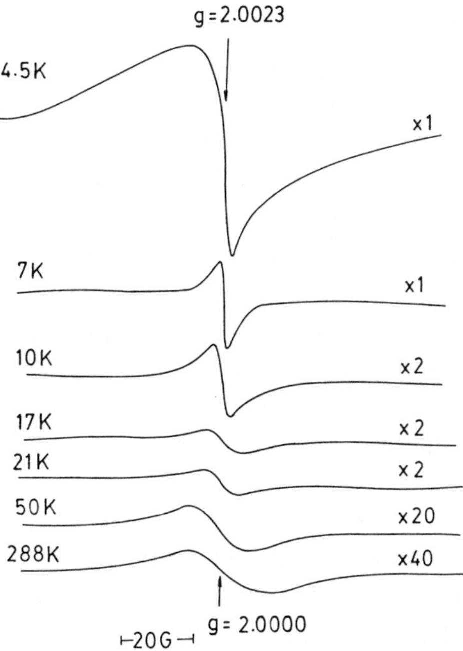

Fig. 18. ESR spectra of C_{60}.TDAE at different temperatures (from Seshadri et al. [47]).

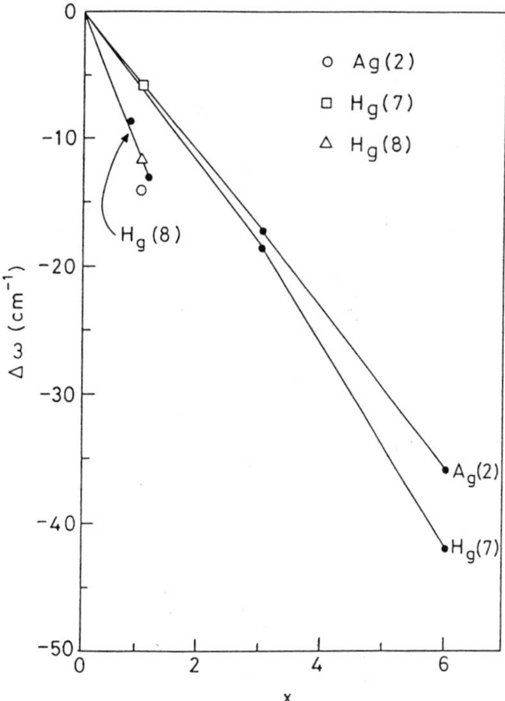

Fig. 19. Raman phonon frequencies of C_{60}.TDAE titrated against x for C_{60}^{x-} (from Muthu et al. [48]).

7. Ferromagnetism in C_{60}.TDAE

C_{60} is an excellent electron donor forming radical anions. With aromatic amines, it forms weak ground state complexes and exciplexes in solution [43]. With a strong donor such as TDAE (*tetrakis*-dimethylaminoethylene), C_{60} forms a 1:1 radical ion salt. C_{60}.TDAE is the best organic ferromagnet known to date with a $T)_c$ of 16 K. It is a soft ferromagnet, i.e. the magnetization shows no remanence (Fig. 17) [44]. The structure of C_{60}.TDAE has been established from powder X-ray diffraction data [45]. The unit cell is monoclinic with the short $C_{60} - C_{60}$ contact along the c-axis. ESR studies have shown that a single electron is doped from TDAE to C_{60} [46, 47]. The linewidths in ESR decrease with decreasing temperature (Fig. 18) suggesting that the sample is metallic. Single electron doping has also been confirmed from our Raman studies [48] by titrating the observed phonon frequency against x, using the data for C_{60}^{x-} from the $A_x C_{60}$ phases. This is shown in Fig. 19. Raman studies also show that the A_g modes of C_{60} in C_{60}.TDAE are split due to the lower symmetry of this system.

By fitting the low temperature susceptibility data to a series expansion for the spin-1/2, 1D Heisenberg system [49], we find a $1/T^2$ dependence of the susceptibility which confirms the quasi-1D behaviour and gives a ferromagnetic coupling constant of 50 K. This plot is shown in Fig. 20a. McConnell's mechanism for ferromagnetic exchange

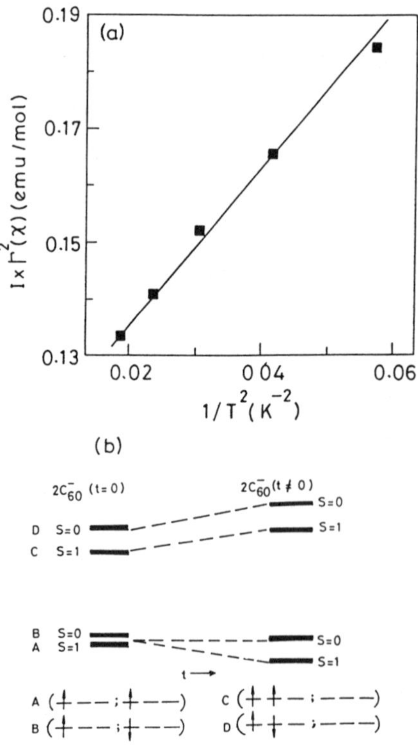

Fig. 20. (a) Low temperature susceptibility of C_{60}.TDAE plotted against $1/T^2$ showing linear behaviour; and (b) Configuration interaction mechanism for the stabilization of triplet state between C_{60}^- (from Seshadri et al. [47]).

has been used to understand this system [50]. A configuration-interaction picture for stabilizing the triplet ground state between two C_{60}^- molecules is shown in Fig. 20b. Comparing C_{60}.TDAE with superconducting A_3C_{60}, it would seem that the lower dimensionality and single electron doping are the key features of the ferromagnet.

8. Carbon nanotubes and onions

The production of macroscopic amounts of fullerenes has given impetus to the study of various forms of carbon. Iijima [3] reported needle-like carbon tubules in the cathode during the arc-preparation of fullerenes. The tubules are composed of concentric graphitic sheets with the dangling bonds at the end of the cylinders capped with fullerene like cages (see Fig. 21). These nanotubes display interesting properties such as capilarity [51]. The capped ends can be oxidized to give open tubes [52, 53]. Calculations indicate that tubules could be metallic at room temperature depending on their diameter and pitch [54, 55]. Scanning tunneling measurements carried out in this laboratory show that there is a conductance gap in the carbon tube, which decreases with increasing diameter or the number of graphitic sheaths in the tube [56].

Hyperfullerenes consisting of smaller fullerenes nested within larger one have been found to form on heating soot and fullerenes to very high temperatures [4]. These

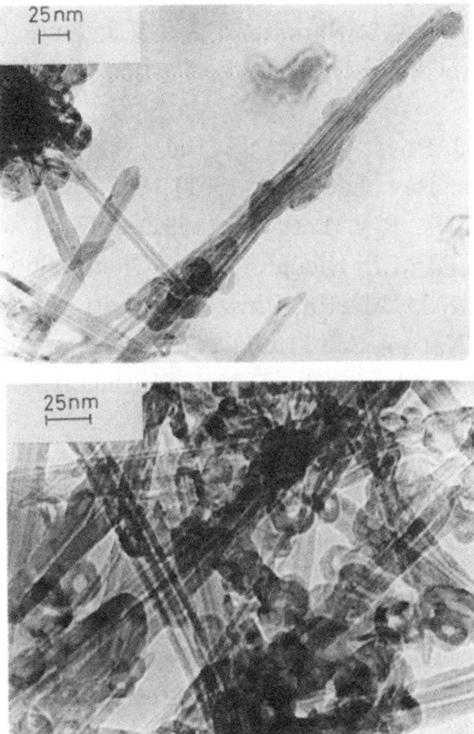

Fig. 21. (a) High resolution electron micrographs of carbon nanotubes; and (b) High resolution electron micrographs of nanotubes and onions (from this laboratory).

nested fullerenes have been christened 'carbon onions'. We have investigated nano-tubes by means of STM, HREM and powder XRD. The last technique shows that the material containing the tubules are not unlike turbostratic graphite in that only hkO and OOl reflections are present but not the general hkl reflections. The hkO reflections have the expected sawtooth line shape [57]. By stuffing the anode with various metals, filled onions and other exotic species are possible. We are able to thus make fine γ-Fe particles which are superparamagnetic, being nanometre-sized, yet stable against oxidation because of the protecting carbon layers. Onions stuffed with iron are however ferromagnetic with a T_c considerably lower than bulk iron [58].

References

1. H. W. Kroto et al., Nature, 318 (1985) 162.
2. W. Krätschmer et al., Nature, 347 (190) 354.
3. S. Iijima, Nature, 359 (1991) 56.
4. D. Ugarte, Nature, 359 (1992) 707.
5. C.N.R. Rao et al., Indian J. Chem., 31 A&B (1992) F5; A. Govindraj and C.N.R. Rao, Fullerene Sci. Technol., 1 9193) 557.
6. A.K. Santra, R. Seshadri, A Govindraj, V. Vijayakrishnan and C.N.R. Rao, Solid State Commun., 85 (1993) 77; J.W. Weaver, J. Phys. Chem. Solids, 53 (192) 1433.
7. R.M. Fleming et al., in G.S. Hammond and V.J. Kuck (eds.), Fullerenes: Synthesis, Properties and Chemistry of Large Carbon Clusters, ACS Symposium Ser. 481 (1991) 25.
8. C.S. Yannoni et al., J. Phys. Chem., 95 (1991) 9.
9. R. Tycko et al., Phys. Rev. Lett., 67 (1991) 1886.
10. P.A. Heiney et al., Phys. Rev. Lett., 66 (1991) 2911.
11. N. Sprik, A. Cheng and M.L. Klein, J. Phys. Chem., 96 (1992) 2027.
12. J.P. Lu, X.P. Li and R.M. Martin, Phys. Rev. Lett., 68 (1992) 1551.
13. W.I.F. David et al., Nature, 353 (1991) 147.
14. N. Chandrabhas, M.N. Shashikala, D.V.S. Muthu, A.K. Sood and C.N.R. Rao, Chem. Phys. Lett., 197 (1992) 319.
15. A. Chakrabarti, S. Yashonath and C.N.R. Rao, Chem. Phys. Lett., 215 (1993) 519.
16. F. Gugenberger et al., Phys. Rev. Lett., 69 (1992) 3774.
17. W.I.F. David et al., Europhys. Lett., 18 (1992) 219.
18. G.B.M. Vaughan et al., Science, 254 (1991) 1350.
19. M.A. Verheijen et al., Chem. Phys., 166 (1992) 287.
20. M. Sprik, A. Cheng and M.L. Klein, Phys. Rev. Lett., 69 (1992) 1660.
21. S.K. Ramasesha, A.K. Singh, R. Seshadri, A.K. Sood and C.N.R. Rao, Chem. Phys. Lett., 220 (1994) 203.
22. V. Varma, R. Seshadri, A. Govindraj, A.K. Sood and C.N.R. Rao, Chem. Phys. Lett., 203 (1993) 545.

23. N. Chandrabhas, K. Jayaraman, D.V.S. Muthu, A.K. Sood, R. Seshadri and C.N.R. Rao, Phys. Rev., B47 (1993) 10963.

24. P.H.M. van Loosdrecht et al., Phys. Rev. Lett., 68 (1992) 1176.

25. G.A. Samara et al., Science, 225 (1992) 1235.

26. E. Grivei et al., Phys. Rev., B47 (1993) 1705.

27. S.J. Duclos et al., Nature, 351 (1991) 380.

28. Y.S. Raptis et al., High Pressure Res., 9 (1992) 41.

29. D.W. Snoke, Y.S. Raptis and K. Syassen, Phys. Rev. B45 (192) 14419.

30. N. Chandrabhas et al. (to be published).

31. A. K. Sood et al., Solid State Commun., 81 (1992) 319.

32. R.C. Haddon et al., Nature, 350 (1991) 320.

33. A.F. Hebard et al., Nature, 350 (1991) 600.

34. K. Holczer et al., Science, 252 (1991) 1154.

35. P.W. Stephens et al., Nature 351 (1991) 632.

36. O. Zhou et al., Nature, 351 (1991) 462.

37. K. Tanigaki et al., Nature, 352 (1991) 222; S.P. Kelty et al., ibid 223.37.

38. P.J. Benning et al., Science, 252 (1991) 1417.

39. D.W. Murphy et al., J. Phys. Chem. Solids, 53 (1992) 1321.

40. C.M. Varma et al., Science, 254 (1991) 989.

41. A.R. Kortan et al., Nature, 355 (1992) 529.

42. A.R. Kortan et al., Nature, 360 (1992) 566.

43. R. Seshadri, C.N.R. Rao, H. Pal, T. Mukherjee and J.P. Mittal, Chem. Phys. Lett., 205 (1993) 395.

44. P.M. Allemand et al., Science, 253 (1991) 301.

45. P.W. Stephens et al., Nature, 355 (1992) 331.

46. K. Tanaka et al., Phys. Lett., A64 (1992) 221.

47. R. Seshadri, A. Rastogi, S.V. Bhat, S. Ramasesha and C.N.R. Rao, Solid State Commun., 85 (1993) 971.

48. D.V.S. Muthu, M. N. Shashikala, A. K. Sood, R. Seshadri and C.N.R. Rao, Chem. Phys. Lett., 217 (1994) 146.

49. M. Takahashi and M. Yamada, Phys. Soc., Japan, 54 (1985) 2808.

50. F. Wudl and J.D. Thomson, J. Phys. Chem. Solids, 53 (1992) 1449.

51. P.M. Ajayan and S. Iijima, Nature, 361 (1993) 333.44.

52. S.C. Tsang et al., Nature, 362 (1993) 520.

53. P.M. Ajayan et al., Nature, 362 (1993) 522.

54. J.M. Mintmire et al., Phys. Rev. Lett., 68 (1992) 631.

55. K. Tanaka et al., Fullerene Science and Technology 1 (1993) 137.

56. R. Seshadri, H.N. Aiyer, A. Govindaraj and C.N.R. Rao, Solid State Commun., 91 (1994) 195.

57. B.E. Warren, Phys. Rev., 59 (1941) 693.

58. R. Seshadri, R. Sen, G.N. Subbanna, K. R. Kannan and C.N.R. Rao, Chem. Phys. Lett., in print.

II. PHASE TRANSITIONS IN SOLIDS

Commentary by C.N.R. Rao

The study of phase transitions is an important aspect of modern physical science. In some of the simple solid state transitions, only changes in primary or secondary coordination occur. Solids undergo a variety of phase transitions, accompanied by significant changes in some of the properties. These transitions with their implications for materials applications are of vital interest in materials chemistry.[1] Typical of such transitions are those involving changes in magnetic, electrical and dielectric properties. Spin-state transitions occur both in molecular and ionic solids. Metal-insulator transitions in solids involving marked changes in electrical properties continue to be of interest. There are transitions such as the glass transition, which have taken a long time to understand. The advent of Monte Carlo and molecular dynamics methods has made a big impact on the study of phase transitions. I have carried out investigations of phase transitions in solids over the last 50 years. In this section, a few articles, dealing with phase transitions are included. These include transitions of TiO_2 and alkali halides as well as metal-insulator transitions and amorphization of fullerenes. Phase transitions in fullerenes are described in article no. 7 in section I of this book.

References

1. C.N.R. Rao and K.J. Rao, *Phase Transitions in Solids – An Approach to the Study of Chemistry and Physics of Solids* (McGraw-Hill, New York, 1978).

Reprinted from Accounts of Chemical Research, 1984, *17*, 83-89
Copyright © 1984 by the American Chemical Society and reprinted by permission of the copyright owner.

Phase Transitions and the Chemistry of Solids[†]

C. N. R. RAO

Solid State and Structural Chemistry Unit, Indian Institute of Science, Bangalore-560012, India

Received January 27, 1983 (Revised Manuscript Received September 1, 1983)

A variety of solids exhibit transformations from one crystal structure to another (polymorphism) as the temperature or pressure is varied. Besides such phase transitions involving changes in atomic configuration, solids also undergo transformations where the electronic or the spin configuration changes. The subject of phase transitions has grown enormously in recent years, with new types of transitions as well as new approaches to explain the phenomena having been reported extensively in the literature.[1,2] Traditionally, metallurgists and physicists have evidenced keen interest in this subject, but it is equally of importance in solid-state chemistry. In this article, I shall discuss some interesting types of phase transitions of relevance to solid-state chemistry investigated by my co-workers and myself in order to illustrate the scope and vitality of the subject.

General Features of Phase Transitions

During a phase transition, the free energy of the solid remains continuous, but thermodynamic quantities such as entropy, volume, and heat capacity exhibit discontinuous changes. Depending on which derivative of the Gibbs free energy, G, shows a discontinuous change at the transiton, phase transitions are generally classified as first order or second order. In a first-order transition where the $G(P,T)$ surfaces of the parent and product phases intersect sharply, the entropy and the volume show singular behavior. In second-order transitions, on the other hand, the heat capacity, compressibility, or thermal expansivity shows singular behavior.

We all know that when a liquid transforms to a crystal, there is a change in order; the crystal has greater order than the liquid. The symmetry also changes in such a transition; the liquid has more symmetry than a crystal since the liquid remains invariant under all rotations and translations. Landau introduced the concept of an order parameter, ξ, which is a measure of the order resulting from a phase transition. In a first-order transition (e.g., liquid–crystal), the change in ξ is discontinuous, but in a second-order transition where the change of state is continuous, the change in ξ is also continuous. Landau proposed that G in a second-order (or structural) phase transition is not only a function of P and T but also of ξ and expanded G as

a series in powers of ξ around the transition point. The order parameter vanishes at the critical temperature, T_c, in such transitions. Landau also considered the symmetry changes across phase transitions. Thus, a transition from a phase of high symmetry to one of low symmetry is accompanied by an order parameter. In a second-order transition, certain elements of symmetry appear or disappear across the transition; for example, when the tetragonal, ferroelectric $BaTiO_3$ in which the dipoles are all ordered, transforms to the cubic, paraelectric phase where the dipoles are randomly oriented, there is an increase in symmetry (appearence of certain syymetry elements) but decrease in order. In a ferroelectric–paraelectric transition, electric polarization is the order parameter; in ferromagnetic–paramagnetic transition, magnetization is the order parameter.

Many physical properties diverge near T_c, i.e., show anomalously large values as T_c is approached from either side. The divergences in different phase transitions are, however, strikingly similar. These divergences can be quantified in terms of critical exponents, λ:

$$\lambda = \lim_{\epsilon \to 0} \left| \frac{\ln f(\epsilon)}{\ln |\epsilon|} \right|$$

where $\epsilon = (T - T_c)/T_c$ and λ is called an exponent since $f(\epsilon)$ is proportional to ϵ^λ. The most important exponents are those associated with the specific heat (α), the order parameter (β), the susceptibility (γ), and the range over which individual constituents like atoms and atomic moments are correlated (ν). It so happens that the individual exponents for many different transitions are roughly similar (e.g., $\beta \approx 0.33$). More interesting is the fact that $\alpha + 2\beta + \gamma = 2$ in most transitions, independent of the detailed nature of the system. In other words, although individual values of exponents may vary from one transition to another, they all add up to 2. Such a universality in critical exponents is understood in the light of Kadanoff's concept[3] of scale invariance associated with the fluctuations near T_c. The exponents themselves can be calculated by employing the renormalization group method developed by Wilson.[4] Thanks to all these developments, we are now able to characterize all higher order phase transitions in terms of the physical dimensionality of the system, d, and the dimensionality of the order parameter, n. This is illustrated in the case of some observed phase transitions in Figure 1. It is noteworthy that there can be no phase transitions in one dimension if short-range forces alone operate.

C. N. R. Rao obtained his Master's degree in chemistry from the Banaras Hindu University, Doctor of Science degree from the University of Mysore, and Ph.D. (and later D.Sc. honoris causa) from Purdue University. He worked as a postdoctoral research associate with Professor K. S. Pitzer at the University of California at Berkeley. Before joining the Indian Institute of Science about 7 years ago to initiate a new department devoted to solid-state and structural chemistry, he was a Professor of Chemistry at the Indian Institute of Technology, Kanpur. Dr. Rao's interests are in solid-state chemistry, surface science, spectroscopy, and molecular structure. He is a Fellow of the Indian National Science Academy and of the Royal Society, London. He was selected a Centennial Foreign Fellow by the American Chemical Society in 1976. He is Vice President/Present-Elect of the International Union of Pure and Applied Chemistry.

[†] Contribution No. 220 from the Solid State and Structural Chemistry Unit.
(1) H. K. Henisch, R. Roy, and L. E. Cross, Ed., "Phase Transitions", Pergamon Press, New York, 1973.
(2) C. N. R. Rao and K. J. Rao, "Phase Transitions in Solids", McGraw-Hill, New York, 1978.
(3) L. P. Kadanoff, *Physics*, **2**, 263 (1966).
(4) K. G. Wilson and J. Kogut, *Phys. Rep.*, **12C**, 77 (1974).

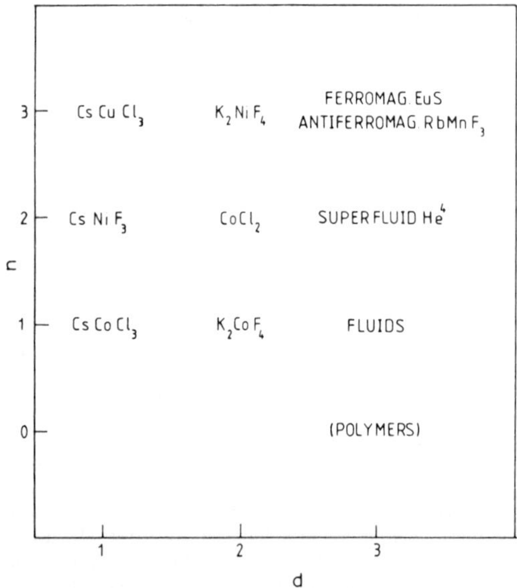

Figure 1. Some observed phase transitions in the d–n plane.

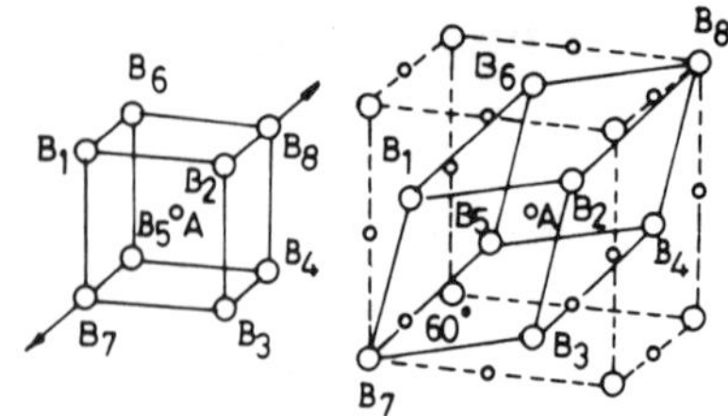

Figure 2. Dialatation transformation from CsCl structure to NaCl structure of an AB-type compound. Symmetry about the the the unique axis of dilatation ($3m$) is preserved.

Another important aspect of phase transitions in solids is the presence of soft modes. Operationally, a soft mode is a collective excitation whose frequency decreases anomalously as the transition point is reached. In second-order transitions, the soft mode frequency goes to zero at T_c, but in first-order transitions, the phase change occurs before the mode frequency goes to zero. Soft modes have been found to accompany a variety of solid-state transitions, including those of superconductors and organic solids.[2,5] Occurrence of soft modes in phase transitions can be inferred from Landau's treatment wherein atomic displacements may themselves be considered to represent an order parameter.

It has been found convenient to classify phase transitions in solids on the basis of the mechanism.[2] Three important kinds of transitions of common occurrence are as follows: (i) nucleation and growth transitions, a typical example being the anatase-rutile transformation of TiO_2, (ii) positional and orientational order-disorder transitions, and (iii) martensitic transitions. A typical example of a positional order–disorder transition is that of AgI; orientational order–disorder transitions are exhibited by many solids such as ammonium halides, plastic (orientationally disordered) crystals, and salts of di- or polyatomic anions. A martensitic transition is a structural change caused by atomic displacements (and not by diffusion) corresponding to a homogeneous deformation wherein the parent and product phases are related by a substitutional lattice correspondence, an irrational habit plane and a pracise orientationl relationship. These transitions that occur with high velocities of the order of sound velocity were originally discovered in steel but are now known to occur in several inorganic solids such as $KTa_{0.65}Nb_{0.35}O_3$ and ZrO_2.

On the basis of our knowledge of crystal chemistry, we can predict the nature of structural changes in the phase transitions of simple ionic solids. Thus, the Born model satisfactorily explains the relative stabilities of structures of simple ionic solids. On the basis of ionicity considerations, we can account for the structures of III–V, II–VI, and such binary compounds.[2] Some years ago, Buerger[6] classified phase transitions in solids on the basis of changes in the primary or higher coordination. Transformations involving primary coordination (e.g., CsCl structure–NaCl structure, aragonite-calcite) can be reconstructive or dilatational. Transformations involving second or higher coordination (e.g., α–β quartz, tetragonal-cubic transition of $BaTiO_3$) can be reconstructive or displacive. Buerger suggested that transformations involving changes in primary coordination such as in the CsCl–NaCl transition can occur more readily by a dilatational or deformational mechanism rather than by a drastic mechanism necessitating the breaking and making of bonds (Figure 2). It is interesting that the CsCl–NaCl transition is now considered to be martensitic with orientational relations between the two phases. It seems likely that phase transitions of many inorganic solids do not require diffusion and involve a deformational mechanism[2] wherein the parent and product phases have orientational relationships. Many such transitions may also exhibit soft mode behavior, an aspect that is worthy of investigation.

Polytypism

Solids such as SiC, ZnS, CdI_2, TaS_2, mica, and perovskite oxides exhibit polytypism wherein the unit cells of the different polytypic forms differ from one another only in the c dimension; the c dimension of polytypes of certain substances vary anywhere between a few angstroms and a few thousand angstroms. This phenomenon arises because of differences in the sequence in which the atomic layers are stacked in different polytypes along the c axis. Although several theories have been put forward to explain the varied characteristics of polytypic substances,[7] it is difficult to understand some of their features. These include the existence of different types of unidimensional order (since one-dimensional systems should not normally show long-range ordering), which is an apparent violation of Gibbs' phase rule (owing to the coexistence of several phases under identical conditions), syntactic coalescence (growth of different polytypic forms in different parts of the same specimen), and the existence of varying extents of disorder.

(5) J. F. Scott, *Rev. Mod. Phys.*, **46**, 83 (1974).

(6) M. J. Buerger, *Forschr. Miner*, **39**, 9 (1961).
(7) A. R. Verma and G. C. Trigunayat in "Solid State Chemistry", C. N. R. Rao, Ed., Marcel Dekker, New York, 1974.

```
A B A B C A C A B C B . B A C A C B A B A C B C B A C A C B A B A C B C B A C A C B A . C A C A B C
B C A B A B C A . B A B A C B C B A C A C B A B A C B C B A C A C . C B A C A C B A B A C B C B A C
A C B A B A C B C B A C A C B A . C A C A B C B C A B A B C A C A B C B C A B A . A C B C B A C A C
B A B A C B . A B A B C A C A B C B C A B A B C A . . . . . . . . . B A C B C B A C A C . . . . . .
C B C B A C A C B A B A C B C B A C A C . C B A C A C B A B A C B C B A C A C . C B C A A C B A B A
C B C B A C A C B A B A C B C B A C . B C B C A B A B C A C A B C B C A B A B C A C A B C B C A B A
B C A C A B C B C . C A B C B C A B A B C A . B A B A C B C B A C A C B A B A C B C B A . . . . . . C
A B C B C A B A B C A C A B C B C A B A B . B C A B A B C A C A B C . . . . . . . . . A C B A B A C
B . . B A C B C B A C A C B A B . B C A C A B C B C A B A B C A C A B C B C A B A B C A C A B C B C
A B A B C A C A B C B C A B A B . B C A B A B C A C A B C B C A B A B C A C A B C B C A B A B C A C
A B C B C A B A B C A C A B C B C . C A B C B C A B A B C A . B A B A C B C B A C A C B A B A C B C
. A C A C B A B A C B . A B A B C A C A . A B C A C A B C B C . C A B C B C A B A B C A C A B C B C
. . . . C B A C A C B A . C A C A B C B C A B A B C A C A B C . A C A C B A B A C B C B A C A C B A
B A C B C B A C A C B A B A C B C B A C A C B A B A C B C B A C A C B A B A C B C B A C A C B A B A
C B C B A C . . . . . . . . . C B A C A C B A . . . . . . C B A C A C B A B A C B C B A C A C B A B
A C B C B A C A C B A B A C B . A B A B C A C A B C B C A B A B . . C A C A B C B C A B . C B C B A
C A C B A B A C B C B A C . . C B A C A C B A B A C B . A C A B C B C A B A B C A C A B C B C . C A
B C B C A B A B C A C A B C B C A B A B C A C A B C B C A B A B . . C A C A B C B C A B A B C A C A
B C B C A B A B C A C A B C B C . C A B C B C A B A B . . . . . . . . B A B A C B C B A C A C . . .
. C A B C B C A B A B C A C A B C B C A B A B C A C A B C B C A B A B C A C A B C B C A B A B C A C
```

Figure 3. Display of long-range order as in 12R polytype (ABACBCBACACB) obtained by computer simulation (taken from ref 8b).

Since polytypism is essentially a one-dimensional phenomenon and the atomic layers exist in either cubic or hexagonal configurations, we can, in principle, treat polytypes as different ordered states of a spin-half Ising chain.[8] Since such an Ising chain should have long-range order at nonzero temperatures, it is essential to have an infinite-range interaction. At the same time, in order to obtain different ordered states of the chain, it is necessary to have a short-range interaction competing with the infinite-range interaction. Spin-half Ising chains with competing short-range and infinite-range interactions have been investigated by Theumann and Høye.[9] The Theumann–Høye Ising chain involves the nearest- and the next-nearest-neighbor antiferromagnetic interaction and an infinite-range ferromagnetic interaction of the Kac type. This chain exhibits different spin orderings at $T = 0$ for different interaction strengths and shows phase transitions as the interaction strengths are varied. In view of the close resemblance between the behavior of this Ising chain and that of polytypes, we have carried out[8] a Monte Carlo simulation of polytypes based on the competing interactions model. The basic Hamiltonian employed was that of Theumann and Høye.

The two states of the spin in an Ising chain can be taken to represent the two lowest energy configurations of a layer in a polytype, viz., the cubic (ABC) and the hexagonal (AB) configurations. Other layer configurations like AAB or ABB are of high energy and are not observed. We should, therefore, expect correspondence between the thermodynamic properties of a polytype and those of an appropriate Ising chain. In order to have an ordered spin arrangement in an Ising chain, it is necessary to introduce an infinite-range interaction among spins, the equivalent-neighbor type being most suitable for Monte Carlo simulations. The contribution to internal energy from this form of infinite-range interaction is given by $-J_{ir}M^2$. An analogous squared term

has been suggested for the major part of the contribution to the elastic energy between atoms of different sizes.[10] It is, however, necessary to add a competing short-range interaction term to the equivalent neighbor interaction term in order that the spin orderings in the Ising chain describe polytypism.

We have found that a general double-layer mechanism (e.g., ABCACB → ABCBCB or ABABCB) connects different states of the polytype chain with about the same probability as the spin–flip mechanism in magnetic Ising chains.[8a] It has been possible to simulate various polytypes with periodicities extending up to 12 layers in this manner (see Figure 3). During the growth of a polytype (whether it be from melt or vapor), it is reasonable to assume that the atoms in the layers are jostling about, leading to fluctuations in interlayer interaction strength. A proper simulation of polytypes should take into account such fluctuations.[8b] We have therefore treated the interaction parameters in the simulation as random variables (varying with time) during any given simulation and carried out a computer simulation of polytype growth from vapor, employing both constant and fluctuating short-range interaction parameters. We have obtained short stretches of fairly long ordered polytypes such as 14H and 33R in such simulations. It is my feeling that the simulations discussed hitherto may be useful in understanding the occurrence of long period structures which include coherent intergrowth phases in systems such as hexagonal barium ferrites and intergrowth bronzes. It appears that long periodicities could originate from more than one cause, but it is likely to be thermodynamic in some systems. This indeed appears to be the case in alloys such as CuAu. Sato and co-workers[11] have shown that the operative factor in such a system is the lowering of the total electronic energy by decreasing the size of the Brillouin zone; the requisite periodicity for creation of

(8) (a) S. Ramasesha and C. N. R. Rao, *Philos. Mag.*, **36**, 827 (1977); (b) M. K. Uppal, S. Ramasesha, C. N. R. Rao, *Acta Crystallogr., Sect. A* **36**, 356 (1980).

(9) W. K. Theumann and J. S. Høye, *J. Chem. Phys.*, **55**, 4159 (1971).

(10) (a) P. W. Anderson and S. T. Chui, *Phys. Rev. B*, **9**, 3229 (1974); (b) J. Friedel in "Solid State Physics", F. Seitz, D. Turnbull, and H. Ehrenreich, Ed., Academic Press, New York, 1956, Vol. 3.

(11) H. Sato, R. S. Toth, and G. Honjo, *J. Phys. Chem. Solids*, **28**, 137 (1967), and the references cited therein.

energy gaps varies in some systems with the electron/atom ratio.

Spin-State Transitions

Transitions of solids from magnetically ordered states to paramagnetic states are well-known. We have been interested in transitions between spin states of solids containing transition metal ions in the d^4–d^8 configuration. These transition metal ions can exist either in the low-spin or the high-spin ground state, depending upon the crystal field strength. In some of the transition-metal compounds, where the crystal field strength is close to the crossover point, interesting magnetic and structural behavior is observed. Spin-state transitions are known to occur in molecular systems of transition-metal complexes as well as in transition-metal oxides and other solids. Typical complexes belonging to the first category are $Fe(phen)_2(NCS)_2$ and $Fe(phen)_2$-$(NCSe)_2$. The latter category consists of solid materials such as MnAs and rare-earth cobaltites, $LnCoO_3$ (Ln = La or rare earth). The low-spin (t_{2g}^6) to high-spin ($t_{2g}^4e_g^2$) transition in $LaCoO_3$ manifests itself as a plateau in the inverse susceptibility–temperature curve, the low- and high-spin ion ordering themselves on unique sites above a particular temperature.[12,13] The other rare-earth cobaltites, on the other hand, only show a maximum in the inverse susceptibility–temperature curve.[14] Spin-state transitions have been studied by several experimental techniques including measurements of heat capacity and magnetic susceptibility and Mössbauer spectroscopy. We have recently examined them by X-ray photoelectron spectroscopy[15] and NMR spectroscopy.[16]

Spin-state transitions have been found to occur in two-dimensional oxides of K_2NiF_4 structure. Thus, La_4LiCoO_8 shows a peak in the inverse susceptibility–temperature curve not unlike $NdCoO_3$ and other rare-earth cobaltites.[14] Trivalent cobalt in this system appears to transform from the low-spin state to the intermediate ($t_{2g}^5e_g^1$) as well as the high- ($t_{2g}^4e_g^2$) spin states.[17,18] Other oxides of cobalt of K_2NiF_4 structure[18] that seem to undergo low intermediate spin transitions are Sr_4TaCoO_8 and Sr_4NbCoO_8. in $LaMNiO_4$ where M is Sr or Ba, Ni^{3+} is supposed to be in the low-spin state[19] when M is Sr and in the high-spin state[20] when M is Ba. We have examined the $LaSr_{1-x}Ba_xNiO_4$ system[18] and found that, with increase in x, the width of the $\sigma^*_{x^2-y^2}$ band (already present when $x = 0$) decreases, accompanying an increase in the unit cell volume; high-spin Ni^{3+} ions are formed to a small extent with increasing x, but there appears to be no spin-state transition.

A successful model of spin-state transitions should be capable of explaining the following observations: (i) smooth as well as abrupt changes in the spin-state population ratio with temperature, (ii) occurrence of thermodynamically second-order transitions and first-order transitions in certain instances and a spin-state population ratio around unity near the transition, (iii) nonzero population of the high-spin state at low temperatures found in some systems, and (iv) a plateau or a peak in the plots of inverse magnetic susceptibility against temperature.

We have examined several models for spin-state transitions.[21] An earlier model described by Bari and Sivardiere[22] is static and can be solved exactly even when the dynamics of the lattice are included; the dynamic model does not, however, show any phase transition. We have investigated a dynamic model of spin-state transitions in which the high-spin and the low-spin states are mixed by a coupling to the lattice. The mode that can bring about such mixing is an ion-cage mode wherein the transition-metal ion moves off-center with respect to the octahedral cage in which it is placed. During such a vibration, the symmetry of the crystal field does not remain octahedral and hence the two spin states mix. This model predicts nonzero population of the high-spin state at low temperature but no spin-state transition. Susceptibility behavior of some Fe^{2+} complexes can be explained by this model.

We have also examined a two-sublattice model, where the displacement on one sublattice is opposite to that on the other, but this model shows only second-order spin-state transitions. In order to explain the occurrence of both first- and second-order spin-state transitions, we have explored a two-sublattice model where the spin states are coupled to the cube of the breathing mode displacement. This model predicts first- or second-order transitions but only zero high-spin-state population at low temperatures. The most general model that predicts nonzero high-spin-state population at low temperatures, a first- or a second-order transition, and other features appears to be one where the coupling of the spin states to a breathing mode is linear and that to an ion-cage mode is quadratic. Nonetheless, spin-state transitions in extended solids need to be further explored to enable us to fully understand the mechanism of these transitions.

Electronic Transitions

Among the phase transitions where electronic factors play a major role, the most well-known are the metal–insulator transitions exhibited by transition-metal oxides, sulfides, and so on. This subject has been discussed at length.[2,23,24] A recent observation[25] of some interest is that the metal–nonmetal transition occurs at a critical electron concentration as given by the particular form of the Mott criterion, $n_c^{1/3}a_H = 0.26 \pm 0.05$. The Verwey transition in Fe_3O_4 is associated with a marked jump in conductivity, but the material remains a semiconductor both above and below the transition temperature (123 K); below 123 K, there is

(12) P. M. Raccah and J. B. Goodenough, *Phys. Rev.*, **155**, 932 (1967).
(13) V. G. Bhide, D. S. Rajoria, G. Rama Rao, and C. N. R. Rao, *Phys. Rev. B*, **6**, 1021 (1972).
(14) W. H. Madhusudan, K. Jagannathan, P. Ganguly, and C. N. R. Rao, *J. Chem. Soc., Dalton Trans.*, 1397 (1980).
(15) S. Vasudevan, H. N. Vasan, and C. N. R. Rao, *Chem. Phys. Lett.*, **65**, 444 (1979).
(16) M. Bose, A. Ghoshray, A. Basu, and C. N. R. Rao, *Phys. Rev. B*, **26**, 4871 (1982).
(17) G. Demazeau, M. Pouchard, M. Thomas, J. F. Colombet, J. Grenier, L. Lournes, J. Souveyroux, and P. Hagenmuller, *Mater. Res. Bull.*, **15**, 451 (1980).
(18) R. Mohan Ram, K. K. Singh, W. H. Madhusudan, P. Ganguly, and C. N. R. Rao, *Mater. Res. Bull.*, **18**, 703 (1983).
(19) G. Demazeau, M. Pouchard, and P. Hagenmuller, *J. Solid State Chem.*, **18**, 159 (1976).
(20) G. Demazeau, J. L. Marty, B. Buffat, J. M. Dance, N. Pouchard, P. Dordor, and B. Chevalier, *Mater. Res. Bull.*, **17**, 37 (1982).

(21) S. Ramasesha, T. V. Ramakrishnan, and C. N. R. Rao, *J. Phys. C*, **12**, 1307 (1979).
(22) R. A. Bari and J. Sivardiere, *Phys. Rev. B*, **5**, 4466 (1972).
(23) N. F. Mott, "Metal–Insulator Transitions", Taylor and Francis, London, 1974.
(24) J. M. Honig, *J. Solid State Chem.*, **45**, 1 (1982), and the references cited therein.
(25) P. P. Edwards and M. J. Sienko, *Acc. Chem. Res.*, **15**, 87 (1982).

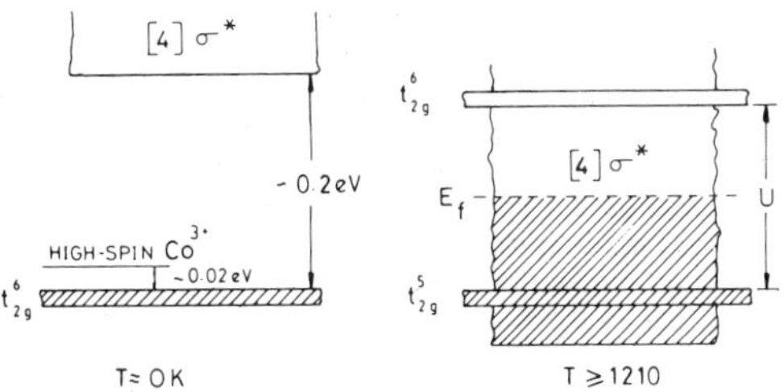

Figure 4. Energy band scheme for $LaCoO_3$ at 0 K and above the first-order electronic transition at 1200 K (taken from ref 13). At 0 K, all the 3d electrons are in the t_{2g} level (low-spin state), and the high-spin Co^{3+} level is empty. Above 1200 K, the e_g electrons form a partially filled σ^* band and the t_{2g} electrons remain localized.

charge ordering of Fe^{2+} and Fe^{3+} ions. The Verwey transition has been a subject of intensive discussion,[24] but there is still some doubt with regard to the mechanism of the transition. In this section, I shall briefly present the features of a few other interesting electronic transitions investigated in this laboratory.

$LaCoO_3$ and other rare-earth cobaltites show first-order phase transitions around 1200 K, which seem to be essentially governed by the change in electronic entropy.[12,13,26] Temperature evolution of the electronic and spin configurations of cobalt in these cobaltites is interesting, and we have investigated this by employing Mössbauer spectroscopy[13,26] and X-ray photoelectron spectroscopy.[27] At low temperatures, cobalt ions are in the diamagnetic low-spin state (t_{2g}^6) and transform to the high-spin state ($t_{2g}^4 e_g^2$) with increase in temperature, the two spin states being clearly distinguished in Mössbauer spectra. Electron hopping between the two spin states gives rise to charge-transfer states ($Co^{2+} + Co^{4+}$) and associated increase in electrical conductivity. The magnitude of charge transfer depends on the acidity of the rare-earth ion. As the temperature is increased further, the e_g electrons tend to form a σ^* band; accordingly, the center shift in the Mössbauer spectra shows a decrease in this temperature region ($\sim$700–1000 K) due to progressive increase in the cation–anion orbital overlap. Mössbauer spectra show a single resonance with a chemical shift close to zero (corresponding to the band state of e_g electrons), as we approach the first-order transition temperature ($\sim$1200 K). Above 1200 K, the cobaltites become metallic due to the change in the nature of d electrons from localized to itinerant behavior (Figure 4). Since no change in crystal symmetry was noticed at the transition, it was considered that the entire entropy change was electronic in origin. However, it seems likely that there is an increase in the symmetry of $LaCoO_3$ (change from rhombohedral to cubic structure) after the transition.[12,13] Even so, the large ΔS of the transition (over 4 J K^{-1} mol^{-1}) can only arise because of a significant electronic contribution. The Lamb–Mössbauer factor (area under the resonance) decreases markedly before this transition, indicative of large ionic vibrations, and increases sharply above the transition, suggesting the

establishment of long-range order.

Another interesting electronic transition that we have been interested in is that of La_2NiO_4, which crystallizes in the two-dimensional K_2NiF_4 structure. Unlike K_2NiF_4, La_2NiO_4 is not known to show any long-range antiferromagnetic order but only shows deviations from Curie–Weiss law due to short-range interactions.[28] What is more interesting is that La_2NiO_4 undergoes a gradual semiconductor–metal transition[29] around 550 K; no structural change accompanying the transition has been reported. The Ni–O–Ni distance in La_2NiO_4 (3.86 Å) is shorter than in NiO (4.18 Å). The transfer integral, b, in NiO is close to the critical transfer integral, b_c, at which the description for localized electron states breaks down. In La_2NiO_4, b should be nearly equal to b_c, so that we obtain partially filled $\sigma^*(x^2 - y^2)$ bands of strongly correlated itinerant d electrons and hence the metallic behavior. The half-filled d_{z^2} orbitals of Ni^{2+}, on the other hand, would be localized. A likely origin of the semiconductor-metal transition is the splitting of the $\sigma^*(x^2 - y^2)$ band on lowering the temperature because of strong electron correlations, which in turn may be accentuated by short-range antiferromagnetic ordering d_{z^2} spins.[30]

In spite of the great interest in the 550 K electronic transition of this two-dimensional system, all the electrical measurements have hitherto been carried out on pellets of polycrystalline material. Preliminary measurements on single crystals[31] indicate the occurrence of a sharp transition (in the ab plane) with at least an order of magnitude jump in conductivity. Furthermore, Ni^{3+} ions that are inevitably present (3–9%) in all preparations of La_2NiO_4 seem to influence the structure and properties quite significantly. Samples of La_2NiO_4 annealed in a CO_2 atmosphere at 1400 K (to removed Ni^{3+}) seem to show evidence for long-range antiferromagnetic ordering. In addition, electron diffraction studies show them to have a monoclinic distortion contrary to published crystallographic data. Careful investigations of the electronic and magnetic properties of well-characterized samples of La_2NiO_4 are therefore warranted.

The last type of electronic transition that I shall discuss is the one exhibited by silver chalcogenides, $Ag_{2+\delta}Ch$ (Ch = S, Se, or Te). These compounds transform to a symmetrical phase on heating wherein Ag^+ are randomly distributed, giving rise to superionic conductivity just as AgI. These materials are small-gap semiconductors at room temperature and exhibit interesting electronic behavior as a function of temperature as well as of composition.[32] Thus, in the high temperature phase ($T > 406$ K), $Ag_{2+\delta}Se$ shows metallic behavior of electronic conductivity for high values of X. With decrease in δ, the electronic conductivity shows evidence for an interesting transition (Figure 5), the ionic conductivity of the high temperature phase being essentially independent of δ. The magnitude of change in electronic conductivity at the phase transition is also

(26) (a) V. G. Jadhao, G. Rama Rao, D. Bahadur, R. M. Singru, and C. N. R. Rao, *J. Chem. Soc., Faraday Trans. 2*, **71**, 1885 (1975); (b) V. G. Jadhao, R. M. Singru, G. Rama Rao, D. Bahadur, and C. N. R. Rao, *J. Phys. Chem. Solids*, **37**, 113 (1976).

(27) G. Thornton, A. F. Orchard, and C. N. R. Rao, *J. Phys. C*, **9**, 1991 (1976).

(28) P. Ganguly, S. Kollali, and C. N. R. Rao, *Magn. Lett.*, **1**, 107 (1978).

(29) P. Ganguly and C. N. R. Rao, *Mater. Res. Bull.*, **8**, 405 (1973).

(30) J. B. Goodenough and S. Ramasesha, *Mater. Res. Bull.*, **17**, 383 (1982).

(31) C. N. R. Rao, D. Buttrey, N. Otsuka, P. Ganguly, H. R. Harrison, C. J. Sandberg, and J. M. Honig, *J. Solid State Chem.*, in press.

(32) A. K. Shukla, H. N. Vasan, and C. N. R. Rao, *Proc. R. Soc. London, Ser. A*, **376**, 619 (1981).

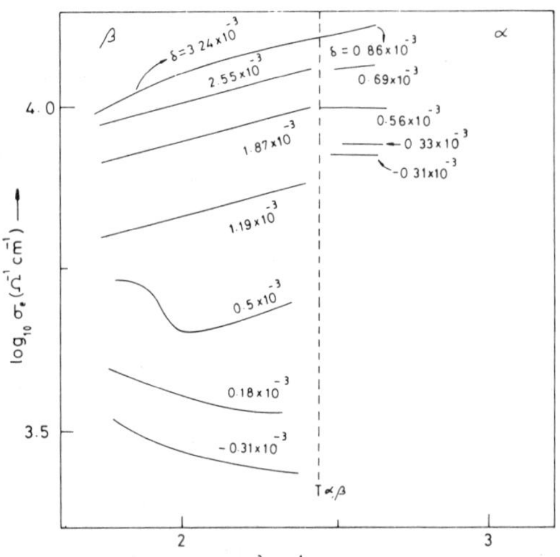

Figure 5. Plot of the logarithm of electronic conductivity against inverse of absolute temperature for various compositions of α- and β-silver selenide. The degree of nonstoichiometry, δ, was determined by solid-state coulometry (taken from ref 32).

determined by stoichiometry. In the low-temperature phase, the material conducts like a semimetal around 400 K and the conductivity decreases substantially at low temperatures, the magnitude of the decrease depending on the value of δ. This behavior as well as the δ-dependent electronic transition in the high-temperature phase is indeed interesting. We are now investigating $Ag_{2+\delta}Te$, which is a p-type material with a small gap (unlike $Ag_{2+\delta}Se$, which is a n-type material), at ordinary temperatures.

Organic Solids and Related Systems

Although phase transitions in organic solids have been reported from time to time, extensive mechanistic studies have not been carried out as in the case of inorganic solids.[2] It was believed until recently by many workers that there are no structural relations between phases in the case of organic solids. Jones et al.,[33] however, have reported that the stress-induced phase transition of 1,8-dichloro-10-methylanthracene proceeds by a diffusionless displacive transition (somewhat similar to a martensitic transition) with definite orientational relationships. The irrational habit plane seems to be composed of close packed planes and the properties of the interface could be formulated in terms of slip dislocations. The reversible topotactic phase transition of 5-methyl-1-thia-5-azoniacyclooctane 1-oxide perchlorate has been explained by Parkinson et al.[34] in terms of recurrent glissile partial dislocations; it was earlier thought[35] that the transition involved a cooperative inversion and rotation of half of the molecular cations. The phase transition of paraterphenyl involving rotational disorder has been elucidated by Ramdas and Thomas[36] by evaluating pairwise interac-

tions between nonbonded atoms.

In this laboratory, we have been studying phase transitions of a variety of organic solids. For example, we have investigated the α–γ–α–β transitions of p-dichlorobenzene by infrared spectroscopy.[37] The γ phase is characterized by unusually high intramolecular vibrations mode frequencies. The α–γ transition shows athermal nucleation behavior as in martensitic transitions; the α–β transition seems to be associated with some disorder. When the asymmetry of the molecules is not too high, organic crystals often exhibit order–disorder transitions (e.g., benzothiophene). We have investigated order–disorder transitions in benzothiophene and other compounds by making use of changes in infrared band intensities and half-widths.[38] Raman spectroscopy would be especially useful in investigating order–disorder transitions.

We have investigated the phase transitions of compounds of the type $(C_nH_{2n+1}NH_3)_2MCl_4$, M = Mn, Fe, Cd, or Cu, which provide interesting model systems to investigate magnetic phenomena in two dimensions. Earlier spectroscopic investigations[39] seemed to indicate that in $(CH_3NH_3)_2MCl_4$, the phase transitions are determined essentially by the motions of the methylammonium groups. We have examined the infrared spectra of several $(C_nH_{2n+1}NH_3)_2MCl_4$ systems through their phase transitions to investigate whether the intramolecular vibration modes show the expected changes.[40] In the high-temperature phase of these solids, the $(CH_3NH_3)^+$ ion has C_{3v} symmetry, but the symmetry goes down to C_1 or C_s in the ordered low-temperature phases. The spectra indeed show the expected site-groups as well as factor-group splittings in the low-temperature phases, the degenerate bending modes of NH_3 and CH_3 being particularly sensitive. The phase transitions of the tetrachlorometallates are similar to those of the corresponding alkylammonium chlorides, $C_nH_{2n+1}NH_3Cl$, thereby establishing that the phase transitions in the former are entirely controlled by the motions of the $(C_nH_{2n+1}NH_3)^+$ group. Accordingly, we find that the $(C_nH_{2n+1}NH_3)_2MBr_4$ system shows transitions similar to those of the chloro compounds.

We have been particularly interested in the study of the plastic states of organic compounds,[41] which are characterized by high values of ΔS of formation from the crystalline state, the ΔS of fusion (plastic–liquid transition) being much smaller. We find that the ΔH as well as the ΔS of the crystal–plastic transition generally decrease as the temperature range of stability of the plastic phase increases; the ΔH and the ΔS of the plastic–liquid transition, on the other hand, increase as the temperature range of stability of the plastic phase increases.[42] Neutron scattering, NMR spectroscopy, and several other techniques have been employed to study molecular reorientation in the plastic state.[41] We

(33) W. Jones, J. M. Thomas, and J. O. Williams, *Philos. Mag.*, **32**, 1 (1975).

(34) G. M. Parkinson, J. M. Thomas, J. O. Williams, M. J. Goringe, and L. W. Hobbs, *J. Chem. Soc., Perkin Trans. 2*, 836 (1976).

(35) I. C. Paul and K. T. Go, *J. Chem. Soc. B*, 33 (1969).

(36) S. Ramdas and J. M. Thomas, *J. Chem. Soc., Faraday Trans. 2*, **72**, 1251 (1976).

(37) S. Ganguly, J. R. Fernandes, G. Bahadur, and C. N. R. Rao, *J. Chem. Soc., Faraday Trans. 2*, **75**, 923 (1979).

(38) C. N. R. Rao, S. Ganguly, and H. R. Swamy, *Croat. Chem. Acta*, **55**, 207 (1982), and the references cited therein.

(39) (a) D. M. Adams and D. C. Stevens, *J. Phys. C*, **11**, 617 (1978); (b) R. Blinc, B. Zeks, and R. Kind, *Phys. Rev. B*, **17**, 3409 (1978).

(40) C. N. R. Rao, S. Ganguly, H. R. Swamy, and I. A. Oxton, *J. Chem. Soc., Faraday Trans. 2*, **77**, 1825 (1981).

(41) J. N. Sherwood, Ed., "The Plastically Crystalline State", Wiley, New York, 1979.

(42) S. Ganguly, J. R. Fernandes, and C. N. R. Rao, *Adv. Mol. Relaxation Interact. Processes*, **20**, 149 (1981).

have employed Raman band shape analysis to obtain rotational correlation function of plastic phases of a few organic compounds.[43] The correlation times are continuous through the plastic–liquid transition.

Molecular dynamics simulation of the plastic state of CH_4 has been carried out by Bounds et al.[44] who have evaluated the static and dynamic structure factors. The crystalline phase of CH_4 described by James and Keenan[45] has been investigated by several techniques. We have carried out a Monte Carlo simulation study[46] on the orientational glasses (or glassy crystalline phases) obtained by annealing or quenching the plastic phase. Different cooling rates lead to different states of the glassy crystalline phase. Temperature variation of the orientational parameter of molecules suggests the presence of a transition between the plastic and glassy crystalline phases.

Phase transitions of hydrogen-bonded solids such as ferroelectric hydrogen phosphates and Rochelle salt have been investigated widely in the literature.[2] We have been recently investigating phase transitions of alkanedioc acids employing vibrational spectroscopy.[38] The phase transition of malonic acid at 360 K is especially interesting. At ordinary temperatures, the unit cell of malonic acid contains two cyclic dimeric rings orthogonal to each other; above 360 K, the two hydrogen-bonded rings become similar as evidenced from IR and Raman spectra.[47,48] Hydrogen bonds in the high-temperature phase are on the average weaker than those in the low-temperature phase. The phase transition occurs at a higher temperature (366 K) in the fully deuterated acid, and the vibrational bands show a positive deuterium isotope effect. It appears that the transition is governed by librational and torsional modes of the hydrogen-bonded rings (around 90 and 50 cm^{-1}, respectively, below the transition temperature), which show a tendency to soften.

Concluding Remarks

The above discussion should serve to indicate the variety of problems in the area of phase transitions that are of relevance to solid-state chemistry. Of course there are other important aspects of phase transitions, such as cooperative Jahn–Teller effect, commensurate–incommensurate structure transitions and transitions in ferroics, that I have not touched upon. Neither have I dealt with technological applications.[2] The main point that I have tried to make is that the study of structural and mechanistic aspects of phase transitions in the solid state constitutes an essential and interesting aspect of the chemistry of solids.[49]

A word about the so-called normal-superionic conducting state transitions would be in order. It seems to be wrong to refer to these transitions as superionic transitions by analogy with the electronic superconducting transitions. While AgI related materials show a marked jump in ionic conductivity at a transition temperature, there are many other superionic materials that do not.[50]

A transition that has eluded a proper understanding and has increasingly become an area of vital interest is the glass transition.[51] Although many models have been proposed, none of them is able to explain all the features of this transition. An interplay of kinetic and thermodynamic effects further complicates the situation. We have recently proposed a cluster model of the glass transition[52] based on the premise that glass is an ensemble of more ordered regions (clusters) embedded in a liquidlike, disordered tissue material. Toward the glass transition, clusters gradually melt and add on to the tissue material. The cluster model making use of the relative size of the cluster as an order parameter seems to explain several features of the transition. Another aspect of the glass transition that is worth noting is that it is not merely characteristic of the normal variety of (positionally disordered) glasses but also of orientationally disordered glasses and dipolar glasses. A satisfactory model would have to account for this feature as well. Some other aspects of phase transitions that deserve attention are transitions in organic solids, deformational mechanisms of phase transitions in complex solids, transitions of orientationally disordered crystals, and transitions involving ordering of defects. Computer simulation studies are bound to be increasingly useful in understanding the nature of many of the solid-state phase transitions.

(43) S. Ganguly, H. R. Swamy, and C. N. R. Rao, _J. Mol. Liq._, 25, 139 (1983).
(44) D. G. Bounds, M. L. Klein, and G. N. Patey, _J. Chem. Phys._, 72, 5348 (1980).
(45) H. M. James and T. A. Keenan, _J. Chem. Phys._, 31, 12 (1959).
(46) S. Yashonath and C. N. R. Rao, _Chem. Phys. Lett._, 101, 524 (1983).
(47) S. Ganguly, J. R. Fernandes, G. R. Desiraju, and C. N. R. Rao, _Chem. Phys. Lett._, 69, 227 (1980).
(48) J. de Villepin, M. H. Limage, A. Novak, M. LePostollec, H. Poulet, S. Ganguly, and C. N. R. Rao, _J. Raman Spectrosc._, in press.

(49) C. N. R. Rao, _Chem. Scr._, 19, 124 (1982).
(50) R. A. Huggins and A. Rabenau, _Mater. Res. Bull._, 13, 1315 (1978).
(51) R. Parthasarathy, K. J. Rao, and C. N. R. Rao, _Chem. Soc. Rev._, in press.
(52) K. J. Rao and C. N. R. Rao, _Mater. Res. Bull._, 17, 1337 (1982).

For a discussion of some aspects of the glass transition, see R. Parthasarathy, K.J. Rao and C.N.R. Rao, The Glass Transition: Salient Facts and Models, <u>Chem. Soc. Revs.</u>, <u>12</u>, 361–385 (1983)

Offprinted from the *Transactions of The Faraday Society*
No. 574, Vol. 66, Part 10, October, 1970

Pm3m-Fm3m Transformations of Alkali Halides

Solid Solutions of CsCl with KCl, CsBr, SrCl$_2$

BY M. NATARAJAN, K. J. RAO AND C. N. R. RAO *

Dept. of Chemistry, Indian Institute of Technology, Kanpur, India

Received 31st December, 1968

Pm3m-Fm3m transformations of solid solutions of CsCl with KCl and CsBr exhibit different behaviours. With increasing percentages of KCl, the NaCl structure gets stabilized in the CsCl+KCl system. In the CsCl+CsBr system, the transformation temperature increases with % CsBr and ΔH essentially remains constant. Both these behaviours can be satisfactorily explained in terms of the Born treatment of ionic solids. The *Pm3m-Fm3m* transformation retains its first-order characteristics in the CsCl+KCl system, but higher-order components seem to be present in the CsCl+CsBr system. Incorporation of vacancies do not affect the transformation of CsCl markedly.

Caesium chloride transforms from the CsCl structure (*Pm3m*) to the NaCl structure (*Fm3m*) around 480°C.[1] The transformation is thermodynamically first order and is associated with considerable thermal hysteresis. The *Pm3m-Fm3m* transformations of CsCl and other alkali halides have been examined by employing the Born treatment of ionic solids.[2,3] The Born-Mayer expression with four repulsive parameters explains the relative stabilities of the *Pm3m* and *Fm3m* structures satisfactorily. Further, this treatment also accounts for the stabilization of solid solutions of CsCl with RbCl in the *Fm3m* structure; the behaviour of CsCl+KCl solid solutions appears to be similar[2] to that of CsCl+RbCl solid solutions. On the other hand, in the transformations of CsCl+CsBr solid solutions, the transformation temperature increases with the percentage of CsBr.[4,5]

We have studied the transformations of the CsCl+KCl and CsCl+CsBr solid solutions in order to find the limitations and applicability of the Born treatment in explaining the two entirely different behaviours of the solid solutions of these two systems. Such a study is of value since theoretical approaches to explain the relative stabilities of structures of ionic solids have not been quite successful, and it is important to explain the relative stabilities of at least the two simplest structure types in ionic solids, viz., the NaCl and CsCl structures. We also wished to find out whether the first order characteristics of *Pm3m-Fm3m* transitions are retained in the solid solutions. We have therefore examined the crystallography of the *Pm3m* and *Fm3m* phases of the solid solutions as functions of temperature; from these data, coefficients of expansion of the two structures have been calculated.

Menary, Ubbelohde and Woodward,[1] have reported that a large increase in vacancies occurs during the transformation of CsCl; and the formation energy of a Schottky pair in CsCl is considerably lower in the *Pm3m* phase (~ 1 eV) compared, to the *Fm3m* phase (~ 2 eV).[6,7] The *Pm3m-Fm3m* transformations of two CsCl+SrCl$_2$ solid solutions have now been examined to find out whether the vacancies produced by the incorporation of Sr^{+2} favour the transformation.

EXPERIMENTAL

All the compounds were better than 99.99 % purity or of spectroscopic grade (J. & M. or Alfa). The preparation of solid solutions as well as the procedure for recording the

* to whom all the correspondence should be addressed.

differential thermal analysis (DTA) curves have been described.[2, 8]   The procedures to obtain enthalpies of transformation ΔH_{tr} (from DTA peak areas) and activation energies E_a are also described earlier.[2, 8]

The X-ray patterns were recorded (with Cu $K\alpha$ radiation) at different temperatures on a General Electric diffractometer fitted with a high-temperature camera and a temperature programmer.   The a_0 values of the $Pm3m$ and $Fm3m$ phases of the CsCl+KCl solid solutions were determined at different temperatures, and that a_0 values of the CsCl+CsBr solid solutions were determined at 25 and 560°C.   All the calculations employing the Born treatment were carried out on the IBM 1620 and IBM 7044 computers in this Institute.

RESULTS AND DISCUSSION

CsCl+KCl SOLID SOLUTIONS

The results of the transformation of various CsCl+KCl solid solutions are summarized in table 1 and typical differential thermograms are shown in fig. 1. The lattice dimensions (at 25°C) of the $Pm3m$ and $Fm3m$ forms of the solid solutions are also given in table 1.   The a_0 values of both the forms decrease with increase in

TABLE 1.—$Pm3m$-$Fm3m$ TRANSITIONS OF CsCl+KCl SOLID SOLUTIONS

% KCl (mol)	T_t/°C forward (DTA peak)	$T/_t$°C reverse (DTA peak)	ΔH_{tr} cal mol^{-1}	E_a kcal mol^{-1}	$a_0(Pm3m)$ 25°C (a)	$a_0(Fm3m)$ 570°C	$a_0(Fm3m)$ 25°C (b) (extra.)	ΔV (25°C) (c)	$\eta Pm3m$ deg.$^{-1}$	$\eta Fm3m$ deg.$^{-1}$ (e)
0	479	444	580	180	4.1210	7.0980	6.9050	7.34	1.25×10^{-4}	1.5×10^{-4}
2.5	470	—	520	114	4.1125	7.0750	6.8960	7.48	2.24×10^{-4} (d)	2.8×10^{-4}
5.0	470	—	465	100	4.1000	7.0630	6.8820	7.49	2.16×10^{-4} (d)	2.7×10^{-4}
10.0	470	—	420	—	4.0760	7.0250	6.8580	7.61	1.2×10^{-4}	1.2×10^{-4}
20.0	470	—	258	—	4.0400	7.0100	6.8100	7.83	2.6×10^{-4}	1.0×10^{-4}
25.0	no transition	—	—	—	—	—	—	—	—	—

(a) The $a_0(Pm3m)$ of pure KCl found by linear extrapolation was 3.65 Å (3.76 Å is the value calc. assuming $a_0(Pm3m)/a_0(Fm3m) = 0.598$).

(b) The $a_0(Fm3m)$ of pure KCl found by linear extrapolation was 6.30 Å (N.B.S. value = 6.294 Å).

(c) change in molar volume.

(d) The $\eta^{(Pm3m)}$ values in these solutions were calculated assuming $a_0^{(Pm3m)}/a_0^{(Fm3m)} = 0.598$; the η^{Pm3m}/η^{Fm3m} values are in the range 0.8-2.6, the value increasing with % KCl in the solid solution.   The η^{Pm3m}/η^{Fm3m} values for pure CsCl and solutions with 10 % and 20% KCl are obtained from experimental a_0 values.

(e) calc. from a_0 values in the temperature range, 500-570°C.

% KCl.   The calculated value of the change ΔV in molar volume accompanying the transformation increases slightly with the percentage of KCl suggesting that the transformation remains first order (thermodynamically) even in solid solutions.

The volume coefficient of expansion of the $Pm3m$ and $Fm3m$ phases of pure CsCl were 1.25×10^{-4} and 1.5×10^{-4} deg.$^{-1}$ respectively in the transition region.   The η^{Fm3m} values for the solid solutions determined experimentally are given in table 1, with the calculated values of the η^{Pm3m}.   The ratio, η^{Pm3m}/η^{Fm3m}, is about 0.80 for pure CsCl, a value which satisfactorily accounts for the melting behaviour of CsCl.[9]   The η^{Pm3m}/η^{Fm3m} ratio for the CsCl+KCl solid solutions increases in the range 0.80-2.60 % KCl.   The melting points of these solid solutions decrease slightly with % KCl.

The results from DTA and X-ray studies (table 1) indicate that the $Pm3m$-$Fm3m$ transformation of CsCl+KCl solid solutions is reversible only up to 3 % KCl. With further increase in the proportion of KCl, the transformation becomes irreversible and at ~ 25 % KCl the $Fm3m$ phase of CsCl is stable.   At, or above, 25 % KCl, the solid solutions showed no indication of the transformation.   Thus, the ΔH of the transformation decreases continuously with % KCl in the solid solutions up to 25 % KCl.   While one may visualise how high percentages of KCl or RbCl stabilize the *Fm3m* form of the solid solutions by causing an appreciable decrease in the interionic distance, it is difficult to understand why the reversibility of the *Pm3m-Fm3m* transformation ceases above 3 % KCl (or 5 % RbCl). The energy of activation also decreases with % KCl indicating that the transformation is favoured by these additives.

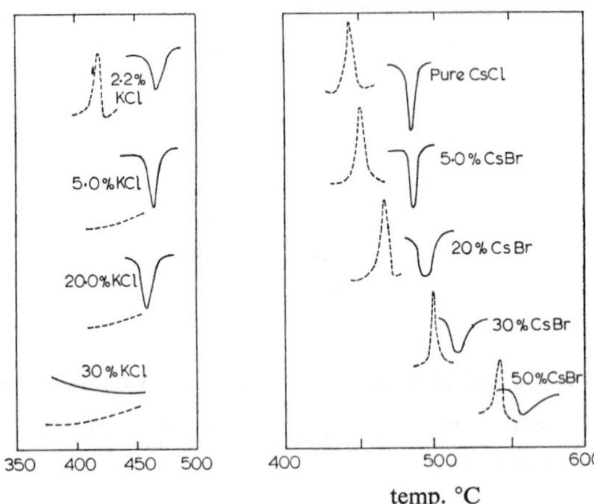

FIG. 1.—DTA curves of CsCl+KCl and CsCl+CsBr solid solutions.

The behaviour of the CsCl+KCl solutions may be understood in terms of the modified Born treatment [2, 3] as for the CsCl+RbCl system.[2] The lattice energy difference between the *Pm3m* and *Fm3m* structures of a solid solution is given by

$$\Delta U^{s.s} = U^{s.s}_{Pm3m} - U^{s.s}_{Fm3m} \tag{1}$$

where $U^{s.s}_{Pm3m}$ and $U^{s.s}_{Fm3m}$ are given by the weighted sum of the lattice energies of CsCl and KCl of the appropriate structures.[2] The lattice energy U has the general form,

$$U = N_0\left[\frac{\alpha e^2}{r_0} + K\left\{\frac{C}{r_0^6} + \frac{D}{r_0^8}\right\} - B(r_0) - \phi_0\right] \tag{2}$$

where $B(r_0) = M_1 b_1 \exp(-r_0/\rho_1) + M_2 b_2 \exp(-ar_0/\rho_2)$. The terms in eqn. (2) have been described earlier.[2, 3] In the repulsive term $B(r_0)$, a is the ratio of the distance between the second nearest neighbour to that between the nearest neighbours. Calculations employing eqn. (1) and (2) provide $\Delta U^{s.s}$ at 25°C since r_0 values are taken from X-ray data at this temperature. The experimental ΔH_{tr} values (table 1) are, however, at the transformation temperatures. Since no transformation was noticed when the fraction f of KCl was 0.25, it may be taken that $\Delta U^{s.s}$ (25°C) approaches zero at this composition.

By choosing various values of K and corresponding values of the repulsive parameters for CsCl and KCl,[2, 3] $\Delta U^{s.s}$ was calculated as a function of f (fig. 2). $\Delta U^{s.s}$ becomes zero at $f \approx 0.25$, when $K = 2.8$ or 3.00. The values of the repulsive parameters when $K = 3.00$ are:

	b_1 (ergs mol^{-1})	b_2 (ergs mol^{-1})	ρ_1 (cm)	ρ_2 (cm)
CsCl	644×10^{-12}	795×10^{-12}	0.367×10^{-8}	0.424×10^{-8}
KCl	3879×10^{-12}	35.6×10^{-12}	0.345×10^{-8}	0.487×10^{-8}

These results indicate that the Born-Mayer expression with four repulsive parameters

2500 *Pm3m-Fm3m* TRANSFORMATIONS OF ALKALI HALIDES

is satisfactory in predicting the general behaviour of the solid solution of CsCl with KCl or RbCl.   Again, the hardness parameters ρ_2 and ρ_1 are related by the structure constant a.

Several workers [10, 11] have proposed revised van der Waals coefficients in alkali halides which eliminate the need for employing high van der Waals terms in the study

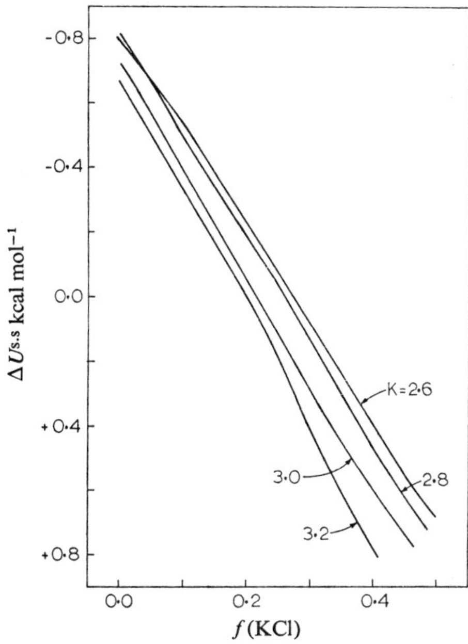

FIG. 2.—Variation of $\Delta U^{s.s}$ of CsCl+KCl solid solutions calculated for various values of K for CsCl and KCl (the graphs correspond to $K_{CsCl} = K_{KCl} =$ the value indicated).

of phase transitions of alkali halides.[12]   By employing these coefficients of Hajj [10] for CsCl (with $K = 1$), we find that $\Delta U^{s.s}$ becomes zero at $f \sim 0.40$; by employing the parameters of Lynch [11] for KCl along with Hajj's values for CsCl, $\Delta U^{s.s}$ becomes zero at $f \approx 0.25$.

TABLE 2.—*Pm3m-Fm3m* TRANSITIONS OF CsCl+CsBr SOLID SOLUTIONS

% CsBr (mol)	T_t/°C forward (DTA peak)	T_t/°C reverse (DTA pek)	ΔH_{tr} cal mol^{-1}	a_0(Pm3m) 25°C	a_0(Fm3m) 560°C	a_0(Fm3m) 25°C (a)	ΔV (25°C) cm^3 mol^{-1}
0	479	444	580	4.1210	7.0010	6.9050	7.34
5	480	450	600	4.1300	7.0130	6.9150	7.33
10	485	455	825	4.1350(b)	7.0320(b)	6.9250	7.31
20	495	462	950	4.1550	7.0600	6.9400	7.15
30	505	480	1025	4.1710	7.1000	6.9550	6.91
40	530	503	1100	4.1900	7.1300	6.9675	6.63
50	555	535	1050	4.2080	7.1515	6.9825	6.37
60	610(c)	—	—	—	—	—	—

(a) calc. on the basis that a_0 (Pm3m)/a_0(Fm3m)≈ 0.598.

(b) The a_0 (Pm3m) and a_0 (Fm3m) values for this composition were determined at several temperatures; the η^{Fm3m} and η^{Pm3m} are 1.17×10^{-4} and 0.68×10^{-4} deg.$^{-1}$ repectively.   The η^{Pm3m}/η^{Fm3m} is thus 0.53

(c) very close to the m.p. (614°C).

CsCl+CsBr SOLID SOLUTIONS

The results from DTA and X-ray analysis are shown in table 2. Typical differential thermograms are given in fig. 2. The transformation temperature increases markedly with % CsBr, the trend continuing up to the melting point of the solid solution (~ 65 % CsBr). The transformation temperatures reported here essentially agree with those of earlier workers.[4,5] The ΔH of the transformation varies little in these solid solutions. The η^{Pm3m}/η^{Fm3m} ratio in the solution with 10 % CsBr is ~ 0.5, a value much less than in pure CsCl.

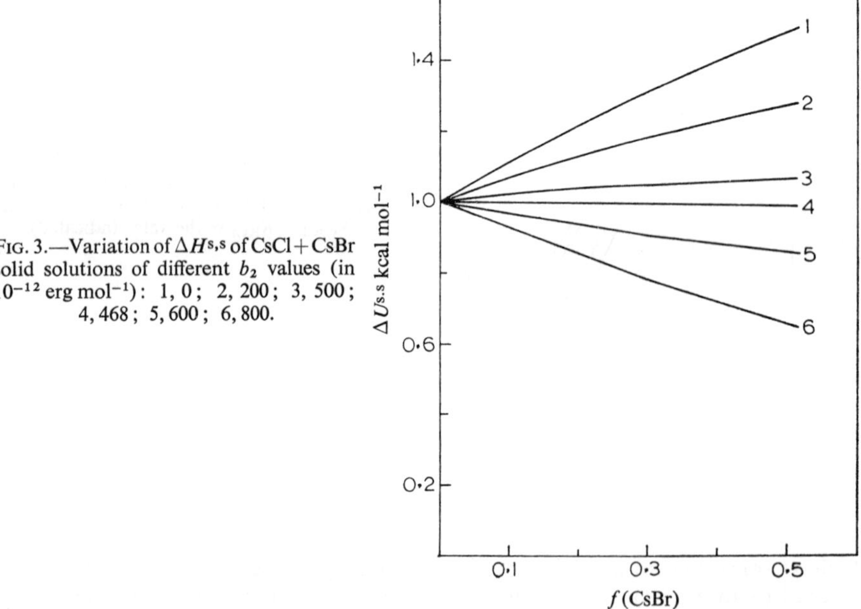

FIG. 3.—Variation of $\Delta H^{s,s}$ of CsCl+CsBr solid solutions of different b_2 values (in 10^{-12} erg mol^{-1}): 1, 0; 2, 200; 3, 500; 4, 468; 5, 600; 6, 800.

Since CsBr has the same $Pm3m$ structure as CsCl, it does not stabilize the $Fm3m$ structure of the olid solutions; accordingly, there is an increase in the a_0 values of the $Pm3m$ and $Fm3m$ structures with increasing % CsBr. On the other hand, in the solid solutions of CsCl with KCl (or RbCl), the a_0 values of both the structures decrease with the increases in the % KCl (or RbCl). The change ΔV in the molar volume accompanying the $Pm3m$-$Fm3m$ transformation (table 2) decreases with increasing CsBr content and at 50 % CsBr the decreases in ΔV is about 14 %; ΔV is expected to decrease with % CsBr on the basis of ionic size considerations.[2] This may mean that the transformation has a component of higher or second order in CsCl+CsBr solid solutions. The DTA peaks at high % CsBr also become (less sharp) compared with those for pure CsCl. The magnitude ΔT of thermal hysteresis, decreases slightly with increasing % CsBr, thus indirectly substantiating that the transformation has components of higher order; ΔT would be expected to decrease with % CsBr since ΔT is proportional to the ΔV of the transformation.[8] Samples containing 30 % CsBr (or > 30 %) showed evidence for the coexistence of the $Pm3m$ and $Fm3m$ phases in the transformation region (~ 500-$550°C$). The coexistence of phases and the formation of hybrid cystals have been noticed by Ubbelohde[13] in continuous or higher-order transformation.

2502   *Pm3m-Fm3m* TRANSFORMATIONS OF ALKALI HALIDES

In order to employ the Born treatment to explain the transformations of CsCl + CsBr solid solutions, we have used the revised van der Waals coefficients of CsBr proposed by Hajj.[10] The repulsive parameters of CsBr can be evaluated as follows (i) In the expression for the lattice energy (eqn. (3)) we have assumed that $\rho_2 = a\rho_1$ as proposed by Rao and Rao.[3] (ii) We calculate $(8b_1 + 6b_2)_{CsBr}$ by substituting the value of ρ_1 obtained from the compressibility relations.[2]

$$U^{Pm3m} = N_0[\alpha e^2/r_0 + C/r_0^6 + D/r_0^8 - (8b_1 + 6b_2)\exp(-r_0/\rho_1) - \phi_0]. \tag{3}$$

The value of $(8b_1 + 6b_2)$ thus calculated is 76867×10^{-12} ergs mol^{-1}. The $\Delta U^{s.s}$ of solid solutions of CsCl and CsBr were evaluated (eqn. (1)) for different values of b_2 (in the range $0\text{-}900 \times 10^{-12}$ erg mol^{-1}) using the r_0 values from table 2. The results are shown diagramatically in fig. 3, the values of b_1 and b_2 are 9257×10^{-12} and 468×10^{-12} erg mol^{-1} respectively when $(6b_2/8b_1 + 6b_2)_{CsCl} \approx (6b_2/8b_1 + 6b_2)_{CsBr}$. The $\Delta U^{s.s}$ remains nearly constant with the % CsBr when b_2 is between 400 and 500×10^{-12} erg mol^{-1}. Thus, curves 3 and 4 of fig. 3 closely represent the experimentally observed situation. Thus, the Born model of ionic solids describes the relative stabilities of structures of alkali halides in widely different situations as manifested here by the CsCl + KCl and CsCl + CsBr systems.

CsCl + SrCl$_2$ SOLID SOLUTIONS

Although there is doubt with regard to the importance of vacancies in the mechanism of the *Pm3m-Fm3m* transformation in CsCl,[1, 14] experimental and theoretical studies on Schottky defects seem to indicate that vacancies could also facilitate the *Pm3m-Fm3m* transformation in CsCl, to some extent. The formation energy of Schottky defects in the *Pm3m* structure in CsCl is about half that in the *Fm3m* structure. The concentration of defects should therefore decrease after the *Pm3m-Fm3m* transformation. Studies on CsCl with different amounts of Sr^{2-} show that the transformation temperature decreases slightly due to the increased vacancy concentrations. Thus, T_t (forward) is 475 and 474°C with 2.5 and 5 % Sr^{2-} respectively; T_t (reverse) is unaffected. The results seem to indicate that *Pm3m-Fm3m* transformation of CsCl may essentially proceed by a dialatational mechanism.[15]

The authors thank the U.S. National Bureau of Standards for a research grant (G-51) under the Special International Programme, and the staff of the IIT Computer Center.

[1] J. W. Menary, A. R. Ubbelohde and I. Woodward, *Proc. Roy. Soc. A*, 1951, **208**, 158.

[2] K. J. Rao ,G. V. Subba Rao and C. N. R. Rao, *Trans. Faraday Soc.*, 1967, **63**, 1013.

[3] K. J. Rao and C. N. R. Rao, *Proc. Phys. Soc.*, 1967, **91**, 754.

[4] L. J. Wood, C. Seeney and M. T. Derbes, *J. Amer. Chem. Soc.*, 1959, **81**, 6148.

[5] L. J. Wood, W. Secunda and C. H. McBride, *J. Amer. Chem. Soc.*, 1958, **80**, 307.

[6] K. J. Rao and C. N. R. Rao, *Physica stat solidi*, 1968, **28**, 158.

[7] (a) M. Natarajan and C. N. R. Rao unpublished results.

[7] (b) Z. Morlin and J. Tremmel, *Acta Phys. Hungary*, 1966, **21**, 129.

[8] K. J. Rao and C. N. R. Rao, *J. Materials Sci.*, 1966, **1**, 238.

[9] K. J. Rao and C. N. R. Rao, *Chem. Phys. Letters*, 1967, **1**, 499.

[10] F. Hajj, *J. Chem. Phys.*, 1966, **44**, 4618.

[11] D. W. Lynch, *J. Phys. Chem. Solids*, 1967, **28**, 1941.

[12] K. J. Rao, G. V. Subba Rao and C. N. R. Rao, *Proc. Phys. Soc.*, 1968, **1**, 134.

[13] A. R. Ubbelohde in *Reactivity in Solids* ed. by J. H. de Boer, (Elsevier Publishing Co., Amsterdam, 1961).

[14] I. M. Hoodless and J. A. Morrison, *J. Phys. Chem.*, 1962, **66**, 557.

[15] M. J. Buerger, in *Phase transformation in Solids*, ed., by R. Smoluchowski, (John Wiley and Sons. Inc., New York, 1951).

See M. Natarajan and C.N.R. Rao, J. Chem. Soc. A 3087 (1970) for a study of phase transitions of silver halides.

For a study of heats of formation of solid solutions of CsCl with other alkali halides, see A.K. Shukla, J.C. Ahluwalia and C.N.R. Rao, JCS Faraday 1 72, 1288 (1976)

Offprinted from the *Transactions of The Faraday Society*,
No. 476, Vol. 58, Part 8, August, 1962

Mechanism of Crystal Structure Transformations

Part 3.—Factors Affecting the Anatase-Rutile Transformation *

By S. R. Yoganarasimhan and C. N. R. Rao[†]

Dept. of Inorganic and Physical Chemistry, Indian Institute of Science,
Bangalore 12, India

Received 20th November, 1961

The particle size and crystallite size of anatase increase markedly in the region of the crystal structure transformation. The unit cell of anatase seems to expand prior to the transformation to rutile. This expansion has been attributed to a displacive transformation of the type defined by Buerger. Smaller particle size and larger surface area seem to favour the transformation. The kinetics of the transformation of anatase prepared by the hydrolysis of titanium sulphate have been studied at different temperatures and are found to be considerably different from the kinetics of the transformation of pure anatase. The transformation becomes immeasurably slow below $\sim 695 \pm 10°C$ compared to $\sim 610°C$ for pure anatase. An induction period is observed in the transformation of anatase obtained from sulphate hydrolysis and the duration decreases with increase in temperature. The activation energy is ~ 120 kcal/mole, a value higher than that for the pure anatase-rutile transformation. The results have been interpreted in terms of the relative rates of nucleation and propagation processes. The activation energy for the nucleation process seems to be much larger than for the propagation process. The kinetics of the transformation of anatase samples doped with different amounts of sulphate ion impurity have also been studied and the transformation is found to be progressively decelerated with increase in the impurity concentration. The energy of activation for the transformation appears to increase progressively with increase in impurity concentration.

The transformation of pure anatase (tetragonal) to rutile (tetragonal) has been systematically investigated by Czanderna, Rao and Honig [1] and Rao,[2] and the transformation of brookite (orthorhombic) to rutile by Rao, Yoganarasimhan and Faeth.[3] These studies prompted us to determine the various parameters that affect crystal structure transformations and the nature and magnitude of their effects. The transformation of anatase to rutile is a reconstructive transformation [4] involving changes in secondary coordination. The rate of transformation of spectroscopically pure anatase [2] is exponential, with an energy of activation ~ 90 kcal/mole. It has been shown that the ratio of the propagation rate constant to the nucleation rate constant is small and that the energy of activation is utilized mainly for the random production of nucleation sites.[5] We have now investigated the changes in particle size, crystallite size and lattice dimensions accompanying this crystal structure transformation and also the effect of particle size and surface area on the transformation. Such a study is considered important for understanding the features and mechanism of crystal-structure transformations. Anatase of the highest purity has been used in this investigation, since impurities are known to have marked effect on the transformation.[6, 7]

It is also known that the transformation characteristics depend on the method of preparation of anatase and the major cause for the wide variation in the transformation temperature reported in the literature $(400\text{-}1000°C)$[8-11] is probably due to impurity

---

* Contains material from the Ph.D. thesis of S. R. Yoganarasimhan to be submitted to Indian Institute of Science under the guidance of C. N. R. Rao. Part 1, ref. (2); part 2, ref. (3).

† To whom all the correspondence should be addressed.

1580 CRYSTAL STRUCTURE TRANSFORMATIONS

effects. Thus the transformation of anatase prepared by the hydrolysis of titanium sulphate occurs at much higher temperature than that of pure anatase possibly due to the presence of sulphate ion impurity in the sample. The effect of impurity atoms is by no means specific to anatase-rutile transformation and is found in several systems such as aragonite + calcite, marcosite + pyrite, sphalerite + wurtzite, etc. Although the impurity effects on phase transformations have been noted qualitatively, no detailed investigations on the nature and magnitude of their effects have hitherto been reported. We have now investigated, in detail, the effect of sulphate ion impurity on the kinetics of the anatase-rutile transformation.

EXPERIMENTAL

Spectroscopically pure peroxy titanium oxide was prepared by the method described in the literature,[12] using spectrographically standardized titanium powder supplied by Johnson and Matthey, London, England. Anatase was then obtained by the thermal decomposition of the peroxy compound. Differential thermal analysis of the peroxy compound was carried out using an apparatus described by Pask and Warner [13] (fig. 1a).

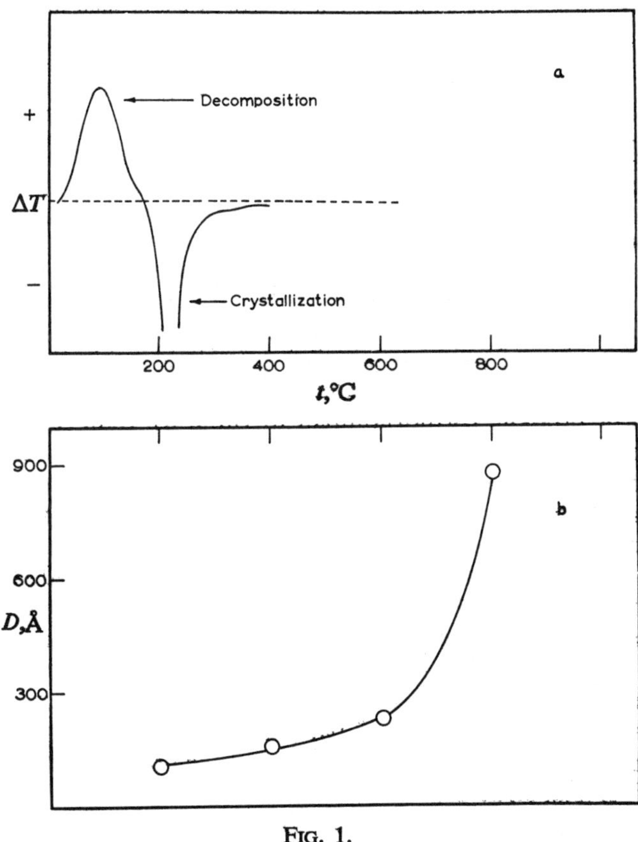

Fig. 1.

(a) Differential thermal analysis curve of peroxy titanium oxide.
(b) Variation of crystallite size with temperature.

Particle size was studied by means of photomicrographs taken with an Earnst Leitz Wetzlar unit with a magnification of 100. Surface areas of the samples were determined by employing a volumetric apparatus described by Srinivasan.[14] A North American Philips X-ray diffractometer equipped with a Geiger-counter recording device was employed for the

X-ray line-width measurements, while a Rich Seifert Debyeflex unit was used for the other powder patterns. A North American Philips back-reflection camera was employed for the precision determination of lattice constants. The uncertainty in the lattice constants varies between ± 0.001 and ± 0.0025 Å. Anatase samples were heated in quartz tubes placed in a vertical tubular furnace fitted with a thermo-regulator. The fraction of anatase in a mixture was determined by the equation,

$$f_A = (1 + 1.26 \, I_R/I_A)^{-1}, \tag{1}$$

where I_R and I_A are the intensities of rutile ($d = 3.245$ Å) and anatase ($d = 3.51$ Å) lines respectively. The approximate uncertainty in the energy of activation of the transformation is ± 10 kcal/mole.

Titanium sulphate was prepared by digesting titanium hydroxide, obtained by hydrolysis of pure titanium tetrachloride, in sulphuric acid. A sample of anatase was prepared by hydrolyzing titanium sulphate solution with ammonia. The precipitate was washed thoroughly till the washings gave no test for sulphate ion. The precipitate was dried at $\sim 100°$C. Such precipitates were, however, found to contain some sulphate impurity.

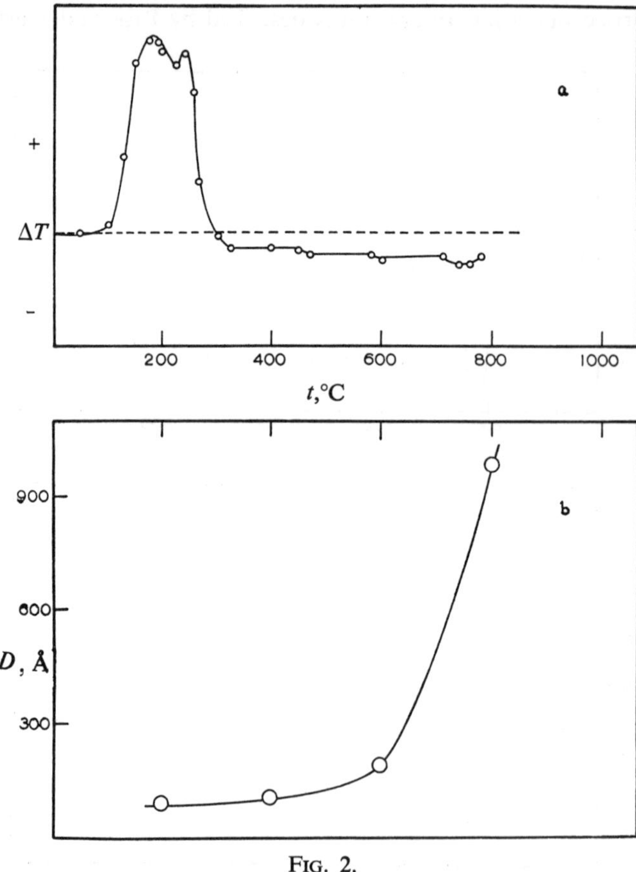

Fig. 2.

(a) Differential thermal analysis curve of titanium dioxide obtained from sulphate hydrolysis.
(b) Crystallite size variation of anatase obtained from sulphate hydrolysis.

Quantitative estimation of the sulphate content by the fusion method was found to be difficult because of the low percentage of the impurity. The anatase, thus prepared, was amorphous. The surface area of this anatase sample (B.E.T.) was ~ 54 m²/g and the differential thermal analysis curve of the anatase sample is shown in fig. 2a. Although no exothermic peak due to crystallization was observed, the endothermic peak shows a definite splitting around

1582   CRYSTAL STRUCTURE TRANSFORMATIONS

250°C which may be due to crystallization.[15]   The sample, in fact, gives an X-ray pattern (though ill-defined) when heated at $\sim 250°C$.   The crystallite size is very low ($\sim 100\,\text{Å}$) and increases markedly when heated to higher temperatures (fig. 2b) just as for pure anatase (fig. 1b).   The kinetics of the transformation of this anatase sample were studied at different temperatures.

Anatase samples doped with 0·1, 1 and 5 atom % sulphate ion impurity were prepared by the method reported earlier.[6]   Spectrographically standardized titanium metal powder supplied by Johnson and Matthey, London, was used for the purpose.   The lattice constants of these anatase samples remained the same within experimental error ($a = 3·771 \pm ·005\,\text{Å}$; and $c = 9·498 \pm ·013\,\text{Å}$).

RESULTS AND DISCUSSION

The amorphous peroxy titanium oxide, on heating, lost the peroxy oxygen and water around 100°C.   The resulting oxide was poorly crystalline.   The differential thermal analysis of peroxy titanium oxide (fig. 1a) showed an endothermic peak

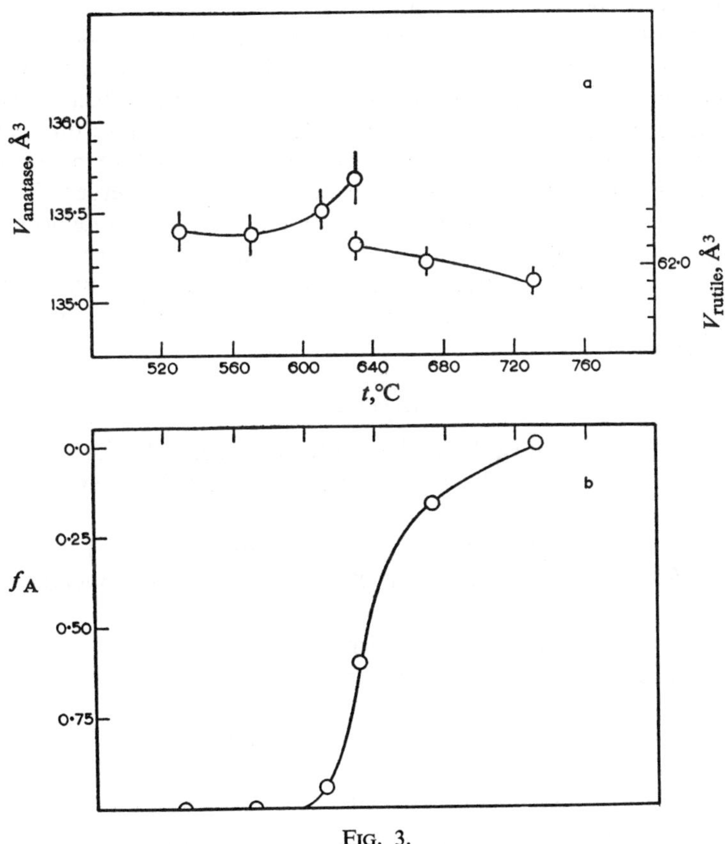

FIG. 3.

(a) Variation of unit cell volumes of anatase and rutile with temperature.

(b) Variation of fraction of anatase at different temperatures.

around 100°C due to the decomposition reaction and a very large exothermic peak around 250°C.   While the oxide was very poorly crystalline before the exothermic peak, it gave a good X-ray diffraction pattern characteristic of crystalline anatase, just after passing the peak.   The exothermic reaction (fig. 1a) is clearly associated with the crystallization reaction.[15]

In order to study the growth of particles with temperature, anatase powder (preheated to 150°C) was heated for a period of 3 h at 400, 600, 800 and 1000°C. Marked increase in particle size was noticed in the 600-1000°C region, as indicated by the photo-micrographs. The specific surface area (B.E.T.) of anatase heated at 400°C was ~ 55 m²/g and decreased markedly for samples heated to higher temperatures. The crystallite size normal to the (101) and (110) reflecting planes of anatase and rutile samples was calculated by measuring the X-ray diffraction line-widths of the samples heated at 200, 400, 600, 800 and 1000°C for 3 h. The Scherrer equation corrected for instrumental line-broadening by Warren's equation was employed for the calculation.[16] The line-width of the sample heated at 1000°C was taken as the reference. The crystallite size increases rapidly after 600°C (fig. 1b). The transformation of pure anatase also starts only above 600°C.

It was considered interesting to see if lattice dimensions change during the transformation of anatase to rutile. To study such an effect, lattice constants of anatase heated for 3 h at different temperatures in the transformation region were determined by taking Debye-Scherrer X-ray patterns at 25°C and subjecting the data to Cohen's extrapolation technique followed by least-square treatment. The results are presented in fig. 3a, where the unit cell, volumes V of anatase and rutile are plotted against temperature. The composition at each temperature is indicated by the curve in fig. 3b. The results in fig. 3a seem to indicate that the anatase unit cell expands before transforming to rutile. An expanded rutile unit cell then contracts with increase in temperature. It is interesting to note that Schossberger[11] had noticed that the density (pyknometric) of anatase underwent a small but finite decrease during transformation to rutile. The expansion of the anatase lattice prior to transformation may be taken as an indirect indication that disordered lattices favour this crystal-structure transformation. The anatase-rutile transformation may be visualized as shown in scheme below.

Scheme

	anatase I; collapsed form (ordered lattice)
~ 620°C		Displacive transformation (Buerger [4]).
	$\downarrow$
	anatase II; open form (disordered lattice)
$\sim 620\text{-}650$°C		Reconstructive transformation (Buerger [4]).
	$\downarrow$
	rutile II; open form (disordered lattice)
above ~ 650°C		Displacive transformation (?)
	$\downarrow$
	rutile I; collapsed form (ordered lattice).

EFFECT OF PARTICLE SIZE ON THE TRANSFORMATION

Anatase with relatively large particle size was prepared by heating the freshly prepared oxide to 400°C and a sample of particle size $74 > P_1 > 53$ microns was obtained by employing Tyler standard sieves ($-200 + 270$ mesh). These particles were crushed to finer particles and sieved (mesh sizes $-270 + 325$ and -325) to provide anatase samples (P_2 and P_3) with particle sizes $53 > P_2 > 43$ and $P_3 < 43$ microns respectively. Since the thermal history of all the three samples is the same, it was assumed that any changes in transformation characteristics can be attributed to particle-size effects. The kinetics of transformation of the three samples were studied at 700°C and the transformation curves shown in fig. 4a, clearly indicate that smaller particle size of

     CRYSTAL STRUCTURE TRANSFORMATIONS

anatase favours the transformation.   The values of the energy of activation of the transformation for P_1 and P_3, however, are found to be about the same within the limits of experimental error (~ 114 and ~ 107 kcal/mole respectively for P_1 and P_3). Recently, larger particle size has been found not to favour quartz-cristobalite transformation.[17]   The effect of particle size on the transformation is possibly due to dislocations, produced during the grinding operation, acting as nucleation sites.

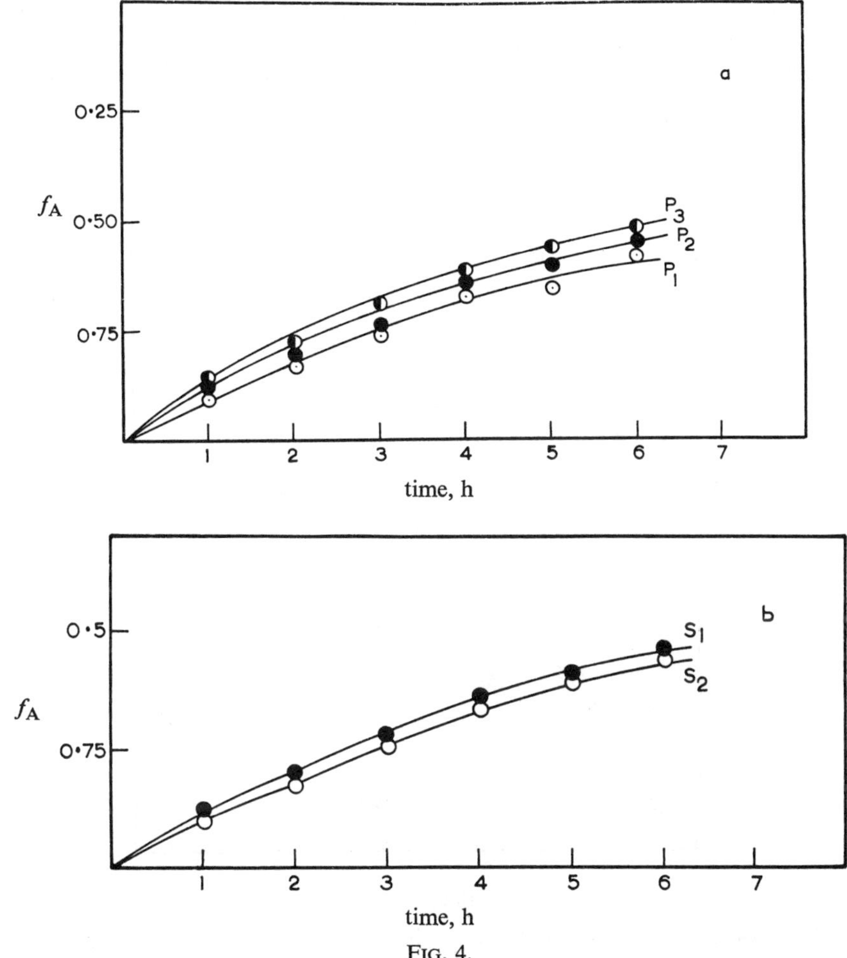

FIG. 4.

(*a*) Transformation curves for anatase samples of different particle sizes at 700°C:
   $P_1 = 74 > P_1 > 53$ microns, $P_2 = 53 > P_2 > 43$ microns, and $P_3 < 43$ microns.

(*b*) Transformation curves of anatase samples of two different surface areas at 700°C: S_1, 55 m²/g and S_2, 25 m²/g.

EFFECT OF SURFACE AREA ON THE TRANSFORMATION

An anatase sample P_2 of specific surface area ~ 55 m²/g was heated at 600°C for 3 h and the specific surface area was found to decrease to ~ 25 m²/g.   The transformation characteristics of these two samples were studied (fig. 4*b*).   Since the growth in the crystallite size is negligible in the 400-600°C region and the particle size is nearly the same for the two samples, the observed effect may be interpreted as due to the effect of surface area.   It can be seen from fig. 4*b* that the effect of surface area is not very marked and, if any, larger surface area favours the transformation. While the energy of activation for the transformation of the sample of high surface area is ~ 110 kcal/mole, that for the lower surface area sample is ~ 102 kcal/mole.

EFFECT OF IMPURITIES ON THE TRANSFORMATION

The kinetics and energetics of the transformation of anatase prepared by the hydrolysis of titanium sulphate are markedly different from those of pure anatase. The transformation of anatase prepared by sulphate-hydrolysis becomes immeasurably slow below $\sim 695 \pm 10°$C, compared to $610 \pm 10°$C [2] for pure anatase (fig. 6). While the transformation of pure anatase follows the exponential rate law, that of anatase

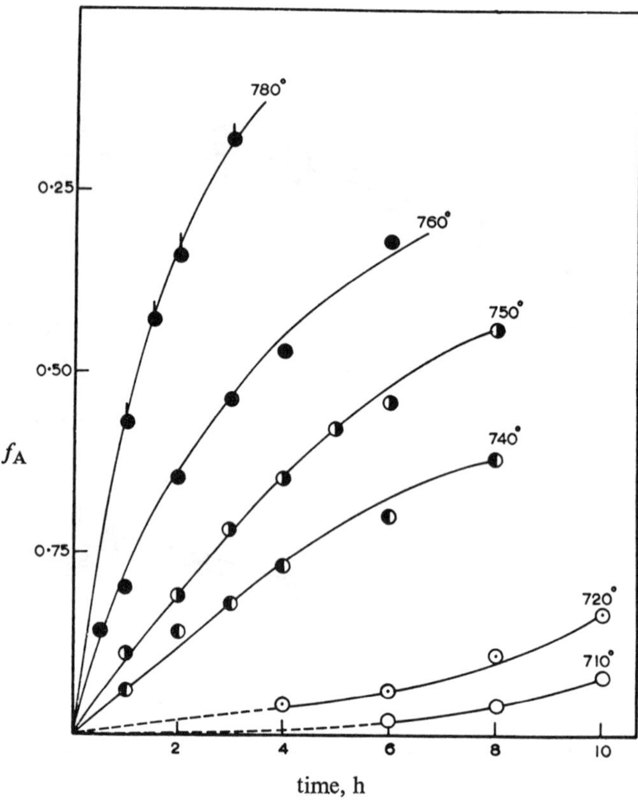

FIG. 5.—Plots of f_A as a function of time at different temperatures (sulphate hydrolysis).

from sulphate hydrolysis shows a considerable induction time at lower temperatures (fig. 5). The induction period decreases with increase in temperature and at sufficiently high temperatures ($> 740°$ for this sample) the rate becomes exponential (fig. 5 and 7).

These observations can be represented as a special case of the general rate equation derived by the application of order-disorder theory to diffusionless transitions in solids.[3] According to this equation, the shape of the rate curve is determined by the relative numerical values of zk_p/k_n and of c. The larger the factor k_p/k_n is relative to c, the more sigmoidal the curves become. This is understandable since the propagation effect which is responsible for the autocatalytic character of the transformation becomes more noticeable when k_p/k_n is large and c small. Under these conditions some time elapses before a sufficient number of nucleation sites are formed ; then the

    CRYSTAL STRUCTURE TRANSFORMATIONS

propagation becomes dominant.   The duration of the induction period itself varies depending on the system.   The observation of induction period in the rate curves for the transformation of anatase prepared by sulphate hydrolysis may thus be interpreted as due to a relatively large ratio of propagation rate constant to nucleation rate constant.

The decrease of induction time with increase in temperature is probably due to the greater sensitivity of the nucleation process to temperature changes.   The nucleation process should, therefore, be associated with a high energy of activation E_a.   This is indeed confirmed by the calculation of E_a from the rate data presented in

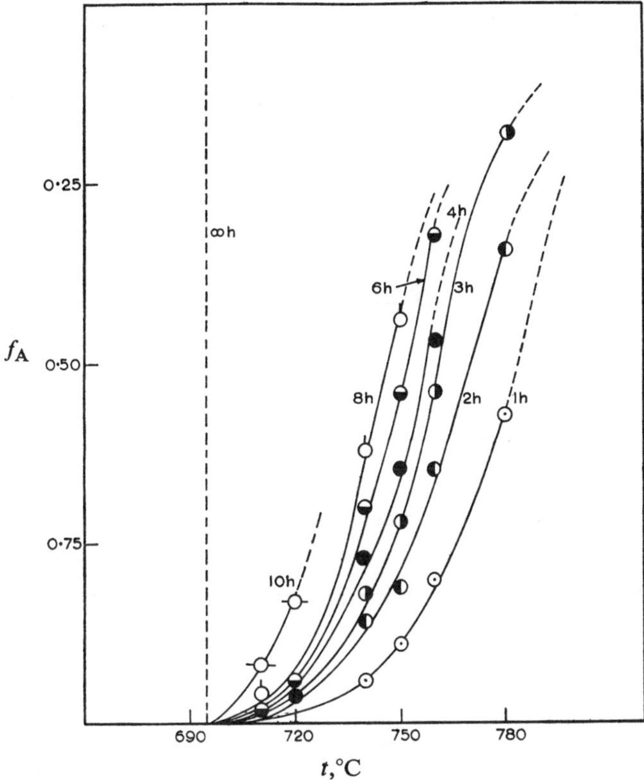

FIG. 6.—Transformation of anatase obtained from sulphate hydrolysis as a function of temperature at different times.

fig. 5.   The E_a of transformation for the rate data at high temperatures (where the rates are exponential) is $\sim 120 \pm 10$ kcal/mole.   From the rate data at lower temperatures, two values of E_a can be calculated; ~ 215 kcal/mole for the induction-time region (fig. 8) and ~ 40 kcal/mole after the induction-time region.   The observation that the E_a for the rate data in the induction-time region is much greater in magnitude than that for the data in the region where the rates are exponential seems to support the earlier postulate [2] that E_a mainly represents the energy for establishing nucleation sites.   The mean of these two values of E_a is nearly equal to the observed energy of activation for the rate data at higher temperatures.   These results are diagrammatically represented in fig. 8.   It is, therefore, thought that the exponential rate curves observed in crystal structure transformations [2,3] represent a combination of two processes, viz., the nucleation process associated with a very large E_a and the propagation process associated with a relatively low E_a.

S. R. YOGANARASIMHAN AND C. N. R. RAO

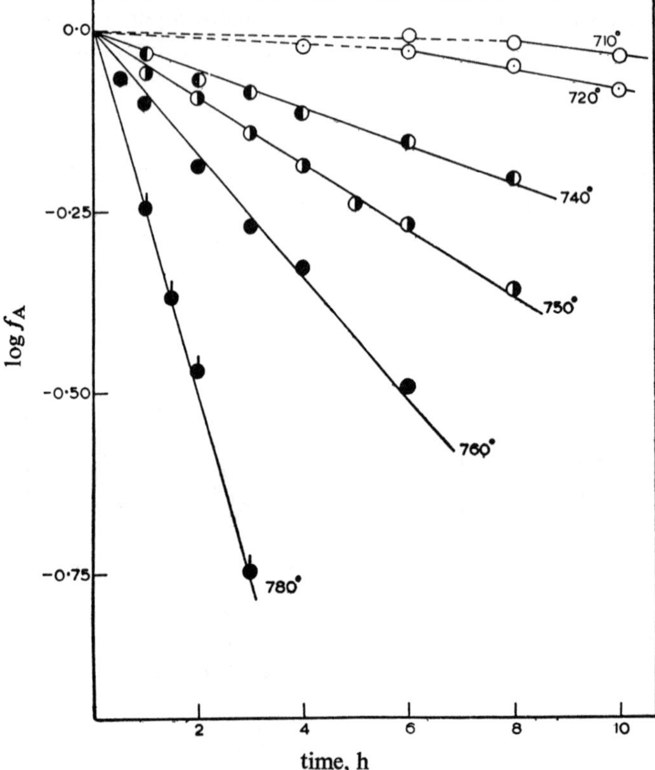

FIG. 7.—First-order plots of the data in fig. 5.

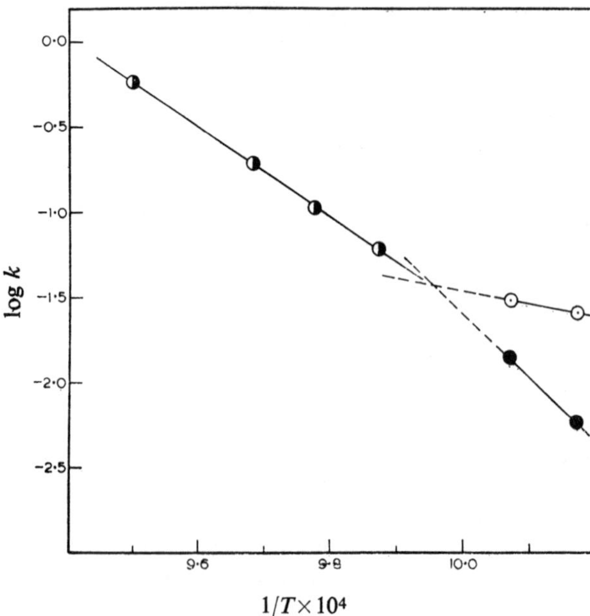

FIG. 8. —Arrhenius plot of the data in fig. 5.

1588   CRYSTAL STRUCTURE TRANSFORMATIONS

The marked difference in the kinetics and energetics of the transformation of anatase prepared by sulphate hydrolysis as compared to that of pure anatase may be ascribed to the decelerating effect of sulphate impurity. It is interesting to note that E_a of the transformation of anatase prepared by sulphate hydrolysis is considerably higher than that of pure anatase ($\sim 90 \pm 10$ kcal/mole). The widely different temperatures reported in the literature [8-11] for the transformation of anatase samples prepared by sulphate hydrolysis are considered to be due to the variation in the extent of washing of the precipitate (resulting in the variation in the concentration of sulphate impurity in the samples). In order to substantiate further that the observed effects are due to the presence of varying amounts of sulphate impurity, transformation of anatase samples doped with sulphate to different extents [6] (0·1, 1 and 5 atom %) has been studied (fig. 9). The sulphate ion progressively inhibits the transformation with

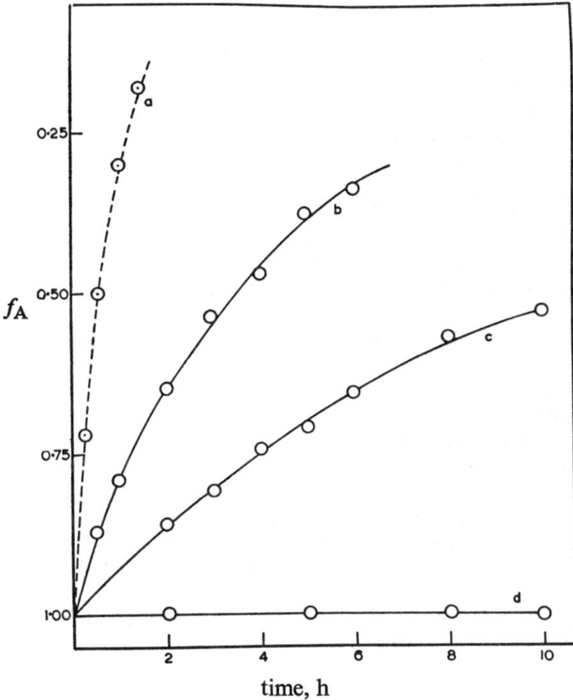

FIG. 9.—Influence of sulphate ion impurity on the anatase-rutile transformation.

(*a*) spectroscopically pure,　　　　　(*b*) 0·1 % sulphate-doped,
(*c*) 1 % sulphate-doped,　　　　　(*d*) 5 % sulphate-doped.

increasing concentration. The E_a also increases with increase in the impurity concentration. Thus, the values of E_a for the transformations of pure 0·1 % sulphate doped and 1 % sulphate doped samples of anatase were ~ 90, ~ 100, and ~ 120 kcal/mole respectively.

The results discussed show that crystal structure transformations are considerably dependent on the thermal history of the samples ; to be more specific, the crystallite size, particle size and surface area have measurable effects on the transformation. It would, therefore, probably be difficult to reproduce strictly transformation data with different samples. The magnitudes of these effects are, however, not too great to result in the wide variability of temperatures of polymorphic transformations. The wide variations in transformation temperature can only be due to other factors such as impurity effects.   Impurities not merely affect the transformation temperatures,[6,7] but also profoundly affect the basic mechanism(s) underlying the transformation.   The sulphate ion impurity decelerates the anatase-rutile transformation and markedly affects the rates of nucleation and propagation processes.   However, the mechanism by which the sulphate ion impurity affects the transformation characteristics is not clear.   In order to understand the mechanism of impurity effects, detailed investigations on the effects of various anionic and cationic impurities are presently in progress in these laboratories.

The authors are thankful to Prof. M. R. A. Rao for his interest and to Dr. P. A. Faeth for his assistance in some of the work.   One of them (S. R. Yoganarasimhan) is indebted to the University Grants Commission of India for the award of a research Fellowship.

[1] Czanderna, Rao and Honig, *Trans. Faraday Soc.*, 1958, **54**, 1069.

[2] Rao, *Can. J. Chem.*, 1961, **39**, 498.

[3] Rao, Yoganarasimhan and Faeth, *Trans. Faraday Soc.*, 1961, **57**, 504.

[4] Buerger, *Phase Transformations in Solids*, ed. Smoluchowski (John Wiley, New York and Chapman and Hall, London, 1957), chap. 6.

[5] Rao and Honig, *Symp. Rate Processes in Physico-chemical Reactions* (Bombay, India, Jan. 1960), to be published in the special volume of the *J. Indian Chem. Soc.*

[6] Rao, Turner and Honig, *J. Physic. Chem. Solids*, 1959, **11**, 173.

[7] Rao and Lewis, *Curr. Sci. (India)*, 1960, **29**, 52.

[8] Barksdale, *Titanium: its Occurrence, Chemistry and Technology* (Ronald Press Co., New York, 1949).

[9] Skinner, Johnston and Beckett, *Titanium and its Compounds* (Herrick L. Johnston Enterprises, 1954), pp. 22-23.

[10] Gmelin, *Handbuch der anorganischen chemie* (Gmelin Institute, Verlag Chemie, Weinheim, Germany, 1951), **41**, pp. 233-34.

[11] Schossberger, *Z. Krist.*, 1942, **104**, 358.

[12] Czanderna, Clifford and Honig, *J. Amer. Chem. Soc.*, 1957, **79**, 5407.

[13] Pask and Warner, *Bull. Amer. Ceram. Soc.*, 1954, **33**, 168.

[14] Srinivasan, *Proc. Indian Acad. Sci. A*, 1957, **46**, 123.

[15] Rao, Yoganarasimhan and Lewis, *Can. J. Chem.*, 1960, **38**, 2359.

[16] Klug and Alexander, *X-Ray Diffraction Procedures* (John Wiley and Sons Inc., New York, 1954), chap. 9.

[17] Roberts, *Kinetics of High Temperature Processes*, ed. Kingery (Technology Press and John Wiley, New York and Chapman and Hall, London, 1959), chap. 27.

For a study of cubic-monoclinic transformation of ZrO_2 (xerogels), see G.N. Subbanna and C.N.R. Rao, <u>Eur. J. Solid State Chem.</u> <u>26</u>, 7 (1989)

JOURNAL OF SOLID STATE CHEMISTRY **68**, 193–213 (1987)

Computer Simulation of Transformations in Solids*

C. N. R. RAO† AND S. YASHONATH

Solid State and Structural Chemistry Unit, Indian Institute of Science, Bangalore 560 012, India

Received August 18, 1986

Recent developments in molecular dynamics (MD) and Monte Carlo (MC) methods enable us to fruitfully investigate transformations in solids by employing appropriate potentials. The possibility of varying both the volume and the shape of the simulation cell in these simulation techniques is especially noteworthy. In this article we briefly describe some of the highlights of the recent MD and MC methods and show how they are useful in the study of transitions in monatomic solids, ionic solids, molecular solids (especially orientationally disordered solids), and glasses. The availability of reliable pair potentials will undoubtedly make these methods more and more useful for studying various aspects of condensed matter in the years to come. © 1987 Academic Press, Inc.

1. Introduction

Computer simulation is being used increasingly in diverse areas of science in the past few years. It has also emerged to become one of the powerful means for investigating condensed matter (*1*). The principal tools employed in computer simulation are the Monte Carlo and the molecular dynamics methods. In these methods, properties of a collection of particles, usually between 30 and 1000 in number, interacting via a potential $\phi(r)$ are obtained numerically. Reliable estimates of equilibrium and transport properties as well as microscopic properties can be obtained from such calculations.

In the Monte Carlo method, one performs stochastic averaging in the configuration space by means of the Metropolis importance sampling scheme (*2*). Monte Carlo calculations are generally performed in the canonical ensemble, the isothermal isobaric ensemble, or the grand canonical ensemble. In the canonical or the NVT ensemble calculations, the number of particles, N, the volume of the simulation cell, V, and the temperature, T, are constant during the course of the simulation. In isothermal isobaric or the NPT ensemble calculations, the number of particles, the pressure, P, and the temperature, T, are held constant. In the molecular dynamics method, Newton's equations of motion are solved numerically (*3*). The energy, E, of the system is conserved over the generated trajectory and the average of any property over this trajectory corresponds to the average in the microcanonical or the NVE ensemble.

There have been some significant advances in the methods of Monte Carlo and

---

* Contribution No. 356 from the Solid State and Structural Chemistry Unit.

† To whom all correspondence should be addressed.

molecular dynamics in the last few years. The pioneering contribution of Andersen (4), provides a method for performing molecular dynamics calculations at constant pressure or constant temperature. The method enables the generation of trajectories corresponding to the isoenthalpic isobaric and isothermal isobaric ensembles. This is made possible by coupling the temperature and the pressure of the system to those of a bath. Another significant development is due to Parrinello and Rahman (5), who allowed for variation not only in the volume but also in the shape of the simulation cell in constant-pressure calculations. The modified molecular dynamics methods have been suitably employed by Nosé and Klein (6) for the study of polyatomic molecules. The Monte Carlo method has been modified by Yashonath and Rao (7) to include the variation in the shape of the simulation cell. These developments have rendered the molecular dynamics and the Monte Carlo methods most useful for the investigation of a variety of phenomena exhibited by condensed matter. The methods are important because they can be employed not only for the study of stable phases but also to examine systems undergoing transformations. The methods are therefore ideally suited to investigate microscopic as well as macroscopic changes accompanying phase transitions in solids.

In this article, we shall discuss studies of phase changes in solids carried out by the application of the generalized Monte Carlo and the molecular dynamics methods. This topic is of particular significance because of the recent modifications of the method to include both variation in size and shape of the simulation cell. What is especially gratifying is that we are now able to make meaningful predictions of phase transitions in real solids by employing reliable pair potentials. Besides phase transitions of molecular solids, we shall examine phase transi-

tions in monatomic and ionic crystals. Some of the molecular systems discussed are CCl_4, CF_4, and adamantane. We shall present recent results on the glass transition obtained by the two methods, a noteworthy feature being the simulation of a real molecular glass formed by isopentane interacting via a realistic potential. The simulation study on isopentane glass has thrown some light on the structure of the glassy state. Andersen's cooling experiments on a Lennard-Jones fluid are also significant in this regard. We shall also indicate in the article the scope for further investigations of solids by computer simulation. Before discussing the various types of phase transitions of solids, we shall briefly review the recent developments in the Monte Carlo and molecular dynamics methods.

2. Recent Developments in Molecular Dynamics and Monte Carlo Methods

There are many excellent reviews on the standard molecular dynamics method dealing with calculations in the microcanonical ensemble as well as on the Monte Carlo method involving calculations in the canonical, isothermal isobaric, and grand canonical ensemble (8). In the present article, we shall limit ourselves exclusively to those developments that have taken place since the work of Andersen (4). In the molecular dynamics method, the developments are the constant-pressure, constant-temperature, constant-temperature–constant-pressure, variable shape simulation cell MD, and isostress calculations; in the Monte Carlo method, it is the variable shape simulation cell calculation.

2.1. Molecular Dynamics Methods

Constant-pressure calculations. In the traditional molecular dynamics calculations, the equations of motion for the coordinates r_1, r_2, . . . , r_N of the N particles confined to a cell of fixed volume V are solved numerically. This corresponds to constant volume and constant energy or the microcanonical ensemble. In the constant-pressure calculations, the volume is considered to be a dynamical variable in addition to these (4). The well-defined lagrangian

$$\alpha = \frac{1}{2} \sum_i m_i \dot{\mathbf{s}}_i \cdot \dot{\mathbf{s}}_i - \sum_{i<j} \phi(r_{ij})$$
$$+ \frac{1}{2} M \dot{U}^2 - PU, \quad (1)$$

first utilized by Haile and Graben (9), is employed for the purpose. Here, m_i is the mass of the ith atom and the particle–particle interaction potential is denoted by $\phi(r)$. The scaled coordinate $\mathbf{s}_i$ of the ith particle and $\mathbf{r}_i$ are related by

$$\mathbf{s} = U^{1/3}\mathbf{r}. \quad (2)$$

If U is taken to be the volume V of the cell, the last two terms represent respectively the kinetic and the potential energies associated with motion of the simulation wall whose mass is M. The parameter P is the uniform hydrostatic external pressure acting on the simulation cell wall.

The lagrangian equations of motion

$$\frac{d}{dt}\left(\frac{\partial \alpha}{\partial \dot{x}}\right) = \frac{\partial \alpha}{\partial x}, \quad (3)$$

where x is a coordinate, are solved numerically to obtain the particle coordinate and momenta as a function of time t.

Andersen (4) has shown that the average of any function, F, over the trajectory thus generated is equal to the ensemble average for the isoenthalpic isobaric ensemble, i.e.,

$$\bar{f} = \bar{f}(N, P, H). \quad (4)$$

The enthalpy H is conserved over this trajectory and the external pressure corresponds to the value of P.

The value for the quantity M is chosen to approximately equal $U^{1/3}$ divided by the speed of sound in the system being simu-

lated. Equilibrium properties of the system are independent of M and dynamical properties are dependent on M through the time scale of the volume fluctuations, the volume changes depending on the imbalance between the internal and the external pressure. The effect of M on single-particle dynamical properties seems to be small as indicated by the work of Nosé and Klein (6).

Constant-temperature calculations. Andersen (4) also proposed a constant-temperature method in which the energy fluctuates during the simulation. The lagrangian in this method is the same as that in the standard MD method. However, it differs from the standard MD method in that the particles undergo random stochastic collisions. The collisions are considered to be instantaneous and bring about a change in the momentum of the atom. The number of collisions and outcome of the collisions is determined by ν, the mean rate at which each particle suffers a collision and T the required temperature at which the simulation is to be performed. Particles are chosen randomly and a new value of momentum is assigned while the momentum and coordinate of all the other particles are unaltered. The new value for the momentum of the particle undergoing collision is chosen at random from the Boltzmann distribution corresponding to the temperature T of the simulation. The average of any property calculated from this trajectory equals the canonical ensemble average of that property for the specified values of (T, V, N) provided the Markov chain generated is irreducible (4). At low temperature and high densities, the Markov chain may not satisfy the irreducibility condition as in the tradiional MD and MC techniques.

Equilibrium properties are independent of the value of ν, but the dynamical properties are not the same for two different values of ν. It is necessary to choose the value of ν within a small range (10). If the value is too high, the diffusion coefficient decreases, and if it is too small, large fluctuations occur in the kinetic energy and hence the temperature of the system. An appropriate choice for ν would be that resulting in a time for the decay of energy fluctuations approximately equal to that expected for a small volume of liquid in a large volume. Andersen (4) has discussed the method to estimate the optimum value of ν.

Nosé (11) has described another alternative formulation for performing calculations under constant-temperature conditions wherein an additional degree of freedom is introduced to permit energy fluctuations. The lagrangian for this system is given by

$$\alpha = \frac{1}{2} \sum_i m_i \dot{\mathbf{r}}^2 g^2 - \sum_{i<j} \phi(r_{ij})$$
$$+ \frac{Q}{2} \dot{g}^2 - (f + 1)kT_0 \ln g, \quad (5)$$

where the last two terms give the kinetic and potential energies associated with g, T_0 is the temperature at which simulation is carried out, and f is the number of degrees of freedom associated with N atoms. The velocities are scaled (11) as

$$\mathbf{v}_i = g\dot{\mathbf{r}}_i, \quad (6)$$

where $\mathbf{v}_i$ is the real velocity of particle i.

The physical system and the scaled system are related by Eq. (6). The scaling of velocities permits the exchange of heat between the simulated and the external heat reservoir. The equations of motion for the N atoms and the scaling factor g are solved numerically and averages calculated from the trajectory. The special choice $(f + 1)$ $kT_0 \ln g$ for the potential energy associated with g guarantees that the averages of equilibrium quantities calculated from the MD trajectory are the same as those in the canonical ensemble.

While Andersen's method for the constant-temperature calculation introduces energy fluctuations in the simulated system stochastically, Nosé's method achieves the same goal by the introduction of an additional degree of freedom; the scaling of velocities by a factor g can be interpreted as the scaling of time. This is similar to the scaling of coordinates in the constant-pressure method. The real time step $\Delta t'$ is given by

$$\Delta t' = \Delta t/g. \quad (7)$$

Thus, the real time steps, $\Delta t'$, are unequal in Nosé's formulation. As in constant-pressure calculations, the value of Q, in units of energy (time)^2, determines the dynamical quantities, although the equilibrium quantities are independent of the value of Q. The procedure for the selection of the optimum value for Q has been discussed by Nosé (11). It is believed that single-particle dynamics are less sensitive to the value of Q.

In the formulation of Nosé, the total hamiltonian, the total momentum, and total angular momentum are conserved. The conserved quantities help to monitor the precision of the calculation and also provide a powerful criterion to check for the correctness of the program.

Other methods for performing constant-temperature molecular dynamics calculations have been proposed recently. Evans (12) has introduced an external damping force in addition to the usual intermolecular force in order to keep the temperature constant in the simulation of a dissipative fluid flow. In another method, Haile and Gupta (13) have imposed the constraint of constant kinetic energy on the lagrangian equations of motion to perform calculations at constant temperature.

Constant-temperature--constant-pressure calculations. The trajectory for the atoms and volume are generated according to the solution of the equations of motion for the lagrangian (1) discussed earlier in connection with constant-pressure calculations. In addition, stochastic collisions are introduced to allow for fluctuations in enthalpy (*4*). The new momentum is sampled randomly from the Boltzmann distribution

$$\exp(-\pi_i \cdot \pi_i / 2m_i V^{2/3} kT), \qquad (8)$$

where π_i is the momentum conjugate to s_i. The average over this trajectory obtained thus equals the ensemble average of the property in the isothermal isobaric ensemble.

Another MD formulation for constant-temperature and -pressure calculations based on the constant temperature formulation of Nosé (*11*) has been described recently. The stochastic collisions which permit energetic fluctuations has been replaced by the dynamic method of scaling of velocities of the atoms, in addition to the scaling of velocities by $V^{1/3}$. The method, which is completely dynamical, requires one to choose appropriate values for Q and M which respectively determine the time scale of the temperature and volume fluctuations.

Parrinello–Rahman method. The constant pressure MD formulation of Andersen (*4*) provides a method for permitting the variation in the volume of the simulation cell. Parrinello and Rahman (*5, 14, 15*) generalized Andersen's method to include variations in the volume as well as the shape of the simulation cell. In the Parrinello–Rahman formulation, the simulation cell is represented by three vectors, **a**, **b**, and **c**. The lagrangian is given by

$$\alpha = \frac{1}{2} \sum_i^N m_i \dot{s}_i G s_i - \sum_{i<j} \phi(r_{ij})$$
$$+ \frac{1}{2} M \, \mathrm{Tr}(\mathbf{h}'\mathbf{h}) - PV, \qquad (9)$$

where the simulation cell is represented by $\mathbf{h} = \{\bar{\mathbf{a}}, \bar{\mathbf{b}}, \bar{\mathbf{c}}\}$ and the scaled coordinates by $s_i = \mathbf{h}^{-1}\mathbf{r}_i$. The metric tensor $G = \mathbf{h}'\mathbf{h}$.

The external hydrostatic pressure on the system is denoted by P. The equations of motion are

$$\ddot{s}_i = m_i^{-1} \sum_{i \neq j}^N \frac{1}{r_{ij}} \frac{d\phi(r_{ij})}{dr_{ij}} (s_i - s_j) - G^{-1}\dot{G}\dot{s}_i,$$
$$i, j = 1, 2 \ldots N$$
$$\ddot{\mathbf{h}} = W^{-1}(\pi - P)\sigma, \qquad (10)$$

where $\sigma_{ij} = \partial V/\partial h_{ij}$. The internal stress tensor π is given by

$$V\pi = \sum_i m_i \mathbf{v}_i \mathbf{v}_i + \sum_{i<j} \frac{1}{r_{ij}} \frac{d\phi(r_{ij})}{dr_{ij}}$$
$$(\mathbf{r}_i - \mathbf{r}_j)(\mathbf{r}_i - \mathbf{r}_j), \qquad (11)$$

the vector $\mathbf{v}_i$ being $\mathbf{h}\dot{s}_i$. The imbalance between the internal stress π and the external pressure P acting on the cell wall given by σ determines the motion of the cell **h**. This method has been appropriately modified for calculations on molecular systems by Nosé and Klein (*6*).

Andersen's formulation is useful in the study of fluid–solid transitions where the change in volume is significant. The generalization of Parrinello and Rahman which allows variations in both the volume and the shape of the simulation cell is most suited for the study of phase transformations in solids. In the case where only the volume is allowed to change, the periodic boundary conditions do not permit the solid to undergo structural transformations involving a change in crystal symmetry. The variable shape MD method has been modified to include external stress rather than uniform hydrostatic pressure (*16*). A generalization to the isothermal isotension ensemble calculation giving a somewhat different interpretation of the additional degree of freedom introduced in the constant-temperature and -pressure formulation of Nosé has been discussed recently by Ray and Rahman (*17*).

2.2. Modified Monte Carlo Method

The traditional MC method in the isothermal isobaric ensemble has been modified to include variations in size as well as the shape of the simulation cell (*7*). The

 RAO AND YASHONATH

average of any quantity, F, in the usual isothermal isobaric calculation is given by

$$F(\mathbf{r}^N, V)$$
$$= \frac{\int_0^\infty dV \int_V d\mathbf{r}^N f(\mathbf{r}^N, V) \exp[-\beta\Phi(\mathbf{r}^N)]}{\int_0^\infty dV \int_V d\mathbf{r}^N \exp[-\beta\Phi(\mathbf{r}^N)]}, \quad (12)$$

where $\mathbf{r}_1, \mathbf{r}_2, \ldots, \mathbf{r}_N$ are the coordinates of the N atoms and are represented by $\mathbf{r}^N$, $\beta = 1/kT$, and Φ is the total potential energy. The simulation cell is a cube with edge length $L = V^{1/3}$. In order to allow both the size and shape of the cell to vary during the simulation, the cell is defined as in the Parrinello–Rahman MD method by the three vectors $\mathbf{a}, \mathbf{b}, \mathbf{c}$ (7). The matrix $\mathbf{h} = \{\bar{\mathbf{a}}, \bar{\mathbf{b}}, \bar{\mathbf{c}}\}$ represents the cell. For purposes of integration, the scaled coordinates $\alpha_i = \mathbf{h}\mathbf{r}_i$ are employed rather than $\mathbf{r}_i$, where the components of α_i vary between 0 and 1. The average of any quantity f in the modified isothermal isobaric ensemble is given by

$$f([\mathbf{h}\alpha]^N, \mathbf{h}) = \frac{\int_0^\infty d\mathbf{h} \int_\mathbf{h} d[\mathbf{h}\alpha]^N f([\mathbf{h}\alpha]^N, \mathbf{h}) \exp(-\beta\Phi([\mathbf{h}\alpha]^N, \mathbf{h}))}{\int_0^\infty d\mathbf{h} \int_\mathbf{h} d[\mathbf{h}\alpha]^N \exp(-\beta\Phi([\mathbf{h}\alpha]^N, \mathbf{h}))}.$$

$$(13)$$

A trial move is made by displacing the particle and also the cell edge:

$$\alpha_i \to \alpha_i \pm \varepsilon_P \bar{\mathbf{R}}$$
$$\mathbf{a}_i \to \mathbf{a}_i \pm \varepsilon_c \bar{\mathbf{A}}, \quad (14)$$

where ε_P and ε_c are random numbers between 0 and 1 and $\bar{\mathbf{R}}$ and $\bar{\mathbf{A}}$ specify respectively the maximum particle and cell displacements.

It is noteworthy that Monte Carlo simulations involve simpler coding than molecular dynamics simulations. The Monte Carlo method therefore provides a convenient alternative to the molecular dynamics method if one is not interested in the dynamics of the system. In recent years many workers have felt the need for employing more complicated intermolecular potentials in order to predict accurately the properties of matter, especially those in the solid state. In such situations, the MC method is to be preferred as it is easier to code programs which incorporate complicated intermolecular potentials. Another advantage of the MC method is the ease with which the number of degrees of freedom can be altered in any calculation.

The above method is easily extended to perform calculations in isostress isothermal ensemble employing an isotropic stress tensor in place of a uniform hydrostatic pressure (18). The extension of the MC method to the case of polyatomic systems is straightforward (19). In the case of molecular systems, however, there are instances where the simulation cell rotates as a whole in space. For example, in the simulation of the low-temperature, high-pressure phase of carbon tetrachloride carried out by us, the simulation cell rotated in space (19). The variation of the magnitudes of $\mathbf{a}$, $\mathbf{b}$, and $\mathbf{c}$ and angles α, β, and γ of the simulation cell, as well as the variation of the cell components, is shown in Fig. 1. It is apparent from the figure that while the cell components change continuously during the course of a run, the magnitudes and angles of the vectors fluctuate around their average values. Similar rotation has been observed in a molecular dynamics study of the high-pressure phase of nitrogen by Nosé and Klein (6). Only six degrees of freedom are required (a, b, c, α, β, and γ) to define the cell. The rotation is a consequence of the three extra degrees of freedom associated with the simulation cell. The cell can be prevented from rotating by imposing a restriction on the cell so as to allow six degrees of freedom to be varied (19). Nosé and Klein (6) have proposed a method wherein the system is rotated to keep the

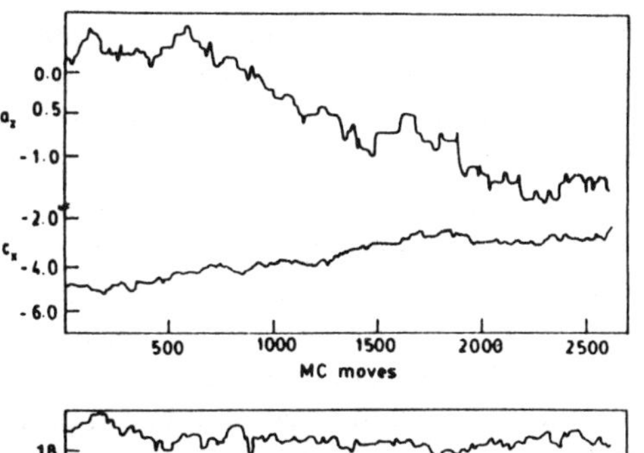

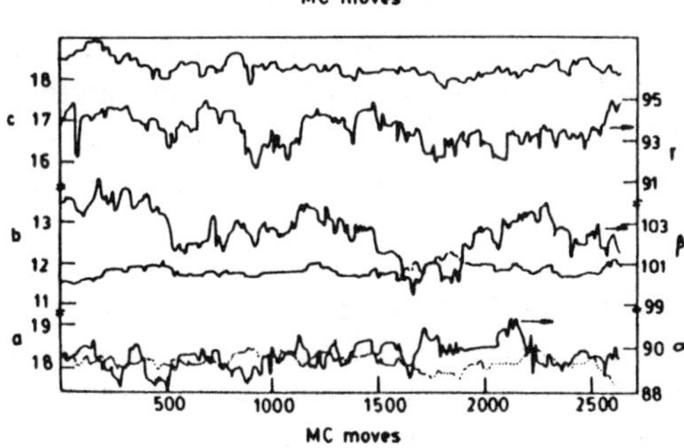

FIG. 1. Variation of the cell components c_x and a_z, the cell edges a, b, and c, and angles α, β, and γ during the first 2500 MC moves in the simulation of carbon tetrachloride, indicating the rotation of the cell. Angles are in degrees and the lengths in Å. (From Yashonath and Rao (19).)

growth of the antisymmetric component to zero after each step. While both methods can be used in the simulation of polyatomic systems, the former has the advantage that it is simpler to implement and also requires significantly less computer time.

2.3. General Remarks Regarding the MD and MC Methods

The MC and MD methods permitting the variation of the shape of the cell are best suited for the study of phase transitions in solids. These methods have been used to study phase transitions of a few solids in the last few years. Among these are monatomic solids such as rare gas solids, ionic solids, and molecular solids. There are, however, some inherent limitations in these methods. While certain transitions are readily investigated by these methods, others are more difficult. The b.c.c. to f.c.c. transformation of monatomic solids is an example of a transition that is readily observed (5, 7) (see Figs. 2 and 3). This transition has been observed as a function

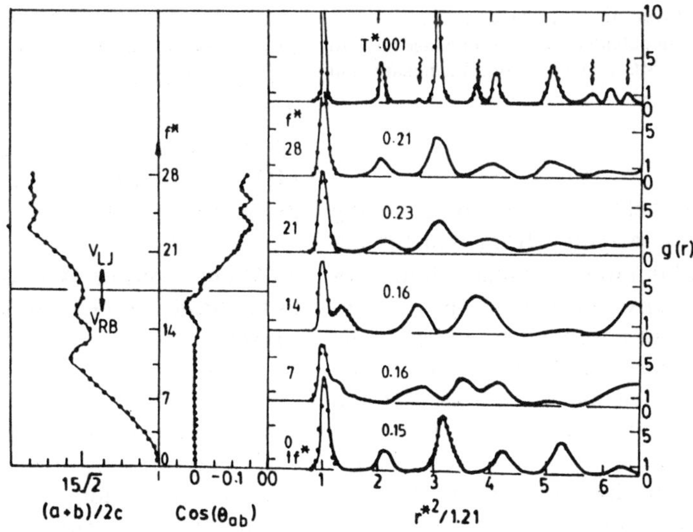

FIG. 2. Variation of the ratio $(a + b)/2c$ and $\cos \theta_{ab}$ (left), where θ_{ab} is the angle between the cell edges a and b, and the r.d.f.'s at various times (right) from the MD study of Parrinello and Rahman (5), showing the transition from f.c.c. to b.c.c. and from b.c.c. to f.c.c. due to a change of interaction potential between the particles.

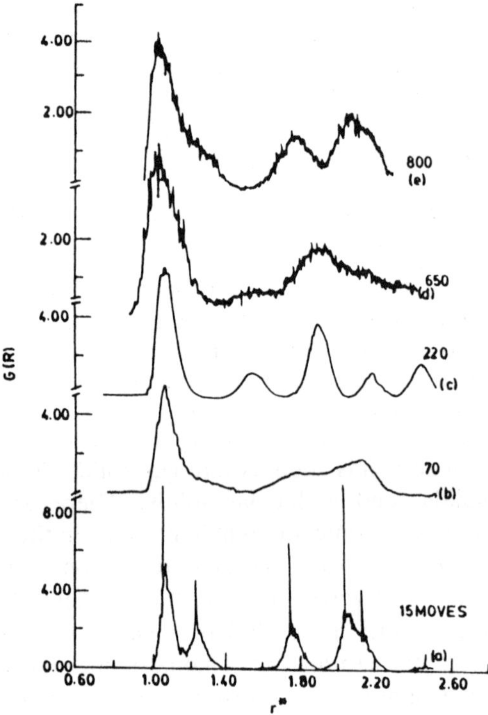

FIG. 3. Pair correlation functions as functions of reduced distance r^*. The b.c.c. arrangement transforms to f.c.c. at 220 MC moves when the Lennard-Jones potential is employed. If at this stage the interaction is changed to cesium potential, there is a change back to the b.c.c. at 450 MC moves. (From Yashonath and Rao (7).)

of pressure in helium and also on changing the interaction potential from that for cesium or rubidium to the Lennard-Jones potential. The transition from h.c.p. to f.c.c. is an example of a difficult transition. This transition requires a modification at the fourth shell of neighbors and both MD and MC calculations have failed to simulate the transformation (20, 21). There are also limitations imposed by the use of periodic boundary conditions and finite size effects (22).

In spite of the aforementioned limitations, the MD and MC methods can be most profitably employed in the investiga-

tions of a variety of systems. These methods provide fairly good approximations to average properties even when a relatively small number of particles (30 to 1000) are employed and are computationally quite economical. This is in contrast to the alternative method described in the literature (22, 23) which requires expensive hardware, considerable memory, and computer time, as it requires a much larger number of particles for the same accuracy. This method, however, has the advantage of being relatively free from effects due to finite size and periodic boundary conditions.

Calculations carried out by us (7) by the MC method where only the nine degrees of freedom were permitted and no particle displacements were allowed could successfully reproduce the f.c.c. to b.c.c. transition (Fig. 3). These and other considerations suggest that the MC method is to be preferred to the MD when only the equilibrium properties are of interest. For example, if one is interested in the determination of a phase diagram, the MC method would be more appropriate than the MD method. The MD method is preferable for obtaining dynamic properties even though it is not clear how one can obtain the exact dynamical behavior (4, 6). The technical details and aspects of programming for the MD and the MC methods are available (6, 21) and we shall not discuss them here.

3. Applications of MD and MC Methods to the Study of Transformations in Crystals

3.1. Monatomic Solids

Parrinello and Rahman (5) in their MD study investigated the relation between structure and interaction potential by employing two different potentials. The (6–12) Lennard-Jones potential which predicts the f.c.c. arrangement as the stable structure and the rubidium potential which has the stable b.c.c. structure were employed for the purpose. Starting with a f.c.c. structure, they found that the solid transformed to a b.c.c. structure on changing the particle interaction from the Lennard-Jones to the rubidium potential. The radial distribution functions (r.d.f.'s) show distinct peaks corresponding to a b.c.c. structure within a few hundred MD steps of imposing the rubidium potential (see Fig. 2). The cubic cell edges transformed into a rectangular parallelopiped by a decrease in the cell edge, $\vec{c}$. The particle interaction was then changed back to the Lennard-Jones. It was found that the system transformed back to the f.c.c. arrangement within a few hundred MD steps. This calculation illustrates the usefulness of the Parrinello–Rahman method for studies involving the structure–potential relationships. We have carried out an isobaric isothermal calculation employing a variable shape MC cell (7) and the results are indeed similar (Fig. 3).

Najafbadi and Yip (18) have investigated the stress–strain relationship in iron under uniaxial loading by means of a MC simulation in the isostress isothermal ensemble. At finite temperatures, a reversible b.c.c. to f.c.c. transformation occurs with hysteresis. They found that the transformation takes place by the Bain mechanism and is accompanied by sudden and uniform changes in local strain. The critical values of stress required to transform from the b.c.c. to the f.c.c. structure or vice versa are lower than those obtained from static calculations. Parrinello and Rahman (14) investigated the behavior of a single crystal of Ni under uniform uniaxial compressive and tensile loads and found that for uniaxial tensile loads less than a critical value, the f.c.c. Ni crystal expanded along the axis of stress reversibly.

3.2. Silver Iodide

Structural transformation in the superionic conductor silver iodide has been investigated by employing the modified molecular dynamics method in the constant-enthalpy–constant-pressure ensemble (24). A potential derived by considering the properties of both the β and the superionic α phases was employed for the purpose. The calculated structure, diffusion coefficient, etc., for α-AgI at 700 K were in agreement with the experiment (25). On cooling from 700 to 350 K, the b.c.c. lattice of iodine transformed to the close-packed h.c.p. structure accompanied by a marked decrease in D_{Ag}; the transformation was reversible on heating. Structural and dynamical properties as well as the transition temperature for the α-β transformation obtained from the MD study were in good agreement with experiment.

3.3. Molecular Crystals

Nitrogen. Nosé and Klein (26) carried out a MD simulation of the various phases of nitrogen using a simple Lennard-Jones 6–12 potential with $\varepsilon = 37.3$ K, $\delta = 3.31$ Å. Starting with the high-pressure room-temperature structure ($Pm3n$), results have been reported for $T = 300$ K and at lower temperatures and a presence of 70 kbar. The cell was allowed to vary in size as well as shape. Molecules in the T_d sites perform free rotation while the remaining three-quarters in D_{2d} sites rotate within a plane. Below 230 K, molecules in the D_{2d} sites align themselves parallel to the direction of the unit cell vectors while those in the T_d sites rotate almost freely. This phase has cubic symmetry ($R3c$) in which all the molecules are aligned along the same direction. The transitions were reversible.

Carbon tetrafluoride. Carbon tetrafluoride, which undergoes a transition to a plastically crystalline (orientationally disordered) phase, has been investigated by the Parrinello–Rahman molecular dynamics method under constant-pressure conditions (6). A simple intermolecular potential model of the Lennard-Jones form was derived by taking into account the experimen-

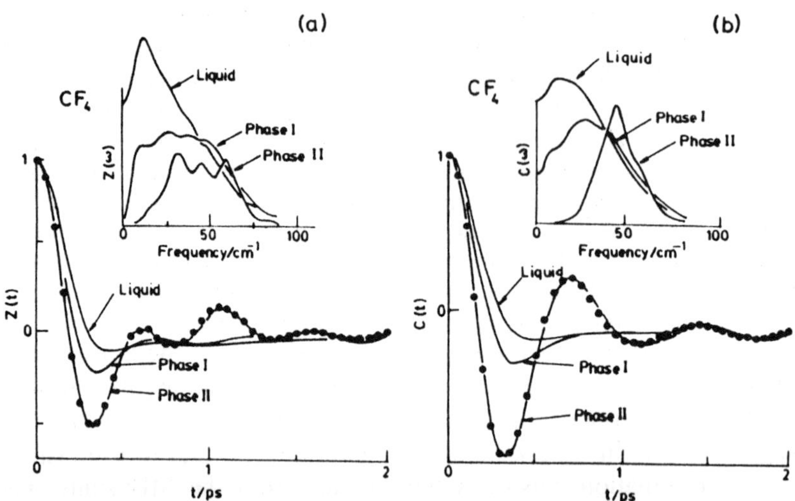

FIG. 4. (a) The velocity autocorrelation function and its power spectrum and (b) the angular velocity autocorrelation function and the power spectrum for different phases of CF_4. (From Nosé and Klein (6).)

tal molar volume and lattice energy. Calculations were carried out at 50.5 K and 1.2 kbar employing this potential with the starting arrangement of the ordered monoclinic phase II ($C2/c$) with $a = 8.43$, $b = 4.32$, $c = 8.48$ Å, $\beta = 120.7°$; calculations were also performed at higher temperatures and pressures. The calculated volume, lattice energy, and the r.d.f.'s of phase II were in agreement with experiment (27). On increasing the temperature, the system showed considerable increase in the molar volume around 75 K and the features in the radial distribution function became broad, suggesting increased molecular motion; the molecules in the lattice were orientationally disordered. The calculated change in enthalpy and molar volume were found to be in good accord with experiment (28). Properties of liquid carbon tetrafluoride have also been reported. Properties of solid carbon tetrafluoride have been studied by the MC method and found to be in agreement with the above results.

Time-dependent quantities have been examined (6) for the orientationally ordered and disordered phases as well as the liquid phases of CF_4. Power spectra for the translational and librational motions are reported (see Fig. 4). The peaks appear in the same region of frequency, suggesting strong translation–rotation coupling. The far-infrared spectrum in the ordered solid (29) has three peaks at 51, 57, and 66 cm^{-1}. On the basis of the significant intensity and temperature dependence of the 51-cm^{-1} band, it has been assigned to the 43-cm^{-1} band of the calculated $Z(\omega)$ and $C(\omega)$. This is a good illustration of the application of the Parrinello–Rahman MD method for the study of a crystal to plastic crystal transition. The study also shows how a simple potential can account for many of the properties exhibited by the solid. The dynamical properties seem to require a rather accurate description of intermolecular interactions. Constant-pressure and constant-volume calculations on carbon tetrafluoride (6) demonstrate that the single-particle dynamical properties are not very sensitive to the choice of the mass of the wall.

Bicyclo[2.2.2]octane. At low tempera-

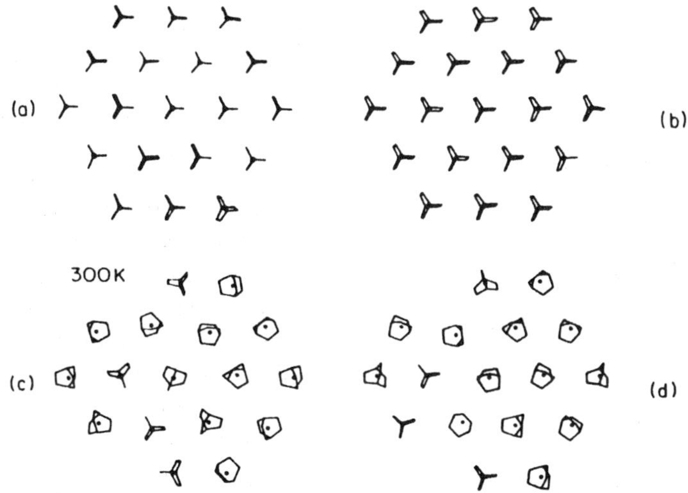

FIG. 5. A view of the bicyclo[2.2.2]octane down the trigonal axis (a) at 50 K for the 8-site model, (b) at 50 K for the 22-site model, (c) at room temperature for the 8-site model, and (d) at room temperature for the 22-site model. (From Neusy et al. (32).)

tures, bicyclo[2.2.2]octane exists as a trigonal phase with the individual molecular C_3 symmetry axes parallel to the crystal [111] directions. At 164.35 K, the ordered crystalline phase transforms to a cubic, orientationally disordered phase in which the quasi-spherical molecules exhibit increased molecular reorientation (30, 31). This order–disorder phase transition has been investigated by Neusy et al. (32) by means of constant-pressure MD employing two different potential models. In the first potential model of the Lennard-Jones form given by Jorgensen (33), each CH_2 or CH group is represented by a single interaction site and hence there are only eight interaction sites per molecule. In the model given by Williams (34), which is of the 6-exp form, each atom is represented by an interaction site that is placed on each atom.

Calculations carried out at 50 and 137.5 K on the ordered phase with an initial arrangement corresponding to the experimental structure (a = 6.4, c = 15.1 Å) resulted in a slightly elongated unique axis (35). The calculated c/a ratio was 2.6 for both the 8-site and the 22-site model as compared to the experimental value of 2.3. The molar volume determined from the 22-site model was in better agreement with experiment although the calculated molar volume was about 5% higher than the experimental value for the 8-site model. On heating to room temperature, the molecules oriented themselves almost randomly with a preference for the [111] crystallographic direction (Fig. 5). The ordered trigonal crystal was found to transform to the disordered cubic phase at about 170 K, in good agreement with experimentally determined transition temperature (164.25 K).

Power spectra of the autocorrelation functions of the linear and angular velocities parallel and perpendicular to the C_3 symmetrical axes have also been examined by Neusy et al. (32). In the rotator phase, there is good agreement with the Raman data (36). The calculated characteristic time (τ_4) for reorientation of the C_3 axes from one [111] direction to another and also the reorientation time (τ_3) for rotation of molecules around the C_3 axes were similar

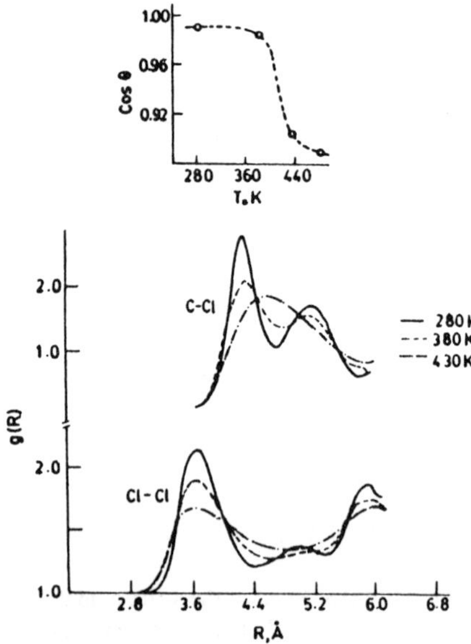

FIG. 6. Radial distribution functions C–C, C–Cl, and Cl–Cl in CCl_4 at 280, 380, and 430 K and 1.0 GPa pressure showing the changes from an ordered to an orientationally ordered phase. Inset shows a plot of $\cos \theta$ against temperature. Here θ is the smallest angle between the $(\vec{a} + \vec{c})$ direction and the four Cl–Cl bonds. The dashed line is drawn to guide the eye. (From Yashonath and Rao (19).)

for both potential models. These results indicate that transferable intermolecular potential functions provide a reasonably good model for solids. Experimentally, from incoherent neutron scattering studies it is found that τ_4 is much slower than τ_3. It appears from these results that while simple potentials can provide a reasonable picture of the phase transitions, they cannot make detailed predictions especially with regard to the dynamics of the system.

Carbon tetrachloride. Carbon tetrachloride exists in at least five phases, Ia, Ib, II, III, and IV, of which two phases, Ia and Ib, are orientationally disordered (37); phases III and IV exist only at high pressures. We have carried out an investigation (19) of the transition from the high-pressure ordered phase III to an orientationally disordered

phase by employing the generalized variable shape MC method in the isothermal isobaric ensemble. A simple Lennard-Jones potential used by McDonald *et al.* (38) in the simulation of liquid and solid phases with $\delta_{CC} = 4.6$ Å, $\varepsilon_{CC} = 51.2$ K, $\delta_{ClCl} = 3.5$ Å, and $\varepsilon_{ClCl} = 102.4$ K was employed in our calculations. Starting with the experimental structure (39) of phase III of CCl_4 with lattice parameters $a = 9.1$, $b = 5.8$, $c = 9.2$ Å and $\beta = 104.3°$, simulation was carried out at 280 K and 1.0 GPa. The calculated values of molar volume, configurational energy, lattice parameters, etc., were in agreement with the available experimental values (39). The r.d.f.'s for phase III are shown in Fig. 6. On increasing the temperature to 380 K, the thermodynamic properties were found to undergo small changes but, however, no significant changes were observed in the structure, the structure remaining ordered at this temperature (Fig. 7). Considerable changes in molar volume and intermolecular energy were observed on increasing the temperature to 430 K. The r.d.f.'s showed broad peaks similar to those of the liquid, suggesting increased molecular motion. The snapshot picture of the molecular arrangement in Fig. 7 shows that the crystal is orientationally disordered. This was con-

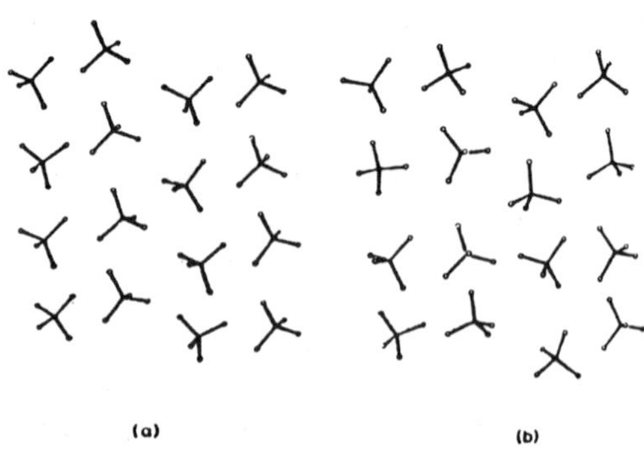

FIG. 7. A snapshot picture of the molecular arrangement viewed down the *b*-axis (a) at 380 K and (b) at 430 K in carbon tetrachloride. (From Yashonath and Rao (19).)

firmed by the drastic change in the minimum of the angle between the four C_3 axes of the molecule and the crystallographic [101] direction (see inset of Fig. 6). Calculations at an intermediate temperature of 405 K suggest that the transition temperature may be between 400 and 410 K, which compares well with the experimental value of 410 K (*37*).

Adamantane. Adamantane is a globular molecule in which the six methylene groups form an octahedron and the four methine groups projecting out form a tetrahedron. Adamantane exists as an ordered body-centered tetragonal solid at low temperatures (space group $P42_1c$; a = 6.6, b = 8.8 Å) (*40–42*). At 208.6 K, the solid undergoes a structural transition from the tetragonal to a cubic phase (*43*). Early X-ray studies of this phase by Nowacki (*40*) and Giocomello and Illuminati (*44*) suggested an ordered structure (space group $F43m$), but later work of Nordman and Schmitkons (*41*) showed the molecules to be orientationally disordered in the cubic phase ($Fm3m$). The nature of the high-temperature phase has, however, remained controversial (*45–47*).

We have carried out a MC simulation of adamantane in the isothermal isobaric ensemble permitting variation of the shape of the simulation cell (*48*). In these calculations, we have employed potential functions described by Jorgensen (*33*) in which there is a single interaction site for groups such as CH or CH_2. The potential parameters are, for CH_2–CH_2; ε = 0.478 kJ mole^{-1}, σ = 3.983 Å; for CH_2–CH: ε = 0.312 kJ mole^{-1}, σ = 4.118 Å; and for CH–CH: ε = 0.203 kJ mole^{-1}, σ = 4.252 Å. Calculations were carried out at 110, 213, and 298 K and atmospheric pressure. The calculated molar volume and the configurational energy are in good agreement with those from experiment (*41*). The molecules are completely ordered at 110 K and perform small librations. The CH_2–CH_2, CH–CH_2, and CH–CH r.d.f.'s (Fig. 8) show

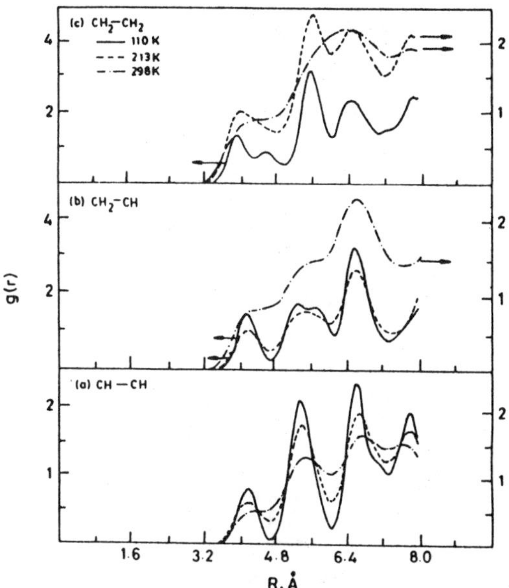

FIG. 8. Radial distribution functions for adamantane between (a) CH_2–CH_2, (b) CH_2–CH, and (c) CH_2–CH_2 groups at 110, 213, and 298 K and 1 atmosphere pressure. (From Yashonath and Rao (*48*).)

well-defined peaks at this temperature. At 213 K, the peaks become somewhat broad, but the arrangement at this temperature is still ordered. On increasing the temperature to 298 K, the r.d.f.'s show broad peaks unlike those of the ordered solid (Fig. 8). The molecules were found to perform large rotational jumps around the C_4 axis (Fig. 9). The calculated unit cell parameters are 9.31, 9.31, 9.25 Å with the angles close to 90°. These compare well with the experimental cubic cell parameter of 9.44 Å (*41*). The elongation of the c-axis on going from 213 to 298 K is also known experimentally. The calculated configurational energy (-58.0 kJ/mole) is in agreement with the heat of sublimation (-58.5 kJ/mole) (*49*). This study supports the view that the room-temperature phase of adamantane is orientationally disordered.

Biphenyl. In biphenyl, the angle between the normal to the two phenyl rings, θ,

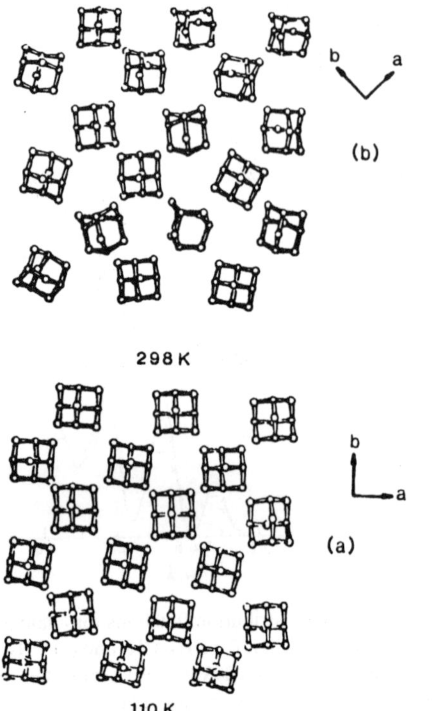

298 K

110 K

FIG. 9. A view of the instantaneous arrangement in adamantane looking down the c-axis: (a) tetragonal ordered phase at 110 K and (b) orientationally disordered cubic phase at 298 K. (From Yashonath and Rao (48).)

which is 42° in the vapor phase, decreases to about 32° in the melt (50). Upon cooling, the melt crystallizes in the space group $P2_1/a$, where the two phenyl rings are essentially planar. Further cooling results in a phase with the two rings deviating from this planar arrangement. Raman spectroscopic and other studies indicate that this solid–solid phase transformation is a second-order transition occurring over a range of temperatures (75 to 40 K) (50, 52). The conformations in the vapor and the liquid phases result from a balance between resonance stabilization and ortho–ortho hydrogen repulsion. In the solid, intermolecular interactions also play a very significant role and hence packing considerations lead to a

planar conformation in the high-temperature solid phase (53). We have carried out some preliminary MC investigations on this interesting solid by employing a 6-exp atom–atom potential derived by Misskaya *et al.* (54) and used by Ramdas and Thomas (55) in their study of the structure of *p*-terphenyl. The same potential was used to calculate the intermolecular interactions as well as the intramolecular interactions arising out of ortho–ortho hydrogen repulsions. The potential parameters were

$$A_{CC} = 1761.9 \text{ kJ A}^6 \text{ mole}^{-1};$$
$$B_{CC} = 299{,}646 \text{ kJ mole}^{-1};$$
$$C_{CC} = 3.68 \text{ Å}^{-1};$$
$$A_{CH} = 493.8 \text{ kJ A}^6 \text{ mole}^{-1};$$
$$B_{CH} = 77{,}841 \text{ kJ mole}^{-1};$$
$$C_{CH} = 3.94 \text{ Å}^{-1};$$
$$A_{HH} = 121.4 \text{ kJ A}^6 \text{ mole}^{-1};$$
$$B_{HH} = 20{,}506.5 \text{ kJ mole}^{-1};$$
$$C_{HH} = 4.29 \text{ Å}^{-1}.$$

The initial arrangement was the experimental structure of the high-temperature phase with $a = 7.82$, $b = 5.58$, $c = 9.44$ Å and $\beta = 94.6°$ (51). The calculated intermolecular energy at 110 K was -68.2 kJ mole^{-1} compared to the sublimation energy of -81.6 kJ mole^{-1}. The average angle between the two phenyl rings was 20° as against the experimentally observed planar conformation (51). The discrepancy in the observed angle could be due to the large repulsion in the region of 1.8–2.2 Å in the hydrogen–hydrogen interaction potential functions. Potential functions were derived by considering the lattice energy of aromatic hydrocarbons such as benzene, naphthalene, etc., where only intermolecular distances larger than 2.2 Å are present (54). It is, therefore, not surprising that the potential may not represent the interactions correctly at lower distances even though it may show excellent agreement at larger distances. A recent calculation on *n*-butane has shown that the conformational angle is much more sensitive to the potential than other properties (56). On cooling the solid to 40 K, we found that θ increased to 25.3° with significant decrease in libration around the long molecular axis. This trend is in agreement with experiment (57). We plan to carry out further calculations using a different potential model where the intermolecular interactions and intramolecular interactions are modelled by different potential functions. These calculations are expected to answer several questions, especially those concerning the distribution of the interplanar angle, the structure of the low-temperature solid, and the nature of the motion in the high-temperature phase.

4. Glasses

Computer simulation has contributed significantly to the understanding of equilibrium and dynamic properties of supercooled liquids and glasses. Hard-sphere, Lennard-Jones, and ionic glasses have been studied (58–61), providing some insights into structural and other aspects (62, 63). For example, it has been found that the split second peak, a characteristic feature of the r.d.f.'s of metallic glasses, is observed in computer-simulated Lennard-Jones glasses (62). The MD method was limited to the microcanonical ensemble in these studies until the constant-temperature and constant-pressure MD was introduced by Andersen. The constant-volume and energy conditions used in the earlier studies did not permit direct comparison with the laboratory glasses, where the conditions of preparation of glasses are very different (64, 65). Also, some of the typical characteristics of laboratory glasses were absent in computer-simulated glasses (64). The smeared nature of the transition at temperatures which are higher than the experimental transition temperature prompted Angell and co-workers (64, 65) to suggest that the transition observed in computer simulation studies bears little resemblance to laboratory glasses. More recent results obtained under isothermal- isobaric conditions by MC and MD methods indicate that there is, in fact, a close parallel between the laboratory and the computer-simulated glasses. The quenching rates employed in computer simulation are, however, orders of magnitude higher than those employed in the laboratory and, as a consequence, computer-simulated glasses differ in certain aspects from those obtained in the laboratory. Frenkel and McTague (66) have reviewed the earlier studies in this area. We shall limit the present discussion to certain significant findings based on some of the recent studies on monatomic, molecular, and orientational glasses.

Lennard-Jones glass. Extensive calculations on a monatomic Lennard-Jones argon-like system have been carried out by Fox and Andersen (67) under conditions of constant temperature and constant pressure with a variety of cooling rates. These workers have reported density, enthalpy, and self-diffusion coefficient at different pressures and temperatures. The glass transition temperature obtained from the density–temperature plot is essentially identical to that obtained from the enthalpy–temperature plots. The glass transition temperature is higher at higher pressures. Also the slower the rate of cooling, the higher was the density of the glass, indicating the existence of slow relaxation processes involving volume changes (see Fig. 10). The apparent self-diffusion coefficient could be fitted equally well to both Arrhenius and Doolittle equations. Interestingly, the pressure dependence of the structural relaxation was different from that of self-diffusion. Hysteresis in the transition was evident from a comparison of the heating and the cooling runs.

The above results indicate that even though the quenching rates employed in

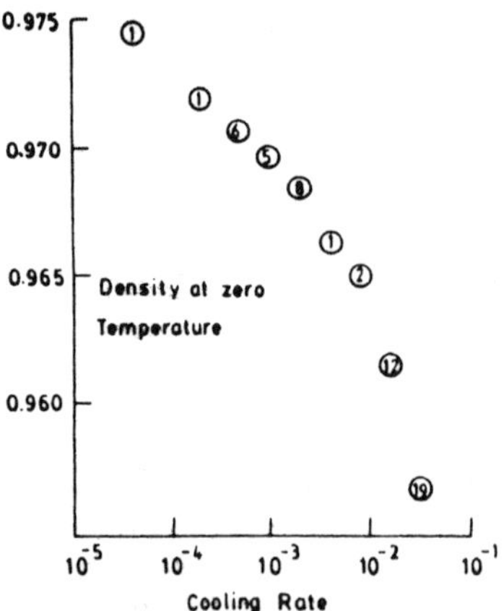

FIG. 10. The densities obtained at different cooling rates at $T = 0$ and $p = 1$ for the Lennard-Jones glass. The cooling rate is the value of the stochastic collision frequency per particle, ν. The number of runs employed to obtain the reported average is shown inside each point. (From Fox and Andersen (67).)

computer simulation are very high, the general features of simulated glasses are quite similar to those of laboratory glasses. The demonstration of the history dependence of the properties and also the presence of hysteresis near the transition are significant. The radial distribution functions for glasses obtained by employing two different cooling rates shown in Fig. 11 are indeed similar; the r.d.f.'s are somewhat insensitive to the structural changes taking place in the glass. Our calculations on methane (68) in which only orientational degrees of freedom were permitted, however, show distinct changes in the r.d.f.'s of the glass before and after annealing (see Fig. 12). This is the only study that we are aware of that shows changes in r.d.f.'s even though it is for an orientationally disordered glass and not for a positionally disordered glass. This is probably because barriers for going from one state to another state are much smaller in the case of orientational glass. The history dependence of the properties of the glass has also been observed in a study on

isopentane glass (69); the intermolecular energy of the glass is lowered by a small amount on annealing near the glass transition.

Isopentane glass. A recent MC simulation of isopentane glass (70) using realistic potentials has yielded some insight into glass structure and the nature of rearrangement occurring during the glass transition. Isothermal isobaric ensemble calculations have been performed on glasses obtained by instantaneous quenching of the liquid. There are four distinct groups in isopentane: CH, CH_2, the methyl group attached to CH, $CHCH_3^*$, and the methyl group attached to CH_2, $CH_2CH_3^*$. These give rise to ten r.d.f.'s. Four of these r.d.f.'s, between the somewhat inaccessible CH and other groups (referred to as the i-type r.d.f.'s), are similar in nature. They have a single peak within the 9-Å region at a somewhat larger distance of 5.3 to 6.2 Å. The remaining six r.d.f.'s (the p-type r.d.f.'s) in the liquid between the peripheral groups CH_2, $CH_2CH_3^*$, and $CHCH_3^*$ show a shoulder around 4 Å and a main peak around 6 Å. A

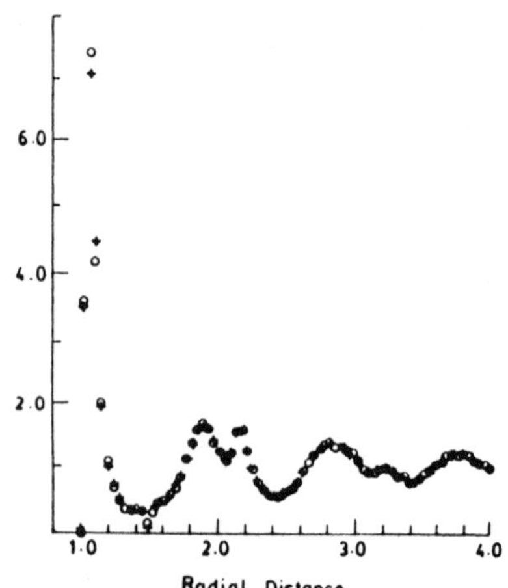

FIG. 11. Radial distribution functions for systems obtained by continuous cooling at $p = 1$. The stochastic collision frequencies were 3.2 for the + and 0.04 for the O points. (From Fox and Andersen (67).)

COMPUTER SIMULATION   209

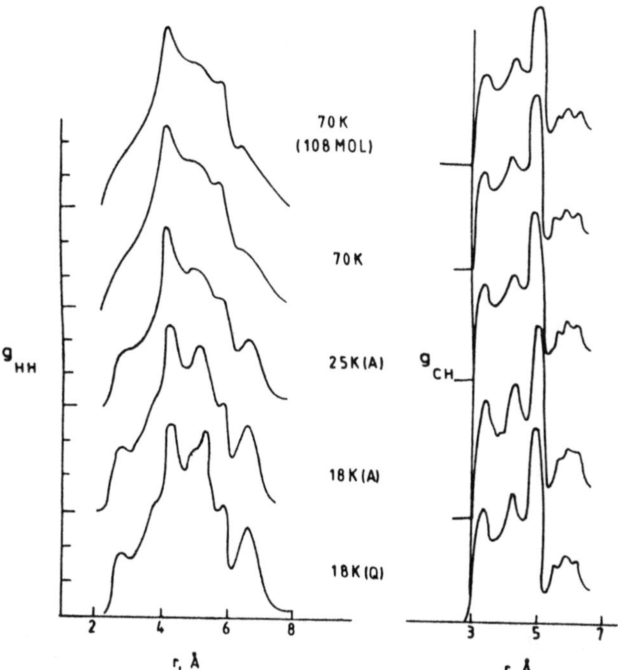

FIG. 12. Pair distribution functions $g_{HH}(r)$ and $g_{CH}(r)$ for the plastic and annealed (A) and quenched phases (Q) of methane. Only the orientational degrees of freedom were considered in the calculation. Note the distinct change in the r.d.f.'s in $g_{HH}(r)$ and to a smaller extent of $g_{CH}(r)$ on annealing. (From Yashonath and Rao (*68*).)

few typical r.d.f.'s are shown in Fig. 13. On cooling below the glass transition temperature, peaks generally become sharper and show more features. However, there are several interesting differences between the i-type and the p-type r.d.f.'s. While the

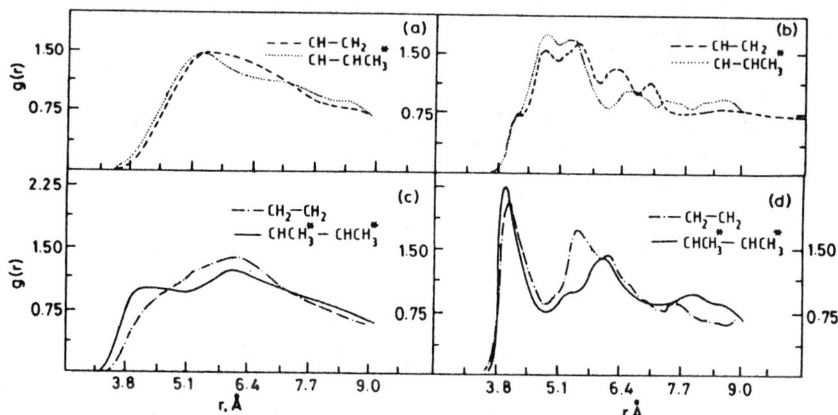

FIG. 13. Typical r.d.f.'s (i-type) between the inaccessible CH and other groups (a) for liquid isopentane at 301 K and (b) for the glass at 30 K. Typical r.d.f.'s (p-type) between the peripheral groups (c) for the liquid at 301 K and (d) for the glass at 30 K. (From Yashonath and Rao (*70*).)

                                          RAO AND YASHONATH

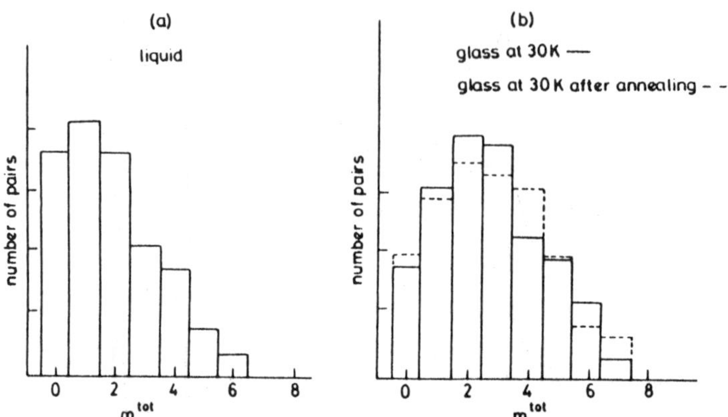

FIG. 14. Histograms for the distribution of the number of pairs of neighbors, in arbitrary units, for different values of m_{tot} for (a) isopentane liquid at 301 K and (b) glass at 30 K, with and without annealing. (From Yashonath and Rao (71).)

For the role of molecular shape on orientational rearrangement during vitrification, see A. Chakrabarti, S. Yashonath and C.N.R. Rao, J. Phys. Chem., 96, 6762 (1992)

peak positions of the i-type r.d.f.'s are shifted towards lower distances, those of p-type r.d.f.'s do not show such shifts. This is surprising as one would have expected a more or less uniform shift of peaks in all the r.d.f.'s towards lower distances on vitrification. Furthermore, on glass formation, the shoulder in the p-type r.d.f.'s of the liquid around 4.2 Å gains significantly in intensity as compared to the main peak. These observations suggest the existence of a somewhat complex rearrangement of the neighbors. This rearrangement seems to involve the shift of the center of mass of the neighbors towards each other. In addition, the disproportionate increase in the first peak intensity on vitrification and the absence of shift towards lower distances for the p-type r.d.f.'s suggests the presence of a molecular reorientational mechanism.

The number of pairs of neighbors specified in arbitrary units is plotted against m_{tot}, where m_{tot} is the sum of the number of groups in the region of the first peak (for the six p-type r.d.f.'s), in Fig. 14 for the liquid at 301 K and the glass at 30 K; the effect of annealing the glass is also shown in the figure. The shift of the histogram towards higher values of m_{tot} indicates the overall increase in the number of neighbors in the region of the first peak. We have been able to estimate the reorientational contribution E_r^g to the increase in energy from the r.d.f.'s and find the reorientational contribution to the increase in the intermolecular energy on vitrification (71) to be nearly 50%. In obtaining an estimate of E_r^g we have neglected the reorientational contribution from the i-type r.d.f.'s and made the assumption that the p-type r.d.f.'s of the hypothetical glass are similar to those of the liquid but for the increased intensity due to the higher density. In other words,

$$g_h(r) = g_l(r)n_t^g/n_t^l, \qquad (15)$$

where the $g_h(r)$ and $g_l(r)$ are the r.d.f.'s of the hypothetical glass and the liquid respectively and n_t^g and n_t^l are the total number of neighbors within a distance of 9 Å for the glass and the liquid.

The total intermolecular energy, E_{inter}, and the total reorientational contribution, E_r, are plotted against temperature in Fig. 15 for isopentane cooled from 301 K to 220, 120, 50, 40, and 30 K. The change in slope observed around 80 K is considerably more

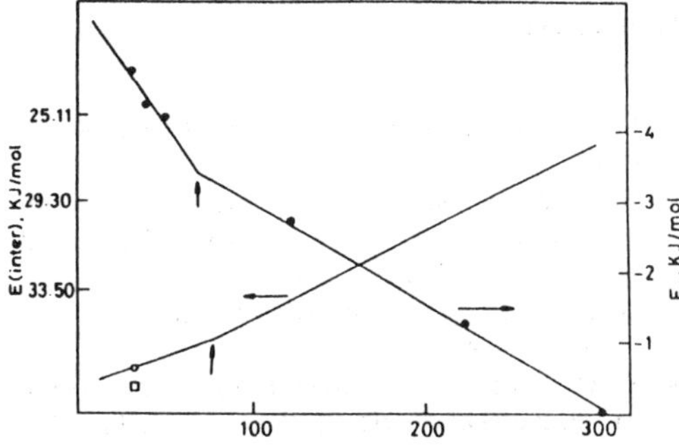

FIG. 15. Variation of the total intermolecular energy, E_{inter}, and the contribution from reorientation, E_r, for isopentane at different temperatures obtained by instantaneous cooling from 301 K. (From Yashonath and Rao (70, 71).)

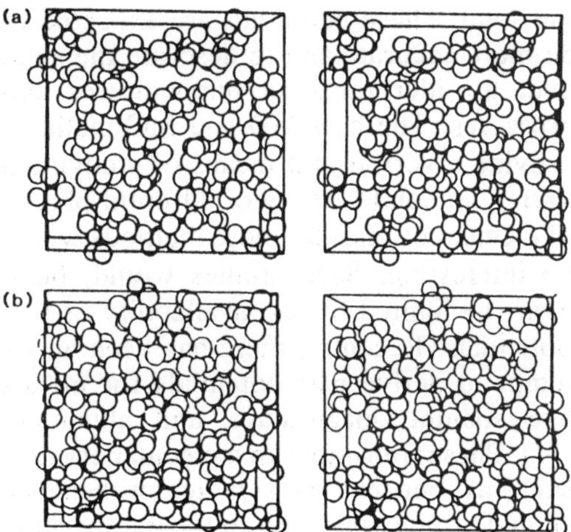

FIG. 16. Stereoplots showing the structure (a) in the liquid at 301 K and (b) in the glass at 30 K for isopentane. Note that the molecules at the surface interact with those at the opposite surface. (From Yashonath and Rao (70).)

marked in the case of E_r as compared to E_{inter}. It appears, therefore, that the reorientational mechanism contributes significantly to the changes in properties near the glass transition. We have also found that significant structural rearrangements involving orientational degrees of freedom occur on annealing the glass. Stereoplots of the molecular arrangement in the liquid and the glass are shown in Fig. 16. The presence of considerable free volume even in the glass is evident from this figure.

Investigations into the properties of glassy water and methanol are presently in progress in our laboratory. Whether amorphous solid water exhibits a glass transition is an aspect of considerable interest (72). Preliminary studies (73) indicate that liquid water quenched to low temperatures does indeed show a glass transition, with the nature of species, intermolecular energy, heat capacity, etc., showing the expected changes at the transition temperature (Fig. 17).

5. Concluding Remarks

The generalized MD and MC methods can be used for studying different classes of materials and phenomena. The method introduced by Parrinello and Rahman (5) permits one to derive interparticle potentials taking into account more than one phase of

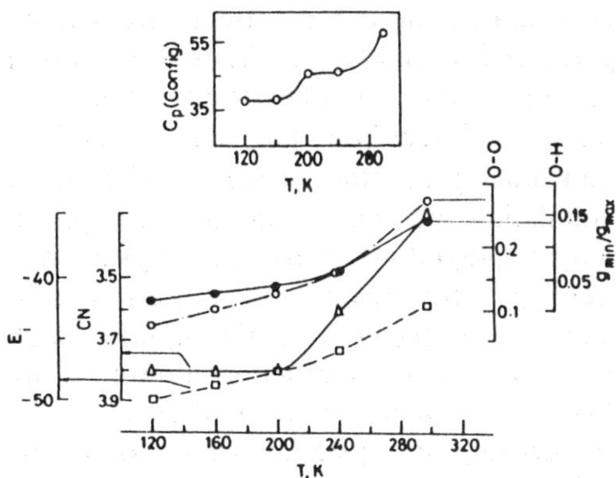

FIG. 17. Temperature variation of internal energy. E_i (kJ mole^{-1}), coordination number (CN), and $g_{\min}/g_{\max}$ ratios of water showing the occurrence of the glass transition in the 200–240 K range. Volume also shows a similar change, but a slightly lower temperature. In the inset, variation of the configurational heat capacity, C_p (J deg^{-1} mole^{-1}), with temperature is shown. (From Chandrasekhar and Rao (73).)

the solid. These potentials are expected to be more accurate and to describe the solid over a wider range of temperatures and pressures. The new MD and MC methods provide a very useful means of studying a variety of solids. Study of different phases of elemental solids by these methods would be interesting. Such studies would, however, be limited by the availability of pair potentials. Recently, there has been an attempt to derive pair potentials for metals like sodium, magnesium, and aluminum (74). Phase transitions in molecular crystals such as benzene and p-dichlorobenzene should provide insight into the microscopic behavior in these solids and be useful in interpreting the large amount of experimental data available in the literature. The mechanism of transitions and the dependence of transition temperature and other properties on the intermolecular potential can be fruitfully investigated. In our opinion, systematic studies of orientationally disordered solids (see Refs. (75–77) for the very recent papers) as well as matrix isolated molecular systems such as CH_4 in Ar matrix should be worthwhile. The liquid crystalline state can be effectively investigated by MC and MD methods. Yet another interesting area would be to investigate the effect of defects on the phase transitions and other properties of solids. The above simulation methods could be very profitably employed in the area of "crystal engineering" and organic solid-state reactions as well as in the study of the liquid crystalline state. If the interaction potentials are known accurately, they could also be used in deriving the crystal structures of simple solids at finite temperatures.

Acknowledgments

The authors thank the Department of Science and Technology and the University Grants Commission for support of this research.

References

1. M. L. KLEIN, *Annu. Rev. Phys. Chem.* **36**, 525 (1985).
2. N. METROPOLIS, A. W. ROSENBLUTH, M. N. ROSENBLUTH, A. H. TELLER, AND E. TELLER, *J. Chem. Phys.* **21**, 1087 (1953).
3. B. J. ALDER AND T. W. WAINWRIGHT, *J. Chem. Phys.* **31**, 459 (1959).
4. H. C. ANDERSEN, *J. Chem. Phys.* **72**, 2384 (1980).
5. M. PARRINELLO AND A. RAHMAN, *Phys. Rev. Lett.* **45**, 1196 (1980).
6. S. NOSÉ AND M. L. KLEIN, *Mol. Phys.* **50**, 1055 (1983); *J. Chem. Phys.* **78**, 6928 (1983).
7. S. YASHONATH AND C. N. R. RAO, *Mol. Phys.* **54**, 245 (1985).
8. (a) W. W. WOOD, *in* "Physics of Simple Liquids" (H. N. V. Temperley, J. S. Rowlinson, and G. S. Rushbrooke, Eds.), North-Holland, Amsterdam (1968); (b) J. P. VALLEAU AND S. G. WHITTINGTON, *in* "Statistical Mechanics, Part A: Equilibrium Techniques" (B. J. Berne, Ed.), p. 137, Plenum, New York (1977); (c) J. P. VALLEAU AND G. M. TORRIE, *ibid.*, p. 169; (d) J. J. ERPENBECK AND W. W. WOOD, *in* "Statistical Mechanics, Part B: Time-Dependent Processes" (B. J. Berne, Ed.), p. 1, Plenum, New York, (1977); (e) J. KUSHICK AND B. J. BERNE, *ibid.*, p. 41; (f) K. BINDER (Ed.), "Monte Carlo Methods," Methuen, London (1964); (g) K. BINDER (Ed.), "Monte Carlo Methods in Statistical Physics," Springer-Verlag, New York (1979).
9. J. M. HAILE AND H. W. GRABEN, *J. Chem. Phys.* **73**, 2412 (1980).
10. H. TANAKA, K. NAKANISHI, AND N. WATANABE, *J. Chem. Phys.* **78**, 2626 (1983).
11. S. NOSÉ, *Mol. Phys.* **52**, 255 (1984); *J. Chem. Phys.* **81**, 511 (1984).
12. D. J. EVANS, *J. Chem. Phys.* **78**, 3297 (1983).
13. J. M. HAILE AND S. GUPTA, *J. Chem. Phys.* **79**, 3067 (1983).
14. M. PARRINELLO AND A. RAHMAN, *J. Appl. Phys.* **52**, 7182 (1981).
15. M. PARRINELLO AND A. RAHMAN, *J. Chem. Phys.* **76**, 2622 (1982).
16. J. R. RAY AND A. RAHMAN, *J. Chem. Phys.* **80**, 4423 (1984).
17. J. R. RAY AND A. RAHMAN, *J. Chem. Phys.* **82**, 4243 (1985).
18. R. NAJAFBADI AND S. YIP, *Scr. Metall.* **17**, 1199 (1983).
19. S. YASHONATH AND C. N. R. RAO, *Chem. Phys. Lett.* **119**, 22 (1985).
20. M. PARRINELLO AND A. RAHMAN, *in* "Melting, Localization and Chaos" (R. K. Kalia and P. Vashishta, Eds.), Elsevier, Amsterdam/New York (1982).

21. S. YASHONATH, Ph.D. thesis, Indian Institute of Science, Bangalore, India (1985).

22. G. S. PAWLEY AND G. W. THOMAS, *Phys. Rev. Lett.* **48**, 410 (1982).

23. G. S. PAWLEY, *J. Mol. Struct.* **30**, 17 (1985).

24. M. PARRINELLO, A. RAHMAN, AND P. VASHISHTA, *Phys. Rev. Lett.* **50**, 1073 (1983).

25. W. ANDREONI AND J. C. PHILLIPS, *Phys. Rev. B* **23**, 6456 (1981).

26. S. NOSÉ AND M. L. KLEIN, *Phys. Rev. Lett.* **50**, 1207 (1983).

27. C. N. BOL'SHUTKIN, V. M. GASAN, A. I. PROKHAVTILOV, AND A. I. ERENBURG, *Acta Crystallogr., Sect. B* **28**, 3542 (1972).

28. J. W. STEWART AND R. I. LAROCK, *J. Chem. Phys.* **28**, 3542 (1972).

29. Y. A. SATATY, A. RON, AND F. H. HERBSTEIN, *J. Chem. Phys.* **62**, 1094 (1975).

30. J. N. SHERWOOD (Ed.), "The Plastically Crystalline State," Wiley, New York (1979); C. N. R. RAO, *Proc. Indian Acad. Sci. (Chem. Sci.)* **94**, 181 (1985).

31. E. F. WESTRUM, JR., W. K. WONG, AND E. MORAWETZ, *J. Phys. Chem.* **74**, 2542 (1970).

32. E. NEUSY, S. NOSÉ, AND M. L. KLEIN, *Mol. Phys.* **52**, 269 (1984).

33. W. L. JORGENSEN, *J. Amer. Chem. Soc.* **103**, 335 (1981).

34. D. E. WILLIAMS, *J. Chem. Phys.* **47**, 4680 (1967).

35. A. J. LEADBETTER, J. PIPER, R. M. RICHARDSON, AND P. G. WRIGHTON, *J. Phys. C* **15**, 5921 (1982).

36. J. L. SAUVAJOL, Thesis, Universite de Lille (1983).

37. R. G. ROSS AND P. ANDERSON, *Mol. Phys.* **36**, 39 (1978).

38. I. R. McDONALD, D. G. BOUNDS, AND M. L. KLEIN, *Mol. Phys.* **45**, 521 (1982).

39. G. J. PIERMARINI AND A. B. BRAUN, *J. Chem. Phys.* **58**, 1974 (1973).

40. W. NOWACKI, *Helv. Chim.* **28**, 1283 (1945).

41. C. E. NORDMAN AND D. L. SCHMITKONS, *Acta Crystallogr.* **18**, 764 (1965).

42. J. DONOHUE AND S. H. GOODMAN, *Acta Crystallogr.* **22**, 352 (1967).

43. S. C. CHANG AND E. F. WESTRUM, *J. Phys. Chem.* **64**, 1547 (1960).

44. G. GIOCOMELLO AND G. ILLUMINATI, *Gazz. Chim. Ital.* **75**, 246 (1945).

45. P. A. REYNOLDS, *Acta Crystallogr., Sect. A* **34**, 362 (1978).

46. J. P. AMOUREUX, M. BEE, AND J-C. DAMIEN, *Acta Crystallogr., Sect. B* **36**, 2633 (1980).

47. P. A. REYNOLDS AND B. R. MARKEY, *Acta Crystallogr., Sect. A* **35**, 627 (1979).

48. S. YASHONATH AND C. N. R. RAO, *J. Phys. Chem.* **90**, 2552 (1986).

49. R. J. JOCHEMS, *Chem. Thermodyn.* **14**, 395 (1982).

50. A. M. PONTE GONCALVES, *Prog. Solid State Chem.* **13**, 1 (1980).

51. G. P. CHARBONNEAU AND Y. DELUGEARD, *Acta Crystallogr., Sect. B* **32**, 1420 (1976).

52. R. M. HOCHSTRASSER, G. W. SCOTT, A. H. ZEWAIL, AND H. F. FUESS, *Chem. Phys.* **11**, 273 (1975); C. A. HUTCHINSON, JR., AND V. H. McCANN, *J. Chem. Phys.* **61**, 820 (1974).

53. G. CASALONE, C. MARIANI, A. MUGNOLI, AND M. SIMONETTA, *Mol. Phys.* **15**, 339 (1968).

54. K. V. MISSKAYA, I. E. KOZLOVA, AND V. F. BEREZNITSKAYA, *Phys. Status Solidi B* **62**, 291 (1974).

55. S. RAMDAS AND J. M. THOMAS, *J. Chem. Soc., Faraday Trans. 2* **72**, 1251 (1976).

56. A. BANBN, F. SERRANO ADAN, AND J. SANTAMERIA, *J. Chem. Phys.* **83**, 297 (1985).

57. A. S. CULLICK AND R. E. GERKIN, *Chem. Phys.* **23**, 217 (1977).

58. L. V. WOODCOCK, *Ann. N.Y. Acad. Sci.* **371**, 274 (1981).

59. J. H. R. CLARKE, *J. Chem. Soc., Faraday Trans. 2* **75**, 1371 (1979).

60. S. BREWER AND M. J. BISHOP, *J. Chem. Phys.* **75**, 3522 (1981).

61. L. V. WOODCOCK, C. A. ANGELL, AND P. CHEESMAN, *J. Chem. Phys.* **65**, 1656 (1976).

62. F. F. ABRAHAM, *J. Chem. Phys.* **72**, 359 (1980).

63. J. R. FOX AND H. C. ANDERSEN, *Ann. N.Y. Acad. Sci.* **371**, 123 (1981).

64. C. A. ANGELL, J-H-R. CLARKE, AND L. V. WOODCOCK, *Adv. Chem. Phys.* **48**, 397 (1981).

65. C. A. ANGELL, *Ann. N.Y. Acad. Sci.* **371**, 136 (1981).

66. D. FRENKEL AND J. P. McTAGUE, *Annu. Rev. Phys. Chem.* **31**, 491 (1980).

67. J. R. FOX AND H. C. ANDERSEN, *J. Phys. Chem.* **88**, 4019 (1984).

68. S. YASHONATH AND C. N. R. RAO, *Chem. Phys. Lett.* **101**, 524 (1983).

69. S. YASHONATH, K. J. RAO, AND C. N. R. RAO, *Phys. Rev. B* **31**, 3196 (1985).

70. S. YASHONATH AND C. N. R. RAO, *Proc. R. Soc. London, Ser. A* **400**, 61 (1985).

71. S. YASHONATH AND C. N. R. RAO, *J. Phys. Chem.* **90**, 2581 (1986).

72. S. R. ELLIOTT, C. N. R. RAO, AND J. M. THOMAS, *Angew Chem. Int. Ed. (English)* **25**, 31 (1986).

73. J. CHANDRASEKHAR AND C. N. R. RAO, *Chem. Phys. Lett.,* in press.

74. D. G. PETTIFER AND M. A. WARD, *Solid State Commun.* **49**, 291 (1984).

75. R. D. MOUNTAIN AND A. C. BROWN, *J. Chem. Phys.* **82**, 4236 (1985).

76. M. FERRORIO, I. R. McDONALD, AND M. L. KLEIN, *J. Chem. Phys.* **83**, 4726 (1985).

77. R. W. IMPEY, M. L. KLEIN, AND I. R. McDONALD, *J. Chem. Phys.* **82**, 4690 (1985).

MOLECULAR PHYSICS, 1995, VOL. 84, NO. 1, 49–68

A Monte Carlo study of the condensed phases of biphenyl†

By APARNA CHAKRABARTI, S. YASHONATH and C. N. R. RAO

Solid State and Structural Chemistry Unit, Indian Institute of Science, Bangalore 560012, India

(Received 12 July 1994; revised version accepted 9 September 1994)

Detailed calculations on the condensed phases of biphenyl have been carried out by the variable shape isothermal–isobaric ensemble Monte Carlo method. The study employs the Williams and the Kitaigorodskii intermolecular potentials with several intramolecular potentials available from the literature. Thermodynamic and structural properties including the dihedral angle distributions for the solid phase at 300 K and 110 K are reported, in addition to those in the liquid phase. In order to get the correct structure it is necessary to carry out calculations in the isothermal–isobaric ensemble. Overall, the Williams model for the intermolecular potential and Williams and Haigh model for the intramolecular potential yield the most satisfactory results. In contrast to the results reported recently by Baranyai and Welberry, the dihedral angle distribution in the solid state is monomodal or weakly bimodal. There are interesting correlations between the molecular planarity, the density and the intermolecular interaction.

1. Introduction

The geometry of biphenyl in the free state as well as in the different condensed phases has been of considerable interest for several decades. Gas phase electron diffraction studies [1–3] suggest that the phenyl–phenyl torsion angle is $\sim 43°$. Molecular force-field calculations [4] suggest that the dihedral angle is $32°$ in solution, but experimental measurements [5, 6] yield a value in the 19–26° range. X-Ray diffraction studies [7, 8] of the solid phase at room temperature, on the other hand, show the molecule to be planar. At room temperature the solid has a monoclinic structure with two molecules per unit cell. The low temperature phase at 110 K also has monoclinic symmetry, with slightly different cell parameters [9]. X-Ray diffraction studies [9] at 110 K and 293 K have shown a large component of the librational motion about the long molecular axis with a mean-square amplitude of $45·7°^2$ at 110 K and $105·9°^2$ at 293 K, suggesting that the zero dihedral angle between the phenyl rings obtained from X-ray diffraction studies [7, 8] is likely to be an average value, and that there is a considerable deviation from the planar conformation. Apart from exhibiting large variations in the dihedral angle in the gaseous and condensed phases, biphenyl undergoes several phase transitions at low temperatures. High-resolution elastic neutron scattering studies [10] have shown that the transition around 38 K, referred to as the 'twist' transition, involves an incommensurate modulation in both the a^* and b^* directions in orientational space and the average dihedral angle changes from 0° to 7–10°. Another transition, called the 'lock-in' transition, is associated with the disappearance of the modulation in the a^* direction below 17–20 K. Raman spectroscopic studies [11] confirm the

† Contribution No. 1003 from the Solid State and Structural Chemistry Unit, Bangalore.

50   A. Chakrabarti *et al.*

existence of these two transitions. Variable-temperature Raman [12] as well as electron paramagnetic resonance and electron–nuclear double resonance studies [13] also support this observation.

Computational approaches have been used with some success in understanding the different properties of biphenyl. Fischer-Hjalmars [14] calculated the dependence of the conjugation energy on dihedral angle θ using a semi-empirical approach. Busing [15] calculated the properties of biphenyl at 22 K, 110 K and 293 K using a temperature-dependent potential. Benkert, Heine and Simmons (BHS) [16] proposed an intramolecular potential which includes both the non-bonded and the conjugation energies . They modelled this system to study the incommensurate phase transition of biphenyl [16]. Baranyai and Welberry [17, 18] have recently reported molecular dynamics calculations at constant volume using the intermolecular potential of Williams and Cox [19] and the BHS intramolecular potential [16]. Their results suggest the presence of two sublattices in the solid state with dihedral angles of 0–2° and 22°. However, on the basis of available experimental results one cannot distinguish between a double minimum in the potential proposed by Charbonneau and Delugeard [9] and the arrangement proposed by Baranyai and Welberry.

We have carried out detailed Monte Carlo (MC) calculations in the isothermal–isobaric ensemble employing the intermolecular potentials of Williams and Cox [19] and Kitaigorodskii [20] along with the intramolecular potentials of Haigh [14], Bartell [21], and BHS [16]. In addition, we have used the potentials of Williams and Kitaigorodskii for the intramolecular contributions. We report thermodynamic properties as well as the crystal and the molecular structures of biphenyl based on these calculations. We also examine the structural aspects of this fascinating molecule in the crystalline (monoclinic) phase at 300 K and 110 K and in the liquid state.

2. Potential models

There are several potentials for hydrocarbons in the literature. One of the most widely used potentials among these is that due to Williams and Cox [19]. This has been derived by fitting to about 30 different aromatic and non-aromatic hydrocarbons. It is an atom–atom potential with 22 sites on each of the C and H atoms and has the following form:

$$\phi_{jk}(r_{jk}) = -A_{jk}/r_{jk}^6 + B_{jk} \exp\left(-C_{jk}r_{jk}\right) + q_j q_k/r_{jk}, \quad j, k = C, H \tag{1}$$

Kitaigorodskii [20] has proposed several potentials for hydrocarbons, and these are used frequently to model aromatic systems. The parametrization of these potentials has been done using structural and enthalpic data for a large number of hydrocarbons. We have used one of these potentials also for the intermolecular interaction. Both potentials have the 6-exp form for the short range interaction but, unlike the Williams potential, no electrostatic interactions are included in the potential given by Kitaigorodskii. In the Williams model, a charge of $+q$ is placed on hydrogen atoms and $-q$ on the carbon atoms. The potential parameters are listed in table 1.

The intramolecular potential for biphenyl consists of two parts. One is the steric interaction of the atoms at the *ortho* position modelled in terms of the non-bonded interactions between hydrogens and carbons of the two rings at the *ortho* position. The second is the variation of the conjugation energy [14]. Several models have been proposed for the former. Of the available potentials, the one proposed by Bartell [21] is the oldest. Most potentials include only the H–H non-bonded interaction, but as

Table 1. Atom–atom interaction parameters for biphenyl. The interaction is assumed to be of the Buckingham form (6-exp).

Model	Atom	$A/\text{kJ mol}^{-1}\,\text{Å}^{-6}$	$B/\text{kJ mol}^{-1}$	$C/\text{Å}^{-1}$	q/e
WW	C	2439·8	369 743·0	3·60	−0·153
	H	136·4	11 971·0	3·74	0·153
KK	C	1498·6	175 812·0	3·58	0·0
	H	238·6	175 812·0	4·86	0·0

Bartell has pointed out the C–H and C–C interactions also play an important role. As we shall see, more important than the inclusion of the C–C, C–H interaction is the overall barrier for the planar conformation. Bartell's potential is of the form:

$$V(C, C) = D/r^{12} - E/r^6 \tag{2}$$

$$V(C, H) = F/r^6(G \exp(-r/0\cdot49) - 1) \tag{3}$$

$$V(H, H) = H \exp(-r/0\cdot245) - I/r^6. \tag{4}$$

This potential yields a barrier height of 21·8 kJ mol^{-1} across $\theta = 0°$, which is rather large. π Electron calculations [22, 23] without geometry optimization have yielded a higher barrier height across the planar conformation (20 kJ mol^{-1}) than across the perpendicular conformation (8·4 kJ mol^{-1}). Barrier heights for the Bartell potential are close to the values obtained from this calculation. More accurate *ab initio* [24] calculations incorporating geometry optimization especially of the inter-ring C–C bond suggests that the barrier heights are in fact 5·0 and 18·8 kJ mol^{-1} for the planar and the perpendicular forms. All barrier heights reported by us here take into account both the conjugation effect and the *ortho–ortho* steric interaction. Table 2 lists the values of barrier heights for the various potential models, and also from π electron calculations [22, 23] and *ab initio* [24] studies. Bartell's potential has been used earlier to predict the properties of solid and gaseous biphenyl [14, 25]. The potential due to Haigh [14] models the non-bonded interaction potential in terms of the H–H interaction term alone. It has the same form as that of Bartell, but the values of the parameters are different:

$$V(H, H) = J \exp(-r/0\cdot234) - K/r^6. \tag{5}$$

More recently, Benkert, Heine and Simmons [16] have proposed a potential of the form

$$V(\theta) = g(L \exp(-N\theta^2) + M \sin^2(\theta)), \tag{6}$$

where θ is the dihedral angle (in radians). This potential includes both the conjugation energy and the energy arising out of the interaction of the H atoms at the *ortho* position. In contrast, all other potentials take into account only the non-bonded interaction, viz., the steric interaction between the atoms at the *ortho* positions. Hence the conjugation energy had to be accounted for separately. The intramolecular potential parameters are listed in table 3. The barrier across the planar conformation is of utmost importance. The intramolecular non-bonded inter-ring interaction between the atoms at the *ortho* positions has also been modelled additionally using the standard Williams and Kitaigorodskii potentials. The parameters are the same as those used in evaluating the intermolecular contribution and all three interactions (H–H, C–H

Table 2.   Approximate barrier heights in kJ mol^{-1} for different calculations and potential models of an isolated biphenyl molecule.

Barrier height at	Williams[a]	Kitaigorodskii[b]	BHS[c]	Bartell[d]	Haigh[e]	Ab initio[f]	π Calculation[g]
$\theta = 0°$	2·8	2·6	8·0	21·8	8·4	5·0	20·0
$\theta = 90°$	9·9	17·1	14·0	10·0	14·0	18·8	8·4

[a] From the Williams intramolecular potential.
[b] From the Kitaigorodskii intramolecular potential.
[c] From the BHS intramolecular potential [16].
[d] From the Bartell intramolecular potential [21].
[e] From the Haigh intramolecular potential [14].
[f] *Ab initio* calculation with geometry optimization [24].
[g] π-Electron calculation [14, 22, 23].

Table 3.   Intramolecular potential parameters for the BHS, Bartell and Haigh potentials.

D/kJ mol^{-1} Å^{-12}	E/kJ mol^{-1} Å^{-6}	F/kJ mol^{-1} Å^{-6}	G	H/kJ mol^{-1}	I/kJ mol^{-1} Å^{-6}	J/kJ mol^{-1}	K/kJ mol^{-1} Å^{-6}	L/kJ mol^{-1}	M/kJ mol^{-1}	N/rad^{-2}	g
12·51 × 10^5	13·59 × 10^2	5·22 × 10^2	358	2·75 × 10^4	2·05 × 10^2	2·86 × 10^4	3·39 × 10^2	35·2	41·05	2·5	2

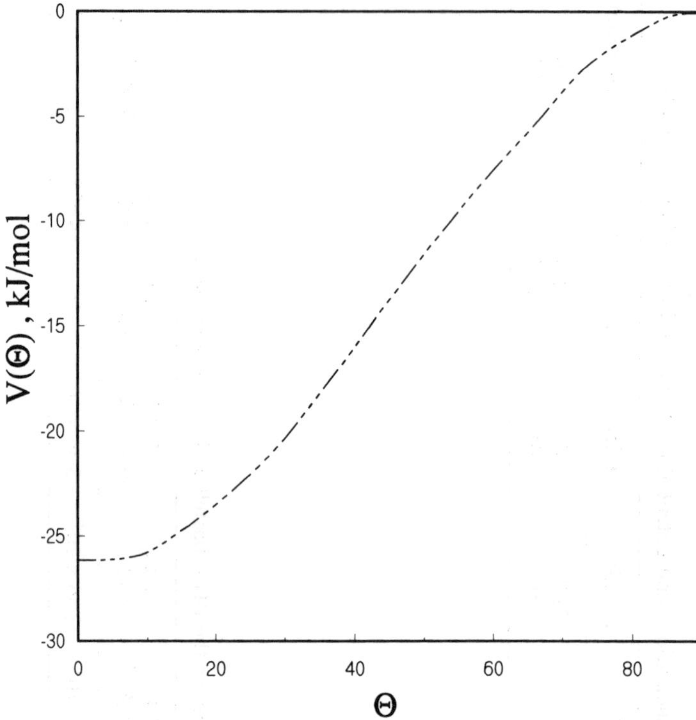

Figure 1. Dependence of conjugation energy on the dihedral angle θ. This is used in all models except the BHS to compute the intramolecular contribution.

and C–C) have been taken into account. The Williams and Kitaigorodskii potentials have a barrier lower than the *ab initio* value for the planar conformation (see table 2).

Fischer-Hjalmars [14] has reported several ways of estimating the conjugation energy. In this work we have used the estimate obtained from semi-empirical quantum chemical calculations [14]. The variation of the conjugation energy as a function of the dihedral angle is shown in figure 1.

Based on the electron diffraction data, Almeningen *et al.* [26] have proposed an intramolecular potential function with barrier heights 6·0 and 6·5 kJ mol^{-1} for the planar and perpendicular conformations, respectively. These values, especially, the latter, are considerably different from the *ab initio* [24] values and hence we preferred not to employ this potential. Similarly we do not employ the intramolecular potential function obtained from the molecular mechanics calculation by Stolevik and Thingstad [27], which predicts barrier heights of 8·4 and 7·5 kJ mol^{-1} for the planar and the perpendicular conformations, respectively. Recent *ab initio* work of Lenstra *et al.* [28] using a 4-21G basis set suggest a barrier of 7·9 kJ mol^{-1} for the planar conformation which compares well with those of the Haigh and BHS potential models. However, they have not given the full potential or the barrier for the perpendicular conformation.

In total, results from eight models were obtained. Of these, four employ the Williams intermolecular potential with the intramolecular potential of Williams, BHS, Bartell and Haigh. These are designated as WW, WBHS, WB and WH models, respectively. The remaining four models, referred to as KK, KBHS, KB and KH,

54   A. Chakrabarti *et al.*

are obtained by replacing the Williams intermolecular interaction with that of Kitaigorodskii.

3. Computational details

In recent molecular dynamics calculations on biphenyl, Baranyai and Welberry [17, 18] have employed the canonical ensemble and carried out simulations of the solid at room temperature using the Williams and the BHS potentials for the intermolecular and the intramolecular interactions. In order to ensure that the results obtained actually correspond to the potential used, we investigated the effect of the cut-off criteria and variable shape simulation on the results. Three sets of calculations were carried out on the Williams–BHS model. Calculations using the atom–atom cut-offs in the canonical ensemble, and the variable shape isothermal–isobaric ensemble are referred to as sets A and B, respectively. Set C uses the conditions of set B except that centre of mass–centre of mass (com–com) cut-off has been used in place of the atom–atom cut-off. The unit cell parameters for the three sets are listed in table 4. The com–com and the C–C radial distribution functions (RDFs) for set B and set C are shown in figure 2. The RDFs for set A are similar to those of set B and hence are not shown. RDFs of set B are sharp and narrow, indicating that the molecules are not performing the thermal motion normally expected of them at room temperature. In contrast, the RDFs of set C are broad and smoothly varying as expected of a room temperature solid. Closer examination revealed that the Coulomb

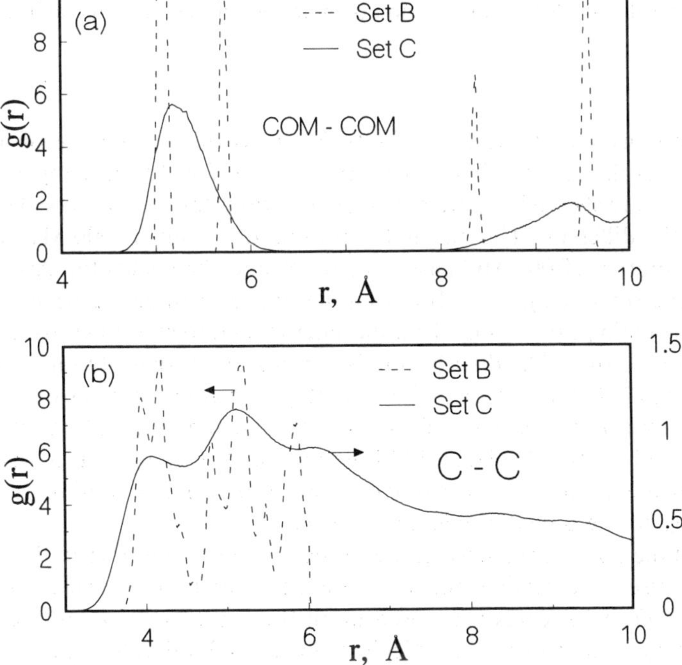

Figure 2.   (a) The centre of mass–centre of mass radial distribution function (RDF) for set B and set C for the room temperature solid phase. (b) C–C RDF for set B and set C for the room temperature solid phase.

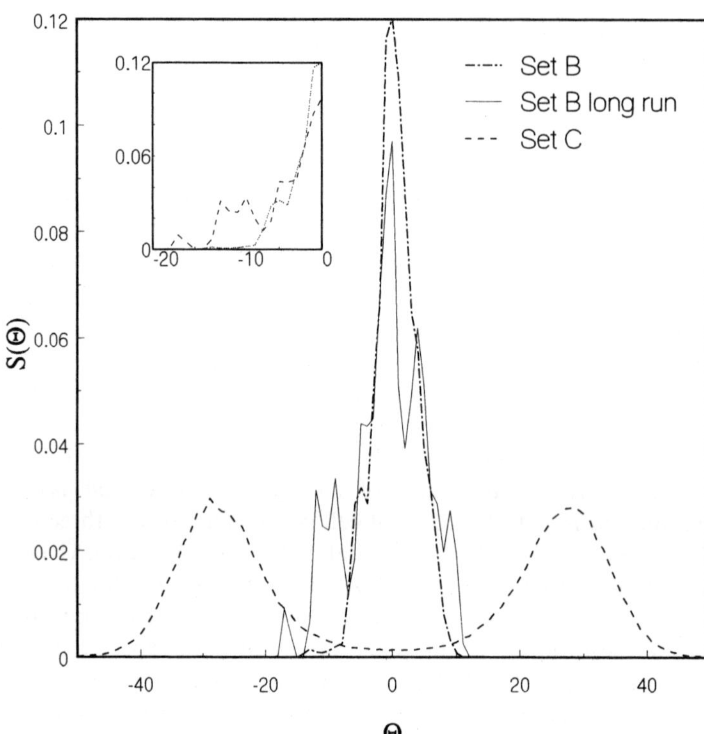

Figure 3. Distribution $S(\theta)$ of the dihedral angle between two phenyl rings from set B, set C, and for a long run of set B. Inset shows $S(\theta)$ from set B and longer run of set B on an expanded scale. $S(\theta)$ is the fraction of molecules per degree interval at a given value of θ.

term did not converge in the case of sets A and B due to the use of atom–atom cut-offs, and that the sum itself was fluctuating widely. The dihedral angle distribution for sets B and C are shown in figure 3. The distributions for the two sets of calculations are completely different. The normal duration of about 10 000 MC passes was extended by another 20 000 MC passes for set B, and the results for this are referred to as set B long run (see figure 3). It is seen that for set B there is a peak near $\theta = 10°$ apart from the main peak near $0°$. This shoulder tended to shift towards higher values of θ for the longer runs. The distribution is reminiscent of that obtained by Baranyai and Welberry [17, 18]. For set C, which employs the com–com cut-off, a symmetric bimodal distribution is observed. This suggests that the atom–atom cut-off should not be employed for the Williams–BHS model. Further, these results suggest that a constant pressure variable shape simulation is essential for obtaining the properties corresponding to the potential being employed.

In view of the non-convergence of the Coulomb contribution when the atom–atom cut-offs were used, we carried out all subsequent simulations using the com–com cut-off. A cut-off of 10 Å has been used in all the runs. The Coulomb contribution was obtained by direct summation. Neglect of the Coulomb interaction beyond 10 Å leads to a $\sim 2\%$ error in the calculation of the total intermolecular energy.

All the results discussed in the paper relate to the isothermal–isobaric ensemble on a system of $N = 72$ biphenyl molecules. The simulation was carried out on

A. Chakrabarti *et al.*

Table 4. The energetics and cell parameters for set A, set B and set C[a].

Set	$\dfrac{\langle E_{tot}\rangle}{\text{kJ mol}^{-1}}$	$\dfrac{\langle E_{c}\rangle}{\text{kJ mol}^{-1}}$	$\dfrac{\langle E_{intra}\rangle}{\text{kJ mol}^{-1}}$	$\dfrac{\langle E_{inter}\rangle}{\text{kJ mol}^{-1}}$	$\dfrac{a}{\text{Å}}$	$\dfrac{b}{\text{Å}}$	$\dfrac{c}{\text{Å}}$	$\dfrac{\alpha}{\text{deg}}$	$\dfrac{\beta}{\text{deg}}$	$\dfrac{\gamma}{\text{deg}}$	$\dfrac{v}{\text{Å}^3}$
A	$-13\cdot98$		$70\cdot45$	$-84\cdot43$	$8\cdot12$	$5\cdot63$	$9\cdot47$	$90\cdot00$	$95\cdot4$	$90\cdot00$	$216\cdot5$
B	$-24\cdot57$	$-34\cdot55$	$70\cdot01$	$-94\cdot58$	$8\cdot33$	$5\cdot73$	$9\cdot63$	$89\cdot88$	$95\cdot4$	$90\cdot09$	$228\cdot94$
C	$-13\cdot46$	$-14\cdot82$	$58\cdot47$	$-71\cdot93$	$8\cdot69$	$5\cdot53$	$9\cdot42$	$90\cdot06$	$97\cdot18$	$89\cdot39$	$227\cdot41$

[a] Experimental cell parameters are $a = 8\cdot12$ Å, $b = 5\cdot63$ Å, $c = 9\cdot47$ Å, $\alpha = \gamma = 90°$, $\beta = 95\cdot4°$.

$3 \times 4 \times 3$ (36 unit cells $\times$ 2 molecules per cell = 72 molecules) crystallographic unit cells. Solid biphenyl was modelled using a variable shape simulation cell. Calculations on solid biphenyl were carried out at a pressure of one atmosphere and at temperatures of 300 and 110 K. The imposition of periodic boundary conditions in the simulation prevent the system from exhibiting incommensurate behaviour. The starting configuration for the room temperature solid was taken from the room temperature crystallographic data of Trotter [7], and for the 110 K run the structure reported by Charbonneau and Delugeard [9] has been used. In modelling liquid biphenyl, a cubic simulation cell was used. Calculations on the liquid are reported at 400 K and atmospheric pressure. For liquid state calculations random starting configurations were used.

One MC pass is defined as N attempted MC steps, once for each of the N molecules. A single MC step consists of an attempted random translational displacement of the com, a random rotational displacement around a randomly chosen axis and a random intramolecular rotation around the central C–C bond joining the two rings. In the simulation of the liquid beiphenyl, prior to equilibration at 400 K, the liquid was kept at 900 K for a few hundred MC passes. For all the condensed phases, equilibration was carried out for about 10 000 MC passes and the averages were calculated over an equivalent number of MC passes.

4. Results and discussion

4.1. *Solid biphenyl*

In table 5, we list the thermodynamic properties of solid biphenyl at 300 K for all eight potential models. The heat of vapourization ΔH_{vap} was obtained from the expression.

$$\Delta H_{\mathrm{vap}} = E_{\mathrm{intra}}^{\mathrm{gas}} - (E_{\mathrm{inter}} + E_{\mathrm{intra}}) + P(V^{\mathrm{gas}} - V) + (H_{\mathrm{o}} - H) \tag{7}$$

$E_{\mathrm{intra}}^{\mathrm{gas}}$ was taken to be the Boltzmann average of intramolecular energy at 800 K. The $(H_{\mathrm{o}} - H)$ term represents the Berthelot correction for the deviation from ideal gas behaviour, and V^{gas} is the volume of ideal gas. It is seen that the best agreement for the heat of vapourization is obtained for models employing the Williams intermolecular interaction potential. For models employing the Kitaigorodskii intermolecular potential, the heat of vapourization is generally lower by 8–18 kJ mol^{-1} compared with the experimental value (81 kJ mol^{-1}) [29, 30] Among the models employing the Williams intermolecular potential, the best agreement is obtained for the WW model. The intramolecular contribution to the total interaction energy is between -8 and -16 kJ mol^{-1} for all potential models with the exception of the BHS model which gives a large and positive intramolecular contribution of nearly 57 kJ mol^{-1}. The densities for the various models listed in table 5, compare favourably with the experimental value of 1.185×10^3 kg m^{-3} [7]. The best agreement is, however, found for the WW model. The density for the KK model is somewhat higher (1.22×10^3 kg m^{-3}).

Table 6 lists the unit cell parameters a, b, c, α, β and γ. Overall, the models predict the unit cell parameter c better than b, the deviations being 1.4% and 9%, respectively. The deviation in the unit cell parameter a is considerably larger being in the range of 6–16%. All the models with the exception of the KK model, underestimate b and overestimate a. Models employing the Williams intermolecular potential predict the

A. Chakrabarti *et al.*

Table 5. Density and energies for solid biphenyl.

Temp.	Model	$\langle E_{tot}\rangle$/kJ mol^{-1}	$\langle E_{inter}\rangle$/kJ mol^{-1}	$\langle E_{intra}\rangle$/kJ mol^{-1}	ΔH_{vap}/kJ mol^{-1}	$\rho \times 10^3$/kg m^{-3}
300 K	Expt	—	—	—	81·00	1·185
	WW	−86·06	−78·18	−7·88	81·93	1·181
	WH	−88·16	−76·21	−11·95	78·24	1·142
	WBHS	−13·46	−71·93	58·47	75·38	1·138
	WB	−79·37	−68·91	−10·46	73·22	1·093
	KK	−84·89	−69·26	−15·63	73·85	1·226
	KH	−76·91	−62·86	−14·05	66·99	1·153
	KBHS	−4·45	−61·26	56·81	66·32	1·138
	KB	−69·88	−58·37	−11·51	63·73	1·127
110 K	Expt	—	—	—	—	1·245
	WW	−92·68	−85·20	7·48	95·49	1·238
	WH	−93·38	−80·22	−13·16	90·39	1·202
	WBHS	−20·85	−77·85	57·00	89·65	1·168
	WB	−86·64	−75·89	−10·75	87·42	1·155
	KK	−93·79	−78·83	−14·96	89·64	1·285
	KH	−87·17	−73·55	−13·62	84·18	1·256
	KBHS	−12·20	−68·93	56·73	81·00	1·226
	KB	−79·74	−68·18	−11·56	80·53	1·232

Table 6. Cell parameters of solid biphenyl at 300 K.

Model	$a/\text{Å}$	$c/\text{Å}$	$c/\text{Å}$	α/deg	β/deg	γ/deg
Expt	8·12	5·63	9·47	90·00	95·40	90·00
WW	8·26	5·62	9·34	90·48	92·75	90·18
WH	8·59	5·57	9·38	90·26	93·59	90·15
WBHS	8·69	5·53	9·42	90·06	97·18	89·39
WB	9·08	5·51	9·39	91·23	95·23	90·18
KK	7·52	5·95	9·39	86·16	95·15	89·46
KH	8·77	5·40	9·40	92·49	95·10	90·64
KBHS	9·40	5·12	9·48	89·30	99·68	88·17
KB	8·62	5·58	9·60	88·60	100·88	90·64

value of α more accurately. A larger deviation is observed in the value of β. Considering the overall performance, the WW model seems to perform better than others. The value of β (92·7°), however, is lower than the crystallographic value of 95·4°. This result is in agreement with the work of Busing [15] on the solid phase of biphenyl; in earlier work on the high pressure monoclinic phase of benzene [31] it was found that the Williams potential underestimated the value of β. But γ is predicted satisfactorily by all the models.

Figure 4 shows a plot of the com–com and the C–C RDFs for different potential models. The positions of the com–com peaks from the X-ray crystal structure [7] are indicated by vertical lines. The relative heights of the lines are proportional to their intensities. The RDFs exhibit the following characteristics. All models except the KK and the WW show an absence of fine structure in the com–com as well as the C–C RDFs. As we shall see shortly, this seems to be related to the nature of the dihedral angle distribution. Model KK exhibits a shift by about 0·25 Å of the first peak towards lower distances; the peak appears at 4·7 Å for model KK compared with 4·95 Å for the X-ray data. The second peak at 5·5 Å shows an outward shift to higher r values for model KK whereas all the other models predict the second peak correctly. The third peak near 8 Å is not predicted correctly by any of the KX (X = BHS, B, H and K) models. The Williams models show the third peak, or a shoulder at a slightly higher distance, around 8·5 Å. The WW model shows a clear peak around 8·2 Å. The principal lacunae in the Kitaigorodskii intermolecular potential appears to be the absence of Coulomb interaction. In particular, recent studies have suggested that a quadrupolar interaction between aromatic rings is of vital importance in modelling any molecule possessing phenyl rings. Thus, models for the benzene molecule not accounting for the quadrupolar term were unsuccessful in predicting the properties of condensed phases of benzene. In particular, the PS model of benzene [31], which underestimated the quadrupolar contribution, exhibited a large inward shift of the first peak in the com–com RDF as compared with the experimental structure; this is similar to the shift of first peak exhibited by the KK model in figure 4. The intramolecular interactions in the KBHS, KB and KH models seems to be responsible for the absence of any such inward shift. Another consequence of the non-inclusion of the quadrupolar interactions is the high density of the KK model. Judging from the relative success of the various models in predicting the heat of vapourization, density, cell parameters and the com–com and C–C RDFs, it appears that only WW, WH and WBHS perform satisfactorily.

60   A. Chakrabarti *et al.*

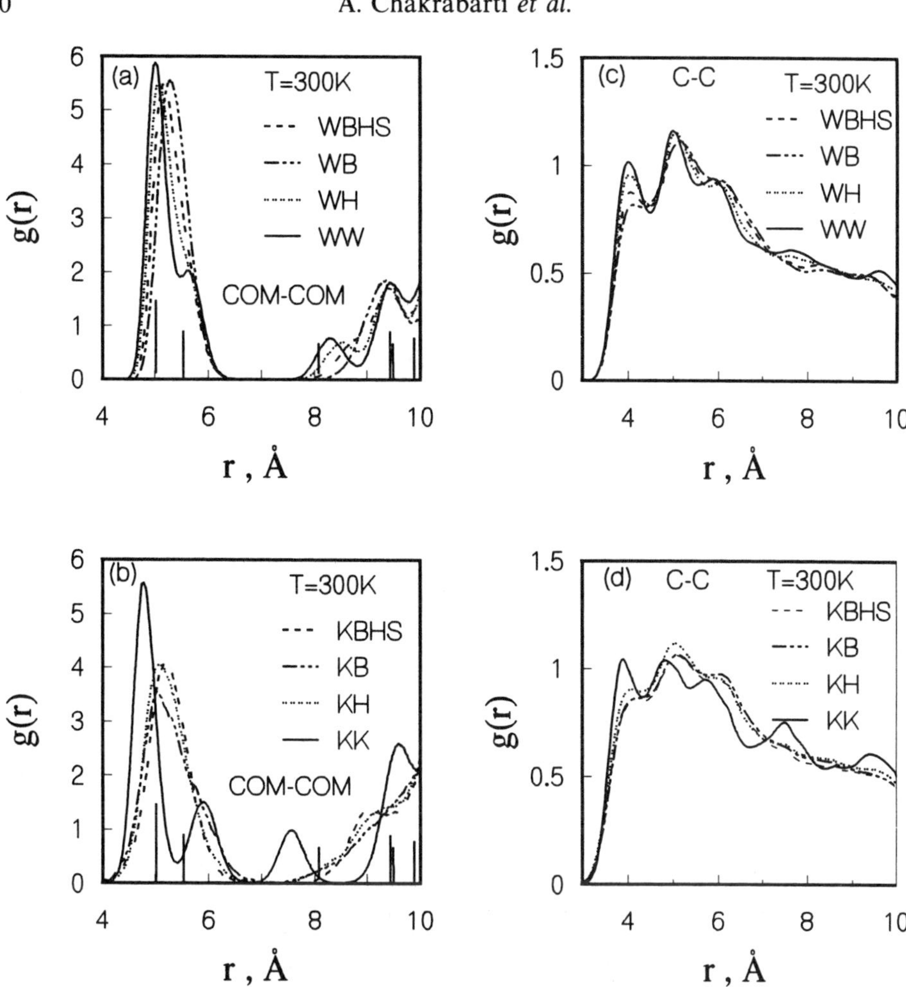

Figure 4. Radial distribution functions for solid biphenyl at 300 K with the Bartell, BHS, and Haigh models and the Williams or Kitaigorodskii intramolecular potential. (*a*) Centre of mass–centre of mass RDF for models employing intermolecular potential given by Williams and Cox; (*b*) same as (*a*) but for models employing the Kitaigorodskii intermolecular potential; (*c*) C–C RDF for the Williams intermolecular potential; and (*d*) C–C RDF for the Kitaigorodskii intermolecular potential. Peak positions in the com–com RDF corresponding to the X-ray structure are marked by vertical lines. The lengths of the lines correspond to the relative peak heights.

Dihedral angle distribution functions for the various models are shown in figure 5. Models using the Bartell and BHS intramolecular potential functions show a clear bimodal distribution. The former shows a zero intensity near $\theta = 0°$. The latter shows a small non-zero intensity near $\theta = 0°$. The Haigh potential shows a distribution which may be described as lying somewhere between bimodal and monomodal. Both the WW and KK models show a monomodal function with a maximum near $\theta = 0°$, suggesting the most probable conformation is the planar conformation in the room temperature solid phase. The RDFs for these two models show well defined features which seem to be correlated with the monomodal $S(\theta)$ exhibited by them.

The average dihedral angles calculated from the $S(\theta)$ for the various models are listed in table 7. Two different types of average can be calculated. (i) The first is the

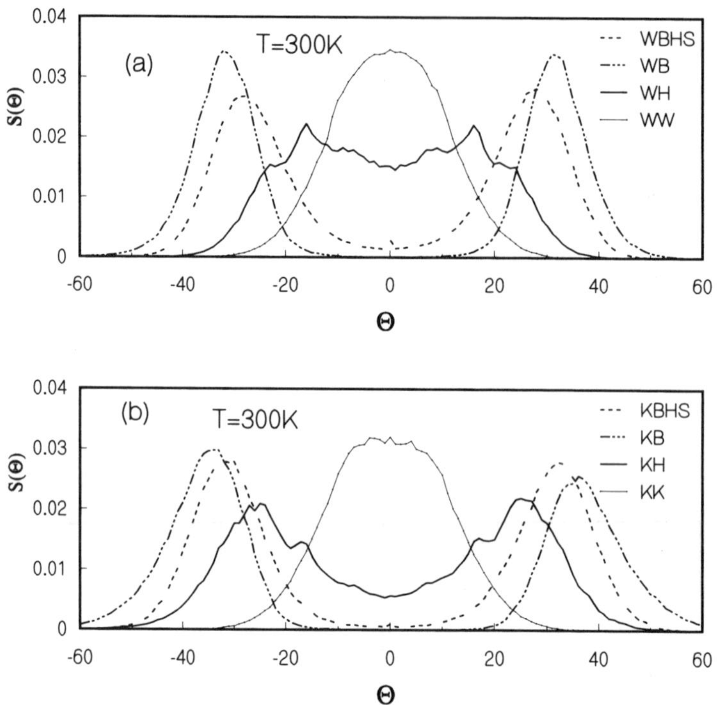

Figure 5. Dihedral angle distribution function for solid biphenyl at 300 K for models containing the intermolecular potential given by (*a*) Williams and Cox, and by (*b*) Kitaigorodskii.

Table 7. Average dihedral angle calculated as described in the text.

Model	$\bar{\theta}$/deg	$\langle\theta\rangle$
WW	8·78	−0·02
WH	14·94	0·02
WBHS	26·20	0·43
WB	31·92	−0·55
KK	9·39	−0·09
KH	22·70	0·27
KBHS	30·90	0·32
KB	37·10	−1·94

average angle as obtained by averaging between $0°$ and $90°$ and between $-90°$ and $0°$ separately. The overall average is then obtained by taking the mean of the modulus of the two values. This is represented by $\bar{\theta}$ in table 7. The average angle that is obtained from X-ray crystallography, however, is different from $\bar{\theta}$. (ii) Another average $\langle\theta\rangle$ has been obtained by calculating the average over the range between $-90°$ and $+90°$. Such an average would correspond to the X-ray crystallographic average. It is seen that $\langle\theta\rangle$ is nearly zero for all of the models investigated. Hence, it is not possible to rule out any of the models here on the basis of X-ray crystallographic data [7, 8]. However, the X-ray data of Charbonneau and Delugeard [9] have also indicated that the libration tensor L_{22} along the long axis has a

                           A. Chakrabarti *et al.*

significantly large amplitude of $105 \cdot 9°^2$. *Ab-initio* [24] calculations and X-ray and electron-diffraction studies indicate a torsional amplitude of $10–15°$ [32]. From figure 5 it is clear that only the WH, WW and KH, KK models have an amplitude in this range. Other models, in particular WBHS, WB, KBHS and KB, have a considerably larger amplitude $(30–35°)$. As has been pointed out, the recent molecular dynamics calculations of Baranyai and Welberry [17, 18] have suggested yet another possible dihedral angle distribution: a main peak near $\theta = \theta°$ and a shoulder around $22°$ which develops into a peak at low temperatures. However, neither experimental nor any of the existing theoretical investigations to date have confirmed such a possibility. Furthermore, as pointed out in the previous section, the dihedral angle distributions obtained by them seems to be an artefact of the possible use of the atom–atom cut off in their calculations. The thermodynamic and unit cell parameters obtained by us also indicate that the constant pressure calculation with variable shape yield properties which are significantly different from the constant volume calculation.

4.2. *Structure–potential correlations*

Here we analyse possible correlations between the structure and the potential employed. We are able to do this because of the large number of potentials employed in the present study. The densities obtained with the different models based on the Kitaigorodskii intermolecular potential (see table 5) are in the order: KK(1·226) > experiment(1·185) [7] > KH(1·153) > KBHS(1·138) > KB(1·127). Let us now look at the probability of the biphenyl molecules being planar in the solid phase, by taking the intensity of the dihedral angle distribution at $\theta = 0°$ as a measure of the planarity. We see that the intensity decreases in the order KK > KH > KBHS > KB (see figure 5), suggesting that there is a strong correlation between the planarity of the molecule and the density of the solid. Similar trends are seen in the models based on Williams intermolecular potential models, where the density as well as the planarity vary in the order WW > experiment > WH > WBHS > WB. These trends suggest that the large barrier for rotation around the inter-ring C–C bond near $\theta = 0°$ is responsible for the low density and demonstrate the relationship between intramolecular potential and thermodynamic properties. Since the experimental value of the density lies close to the values given by the WW and WH models, and since the torsional amplitudes for these two models are comparable with the experimental value, it appears that the dihedral angle distribution in the real crystal could be closer to that exhibited by the WW and WH models.

We now compare two potential models with the same intramolecular potential but different intermolecular potentials. For example, comparing the results from KX (X = BHS, Bartell or Haigh) with those from the corresponding WX models indicates that the WX models always have a higher intermolecular contribution to the total energy (see table 5). Figure 5 indicates that, for a given intramolecular potential, the molecules in the solid are more planar if the intermolecular contribution is larger. This finding is in agreement with that of Casalone *et al.* [25], who found that the planarity of the phenyl rings in the solid phase may be related to the intermolecular energy. The present study demonstrates that there is indeed a one-to-one correspondence.

Overall, it appears that the WW and WH models give the best description of the room temperature solid. Both of them, however, underestimate β. In the present

Table 8. Cell parameters of solid biphenyl at 110 K.

Model	$a/\text{Å}$	$b/\text{Å}$	$c/\text{Å}$	α/deg	β/deg	γ/deg
Expt	7·82	5·58	9·44	90·00	94·62	90·00
WW	8·00	5·61	9·20	89·59	92·24	89·86
WH	8·35	5·51	9·28	89·14	95·00	89·25
WBHS	8·55	5·50	9·36	90·72	96·44	90·29
WB	8·62	5·50	9·41	90·47	97·70	90·35
KK	7·17	5·98	9·32	88·26	94·30	89·14
KH	7·81	5·61	9·34	89·94	95·37	90·00
KBHS	7·98	5·59	9·42	90·61	97·67	90·96
KB	8·08	5·50	9·45	90·38	98·80	90·28

study no dynamical properties have been calculated and hence it is not possible to comment on the performance of the Williams model in predicting dynamical properties accurately. However, it is worthwhile to note that an earlier study of solid benzene using the Williams intermolecular potential found that it leads to a rather tightly packed structure [31].

4.3. *The solid at 110 K*

Table 8 lists the values of the cell parameters for the various models along with the values obtained from X-ray diffraction measurements [9]. It is seen that the deviations of the cell parameters from the experimental values of Charbonneau and Delugeard [9] are similar to those seen at 300 K. Thus, all of the cell parameters are predicted satisfactorily by the WW and the WH models with the exception of β and a, respectively. The KK model predicts β well but the density is too large, a is too small and α deviates significantly from 90°. The dihedral angle distribution functions for different models are shown in figure 6. The general features for the various models are similar to those of the room temperature solid. Models employing the Haigh intramolecular potential show a slight change in the shape of the curve with respect to the room temperature solid; the peaks near $\pm 20°$ appear to be split peaks. At sufficiently low temperatures, the dihedral angle in the solid is non-zero (10° at 22 K from neutron diffraction studies of Cailleau and Baudour [33]).

4.4. *The liquid phase*

Only the WW and WH models perform satisfactorily in predicting the properties of solid biphenyl, and since the results for these two models are on the whole only slightly different, we choose only the WW model to study the liquid phase of biphenyl. In addition, for purposes of comparison and to understand the effect of neglect of quadrupolar interaction on the thermodynamic and structural properties of liquid biphenyl, we have chosen the KK model. The thermodynamic properties at 400 K are listed in table 9. The calculated heat of vapourization for both models lies within about 4% of the experimental value [34], the densities being 0·879 and 0·978, respectively, for the two models (the experimental value is $0·866 \times 10^3 \text{ kg m}^{-3}$ [34]).

In figure 7 we show a plot of the com–com and the C–C RDFs for the KK and

Table 9. Density and energies for liquid biphenyl at 400 K.

Model	$\langle E_{tot}\rangle$/kJ mol^{-1}	$\langle E_{inter}\rangle$/kJ mol^{-1}	$\langle E_{intra}\rangle$/kJ mol^{-1}	ΔH_{vap}/kJ mol^{-1}	$\rho \times 10^3$/kg m^{-3}
Expt[a]	—	—	—	54·00	0·866[a]
WW	−60·79	−52·31	−8·48	56·18	0·879
KK	−67·85	−51·89	−15·96	56·29	0·978

[a] Studied at 293 K.

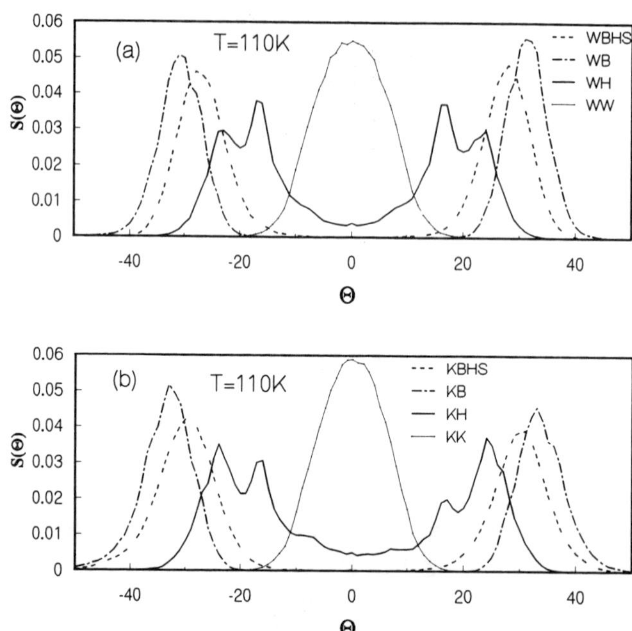

Figure 6. Dihedral angle distribution function for solid biphenyl at 110 K for models containing the intermolecular potential given by (*a*) Williams and Cox, and by (*b*) Kitaigorodskii.

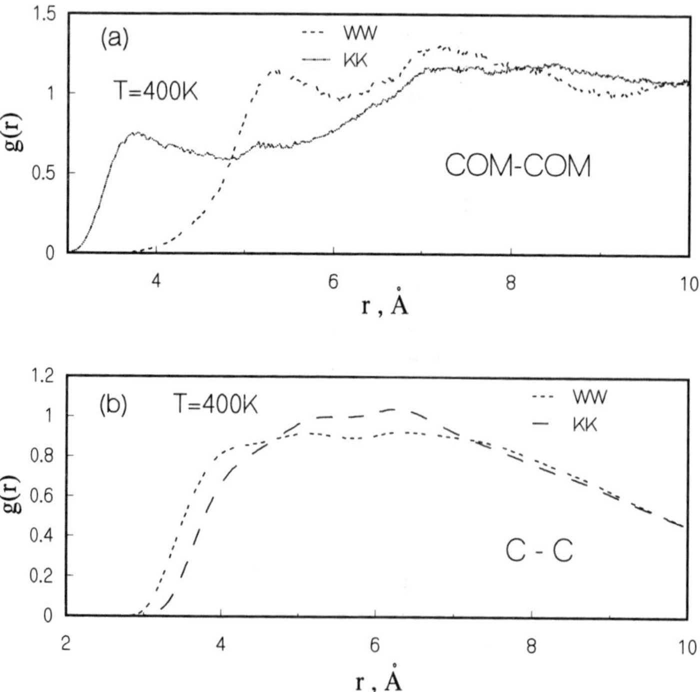

Figure 7. Radial distribution function for liquid biphenyl at 400 K for the WW and KK models: (*a*) com–com; and (*b*) C–C.

66   A. Chakrabarti *et al.*

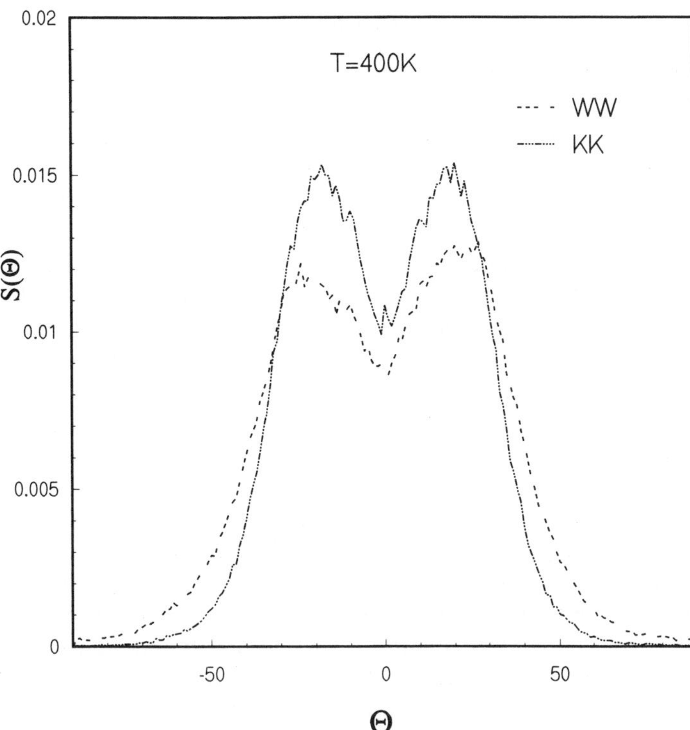

Figure 8.   Dihedral angle distribution function for liquid biphenyl at 400 K for the WW and KK models.

WW models. The KK model starts at a significantly lower distance (3 Å) compared with the WW model. This distance seems to be too small in view of the fact that two benzene molecules in liquid benzene do not approach closer than 4 Å [31]. The Kitaigorodskii intermolecular potential seems to give a wrong packing of the nearest neighbours in the liquid as well. Thus, it appears that the absence of quadrupolar interactions leads to the inward shift of the first peak of the com–com RDF and to a significantly higher density. The first peak and the main peak in the com–com $g(r)$ appear around 5·2 Å and 7·2 Å, respectively. The C–C RDF shows a broad peak with a maximum around 6·4 Å and a shoulder around 4·2 Å. This shoulder is characteristic of all hydrocarbons and corresponds to the peripheral groups between neighbouring molecules [31]. Thus, the shoulder around 4·2 Å in liquid biphenyl arises from the nearest CH groups.

The dihedral angle distribution for the liquid is shown in figure 8. Both the WW and the KK models show a bimodal distribution. This is in contrast to the solid where a monomodal distribution was observed for these potential models. The trends are in agreement with experiment where it is known that the rings have a non-zero dihedral angle in the solution phase. The average dihedral angle is found to be 24·6° and 20·4° for WW and KK models, respectively. This may be compared with the experimental value of 19–26° for the solution phase [5, 6]. The distribution has a larger width for the WW model, extending almost over the whole range of θ ($-90°$ to $90°$).

5. Conclusion

Of the eight models investigated, the overall agreement with the experimental structure of biphenyl could be reproduced most satisfactorily only by the WW and WH models. Structurally, the WW and WH models predict all the unit cell parameters satisfactorily except β. An inward shift of the nearest neighbours is observed for the room temperature solid phase as well as the liquid phase for the KK model. This is attributed to the absence of quadrupole interaction between the phenyl rings. The dihedral angle distribution and the torsional amplitude of 10–$15°$ observed experimentally are again best reproduced by the WW and WH models. Consequently, $S(\theta)$ in real crystals may be described as somewhere in between monomodal and weakly bimodal with a significant non-zero intensity near $\theta = 0°$. The results obtained here suggest that the distribution obtained by Baranyai and Welberry [17, 18] where some biphenyls have near zero dihedral angle while others have a value of $\theta = 22°$, may be an artefact due to the possible use of atom–atom cut-off in their simulations.

Important correlations have been observed in solid biphenyl. The planarity of the molecules in the crystal is higher, the higher the density. In addition, the intermolecular contribution to the total interaction energy is larger, the larger is the planarity. Potentials employing the Williams intermolecular function generally lead to higher planarity. The calculated dihedral angle for the liquid phase is in good agreement with available experimental data for the solution phase. Table 2 and the results obtained here suggest that actual barrier height for the planar conformation are between $2\cdot8$ and $8\cdot4$ kJ mol^{-1}, which are the heights for the WW and WH models, respectively. The range of barrier heights is comparable to the *ab initio* result ($5\cdot0$ kJ mol^{-1}). Any proposed potential should improve the prediction of unit cell parameters β and a.

References

[1] KARLE, I. L., and BROCKWAY, L. O., 1944, *J. Am. chem. Soc.*, **66**, 1974.
[2] BASTIANSEN, O., 1949, *Acta Chem. Scand.*, **3**, 408.
[3] ALMENINGEN, A., and BASTIANSEN, O., 1958, *K. Nors. Vidensk. Seisk. Skr.*, **4**, 1.
[4] EATON, V. J., and STEELE, D., 1973, *J. chem. Soc. Faraday Trans. ii*, **69**, 1601.
[5] ALMASY, F., and LAEMMEL, H., 1950, *Helv. Chim. Acta*, **33**, 2092.
[6] SUZUKI, H., 1959, *Bull. chem. Soc. Jap.*, **32**, 1340.
[7] TROTTER, J., 1961, *Acta Crystallogr.*, **14**, 1135.
[8] HARGREAVES, A., and RIZVI, S. H., 1962, *Acta Crystallogr.*, **15**, 365.
[9] CHARBONNEAU, G. P., and DELUGEARD, Y., 1976, *Acta Crystallogr. B*, **32**, 1420.
[10] CAILLEAU, H., MOUSSA, F., and MONS, J., 1979, *Solid State Commun.*, **31**, 521.
[11] BREE, A., and EDELSON, M., 1977, *Chem. Phys. Lett*, **46**, 500
[12] FRIEDMAN, P. S., KOPELMAN, R., and PRASAD, P. N., 1974, *Chem. Phys. Lett.*, **24**, 15.
[13] BRENNER, H. C., HUTCHISON, C. A. and KEMPLE, M. D., 1974, *J. chem. Phys.*, **60**, 2180.
[14] FISCHER-HJALMARS, I., 1963, *Tetrahedron*, **19**, 1805.
[15] BUSING, W. R., 1983, *Acta Crystallogr. A*, **39**, 340.
[16] BENKERT, C., HEINE, V., and SIMMONS, E. H., 1987, *J. Phys. C*, **20**, 3337.
[17] BARANYAI, A., and WELBERRY, T. R., 1991, *Molec. Phys.*, **73**, 1317.
[18] BARANYAI, A., and WELBERRY, T. R., 1992, *Molec. Phys.*, **75**, 867.
[19] WILLIAMS, D. E., and COX, S. R., 1984, *Acta Crystallogr. B*, **40**, 404.
[20] KITAIGORODSKII, A. I., *Molecular Crystals and Molecules* (New York: Academic Press).
[21] BARTELL, L. S., 1960, *J. chem. Phys.*, **32**, 827.
[22] GONDO, Y., 1964 *J. chem. Phys.*, **41**, 3928.
[23] GOLEBRIEWSKI, A., and PARCZEWSKI, A., 1967, *Theoret. Chim. Acta*, **7**, 171.

68   A. Chakrabarti *et al.*

[24] ALMLOF, J., 1974, *Chem. Phys.* **6,** 135.
[25] CASALONE, G., MARINARI, C., MUGNOLI, A., and SIMONETTA, M., 1968, *Molec. Phys.*, **15,** 339.
[26] ALMENINGEN, A., BASTIANSEN, O., FERNHOLT, L., CYVIN, B. N., CYVIN, S. J., and SAMDAL, S., 1985, *J. molec. Struct,* **128,** 59.
[27] STOLEVIK, R., and THINGSTAD, O., 1984, *J. molec. Struct.,* **106,** 333.
[28] LENSTRA, A. T. H., VAN ALSENOY, C., VERHULST, K., and GEISE, H. J., 1994, *Acta Crystallogr. B,* **50,** 96.
[29] SAITO, K., ATAKE, T., and CHIHARA, H., 1988, *Bull. chem. Soc. Jap.,* **61,** 679.
[30] MORAWETZ, E., 1972, *J. chem. Thermodyn.,* **4,** 455.
[31] YASHONATH, S., PRICE, S. L., and McDONALD, I. R., 1988, *Molec. Phys.,* **64,** 361.
[32] BASTIANSEN, O., KVESETH, K., and MOLLENDAL, H., 1979, *Topics current Chem.,* **81,** 99.
[33] CAILLEAU, H., and BAUDOUR, J. L., 1979, *Acta Crystallogr. B,* **45,** 426.
[34] *CRC Handbook of Physics and Chemistry,* 1985 (Boca Raton: CRC Press).

5228 Reprinted from **The Journal of Physical Chemistry, 1995, _99_.**
Copyright © 1995 by the American Chemical Society and reprinted by permission of the copyright owner.

The Metal–Nonmetal Transition: A Global Perspective

P. P. Edwards

The School of Chemistry, University of Birmingham, Birmingham, B 15 2TT, U.K.

T. V. Ramakrishnan

Department of Physics, Indian Institute of Science, Bangalore 560 012, India

C. N. R. Rao*

CSIR Centre of Excellence in Chemistry, Indian Institute of Science, Bangalore 560 012, India, and Department of Chemistry, University of Wales, Cardiff CF1 3TB, U.K.

Received: October 24, 1994; In Final Form: January 26, 1995

A wide range of condensed matter systems traverse the metal–nonmetal transition. These include doped semiconductors, metal–ammonia solutions, metal clusters, metal alloys, transition metal oxides, and superconducting cuprates. Certain simple criteria, such as those due to Herzfeld and Mott, have been highly successful in explaining the metallicity of materials. In this article, we demonstrate the amazing effectiveness of these criteria and examine them in the light of recent experimental findings. We then discuss the limitations in our understanding of the phenomenon of the metal–nonmetal transition.

1. Introduction

Of all the physical properties of condensed materials, the electrical conductivity exhibits the widest range, anywhere from 10^{-22} ohm^{-1} cm^{-1} in the best nonmetals to around 10^{10} ohm^{-1} cm^{-1} in pure metals (not in the superconducting state). There are several situations where condensed phases transform from the metallic to the nonmetallic state on changing thermodynamic parameters such as temperature, pressure, and composition, with the electrical conductivity changing by factors of 10^3–10^{14} over a small range of the thermodynamic parameter.[1-3] In Figure 1, we show the temperature–composition plane with the electron density varying between 10^{12} and 10^{30}, with the temperature going up to 10^{10} K. The normal experimental conditions where we find metals and semiconductors are indicated in the figure, as are also the conditions appropriate to stars. The elements H, Xe, Cs, and Hg shown by asterisks are at conditions close to the critical points attained by one of many routes such as shock waves, wire explosions, or MHD (magnetohydrodynamic implosion). The figure also shows the regime for a degenerate strongly coupled plasma, besides the values of the Wigner–Seitz radius, r_s, given by

$$r_s = \left(\frac{3}{4\pi n a_0^3}\right)^{1/3} \tag{1}$$

where n is the conduction electron density and a_0 is the Bohr radius. The largest value of r_s in metals[5] is found for Cs. In all the systems in Figure 1, the possibility of a thermodynamically induced transition from the metallic to the nonmetallic regime exists. Clearly, the metal–nonmetal transition is a

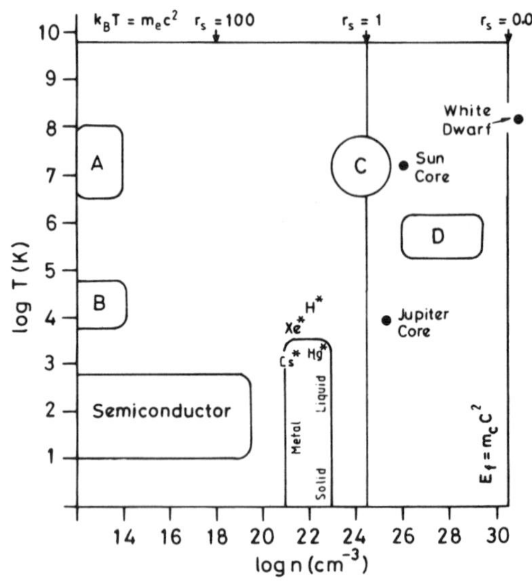

Figure 1. A nonrelativistic window of the temperature–composition plane, showing electron density (n) and temperature (T). Normal conditions (on earth) for semiconductors and elemental metals and conditions on the Sun, Jupiter, and the White Dwarf are shown. Experimental methods in A, B, C, and D are Tokamak, glow-discharge, laser fusion, and degenerate strongly coupled plasma, respectively. Wigner–Seitz radii, r_s, are also shown (adapted from Redmer[4]).

problem of vital interest, being concerned with a wide range of issues from the metallization of stars to the size-induced transition in small clusters of metals.[3,6,7] Between these two extrema is a myriad of examples in a variety of condensed matter systems.[1,2] The range of systems traversing the metal–nonmetal

* To whom correspondence should be addressed at CSIR Centre of Excellence in Chemistry, Indian Institute of Science, Bangalore 560 012, India.
® Abstract published in *Advance ACS Abstracts*, March 15, 1995.

Feature Article

J. Phys. Chem., Vol. 99, No. 15, 1995 **5229**

transition is continually increasing. Thus, oxides exhibiting high-temperature superconductivity are close to or at the metal—nonmetal boundary,[8,9] and some of them actually exhibit metal to nonmetal transitions with a change in composition.

In spite of the plethora of experimental findings, the status of our theoretical understanding of metal—nonmetal transitions is far from satisfactory. The difficulty is largely intrinsic to the phenomenon. Electronic states involved in charge transport(i.e., those near the Fermi energy) are spatially extended in the metal and are localized in the insulator. The localization may be due to static disorder (Anderson localization), to strong local electron—electron correlations which "freeze" the local electron number (Mott transition), or to strong electron—lattice coupling which traps the electron locally.[1,2,10,11] In all these cases, the natural modes of description of the electronic states in the different phases are diametrically opposite; it is difficult to find an approach which does both. Secondly, in many cases, more than one mechanism is operative, and one may reinforce the other. For example, in a disordered, strongly correlated oxide, Anderson localization due to disorder tends to increase the local correlation effect. In all electronic systems, the Coulomb interaction which is relatively weak and short ranged in a metal but strong and long ranged in an insulator (giving rise to bound electron—hole states) is present and can be important in promoting the insulating state. There is one mechanism for the metal—nonmetal (M—NM) transition, in a crystalline solid, that does not involve localized states. This is the transition of electrons from a fully filled band (insulator) to a partially filled band (metal) under pressure or structural change. This transition, however, appears to be uncommon. We discuss these mechanisms in some detail in section 6 (see also refs 1, 2, 10, and 11).

There are certain simple criteria for the occurrence of the metal—nonmetal transition, based on powerful physical concepts which turn out to be surprisingly successful. One such criterion is the idea due to Mott that in a low carrier density metal, the screened Coulomb attraction may be strong enough to bind an electron hole pair, thus destabilizing the metal.[2,12] Another useful criterion due to Mott[13] is the idea of a minimum conductivity, σ_{min}, that a metal can support, corresponding to the mean free path being equal to the de Broglie wavelength of the electron at the Fermi energy. Then, there is the Herzfeld metallization criterion of the dielectric catastrophe[14] which could occur for a dense collection of polarizable atoms. We discuss these criteria further in sections 3—5. We shall first present a brief overview of phenomena and systems associated with the M—NM transition.

2. Diverse Systems Exhibiting Metal—Nonmetal Transitions

As mentioned earlier, a large variety of systems exhibit M—NM transitions. The systems include[1-3] metal—ammonia (or amine) solutions, expanded metals, doped semiconductors, metal—noble gas films, metal—metal halide melts, alloys of gold with metals such as cesium, transition metal oxides and sulfides, and other inorganic and organic solids. These systems have been adequately reviewed, and we shall briefly examine only those findings that are new or directly relevant to the later discussion. Transition metal oxides[15] are especially noteworthy in that the M—NM transition in them can arise from one of many causes. Typical of the transitions found in oxide systems are the following: (i) pressure-induced transitions as in NiO, (ii) transitions as in Fe_3O_4 involving charge ordering, (iii) transitions as in $LaCoO_3$ that are initially induced because of the different spin configurations of the transition metal ion, (iv)

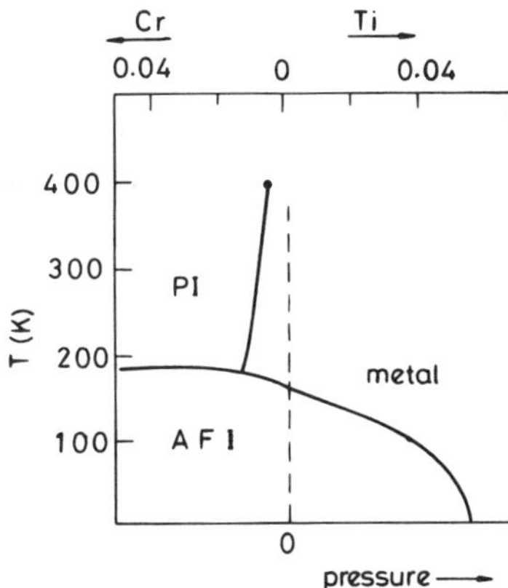

Figure 2. Phase diagram of the metal insulator transition in V_2O_3 as a function of doping with Cr or Ti (regarded as chemical pressure) as well as actual pressure, showing a critical point (after McWhan, D. B.; et al. *Phys. Rev. Lett.* **1971**, 27, 941). The three phases are paramagnetic insulator (PI), metal, and antiferromagnetic insulator (AFI). The different kinds of resistivity behavior in Figure 3 correspond to different constant pressure (or Cr, Ti content) cuts in this phase diagram.

transitions as in EuO arising from the disappearance of spin polarization band-splitting effects when the ferromagnetic Curie temperature is reached, (v) compositionally induced transitions, as in $La_{1-x}Sr_xCoO_3$ and $LaNi_{1-x}Mn_xO_3$, in which changes of band structure in the vicinity of the Fermi level are brought about by a change in composition or are due to disorder-induced localization, (vi) transitions as in $K_{0.3}MoO_3$ due to charge-density waves, and (vii) temperature-induced transitions in a large class of oxides such as Ti_2O_3, VO_2, and V_2O_3 involving more than one mechanism.

The last category of M—NM transitions has attracted considerable attention. In Ti_2O_3, a second-order transition occurs around 410 K, accompanied by a gradual change in the rhombohedral c/a ratio and a 100-fold jump in conductivity; the oxide remains paramagnetic throughout. A simple band-crossing mechanism occurring with the change in the c/a ratio can explain this transition. Accordingly, substitution of Ti by V up to 10% in Ti_2O_3 makes the system metallic; the c/a ratio of this metallic solid solution and the high-temperature phase of Ti_2O_3 are similar. In VO_2, a first-order transition occurs around 340 K, accompanied by a change in structure (monoclinic to tetragonal) and a 10^4-fold jump in conductivity; the material remains paramagnetic throughout. A crystal distortion model wherein a gap opens up in the low-temperature low-symmetry structure adequately explains the transition. Substitution of trivalent ions such as Cr^{3+} and Al^{3+} for vanadium in VO_2 leads to a complex phase diagram with at least two insulating phases whose properties are significantly different from those of the insulating phase of pure VO_2. These phases are now fairly well understood.

The M—NM transition in V_2O_3 and its alloys has been the subject of a large number of publications.[1,2,15-19] Pure V_2O_3 undergoes a first-order transition (monoclinic—rhombohedral) at 150 K accompanied by a 10^7-fold jump in conductivity and an antiferromagnetical—paramagnetic transition. The carrier effective mass and other properties also show large changes at this transition. Application of pressure makes V_2O_3 increasingly metallic, thus suggesting that it is near a critical point. Accordingly, doping with Ti or Cr has a marked effect on the transition; the former has a positive pressure effect and the latter a negative pressure effect (Figure 2). V_2O_3 also shows a second-

5230 *J. Phys. Chem., Vol. 99, No. 15, 1995*

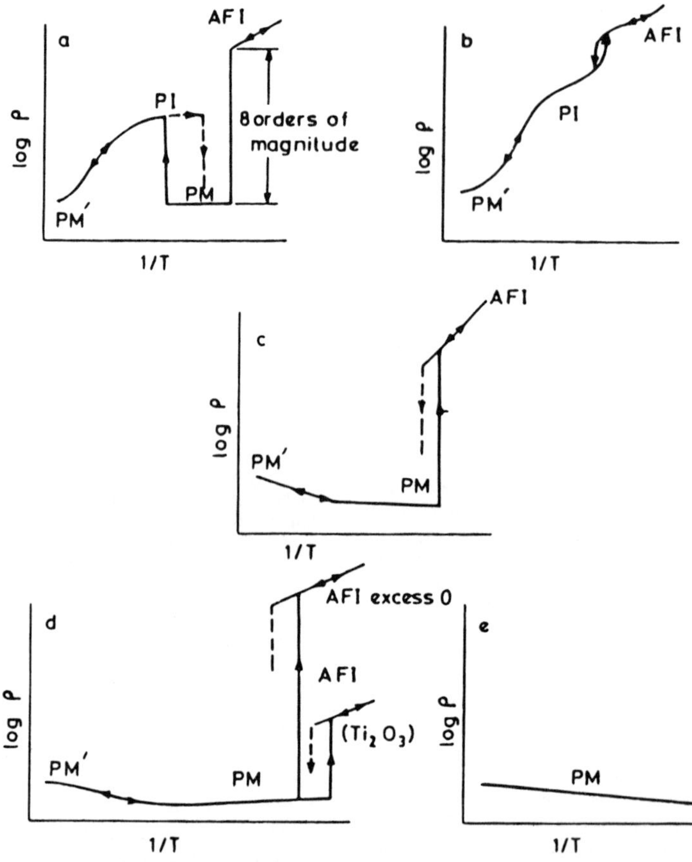

Figure 3. Schematic diagram depicting the changes of resistivity, ϱ, with temperature, T, in the V_2O_3 system: (a) V_2O_3 doped with 1.5% Cr_2O_3, (b) V_2O_3 doped with 3% Cr_2O_3, (c) pure V_2O_3, (d) V_2O_3 doped with 1% Ti_2O_3 and nonstoichiometric V_2O_3, and (e) 5.5% Ti_2O_3 doped V_2O_3. AFI, antiferromagnetic insulator; PM (PM′), paramagnetic metal; PI, paramagnetic insulator (after Honig and Spalek[16]).

order transition around 400 K with a small conductivity anomaly. Mere crystal distortion or magnetic ordering cannot explain the large connectivity jump at 150 K. The current status of the V_2O_3 transition can be represented in terms of Figure 3. This figure also serves to indicate the complexity of the metal—nonmetal transition in a relatively simple oxide system. There are many recent findings on V_2O_3 which are interesting. Thus, Carter et al.[17] have carefully measured the electrical and magnetic properties of single crystals of pure and doped V_2O_3 near the M—NM transition. Bao et al.[18] have examined the phase diagram of $V_{2-x}O_3$ in the x, P, T space and identified an incommensurate spin density wave (SDW) in metallic V_2O_3 close to the transition. The optical conductivity of V_2O_3, α (ω), has been investigated by Thomas et al.[19] In spite of extensive experimental and theoretical effort, a complete understanding of the transition in the V_2O_3 system is yet to emerge. We shall examine some of the factors responsible for this situation in section 6.

Compositionally controlled M—NM transitions in oxides are worthy of special mention. We shall examine two types of compositionally controlled transitions[20] as typified by $La_{1-x}A_xMO_3$ (A = Ca or Sr and M = V, Mn, or Co) and $LaNi_{1-x}M_xO_3$ (M = Mn or Fe). In $La_{1-x}A_xMO_3$, progressive substitution of trivalent La by divalent A brings about itinerant behavior of the d electrons, because every A ion creates an M^{4+} ion and promotes electron hopping from M^{3+} to M^{4+} ions (impurity-

band formation). This is to be contrasted with $LaNi_{1-x}M_xO_3$ where $LaNiO_3$ ($x = 0$), which is a correlated metal, transforms to an insulator on progressive substitution of Ni by M (somewhat like a deep-impurity situation). In Figure 4 we show typical electrical resistivity data in the two types of transitions. Unlike the above two systems, $A_{0.3}MoO_3$ (M = K or Rb) shows a metal—nonmetal transition associated with charge-density waves.[21]

Many of the high-temperature superconducting cuprates show compositionally controlled M—NM transitions.[8] Thus, in $TlCa_{1-x}Ln_xSr_2Cu_2O_y$ (Ln = Y or rare earth), the superconducting T_c shows a maximum at an optimal value of x (corresponding to the optimal value of the hole concentration). The system also shows a M—NM transition in the normal state as x is varied (Figure 5). Thus, the cuprate is metallic when $x = 0.25$ and insulating when $x = 1.0$. Rather interesting behavior occurs at $x = 0.75$ when the superconducting transition occurs from a seemingly semiconducting state. A compositionally controlled M—NM transition is also exhibited by $Bi_2Ca_{1-x}Ln_xSr_2Cu_2O_{8+\delta}$ (Figure 5) which has a maximum T_c at an optimal x value of 0.25. $La_{2-x}Sr_xCuO_4$ shows a similar metal—nonmetal transition ($x = 0$ is an insulator and $x = 0.3$ is a metal) with change in x and the maximum T_c is at $x = 0.20$. $La_{2-x}Sr_2CuO_4$ and a few other systems traverse the insulator—superconductor—metallic regimes with change in composition (increase in x from 0.0 to 0.3 in $La_{2-x}Sr_xCuO_4$). This suggests that the high-temperature

J. Phys. Chem., Vol. 99, No. 15, 1995 **5231**

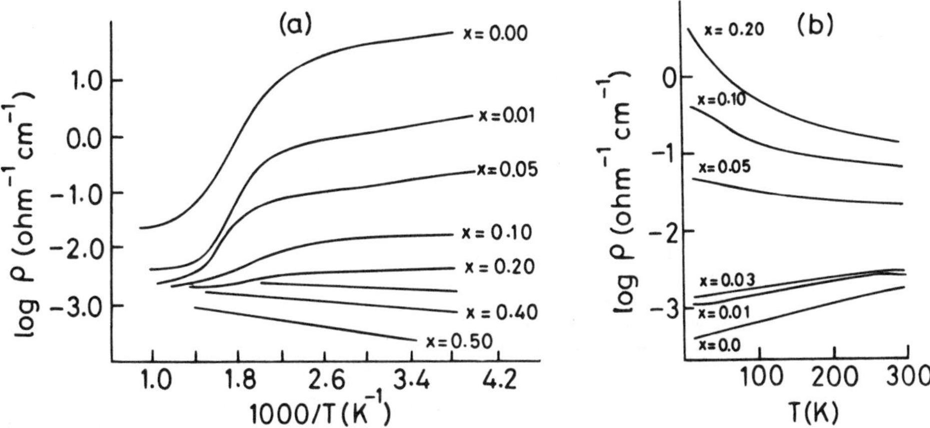

Figure 4. Compositionally controlled metal−nonmetal transitions in (a) $La_{1-x}Sr_xCoO_3$ and (b) $LaNi_{1-x}Mn_xO_3$ (from Rao and Ganguly[20]).

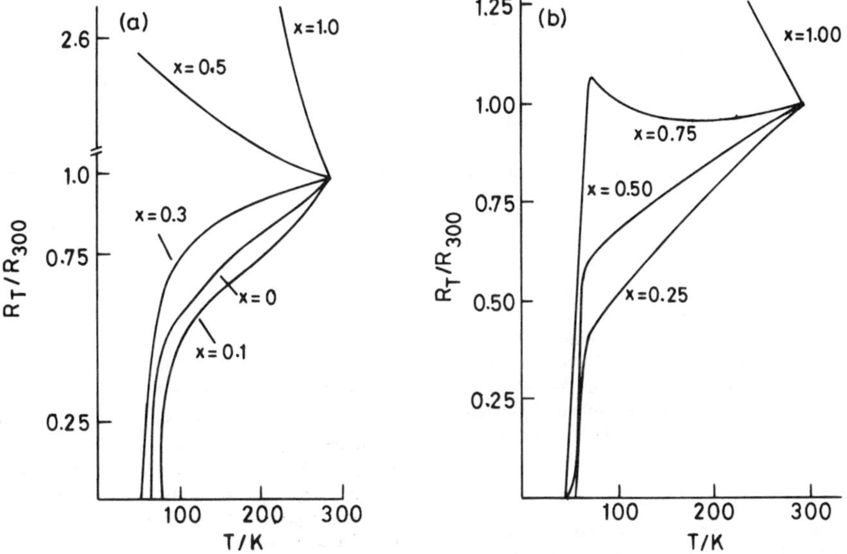

Figure 5. Compositionally controlled metal−nonmetal transition in superconducting cuprates: (a) $Bi_2Sr_2Ca_{1-x}Nd_xCu_2O_8$ and (b) $TlCa_{1-x}Nd_xSr_2CuO_4$ (after Rao[8]).

superconducting cuprates are at the boundary between metals and insulators.

One of the fundamental questions that has intrigued scientists is, "how many atoms maketh a metal?". The question is concerned with the possibility of a transition from the metallic to the nonmetallic state as the bulk metal is divided into finer particles. Recent studies of metal clusters[22,23] have attempted to answer the question. Careful investigations of gold clusters have shown that the binding energy of the core levels increases markedly (relative to the bulk metal value) when the cluster size decreases below 1 nm. That the effect is not merely due to final state effects but due to the occurrence of metal to nonmetal transitions as the cluster size decreases is reinforced by valence band and Bremstraahlung isochromat spectroscopic studies. Furthermore, tunneling conductance measurements show the existence of a gap in clusters smaller than 1 nm containing around 50 atoms or less; in Hg atom clusters, the 6s−6p atomic transition gives way to a collective metal-like plasmon absorption for a cluster size between 7 and 20

atoms.[23] The size-dependent transition from the metallic to the nonmetallic state does not occur abruptly. The possibility of a matrix-bound insulator−metal transition in alkali metal-doped zeolites at a critical stage of loading has been suggested by Edwards et al.[24] The dependence of the energetics and charge distribution of electron states on the cluster size and the dielectric constant has been examined by Rosenblit and Jortner,[7] to shed light on the cluster size-induced metal−nonmetal transition.

3. The Herzfeld Criterion

The earlier theoretical prediction of a M−NM transition is that derived from the work of Goldhammer[25] and Herzfeld.[14] These authors considered the effect of increasing density on the atomic polarizability and suggested that there would be a divergence in the polarizability or the dielectric constant causing the release of bound electrons. The Herzfeld criterion for dielectric catastrophe is given by,

5232 *J. Phys. Chem., Vol. 99, No. 15, 1995*

$$\frac{4\pi N \alpha_0}{3V} = \frac{R}{V} = 1 \qquad (2)$$

where α_0 is the low-density static polarizability of the atom, N Avogadro's number, V the molar volume, and R the molar polarizability. As a result of cooperative polarization effects, valence electrons get delocalized from the lattice sites at very high (metallic) densities[26] and the Drude free electron model becomes applicable. In the metallic regime, $(R/V) > 1$. In Figure 6 we show how this criterion excellently delineates metals from nonmetals in the periodic table.[27] Rao and Ganguly[28] have pointed out that the latent heat of vaporization, ΔH_v, of elements with metallic cohesion is larger than that of elements which are insulating because of relatively weak bonding (Figure 7). There are, however, exceptions such as strongly covalently bonded solids e.g., carbon.

A recent development is the realization of a link between the Herzfeld metallization view and the stress-induced transformations in solids, notably in semiconductors.[29] Under a diamond pressure indentor, ordinary (semiconducting) silicon transforms to the much denser β-tin (metallic) structure, the critical pressure being in the range 11–12 GPa. This is consistent with the experiments which reveal a large, reversible drop in the electrical resistivity in a thin layer surrounding a microindentation in Si, as would be anticipated because of the metallic characteristics of the β-tin structure. Good correlations are found between the experimental metallization pressure and the values calculated from the Herzfeld polarization catastrophe criterion. Experimental transition pressures also correlate with Vickers hardness numbers and activation energies for dislocation motion. We show in Figure 8, a comparison of the calculated critical pressure (Herzfeld model) with the measured Vickers hardness (expressed in kilobars) for the group IV elements and SiC. Equally good correlations are obtained for all tetrahedrally bounded semiconductors and alkali and alkaline earth oxides.

Fujii et al.[30] have reported experimental evidence for the molecular dissociation process in Br_2 near 80 GPa. This transition, which is coincident with the onset of pressure-induced metallization, was first discovered in molecular/metallic iodine.[31] A diatomic molecular crystal loses its molecular character in the limit when the intermolecular distance becomes equal to the intramolecular bond length. Fujii et al.[30] applied the Herzfeld criterion to I_2 and Br_2 and estimated that the molar refractivity reaches the atomic limit around 20 GPa in I_2 and 80 GPa in Br_2. In both cases, the computed pressure coincides with that for molecular dissociation accompanied by metallization.

4. The Mott Criterion

The Herzfeld criterion considers the M–NM transition as viewed from the nonmetallic side. Over 30 years ago, Mott[12] proposed a simple model of the M–NM transition which considers how electron localization occurs as the transition is approached from the metallic side. In Figure 9, we show a schematic representation of a lattice of one-electron hydrogenic centers (P donors in Si) in the two limiting electronic regimes of high and low donor densities. At a sufficiently high density (small interparticle distance), the system would be a metal;[32] in this state, the system has a finite electrical conductivity at the absolute zero of temperature, i.e., $\sigma(T = 0) \neq 0$. At large interparticle distances, the system must surely become nonmetallic, or insulating, having a conductivity of 0 at $T = 0$ K, viz., $\sigma(T = 0) \rightarrow 0$. Mott argued that at a critical interdonor distance (d_c) a first-order (discontinuous) transition from metal to nonmetal would occur (Mott transition). A discontinuous

transition at the absolute zero of temperature will always remain a tantalizing theoretical prediction. Experimentally, however, even very close to $T = 0$ K (down to 0.03 K), the experimental situation is equivocal.[32–35] The physics of the problem is that, on the metallic side of the transition, the effective valence electron–cation potential in an atom is completely screened by the conduction electron gas and bound states are therefore nonexistent. However, if the conduction electron density, or equivalently the average separation between donor centers is changed, there comes a point at which bound levels (i.e., nonmetallic states) appear at some critical concentration of centers such that the following condition is satisfied.

$$n_c^{1/3} a_H^* \leq 0.25 \qquad (3)$$

Here, a_H^* is the Bohr orbit radius of the isolated center and n_c is the critical carrier density at the M–NM transition. Another way of viewing the transition is that of an electronic instability which ensues when the trapping of an electron into a localized level also removes one electron from the Fermi gas of electrons. This must clearly lead to a further reduction in the screening properties (which are themselves directly related to the conduction electron density) and a catastrophic situation then ensures the localization of electrons from the previously metallic electron gas.

There appears to be little doubt that the Mott criterion given by eq 3 is an effective indicator of the critical condition at the M–NM transition itself. At the least, this simple criterion provides a numerical prediction for the metal–nonmetal transition in many situations. Figure 10 summarizes some of the experimental data.[34,36] Interestingly, besides doped semiconductors, metal–ammonia and metal–noble gas systems and superconducting cuprates all follow the linear relation given by eq 3. This is truly remarkable.

5. Minimum Metallic Conductivity at the Metal–Nonmetal Transition

Mott[13] has argued that the M–NM transition in a perfect crystalline material at $T = 0$ K is discontinuous (Figure 8) and proposed that, at the transition, there exists a minimum conductivity, σ_{min}, for which the material could still be viewed as metallic, prior to the localization of electrons.[2] Mott's ideas were based on arguments developed earlier by Ioffe and Regel[37] for the breakdown of the theory of electronic conduction in semiconductors. The conventional Boltzmann transport theory becomes meaningless when the mean-free path, l, of the itinerant conduction electrons becomes comparable to, or less than, the interatomic spacing, d. The Ioffe–Regel mean free path, l_{IR}, at the minimum metallic conductivity is equal to d. Abrahams et al.[38] have, however, predicted a continuous M–NM transition on the basis of a scaling theory of noninteracting electrons in a disordered system,[39] and their results question the existence of σ_{min} in both two and three dimensions.[32,39,40] The two possible scenarios of the transition are compared in Figure 11.

5.1. The Situation in Doped Semiconductors. There is an increasing belief amongst workers in the field that the M–NM transition is continuous, based on experimental measurements carried out at low temperatures down to 3 mK. In Figure 12, we show the experimental evidence in P-doped Si. Note that at a fixed (very low) temperature, the conductivity changes continuously with, for example, donor concentration. In addition, the extrapolated zero-temperature value of the conductivity $(\sigma(0))$ varies continuously with impurity concentration. An example showing the variation of the extrapolated "zero-temperature" conductivity[41] in the case of B-doped Si is

Feature Article

J. Phys. Chem., Vol. 99, No. 15, 1995 **5233**

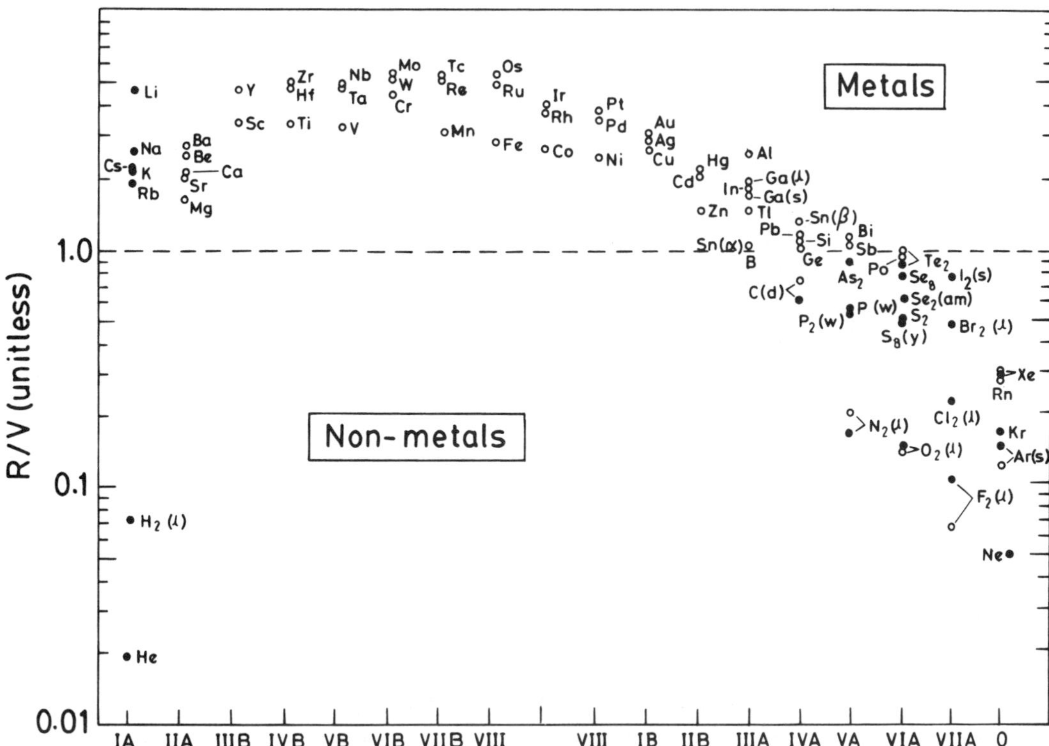

Figure 6. The occurrence of metallic vs nonmetallic character in the periodic table of the elements: the variation of R/V for naturally occurring elements of the s, p, and d blocks (adapted from Edwards and Sienko[27]).

shown in Figure 13. This could be taken as strong experimental evidence for a continuous M−NM transition in doped semiconductors at $T = 0$ K. Möbius,[35,42] however, questions the reliability of such 0 K extrapolations and suggests that these findings do not disprove the existence of finite σ_{min} at the transition. He argues that the data can be explained by a combination of σ_{min} on the metallic side and a Coulomb interaction dominated $\sigma(T) \simeq \sigma_{min} \exp-\{(T_0/T)^{1/2}\}$ on the insulating side, with $T_0 \rightarrow 0$ as the disorder decreases to the critical value. As noticed by him,[35,42] not all the measurements are consistent with this suggestion; furthermore, the clear observation of characteristic weak localization or precursor effects in transport and magnetotransport weakens the argument for a σ_{min}. However, this analysis points out the fact that while the continuous conductivity transition prediction is for *noninteracting* disordered electrons, just on the insulating side the Coulomb interaction is by definition of long range and hence qualitatively important. It is possible that 3 mK may not be a low enough temperature since $k_B T$ may exceed the activation energy for conduction even at this temperature.

While the presence of a high degree of disorder may wipe out the discontinuous nature of the M−NM transition,[43,44] inclusion of strong correlation in scaling models could conversely change a continuous transition to a discontinuous one. In spite of such difficulties, however, σ_{min} continues to be a useful experimental criterion[17,20,45] at least at the "high-temperature" limit. Earlier results providing experimental evidence for σ_{min} have been reviewed by several authors. In Figure 14, we show some of the results. We note that σ_{min} scales with n_c. As pointed out by Fritzsche,[46] σ_{min} appears to satisfactorily represent the value of conductivity where the

activation energy for conduction disappears. This aspect is specially borne out by investigations of transition metal oxide systems.

5.2. The Situation in Transition Metal Oxides. In many of the oxide systems,[15,20] especially those exhibiting compositionally controlled M−NM transitions (Figures 4 and 5), the temperature coefficient of the conductivity changes sign around σ_{min} ($\sim 10^3$ ohm^{-1} cm^{-1}). Most of these oxides, including the superconducting cuprates, follow the relation shown in Figure 14. The points corresponding to these oxides fall somewhere between those of fluid alkali metals and $La_{1-x}Sr_xVO_3$. The critical carrier concentrations in these materials from Figure 10 also give σ_{min} values close to the observed values. Accordingly, σ_{min} is often taken to represent the separation of localized and itinerant electron regimes.

Recent work of Raychaudhuri and co-workers[47,48] suggests the need to reevaluate the status of transition metal oxides with regard to σ_{min}. It appears that in these disordered oxide systems, genuine metallic states (with $\sigma(T = 0) \neq 0$) exist even when the conductivity is activated ($\sigma < \sigma_{min}$). This implies that the earlier values of the critical electron density at the transition in such oxides may be overestimates. In Figure 15, we show the behavior of $Na_xW_{1-y}Ta_yO_3$ where the temperature coefficient of the conductivity changes sign when $(x - y) \approx 0.20$, but the transition actually occurs at $(x - y) = 0.19$. The curves for $x = 0.35$ and 0.34 compositions both show "activated conductivity" at $T > 10$ K (i.e., a negative temperature coefficient of the resistivity), but the $\sigma(T)$ *saturates* at a fairly high residual value in the case of the $x = 0.35$ sample as expected of a metal ($\sigma(T = 0) \neq 0$). The $x = 0.34$ sample, however, shows a very much lower σ value at low T, tending to infinity, and fits an activated

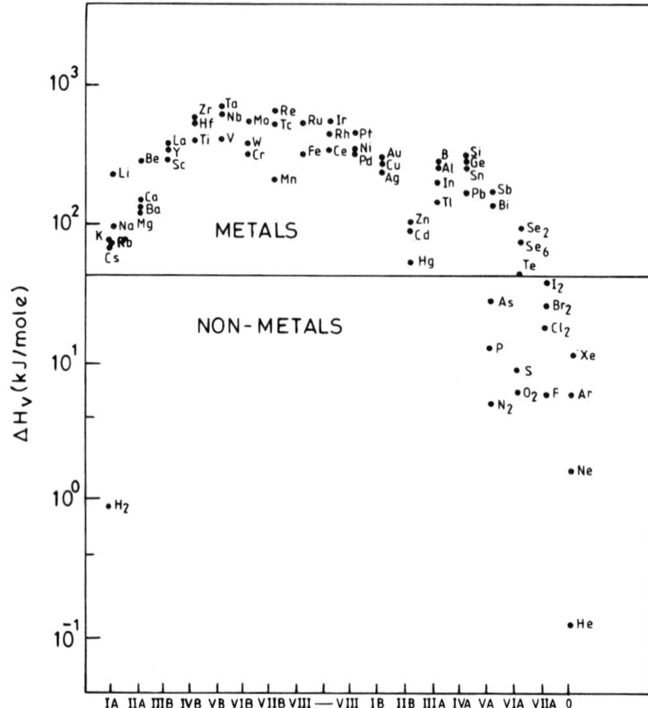

Figure 7. Plot of the latent heat of vaporization, ΔH_v, for metals and nonmetals of the periodic table (from Rao and Ganguly[28]).

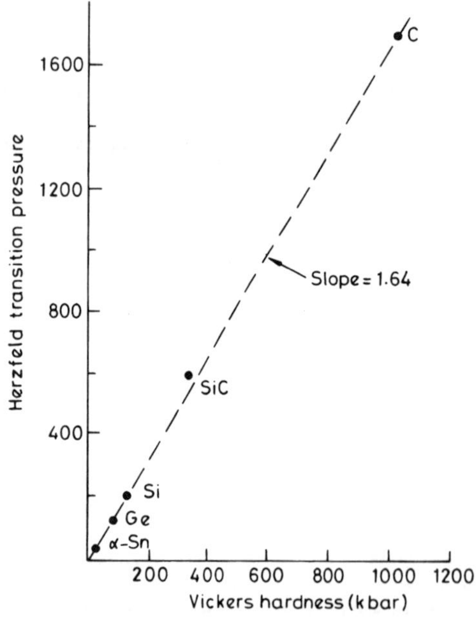

Figure 8. Correlation of the metallization pressure as calculated from the Herzfeld model and Vicker's indentation hardness number (from Gilman[29]).

temperature dependence, with the behavior of a nonmetal ($\sigma(T = 0) = 0$). These results show that it is difficult to differentiate metallic *vs* nonmetallic behavior in oxides based on high-temperature measurements. What emerges, however, is that the

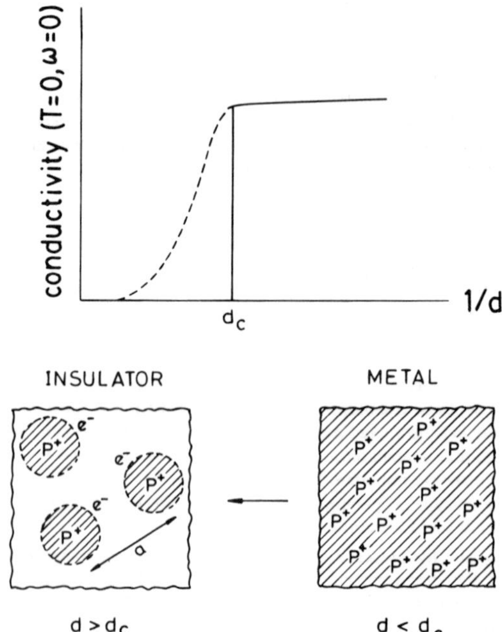

Figure 9. A schematic representation of the situation ($T = 0$ K) for P doped Si at both high and low donor densities. Also shown are two scenarios for the composition dependence of the electrical conductivity, showing the metal−nonmetal transition.

"high-temperature" σ_{min} value differentiates the (metallic *vs* nonmetallic) conductivity behavior in that temperature regime.

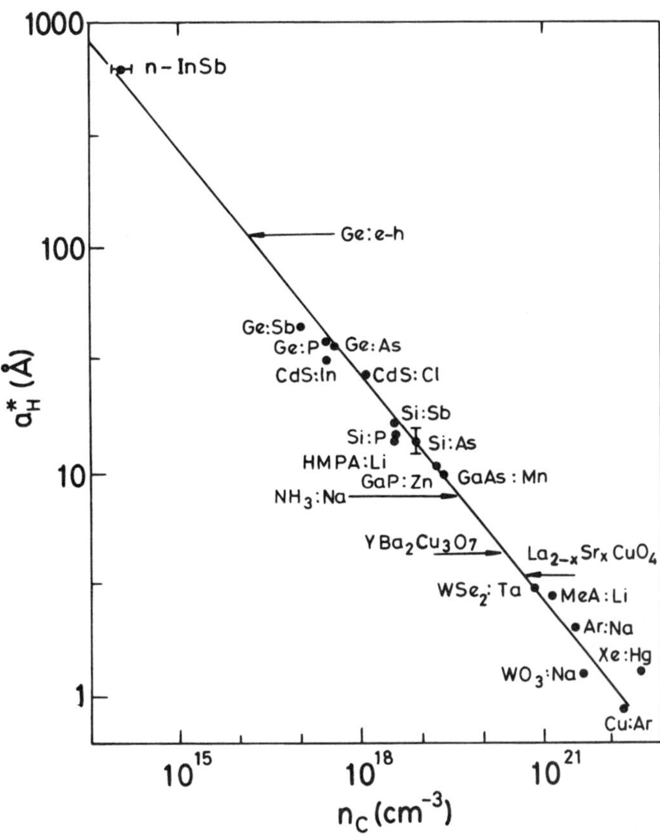

Figure 10. An experimental correlation of the critical density, n_c, with the Bohr radius, a_H^*. The solid line represents $n_c^{1/3} a_H^* = 0.26$ (from Edwards and Sienko[36] and Thomas[34]). Here, InSb refers to lightly doped n-InSb, Ge:e-h means germanium with photoexcited electrons and holes; Cu:Ar is a codeposited Cu and Ar mixture on a cold substrate.

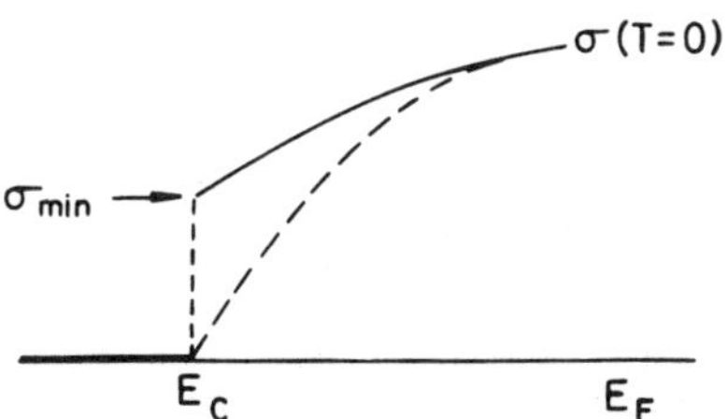

Figure 11. Schematic illustration of the two possibilities of a continuous or discontinuous metal−nonmetal transition at $T = 0$ K. The minimum metallic conductivity, σ_{min}, at the transition is also shown (from Lee and Ramakrishnan[39]). The conductivity at zero temperature (ordinate) and Fermi energy (abscissa) are shown. The discontinuous conductivity transition suggested by Mott is the full curve, with σ_{min} occurring as E_F crosses the mobility edge energy E_c. The dotted curve is the continuous conductivity transition predicted by the scaling theory.

6. Our Theoretical Understanding

The metal to nonmetal transition, a basic electronic change, has proven surprisingly difficult to understand in detail. The underlying reason is the diametrically opposite modes of description natural for the metal (extended electronic states) and for the insulator (localized states, often with local constraints on electron number). The observed diversity of systems and phenomena indicates that a number of causes may be at work, e.g., disorder, short-range electron correlation, long-range Coulomb interaction, and electron lattice coupling. The effects are often present simultaneously and interactively. Many reasonable scenarios as well as detailed theories of model systems do exist, but the distance between them and experiments is still large. In this section, we first look at the implications of the successful criteria for the M−NM transition. We then describe some theoretical models and approaches.

6.1. Criteria for the Transition. The three criteria discussed earlier are the dielectric catastrophe of Herzfeld, the Mott density criterion, and the idea (again due to Mott) that a metal is characterized by a minimum value of the electrical conductivity at the transition. The first two are related and consider how the Coulomb interaction and electron or atom density together can lead to the metal−insulator transition. The polarizability of an isolated atom depends on its size, being roughly proportional to its volume (a_H^{*3}). When such atoms are put together, the dielectric constant is enhanced because of local field effects, i.e., the large electric field at each atom due the collection of polarized atoms (or dipoles). At a density such that the polarizability proportional to a_H^{*3} is comparable to the volume per atom, i.e., when $a_H^* n^{1/3} \simeq 1$, the dielectric constant diverges. It is difficult to develop a detailed theory along these lines, because at such high densities electron overlap effects are large. The electron energy level structure (band structure) and wave functions are very different from those at the atomic limit, so that the dielectric constant is not easily related to properties of isolated atoms. Further, disorder and lattice type can also have a large quantitative effect.

5236 *J. Phys. Chem., Vol. 99, No. 15, 1995*

Edwards et al.

The Mott criterion, namely $a_H^* n^{1/3} \simeq 0.25$, is dimensionally identical to that of Herzfeld and is physically related. It considers the transition from the metallic side. If an electron—hole excitation is created in a metal, the attraction between them is screened, and they may not form a bound state unless the screened interaction is strong enough. This will happen at a low enough carrier density estimated by Mott to be such that $a_H^* n^{1/3} \lesssim 0.25$. The metal is then unstable against the formation of a bound electron—hole pair; a collection of these is an insulator or a nonmetal. Both the criteria thus point to the Coulomb interaction which is screened and effectively weak in the metal and is strong and long ranged in a nonmetal. A theory describing this crossover and its consequences does not exist. These approaches neglect electron spin and associated magnetic effects, *local* electron correlation, as well as disorder.

The minimum metallic conductivity criterion of Mott emphasizes the role of disorder. It was first realized by Anderson[49] that electronic states originally extended can become spatially localized if the disorder is strong enough. Mott argued that in a disordered metal, the minimum possible value of the mean free path is the interatomic spacing. For stronger disorder, the picture of electron waves scattered randomly is inappropriate; the electron state is localized, and one has a nonmetal. Thus, a minimum electrical conductivity, σ_{min}, characterizes the transition. The conductivity discontinuously jumps from this value to zero on the nonmetallic side. As mentioned earlier, this is a very good first approximation; very close to the critical disorder *and* at very low temperatures, $\sigma \rightarrow 0$ and values much less than σ_{min} (but tending to a nonzero limit at $T = 0$) have been observed. A detailed theory for the disorder caused continuous M—NM transition exists.[38,39] The minimum metallic conductivity criterion emphasizes disorder and is silent about the effect of interactions. The Mott density criterion focuses on the Coulomb interaction and completely neglects disorder effects. The success of both these criteria, often in the same system, implies that both interactions and disorder are significant in real systems.

6.2. Detailed Models. *6.2.1. Anderson Localization.* It was pointed out by Anderson[49] that if the randomness in electronic state energies at different sites is large enough, electrons do not diffuse from one end of the system to another but are localized at potential fluctuations. Since for weak disorder electrons do diffuse, there must be a critical disorder at which an electron of a particular energy (say, the Fermi energy) does not diffuse, i.e., the dc electrical conductivity is zero. The energy separating the band of extended states from localized states is called the mobility edge. As disorder or electron density changes, the Fermi energy and the mobility edge coincide, and the system makes a transition from a metal to a nonmetal; this is the disorder-driven Anderson transition.

Predictions for physical properties near the transition were first made by Mott. He argued for a nonzero minimum conductivity of the metal at the transition point and for a divergence of localization length in the nonmetal with a consequent divergence in the dielectric constant.

A detailed theory of the transition, which identified singular backscattering as the mechanism of localization, was developed by Abrahams et al.[38] Quantum mechanically, an electron in a random medium can go from one point to another via infinitely many paths with appropriate amplitudes. Of these, self intersecting paths in which the loop is traversed in opposite directions have the same amplitude. Thus the probability (the square of the amplitude) for return to the origin is enhanced because of constructive interference. Nearby paths also contribute, depending on dimensionality. This process thus tends to localize

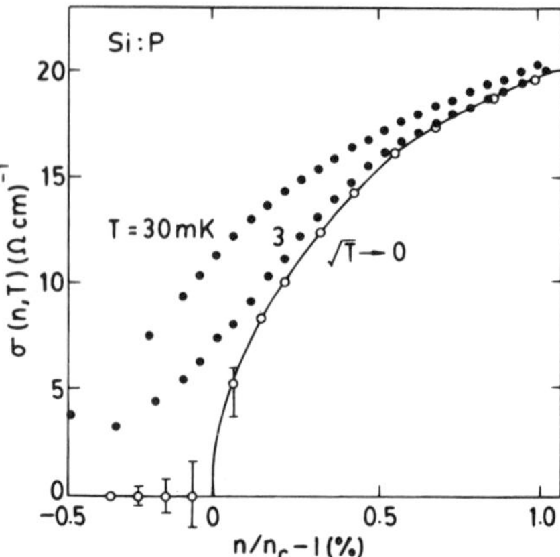

Figure 12. Electrical conductivity as a function of relative electron (donor) concentration in P doped Si to illustrate the metal—nonmetal transition at 0 K. The relative P density is varied by changing n_c with uniaxial stress on a single sample. The open circles are extrapolated to $T = 0$ K (from Thomas[33]).

electrons, depending on disorder and dimensionality. Detailed calculations show for example that the conductivity decreases from its value σ at atomic length scales (of order the mean free path l) to smaller values at a larger length scale, L, on account of interference effects. In two dimensions, the decrease is logarithmic in L.

$$\sigma(L) = \sigma_0 - \left(\frac{e^2}{\hbar\pi^2}\right) \ln(L/l) \tag{4}$$

This means, from a scaling analysis, that for large enough systems, the conductivity (or conductance) is zero. There are no disordered *metal* films at absolute zero. In three dimensions, $\sigma(L)$ has the form

$$\sigma_{3d}(L) = \sigma_0 - \frac{e^2}{\hbar\pi^3}\left(\frac{1}{l} - \frac{1}{L}\right) \tag{5}$$

This implies for example that when $\sigma(l)$ has the Mott minimum value $(e^2/\hbar\pi^3 l)$, the dc or large length scale $(L \rightarrow \infty)$ conductivity is zero. Thus, the conductivity goes continuously to zero at the Anderson transition, due to interference effects not envisaged in the Mott approximation. The exponent with which σ vanishes is the same as that with which the localization length diverges.

The scaling picture, starting from weak localization due to interference, puts in perspective both the minimum metallic conductivity σ_{min} and the possibility of a continuous conductivity transition. At nonzero temperatures, inelastic scattering limits the length scale L to $L_{in} \sim T^{-\alpha}$ up to which interference effects take place. If L_{in} is small, the dc conductivity even near critical disorder will be close to σ_{min}. As T decreases, L_{in} increases, the interference effects are stronger, and the conductivity decreases. This is considered "nonmetallic" behavior. However $\sigma(T = 0)$ is nonzero, so that one can naturally have strongly disordered metals with a negative temperature coefficient of the resistivity. The purely disorder driven transition is a continuous quantum $(T = 0)$ transition. A number of predictions for quantities such as magnetoresistance, the Hall effect, and resistivity in the presence of magnetic and other scatterers have been made. These agree with experiment.[39,40] It is also realized that many of these phenomena are common to *all* waves in random media, e.g., light and sound waves.

Feature Article

J. Phys. Chem., Vol. 99, No. 15, 1995 **5237**

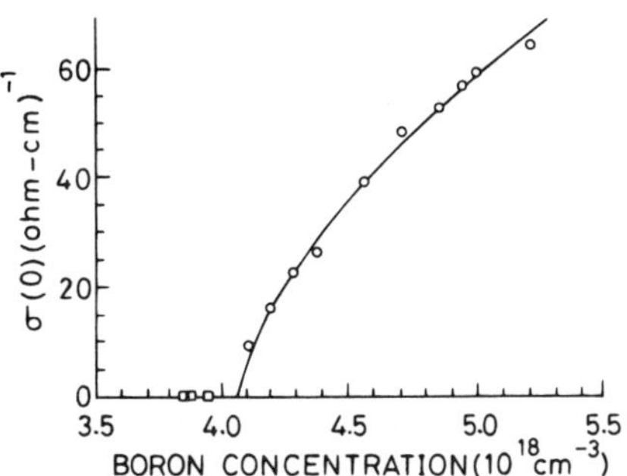

Figure 13. Zero-temperature conductivity $\sigma(T = 0)$ vs boron concentration in B doped Si. The circles represent metallic samples, while the three squares indicate samples which are nonmetallic (from Dai et al.[41]).

Real disordered electronic systems cannot be thought of as a collection of noninteracting particles moving independently in a random potential. With increasing disorder, electrons diffuse very slowly and so in effect interact strongly. The range of interaction increases until in the nonmetal it decreases only inversely with distance. It has proven very difficult to describe the passage of a disordered interacting electron gas to the insulating state. In the metallic regime, for effectively weak interaction, there are predictions for characteristic effects such as a dip in the density of states proportional to $(|E - E_F|)^{1/2}$ where E is the energy and E_F the Fermi energy. Scaling approaches to the transition starting from the perturbative regime show that there are several kinds of M−NM transitions, characterized by different critical behavior for density of states, conductivity, magnetic susceptibility, low-temperature specific heat, etc.[39,50] Unfortunately, the actual transition is often in the regime of strong coupling and so the perturbative scaling analysis is not reliable. Also, there is no theory which describes the Coulomb interaction realistically in the entire regime (weakly disordered metal−critical region−Coulomb glass). Experimentally, this transition has been recently explored in several systems.[47,48] The square root dip in the density of electronic states near the Fermi energy steepens, changes to a linear form, and becomes a soft quadratic gap just on the

insulating side (i.e., the density of states goes as $(E - E_F)^2$). There are characteristic temperature and field dependences for the resistivity. These are not yet comprehensively understood.

6.2.2. Correlation Effects. The effects of electron correlation are strong in narrow band systems and can cause them to be insulating when they should be metallic according to band theory. Well-known examples are NiO and La₂CuO₄, which, with an odd number of electrons per site in the highest unfilled band, ought to be metallic. In the latter, for example, only the Cu−O configurations d^9p^6 and $d^{10}p^5$ are close in energy (the configuration d^8p^6 has very high energy). The former, containing one d hole per site, is more stable so that the system has a gap against transfer of one hole from Cu to O. This is a charge transfer insulator. An even simpler correlation driven one-band (d band) insulator requires the energy cost U of local configurational changes involving sites l, $l + 1$, i.e., the change ($d_l^n - d_{l+1}^n$ to $d_l^{n-1} - d_{l+1}^{n+1}$) to be greater than the kinetic energy gain, KE, due to electron hopping. Thus, as U decreases or increases (e.g., due to changes in physical conditions such as pressure) in relation to KE, the system can make a transition from insulator to metal. These ideas were extensively discussed by Mott. The one-band model of electrons on a lattice, with local correlations, is the well-known Hubbard model, and the correlation driven transition for one (or commensurate) electron per site, from a paramagnetic metal to a paramagnetic insulator is the Mott transition, mentioned earlier in section 4.

The Mott transition has proven difficult to describe in detail, though there are several plausible conclusions and results for spatial dimensionality $d = 1$ and ∞. On the metallic side close to the transition, for the case of one electron per site, the probability of two electrons being on the same site is small because of correlation. Using this as a variational parameter in a ground state wave function as was first proposed by Gutzwiller (see ref 40 for details) a number of conclusions have been drawn. As $U \rightarrow U_c$ (the critical correlation), electron hopping becomes infrequent so that the electron effective mass is large. Since each site has mostly one electron, local magnetic moments are very long lived, and the Pauli spin susceptibility is large, diverging as $U \rightarrow U_c$. On the insulating side, a gap (the Mott−Hubbard gap) opens up for charge excitations. This gap is different from that in a semiconductor. The former is

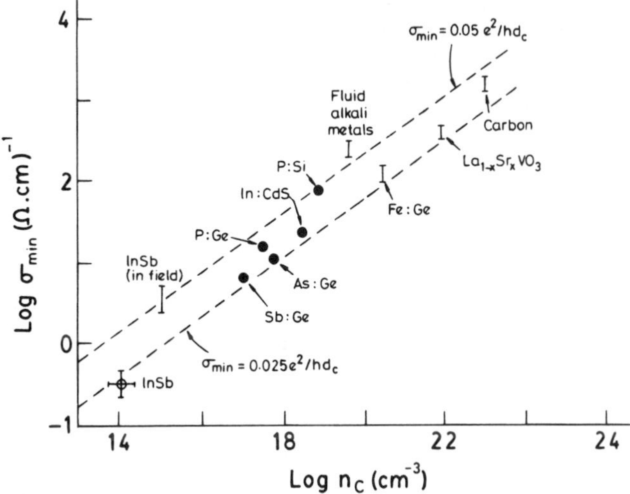

Figure 14. A plot of the minimum metallic conductivity, σ_{min}, against log n_c, the critical electron density at the metal−nonmetal transition. The two straight lines correspond to the values of the constant 0.05 and 0.026 in the equation relating σ_{min} and $n_c^{1/3}$ (adapted from Mott[44] and Fritzche[46]).

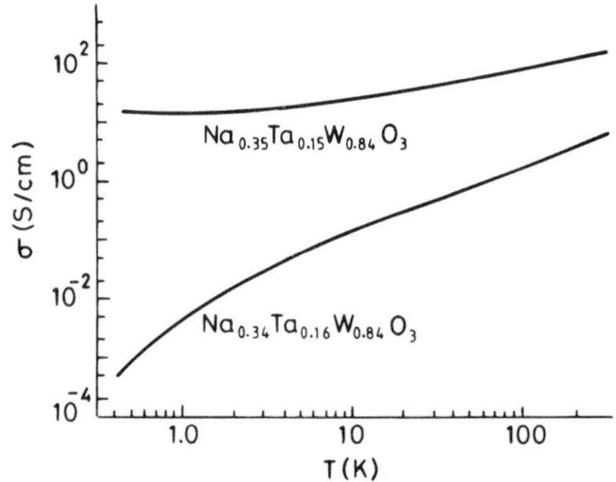

Figure 15. Electrical conductivity versus temperature for two samples of Na$_x$Ta$_y$W$_{1-y}$O$_3$, one having metallic properties for $T \rightarrow 0$ (top curve) and one having nonmetallic status for $T \rightarrow 0$ (bottom curve). Notice the extremely small differences in concentration for the metallic and nonmetallic samples (from Raychaudhuri[47]).

due entirely to local correlations and the latter to periodicity. On the insulating side, local moments are well defined and interact antiferromagnetically. There is experimental evidence in metallic V$_2$O$_3$, near the M−NM transition, for mass enhancement, increased Pauli paramagnetism, a large T^2 term in the resistivity etc.[17−19,40]

Recently, there has been considerable progress in developing a theory of the Hubbard model in high spatial dimensions.[51] There are reasons to believe that the results can be usefully extrapolated to $d = 3$. The development of a Mott−Hubbard gap is due to the size and time dependence of the field on an electron at a given site. This field is calculated self-consistently. Some results are as follows: In the (U,T) plane, there are three phases, metallic (PM), insulating paramagnetic (PI), and insulating antiferromagnetic (AFI). By suppressing the intermoment exchange coupling, the Mott transition (PM−PI) can be located and followed. It is discontinuous at $T \neq 0$ and is believed to become continuous as $T \rightarrow 0$. The physical properties of the heavy fermion metal for $U \lesssim U_c$ can also be calculated and agree with expectations. Electron−lattice coupling can drive the Mott transition discontinuous, with a $T \neq 0$ critical point.[52] Many of these results are similar to those found in V$_2$O$_3$.

The large d approach developed so far concentrates exclusively on the single-site time-dependent electron potential. The other low-lying fluctuations (charge and spin) are integrated out so that we do not have a detailed picture of the quantum dynamics. There is little understanding of the effects of deviations from integer site occupation, i.e., of doping. This lacuna is serious because cuprate superconductors can be thought of as doped Mott insulators. The effect of disorder is also not known.

7. Concluding Remarks

The transition of a metal to an insulator is caused and accompanied by changes in the nature of chemical bonding as well as by changes in physical properties. The transition occurs in a wide variety of systems, as we have tried to indicate. In spite of this variety, simple criteria reflecting a basic competition between potential and kinetic energies have been remarkably successful in locating broad conditions under which such transitions may occur. A detailed understanding of transport, magnetic, optical, and other properties in the transition regime has proven difficult, since several causes seem to be at work

simultaneously. However, a number of theoretical models valid for various limiting situations have given us valuable qualitative and quantitative insights.

The entire area has experienced a rebirth in the last few years, for several reasons. The most important perhaps is the realization that cuprate superconductors are not far from a largely correlation driven metal−insulator transition. Another is the appreciation that the transition is of wide occurrence, some recent examples being the superconductor−insulator, the quantum Hall fluid−insulator, and the stable quasicrystal metal−insulator transitions. Hopefully, the next few years should see considerable progress in the experimental and theoretical description of this basic electronic transition.

References and Notes

(1) Edwards, P. P., Rao, C. N. R., Eds. *The Metallic and the Nonmetallic States of Matter*; Taylor and Francis: London, 1985.

(2) Mott, N. F. *Metal-Insulator Transitions*, 2nd ed.; Taylor and Francis: London, 1990.

(3) Rao, C. N. R.; Edwards, P. P. *Proc. Indian Acad. Sci. Chem. Sci.* **1986**, *96*, 473.

(4) Redmer, R. Thermodynamische und Transporteigenschafter dicter Alkali plasmen. Habilitationsschrift, Universitat Rostock, 1991.

(5) Ashcroft, N. W.; Mermin, N. D. *Solid State Physics*; Saunders College: Philadelphia, 1976.

(6) See articles of K. Kimura and P. P. Edwards in *Chemical Processes in Inorganic Materials: Metal and Semi-conductor Clusters and Colloids*; Persans, P. D., Bradley, J. S., Chianelli, R. R., Schmid, S., Eds.; *Mat. Res. Soc. Symp. Proc.* **1992**, *272*, 193, 311.

(7) Rosenblit, M.; Jortner, J. *J. Phys. Chem.* **1994**, *98*, 9365.

(8) Rao, C. N. R. *Philos. Trans. R. Soc. London* **1991**, *A336*, 595.

(9) Iye, Y. In *Physical Properties of High Temperature Superconductors*; Ginsberg, D. M., Ed.; World Scientific: Singapore, 1992; Vol. II.

(10) Logan, D. E.; Tusch, M. A. *J. Non-Cryst. Solids* **1993**, *156−158*, 639.

(11) Ramakrishnan, T. V. *J. Solid State Chem.* **1994**, *111*, 4.

(12) Mott, N. F. *Philos. Mag.* **1961**, *6*, 287.

(13) Mott, N. F. *Philos. Mag.* **1972**, *26*, 1015.

(14) Herzfeld, K. F. *Phys. Rev.* **1927**, *29*, 701.

(15) Rao, C. N. R. *Annu. Rev. Phys. Chem.* **1989**, *40*, 291.

(16) Honig, J. M.; Van Zandt, L. L. *Annu. Rev. Mater. Sci.* **1975**, *5*, 225. Honig, J. M.; Spalek, J. *Proc. Indian Natl. Sci. Acad.* **1986**, *A52*, 232.

(17) Carter, S. A.; Rosenbaum, T. F.; Honig, J. M.; Spalek, J. *Phys. Rev. Lett.* **1991**, *67*, 3440. Carter, S. A.; Rosenbaum, T. F.; Metcalf, P.; Honig, J. M.; Spalek, J. *Phys. Rev.* **1993**, *B48*, 16841.

(18) Bao, W.; Broholm, C.; Carter, S. A.; Rosenbaum, T. F.; Aeppli, G.; Trevino, S. F.; Metcalf, P.; Honig, J. M.; Spalek, J. *Phys. Rev. Lett.* **1993**, *71*, 766.

(19) Thomas, G. A.; et al. *Phys. Rev. Lett.* **1994**, *73*, 1529.

(20) Rao, C. N. R.; Ganguly, P. In *The Metallic and the Non-metallic States of Matter*; Edwards, P. P., Rao, C. N. R., Eds.; Taylor and Francis: London, 1985.

(21) Schlenker, C.; Dumas, J.; Escribe-Filippini, G. H.; Marcus, J.; Fourcandot, G. *Philos. Mag.* **1985**, *B52*, 643.

(22) Vijayakrishnan, V.; Chainani, A.; Sarma, D. D.; Rao, C. N. R. *J. Phys. Chem.* **1992**, *96*, 8679.

(23) Rao, C. N. R.; Vijayakrishnan, V.; Aiyer, H. N.; Kulkarni, G. U.; Subbanna, G. N. *J. Phys. Chem.* **1993**, *97*, 11157. Haberland, H.; von Issendorff, B.; Yufeng, Ji; Kolar, T. *Phys. Rev. Lett.* **1992**, *69*, 3212.

(24) Edwards, P. P.; Anderson, P. A.; Armstrong, A. R.; Slaski, M.; Woodall, L. J. *Chem. Soc. Rev.* **1993**, *22*, 305.

(25) Goldhammer, D. A. *Dispersion via Absorption des Lichtes*; Teubner: Leipzig, 1911.

(26) Sommerfeld, A. In *Gesammelten Schriften*; Vieweg: Braunschweig, 1968; Vol. 4, as cited by Ehrenreich, H. E. *Science* **1987**, *235*, 1029.

(27) Edwards, P. P.; Sienko, M. J. *J. Chem. Educ.* **1983**, *60*, 691.

(28) Rao, C. N. R.; Ganguly, P. *Solid State Commun.* **1986**, *57*, 5.

(29) Gilman, J. J. *J. Mater. Res.* **1992**, *7*, 535; see also *Philos. Mag.* **1993**, *67*, 207.

(30) Fujii, Y.; Hase, K.; Ohishi, Y.; Fujihisa, H.; Hamaya, N. H.; Takemura, K.; Shimomura, O.; Kikegawa, T.; Amemiya, Y.; Matsushita, T. *Phys. Rev. Lett.* **1989**, *63*, 536.

(31) Balchan, A. S.; Drickamer, H. G. *J. Chem. Phys.* **1961**, *34*, 1948.

(32) For a review see: Milligan, R. F.; Thomas, G. A. *Annu. Rev. Phys. Chem.* **1985**, *36*, 139.

(33) Thomas, G. A. *J. Phys. Chem.* **1983**, *88*, 3749.

(34) Thomas, G. A. In *Proceedings of the 39th Scottish Universities Summer School in Physics*; Tunstall, D. P., Barford, W., Eds.; Adam Hilger: Edinburgh, 1991.

(35) Möbius, A. *J. Phys. C: Solid State Phys.* **1985**, *18*, 4639.

(36) Edwards, P. P.; Sienko, M. J. *Phys. Rev.* **1978**, *B17*, 2575.

(37) Ioffe, A. F.; Regel, A. R. *Prog. Semicond.* **1960**, *4*, 237.

(38) Abrahams, E.; Anderson, P. W.; Liccardello, D. C.; Ramakrishnan, T. V. *Phys. Rev. Lett.* **1979**, *42*, 693.

(39) Lee, P. A.; Ramakrishnan, T. V. *Rev. Mod. Phys.* **1985**, *57*, 287.

(40) Ramakrishnan, T. V. In *The Metallic and Non-metallic States of Matter*; Edwards, P. P., Rao, C. N. R., Eds.; Taylor and Francis: London, 1985.

(41) Dai, P.; Zhang, Y.; Sarachik, M. P. *Phys. Rev. Lett.* **1991**, *66*, 1914.

(42) Möbius, A. *Phys. Rev.* **1989**, *B40*, 4194.

(43) Mott, N. F. *Philos. Mag.* **1978**, *B37*, 377.

(44) Mott, N. F. *Proc. R. Soc.* **1982**, *A382*, 1.

(45) Edwards, P. P.; Sienko, M. J. *Int. Rev. Phys. Chem.* **1983**, *3*, 83.

(46) Fritzche, H. In *The Metal-Nonmetal Transitions in Disordered Systems*; Friedman, L. R., Tunstall, D. P., Eds.; Scottish Universities Summer School in Physics, 1978.

(47) Raychaudhuri, A. K. *Phys. Rev.* **1991**, *B44*, 8572.

(48) Raychaudhuri, A. K.; Rajeev, K. P.; Srikanth, H.; Mahendiran, R. *Physica B* **1994**, *197*, 124.

(49) Anderson, P. W. *Phys. Rev.* **1958**, *109*, 1492.

(50) Belitz, D.; Kirkpactrick, T. R. *Rev. Mod. Phys.* **1994**, *66*, 261.

(51) Georges, A.; Krauth, W. *Phys. Rev. B* **1993**, *48*, 7167. Rozenberg, M. J.; Zhang, X. Y.; Kotliar, G. *Phys. Rev. B* **1994**, *49*, 10181.

(52) Majumdar, P.; Krishnamurthy, H. R. *Phys. Rev. Lett.* **1994**, *73*, 1525.

JP9428496

16814 *J. Phys. Chem.* **1995,** *99,* 16814−16816

Polymerization and Pressure-Induced Amorphization of C_{60} and C_{70}[†]

C. N. R. Rao,* A. Govindaraj, Hemantkumar N. Aiyer, and Ram Seshadri

CSIR Centre of Excellence in Chemistry, Solid State & Structural Chemistry Unit and Materials Research Centre, Indian Institute of Science, Bangalore 560 012, India

Received: July 13, 1995; In Final Form: September 18, 1995[®]

Scanning tunneling microscopy of solid films of C_{60} and C_{70} clearly demonstrate the occurrence of photochemical polymerization of these fullerenes in the solid state. X-ray diffraction studies show that such a polymerization is accompanied by contraction of the unit-cell volume in the case of C_{60} and expansion in the case of C_{70}. This is also evidenced from the STM images. These observations help to understand the differences in the amorphization behavior of C_{60} and C_{70} under pressure. Amorphization of C_{60} under pressure is irreversible because it is accompanied by polymerization associated with a contraction of the unit cell volume. Monte Carlo simulations show how pressure-induced polymerization is favored in C_{60} because of proper orientation as well as the required proximity of the molecules. Amorphization of C_{70}, on the other hand, is reversible because C_{70} is less compressible and polymerization is not favored under pressure.

It has been reported recently that C_{60} undergoes polymerization in the solid state involving 2+2 cycloaddition induced photochemically[1] or at high pressures and temperatures.[2] C_{60} is also known to undergo irreversible amorphization at high pressures, and the irreversibility has been attributed to the formation of cycloaddition products.[3] C_{70} seems to undergo photopolymerization in the solid state involving 2+2 cycloaddition,[4] but unlike C_{60}, it undergoes reversible amorphization at high pressures.[5] We have been interested in understanding the polymerization of C_{60} and C_{70} in the solid state[6] and its relation to the nature of amorphization of these fullerenes. For this purpose, we have investigated photopolymerization of solid films of the two fullerenes by scanning tunneling microscopy (STM) and X-ray diffraction and carried out certain simulation studies on the effects of pressure on these solids. The study has indeed revealed the subtle differences between C_{60} and C_{70} with regard to features of their polymerization and amorphization.

Photopolymerization of solid films of C_{60} and C_{70} was studied by recording the STM images before and after UV irradiation. Typical STM images of C_{60} and C_{70} shown in Figure 1 bear clear evidence for the occurrence of photopolymerization. Photopolymerization of C_{60} (Figure 1a) does not change the symmetry of the top surface. The internal features of the molecules become visible after photopolymerization because of the complete freezing of molecular motion.[7] Internal features become visible after photopolymerization of C_{70} as well (Figure 1b), but the lattice symmetry of the top layer appears to change from fcc(111)-like to fcc(100)-like. STM results indicate that in C_{60}, a lattice contraction of ≈13% occurs on polymerization while in C_{70}, an expansion of ≈10% is observed. X-ray diffraction measurements on C_{60} and C_{70} films confirm similar changes in the unit-cell volume. Accordingly, we see from Figure 2 that the 2θ values decrease on photopolymerization of C_{70} while the opposite holds in the case of C_{60}. The changes in the unit cell volume can be understood by means of the model for the dimerization of C_{70} and C_{60} shown in Figure 2. The volume expansion in C_{70} arises due to the ellipsoidal shape of

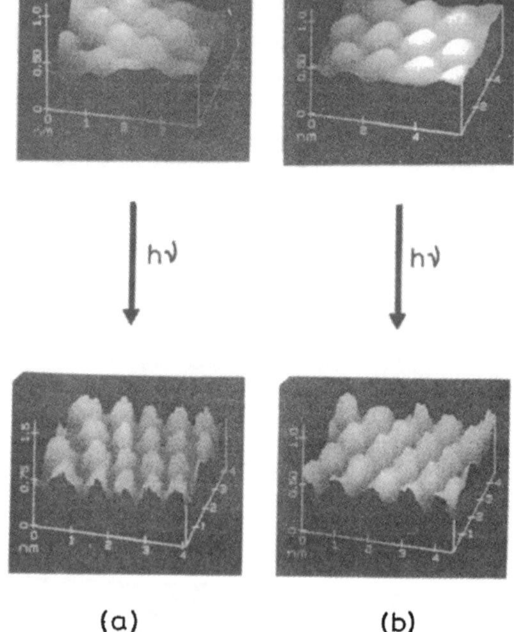

Figure 1. (a) STM images of C_{60} films on highly oriented pyrolytic graphite (HOPG) before and after UV irradiation. (b) STM images of C_{70} films on HOPG before and after UV irradiation. The conditions were 254 nm UV, 12 h, 10^{-5} Torr vacuum.

the molecule as well as the high reactivity of the bonds radial to the capping pentagons toward 2+2 addition, i.e., the pentagons defining the 5-fold axis.[8] Since the molecules do not have to reorient for the polymerization of C_{60} to occur, the only effect is due to intermolecular bonding resulting in the observed contraction. The observation of unit-cell expansion in the case of C_{70} and contraction in the case of C_{60} is indeed noteworthy and has direct bearing on the amorphization of these fullerenes.

[†] Supported by the Jawaharlal Nehru Centre for Advanced Scientific Research.

* Correspondence to be addressed at the CSIR Centre of Excellence in Chemistry, Indian Institute of Science, Bangalore 560 012, India.

[®] Abstract published in *Advance ACS Abstracts,* October 15, 1995.

Letters

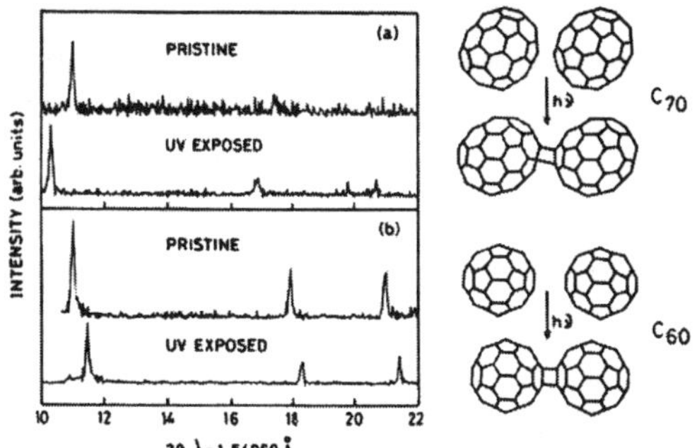

Figure 2. XRD patterns of (a) C_{70} and (b) C_{60} on glass substrates before and after UV irradiation. Whereas the C_{60} lattice contracts on photopolymerization, for C_{70}, the lattice is seen to expand. The 2+2 addition of C_{60} and C_{70} (resulting in photopolymerized products) is shown schematically.

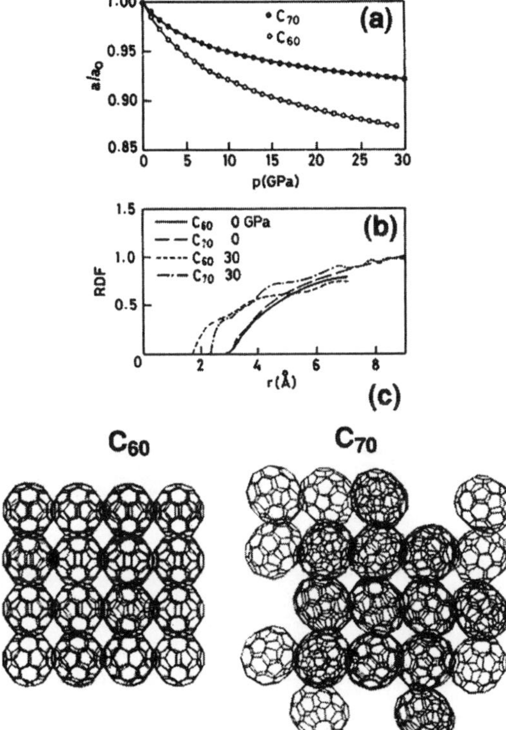

Figure 3. (a) Pressure dependence of the ratio of the pseudocubic lattice parameter a of C_{60} and C_{70} to the parameter a_0 at ambient pressure as given by the Murnaghan equation of state. Experimental compressibilities were employed to produce these plots. It is assumed that there are no intervening phase changes and that C_{60} remains cubic and C_{70} rhombohedral throughout. (b) Atom–atom radial distribution functions of C_{60} and C_{70} molecules at 0 and 30 GPa and 300 K from NVT Monte Carlo studies. The molecules of C_{60} approach closer at a given pressure than do C_{70}. (c) Snapshots of the C_{60} lattice at 20 GPa and C_{70} lattice at 25 GPa obtained from NVT-MC simulations, looking down the (100) direction.

What is also interesting about C_{60} and C_{70} is that their compressibilities are quite different.[9] The ratios of the pseudocubic lattice parameters of C_{60} and C_{70} with respect to their corresponding room-pressure values are shown as a function of pressure in Figure 3a. These plots were made using the Murnaghan equation of state and experimental compressibilities. We see that solid C_{60} is more compressible than C_{70}. Our NVT Monte Carlo studies[10] also throw light on the nature of changes brought about on application of pressure. Among the various results obtained from such studies, the one most relevant to the context of polymerization under pressure is the following: On application of pressure, the C_{60} molecules approach one another much more closely than C_{70} molecules. The calculated closest approach between two C_{60} molecules at 300 K and 30 GPa is ≈ 1.8 Å, while it is ≈ 2.2 Å in C_{70} (Figure 3b). Although the C_{60} molecules are randomly oriented in the solid up to moderate pressures, the molecules align themselves at higher pressures providing so as to be favorable for the 2+2 addition to occur. In the case of C_{70} on the other hand, the alignment of the molecules under pressure occurs with their long axes along the face and body diagonals of the unit cell so as to render the reactive end caps of the molecule to be relatively farther apart. Snapshots of the simulated C_{60} and C_{70} lattices under pressure are shown in Figure 3c.

Clearly, polymerization under pressure is favored in the case of C_{60} but not in C_{70}. Since the polymerization is also accompanied by volume contraction in the case of C_{60}, pressure would indeed favor such a process. This is not so in C_{70}, where polymerization is accompanied by volume expansion. On the basis of these observations, we are able to understand why the amorphization of C_{60} under pressure is irreversible because of polymerization.

References and Notes

(1) (a) Rao, A. M.; Zhou, P.; Wang, K. A.; Hager, G. T.; Holden, J. M.; Wang, Y.; Lee, W. T.; Bi, X. X.; Ecklund, P. C.; Cornett, D. S.; Duncan, M. A.; Amster, I. J. *Science* **1993**, *259*, 955. (b) Weaver, J. H.; Poirier, D. M. *Solid State Phys.* **1994**, *48*, 1. (c) Zhou, P.; Rao, A. M.; Wang, K. A.; Robertson, J. D.; Eloi, C.; Meier, M. S.; Rin, S. L.; Bi, X. X.; Ecklund, P. C.; Dresselhaus, M. S. *Appl. Phys. Lett.* **1992**, *60*, 2871. (d) Bacsa, W. S.; Lannin, J. S. *Phys. Rev.* **1994**, *B49* 14750. (e) Menon, M.; Subbaswamy, K. R. *Phys. Rev.* **1994**, *B49* 13966.

(2) (a) Nunez-Reguero, M.; Marques, L.; Hodeau, J. L.; Bethoux, O.; Perroux, M. *Phys. Rev. Lett.* **1995**, *74*, 278. (b) Xu, C. H.; Scuseria, G. E. *Phys. Rev. Lett* **1995**, *74*, 274.

(3) Rao, A. M.; Menon, M.; Wang, K. A.; Ecklund, P. C.; Subbaswamy, K. R.; Cornett, D. S.; Duncan, M. A.; Amster, I. J. *Chem. Phys. Lett.* **1994**, *224*, 106.

(4) (a) Yoo, C. S.; Nellis, W. J. *Chem. Phys. Lett.* **1992**, *198*, 379. (b) Moshary, F.; Chen, N. H.; Silvera, I. F.; Brown, C. A.; Dorn, H. C.; deVries, M. S.; Bethune, D. S. *Phys. Rev. Lett.* **1992**, *69*, 466. (c) Snoke, D. W.; Syassen, K.; Mittelbach, A. *Phys. Rev.* **1993**, *B47*, 4146.

(5) Chandrabhas, N.; Sood, A. K.; Muthu, D. V. S.; Sundar, C. S.; Bharathi, A.; Hariharan, Y.; Rao, C. N. R. *Phys. Rev. Lett.* **1994**, *73*, 3411.

(6) Since 2+2 cycloaddition is symmetry allowed under photochemical conditions but not under thermal conditions, under the application of pressure, it must take place by routes other than concerted, for example, via polar intermediates.

(7) Both C_{60} and C_{70} are orientationally disordered in the solid state at ordinary temperatures. Moreover, the disorder is dynamic. This prevents the probing of intramolecular features by techniques such as STM. However, when the fullerenes are deposited as films on surfaces such as single crystal metals, ordered structures can be obtained because the molecules are attached to the surface relatively strongly because of charge transfer from the metal to the fullerenes. When the molecules polymerize, they cannot rotate freely and the internal structure becomes visible. For details see: (a) Wang, X. D.; Hashizume, T.; Sakurai, T. *Mod. Phys. Lett.* **1994**, *8*, 1397. (b) Aiyer, H. N.; Govindaraj, A.; Rao, C. N. R. *Bull. Mater. Sci.* **1994**, *17*, 563. (c) Joachim, C.; Gimzewski, J.; Schittler, R.; Chavy, C. *Phys. Rev. Lett.* **1995**, *74*, 2102. (d) Aiyer, H. N.; Govindaraj, A.; Rao, C. N. R. *Philos. Mag. Lett.*, in print.

(8) (a) Henderson, C. C.; Cahill, P. A. *Science* **1994**, *263*, 397. (b) Karfunkel, H. R.; Hrisch, A. *Angew. Chem., Int. Ed. Engl.* **1992**, *31*, 1468. (c) Rathna, A.; Chandrasekhar, J. *Curr. Sci. (India)* **1993**, *65*, 768. (d) Govindaraj, A.; Rathna, A.; Chandrasekhar, J.; Rao, C. N. R. *Proc. Indian Acad. Sci. (Chem. Sci.)* **1993**, *105*, 303.

(9) (a) Duclos, S. J.; Brister, K.; Haddon, R. C.; Kortan, A. R.; Thiel, F. A. *Nature* **1991**, *351*, 380. (b) Christides, C.; Thomas, I. M.; Dennis, T. J. S.; Prassides, K. *Europhys. Lett.* **1993**, *22*, 545.

(10) (a) NVT Monte Carlo simulations were performed on cells containing 32 rigid C_{60} or C_{70} molecules. The intermolecular potential was of the Lennard-Jones form. The unit cells for the simulation were taken from experiments referred to in the text. Visualization was done using the InsightII software from BIOSYM Technologies, San Diego on Indigoll workstations. For details of the simulation as well as the potential parameters used, see: (a) Cheng, A.; Klein, M. L. *J. Phys. Chem.* **1991**, *95*, 6750. (b) Girifalco, L. A. *J. Phys. Chem.* **1992**, *98*, 858. (c) Chakrabarti, A.; Yashonath, S.; Rao, C. N. R. *Chem. Phys. Lett.* **1993**, *215*, 519.

JP951970P

Phase transformations of mesoporous zirconia

Neeraj and C. N. R. Rao*

Chemistry and Physics of Materials Unit, Jawaharlal Nehru Center for Advanced Scientific Research, Jakkur, Bangalore 560064, India

The kinetics of the lamellar→hexagonal→cubic transformations of mesoporous zirconia prepared by using a neutral organic amine as the amphiphile have been studied in phosphoric acid medium. The lamellar→hexagonal transformation is preceded by a loss of the template molecules and the hexagonal→cubic transformation proceeds only when the lamellar form has entirely transformed to the hexagonal phase. The kinetics of the thermal transformation of lamellar zirconia to the hexagonal form have also been examined; this transformation also occurs and is accompanied by a loss of template molecules. Accordingly, the activation energy of the transformation is comparable to the hydrogen bond energy between the amine and the oxo–zirconium species. The phase transformations of mesoporous zirconia can be understood in terms of minimum energy surfaces.

Mesoporous solids are generally prepared by making use of self-assembled ordered aggregates of surfactants as templates, thus rendering the structures exhibited by the mesoporous solids to be similar to those of the self-assembled[1-3] surfactants. Accordingly, mesoporous solids occur in lamellar, hexagonal and cubic forms, just as the surfactant aggregates. Phase transitions among these forms occur in mesoporous materials in the solution phase. For example, changing the pH of the medium or ageing, leads to transformation of the lamellar phase of silica to the hexagonal phase.[4] Lamellar, hexagonal and cubic forms of aluminoborates have been prepared by changing the pH.[5] Other factors such as temperature, presence of counter ions and concentration of the surfactant also affect the phase transitions.[4-10] The phase transitions in mesoporous solids bear some similarity to those in surfactant assemblies.

In surfactant systems, decreasing the concentration of the surfactant instantaneously transforms the lamellar phase to the hexagonal phase.[11,12] The transitions in the surfactant self-assemblies are commonly understood in terms of the surface/interface energies of the ordered aggregates.[13] The effect of ageing on the lamellar to hexagonal phase transformation in mesoporous silica has been related to the extent of polymerization of the silica framework and the balancing of charges.[4] High pH and a low degree of polymerization favour the lamellar phase whereas low pH and highly condensed silica favour the hexagonal phase of silica.[4] The transformation from the hexagonal to the cubic phase in silica has been explained in terms of the formation of a periodic minimal surface governed by a competition between the curvature and packing and the transformation is associated with kinetic barriers.[14] Since there is limited quantitative information on the nature of the phase transitions in mesoporous solids in the literature, we considered it important to investigate the transitions in some detail. For this purpose, we have chosen to investigate mesoporous zirconia which exhibits interesting phase transitions both in solution and in the solid state. Here, we report the results of a kinetic study of the lamellar→hexagonal→cubic transitions of mesoporous zirconia in the presence of phosphoric acid and of the thermal lamellar → hexagonal transition of zirconia in the solid state.

Experimental

The lamellar form of mesoporous zirconium oxide was prepared using dodecylamine (DA) as the surfactant. In a typical synthesis zirconium isopropoxide (0.01 mol) was added to a solution of the dodecylamine (DA) (0.03 mol) in propan-l-ol (0.1 mol) to which ammonium sulfate (0.12 mol) was added under stirring. The pH of the gel was adjusted to 1.5–2.0 by using dilute hydrochloric acid (HCl). The gel was subjected to hydrothermal treatment at 373 K for 20 h, filtered, washed with acetone and dried at 373 K for 2 h. The lamellar nature of the product was verified by X-ray diffraction (XRD) and transmission electron microscopy (TEM). To study the transformation of lamellar ZrO_2 in solution phase, *ca.* 100 mg of the as-synthesized sample was added to 50 ml of 0.87 M phosphoric acid and stirred for different times, filtered, washed with water, acetone and dried at ambient temperature. XRD patterns and TEM images of the samples were recorded for each time.

The kinetics of the lamellar→hexagonal→cubic transition of zirconia in phosphoric acid solution was studied as follows. The (100) reflection of the lamellar and hexagonal phases differ both in intensity and position. Thus, as the lamellar phase ($d_{100} = 3.34$ nm) transforms to the hexagonal phase the intensity of the (100) reflection decreases until it attains a much smaller value, characteristic of that of a disordered hexagonal mesophase (Fig. 1). In addition, the *d*-spacing of the (100) reflection decreases until it reaches a minimum value after the transformation to the hexagonal phase is complete. The intensities of the (200) and (300) reflections of the lamellar phase also decrease continuously with time, as the lamellar→hexagonal transformation proceeds. We have employed both the intensity and position of the (100) reflection to follow the kinetics of the lamellar–hexagonal transformation. These two measurements give slightly different estimates of the phase compositions and we have taken the average value in the kinetic study. The hexagonal to cubic transformation of mesoporous zirconia was followed by the increase in the intensity of the (220) reflection of the cubic phase with time (Fig. 1).

The lamellar→hexagonal thermal transformation of mesoporous zirconia in the solid state was studied by heating the lamellar form at a fixed temperature for different periods of time. The phase composition of the sample subjected to heat treatment was estimated on the basis of the intensity of the (100) reflection, the *d*-spacing showing only small changes in the thermal transformation (Fig. 2). The kinetics of the transformation was studied as a function of time at 373, 403 and 413 K. The transformation was also studied by heating the lamellar sample for a fixed period of 2 h at different temperatures in the range 360–430 K.

Results and Discussion

The lamellar form of mesoporous zirconia, on contact with 0.87 M phosphoric acid, first transforms to the hexagonal form.

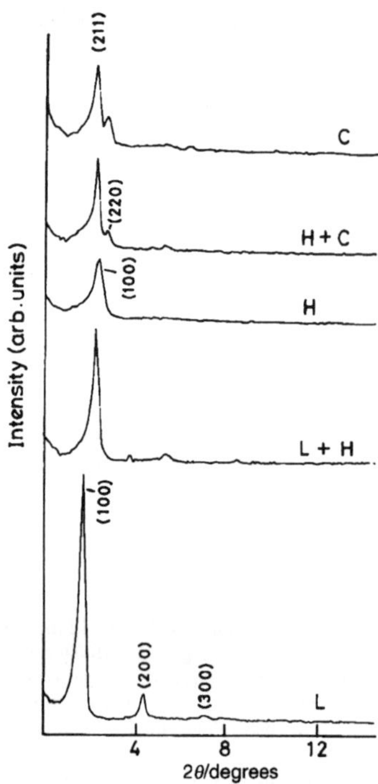

Fig. 1 XRD patterns showing the transformation of lamellar (L) zirconia to hexagonal (H) and then cubic (C) phases in phosphoric acid solution. Intermediate stages during the L–H and H–C transformations are shown.

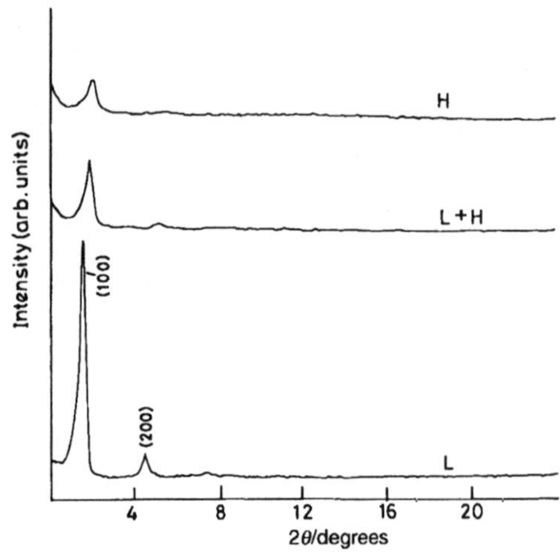

Fig. 2 XRD patterns showing the thermal transformation of lamellar (L) zirconia to the hexagonal (H) form. An intermediate stage during the transformation is shown.

This transformation is complete in 7 h. We have followed the lamellar→hexagonal transformation as a function of time. Fig. 3 shows the progress of the lamellar→hexagonal transformation by plotting the percentages of the lamellar and hexagonal phases *vs.* time. We see that the proportion of the lamellar form decreases while that of the hexagonal form increases. It is noteworthy that the transformation of the hexagonal form to the cubic form starts only after the lamellar phase has completely

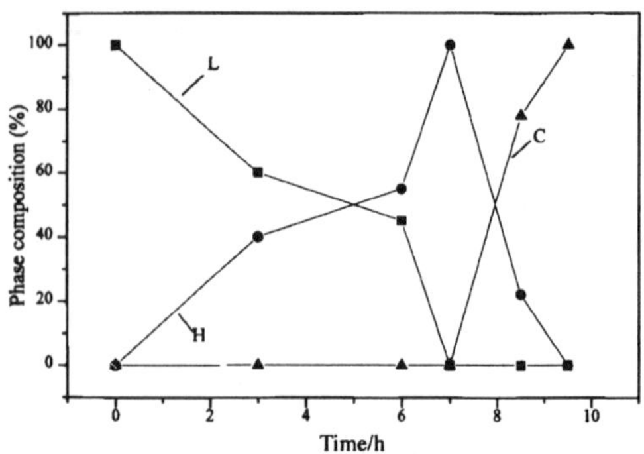

Fig. 3 Time variation in the phase composition of zirconia in phosphoric acid solution; L, lamellar; H, hexagonal; C, cubic

transformed to the hexagonal phase. The hexagonal to cubic transformation occurs over a short time (<3 h).

The lamellar→hexagonal transformation of ZrO_2 is likely to be initiated first by the removal of some of the surfactant species, followed by the curling of the surfactant bilayer in order to minimize the surface/interface energy as shown in Fig. 4(a) and (b).[13,15] The curled bilayers transform to cylindrical rods to further minimize the surface energy as shown in Fig. 4(c) and the cylindrical rods assemble to give the ordered hexagonal structure shown in Fig. 4(d). In order to examine

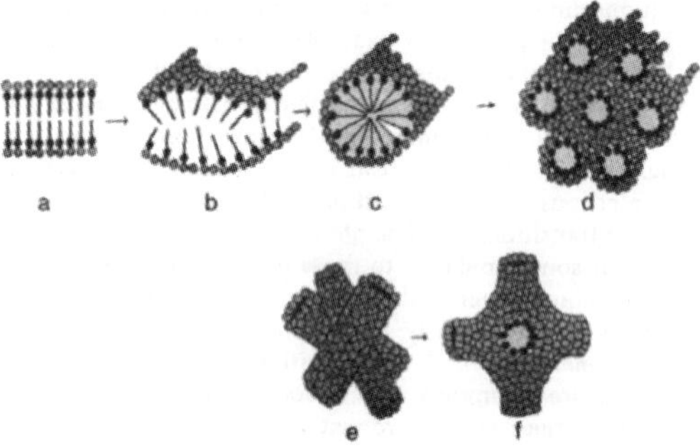

Fig. 4 Schematic representation of lamellar→hexagonal phase transformation (a through d) and the hexagonal→cubic transformation (e and f). The shaded circles around the surfactant aggregates represent the inorganic species (generally metal alkoxides or other metal–oxo species).

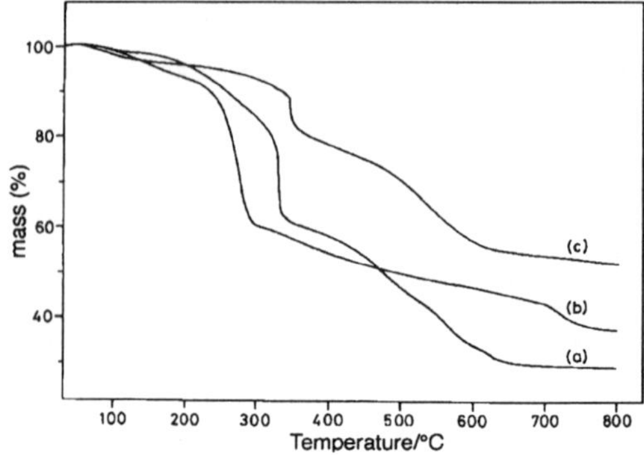

Fig. 5 Thermogravimetric curves of lamellar zirconia maintained for different periods in phosphoric acid solution: (a) as-prepared lamellar zirconia, (b) after 3 h and (c) after 7 h

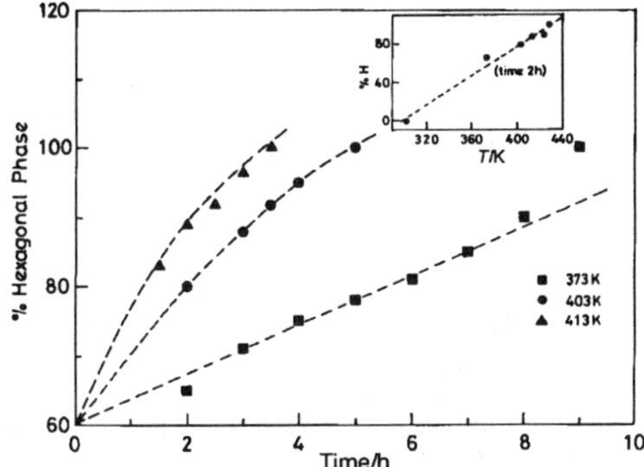

Fig. 6 Kinetics of the lamellar to hexagonal transformation of zirconia at different temperatures. Inset shows the temperature variation of the percentage of the hexagonal (H) phase in a fixed period of 2 h.

whether the loss of the surfactant species precedes the lamellar→hexagonal transformation, we have carried out thermogravimetric analysis (TGA) studies. We find that there is a significant loss of the surfactant in the lamellar→hexagonal→cubic transformation in phosphoric acid solution.

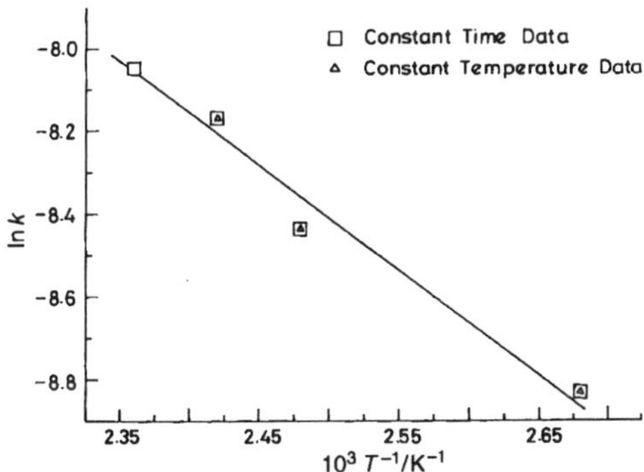

Fig. 7 Arrhenius plots of the kinetic data of the lamellar→hexagonal transformation of zirconia in the solid state

TGA of the lamellar zirconia [Fig. 5(a)] shows a marked mass loss at around 593 K and gradual loss of mass from 593 to 973 K. There is negligible loss of mass below 430 K due to water and other species. The mass loss at 593 K and above appears to be entirely due to the loss of the surfactant. The lamellar ZrO_2 sample treated for 3 h in phosphoric acid [Fig. 5(b)] showed a sharp mass loss around 543 K, followed by a gradual loss up to 973 K. Since the entire mass loss is only due to the surfactant, we can take the difference in the mass loss, say at 973 K, between the curves Fig. 5(a) and (b) as equal to the amount of the surfactant lost by phosphoric acid treatment. We see that around 8% of the surfactant is lost after 3 h of treatment. The ZrO_2 sample treated in phosphoric acid for 7 h was entirely hexagonal and showed a sharp mass loss around 613 K and a gradual loss thereafter, up to 973 K [Fig. 5(c)]. The difference between curves Fig. 5(a) and (c) shows that 23% of the surfactant has been removed after 7 h of treatment. Phosphoric acid treatment for 9.5 h gave the cubic phase, and *ca.* 30% of the template had been removed at this stage. The TGA studies suggest that the loss of the surfactant is a necessary initial step in the lamellar→hexagonal transformation.

The transformation from the hexagonal to the cubic phase is driven by the tendency of the cylindrical rods of the hexagonal phase to minimize their energy by forming a three-dimensional network of rods, forming bicontinuous cubic phases as shown in Fig. 4(e) and (f). The cubic phase can be described using the concept of periodic minimal surface,[15,16] with the space groups *Pn3m, Pm3n, P4₃32, Im3m, Ia3d* or *Fd3m*.[16] The XRD pattern of the cubic phase obtained after the complete transformation of the hexagonal ZrO_2 phase in phosphoric acid, appears to be consistent with the space group *Ia3d*.

The thermal transformation of the lamellar form of mesoporous ZrO_2 to the hexagonal form occurs in the solid state. Thus, on heating at 428 K for 2 h, the lamellar phase completely transforms to the hexagonal phase. We have followed the kinetics of the lamellar→hexagonal transformation at three fixed temperatures. Fig. 6 shows the kinetics of the lamellar→hexagonal transformation at different temperatures. We have also examined the kinetics of this transformation by heating lamellar zirconia at different temperatures for a fixed time of 2 h. The inset of Fig. 6 shows how the proportion of the hexagonal form increases with temperature at a given time (2 h). We were able to fit these kinetic data to a first order

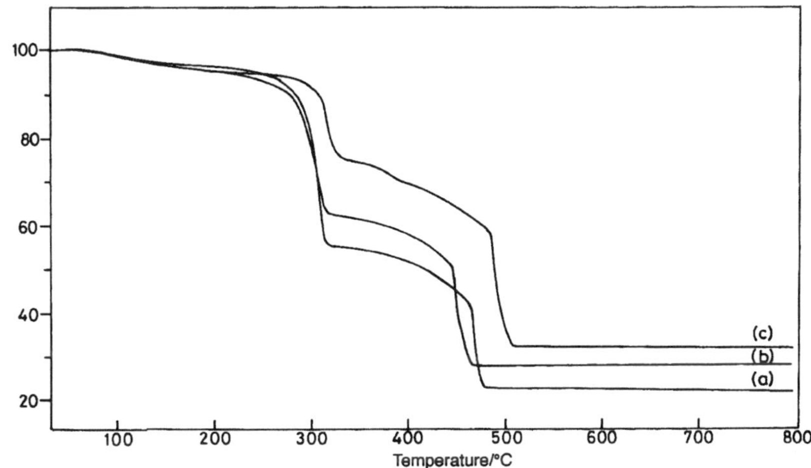

Fig. 8 Thermogravimetric curves of lamellar form heated at different temperatures: (a) as-prepared lamellar zirconia, (b) 403 K, 2 h and (c) 428 K, 2 h.

rate equation. The rate data follow the Arrhenius equation as shown in Fig. 7. The data give an activation energy of *ca.* 22 kJ mol^{-1} for the lamellar→hexagonal transformation in the solid state. This value of the activation energy is comparable to the energy of a medium strength hydrogen bond. This is understandable since the removal of surfactant molecules from the lamellar phase is necessary for the transformation to occur. The surfactants interact with the oxo–zirconium species primarily through hydrogen bonding.

TGA studies show that on heating the lamellar phase to 403 K for 2 h, around 6% of the surfactant is lost (Fig. 8). On heating to 428 K the sample loses 11% of the surfactant. These data demonstrate that during the thermal transformation of the lamellar phase to the hexagonal phase, the amine template is partially removed, leading to the reorganization of the self-assembled surfactant aggregate. The magnitude of the loss of the surfactant in the thermal transformation is somewhat smaller than that accompanying in the transformation in phosphoric acid solution.

In conclusion, the present study of the kinetics of the lamellar to hexagonal transformation of mesoporous zirconia shows that a loss of surfactant molecules accompanies the transformation. Transformation to the cubic form seems to require that all the starting material be in the hexagonal form. The thermally induced lamellar→hexagonal transformation is associated with an activation energy comparable to the hydrogen bond energy.

References

1 P. Mariani, V. Luzatti and H. Delacroix, *J. Mol. Biol.*, 1988, **204**, 165.
2 P. O. Eriksson, G. Lindblom and G. Arvidson, *J. Phys. Chem.*, 1985, **89**, 1050.
3 J. C. Vartuli, K. D. Schmitt, C. T. Kresge, W. J. Roth, M. E. Leonowicz, S. B. McCullen, S. D. Hellring, J. S. Beck, J. L. Schlenker, D. H. Olson and E. W. Sheppard, *Chem. Mater.*, 1994, **6**, 2317.
4 A. Monnier, F. Schuth, Q. Huo, D. Kumar, D. Margolese, R. S. Maxwell, G. D. Stucky, M. Krishnamurthy, P. Petroff, A. Firouzi, M. Janicke and B. F. Chmelka, *Science*, 1993, **261**, 1299.
5 S. Ayyappan and C. N. R. Rao, *Chem. Commun.*, 1997, 575.
6 J. Luo and S. L. Suib, *Chem. Commun.*, 1997, 1031.
7 M. Ogawa, *J. Am. Chem. Soc.*, 1994, **116**, 7941.
8 S. Oliver, A. Kuperman, N. Coombs, A. Laugh and G. A. Ozin, *Nature (London)*, 1995, **378**, 47.
9 C. A. Fyfe and G. Fu, *J. Am. Chem. Soc.*, 1995, **117**, 9709.
10 H. P. Lin and C. Y. Mou, *Science*, 1996, **273**, 765.
11 E. S. Blackmore and G. J. T. Tiddy, *J. Chem. Soc., Faraday Trans. 2*, 1988, **84**, 1115.
12 A. Sein and J. B. F. N. Engberts, *Langmuir*, 1995, **11**, 455.
13 J. M. Seddon, *Biochim. Biophys. Acta.*, 1990, **1031**, 1.
14 S. M. Gruner, *J. Phys. Chem.*, 1989, **93**, 7562.
15 G. Porte in *Micelles, Membranes, Microemulsions and Monolayers*, ed. W. M. Gelbart, A. Ben-Shaul and D. Roux, Springer-Verlag, New York, 1994, p. 105.
16 S. T. Hyde, *Curr. Opin. Solid State Mater. Sci.*, 1996, **1**, 653.

Paper 8/01419A; Received 19th February, 1998

III. TRANSITION METAL OXIDES

Commentary by C.N.R. Rao

Transition metal oxides constitute probably one of the most interesting classes of solids, exhibiting a variety of structures and properties.[1,2] The nature of metal-oxygen bonding can vary between nearly ionic to highly covalent or metallic. The unusual properties of transition metal oxides are clearly due to the unique nature of the outer *d*-electrons. The phenomenal range of electronic and magnetic properties exhibited by transition metal oxides is noteworthy. Thus, we find oxides with metallic properties (e.g., RuO_2, ReO_3, $LaNiO_3$) at one end and insulating oxides (e.g., MnO, $BaTiO_3$) at the other. There are also oxides that traverse both these regimes with change of temperature, pressure, or composition (e.g., V_2O_3, $La_{1-x}Sr_xCoO_3$). Interesting electronic properties also arise from charge density waves (e.g., $K_{0.3}MoO_3$), charge ordering (e.g., Fe_3O_4), and defect ordering (e.g., $Ca_2Fe_2O_5$). Oxides with diverse magnetic properties anywhere from ferromagnetism (e.g. CrO_2, $La_{0.5}Sr_{0.5}MnO_3$) to antiferromagnetism (e.g., NiO, $LaCrO_3$) are known. Many oxides possess switchable orientation states, as in ferroelectric (e.g., $BaTiO_3$, $KNbO_3$) and ferroelastic (e.g., $Gd_2(MoO_4)_3$) materials. No discovery in solid state science has created as much sensation as that of high-temperature superconductivity in cuprates. Although superconductivity in transition metal oxides has been known for some time, the highest T_c reached was around 13K; we now have oxides with T_c's approaching 160K. The discovery of high T_c oxides has focused worldwide attention on the chemistry of metal oxides and at the same time revealed how inadequate our understanding is of these fascinating materials. This section has a few representative articles on metal oxides, covering structure and properties, as well as high-temperature superconductivity.[3,4]

An important new addition to the fascinating phenomena exhibited by transition metal oxides is colossal magnetoresistance (CMR), discovered in 1993 in rare earth manganates of the formula $Ln_{1-x}A_xMnO_3$ (Ln = rare earth, A = Alkaline earth or other divalent ions).[5] These manganates containing Mn in the +3 and +4 states become ferromagnetic and metallic due to the double exchange mechanism of electron hopping. At the ferromagnetic T_c, they show a large decrease in resistance on application of magnetic fields. The T_c and other properties of the manganates are sensitive to the average radius of the A-site cations, $<r_A>$, as well as size disorder (mismatch). Besides CMR, the manganates undergo charge-ordering of the Mn^{3+} and Mn^{4+} ions depending on the $<r_A>$, small values favoring charge-ordering.[5] Charge-ordering competes with double exchange and gives rise to anti-ferromagnetic insulating states. Besides the ordering of charges and spins, the ordering of the e_g orbitals of Mn^{3+} ions in the rare earth manganates plays a significant role. In this section, articles dealing with CMR, charge-ordering and related aspects of rare earth manganates is presented. Other interesting aspects such as electron hole asymmetry and electronic phase separation are also examined in the papers. Electronic phase separation in solids is a phenomenon of importance being understood more clearly in the last few years.[6] It is likely that many transition metal oxides with correlated electrons will exhibit phase separation.

Multiferroics constitute an important area of materials research.[7] Multiferroics are materials which are magnetic as well as ferroelectric. In principle, these two properties cannot occur together since one requires the presence of d-electrons and the other does not. Strategies to attain multiferroic properties in oxides are presented in an article.

References

1. C.N.R. Rao, *Ann. Rev. Phys. Chem.* **40**, 291 (1989).
2. C.N.R. Rao and B. Raveau, *Transition Metal Oxides*, Wiley-VCH, Second Edition, 1999.
3. T.V. Ramakrishnan and C.N.R. Rao, *Superconductivity Today*, Second Edition, University Press, Hyderabad, India, 1999.
4. C.N.R. Rao (Ed.), *Chemistry of High Temperature Superconductors*, World Scientific, Singapore, 1991.
5. C.N.R. Rao and B. Raveau (Eds.), *Colossal Magnetoresistance, Charge Ordering and Related Properties of Manganese Oxides*, World Scientific, Singapore, 1998.
6. V.B. Shenoy and C.N.R. Rao, *Phil. Trans. Royal. Soc. London*, **A366**, 63 (2008).
7. C.N.R. Rao and C.R. Serrao, *J. Mater. Chem.* **17**, 4931 (2008).

Proc. R. Soc. Lond. A. **355**, 301–312 (1977)
Printed in Great Britain

Electron microscopy of ferroelectric bismuth oxides containing perovskite layers

By J. L. HUTCHISON[†], J. S. ANDERSON, F.R.S.[†] AND C. N. R. RAO[‡]

Inorganic Chemistry Laboratory, University of Oxford

(*Received* 12 *January* 1976 – *Revised* 23 *December* 1976)

[Plates 1–4]

Ferroelectric bismuth oxides of the general formula

$$(Bi_2O_2)^{2+}(A_{n-1}B_nO_{3n+1})^{2-}$$

(A = Ba, etc., B = Ti or other transition metal) have been examined by high-resolution lattice imaging electron microscopy. The lattice images show dark bands at the positions of the Bi_2O_2 layers, with $n-1$ lines between them due to the layers of the perovskite A cations or, in favourable circumstances, the fully resolved 0.4 nm square perovskite grid. Dislocations and domain boundaries have been imaged for the first time in ferroelectric crystals. The structure of the dislocations and domain walls is discussed in the light of the microstructural evidence.

1. INTRODUCTION

Lattice imaging methods of electron microscopy provide a powerful tool for investigating the local structure of crystals, in real space. They have been applied, to good effect, to a number of complex oxide systems and have yielded information, that would hardly be accessible to other methods, about stacking sequences, the regularity and ordering of layer structures, the existence of new intergrowth phases, superstructures and extended defects. Our study of the barium ferrite layer structures (McConnell, Hutchison & Anderson 1974) showed the value of examining structures in which a variety of large crystallographic units could be built up by the ordered repetition of a small number of distinct sub-units.

In this paper we describe some observations on a somewhat analogous family of layered structures, first described by Aurivillius (1949) as having the general chemical formula $Bi_2A_{m-2}B_{m-1}O_{3m-1}$, where A is a bulky cation (Bi^{3+}. Ba^{2+}, Pb^{2+} etc.) and B is a transition metal cation normally found in 6-coordination with oxygen (Ti^{4+}, Nb^{5+} etc.). The structural principle common to all the family is better expressed by the formulation $(Bi_2O_2)^{2+}(A_{n-1}B_nO_{3n+1})^{2-}$ (in which $n = m - 1$); this emphasizes their construction from finite (and therefore anionically charged) slabs of

---

† Now at Edward Davies Chemical Laboratory, University College of Wales, Aberystwyth, SY23 1NE.

‡ Permanent address: Indian Institute of Science, Bangalore 560012, India.

302   J. L. Hutchison, J. S. Anderson and C. N. R. Rao

perovskite structure, $n-1$ unit cells thick and thus having n layers of B-site cations in corner-sharing octahedral coordination, enclosing $n-1$ layers of 12-coordinate A sites (Goodenough & Longo 1970). Between these slabs, and having their oxygen atoms in common with them, are double-sided $Bi_2O_2^{2+}$ sheets, of the kind frequently

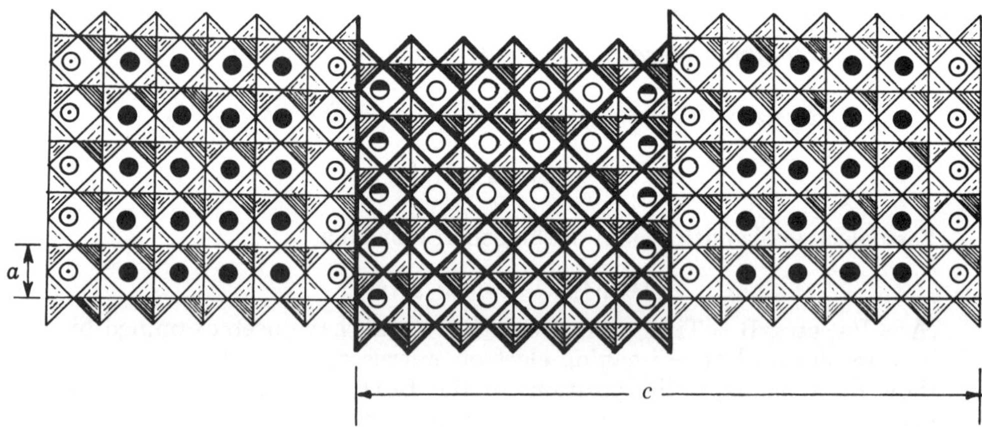

FIGURE 1. Structure of $Ba_2Bi_4Ti_5O_{18}$ (schematic), projected on (101) of the pseudotetragonal cell. Lightly drawn octahedra: B cations at $y = 0$; heavily drawn octahedra: B cations at $y = \frac{1}{2}$; ⊖, Bi (in Bi_2O_2 layers) at $y = 0$; ⊙, Bi (in Bi_2O_2 layers) at $y = \frac{1}{2}$; ○, A cations at $y = 0$; ●, A cations at $y = \frac{1}{2}$.

found in bismuth oxy-compounds, made up of square pyramidal $[BiO_4]$ groups sharing their basal edges. Thus, in $Bi_4Ti_3O_{12}$ ($n = 3$) and $Bi_4BaTi_4O_{15}$ ($n = 4$) the perovskite units are slabs with the composition $BiTiO_3$ and $(Ba_{0.33}Bi_{0.67})TiO_3$ per perovskite cell respectively. An apparent variability of composition, as appeared from Aurivillius' papers, could arise if the perovskite slabs included between the Bi_2O_2 layers varied in thickness from layer to layer. All the compounds are tetragonal (or nearly so, with slight orthorhombic distortion); their a and b axes lie along $[110]_c$ (where the suffix denotes the cubic perovskite subcell), so that

$$a \approx b \approx \sqrt{2}a_c \approx 0.54\,\mathrm{nm}.$$

Their c axes are inherently long: $\frac{1}{2}c_{ss} = (n-1)a_c + $ thickness of the Bi_2O_2 layer $\approx na_c$ (the suffix ss denoting dimensions, etc., of the large superstructure cell). For compounds with $n = 1$ to 5, c_{ss} is in the range 1.54–5.04 nm.

Figure 1 shows a typical member of the series, $Ba_2Bi_4Ti_5O_{18}$ ($n = 5$), drawn as a polyhedron network structure. It may be noted that the bismuth atoms of the Bi_2O_2 sheets occupy positions close to virtual A sites, and also that the pyramidal BiO_4 groups (which could alternatively be regarded as half octahedra terminated by a Bi atom in place of an oxygen atom) share edges in such a way that successive slabs of perovskite structure are in antiphase, displaced by $\frac{1}{2}[110]_c$.

These layered perovskite compounds are ferroelectric, with fairly high Curie temperatures (Schmid 1975; Subbarao 1974). Some ambiguity still attaches to the interpretation of paraelectric/ferroelectric transitions, but it is clear that ionic polarization makes an important contribution to the permanent dipole in the ferro-electric state and that, in the perovskites and related structures, this ionic polarization involves a displacement of the highly charged B-site cations away from the centres of their coordination octahedra. Whether there is also a displacement of the oxide ions appears less certain. The displacements are small; in the case of $BaTiO_3$ the shift of the Ti^{4+} ions has been estimated as being of the order of 0.006–0.010 nm, relative to the A sites. The structural consequence of this polarization is to distort the environment of the B cations from nearly regular octahedral to irregular octahedral coordination, and thereby to lower the crystal symmetry. The topology of the structure, however, as a network of vertex-linked octahedra, remains quite unaltered. It is permissible to extend these considerations from the perovskites proper to the Aurivillius phases considered here. It can be concluded that the only effects of ferroelectric polarization likely to show up in lattice images are distortions from orthogonality in the ferroelectric state; displacements of heavy atoms are small, as compared with the attainable resolution, and electron scattering (hence image contrast) will be dominated by the heavy atoms in the Bi_2O_2 layers and the A sites.

Regularity of microstructure is readily investigated, but it was of interest to see whether ferroelectric domains and boundaries could be revealed in lattice images. Evidence on this is given below. In addition, the materials afforded opportunity for observations of structural relaxation in the core regions of large dislocations.

2. Experimental

Materials

Compounds with $n = 1$ to 5 were prepared from appropriate mixtures of the constituent oxides (Bi_2O_3, WO_3, Nb_2O_5, TiO_2; Johnson Matthey Specpure) and barium carbonate. The mixtures were ground thoroughly and precalcined at about 100 °C below the final firing temperature. They were then re-ground and heated at the temperatures shown in Table 1. The weight loss during the preparation was less than 0.5 %. Lattice parameters and ferroelectric Curie temperatures of the compounds thus prepared are listed in table 1. According to Ismailzade (1967) the $n = 8$ compound $Bi_9Ti_3Fe_5O_{27}$ is formed by heating an appropriate oxide mixture at 1300 K; a preparation following his procedure proved, however, to be structurally inhomogeneous (see below).

Electron microscopy

Procedures for electron microscopy were those now customary in many-beam lattice imaging work. Crystals were finely crushed in an agate mortar and the finer fragments slurried on to carbon-coated grids, for examination in a Siemens Elmiskop 102 electron microscope, equipped with a high resolution, double tilt stage and a z-translation. Micrographs were recorded at 100 kV, at magnifications from 300 000 to 700 000, in through-focal series spanning the optimum defocus condition (about

304 J. L. Hutchison, J. S. Anderson and C. N. R. Rao

70 nm under focus). As is well known, it is necessary to orient specimens with great accuracy, in order that corresponding structural units should be exactly lined up in projection. Suitably thin fragments were therefore tilted to bring the projection axis, usually $[100]_c$ or $[110]_c$, parallel to the electron beam, within $0.1°$. A 40 µm objective aperture was used to combine the primary beam with all diffracted beams out to about 3.3 nm^{-1}; about 70 beams were thereby included for the $n = 5$ compound. Experience has shown that, in these conditions, the effective resolving power is about 0.35 nm.

TABLE 1. BISMUTH OXIDE LAYERED COMPOUNDS STUDIED

		final heating temperature/K	lattice parameters nm			ferroelectric T_c
n	compound		a	b	c	K
1	Bi_2WO_6	1040	0.5458	0.5438	1.5434	1223
2	$BaBi_2Nb_2O_9$	1370	0.5533	0.5333	2.5550	473
3	$Bi_4Ti_3O_{12}$	1280	0.5448	0.5411	3.283	498
4	$BaBi_4Ti_4O_{15}$	1450	0.5461	0.5461	4.185	668
5	$Ba_2Bi_4Ti_5O_{18}$	1450	0.5514	0.5526	5.0370	590
8	$Bi_9Ti_3Fe_5O_{27}$	1450	0.5491	0.5502	7.62	—

Work on the calculation of N-beam lattice image contrast (Allpress & Sanders 1973; O'Keefe 1973; Lynch, Moodie & O'Keefe 1975) has shown that, for materials like the heavy metal oxides, in zone axis orientations, specimens less than 7–10 nm thick furnish images that approximate fairly closely to the projected potential, or the projected charge density. All the observations in this paper relate to crystal flakes or thin edges that fulfilled this condition. In terms of the projected charge density approximation, such images can therefore be interpreted in chemical, structural terms, bearing in mind that electron scattering is dominated by the heaviest atoms present (Bi atoms and A site cations). There is ample evidence (see, for example, the calculations and critical review of O'Keefe (1976)) that this approach is valid for regular structures, and empirical evidence from our own and other work, now being supplemented by calculations, that it can be extended to local modification of otherwise regular structures. We have therefore used the 'intuitive-structural' approach, bringing in what is already known of the crystal chemistry of these and related complex oxides, to interpret the micrographs. It would not be practicable to carry out a full N-beam multislice image contrast calculation for such a large image field as figure 3, but appropriate diffraction contrast calculations for crystals with dislocations would be useful. They would provide an assessment of the validity of inferences drawn from the intuitive approach.

3. Results and discussion

Lattice images

The lattice images generally showed dark bands of contrast corresponding to the Bi_2O_2 layers. Between these were $n-1$ lines, arising from the layers of perovskite A-site cations (Bi and Ba), 0.38 nm apart. Under favourable conditions, a second set of fringes, perpendicular to the direction of the Bi_2O_2 bands and also (in $[010]_c$ projection) 0.38 nm apart fully delineated the square grid of the perovskite structure. These features are illustrated by the two-dimensional image in figure 2, plate 1, which can be understood in terms of the projected structure diagram, figure 1.

When a family of compounds can be generated by stacking of similar subunits, it is possible for a single crystal (in the macroscopic sense of a morphological entity) to vary in composition and structure at different points taken along the stacking direction. The lattice image (figure 2) shows that this is the case for the preparation which, replicating Russian work, should have yielded the $n = 8$ compound $Bi_9Ti_3Fe_5O_{27}$. No lattice images were found which showed the expected seven lines of A sites between Bi_2O_2 contrast bands. Instead, sequences corresponding to $n = 4$ and $n = 5$, ordered in short arrays of each kind, occurred randomly. The $n = 5$ layers predominated and the average multiplicity was probably around 4.7–4.8, but there were no long, homogeneous sequences. In X-ray diffraction, the superlattice lines would probably simulate an averaged multiplicity, but the electron diffraction pattern was of interest in view of the relatively small extent (15–20 nm) of any one uniform lamella. Although streaked, the diffraction spots corresponded to a superposition of the $n = 4$ and $n = 5$ patterns, not to an averaged multiplicity. It is also of interest – and relevant to the validity of the intuitive-structural interpretation of the lattice images – that although the sequence of Bi_2O_2 layers is faulted, the two-dimensional structure of the perovskite slabs is fully resolved.

Dislocations

Several of the ferroelectric bismuth oxide perovskites showed the presence of dislocations in their lattice images: these were especially frequent in crystals of the $n = 2$ and $n = 5$ compounds. It may be relevant that these have the lowest Curie temperatures of the systems studied. It has been considered that dislocations play a role in the motion of domains in perovskite ferroelectrics and it is possible that, during microscopy, the electron beam raised the temperature of the samples close to T_c. On prolonged irradiation, electron beam damage gave rise to dislocation loops, probably as a result of chemical decomposition. Van Landuyt, Remaut and Amelinckx (1969) earlier reported observing dislocation ribbons in diffraction contrast micrographs of $Bi_4Ti_3O_{12}$ ($n = 3$); lattice imaging of dislocated regions in these and other polyhedron network structures affords, for the first time, the possibility of examining their intimate structure. In another paper (Anderson, Hutchison and Lincoln 1976) we have considered dislocations in the complex niobium oxide structures, which are also networks based on a simple cubic framework of vertex-

306   J. L. Hutchison, J. S. Anderson and C. N. R. Rao

linked octahedral groups around the cations. We there point out that (unlike metals and some simple binary compounds) the local chemistry imposed by dislocations cannot be neglected. In particular, strong near-neighbour atomic interactions are important in determining the local structure adopted as a result of atomic displacements: the maintenance of local charge balance around cations of high formal charge; atomic sizes, determining the packing around the cations; and the not negligible covalent anion–cation interactions. The effect of these constraints is that, except in the most highly distorted core region, the general topological characteristics of the structure must be conserved. This requirement restricts the most likely values for the displacement vectors of dislocations.

Figure 3, plate 1, shows a dislocation in orthorhombic $Ba_2Bi_4Ti_5O_{18}$. The clarity of the image, particularly in the core region, justifies an attempt to analyse the distorted region in some detail; for reasons discussed earlier, the lattice image will be treated as a projection of the charge density distribution, and the structure deduced accordingly. The immediate character of the dislocation and its strain field can be seen by comparison with a superimposed grid representing the Bi_2O_2 layers in an ideal, unperturbed structure. By using the 0.38 nm separation between rows of A sites as an internal calibration, the dimensional correctness of the grid is ensured, and the distortions in the defect itself, and in the surrounding good crystal, can be measured at every point.

In the upper half of the micrograph, and terminating in the less well resolved core, two extra slabs of perovskite, with their Bi_2O_2 layers, are present. Since these have the same thickness as in the bulk of the crystal, they represent the introduction of a half-slice of $n = 5$ structure, one complete unit cell in thickness. Strain due to this insertion appears to be accommodated in two ways. The first, clearly seen in regions a few nanometres away from the core, is by elastic deformation: a small change in the spacing between Bi_2O_2 layers (a small change in the local c axis dimension). On the upper (compressed) side, this spacing is clearly diminished, to 2.40 nm as measured across five unit cells, a compression of 4–5 %. Below the dislocation the structure is slightly dilated, the average spacing of Bi_2O_2 layers, measured across several unit cells, being 2.54 nm; the stretching by about 1 % is just significant. These deformations imply a change in the spacing between the B-site cations of the framework, but not necessarily a corresponding change in cation–anion distances. The linear –Ti–O–Ti– chains must necessarily be stretched to produce the small dilation, but the shape of next-neighbour potential energy curves makes it unlikely that the larger compressional distortion involves a substantial decrease in Ti–O distances. It is more likely that the octahedral groups twist, to give puckered sheets of oxygen atoms. In terms of a local mode description of the lattice vibrations, the strain energy of dilation may be correlated with the restoring force of a local stretching mode, that of compression with a local bending mode of the $[TiO_6]$ groups. It is known that the former are usually the stiffer; in $BaTiO_3$ the frequency of the symmetric stretch is about 500 cm^{-1} as compared with about 340 cm^{-1} for the local bending modes. Thus a difference between the compressed and dilated sides of the

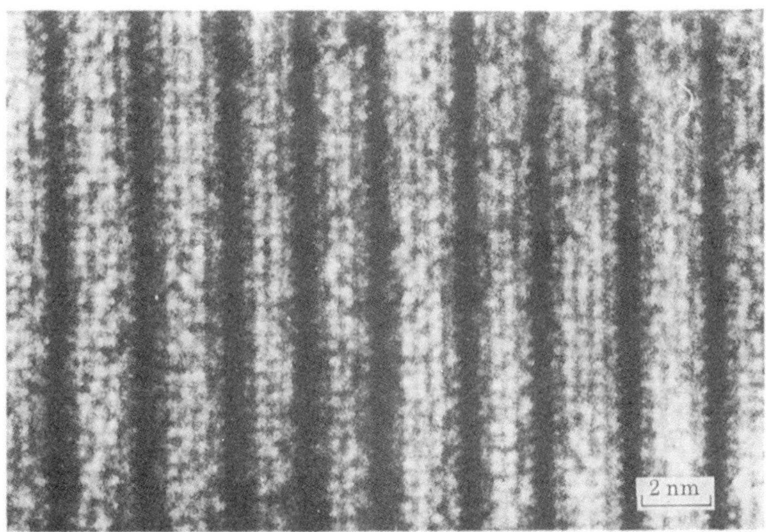

FIGURE 2. Lattice image showing faulted sequence in nominal $Bi_9Ti_3Fe_5O_{27}$ ($n = 8$), showing random occurrence of $n = 4$ and $n = 5$ sequences. The square grid of the perovskite structure can be seen.

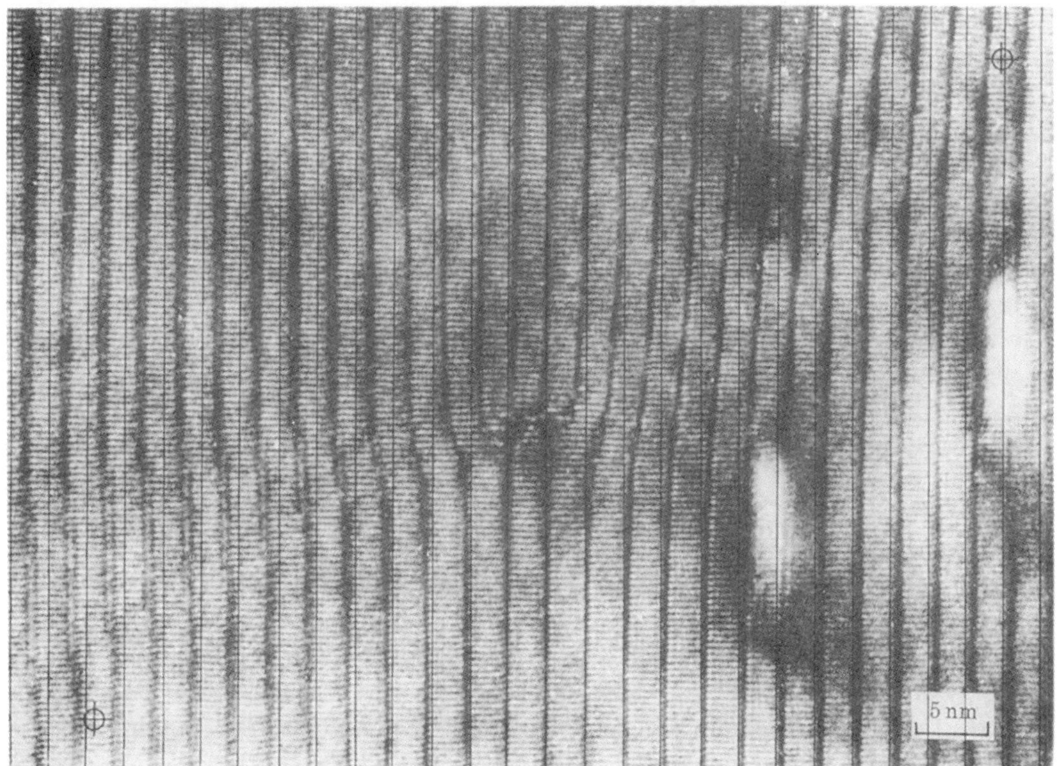

FIGURE 3. Lattice image of $Ba_2Bi_4Ti_5O_{18}$ ($n = 5$) containing a dislocation. A grid showing positions of Bi_2O_2 layers in unperturbed structure is superimposed for comparison.

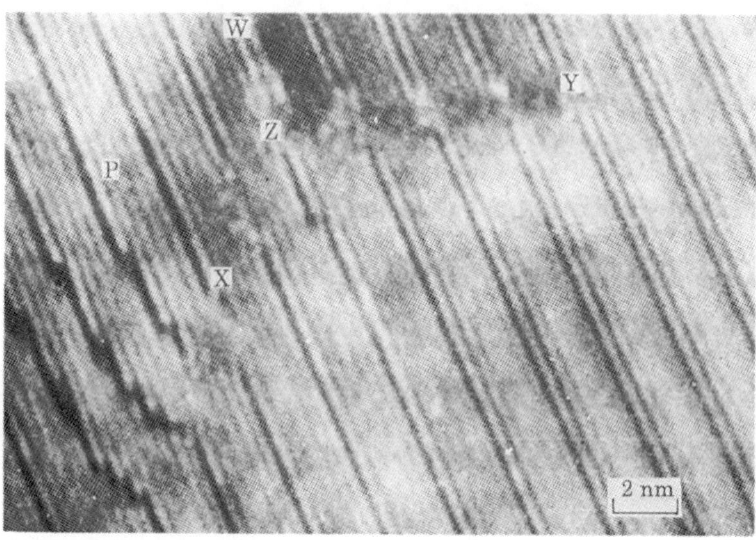

FIGURE 5. Fault in $Ba_2Bi_4Ti_5O_1$, with one extra half-slab of perovskite structure introduced, showing resultant displacement of Bi_2O_2 layers by several discrete steps.

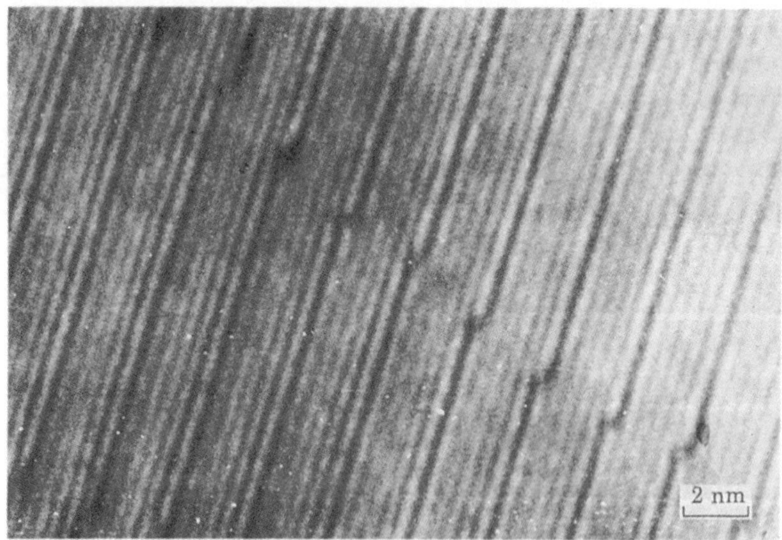

FIGURE 7. Stacking fault in $Ba_2Bi_4Ti_5O_{18}$, across which Bi_2O_2 layers are out of step by the width of one perovskite cell.

defect is explicable. The expansion of the structure produces a marked warping of the layers over considerable distances, as seen in the bending of the Bi_2O_2 bands over the whole field shown; this is particularly marked on the right-hand side.

There is a marked difference between the left-hand and right-hand sides of the dislocation. On the left, the Bi_2O_2 layers are much less severely bent in the a_c–c_c plane, but the dark fringes showing their location are broadened, and this broadening increases with distance from the dislocation. It is possible that this broadening can be attributed to buckling in the b_c–c_c plane: a general warping that destroys the exact alignment of bismuth atoms along the projection direction. Running left from the dislocation core is an extended fault of some kind, within which the Bi_2O_2 layers are seen as doubled fringes, each of the same width as in good crystal and separated by about 0.4 nm. The broadened contrast bands are offset by the same amount in crossing the fault. Such a doubling of the projected image would be observed if well aligned Bi_2O_2 layers were located at different positions, spaced one perovskite cell apart, near the upper surface and the lower surface of the crystal flake respectively. It is possible that buckling has been locally relieved by side-stepping of the Bi_2O_2 sheets, giving a strip in which two Bi_2O_2 layers are then seen in projection.

Distribution of strain about the dislocation is thus highly asymmetric. One contribution to this asymmetry, as between compressed and dilated sides, has been considered. There is little evidence on which to interpret the considerable differences between the strain on the right and left sides. Whether some ferroelectric distortion – bringing small changes in cell dimensions and a polar axis no longer collinear with the crystallographic c axis – could contribute to the asymmetry is speculative.

The extent to which a polyhedron network can be elongated or compressed is very limited. Dimensional changes in the core region are abrupt, and the image shows evidence of a second way in which they can be locally accommodated. In several places, most clearly on the left-hand side, the Bi_2O_2 layers side-step by about 0.4 nm, a unit step of one perovskite cell, corresponding to a change of ± 1 in the number of $[TiO_6]$ octahedra between layers, as is shown schematically in figure 4. In effect, elastic strain in the most perturbed part of the transition region can be relieved by introducing elements of other members of the homologous series of compounds. Their presence implies a localized change of chemical composition.

Dislocations in these layered perovskite derivatives need structural readjustments very similar to those that we have analysed in the niobium oxide block structures. In each case, the basic structure is a simple cubic array of octahedrally coordinated B site cations; in order that the structure should join up as good crystal around the dislocation, all displacements must result in configurations that are compatible with that structure and with the superlattice, which is imposed, in the present case, by the Bi_2O_2 layers. The significant feature of this superlattice is the antiphase displacement of successive perovskite slabs by $\frac{1}{2}a_{ss}[100]_{ss}$ ($= \frac{1}{2}a_c[110]_c$) across each Bi_2O_2 layer. Since dislocations such as that in figure 3 generate two extra perovskite slabs, with their bounding Bi_2O_2 layers, they interpolate one complete crystallographic unit cell; this preserves the matching of atomic positions at the

308   J. L. Hutchison, J. S. Anderson, and C. N. R. Rao

Bi_2O_2 layers. These are perfect dislocations of the superstructure, with the very large Burgers vector $b = c_{ss}$, and they can have pure edge character.

Most of the dislocations found have been of this type. In principle, dislocations introducing only a single perovskite slab are possible. In these, the antiphase displacement across the Bi_2O_2 layers, i.e. $\frac{1}{2}a_{ss}$ must still be accommodated, and the smallest Burgers vector for such dislocations would appear to be $\frac{1}{2}[a\,0\,c]_{ss}$.

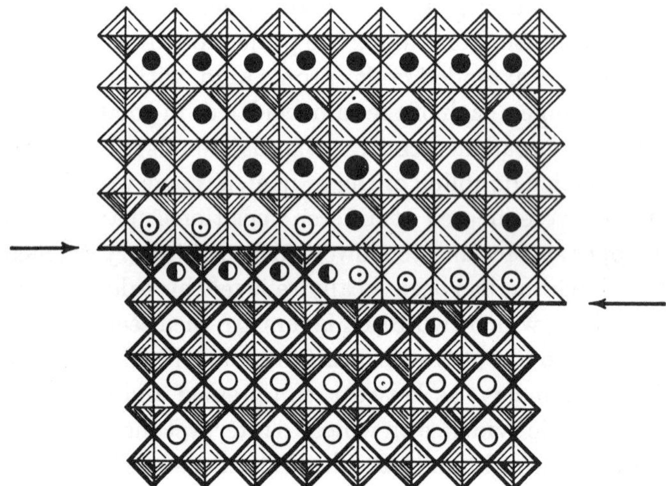

FIGURE 4. Idealized drawing of a unit shift of *ca.* 0.4 nm (one perovskite cell) in the Bi_2O_2 layer.

Dislocations of this type have not been identified with certainty. It can be shown that combinations of smaller faults enable an extra half slab of perovskite structure, with its Bi_2O_2 layer, to be accommodated with minimal strain. A fault that may be of this type is shown in figure 5, plate 2, with an extra half slab inserted at P into the upper half of the field shown. In this, unlike figure 3, the sheets of perovskite sites parallel to the Bi_2O_2 layers run through the structure without bending or interruption; the Bi_2O_2 layers on the left are laterally displaced by a succession of unit steps. The insertion of the extra half-slab is crossed, from a point below X–Y, by a narrow strip within which the resolution is poor, and at Z there is a strip of enhanced contrast, parallel to the Bi_2O_2 layers (i.e. to a_c) and with its lower edge almost parallel to $[101]_c$. With one extra Bi_2O_2 layer inserted, the topological requirements considered above could be fulfilled by invoking a single dislocation at X, with the displacement vector $\frac{1}{2}a_c[110]_c$. However, this would leave the other features of the micrograph unexplained. Consideration of the nature of antiphase boundaries (APBs), and the lattice images expected from them, suggests a possible interpretation. Permitted APBs in the perovskite structure lie on $\{110\}_c$; they have the structure shown in figure 6a. In a projection down $[010]_c$, APBs of the types $(110)_c$ and $(011)_c$ will be parallel to lines of perovskite sites and run into the crystal at 45° to the plane of projection. In a micrograph, their projection should be a band of changed contrast. There is no change in the spacing between lines of perovskite sites crossing the APB, although there is a spacing anomaly along each line of A sites; this would not be visible in the a_c direction of figure 5, which affords only one-dimensional resolution. Within the band covered by the projection of the oblique APB, lines of perovskite sites parallel to it are spaced only 0.2 nm apart and would not be resolved.

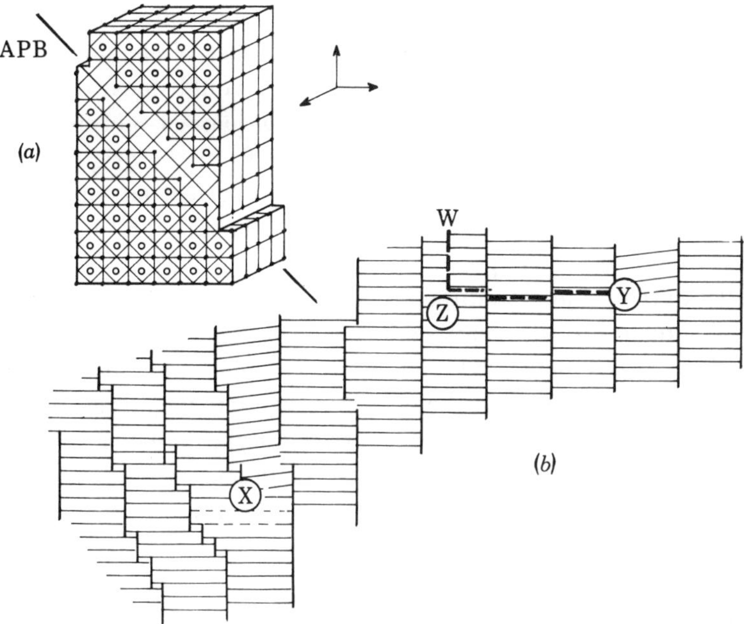

FIGURE 6. (a) Antiphase boundary in the perovskite structure. (b) Schematic representation of the possible structure of the faulted region in figure 5, with dislocations of $\frac{1}{2} aa0_{ss}$ at X and Y and antiphase boundary WZY. Lines of perovskite sites perpendicular to the Bi_2O_2 layers (not resolved in the micrograph) are marked to show distortions imposed by the introduction of an odd number of Bi_2O_2 layers.

Hence, in figure 5, an APB on $(011)_c$ should be seen as a band of darker contrast. It is therefore possible that the feature YZ represents an APB on $(110)_c$, initiated by a dislocation of $\frac{1}{2}a_c[110]_c$ at Y, which at Z intersects another APB on $(011)_c$ (the dark band ZW). The structure of this faulted region, on that hypothesis, is schematically represented in figure 6b.

Ths dislocations considered above operate upon the superstructure; imperfect partials could introduce additional perovskite layers between Bi_2O_2 layers, which would thereby be displaced by about 0·4 nm. Stepped displacements of this kind, in the core region, have been referred to. Figure 7, plate 2, shows an area with a stacking fault, across which four of the five layers in each slab of $n = 5$ structure run without interruption, but every Bi_2O_2 layer is displaced by the width of one perovskite cell. This fault originates outside the field of the micrograph, but could be accounted for if a half-sheet of octahedra – a single layer of $BaTiO_3$ – had been interpolated at some point (to form an element of the $n = 6$ structure) or eliminated (an element of $n = 4$ structure) by a partial dislocation.

Tentative inferences about the direction of the dislocation lines can be drawn, on the basis that lattice imaging resolution is attained only when equivalent elements of structure are rather exactly aligned throughout the thickness of the crystal. Figure 3 is a projection down $[010]_c$. If the dislocation line were tilted in the b_c–c_c plane, the termination of the added Bi_2O_2 layers would be offset, as between the top and bottom of the crystal. This should have the effect of changing the extent of distortion in the core region from level to level, and would smear out the projected image of the Bi_2O_2 layers laterally. A tilt in the a_c–b_c plane would similarly spoil the alignment through the layers where they deviate from the original a_c direction. Whilst ordering along the projection direction is imperfect near the core, there is little evidence for a progressive shift of the strain field through the thickness of the crystal. No firm assignment is possible, but it is likely that the dislocation line is nearly parallel to the projection direction.

The foregoing relates to the components of the Burgers vector in the a_c–c_c plane. In principle, the lattice image would not detect a displacement along b_c. This would hold good in regions remote from the core, but any screw character would impose very severe distortions of the polyhedron network in the core itself and these would almost certainly impair the resolution of the image. Since resolution is preserved rather well into the core region, it is likely that dislocations such as that in figure 3 have a Burgers vector c_{ss}.

Domain walls

Some insight into the structure of domain walls may be expected if domain boundaries could be directly observed by lattice imaging. We have, for the first time, made such observations on these bismuth oxide ferroelectrics.

When a single crystal in the paraelectric state transforms to a ferroelectric state of lower symmetry, stresses induced by dimensional changes can be relieved by domain twinning. For the tetragonal, pseudocubic $BaTiO_3$, there are two possible modes of twinning: $180°$ twins and $90°$ twins, with (100) and (101) respectively as composition planes. For the highly elongated structures considered in this paper, it may be expected that $180°$ twins could have $(100)_{ss}$, $(010)_{ss}$ or $(001)_{ss}$ as composition planes, but that $90°$ twins would be restricted to $(110)_{ss}$ as composition plane.

These compounds are generally pseudotetragonal or pseudoorthorhombic in the ferroelectric state. Cummins and Cross (1967) found a slight monoclinic distortion in $Bi_4Ti_3O_{12}$, where the monoclinic a_m–c_m plane coincided with the orthorhombic (paraelectric) b_o–c_o plane. The polar axis in these bismuth oxides has been considered to lie along, or close to, the crystallographic c axis (Fang, Robbins & Aurivillius 1962; Ismailzade 1964); Cummins and Cross found that it was slightly tilted, by $7°$ or less, from the b_o–c_o plane. By analogy with $BaTiO_3$ we would expect to find $180°$ domain walls parallel to the a_o–c_o or b_o–c_o planes and $90°$ domain walls with (110) contact planes. Whilst structural distortion along a $180°$ wall should be small, that along a $90°$ wall would be considerable.

Proc. R. Soc. Lond. A, volume 355   *Hutchinson et al., plate 3*

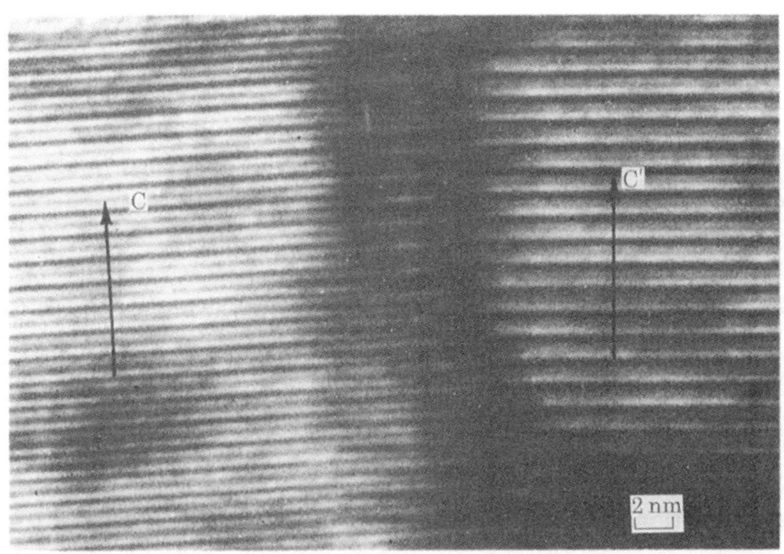

FIGURE 8. Lattice image of $BaBi_2Nb_2O_9$ ($n = 2$) showing a 180° domain wall.

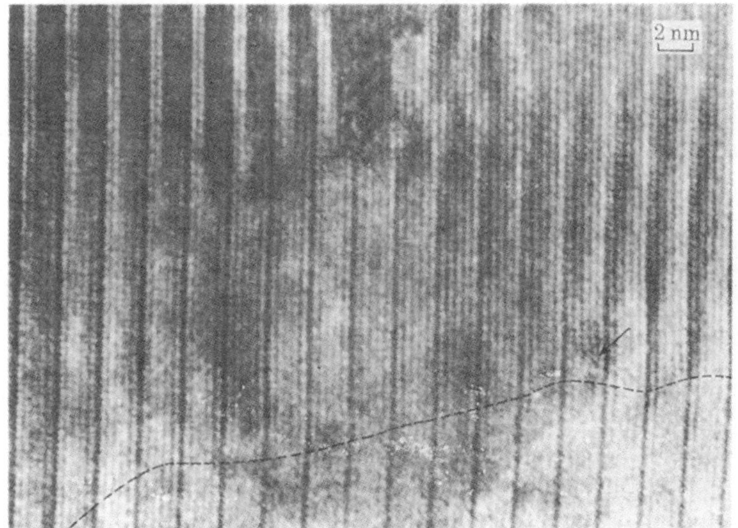

FIGURE 9. Lattice image of $Ba_2Bi_4Ti_5O_{18}$ showing a 180° domain wall.

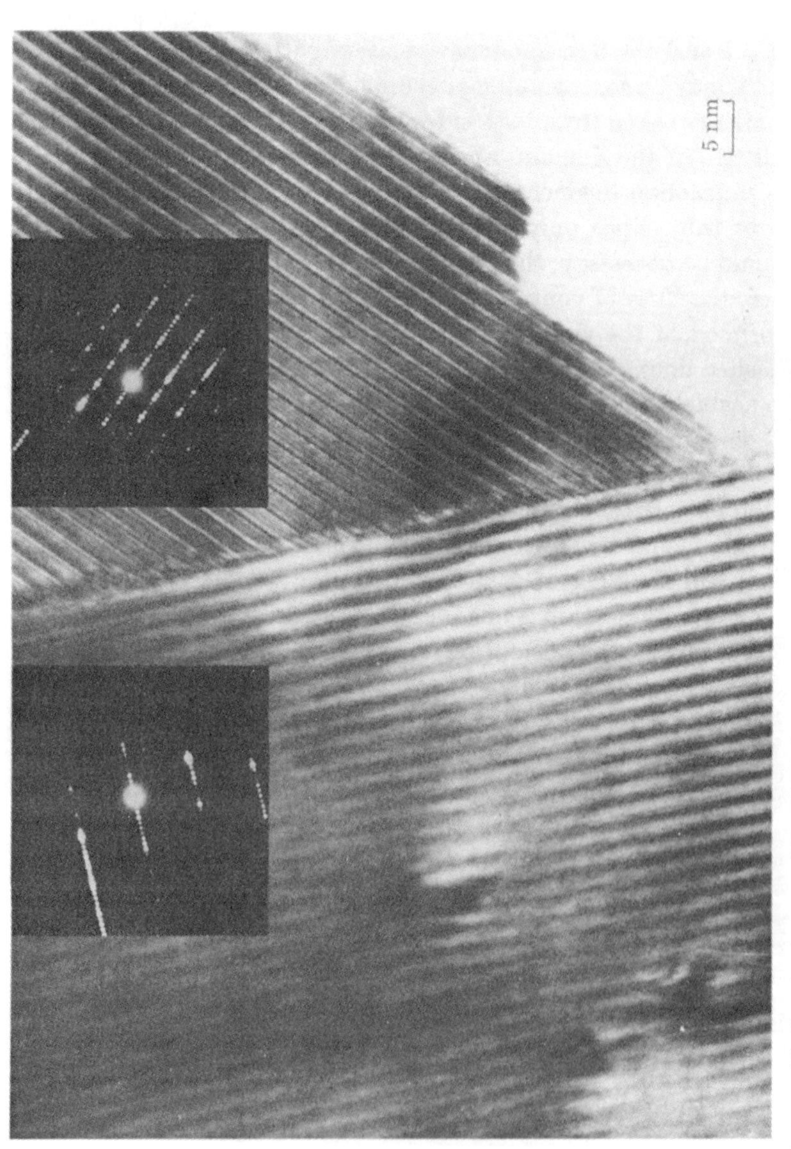

FIGURE 10. Lattice image of $BaBi_4Ti_4O_{15}$ ($n = 4$) showing an unusual domain wall, possibly of 90° type. Inserts: electron diffraction patterns from the two domains.

Electron microscopy of ferroelectric bismuth oxides   **311**

Figures 8 and 9, plate 3, are lattice images of two compounds ($n = 2$ and $n = 5$ respectively) showing evidence of 180° domain walls. The c axes, normal to the dark Bi_2O_2 sheets in the plane of the image, are nearly, but not exactly, parallel in the two domains; the two component c axes, C and C′, are inclined at about 3°. These observations suggest that the domain walls are of 180° type and that the polar axes of both the $n = 2$ and $n = 5$ compounds deviate slightly from the c_t or c_o axes. As in $Bi_4Ti_3O_{12}$, there may be some small monoclinic distortion. Selected area diffraction patterns, nominally taken from both sides of the domain boundary, were identical, but the small size of the domain, about 50 nm across, makes the evidence from selected area diffraction inconclusive. As expected for 180° domains, the domain walls are quite thin; since only a small adjustment or relaxation of the $[BO_6]$ octahedra would be necessary, the structure of the wall is undoubtedly simple. As a consequence, the lines of contrast due to the A site cations are clearly visible in the neighbourhood of the wall (figure 9), with only minor distortions around the boundary (dashed line).

A boundary which may represent a domain wall of a different type, in the tetragonal $n = 4$ compound, is shown in figure 10; plate 4; this presents some problems that cannot be resolved. In projection, the Bi_2O_2 layers are inclined at 45°; selected area diffraction patterns show that the two domains present different orientations. $[110]_{ss}$ of the tetragonal structure on the right hand side is exactly parallel to the electron beam, and the lines of A site cations between the Bi_2O_2 sheets are fully resolved. The left-hand domain is not thereby brought into an orientation in which its structure is resolved; the beam direction is approximately along $[210]_c$, but off-centre Laue zones show that it is tilted a few degrees away from this direction. The boundary between the domains, if it can be regarded as a domain wall, appears to be complex and thick; the interface may, indeed, be slightly inclined to the image plane, thereby giving rise to overlap contrast. Although this interface, which appears to be close to $(320)_c$, seems to be coherent, it is difficult to understand how that orientation could meet the prime requirement both for a physical composition plane and for a ferroelectric contact plane – continuity of structure – without severe distortions.

The authors thank the Science Research Council for support of this research and one of the authors (C.N.R.R.) is grateful to the Association of the Commonwealth Universities and the University of Oxford for the award of a Commonwealth Visiting Professorship during 1974–5.

312   J. L. Hutchison, J. S. Anderson and C. N. R. Rao

REFERENCES

Allpress, J. G. & Sanders, J. V. 1973 *J. Applied Cryst.* **6**, 165.
Anderson, J. S., Hutchison, J. & Lincoln, F. J. 1976 *Proc. R. Soc. Lond.* A **352**, 303.
Aurivillius, B. 1949 *Arkiv Kemi* **1**, 463.
Cummins, G. E. & Cross, L. E. 1967 *Appl. Phys. Lett.* **10**, 14.
Fang, P., Robbins, C. R. & Aurivillius, B. 1962 *Phys. Rev.* **126**, 892.
Goodenough, J. G. & Longo, J. M. 1970 *Landolt-Bornstein, New Series Group III*, vol. 4*a*. New York: Springer.
Ismailzade, I. H. 1964 *Soviet Phys. Cryst.* **8**, 686.
Ismailzade, I. H. 1967 *Soviet Phys. Cryst.* **12**, 400.
Lynch, D. F., Moodie A. F. & O'Keefe, M. A. 1975 *Acta Cryst.* A **31**, 300.
McConnell, J. D. M., ,Hutchison, J. L. & Anderson, J. S. 1974 *Proc. R. Soc.* A **339**, 1.
O'Keefe, M. A. 1973 *Acta Cryst.* A **29**, 389.
O'Keefe, M. A. 1976 Thesis, University of Melbourne.
Schmid, H. 1975 *Magnetoelectric Interaction Phenomena in Crystals* (ed. H. J. Freeman & H. Schmid) New York: Gordon and Breach.
Subbarao, E. C. 1974 *Solid state chemistry* (ed. C. N. R. Rao). New York: Marcel Dekker.
Van Landuyt, J., Remaut, G. & Amelinckx, S. 1969 *Mat. Res. Bull.* **4**, 329.

JOURNAL OF SOLID STATE CHEMISTRY **53**, 193–216 (1984)

Crystal Chemistry and Magnetic Properties of Layered Metal Oxides Possessing the K₂NiF₄ or Related Structures*

P. GANGULY AND C. N. R. RAO†

Solid State and Structural Chemistry Unit, Indian Institute of Science, Bangalore-560012, India

Received June 30, 1983; in revised form January 20, 1984

There is increasing interest in recent years in the structural chemistry and properties of layered metal oxides possessing the K_2NiF_4 or related structures. Many new oxides of this structure exhibiting novel properties are being reported from time to time in the literature. The crystal chemistry of the oxides of the general formula A_2BO_4 with particular reference to the stability of the K_2NiF_4 structure and the relations between the different structures exhibited by this family of oxides is discussed. Non-stoichiometry in these oxides is another aspect of interest discussed in the article. While K_2NiF_4 itself is a well-known two-dimensional antiferromagnet, oxides of this structure with a variety of magnetic properties are examined in some detail. Besides the ternary A_2BO_4 oxides, the structure and magnetic properties of complex oxides, where the A or/and the B ions are partly substituted by other cations, is discussed. Some of the problems related to this family of oxides that are worth investigating are indicated. Much of the discussion in this article would have relevance in understanding the structure and properties of layered materials.

1. Introduction

K_2NiF_4 is a prototype, two-dimensional antiferromagnetic material (*1*). The tetragonal K_2NiF_4 structure consists of alternating layers of $KNiF_3$ perovskite layers and KF rock-salt layers. Antiferromagnetic interactions between the transition metal ions occur only in the planes containing the $KNiF_3$ perovskite layers. There is no Ni–F–Ni interaction in the direction parallel to the c axis. A variety of oxides are known to crystallize in structures related to K_2NiF_4 (*2–*

14). Besides the tetragonal K_2NiF_4 structure, A_2BO_4 type oxides with monoclinic, orthorhombic, and other tetragonal structures have been characterized in recent years (*6–9, 13–15*). The main objective of this article is to examine the stability of the K_2NiF_4 structure and the relations between the different structures of metal oxides in this family and to rationalize their properties in terms of the structures. We have discussed at length, the crystal chemistry of various oxides of the A_2BO_4 type in terms of the nature of A–O and B–O bonding and the consequences of the bonding on the structure and properties of the oxides. Non-stoichiometry in these oxides has also been discussed briefly.

Oxides with K_2NiF_4 or related structures exhibit interesting magnetic and electronic

---

* Contribution No. 213 from the Solid State and Structural Chemistry Unit.

† To whom all correspondence should be addressed at: Department of Physical Chemistry, University of Cambridge, Cambridge, CB2 1EP, United Kingdom.

properties. It was reported a few years ago
(2) that La_2NiO_4 shows an activated electrical conductivity behavior at low temperatures, but the resistivity has a positive temperature coefficient above 550 K. Such a
semiconductor–metal transition is indeed
of great interest. La_2CuO_4 is reported to exhibit a temperature-independent electrical
resistivity ($\sim 10^{-1}$ ohm cm), but other rare
earth derivatives of the general formula
Ln_2CuO_4 are semiconductors (2). Electron
transport properties of more complex oxides of the K_2NiF_4 family such as $LaSr_3$
Co_2O_8 are quite different from those of the
corresponding perovskite oxides.

Magnetic properties of oxides possessing
K_2NiF_4 structure have been investigated
more extensively than electron transport
properties. Some of the oxides such as
Ca_2MnO_4 show long-range antiferromagnetic ordering while others such as La_2NiO_4
do not. Ferromagnetism is known to occur
in oxides such as $LaSr_3Co_2O_8$. The K_2NiF_4
structure appears to preferentially stabilize
certain spin states of transitional metal ions
and accordingly, some of the transition
metal oxides of this family exhibit spin-
state transitions. We shall discuss magnetic
properties of a variety of oxides possessing
K_2NiF_4 type structure in the light of their
crystal chemistry.

2. Crystal Chemistry

Compounds of the general formula
A_2BX_4 with the K_2NiF_4 structure (Fig. 1)
may be considered to be built up of alternating layers of perovskite (ABX_3) and rock-
salt (AX) structures. There are no close-
packed A_2X_4 layers in A_2BX_4 similar to the
close-packed AX_3 layers (Fig. 2a) present in
perovskites. In a close-packed A_2X_4 layer,
there would be considerable electrostatic
repulsion between the two A ions since
they are forced to be adjacent to each other
if the perovskite AX_3 layers are to be retained within the structure (see Fig. 2b).

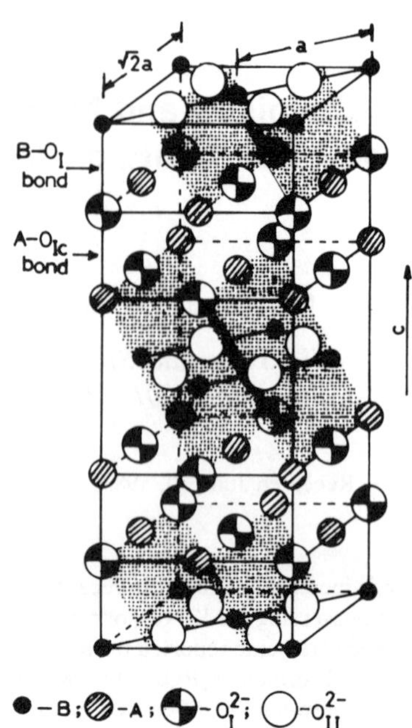

FIG. 1. The K_2NiF_4 structure of A_2BO_4 oxides showing O_I and O_{II} ions. B–O_I and A–O_{Ic} bonds are also
shown. Shaded portion shows close-packed AO_3 lay-
ers.

The two AX_3 layers can, however, be displaced (Fig. 2c) to give alternating layers of
rock-salt and perovskite structures. It is evident from Fig. 1 that along the {110} planes

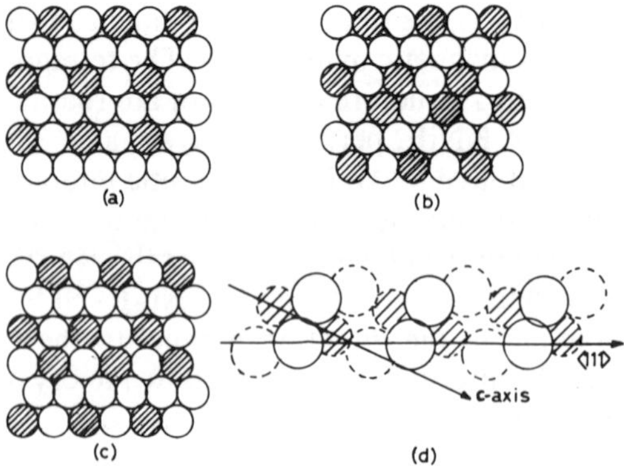

FIG. 2. (a) AX_3 close-packing: hatched circles are A
ions and open circles are X ions; (b) hypothetical A_2X_4
close-packing built up from AX_3 layers; (c) displacement of AX_3 layers to give A_2X_4 packing as found in the
ideal K_2NiF_4 structures; (d) corrugated packing of
A_2X_4 layers as in La_2NiO_4.

of the K_2NiF_4 structure, the {110} planes of the perovskite layers are stacked alternately with the {100} planes of the rock-salt layer. In order to obtain the A_2X_4 packing shown in Fig. 2c, the {111} planes of the perovskite layer should be coplanar with the {100} planes of the rock-salt layer. This is roughly what is achieved along the {111} planes of the K_2NiF_4 structure (see Figs. 1 and 2d), the two planes being corrugated instead of being coplanar. This results in a better packing efficiency. The low c/a ratio (3.25–3.30) found in several A_2BO_4 oxides (compared to the theoretical value of 3.414 ($c = (2 + \sqrt{2})a$) if the BO_6 octahedra were regular and the $A–O$ distances were identical) is perhaps due to the corrugated nature of the packing in the {111} planes. The perovskite (AX_3) layers (Fig. 3a) are displaced in the K_2NiF_4 structure as represented in Fig. 3b. Structures of the related Ruddlesdon–Popper type compounds of the general formula $AX(ABX_3)_n$ with $n = 2$ and 3 are shown in Figs. 3c and d where 2 and 3 layers of perovskites are displaced, respectively. Insertion of an AX layer in the $n = 2$ case (Fig. 3c) between the two perovskite bilayers and a reshuffling of the perovskite layers could give rise to the hypothetical compound $(AX)_2(ABX_3)_2$ with alternating bilayers of rock-salt and perovskite (Fig. 3e) instead of the monolayers found in the K_2NiF_4 structure. We shall discuss later why such compounds are not readily formed.

2.1. Stability of the K_2NiF_4 structure in oxides. Just as for perovskite oxides, a tolerance factor t may be defined for A_2BO_4 oxides as

$$t = \frac{r(A–O)}{\sqrt{2}r(B–O)} \tag{1}$$

Here, $r(A–O)$ and $r(B–O)$ are distances obtained from ionic radii. Poix (*3*) has defined the tolerance factor for oxides as

$$t = \psi_A/\sqrt{2}\,\beta_B \tag{2}$$

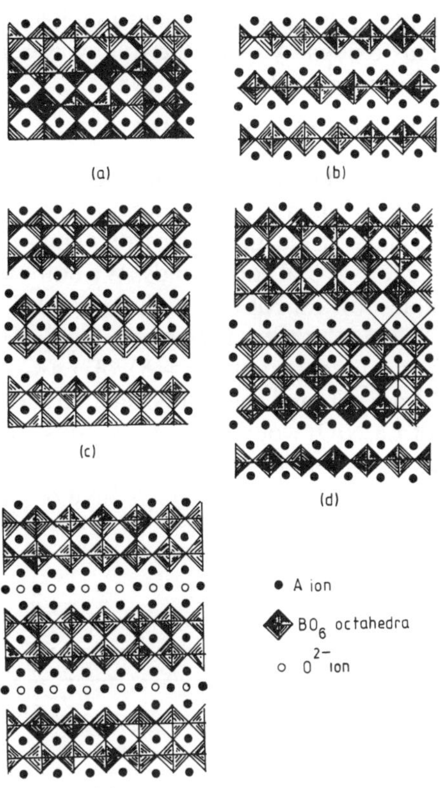

FIG. 3. Projection drawings of (a) ABO_3 perovskites and (b) A_2BO_4 oxides with K_2NiF_4 structure. Projection drawings of $n = 2$ and $n = 3$ members of the series $AO(ABO_3)_n$ are shown in (c) and (d). Projection drawing of hypothetical $(AO)_2(ABO_3)_2$ is shown in (e). After Tilley, Ref. (*45*).

where ψ_A and β_B are invariant values (*3, 4*) associated with $B–O$ and $A–O$ distances in six- and ninefold coordinations, respectively. The tetragonal K_2NiF_4 structure is assumed to be stable (*3*) within the limits $1.02 > t > 0.85$. This criterion is conceptually more appealing than Ganguli's criterion (*5*) according to which the tetragonal structure is stable when r_A/r_B is between the limits 2.4 and 1.7 when one employs the ionic radii of Shannon (*16*). Such a criterion in terms of $A–O$ and $B–O$ distances is indeed equivalent to Poix's criterion. In this article, we shall adopt Poix's criterion while examining the stability of the tetragonal structure. We have calculated t and ψ_A for various oxides by using the following relationship proposed by Poix (3):

$$\beta_B + \sqrt{2}\,\psi_A = 0.996 V^{1/3} \qquad (3)$$

where V is the volume of the unit cell and β_B is available in the literature (4). In Table I, we have listed the values of t thus obtained for several compounds. We note that t values of La_2NiO_4, La_2CuO_4, and La_2CoO_4 are close to the lower limit. The orthorhombic structure (6–8) of La_2CoO_4 and La_2CuO_4 and the tetragronal structure (6) of La_2NiO_4 would seem to be consistent with β_{Ni} being less than β_{Cu} or β_{Co}. The monoclinic distortion in Nd_2NiO_4 and Pr_2NiO_4 (9) can be attributed to the low tolerance factor in these two oxides.

In compounds where t is close to the lower limit of stability of the tetragonal K_2NiF_4 structure, it is possible to increase t by several possible mechanisms. In compounds of the type La_2BO_4, β_B can be reduced by the incorporation of a small proportion of B^{3+} ions. Thus, La_2NiO_4 possessing the tetragonal structure almost always contains a finite proportion of Ni^{3+} ions. Low β_B can also arise if the B ions are in the low-spin state instead of in the high-spin state. In Sm_2CoO_4, there is crystal structure evidence (8) for two different environments for the Co^{2+} ions.

Investigations on polycrystalline as well as single crystal samples (6, 10, 11) of La_2NiO_4 have established the structure of this oxide to be tetragonal. Indeed, it was the first oxide to be shown to have the K_2NiF_4 structure. Recently (12), supercells corresponding to $\sqrt{2}\,a$ unit cells (with crystals grown from a skull melter) have been observed in the electron diffraction patterns. Subsequently, it has been found that electron diffraction patterns of La_2NiO_4 prepared by the ceramic method also show such superlattice reflections. These reflections are similar to those in the neutron diffraction pattern of three-dimensional, antiferromagnetically ordered K_2NiF_4. Such superlattice reflections are also observed in La_2CuO_4. Compounds such as La_4LiBO_8 which have similar tolerance factors show superlattice reflections similar to those observed in La_2NiO_4. In Fig. 4a we show electron diffraction patterns corresponding to the tetragonal [111] zone axis of La_4LiCoO_8 and to the tetragonal [110] zone axis of La_2NiO_4 (Fig. 4b) as well as of La_4LiCoO_8 (Fig. 4c). Such superlattice reflections have not been seen in the case of other $LaSrBO_4$ compounds including $La_2Sr_2NiTiO_8$; these oxides have higher tolerance factors. The superlattice reflections may be associated with two different envi-

TABLE I

LATTICE PARAMETERS, $\psi_{Ln^{3+}}$ AND t VALUES FOR SOME OXIDES WITH K_2NiF_4 STRUCTURE

Compounds	a (Å)	c (Å)	c/a	$\psi_{Ln^{3+}}$	t
$LaSrAlO_4$[a]	3.761	12.649	3.363	2.586	0.978
$LaSrVO_4$	3.87	12.65	3.270	2.563	0.914
$LaSrCrO_4$	3.85	12.50	3.25	2.549	0.928
$LaSrMnO_4$	3.804	13.10	3.44	2.537	0.901
$LaSrFeO_4$[b]	3.88	12.76	3.29	2.602	0.922
$LaSrCoO_4$	3.806	12.503	3.285	2.626	0.989
$LaSrNiO_4$	3.80	12.51	3.292	2.521	0.935
La_2CoO_4	3.896[c]	12.66	3.249	2.561	0.852
La_2NiO_4	3.855	12.652	3.282	2.558	0.867
La_2CuO_4	3.807[c]	13.17	3.459	2.53	0.834
$La_2Li_{0.5}Co_{0.5}O_4$[d]	3.784	12.624	3.336	2.519	0.861
$La_2Li_{0.5}Ni_{0.5}O_4$[d]	3.756	12.87	3.426	2.54	0.877
$PrSrAlO_4$	3.732	12.54	3.36	2.523	0.966
$PrSrCrO_4$	3.836	12.377	3.226	2.485	0.909
$PrSrFeO_4$	3.838	12.597	3.282	2.511	0.906
Pr_2NiO_4	3.845[c]	12.44	3.235	2.53	0.857
$NdSrAlO_4$	3.726	12.49	3.352	2.504	0.963
$NdSrCrO_4$	3.834	12.36	3.223	2.501	0.919
$NdSrMnO_4$	3.768	12.98	3.445	2.461	0.888
$NdSrFeO_4$	3.846	12.594	3.274	2.521	0.908
$NdSrNiO_4$	3.786	12.26	3.238	2.448	0.922
Nd_2NiO_4	3.810	12.31	3.231	2.491	0.844
$GdSrAlO_4$[a]	3.701	12.362	3.340	2.442	0.952
$GdSrCrO_4$	3.823	12.263	3.207	2.442	0.902
$GdSrMnO_4$	3.754	12.87	3.428	2.418	0.881
$GdSrFeO_4$	3.853	12.554	3.258	2.523	0.908
$GdSrNiO_4$	3.768	12.23	3.245	2.416	0.916
Gd_2CuO_4	3.89	11.85	3.046	2.452	0.806

[a] Based on the present study.
[b] Lattice parameter data from Ref. (54).
[c] Pseudo-tetragonal lattice parameters.
[d] Lattice parameter data from Ref. (25). Rest of the lattice parameter data from Ref. (35, 36).

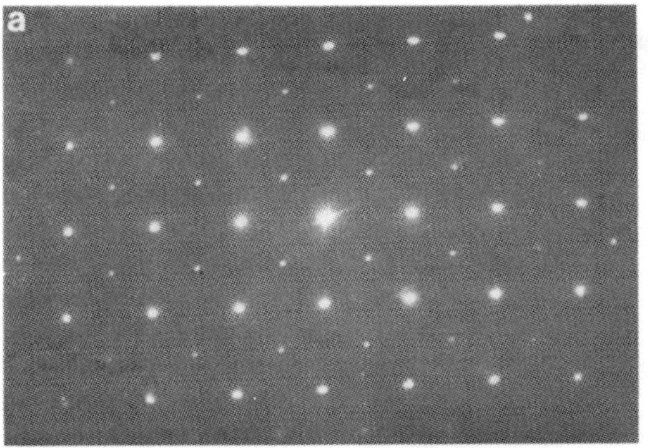

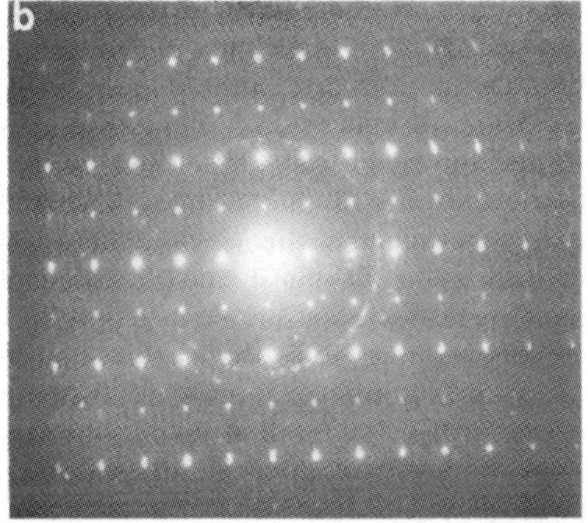

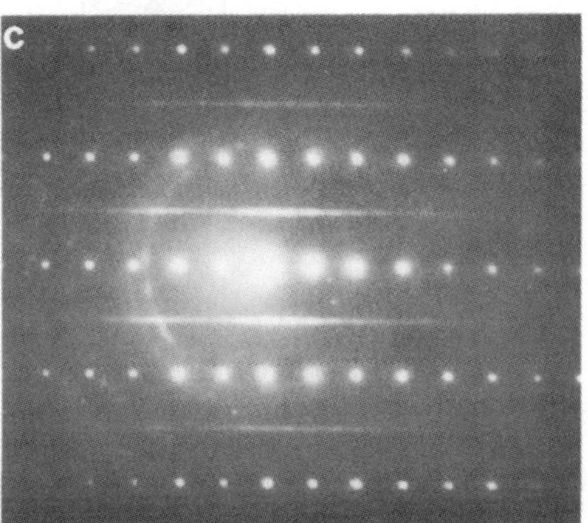

FIG. 4. Electron diffraction patterns corresponding to (a) the [111] zone axis of La₄LiCoO₈, (b) [110] zone axis of La₂NiO₄, and (c) [110] zone axis of La₄LiCoO₈.

ronments of the B ions arising from the deviation of the B–O_{II}–B angle from 180° or from changes in the B–O distance. The former can happen if $t \approx 0.85$ and the BO_6 octahedra get rotated about the c axis (Fig. 5a) or if the octahedra are tilted in the $\langle 100 \rangle$ or the $\langle 110 \rangle$ directions (Figs. 5b and c). The latter is likely to happen in compounds such as La₄LiBO₈ where the nature of Li–O and B–O bonds is quite different. The distortions (and hence the superlattice reflections) in many of the K₂NiF₄ related compounds may be similar to those depicted in Fig. 5.

Several structures can result from a rear-

rangement of the oxygen ion positions in the K₂NiF₄ structure. Three of these arise from monoclinic distortions with $\gamma \neq 90°$. Of these, there are two structures with $a = b$ which can be indexed on the basis of a orthorhombic unit cell; two types of orthorhombic structures O and O' have been distinguished (13). For the O structure, the conditions for the allowed reflections are $h + k = 2n$, $k + l = 2n$ and $h + l = 2n$ while for the O' structure, the condition is $k + l = 2n$. In the O' structure ($8, 13, 14$) found in La₂CuO₄, La₂CoO₄, and CaY CrO₄, the BO_6 octahedra are tilted as in Fig. 5c. The O structure was established for

                              GANGULY AND RAO

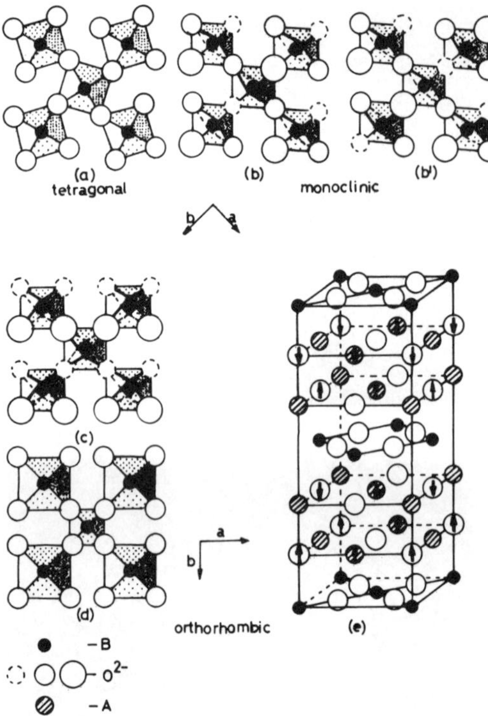

FIG. 5. (a) Rotation of BO_6 octahedra along the c axis; (b, b') Tilting of the octahedra in the $\langle 100 \rangle$ direction by rotation along the b axis, the sense of rotation for adjacent B ions along the b axis being different in (b), and the same in (b'); (c) Tilting of octahedra in the $\langle 110 \rangle$ direction. Larger circles are above the plane of the paper and broken circles are below the plane. (d) BO_6 octahedra with different $B–O_{II}$ and possibly $B–O_I$ distances (e) Movement of $A–O_I–A$ linkages (Ref. (13)) to give rise to O or O' structures.

Sm_2CoO_4 which has the $Fmm2$ symmetry with two $Sm–O_I$ and $Co–O_I$ distances (8); the BO_6 octahedra in this oxide may be arranged as in Fig. 5d. Although the orthorhombic nature of the unit cell is apparent in Fig. 5c (with the a' parameter greater than the b' parameter), the orthorhombic nature of the distortion in the O structure is not apparent from Fig. 5d. The O structure has been considered (13) to be due to a change in the $A–O_{Ib}–A$ angle from 180° along the b direction by a shift of the $A–O_{Ia}–A$ linkages along the a direction as shown in Fig. 5e ($a > b$).

In a study of the evolution of the tetragonal (T) K_2NiF_4 structure as a function of x in the series $La_{1-x}Y_xCaCrO_4$ and as a function of temperature in $CaYCrO_4$, it has been found that the sequence $O' \rightarrow O \rightarrow T$ occurs in both cases (13). Furthermore, as the O phase approaches the O' phase, the b parameter decreases, but the a parameter remains constant. This is consistent with a continuous decrease in the $A–O_{Ib}–A$ angle. In the O' phase, decreasing t or temperature, has the opposite effect (b remaining nearly constant and a decreasing rapidly). This can be understood by assuming that in the O' phase, the $A–O_{Ib}–A$ angle resists further reduction and that the decrease in the tolerance factor imposes a strain on the $A–O_{Ia}–A$ linkage (15). Since the $A–O_{Ia}–A$ angle deviates from 180° in this phase, a approaches b with decreasing t. Accompanying these changes the $B–O_{II}–B$ angle also changes in the O' phase and the BO_6 octahedra get tilted as shown in Fig. 5c. It would thus appear as though the O structure is superimposed on the O' structure shown in Fig. 5c so that one may expect two different $B–O_I$ distances even in the O' phase (15). This has, however, not been observed in the crystal structures of La_2CoO_4 or La_2CuO_4 (8, 14).

The truly monoclinic (M) distortion with $a \neq b \neq c$ and $\gamma \neq 90°$ has not been reported so far. Pr_2NiO_4 and Nd_2NiO_4 were reported to be monoclinic (9), but with $a = b$, so that the unit cell can be indexed on the basis of an orthorhombic cell. On rechecking the lattice parameters of Pr_2NiO_4 (15), it is found that the structure is indeed truly monoclinic with $a \neq b$. We have observed that $La_2NiO_{4+\delta}$ on heating for long periods in CO_2 at 1150°C gives rise to a monoclinic structure with $a \neq b$; in such a sample, δ would be zero or even slightly negative. As mentioned earlier, small amounts of Ni^{3+} seems to stabilize the tetragonal structure of La_2NiO_4.

The tetragonal to monoclinic distortion

could be associated with changes in the tolerance factor. The monoclinic structure would be associated with the tilting of the BO_6 octahedra of the type shown in Fig. 5b; the sense of tilting of the octahedra between nearest neighbor B ions may be assumed to be the same for a row of octahedra along the b or a axes (Fig. 5b′).

Another modification of the tetragonal K$_2$NiF$_4$ structure is the tetragonal T' structure (Fig. 6) found in the copper oxides $(17, 18)$ Ln_2CuO_4 (Ln = Pr, Nd, Sm, Eu, Gd). The small c/a ratio of the Ln_2CuO_4 compounds was initially associated with compressed CuO$_6$ octahedra (7) in the same manner as compressed CuF$_6$ octahedra were initially postulated in K$_2$CuF$_4$ (19). In the oxides, the a parameter is ~3.97 Å and in order to obtain compressed octahedra, the Cu–O$_I$ distance should be less than 1.98 Å. The suggestion of Longo and Raccah (7) that CuO$_6$ octahedra are compressed is therefore subject to some doubt. The T' structure is derived form the T structure by a shift of the oxide ions from the $(0, 0, z)$ positions in the T structure to $(0.5, 0.0,$

0.25) position in the T' structure. As a consequence, the oxide ion changes its coordination from 6 to 4 in the T' structure. In such a structure, the rare earth and Cu ions have no intervening anion along the c axis. The structure appears to be specific to Cu^{2+} ions (15) as the $d^2_{z^2}$ electrons can provide sufficient screening to minimize repulsion between the Cu^{2+} and Ln^{3+} ions. The relationship between K$_2$NiF$_4$ related structures and the T' structure has been examined by Singh $et\ al.$ (20) in the case of the solid solutions, La$_{2-x}Ln_x$CuO$_4$ (Ln = Pr, Nd). A first-order transition occurs between the two structures as a function of x accompanied by a marked increase in the volume of the unit cell of the T' phase across the critical value x_c; x_c decreases with the decreasing size of the Ln^{3+} ion. The T' structure may be considered to be composed of alternating layers of (CuO$_2$)$^{2-}$ layers with the Cu^{2+} ions in the square-planar, fourfold coordination and (Ln_2O$_2$)$^{2+}$ layers with the fluorite structure containing Ln^{3+} ions in the eightfold coordination. The increase in volume in the T' phase is attributed to the lower packing efficiency while the collapse in the c/a ratio is attributed to the change from the rock-salt-like packing of the (Ln_2O$_2$)$^{2+}$ layers in the K$_2$NiF$_4$ structure to the fluorite type of packing in the T' structure.

The driving force for the transition from the K$_2$NiF$_4$ related structure to the T' structure in the La$_{2-x}Ln_x$CuO$_4$ compounds can perhaps be understood in terms of the competition between the A and B ions for covalency with the O$_I$ ion in the A–O$_I$–B linkages. The higher acidity of the smaller Ln^{3+} ion could further elongate the Cu–O$_I$ bond (compared to that in La$_2$CuO$_4$) and drive the Cu^{2+} ions to a square-planar coordination.

It is interesting that the relationship between Ln_2NiO_4 and Ln_2CuO_4 (Ln = Pr, Nd) with respect to the unit-cell volume and c/a ratio is similar to that between K$_2$NiF$_4$ and

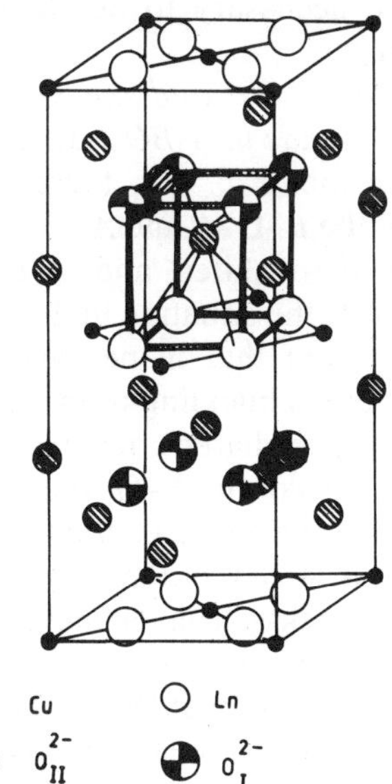

Fig. 6. The T' structure of Ln_2CuO_4 compounds.

K_2CuF_4. This would suggest an antiferrodistortive ordering (Fig. 7a) of CuO_6 octahedra enhancing the a parameter to be the driving force for the transition to the T' structure in Ln_2CuO_4 compounds. An antiferrodistortive type of ordering of elongated BO_6 octahedra of the type found in K_2CuF_4 (21, 22) would be unstable in A_2BO_4 oxides. In the fluorides of K_2NiF_4 structures ($t \gtrsim 1$), a mechanism which enhances the a parameter would be favored. In A_2BO_4 oxides, since $t < 1$, mechanisms that reduce the a parameter should be operative and it is perhaps for this reason that La_2CuO_4 has a c/a ratio suggestive of ferrodistortive ordering (18, 22) of elongated CuO_6 octahedra (Fig. 7b).

The relationship between the O (or O'), T, M, and T' structures may be obtained from a study of the solid solutions of La_2NiO_4 (T) and La_2CuO_4 (O) and that between Pr_2NiO_4 (M) and Pr_2CuO_4 (T'). In the series $La_2Ni_{1-x}Cu_xO_4$, the O structure

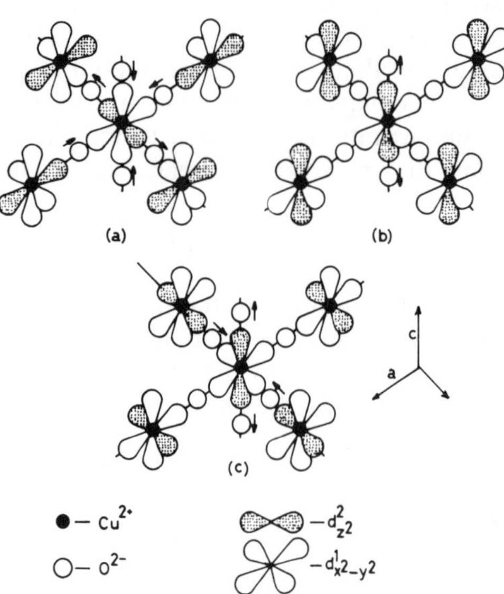

FIG. 7. (a) Antiferrodistortive ordering of elongated CuO_6 octahedra (arrows indicate direction of displacement of oxide ions). (b) Ferrodistortive ordering of elongated CuO_6 octahedra. (c) Ordering of elongated CuO_6 octahedra as in the ac or the bc planes of $KCuF_3$.

is obtained (20) from the T structure for $x \geq 0.9$. Since $\beta_{Cu} > \beta_{Ni}$, the changes may be considered to be due to changes in the tolerance factor. However, in the $Pr_2Ni_{1-x}Cu_xO_4$ series (15), the M structure which is stable in the range $0 < x \leq 0.1$ goes over to the T structure for $0.25 \leq x \leq 0.50$. In the range $0.50 < x < 0.90$, a mixture of T and T' phases is obtained. For $0.9 \leq x \leq 1.0$, the T' structure is obtained. The $M \rightarrow T$ transition in $Pr_2Ni_{1-x}Cu_xO_4$ is in the opposite direction to that expected from tolerance factor effects. Both the O and O' structures are found (13) in oxides of the formula $La_{1-x}Y_xCaCrO_4$ which have high tolerance factors. The fact that such a distortion has not been observed (23) in $CaYAlO_4$ probably implies that distortions in $La_{1-x}Y_xCaCrO_4$ are associated with the high energy required to distort the CrO_6 octahedra so that the $Ca-O_{Ic}$ and $Y-O_{Ic}$ distances are constrained to be different. This is in effect reduces the tolerance factor just as Jahn–Teller effects associated with Cu^{2+} ions could enhance it. The immiscibility range in the $Pr_2Ni_{1-x}Cu_xO_4$ solid solutions is consistent with the requirement that two d_{z^2} electrons are necessary to be stabilize the T' structure.

2.2. Ratios c/a and the nature of $A–O$ and $B–O$ bonds in A_2BO_4 oxides. When the tolerance factor $t = 1$, there is a perfect match of the $B–O–B$ and $A–O–A$ distances, in both perovskites and A_2BO_4 oxides. When $t < 1$, the situation in A_2BO_4 oxides is different from that in perovskites. In the latter, there is buckling of the three-dimensional corner-shared octahedral network tending to make the $B–O–B$ angle less than 180° so that the effective $B–O–B$ distance is reduced. In A_2BO_4 oxides, however, the intervening rock-salt layer imparts a rigidity to the two-dimensional octahedral network and prevents the buckling of the octahedra. Instead, there is a pressure effect on the $B–O_{II}–B$ bond tending to reduce the distance while the $A–O_I–A$ distance is stretched in order that the two distances match in the tetragonal K_2NiF_4 structure. In most oxides including Sr_2TiO_4, $t < 1$ and hence the B–O_{II}–B distance would be expected to be smaller than that computed from the ionic radii. As a consequence, we would expect the B–O_I bond to be lengthened and the A–O_I bond to be shortened. Accordingly, fluoride ions in Sr_2FeO_3F substitutes for the O_I ions in keeping with the weaker B–O_I bond strength (24).

Evidence for covalent Ln–O_I bonding in $LnSrBO_4$ and Ln_2BO_4 compounds may also be obtained indirectly by the application of the method of invariants. Poix (3) has applied successfully the method of invariants (Eq. (3)) to oxides of the formula Sr_2BO_4 and to other oxides where the B ions are in the 4+ state. We have evaluated the value of ψ_{Ln} for a series of compounds of the type Ln_2BO_4 and $LnSrBO_4$ (Table I) and find that ψ_{Ln} is not really an invariant. The reason for this may lie in the covalency of the Ln–O_I bonds and the possible competition between Ln–O_I and B–O_I bonds.

Elongation of the BO_6 octahedra in A_2BO_4 type oxides can lead to the stabilization of unusual electronic configurations of the B ions. Thus, intermediate-spin Co^{3+} ions ($t_{2g}^5 d_{z^2}^1 d_{x^2-y^2}^0$) are found to exist at low temperatures in La_4LiCoO_8 and $LaSrCoO_4$ (25–27). At high temperatures, they are transformed to the high-spin configuration ($t_{2g}^4 e_g^2$). High-spin Fe^{4+} ions are found in $La_3SrLiFeO_8$ (28). In La_4LiNiO_8, ESR and magnetic susceptibility studies (29) have shown that the Ni^{3+} ions are in the low-spin ($t_{2g}^6 d_{z^2}^1 d_{x^2-y^2}^0$) configuration. In Sr_2FeO_3F, the Fe^{3+} ions are in the low-spin configuration due to the apical positioning of the fluorines (24); Sr_4FeTaO_8 also seems to show Fe^{3+} ions in the low-spin configuration (16).

Elongated BO_6 octahedra, short B–O_{II}–B bonds as well as short A–O_{Ic} bonds (A–O_I bonds along the c axis), are commonly found in many of the A_2BO_4 oxides. Thus, in $LaSrFeO_4$ and $LaSrCrO_4$, the values of

the a parameter (3.86 and 3.84 Å, respectively) are considerably smaller than the pseudo-cubic unit cell parameters of $LaFeO_3$ and $LaCrO_3$ (3.931 and 3.883 Å, respectively); similarly, the a parameter of Sr_2TiO_4 (3.88 Å) is smaller than that of $SrTiO_3$ (3.90 Å). In general, pressure on the B–O_{II}–B distance increases as t decreases from unity or as the size of the B ion increases (or the formal charge decreases). Accordingly, the B–O_{II} distance in La_2NiO_4 is 1.93 Å (6) compared to the value of 2.09 Å in NiO (30). The A–O_{Ic} distance would be expected to decrease as the charge of the A ion increases. In La_2NiO_4, the La–O_{Ic} distance (2, 11) is around 2.36 Å compared to the value of 2.616 Å computed from ionic radii (16).

Pressure effect on the B–O_{II}–B bond can be viewed in another manner. Along the c axis, there are A–O_I–B–O_I–A $\cdots$ A–O_I–B–O_I–A linkages. There would be strong electrostatic repulsion between $A \cdots A$ ions with no intervening anions between them. The O_{II} ions in the basal plane could come closer and screen the charge of the A ion and thereby reduce the a parameter. Electrostatic repulsion between the A ions could push the A ions closer toward the O_{Ic} ions. Strong A–O_I bonding would reduce the effective charge on the A ion and hence the $A \cdots A$ electrostatic repulsion. If we consider the ionic potential of A ions, we would expect the smaller ions to increase the $A \cdots A$ electrostatic repulsion. Furthermore, because of the competition between A and B ions for bonding with O_I ions, the A–O_{Ic} bond will get shorter (and the B–O_I bond longer or BO_6 octahedra elongated) as the charge on the A ion increases. Accordingly La–O_{Ic} bond distances in $LaSrAlO_4$ and La_2NiO_4 are 2.53 and 2.40 Å, respectively (6, 15). A compression of the BO_6 octahedra may therefore be taken to indicate a weakening of the A–O_{Ic} interaction.

It is interesting to examine the variation of the lattice parameters and the c/a ratios in $LnSrBO_4$ (Ln = rare earth) type compounds with the size of the rare earth ion. In general, the a parameter shows a linear dependence on the size of the rare earth ion. In Fig. 8, we have shown the variation of the c/a ratios in a series of compounds of the formula $LnSrBO_4$ (B = Al, Cr, Fe, Ni) with the size of the rare earth ion. In $LnSrAlO_4$ (*15, 31*), there is a linear variation of the a parameter with the size of the Ln ion and a small change in the c/a ratio between Pr and Nd (Fig. 8a). Refinement of the positional parameters has shown (*15*) that the $Al-O_I$ distance in $LaSrAlO_4$ (2.01 Å) is larger than that in $GdSrAlO_4$ (1.95 Å); the Al–O distance computed from ionic radii (*16, 32*) is 1.935 Å. A compression of the

AlO_6 octahedra implies that the $Gd-O_{Ic}$ interaction is weaker than the $La-O_{Ic}$ interaction. The average Al–O distance in $LaSrAlO_4$ is 1.92 Å while that in $GdSrAlO_4$ is 1.885 Å. The average $(Ln,Sr)-O_{Ic}$ distance is 2.53 Å in $LaSrAlO_4$ and 2.48 Å in $GdSrAlO_4$ compared to the computed values of 2.66 and 2.61 Å for the (La,Sr)–O and (Gd,Sr)–O distances and 2.62 and 2.51 Å for the La–O and Gd–O distances. While the $Gd-O_{Ic}$ distance in $GdSrAlO_4$ is close to that predicted by ionic radii, the $La-O_{Ic}$ distance in $LaSrAlO_4$ is considerably smaller. This would again imply that $La-O_{Ic}$ interaction is stronger than the $Gd-O_{Ic}$ interaction.

In $LnSrNiO_4$ (*33*), the a parameter decreases linearly with the size of the Ln ion, but there is an abrupt change in the c parameter and in the c/a ratio (Fig. 8c). Since the size of low-spin Ni^{3+} is comparable to that of Al^{3+}, the markedly different behavior of $LnSrNiO_4$ compounds cannot be due to ionic size effects; electronic factors seem to be important. A similar behavior is observed in $LnSrFeO_4$ and $LnSrCrO_4$ (*34*) as can be seen from Fig. 8.

When the A ion is kept constant and the B ion is varied, systematics in lattice parameters are not obvious. In Fig. 9, the lattice parameters and c/a ratios (*35, 36*) of some Sr_2BO_4 compounds are plotted against the ionic radius of the B ion. The a parameter varies linearly with the radius of the B ion provided that it has partially filled d orbitals. Thus, ions such as Sn^{4+}, Hf^{4+}, and Zr^{4+} do not fall on this straight line. Poix (*3*) has, however, found a linear relationship using the β_B parameters. What is important is that there is no linear relationship between the c parameters or the c/a ratios and the size of the B ion in these compounds. Furthermore, compounds containing B ions with partially filled d orbitals exhibit larger c/a ratios than those with filled or empty d orbitals. When the B ions have partially filled d orbitals, the c/a ratio

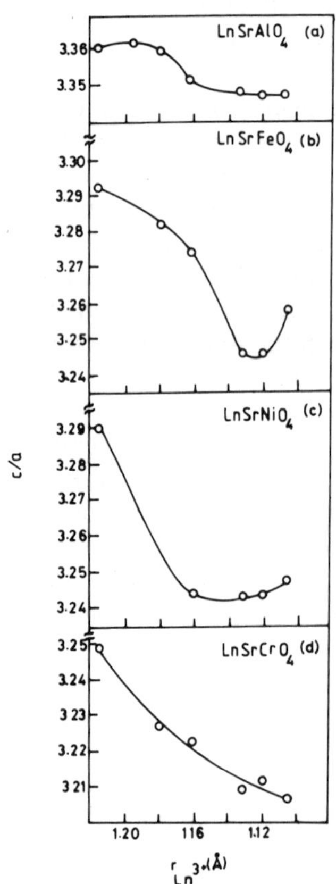

FIG. 8. Variation of the c/a ratio in $LnSrBO_4$ as function of the radius of the Ln ion.

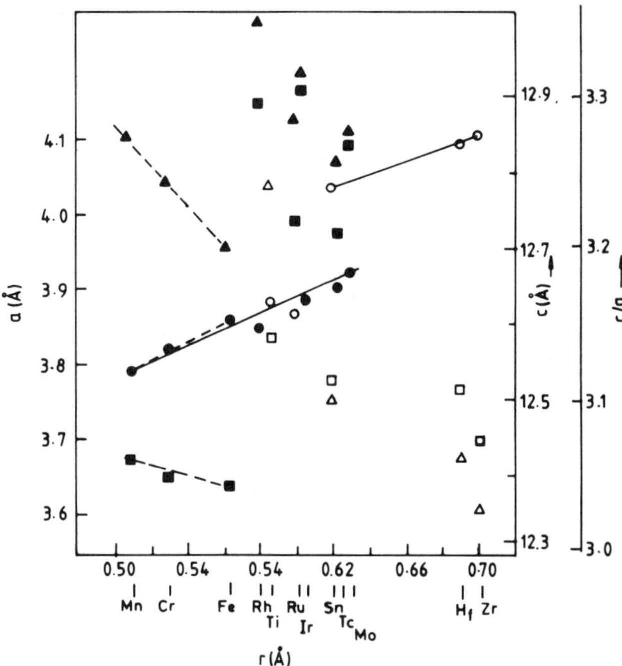

FIG. 9. Variation of the a and c parameters and the c/a ratio of several Sr_2BO_4 compounds with the radius of the B ion: circles, a parameter; squares, c parameter; triangles, c/a ratio; closed symbols represent partly filled d orbitals.

seems to increase with the decreasing size of the B ion (of the same period). In the case of $LaSrBO_4$ compounds (*35, 36*) even the a parameter does not show systematic changes with the radius of the B ion. In Sr_2BO_4 compounds, the high charge of the B^{4+} ion compared to that of Sr^{2+} ensures that the BO_6 octahedra are regular since the B–O_I bonds would be much stronger than the Sr–O_{Ic} bonds. In $LaSrBO_4$ compounds, however, competition between La and B ions for covalent bonding with the oxygen ions could complicate the situation. When the B ion is a $3d$ transition metal ion such as Mn^{3+} or low-spin Ni^{3+}, further complications enter because of the possibility of static Jahn–Teller distortions.

In Fig. 10, we have plotted the c/a ratios in some $LaSrBO_4$ compounds where the B ion (Fe, Cr, V) is neither a Jahn–Teller ion nor expected to occur in the low-spin state,

against the octahedral crystal field stabilization energy (*37*) and the optical electronegativity of the B ion in B_2O_3 compounds (*38*). The linear relationships found here indicate that B ions which do not form strong cova-

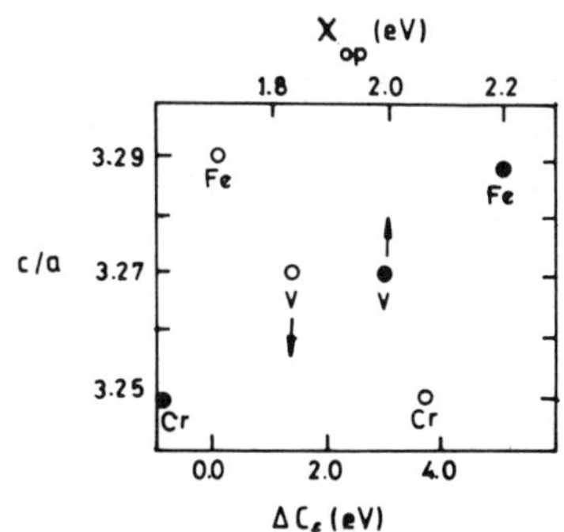

FIG. 10. Variation of the c/a ratio of $LaSrBO_4$ ($B =$ Cr, V, Fe) with the octahedral crystal field stabilization energy, ΔC_f, and the optical electronegativity of the B ions in B_2O_3 compounds.

lent bonds (or which favor an octahedral environment) also favor a low c/a ratio. Thus, we may conclude that when $t < 1$, increasing covalency of the $B-O_{II}$ bond favors an elongation of the BO_6 octahedra. This elongation is a consequence of the pressure on the $B-O_{II}$ bond in the ab plane. In Sr_2BO_4 compounds, $t \approx 1$ and the pressure on the $B-O_{II}$ bond is considerably reduced. Nevertheless, the above considerations account for the high c/a ratios in Sr_2BO_4 compounds containing B ions with partially filled d orbitals. Elongation of the BO_6 octahedra associated with B ions which are Jahn–Teller ions, may favor a cooperative, ferrodistortive ordering of octahedra elongated parallel to the c axis. The unusually high c/a ratios (≈ 3.40) in La_2CuO_4 and $LaSrMnO_4$ (7, 39) could be indicative of such ordering.

Another feature of A_2BO_4 oxides with the K_2NiF_4 structure is that the average $B-O$ distance is less than that computed from ionic radii especially when t is considerably less than unity. Thus, in La_2NiO_4 and La_2CuO_4, the average Ni–O and Cu–O distances are 2.03 and 2.07 Å, respectively (6, 7), while the values from ionic radii are 2.09 and 2.13 Å, respectively (16, 32). When the average $B-O$ distance is very small, the question arises as to whether a disproportion of B ions can occur. Thus, Cu^{2+} can disproportionate to Cu^{1+} and Cu^{3+}. The average (Cu^{1+}, Cu^{3+})–O distance is coincidentally 2.07 Å, which is the average Cu–O distance in La_2CuO_4 (7). Significantly again, the lattice parameters of La_2CuO_4 are close to that of $La_4Li^+Cu^{3+}O_8$ (40) and the radius of the Li^+ ion (0.76 Å) is only slightly smaller than that of Cu^{2+} ion (0.77 Å). Although we do not propose that there is actually a static disproportionation of Cu^{2+} ions in La_2CuO_4, the above arguments suggest the possible presence of charge-density waves. Transition metal ions with incompletely filled d orbitals have a mechanism to adjust their ionic radii toward a more favorable value of t by disproportionation or by forming charge-density waves. For example, disproportionation of Fe^{4+} ions into Fe^{3+} and Fe^{5+} ions is known to occur in $CaFeO_3$ (41) but not in $SrFeO_3$ with a higher tolerance factor. Studies on Ca_2FeO_4 and Sr_2FeO_4 would be interesting to establish whether the disproportionation is associated with the tolerance factor.

It is clear from the above discussion that the c/a ratio in A_2BO_4 oxides is determined by several factors. Besides the covalency of the $B-O_{II}-B$ bond and the competition between $A-O_I$ and $B-O_I$ linkages, other factors may also be important. Thus, $LaSrCuO_4$ in which the Cu^{3+} ion is in the low-spin state has an unusually high c/a ratio (40); the preference for square-planar coordination of low-spin Cu^{3+} ion is possibly an important factor here. $LaNaTiO_4$ and compounds of the formula La_4LiBO_8 ($B = $ Co, Cu, Ni) also have unusually high c/a ratios (35, 36). In the former, the Ti–O–Ti distance is unusually short (3.77 Å) compared to 3.88 Å in Sr_2TiO_4; in line with our earlier arguments, we expect the TiO_6 octahedra to be elongated. Since the LiO_6 octahedra would be elongated because of the short $Li-O_{II}$ distances we would expect increased c/a ratios in La_4LiBO_8 compounds.

2.3. One-dimensional antiferromagnet model of K_2NiF_4 structure. It is instructive to visualize the K_2NiF_4 structure as having a stacking sequence . . . *RPRPRPRPRPRP* . . . where R is a rocksalt layer and P is a perovskite layer. In so doing, we ignore the displacement of alternate perovskite layers mentioned earlier. Such a sequence is formally analogous to a one-dimensional antiferromagnetic Ising chain where R and P are treated as pseudospins with anisotropy. Long-range order is provided by an ordering field which can, in principle, arise from several mechanisms.

When t is less than unity, compression of the $B-O_{II}-B$ bond (and expansion of the $A-O_I-A$ bond) is necessitated. Long-range one-dimensional *RPRPRP* order would be favored because of the higher energy associated with interactions such as *RR* or *PP*.

In compounds with the Ruddlesdon–Popper structures (*42*) such as (SrO) (Sr $TiO_3)_n$, compounds with $n = 1$ have an a parameter of 3.88 Å (compared to 3.906 Å in $SrTiO_3$) while the a parameter of compounds with $n = 2$ and 3 is 3.90 Å, implying that in these compounds, the perovskite layer cannot be compressed. Furthermore, since the Ti–O–Ti distance in these oxides is larger than the $(1/\sqrt{2})$ A–O–A distance, it would require considerable energy to stretch the Sr–O bond in a bilayer of SrO. Consequently, compounds of the type $(SrO)_m$ $(SrTiO_3)_n$ are not known including the type $m = n = 2$ shown in Fig. 3e. The $(AX)_m$ $(ABX_3)_n$ system may be possible, however, with larger A ions and it would be interesting to investigate systems such as LaMBO_4 (M = monovalent ion) and Ba_2TiO_4–Sr_2TiO_4 where Ba_2TiO_4 has the K_2SO_4 structure with Ti in the tetrahedral site (*43*). In compounds such as La_2BO_4, the $(LaO)^+$ layer is positively charged while the $(LaBO_3)^-$ layer is negatively charged. The stacking sequence in such oxides is therefore a favorable one for ordering along the c axis, $\ldots R^+ P^- R^+ P^- R^+ P^- R^+$ $\ldots$ In oxygen excess $La_2BO_{4+\delta}$ compounds (e.g., La_2NiO_4), if the δB^{3+} ions are randomly distributed, the $LaBO_3$ layers would be electricaly neutral (P^0). We can then have intergrowths of $P^- P^0$ and $P^- P^0 P^0$ type of layers; intergrowths of the composition $La_3Ni_2O_7$ and $La_4Ni_3O_{10}$ have been observed (*44*) in oxygen-excess La_2NiO_4. Tilley (*45*) has carried out an electron microscopic investigation of the SrO–TiO_2 system with special reference to Sr_2TiO_4. Besides the intergrowth of Ruddlesdon–Popper phases such as $Sr_3Ti_2O_7$ and Sr_4 Ti_3O_{10}, intergrowth of SrO-rich phases has been observed by Tilley. In the La_2BO_4 phases, intergrowth of two $(LaO)^+$ layers adjacent to each other would not be likely

unless at least one of the LaO layers adjusts its oxygen and La content in such a manner as to make the layer neutral. Since this is unlikely, some La_2O_3 should be precipitated out in oxygen-excess La_2BO_4 compounds. Another possibility is that the excess oxygen is accommodated by A-site deficiency. In compounds such as Sr_2BO_4, R and P layers are both neutral and intergrowth of RR and PP sequence would be possible.

2.4. Superlattice ordering in A_2BO_4 compounds. Ordering of B ions in compounds of the type $A_2BB'O_4$ due to charge difference potential between B and B' ions in different oxidation states would be analogous to antiferromagnetic ordering of spins in K_2NiF_4 due to an exchange potential. Since a three-dimensionally ordered antiferromagnetic structure of K_2NiF_4 is known (*1*), it seems reasonable to expect that B and B' ions may similarly get ordered. However, not all possible antiferromagnetic interactions can be satisfied in the K_2NiF_4 structure (*46*) and this frustration leads to a two-dimensional order.

In compounds such as La_4LiBO_8 (B = Co, Ni, or Cu) or $Sr_4BB'O_8$ (B = Co, Fe, Ni, etc.; B' = Nb, Ta), X-ray diffraction studies do not reveal any evidence for an ordered superlattice (*47, 48*). Demazeau *et al.* (*25*) have found evidence for a $\sqrt{2}$ increase in the tetragonal a parameter by employing neutron diffraction and X-ray diffraction (with monochromatized Cu $K_{\alpha l}$ radiation) in La_2LiCoO_8. Our electron diffraction studies also reveal such ordering as mentioned earlier; the electron diffraction patterns of La_4LiCoO_8 show streaking parallel to the c^* axis (Fig. 4c) similar to the ridges found in the neutron diffraction pattern of K_2NiF_4 in the temperature range where there is only two-dimensional antiferromagnetic ordering (*1*). We feel that the streaking in Fig. 4c may indeed be associated with two-dimensional ordering of Li^+ and Co^{3+} ions.

Three-dimensional antiferromagnetic ordering in K_2NiF_4 is associated with an orthorhombic distortion. It is interesting that the distortion which gives rise to M, O, and O' structures involves the movement of O_I ions along the c axis (Fig. 4) besides that of O_{II} ions in the ab plane. By analogy with the three-dimensional antiferromagnetic ordering in K_2NiF_4, we would expect three-dimensional ordering of B ions in A_2BO_4 only in distorted structures or when t is close to 0.85.

In A_2BO_4 oxides, there are eight A ions surrounding the B ions as in the perovskite structure and two A ions linked through O_I ions with the B ions along the c axis. Random occupation of these sites by La and Sr ions in $LaSrBO_4$ (33) could give rise to a distribution of crystal fields. Of these, the most important are those involving the O_I ions. La–O_I–B–O_I–La arrangement would give rise to the lowest crystal field while Sr–O–B–O–Sr arrangement would give rise to the highest crystal field. Evidence for a distribution of sites is seen in the Fe^{3+} ESR spectra of $LaSrAl_{0.98}$ $Fe_{0.02}O_4$ (Fig. 11). There are two prominent lines at $g \approx 6$ and $g \approx 4.25$ which may be associated with Fe^{3+} ions in axial and orthorhombic sym-

metry (49–51); the $g \approx 6$ line is also found in oxide glasses containing Fe^{3+} (51). There is also a line at $g \approx 2.1$. The deviation from $g = 2$ may be attributed to spin-orbit coupling effects in Fe^{3+} ions in distorted octahedra. Similarly, $LaSrAlO_4$ containing small amounts of Ni^{3+} ions clearly shows evidence for the simultaneous existence of low-spin and high-spin Ni^{3+} ions (52). Random ordering of B ions in LaSr $(B,B')O_4$ has been used to explain percolation effects in electrical and magnetic properties of these compounds (16, 52).

The only $AA'BO_4$ compound which shows ordering is $LaNaTiO_4$ (53). Ordering in this compound interestingly involves $(La_2O_2)^{2+}$ and $(Na_2O_2)^{2-}$ layers (53). There is considerable pressure on the $Ti–O_{II}–Ti$ bonds in $LnNaTiO_4$ ($a \approx 3.78$ Å compared to 3.88 Å for Sr_2TiO_4), consistent with the low tolerance factor. As discussed earlier, a structure containing bilayers of rock-salt and perovskite (Fig. 3e) such as $(Ln_{0.5} Na_{0.5}O)_2$ $(Ln_{0.5}Na_{0.5}TiO_3)_2$ would also impose considerable pressure on the $Ti–O_{II}–Ti$ bond. We have observed that X-ray diffraction line intensities calculated for such a structure by assuming a random distribution of Ln and Na ions are comparable to those reported by Blasse (53).

2.5. Non-stoichiometry. Oxides with the K_2NiF_4 structure can accommodate considerable non-stoichiometry. This could lead to a significant variation in lattice parameters as indeed found in oxides such as LaSr FeO_4, Pr_2NiO_4, or Nd_2NiO_4, (9, 10, 15, 54). In the $La_2NiO_{4+\delta}$ system, Drennan *et al.* (44) have shown that intergrowth of Ruddlesdon–Popper type phases such as $La_3Ni_2O_7$ and $La_4Ni_3O_{10}$ would account for anion-excess non-stoichiometry. Lewandowski *et al.* (55) have found A-site deficiency in lanthanum cobalt oxide; such a defect structure would be favored by δB^{3+} ions.

Oxygen-deficient non-stoichiometry is more difficult to account for. Poeppelmeier

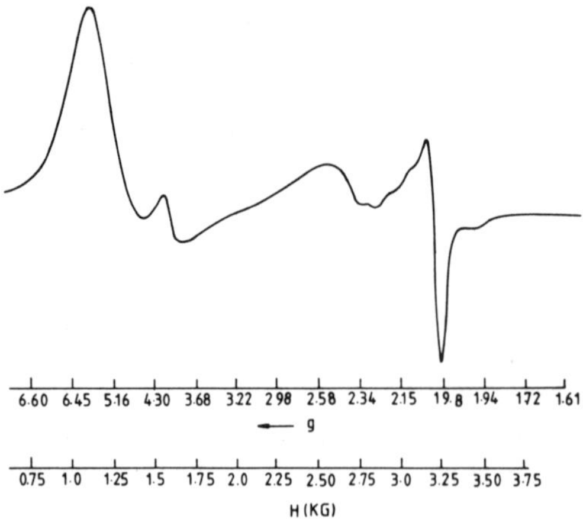

FIG. 11. ESR signal from Fe^{3+} ions in $LaSrAl_{0.98}$ $Fe_{0.02}O_4$.

et al. (*56*) have observed that Ca$_2$MnO$_{3.5}$ can be obtained topotactically from Ca$_2$MnO$_4$ by reduction, just as CaMnO$_{2.5}$ can be obtained from CaMnO$_3$. In CaMn O$_{2.5}$, Mn^{3+} ions are in five-coordinated square-pyramidal coordination. It is assumed that the same situation may be present in Ca$_2$MnO$_{3.5}$. The idealized structure proposed for Ca$_2$MnO$_{3.5}$ is shown in Fig. 12 with the O$_{II}$ atoms being labile. The loss of O$_{II}$ atoms instead of O$_I$ atoms is consistent with $t < 1$; since the B–□–B (□ = vacancy) distance would be smaller than the B—O$_{II}$—B distance, it would favor the K$_2$NiF$_4$ structure. In Sr$_2$CuO$_3$, half the O$_{II}$ oxygen sites are vacant (*18*) and this is possible with Cu^{2+} because of the presence of $d_{z^2}^2$ electrons. In this laboratory, a layered brownmillerite phase of the formula Ca$_2$Fe O$_{3.5}$(a = 14.768, b = 13.724, and c = 12.20 Å) has been recently synthesized (*57*). This structure seems to have alternate columns of octahedra and tetrahedra in the ab plane with oxygen vacancies in both O$_I$ and O$_{II}$ positions. It is possible that in Ca$_2$MnO$_{3.5}$, fivefold coordination of Mn is achieved by the loss of O$_I$ oxygens. In any case, it is important to to note that anion-deficient non-stoichiometry can be achieved by the loss of O$_I$ or O$_{II}$ oxygens.

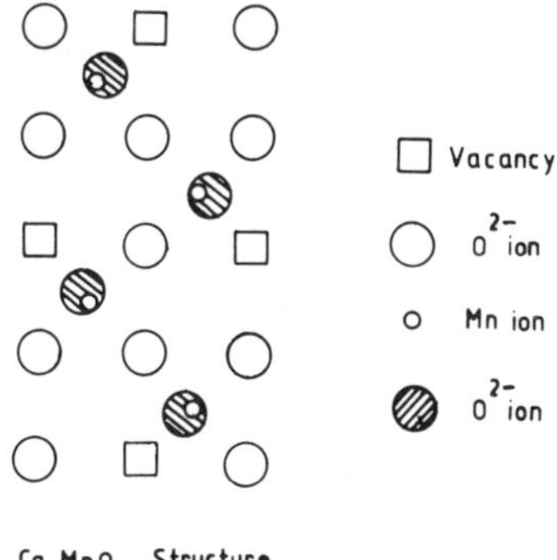

FIG. 12. Proposed ordering scheme in the ab plane of fivefold coordinated Mn^{3+} ions in Ca$_2$MnO$_{3.5}$. Oxide ions above the plane of the paper and Ca ions are not shown (from Ref. (*56*)).

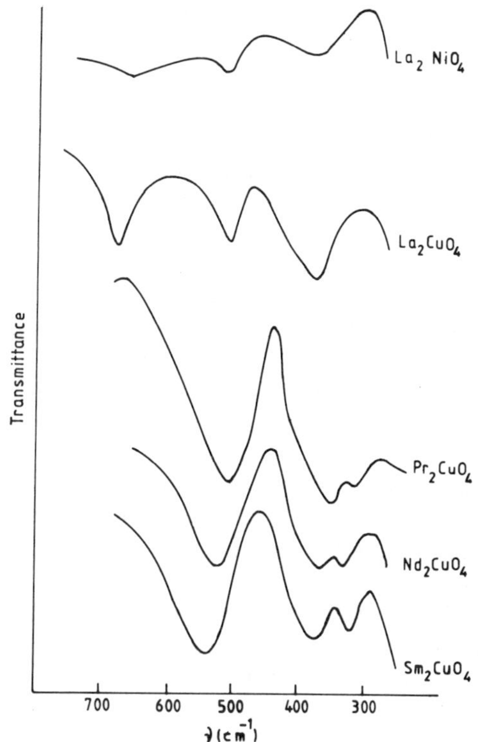

FIG. 13. Infrared spectra of some Ln_2BO_4 compounds.

3. Infrared Spectra

Infrared spectra of A_2BO_4 type oxides provide useful, diagnostic information on the nature of the BO_n polyhedra. Thus, the spectra clearly distinguish Ln_2CuO_4 compounds of orthorhombic structure containing distorted CuO$_6$ octahedra from those of tetragonal structure containing square-planar CuO$_4$ polyhedra. We see from Fig. 13 that the Cu–O asymmetric stretching vibration band is split into a doublet (~690 and 520 cm^{-1}) in La$_2$CuO$_4$ (suggesting D_{4h} symmetry of the isolated octahedron), while it appears as a single band in other Ln_2CuO_4 compounds with square-planar coordination. The spectrum of La$_2$NiO$_4$ possessing distorted NiO$_6$ octahedra is similar to that of La$_2$CuO$_4$. In the spectra of La$_{2-x}$Pr$_x$CuO$_4$, Cu–O asymmetric stretching is a doublet when $x < 0.75$ and a single band when $x \geq 0.75$ and Cu coordination is square-planar. In $LaSrBO_4$ (B = Al or Fe), there seem to be three bands in the B–O stretching region suggesting a low symmetry of the BO_6 octahedra. The B–O stretching mode in $LaSrBO_4$ shifts to higher frequencies compared to that in the corresponding La_2BO_4. It is interesting that the B–O stretching frequency in $AA'BO_4$ oxides is higher compared to that in the corresponding ABO_3 perovskites. In $GdSrAlO_4$, for instance, the Al–O_{II} stretching frequency increases by as much as 75 cm^{-1} compared to that in $GdAlO_3$ and by about 50 cm^{-1} compared to that in $LaSrAlO_4$. The Al–O stretching frequencies in $LaAlO_3$ and $GdAlO_3$ are similar, consistent with our earlier observation that in ABO_3 perovskites, lowering of t leads to a buckling of the octahedra, the B–O distance remaining roughly the same. In A_2BO_4 oxides, the layered structure does not allow buckling and there is greater pressure on the B–O_{II}–B bond with decreasing t.

4. Magnetic Properties

4.1. Spin-state equilibria of transition metal ions. Transitions between low-spin and high-spin states of transition metal ions have been found in perovskite oxides (*58, 59*) such as $LaCoO_3$. There is evidence for such transitions between spin states in oxides of K_2NiF_4 structure as well. The earliest evidence for such a transition was obtained with La_4LiCoO_8 by Blasse (*60*). Because of the elongated nature of the BO_6 octahedra in this oxide, the degeneracy of the e_g orbital is expected to be lifted, shifting the d_{x2-y2} orbital to higher energies. This orbital can remain unoccupied under certain conditions and this indeed appears to be the case with Co^{3+} ions in La_4Li CoO_8. In Fig. 14, we have shown the in-

verse susceptibility–temperature curve of La_4LiCoO_8 and this is best interpreted in terms of a low-spin to intermediate-spin state transition. The intermediate-spin state with the configuration $t_{2g}^5 d_{z2}^1 d_{x2-y2}^0$ seems to be stabilized (at intermediate temperatures) in compounds such as La_4LiCoO_8 (*25*) and also perhaps in Sr_4CoNbO_8 and Sr_4CoTaO_8 (*52*). In the last two compounds, there is some evidence of ordering of the two spin state (just as in $LaCoO_3$), the inverse susceptibility–temperature curve showing a plateau.

An interesting example of spin-state equilibrium between low- and high-spin Ni^{3+} ions has been reported in $LaBaNiO_4$ on the basis of ESR evidence (*61*). The average Ni–O distance of 2.03 Å in this compound is consistent with the Ni–O distance expected from the ionic radius of high-spin Ni^{3+}. Magnetic susceptibility studies (*52*) on $LaSr_{1-x}Ba_xNiO_4$, however, reveal that the susceptibility can be entirely described on the basis of an equation of the form $\chi = [C/(T + \theta)] + \alpha$ where α is of the order of 6 $\times 10^{-4}$ emu and C is of the order of 0.01–0.1 emu/K; C increases with increasing x while θ (in the range 10–20 K) decreases with increasing x. Magnetic susceptibility measurements show no evidence for an activated behavior. It would therefore seem that most of the e_g electrons of the Ni^{3+} ions are in extended states (as in $LaSrNiO_4$,

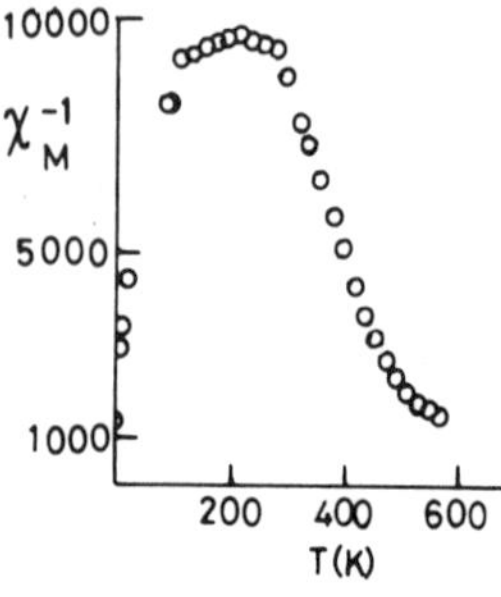

FIG. 14. χ_M^{-1} vs T plot of La_4LiCoO_8 (after Ref. (*25*)).

with $x = 0$, where the $d_{x^2-y^2}$ electrons from a $\sigma^*_{x^2-y^2}$ band) with only a few of the electrons being localized. Localized electrons in $LaBaNiO_4$ may be associated with Ni sites with high crystal field, introduction of Ba^{2+} in $LaSrNiO_4$ increasing the number of localized Ni^{3+} states along with the unit-cell volume. It is possible that the ESR evidence of Demazeau *et al.* (*61*) is associated with such localized states.

An unusual configuration change of low-spin Ni^{3+} has been observed (*52*) in the solid solution $LaSrAl_{1-x}Ni_xO_4$. For small values of x, the Ni^{3+} ions are predominantly in the low-spin state showing an ESR signal similar to that observed in La_4LiNiO_8 (*29*) in which the Ni^{3+} ions have the configuration $t_{2g}^6 d_{z^2}^1 d_{x^2-y^2}^0$. For $x \geq 0.75$, the ESR signal disappears and the magnetic susceptibility decreases sharply. The change in magnetic properties is accompanied by a significant decrease in the c/a ratio. This points to a change to the configuration $t_{2g}^6 d_{z^2}^0 d_{x^2-y^2}^1$ with the $d_{x^2-y^2}$ electrons forming $\sigma^*_{x^2-y^2}$ band-like states. It is significant that the magnitude as well as the temperature dependence of the susceptibility of the $x = 0.75$ sample (*52*) is close to that of pure $LaBaNiO_4$. It seems therefore extremely unlikely that the bulk of the Ni^{3+} ions in the latter compound are involved in a low-spin to high-spin transition.

4.2. Antiferromagnetic ordering. When the A ion is nonmagnetic and the B ion is magnetic, A_2BO_4 oxides may be expected to behave as the fluorides with strong intralayer coupling and weak interlayer coupling. The earlier study (*62*) of such an oxide was on Ca_2MnO_4. In this oxide, Mn^{4+} ions are coupled antiferromagnetically with the spins parallel to the c axis. The magnetic structure determined by neutron diffraction (*62*) is shown in Fig. 15. Interplanar magnetic ordering leads to a doubling of the unit-cell parameter. Poeppelmeier *et al.* (*56*) have recently reported that the c parameter of Ca_2MnO_4 is actually twice that of the value reported earlier. Considerable work has been carried out at Bordeaux on magnetic ordering in insulating A_2BO_4 oxides in which the spins couple antiferromagnetically. In Fig. 15, we show the magnetic structure of β-Sr_2MnO_4 (*63*) and $LaCaFeO_4$. In the latter, the magnetic moments are aligned along the a axis.

Le Flem *et al.* (*64*) have pointed out that β-Sr_2MnO_4 (*63*), $LaSrCrO_4$ (*65*), and La-rich $La_{1-x}Y_xCaCrO_4$, show strong two-dimensional behavior as indicated by the critical exponents in the vicinity of the ordering

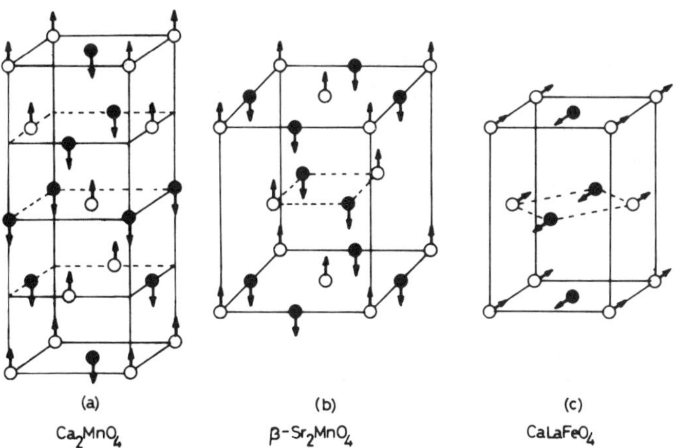

FIG. 15. Magnetically ordered structure of (a) Ca_2MnO_4, (b) β-Sr_2MnO_4, and (c) $LaCaFeO_4$ (from Ref. (*64*)).

temperatures as well as by neutron scattering investigations. Compounds such as Ca_2MnO_4 and $CaYCrO_4$ have essentially three-dimensional magnetic correlations. These authors also point out that the change from two-dimensional to three-dimensional correlation depends on the extent of covalency of the $A–O$ bond. Making use of Goodenough's arguments in the case of perovskites (66), these authors suggest that there is competition between $t_{2g}–O_{II}–t_{2g}$ interaction (involving the t_{2g} orbitals of the B ions) and $\sigma_{A–O_{II}}$ bonding (involving the A ions in the perovskite layer with a 90° $A–O_{II}–B$ linkage). The greater the covalency of the $A–O_{II}$ bonding, the weaker is the $B–O_{II}–B$ interaction and the stronger is the three-dimensional coupling. Thus, ions such as Cr^{3+} and the isoelectronic Mn^{4+} with only t_{2g} electrons, have their magnetic interactions strongly determined by the nature of $A–O$ interactions. The smaller the A cation, the stronger is the $A–O_{II}$ linkage (and the three-dimensional correlations) and the lower is the antiferromagnetic ordering temperature. Applying the same arguments to oxides containing Fe^{3+} ions, it is stated that since the magnetic interactions are strongly determined by the e_g electrons, the influence of the A ion is considerably diminished. Accordingly, there is not much difference in the ordering temperatures of $LaCaFeO_4$ (373 K) and $LaSrFeO_4$ (380 K). However, compounds such as $LnSrFeO_4$ show a marked dependence of the magnetic ordering temperature on the size of the Ln ion (67). Although the above arguments seem plausible, we should note that we have ignored the $A–O_I–B$ linkage which would be exected to play a role. The collapse in the c/a ratio of $LnSrBO_4$ compounds with the decreasing size of Ln ion (Fig. 4) is consistent with a decreased $B–O_I$ distance. Since the $B–O_I–O_I–B$ interaction determines the interlayer coupling, we may expect the three-dimensional character to increase with decreasing $B–O_I$ bond length.

In $LaSrAl_{1-x}Fe_xO_4$ solid solutions (15), the Néel temperature drops to zero when $x > x_c$ where x_c (0.59) is the critical percolation threshold for nearest-neighbor interactions in a square-planar array (68). The results are similar to those found in the $Rb_2Mg_{1-x}Mn_xF_4$ system (69). The surprising conclusion, therefore, is that even in oxides, long-range magnetic ordering is dominated by nearest-neighbor interactions.

4.3. Unusual Behavior of La_2NiO_4, La_2CuO_4, and La_2CoO_4. La_2NiO_4 shows a Curie–Weiss behavior at high temperatures with high θ (~ -500 K) and μ_{eff} ($\sim 3.00\ \mu_B$) values (70). Below 200 K, there is deviation from the Curie–Weiss law, but neither neutron diffraction nor magnetic susceptibility studies down to the lowest temperature show any evidence for long-range antiferromagnetic ordering (70, 71). It has been found recently that below 100 K, the magnetic susceptibility again conforms to a Curie–Weiss behavior of (Fig. 16) with a μ_{eff} of

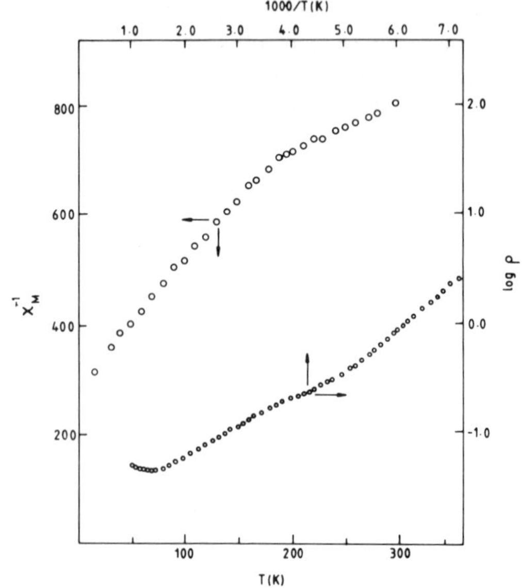

Fig. 16. Plots log ρ vs $1/T$ and χ_M^{-1} vs T of La_2NiO_4 (from Ref. (2) and unpublished results from this labratry).

1.8 μ_B indicating the presence of one unpaired electron (15). The θ value obtained from the slope of the χ_M^{-1} vs T plot is consistent with the expression $\theta = CW$ with a nearly temperature-independent W and with a C corresponding to two unpaired electrons above 200 K and one unpaired electron below 100 K. Such a behavior is also observed in the $La_2Ni_{1-x}Cu_xO_4$ system in the range $0.75 \geq x \geq 0.0$ below 100 K with a μ_{eff} value corresponding to one unpaired electron (15) per g. atom of Ni. Below 200 K, there is also an increase in the activation energy for electrical conduction (Fig. 16) suggesting that there is spin-pairing of the $d_{x^2-y^2}$ electron around 200 K. It is rather surprising that in La_2NiO_4, below 200 K, the $d_{x^2-y^2}$ electrons are coupled into nearly dimagnetic spin-paired states while the d_{z^2} electrons are localized and coupled antiferromagnetically without the onset of long-range order. Since La_2NiO_4 prepared by ceramic techniques always has an excess of oxygen due to the presence of Ni^{3+} ions, it is not clear whether long-range order is frustrated by the presence of such ions. Singh *et al.* (15) propose that a disproportionation of the Ni^{2+} ions to Ni^{1+} and low-spin Ni^{3+} (which could result in the formation of charge-density waves) would stabilize the diamagnetic nature of the spin-paired states involving the $d_{x^2-y^2}$ electrons.

La_2CuO_4 was first reported to show nearly temperature independent magnetic susceptibility which was attributed to antiferromagnetism (7). Because of the low value the electrical resistivity, it was assumed that the oxide exhibited broad-band Pauli paramagnetism. It was subsequently shown (72) that there is an enhancement in the susceptibility at the lowest temperature with an anomaly around 200 K (Fig. 17). The anomaly in the susceptibility at low temperatures can be attributed to paramagnetic impurities and it has been shown by Saez-Puche *et al.* (73) that starting with high purity oxides, the paramagnetic be-

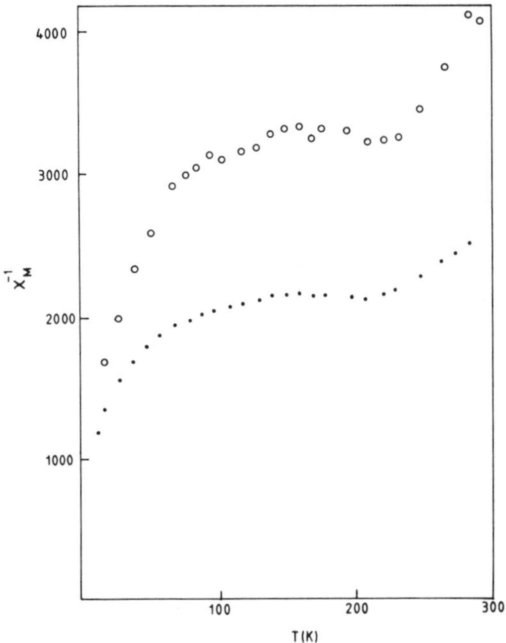

FIG. 17. χ_M^{-1} vs T plot of La_2CuO_4 at 4000 G (dotted curve); after subtraction of the contribution from the ferromagnetic component (line of circles) (from Ref. (15)).

havior at low temperature is suppressed although there is a tendency toward an increase in the susceptibility below 100 K. Singh *et al.* (15) have since found that at low temperatures, there is a marked increase in the resistivity with $\rho(12\ K)/\rho(300\ K) \approx 10^2$. Careful magnetic susceptibility studies show that above 200 K, the susceptibility obeys a Curie law with a μ_{eff} corresponding to the contribution from about 16% of Cu^{2+} ions. At low temperatures, point by point calculation of μ_{eff} (assuming a Curie law) shows a linear decrease in the concentration of paramagnetic Cu^{2+} ions with decreasing temperature. Such a behavior seems to be consistent with the disproportionation of Cu^{2+} to Cu^{1+} and Cu^{3+} as mentioned earlier (15) and the formation of charge-density waves. The two B–O_1 distances in the O (or O') structure of La_2CuO_4 would be consistent with orbital ordering of the type shown in Fig. 7c and it is interesting to speculate on the consequences of such ordering.

There are several intriguing features associated with Cu^{2+} ions in such oxides which are not readily understood. A comparison of the high-temperature susceptibilities of the oxides in the series $La_2Ni_{1-x}Cu_xO_4$ and $La_{2-2x}Sr_{2x}Ni_{1-x}Ti_xO_4$ shows that there is no contribution to the magnetic susceptibility at high temperatures from Cu^{2+} ions (15). It has been observed that Cu^{2+} ions do not contribute to the susceptibility of Ln_2CuO_4 compounds (Ln = rare earth) at low temperatures and this has been taken to indicate antiferromagnetic ordering (72–74).

It is extremely difficult to prepare stoichiometric La_2CoO_4. Magnetic susceptibility of $La_2CoO_{4.04}$ measured in this laboratory (75) shows a broad maximum around 500 K and a sharp maximum around 400 K. DSC studies show a large number of the phase transitions in this temperature region. The results have been interpreted in terms of two-dimensional magnetic ordering with the onset of three-dimensional order at low temperatures. Further work on stoichiometric samples of La_2CoO_4 is necessary to establish the magnetic behavior of this oxide.

4.4. Ferromagnetic oxides. One of the earliest oxides to be studied was $LaSr_3Mn_2O_8$ and measurements on this oxide were reported in the same paper (76) dealing with the low-dimensional magnetic susceptibility behaviour of K_2NiF_4. The peculiarity of $LaSr_3Mn_2O_8$ is that although the high-temperature behavior is typical of ferromagnets (the susceptibility showing a Curie–Weiss plot with a high negative values of θ), there is no spontaneous magnetization at low temperatures. This has been attributed (77) to the presence of small superparamagnetic clusters due to the absence of ordering of Mn^{3+} and Mn^{4+} ions (since only Mn^{3+}–O–Mn^{4+}

interactions are ferromagnetic). However, the corresponding cobalt analog $LaSr_3Co_2O_8$ is a true ferromagnet with a well-defined Curie temperature (75). In this oxide, Co^{3+} ions could have the intermediate-spin configuration ($t_{2g}^5 e_g^1$) while the Co^{4+} ions are in the low-spin configuration (t_{2g}^5). Co^{3+}–O–Co^{4+} would then be a Zener double exchange pair since the transfer of an electron from intermediate-spin Co^{3+} to low-spin Co^{4+} would have the initial and final state degenerate. Although this oxide is semiconducting, changes in the electron transport properties observed at the Curie temperature (E_a decreases below T_c) is consistent with a double exchange mechanism. Such changes in transport properties at T_c have not been observed in the corresponding three-dimensional perovskite compound $LaSrCo_2O_6$. The μ_{eff} value calculated from the χ_M^{-1} vs T plot (Fig. 18) above the Curie temperature also supports the existence of intermediate-spin Co^{3+} and low-spin Co^{4+}. At high temperatures (Fig. 18), the χ_M^{-1} vs T plot shows a Curie behavior with the μ_{eff} corresponding to high-spin Co^{3+} ions, and low-spin Co^{4+} ions. This could be associated with a low-spin to high-spin transition. It should be mentioned that the nature of the χ_M^{-1} vs T plot is similar to that predicted by Anderson and Hasegawa (78) for double exchange systems.

$LaSrMnO_4$ has been reported to be ferro-

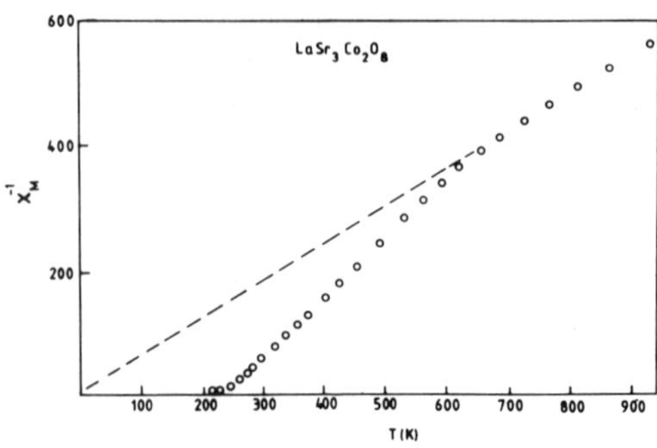

FIG. 18. χ_M^{-1} vs T plot of $LaSr_3Co_2O_8$ above 300 K.

magnetic (*79*). The electronic configuration of Mn^{3+} being the same as that of Cr^{2+}, we would expect ferromagnetism in LaSrMnO$_4$ by analogy with halides of the formula A_2CrX_4 which are transparent ferromagnets (*80*). Ferromagnetic LaSrMnO$_4$ is reported to have a tetragonal structure with a small c/a ratio while the A_2CrX_4 compounds are orthorhombic with unequal Cr–X distances in the basal plane indicative of antiferrodistortive ordering of CrX_6 octahedra. Recently, stoichiometric LaSrMnO$_4$ with a large c/a ratio (value) has been reported (*39*); we would expect ferrodistortive ordering of the elongated MnO$_6$ octahedra in such an oxide with the e_g electron in the d_{z^2} orbital. Considerations based on the Goodenough–Kanamori rules (*81*) indicate that the interaction would be ferromagnetic if the transfer of the $d_{z^2}^1$ electron is to the empty $d_{x^2-y^2}^0$ orbital. Magnetic measurements on stoichiometric LaSrMnO$_4$ have not been reported. The small c/a ratio of the earlier sample of LaSrMnO$_4$ suggests considerable oxygen excess non-stoichiometry and it is indeed known that the c/a ratio of La$_{1-x}$Sr$_{1+x}$MnO$_4$ decreases with increasing x.

Two oxides that have been recently studied in this laboratory (*15*) which show evidence for ferromagnetic interactions are La$_2$Sr$_2$MnNiO$_8$ and La$_2$Sr$_2$MnCoO$_8$ which are the two-dimensional analogs of the three-dimensional ferromagnets La$_2$MnNiO$_6$ and La$_2$MnCoO$_6$, respectively (*82, 83*). La$_2$Sr$_2$MnNiO$_8$ shows a large negative value of the Weiss constant, but is like LaSr$_3$Mn$_2$O$_8$ in that it does not show spontaneous magnetization at low temperatures. There is no ion ordering in either of these oxides. The c/a ratio of the former compound (3.27) is much less than that of stoichiometric LaSrMnO$_4$ (*39*), but similar to that of LaSrNiO$_4$ (*33*). This suggests that the electronic configurations of the Mn and Ni ions are $t_{2g}^3 d_{x^2-y^2}^1$ and $t_{2g}^6 d_{x^2-y^2}^1$, respectively, the d_{z^2} orbitals being empty in both

cases. La$_2$Sr$_2$MnCoO$_8$ shows the behavior of a true ferromagnet with a well-defined Curie temperature. The c/a ratio in this compound is similar (3.28) to that of LaSrCoO$_4$ (*26, 27*), but much less than that of LaSrMnO$_4$ (*39*). Electronic configurations of the Mn^{3+} and Co^{3+} ions in this oxide appear to be $t_{2g}^3 d_{x^2-y^2}^1$ and $t_{2g}^5 d_{z^2}^1$, respectively. The presence of a localized d_{z^2} electron is significant. It is possible that the presence of a localized d_{z^2} electron enhances the three-dimensional B–O$_1$–O$_1$–B interlayer coupling. A mechanism that could be of relevance to compounds such as LaSr$_3$Mn$_2$O$_8$ and La$_2$Sr$_2$MnNiO$_8$ is one where the magnetic moments in the ferromagnetic state may be aligned perpendicular to the ab plane. In the absence of intralayer coupling, demagnetization effects acting on spins aligned perpendicular to a layer would be considerable.

5. Concluding Remarks

It should be clear from the above discussion that oxides with the K$_2$NiF$_4$ structure offer considerable scope for research. The anisotropic bonding coupled with strong covalency effects in these oxides could give rise to unique properties which have not been adequately investigated. For example, we are yet to understand the nature of the semiconductor–metal transition in La$_2$NiO$_4$ and other rare earth nickelates. Although many models have been suggested (*84*) to explain the transition in La$_2$NiO$_4$ (*2*), all the measurements reported hitherto are on polycrystalline samples. Preliminary studies (*12*) on single crystals of La$_2$NiO$_4$ indicate that the transition occurs sharply (550 K) with about an order of magnitude jump in conductivity along the ab plane. Anisotropic magnetic susceptibility of La$_2$NiO$_4$ needs to be investigated as also the effect of Ni^{3+} ions on these properties. The structure of La$_2$NiO$_4$ itself seems to require a revision in the light of the superlattice spots found in the diffraction patterns. Electron transport properties of single crystals of La_2CuO_4 and related rare earth compounds are yet to be investigated. The possible occurrence of two-dimensional antiferromagnetic ordering in La_2CoO_4 is worth exploring.

An interesting feature of some of the A_2BO_4 oxides is that their electrical properties are considerably different from the corresponding perovskites, even though the magnetic properties are similar. For example, $LaSr_3Co_2O_8$ is a semiconducting ferromagnet while $LaSrCo_2O_6$ is an itinerant-electron ferromagnet. The corresponding manganese compounds also show a similar behavior. Another system showing such a behavior is that of vanadium, $La_{1-x}Sr_xVO_3$ and $La_{1-x}Sr_{1+x}VO_4$, where the latter is insulating while the former is metallic for $0.3 > x > 0.05$.

None of the A_2BO_4 oxides seems to exhibit a true metallic behavior down to the lowest temperatures. Most of these oxides show activated conduction and even those phases that have been considered to be metallic (e.g., La_2CuO_4) exhibit conductivities of the order of 10 ohm^{-1} cm^{-1}. This is much less than the conductivities found in metallic oxides of perovskite structure (e.g., La NiO$_3$ or $LaSrCo_2O_6$ with σ of ~10^3 ohm^{-1} cm^{-1}). It is not clear whether the absence of true metallic conductivity in A_2BO_4 oxides has something to do with localization in two dimensions (85, 86).

In systems where the electrical properties are determined by the concentration of the component ions, the A_2BO_4 system shows an unusual concentration dependence of resistivity. For example, delocalization of e_g electrons is found in systems like $LaSrAl_{1-x}Ni_xO_4$ when $x > 0.6$. In perovskite systems such as $La_{1-x}Sr_xCoO_3$ and $LaFe_{1-x}Ni_xO_3$, the oxides become metallic when $x = 0.25-0.30$. It is interesting to ponder whether such concentration limits are related to percolation limits in two-

dimensional and cubic systems.

Some of the well-known ferroelectric materials are perovskite oxides. No ferroelectric oxide of K_2NiF_4 structure has been reported until now; similarly, other ferroic properties (87) are yet to be explored.

Acknowledgments

The authors thank the Department of Science and Technology, Government of India and the University Grants Commission for support of this research.

References

1. R. J. BIRGENEAU, H. J. GUGGENHEIM, AND G. SHIRANE, *Phys. Rev. B* **1**, 2211 (1970).
2. P. GANGULY AND C. N. R. RAO, *Mater. Res. Bull.* **8**, 405 (1973).
3. P. POIX, *J. Solid State Chem.* **31**, 95 (1980).
4. P. POIX, *C.R. Acad. Sci. (Paris) C* **268**, 1139 (1969).
5. D. GANGULI, *J. Solid State Chem.* **30**, 353 (1979).
6. A. RABENEAU AND P. ECKERLIN, *Acta Crystallogr.* **11**, 304 (1958).
7. J. M. LONGO AND P. RACCAH, *J. Solid State Chem.* **6**, 526 (1973).
8. V. U. LEHMANN AND H. R. MÜLLER-BUSCHBAUM, *Z. Anorg. Allg. Chem.* **470**, 59 (1980).
9. B. WILLER AND M. DAIRRE, *C.R. Acad. Sci. (Paris) C* **267**, 1482 (1968).
10. M. FÖEX, *Bull. Soc. Chim. Fr.* 109 (1961).
11. B. GRANDE AND H. R. MÜLLER-BUSCHBAUM, *Z. Anorg. Allg. Chem.* **433**, 152 (1977).
12. C. N. R. RAO, D. BUTTREY, N. OTSUKA, P. GANGULY, H. R. HARRISON, C. J. SANDBERG, AND J. M. HONIG, *J. Solid State Chem.* **51**, 266 (1984).
13. R. BERJOAN, J. P. COUTURES, G. LE FLEM, AND S. SAUX, *J. Solid State Chem.* **42**, 75 (1982), and references therein.
14. B. GRANDE, H. R. MÜLLER-BUSCHBAUM, AND M. SCHWEIZER, *Z. Anorg. Allg. Chem.* **428**, 120 (1977).
15. K. K. SINGH, P. GANGULY, AND J. B. GOODENOUGH, *J. Solid State Chem.*, in press; also see K. K. SINGH, Ph.D. thesis, Indian Institute of Science, Bangalore, India (1983).
16. R. D. SHANNON, *Acta Crystallogr. Sect. A* **32**, 751 (1976).
17. H. R. MÜLLER-BUSCHBAUM AND W. WOLLSCHLAGER, *Z. Anorg. Allg. Chem.* **414**, 76 (1975).
18. H. R. MÜLLER-BUSCHBAUM, *Angew. Chem. (English transl.)* **16**, 674 (1977).
19. K. KNOX, *J. Chem. Phys.* **30**, 991 (1959).

20. K. K. SINGH, P. GANGULY, AND C. N. R. RAO, *Mater. Res. Bull.* **17**, 493 (1982).

21. V. R. HAEGELE AND D. BABEL, *Z. Anorg. Allg. Chem.* **409**, 11 (1974).

22. D. REINEN AND C. FRIEBEL, *Struct. Bonding* **37**, 1 (1979).

23. J. P. OUDALOV, A. DAOUDI, J. C. JOUBERT, G. LE FLEM, AND P. HAGENMULLER, *Bull. Soc. Chim. Fr.* 3408 (1970).

24. L. FOURNES, N. KINOMURA, AND F. MENIL, *C.R. Acad. Sci. (Paris) C* **291**, 235 (1980).

25. G. DEMAZEAU, M. POUCHARD, M. THOMAS, J. F. COLOMBET, J. C. GRENIER, L. FOURNES, J. L. SOUBEYROUX, AND P. HAGENMULLER, *Mater. Res. Bull.* **15**, 451 (1980).

26. G. DEMAZEAU, P. COURBIN, G. LE FLEM, M. POUCHARD, P. HAGENMULLER, J. L. SOUBEYROUX, I. G. MAIN, AND G. A. ROBINS, *Nouv. J. Chim.* **3**, 171 (1979).

27. G. A. ROBINS, M. F. THOMAS, J. D. RUSH, G. DEMAZEAU, AND I. G. MAIN, *J. Phys. C* **15**, 233 (1982).

28. G. DEMAZEAU, M. POUCHARD, N. CHEVREAU, M. THOMAS, F. MENIL, AND P. HAGENMULLER, *Mater. Res. Bull.* **16**, 689 (1981).

29. G. DEMAZEAU, J. L. MARTY, M. POUCHARD, T. ROJO, J. M. DANCE, AND P. HAGENMULLER, *Mater. Res. Bull.* **16**, 47 (1981).

30. H. P. ROOKSBY, *Acta Crystallogr.* **1**, 266 (1948).

31. M. M. J. FAVA, Y. OUDALOV, J. M. REAU, G. LE FLEM, AND P. HAGENMULLER, *C.R. Acad. Sci. (Paris) C* **274**, 183 (1972).

32. R. D. SHANNON AND C. T. PREWITT, *Acta Crystallogr. Sect. B* **25**, 925 (1969).

33. G. DEMAZEAU, M. POUCHARD, AND P. HAGENMULLER, *J. Solid State Chem.* **18**, 159 (1976).

34. J. C. JOUBERT, A. COLLOMB, D. ELMALEH, G. LE FLEM, A. DAOUDI, AND G. OLLIVER, *J. Solid State Chem.* **2**, 343 (1970).

35. J. B. GOODENOUGH AND J. M. LONGO, "Landolt-Bornstein" (K. H. Hellwege, Ed.), Group III/4a, p. 126, Springer-Verlag, Berlin (1970).

36. S. NOMURA, "Landolt-Bornstein" (K. H. Hellwege and A. M. Hellwege, Eds.), Group III/12a, p. 425, Springer-Verlag, Berlin (1978).

37. D. S. MCCLURE, *J. Phys. Chem. Solids* **3**, 311 (1957).

38. J. A. DUFFY, *Struct. Bonding* **32**, 147 (1977).

39. A. BENABAD, A. DAOUDI, R. SALMON, AND G. LE FLEM, *J. Solid State Chem.* **22**, 121 (1977).

40. J. B. GOODENOUGH, G. DEMAZEAU, M. POUCHARD, AND P. HAGENMULLER, *J. Solid State Chem.* **8**, 325 (1973).

41. M. TAKANO, N. NAKANISHI, Y. TAKADA, S. NAKA, AND T. TAKADA, *Mater. Res. Bull.* **12**, 923 (1977).

42. S. N. RUDDLESDON AND P. POPPER, *Acta Crystallogr.* **10**, 538 (1957); **11**, 54 (1958).

43. J. A. BLAND, *Acta Crystallogr.* **14**, 875 (1961).

44. J. DRENNAN, C. P. TAVARES, AND B. C. H. STEELE, *Mater. Res. Bull.* **17**, 621 (1982).

45. R. J. D. TILLEY, *J. Solid State Chem.* **21**, 293 (1977).

46. R. PLUMIER, *J. Appl. Phys.* **35**, 950 (1964).

47. G. BLASSE, *J. Inorg. Nucl. Chem.* **27**, 2683 (1965).

48. J. F. ACKERMANN, *Mater. Res. Bull.* **14**, 487 (1979).

49. J. S. GRIFFITHS, *Mol. Phys.* **8**, 213, 217 (1964).

50. M. S. HADDED, M. W. LYNCH, W. D. FEDERER, AND D. N. HENDRICKSON, *Inorg. Chem.* **20**, 123, 131 (1981).

51. J. T. CASTNER, G. S. NEWELL, W. C. HOLTON, AND C. P. SLICHTER, *J. Chem. Phys.* **32**, 668 (1960).

52. R. MOHAN RAM, K. K. SINGH, W. H. MADHUSUDHAN, P. GANGULY, AND C. N. R. RAO, *Mater. Res. Bull.* **18**, 703 (1983).

53. G. BLASSE, *J. Inorg. Nucl. Chem.* **30**, 656 (1968).

54. J. L. SOUBEYROUX, P. COURBIN, L. FORUNES, D. FRUCHART, AND G. LE FLEM, *J. Solid State Chem.* **31**, 313 (1980).

55. J. T. LEWANDOWSKI, J. M. LONGO, AND R. A. MCCAULEY, *Amer. Ceram. Soc. Bull.* **61**, 333 (1982).

56. K. POEPPELMEIER, M. E. LEONWICZ, AND J. M. LONGO, *J. Solid State Chem.* **44**, 89 (1982).

57. K. VIDYASAGAR, J. GOPALAKRISHNAN, AND C. N. R. RAO, *Inorg. Chem.*, in press.

58. W. H. MADHUSUDHAN, K. JAGANNATHAN, P. GANGULY, AND C. N. R. RAO, *J. Chem. Soc. Dalton Trans.* 1397 (1980).

59. S. RAMASESHA, T. V. RAMAKRISHNAN, AND C. N. R. RAO, *J. Phys. C* **12**, 1307 (1979).

60. G. BLASSE, *J. Appl. Phys.* **36**, 879 (1965).

61. G. DEMAZEAU, J. L. MARTY, B. BUFFAT, J. M. DANCE, N. POUCHARD, P. DORDOR, AND B. CHEVALIER, *Mater. Res. Bull.* **17**, 37 (1982).

62. D. E. COX, G. SHIRANE, R. J. BIRGENEAU, AND J. B. MACCHESNEY, *Phys. Rev.* **188**, 930 (1969).

63. J. C. BOULOUX, J. L. SOUBEYROUX, M. PERRIN, AND G. LE FLEM, *J. Solid State Chem.* **38**, 34 (1981).

64. G. LE FLEM, G. DEMAZEAU, AND P. HAGENMULLER, *J. Solid State Chem.* **44**, 82 (1982).

65. G. OLLIVER, These de Doctorates Sciences Physiques, Univ. Scientifique et Medicale, Grenoble (1973).

66. J. B. GOODENOUGH, *Progr. Solid State Chem.* **5**, 145 (1972); "Solid State Chemistry," (C. N. R. Rao, Ed.), Dekker, New York (1973).

67. M. SHIMADA AND M. KOIZUMI, *Mater. Res. Bull.* **11**, 1237 (1976).

68. B. K. Shante and S. Kirkpatrick, *Advan. Phys.* **20**, 325 (1971).

69. W. M. Walsh, R. J. Birgeneau, L. W. Rupp, and H. J. Guggenheim, *Phys. Rev. B* **20**, 4645 (1979).

70. G. A. Smolenskii, V. N. Yudin, and E. Sher, *Sov. Phys. Solid State* **4**, 2452 (1962).

71. G. A. Smolenskii, V. A. Bokov, S. A. Kizaev, E. I, Mal'teev, G. M. Nedhir, V. P. P. Plaknty, A. G. Tutov, and V. N. Yudin, "Proceedings International Conference on Magnetism, Nottingham, 1964," p. 354, Inst. Phys. and Phys. Soc. London (1965).

72. P. Ganaguly, S. Kollali, C. N. R. Rao, and S. Kern, *Magn. Lett.* **1**, 107 (1980).

73. R. Saez-Puche, M. Norton, and W. S. Glaunsinger, *Mater. Res. Bull.* **17**, 1523 (1982).

74. R. Saez-Puche, M. Norton, and W. S. Glaunsinger, *Mater. Res. Bull.* **17**, 1539 (1982).

75. P. Ganguly and S. Ramasesha, *Magn. Lett.* **1**, 131 (1980).

76. K. G. Srivatsava, *Phys. Lett.* **4**, 55 (1963).

77. J. C. Bouloux, J. L. Soubeyroux, A. Daoudi, and G. Le Flem, *Mater. Res. Bull.* **16**, 855 (1981).

78. P. W. Anderson and H. Hasegawa, *Phys. Rev.* **100**, 675 (1955).

79. J. B. MacChesney, J. F. Potter, and R. C. Sherwood, *J. Appl. Phys.* **40**, 1243 (1969).

80. P. Day, *Acc. Chem. Res.* **12**, 236 (1973), and references therein.

81. J. B. Goodenough, "Magnetism and the Chemical Bond," Wiley–Interscience, New York (1963).

82. J. B. Goodenough, A. Wold, R. J. Arnott, and N. Menyuk, *Phys. Rev.* **124**, 373 (1961).

83. N. Y. Vasanthacharya, K. K. Singh, and P. Ganguly, *Rev. Chim. Miner.* **18**, 333 (1981), and references therein.

84. J. B. Goodenough and S. Ramasesha, *Mater. Res. Bull.* **17**, 383 (1982).

85. E. Abraham, P. W. Anderson, D. C. Liccardello, and T. V. Ramakrishnan, *Phys. Rev. Lett.* **42**, 673 (1979).

86. E. Abraham and T. V. Ramakrishnan, *J. Non-Cryst. Solids* **35**, 15 (1980); *Philos. Mag. [Part] B* **42**, 827 (1980).

87. R. E. Newnham and L. E. Cross, *in* "Preparation and Characterization of Materials" (J. M. Honig and C. N. R. Rao, Eds.), Academic Press, New York (1981).

The Blackett Memorial Lecture, 1991

Chemical insights into high-temperature superconductors

BY C. N. R. RAO

Solid State and Structural Chemistry Unit and Jawaharlal Nehru Centre for Advanced Scientific Research, Indian Institute of Science, Bangalore 560 012, India

Contents

The high-temperature superconductors are complex oxides, generally containing two-dimensional CuO_2 sheets. Various families of the cuprate superconductors are described, paying special attention to aspects related to oxygen stoichiometry, phase stability, synthesis and chemical manipulation of charge carriers. Other aspects discussed are chemical applications of cuprates, possibly as gas sensors and copper-free oxide superconductors. All but the substituted Nd and Pr cuprates are hole-superconductors. Several families of cuprates show a nearly constant n_h at maximum T_c. Besides this universality, the cuprates exhibit a number of striking common features. Based on Cu(2p) photoemission studies, it is found that the Cu–O charge-transfer energy, Δ, and the Cu(3d)–O(2p) hybridization strength, t_{pd}, are key factors in the superconductivity of cuprates. The relative intensity of the satellite in the Cu(2p) core-level spectra, the polarizability of the CuO_2 sheets as well as the hole concentration are related to Δ/t_{pd}. These chemical bonding factors have to be explicitly taken into account in any model for superconductivity of the cuprates.

1. Introduction

I am delighted that I have been asked to deliver the Blackett Memorial Lecture this year. Professor Blackett was a man of many parts and was keenly interested in India,

Phil. Trans. R. Soc. Lond. A (1991) **336**, 595–624

Printed in Great Britain

21-2

596 *C. N. R. Rao*

especially with regard to the policy for science and development. I have been myself involved in planning science and technology for development in the past few years and have greatly appreciated the concerns of Professor Blackett. Although I am not certain whether Professor Blackett would have predicted the science and technology scenario prevalent today, I feel that he would have been amused by a chemist from India talking about recent developments in warm superconductors. I have chosen this topic not only to demonstrate how chemists have much to do in this frontline area of condensed matter science, but also to show how certain classes of transition metal oxides – on which I have been working for many years (Rao & Subbarao 1970; Rao 1989) in a corner of the globe – have become so prominent because of high-temperature superconductivity. This area is no longer the one I pursued years ago out of curiosity, but one which has become so competitive that it is foolhardy to try to read all the literature. It has been really an exciting experience to witness this area develop explosively in so short a time.

Metal oxides themselves are not new to superconductivity. Superconducting transition temperatures of around 13 K were obtained some years ago in $Li_{1+x}Ti_{2-x}O_4$ (Johnston *et al.* 1973) and $BaBi_{1-x}PbO_3$ (Sleight *et al.* 1975). The discovery of superconductivity in the La–Ba–Cu–O system (Bednordz & Müller 1986) pushed the upper limit of the transition temperature from a stagnant value of 23 K to around 30 K and initiated an unprecedented pace of search for high-temperature superconductivity. The superconducting phase in the La–Ba–Cu–O system had the quasi two-dimensional K_2NiF_4 structure containing a perovskite layer, the parent compound being La_2CuO_4. I may recall here that my own interest in the structure and properties of this class of transition metal oxides goes back to several years (Ganguly & Rao 1973, 1984). When superconductivity above the liquid nitrogen temperature was reported in the Y–Ba–Cu–O system in early 1987 (Wu *et al.* 1987), we independently identified the phase responsible for superconductivity to be $YBa_2Cu_3O_7$ with a defect perovskite structure (Rao *et al.* 1987). We have since worked on several other families of superconducting cuprates, all containing perovskite layers, the highest T_c till to date being 125 K. High-temperature superconductivity is not restricted to cuprates alone, but what is interesting is that all the high T_c materials discovered up to now are metal oxides. It is not entirely unlikely that the ability of the metal–oxygen bond to traverse the entire range from the extreme ionic limit to the highly covalent limit is related to this feature. We cannot forget that transition metal oxides are versatile materials (Rao & Subbarao 1970; Goodenough 1971; Rao 1989) showing metallic behaviour at one end (e.g. ReO_3, $LaNiO_3$) and insulating behaviour at the other (e.g. $BaTiO_3$); then, we have oxides exhibiting metal–insulator transitions (e.g. V_2O_3, $LaNi_{1-x}Mn_xO_3$, $La_{1-x}Sr_xCoO_3$). Oxides can be ferromagnetic (e.g. $La_{0.5}Sr_{0.5}MnO_3$) or antiferromagnetic (e.g. $LaCrO_3$). Metal oxides exhibit interesting electronic properties arising from valence ordering (e.g. Fe_3O_4), charge-density-wave transitions (e.g. Na_xWO_3) and defect ordering (e.g. $Ca_2Fe_2O_5$). Properties of transition metal oxides depend on the dimensionality as well. Thus two-dimensional oxides do not exhibit ferromagnetism or real metallicity (Rao *et al.* 1988). The various cuprate families exhibiting superconductivity possess two-dimensional CuO_2 sheets just like La_2CuO_4.

I present here some of the highlights of the structure–property relations in the various families of high T_c cuprates and illustrate how chemistry plays a major role in the development of such important materials and how solid state chemistry constitutes a fascinating and important branch of chemical science in its own right.

Chemical literature on high-temperature superconductors has become voluminous and I shall therefore cite only some of the very recent references, primarily those based on the work carried out in my laboratory. The earlier results have been adequately covered in many reviews and conference reports (Cava 1990; Goodenough & Manthiram 1990; Joshi *et al.* 1990; Kitazawa & Ishiguro 1989; Nelson *et al.* 1987; Ramakrishnan & Rao 1989; Rao 1988*a*, *b*; Rao & Raveau 1989; Sleight 1988). I shall attempt to point out the important commonalities among the various families of cuprate superconductors and examine such chemical factors as stoichiometry, oxygen disorder, oxidation states, phase stability and chemical manipulation of charge carriers as well as the crucial role of chemical bonding in understanding the phenomenon of high-temperature superconductivity. I shall also briefly touch upon certain synthetic aspects and possible chemical applications of these materials.

2. Cuprate families

The first family of high-temperature oxide superconductors are derived from La_2CuO_4 possessing the K_2NiF_4 structure (figure 1). Stoichiometric La_2CuO_4 is an antiferromagnetic insulator which when doped with holes (by formally creating trivalent Cu species through the substitution of La^{3+} by divalent ions such as Sr^{2+} or by incorporating excess oxygen) becomes superconducting. While La_2CuO_4 is orthorhombic at 300 K and becomes tetragonal at higher temperatures, super-conducting $La_{2-x}M_xCuO_4$ ($x \approx 0.2$ when $M = Sr$) is tetragonal at 300 K and becomes orthorhombic around 180 K, well above the superconducting transition temperature (*ca.* 35 K). Accordingly, the Cu–O–Cu angle in the superconducting oxides of K_2NiF_4 structure are slightly bent (less than 180°) causing a buckling of the Cu–O sheets. In figure 2, the phase diagram of $La_{2-x}Sr_xCuO_4$ is shown to indicate the narrow range of the antiferromagnetic phase and the maximum in T_c at a specific value of x where the hole concentration is also a maximum (Torrance *et al.* 1988). Some doubt has been raised as to whether the maximum T_c at $x \approx 0.2$ exhibited by $La_{2-x}Sr_xCuO_4$ is due to the presence of inhomogeneities in the compositions other than the one showing maximum T_c. This seems unlikely; as we will show later, the T_c maximum occurs at an optimal hole concentration in all the families of cuprate superconductors (Rao *et al.* 1991*a*). In oxygen-excess $La_2CuO_{4+\delta}$, however, separation into superconducting and antiferromagnetic phases seems to occur (Chaillout *et al.* 1990; Jorgensen *et al.* 1988).

The La^{3+} ion in $La_{2-x}M_xCuO_4$ can be substituted to some extent by Pr^{3+} and other rare earth ions without losing superconductivity. It is to be noted that Pr_2CuO_4 and Nd_2CuO_4 themselves possess the so-called T′-structure with square-planar CuO_4 units unlike the T-structure of La_2CuO_4. The T and T′ cuprates form solid solutions over a reasonable range of compositions (Goodenough & Manthiram 1990; Singh *et al.* 1982).

The next homologue of the $La_{2-x}Sr_xCuO_4$ family containing two CuO_2 sheets had earlier been reported to be an insulator. However, recently Cava *et al.* (1990*a*) have synthesized $(La,Sr)_2CaCu_2O_6$ under a high oxygen pressure and found it to be superconducting with a T_c of 60 K. This removes the discomfort one had related to the absence of superconductivity in the two-layer cuprate of this family and also underscores the importance of oxygen stoichiometry. Vijayaraghavan has recently synthesized analogous double-layer compounds of the general formula $LaSrLnCu_2O_6$ (Ln = Nd, Gd or Y) in my laboratory.

Phil. Trans. R. Soc. Lond. A (1991)

C. N. R. Rao

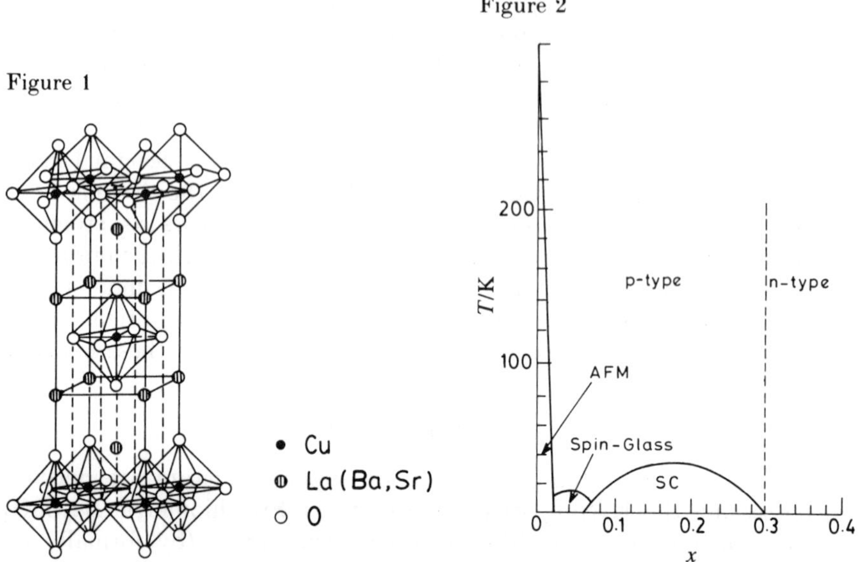

Figure 1. Structure of $La_{2-x}Sr_xCuO_4$.

Figure 2. Phase diagram of hole-doped $La_{2-x}Sr_xCuO_4$.

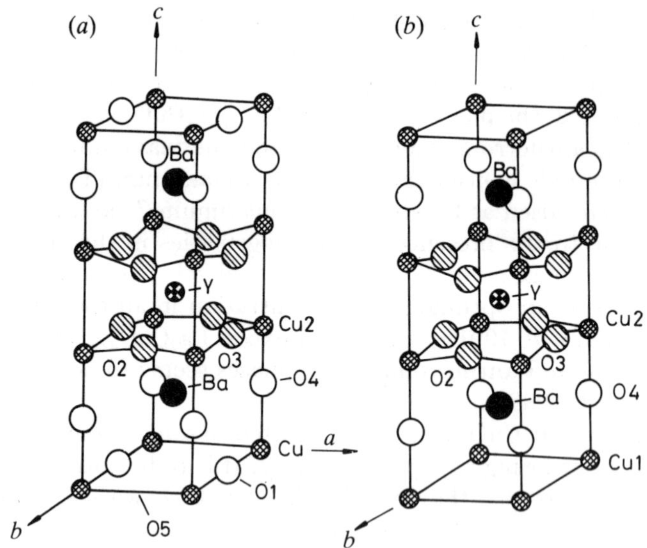

Figure 3. Structures of (a) $YBa_2Cu_3O_7$ and (b) $YBa_2Cu_3O_6$.

$YBa_2Cu_3O_{7-\delta}$ and the other 123 cuprates of the general formula $LnBa_2Cu_3O_7$ (Ln is a rare earth other than Ce, Pr and Tb) show superconductivity with a T_c of *ca.* 90 K in the near stoichiometric compositions (not more than 0.2). These compounds have an orthorhombic structure. The structure as well as the superconductivity are sensitive to oxygen stoichiometry. Stoichiometric 123 cuprates contain Cu–O chains along the *b*-axis in addition to the CuO_2 sheets (figure 3). When the Cu–O chains are fully depleted of oxygen, we get the non-superconducting, tetragonal $YBa_2Cu_3O_6$.

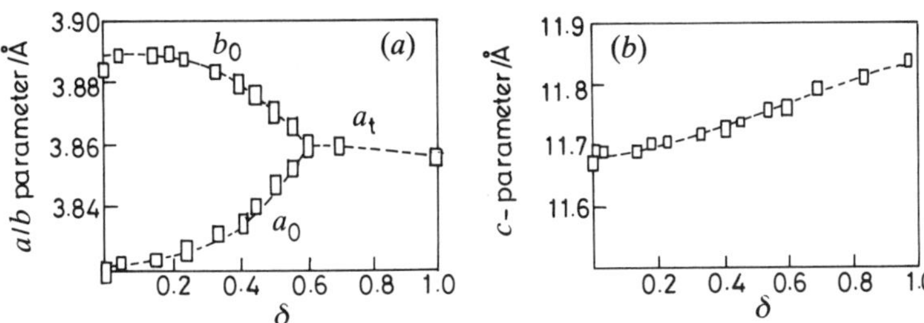

Figure 4. Variation of the lattice parameters of $YBa_2Cu_3O_{7-\delta}$ with δ.

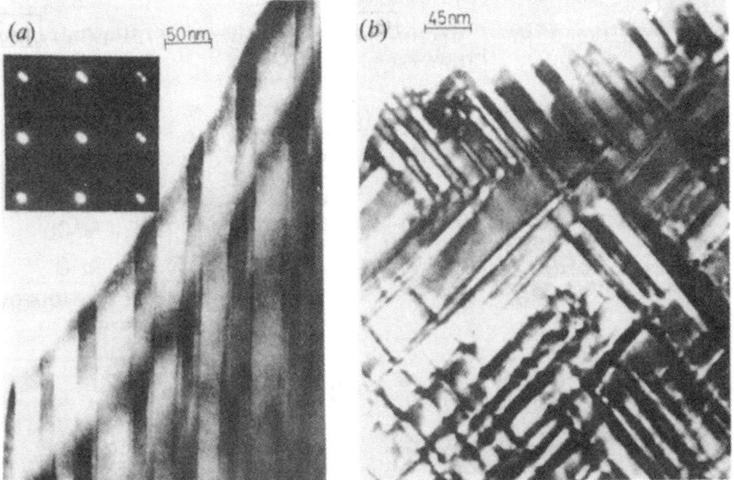

Figure 5. (a) Bright field electron microscopic images of twins in $YBa_2Cu_3O_7$. The splitting of the {110} Brugg spots is clearly seen in SAED pattern given in the inset. (b) 90° twins in $YBa_2Cu_3O_7$.

The variation of the unit cell parameters of $YBa_2Cu_3O_{7-\delta}$ with the oxygen stoichiometry is shown in figure 4 to illustrate how the orthorhombic structure gives way to the tetragonal structure around $\delta \approx 0.6$. All the $LnBa_2Cu_3O_7$ compounds undergo the orthorhombic tetragonal transition, the temperature of the transition depending on the Ln ion (Y = 970 K, La = 590 K). The orthorhombic phases show extensive twinning with a rotation of the a and b axes across the twin boundary. In figure 5 we show typical twins recorded by us in early March 1987. The twinning has no direct bearing on superconductivity; accordingly non-superconducting $PrBa_2Cu_3O_7$ shows twinning because of the orthorhombic structure.

The variation of T_c with δ in $YBa_2Cu_3O_{7-\delta}$ is shown in figure 6. We see that T_c is nearly constant (*ca.* 90 K) upto $\delta = 0.2$ and then drops sharply showing a sort of a plateau around 60 K for $\delta = 0.3$–0.4; the T_c value reaches 45 K when $\delta = 0.5$. The formal valence of Cu in the sheets also shows a plateau in the 60 K region as shown in figure 6 (Cava *et al.* 1990 *b*).

$YBa_2Cu_3O_{6.5}$ ($\delta = 0.5$) is a oxygen vacancy-ordered structure with fully oxidized Cu–O chains (O_7 units) alternating empty chain (O_6 units) as shown in figure 7. In $YBa_2Cu_3O_7$ and related 123 compounds, the orthorhombic c parameter is exactly equal to $3b$ (figure 8). $YBa_2Cu_3O_{6.75}$ also seems to have a vacancy-ordered structure (figure 7), but compositions with δ between 0.3 and 0.4 ($T_c \approx 60$ K) show no such

600   *C. N. R. Rao*

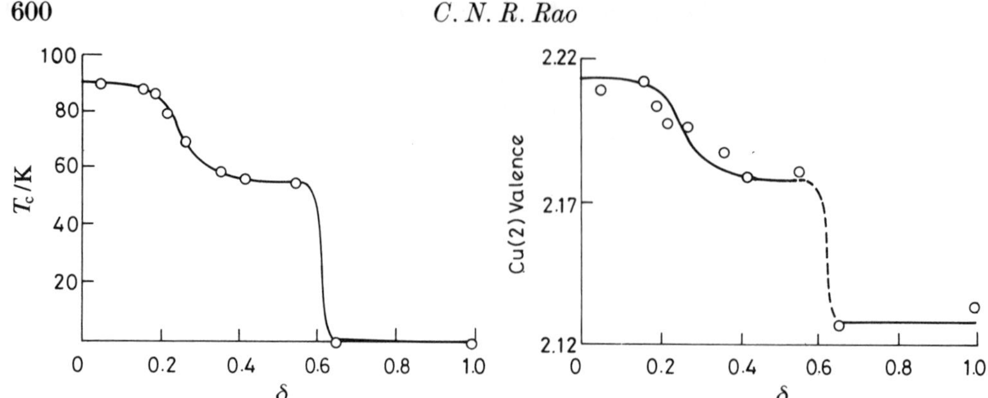

Figure 6. Variation of the T_c and in-plane Cu(2) valence of $YBa_2Cu_3O_{7-\delta}$ with δ (From Cava *et al.* 1990*b*).

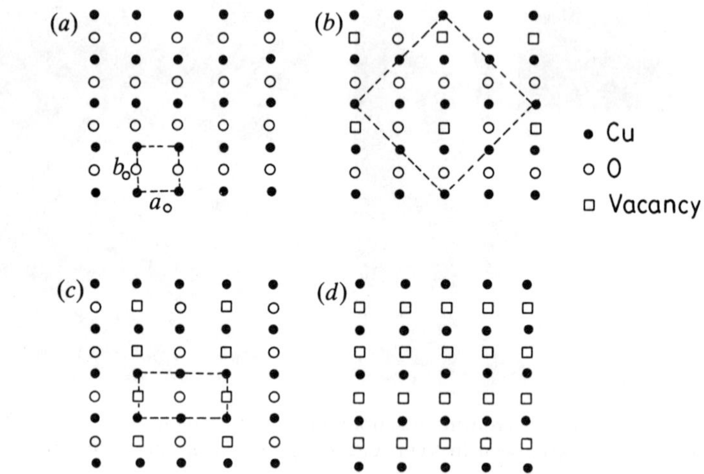

Figure 7. Structure of the basal (*ab*) plane of $YBa_2Cu_3O_{7-\delta}$: (*a*) $\delta = 0$, (*b*) $\delta = 0.25$, (*c*) $\delta = 0.5$, (*d*) $\delta = 1.0$. (From Rao *et al.* 1990*d*.)

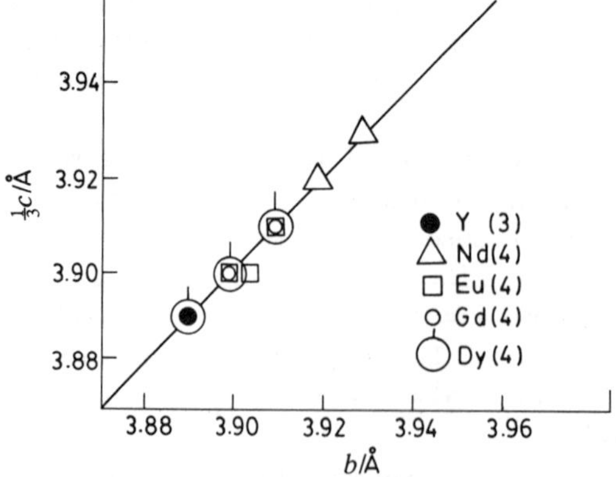

Figure 8. Relation between the c and b parameters in orthorhombic $LnBa_2Cu_3O_{7-\delta}$ (Ln = rare earth or Y, $\delta < 0.2$). (From Rao *et al.* 1990*d*.)

ordering. It has indeed been shown that this composition range is metastable, disproportionating to orthorhombic and tetragonal phases on annealing at low temperatures of the order of 470 K (Rao *et al.* 1990*a*). Thus, on annealing $YBa_2Cu_3O_{6.7}$ at 470 K for a few hours, the X-ray diffraction pattern changes drastically while the electron diffraction pattern shows considerable streaking. Every oxygen added in this δ range oxidizes Cu^+ in the chains without adding extra holes to the CuO_2 sheets, thereby giving rise to the T_c (as well as the Cu valence) plateau around 60 K (figure 6). It is noteworthy that tetragonal $YBa_2Cu_3O_{6.8}$ even though oxygen-rich and contains a fair proportion of Cu^{III} species, is not superconducting since the different Cu layers get connected three dimensionally through the presence of orthorhombic microdomains (Caignaert *et al.* 1990). In $YBa_2Cu_3O_{6.4}$, local oxygen ordering seems to increase the T_c from 0 to 20 K (Jorgensen *et al.* 1990).

The chemistry of $YBa_2Cu_3O_{7-\delta}$ is best understood starting with $YBa_2Cu_3O_6$ ($\delta = 1.0$) which is an antiferromagnetic insulator containing Cu^I ions in the chain. As oxygen diffuses into tetragonal $YBa_2Cu_3O_6$, the linear $O-Cu^I-O$ units are progressively converted into square-planar CuO_4 that share corners to form chains (figure 3); when *ca.* 40% of the Cu–O chains are thus oxidized, we get the orthorhombic structure. Fully oxidized $YBa_2Cu_3O_7$ itself can be formally considered to be $YBa_2Cu_2^{II}Cu^{III}O_7$. Substitution chemistry of $YBa_2Cu_3O_{7-\delta}$ has been examined by several workers who have partly substituted Cu by Fe, Ni or Zn; such substitutions are generally unfavourable to superconductivity just as in the case of $La_{2-x}Sr_xCuO_4$. By substitution of Ba by La or Y by Ca in $YBa_2Cu_3O_7$, one changes the oxygen stoichiometry and also brings about a decrease or an increase in the carrier concentration (Manthiram & Goodenough 1989; Tokura *et al.* 1988). We should note here that only the holes in the CuO_2 sheets are pertinent to superconductivity. It is therefore necessary to subtract out the holes in the chains from the total hole concentration as given by n_h (plane) $= \frac{3}{2}$ (n_h total $- n_h$ chain) (Shafer & Penney 1990).

As mentioned earlier, $PrBa_2Cu_3O_7$ is not superconducting (Ganguli *et al.* 1989*c*). This was first considered to be due to the presence of a small proportion of the Pr ions in the 4+ state. It now seems that this is not the case. It has been suggested that the presence of Pr (4f) levels in the vicinity of the Cu–O band could be responsible for the absence of superconductivity (Sarma *et al.* 1991). The recent observation (Norton *et al.* 1991) of superconductivity ($T_c \approx 43$ K) in films of $Pr_{0.5}Ca_{0.5}Ba_2Cu_3O_{7-\delta}$, however, suggests that absence of superconductivity in Pr cuprate $PrBa_2Cu_3O_7$ may be related to the small hole concentration. This aspect needs to be examined further.

Basically, $YBa_2Cu_3O_7$ is not thermodynamically stable. The more stable $YBa_2Cu_4O_8$ containing two Cu–O chains (figure 9) formed by CuO_4 units sharing edges has been prepared. This cuprate has a T_c of 80 K. Other rare earth analogues of the general formula $LnBa_2Cu_4O_8$ are also known. Although the so-called 124 compounds were first prepared under high oxygen pressures (Karpinski *et al.* 1988), they can be prepared under ambient conditions (Cava *et al.* 1989; Liu *et al.* 1990; Rao *et al.* 1990*d*). Superconducting 247 compounds of the general formula $Ln_2Ba_4Cu_7O_{15}$ with a T_c of *ca.* 90 K have been characterized. The structure of these compounds may be considered to be composed of an 1:1 ordered intergrowth of $LnBa_2Cu_3O_7$ and $LnBa_2Cu_4O_8$ (figure 9). Recall $LnBa_2Cu_3O_{6.5}$ is an 1:1 intergrowth of $LnBa_2Cu_3O_7$ and $LnBa_2Cu_3O_6$. The 247 as well as the 124 cuprates on thermal decomposition give 123 throwing out excess CuO. Effects of temperature and pressure on the structures of the 124 and 247 cuprates have been examined in detail

C. N. R. Rao

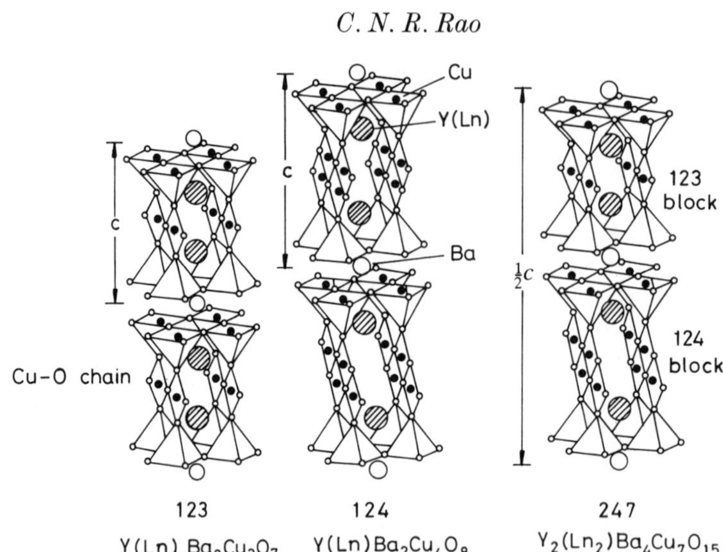

Figure 9. Comparison of the structures of 123, 124 and 247 cuprates. Note the presence of two chains in 124.

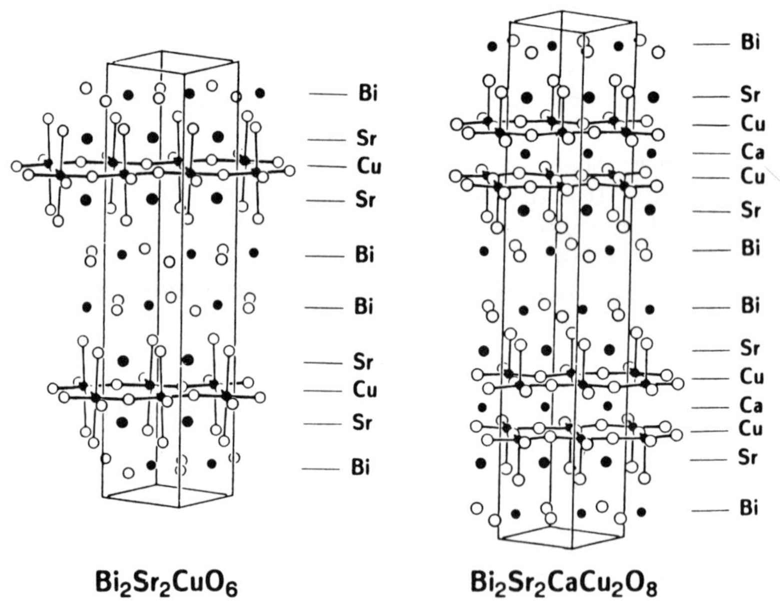

Figure 10. Structures of Bi cuprates. (From Torardi *et al.* 1989.)

by Hewat *et al.* (1990). Properties of the 124 cuprate have been reviewed by Miyatake *et al.* (1990).

Bismuth cuprates of the general formula $Bi_2(Ca,Sr)_{n+1}Cu_nO_{2n+4}$ possessing an orthorhombic structure and containing two rock-salt type layers of BiO constitute an important family of superconductors, with the $n = 2$ and the $n = 3$ members showing T_cs of 90 K and 110 K respectively (figure 10). The $n = 1$ member of the formula $Bi_{2+x}Sr_{2-x}CuO_6$ (without Ca) shows a maximum T_c of around 20 K. The $n = 1$ member containing calcium has been reported, but it does not appear to be

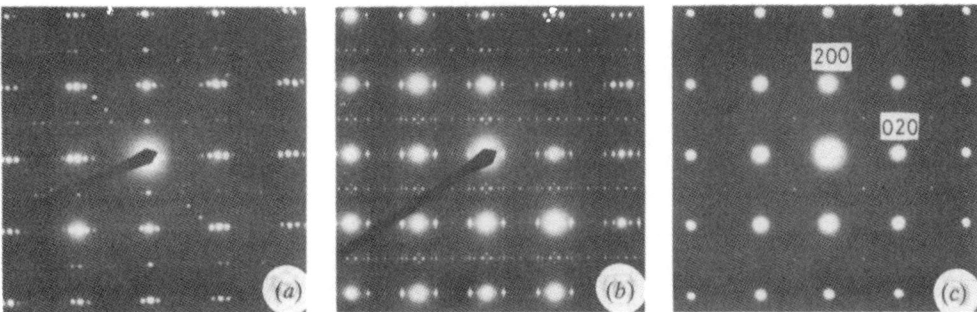

Figure 11. Electron diffraction patterns showing (a) superlattice modulation in superconducting $Bi_2CaSr_2Cu_2O_8$ recorded along the (001) direction, (b) superlattice modulation in non-superconducting $Bi_2YSr_2Cu_2O_8$, and (c) absence of superlattice modulation in superconducting $BiPbSr_2Y_{0.5}Ca_{0.5}Cu_2O_8$.

superconducting. One of the difficulties with the bismuth cuprates is that it is difficult to make them absolutely monophasic because of the presence of disordered intergrowths of different sequences. A unique feature of the bismuth cuprate superconductors is that they exhibit $4b$-type superlattice modulation along the b-direction (figure 11) due to the insertion of extra oxygens in the BiO layers (LePage *et al.* 1989). It was first thought that superconductivity in these compounds had some relation to the superlattice modulation. However, it was possible to prepare compounds of the type $Bi_2Sr_2LnCu_2O_8$ (Ln = Y or rare earth) which are not superconducting but exhibited superlattice modulation (figure 11) of the $4b$ or the $8b$ type (Rao *et al.* 1990b). Similarly, non-superconducting modulation-free oxides of the type $BiPbSr_2MO_y$ (M = Mn, Fe or Co) have been prepared (Tarascon *et al.* 1990). Recently, we have been able to prepare two series of superconductors of the general formula $BiPbSr_{1+x}Ln_{1-x}CuO_6$ (maximum $T_c \approx 80$ K) both of which are modulation-free (Manivannan *et al.* 1991) as can be seen from figure 11. Since Pb in these cuprates is in the $2+$ state, it helps to decrease the oxygen content in the BiO layers, thereby eliminating the modulation. In both these new series of superconductors, the hole concentration, n_h, varies with x and the T_c reaches a maximum value at an optimal value at n_h.

An interesting series of bismuth cuprates in terms of the variation of the T_c as well as the hole concentration with composition is provided by $Bi_2Sr_2Ca_{1-x}Ln_xCu_2O_8$ where Ln = Y or rare earth (Rao *et al.* 1990b). The electrical resistivity data show a metal–insulator transition in the normal state with change in x (figure 12). The T_c as well as the n_h show a maximum at a composition of $x = 0.25$ (figure 13). Note that when Ca is fully substituted by Ln, the material becomes a non-superconducting insulator. Hole concentration in these bismuth cuprates is readily determined by Fe^{II}–Fe^{III} redox titrations.

Thallium cuprates of the general formula $Tl_2Ca_{n-1}Ba_2Cu_nO_{2n+4}$ with two Tl–O layers possessing a tetragonal structure (figure 14) show superconductivity with T_cs of 80, 110, and 125 K when $n = 1$, 2, 3 respectively. The corresponding $TlCa_{n-1}Ba_2Cu_nO_y$ series of cuprates with only a single Tl–O layer are also tetragonal (figure 14) with T_cs of 90 and 115 K respectively for $n = 2$ and 3. Higher members of the thallium cuprate families with $n > 3$ have been prepared, but the maximum T_c is generally found for $n = 3$. The thallium cuprates also exhibit disordered intergrowths of different members in most of the preparations. Oxygen stoichiometry

C. N. R. Rao

Figure 13

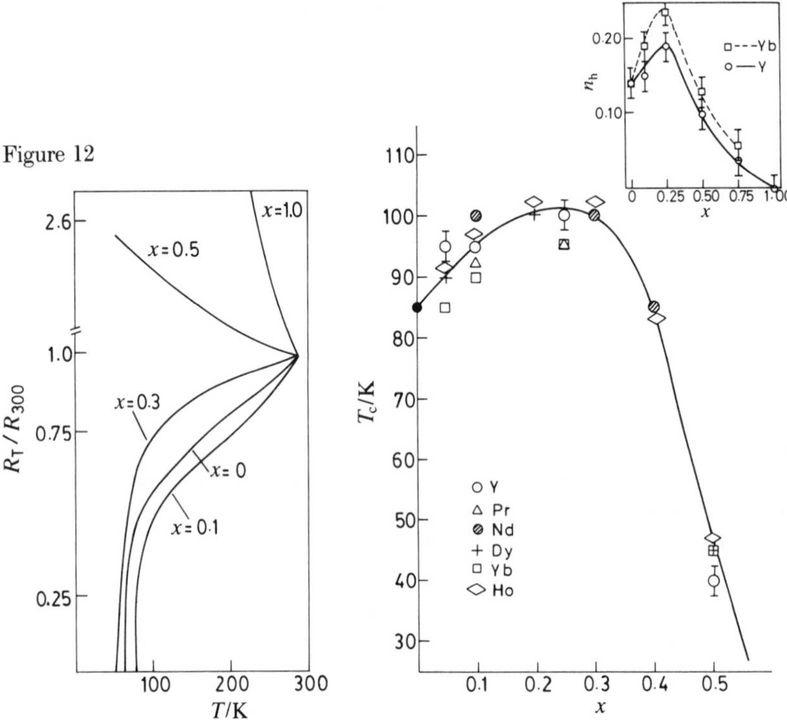

Figure 12. Resistivity data of superconducting $Bi_2Sr_2Ca_{1-x}Nd_xCu_2O_8$ showing the occurrence of a metal–insulator transition in the normal state. (From Rao *et al.* 1990*b*.)

Figure 13. Variation of T_c with composition in $Bi_2Ca_{1-x}Ln_xSr_2Cu_2O_{8+\delta}$. Inset shows variation of hole concentration n_h with DC. (From Rao *et al.* 1990*b*.)

plays an important role in the superconductivity of Tl cuprates as well. Thus hydrogen annealing of some of these materials increases the T_c (Maignan *et al.* 1990). A serious problem with these Tl cuprates is the difficulty in exactly determining the hole concentration. A method has been suggested recently in the literature (Manthiram *et al.* 1990), but it does not appear to be entirely satisfactory. A more reliable method has been developed by Gopalakrishnan and others in this laboratory by making use of a reducing agent (HBr) which selectively reacts with Cu^{III} giving Br_2 while thallium forms $TlBr_3$. The Tl cuprates are somewhat different from the corresponding bismuth cuprates although both Tl and Bi are essentially in the $3+$ state. (Vijayakrishnan *et al.* 1990). The source of holes in the Tl cuprates is yet to be fully understood unlike in the bismuth cuprates (Goodenough & Manthiram 1990). It appears that one has to consider the presence of Tl vacancies as well as the possible overlap of the Tl(6s) band with the conduction band.

Although members of the Tl–Ca–Ba–Cu–O system of superconductors are readily prepared, the corresponding members of Tl–Ca–Sr–Cu–O cannot be prepared in pure form. It is, however, possible to prepare $Tl_{1-x}Pb_x(Ca,Sr)_{n+1}Cu_nO_y$ which are superconducting with the $n = 2$ member showing a T_c of 90 K and the $n = 3$ member a T_c of 120 K (Ganguli *et al.* 1988, 1989*b*; Subramanian *et al.* 1988). In these cuprates, Pb is in the $4+$ state (Kulkarni *et al.* 1989), unlike in the Bi cuprates where it is in the $2+$ state. A novel, analogous series of cuprates with the general formula,

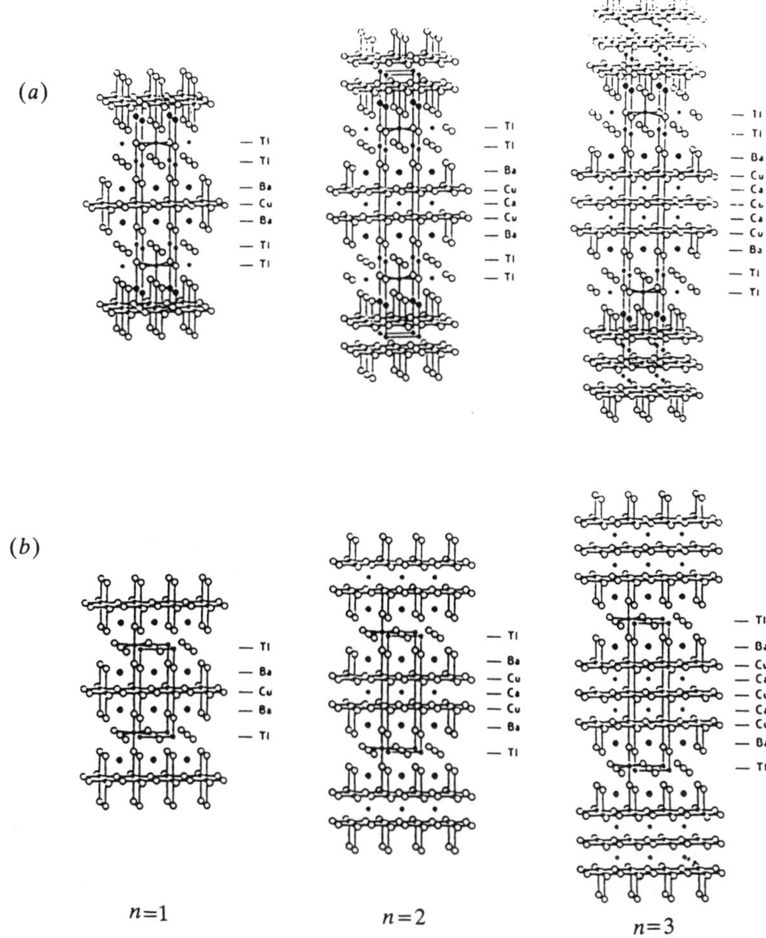

Figure 14. Structures of Tl cuprates of the type $Tl_mCa_{n-1}Ba_2Cu_nO_y$, where (a) $m = 1$ or (b) $m = 2$. (From Sleight *et al.* 1990.)

$TlCa_{1-x}Ln_xSr_2Cu_2O_y$ (Ln = Y or rare earth) has been prepared and characterized recently (Rao *et al.* 1989 a, c). Here, Ca^{2+} is replaced by Ln^{3+} instead of Tl^{3+} by Pb^{4+} as in the earlier series. These cuprates also show a maximum T_c of 90 K. In figure 15 we show the electrical resistivity behaviour of a series of these cuprates; note the occurrence of the metal–insulator transition in the normal state as x is varied. The effect of substitution of Tl by Pb or of Ca by Ln is associated with the variation of the hole concentration. The interplay of chemical substitution with hole concentration can be nicely visualized in the $Tl_{1-y}Pb_yY_{1-x}Ca_xCu_2O_y$ system. This system may be considered as derived from the parent insulator $TlYSr_2Cu_2O_y$ wherein the substitution of Ca in place of Y increases the number of holes, while that of Pb in place of Tl does the opposite. Accordingly, T_c becomes maximum at a higher value of x (Ca concentration) as y (Pb concentration) increases as shown in figure 16. We notice that both T_c and the concentration of doped holes, x, reach a maximum at $y = 0.5$. The T_c goes up to 105 K by the chemical manipulation of hole concentration.

All the cuprate superconductors discussed hitherto contain Ca, Ba or/and rare earth. Recently, superconducting thallium cuprates not containing Ca, Ba or even a

Figure 15

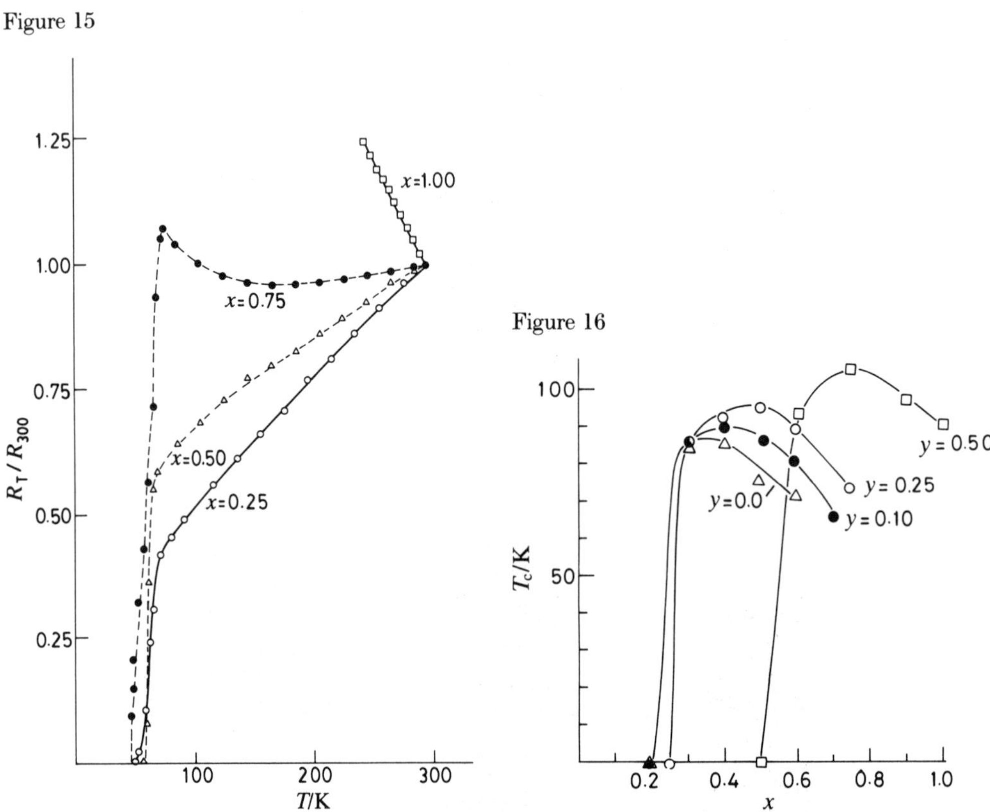

Figure 15. Resistivity data of superconducting $TlCa_{1-x}Nd_xSr_2Cu_2O_y$ showing composition-dependent metal–insulator transition in the normal state. (From Rao *et al.* 1989*c*.)

Figure 16. Variation of T_c with composition in $Tl_{1-y}Pb_yY_{1-x}Ca_xSr_2Cu_2O_y$. (Unpublished results of Vijayaraghavan, Grantscharova and Rao.)

rare earth have been characterized. Thus, in the series, $TlSr_{n+1-x}Ln_xCu_nO_{2n+3+\delta}$, the $n = 1$ and 2 members with T_cs of 40 and 90 K have been prepared (Ganguli *et al.* 1989*a*). Similarly, the $n = 3$ member in the series $(Tl,Pb)Sr_{n+1-x}Ln_xCu_nO_{2n+3}$ with a T_c of 60 K has been prepared though not in pure form (Manivannan *et al.* 1990). The $n = 1$ derivatives of these families may be considered to be derived from $TlSr_2CuO_5$. Substitution of La^{3+} for Sr^{2+} stabilizes the structure and reduces Cu^{III}, permitting superconductivity (Kovatcheva *et al.* 1991).

Lead cuprates of the general formula $Pb_2Sr_2(Ln,Ca)Cu_3O_8$ containing PbO layers and $O–Cu^I–O$ sticks (figure 17) with a T_c of about 60 K have been prepared (Cava *et al.* 1988*b*). In this system the average oxidation state of Cu is less than 2. The normal and superconducting state properties of $Pb_2Sr_2Y_{1-x}Ca_xCu_3O_{8+\delta}$ (note that the $x = 0.0$ composition is an insulator) have been examined in some detail (Koike *et al.* 1990). The superconducting lead cuprates have to be synthesized in an atmosphere deficient in oxygen (N_2 with 1 % O_2) to prevent oxidation of Cu and Pb. Analogous to the single Tl–O layer compounds, lead cuprates of the formula $(Pb,Cu)Sr_2(Y,Ca)Cu_2O_y$ are found to exhibit a T_c of 52 K (Maeda *et al.* 1990). Lower members of the Pb cuprate family of the type $(Pb,Cu)(SrLa)_2CuO_5$ (T_c of 34 K) have also been synthesized. Recently, a lead cuprate of the formula

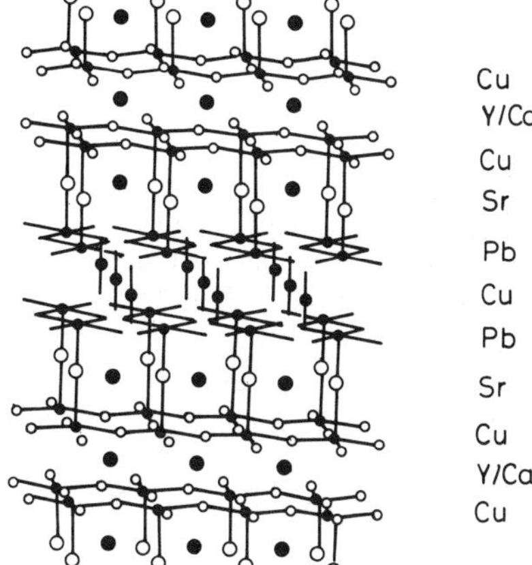

Figure 17. Structure of $PbSr_2(Ca,Y)Cu_3O_8$ (after Sleight 1988).

$(Pb,Cu)(Eu,Ce)_2(Sr,Eu)_2Cu_2O_9$ containing a fluorite layer has been prepared with a T_c of 25 K (Maeda *et al.* 1990).

3. Nature of holes in the cuprate superconductors

All the cuprates described till now are hole superconductors. The nature of holes has been subject of considerable discussion (Chakraverty *et al.* 1988; Rao *et al.* 1989*b*; Sarma & Rao 1989). There has been no experimental evidence for the presence of Cu^{III} type species in the doped cuprates. Instead, there is considerable evidence from electron and X-ray spectroscopies for the presence of hybridized oxygen holes which can be represented as O^-. The detailed description of the holes in terms of the d and p characters has been investigated (Bianconi 1990). Essentially, the mobile holes in the cuprates are present in the in-plane π^* band which has O–2p character. The concentration of holes (in all but the Tl cuprates) are easily determined by iodometry or Fe^{II}–Fe^{III} titrations (Rao *et al.* 1991*a*; Shafer & Penney 1990). Since the Hall coefficients are temperature dependent, the chemical titration method becomes invaluable.

In figure 18, we show the variation of T_c with n_h (obtained by chemical titrations) in a number of cuprate families. We see that in all these families, the T_c goes through a maximum around the same hole concentration in the different series of cuprates containing the same number or CuO_2 sheets. Accordingly, $n_h \approx 0.2$ at the maximum T_c in all the cuprates containing two CuO_2 sheets. This universality is noteworthy. In figure 19 we show the variation of T_c in $Tl_{1-y}Pb_yY_{1-x}Ca_xSr_2Cu_2O_y$ against $(x-y)$ which is a direct measure of the hole concentration. We see that the maximum T_c (105 K) is found for $y = 0.25$ and 0.5 at a $(x-y)$ value of *ca.* 0.22. This value is close to the n_h value at maximum T_c in the other cuprates; the maximum T_c is lower when $(x-y)$ is larger as in the cases $y = 0.0$ and 0.1.

4. Electron-superconducting cuprates

Unlike the various cuprates discussed in §2 where the CuO_2 sheets could be doped with holes, Pr_2CuO_4 or Nd_2CuO_4 possessing the T′ tetragonal structure (figure 20) can be doped with electrons by partly substituting Nd by Ce or Th or oxygen by fluorine

608 *C. N. R. Rao*

Figure 18

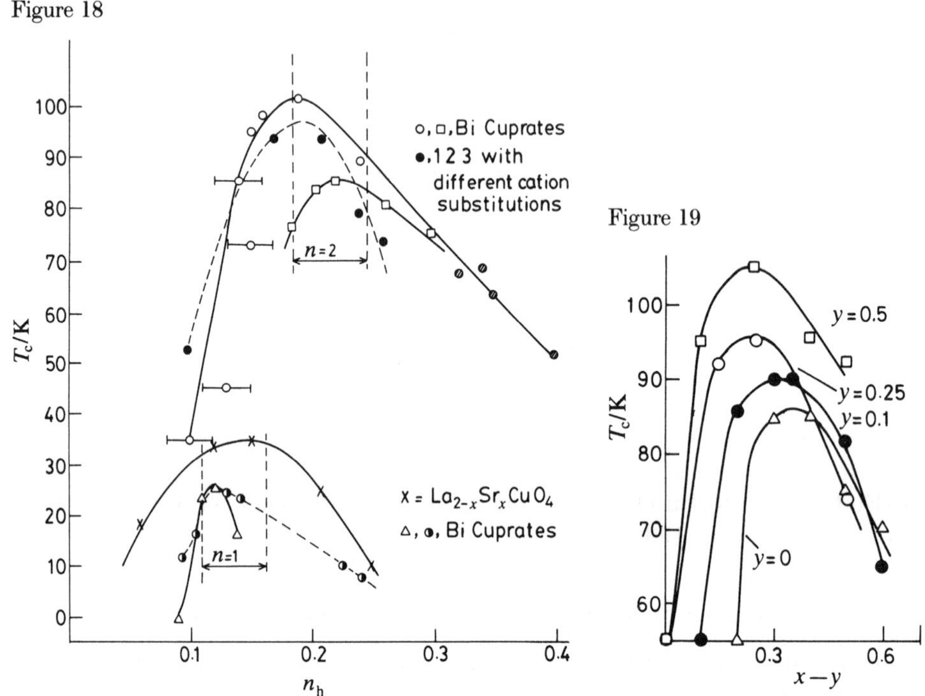

Figure 18. Variation of T_c with hole concentration in cuprates (from Rao *et al.* 1991*a*).

Figure 19. Variation of T_c with effective hole concentration, $(x-y)$, in $Tl_{1-y}Pb_yY_{1-x}Ca_xSr_2CuO_y$.

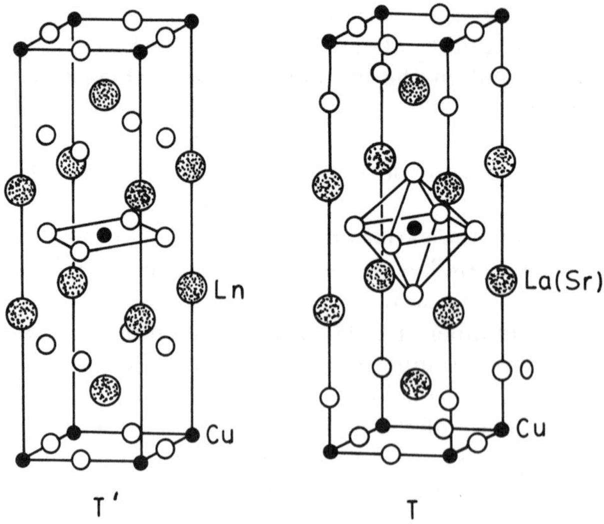

Figure 20. The T′ and T structure of cuprates (T, Ln = Nd, Pr, Ce).

(Maple 1990; Tokura *et al.* 1989*b*). The maximum T_c that these compounds exhibit is around 25 K. Even though these cuprates are formally considered to be electron superconductors, the exact nature of the charge carriers is not fully certain. Some holes could also be present in such materials. In many ways, electron-superconducting cuprates are similar to the La_2CuO_4 family of hole superconductors. In $Nd_{2-x}Ce_xCuO_4$, it appears that the compositions are monophasic only for $x = 0.0$ and for the optimal x value where T_c is maximum (Lightfoot *et al.* 1990). Such inhomogeneity has been found in oxygen-excess Ln_2CuO_4 (Jorgensen *et al.* 1988). We do not have electron-superconducting oxides with T_cs comparable with those of the hole superconductors.

5. Synthetic aspects

In this section, I shall briefly present some preparative aspects of the cuprate superconductors based on the experience gained by us in the past four years. The cuprates are ordinarily made by the traditional ceramic method (mix, grind and heat), which involves thoroughly mixing the various oxides or/and carbonates (or any other salt) in the desired proportion and heating the mixture (preferably in pellet form) at a high temperature. The mixture is ground again after some time and reheated until the desired product is formed as indicated by X-ray diffraction. This method may not always yield the product with the desired structure purity or in oxygen stoichiometry. Variants of this method are often used. For example, decomposing a mixture of nitrates has been found to yield a better product in the case of the 123 compounds by some workers; some others prefer to use BaO_2 in place of $BaCO_3$ for the synthesis.

Coprecipitation and sol-gel methods are conveniently employed for the synthesis of 123 compounds and other cuprates. The sol-gel method provides a homogeneous dispersion of the various component metals when a solution containing the metal ions is transformed into a gel by adding an organic solvent such as a glycol or an alcohol often in the presence of other chemicals such as organic amines. The gel is then decomposed at relatively low temperatures to obtain the desired oxide, generally in fine particulate form. Materials prepared by such low-temperature methods may need to be annealed or heated under suitable conditions to obtain the desired oxygen stoichiometry as well as the characteristic high T_c. 124 cuprates, lead cuprates and bismuth cuprates have all been made by this method; the first two are particularly difficult to make by the ceramic method.

One of the problems with the bismuth cuprates is the difficulty in obtaining phasic purity (minimizing the intergrowth of the different layered phases). The glass or the melt route has been used to obtain better samples. The method involves preparing a glass by quenching the melt; the glass is then crystallized by heating it above the crystallization temperature. Thallium cuprates are best prepared in sealed tubes (gold or silver). Heating Tl_2O_3 with a matrix of the other oxides (already heated to 1100–1200 K) in a sealed tube is preferred by some workers. It is important that thallium cuprates are not prepared in open furnaces since Tl_2O_3 which readily sublimes is highly toxic. To obtain superconducting compositions corresponding to a particular copper content (number of CuO_2 sheets) by the ceramic method, one often has to start with various arbitrary compositions especially in the case of the Tl cuprates. The real composition of a bismuth or a thallium cuprate superconductor is not likely to be anywhere near the starting composition. The actual composition can be determined by analytical electron microscopy and other methods.

Heating oxidic materials under high oxygen pressures or in flowing oxygen often becomes necessary to attain the desired oxygen stoichiometry. Thus, La_2CuO_4 and $La_2Ca_{1-x}Sr_2Cu_2O_6$ heated under high oxygen pressure become superconducting with T_cs of 40 and 60 K respectively. In the case of the 123 compounds, one of the

 C. N. R. Rao

problems is that it loses oxygen easily. It therefore becomes necessary to heat the material in an oxygen atmosphere at an appropriate temperature below the orthorhombic–tetragonal transition temperature. Oxygen stoichiometry is, however, not a problem in the bismuth cuprates. The 124 superconductors were first prepared under high oxygen pressures. It was later found out that heating the oxide or nitrate mixture in the presence of Na_2O_2 in flowing oxygen is sufficient to obtain 124 compounds. Superconducting Pb cuprates, on the other hand, can only be prepared in presence of very little oxygen (N_2 with a small percentage of O_2). In the case of the electron superconductor, $Nd_{2-x}Ce_xCuO_4$, it is necessary to heat the material in an oxygen-deficient atmosphere; otherwise, the electron given by Ce will merely go into giving an oxygen excess material. It may be best to prepare $Nd_{2-x}Ce_xCuO_4$ by a suitable method (say decomposition of mixed oxalates or nitrates) and then reduce it with hydrogen.

6. Commonalities in the cuprates

There are many striking commonalities in the structure and properties of the high T_c cuprates. All the cuprates can be considered to be a result of the intergrowth of defect perovskite layers of $ACuO_{3-x}$ with AO-type rock-salt layers leading to the general formula $[ACuO_{3-x}]_n [AO]_{n'}$ as shown in figure 21. The 123 compounds, however, do not have rock-salt layers and may be considered as the $n' = 0$ member of this general family. The more important common features are the following:

(i) All the cuprates possess CuO_2 layers sandwiched between certain M–O layers (e.g. TlO, BiO) acting as charge reservoirs or spacers. The seat of superconductivity is in the CuO_2 layers. The T_c in Bi and Tl cuprates increases up to $n = 3$ and then decreases, the cuprate with an infinite number of Cu–O layers being an antiferromagnetic insulator (figure 22). Interaction or spacing between the Cu–O layers is crucial. This is demonstrated by recent experiments where the introduction of a fluorite layer, $[Ln_{1-x}Ce_x]_2O_2$, between two CuO_2 sheets in Bi cuprates lowers the T_c markedly (Tokura *et al.* 1989*a*). However, intercalation of iodine between the BiO layers in $Bi_2CaSr_2Cu_2O_8$ (causing a substantial increase in the *c*-parameter) does not affect the T_c (Xiang *et al.* 1990).

(ii) The Cu–O bonds in the cuprates are highly covalent.

(iii) There is an interesting comparison between the Cu–O sheets in the hole and the electron superconductors. Cuprates with the T′-structure where Cu has a square-planar coordination can be doped with electrons while those with the T-structure as in $La_{2-x}Sr_xCuO_4$ can be doped with holes (figure 20). There is an interesting symmetry between these two situations as shown in figure 23. The Cu–O–Cu angle is less than 180° in the hole superconductors while it is close to 180° in the electron superconductors. The position of the apical oxygen in the Cu–O square-pyramids or octahedra in the hole superconductors seems to modulate the width of the conduction band.

(iv) The parent cuprates in all the superconductors are antiferromagnetic insulators. For example, La_2CuO_4, $YBa_2Cu_3O_6$, $Bi_2Sr_2LnCu_2O_8$ and $Pb_2Sr_2LnCu_3O_8$ are the antiferromagnetic insulators corresponding to the superconductors $La_{2-x}Sr_xCuO_4$, $YBa_2Cu_3O_7$, $Bi_2Ca_{1-x}Ln_xSr_2Cu_2O_8$ and $Pb_2Sr_2Ca_{1-x}Ln_xCu_3O_8$ respectively. In the case of the electron superconductor $Nd_{2-x}Ce_xCuO_4$, parent Nd_2CuO_4 is the antiferromagnetic insulator.

(v) All the cuprates nominally contain mixed valent copper which can disproportionate ($Cu^{II} \rightarrow Cu^{III} + Cu^{I}$). In other words, the phenomenon is associated

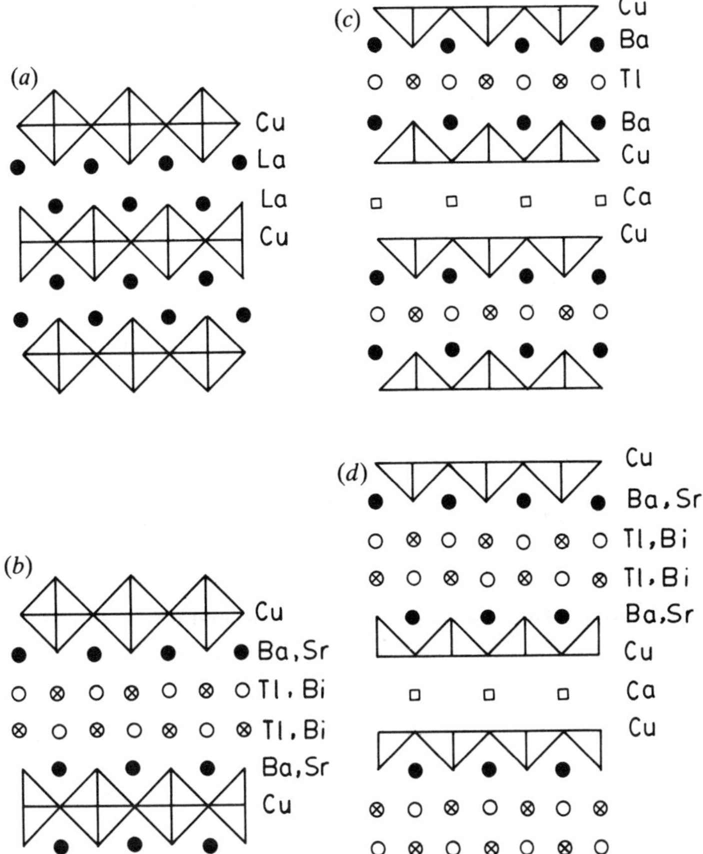

Figure 21. Schematic representation of the structures of (*a*) La_2CuO_4, (*b*) $Bi_2Sr_2CuO_6$ and $Tl_2Ba_2CuO_6$, (*c*) $TlCaBa_2Cu_2O_7$ and (*d*) $Bi_2CaSr_2Cu_2O_8$ and $Tl_2CaBa_2Cu_2O_8$. Oxygens are shown by open circles and Bi and Tl by circles with a cross. (From Rao & Raveau 1989.)

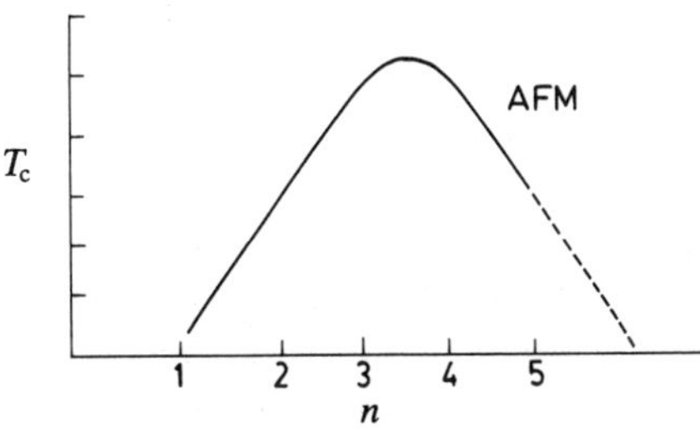

Figure 22. Variation of T_c with the number of CuO_2 sheets, n (schematic).

<u>NOTE</u>: Infinitely layered cuprates are since found to be superconducting

C. N. R. Rao

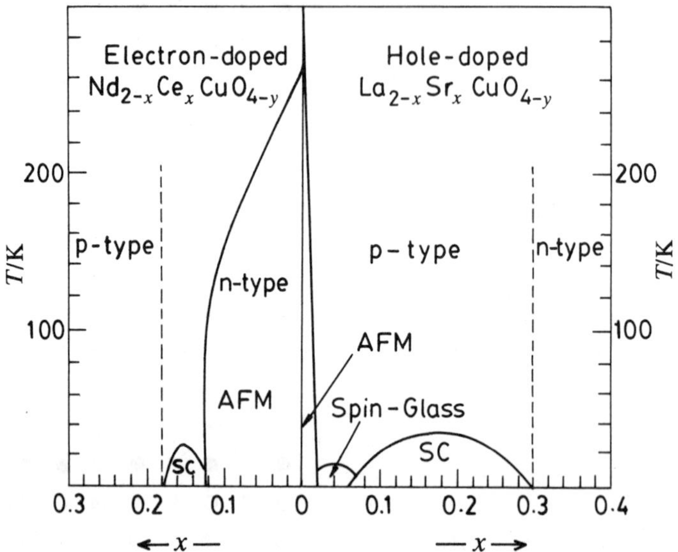

Figure 23. Symmetry in the phase diagrams of electron- and hole-superconductors, $Nd_{2-x}Ce_xCuO_4$ and $La_{2-x}Sr_xCuO_4$ (after Maple 1990.)

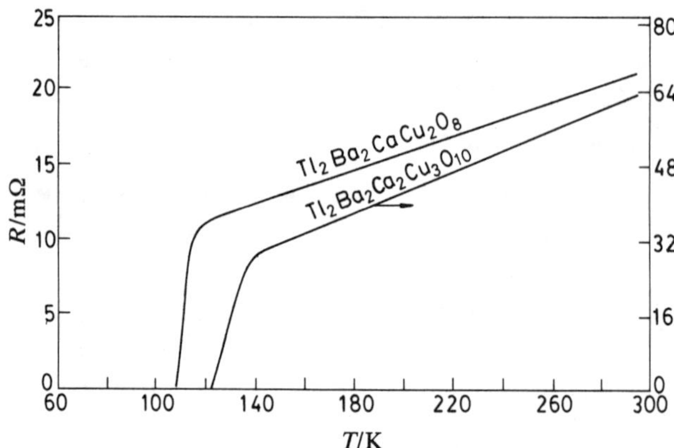

Figure 24. Linearity of the resistivity with temperature in the normal state of Tl cuprates (after Torardi *et al.* 1989).

with a band that gets filled with two electrons with stability associated with empty, half-filled and filled states.

(vi) Local charge distribution provides a basis to understand superconductivity in the cuprates. This is well demonstrated in $Pb_2Sr_2Ln_{1-x}Ca_xCu_3O_8$ where excess oxygen oxidizes Pb^{2+} and Cu^+ without affecting the CuO_2 sheets. In this system, it is necessary to replace the yttrium by calcium in between the CuO_2 sheets to render it superconducting (Cava 1990).

(vii) Oxygen stoichiometry, homogeneity and disorder play an important role in the superconductivity of the cuprates (Hewat *et al.* 1989; Rao *et al.* 1990*d*; Raveau *et al.* 1990) as exemplified in the discussion of the various families.

(viii) All the superconducting cuprates are marginally metallic in the normal state, sitting on a metal–insulator boundary. We would, therefore, expect abnormal properties in the normal state. One of the striking abnormal normal-state properties of these materials is the linearity of resistivity over a wide range of temperatures as shown in figure 24 in the case of thallium cuprates.

7. Relation between the electronic structure and the superconductivity of cuprates

Photoemission spectroscopic studies (Allen & Olson 1990) of the high T_c cuprates show superconducting gap formation and establish that the normal state is strongly correlated with atomic spectral features at high energy. There is strong Cu(3d)–O(2p) hybridization and the E_F has nearly the same value for both hole and electron doping, lying in states filling the gap of the parent antiferromagnetic insulator. The gap filling states near E_F obey the Luttinger counting theorem.

It is important to understand the relation between the electronic structure and superconductivity of the cuprates. The problem, however, is that it is not easy to exactly describe the electronic structure of such complex oxides or to develop meaningful models to describe the properties in the superconducting and normal states. Empirical relations and chemical intuition, however, continue to be useful. Accordingly, it has been shown that the difference in the Madelung site potential between a hole on copper and one on oxygen can control their site preference and the ability to delocalize (Torrance & Metzger 1989). A relation between the oxygen content and the average [Cu–O] charge of Cu valence shows the presence of a boundary between the insulators and superconductors. The [Cu–O] charge in the sheets has been shown by an large to determine T_c (Tokura *et al.* 1988). The formal valence of Cu and O ions in the central CuO_2 planes has been calculated by de Leeuw *et al.* (1990) from experimental bond lengths following Zachariasen rules. The results show that in all the structures, the values of formal valence correlate well with the T_c; the T_c(max) increases when the holes prefer the oxygen sites over the copper sites in the CuO_2 sheets. This implies a higher value for $U - \Delta + \frac{1}{2}W$ where U is the correlation energy, Δ, the charge-transfer energy and W the band width.

In the superconducting cuprates, the correlation energy, U_{dd} within the Cu 3d manifold is considerably larger than the Cu–O charge-transfer energy, Δ, or the Cu(3d)–O(2p) hybridization strength, t_{pd}. Hybridized oxygen holes in the CuO_2 sheets are the charge carriers responsible for the superconductivity. Although many of the models for superconductivity in the cuprates consider the parent compounds to be charge-transfer gap insulators and the t_{pd} to be substantial, they do not explicitly take Δ and t_{pd} as crucial parameters. Some of the phenomenological models consider the electronic polarizability, α, to be important, but do not evaluate or relate α to Δ, t_{pd} or carrier concentration. Although it seems clear that chemical bonding factors such as Δ and t_{pd} have to be given greater attention, there has been hitherto no experimental proof to show that these are indeed as important as they appear to be. Recent Cu core-level photoemission studies of the cuprates supported by theoretical calculations have, however, changed the picture (Rao *et al.* 1990*c*, 1991*b*; Rao & Sarma 1991; Santra *et al.* 1991).

The Cu $2p_{\frac{3}{2}}$ core-level spectra of cuprate superconductors show a main feature, M, around 933 eV due to the well-screened core-hole state of $2p^5 3d^{10}$ configuration and a broad satellite, S, centred around 942 eV due to the poorly screened state of $2p^5 3d^9$

C. N. R. Rao

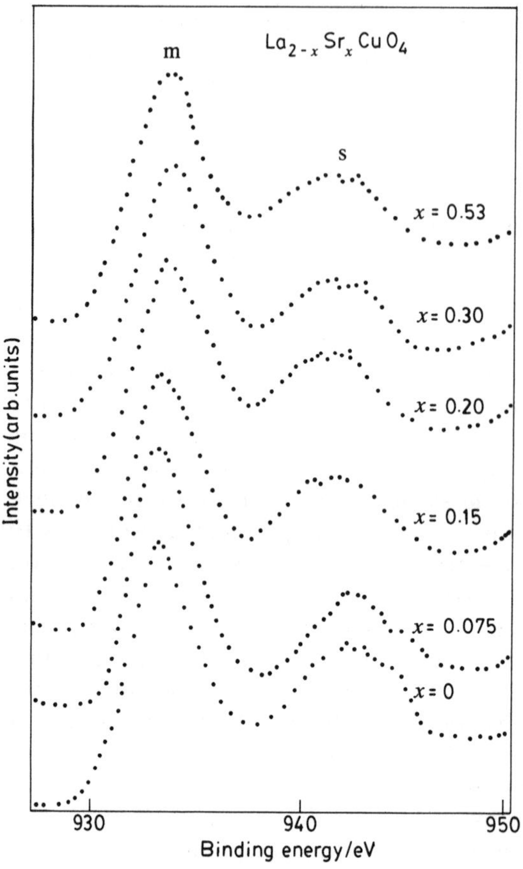

Figure 25. Cu 2p core-level spectra of La$_{2-x}$Sr$_x$CuO$_4$.

configuration. The relative intensity of the satellite with respect to that of the main feature, I_s/I_m, is determined by the charge-transfer energy, Δ, and the Cu–O hybridization strength, t_{pd}. Making use of the I_s/I_m ratio as an experimental handle, we have investigated the role of Δ and t_{pd}. For this purpose I_s/I_m has been carefully measured in several series of cuprates with known hole concentration, n_h. To explain the nature of variation of I_s/I_m with the composition of n_h, model calculations have been carried out on a CuO$_4$ cluster including configuration interaction. These calculations indeed reveal how I_s/I_m depends sensitively on the Δ/t_{pd} ratio. We shall briefly examine the results of our Cu 2p core-level photoemission studies and theoretical calculations to appreciate how the Cu–O charge-transfer energy and the Cu(3d)–O(2p) hybridization strength play a crucial role in the superconductivity of the layered cuprates.

In figure 25 we show the X-ray photoemission spectra of La$_{2-x}$Sr$_x$CuO$_4$ for various values of x in the Cu 2p$_{\frac{3}{2}}$ region. The spectra exhibit the well-known two-peak structure with a peak at about 933 eV (the main peak) and the other at 941 eV (the satellite) binding energies. The ratio of the satellite to the main peak intensity, I_s/I_m, in figure 25 exhibits a systematic variation with x. We have quantitatively estimated I_s/I_m as the ratio of the integrated areas under the main peak and the satellite after background subtraction of the spectra. The resulting I_s/I_m is plotted as a function of x in figure 26 for the La$_{2-x}$Sr$_x$CuO$_4$ series. In the same figure, we have also shown the

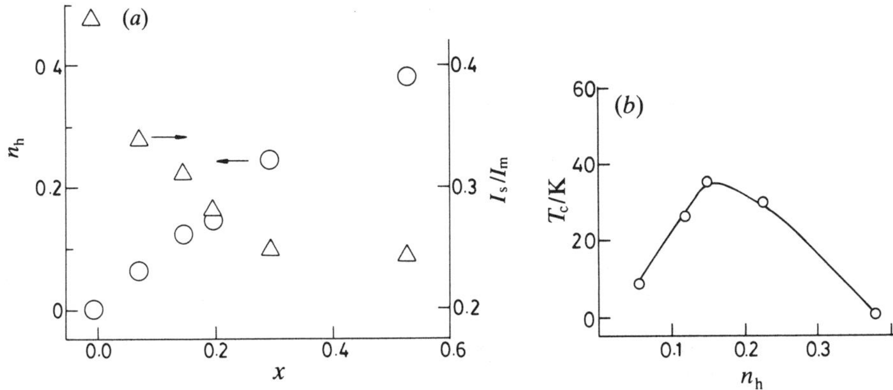

Figure 26. (*a*) Variation of I_s/I_m and n_h with x in $La_{2-x}Sr_xCuO_4$. (*b*) Variation of T_c with n_h.

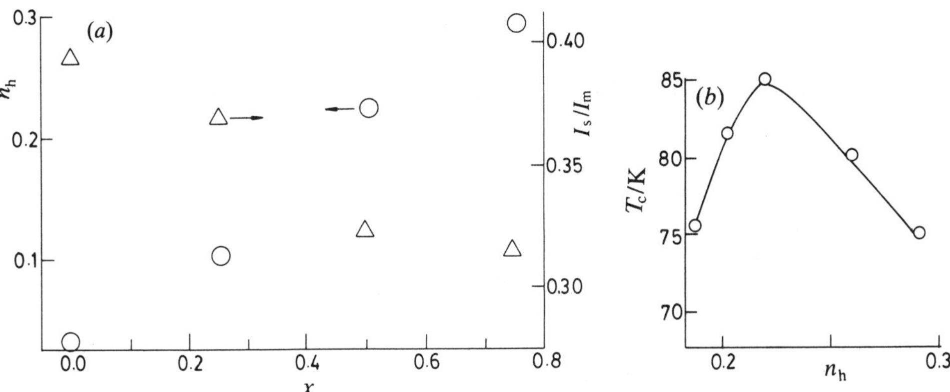

Figure 27. (*a*) Variation of I_s/I_m and n_h with x in $BiPbSr_2Y_{1-x}Ca_xCu_2O_8$. (*b*) Variation of T_c with n_h.

dependence of the experimentally obtained (by iodometric titrations) hole concentration, n_h, on x in these compounds. The inset shows the variation of the superconducting transition temperature T_c on n_h. This system traverses through the insulator–superconductor–metal régimes with increase in x. The I_s/I_m decreases continuously though these régimes, while n_h increases.

This behaviour seems to be common to all the series of superconducting cuprates that we have investigated. In figure 27 we plot the variation of I_s/I_m and n_h with x in the $BiPbSr_2Y_{1-x}Ca_xCu_2O_8$ series. We also show the variation of T_c with n_h in this series in figure 27*b*. The T_c reaches a maximum of about 85 K around $n_h = 0.22$; n_h once again exhibits a linear dependence on x with a slope of less than unity, while I_s/I_m monotonically decreases with x.

We show the dependence of n_h and I_s/I_m on x in the $Bi_2Ca_{1-x}Ln_xSr_2Cu_2O_8$ (Ln = Y or rare earth) series of compounds in figure 28. The inset shows the variation of T_c with n_h in this series. The T_c appears to exhibit a broad maximum at about $n_h = 0.2$ in these compounds. Interestingly in this series, we obtain a non-monotonic dependence of n_h on x. The n_h increases with x for small values of x up to about 0.25 and then decreases continuously up to $x = 1.0$. This is in contrast to the dependence of n_h on x in the other two series shown in figures 26 and 27. It is significant that in this series, I_s/I_m also exhibits a non-monotonic behaviour, showing a decrease up to

C. N. R. Rao

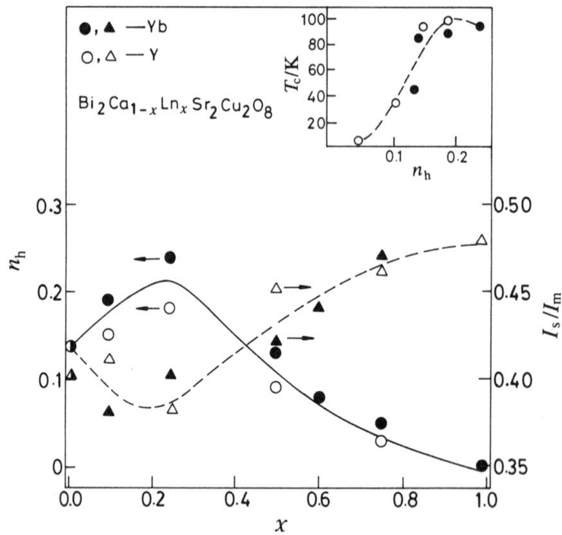

Figure 28. Variation of I_s/I_m and n_h with x in Bi$_2$Ca$_{1-x}$Ln$_x$Sr$_2$Cu$_2$O$_8$.

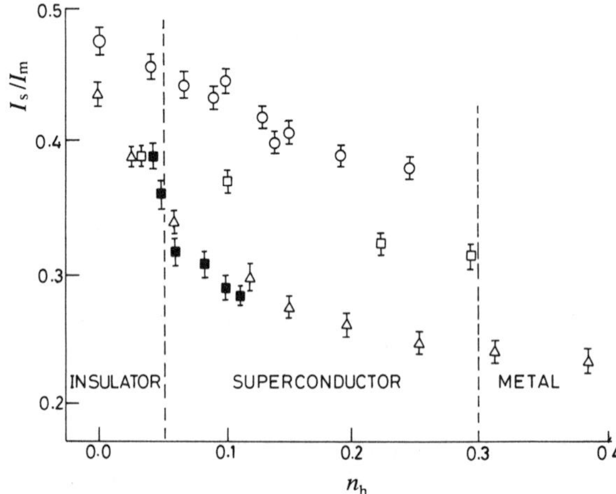

Figure 29. Variation of I_s/I_m with n_h. ■, BiPbSr$_{1+x}$Pr$_{1-x}$CuO$_6$; ○, Bi$_2$Ca$_{1-x}$Ln$_x$Sr$_2$Cu$_2$O$_8$ (Ln, rare earth); △, La$_{2-x}$Sr$_x$CuO$_4$; □, Bi$_2$PbSr$_2$Y$_{1-x}$Ca$_x$Cu$_2$O$_8$.

$x = 0.2$ and then a monotonic increase up to $x = 1.0$. This provides a crucial test of the fact that I_s/I_m appears to have the complimentary dependence on x compared with n_h. This implies that I_s/I_m will monotonically decrease in all the series with increasing n_h.

There is a relation between the experimentally obtained I_s/I_m and n_h values as shown in figure 29 where we have plotted the I_s/I_m ratios against the n_h values for the three series, La$_{2-x}$Sr$_x$CuO$_4$, BiPbSr$_2$Y$_{1-x}$Ca$_x$Cu$_2$O$_8$ and Bi$_2$Ca$_{1-x}$Ln$_x$Sr$_2$Cu$_2$O$_8$ (Ln = Y or Yb) as well as for the series, BiPbSr$_{1+x}$Pr$_{1-x}$CuO$_6$. It becomes absolutely clear from this figure that in each of the three series, I_s/I_m decreases monotonically with increased hole doping. The La$_{2-x}$Sr$_x$CuO$_4$ and the BiPb Sr$_{1+x}$Pr$_{1+x}$CuO$_6$ series exhibit the most pronounced dependence of I_s/I_m on n_h (over a narrow range of n_h), while the Bi$_2$Ca$_{1-x}$Ln$_x$Sr$_2$Cu$_2$O$_8$ and Bi$_2$PbSr$_2$Y$_{1-x}$Ca$_x$Cu$_2$O$_8$ series have a weaker dependence on n_h. In figure 29 we have also marked the

Figure 31

Figure 30

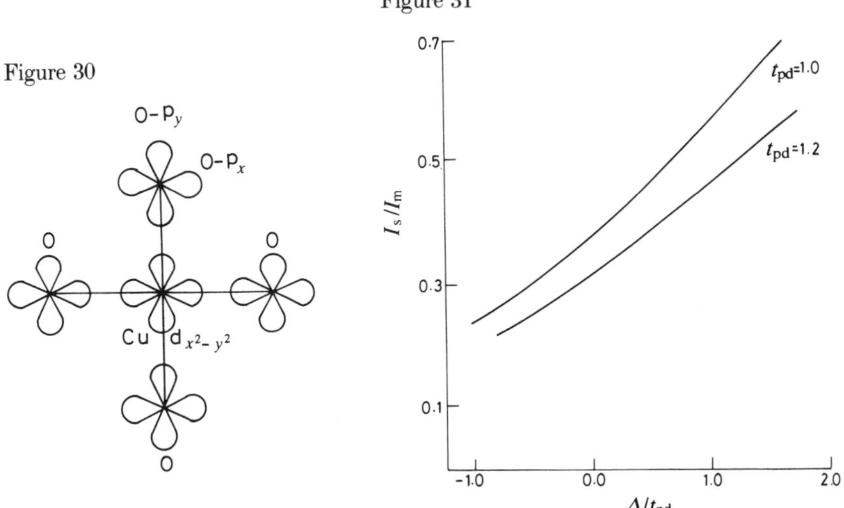

Figure 30. Bonding in a CuO_4 unit.

Figure 31. Variation of I_s/I_m with Δ/t_{pd}.

insulating, superconducting and metallic regions of the cuprates as n_h is varied by chemical doping to demonstrate how I_s/I_m varies continuously through these different régimes.

To understand the variation of I_s/I_m in the cuprates, we have performed model calculations of the Cu $2p_{\frac{3}{2}}$ core level photoemission spectral features (Rao *et al.* 1991*b*; Rao & Sarma 1991; Santra *et al.* 1991). For this purpose, we consider a CuO_4 cluster as shown in figure 30. We include the Cu $3d_{x^2-y^2}$ and the oxygen in-plane $2p_x$, $2p_y$ orbitals. In the D_{4h} symmetry of the CuO_4 cluster, the Cu $3d_{x^2-y^2}$ orbital transforms according to the b_{1g} irreducible representation. Out of the eight oxygen $2p_{x,y}$ orbitals shown in the figure, only one linear combination of these (namely $p_x^1 - p_y^2 - p_x^3 + p_y^4$) has the right symmetry (b_{1g}) to mix with the $3d_{x^2-y^2}$ orbital of Cu. Within this approximation the hybridization of the various oxygen orbitals do not play any important role as far as the spectral features are concerned, since these hybridization interactions will merely shift the b_{1g} combination of the O 2p orbitals in energy and renormalize the charge-transfer excitation energy (Sarma & Ovchinnikov 1990). We consider only one hole per Cu, so that the Coulomb interaction strengths, U_{dd}, U_{pp} and U_{pd} do not play any role. Thus, the spectral features depend only on the charge-transfer energy, Δ, between the Cu $3d_{x^2-y^2}$ and the oxygen-derived b_{1g} orbitals, the hybridization interaction strength, t, between the 3d and ligand b_{1g} level, and the 2p core-hole–3d-hole Coulomb repulsion, U. This problem can be exactly solved within the sudden approximation for the spectral features (Rao & Sarma 1991). In all our calculations we fix the value of the Coulomb repulsion, U, between the Cu 2p-core hole in the Cu 3d valence-hole at 8.0 eV.

In figure 31, we show the variation of the calculated I_s/I_m with Δ/t_{pd}, where $t_{pd} = \frac{1}{2}t$ is the hybridization strength between the Cu 3d and the O 2p orbitals for two different values of t_{pd}. We find from this figure that I_s/I_m increases with increasing Δ/t_{pd}. At large values of t_{pd}, the variation of I_s/I_m with Δ/t_{pd} becomes less pronounced; the dependence of I_s/I_m on Δ appears to be more pronounced when t_{pd} is small. In this context, we note that the experimental I_s/I_m values for $La_{2-x}Sr_xCuO_4$

618   *C. N. R. Rao*

Figure 32   Figure 33

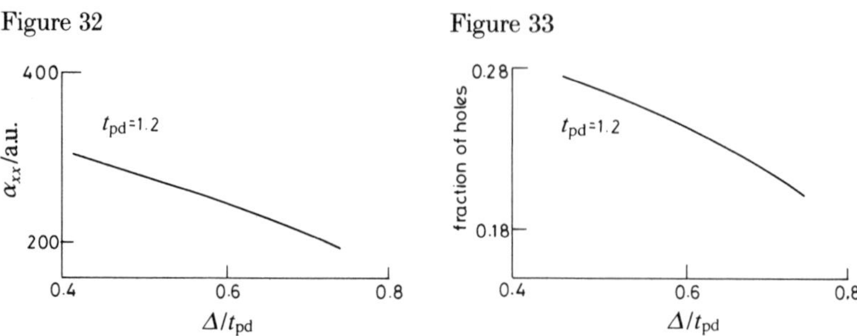

Figure 32. Variation of the polarizability of the CuO_2 sheets, α_{xx}, with Δ/t_{pd} (from Rao *et al.* 1991).

Figure 33. Variation of the fraction of holes with Δ/t_{pd} (from Rao *et al.* 1991).

and $BiPbSr_{1-x}Pr_{1+x}CuO_6$ show a marked variation with n_h and x, while I_s/I_m for $Bi_2Ca_{1-x}Ln_xSr_2Cu_2O_8$ and $Bi_2PbSr_2Y_{1-x}Ca_xCu_2O_8$ exhibit a less pronounced variation. It, therefore, appears that the decrease in I_s/I_m with x or n_h observed in these series of cuprates as well as the slopes of these variations are related to the magnitude of Δ/t_{pd}.

Our calculations of I_s/I_m in the cuprates indicate that the decrease in the charge-transfer energy is primarily responsible for the variation of the Cu 2p satellite intensity with increasing n_h. This is not an unreasonable expectation since increasing n_h tends to renormalize the charge-transfer energy to smaller values in the presence of a finite interatomic Coulomb interaction strength (U_{pd}). Since we do not explicitly take into account U_{dd} in our model, the renormalized Δ value should appear to be decreasing with increasing n_h. An earlier model (Sarma & Taraphder 1989) that includes U_{pd} interactions explicitly has indeed shown that the I_s/I_m is expected to decrease on hole-doping (increasing n_h) primarily due to the renormalization of Δ to a smaller value.

The T_c in the various families of cuprates generally show a maximum at a certain n_h value (figure 18). Accordingly, at a given n_h-value where the T_c is maximum (say $n_h \approx 0.15$), increasing I_s/I_m (going vertically in figure 29) is accompanied by an increase in T_c. Thus, $Bi_2Ca_{1-x}Ln_xSr_2Cu_2O_8$ exhibits the highest T_c of around 100 K, while $La_{2-x}Sr_xCuO_4$ is associated with the lowest T_c; $BiPbSr_2Y_{1-x}Ca_xCu_2O_8$ falls in between, $YBa_2Cu_3O_7$ ($n_h = 0.2$ at maximum T_c of *ca.* 90 K) shows an I_s/I_m value (0.45) close to that of $Bi_2Ca_{1-x}Ln_xSr_2Cu_2O_8$. As we have already shown, an increase in the I_s/I_m ratio is associated with an increase in Δ/t_{pd} (figure 31) which can arise from either a decrease in the t_{pd} value or an increase in the Δ value. It therefore appears that T_c in the different series of the cuprates are tuned by changing the value of Δ/t_{pd} *via* a change in the hole-doping level.

We show in figure 32 the static electronic polarizability, α_{xx}, calculated (Rao *et al.* 1991*b*) for a Cu_4O_8 cluster as a function of Δ/t_{pd}, while we show the fraction of the oxygen holes in the ground state wave function as a function of Δ/t_{pd} in figure 33. We find that α is rather large compared with the value for a single-band Hubbard model at $U = 4t$ of a similar size system at half-filling. The large static polarizability indicates a large dynamic polarizability as well, which would favour hole pairing in these systems. The polarizability increases with decreasing Δ/t_{pd}, concomitant with the decrease of the I_s/I_m ratio with decreasing Δ/t_{pd}. At the same time, there is an increase in the weightage of the fraction of oxygen holes in the ground state wave function.

Phil. Trans. R. Soc. Lond. A (1991)

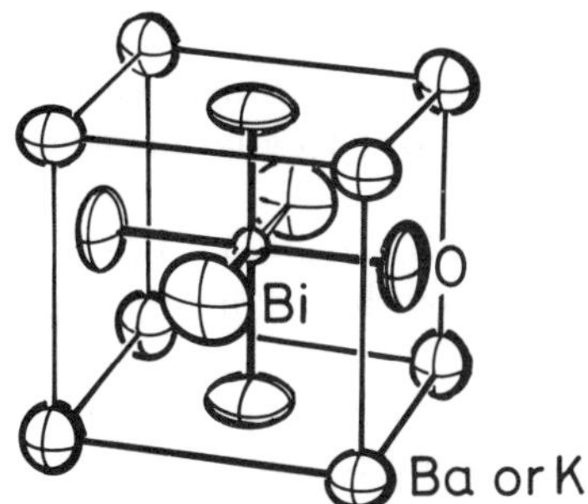

Figure 34. Structure of $Ba_{1-x}K_xBiO_3$.

Our studies of the Cu(2p) photoemission in cuprates combined with theoretical calculations enable us to come to the following conclusions.

(i) The intensity of the Cu 2p satellite, relative to the main feature, I_s/I_m, in the core-level spectra of the cuprates provides an experimental handle to investigate the role of important chemical bonding factors such as the Cu–O charge-transfer energy, Δ, and the Cu(3d)–O(2p) hybridization strength, t_{pd}.

(ii) The I_s/I_m ratio varies continuously with composition through the insulator–superconductor–metal régimes in the various series of cuprates. The experimental hole concentration n_h and I_s/I_m vary in opposite directions with composition suggesting that these two quantities are inversely related.

(iii) Theoretical calculations show that I_s/I_m increases monotonically with increasing Δ/t_{pd}. A relatively small Δ/t_{pd} gives rise to a large n_h (and small I_s/I_m). The value of n_h itself is nearly the same at maximum T_c in all the cuprate superconductors containing the same number of CuO_2 sheets.

(iv) At small Δ/t_{pd} values, the electronic polarizability, α, of the CuO_2 sheets will be large, favouring hole-pairing; α increases with increasing n_h or decreasing Δ/t_{pd}. Furthermore, Bose condensation of such pairs would be favoured by the two dimensionality of the CuO_2 sheets.

(v) Since all the cuprates containing the same number of CuO_2 sheets have roughly the same n_h values at maximum T_c, it appears that they will also be associated with similar, low Δ/t_{pd} and high α values.

8. Copper-free oxide superconductors

Historically, the two oxide systems not containing copper which showed relatively high T_cs in the range of 13 K are $BaBi_{1-x}Pb_xO_3$ (Sleight *et al.* 1975) and $Li_{1+x}Ti_{2-x}O_4$ (Johnston *et al.* 1973). Both these oxides have mixed valent cations. In the bismuthate system, which has the perovskite structure, the nominal Bi^{IV} disproportionates into Bi^{III} and Bi^{V} in the insulting phase ($x < 0.75$); in other words, there is a charge-density-wave (CDW) gap. There is a sharp insulator-metal transition at $x = 0.75$, when the CDW gap disappears and superconductivity manifests itself; Bi^{IV} ions are delocalized in the metallic/superconducting phase (Kulkarni *et al.* 1990). In superconducting $Ba_{1-x}K_xBiO_3$ (figure 34) with a T_c of *ca.* 30 K (Cava *et al.* 1988*a*), introduction of mixed valency by substitution of Ba by K competes with the CDW. These bismuthates are considered to be negative U cases with $\Delta \gg U$. Although some of the features of bismuthates are similar to those of the cuprates, they show some properties which are quite different (Hinks 1990). Thus $Ba_{1-x}K_xBiO_3$ shows a

 C. N. R. Rao

large ^{18}O isotope effect (unlike the two-dimensional cuprates), but no static magnetic order. $Ba_{1-x}K_xBiO_3$ appears to be a superconductor in the weak to moderate coupling limit and the high T_c is due to a large electron–phonon coupling constant. The mechanism of superconductivity in these materials seems to be different from that of cuprates. Other than $Ba_{1-x}K_xBiO_3$, there has not been much success in synthesizing three-dimensional oxides with high T_cs (see, for example, Nagarajan *et al.* 1991).

Among the other copper-free oxide superconductors, the lanthanum nickelates were suspected to show superconductivity because of the diamagnetic behaviour found in some samples of $La_{2-x}Sr_xNiO_4$ and related compounds (Nanjundaswamy *et al.* 1990). It has, however, not been possible to reproduce these findings universally and the origin of diamagnetism observed in some of the samples is not clear (Sreedhar & Rao 1990). Hopes were raised by a recent Japanese report of T_cs near 200 K in the Tl–Sr–V–O system, but we have found that these results are not reproducible. There was a Russian report that $LaCa_2Co_3O_y$ was superconducting with a T_c of around 227 K, but we have not been able to reproduce this result as well. Clearly there must be other interesting oxides without copper which should exhibit high T_cs. Future investigations may bring into light such metal oxides.

9. Chemical applications

Potential applications of superconducting cuprates in electronics and other technologies are commonly known. These cuprates also exhibit significant catalytic activity. Thus, $YBa_2Cu_3O_{7-\delta}$ and related cuprates act as catalysts in oxidation or dehydrogenation reactions (Hansen *et al.* 1988; Halasz 1989; Mizuno *et al.* 1988). Carbon monoxide and alcohol are readily oxidized over the cuprates. NH_3 is oxidized to N_2 and H_2O on these surfaces. Ammoxidation of toluene to benzonitrile has been found to occur on $YBa_2Cu_3O_7$ (Hansen *et al.* 1990).

The catalytic activity of the cuprate superconductors prompted us to examine their possible use as gas sensors. It should be noted that a good gas sensor would require not only the catalytic property of the oxide surface but also high sensitivity. For example, the electrical resistivity of the material should change sharply on contact with the gas or vapour in question. The superconducting compositions of the cuprates being metallic in the normal state, would therefore not be best suited for sensing while they may be good catalysts. With suitable compositional variation wherein the resistivity of the material is increased in order to have the right régime for sensing, it is possible to effectively use these materials. We have indeed found $La_{2-x}Sr_xCuO_4$ ($0 < x < 0.2$) and $Bi_2Ca_{1-x}Y_xSr_2Cu_2O_{8+\delta}$ ($x > 0.5$) are good sensors for alcohol and other vapours; the superconducting compositions are not. In figure 35, we show some of the recent results obtained by Grantscharova and Raju in this laboratory. We see that these materials are good sensors for alcohol and ether. Further studies on the gas sensor characteristics of superconducting materials would be worthwhile.

10. Concluding remarks

Structure–property relations and other aspects of the oxide superconductors that I have described so far should clearly indicate how chemistry becomes important in not only synthesizing novel materials of desired structures and properties, but also in understanding the phenomenon of high-temperature superconductivity. Our

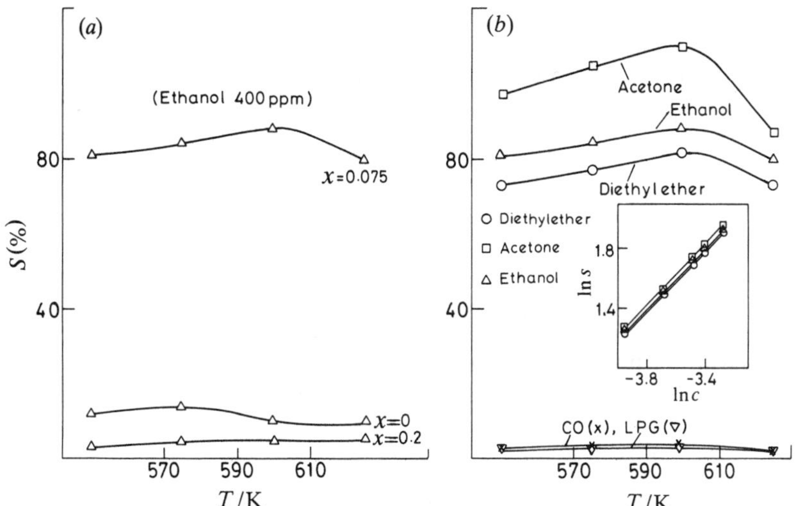

Figure 35. Gas sensing ability of $La_{2-x}Sr_xCuO_4$ (S is sensitivity). Inset shows log sensitivity against log concentration plot (unpublished results of Grantscharova, Raju and Rao).

search for newer and better high T_c materials is far from over. We are yet to investigate many other oxide and related systems, besides improving on the already known ones. Some of the important objectives of research in superconductivity today would be the following.

(i) To acquire better and more experimental data on the known high T_c materials in order to improve our understanding of the properties.

(ii) To develop suitable models to understand the mechanism responsible for high T_c superconductivity and carry out those experiments which would clearly delineate the various factors responsible for superconductivity.

(iii) To search for new materials exhibiting high T_c and especially those not containing Cu (since most of present models require the d orbitals of Cu).

(iv) To improve our understanding of the chemistry of materials processing as well as the ceramic properties of the oxide superconductors.

(v) To prepare high quality films, tapes, wires, etc. with desired J_c and other properties for applications.

(vi) To investigate catalytic, gas-sensing and other chemical applications, besides the well-known applications in electronics and other areas.

I thank the National Superconductivity Programme, the Department of Science and Technology, the University Grants Commission and the US National Science Foundation for support of this research. I would like to place on record my sincere appreciation of the dedicated effort made by my students and other coworkers in carrying out this research under difficult circumstances. This paper is contribution no. 771 from the Solid State and Structural Chemistry Unit.

References

Allen, J. W. & Olson, C. G. 1990 *MRS Bull.* **15**, 34, and the references cited therein.

Bednorz, J. G. & Müller, K. A. 1986 *Z. Phys.* B **64**, 189

Bianconi, A. 1990 In *Superconductivity* (ed. S. K. Joshi, C. N. R. Rao & S. V. Subramanyam), p. 448. Singapore: World Scientific.

Caignaert, V., Hervieu, M., Wang, J., Desgardin, G., Raveau, B., Boterel, F. & Haussone, J. M. 1990 *Physica* C **170**, 139.

622   *C. N. R. Rao*

Cava, R. J. 1990 *Science, Wash.* **247**, 656.

Cava, R. J., Batlogg, B., Krajewski, J. J., Farrow, R., Rupp, L. W. Jr, White, A. E., Short, K., Peck, W. F. Jr & Kometani, T. 1988a *Nature, Lond.* **332**, 814.

Cava, R. J., Batlogg, B., Krajewski, J. J., Rupp, L. W., Schneemeyer, L. F., Siegrest, T., Van Dover, R. B., Marsh, P., Peok, W. F. Jr, Gallagher, P. K., Glarium, S. H., Marshall, J. H., Farrow, R. C., Waszezak, J. W., Hull, R. & Trevor, R. 1988b *Nature, Lond.* **336**, 221.

Cava, R. J., Batlogg, B., Van Dover, R. B., Krajewski, J. J., Waszezak, J. W., Fleming, R. M., Peck, W. F. Jr, Rupp, L. W. Jr, Marsh, P., James, A. C. W. P. & Schreemeyer, L. F. 1990a *Nature, Lond.* **345**, 602.

Cava, R. J., Hewat, A. W., Hewat, E. A., Batlogg, B., Marezio, M., Rabe, K. M., Krajewski, J. J., Peck, W. F. Jr & Rupp, L. W. Jr 1990b *Physica* C **165**, 419.

Cava, R. J., Krajewski, J. J., Peck, W. F. Jr, Batlogg, B., Rupp, L. W., Fleming, R. M., James, A. C. W. P. & Marsh, P. 1989 *Nature, Lond.* **338**, 328.

Chaillout, C., Chenavas, J., Cehong, S. W., Fisk, Z., Marezio, M., Morosin, B. & Schriber, J. E. 1990 *Physica* C **170**, 87.

Chakraverty, B., Sarma, D. D. & Rao, C. N. R. 1988 *Physica* C **156**, 413.

de Leeuw, D. M., Groen, W. A., Feiner, L. F. & Havinga, E. E. 1990 *Physica* C **166**, 133.

Ganguli, A. K., Nanjundaswamy, K. S. & Rao, C. N. R. 1988 *Physica* C **156**, 788.

Ganguli, A. K., Manivannan, V., Sood, A. K. & Rao, C. N. R. 1989a *Appl. Phys. Lett.* **55**, 2664.

Ganguli, A. K., Nanjundaswamy, K. S., Rao, C. N. R., Sequeira, A. & Rajagopal, H. 1989b *Mat. Res. Bull.* **24**, 883.

Ganguli, A. K., Rao, C. N. R., Sequeira, A. & Rajagopal, H. 1989c *Z. Phys.* B **74**, 215.

Ganguly, P. & Rao, C. N. R. 1973 *Mat. Res. Bull.* **17**, 493.

Ganguly, P. & Rao, C. N. R. 1984 *J. Solid State Chem.* **53**, 193, and the references cited therein.

Goodenough, J. B. 1971 *Prog. Solid State Chem.* **5**, 149.

Goodenough, J. B. & Manthiram, A. 1990 *J. Solid State Chem.* **88**, 115, and the references cited therein.

Halasz, I. 1989 *Appl. Catal.* **47**, L17.

Hansen, S., Otamiri, J. & Andersson, A. 1990 *Catal. Lett.* **6**, 33.

Hansen, S., Otamiri, J., Bovin, J. & Andersson, A. 1988 *Nature, Lond.* **334**, 143.

Hewat, A. W., Fischer, P., Kaldis, E., Karpinski, J., Physiecki, S. & Julek, E. 1990 *Physica* C **167**, 579.

Hewat, A. W., Hewat, E. A., Bordet, P., Capponi, J. J., Chaillout, C., Chenavas, J., Hodeau, J. L., Marezio, M., Strobel, P., Francois, M., Yoon, K., Fischer, P. & Tholence, J. L. 1989 *IBM J. Res. Dev.* **33**, 220.

Hinks, D. G. 1990 *MRS Bull.* **15**, 60, and the references cited therein.

Johnston, D. C., Prakash, H., Zachariasen, W. H. & Viswanathan, R. 1973 *Mat. Res. Bull.* **8**, 777.

Jorgensen, J. D., Dabrowski, B., Pei, S., Hinks, D. G., Soderholm, L., Morosin, B., Schriber, J. E., Venturini, E. L. & Ginley, D. S. 1988 *Phys. Rev.* B **42**, 6765.

Jorgensen, J. D., Pei, S., Lightfoot, P., Shi, H., Paulikas, A. P. & Veal, B. W. 1990 *Physica* C **167**, 571.

Joshi, S. K., Subramanyam, S. V. & Rao, C. N. R. 1990 *Superconductivity (ICSC).* Singapore: World Scientific.

Koike, Y., Masuzawa, M., Noji, T., Sunagawa, H., Kawabe, H., Kobayashi, N. & Saito, Y. 1990 *Physica* C **170**, 130.

Karpinski, J., Kaldis, E., Jilek, E. & Bucher, B. 1988 *Nature, Lond.* **336**, 660.

Kitazawa, K. & Ishiguro, T. (eds) 1989 *Advances in superconductivity.* Tokyo: Springer-Verlag.

Kovatcheva, D., Hewat, A. W., Rangavittal, N., Manivannan, V., Guru Row, T. N. & Rao, C. N. R. 1991 *Physica* C **173**, 444.

Kulkarni, G. U., Sankar, G. & Rao, C. N. R. 1989 *Appl. Phys. Lett.* **55**, 388.

Kulkarni, G. U., Vijayakrishnan, V., Rao, G. R., Seshadri, R. & Rao, C. N. R. 1990 *Appl. Phys. Lett.* **57**, 1823, and the references cited therein.

Le Page, Y., McKinnon, W. R., Taraseon, J. M. & Barboux, P. 1989 *Phys. Rev.* B **40**, 4810.

Lightfoot, P., Richards, D. R., Dabroowski, B., Hinks, D. G., Pei, S., Marx, D. T., Mitchell, A. W., Zing, Y. & Jorgensen, J. D. 1990 *Physica* C **168**, 627.

Liu, R. S., Janes, R., Bennett, M. J. & Edwards, P. P. 1990 *Appl. Phys. Lett.* **57**, 920.

Maeda, T., Sakyuama, K., Koriyama, S. & Yamanchi, H. 1990 *ISTEC J.* **3**, 16, and the references cited therein.

Maignan, A., Martin, C., Huve, M., Provost, J., Hervieu, M., Michel, C. & Raveau, B. 1990 *Physica* C **170**, 350.

Manivannan, V., Ganguli, A. K., Subbanna, G. N. & Rao, C. N. R. 1990 *Solid State Commun.* **74**, 87.

Manivannan, V., Gopalakrishnan, J. & Rao, C. N. R. 1991 *Phys. Rev.* B **43**, 8686.

Manthiram, A. & Goodenough, J. B. 1989 *Physica* C **159**, 760.

Manthiram, A., Paranthaman, M. & Goodenough, J. B. 1990 *Physica* C **171**, 35.

Maple, M. B. 1990 *MRS Bull.* **15**, 60, and the references cited therein.

Miyatake, T., Itti, R. & Yamguchi, Y. 1990 *ISTEC J.* **3**, 23.

Mizuno, N., Yamato, M. & Misono, M. 1988 *J. chem. Soc. chem. Commun.* 887.

Nagarajan, R., Vasantacharya, N. Y., Gopalakrishnan, J. & Rao, C. N. R. 1991 *Solid State Commun.* **77**, 373.

Nanjundaswamy, K. S., Lewicki, A., Karol, Z., Gopalan, P., Metcalf, P., Honig, J. M., Rao, C. N. R. & Spalek, J. 1990 *Physica* C **166**, 361, and the references cited therein.

Nelson, D. L., Whittingham, M. S. & George, T. F. (eds) 1987 *Chemistry of high-temperature superconductors. ACS Symposium Series 351.* Washington, D.C.: American Chemical Society.

Norton, D. P., Lowndes, D. H., Sales, B. C., Budai, J. D., Chakoumakos, B. C. & Kerchner, H. R. 1991 *Phys. Rev. Lett.* **66**, 1537.

Ramakrishnan, T. V. & Rao, C. N. R. 1989 *J. Phys. Chem.* **93**, 4414, and the references cited therein.

Rao, C. N. R. 1988*a* *J. Solid State Chem.* **74**, 147, and the references cited therein.

Rao, C. N. R. (ed.) 1988*b* *Chemical and structural aspects of high-temperature superconductors.* Singapore: World Scientific.

Rao, C. N. R. 1989 *A. Rev. Phys. Chem.* **40**, 291.

Rao, C. N. R. & Raveau, B. 1989 *Acc Chem. Res.* **22**, 106, and the references cited therein.

Rao, C. N. R. & Sarma, D. D. 1991 In *Studies of high-temperature superconductors*, vol. 9. New York: Nova.

Rao, C. N. R. & Subbarao, G. V. 1970 *Physica Status Solidi* A **1**, 597.

Rao, C. N. R., Ganguli, A. K. & Vijayaraghavan R. 1989*a* *Phys. Rev.* B **40**, 2565.

Rao, C. N. R., Ganguly, P. & Mohan Ram, R. A. 1988 *J. Solid State Chem.* **72**, 14.

Rao, C. N. R., Sarma, D. D. & Rao, G. R. 1989*b* *Phase Transitions* **19**, 69, and the references cited therein.

Rao, C. N. R., Gopalakrishnan, J., Santra, A. K. & Manivannan, V. 1991*a* *Physica* C **174**, 11.

Rao, C. N. R., Ramasesha, S., Sarma, D. D. & Santra, A. K. 1991*b* *Solid State Commun.* **77**, 709.

Rao, C. N. R., Rao, G. R., Rajumon, M. K. & Sarma, D. D. 1990*c* *Phys. Rev.* B **42**, 1026.

Rao, C. N. R., Ganguly, P., Raychaudhuri, A. K., Mohan Ram, R. A. & Sreedhar, K. 1987 *Nature, Lond.* **326**, 856.

Rao, C. N. R., Nagarajan, R., Ganguli, A. K., Subbanna, G. N. & Bhat, S. V. 1990*a* *Phys. Rev.* B **42**, 6765.

Rao, C. N. R., Vijayaraghavan, R., Ganguli, A. K., Manivannan, V. & Vasanthacharya, N. Y. 1989*c* In *Studies of high-temperature superconductors*, vol. 4. New York: Nova.

Rao, C. N. R., Subbanna, G. N., Nagarajan, R., Ganguli, A. K., Ganapathi, L., Vijayaraghavan, R., Bhat, S. V. & Raju, A. R. 1990*d* *J. Solid State Chem.* **88**, 163.

Rao, C. N. R., Nagarajan, R., Vijayaraghavan, R., Vasanthacharya, N. Y., Kulkarni, G. U., Rao, G. R., Umarji, A. M., Somasundaram, P., Subbanna, G. N., Raju, A. R., Sood, A. K. & Chadrabhas, N. 1990*b* *Superconductor Sci. Technol.* **3**, 242.

624   *C. N. R. Rao*

Raveau, B., Michel, C. & Hervieu, B. 1990 *J. Solid State Chem.* **88**, 140.

Santra, A. K., Sarma, D. D. & Rao, C. N. R. 1991 *Phys. Rev.* B **43**, 5612.

Sarma, D. D. & Ovchinnikov, S. G. 1990 *Phys. Rev.* B **42**, 6817.

Sarma, D. D. & Rao, C. N. R. 1989 *Synthetic Metals* **33**, 131.

Sarma, D. D. & Taraphder, A. 1989 *Phys. Rev.* B **39**, 11570.

Sarma, D. D., Sen, P., Cimino, R., Carbone, C., Gudat, W., Sampathkumaran, E. V. & Das, I. 1991 *Solid State Commun.* **77**, 377.

Shafer, M. W. & Penney, T. 1990 *Eur. J. Solid State inorg. Chem.* **27**, 191.

Singh, K. K., Ganguly, P. & Rao, C. N. R. 1982 *Mat. Res. Bull.* **17**, 493.

Sleight, A. W. 1988 *Science, Wash.* **242**, 1519, and the references cited therein.

Sleight, A. W., Gillson, J. L. & Bierstedt, P. E. 1975 *Solid State Commun.* **17**, 27.

Sleight, A. W., Gopalakrishnan, J., Torardi, C. C. & Subramanian, M. A. 1989 *Phase Transitions* **19**, 149.

Sreedhar, K. & Rao, C. N. R. 1990 *Mat. Res. Bull.* **25**, 1235.

Subramanian, M. A., Torardi, C. C., Gopalakrishnan, J., Gai, P. L., Calabrese, J. C., Askew, T. R., Flippen, R. B. & Sleight, A. W. 1988 *Science, Wash.* **242**, 249.

Tarascon, J. M., Le Page, Y., McKinnon, W. R., Ramesh, R., Eibuschutz, M., Tselepis, E., Wang, E. & Hull, G. W. 1990 *Physica* C **167**, 20.

Tokura, Y., Arima, T., Takagi, H., Uchida, S., Ishigaki, T., Asamo, H., Beyers, R., Nazzal, A. I., Lacorre, P. & Torrance, J. B. 1989*a Nature, Lond.* **342**, 890.

Tokura, Y., Takagi, H. & Uchida, S. 1989*b Nature, Lond.* **337**, 345.

Tokura, Y., Torrance, J. B., Huang, T. C. & Wazzal, A. I. 1988 *Phys. Rev.* B **38**, 7156.

Torardi, C. C., Subramanian, M. A., Gopalakrishnan, J., McCarron, E., Calabrese, J. C., Morrissey, K. J., Askew, T. R., Flippen, R. B., Chowdhry, U., Sleight, A. W. & Cox, D. E. 1989 In *High-temperature superconductivity. Proc. Alabama Conf.* (ed. R. M. Metzger). New York: Gordon Breach..

Torrance, J. B. & Metzger, R. M. 1989 *Phys. Rev. Lett.* **63**, 1515.

Torrance, J. B., Tokura, Y., Nazzal, A. I., Bezinge, A., Huang, T. C. & Parkin, S. S. P. 1988 *Phys. Rev. Lett.* **61**, 1127.

Vijayakrishnan, V., Kulkarni, G. U. & Rao, C. N. R. 1990 *Mod. Phys. Lett.* B **4**, 451.

Wu, M. K., Ashburn, J. R., Torng, C. J., Hor, P. H., Meng, R. L., Gao, L. L., Huang, Z. J., Wang, Y. Q. & Chu, C. W. 1987 *Phys. Rev. Lett.* **58**, 908.

Xiang, X. D., McKernan, S., Vareka, W. A., Zettle, A., Corkill, J. L., Barbee, T. W. & Cohen, M. L. 1990 *Nature, Lond.* **348**, 145.

Lecture delivered 25 March 1991; typescript received 15 April 1991

Note

(a) For a review of synthesis of cuprate superconductors, see C.N.R. Rao et al, Supercond. Sci. Tech., **6**, 1 (1993)

(b) For tunneling studies of $Bi_2Ca_{1-x}Y_xSr_2Cu_2O_{8+\delta}$ see: H. Srikant, A.K. Raychaudhuri, C.R.V. Rao, P. Ramasamy, H.N. Aiyer and C.N.R. Rao, Physica C, **200**, 372 (1992)

(c) For analysis of thermopower data of superconducting Bi and Tl cuprates, see C.N.R. Rao, T.V. Ramakrishnan and N. Kumar, Physica C, **165**, 183 (1990)

(d) For a discussion of polarizability of CuO_2 sheets, see S. Ramasesha and C.N.R. Rao, Phys. Rev. B, **44**, 7046 (1991)

Structure–Property Relationships in Superconducting Cuprates

C. N. R. Rao*
Department of Chemistry, University of Wales, Cardiff CF1 3TB, U.K., and Solid State and Structural Chemistry Unit, Indian Institute of Science, Bangalore 560 012, India

A. K. Ganguli
CSIR Centre of Excellence in Chemistry, Indian Institute of Science, Bangalore 560 012, India

1 Introduction

The highest superconducting transition temperature known till 1986 was 23K and it seemed as though this barrier would not be broken.[1] The discovery of 30K superconductivity in an oxide of the La–Ba–Cu–O system by Bednorz and Müller[2] changed the picture. A variety of superconducting oxides, especially cuprates, have since been synthesized and characterized,[3—6] the highest transition temperature as of today being 155K in $HgBa_2Ca_2Cu_3O_{8+\delta}$. Studies of the various families of cuprates have shown many commonalities and unifying features.[6,7] Properties of the cuprates have been related to certain structural and electronic parameters. Although there is no simple relationship between the superconducting transition temperature and any specific structural feature of the cuprates, the various correlations help us to understand these materials better and to design newer ones. In this article, we shall briefly examine some of the significant structure–property relationships in superconducting cuprates along with the structural commonalities.

2 Common Structural Features

Many cuprate families have been discovered in the past seven years. The general features of the cuprates are shown schematically in Figure 1. The major families of cuprates are:
(a) $La_{2-x}A_xCuO_4$ (A = alkaline earth) possessing the K_2NiF_4(T) structure; (b) $LnBa_2Cu_3O_{7-\delta}$ (Ln = Y or rare-earth other than Ce, Pr, and Tb) referred to as the 123 type (Figure 2) and the related $LnBa_2Cu_4O_8$ (124) and $Ln_2Ba_4Cu_7O_{15}$ (247) cuprates containing perovskite layers with CuO_2 sheets as well as Cu–O chains; (c) $Bi_2(Ca,Sr)_{n+1}Cu_nO_{2n+4}$ containing two BiO layers and perovskite layers with CuO_2 sheets (Figure 2); (d) $Tl_2A_{n+1}Cu_nO_{2n+4}$ and $TlA_{n+1}Cu_nO_{2n+3}$ (A = Ca, Ba, Sr etc.) containing Tl–O layers and perovskite layers with CuO_2 sheets (Figure 2); (e) lead-based superconducting cuprates such as $Pb_2Sr_2LnCu_3O_8$ containing CuO_2 sheets and CuI–O sticks; (f) Tl, Bi, and Pb cuprates containing fluorite

* For correspondence

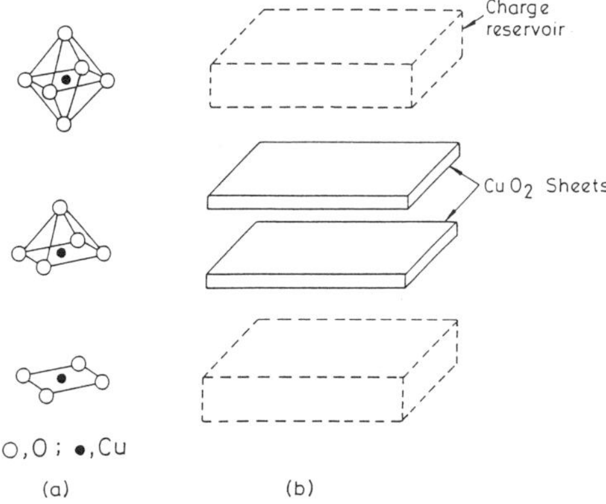

Figure 1 (a) The three types of Cu—O polyhedra found in the superconducting cuprates (b) Schematic representation of cuprates.

layers and CuO_2 sheets; (g) mercury cuprates of the type $HgCa_{n-1}Ba_2Cu_nO_{2n+2+\delta}$; and (h) infinite layer cuprates such as $ACuO_2$ (A = Ca, Sr, Ba) and $Ln_{1-x}A_xCuO_2$ (Ln = Nd, Pr; A = Sr, Ba). All these cuprates have holes as charge carriers. The only well established superconducting cuprates with electrons as charge carriers are $Nd_{2-x}M_xCuO_4$ (M = Ce, Th) and related compounds with the T′ structure possessing square-planar CuO_4 units instead of the octahedra in the T structure (Figure 1).

A. K. Ganguli is a scientific officer at the Centre of Excellence in Chemistry at the Indian Institute of Science. He has an M.Sc. degree from the University of Delhi and a Ph.D. degree from the Indian Institute of Science and has carried out post-doctoral work at du Pont and Iowa State University.

C. N. R. Rao is a Professor of Chemical Science at the Indian Institute of Science, and President of the Jawaharalal Nehru Centre for Advanced Scientific Research, Bangalore, India. He is an Honorary Professor of Chemistry at the University of Wales, Cardiff. He was Commonwealth Visiting Professor at the University of Oxford and Nehru Professor at the University of Cambridge. He is a Fellow of the Royal Society, London, Foreign Associate of the U.S. National Academy of Sciences, and Foreign member of several other academies. He is a member of the Pontifical Academy of Sciences and an Honorary Fellow of the Royal Society of Chemistry. He is a recipient of the Marlow Medal of the Faraday Society and the RSC Medal for solid-state chemistry. His main research interests are in solid-state chemistry, spectroscopy, and molecular structure and surface science. He is the author of over 500 research papers and several books in solid state chemistry.

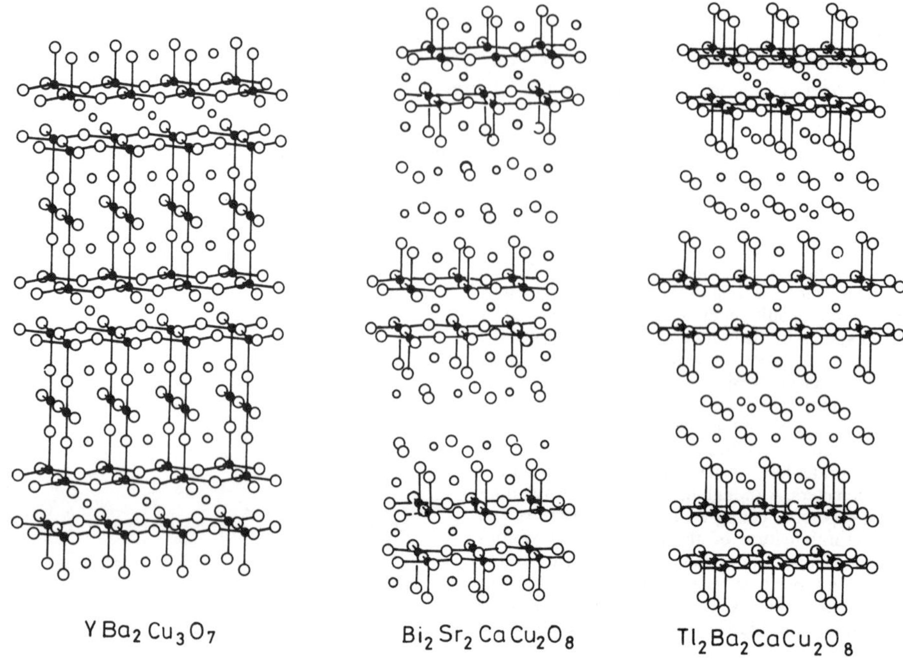

Figure 2 Structures of $YBa_2Cu_3O_7$, $Bi_2CaSr_2Cu_2O_8$, and $Tl_2CaBa_2Cu_2O_8$. Notice the presence of CuO_2 sheets containing CuO_5 square-pyramids. Additionally, there are Cu–O chains in $YBa_2Cu_3O_7$.

Many of the cuprate families have an antiferromagnetic insulator member at one end of the composition (*e.g.* La_2CuO_4 in $La_{2-x}A_xCuO_4$, $YBa_2Cu_3O_6$ in $YBa_2Cu_3O_{7-\delta}$, and $TlYSr_2Cu_2O_7$ in $TlY_{1-x}Ca_xSr_2Cu_2O_{7-\delta}$. What is more interesting is that all the cuprates are at a metal–insulator boundary. Some of them undergo a metal–non-metal transition as a function of composition (*e.g* $Bi_2Ca_{1-x}Y_xSr_2Cu_2O_8$ and $TlY_{1-x}Ca_xSr_2Cu_2O_7$ with change in x).

The cuprates can be described on the basis of certain structural features common to many of them. For example, the structures of $La_{2-x}A_xCaO_4$, $Bi_2(Ca, Sr)_{n+1}Cu_nO_{2n+4}$ and the thallium cuprates can be considered to be intergrowths of oxygen-deficient perovskite layers, $ACuO_{3-x}$, with AO-type rock-salt layers.[7] The cuprates contain different types of Cu–O polyhedra with the hole superconductors necessarily having CuO_5 or CuO_6 units and the electron-superconducting $Nd_{2-x}M_xCuO_4$ containing only CuO_4 square-planar units (Figures 1 and 2). Thus, the essential feature of the cuprates is the presence of CuO_2 sheets with or without apical oxygens. The mobile charge carriers in the cuprates are in the CuO_2 sheets. All the cuprates have charge reservoirs as exemplified by the Cu–O chains in the 123 and 124 cuprates and the TlO, BiO, and HgO layers in the other cuprates. That the CuO_2 sheets are the seat of high-temperature superconductivity is demonstrated by the fact that intercalation of iodine between BiO layers in the bismuth cuprates does not affect the superconducting transition temperature while introduction of fluorite layers between the CuO_2 sheets adversely affects superconductivity. In the different series of cuprates with varying number of CuO_2 sheets studied hitherto, the T_c reaches a maximum when $n = 3$ except in single thallium layer cuprates where the maximum is at $n = 4$. The infinite layered cuprates, where the CuO_2 sheets are separated by alkaline earth and other cations, show T_c's in the 40—110K range.[8] Superconductivity in these materials appears to be due to the presence of Sr–O defect layers corresponding to the insertion of $Sr_3O_{2\pm x}$ blocks.[8]

Based on the interplanar Cu–Cu distances, one can classify cuprates into two categories.[9] In one category, r(Cu–Cu) lies between 3.0 and 3.6Å with T_c's varying between 50 and 133K and in another it is between ~6 and 12.5Å encompassing superconductors with lower T_c (< 50K), except $Tl_2Ba_2CuO_6$ and $HgBa_2CuO_{4.065}$ with T_c's of ~90K. In the first category with r(Cu–Cu) < 3.6Å, the T_c increases as the Cu–Cu distance decreases. In the 2222-type fluorite-based superconductors, there are three copper oxygen sheets, [CuO_2–CuO_δ–CuO_2–fluorite–CuO_2–CuO_δ–CuO_2], each block of three sheets separated by a Ln_2O_2 fluorite layer. The r(Cu–Cu) relevant to these compounds would be the distance between the CuO_2 sheets across the fluorite layer (~ 6.2Å) and not the distance between two neighbouring sheets. Accordingly, these cuprates exhibit a low T_c(45—50K). It therefore appears that the distance between the CuO_2 sheets is a factor in determining the value of T_c, indicating that there is some interaction between the closely spaced CuO_2 sheets although the cuprates have quasi two-dimensional character.

Oxygen stoichiometry and ordering play a crucial role in determining the structure and properties of cuprates. The dependence of the structure and properties of $YBa_2Cu_3O_{7-\delta}$ on oxygen content has been studied in detail. Thus, $YBa_2Cu_3O_{7-\delta}$ which is orthorhombic ($c \simeq 3 b$) with a T_c of 90K when $0.0 \lesssim \delta \lesssim 0.25$, assumes another orthorhombic structure ($c \neq 3b$) when $0.3 \lesssim \delta \lesssim 0.4$ with a T_c of 60K. When $\delta = 1.0$, all the oxygens in the CuO chains are depleted and the structure becomes tetragonal and the material is non-superconducting. When $\delta = 0.5$, there is an ordered arrangement of oxygen vacancies with the presence of fully oxidized (O_7) and fully reduced (O_6) chains alternately. The compositions showing 60K superconductivity are metastable and transform to a 124-type phase on heating at low temperatures.[10]

Oxygen-excess La_2CuO_4 is biphasic, consisting of the stoichiometric antiferromagnetic phase and an oxygen-excess superconducting phase.[11] In bismuth cuprates, excess oxygen in the BiO layers gives rise to incommensurate modulation. Modulation-free superconducting bismuth cuprates have been made[12] by replacing one Bi^{3+} by Pb^{2+}. In $HgBa_2CuO_{4+\delta}$ oxygen excess in the Hg plane is necessary to render it superconducting.

3 The Relationship between T_c and the Hole Concentration

As mentioned earlier, a majority of the cuprates have holes as charge carriers. These holes are created by the extra positive charge on copper (*e.g.* Cu^{3+}) or on oxygen (*e.g.* O^{1-}). The excess positive charge can be represented in terms of the formal valence of copper, which in the absence of holes will be $+2$ in the CuO_2 sheets. In hole superconducting cuprates, it is generally around $+2.2$. In electron superconductors, it would be less than $+2$ as expected. The actual concentration of holes, n_h, in the CuO_2 sheets in $La_{2-x}A_xCuO_4$, $YBa_2Cu_3O_7$, and Bi cuprates is readily determined by redox titrations. In the 123 cuprates, the concentration of mobile holes in the CuO_2 sheets can be delineated from that in the Cu–O chains.[13] Determination of n_h in thallium cuprates poses some problems, but in single Tl–O layer cuprates, chemical methods have been developed to obtain reasonable estimates.[14] Generally, T_c in a given family of cuprates reaches a maximum value at an optimal value of n_h as shown in Figure 3; the maximum is around $n_h \sim 0.2$ in most cuprates.[15] Notice that the points in the underdoped region in Figure 3 fall close to a straight line. Deviations occur in the overdoped region. Single layer thallium cuprates also show this behaviour. In $Tl_{1-y}Pb_yY_{1-x}Ca_xSr_2Cu_2O_7$ where the substitution of Tl^{3+} by Pb^{4+} has an effect opposite to that due to the substitution of Y^{3+} by Ca^{2+}, the T_c becomes a maximum at an optimal value of $(x-y)$, which is a measure of the hole concentration[16] (Figure 4). By suitably manipulating x and y, the T_c of this system can be increased from 85—90K up to 110K.

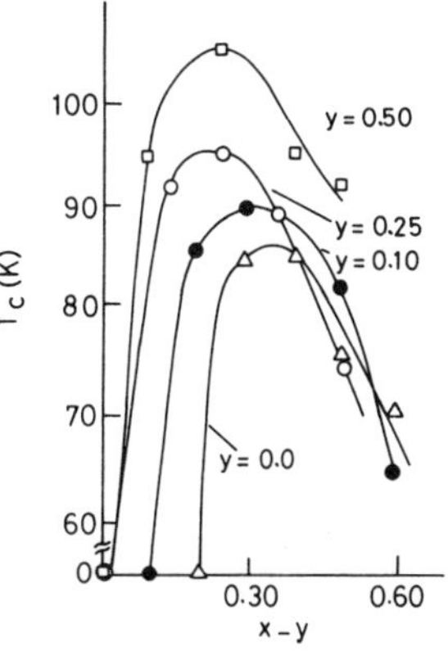

Figure 4 Plots of T_c against the effective hole concentration $(x-y)$ in $Tl_{1-y}Pb_yY_{1-x}Ca_xSr_2Cu_2O_7$ (from reference 16).

Another way of representing the variation[17] of T_c with n_h is to plot the reduced T_c (T_c observed/maximum value of T_c) against n_h as shown in the inset of Figure 3. The ends of the plateau region in this curve correspond to insulating (possibly antiferromagnetic) and metallic regimes of the materials.

4 Relation between T_c and the In-plane Cu–O Distance

The Cu–O bonds in the CuO_2 sheets involve an antibonding π interaction and doping with holes reduces the bond distance. The in-plane Cu–O bond distance r(Cu–O), therefore reflects the hole concentration and a variation of T_c with r(Cu–O) represents an alternative way of examining the T_c–n_h relationship. In cuprates where n_h cannot be determined, as for example in Tl cuprates, the T_c *vs.* r(Cu–O) plots show maxima at an optimal distance. The value r(Cu–O) is around half that of the a-parameter in most cuprates. Whangbo *et al.*[18] find three distinct T_c–r(Cu–O) relationships depending on the cation located above and below the CuO_2 sheets, with each exhibiting a T_c maximum at an optimal value of the distance (Figure 5). If we plot the reduced T_c against r(Cu–O), we get the curves shown in Figure 6 where the highest T_c values occur in the 1.89—1.94 Å

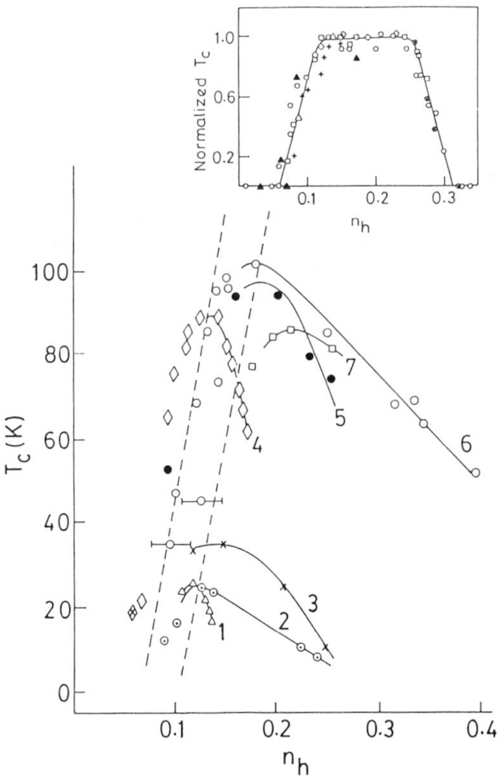

Figure 3 Variation of T_c with hole concentration n_h in superconducting cuprate families (adapted from reference 15). Deviations in overdoped region are similar to those in the plots of muon spin relaxation rate (n_h/m^*) against T_c of Uemura *et al.* (*Phys. Rev. Lett.*, 1991, **66**, 2665). 1, 2, 6, and 7, Bi cuprates; 5, 123 cuprates; 3, $La_{2-x}Sr_xCuO_4$, 4, T_c cuprates. The variation of normalized T_c with n_h is shown in the inset (from reference 17).

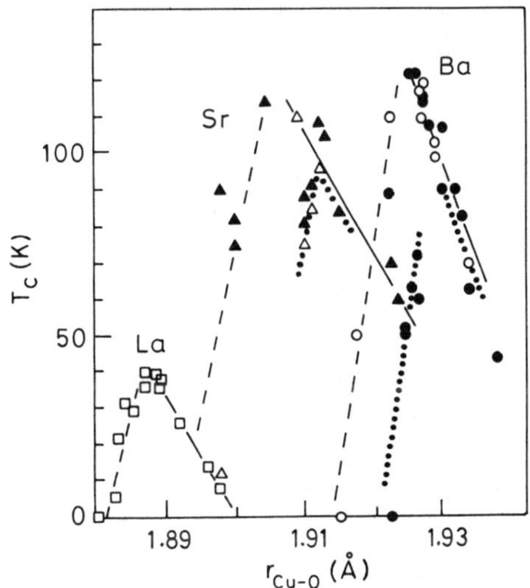

Figure 5 Variation of T_c with in-plane Cu–O distances in various families of cuprates (from reference 18).

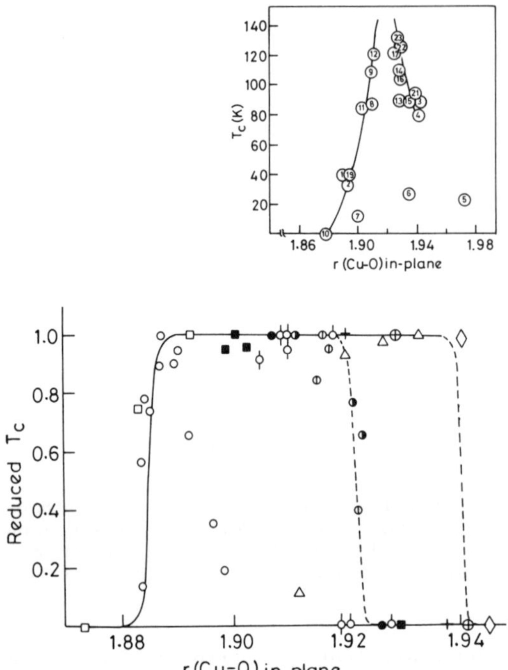

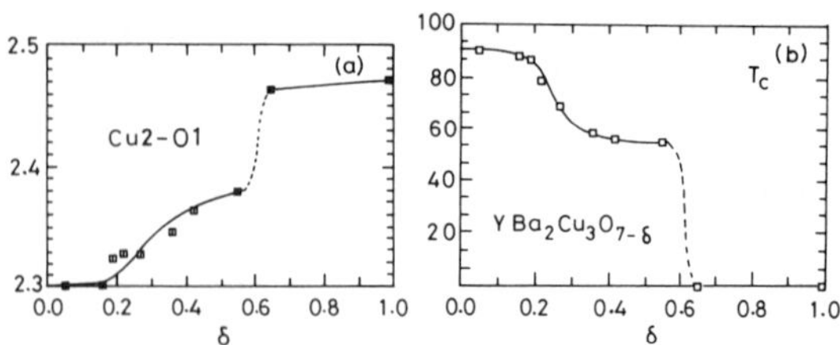

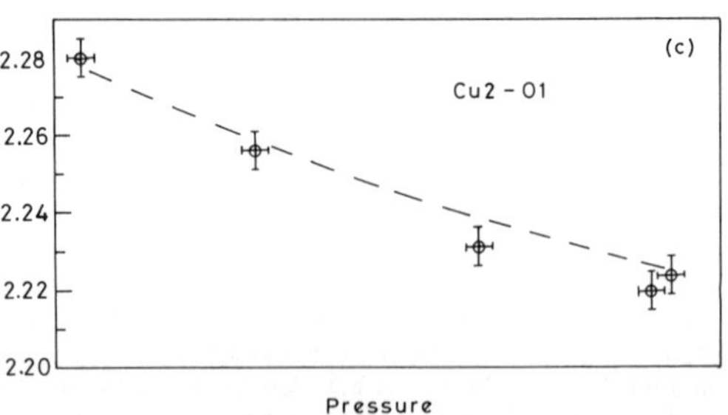

Figure 7 Variation of (a) the apical Cu–O distance with δ in $YBa_2Cu_3O_{7-\delta}$ (b) T_c with δ and (c) variation of the Cu–O apical distance with pressure in $YBa_2Cu_4O_8$ as shown by the points (the dashed line is obtained by the oxidation of $YBa_2Cu_3O_{7-\delta}$) (from references 19 and 20).

Figure 6 Plot of reduced T_c with in-plane r(Cu–O) distance: $La_{2-x}Sr_x$-CuO_4 (open circles), $YBa_2Cu_3O_{7-\delta}$ (circles with cross), Bi_2 (Ca, Y, Sr)$_3Cu_2O_8$ (half-shaded circles), $Bi_2Ca_{1-x}Y_xSr_2Cu_2O_8$ (circles with a line in the centre), modulation-free Bi-cuprates (filled squares), $TlSr_{2-x}La_xCuO_5$ (open squares), $Tl_2Ba_{2-x}Sr_xCuO_6$ (open triangles), $TlCa_{1-x}Y_xBa_2Cu_2O_7$ (crosses), $TlCa_{1-x}Ln_xSr_2Cu_2O_7$ (circles with one line on top), and $Tl_{0.5}Pb_{0.5}Ca_{1-x}Y_xSr_2Cu_2O_7$ (circles with lines above and below), $Pb_2Sr_2Y_{1-x}Ca_xCu_3O_{8+\delta}$ (filled circles), $HgBa_2CuO_{4+\delta}$ (open diamonds). Inset shows the variation of T_c with the in-plane Cu–O distance where only the cuprate compositions showing the maximum T_c in each family are taken into account. (1) $La_{1.85}Sr_{0.15}CuO_4$. (2) $La_{1.85}Ba_{0.15}CuO_4$. (3) $YBa_2Cu_3O_{6.91}$. (4) $YBa_2Cu_4O_8$. (5) $Nd_{1.85}Ce_{0.15}CuO_4$. (6) $Nd_{1.4}Ce_{0.2}Sr_{0.4}CuO_4$. (7) $Bi_2Sr_2CuO_6$. (8) $Bi_2CaSr_2Cu_2O_8$. (9) $Bi_2Ca_2Sr_2Cu_3O_{10}$. (10) $Tl_{0.5}Pb_{0.5}Sr_2CuO_5$. (11) $Tl_{0.5}Pb_{0.5}CaSr_2Cu_2O_7$. (12) $Tl_{0.5}Pb_{0.5}Ca_2Sr_2Cu_3O_9$. (13) $TlCaBa_2Cu_2O_7$. (14) $TlCa_2Ba_2Cu_3O_9$. (15) $Tl_2Ba_2CuO_6$. (16) $Tl_2CaBa_2Cu_2O_8$. (17) $Tl_2Ca_2Ba_2Cu_3O_{10}$. (18) $Tl_2Ca_3Ba_2Cu_4O_{12}$. (19) $TlSrLaCuO_5$. (2) $TlCa_{0.5}La_{0.5}Sr_2Cu_2O_7$. (21) $HgBa_2CuO_{4.065}$. (22) $HgCaBa_2Cu_2O_{6.22}$. (23) $HgCa_2Ba_2Cu_3O_{8.41}$.

range. When r(Cu–O) < 1.88 Å, the material is metallic; those with r(Cu–O) > 1.94 Å are certainly insulating, but there are different insulating boundaries for the different cation families, somewhat like in Figure 5. However, if we consider only the maximum T_c value in each cuprate family and the corresponding r(Cu–O), we get the curve shown in the inset of Figure 6 which peaks at $r \approx 1.92$ Å. $Bi_2Sr_2CuO_6$ with a T_c of 12K would not fall on the curve, but $Bi_2Sr_{2-x}La_xCuO_{6+\delta}$ with a T_c of 30K would.

5 The Relationship between T_c and the Apical Cu–O Distance

All the cuprates which are hole superconductors have apical oxygens which act as the link between the charge reservoirs and the CuO_2 sheets. (Note that electron superconductors such as $Nd_{2-x}Ce_xCuO_4$ contain only CuO_4 units without apical oxygens.) In $YBa_2Cu_3O_{7-\delta}$, the T_c–r(Cu2–O1) relationship (Figures 7a and b) mirrors the T_c–δ relationship.[19] In $YBa_2Cu_4O_8$, the T_c increases with pressure from 80K to 90K, as the apical Cu–O distance decreases[20] (Figure 7c). We have sought to find relationships between T_c and apical Cu–O distance in

different cuprate families (Figure 8). Within a series of cuprates with varying number of CuO_2 sheets, (e.g., $TlCa_{n-1}Ba_2Cu_n O_{2n+3}$), the apical (Cu–O) distance decreases with the increase in n while T_c increases linearly with the decrease in the apical distance. The slope of the T_c vs. the apical Cu–O distance plot is nearly the same in $TlCa_{n-1}Ba_2Cu_nO_{2n+3}$, $Tl_{0.5}Pb_{0.5}Ca_{n-1}Sr_2Cu_nO_{2n+3}$, and $HgCa_{n-1}Ba_2Cu_nO_{2n+3}$, all of them having a single rock-salt layer (TlO, $Tl_{0.5}Pb_{0.5}O$, or HgO_δ). The $Tl_2Ca_{n-1}Ba_2Cu_nO_{2n+4}$ family also shows increasing T_c with the decrease in the apical Cu–O distance, but with a different slope. The mercury-based superconductors have larger apical distances compared to the other cuprates and they also show a large pressure dependence of T_c. Interestingly, if we consider the maximum T_c points at the top of the plots for the different

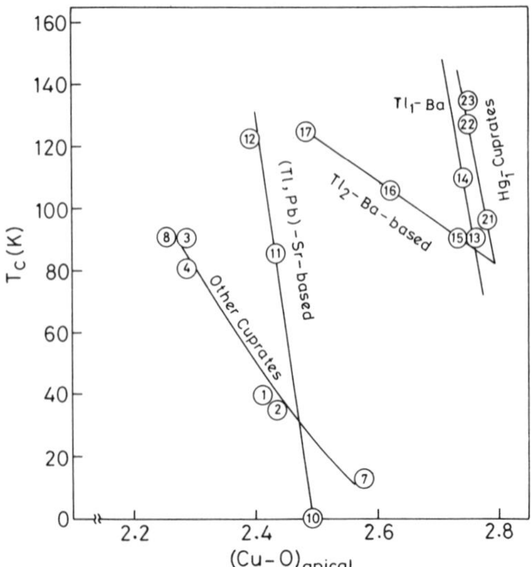

Figure 8 Variation of T_c with the apical Cu–O distance in cuprate superconductors.

groups of cuprates in Figure 8, we see that the T_c increases with the increase in the apical Cu—O distance.

6 Covalency of the Charge Reservoir

All the superconducting cuprates have charge reservoirs and any damage to these reservoirs adversely affects superconductivity. The nature of the charge reservoir determines the carrier concentration and the ease of charge-transfer to the CuO_2 sheets. Covalent charge reservoirs can redistribute charge effectively through the apical oxygen of the CuO_5 square pyramids giving rise to high T_c's. Ionic charge reservoirs, on the other hand, would be less flexible with regard to the charge states and do not favour high T_c's. Structural mismatch as well as disorder in the reservoirs also adversely affect the superconducting properties. The covalency of the Hg—O bond could be related to the high T_c of Hg cuprates.

The effect of covalency of the charge reservoir is clearly seen in $Tl_{0.5}Pb_{0.5}Sr_2Y_{1-x}Ca_xCu_2O_7$ which shows a T_c of 110K at an optimal x value, the material being an insulator when $x = 0.0$. The a parameter decreases with the increase in x, because the population of the antibonding Cu $3d_{x^2-y^2}$ orbitals decreases with the increase in x, causing a strengthening of the Cu—O bond. The puckering of the CuO_2 sheets decreases with increasing hole concentration. The displacement of the apical oxygen is around 0.06 Å when $x = 1.0$ and 0.20 Å when $x = 0.0$. An increase in Y content (increased electron population), however, increases the puckering and pulls the apical oxygen away from the base of the pyramid.

7 The Relationship between T_c and Madelung Potentials

The role of the Madelung site potential in the hole conductivity of the cuprate superconductors was first pointed out by Torrance and Metzger.[21] Two classes of cuprates can be delineated depending on the value of ΔV_M which is the difference in Madelung site potential for a hole on a Cu site and that on an oxygen site. Those with high ΔV_M ($\geq 47eV$) are metallic and superconducting; those with lower ΔV_M are semiconducting with localized holes. It is possible to define a term ΔV_A which is the difference in the Madelung site potentials for a hole between the apex and the in-plane oxygen atoms and provides a measure of the position of the energy level of the p_z-orbital on the apical oxygens.[22] When the maximum values of T_c of hole-doped superconductors are plotted against ΔV_A (Figure 9), one finds that nearly all the cuprates are located on a curve (with some width), the cuprates with large ΔV_A exhibiting high T_c's. It appears that the energy level of the apical oxygen plays a significant role in the electronic states of the doped holes,

thereby affecting the T_c. The correlation probably owes its origin to the stability of local singlet states made up of two holes in the Cu $3d_{x^2-y^2}$ and O $2p$ orbitals in the CuO_2 sheet. The local singlet is well defined and stable when the energy level of the apical oxygen atom is sufficiently high. A comparison of the correlations of T_c with ΔV_A and ΔV_M indicates that ΔV_A scales better with T_c.

It is instructive to correlate T_c simultaneously with ΔV_M and the in-plane Cu—O bond length, d_p, in the CuO_2 planes. The maximum T_c for each cuprate is shown in the d_p vs. ΔV_M plot[23] in Figure 10. The data points are confined to a narrow strip running from the top left (high d_p and low ΔV_M) to the bottom right corner (low d_p and high ΔV_M). The T_c value increases as we go from right to left (larger to smaller value of ΔV_M) or from top to bottom (from higher d_p to lower d_p). An important observation is that T_c changes to a small extent as one goes from the top left to bottom right (d_p decreasing and ΔV_M increasing) while T_c increases drastically when traversing from the top right to bottom left of the map (both d_p and ΔV_M decreasing). Clearly, the T_c is governed by both ΔV_M and d_p. It appears that the difference in T_c is a result of the difference in internal stress of the crystal (high T_c when the CuO_2 planes are under compression and low T_c when they are strained).

8 The Relationship between Bond Valence Sums and T_c

The bond-valence sum is a measure of the total charge on an atom in a structure. Its value changes with oxygen doping, cation substitution, or applied pressure indicating the occurrence of charge transfer within the structure. One defines, $V_- = 2 + V_{Cu2} - V_{O2} - V_{O3}$ where V_- is the total excesss charge in the planes (O2 and O3 are the oxygens in the plane) and $V_+ = 6 - V_{Cu2} - V_{O2} - V_{O3}$. The T_c vs. V_- plot for $YBa_2Cu_3O_{7-\delta}$ is similar to the T_c vs. δ plot; the plot of T_c against V_+ is linear[24] (Figure 11).

9 The Importance of the Cu—O Charge-Transfer Energy

One of the unique features of the cuprates is the relatively small Cu—O charge-transfer energy. It is therefore of significance to relate this property with superconductivity. X-Ray photoemission spectra of cuprates in the Cu $2p_{3/2}$ region show a characteristic two-peak feature. The peak around 933 eV (main peak) is due to a final state with primarily $3d^{10}$ character, while the peak of weaker intensity at about 941 eV binding energy is mainly due to a $3d^9$ final state (satellite). Model calculations of the Cu $2p$ core-level photoemission within a CuO_4 square-planar cluster show that the I_s/I_m ratio (relative intensity of the satellite to the main peak) is related to Δ/t_{pd} where Δ is the charge-transfer excitation energy and t_{pd} is the hybridization strength between the Cu $3d$ and O $2p$ orbitals. Thus, any relationship between the experimentally observed I_s/I_m ratio with n_h would suggest a link between n_h (T_c) and the Δ/t_{pd} ratio. It is indeed found that the I_s/I_m ratio is related to the hole concentration n_h in many of the superconducting cuprates.[25]

In Figure 12(a), we show the relative intensity of the satellite (I_s/I_m) as a function of x in $La_{2-x}Sr_xCuO_4$. In the same plot, we also show the variation of the experimentally determined hole concentration with x in this series. The inset of Figure 12(a) shows the dependence of T_c on n_h exhibiting a maximum around $n_h = 0.15$. We see that the value of I_s/I_m decreases markedly with increasing x until about 0.3, after which it changes slope. What is significant is that I_s/I_m decreases as the hole concentration increases. In Figure 12(b) we show the variation of n_h and I_s/I_m with x in $BiPbSr_2Y_{1-x}Cu_2O_8$. In the inset we show the variation of T_c with n_h. Here also we see that the I_s/I_m ratio decreases as the hole concentration increases. Figure 13 shows how in three series of high T_c cuprates, the I_s/I_m ratio decreases monotonically with the increase in magnitude of hole doping. The

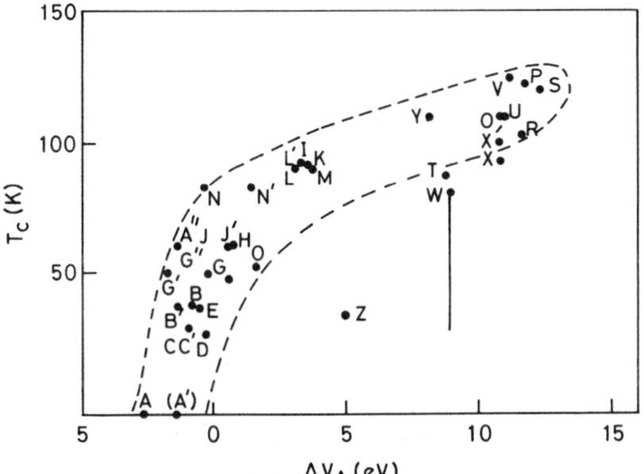

Figure 9 Variation of T_c with ΔV_A. See reference 22 for the explanation of the letter symbols.

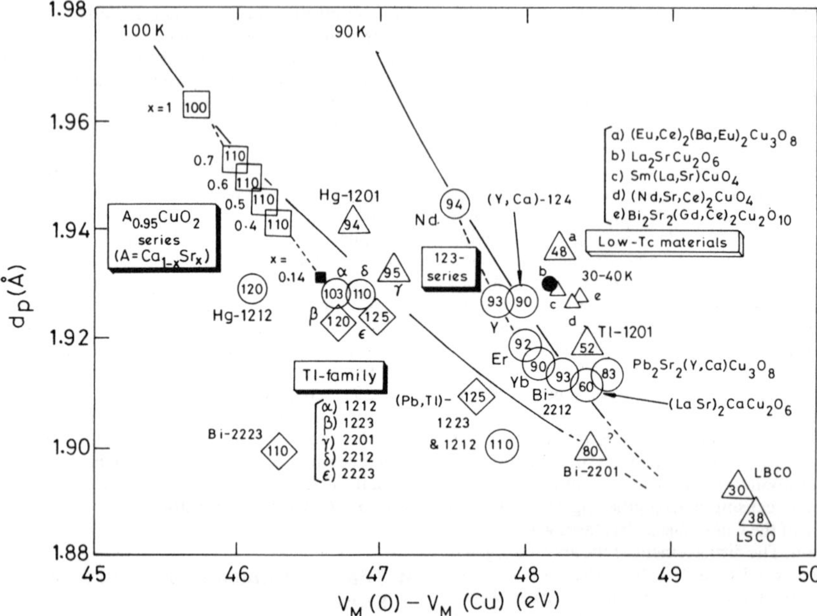

Figure 10 Plot of d_p *vs.* ΔV_M showing region of high T_c and low T_c superconductors (from reference 23).

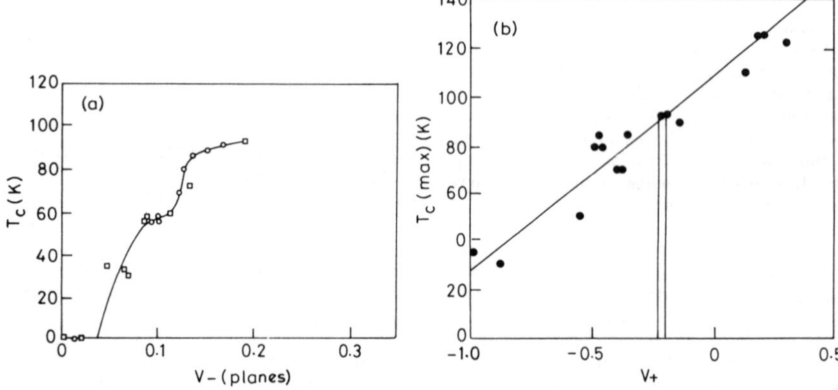

Figure 11 Variation of (a) T_c with V_- (see text) and (b) T_c with V_+ for $YBa_2Cu_3O_{7-\delta}$ (from reference 24).

variation is smooth across the insulator–superconductor–metal boundaries. Calculations show that the I_s/I_m ratio increases with the Δ/t_{pd} ratio in the cuprates. Accordingly, at a given n_h value where the T_c is maximum, increasing I_s/I_m is accompanied by an increase in T_c. Thus, $Bi_2Ca_{1-x}Ln_xSr_2Cu_2O_{8+\delta}$ exhibits the highest T_c of around 100K while $La_{2-x}Sr_xCuO_4$ has the lowest T_c value. The observation of a maximum T_c around an optimal value of n_h or Cu–O distance can therefore be traced to the optimal Δ/t_{pd} value or of the Cu–O charge-transfer energy.

10 Concluding Remarks

We have discussed several interesting correlations between the superconducting transition temperature and some of the crucial structural and electronic parameters. These correlations along with the structural commonalities in the cuprates show how superconductivity is intimately connected with the structural

chemistry of these materials. They may also suggest many other relationships, some of which may be even more significant.

Acknowledgement. The authors thank the Science Office of the European Union, the Department of Science and Technology, and the University Grants Commission for support of this research.

7 References

1 C. N. R. Rao, K. J. Rao, and J. Gopalakrishnan, *Ann. Rep. Prog. Chem., Sect C*, 1985, **82**, 193.
2 J. Bednorz and K. A. Müller, *Z. Phys.*, 1986, **B64**, 189.
3 B. Raveau, C. Michel, M. Hervieu, and D. Groult, 'Crystal Chemistry of High T_c Superconducting Copper Oxides', Springer-Verlag, Berlin, 1991.
4 'Chemistry of High-Temperature Superconductors', ed. C. N. R. Rao, World Scientific, Singapore, 1991.

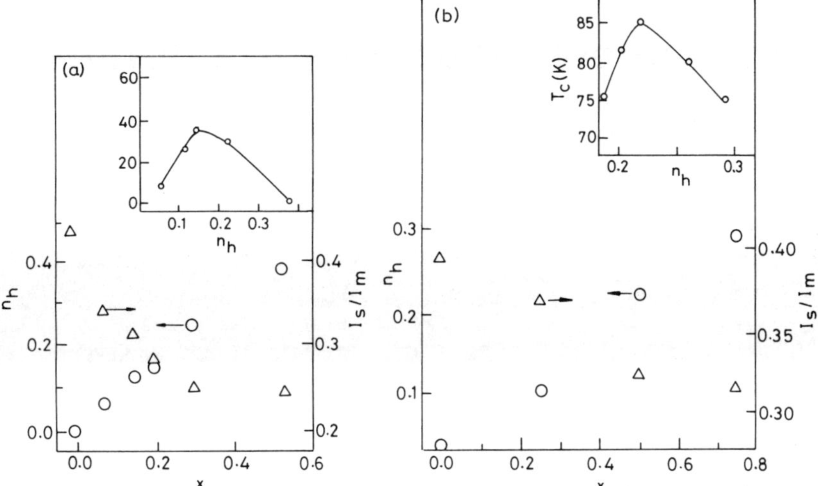

Figure 12 (a) Variation of I_S/I_M ratio with x in $La_{2-x}Sr_xCuO_4$. (b) Also shown is the variation of n_h with x and I_S/I_M and n_h with x in $BiPbSr_2Y_{1-x}Ca_xCu_2O_8$ (from reference 25).

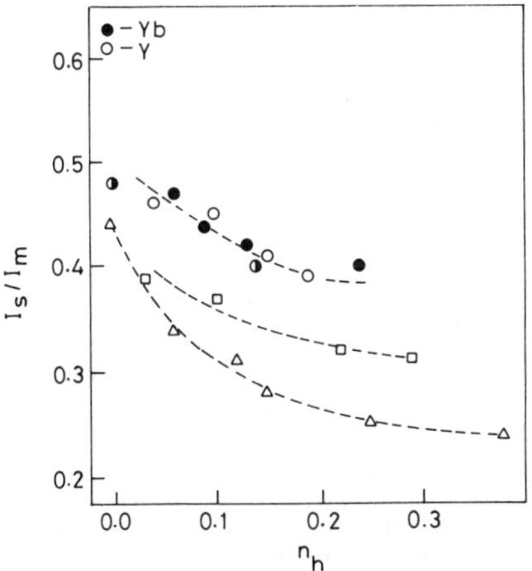

Figure 13 Variation of the I_S/I_M ratio with hole concentration (n_h): $La_{2-x}Sr_xCuO_4$ (triangles), $BiPbSr_2Y_{1-x}Ca_xCu_2O_8$ (squares), and $Bi_2Ca_{1-x}L_xSr_2CuO_2O_8$ (circles) (from reference 25).

5 M. Nunez-Regueiro, J. L. Tholence, E. V. Antipov, J. J. Capponi, and M. Marezio, *Science*, 1993, **262**, 976.
6 A. R. Armstrong and P. P. Edwards, *Ann. Rep. Prog. Chem., Sect. C*, 1991, **88**, 259.
7 C. N. R. Rao, *Philos. Trans. R. Soc. London*, 1991, **A336**, 595, 106.
8 H. Zhang, Y. Y. Wang, V. P. Dravid, L. D. Marks, P. D. Han, D. A. Payne, P. G. Radaelli, and J. D. Jorgensen, *Nature*, 1994, **370**, 352, and references therein.
9 H. Nobumasa, K. Shimizu, and T. Kawai, *Physica C*, 1990, **167**, 515.
10 R. Nagarajan and C. N. R. Rao, *J. Solid State Chem.*, 1993, **103**, 533.
11 J. D. Jorgensen, *Physics Today*, 1991, **44**, 34
12 V. Manivannan, J. Gopalakrishnan, and C. N. R. Rao, *Phys. Rev. B*, 1991, **43**, 8686.
13 R. Nagarajan and C. N. R. Rao, *J. Mater. Chem.*, 1993, **3**, 969.
14 C. N. R. Rao in 'Thallium-based High Temperature Superconductors , ed. A. M. Hermann and J. V. Yakhmı, Marcel Dekker, New York, 1994.
15 C. N. R. Rao, J. Gopalakrishnan, A. K. Santra, and V. Manivannan, *Physica C*, 1991, **174**, 11.
16 R. Vijayaraghavan, N. Rangavittal, G. U. Kulkarni, E. Grantscharova, T. N. Guru Row, and C. N. R. Rao, *Physica C*, 1991, **179**, 183.
17 H. Zhang and H. Sato, *Phys. Rev. Lett.*, 1993, **70**, 1697.
18 M. H. Whangbo, D. B. Kang, and C. C. Torardi, *Physica C*, 1989, **158**, 371.
19 R. J. Cava, A. W. Hewat, E. A. Hewat, B. Batlogg, M. Marezio, K. B. Rabe, J. J. Krajewski, W. F. Peck, and L. W. Rupp, *Physica C*, 1990, **165**, 419.
20 R. J. Nelmses, E. Loveday, E. Kaldis, and J. Karpinski, *Physica C*, 1992, **172**, 311.
21 J. B. Torrance and R. M. Metzger, *Phys. Rev. Lett.*, 1989, **63**, 1515.
22 Y. Ohta, T. Tohyama, and S. Maekawa, *Physica C*, 1990, **167**, 515.
23 M. Muroi, *Physica C*, 1994, **219**, 129.
24 J. L. Tallon and G. V. M. Williams, *J. Less Common Metals*, 1990, **164—165**, 60.
25 A. K. Santra, D. D. Sarma, and C. N. R. Rao, *Phys. Rev. B*, 1991, **43**, 5612.

Stripes and Superconductivity in Cuprates – Is there a Connection?

N. Kumar[a] and C. N. R. Rao*[b]

KEYWORDS:

cuprates · scanning probe microscopy · soft X-ray scattering · stripes · superconductors

Charge stripes can be traced back to the 1990s or even earlier.[1] Stripes are a special case of the general phenomenon of electronic phase separation in strongly correlated transition metal oxide systems wherein regions with a high concentration of mobile carriers are separated from those with a low concentration due to Coulomb interactions. The carrier-rich regions are generally metallic and/or ferromagnetic, while the carrier-poor regions are antiferromagnetic and insulating. Such phase separation manifests itself in the form of metallic droplets in an antiferromagnetic insulating matrix or self-organization, for example, in one dimension as stripes.[1, 2] The phase-separated regions tend to be of nanometric dimensions.

Over the years, experimental evidence[1–12] has been steadily mounting in support of the theoretical[13–19] proposition that the cuprate superconductors may be intrinsically inhomogeneous in the distribution of their spins and charges in the CuO_2 sheets, and that this inhomogeneity may hold a clue to the mechanism of high-T_c (high critical temperature) superconductivity in these materials. The growing realization that stripes matter has enlivened the debate on high temperature superconductivity in these ceramic materials, and emphasizes the incompleteness in our understanding of the phenomenon discovered in 1986 by Bednorz and Müller.[20]

Stripes have been reported in at least three families of hole-doped high-T_c cuprates, all of which contain the CuO_2 sheets. These are the single-layered cation-doped 214 compound LSCO ($La_{2-x}Sr_xCuO_4$) or the oxygenated (anion-doped) LCO214 (La_2CuO_{4+x}), the 123 compound YBCO ($YBa_2Cu_3O_{7-x}$), which contains CuO chains in addition to the CuO_2 sheets, and the two-layered BSCCO-2212 ($Bi_2Sr_2CaCu_2O_{8+x}$) with two CuO_2 layers per unit cell. The strongest case for the stripes so far has been made for the model system NdLSCO ($La_{1.6-x}Nd_{0.4}Sr_xCuO_4$),[2] in which the stripes are structurally pinned, nearly static. The commonality of the all-important CuO_2 sheets, and the associated low-dimensional physics,[21–23] which is known to favor solitons and the solitonic doping, has been the compelling reason to believe that stripes may be generic to the layered cuprates—at least in the underdoped region of the phase diagram, which is shown schematically in Figure 1.

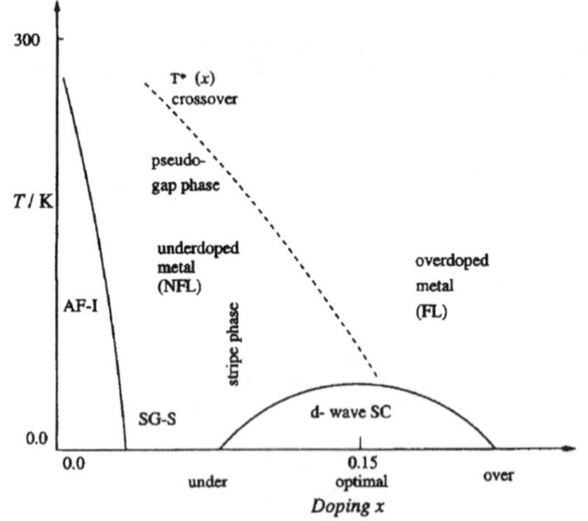

Figure 1. Schematic temperature (T) – doping (x) phase diagram typical of hole-doped high-T_c layered cuprate superconductors. AF-I denotes antiferromagnetically ordered insulator; SG-S, the spin-glass semiconductor; d-wave SC, the $d_{x^2-y^2}$ ordered superconductor; and T*(x), the crossover temperature to the pseudo-gapped, abnormal metallic underdoped phase. Stripe phase occupies an underdoped pseudogapped region, possibly extending into the superconducting phase. The overdoped region is a normal metal.

Charge stripes have also been found in a few other transition metal oxides and were first reported in the isostructural hole-doped nickelates, $La_{2-x}Sr_xNiO_4$ and La_2NiO_{4+x} which contain modulated NiO_2 sheets.[24] But there is a striking difference between the nickelates and the cuprates in that the stripe phase in the former is insulating, much less superconducting, unlike in

[a] Prof. Dr. N. Kumar
Raman Research Institute
Bangalore 560 080 (India)

[b] Prof. Dr. C. N. R. Rao
CSIR Center of Excellence in Chemistry and Chemistry & Physics of Materials Unit
Jawaharlal Nehru Center for Advanced Scientific Research
Bangalore 560 064 (India)
Fax: (+ 91) 80 846 2760
E-mail: cnrrao@jncasr.ac.in

CHEMPHYSCHEM

the cuprates where it is conducting or superconducting, thus providing a case for a study in contrast between two related layered oxides systems with the spin-$1/2$ containing Cu^{2+}:$3d^9$ and the spin-1 containing Ni^{2+}:$3d^8$. Although cuprates are superconducting, while the nickelates are insulators, the characteristics of the stripes in the two materials are comparable in that for example, the stripes in both the systems are fluctuating, strongly correlated fluids. Interestingly, stripes have been reported in the rare earth manganates of the general formula, $Ln_{1-x}A_xMnO_3$ (Ln = rare earth metal, A = alkaline earth metal) as well.[25] Here, ferromagnetic metallic stripes are bounded by antiferromagnetic insulating stripes.

While the existence of charged stripes is not much in doubt, their relevance to high-temperature superconductivity (HTSC) is by no means established, particularly in the light of the recent anomalous X-ray scattering measurements on $La_2CuO_{4+\delta}$[26] (see, however, refs.[27, 28]). The presence of stripes raises many questions. Thus, we may ask whether they constitute a truly thermodynamic phase. Does the stripe order compete with superconductivity in the layered cuprates? Does it promote HTSC? Does it coexist with superconductivity—trivially, spatially separated from it, or nontrivially, overlapping with it? Or, as is likely, do the stripe order and the HTSC have a common underlying physics? How can the charged stripes, with 1D topology, subtend a 2D electrical conductivity, let alone superconductivity? Or, is it possible that the 1D stripes merely provide a solitonic route to hole doping into the 2D CuO_2 sheets at low levels of doping, as in its low-dimensional analogs in which point defects (zero-dimensional domain walls) provide solitonic doping, for example, polyacetylene.[21-23] Herein, we address some of these and related issues, including the provenance of stripes.

Figure 2 depicts schematically the geometry and the spin-charge order for an in-plane stripe as derived from low-energy neutron diffraction/scattering in a typical hole-doped cuprate

superconductor, for example, $La_{2-x}Sr_xCuO_4$ (LSCO), or NdLSCO ($La_{2x-y}Nd_ySr_xCuO_4$) for $x = 1/8$ per CuO_2 unit.[2] The charge carriers (holes), which are introduced by doping, are segregated into linear (1D) domain walls—the charged stripes—that are barely a few atoms wide and separate the antiferromagnetically (AFM) ordered planar (2D) domains in all the CuO_2 sheets. Furthermore, the domain walls are antiphase, so that the AFM domain order is phase shifted by π across the charged domain wall. This shift has the consequence of making the spatial period of the magnetic stripes twice that of the charge stripes. This factor of two follows from a general consideration of the ground state of the interacting spin–hole system. Such a real-space static stripe modulation gives the incommensurate elastic magnetic peaks, which are positioned in the reciprocal space at $(2\pi/a)$. The main Bragg diffraction peak for the parent AFM order is (0.5, 0.5). The incommensuration δ is a function of doping x (at low doping, $x \approx \delta$) and is related to the real interstripe spacing ($= a/2\delta$). The tetragonal zone indexing (h,k) is appropriate to the planar geometry of the CuO_2 sheets.

In simplistic terms, the doped holes in the cuprates are mostly on the oxygen in the CuO_2 sheets creating oxygen 2p-holes, while the spins reside on the copper (the Cu^{2+}:$3d^9$-spins). In Figure 2 we have only shown the copper atoms. The charge stripes are aligned parallel to the Cu—O bonds, but may be centered on the sites (on the Cu—O—Cu legs), on the bonds (between two Cu—O—Cu legs), or in between. The stripe phase is singly modulated, but the stripes in the adjacent CuO_2 planes are at 90° because of the symmetry-lowering lattice distortions. Hence, no incommensurate peaks occur at $(2\pi/a)$ $(1/2 + \delta, 1/2 + \delta)$, or $(2\pi/a)$ $(1/2 - \delta, 1/2 - \delta)$. Figure 2 assumes the stripes to be static; this is the case for low levels of doping (for the underdoped samples), where one observes narrow diffraction (elastic) peaks. In the optimally doped superconducting samples, on the other hand, the stripes fluctuate, giving relatively broad inelastic peaks, still positioned incommensurately about the AFM diffraction peaks. In the overdoped samples the stripes simply fade away. We may recall that the strongest evidence for the stripes comes from NdLSCO ($La_{2-x-y}Nd_ySr_xCuO_4$, with $x = 0.12$, $y = 0.4$),[2] in which the stripes are immobilized by commensuration pinning to the lattice distortion of the low-temperature tetragonal (LTT) phase. It is the modulation of the lattice distortion induced by the stripe charge order that is picked up in neutron scattering as an indirect signature of the charged stripes.

Let us briefly examine the nature of the stripes in cuprates. Several of the reciprocal-space bulk scattering probes and real-space surface-imaging tools have been used to study the stripes. The two methods are somewhat complementary. The scattering probes can detect fluctuating stripes as well, while surface imaging can only probe the static stripes at atomic resolution. The geometry of the stripe order in the bulk is best seen by neutron diffraction. If the stripes can be immobilized through pinning due to impurity atoms and structural commensuration as in NdLSCO, the applicability of elastic scattering allows a more thorough investigation. In order to probe spatially modulated dynamical correlations, low-energy ($\approx$few meV) inelastic neutron scattering is employed. The problem, however, is that neutrons couple directly to the electron spin and to the positions

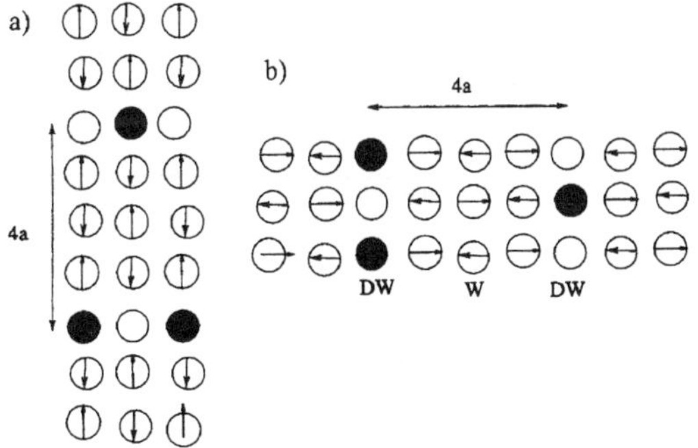

Figure 2. A schematic diagram of the stripe showing spin-charge order in CuO_2 plane for hole doping $x = 1/8$. AFM ordered planar domains separated by charged linear domain walls (DW)—the charge stripes. The domain walls are antiphase: with AFM order phase shifted by π across it, making the spacing of magnetic stripe (8a) twice that of the charged stripes (4a). The open circle with arrow denotes a spin $1/2$ on copper, and the filled circle denotes a hole nominally on copper (actually on oxygen, not shown) with one hole per two copper atoms in the linear DW. The stripes in the adjacent planes (a) and (b) are orthogonal to each other.

of the nuclei, whose forced displacement in response to any charge-ordering gives only an indirect evidence for charge order. Thus, the incommensurate, split superlattice peaks at $(2\pi/a)$ $(2 \pm 2\delta)$ and their near absence at $(2\pi/a)$ $(2 \pm 2\delta)$ constitute indirect evidence for the charged stripes (with a spacing half of that for the magnetic stripes). For direct evidence, one must turn to X-rays that directly couple to the charge. X-rays, however, couple nonspecifically to all the charges, including those of the core electrons that outnumber the mobile charge carriers (doped holes) by a factor of ≈ 500 to 1, for example, in optimally doped LSCO. The normal X-ray diffraction cannot resolve this small proportion. This problem has been overcome recently by Abbamonte et al.[26] by the use of multiwavelength anomalous soft X-ray scattering, wherein the elastic (or the low-energy inelastic) scattering is resonantly and selectively enhanced many-fold (by over a factor of $\approx 10^3$) by operating close to an absorption pre-edge (pre-edge of the OK shell) because of the near-vanishing of the energy denominator that is associated with the virtual excitation of the O (1s) electron to the Fermi level vacated due to hole doping. Even this enhanced sensitivity and specificity have failed to detect incommensurate X-ray diffraction peaks corresponding to a charged-stripe modulation in La_2CuO_{4+x} films of atomic perfection deposited on $SrTiO_3$ substrates. But, this new technique clearly needs to be tested out on samples proven to give the incommensurate magnetic peaks in neutron scattering. Also, Abbamonte et al.[26] studied the sample at a high temperature ($> T_c$), and the stripes may well appear below T_c. The issue is, therefore, not fully resolved.

Complementary to the soft X-ray scattering probe, which is a bulk probe of mesoscopic resolution, is the real-space surface imaging by scanning tunneling microscopy (STM) at atomic resolution. STM measures the differential tunneling conductance, which is in turn proportional to the local density of states (LDOS), $\alpha \sum_k |\varphi_k(r)|^2 \delta[E - e(k)]$, at the position (r) of the STM tip and at the energy E corresponding to the bias voltage for the charge-carrier injection. An earlier study of a single crystal of BSCCO 2212, cleaved along the BiO planes, reported charge-stripe modulations for optimal doping in the superconducting phase.[27] This result is subject to some doubt since STM studies find that the spatial modulation wavelength varies with the STM bias voltage, which suggests an extrinsic interpretation involving interference of the electron waves, which are scattered from defects/impurities—akin to Friedel oscillations or more accurately, von Laue-theoretic oscillations. Stripes should have given intrinsic voltage-independent spatial oscillations. It is, however, possible that the latter are masked by the extrinsic effect. It is important to note that, over and above the modulations due to any possible stripe order, there are microscopic inhomo-

geneities arising from local doping concentration (LDC) effects, because of the poor screening of the dopant potentials in these doped Mott insulators. The LDC effects have been isolated[29] by Fourier-filtering out of the stripe modulations, and remarkable correlations of, for example, the LDOS, the superconducting gap, and the doping, have been found; this is indeed intriguing.

The topological issue of one-dimensionality of the charge stripes, namely, that the mobile charge carriers must move along the linear tracks, defined laterally by the Mott-insulating AFM domains on the sides, is best probed through studies of charge motion in real space (for example, Hall effect measurements) and in the reciprocal (momentum) space (for example, through angle resolved photo electron spectroscopy; ARPES).

The 1D topology should block any off-diagonal Hall response making the Hall coefficient vanish, even for an arbitrary collection of nonpercolating stripes containing mobile holes. This is found to be the case in NdLSCO ($La_{1.4-x}Nd_{0.6}Sr_xCuO_4$).[6] At low doping ($x < 1/8$, the magic number at which the stripes are known to be structurally pinned in the LTT phase) the Hall coefficient decreases rapidly with decreasing temperature. For $x > 1/8$, however, where the stripes are known to be fluctuating, the Hall coefficient is large and featureless despite the charge-stripe order. The magic number $x = 1/8$ seems to mark a cross over from 1D to 2D-hole motion. The 1D electronic nature of the stripes with mobile carriers is strongly supported by the ARPES studies.[4]

ARPES measures the single-particle spectral function $A(\mathbf{k},w)$, within the photoionization weight factor (w), and gives, on frequency integration the in-plane momentum $(\mathbf{k})$ distribution, function

$$n(\mathbf{k}) = \int_{-\infty}^{+\infty} A(\mathbf{k},w)f(w)dw/2\pi,$$

from which one constructs the Fermi surface that contains the high spectral weight region. Figure 3a shows such a Fermi surface, which is redrawn schematically from ARPES measurements on NdLSCO.[4] Figure 3b shows the Fermi surface expected of the orthogonal 1D charge stripes of the type shown in Figure 2, where the vertical dotted lines are along [100] and the horizontal lines (solid lines) are along [010]. (Recall that the

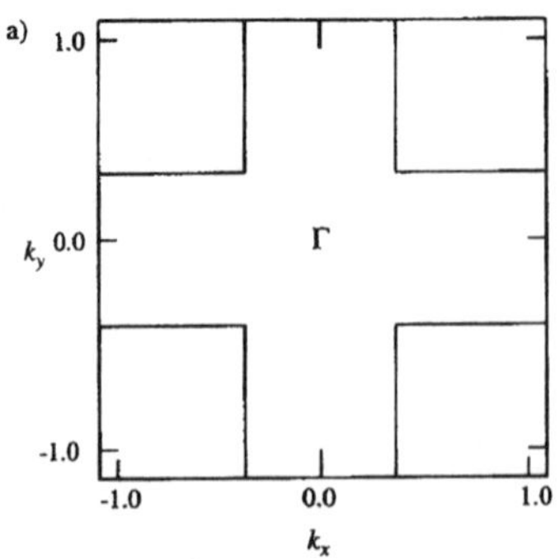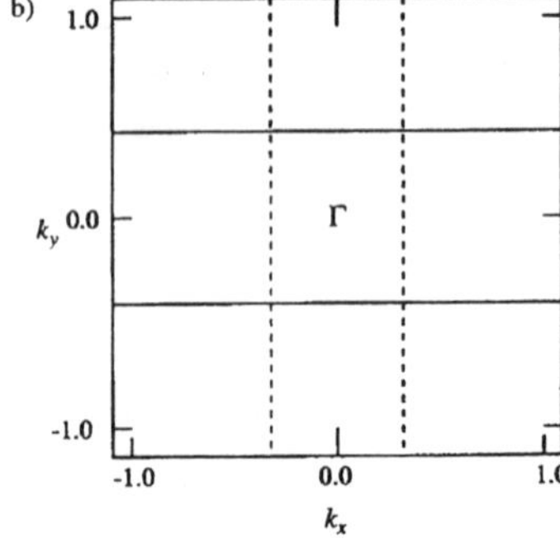

Figure 3. a) A schematic of the ARPES results for the Fermi surface of NdLSCO for $x = 1/8$, taken and redrawn from ref.[4]. b) Fermi surfaces expected for $1/4$-filled 1D bands, for comparison. Dashed lines (----) are for the horizontal stripes, and solid lines (——) for the vertical stripes.

stripes in the adjacent singly modulated CuO_2 planes are orthogonal). The striking similarity suggests a $1/4$-filled 1D band for a $1/2$-filled charge stripe corresponding to one hole per two copper atoms in the linear stripe and satisfying the Luttinger theorem. The Fermi surfaces, or rather lines, are $(\pm\pi/4a, k_y)$ and $(k_x, \pm\pi/4a)$. A 2D homogeneous electronic structure clearly appears to be discredited, which thus suggests the striped nature of high-T_c cuprates.

Whether there is any relation between stripes and cuprate superconductivity is the important question facing us today. Systematics of the dependence of the magnetic incommensur-ation δ and the superconductivity transition temperature T_c on the hole concentration (the value of x) in LSCO and NdLSCO show striking correlations.[2, 3] First, superconductivity is anom-alously suppressed (T_c goes to almost zero) for $x = 1/8$, the magic number at which the stripes are structurally pinned, showing up in low-energy neutron scattering as incommensurate peaks of small momentum width. Second, away from this magic number, the dependence of δ and of T_c on x from underdoping to optimal doping closely track each other. Thus, the critical doping for the onset of superconductivity and for the appearance of the magnetic stripes are the same; signalled by the shifting of the neutron diffraction peaks away from the AFM position to the incommensurate split-peak positions. The incommensurate, low-energy inelastic magnetic peaks remain narrow in momentum in the superconducting phase, except in the weakly superconduct-ing overdoped region, where they become broad and incoher-ent. Thus, fluctuating stripes are correlated with, and certainly can coexist with, superconductivity. Static stripes, on the other hand, appear to suppress superconductivity. ^{63}Cu NQR (Nuclear Quadrupole Resonance) measurements that detect the charge dynamics in the proximate charge stripes through the electric field gradient they produce, however, indicate that static stripes and superconductivity can coexist.[30]

An intriguing aspect of the doping dependence of incom-mensuration δ is that the linear variation of δ with x saturates at $x = 1/8$, while T_c continues to increase up to optimal doping. While the linearity of δ with x implies that the added holes are accommodated in the linear stripes, occurrence of saturation suggests an alternative mode of accommodation of the holes. This is reminiscent of the low-dimensional polyacetylene-like system,[22] where solitonic doping gives way to band-type doping beyond a critical doping. This would suggest that the stripes may well be a solitonic route to hole doping at low levels.

We now turn to microscopic considerations of the stripes and their implication for a theory of HTSC. Incommensurate modu-lations, structural or electronic (charge and spin), can result from competing length scales, for example, the length scale set by the nested Fermi surface and that of the lattice periodicity as in the Peierls or spin-Peierls instability that leading to spatial modu-lations (and gapping of the Fermi surface), well known in low-dimensional systems as the CDW (charge density wave) and the SDW (spin density wave) orders.[21, 23] In the case of cuprates, however, it is known that charge ordering occurs at a higher temperature than spin ordering, and so the stripe phase is charge driven[2]—as in $La_{2-x}Sr_xCuO_4$ as well as in the nickelates. This is not consistent with the usual Fermi surface instabilities,

where the spins would order first. Incommensuration can also arise from competing interactions such as short-range attraction and the long-range repulsion, or short-range FM and the long-range AFM interactions in a spin-1 lattice gas that frustrate any macroscopic phase separation. Charge stripes, and the magnetic stripes driven by the charge order, ultimately result from the fact that the cuprates are essentially hole-doped Mott (or rather charge-transfer) insulators and the doped holes are energetically expelled from the AFM ordered domains, which they would otherwise disrupt through their motion. This basic physics is well captured by a short-range two-band 2D Hubbard model with strong on-site repulsion (the Hubbard correlation energy) on the copper atoms, and the $3d_{x^2-y^2}$-$2p_{x,y}$ hybridization between copper and oxygen, and the energy of charge transfer, Cu^{2+}:$3d^9 \rightarrow Cu^{1+}$:$3d^{10}$ + oxygen 2p-hole, in these strongly corre-lated insulators.[13] A mean-field treatment indeed gives charge-stripes, but they turn out to be insulating, as in the nickelates (unlike the conducting cuprate stripes). Antiphase spin-charge stripes of the form shown in Figure 2 result from a short-range one-band $t-J$ model.[19] Thus, a long-range repulsion is not a necessary condition for the occurrence of the stripes.

Next, we shall look at the possible connection between stripes and the pairing mechanism. Almost all the nonphonon pairing mechanisms proposed so far invoke interaction between spin fluctuations (magnetism) and hole motion (charge) in some form.[31] For a pairing mechanism involving spin–hole interac-tion, the electronic specific heat near the superconducting transition must track the temperature derivative of the average exchange energy of the spins. This follows from the thermody-namics of the superconducting phase transition, derived first in the context of the conventional electron–phonon pairing mechanism,[32] and generalized later to the spin-fluctuation mechanism.[33] This mechanism is really much the same as the temperature derivative of the spin-disorder resistivity that shows the critical behavior of the electronic specific heat in bad itinerant metals.[34] Recent inelastic neutron scattering measure-ments[35] of the spin-fluctuation spectrum of YBCO123, which are enhanced by the so-called π-resonance at the wavevector (π/a, π/b, π/c), verify the above relation between the temperature dependence of the AFM exchange energy and the electronic specific heat.[36] While this does implicate magnetic fluctuations as a hole-pairing mechanism, it is too nonspecific to support any stripe-based theory.

A minimal model for the layered cuprates is the 2D short-range single-band Hubbard model that effectively describes the strongly correlated system.[37, 38] The undoped parent material, for example, La_2CuO_4, is a Mott insulator, as can be seen from the electron count, for example, $(La^{3+})_2(Cu^{2+})(O^{2-})_4$, that would give a half-filled band and is therefore a metal. It becomes conducting when doped away from stoichiometry as in, $La_{2-x}Sr_xCuO_4$. When the doping is not too small, the holes get unbound from the charged dopants as their potentials begin to overlap. The hole-doped Mott insulator is then well-described in the space of nondouble occupancy by the so-called $t-J$ Hamiltonian [Equa-tion (1) and (2)][38]

$$H = -t \sum_{\langle ij \rangle, \sigma} (C_{i\sigma}^+ C_{j\sigma} + Hc) + J \sum_{\langle ij \rangle} \left(S_i S_j - \frac{n_i n_j}{4} \right) \qquad (1)$$

MINIREVIEWS

$$S_i = \frac{1}{2} \sum_{\alpha,\beta} C_{i\beta} \sigma_{\beta\alpha} C_{i\alpha} \qquad (2)$$

that describes the motion of the holes in the active background of spins. This spin-hole Hamiltonian suggests a kinetic hole-pairing mechanism driven by the lowering of the kinetic energy due to delocalization, as described below.

A long-range or even a mesoscopic AFM order (Néel ordered) encumbers the hole motion. (As a hole moves to the nearest-neighbouring site, the spin at that site concomitantly moves to the site previously occupied by the hole). A succession of such moves shift-registers a line of spins, which become laterally misaligned with respect to the spins on the sides (Figure 4). The broken AFM bonds shown by the dashed lines create a linearly rising potential that impedes free hole motion. A compact

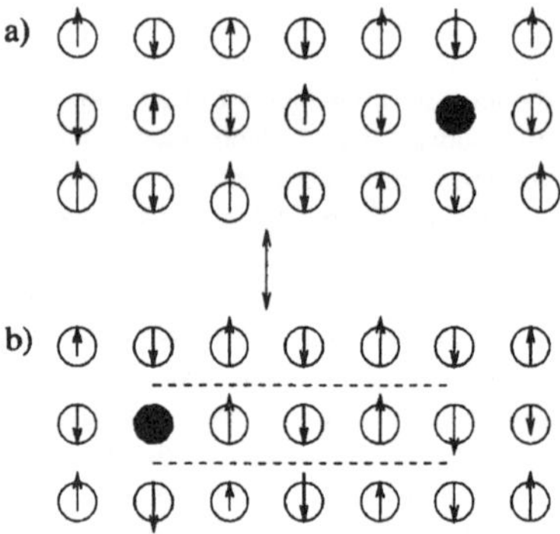

Figure 4. *a) A hole at the initial site in a Néel-ordered antiferromagnet; b) a hole at the final site after four hops. - - - - show disrupted bonds between misaligned spins giving a linearly rising potential.*

pair of holes can, however, toboggan together, and thus delocalize quite freely. The trailing hole undoes what the leading hole does, thus healing the lateral misalignment of the spins. The lowering of the pair kinetic energy by the ease of pair delocalization then provides the kinetic energy mechanism for pairing.[39] (The quantum fluctuation of spins away from the Néel order must modify the above naive visualization, but one can still talk meaningfully about the hole-pair moving unencumbered by entanglement with the background spins). Clearly, for low doping the hole-hole repulsion will dominate over the kinetic mechanism that favors compact pairing. It is only for higher doping, when the mean hole separation matches the pair size that the kinetic pairing wins. For such a kinetic mechanism for pairing, stripe modulation in the zeroth approximation merely provides a modified spin background for the hole motion—not necessarily conducive to pairing. This basic idea of kinetic pairing seems to suggest a novel, highly elaborate quantum liquid-crystalline model for HTSC, in which the fluctuating stripes play a crucial role.[14–18]

The charge stripe is normally viewed as a 1D lane, along which the holes can freely move longitudinally, but are constrained in the transverse direction by the antiphase AFM-ordered magnetic domains. Indeed, the pinning at the magic number doping ($x =$

$\frac{1}{8}$) gives such a stripe which is static and gives a resolution-limited elastic neutron scattering. It is well known that an array of such static charge stripes will be insulating through a CDW instability. In the electronic quantum liquid-crystalline phase model of the doped Mott insulator the stripes show transverse quantum fluctuations that affect the motion of the charged particles which is guided by the fluctuating stripes. Thus, the charge stripes are viewed as a quantum mechanical deformable object with a topological integrity, and as having many degrees of freedom. The role of zero-point fluctuations was recognized quite early.[40] The mechanism for the mobile hole pairing here involves a magnetic analog of the superconducting proximity effect between the metallic stripe and the Mott insulators that bound it transversely. The mobile hole pairs virtually hop between the Mott-insulator domains and acquire a spin gap. This is in an underdoped system. At higher doping the hole pairs can tunnel between the stripes giving a global 3D phase stiffness to the local pairs. Several known features of HTSC such as the normal-state pseudogap in the underdoped samples, which have the same d-wave symmetry as the d-wave superconducting gap at lower temperatures, come out of this model. The electronic quantum liquid-crystalline model has many topological features of its underlying one-dimensionality that affect its charge and spin degrees of freedom. It certainly warrants a detailed examination—experimentally and theoretically, as something interesting in its own right.

Any discussion of the role of stripes in the high-T_c layered cuprates is necessarily inconclusive at present. It is possible, however, that stripes may well be a solitonic route to hole doping. In any case, as of now the stripes cannot confidently signify superconductivity.

[1] For a critical discussion of stripes and electronic phase separation see a) J. Zaanen, *J. Phys. Chem. Solids* **1998**, *59*, 1769; b) J. M. Tranquada, *Physica B* **1998**, *241–243*, 745; c) J. M. Tranquada, *J. Phys. Chem. Solids* **1998**, *59*, 2150; d) *Proceedings of the workshop on Phase Separation in Cuprate Superconductors* (Eds.: K. A. Müller, G. Benedek), World Scientific, Singapore, **1993**; e) *Phase Separation in Cuprate Superconductors* (Eds.: E. Sigmund, K. A. Müller), Springer-Verlag, Heidelberg, **1994**; f) E. Dagotto, T. Hotta, A. Moreo, *Phys. Rep.* **2001**, *344*, 1; g) C. N. R. Rao, P. V. Vanitha, A. K. Cheetham, *Chem. Eur. J.* **2003**, *9*, 828..

[2] a) J. M. Tranquada, S. J. Sternlieb, J. D. Axe, Y. Nakamura, S. Uchida, *Nature* **1995**, *375*, 561; b) J. M. Tranquada, J. D. Axe, N. Ichikawa, Y. Nakamura, S. Uchida, B. Nachumi, *Phys. Rev. B* **1996**, *54*, 7489; c) J. M. Tranquada, J. D. Axe, N. Ichikawa, A. R. Moodenbaugh, Y. Nakamura, S. Uchida, *Phys. Rev. Lett.* **1997**, *78*, 338; d) *Stripes and Related Phenomena* (Eds.: A. Bianconi, N. L. Saini), Kluwer Academic Publishers, Dordrecht, **2000**.

[3] K. Yamada, C. H. Lee, K. Kurahashi, J. Wada, S. Wakimoto, S. Ueki, H. Kimura, X. Endoh, S. Hosaya, G. Shirane, R. J. Birgeneau, M. Greven, M. A. Kastner, Y. J. Kim, *Phys. Rev. B* **1998**, *57*, 6165.

[4] X. J. Zhou, P. Bogdanov, S. A. Kellar, T. Noda, H. Eisaki, S. Uchida, Z. Hussain, Z.-X. Shen, *Science* **1999**, *286*, 268.

[5] X. J. Zhou, T. Yoshida, S. A. Kellar, P. V. Bogdanov, E. D. Lu, A. Larzava, M. Nakamura, T. Noda, T. Takeshita, H. Eisaki, S. Uchida, A. Fujimori, Z. Hussain, Z.-X. Shen, *Phys. Rev. Lett.* **2001**, *86*, 5578.

[6] T. Noda, H. Eisaki, S.-I. Uchida, *Science* **1999**, *286*, 265.

[7] H. A. Mook, P. Dai, F. Dogan, *Phys. Rev. Lett.* **2002**, *88*, 097004.

[8] B. Lake, G. Aeppli, K. N. Clausen, D. F. McMorrow, K. Letmann, N. E. Hussey, N. Manykorntong, M. Nohava, H. Takagi, T. E. Mason, A. Schröder, *Science* **2001**, *291*, 1759.

CHEMPHYSCHEM

[9] B. Lake, H. M. Ronnow, N. B. Christensen, G. Aeppli, K. Lefmann, D. F. McMorrow, P. Verderwisch, P. Smeibidl, N. Mangkorntong, T. Susagawa, M. Nohara, H. Takagi, T. E. Mason, *Nature* **2002**, *415*, 299.

[10] F. Mitrovic, E. E. Sigmund, M. Eschrig, H. N. Bachman, W. P. Halperin, A. P. Reyas, P. Kuhns, W. G. Moulton, *Nature* **2001**, *413*, 501.

[11] Y. S. Lee, R. J. Birgeneau, M. A. Kastner, Y. Endoh, S. Wakimoto, K. Yamada, R. W. Erwin, S. H. Lee, G. Shirane, *Phys. Rev. B* **1999**, *60*, 3643.

[12] C. Howald, P. Fournier, A. Kapitulnik, *Phys. Rev. B* **2001**, *64*, R100504.

[13] J. Zaanen, O. Gunnarsson, *Phys. Rev. B* **1989**, *40*, 7391.

[14] V. J. Emery, A. Kivelson, *Strongly Correlated Electronic Materials: The Los Alamos Symposium 1993* (Eds.: K. S. Bedell, Z. Wang, D. E. Meltzer, A. V. Balatsky, E. Abrahams), Addision-Wesley, Redwood City, **1994**, p. 619.

[15] A. Kivelson, E. Fradkin, V. J. Emery, *Nature* **1998**, *393*, 550.

[16] V. J. Emery, S. A. Kivelson, O. Zachar, *Phys. Rev. B* **1997**, *56*, 6120.

[17] a) V. J. Emery, A. Kivelson, *Phys. C* **1993**, *209*, 597; b) V. Löw, V. J. Emery, K. Fabricius, S. A. Kivelson, *Phys. Rev. Lett.* **1994**, *72*, 1918.

[18] a) M. Salkola, V. J. Emery, S. A. Kivelson, *Phys. Rev. Lett.* **1996**, *77*, 155; b) V. J. Emery, S. A. Kivelson, J. M. Tranquada, *Proc. Natl. Acad. Sci. USA* **1999**, *96*, 8814.

[19] a) S. R. White, D. J. Scalapino, *Phys. Rev. Lett.* **1998**, *81*, 3227; b) S. R. White, D. J. Scalapino, *Phys. Rev. Lett.* **1998**, *80*, 1272.

[20] J. G. Bednorz, K. A. Müller, *Z. Phys. B* **1986**, *64*, 189.

[21] W. P. Su, J. R. Schrieffer, A. J. Heeger, *Phys. Rev. B* **1980**, *22*, 2099.

[22] E. M. Conwell, *Phys. Today*, **1985**, 48 (June).

[23] V. L. Pokhrovsky, A. L. Talapov, *Theory of Incommensurate Crystals, Soviet Scientific Reviews Suppl. Sec. A, Vol. 1*, (Ed.: I. M. Khalatnikov), Harwood Academics, Switzerland, **1984**.

[24] a) J. M. Tranquada, J. E. Lorenzo, D. J. Buttrey, V. Sachan, *Phys. Rev. B* **1995**, *52*, 3581; b) S. W. Cheong, H. Y. Hwang, C. H. Chen, B. Batlogg, L. W. Rupp, Jr., S. A. Carter, *Phys. Rev. B* **1994**, *49*, 7088; c) H. Kuwahara, Y. Tomioka, Y. Moritomo, A. Asamitsu, M. Kasai, R. Kumai, Y. Tokura, *Science* **1996**, *272*, 80; d) S. H. Lee, J. M. Tranquada, K. Yamada, D. J. Buttrey, Q. Li, S. W. Cheong,24935 *Phys. Rev. Lett.* **2002**, *88*, 126401; e) M. Abu-Sheikah, O. Bakharev, H. B. Bram, J. Zaanen, *Phys. Rev. Lett.* **2001**, *87*, 237201.

[25] a) *Colossal magnetoresistance, Charge-ordering and Related Properties of Manganese Oxides* (Eds.: C. N. R. Rao, B. Raveau), World Scientific, Singapore, 1999; b) C. N. R. Rao, A. Arulraj, A. K. Cheetham, B. Raveau, *J. Phys. Condens. Matter* **2000**, *12*, R83; c) C. N. R. Rao, P. V. Vanitha, *Curr. Opin. Solid State Mater. Sci.* **2002**, *6*, 97.

[26] P. Abbamonte, L. Venema, A. Rusydi, G. A. Sawatzky, G. Logvenov, I. Bozovic, *Science* **2002**, *297*, 581.

[27] C. Howald, H. Eisaki, N. Kaneko, A. Kapitulnik, *cond-mat/0201546*.

[28] J. E. Hoffman, K. McElroy, D.-H. Lee, K. M. Lang, H. Eisaki, S. Uchida, J. C. Davis, *Science* **2002**, *297*, 1148.

[29] S. H. Pan, J. P. O'Neal, R. D. Badzey, C. Chamon, N. Ding, J. R. Engelbracht, Z. Wang, N. Eisaki, S. Uchida, A. K. Gupta, K.-W. Ng, E. W. Hudson, K. M. Lang, J. C. Davis, *Nature* **2001**, *413*, 282.

[30] P. M. Singer, A. W. Hunt, A. F. Cederstrom, T. Imai, *cond-mat/9906140*.

[31] J. R. Schrieffer, *J. Low Temp. Phys.* **1995**, *99*, 97.

[32] G. V. Chester, *Phys. Rev.* **1956**, *103*, 1693.

[33] D. J. Scalapino, S. R. White, *Phys. Rev. B* **1998**, *58*, 8222.

[34] L. Klein, J. S. Dodge, C. H. Ahn, G. J. Snyder, T. H. Geballe, M. R. Beasley, A. Kapitulnik, *Phys. Rev. Lett.* **1996**, *77*, 2774.

[35] a) P. Dai, H. A. Mook, S. M. Hayden, G. Aeppli, T. G. Perring, R. D. Hunt, F. Dogan, *Science* **1999**, *284*, 1344; b) E. Demler, S. C. Zhang, *Nature* **1998**, *396*, 733.

[36] J. W. Loram, K. A. Mirza, P. F. Freeman, *Phys. C* **1990**, *171*, 243.

[37] F. C. Zhang, T. M. Rice, *Phys. Rev. B* **1988**, *37*, 3759.

[38] E. Dagotto, *Rev. Mod. Phys.* **1994**, *66*, 763.

[39] a) J. E. Hirsch, *Phys. Rev. Lett.* **1987**, *59*, 228; b) M. M. Mohan, N. Kumar, *J. Phys. C* **1987**, *20*, L527; c) N. Kumar, *Phys. Rev. B* **1990**, *42*, 6138; d) Yu. A. Dimashko, *JETP Lett.* **1993**, *76*, 267.

[40] a) P. Prelovšek, X. Zotos, *Phys. Rev. B* **1993**, *47*, 5984; b) P. Prelovšek, I. Sega, *Phys. Rev. B* **1994**, *49*, 15241.

Received: November 28, 2002 [M601]

J. Phys. Chem. B **2000**, *104*, 5877−5889

FEATURE ARTICLE

Charge, Spin, and Orbital Ordering in the Perovskite Manganates, $Ln_{1-x}A_xMnO_3$ (Ln = Rare Earth, A = Ca or Sr)

C. N. R. Rao*

Chemistry & Physics of Materials Unit and the CSIR Centre of Excellence in Chemistry, Jawaharlal Nehru Centre for Advanced Scientific Research, Jakkur P.O, Bangalore 560 064, India

Received: February 7, 2000; In Final Form: March 29, 2000

Charge ordering occurs in some mixed-valent transition metal oxides. The perovskite manganates of the formula $Ln_{1-x}A_xMnO_3$ (Ln = rare earth; A = Ca, Sr) are especially interesting because long-range ordering of the Mn^{3+} ($t_{2g}^3\ e_g^1$) and Mn^{4+} ($t_{2g}^3\ e_g^0$) ions in these materials is linked to antiferromagnetic spin ordering, and also to the long-range ordering of the Mn^{3+} (e_g) orbitals and the associated lattice distortions. Charge ordering occurs at a higher temperature than spin ordering in some of the manganates ($T_{CO} > T_N$), whereas in some others $T_{CO} = T_N$. Orbital ordering occurs without charge ordering in the A-type antiferromagnetic manganates, but in the manganates where charge ordering occurs, antiferromagnetism of CE-type is found along with orbital ordering. The subtle relations between charge, spin, and orbital ordering are discussed in the article, with special attention to the effects of cation size, chemical substitution, dimensionality, pressure, and magnetic and electric fields. Unusual features such as phase separation and electron−hole asymmetry are also examined.

Introduction

Charge ordering is a phenomenon generally observed in mixed-valent transition metal oxides. When differently charged cations (i.e., 2+ and 3+) in an oxide order on specific lattice sites, the hopping of electrons between the cations is no longer favored. One therefore observes an increase in the electrical resistivity at the charge-ordering transition, often accompanied by a change in crystal symmetry. Because transition metal ions also carry spins, it is interesting to examine the magnetic (spin) ordering in the solids in relation to charge ordering. A well-known example of charge ordering is found in Fe_3O_4 (magnetite) where it occurs at a temperature well below spin ordering.[1,2] Charge and spin ordering in real space in metal oxides received renewed attention because of their role in cuprate superconductors.[3] In recent years, charge and spin ordering have been discovered in a few other transition metal oxides,[4] typical examples being $La_{1-x}Sr_xFeO_3$, $La_{2-x}Sr_xNiO_4$, and $LiMn_2O_4$. Charge ordering in the rare earth manganates of the perovskite structure, with the general composition $La_{1-x}A_xMnO_3$ (Ln = rare earth; A = alkaline earth), is considerably more interesting, because it is closely associated with spin and orbital ordering, giving rise to fascinating properties.[5−7]

The perovskite manganates originally became popular because of the discovery of colossal magnetoresistance (CMR).[5,6] CMR and related properties essentially arise from the double-exchange mechanism of electron hopping between the Mn^{3+} ($t_{2g}^3\ e_g^1$) and Mn^{4+} ($t_{2g}^3\ e_g^0$) ions.[8] In this mechanism, lining up of the spins (ferromagnetic alignment) of the incomplete e_g orbitals of the adjacent Mn ions is directly related to the rate of hopping of the electrons, giving rise to an insulator−metal transition in the material at the ferromagnetic Curie temperature, T_c. In the ferromagnetic phase ($T < T_c$), the material becomes metallic, but is an insulator in the paramagnetic phase ($T > T_c$). In the insulating phase, a Jahn−Teller distortion associated with the Mn^{3+} ions favors the localization of electrons. Charge ordering of the Mn^{3+} and Mn^{4+} ions competes with double-exchange and promotes insulating behavior and antiferromagnetism. It may be recalled that the $Mn^{3+}-O-Mn^{3+}$ and $Mn^{4+}-O-Mn^{4+}$ superexchange interaction, through the e_g orbitals, is antiferromagnetic. Although charge ordering in the rare earth manganates was investigated by Jirak et al.[9] as early as 1985, the subject has received renewed attention only in the past 5 years, for reasons described below.

Charge ordering occurs through a fairly wide range of compositions of $Ln_{1-x}A_xMnO_3$, provided the Ln and A ions are not too large. Large Ln and A ions (e.g., La, Sr) favor ferromagnetism and metallicity, whereas the smaller ones (e.g., La, Ca, or Pr, Ca) favor charge ordering. Charge ordering and spin (antiferromagnetic) ordering may or may not occur at the same temperature. Besides, the Mn^{3+} (d_{z^2}) orbitals and the associated lattice distortions develop long-range order (as illustrated later in this article). Such orbital ordering may or may not occur with charge ordering in the manganates, but it generally accompanies antiferromagnetic (spin) ordering. In this article, we discuss the interplay of charge, spin, and orbital ordering in the rare earth manganates in some detail, and highlight some of the important recent results. Before discussing the manganates, we shall briefly review the properties of a few of the other transition metal oxides that exhibit charge ordering.

Examples of Charge and Spin Ordering in Oxides. Fe_3O_4 has a spinel structure, represented as $Fe^{3+}[Fe^{2+},\ Fe^{3+}]O_4$ in which one third of the cations are tetrahedrally coordinated (A-sites) and two thirds of the cations are octahedrally coordinated (B-sites). It is ferrimagnetic below 858 K (T_N), with the spins

* For correspondence: e-mail: cnrrao@jncasr.ac.in.

10.1021/jp0004866 CCC: $19.00 © 2000 American Chemical Society
Published on Web 06/02/2000

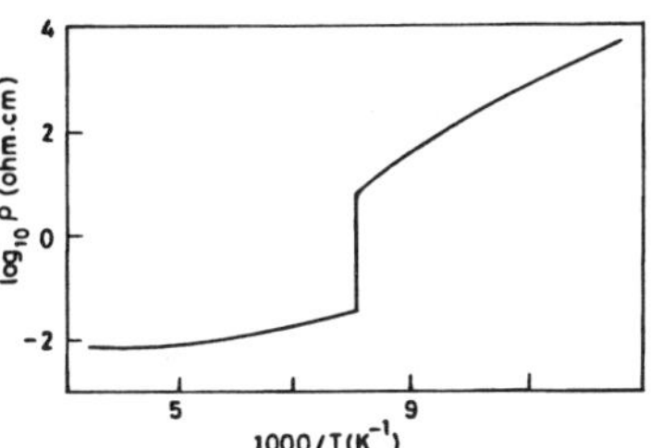

Figure 1. Verwey transition in Fe_3O_4 caused by charge ordering (from Honig[1b]).

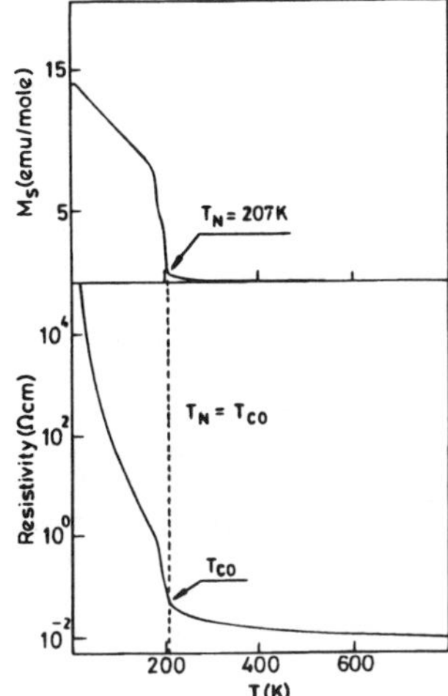

Figure 2. Temperature variation of electrical resistivity and magnetization of $La_{0.33}Sr_{0.67}FeO_3$ showing marked changes on charge ordering ($T_{CO} = T_N$) (from Park, Yamaguchi and Tokura, as quoted by Imada et al.[4]).

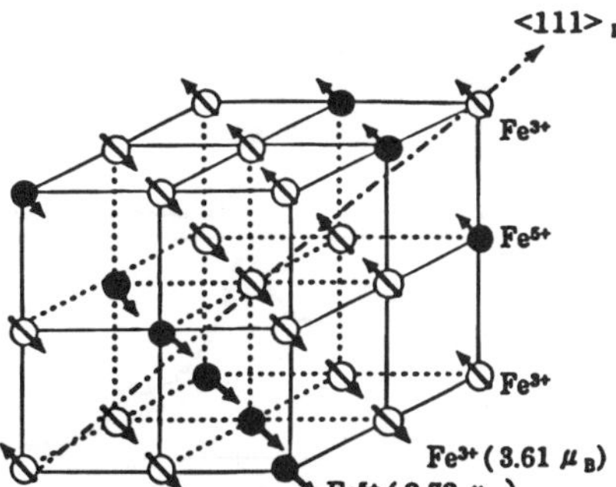

Figure 3. Structure showing charge and spin ordering in $La_{0.33}Sr_{0.67}$-FeO_3 (from Battle et al.[10b]).

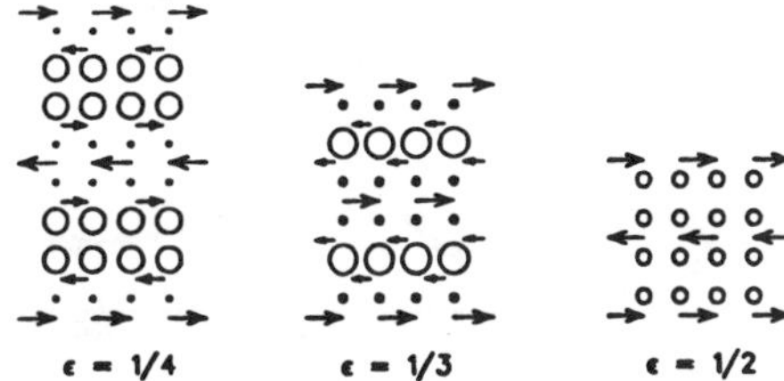

Figure 4. Spin and charge modulation in the NiO_2 planes of $La_{2-x}Sr_xNiO_{4+y}$ with increasing hole concentration. Here, the modulation wavenumber, ϵ, is equivalent to the hole concentration, $p = x + 2y$ (from Tranquada et al.[11a]). Spin density on nickel ions is shown by arrows, and charge density on oxygen ions is shown by circles.

of the A- and B-sublattices being antiparallel. Around 120 K (T_V), Fe_3O_4 shows a sharp increase in resistivity, commonly referred to as the Verwey transition (Figure 1). Below T_V, the Fe^{2+} and Fe^{3+} ions are considered to be ordered, thereby giving rise to high electrical resistivity. Above T_V, conduction occurs on the B site sublattice. It actually turns out, however, that charge ordering in Fe_3O_4 is much more complicated, with some short-range ordering present even above T_V. Even today, there is argument about the symmetry of the low-temperature ordered phase. The Verwey transition in this oxide has been reviewed excellently by Honig[1] and Tsuda et al.[2]

$La_{1-x}Sr_xFeO_3$ ($x = 0.67$) undergoes charge ordering at approximately 207 K (T_{CO}), at which temperature it also exhibits antiferromagnetic spin ordering. Thus, T_{CO} is also the Néel temperature (T_N) in this oxide. The resistivity shows a marked increase at $T_{CO} = T_N$ (Figure 2). Although the formula requires the presence of Fe^{3+} and Fe^{4+} ions, the Fe^{4+} ions disproportionate to give Fe^{3+} and Fe^{5+} ions.[10] The charge- and spin-ordered structure of $La_{0.33}Sr_{0.67}FeO_3$ is shown in Figure 3.

Quasi two-dimensional $La_{2-x}Sr_xNiO_{4+y}$ undergoes cooperative ordering of the Ni spins and of the charge carriers below a temperature (T_{CO}). Charge ordering occurs at nearly all doping levels ($p = x + 2y$), in the insulating regime of the material ($p < 0.7$). Spin and charge modulations are observed in the NiO_2 plane and these vary with the hole concentration (p value). This nickelate system is a typical instance of microscopic phase separation wherein the charge carriers (holes) localize in the domain walls in an antiferromagnetic system. Such phase

separation causes stripe modulations. Here, the ordering may be viewed as alternating stripes of antiferromagnetic and hole-rich regions.[11] In Figure 4, we show the spin and charge modulations in the NiO_2 planes at two hole concentrations. The $\epsilon = 1/4$ case corresponds to $La_2NiO_{4.125}$ ($p = 0.25$), where the interstitial oxygens form a superlattice with a unit cell of $3a \times 5b \times 5c$. The compositional dependence of resistivity of $La_{2-x}Sr_xNiO_4$ shows peaks at $x = 0.33$ and 0.25, owing to charge ordering, which in the $x = 0.33$ composition occurs at 240 K. At 240 K, various properties show anomalies, as depicted in Figure 5. Superlattice peaks show up in the electron diffraction pattern at this temperature. Antiferromagnetic spin ordering in the nickelate occurs at a lower temperature (180 K). It appears that ordering is driven by charge. Charge ordering of holes, accompanied by spin ordering or the segregation of holes and spins in the stripe form, also occurs in $La_{2-x}A_xCuO_4$ (A = Sr, Ba, $x = 0.125$), causing anomalous suppression of superconductivity.[3]

$LiMn_2O_4$ is a spinel with equal proportions of 3+ and 4+ Mn cations. It has been shown recently to undergo a Verwey-type transition with a resistivity anomaly at approximately 290 K where there is a structural (cubic-tetragonal) transition. Around 65 K, the material exhibits a complex long-range magnetic order.[13] Clearly, this charge-ordering transition requires further study.

A Brief Description of the Different Types of Ordering in the Manganates. $LaMnO_3$ has a layered antiferromagnetic structure, referred to as A-type antiferromagnetic ordering[14] (Figure 6a). Because there is a doubly degenerate e_g orbital in each Mn^{3+} ion ($t_{2g}^3 e_g^1$), $LaMnO_3$ and the other analogous rare earth compounds such as $NdMnO_3$ show a fine interplay between spin and orbital ordering. The orbital ordering is coupled to the Jahn−Teller (JT) distortion. Figure 6b describes the JT distortion in $LaMnO_3$. The distortion disappears above

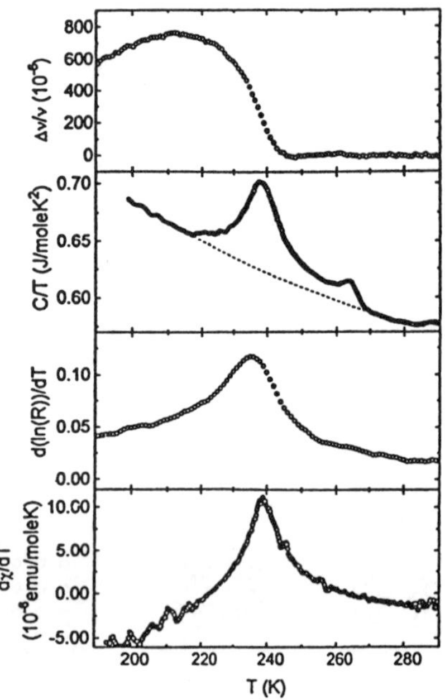

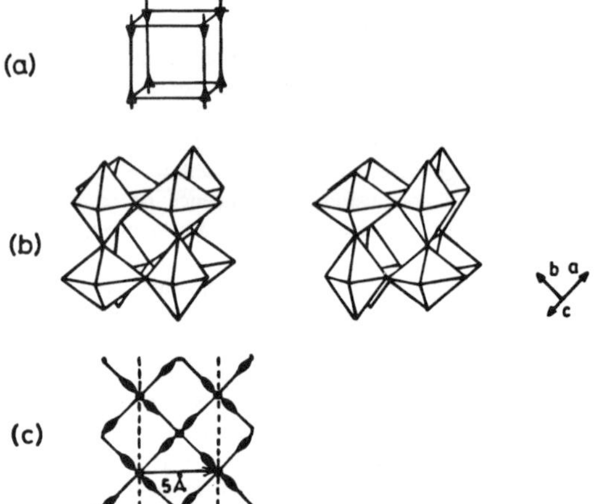

Figure 5. Normalized sound velocity ($\Delta v/v$), specific heat (c), as well as temperature derivatives of resistivity (ρ) and magnetic susceptibility (χ) showing anomalies at the charge ordering transition (240 K) in $La_{1.67}$-$Sr_{0.33}NiO_4$ (from Ramirez et al.[12]).

Figure 6. (a) A-type AFM ordering, (b) JT distortion, and (c) orbital ordering in $LaMnO_3$. The orientation of e_g orbitals shown in panel c is consistent with the 5 Å lattice spacing.

750 K. Orbital ordering of $3x^2 - r^2$ or $3y^2 - r^2$ type accompanied by the JT distortion leads to a superexchange coupling in $LaMnO_3$, which is ferromagnetic (FM) in the planes and antiferromagnetic (AFM) between the planes (Figure 6a).[15] Orbital ordering in $LaMnO_3$ is shown in Figure 6c. Without the JT distortion, $LaMnO_3$ would have been a FM insulator; it is an A-type AFM insulator instead.

In $Ln_{1-x}A_xMnO_3$, besides orbital and spin ordering, we can have charge ordering because of the presence of Mn^{3+} and Mn^{4+} ions. Small Ln and A ions stabilize the charge-ordered (CO) state. Thus, $Pr_{0.7}Ca_{0.3}MnO_3$, charge-orders around 230 K in the paramagnetic state, becoming AFM at 170 K; it is an insulator and does not show ferromagnetism in the absence of a strong

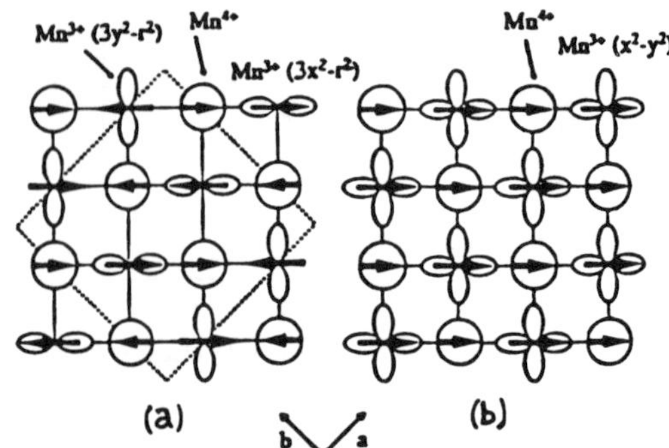

Figure 7. Charge, spin, and orbital ordering in (a) CE-type and (b) A-type AFM $Ln_{1-x}A_xMnO_3$. In part a, the broken line shows the unit cell for the CE-type AFM CO order. Mn^{4+} ions are shown by circles.

magnetic field. $La_{0.7}Ca_{0.3}MnO_3$, on the other hand, is a FM metal below the Curie temperature ($T_c \approx 230$ K) and a paramagnetic insulator above T_c. $La_{0.5}Sr_{0.5}MnO_3$ is metallic both in the FM and paramagnetic states, whereas $Nd_{0.5}Sr_{0.5}MnO_3$ shows a transition from a ferromagnetic metallic (FMM) state to an AFM charge-ordered state on cooling. The CO states in these manganates are associated with CE-type AFM ordering.[14] In the CE-type ordering, Mn^{3+} and Mn^{4+} ions are arranged as in a checker board and the Mn^{3+} sites are JT distorted.[15] Along the c-axis, the in-plane arrangement mentioned above gets stacked and the neighboring planes are coupled antiferromagnetically. The exchange coupling between the Mn^{3+} and Mn^{4+} ions depends on the type of e_g orbital at the Mn^{3+} site, and hence the nature of orbital ordering becomes important. The CE-type AFM CO state in $Ln_{1-x}A_xMnO_3$ is associated with the ordering of $3x^2 - r^2$ or $3y^2 - r^2$ type orbitals at the Mn^{3+} site. The JT distortion that follows such orbital ordering stabilizes the CE-type AFM state (relative to the FMM state). In Figure 7a, we show the spin, charge, and orbital ordering in the CE-type AFM state. The CO states in the manganates exhibit CE-type AFM ordering at the same temperature as the charge-ordering transition or at a lower temperature ($T_{CO} \geq T_N$). That is, spin ordering occurs concurrently or after charge ordering. Complete orbital ordering is achieved when there is both charge and spin ordering. Orbital and spin ordering occur without charge ordering in some of the manganates showing A-type antiferromagnetism.

The A-type AFM state described earlier in relation to $LaMnO_3$ (Figure 6a) is also encountered in the $Ln_{1-x}A_xMnO_3$ system. This state generally is not accompanied by charge ordering, because some electron transfer can occur between the Mn cations in the ab plane. Orbital ordering in A-type AFM ordering is depicted in Figure 7b. Here, the $x^2 - y^2$ type orbital is present at the Mn^{3+} site. $Pr_{0.5}Sr_{0.5}MnO_3$ transforms from a FMM state to a A-type AFM state on cooling. $Nd_{0.45}Sr_{0.55}MnO_3$ is a A-type antiferromagnet, unlike $Nd_{0.5}Sr_{0.5}MnO_3$, which is a CE-type antiferromagnet exhibiting charge ordering at low temperatures; the former shows conductivity in the ab plane and is not charge-ordered.

Evidence for charge ordering can be observed in crystal structures at low temperatures. Thus, in $La_{0.5}Ca_{0.5}MnO_3$, the Mn^{4+} environment is nearly isotropic, with all the Mn−O distances being nearly equal ($\sim$1.92 Å). In the $Mn^{3+}O_6$ octahedra, one of the Mn−O bonds is much longer ($\sim$2.07 Å) than the other bonds ($\sim$1.92 Å), consistent with orbital ordering. It must be noted that the Mn−O distances in the ab plane of the manganates are much longer than in the c-direction,

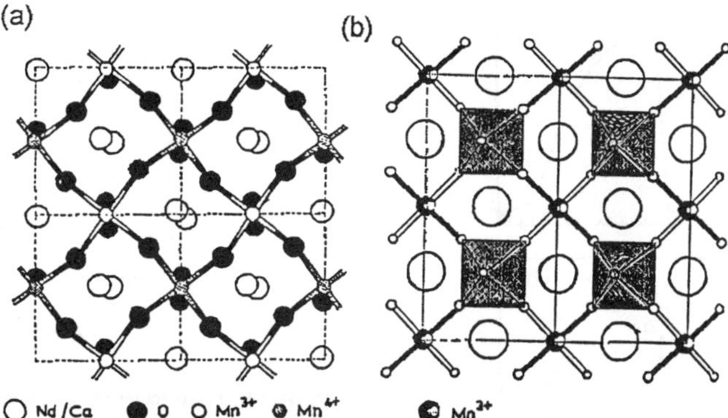

Figure 8. (a) Structure of charge-ordered $Nd_{0.5}Ca_{0.5}MnO_3$ in the *ab* plane at 10 K. Mn^{4+} (black) located at (1/2, 0, 0) and Mn^{3+} (grey) located (0, 1/2, 0) can be distinguished. The structure has zigzag chains with alternate long and short Mn–O distances. The distances are 1.921 and 2.021 Å $\langle 110 \rangle$ and 1.881 and 2.020 Å along $\langle -110 \rangle$ (from Vogt et al.[16]). (b) Charge-ordered structure of $Nd_{0.5}Sr_{0.5}MnO_3$. The Mn^{3+} ions are the filled circles, and Nd/Sr ions are large open circles. Oxygens are small open circles. The long Mn^{3+}–O bonds shown in black also represent the orientation of the e_g orbitals. Mn^{4+}–O octahedra are shown in polyhedral representation (from Woodward et al. [50]).

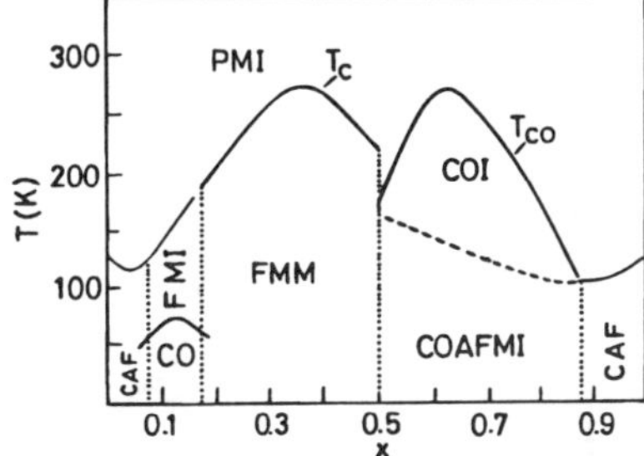

Figure 9. Phase diagram of $La_{1-x}Ca_xMnO_3$ (following Cheong). CAF, canted antiferromagnet; CO, charge-ordered phase; FMI, ferromagnetic insulator; PMI, paramagnetic insulator; FMM, ferromagnetic metallic state; CO AFMI, charge-ordered antiferromagnetic insulator; COI, charge-ordered insulator. Notice electron hole asymmetry by comparing the $x < 0.5$ and $x > 0.5$ regimes.

particularly in the AFM state. In Figure 8a, we show the projection of the structure of charge-ordered $Nd_{0.5}Ca_{0.5}MnO_3$ down the *c*-axis to illustrate the definitive identification of the unique sites occupied by Mn^{3+} and Mn^{4+} ions in the CO state. The structure of charge-ordered $Nd_{0.5}Sr_{0.5}MnO_3$, where the Mn^{4+}-centered octahedra are represented in polyhedral notation, is shown in Figure 8b.

Representative Phase Diagrams of the Manganates. In Figure 9 we show the phase diagram of $La_{1-x}Ca_xMnO_3$. In this system, charge ordering occurs in the $x \approx 0.5$–0.8 range. In Figure 10 we show the phase diagrams of $Pr_{1-x}Ca_xMnO_3$ and $Pr_{1-x}Sr_xMnO_3$. The latter system shows no charge ordering, but $Pr_{1-x}Ca_xMnO_3$ exhibits charge ordering over the $x \approx 0.3$–0.8 range. Note that $Pr_{0.7}Ca_{0.3}MnO_3$ exhibits charge ordering but $La_{0.7}Ca_{0.3}MnO_3$ does not. All such variations are essentially due to the effect of the size of the A-site cations, the smaller size favoring charge ordering. From Figures 9 and 10, we also see that the CO regime is prominent at large x. In fact, the $x > 0.5$ compositions in $Ln_{1-x}Ca_xMnO_3$ are almost entirely in the CO regime both when Ln = La and Pr. This regime can be considered as the electron-doped regime (substitution of trivalent rare earth in $CaMnO_3$), whereas the $x < 0.5$ compositions may be considered as the hole-doped regime (substitution of divalent Ca in $LnMnO_3$). Clearly, there is electron–hole asymmetry in these manganates.

It is surprising that ferromagnetism is not encountered in the electron-doped regime ($x > 0.5$). Effects of magnetic and electric

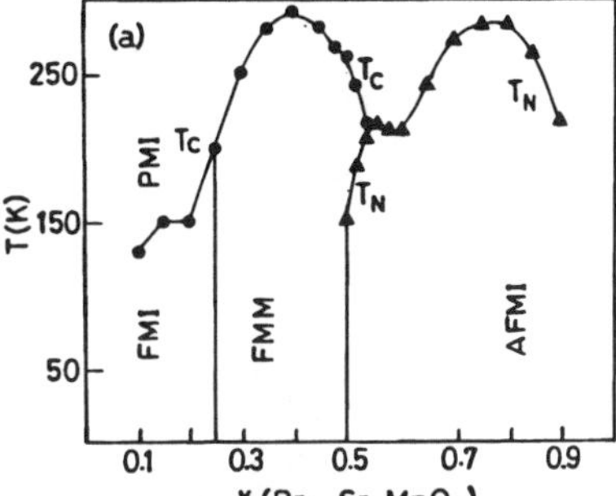

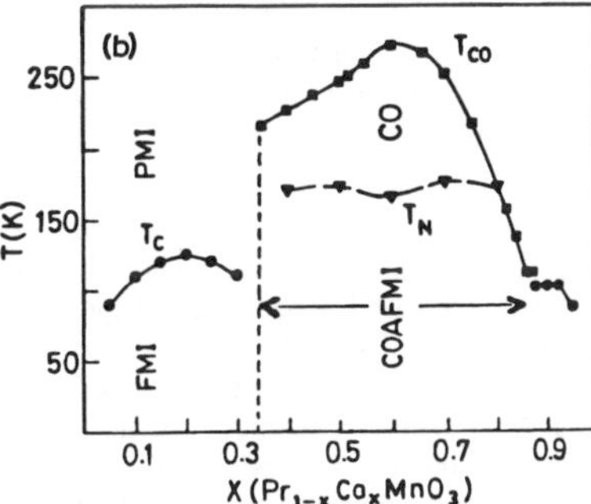

Figure 10. Phase diagrams of (a) $Pr_{1-x}Sr_xMnO_3$ and (b) $Pr_{1-x}Ca_xMnO_3$. Notice the wide charge ordering regime and electron–hole asymmetry in part b and the absence of charge ordering in part a. In part b, there is spin-glass or CAF behavior when $x \geq 0.8$.

fields on the hole- and electron-doped manganates (e.g., $Pr_{0.7}$-$Ca_{0.3}MnO_3$ and $Pr_{0.3}Ca_{0.7}MnO_3$) are also different. There are some similarities between the hole- and electron-doped regimes. For example, in $Pr_{1-x}Ca_xMnO_3$, the charge-ordering transition temperature increases with hole concentration in the $0.3 < x \leq 0.5$ regime and with electron concentration in the $0.5 \leq x \leq 0.8$ regime.

Case Studies. To understand typical scenarios of charge-ordered manganates, it is instructive to examine the properties of two manganates with different sizes of the A-site cations. For this purpose, we choose $Nd_{0.5}Sr_{0.5}MnO_3$ with a weighted average radius of the A-site cations, $\langle r_A \rangle$, of 1.24 Å and $Pr_{0.6}$-$Ca_{0.4}MnO_3$ with an $\langle r_A \rangle$ of 1.17 Å (Shannon radii are used here).

Feature Article

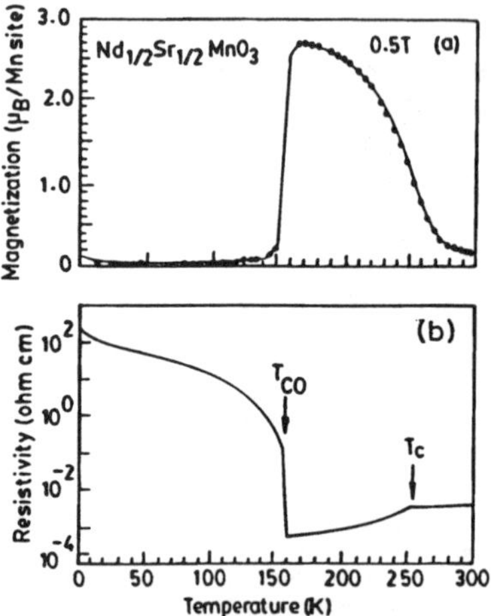

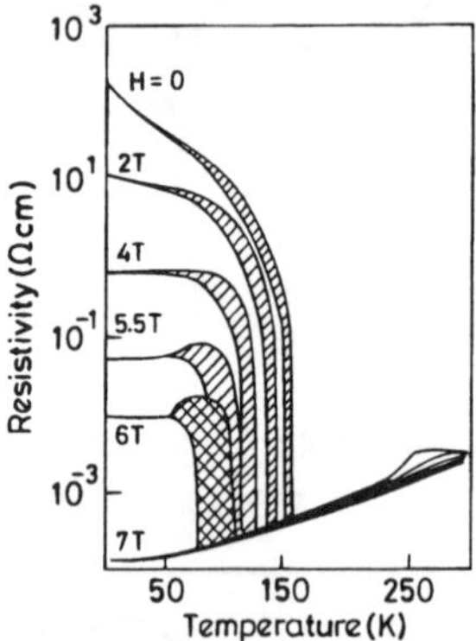

Figure 11. Temperature variation of (a) the magnetization and (b) the resistivity of $Nd_{0.5}Sr_{0.5}MnO_3$ (from Kuwahara et al.[17]).

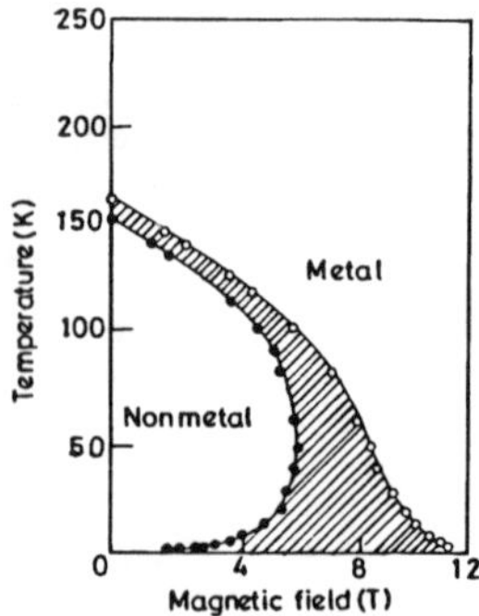

Figure 13. Temperature-magnetic field phase diagram for $Nd_{0.5}Sr_{0.5}$-MnO_3 (from Tokura et al.[18]).

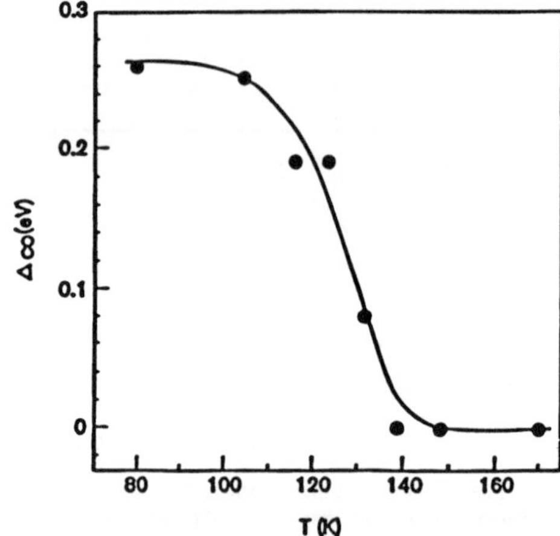

Figure 14. CO gap in $Nd_{0.5}Sr_{0.5}MnO_3$ as revealed by vacuum tunneling measurements (from Biswas et al.[19]).

Figure 12. Effect of magnetic fields on the charge ordering transition of $Nd_{0.5}Sr_{0.5}MnO_3$ (from Kuwahara et al.[17]).

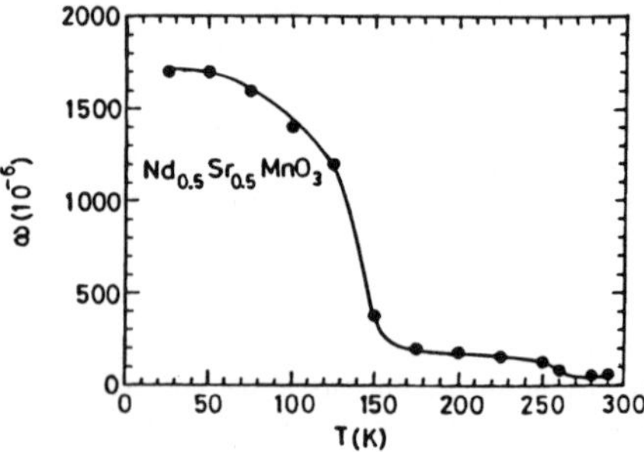

Figure 15. Temperature variation of the maximum volume magnetostriction in $Nd_{0.5}Sr_{0.5}MnO_3$ at 13.7T (from Mahendiran et al.[21]).

$Nd_{0.5}Sr_{0.5}MnO_3$ is a ferromagnetic metal with a T_c of ~250 K and transforms to an insulating CO state at about 150 K (Figure 11). The CO transition is accompanied by spin ordering, and the CO insulator is a CE-type antiferromagnet.[17] The orbital ordering therefore involves $3x^2 - r^2/3y^2 - r^2$ orbitals as in Figure 7a. Application of a magnetic field of 7T destroys the CO state, and the material becomes metallic (Figure 12), the sharpness of the transition decreasing with increasing strength of the field. The transition is first-order, showing hysteresis, and is associated with changes in unit cell parameters. The unit cell volume of the CO state is considerably smaller than that of the FMM state. The properties of $Nd_{0.5}Sr_{0.5}MnO_3$ can be described in terms of the phase diagram shown in Figure 13. The *Imma* space group of this manganate renders the Mn−O−Mn angle in the *ab* plane closer to 180°, promoting the overlap of the Mn(e_g) and O(2p) orbitals. Vacuum tunneling measurements[19] show that a gap of 250 meV opens up below T_{CO} (Figure 14). The gap collapses on applying a magnetic field, suggesting that a gap in the density of states at E_F is necessary for the stability of the CO state. Photoemission studies indicate a sudden change in electron states at the transition and give an estimate of 100 meV for the gap.[20] These estimates of the gap are considerably larger than the T_{co} (12 meV); it is not clear how a magnetic field of 6T (1.2 meV) can destroy the CO state. $Nd_{0.5}Sr_{0.5}MnO_3$ shows anomalous magnetostriction behavior, with a large positive magnetovolume effect (Figure 15), owing

to the magnetic field-induced structural transition accompanying a change from the AFM CO state to the FMM state.[21]

$Pr_{0.6}Ca_{0.4}MnO_3$ is an insulator at all temperatures and becomes charge-ordered at about 230 K (T_{CO}). At this temperature, anomalies are found in the magnetic susceptibility, as well as in the resistivity, as shown in Figure 16. In the CO state, the Mn^{3+} and Mn^{4+} ions are regularly arranged in the *ab* plane with the associated ordering of the $3x^2 - r^2/3y^2 - r^2$ orbitals. On cooling, AFM ordering (CE-type) occurs at 170 K (T_N). At about 40 K, $Pr_{0.6}Ca_{0.4}MnO_3$ exhibits canted AFM ordering. Application of an external magnetic field transforms the CO state to a FMM state, as shown in Figure 16, but the field required is much larger than in $Nd_{0.5}Sr_{0.5}MnO_3$. The transition

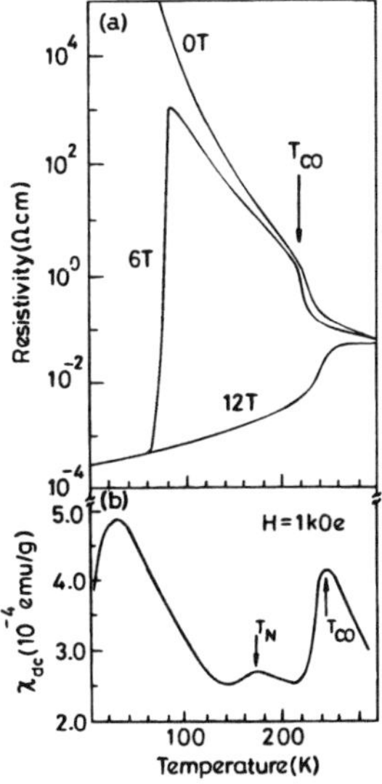

Figure 16. Temperature variation of (a) resistivity and (b) magnetic susceptibility of $Pr_{0.6}Ca_{0.4}MnO_3$ (from Tomioka et al.[22a] and Lees et al.[22b]).

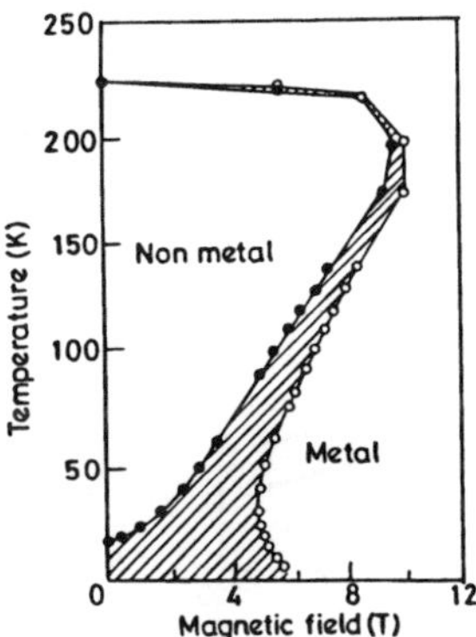

Figure 17. Temperature-magnetic field phase diagram for $Pr_{0.6}Ca_{0.4}$-MnO_3 (from Tokura et al.[18]).

is associated with hysteresis. The properties of $Pr_{0.6}Ca_{0.4}MnO_3$ can be represented by the phase diagram in Figure 17. The basic features of the CO state in $Pr_{0.6}Ca_{0.4}MnO_3$ are exhibited by several other rare earth manganates with relatively small A-size cations, in that the CO state is the ground state. Thus, $Nd_{0.5}$-$Ca_{0.5}MnO_3$ ($\langle r_A \rangle = 1.17$ Å) is a paramagnetic insulator with a charge-ordering transition at about 240 K.

Charge ordering occurs in the paramagnetic state in $Pr_{1-x}Ca_xMnO_3$ ($0.35 \leq x \leq 0.5$) with the T_{CO} increasing with x. The paramagnetic state is characterized by FM spin fluctuations with a small energy scale.[23] At T_{CO}, these fluctuations decrease and disappear at T_N. Electron diffraction and dark-field transmission electron microscopy (TEM) images show the presence of incommensurate charge ordering in the paramagnetic insulating state (180−260 K) of the $x = 0.5$ composition.[24] At T_N, there is an incommensurate−commensurate CO transition. In the incommensurate CO structure, partial orbital ordering is likely to be present. Similar charge, orbital, and spin ordering has been found in the 0.3 composition as well.[25] Optical conductivity spectra of the $x = 0.4$ composition show evidence of spatial charge and orbital ordering at 10 K.[26] The CO state

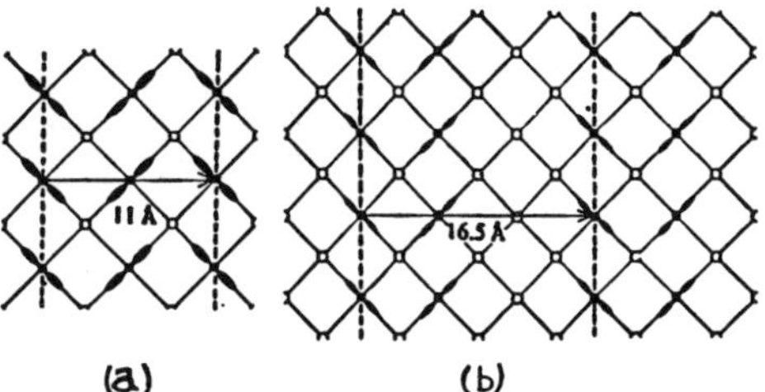

Figure 18. Charge and orbital ordering in $La_{1-x}Ca_xMnO_3$: (a) $x = 0.50$, (b) $x = 0.67$. Open circles represent Mn^{4+} and lobes show e_g orbitals of Mn^{3+}. Charge modulation wavelengths are ~11 and 16.5 Å for $x = 0.50$ and 0.67, respectively (following Cheong).

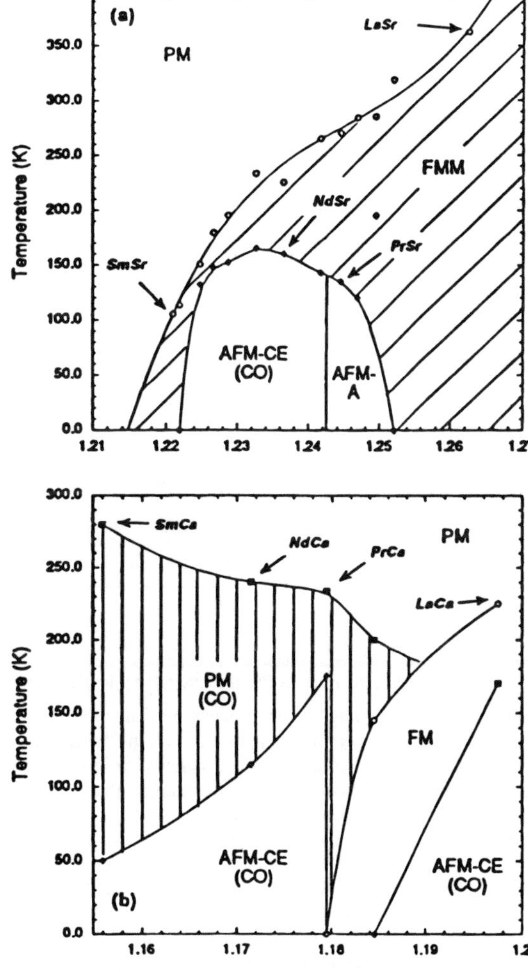

Figure 19. Temperature−$\langle r_A \rangle$ phase diagrams for (a) $Ln_{0.5}Sr_{0.5}MnO_3$ and (b) $Ln_{0.5}Ca_{0.5}MnO_3$ (from Woodward et al.[31]).

has a gap of ~0.2 eV, and the gap remains up to 4.5T. The gap value is the order parameter of the CO state and couples with spin ordering. $Pr_{0.67}Ca_{0.33}MnO_3$ shows thermal relaxation effects from the metastable FMM state (produced by the application of 10T magnetic field) to the CO state.[27] A metal−insulator transition is observed as an abrupt jump in resistivity at a well-

J. Phys. Chem. B, Vol. 104, No. 25, 2000 **5883**

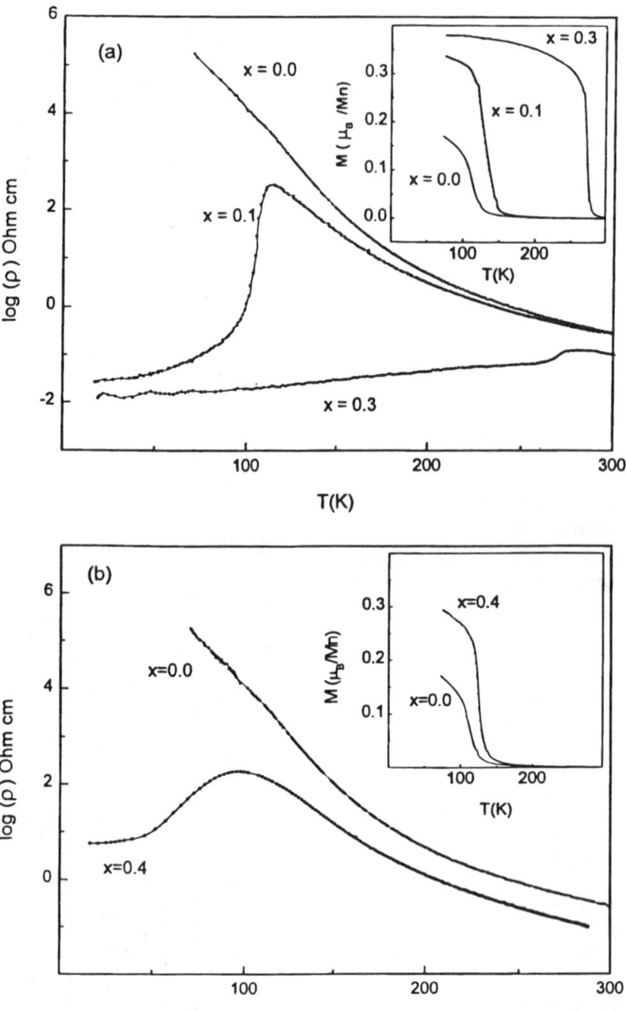

Figure 20. Effect of internal pressure on the properties of (a) $Pr_{0.7}Ca_{0.3-x}Sr_xMnO_3$ and (b) $Pr_{0.7-x}La_xCa_{0.3}MnO_3$. Note that substitution by a larger A-site cation renders the material FM (see insets). An insulator–metal transition is also observed (from Rao et al.[33]).

defined time, depending on the temperature. This observation seems to indicate a percolative nature of current transport.

Charge and orbital ordering in the manganates give rise to stripes.[28] In Figure 18 we show the modulation that can arise in $La_{1-x}Ca_xMnO_3$ with $x = 0.50$ and 0.67. At $x = 0.50$, the same number of Mn^{3+} and Mn^{4+} ions exist and diagonal stripes with a spacing of 11 Å are, therefore, to be expected (Figure 18a). When $x = 0.67$, there are twice as many Mn^{4+} ions as Mn^{3+} ions. Ordering of diagonal rows of Mn^{3+} and Mn^{4+} ions, besides the orientational ordering of the orbitals, would give rise to the striped pattern in Figure 18b, with a periodicity of 16.5 Å.

Paired JT distorted stripes or bistripes are believed to be present in $La_{0.33}Ca_{0.67}MnO_3$.[29] A supercell based on such ordering has been proposed, but a detailed structural investigation[30] has shown that the diffraction data can be explained satisfactorily without bistripes. Apparently, owing to orbital disordering, a mixture of paired and unpaired stripes seems to occur in $La_{0.5}Ca_{0.5}MnO_3$. Because such charge stripes generally are observed by electron microscopy, it is not clear whether these structures truly represent the bulk composition.

Cation Size Effects. In the manganates exhibiting CMR, the ferromagnetic T_c increases with the average radius of the A-site cations, $\langle r_A \rangle$. Increasing $\langle r_A \rangle$ is equivalent to increasing the hydrostatic pressure and is therefore accompanied by an increase in the $Mn-O-Mn$ angle and the e_g bandwidth. If there is considerable mismatch in the radii of the different A-site cations, however, the T_c does not increase with $\langle r_A \rangle$. Charge ordering is also highly sensitive to $\langle r_A \rangle$ and the T_{CO} generally increases with decrease in $\langle r_A \rangle$. The sensitivity of T_{CO} to $\langle r_A \rangle$ has been examined[7,31,32] and is generally attributed to an increased tilting of the MnO_6 octahedra as the $\langle r_A \rangle$ decreases. In Figure 19 we show that the phase diagrams[31] of $Ln_{0.5}Sr_{0.5}MnO_3$ and $Ln_{0.5}$-$Ca_{0.5}MnO_3$ illustrate some of the important features. (Here, the $\langle r_A \rangle$ is varied by changing the Ln ion.) The $Mn-O(eq)-Mn$ and $Mn-O(ax)-Mn$ bonds are identical in $Ln_{0.5}Ca_{0.5}MnO_3$ (excect when Ln = La). For $La_{0.5}Ca_{0.5}MnO_3$ and all the $Ln_{0.5}$-$Sr_{0.5}MnO_3$ compounds, the $Mn-O(eq)-Mn$ angle is significantly larger (by 2–6°) than the $Mn-O(ax)-Mn$ angle. Although the $Ln_{0.5}Ca_{0.5}MnO_3$ manganates crystallize in the *Pnma* symmetry, there is an evolution from *Pnma* to *I4/mcm* through *Imma* in the $Ln_{0.5}Sr_{0.5}MnO_3$ manganates, with increase

5884 *J. Phys. Chem. B, Vol. 104, No. 25, 2000*   Rao

in $\langle r_A \rangle$. The changes in the octahedral tilt system have consequences on the low-temperature magnetic structure. This is seen in $Nd_{0.5}Sr_{0.5}MnO_3$ where the charge ordering in the CE-type AFM state is associated with the *Imma* structure.

As pointed out earlier, the CO states of $Nd_{0.5}Sr_{0.5}MnO_3$ ($\langle r_A \rangle = 1.24$ Å) and $Pr_{0.6}Ca_{0.4}MnO_4$ ($\langle r_A \rangle = 1.17$ Å) are destroyed by magnetic fields. The field required to melt the CO state varies with $\langle r_A \rangle$ and the manganates with very small $\langle r_A \rangle$ remain charge-ordered even on the application of high magnetic fields.[32] Thus, $Y_{0.5}Ca_{0.5}MnO_3$ ($\langle r_A \rangle = 1.13$ Å) has a robust CO state that is not affected by very high magnetic fields (>25T). We can thus distinguish three different categories of manganates with respect to their sensitivity to magnetic fields: (a) manganates that are FM and become charge-ordered at low temperatures (e.g., $Nd_{0.5}Sr_{0.5}MnO_3$ when $T_{CO} = T_N$), with the CO state transforming to a FMM state on the application of a moderate magnetic field; (b) manganates that are charge-ordered in the paramagnetic state ($T_N < T_{CO}$), and do not exhibit an FMM state, but transform to a FMM state under a magnetic field (e.g., $Pr_{1-x}Ca_xMnO_3$); and (c) those that are charge-ordered in the paramagnetic state ($T_N < T_{CO}$) as in b, but are not affected by magnetic fields up to 15T or greater (e.g., $Y_{0.5}Ca_{0.5}MnO_3$). Category c is encountered when $\langle r_A \rangle \leq 1.17$ Å. The apparent one-electron bandwidth estimated on the basis of the experimental Mn$-$O$-$Mn angle and the average Mn$-$O distance in $Ln_{0.5}A_{0.5}MnO_3$ does not vary significantly with $\langle r_A \rangle$, which suggests that other factors may be responsible for the sensitivity of the CO state to $\langle r_A \rangle$. One possibility is a competition between the A- and B-site cations for covalent mixing with the O(2p) orbitals.[32]

By increasing the size of the A-site cations or by the application of pressure, the CO state in the manganates can be transformed to the FMM state.[33,34] In Figure 20, we show the effect of internal pressure on the CO state of $Pr_{0.7}Ca_{0.3}MnO_3$ wherein Ca is substituted by the larger Sr or Pr is substituted by La. The T_c in the $Pr_{0.5}Sr_{0.5-x}Ca_xMnO_3$ system, decreases with an increase in x or a decrease in $\langle r_A \rangle$ up to $x = 0.25$; $T_{CO} = T_N$ from $x = 0.09$ to 0.25. When $\langle r_A \rangle$ is decreased further, T_{CO} increases from 180 K for $x = 0.25$ to 250 K for $x = 0.30$; for $0.30 \leq x \leq 0.50$, $T_{CO} > T_N$.[35]

The effect of cation size disorder on the ferromagnetic T_c of rare earth manganates exhibiting CMR has been investigated in detail. The disorder is quantified in terms of the variance in the distribution $\langle r_A \rangle$, as defined by Attfield.[36] The variance σ^2 is defined by, $\sigma^2 = \Sigma x_i r_i^2 - \langle r_A \rangle^2$, where x_i is the fractional occupancy of A-site ions and r_i is the ionic radius. The ferromagnetic T_c decreases significantly with the variance, σ^2, based on the studies of rare earth manganates with fixed $\langle r_A \rangle$. A similar study of the variation of T_{CO} with σ^2 in $Ln_{0.5}A_{0.5}MnO_3$, for fixed $\langle r_A \rangle$ values of 1.17 and 1.24 Å, has shown that T_{CO} is not very sensitive to the size mismatch.[37] It appears that JT distortion and Coulomb interactions play a prominent role in determining the nature of the CO state in these materials.

Considering that the rare earth manganates with large $\langle r_A \rangle$, as exemplified by $Nd_{0.5}Sr_{0.5}MnO_3$ ($\langle r_A \rangle = 1.17$ Å), exhibit entirely different characteristics of the CO state, and that $\langle r_A \rangle = 1.17$ Å categorizes the manganates with respect to their insensitivity to magnetic fields, we would expect interesting and unusual properties in the intermediate range of $\langle r_A \rangle$ of 1.20 ± 0.20 Å. In this regime T_{CO} approaches T_c, leading to a competition between charge ordering and ferromagnetism. Thus, $La_{0.5}Ca_{0.5}MnO_3$ ($\langle r_A \rangle = 1.20$ Å) exhibits a region of co-existence of ferromagnetism and charge ordering ($T_c = 225$ K, $T_{CO} = 135$ K). At 135 K, the material also becomes AFM (CE-

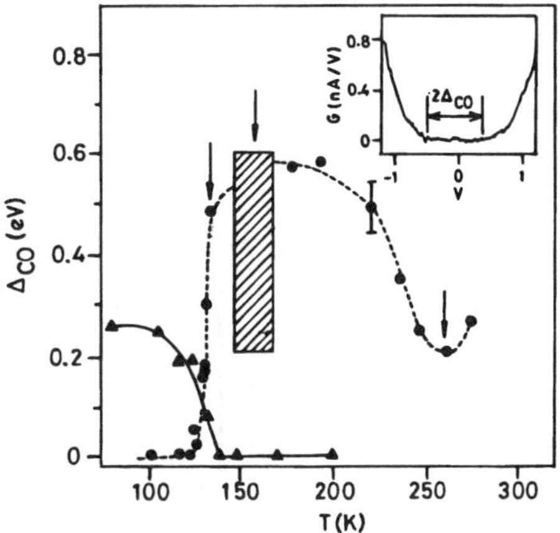

Figure 21. Temperature variation of the CO gap in $Nd_{0.25}La_{0.25}Ca_{0.5}MnO_3$ (closed circles) and $Nd_{0.5}Sr_{0.5}MnO_3$ (closed triangles). Inset shows a typical tunneling conductance curve (from Arulraj et al.[40]). Shaded region represents the coexistence regime.

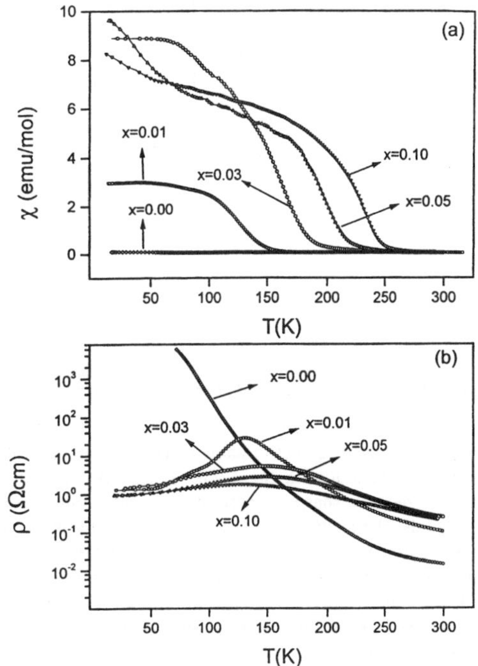

Figure 22. Temperature variation of the (a) magnetic susceptibility and (b) resistivity of $Nd_{0.5}Ca_{0.5}Mn_{1-x}Ru_xO_4$ (from Vanitha et al.[46b]).

type) and the orbital ordering becomes commensurate.[38] In $(Nd_{1-x}Sm_x)_{0.5}Sr_{0.5}MnO_3$, T_c decreases from 255 K to 115 K as x increases from 0.0 to 0.875, as expected of a decrease in $\langle r_A \rangle$.[39] The T_{CO} also decreases from 158 K to 0 K as x changes from 0.0 to 0.875. This unusual behavior wherein T_{CO} is suppressed as T_c approaches T_{CO} is interesting.

$Nd_{0.25}La_{0.25}Ca_{0.5}MnO_3$ ($\langle r_A \rangle = 1.19$ Å) reveals an intriguing sequence of phase transitions.[40] On cooling, this manganate develops an incipient CO state below 220 K. The formation of this state is accompanied by an increase in electrical resistivity and the opening up of a gap in the density of states near E_F. The orthorhombic distortion also increases, as a consequence of cooperative JT distortion of the lattice and short-range

J. Phys. Chem. B, Vol. 104, No. 25, 2000 **5885**

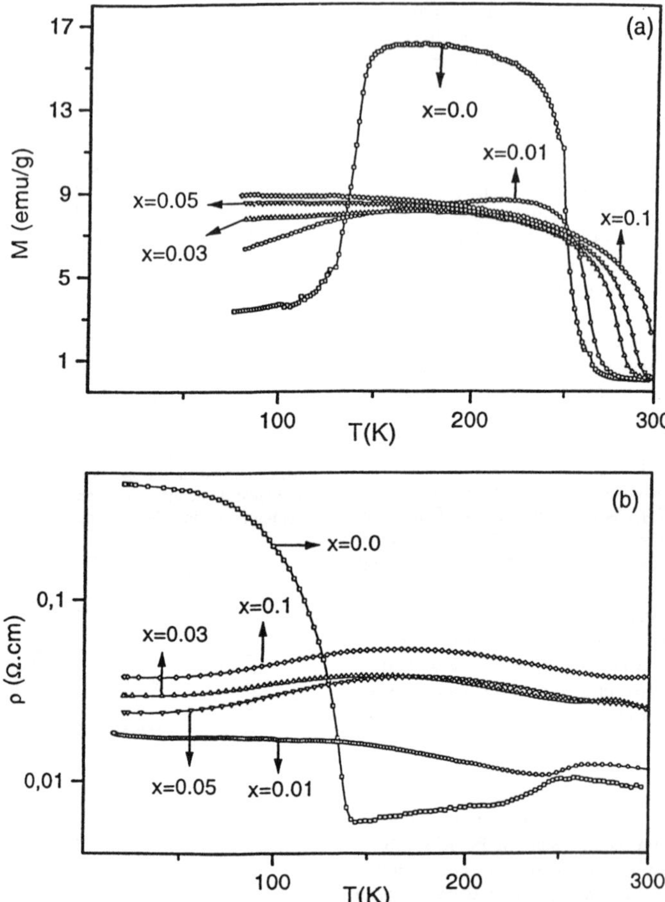

Figure 23. Temperature variation of the (a) magnetization and (b) resistivity of $Nd_{0.5}Sr_{0.5}Mn_{1-x}Ru_xO_4$ (from Vanitha et al.[46b]).

ordering of the Mn^{3+} and Mn^{4+} ions. At about 150 K, the incipient CO state becomes unstable and the material undergoes a reentrant transition to the FMM state. The transition is characterized by a sharp decrease in resistivity, collapse of the CO gap, development of FM ordering, and an abrupt decrease in the orthorhombic distortion. There is a two-phase coexistence region (150–220 K) around the CO–FMM transition. We show the fascinating reentrant transition in Figure 21 where the CO gap, obtained by vacuum tunneling measurements,[40] is plotted against temperature.

Effects of Substitution in the Mn Site and by ^{18}O. Substitution of ^{16}O by ^{18}O (generally up to 85–90%) has a marked effect on the magnetic properties and CMR of the manganates, indicating the important role of electron–phonon coupling. Substituting ^{18}O for ^{16}O in $La_{0.175}Pr_{0.525}Ca_{0.3}MnO_3$ destroys the insulator–metal transition and renders the material insulating down to 4 K.[41] The CO transition in $La_{0.5}Ca_{0.5}MnO_3$ increases by 9 K upon replacing ^{16}O by ^{18}O; the isotope shift increases with the magnetic field in $La_{0.5}Ca_{0.5}MnO_3$ and in $Nd_{0.5}$-$Sr_{0.5}MnO_3$.[42] In $Pr_{0.67}Ca_{0.33}MnO_3$, the magnetic field–induced insulator–metal transition occurs at a higher field on ^{18}O substitution; the heavier isotope favors the insulating state.[43] The isotope effect on T_{CO} is greater in $Nd_{0.5}Sr_{0.5}MnO_4$ than $Pr_{0.5}$-$Ca_{0.5}MnO_3$, indicating the role of $\langle r_A \rangle$ as well.[44]

Substitution of Mn by cations such as Al^{3+} and Fe^{3+} in charge-ordered manganates destroys charge ordering at moderate doping ($x \geq 0.03$), but the materials remain insulating. However, substitution by Cr^{3+} readily destroys charge ordering and renders the material FM and metallic.[45,46] In $Sm_{0.5}Ca_{0.5}Mn_{1-x}Cr_xO_3$, the T_{CO} (275 K when $x = 0.0$) decreases with increasing x and charge ordering disappears at $x = 0.05$. The $x = 0.05$ composition shows an insulator–metal transition when the material becomes FM.[47] The effectiveness of Cr^{3+} in destroying the CO state is considered to be due to its favorable electronic configuration (t_{2g}^3), which is the same as that of Mn^{4+}. However, Cr^{3+} in the Mn^{3+} site would be surrounded immediately by Mn^{4+} ions, which would not allow for near-neighbor electron hopping. Hopping would be more favored if Mn^{4+} were substituted by an appropriate quadrivalent cation such as Ru^{4+} $(t_{2g}^4 e_g^0)$. Recent studies have shown that substitution of Mn by Ru in $Nd_{0.5}Ca_{0.5}MnO_3$ destroys charge ordering and renders the material FM, with the T_c increasing with Ru content; the material also shows an insulator–metal transition[46] (Figure 22). The marked effect of Ru substitution is also seen in $Nd_{0.5}Sr_{0.5}MnO_3$ where the T_c increases with the Ru content to well above 300 K, but charge ordering is destroyed (Figure 23).

Phase Separation. The formation of FM clusters in an AFM host matrix in the rare earth manganates has been noticed by many workers. Thus, spin glass behavior has been encountered in the $Ln_{1-x}A_xMnO_3$ system at either extreme, corresponding to large or small x. Electronic phase separation is also evidenced

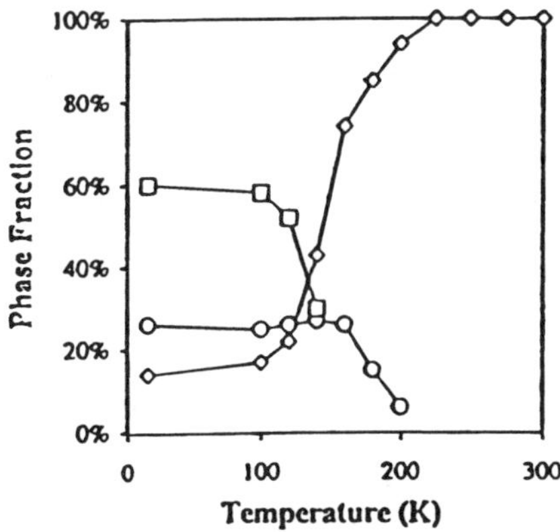

Figure 24. Variation in the percentage of the different phases of $Nd_{0.5}Sr_{0.5}MnO_3$ with temperature: FMM phase (diamonds); orbitally ordered A-type AFM phase (circles); charge-ordered CE-type AFM phase (squares) (from Woodward et al.[50]).

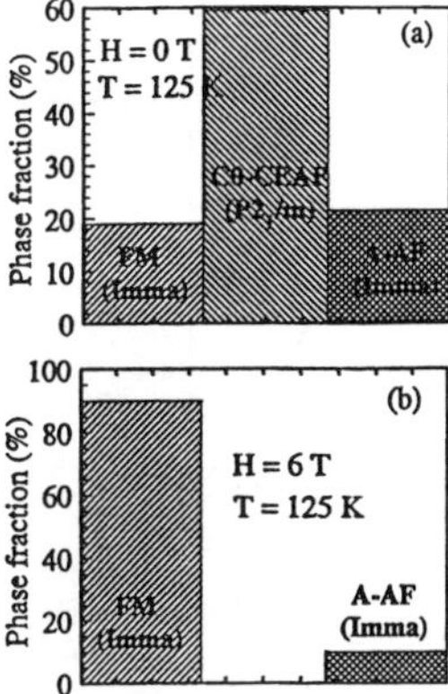

Figure 25. Effect of magnetic field (6T) on phase-separated $Nd_{0.5}Sr_{0.5}MnO_3$ at 125 K (from Ritter et al.[52]).

in the manganates. Thus, an electron microscopic study of $La_{1-x-y}Pr_yCa_xMnO_3$ ($x = 0.375$) has shown electronic phase separation into a submicrometer-scale mixture of CO insulator regions and FMM domains. Perculative transport could occur between the two states.[48] The coexistence of FM and CO states has been observed in $La_{0.5}Ca_{0.5}MnO_3$ and $Nd_{0.25}La_{0.25}Ca_{0.5}MnO_3$.[38,40] Some phase separation is likely in many of the charge-ordered manganates. In Cr-doped $Nd_{0.5}Ca_{0.5}MnO_3$, submicrometer FMM domains are embedded in an AFM CO state, so the material shows a relaxor behavior.[49]

$Nd_{0.5}Sr_{0.5}MnO_3$, which shows evidence for separation into three macroscopic phases, is particularly interesting. These are the high-temperature FMM phase (*Imma*), the A-type AFM intermediate-temperature phase (*Imma*), and the CE-type AFM CO low-temperature phase ($P2_1/m$).[50] The A-type AFM phase starts manifesting itself around 220 K, whereas the CE-type CO phase first appears at 150 K. In Figure 24, we show the phase compositions at different temperatures. There are three phases at the so-called CO transition at 150 K. The presence of the high-temperature FMM phase, even at very low temperatures, is noteworthy. These results are of significance in interpreting many of the properties of this manganate. The fact that such phase separation is seen by X-ray/neutron diffraction implies that even the minority phases have large domains (≥ 100 nm). The FMM phase of $Nd_{0.5}Sr_{0.5}MnO_3$ has a larger volume than the average volume or the volume of the low-temperature CO phase. The phase-separation behavior of this system and the relative stabilities of the structures seem to depend crucially on the Mn^{4+}/Mn^{3+} ratio. Thus, the regime $Mn^{4+}/Mn^{3+} > 1$ seems to stabilize the orbitally ordered AFM (A-type) phase. Unlike $Nd_{0.5}Sr_{0.5}MnO_3$, $Nd_{0.45}Sr_{0.55}MnO_3$ has an A-type AFM state below T_N (220 K) and metallicity is confined to the ferromagnetic ab plane below T_N, whereas the material is insulating along the c-axis. The large anisotropy in this material implies that the carriers are confined to the FM layer by magnetic and orbital ordering.[51] An interesting experiment on phase-separated $Nd_{0.5}Sr_{0.5}MnO_3$ at low temperatures was performed recently wherein the phase composition was determined by neutron diffraction in the absence and presence of a magnetic field.[52] The results show how on applying a magnetic field, the CO state at low temperatures, by and large, transforms to the FMM state (Figure 25). For some reason, a small proportion of the intermediate-temperature A-type AFM phase persists.

Effect of Electric Fields. Charge-ordered manganates such as $Pr_{1-x}Ca_xMnO_3$ ($x \approx 0.3-0.4$) undergo a CO–FMM transition

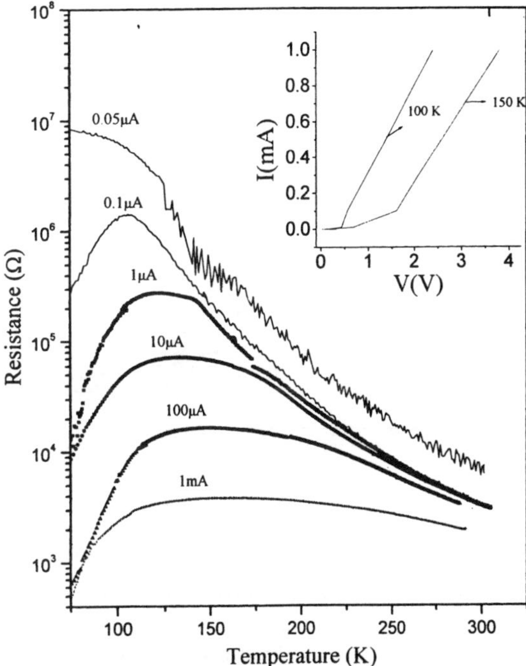

Figure 26. Electric current induced insulator–metal transition in $Nd_{0.5}Ca_{0.5}MnO_3$ films deposited on Si(100) at different values of the current. Inset show I–V curves at different temperatures (from Rao et al.[54]).

on application of moderate magnetic fields. However, manganates with small $\langle r_A \rangle$ such as $Y_{0.5}Ca_{0.5}MnO_3$ are, for all practical purposes, unaffected by magnetic fields. Laser irradiation has been reported to cause an insulator–metal transition in $Pr_{0.7}Ca_{0.3}MnO_3$, generating a localized conduction path, although the bulk of the sample is insulating.[53] It has been found recently that small d.c. currents induce insulator–metal transitions in thin films of several charge-ordered rare earth manganates, including $Y_{0.5}Ca_{0.5}MnO_3$, $Nd_{0.5}Ca_{0.5}MnO_3$, and $Pr_{0.7}Ca_{0.3}MnO_3$.[54] The current–voltage characteristics are nonohmic and show hysteresis. The I–M transition temperature decreases with increasing current (Figure 26). The hysteretic I–M transition in Figure 27 is specially noteworthy in that there is a reproducible memory effect in the cooling and heating cycles. The current-induced I–M transition occurs even in $Y_{0.5}Ca_{0.5}MnO_3$,

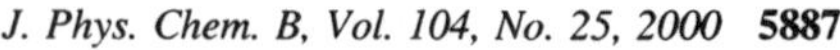

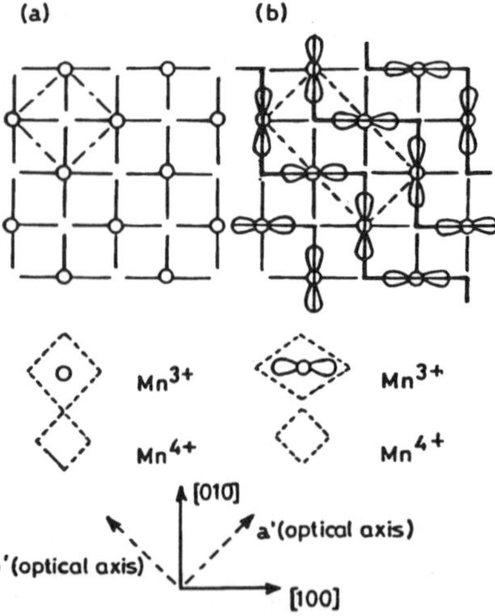

Figure 28. Charge ordering in La$_{0.5}$Sr$_{1.5}$MnO$_4$ showing oxide ion displacements. Mn^{3+} (e$_g$) orbitals are shown (from Ishikawa et al.[56a]).

when $x = 0.7$. La$_{0.5}$Sr$_{1.5}$MnO$_4$ shows ordering of the Mn^{3+} and Mn^{4+} ions at about 220 K, accompanied by the ordering of the e$_g$ orbitals (Figure 28). The material becomes antiferromagnetic at 110 K and exhibits anisotropic properties arising from orbital ordering.[56] At the 220 K CO transition, a significant change occurs in the conductivity spectrum. The order parameter for orbital ordering increases at T_{CO}, grows further with decreasing temperature, and decreases on spin ordering at T_N. The order parameters for charge and orbital ordering seem to evolve together. Application of high magnetic fields gives rise to metamagnetic transitions below T_{CO}, and charge and orbital ordering are destroyed by the magnetic field. Charge ordering in LaSr$_2$Mn$_2$O$_7$ ($n = 2$) has been observed by electron diffraction.[57] There is a need to study charge ordering in this as well as in other rare earth analogues, including the calcium derivatives.

Concluding Remarks

The rare earth manganates exhibit a variety of properties and phenomena, with an extraordinary sensitivity to various factors such as cation size, pressure, and magnetic and electric fields. In particular, the mutual relations between orbital ordering, charge ordering, and spin ordering in the rare earth manganates is truly fascinating. It is because of this interplay that charge ordering in these materials has turned out to be such an attractive area of research. In manganates such as Pr$_{0.6}$Ca$_{0.4}$MnO$_3$ and Nd$_{0.5}$Ca$_{0.5}$MnO$_3$ ($\langle r_A \rangle \approx 1.17$ Å), charge ordering occurs in the paramagnetic state, and commensurate orbital ordering accompanies a transition to the CE-type AFM state. In the paramagnetic CO state, there could be some orbital ordering that would be incommensurate, and FM correlations or clusters are generally present. In Nd$_{0.5}$Sr$_{0.5}$MnO$_3$ ($\langle r_A \rangle = 1.24$ Å), where charge ordering occurs on cooling a FMM state, however, it is possible to delineate transitions associated with orbital and charge ordering. In La$_{0.5}$Ca$_{0.5}$MnO$_3$ ($\langle r_A \rangle = 1.20$ Å), charge ordering and orbital ordering occur when the FM state transforms to an AFM state (CE type) on cooling. In the regime where the FM and CO states coexist, there is no long-range orbital ordering. As expected, the La$_{0.5}$Ca$_{0.5}$MnO$_3$ system gives rise to complex lattice images. In La$_{0.33}$Ca$_{0.67}$MnO$_3$, however, there is a clear-cut charge ordering transition (260 K) accompanied by orbital ordering, and followed by a transition to

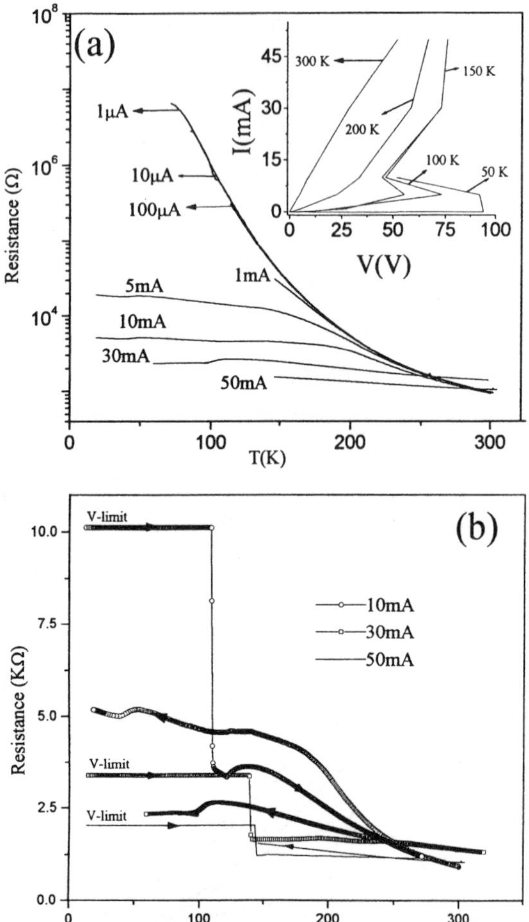

Figure 27. (a) Temperature variation of resistance of an oriented Nd$_{0.5}$Ca$_{0.5}$MnO$_3$ film deposited on LaAlO$_3$(001) for different values of the current; (b) resistance–temperature plots for three current values recorded over cooling and heating cycles showing memory effect. Inset in part a shows I–V curves at different temperatures (from Rao et al.[54]).

which is not affected by large magnetic fields. Furthermore, there is no need for prior laser irradiation to observe the current-induced I–M transitions. It is proposed that electric fields cause depinning of the randomly pinned charge solid. There appears to be a threshold field in the CO regime beyond which nonlinear conduction sets in along with a large broad-band conductivity noise.[55] Threshold-dependent conduction disappears around T_{CO}, which suggests that the CO state gets depinned at the onset of nonlinear conduction. At small currents or low magnetic fields, resistance oscillations occur because of temporal fluctuations between resistive states.

Layered Manganates. The interplay of spin, orbital, and charge ordering in the rare earth manganates in determining their properties becomes even more prominent in the layered manganates. In the Ruddlesden–Popper series of manganates, (Ln, A)$_{n+1}$Mn$_n$O$_{3n+1}$, the $n = \infty$ phases are the three-dimensional perovskites. The $n = 2$ phases, such as La$_{2-2x}$Sr$_{1+2x}$Mn$_2$O$_7$, show a metallic ground state for $x \geq 0.17$ and high CMR; they also exhibit interesting properties arising from the layered nature of the materials. The $n = 1$ member, Ln$_{1-x}$Sr$_{1+x}$MnO$_4$, with the quasi two-dimensional K$_2$NiF$_4$ structure, becomes metallic only

Rao

a CE-type AFM state, and paired stripes of $Mn^{3+}O_6$ octahedra have been observed in the lattice images. In manganates such as $Nd_{0.5}Ca_{0.5}MnO_3$ and $Y_{0.5}Ca_{0.5}MnO_3$ with a small $\langle r_A \rangle$, it is not clear whether ordered single or bistripes would be present. It is important to ensure that bistripes and other features found by electron miscroscopy truly represent the bulk structure. Low-temperature lattice imaging of these materials with high-resolution electron microscopy would be of great value.

Although we are able to distinguish spin, charge, and orbital ordering transitions, we do not understand some aspects. It is not entirely clear whether commensurate orbital ordering occurs only in the AFM (CE) state. The electron−hole asymmetry in the manganates, with respect to the FM and CO states, needs to be explained properly. Is it certain that we can never encounter ferromagnetism in the electron-doped materials? Also, what is the type of AFM ordering in the electron-doped CO manganates?

Phase separation in the manganates is especially interesting. Phase separation generally involves the formation of clusters of one phase in another (e.g., FM clusters in a CO or AFM phase). Macroscopic phase separation involving the presence of distinct domains of fairly large sizes (≥ 100 nm) with Bragg peaks in the diffraction patterns is, however, different from the formation of clusters or stripes. Such inhomogeneous distributions of CO or FM (charge and spin) phases in a material requires understanding. Coulomb forces would prevent accumulation of charge in a phase-separated regime in the absence of the means to compensate the charge. Electronic phase separation is known in $La_2NiO_{4+\delta}$ and $La_2CuO_{4+\delta}$.[3,11,28] In the rare earth manganates, there is increasing evidence for the coexistence of FM and AFM (CO) phases. Although it is possible that small changes in the Mn^{3+}/Mn^{4+} ratio could affect phase separation, we do not understand the mechanism. Although one can consider the presence of large domains (e.g., CO and FM) of different phases of a given composition as a criterion for phase separation, the effects of phase separation, in contrast to those of chemical inhomogenieties or cluster formation, need to be investigated in greater detail.

The $\langle r_A \rangle$ regime of 1.20 ± 0.02 Å in $Ln_{0.5}A_{0.5}MnO_3$, which exhibits complex phenomena and properties, including reentrant transitions, deserves further study. It is useful to examine such systems with fixed $\langle r_A \rangle$ and variable cation size mismatch. Charge ordering in the layered manganates has to be investigated in detail. We do not yet have a full understanding of the extraordinary effect of electric fields on the CO state.

Acknowledgment. The author is thankful to the Department of Science and Technology and the Science Office of the European Union for support of this work.

References and Notes

(1) (a) Honig, J. M. In *The Metallic and the Nonmetallic States of Matter*; Edwards, P. P., Rao, C. N. R., Eds.; Taylor and Francis: London, 1985. (b) Honig, J. M. *Proc. Indian Acad. Sci., Chem. Sci.* **1986**, *96*, 391.

(2) Tsuda, N.; Nasu, K.; Yanase, A.; Siratori, K. *Electronic Conduction in Oxides*; Springer: Berlin, 1991.

(3) (a) Kivelson, S. A.; Fradkin, E.; Emery, V. J. *Nature (London)* **1998**, *393*, 550. (b) Tranquada, J. M.; Sternleib, B. J.; Axe, J. D.; Nakamura, Y.; Uchida, S. *Nature (London)* **1995**, *375*, 561.

(4) Imada, M.; Fujimori, A.; Tokura, Y. *Rev. Mod. Phys.* **1998**, *70*, 1039.

(5) (a) von Helmolt, R.; Holzapfel, B.; Schultz, L.; Samwer, K. *Phys. Rev. Lett.* **1993**, *73*, 2331. (b) Chahara, K.; Ohno, T.; Kasai, M.; Kozono, Y. *Appl. Phys. Lett.* **1993**, *63*, 1990.

(6) (a) Rao, C. N. R.; Raveau, B., Eds. *Colossal Magnetoresistance, Charge Ordering and Related Properties of Manganese Oxides*; World Scientific: Singapore, 1998. (b) Ramirez, A. P. *J. Phys: Condens. Matter* **1997**, *9*, 8171.

(7) Rao, C. N. R.; Arulraj, A.; Santosh, P. N.; Cheetham, A. K. *Chem. Mater.* **1998**, *10*, 2714.

(8) Zener, C. *Phys. Rev.* **1951**, *82*, 403.

(9) Jirak, Z.; Krupicka, S.; Simsa, Z.; Dlouha, M.; Vratislav, S. *J. Magn. Magn. Mater.* **1985**, *53*, 153.

(10) (a) Takano, M.; Takeda, Y. *Bull. Inst. Chem. Res.* **1983**, *61*, 406. (b) Battle, P. D.; Gibb, T. C.; Lightfoot, P. *J. Solid State Chem.* **1990**, *84*, 271.

(11) (a) Tranquada J. M.; Loronzo, J. E.; Buttrey, D. J.; Sachan, V. *Phys. Rev.* **1995**, *B52*, 3581. (b) Tranquada, J. M.; Loronzo, J. E.; Buttrey, D. J.; Sachan, V. *Phys. Rev.* **1996**, *B54*, 12318.

(12) Ramirez, A. P.; Gammel, P. L.; Cheong, S. W.; Bishop, D. J.; Chandra, P. *Phys. Rev. Lett.* **1996**, *76*, 447.

(13) (a) Rodriguez-Carvajal, J.; Rousse, G.; Masquelier, C.; Hervieu, M. *Phys. Rev. Lett.* **1998**, *81*, 4660. (b) Willis, A. S.; Raju, N. P.; Greedan, J. E. *Chem. Mater.* **1999**, *11*, 1510.

(14) Wollan, E. O.; Koehler, W. C. *Phys. Rev.* **1955**, *100*, 545.

(15) (a) Goodenough, J. B. *Phys. Rev.* **1955**, *100*, 564. (b) Goodenough, J. B.; Wold, A.; Arnolt, R. J.; Menyuk, N. *Phys. Rev.* **1961**, *124*, 373.

(16) Vogt, T.; Cheetham, A. K.; Mahendiran, R.; Raychaudhuri, A. K.; Mahesh, R.; Rao, C. N. R. *Phys. Rev.* **1996**, *B54*, 15303.

(17) Kuwahara, H.; Tomioka, Y.; Asamitsu, A.; Moritomo, Y.; Tokura, Y. *Science* **1995**, *270*, 961.

(18) Tokura, Y.; Tomioka, Y.; Asamitsu, A.; Moritomo, Y.; Kasai, M. *J. Appl. Phys.* **1996**, *79*, 5288.

(19) Biswas, A.; Raychaudhuri, A. K.; Mahendiran, R.; Guha, A.; Mahesh, R.; Rao, C. N. R. *J. Phys. Condens. Matter* **1997**, *9*, L355.

(20) Sekiyama, A.; Suga, S.; Fujikawa; Imada, S.; Iwasaki, T.; Matsuda, K.; Matsushita, T.; Kaznacheyev, K. V.; Fujimori, A.; Kuwahara, H.; Tokura, Y. *Phys. Rev.* **1999**, *B59*, 15528.

(21) Mahendiran, R.; Ibarra M. R.; Maignan, A.; Millange, F.; Arulraj, A.; Mahesh, R.; Raveau, B.; Rao, C. N. R. *Phys. Rev. Lett.* **1999**, *82*, 2191.

(22) (a) Tomioka, Y.; Asamitsu, A.; Kuwahara, H.; Moritomo, Y.; Tokura, Y. *Phys. Rev.* **1996**, *B53*, 1689. (b) Lees, M. R.; Barrat, J.; Balakrishnan, G.; Paul D, M. C. K.; Yethiraj, M. *Phys. Rev.* **1995**, *B52*, 14303.

(23) Kajimoto R.; Kakeshita T.; Oohara Y.; Yoshizawa H.; Tomioka Y.; Tokura, Y. *Phys. Rev.* **1998**, *B58*, R11837.

(24) Mori, S.; Katsufuji, T.; Yamamoto, N.; Chen, C. H.; and Cheong, S. W. *Phys. Rev.* **1999**, *B59*, 13573.

(25) Cox, D. E.; Radaelli, P. G.; Marezio, M.; Cheong, S. W. *Phys. Rev.* **1998**, *B57*, 3305.

(26) Okimoto Y.; Tomioka, Y.; Onose, Y.; Otsuka, Y.; Tokura, Y. *Phys. Rev.* **1999**, *B59*, 7401.

(27) Anane, A.; Renard, J. P.; Reversat, L.; Dupas, C.; Veillet P.; Viret M.; Pinsard, L.; Revcolevsci, A. *Phys. Rev.* **1999**, *B59*, 77.

(28) Cheong, S. W.; Chen, C. H. In *Colossal Magnetoresistance, Charge Ordering and Related Properties of Manganese Oxides*; Rao, C. N. R., Raveau, B, Eds. World Scientific: Singapore, 1998.

(29) Mori, S.; Chen, C. H.; Cheong, S. W. *Nature (London)* **1998**, *392*, 473.

(30) Radaelli, P. G.; Cox, D. E.; Capogna, L.; Cheong, S. W.; Marezio, M. *Phys. Rev.* **1999**, *B59*, 14440.

(31) Woodward, P. M.; Vogt, T.; Cox, D. E.; Arulraj, A.; Rao, C. N. R.; Karen, P.; Cheetham, A. K. *Chem. Mater.* **1998**, *10*, 3652.

(32) Arulraj, A.; Santosh, P. N.; Gopalan, R. S.; Guha, A.; Raychaudhuri, A. K.; Kumar, N.; Rao, C. N. R. *J. Phys: Condens. Matter* **1998**, *10*, 8497.

(33) Rao, C. N. R.; Santosh, P. N.; Singh, R. S.; Arulraj, A. *J. Solid State Chem.* **1998**, *135*, 169.

(34) Moritomo, Y.; Kuwahara, H.; Tomioka, Y.; Tokura, Y. *Phys. Rev.* **1997**, *B55*, 7549.

(35) Damay, F.; Martin, C.; Maignan, A.; Hervieu, M.; Raveau, B.; Jirak, Z.; Andre, G.; Bouree, F. *Chem. Mater.* **1999**, *11*, 536.

(36) Attfield, J. P. *Chem. Mater.* **1998**, *10*, 3239.

(37) Vanitha, P. V.; Santosh, P. N.; Singh, R. S.; Rao, C. N. R.; Attfield, J. P. *Phys. Rev.* **1999**, *B59*, 13539.

(38) Radaelli, P. G.; Cox, D. E.; Marezio, M.; Cheong, S. W. *Phys. Rev.* **1997**, *B55*, 3015.

(39) Tokura, Y.; Kuwahara, H.; Moritomo, Y.; Tomioka, Y.; Asamitsu, A. *Phys. Rev. Lett.* **1996**, *76*, 3184.

(40) Arulraj, A.; Biswas, A.; Raychaudhuri, A. K.; Rao, C. N. R.; Woodward, P. M.; Vogt, T.; Cox, D. E.; Cheetham, A. K. *Phys. Rev.* **1998**, *B57*, R8115.

(41) Babushkina, N. A.; Belova, L. M.; Gorbenko, O. Yu; Kaul, A. R.; Bosak, A. A.; Ozhogin, V. I.; Kugel, KI. *Nature (London)* **1998**, *391*, 159.

(42) Zhao, G.; Ghosh, K.; Keller, H.; Greene, R. L. *Phys. Rev.* **1999**, *B59*, 81.

(43) Garcia-Landa, B.; Ibarra, M. R.; De Teresa, J. M.; Zhao, G.; Conder, K.; Keller, H. *Solid State Commun.* **1998**, *105*, 567.

(44) Mahesh, R.; Itoh, M. *J. Solid State Chem.* **1999**, *144*, 232.

(45) Raveau, B.; Maignan, A.; Martin, C.; Hervieu, M. *Chem. Mater.* **1998**, *10*, 2641.

(46) (a) Vanitha, P. V.; Singh, R. S.; Natarajan, S.; Rao, C. N. R. *Solid State Commun.* **1999**, *109*, 135. (b) Vanitha, P. V.; Arulraj, A.; Raju, A. R.; Rao, C. N. R. *C. R. Acad. Sci. Ser. II*, t.2 **1999**, 595.

(47) Barnabe, A.; Hervieu, M.; Martin, C.; Maignan, A.; Raveau, B. *J. Mater. Chem.* **1998**, *8*, 1405.

(48) Uehara M., Mori, S.; Chen, C. H.; Cheong, S. W. *Nature (London)* **1999**, *399*, 560.

(49) Kimura, T.; Tomioka, Y.; Kumai, R.; Okinoto, Y.; Tokura, Y. *Phys. Rev. Lett.* **1999**, *83*, 3940.

(50) Woodward, P. M.; Cox, D. E.; Vogt, T., Rao, C. N. R.; Cheetham, A. K. *Chem. Mater.* **1999**, *11*, 3528.

(51) Kuwahara, H.; Okuda, T.; Tomioka, Y.; Asamitsu, A.; Tokura, Y. *Phys. Rev. Lett.* **1999**, *82*, 4316.

(52) Ritter, C.; Mahendiran, R.; Ibarra, M. R.; Morellon, L.; Maignan, A.; Raveau, B.; Rao, C. N. R. *Phys. Rev.* **2000**, *B61*, R9229.

(53) Feibig, M.; Miyano, K.; Tomioka, Y.; Tokura, Y. *Science* **1998**, *280*, 1925.

(54) Rao, C. N. R.; Raju, A. R.; Ponnambalam. V.; Parashar, S.; Kumar, N. *Phys. Rev.* **2000**, *B61*, 591.

(55) Guha, A.; Ghosh, A.; Raychaudhuri, A. K.; Parashar, S.; Raju, A. R.; Rao, C. N. R. *Appl. Phys. Lett.* **1999**, *75*, 3381.

(56) (a) Ishikawa, T.; Ookura, K.; Tokura, Y. *Phys. Rev.* **1999**, *B59*, 8367. (b) Tokunaga, M.; Miura, N.; Moritomo, Y.; Tokura, Y. *Phys. Rev.* **1999**, *B59*, 11151.

(57) Li, J. Q.; Matsui, Y.; Kimura, T.; Tokura, Y. *Phys. Rev.* **1998**, *B57*, 3205.

Chem. Mater. **2001**, *13*, 787–795

Electron–Hole Asymmetry in the Rare-Earth Manganates: A Comparative Study of the Hole- and the Electron-Doped Materials

K. Vijaya Sarathy,[†] P. V. Vanitha,[†] Ram Seshadri,[‡] A. K. Cheetham,[§] and C. N. R. Rao*,[†,‡,§]

Chemistry and Physics of Materials Unit, Jawaharlal Nehru Centre for Advanced Scientific Research, Jakkur P.O., Bangalore 560 064, India, and Solid State and Structural Chemistry Unit, Indian Institute of Science, Bangalore 560 012, India, and Materials Research Laboratory, University of California, Santa Barbara, California 93106

Received June 6, 2000. Revised Manuscript Received December 18, 2000

Properties of the hole-doped $Ln_{1-x}A_xMnO_3$ (Ln = rare earth, A = alkaline earth, $x < 0.5$) are compared with those of the electron-doped compositions ($x > 0.5$). Charge ordering is the dominant interaction in the latter class of manganates unlike ferromagnetism and metallicity in the hole-doped materials. Properties of charge-ordered (CO) compositions in the hole- and electron-doped regimes, $Pr_{0.64}Ca_{0.36}MnO_3$ and $Pr_{0.36}Ca_{0.64}MnO_3$, differ markedly. Thus, the CO state in the hole-doped $Pr_{0.64}Ca_{0.36}MnO_3$ is destroyed by magnetic fields and by substitution of Cr^{3+} or Ru^{4+} (3%) in the Mn site, while the CO state in the electron-doped $Pr_{0.36}Ca_{0.64}MnO_3$ is essentially unaffected. It is not possible to induce long-range ferromagnetism in the electron-doped manganates by increasing the Mn–O–Mn angles up to 165 and 180 °as in $La_{0.33}Ca_{0.33}Sr_{0.34}MnO_3$; application of magnetic fields and Cr/Ru substitution (3%) do not result in long-range ferromagnetism and metallicity. Application of magnetic fields on the Cr/Ru-doped, electron-doped manganates also fails to induce metallicity. These unusual features of the electron-doped manganates suggest that the electronic structure of these materials is likely to be entirely different from that of the hole-doped ones, as verified by first-principles linearized muffin-tin orbital calculations.

Introduction

Rare-earth manganates of the formula $Ln_{1-x}A_xMnO_3$ (Ln = rare earth and A = alkaline earth) exhibit extraordinary properties such as colossal magnetoresistance (CMR) and charge ordering.[1] The manganates with $x < 0.5$, where the trivalent Ln ions are substituted by divalent A ions, have come to be designated as hole-doped. Accordingly, the manganates with $x > 0.5$, where the Ln ion substitutes a divalent A ion, are being referred to as electron-doped.[2] The manganates exhibiting CMR, by and large, have compositions in the range $0.1 < x < 0.5$, wherein the average radius of the A-site cations, $\langle r_A \rangle$, is fairly large. These manganates become ferromagnetic because of the double-exchange mechanism of electron hopping between Mn^{3+} ($t_{2g}^3 e_g^1$) and Mn^{4+} ($t_{2g}^3 e_g^0$) via the oxygen and undergo an insulator–metal transition around the ferromagnetic T_C. CMR is

generally highest around T_C, and the T_C is sensitive to $\langle r_A \rangle$. When $\langle r_A \rangle$ is sufficiently small, the materials do not ordinarily exhibit ferromagnetism but instead become charge-ordered insulators. Thus, charge ordering and double exchange are competing interactions in the manganates. As x in $Ln_{1-x}A_xMnO_3$ increases, crossing over from the hole-doped regime to the electron-doped regime ($x > 0.5$), charge ordering becomes the dominant interaction and ferromagnetism does not appear to occur in any of the compositions. In this regime, CMR occurs over a narrow range of compositions, $0.80 < x < 1.0$, but there is no long-range ferromagnetism or metallicity associated with the materials.[2] The various features of the manganates are nicely borne out by the approximate phase diagrams shown in Figure 1, prepared on the basis of available data. What is noteworthy is the marked absence of symmetry in these phase diagrams. While the presence of electron–hole asymmetry in the manganates is not surprising, considering that the introduction of e_g electrons increases the lattice distortion and their removal would have the opposite effect, the asymmetry has some unusual features.

Electron–hole asymmetry is encountered in cuprate superconductors.[3] In the cuprates, superconductivity occurs in the electron-doped regime, although not as prominently as in the hole-doped regime. The electron–

---

* To whom correspondence should be addressed. E-mail: cnrrao@jncasr.ac.in.
† Jawaharlal Nehru Centre for Advanced Scientific Research.
‡ Indian Institute of Science.
§ University of California.

(1) (a) Ramirez, A. P. *J. Phys.: Condens. Matter* **1997**, *9*, 8171. (b) Rao, C. N. R., Raveau, B., Eds. *Colossal Magnetoresistance, Charge ordering and Related Properties of Manganese Oxides*; World Scientific: Singapore, 1998. (c) Rao, C. N. R.; Arulraj, A.; Cheetham, A. K.; Raveau, B. *J. Phys.: Condens. Matter* **2000**, *12*, R83.
(2) (a) Maignan, A.; Martin, C.; Damay, F.; Raveau, B. *Chem. Mater.* **1998**, *10*, 950. (b) Santhosh, P. N.; Arulraj, A.; Vanitha, P. V.; Singh, R. S.; Sooryanarayana, K.; Rao, C. N. R. *J. Phys.: Condens. Matter* **1999**, *11*, L27.

(3) Ramakrishnan, T. V.; Rao, C. N. R. *Superconductivity Today*, 2nd ed.; University of India Press: India, 1999.

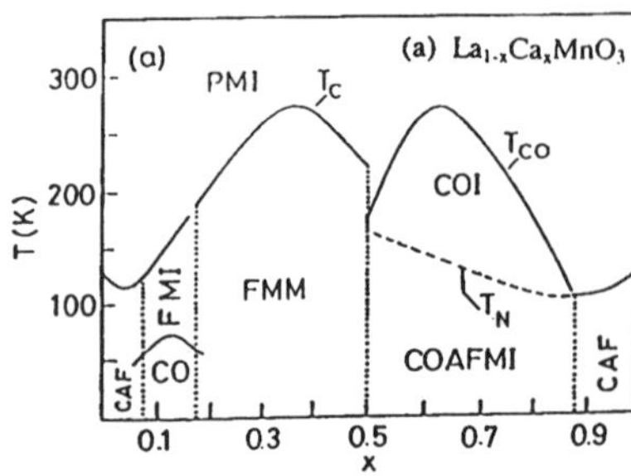

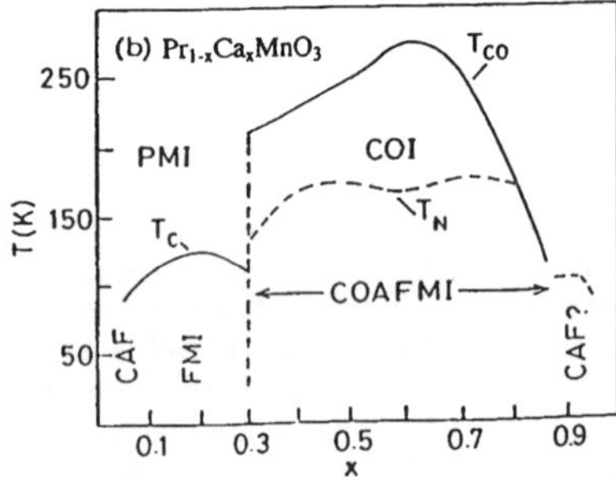

Figure 1. Phase diagrams of (a) $La_{1-x}Ca_xMnO_3$ and (b) $Pr_{1-x}Ca_xMnO_3$: CAF, canted antiferromagnet; CO, charge-ordered state; FMI, ferromagnetic insulator; FMM, ferromagnetic metal; PMI, paramagnetic insulator; COI, charge-ordered insulator (paramagnetic); COAFMI, charge-ordered antiferromagnetic insulator. These diagrams have been prepared based on the available data in the literature and reflect the properties of the systems fairly satisfactorily. The source material for diagram a can be found from ref 1c and from Cheong and Chen in ref 1b. The source material for diagram b can be found in ref 1b,c and in Maignan et al. *Phys. Rev. B* **1999**, *B60*, 12191.

hole asymmetry in the rare-earth manganates, involving the total absence of the ferromagnetic metallic (FMM) state in the electron-doped regime, is therefore worthy of investigation. We have studied this interesting problem by comparing the structural properties of the hole- and electron-doped manganates of similar compositions in the $Pr_{1-x}Ca_xMnO_3$ system ($x = 0.36$ and 0.64). Although there have been several studies on hole-doped compositions of this system,[1] a careful comparison of hole- and electron-doped materials seemed to be necessary. In particular, we have examined whether ferromagnetism can be induced in the electron-doped material by appropriate cation substitution in the B-site and/or application of magnetic fields. The cations chosen for this purpose are Cr^{3+} ($t_{2g}^3 e_g^0$) and Ru^{4+} ($t_{2g}^4 e_g^0$), which have been found to be effective in destroying charge ordering in materials such as $Nd_{0.5}Ca_{0.5}MnO_3$ and $Sm_{0.5}Ca_{0.5}MnO_3$.[4] To ensure that the Mn−O−Mn angle is not the limiting factor, we have prepared a manganate of the composition $La_{0.33}Ca_{0.33}Sr_{0.34}MnO_3$ with significantly large Mn−O−Mn angles and studied the effects of B-site substitution and magnetic fields on the properties of this material. We have also carried out first principles

electronic structure calculations to understand what makes the hole- and electron-doped manganates different.

Experimental Section

Polycrystalline samples of the manganates were prepared by the ceramic method. Stoichiometric quantities of the respective rare-earth oxides, alkaline-earth carbonates, MnO_2 or Mn_3O_4, and the dopant transition-metal oxide (Cr_2O_3 or RuO_2) were mixed and preheated at 1173 K for 12 h in air. They were subsequently ground and heated at 1473 K for another 12 h in air. The mixture so obtained was pelletized and heated at 1673 K. The X-ray diffraction pattern recorded (using a Seifert XRD 3000TT instrument) showed a single phase for all of the compositions prepared. Rietveld analysis was carried using the structure refinement program GSAS. Data were collected between $2\theta = 10°$ and $100°$ with a scan step of $0.02°$.

Electrical resistivity measurements were carried out on pressed pellets of polycrystalline materials by the four-probe method between 300 and 20 K. Magnetic measurements were carried out using a vibrating sample magnetometer (VSM-7300, Lakeshore Inc.) between 300 and 50 K employing a field of 0.01 T. Magnetoresistivity measurements were carried out using a cryocooled closed-cycle superconducting magnet designed by us along with Cryo Industries of America, Manchester, MA.

Results and Discussion

The phase diagrams of two rare-earth manganates, $Ln_{1-x}Ca_xMnO_3$ (Ln = La and Pr) in Figure 1, clearly demonstrate the electron−hole asymmetry present in these materials and also the preponderance of the charge-ordered (CO) state in the electron-doped regime ($x > 0.5$). The FMM state, generally found in the hole-doped regime ($x \leq 0.5$), is favored by the large size of the A-site cations.[5] Thus, the FMM state occurs up to $x = 0.5$ in $La_{1-x}Ca_xMnO_3$ (Figure 1a). In $Pr_{1-x}Ca_xMnO_3$, the FMM state is not found at any composition, and there is only a ferromagnetic insulating (FMI) state when $0.1 \leq x \leq 0.3$. The charge-ordering regime in $La_{1-x}Ca_xMnO_3$ is $0.5 \leq x \leq 0.85$ but is considerably wider ($0.3 \leq x \leq 0.85$) in $Pr_{1-x}Ca_xMnO_3$. Accordingly, charge ordering occurs in both the hole- and electron-doped compositions of the latter system. Charge ordering in these systems is ascertained by the appearance of superlattice reflections in the diffraction patterns and also by the occurrence of anomalies (observation of maxima) in the temperature variation of magnetic susceptibility and the activation energy for conduction.[1c]

Effects of Cation Size and Size Disorder. In Figure 2, we plot the charge-ordering transition temperatures, T_{CO}, in $Ln_{0.5}Ca_{0.5}MnO_3$ and $Ln_{0.36}Ca_{0.64}MnO_3$ against the average radius of the A-site cations $\langle r_A \rangle$. Here, $\langle r_A \rangle$ is varied by changing the Ln ion. While T_{CO} increases with a decrease in $\langle r_A \rangle$ in the case of $Ln_{0.5}Ca_{0.5}MnO_3$, it is not very sensitive to $\langle r_A \rangle$ in electron-doped $Ln_{0.36}Ca_{0.64}MnO_3$. In Table 1, we compare the T_{CO} values and other properties of $Ln_{0.64}Ca_{0.36}MnO_3$ and $Ln_{0.36}Ca_{0.64}MnO_3$. The T_{CO} value is generally higher in the latter system compared to that of the hole-doped materials, but there is little variation with $\langle r_A \rangle$ in both

(4) (a) Barnabe, A.; Maignan, A.; Hervieu, M.; Raveau, B. *Eur. Phys. J.* **1998**, *B1*, 145. (b) Barnabe, A.; Hervieu, M.; Martin, C.; Maignan, A.; Raveau, B. *J. Mater. Chem.* **1998**, *8*, 1405. (c) Martin, C.; Maignan, A.; Damay, F.; Hervieu, M.; Raveau, B. *J. Solid State Chem.* **1997**, *134*, 198. (d) Vanitha, P. V.; Arulraj, A.; Raju, A. R.; Rao, C. N. R. *C. R. Acad. Sci. Paris* **1999**, *2*, 595.

(5) (a) Hwang, H. Y.; Cheong, S. W.; Radaelli, P. G.; Marezio, M.; Batlogg, B. *Phys. Rev. Lett.* **1995**, *75*, 914. (b) Mahesh, R.; Mahendiran, R.; Raychaudhuri, A. K.; Rao, C. N. R. *J. Solid State Chem.* **1995**, *120*, 204.

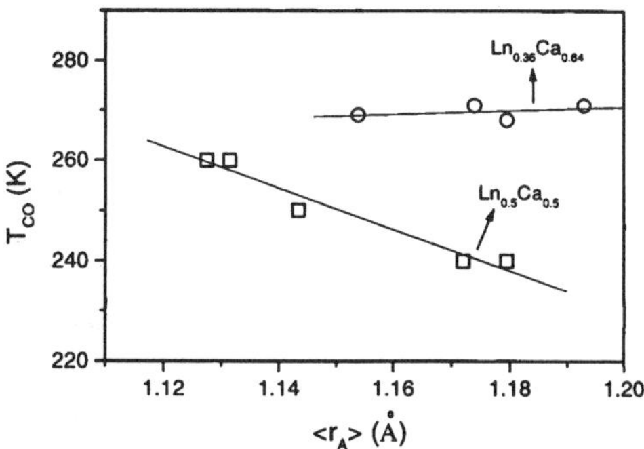

Figure 2. Variation of the charge-ordering transition temperature, T_{CO}, with the average size of the A-site cation, $\langle r_A \rangle$.

Table 1. Effect of $\langle r_A \rangle$ on Charge Ordering in Ln$_{1-x}$Ca$_x$MnO$_3$ (x = 0.36 and 0.64)

Ln	$\langle r_A \rangle$ (Å)	σ^2 (Å^2)	lattice parameter (Å)			$T_{CO}{}^a$ (K)
			a	b	c	
		Ln$_{0.64}$Ca$_{0.36}$MnO$_3$				
La	1.203	0.0003	5.454	5.468	7.704	b
Pr	1.179	0.0000	5.413	5.442	7.676	210
Nd	1.169	0.0001	5.407	5.458	7.646	212
		Ln$_{0.36}$Ca$_{0.64}$MnO$_3$				
La	1.193	0.0013	5.390	5.391	7.588	271
Pr	1.179	0.0000	5.374	5.369	7.576	268
Nd	1.174	0.0001	5.388	5.361	7.570	271

^{a} From magnetization measurements (H = 100 G). ^{b} This compound exhibits ferromagnetism and an insultor−metal transition around the T_C (~250 K).

series of compounds. In the series of manganates listed in Table 1, the cation size disorder,[6a] as measured by the variance, σ^2, is quite small. It may be noted that in the electron-doped manganates T_{CO} increases with electron concentration, x, but the ferromagnetic component[2a] (in the cluster regime $0.0 < x < 0.2$ in Ca$_{1-x}$Ln$_x$MnO$_3$) is only slightly affected by $\langle r_A \rangle$ for a fixed value of x.

The effect of cation size disorder on the charge-ordering transition in Ln$_{0.5}$Ca$_{0.5}$MnO$_3$ has been investigated by keeping the average radius of the A-site cation fixed and varying σ^2. The value of σ^2 is varied by making different combinations of the Ln and alkaline-earth ions.[6b] The slope of the linear $T_{CO}-\sigma^2$ plot for Ln$_{0.5}$Ca$_{0.5}$MnO$_3$ is 10 975 K Å^{-2} and the intercept of the plot, $T_{CO}{}^0$, is 236 K. The value of $T_{CO}{}^0$ corresponds to that of the disorder-free manganate. We have not been able to obtain sufficient reliable data on the variation of T_{CO} with σ^2 (at fixed $\langle r_A \rangle$) in hole-doped compositions of the type Ln$_{0.64}$Ca$_{0.36}$MnO$_3$, but the limited data available show only small changes. In the case of the electron-doped Ln$_{0.36}$Ca$_{0.64}$MnO$_3$, however, we have obtained reliable data for two series of manganates with fixed $\langle r_A \rangle$ values of 1.180 and 1.174 Å, respectively, corresponding to Pr$_{0.36}$Ca$_{0.64}$MnO$_3$ and Nd$_{0.36}$Ca$_{0.64}$MnO$_3$. These compounds, along with their structural data and T_{CO} values, are listed in Table 2. We show the plots of T_{CO} against σ^2 for the two series of manganates in Figure 3. The plots are linear, giving slopes 6408 and 5813 K Å^{-2} for fixed $\langle r_A \rangle$ of 1.180 and 1.174 Å, respec-

tively. The intercept, $T_{CO}{}^0$, is around 266 K in both of the cases, a value considerably higher than that in Ln$_{0.5}$A$_{0.5}$MnO$_3$.[6b] An examination of the phase diagram in Figure 1a shows that in La$_{1-x}$Ca$_x$MnO$_3$ T_{CO} reaches a maximum value around x = 0.65. We see a similar maximum in the phase diagram of Pr$_{1-x}$Ca$_x$MnO$_3$ as well (Figure 1b). In the latter system, however, T_{CO} increases with the hole concentration, x, in the hole-doped regime ($0.3 < x \leq 0.5$) and with the electron concentration, $1 - x$, in the electron-doped regime ($0.6 < x \leq 0.85$). Despite some of these apparent similarities, the hole-doped and electron-doped compositions exhibit significant differences in their electronic and magnetic properties, as detailed in the following sections.

Comparison of the Hole- and Electron-Doped Pr$_{1-x}$Ca$_x$MnO$_3$ (x = 0.36 and 0.64). To understand the nature of electron−hole asymmetry in the rare-earth manganates, it is useful to compare the electronic and magnetic properties of comparable compositions of the hole- and electron-doped materials. Thus, the hole-doped La$_{0.7}$Ca$_{0.3}$MnO$_3$ becomes ferromagnetic around 250 K, at which temperature it exhibits an insulator−metal transition. The electron-doped La$_{0.3}$Ca$_{0.7}$MnO$_3$, on the other hand, gets charge-ordered at 271 K and does not exhibit the FMM state at any temperature. A better appreciation of the differences in the properties of the hole- and electron-doped manganates is obtained by comparing the properties of Pr$_{1-x}$Ca$_x$MnO$_3$ at the same carrier concentration (equal values of $1 - x$ and x). We have carried out detailed studies on Pr$_{0.64}$Ca$_{0.36}$MnO$_3$ and Pr$_{0.36}$Ca$_{0.64}$MnO$_3$, both of which are charge-ordered.

Pr$_{0.64}$Ca$_{0.36}$MnO$_3$, I, and Pr$_{0.36}$Ca$_{0.64}$MnO$_3$, II, are both orthorhombic (*Pbnm*), but the unit cell is larger in the former as expected on the basis of the relative sizes of Mn^{3+} and Mn^{4+}. In Table 3, we list the atomic coordinates and the lattice parameters of the two manganates obtained from the Rietveld analysis. The Mn−O distances in I are longer than those in II, but the Mn−O−Mn angles in the two are comparable.

Both Pr$_{0.64}$Ca$_{0.36}$MnO$_3$ and Pr$_{0.36}$Ca$_{0.64}$MnO$_3$ get charge-ordered in the paramagnetic state, with transition temperatures (T_{CO}) of 210 and 268 K, respectively. They show maxima in the magnetization curves at the charge-ordering transition temperatures (Figure 4). Pr$_{0.64}$Ca$_{0.36}$MnO$_3$ also shows an antiferromagnetic transition around 140 K. The nature of the transitions has been ascertained independently by diffraction studies as well as EPR and other measurements.[1,7] Both Pr$_{0.64}$Ca$_{0.36}$MnO$_3$ and Pr$_{0.36}$Ca$_{0.64}$MnO$_3$ are insulators down to low temperatures, as is expected of charge-ordered compositions, but the electron-doped composition shows a more marked change in resistivity at T_{CO} (Figure 5). The difference between the two lies in the effect of magnetic fields. Application of a magnetic field of 12 T melts the CO state to a metallic state in the case of Pr$_{0.64}$Ca$_{0.36}$MnO$_3$ (Figure 5a). A 12 T magnetic field has no effect whatsoever on the resistivity of Pr$_{0.36}$Ca$_{0.64}$MnO$_3$ (Figure 5b).

Substitution of Cr^{3+} or Ru^{4+} in the Mn site of certain charge-ordered manganates is known to destroy the CO state, rendering them ferromagnetic and metallic.[1c,4,8] The effect of Cr^{3+} doping on Pr$_{1-x}$Ca$_x$MnO$_3$ has been

(6) (a) Rodriguez-Martinez, L. M.; Attfield, J. P. *Phys. Rev.* **1996**, *B54*, R15622; **1998**, *B58*, 2426. (b) Vanitha, P. V.; Santhosh, P. N.; Singh, R. S.; Rao, C. N. R.; Attfield, J. P. *Phys. Rev.* **1999**, *B59*, 13539.

(7) Gupta, R.; Joshi, J. P.; Bhat, S. V.; Sood, A. K.; Rao, C. N. R. *J. Phys.: Condens. Matter* **2000**, *12*, 6919.

Table 2. Structure and Properties of $Ln_{0.36-x}Ln_xCa_{0.64-y}Sr_yMnO_3$ with Fixed $\langle r_A \rangle$ Values[a]

composition	σ^2 (Å^2)	lattice parameter (Å)			% D (300 K)	T_{CO} (K)
		a	b	c		
$\langle r_A \rangle = 1.174$ Å						
$Nd_{0.36}Ca_{0.64}MnO_3$	0.0001	5.388	5.361	7.570	0.33	271 (261)
$Pr_{0.28}Gd_{0.08}Ca_{0.64}MnO_3$	0.0003	5.353	5.366	7.548	0.18	279 (270)
$La_{0.185}Gd_{0.175}Ca_{0.64}MnO_3$	0.0011	5.354	5.369	7.537	0.27	257 (253)
$La_{0.225}Y_{0.135}Ca_{0.64}MnO_3$	0.0017	5.353	5.370	7.563	0.15	249 (248)
$Sm_{0.36}Ca_{0.554}Sr_{0.086}MnO_3$	0.0022	5.355	5.372	7.569	0.15	263 (258)
$Nd_{0.1}Gd_{0.26}Ca_{0.528}Sr_{0.112}MnO_3$	0.0033	5.355	5.364	7.580	0.06	245 (243)
$Gd_{0.15}Y_{0.21}Ca_{0.433}Sr_{0.207}MnO_3$	0.0066	5.346	5.374	7.567	0.21	~229 (223)
$\langle r_A \rangle = 1.18$ Å						
$Pr_{0.36}Ca_{0.64}MnO_3$	0.0000	5.374	5.369	7.576	0.11	267 (267)
$Nd_{0.18}Sm_{0.18}Ca_{0.553}Sr_{0.087}MnO_3$	0.0019	5.369	5.370	7.580	0.07	~256 (251)
$Nd_{0.18}Gd_{0.18}Ca_{0.518}Sr_{0.122}MnO_3$	0.0031	5.363	5.376	7.573	0.14	250 (245)
$La_{0.16}Y_{0.2}Ca_{0.526}Sr_{0.114}MnO_3$	0.0043	5.364	5.370	7.578	0.07	~241 (~238)
$La_{0.1}Y_{0.26}Ca_{0.46}Sr_{0.18}MnO_3$	0.0060	5.386	5.369	7.550	0.32	234 (222)

[a] The values in parentheses are obtained from resistivity data; % D is the orthorhombic distortion.

Table 3. Atomic Coordinates and Structural Parameters of I, $Pr_{0.64}Ca_{0.36}MnO_3$,[a] and II, $Pr_{0.36}Ca_{0.64}MnO_3$[b]

atom	site	x	y	z	frac	U_{ISO}
Pr	4c	−0.0083 (0.5027)	0.0315 (0.4746)	0.2500 (0.2500)	0.6400 (0.3600)	0.0049 (−0.0033)
Ca	4c	−0.0083 (0.5027)	0.0315 (0.4746)	0.2500 (0.2500)	0.3600 (0.6400)	0.0049 (−0.0033)
Mn	4b	0.5000 (0.5000)	0.0000 (0.0000)	0.0000 (0.0000)	1.0000 (1.0000)	0.0040 (0.0063)
O	8d	0.0529 (0.0544)	0.4939 (0.5213)	0.2500 (0.2500)	1.0000 (1.0000)	0.0405 (0.0901)
O	4c	−0.7115 (0.2854)	0.2809 (0.2717)	0.0354 (−0.0460)	1.0000 (1.0000)	0.0122 (−0.0367)

bond	distance (Å)	bond	angle (deg)
Mn−O	2 × 1.935 (2 × 1.892)	Mn−O−Mn	4 × 157.6 (4 × 155.6)
	2 × 1.941 (2 × 1.993)		2 × 162.9 (2 × 161.2)
	2 × 1.989 (2 × 1.916)		

[a] $a = 5.4310$ Å, $b = 5.4573$ Å, $c = 7.6761$ Å; *Pbnm*; $R_{wp} = 3.45\%$. [b] $a = 5.3664$ Å, $b = 5.3746$ Å, $c = 7.5603$ Å; *Pbnm*; $R_{wp} = 3.28\%$. The values in brackets correspond to those of II.

investigated in the composition region $0.6 \leq x \leq 0.7$, with the Cr^{3+} content going up to 12%.[4a] These workers find a marked effect when Cr^{3+} is around 10%, at which composition one would expect clustering of the dopant ions leading to superexchange-induced ferromagnetism. We have substituted Mn by Cr^{3+} or Ru^{4+}, keeping the dopant concentration at 3% to avoid clustering. On doping with 3% Cr^{3+}, $Pr_{0.64}Ca_{0.36}MnO_3$ becomes ferromagnetic with a T_C of 130 K, but $Pr_{0.36}Ca_{0.64}MnO_3$ remains paramagnetic and charge-ordered, albeit with a slightly lower T_{CO} (215 K) as shown in Figure 6a. The 3% Ru doping shows similar differences between the two manganates (Figure 6b).

In Figure 7, we show the effect of 3% Cr^{3+} doping on the resistivity of $Pr_{0.64}Ca_{0.36}MnO_3$ and $Pr_{0.36}Ca_{0.64}MnO_3$. The former exhibits an insulator−metal (I−M) type transition around 80 K, but the latter remains an insulator. The I−M transition in the hole-doped system is shifted to higher temperatures on applying magnetic fields. Application of magnetic fields does not render the Cr^{3+}-doped $Pr_{0.36}Ca_{0.64}MnO_3$ metallic (Figure 7) at any temperature, indicating that it may not be possible to induce the FMM state in this electron-doped material even under favorable conditions. Results with 3% Ru^{4+} doping in the two manganates are similar, in that the hole-doped material becomes FMM while the CO state in the electron-doped material is essentially unaffected. In the insets of Figure 7, we show the results obtained with 3% Ru^{4+}-doped $Pr_{0.64}Ca_{0.36}MnO_3$ and $Pr_{0.36}Ca_{0.64}$-MnO_3. The failure to destroy the CO state of $Pr_{0.36}Ca_{0.64}$-MnO_3 with Cr/Ru doping as well as with a magnetic

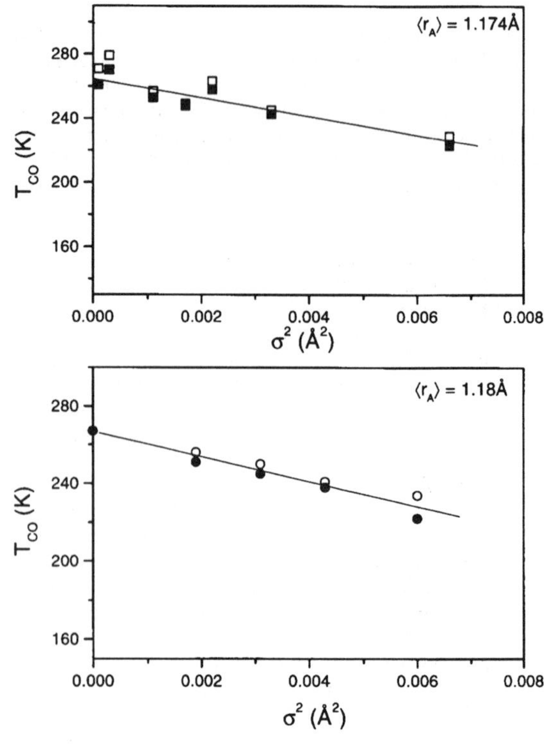

(8) (a) Raveau, B.; Maignan, A.; Martin, C.; Hervieu, M. *Chem. Mater.* **1998**, *10*, 2641. (b) Vanitha, P. V.; Singh, R. S.; Natarajan, S.; Rao, C. N. R. *J. Solid State Chem.* **1998**, *137*, 365.

Figure 3. Variation of the charge-ordering temperature in $Ln_{0.36}Ca_{0.64}MnO_3$ with σ^2 for fixed $\langle r_A \rangle$ values. Open symbols represent data obtained from magnetic measurements, and the corresponding closed symbols are from resistivity measurements.

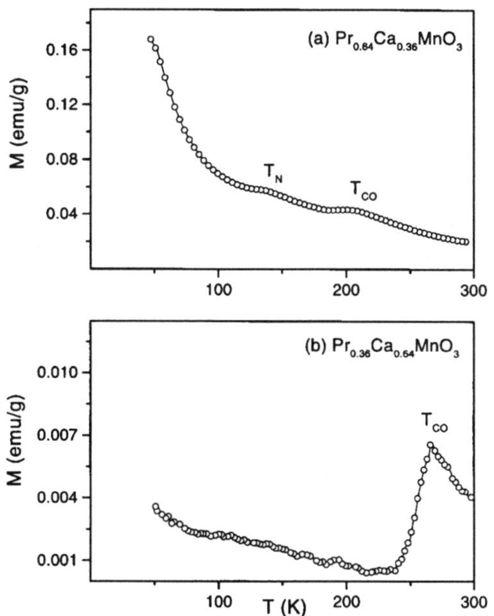

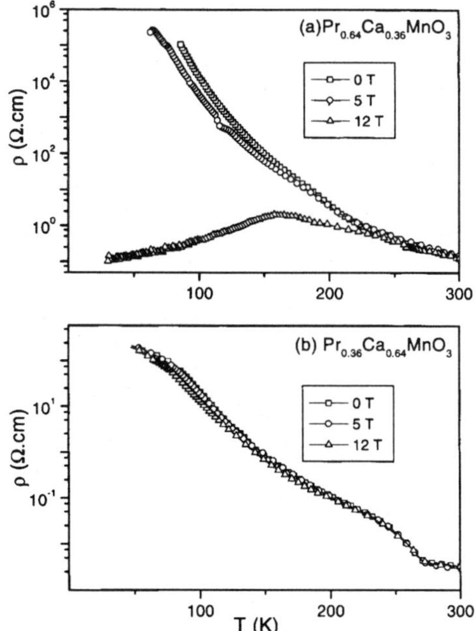

Figure 4. Temperature variation of the magnetization of (a) $Pr_{0.64}Ca_{0.36}MnO_3$ and (b) $Pr_{0.36}Ca_{0.64}MnO_3$.

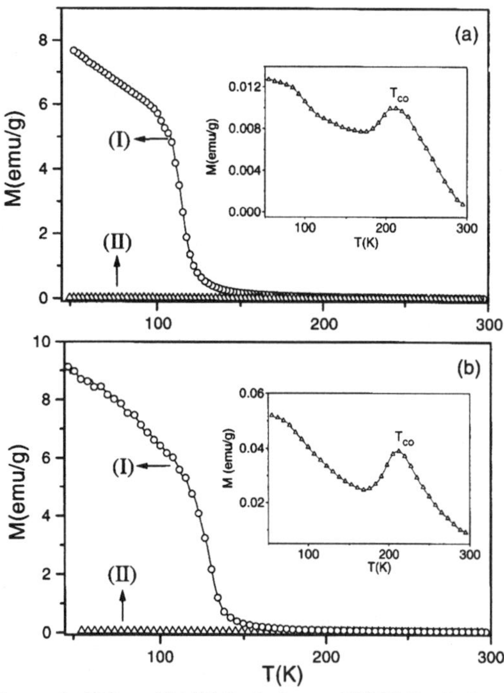

Figure 6. Effect of (a) 3% Cr doping and (b) 3% Ru doping in the Mn site of $Pr_{0.64}Ca_{0.36}MnO_3$, I, and $Pr_{0.36}Ca_{0.64}MnO_3$, II, on the magnetization. The inset gives the magnetization data of II on an enlarged scale to show T_{CO}.

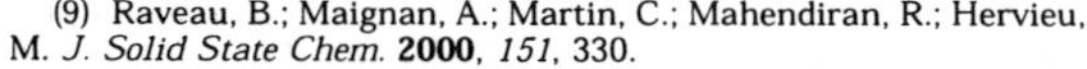

Figure 5. Temperature variation of the electrical reisistivity of (a) $Pr_{0.64}Ca_{0.36}MnO_3$ and (b) $Pr_{0.36}Ca_{0.64}MnO_3$. The effect of magnetic fields is shown.

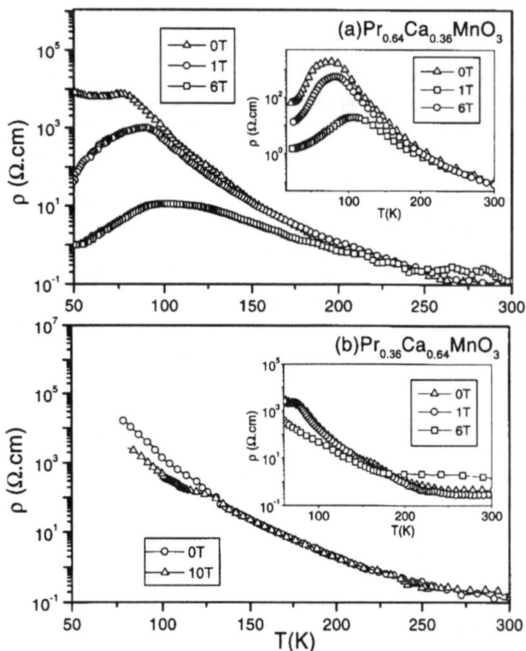

Figure 7. Effect of 3% Cr^{3-} doping on the electrical resistivity of (a) $Pr_{0.64}Ca_{0.36}MnO_3$ and (b) $Pr_{0.36}Ca_{0.64}MnO_3$. The effect of magnetic fields is also shown. Insets in a and b show the effect of 3% Ru doping.

field of 12 T is noteworthy. Raveau et al.[9] have recently observed that Ru substitution in $Ln_{0.4}Ca_{0.6}MnO_3$ gives rise to FMM clusters, but the effect is prominent at high Ru concentrations (>3%). The observed effect is probably due to clusters or domains containing Ru ions, as suspected by these authors. Furthermore, the observed

magnetization in these samples is low. This raises the question as to whether long-range ferromagnetism can ever occur in the electron-doped manganates. We note

(9) Raveau, B.; Maignan, A.; Martin, C.; Mahendiran, R.; Hervieu, M. *J. Solid State Chem.* **2000**, *151*, 330.

Table 4. Atomic Coordinates and Structural Parameters of La$_{0.33}$Ca$_{0.33}$Sr$_{0.34}$MnO$_3$[a]

atom	site	x	y	z	frac	U_{iso}
La	4b	0.0000	0.5000	0.2500	0.3300	−0.0134
Ca	4b	0.0000	0.5000	0.2500	0.3300	0.0245
Sr	4b	0.0000	0.5000	0.2500	0.3400	0.0212
Mn	4c	0.0000	0.0000	0.0000	1.0000	0.0007
O	4a	0.0000	0.0000	0.2500	1.0000	0.0051
O	8h	0.2799	0.7799	0.0000	1.0000	0.0051

bond	distance (Å)	bond	angle (deg)
Mn−O	4 × 1.909	Mn−O−Mn	4 × 166.4
	2 × 1.949		2 × 180.0

[a] $a = b = 5.3608$ Å, $c = 7.7857$ Å; $I4/mcm$; $R_{wp} = 11.82\%$.

here that Maignan et al.[10] find a cluster glass state (but no long-range ferromagnetism) in Ca$_{1-x}$Sm$_x$MnO$_3$ ($0 \leq x \leq 0.12$). Neumeier and Cohn[11] observe only local ferromagnetic regions within an antiferromagnetic host in Ca$_{1-x}$La$_x$MnO$_3$ ($0 \leq x \leq 0.2$), as was indeed observed earlier by Mahendiran et al.[12]

Absence of Long-Range Ferromagnetism in Electron-Doped Manganates. To answer the above question, it is important to ensure that the Mn−O−Mn angle is not a limiting factor. This is because the Mn−O−Mn angle in Pr$_{1-x}$Ca$_x$MnO$_3$ is generally in the 156−162 ° range (Table 3), which may be considered to be somewhat small. For ferromagnetism to be favored in the electron-doped manganates, it is important to have a material with a much larger Mn−O−Mn angle. To decide on the composition of such a material, the following considerations are relevant. The Mn−O−Mn angles in CaMnO$_3$ and La$_{0.5}$Ca$_{0.5}$MnO$_3$ are around 158 ° and 160 °, respectively.[13,14] Therefore, substitution of Ca by any of the rare earths would not increase the Mn−O−Mn angle beyond 160 °. It is, however, possible to increase the angle by Sr substitution, as in Ca$_{1-x}$Sr$_x$MnO$_3$ and La$_{0.5}$Ca$_{0.5-x}$Sr$_x$MnO$_3$.[15,16] We therefore prepared a manganate of the composition La$_{0.33}$Ca$_{0.33}$Sr$_{0.34}$MnO$_3$. The structure of the manganate is tetragonal ($I4/mcm$). On the basis of a Rietveld analysis of the powder X-ray diffraction data, we have obtained the atomic coordinates and structural parameters listed in Table 4. The two Mn−O−Mn distances are around 1.91 and 1.95 Å, while the angles are 166.4 ° and 180 °. These values of the angles are comparable to those found in some of the ferromagnetic and metallic compositions of the hole-doped manganates.

In Figure 8 we show the magnetization and electrical resistivity data of La$_{0.33}$Ca$_{0.33}$Sr$_{0.34}$MnO$_3$. The material is a paramagnetic insulator down to 25 K, with a charge-ordering transition of around 220 K. Application of magnetic fields up to 12 T has no effect on the electrical

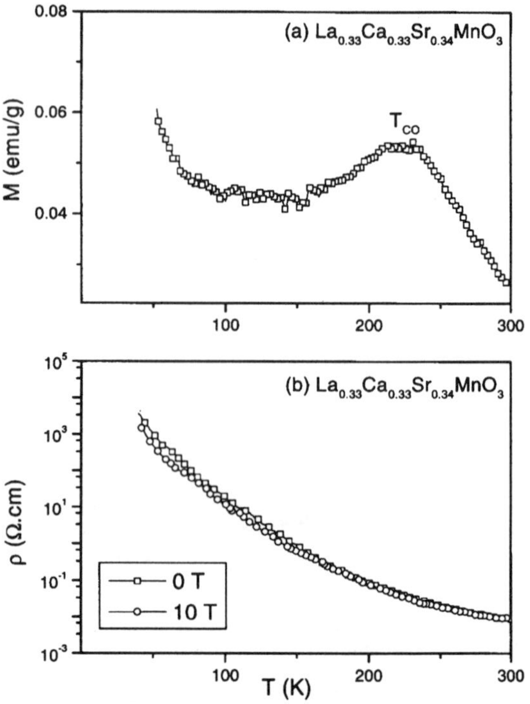

Figure 8. Temperature variation of (a) the magnetization and (b) the resistivity of La$_{0.33}$Ca$_{0.33}$Sr$_{0.34}$MnO$_3$. The effect of magnetic fields on the resistivity is shown.

resistivity (Figure 8b). Clearly, the large Mn−O−Mn angles do not help to make this manganate ferromagnetic. Furthermore, 3% Cr^{3+} doping of La$_{0.33}$Ca$_{0.33}$Sr$_{0.34}$MnO$_3$ does not transform it to a ferromagnetic metal (Figure 9). What is surprising is that the application of magnetic fields up to 10 T does not induce an I−M transition in the Cr-doped material. The same holds for the 3% Ru^{4+}-doped material, as shown in the insets of Figure 9a,b. It must be remembered that the analogous hole-doped composition La$_{0.67}$A$_{0.33}$MnO$_3$ (A = Ca/Sr) becomes a ferromagnetic metal at fairly high temperatures (230−300 K).[1] These results suggest that it may not be possible to make the electron-doped rare-earth manganates exhibit long-range ferromagnetism and metallicity. As observed earlier, other studies also show at best local ferromagnetic interactions or the presence of FM clusters.[9−12]

The absence of long-range ferromagnetism and metallicity in the electron-doped manganates is difficult to comprehend. One difference that is noteworthy is that the hole-doped manganates possess a higher proportion of e$_g$ electrons relative to the degenerate e$_g$ orbitals. There are also some intrinsic differences between the Mn^{3+} (d^4) and Mn^{4+} (d^3) ions. Although the structure of the parent LnMnO$_3$ compounds is influenced by the Jahn−Teller distorted Mn^{3+}, the stability of this state is not sufficient to inhibit the electron-transfer process when Mn^{4+} ions are introduced. The facile electron transfer required for double exchange, and hence ferromagnetism, can be sustained in such a hole-doped system. By contrast, the high ligand-field stabilization of the preponderant Mn^{4+} ions in the electron-doped materials can inhibit electron transfer. To unravel the

(10) Maignan, A.; Martin, C.; Damay, F.; Raveau, B.; Hejtmanek, J. *Phys. Rev.* **1998**, *B58*, 2758.
(11) Neumeier, J. J.; Cohn, J. L. *Phys. Rev.* **2000**, *B61*, 14319.
(12) Mahendiran, R.; Tiwary, S. K.; Raychaudhuri, A. K.; Ramakrishnan, T. V.; Mahesh, R.; Rangavittal, N.; Rao, C. N. R. *Phys. Rev.* **1996**, *B53*, 3348.
(13) Poeppelmeirer, K. R.; Leonowicz, M. E.; Scanlon, J. C.; Longo, J. M.; Yellon, W. B. *J. Solid State Chem.* **1982**, *45*, 71.
(14) Radaelli, P. G.; Cox, D. E.; Marezio, M.; Cheong, S. W.; Schiffer, P. E.; Ramirez, A. P. *Phys. Rev. Lett.* **1995**, *75*, 4488. (b) Radaelli, P. G.; Cox, D. E.; Capogna, L.; Cheong, S.-W.; Marezio, M. *Phys. Rev.* **1999**, *B59*, 14440.
(15) Taguchi, H.; Sonoda, M.; Nagao, M. *J. Solid State Chem.* **1998**, *137*, 82.
(16) Sundaresan, A.; Paulose, P. L.; Mallik, R.; Sampathkumaran, E. V. *Phys. Rev.* **1998**, *B57*, 2690.

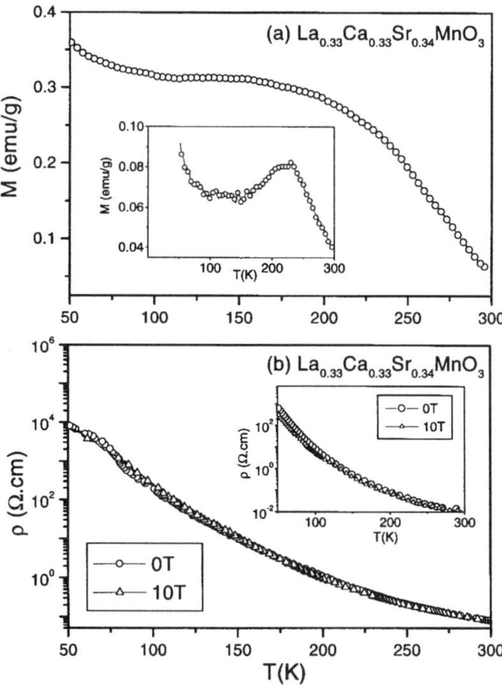

Figure 9. Effect of 3% Cr^{3+} doping on (a) the magnetization and (b) the resistivity of $La_{0.33}Ca_{0.33}Sr_{0.34}MnO_3$. The effect of magnetic fields on the resistivity is also shown. Insets in a and b show the effect of 3% Ru doping.

cause(s) for the absence of metallicity in the electron-doped managanates, we have carried out some electronic structure calculations.

Electronic Structure Calculations. We have used first-principles electronic structure calculations as manifest in the (spin) density functional linearized muffin-tin orbital method to examine whether the asymmetry in properties is reflected in a corresponding asymmetry in the one-electron band structure. While in a more complete analysis explicit electron correlation of the Hubbard U type would be intrinsic to the calculation,[17] we have taken the view that one-electron bandwidths point to the possible role that correlation might play and that correlation could be a consequence of the one-electron band structure rather than an integral part of the electronic structure. We have chosen the $La_{1-x}Ca_xMnO_3$ system for our calculations to avoid complications due to 4f electrons in the corresponding Pr system.

For our calculations on $La_3CaMn_4O_{12}$ ($La_{0.75}Ca_{0.25}$-MnO_3), we have used the structure reported by Radaelli et al.[14] refined from powder neutron diffraction data on 20 K. Ordering La and Ca in a supercell yields a structure with the same lattice parameters but in the space group *Pm* with 14 atoms (rather than 4) in the asymmetric unit. This structure is displayed in Figure 10. La and Ca have closely similar radii, and ordering them over crystallographically distinct sites is unphysical. However, the bond lengths and angles in the structures used in our calculations closely follow the experiments. For the electron-doped $La_{0.25}Ca_{0.75}MnO_3$

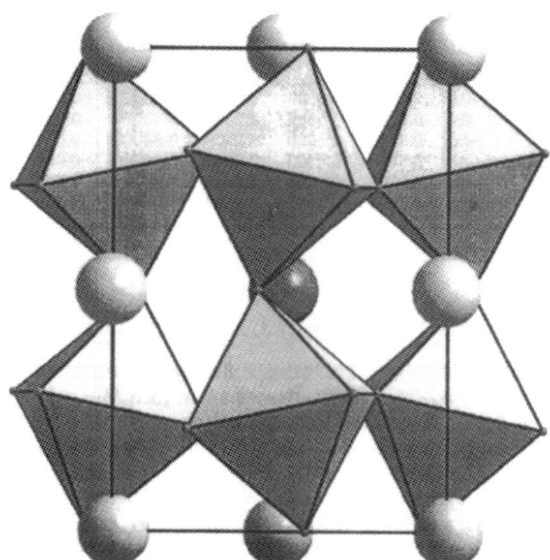

Figure 10. Monoclinic (space group *Pm*) structure of La_3-$CaMn_4O_{12}$. The dark spheres are La, and the light sphere is Ca. The view is looking down the short $a = a_P\sqrt{2}$ axis.

($LaCa_3Mn_4O_{12}$), we have used the cell and positional parameters of the refined 300 K neutron powder diffraction structure of $La_{0.33}Ca_{0.67}MnO_3$ from Radaelli et al.[14] Again, the supercell is in the space group *Pm* rather than in the orthorhombic *Pnma* space group. Calculations were performed using the Stuttgart TB-LMTO-ASA program.[18] The basis sets consisted of 6s, 5d, and 4f orbitals for La, 4s and 3d orbitals for Ca, 4s, 4p, and 3d orbitals for Mn, and 2p orbitals for O. The atomic sphere approximation (ASA) relies on the partitioning of space into atom-centered spheres as well as empty spheres, with the latter being critical in structures that are not closely packed. The basis for the empty spheres is 1s orbitals, with the 2p component treated using downfolding. The spheres are chosen so that the atom-centered spheres do not have a volume overlap of more than 16%. The calculations used 108K points in the primitive Brillouin zone for achieving convergence. Because of the *Pm* supercell employed, the nature of the magnetism becomes a little more complex and perhaps artificial for both of the systems studied. Indeed, there are now two types of somewhat indistinct Mn atoms in the unit cell. We have found a tendency for a ferrimagnetic ground state in both of the manganates, with the two Mn having opposite spins. For simplifying the comparison of the two electronic structures, the Mn atoms were provided similar polarization at the start of the calculations; self-consistency yielded a ferromagnet in both of the cases.

The calculations yielded a ferromagnetic ground state with a magnetic moment of 3.2 μ_B per Mn for La_3Ca-Mn_4O_{12} (spin-only value 3.75 μ_B). Such a reduction in the magnetic moment from the expected value can arise because of an infelicitous choice of sphere radii and does not merit interpretation. The refined neutron moment

(17) Anisimov, V. I.; Zaanen, J.; Andersen, O. K. *Phys. Rev.* **1991**, *B48*, 943.

(18) Andersen, O. K.; Jepsen, O.; et al. *The Stuttgart TB-LMTO-ASA-47 Program*; MPI für Feukörperforschung: Stuttgart, 1998. (b) Skriver, H. L. *The LMTO Method*; Springer: Berlin, 1984.

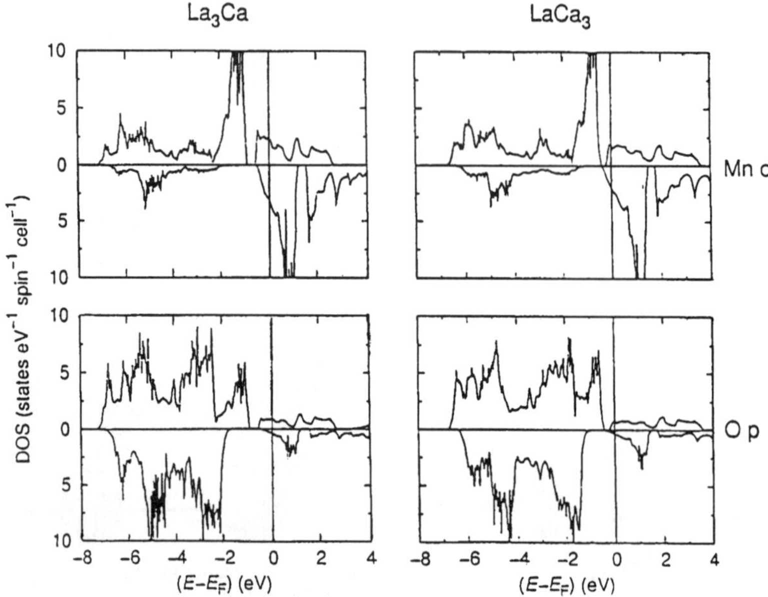

Figure 11. LMTO DOS for Mn and O in $La_3CaMn_4O_{12}$ and $LaCa_3Mn_4O_{12}$ near the Fermi energy. The upper halves of each panel display up-spin states, and the lower halves, down-spin states.

on Mn is 3.5 μ_B.[14] For ferromagnetic $LaCa_3Mn_4O_{12}$, the calculated magnetic moment was 3 μ_B (spin-only value 3.25). Figure 11 compares the spin-polarized densities of state (DOS) in the two spin directions for Mn and O in the two compositions. In each panel, the majority up-spin states are depicted in the upper half and the minority down-spin states in the lower half.

In both of the manganates, we find filled localized t_{2g}^3 (↑) states with a small bandwidth, of the order of 1.5 eV, polarizing a broader conduction band that is essentially e_g (↑) derived, although some t_{2g} (↓) states also occur at the Fermi energy. In $La_3CaMn_4O_{12}$, the Fermi energy E_F lies in a relatively broad conduction band. Both Mn and O states are present at E_F, indicating some covalency. Of note is the spin differentiation at the Fermi energy, with there being significantly more down-spin Mn states at the E_F than up-spin states. In the case of $LaCa_3Mn_4O_{12}$, E_F again lies in an e_g (↑) derived conduction band, but because of the smaller number of e_g electrons, E_F is at the band edge. Through examination of the so-called "fat-bands" (energy bands that have been decorated with the character of the corresponding orthonormal orbitals), we know that the e_g states in the perovskite systems[19] are derived from the combination of the narrow d_{z^2} bands and the broader bands formed by the strong covalent overlap between O p_x and p_y and metal $d_{x^2-y^2}$ orbitals. A scheme depicting the nature of the overlap is displayed in Figure 12. In $LaCa_3Mn_4O_{12}$, where there is a much smaller filling of the e_g, it is the relatively narrow d_{z^2}-derived band that is mostly occupied. The finding that E_F lies on a band edge suggests that this oxide would be susceptible to the opening of a gap in the DOS at E_F, through correlations of the Hubbard type. At the same time, the propensity for the

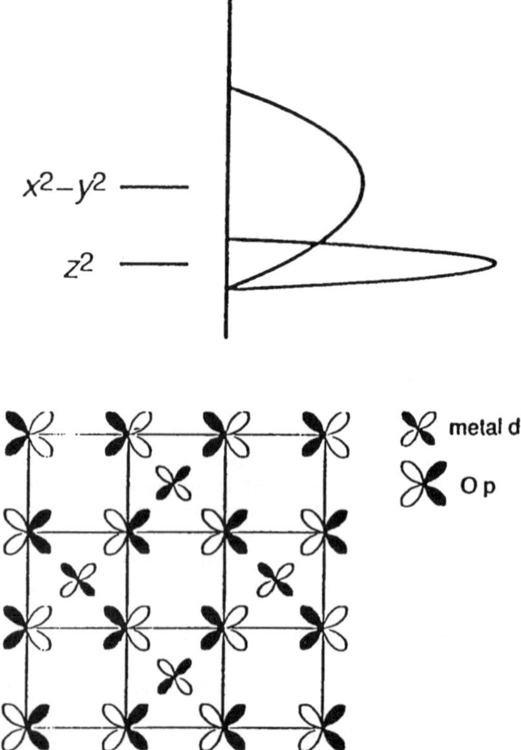

Figure 12. Scheme displaying the Jahn–Teller distorted e_g states becoming bands in solids such as perovskite manganese oxides. Because of overlap between metal $d_{x^2-y^2}$ and O p_x and p_y, the $d_{x^2-y^2}$-derived bands are significantly broader. A scheme for such overlap is displayed along the *ab* plane.

(19) Felser, C.; Seshadri, R.; Leist, A.; Tremel, W. *J. Mater. Chem.* **1998**, *8*, 787.

localized and delocalized states to be separated through a mobility edge (the formation of an Anderson insulator) is also increased by E_F lying on a band edge. In systems such as the present ones, disorder due to disparate ions occupying the A site of the perovskite structure cannot be avoided. A point of interest is that in $LaCa_3Mn_4O_{12}$ the DOS at E_F shows a smaller spin differentiation than that in $La_3CaMn_4O_{12}$. Spin differentiation is believed to be the key to the unusual magnetic field dependence of the electrical transport properties such as colossal magnetoresistance.[20] This suggests that the electron-doped manganates may be less interesting with respect to the CMR properties.

Conclusions

The electron-doped regime ($x > 0.5$) of the rare_earth manganates of the general formula $Ln_{1-x}Ca_xMnO_3$ is dominated by charge-ordering effects. While the effects of cation size and size disorder on the charge-ordered states of the hole- and electron-doped compositions are similar, their properties vary markedly as revealed by the comparison of the properties of the hole-doped $Pr_{0.64}Ca_{0.36}MnO_3$ and the electron-doped $Pr_{0.36}Ca_{0.64}MnO_3$. The CO state of the former is transformed to the FMM state by magnetic fields as well as by 3% Cr^{3+} or Ru^{4+} substitution in the Mn site, but none of these affects the CO state of the electron-doped material, which remains a paramagnetic insulator under all conditions. Increasing the Mn$-$O$-$Mn angle from 158 to 162 °, as in the above two manganates, to 165$-$180 ° in $La_{0.33}Ca_{0.33}Sr_{0.34}MnO_3$ does not result in ferromagnetism and metallicity. The CO state in this material is also unaffected by magnetic fields and Cr^{3+}/Ru^{4+} substitution in the Mn site. We are, therefore, prompted to conclude that it is not possible to induce long-range ferromagnetism in the electron-doped manganates by any means. First-principles calculations suggest that this may be because the Fermi level lies on a band edge in these materials.

Acknowledgment. This work was partially supported by the MRSEC Program of the NSF under Grant DMR 96-32716.

(20) Pickett, W. E.; Singh, D. J. *Phys. Rev.* **1996**, *B53*, 1146.

CM000464W

Phil. Trans. R. Soc. A (2008) **366**, 63–82
doi:10.1098/rsta.2007.2140
Published online 7 September 2007

Electronic phase separation and other novel phenomena and properties exhibited by mixed-valent rare-earth manganites and related materials

By Vijay B. Shenoy[1] and C. N. R. Rao[2],*

[1]*Centre for Condensed Matter Theory, Department of Physics, Indian Institute of Science, Bangalore 560 012, India*
[2]*Chemistry and Physics of Materials Unit and CSIR Centre of Excellence in Chemistry, Jawaharlal Nehru Centre for Advanced Scientific Research, Jakkur Post, Bangalore 560 064, India*

Transition metal oxides, such as the mixed-valent rare-earth manganites $Ln_{(1-x)}A_xMnO_3$ (Ln, rare-earth ion, and A, alkaline-earth ion), show a variety of electronic orders with spatially correlated charge, spin and orbital arrangements, which in turn give rise to many fascinating phenomena and properties. These materials are also electronically inhomogeneous, i.e. they contain disjoint spatial regions with different electronic orders. Not only do we observe signatures of such electronic phase separation in a variety of properties, but we can also observe the different 'phases' visually through different types of imaging. We discuss various experiments pertaining to electronic orders and electronic inhomogeneities in the manganites and present a discussion of theoretical approaches to their understanding. It is noteworthy that the mixed-valent rare-earth cobaltates of the type $Ln_{(1-x)}A_xCoO_3$ also exhibit electronic inhomogeneities just as the manganites.

Keywords: rare-earth cobaltates; rare-earth manganites; electronic phase separation; charge-ordering

1. Introduction

Transition metal oxides such as rare-earth manganites, cuprates and cobaltates exhibit a plethora of properties and phenomena (Imada *et al.* 1998; Rao 2000) that have stimulated sustained investigations by materials chemistry and physics communities. Many of the fascinating properties of these oxides arise from the fact that the transition metal ions can exhibit mixed valence with strongly correlated electrons giving rise to many 'electronic orders'. By electronic orders, we mean spatially correlated arrangement of charge, spin, orbital and/or phase (as in a superconductor). In addition, there is overwhelming evidence that these oxides with high chemical homogeneity can show spatially inhomogeneous structures, i.e. regions showing different electronic orders (Mathur & Littlewood 2003;

* Author for correspondence (cnrrao@jncasr.ac.in).

One contribution of 15 to a Discussion Meeting Issue 'Mixed valency'.

V. B. Shenoy and C. N. R. Rao

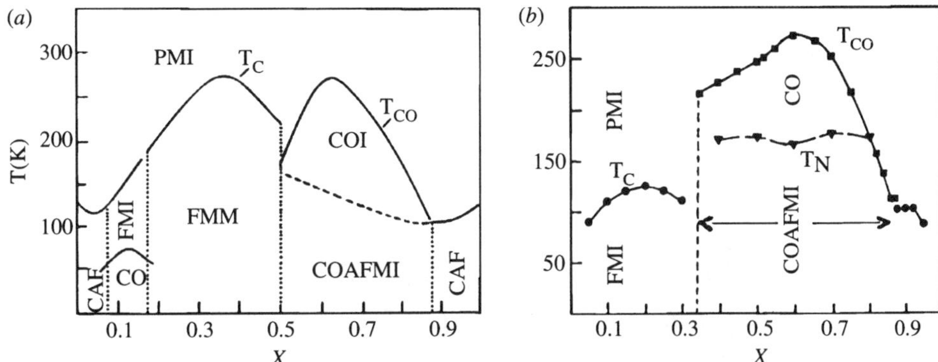

Figure 1. Phase diagram of (a) $La_{1-x}Ca_xMnO_3$ and (b) $Pr_{1-x}Ca_xMnO_3$ (adapted from Rao *et al.* (2004)).

Rao *et al.* 2004; Dagotto 2005; Shenoy *et al.* 2006). While many issues pertaining to these oxides have been understood, their science has many puzzles and challenges in addition to the promise of applications. The development of new oxide materials via the methods of materials chemistry with controlled electronic orders, and their theoretical understanding, therefore, continue to be at the forefront of research in condensed matter science.

In this article, we discuss spin, charge and orbital orders in transition metal oxides, in particular, in the rare-earth manganites which have become famous since 1993 owing to the phenomenon of colossal magnetoresistance (CMR) exhibited by them (Rao & Raveau 1998). We also touch upon the issue of electronic inhomogeneities; we discuss the signatures of the presence of such inhomogeneities and theoretical approaches to understand the same.

2. Electronic orders in rare-earth manganites

Rare-earth manganites of the type $Ln_{(1-x)}A_xMnO_3$ (Ln, rare-earth ion, and A, alkaline-earth ion) which crystallize in the perovskite structure were investigated several years ago by Wollan & Koehler (1955), but it was the discovery of CMR that stimulated their resurgence (Rao & Raveau 1998; Dagotto *et al.* 2001; Salamon & Jaime 2001; Tokura 2006). In the parent undoped compound $LaMnO_3$, the manganese ions exist in a $+3$ state, while the doped system such as $La_{(1-x)}Ca_xMnO_3$ contains (nominally) a fraction $(1-x)$ of Mn^{3+} and x of Mn^{4+} ions. This mixed-valent character of Mn underlies the rich phase diagrams (figure 1) exhibited by the manganites. A study of the phase diagram reveals a large number of electronic orders present in these systems. In what follows, we discuss some of these electronic orders.

$LaMnO_3$ exhibits spin, orbital and lattice orders. The spin order is that of an A-type antiferromagnet (figure 2). This is accompanied by the orbital order, which couples strongly with the lattice distortion due to the Jahn–Teller (JT) effect, giving rise to the lattice distortion pattern shown in figure 2. Interestingly, the JT distortion pattern loses long-range order at approximately 750 K. This transition is also associated with a symmetry change of the crystal from an

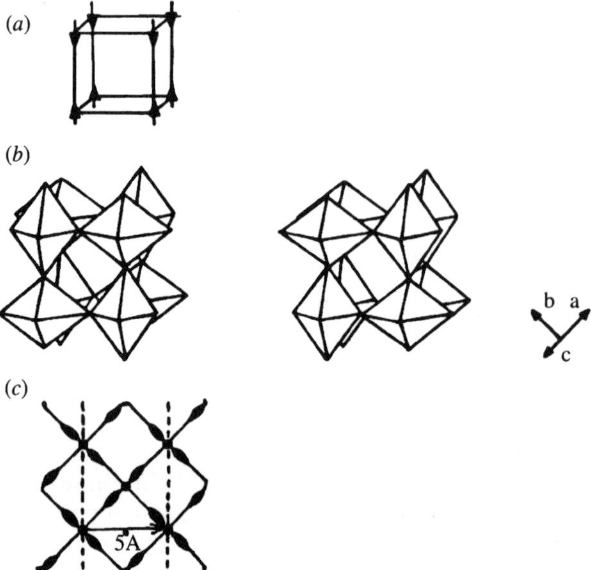

Figure 2. (*a*) A-type antiferromagnetic spin order, (*b*) Jahn–Teller distortion pattern and (*c*) orbital ordering in LaMnO$_3$ (after Rao (2000)).

orthorhombic structure at lower temperatures (due to the cooperative JT distortion) to a cubic structure at higher temperatures. At low temperatures, the combination of orbital ordering and JT distortion leads to an effective ferromagnetic (FM) exchange interaction between the spins on the plane, and a strong antiferromagnetic (AFM) exchange between spins between the planes, resulting in an A-type AFM arrangement of the spins. Other undoped manganites such as NdMnO$_3$ also show a similar electronic order.

Doping of manganites with alkaline-earth ions such as Ca^{2+} and Sr^{2+} results in a fraction of the manganese ions occurring in the +4 state, giving rise a new electronic order, charge order, wherein the +3 and +4 charges of the Mn ions are arranged in a periodic fashion (figure 3). Charge order is known to occur in other metal oxides as well, Fe$_3$O$_4$ being a well-known example (Rao 2000). The charge-ordered state is found predominantly in manganites with small Ln and A ions. Further, charge order disappears at high temperatures. Thus, in Pr$_{0.7}$Ca$_{0.3}$MnO$_3$, charge order appears only below 230 K. Interestingly, Pr$_{0.7}$Ca$_{0.3}$MnO$_3$ also shows AFM order below 170 K, where both antiferromagnetism and charge order coexist. At low temperatures, this manganite is electrically insulating. In contrast, La$_{0.7}$Ca$_{0.3}$MnO$_3$ does not show charge order, although it enters a FM metallic phase below a Curie temperature of 230 K, showing a PM insulating behaviour above this temperature. In the Sr-based La$_{0.7}$Sr$_{0.3}$MnO$_3$, the high-temperature PM phase is metallic. Note that the average radius of the A-site cations, $\langle r_A \rangle$, is larger in these two manganites than in Pr$_{0.7}$Ca$_{0.3}$MnO$_3$. CMR is observed at relatively moderate fields in La$_{0.7}$Ca$_{0.3}$MnO$_3$. Another important aspect of the doped manganites is that, when the size of the A-site cation (or the e$_g$ bandwidth) is large enough, electrons of the Mn^{3+} hop to Mn^{4+} by the double-exchange (DE) mechanism. The DE mechanism plays a crucial role in the occurrence of

66   *V. B. Shenoy and C. N. R. Rao*

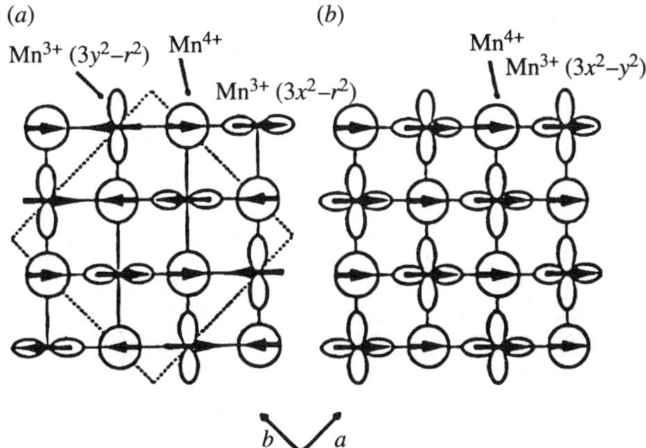

Figure 3. Ordering of charge, spin and orbital in doped manganites. (*a*) CE-type order, the dashed line shows the unit cell and (*b*) A-type order. In both figures, circles correspond to Mn^{4+} (after Rao (2000)).

metallicity and ferromagnetism in these materials. The phenomenon of CMR is explained by invoking the DE mechanism along with certain competing interactions involving phonons. Charge ordering is exactly the opposite of DE, since it localizes electrons creating an AFM insulating state.

Half-doped manganites show many interesting electronic orders. For example, $Nd_{0.5}Sr_{0.5}MnO_3$, which is a FM metal at high temperatures, shows a transition to a charge-ordered AFM state at approximately 150 K. The charge-ordering temperature is usually somewhat higher than the magnetic-ordering temperature ($T_{CO} \gtrsim T_N$). Interestingly, the charge order can be 'melted' by a small magnetic field (Kuwahara *et al.* 1995). At low temperature, the charge-ordered state is associated with a CE-type AFM order (figure 3), orbital order and a regular pattern of JT distortions. The magnetic order consists of *ab* planes stacked along the *c*-axis (figure 3) with an AFM coupling between spins on neighbouring planes. This may be contrasted with A-type order (figure 2). Note that there is no charge ordering in an A-type antiferromagnet unlike in a CE-type antiferromagnet. Usually, orbital and spin orders that are not accompanied by charge order show A-type antiferromagnetism.

The presence of charge and orbital orders can be observed experimentally by a study of the structures of the manganites. For example, the charge-ordered structure of $Nd_{0.5}Sr_{0.5}MnO_3$ consists of distorted oxygen octahedra (figure 4) containing zigzag chains with alternate long and short Mn–O bonds. The bond lengths are 1.921 and 2.021 Å along $\langle 110 \rangle$ and 1.881 and 2.020 Å along $\langle -110 \rangle$ (figure 4). The emergence of a charge gap is also indicated by vacuum tunnelling measurements (Biswas *et al.* 1997). Intriguingly, the low-temperature gap (approx. 250 meV) is much larger than $T_{CO} \sim 12$ meV. Moreover, a magnetic field of 6 T (approx. 1.2 meV) can destroy the charge-ordered state. A consequence of this is the large magnetostrictive effect arising from the magnetic field-induced structural transition that accompanies the AFM charge-ordered insulator to FM-metal transition.

An interesting fundamental question that arises is, what controls the different electronic orders? At the level of the chemistry of the constituents, the mean A-site cation radius $\langle r_A \rangle$ is expected to play an important role. This arises from the

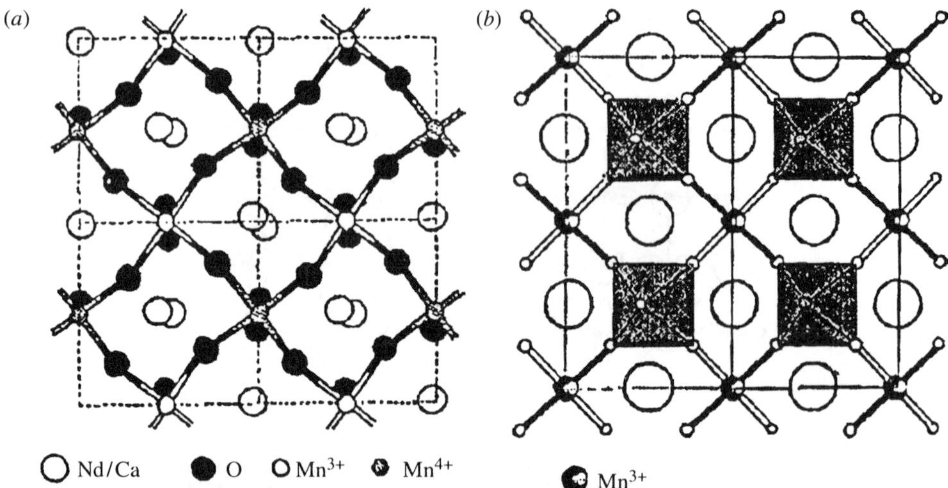

Figure 4. (*a*) Structure of charge-ordered $Nd_{0.5}Sr_{0.5}MnO_3$ in the *ab* plane at 10 K. Mn^{4+} ions are located at $(1/2, 0, 0)$ and Mn^{3+} are located at $(0, 1/2, 0)$. The structure contains zigzag chains with alternate long and short Mn–O bonds (see text). (*b*) The same structure as given in (*a*), shown in polyhedral representation ((*a*) after Vogt *et al.* (1996) and (*b*) after Woodward *et al.* (1999)).

fact that the bandwidth of the e_g electrons of Mn ions is affected by the size of the A-site cation, i.e. increasing $\langle r_A \rangle$ is equivalent to increasing the hydrostatic pressure, which increases the Mn–O–Mn bond angle and consequently the bandwidth. Detailed studies of the effect of $\langle r_A \rangle$ on the electronic order have been reported by Woodward *et al.* (1998). Systems with very large $\langle r_A \rangle$ tend to be FM and metallic, with a Curie temperature T_c that increases with $\langle r_A \rangle$. Charge ordering is also strongly affected by $\langle r_A \rangle$, with a smaller $\langle r_A \rangle$ stabilizing charge-ordered phase at low temperatures. It is evident that the tuning of $\langle r_A \rangle$ via chemical means provides an important means of controlling the phase of the manganite.

3. Electronic inhomogeneities in rare-earth manganites

As noted in §1, there is strong experimental evidence to indicate that many correlated oxides are electronically inhomogeneous (figure 5). Nowhere is it manifested better than the rare-earth manganites. Thus, these materials of high chemical homogeneity consist of different spatial regions with different electronic order (Mathur & Littlewood 2003; Rao *et al.* 2004; Dagotto 2005; Shenoy *et al.* 2006), a phenomenon that has come to be known as 'phase separation'. These regions can be static or dynamic and can be tuned by the application of external stimuli like a magnetic field. Moreover, the size scale of these inhomogeneities can vary from nanometres to as large as micrometres. The presence of electronic inhomogeneities raises many intriguing questions. What is the microscopic origin of these inhomogeneities? Why do they possess such a large range of length-scales? More importantly, are they responsible for the CMR and such responses exhibited by the manganites and related oxide materials? Indeed, there are suggestions in the literature that these features observed in correlated oxides indicate that they

                    V. B. Shenoy and C. N. R. Rao

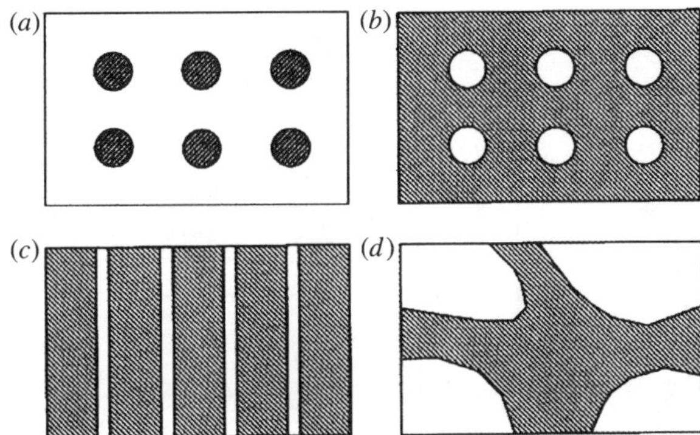

Figure 5. Schematic of electronic phase separation. Shaded portions indicate FM metallic regions; the unshaded portions correspond to AFM insulating regions. (*a*) FM metallic puddles in an insulating AFM background, (*b*) metallic regions with insulating droplets, (*c*) charged stripes and (*d*) phase separation on the mesoscopic scale.

are 'electronically soft' providing the possibility of spatial tuning of electronic properties, not unlike the liquid crystalline materials (Mathur & Littlewood 2003; Dagotto 2005; Milward *et al.* 2005). In this section, we discuss the phase separation phenomena especially in the rare-earth manganites. Following a description of the experimental signatures of electronic inhomogeneities, we discuss the theoretical ideas put forth to understand this phenomenon in §3*a*.

(*a*) *Signatures of electronic inhomogeneities*

Investigations employing a variety of experimental probes indicate the presence of electronic inhomogeneities. When the length-scale of the inhomogeneities is large (a few hundred nanometres), structural and magnetic information from the X-ray and neutron diffraction patterns show clear signatures. Specialized local probes such as NMR and Mössbauer spectroscopies can be used to probe inhomogeneities at a smaller length-scale. Thermodynamic (e.g. magnetization) and transport measurements also indicate the presence of inhomogeneities. Direct observational evidence provided by images employing transmission electron microscopy, scanning probe microscopy and photoemission micro-spectroscopy makes for a convincing case for the presence of the electronic inhomogeneities.

The pioneering work on manganites by Wollan & Koehler (1955) revealed the presence of electronic inhomogeneities. They observed both FM and AFM peaks in the magnetic structure of $La_{1-x}Ca_xMnO_3$ obtained from neutron scattering. They concluded that there is the simultaneous presence of ferromagnetism and antiferromagnetism in this material. A more recent neutron diffraction study is that of Woodward *et al.* (1999) on $Nd_{0.5}Sr_{0.5}MnO_3$. This material first becomes FM at 250 K, partially transforming to an A-type AFM phase at approximately 220 K, followed by a transformation of a substantial fraction to a CE-type AFM phase at approximately 150 K. The CE-type AFM (abbreviated as CO-CEAF) phase has a simultaneous ordering of spins, orbitals and charge in a complex spatial arrangement as shown in figure 3. It is evident from the studies of Woodward *et al.* (1999) that the three phases coexist at low temperatures. The size scale of the inhomogeneities is at least in the mesoscopic range

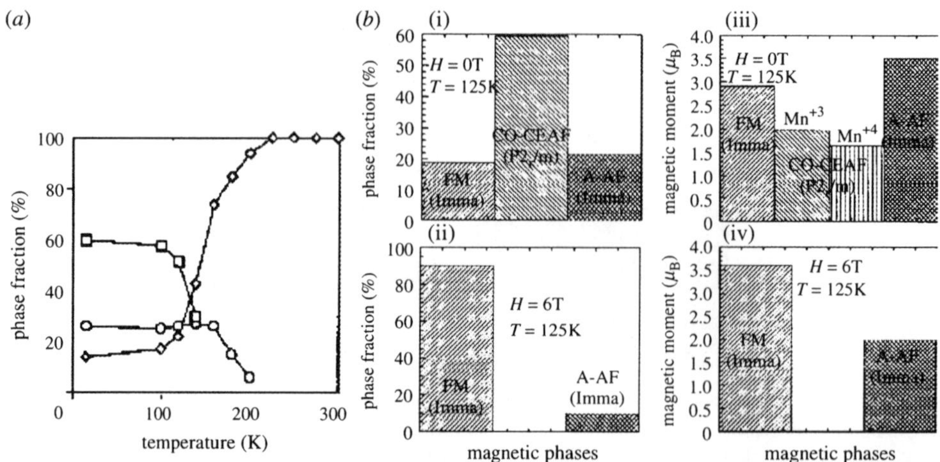

Figure 6. (*a*) Volume fraction of phases as a function of temperature in $Nd_{0.5}Sr_{0.5}MnO_3$. Open diamonds, open circles and open squares represent ferromagnet (FM), antiferromagnet (A-type, A-AF) and antiferromagnet (CE-type, CO-CEAF), respectively. (*b*) Phase fractions at 125 K at different magnetic fields (i) 0 T and (ii) 6 T; (iii) and (iv) show the magnetic moment in each of the phases ((*a*) after Woodward *et al.* (1999) and (*b*) after Ritter *et al.* (2000)).

(a few hundred nanometres or more), since they are large enough to produce well-defined reflections in neutron and X-ray diffraction patterns. The variation of the volume fraction of the three phases with temperature is shown in figure 6*a*. This phase diagram has an unusual feature with the three phases existing at the 150 K transition and below. The effect of the volume fractions of the electronic inhomogeneities due to an external magnetic field was studied by Ritter *et al.* (2000). The results shown in figure 6*b* demonstrate the evolution of the inhomogeneities where the FM fraction grows at the expense of the AFM regions in the presence of the applied magnetic field. Note the growth of one phase at the cost of another in figure 6. Clearly, the phase separation observed here is of a non-trivial type, in contrast to a situation where the sample is not phase-pure containing multiple phases that are both structurally and chemically distinct. Another interesting example (Radaelli *et al.* 2001) of phase separation is in $Pr_{0.7}Ca_{0.3}MnO_3$, which shows the presence of two distinct phases below the charge-ordering transition at 80 K. A charge-ordered AFM phase and a charge-delocalized phase have been observed by neutron diffraction.

Signatures of electronic inhomogeneities may be inferred from magnetic and electron transport measurements as well. Sudheendra & Rao (2003) studied these properties of $(La_{1-x}Ln_x)_{0.7}Ca_{0.3}MnO_3$, where $Ln = Nd$, Gd or Y. The lanthanide element was varied to control the average A-site cation radius $\langle r_A \rangle$. In the case of the La compound ($x = 0.0$) with the largest value of $\langle r_A \rangle$, a clear FM transition with a saturation magnetization of $3\mu_B$ is observed (figure 7*a*) up to a critical Nd doping of $x \lesssim x_c \approx 0.5$. At higher values of Nd doping ($x > x_c$), there is a monotonic increase of the magnetization with decrease in temperature, although the magnetization never attains the maximum value of $3\mu_B$. Interestingly, reducing the value of $\langle r_A \rangle$ by Gd (intermediate value of $\langle r_A \rangle$) or Y (smallest value of $\langle r_A \rangle$) substitution decreases the x_c, i.e. the critical doping value x above which the sharp FM transition vanishes,

70   V. B. Shenoy and C. N. R. Rao

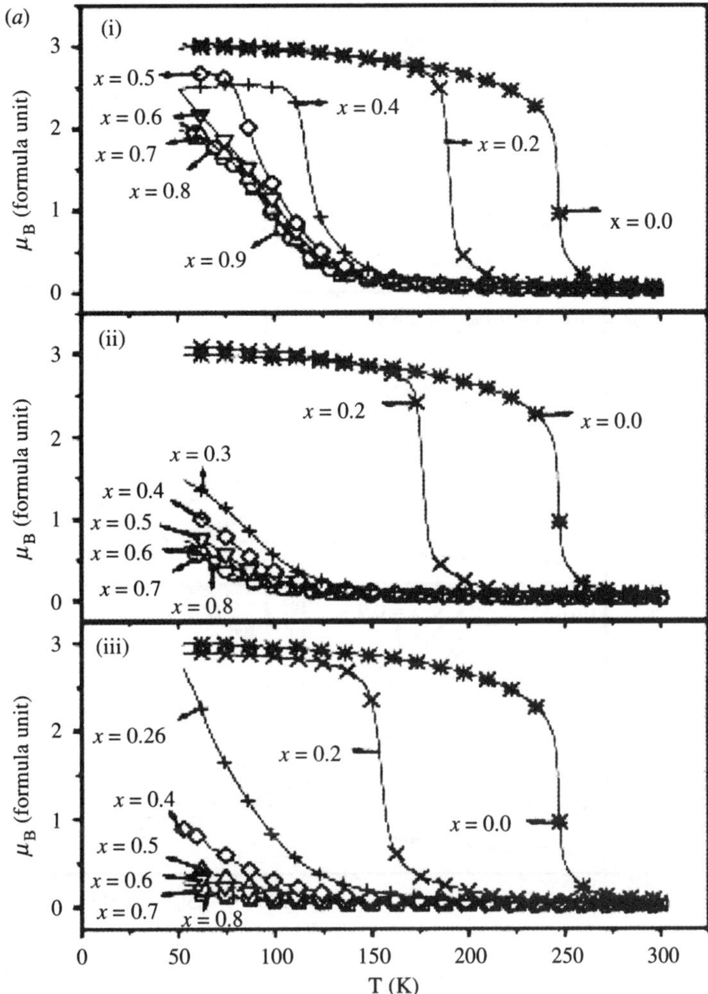

Figure 7. (*a*) Temperature dependence of the effective magnetic moment per formula unit in $(La_{1-x}Ln_x)_{0.7}Ca_{0.3}MnO_3$. (*b*) Temperature dependence of the resistivity in $(La_{1-x}Ln_x)_{0.7}Ca_{0.3}MnO_3$. The insets show the 'resistivity hysteresis' upon warming of the sample. In graphs in (*a*) and (*b*), (i), (ii) and (iii) correspond to Ln=Nd, Gd and Y, respectively. (*c*) Variation of (i) magnetization and (ii) resistivity as with temperature. The data shown are in $Ln_{0.7-x}Ln'_xA_{0.3-y}A'_y MnO_3$ with mean radius of the A-site cation are fixed at 1.216 Å. The different curves show results for different σ^2 ((*a,b*) after Sudheendra & Rao (2003) and (*c*) after Kundu *et al.* (2005*b*)).

reduces with the mean A-site cation radius $\langle r_A \rangle$. Moreover, the FM transition temperature T_c for $x < x_c$ increases with $\langle r_A \rangle$ as expected. The precipitous reduction of the saturation magnetization at $x \geq x_c$ is a signature of electronic and magnetic inhomogeneities induced by the A-site cation disorder. Transport measurements also show the signature of electronic inhomogeneities. In figure 7*b*, we show the variation of resistivity with temperature to demonstrate that all the three substitutions of manganites with Ln=Nd, Gd and Y exhibit a metal–insulator transition for $x < x_c$. The insets in figure 7*b* show the resistivity hysteresis across the

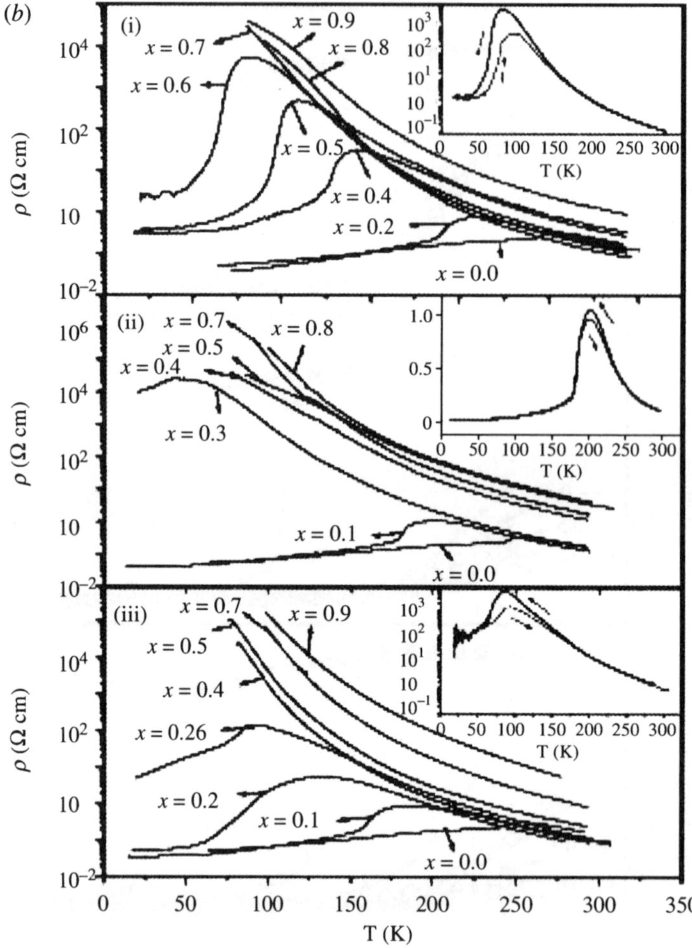

Figure 7. (*Continued.*)

transition. The insulator–metal transition temperature decreases linearly on doping and increases nearly linearly with $\langle r_A \rangle$. Electronic phase separation in such manganites generally occurs when $\langle r_A \rangle$ is relatively small (<1.18 Å) and is favoured by size disorder arising from the size mismatch between the A-site cations. This disorder is quantified (Rodríguez-Martínez & Attfield 2001) by the parameter $\sigma^2 = \sum_i x_i r_i^2 - \langle r_A \rangle^2$, where x_i is the fractional A-site occupancy of species i with radius r_i. A study of several series of manganites with fixed $\langle r_A \rangle$ and varying σ^2 shows that, as σ^2 is lowered, the phase-separated system transforms to a FM metallic state. It has been demonstrated recently (Kundu *et al.* 2005*b*) that in a series of manganites of the type $\mathrm{Ln}_{0.7-x}\mathrm{Ln}'_x\mathrm{A}_{0.3-y}\mathrm{A}'_y\mathrm{MnO}_3$, where $\langle r_A \rangle$ always remains large (thereby avoiding effects due to bandwidth), a decrease in σ^2 transforms the insulating non-magnetic state to a FM metallic state (figure 7*c*).

The above examples, while being indicative of the existence of multiple electronic and magnetic phases within a single apparently structurally and chemically pure phase, do not provide direct evidence with regard to the length-scale of the

72   *V. B. Shenoy and C. N. R. Rao*

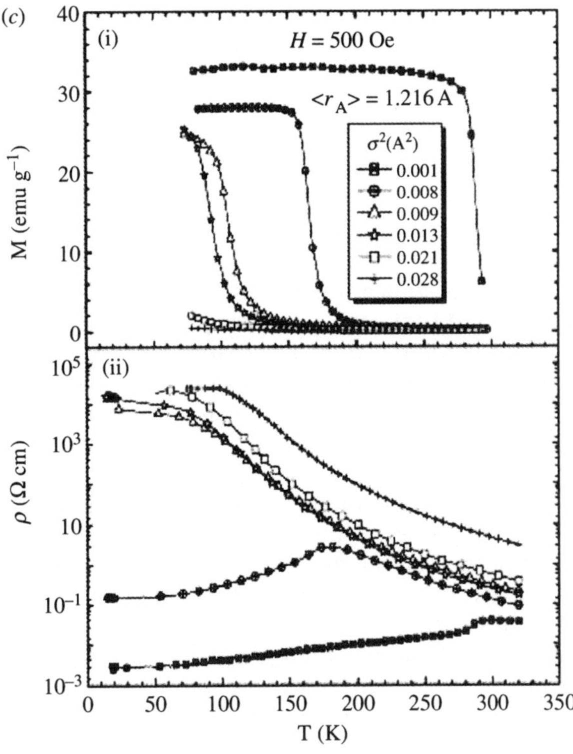

Figure 7. (*Continued.*)

inhomogeneities. While diffraction provides evidence of electronic phase separation on a mesoscale, NMR spectroscopy has revealed the possibility of phase separation at nanoscopic scales (Kuhns *et al.* 2003).

(*b*) *Visual evidence for electronic phase separation*

We now turn to the experimental work that shows direct visual evidence of electronic inhomogeneities, thereby allowing one to probe the length-scale of the inhomogeneity. One of the earliest reports of the direct evidence of electronic patterns in manganites was in the hole-doped side of (La, Ca)MnO_3. Mori *et al.* (1998) observed charged stripes in thin films of $La_{1-x}Ca_xMnO_3$ with $0.5 \leq x \leq 0.75$ by high-resolution lattice images obtained by transmission electron microscopy. The spatial structure consisted of paired JT distorted oxygen octahedra surrounding the Mn^{3+} ions, separated by stripes of $Mn^{4+}O_6$ octahedra. The spacing of the stripes, usually 5–10 lattice spacings, was found to be governed by the doping. The electronic inhomogeneity in this case was of a nanoscopic length-scale. Further evidence of nanoscopic electronic inhomogeneity in manganites was obtained by a scanning tunnelling microscopy study of $Bi_{1-x}Ca_xMnO_3$ ($x \approx 0.75$) by Renner *et al.* (2002). They found nanoscopic charge-ordered and metallic domains which correlated with the structural distortions.

The first direct evidence of mesoscale inhomogeneities in manganites was provided by Uehara *et al.* (1999), who observed electronic inhomogeneities in $La_{0.0625-y}Pr_yCa_{0.375}MnO_3$ by transmission electron microscopy. This work demonstrated the coexistence of charge-ordered (insulating) regions with interspersed FM metallic domains with a typical size of approximately 0.2 µm, bringing these features of inhomogeneity up to the mesoscopic scale. The authors suggested a possible mechanism for the colossal magnetoresistive response arising from such an electronic and magnetic inhomogeneity. The spin alignment of each of the FM metallic domains is random and the spin-polarized conduction electron of one domain cannot hop effectively via other domains due to the absence of available electronic states with the correct spin configuration on other randomly oriented magnetic domains. On application of a magnetic field, the spins of different magnetic domains align, enabling an unhindered hopping of electrons, thereby resulting in a drastic reduction of the resistance, known as the colossal negative magnetoresistance. This study suggested that such a mesoscopic phase separation is essential for colossal responses. Approximately at the same time, Fäth *et al.* (1999) used scanning tunnelling spectroscopy to study metal–insulator transitions in $La_{0.7}Ca_{0.3}MnO_3$ and found evidence of electronic inhomogeneities with a mesoscopic scale of approximately 0.2 µm, below the FM transition temperature, with FM metallic domains interspersed in insulating regions, just as in the work by Uehara *et al.* (1999). Most interestingly, the domains evolved as a function of the applied magnetic field, with the volume fraction of the magnetic (metallic) domains increasing at the cost of the non-magnetic (insulating) parts, closely accompanying the decrease in the resistivity with the magnetic field. This scenario contrasts with the spatially static phase separation scenario suggested by Uehara *et al.* (1999). A magnetic force microscope study of $La_{0.33}Pr_{0.34}Ca_{0.33}MnO_3$ by Zhang *et al.* (2002) has indicated that magnetic domains of mesoscopic scale evolved with temperature showing magnetic hysteresis coinciding with the resistivity hysteresis. This observation again suggested that mesoscale inhomogeneities are possibly crucial for colossal responses in magnetoresistance. Loudon *et al.* (2002) have studied $La_{0.5}Ca_{0.5}MnO_3$ using transmission electron microscopy and electron holography and found mesoscopic domains of FM regions interspersed in insulating regions. They also found evidence that some of the FM regions were charge ordered. A possible interpretation of the result is that the mesoscopic FM region is itself inhomogeneous at the nanoscale with coexisting metallic and charge-ordered regions.

A final example of mesoscale electronic inhomogeneities is from the experiments of Sarma *et al.* (2004), who used photoemission spectromicroscopy which has the advantage of spatially resolving the local metallic/non-metallic nature and simultaneously determining the local chemical composition with a resolution of approximately 0.5 µm. Using this technique, they studied $La_{0.25}Pr_{0.375}Ca_{0.375}MnO_3$. The key finding was that the material showed very large (10×5 µm^2) domains of insulating regions interspersed in a metallic background. These regions evolved on increasing temperature, with the metallic regions undergoing a metal-to-insulator transition at higher temperatures. On cooling the sample back to low temperatures, the insulating regions appeared essentially at the same locations, indicating, for the first time, a memory effect associated with the electronic inhomogeneities.

4. Electronic phase separation in rare-earth cobaltates

Electronic phase separation in rare-earth cobaltates of the type $Ln_{1-x}A_xCoO_3$ has come to the fore in the last 2 years. $La_{0.5}Sr_{0.5}CoO_3$ and other members of this family were once considered to be itinerant electron ferromagnets

V. B. Shenoy and C. N. R. Rao

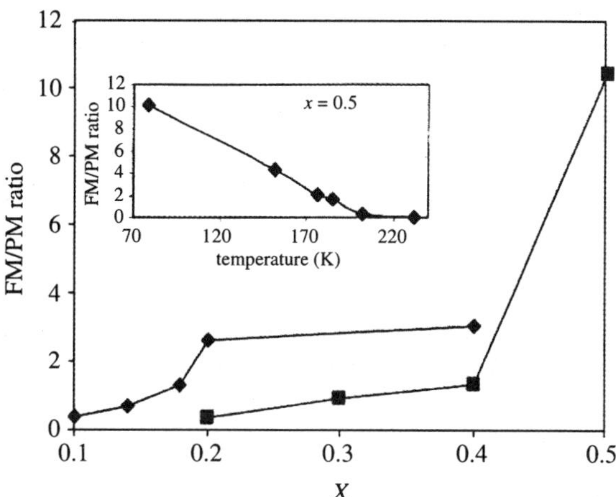

Figure 8. Variation of the ratio of the FM to PM species with composition in $La_{(1-x)}Sr_xCoO_3$ (filled squares, Mössbauer data at 78 K from Bhide *et al.* (1975); filled diamonds, NMR data at 1.9 K from Kuhns *et al.* (2003); inset shows the temperature variation of the FM/PM ratio in $La_{0.5}Sr_{0.5}CoO_3$ from the Mössbauer data).

showing metallic behaviour, but it was recognized later that the FM ordering was not long range (Rao *et al.* 1977; Señaris Rodríguez & Goodenough 1995). It is now established that $La_{0.5}Sr_{0.5}CoO_3$ exhibits a glassy magnetic behaviour (Wu *et al.* 2005). Recent studies of the magnetic and electrical properties of several members of the $Ln_{1-x}A_xCoO_3$ family have revealed that they are electronically phase-separated (Burley *et al.* 2004; Kundu *et al.* 2005a; Wu *et al.* 2005). The phase separation is favoured by small $\langle r_A \rangle$ as well as size disorder (Kundu *et al.* 2006). Measurements of magnetic relaxation and memory effects have shown that glassy behaviour is common among the cobaltates (Kundu *et al.* 2005a, 2006). Studies employing local probes such as NMR and Mössbauer spectroscopy have provided direct evidence for different distinct species involving large magnetic/metallic clusters and small non-magnetic clusters (Bhide *et al.* 1975; Kuhns *et al.* 2003). Thus, the NMR and Mössbauer studies of $La_{(1-x)}Sr_xCoO_3$ show distinct signals corresponding to FM and PM species. Cobalt Mössbauer spectra show a signal due to the CM species, but are always accompanied by a signal due to the PM species even at low temperatures $(T < T_c)$. In figure 8, we show the variation of the FM to PM ratio with composition in $La_{(1-x)}Sr_xCoO_3$. In the inset, the variation of the FM to PM ratio of $La_{0.5}Sr_{0.5}CoO_3$ with temperature is shown (Kundu *et al.* 2007). Cobalt NMR studies of a cobaltate with small $\langle r_A \rangle$ has shown the presence of two distinct signals varying in intensity with temperature (Kundu *et al.* 2007).

5. Theoretical approaches

We now turn to a discussion of theoretical ideas that have been proposed to explain the origin of the electronic inhomogeneities, with specific focus on manganites. The simplest possible description of manganites at a microscopic level involves the Mn d-orbitals and the lattice distortion of the oxygen octahedron surrounding the Mn ion (Dagotto 2003). The octahedral crystal field splits the degeneracy of the five Mn d-orbitals into three degenerate t_{2g} orbitals and two degenerate e_g orbitals. In the doped manganites, both Mn^{3+} and Mn^{4+} configurations are present; both have three electrons in the t_{2g} orbitals, and Mn^{3+} has, in addition, a lone electron in one of the e_g orbitals. The spins of the t_{2g} electrons are aligned parallel due to strong Hund coupling, and it is only the resulting 'core spins' ($S=3/2$) of the t_{2g} electrons that affect the low-energy physics of manganites. Thus, the relevant degrees of freedom at each manganese site are: an average of $(1-x)$ electrons per site populating the two e_g orbitals, the t_{2g} core spins and lattice (phonon) degrees of freedom corresponding to the distortion of the oxygen octahedra surrounding the manganese ion. The e_g electrons hop from a Mn site to neighbouring Mn sites with an amplitude t (approx. 0.2 eV). The spin of the e_g electron has a strong FM Hund coupling J_H (approx. 2.0 eV) with the local t_{2g} core spin. Another important energy scale is the on-site Mott–Hubbard repulsion U (approx. 5.0 eV) which forbids double occupancy of the local e_g sector. Neighbouring t_{2g} spins interact with each other via an AFM superexchange coupling J_{SE} (approx. 0.02 eV). Finally, the energy gained by the JT distortion of the oxygen octahedron is given by E_{JT} (approx. 0.5 eV). Competition between these different interactions, leading to a variety of states very close in energy, is responsible for the complex phase diagram of manganites with many electronic orders, and their extreme sensitivity to external perturbations such as temperature, magnetic field and strain (Sarma *et al.* 1995; Satpathy *et al.* 1996; Millis 1998; Ramakrishnan *et al.* 2003; Cepas *et al.* 2006).

Owing to the difficulties of dealing with all the degrees of freedom and competing interactions mentioned above, much of the early work aimed at understanding colossal responses and electronic inhomogeneities in manganites has been based on simplified models which neglect one or more of them. A prominent example is the work of Dagotto and coworkers (Dagotto 2003), who studied simple *magnetic Hamiltonians* with competing phases that are separated by a first-order transition. Based on these studies, they suggested that the system is prone to macroscopic phase separation, which is frustrated by disorder leading to the electronic inhomogeneities at various scales, the magnitude of the disorder determining the scale. From the real-space structure obtained from simulations, they constructed a random resistor network to explain the colossal responses by a mechanism similar to the one proposed by Fäth *et al.* (1999), again suggesting that the phase separation is key to colossal responses. Other simplified models have been studied, and alternate scenarios have been proposed. There are also suggestions that manganites close to half doping are near a multicritical point with competing phases affected by disorder (Motome *et al.* 2003; Tokura 2006). Ahn *et al.* (2004) consider a model Hamiltonian including electron–phonon interaction, and long-range elastic coupling between local lattice distortions. They present a scenario for mesoscopic/microscopic inhomogeneities and suggest that these are responsible for the colossal responses.

We now turn to a recent study of electronic inhomogeneities in manganites (Shenoy *et al.* 2007), which attempts to take into account all the degrees of freedom and their interactions based on the ℓb model (Ramakrishnan *et al.* 2003, 2004; Krishnamurthy 2005).

(a) The ℓb model for manganites

The ℓb model (Pai *et al.* 2003; Ramakrishnan *et al.* 2004) uses the idea that, under the conditions prevailing in doped manganites, the electrons populating the doubly degenerate e_g states centred at the Mn sites spontaneously reorganize themselves into two types of coexisting electron fluids. One is obtained by populating essentially site-localized states labelled ℓ which are polaronic, with strong local JT distortions of the oxygen octahedra, an energy gain E_{JT} and exponentially reduced intersite hopping. The other, labelled b, is a fluid of broadband, non-polaronic electrons, with no associated lattice distortions and undiminished hopping amplitudes. There is a strong local repulsion between the two fluids, as double occupancy on a site costs an extra Coulomb of energy U. The spins of ℓb and b are enslaved to the Mn-t_{2g} spins on site due to the large FM Hund's coupling J_H. In the simplest picture, assuming all the t_{2g} spins and the e_g spins to be aligned parallel, the ℓ and b electrons can be regarded as spinless, leading to the Falicov–Kimball (Freericks & Zlatić 2003) like ℓb Hamiltonian

$$H_{\ell b} = -E_{JT} \sum_i n_{\ell i} - t \sum_{\langle ij \rangle} (b_i^\dagger b_j + \text{h.c.}) + U \sum_i n_{\ell i} n_{bi}. \qquad (5.1)$$

Here, $\ell_i^\dagger$ and $b_i^\dagger$ create ℓ polarons and b electrons, respectively, at the sites i of a cubic Mn lattice, and $n_{\ell i} \equiv \ell_i^\dagger \ell_i$ and $n_{bi} \equiv b_i^\dagger b_i$ are the corresponding number operators.

(b) The extended ℓb model and the effects of long-range Coulomb interactions

This model assumes that the Mn ions occupy the sites of a cubic lattice (taken to be of unit lattice parameters), while the dopant A ions occupy an x fraction of the 'body centre' sites of each unit cube formed by the Mn ions. Since the aim is to study the effect of long-range Coulomb interactions on phase separation, we make further simplifying assumptions. It is assumed that the t_{2g} core spins are aligned ferromagnetically and that $J_H \to \infty$; this effectively projects out ℓ or b electron spin opposite to that of the t_{2g} core spins—we obtain an effectively spinless model. The above considerations lead us to the following *extended ℓb Hamiltonian*

$$H = H_{\ell b} + H_C \quad \text{and} \quad H_C = \sum_i \Phi_i q_i + \frac{V_0}{2} \sum_{i \neq j} \frac{q_i q_j}{r_{ij}}. \qquad (5.2)$$

Here $\ell_i^\dagger$ and $b_i^\dagger$ create ℓ and b electrons, respectively, at site i, and $n_{\ell i} \equiv \ell_i^\dagger \ell_i$ and $n_{bi} \equiv b_i^\dagger b_i$ are the corresponding number operators. In terms of the hole operator $(h_i^\dagger \equiv \ell_i$ which removes an ℓ polaron at site $i)$, the electron charge operator $q_i \equiv h_i^\dagger h_i - b_i^\dagger b_i$ and has the average value x per site owing to overall charge neutrality. The Coulomb term H_C has two parts; the charge at site i has energy $q_i \Phi_i$, where Φ_i is the electrostatic potential there due to A^{2+} ions, and the interaction between the charges at sites i and j leads to an energy $V_0((q_i q_j)/r_{ij})$. In what follows, we take the short-range Coulomb correlation U to be large (∞).

We now discuss the results of full-scale numerical simulations of the Hamiltonian (5.2) on finite three-dimensional periodic lattices, allowing for a random distribution of the A ions following Shenoy *et al.* (2007). Here, all the energy scales are normalized by the bare intersite hopping amplitude t. Systems as large as $20 \times 20 \times 20$ have been considered. The numerical determination of the groundstate of (5.2) requires further simplifying approximations. The most important approximation is the Hartree approximation, i.e. the charge operator q_i

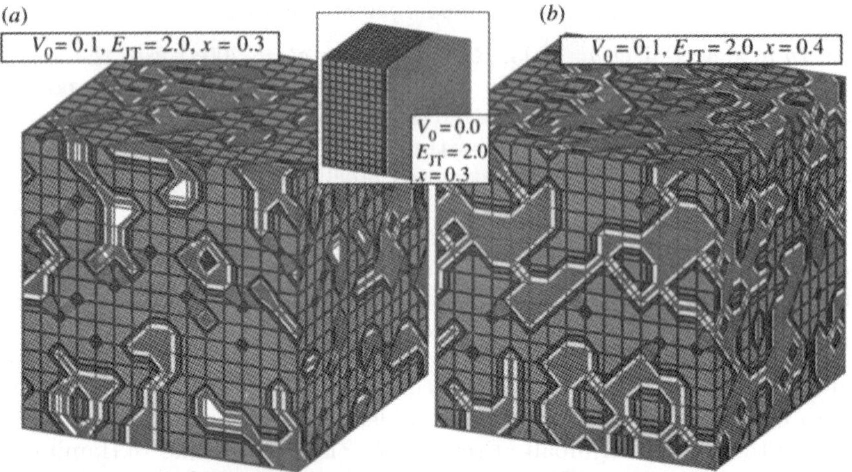

Figure 9. Real-space electronic distribution obtained from simulations on a 16^3 cube. Magenta (darkest) denotes hole clumps with occupied b electrons, white (lightest) denotes hole clumps with no b electrons, cyan (2nd lightest) denote singleton holes, and light blue (2nd darkest) represents regions with ℓ polarons. (a) Isolated clumps with occupied b-electrons (b-electron puddles). (b) Larger doping; percolating clumps. Inset: 'macroscopic phase separation' *absence* of long-range Coulomb interaction ($V_0 = 0.0$) (after Shenoy *et al.* (2007)).

is replaced by its expectation value in the groundstate $\langle q_i \rangle = h_i^\dagger h_i - \langle b_i^\dagger b_i \rangle$. Since we have assumed that $U \to \infty$, the b electrons do not hop to sites where an ℓ polaron is present. This leads to the segregation of the two types of electron into disjoint clusters. In a cluster of hole sites (which has at least two nearest neighbour hole sites), which we call a 'clump', the b electron states are determined by solving the quantum kinetic energy Hamiltonian exactly. This is a new generalization of the common Coulomb glass simulation (Baranovskii *et al.* 1979; Davies *et al.* 1984; Vojta & Schreiber 2001), which includes the quantum mechanically obtained b states within their clump or puddle. The electrostatic energy is calculated accurately using the Ewald technique and fast Fourier transform routines.

In the absence of the long-range Coulomb interaction ($V_0 = 0$), this procedure leads to the macroscopically phase-separated state (see inset of figure 9). The holes aggregate to one side of the simulation box and a fraction of the electrons occupying the ℓ states is promoted to the b states in the one large hole clump— with their concentrations determined by the equality of the chemical potential in the two regions (i.e. the highest occupied b level equals $-E_{JT}$). This phase separation is due to the strong Coulomb repulsion between the two types of electron fluids and is in agreement with known results on the Falicov–Kimball model (Freericks *et al.* 2002; Freericks & Zlatić 2003).

Long-range Coulomb interaction 'frustrates' phase separation, i.e. macroscopic phase separation costs prohibitive energy in the presence of long-range Coulomb interactions. The precise nature of the resulting groundstate electronic configuration depends on the JT energy E_{JT} and the doping level x. There are two critical doping levels x_{c1} and x_{c2} for any given JT energy E_{JT}. When the doping level is less than x_{c1}, i.e. $x < x_{c1}$, no b states are occupied—the holes (and hence also the ℓ polarons) form a Coulomb glass (Efros & Shklovskii 1975). When the doping exceeds x_{c1}, some of

V. B. Shenoy and C. N. R. Rao

the hole clumps are occupied and the groundstate consists of isolated b-electron puddles dispersed in a polaronic background (figure 9). There is a second critical doping level x_{c2} (see Shenoy *et al.* 2007); when $x \gtrsim x_{c2}$ the occupied b-electron puddles percolate and the system attains metallicity (figure 9).

The size scale of the electronic inhomogeneity is nanometric. However, its dependence on V_0 differs fundamentally from the analytical results of the same model, which assumed a homogeneous distribution (jellium) of the dopant A ions. For the more realistic, random distribution of the A ions, the size of the electronic inhomogeneity is almost independent of the long-range Coulomb parameter V_0, in that even an extremely small V_0 produces clump sizes that are of the size scale of a few lattice spacings. The dependence of the average size of the electronic inhomogeneity is thus due to the long-range Coulomb interaction which *along with* the random distribution of A ions acts as a 'singular perturbation' that frustrates macroscopic phase separation. However, the sizes and the distribution of the clumps are determined by the random distribution of the A ions, and thus we conclude that doped manganites (and similarly, possibly many other correlated oxides) are necessarily and intrinsically electronically inhomogeneous, on a *nanometric scale.*

The results of the extended ℓb model with realistic energy parameters presented here provide several new insights into the complex electronic inhomogeneities seen in correlated oxides, and in particular the low-bandwidth manganites with a large FM region in their phase diagram. These arguments may be sharply contrasted with earlier work. In particular, this work suggests that the nanoscale inhomogeneities in manganites arise not out of 'phase competition'-induced phase separation frustrated by disorder as argued from the studies of model spin Hamiltonians (Dagotto 2003), but from short-range Coulomb correlation-induced phase separation frustrated by long-range Coulomb interaction, similar to what has been suggested in cuprates (Emery *et al.* 1990; Emery & Kivelson 1993). Most importantly, nanoscale electronic inhomogeneities are present in *both* the insulating and metallic phases of doped manganites. As is evident from figure 9, each of these constitute a single thermodynamic phase that is homogeneous at mesoscales. One has a metal–insulator transition between two such nanoscopically inhomogeneous phases at x_{c2} as a function of doping. These observations are in agreement with experiments. Indeed, the electron holography results of Loudon *et al.* (2002) show that even the ferro-metallic regions have interspersed in them charge-ordered insulating regions which can be interpreted as the cluster of ℓ states.

6. Concluding remarks

A number of competing energy scales are operative in mixed-valent correlated oxides giving rise to a large number of electronic orders. In the $Ln_{(1-x)}A_xMnO_3$ manganites, a variety of spin, charge and orbital (and associated lattice order) is found. The emergence of these orders can be understood based on a simple parametrization, i.e. the mean A-site cation radius which controls the bandwidth of the e_g electrons. An intriguing aspect of these materials is that small perturbations can drive a transition from one type of electronic order to another, as is evident from the effect of a small magnetic field on the charge-ordered state.

In addition to the large number of electronic orders, it is found experimentally that there is coexistence of *different* electronic orders in disjoint spatial regions, i.e. these materials are electronically inhomogeneous showing signatures in thermodynamic and transport properties and can be seen directly in several experimental probes. These inhomogeneities can be of varied scales (nm to μm) and can be static or dynamic. Although the phenomenon has been called phase separation, it is clear that the regions of distinct electronic order may not necessarily be 'thermodynamic phases', particularly in the case where they are of nanometric scale. Indeed, the theoretical model discussed in the paper shows that a single 'phase' of the manganite itself is inhomogeneous on a nanometric scale.

Although, we have focused on manganites and cobaltates, electronic inhomogeneities are important in many other oxides. In the high T_c cuprate $La_2CuO_{4-\delta}$, the electronic phase separation is caused by compositional variations with superconducting clusters of $La_2CuO_{3.06}$ and AFM clusters of $La_2CuO_{3.01}$ (Sigmund & Müller 1994). Even in the cuprates that possess chemical homogeneity on a mesoscale, it has been suggested that the unusual variation of zero-temperature superfluid density with the superconducting transition temperature is due to the possible coexistence of AFM and superconducting domains (Broun *et al.* 2005). Neutron-scattering measurements suggest the presence of charge stripes (Tranquada *et al.* 2004), while scanning probe microscopy indicates a spatially inhomogeneous superconducting gap (McElroy *et al.* 2005).

Theoretical understanding of the phenomenon of electronic inhomogeneities is still incomplete. Although a large body of previous work based on *model Hamiltonians* and phenomenological models exists, the microscopic mechanisms and the length-scales that arise in these systems remain poorly understood. The microscopically motivated theoretical model that we have discussed here clearly indicates that these materials are electronically inhomogeneous at nanoscales due to strong correlation physics. Naturally, the question of the origin of mesoscale inhomogeneities remains. There are several scenarios that could explain the origin of the mesoscale patterns. There is a possibility that, since many of the transitions are of first order, many of the inhomogeneities are due to an 'incomplete transition', i.e. they are metastable states. The second possibility is the presence of long-range elastic strain fields which are unscreened. Indeed, it is well known that pressure (strain) can drive transitions in the manganites (Soh *et al.* 2002; Postorino *et al.* 2003). The key question is whether strain effects are extrinsic (caused by defects, etc.) or intrinsic, i.e. the system reorganizes spontaneously into domains (much like a martensite).

The need for further exploration of electronic orders and inhomogeneities in oxides is evident. Experiments need to be designed so as to isolate long-range strain effects, with theoretical efforts towards building coarse-grained models to understand the mesoscale physics. This issue is particularly important for exploiting the electronic softness of these materials for applications.

V.B.S. gratefully acknowledges illuminating discussions and collaborative work with H. R. Krishnamurthy and T. V. Ramakrishnan.

References

Ahn, K. H., Lookman, T. & Bishop, A. R. 2004 Strain-induced metal–insulator phase coexistence in perovskite manganites. *Nature* **428**, 401–404. (doi:10.1038/nature02364)

Baranovskii, S. D., Efros, A. L., Gelmont, B. L. & Shklovskii, B. I. 1979 Coulomb gap in disordered systems: computer simulation. *J. Phys. C: Solid State Phys.* **12**, 1023–1034. (doi:10.1088/0022-3719/12/6/015)

                    V. B. Shenoy and C. N. R. Rao

Bhide, V. G., Rajoria, D. S., Rao, C. N. R., Rama Rao, G. & Jadhao, V. G. 1975 Itinerant-electron ferromagnetism in $La_{1-x}Sr_xCoO_3$: mossbauer study. *Phys. Rev. B* **12**, 2832–2843. (doi:10.1103/PhysRevB.12.2832)

Biswas, A., Raychaudhuri, A. K., Mahendiran, R., Guha, A., Mahesh, R. & Rao, C. N. R. 1997 Direct measurement of the charge ordering gap in $Nd_{0.5}Sr_{0.5}MnO_3$ using vacuum tunnelling. *J. Phys. Condens. Matter* **9**, L355–L360. (doi:10.1088/0953-8984/9/24/002)

Broun, D. M., Turner, P. J., Huttema, W. A., Ozcan, S., Morgan, B., Liang, R., Hardy, W. N. & Bonn, D. A. 2005 In-plane superfluid density of highly underdoped $YBa_2Cu_3O_{6+x}$. arxiv.org:cond-mat/0509223.

Burley, J. C., Mitchell, J. F. & Short, S. 2004 Competing electronic ground states in $La_{1-x}Ca_xCoO_3$. *Phys. Rev. B* **69**, 054401. (doi:10.1103/PhysRevB.69.054401)

Cepas, O., Krishnamurthy, H. R. & Ramakrishnan, T. V. 2006 Instabilities and insulator–metal transitions in half-doped manganites induced by magnetic-field and doping. *Phys. Rev. B* **73**, 035218. (doi:10.1103/PhysRevB.73.035218)

Dagotto, E. 2003 *Nanoscale phase separation and colossal magnetoresistance.* Berlin, Germany: Springer.

Dagotto, E. 2005 Complexity in strongly correlated electronic systems. *Science* **309**, 257–362. (doi:10.1126/science.1107559)

Dagotto, E., Hotta, T. & Moreo, A. 2001 Colossal magnetoresistant materials: the key role of phase separation. *Phys. Rep.* **344**, 1–153. (doi:10.1016/S0370-1573(00)00121-6)

Davies, J. H., Lee, P. A. & Rice, T. M. 1984 Properties of the electron glass. *Phys. Rev. B* **29**, 4260–4271. (doi:10.1103/PhysRevB.29.4260)

Efros, A. L. & Shklovskii, B. I. 1975 Coulomb gap and low-temperature conductivity of disordered systems. *J. Phys. C: Solid State Phys.* **8**, L49–L51. (doi:10.1088/0022-3719/8/4/003)

Emery, V. J. & Kivelson, S. A. 1993 Frustrated electronic phase separation and high-temperature superconductors. *Physica C* **209**, 597–621. (doi:10.1016/0921-4534(93)90581-A)

Emery, V. J., Kivelson, S. A. & Lin, H. Q. 1990 Phase-separation in the t–J model. *Phys. Rev. Lett.* **64**, 475–478. (doi:10.1103/PhysRevLett.64.475)

Fäth, M., Freisem, S., Menovsky, A. A., Tomioka, Y., Aarts, J. & Mydosh, J. A. 1999 Spatially inhomogeneous metal–insulator transition in doped manganites. *Science* **285**, 1540–1542. (doi:10.1126/science.285.5433.1540)

Freericks, J. K. & Zlatić, V. 2003 Exact dynamical mean field theory of the Falicov–Kimball model. *Rev. Mod. Phys.* **75**, 1333–1382. (doi:10.1103/RevModPhys.75.1333)

Freericks, J. K., Leib, E. H. & Ueltschi, D. 2002 Phase separation due to quantum mechanical correlations. *Phys. Rev. Lett.* **88**, 106401. (doi:10.1103/PhysRevLett.88.106401)

Imada, M., Fujimori, A. & Tokura, Y. 1998 Metal–insulator transitions. *Rev. Mod. Phys.* **70**, 1039–1263. (doi:10.1103/RevModPhys.70.1039)

Krishnamurthy, H. R. 2005 A new theory of doped manganites exhibiting colossal magnetoresistance. *Pramana* **64**, 1063–1074. (Invited talk published in the *Proc. 22nd IUPAP Int. Conf. on Statistical Physics* (eds S. Dattagupta, H. R. Krishnmurthy, R. Pandit, T. V. Ramakrishanan & D. Sen).)

Kuhns, P. L., Hoch, M. J. R., Moulton, W. G., Reyes, A. P., Wu, J. & Leighton, C. 2003 Magnetic phase separation in $La_{1-x}Sr_xCoO_3$ by ^{59}Co nuclear magnetic resonance. *Phys. Rev. Lett.* **91**, 127202. (doi:10.1103/PhysRevLett.91.127202)

Kundu, A. K., Nordblad, P. & Rao, C. N. R. 2005*a* Nonequilibrium magnetic properties of single-crystalline $La_{0.7}Ca_{0.3}CoO_3$. *Phys. Rev. B* **72**, 144423. (doi:10.1103/PhysRevB.72.144423)

Kundu, A. K., Seikh, M. M., Ramesha, K. & Rao, C. N. R. 2005*b* Novel effects of size disorder on the electronic and magnetic properties of rare earth manganates of the type $La_{0.7-x}Ln_xBa_{0.3}Mn_{0.3}$ (Ln = Pr, Nd, Gd or Dy) with large average radius of the A-site cations. *J. Phys. Condens. Matter* **17**, 4171–4180. (doi:10.1088/0953-8984/17/26/015)

Kundu, A. K., Nordblad, P. & Rao, C. N. R. 2006 Spin-glass behavior in $Pr_{0.7}Ca_{0.3}CoO_3$ and $Nd_{0.7}Ca_{0.3}CoO_3$. *J. Solid State Chem.* **179**, 923–927. (doi:10.1016/j.jssc.2005.12.014)

Kundu, A., Sarkar, R., Pahari, B., Ghoshray, A. & Rao, C. N. R. 2007 A comparative study of the magnetic properties and phase separation behavior of the rare earth cobaltates, $Ln_{0.5}Sr_{0.5}CoO_3$ (*Ln*=rare earth). *J. Solid State Chem.* **180**, 1318–1324. (doi:10.1016/j.jssc.2007.02.003)

Kuwahara, H., Tomioka, Y., Asamitsu, A., Moritomo, Y. & Tokura, Y. 1995 A first-order phase transition induced by a magnetic field. *Science* **270**, 961–963. (doi:10.1126/science.270.5238.961)

Loudon, J. C., Mathur, N. D. & Midgley, P. A. 2002 Charge-ordered ferromagnetic phase in $La_{0.5}Ca_{0.5}MnO_3$. *Nature* **420**, 797–800. (doi:10.1038/nature01299)

Mathur, N. D. & Littlewood, P. B. 2003 Mesoscopic texture in manganites. *Phys. Today* **56**, 25–30. (doi:10.1063/1.1554133)

McElroy, K., Lee, J., Slezak, J. A., Lee, D.-H., Eisaki, H., Uchida, S. & Davis, J. 2005 Atomic-scale sources and mechanism of nanoscale electronic disorder in $Bi_2Sr_2CaCu_2O_{8+\delta}$. *Science* **309**, 1048–1052. (doi:10.1126/science.1113095)

Millis, A. J. 1998 Colossal magnetoresistance manganites: a laboratory for electron–phonon physics. *Phil. Trans. R. Soc. A* **356**, 1473–1480. (doi:10.1098/rsta.1998.0230)

Milward, G. C., Calderón, M. J. & Littlewood, P. B. 2005 Coexisting charge modulation and ferromagnetism produces long period phases in manganites: new example of electronic soft matter. *Nature* **433**, 607–610. (doi:10.1038/nature03300)

Mori, S., Chen, C. H. & Cheong, S.-W. 1998 Pairing of charge-ordered stripes in (La, Ca)MnO_3. *Nature* **392**, 473–476. (doi:10.1038/33105)

Motome, Y., Furukawa, N. & Nagaosa, N. 2003 Competing orders and disorder-induced insulator to metal transition in manganites. *Phys. Rev. Lett.* **91**, 167204. (doi:10.1103/PhysRevLett.91.167204)

Pai, G. V., Hassan, S. R., Krishnamurthy, H. R. & Ramakrishnan, T. V. 2003 Zero-temperature insulator–metal transition in doped manganites. *Europhys. Lett.* **64**, 696–702. (doi:10.1209/epl/i2003-00282-0)

Postorino, P., Congeduti, A., Dore, P., Sacchetti, A., Gorelli, F., Ulivi, L., Kumar, A. & Sarma, D. D. 2003 Pressure tuning of electron–phonon coupling: the insulator to metal transition in manganites. *Phys. Rev. Lett.* **91**, 175501. (doi:10.1103/PhysRevLett.91.175501)

Radaelli, P. G., Ibberson, R. M., Argyriou, D. N., Casalta, H., Andersen, K. H., Cheong, S.-W. & Mitchell, J. F. 2001 Mesoscopic and microscopic phase segregation in manganese perovskites. *Phys. Rev. B* **63**, 172419. (doi:10.1103/PhysRevB.63.172419)

Ramakrishnan, T. V., Krishnamurthy, H. R., Hassan, S. R. & Pai, G. V. 2003 Theory of manganites exhibiting colossal magnetoresistance. arxiv.org:cond-mat/0308396.

Ramakrishnan, T. V., Krishnamurthy, H. R., Hassan, S. R. & Pai, G. V. 2004 Theory of insulator metal transition and colossal magnetoresistance in doped manganites. *Phys. Rev. Lett.* **92**, 157203. (doi:10.1103/PhysRevLett.92.157203)

Rao, C. N. R. 2000 Charge, spin, and orbital ordering in the perovskite manganates, $Ln_{(1-x)}A_{(x)}MnO_{(3)}$ (Ln = rare earth, A = Ca or Sr). *J. Phys. Chem. B* **104**, 5877–5889. (doi:10.1021/jp0004866)

Rao, C. N. R. & Raveau, B. (eds) 1998 *Colossal magnetoresistance, charge ordering and related properties of manganese oxides*. Singapore: World Scientific.

Rao, C. N. R., Parkash, O., Bahadur, D., Ganguly, P. & Nagabhushana, S. 1977 Itinerant electron ferromagnetism in SR2+-doped rare-earth, CA2+-doped rare-earth, and BA2+-doped rare-earth orthocobaltites $(LN_{1-x(3+)}M_{x(2+)}CoO_3)$. *J. Solid State Chem.* **22**, 353–360. (doi:10.1016/0022-4596(77)90011-1)

Rao, C. N. R., Kundu, A. K., Seikh, M. M. & Sudheendra, L. 2004 Electronic phase separation in transition metal oxides. *Dalton Trans.* 3003–3011.

Renner, C., Aeppli, G., Kim, B. G., Soh, Y.-A. & Cheong, S.-W. 2002 Atomic-scale images of charge ordering in a mixed-valence manganite. *Nature* **416**, 518–521. (doi:10.1038/416518a)

Ritter, C., Mahendiran, R., Ibarra, M. R., Morellon, L., Maignan, A., Raveau, B. & Rao, C. N. R. 2000 Direct evidence of phase segregation and magnetic-field-induced structural transition in $Nd_{0.5}Sr_{0.5}MnO_3$ by neutron diffraction. *Phys. Rev. B* **61**, R9229–R9232. (doi:10.1103/PhysRevB.61.R9229)

Rodríguez-Martínez, L. M. & Attfield, J. P. 2001 Disorder-induced orbital ordering in $L_{0.7}M_{0.3}MnO_3$ perovskites. *Phys. Rev. B* **63**, 024424. (doi:10.1103/PhysRevB.63.024424)

Salamon, M. B. & Jaime, M. 2001 The physics of manganites: structure and transport. *Rev. Mod. Phys.* **73**, 583–628. (doi:10.1103/RevModPhys.73.583)

Sarma, D. D., Shanthi, N., Barman, S. R., Hamada, N., Sawada, H. & Terakura, K. 1995 Band theory for ground-state properties and excitation spectra of perovskite LaMO$_3$ (M = Mn, Fe, Co, Ni). *Phys. Rev. Lett.* **75**, 1126–1129. (doi:10.1103/PhysRevLett.75.1126)

Sarma, D. D. *et al.* 2004 Direct observation of large electronic domains with memory effect in doped manganites. *Phys. Rev. Lett.* **93**, 097202. (doi:10.1103/PhysRevLett.93.097202)

Satpathy, S., Popović, Z. S. & Vukajlović, F. R. 1996 Electronic structure of the perovskite oxides: La$_{1-x}$Ca$_x$MnO$_3$. *Phys. Rev. Lett.* **76**, 960–963. (doi:10.1103/PhysRevLett.76.960)

Señaris Rodríguez, M. A. & Goodenough, J. B. 1995 Magnetic and transport properties of the system La$_{1-x}$Sr$_x$CoO$_{3-\delta}$ ($0 < x \le 0.50$). *J. Solid State Chem.* **118**, 323–336. (doi:10.1006/jssc.1995.1351)

Shenoy, V. B., Sarma, D. D. & Rao, C. N. R. 2006 Electronic phase separation in correlated oxides: the phenomenon, its present status and future prospects. *Chemphyschem* **7**, 2053–2059. (doi:10. 1002/cphc.200600188)

Shenoy, V. B., Gupta, T., Krishnamurthy, H. R. & Ramakrishnan, T. V. 2007 Coulomb interactions and nanoscale electronic inhomogeneities in manganites. *Phys. Rev. Lett.* **98**, 097201. (doi:10.1103/PhysRevLett.98.097201)

Sigmund, E. & Müller, K. A. (eds) 1994 *Phase separation in cuprate superconductors.* Heidelberg, Germany: Springer.

Soh, Y.-A., Evans, P., Cai, Z., Lai, B., Kim, C.-Y., Aeppli, G., Mathur, N., Blamire, M. & Isaacs, E. 2002 Local mapping of strain at grain boundaries in colossal magnetoresistive films using x-ray microdiffraction. *J. Appl. Phys.* **91**, 7742–7744. (doi:10.1063/1.1455609)

Sudheendra, L. & Rao, C. N. R. 2003 Electronic phase separation in the rare-earth manganates (La$_{1-x}$Ln$_x$)$_{0.7}$Ca$_{0.3}$MnO$_3$ (Ln = Nd, Gd and Y). *J. Phys. Condens. Matter* **15**, 3029–3040. (doi:10.1088/0953-8984/15/19/306)

Tokura, Y. 2006 Critical features of colossal magnetoresistive manganites. *Rep. Prog. Phys.* **69**, 797–851. (doi:10.1088/0034-4885/69/3/R06)

Tranquada, J. M., Woo, H., Perring, T. G., Goka, H., Gu, G. D., Xu, G., Fujita, M. & Yamada, K. 2004 Quantum magnetic excitations from stripes in copper oxide superconductors. *Nature* **429**, 534–538. (doi:10.1038/nature02574)

Uehara, M., Mori, S., Chen, C. H. & Cheong, S. W. 1999 Percolative phase separation underlies colossal magnetoresistance in mixed-valent manganites. *Nature* **399**, 560–563. (doi:10.1038/21142)

Vogt, T., Cheetham, A. K., Mahendiran, R., Raychaudhuri, A. K., Mahesh, R. & Rao, C. N. R. 1996 Structural changes and related effects due to charge ordering in Nd$_{0.5}$Ca$_{0.5}$MnO$_3$. *Phys. Rev. B* **54**, 15 303–15 306. (doi:10.1103/PhysRevB.54.15303)

Vojta, T. & Schreiber, M. 2001 Localization and conductance in the quantum Coulomb glass. *Philos. Mag. B* **81**, 1117–1129.

Wollan, E. O. & Koehler, W. C. 1955 Neutron diffraction study of the magnetic properties of the series of perovskite-type compounds [La$_{(1-x)}$, Ca$_x$]MnO$_3$. *Phys. Rev.* **100**, 545–563. (doi:10. 1103/PhysRev.100.545)

Woodward, P. M., Vogt, T., Cox, D. E., Arulraj, A., Rao, C. N. R., Karen, P. & Cheetham, A. K. 1998 The influence of cation size on the structural features of Ln$_{1/2}$A$_{1/2}$MnO$_3$ perovskites at room temperature. *Chem. Mater.* **10**, 3652–3665. (doi:10.1021/cm980397u)

Woodward, P. M., Cox, D. E., Vogt, T., Rao, C. N. R. & Cheetham, A. K. 1999 Effect of compositional fluctuations on the phase transitions in (Nd$_{1/2}$Sr$_{1/2}$)MnO$_3$. *Chem. Mater.* **11**, 3528–3538. (doi:10.1021/cm990281d)

Wu, J., Lynn, J. W., Glinka, C. J., Burley, J., Zheng, H., Mitchell, J. F. & Leighton, C. 2005 Intergranular giant magnetoresistance in a spontaneously phase separated perovskite oxide. *Phys. Rev. Lett.* **94**, 037201. (doi:10.1103/PhysRevLett.94.037201)

Zhang, L., Israel, C., Biswas, A., Greene, R. L. & de Lozanne, A. 2002 Direct observation of percolation in a manganite thin film. *Science* **298**, 805–807. (doi:10.1126/science.1077346)

HIGHLIGHT   www.rsc.org/materials | Journal of Materials Chemistry

New routes to multiferroics

C. N. R. Rao*[ab] and Claudy Rayan Serrao[ab]

DOI: 10.1039/b709126e

Multiferroic materials are those which possess both ferroelectric and ferromagnetic properties. Clearly, there is a contradiction here since ferromagnetism requires d-electrons while ferroelectricity generally occurs only in the absence of d-electrons. Several multiferroics demonstrating magnetoelectric coupling effects have, however, been discovered in the past few years, but they generally make use of alternative mechanisms in attaining these properties. Several new ideas and concepts have emerged in the past two years, typical of them being magnetic ferroelectricity induced by frustrated magnetism, lone pair effect, charge-ordering and local non-centrosymmetry. Charge-order driven magnetic ferroelectricity is interesting in that it would be expected to occur in a large number of rare earth manganites, $Ln_{1-x}A_xMnO_3$ (A = alkaline earth), well known for colossal magnetoresistance, electronic phase separation and other properties. In this article, we discuss novel routes to multiferroics, giving specific examples of materials along with their characteristics.

Introduction

Materials which exhibit both magnetic and electrical ordering have attracted great interest in the past few years, partly because of their technological potential. Besides a range of possible device applications, the science of these materials is truly fascinating.[1-6] In the most general terms, multiferroics are materials in which ferromagnetism, ferroelectricity and ferroelasticity occur in the same phase. This implies that they possess spontaneous magnetization which can be reoriented by an applied magnetic field, spontaneous polarization which can be reoriented by an applied electric field and spontaneous deformation which can be reoriented by an applied stress. It is, however, customary to exclude ferro-elasticity and only consider magnetic and ferroelectric characteristics. Multiferroics possess neither spatial inversion symmetry nor time-reversal symmetry.

[a]Chemistry and Physics of Materials Unit and CSIR Centre of Excellence in Chemistry, Jawaharlal Nehru Centre for Advanced Scientific Research, Jakkur P. O., Bangalore - 560064, India. E-mail: cnrrao@jncasr.ac.in; Fax: +91-80-22082760; Tel: +91-80-22082761
[b]Material Research Centre, Indian Institute of Science, Bangalore - 560012, India

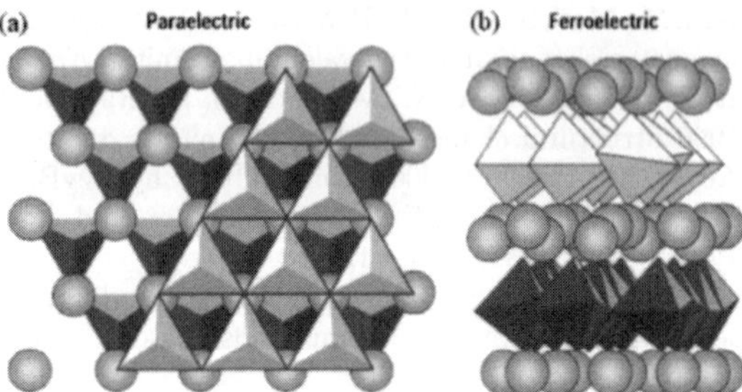

Fig. 1 Crystal structure of $YMnO_3$ in (a) the paraelectric and (b) the ferroelectric phases.

C. N. R. Rao

C. N. R. Rao obtained his PhD degree from Purdue University and DSc degree from the University of Mysore. He is the National Research Professor of India, Linus Pauling Research Professor at the Jawaharlal Nehru Centre for Advanced Scientific Research and Honorary Professor at the Indian Institute of Science. His research interests are in the chemistry of materials. A member of several academies including the Royal Society,

Claudy Rayan Serrao

the US National Academy of Sciences and the French Academy, he is an honorary fellow of the RSC. He is the recipient of the Hughes Medal of the Royal Society and the Dan David Prize for materials research.

Claudy Rayan Serrao obtained his MSc degree in Physics from Mangalore University and is carrying out his PhD studies in Materials Science at the Indian Institute of Science.

It is generally difficult to find materials that are magnetic as well as ferroelectric, since ferroelectricity occurs when the metal ions have empty d-orbitals. Magnetism, on the other hand, occurs in materials containing cations with partially filled d-orbitals. Magneto-electrics are simultaneously ferromagnetic and ferroelectric in the same phase, with coupling between the two orders. Magnetoelectric coupling describes the influence of a magnetic (electric) field on the polarization (magnetization) of a material and vice versa. Magnetoelectric coupling can exist independent of the nature of the magnetic and electrical order parameters. It is an independent phenomenon which may not necessarily arise in materials that are both magnetically and electrically polarizable. Since the coexistence of magnetism and ferroelectric ordering is not favoured, materials with such properties arising from alternative mechanisms have been sought in recent years.

Many years ago, some mixed perovskites were shown to be weakly ferromagnetic and ferroelectric by Smolensky *et al.*,[7] a typical example being $(1 - x)Pb(Fe_{2/3}W_{1/3})O_3 - xPb(Mg_{1/2}W_{1/2})O_3$. Magnetoelectric switching has been known to occur in boracite, $Ni_3B_7O_{13}I$.[8] The ferromagnetic spinel $CdCr_2S_4$ exhibits relaxor ferroelectricity below 135 K.[9] $YMnO_3$ is antiferromagnetic below the Néel temperature (T_N) 80 K and ferroelectric below the Curie temperature (T_{CE}) 914 K.[10] In this material, ferroelectricity is associated with the tilting of the MnO_5 trigonal bypyramids. Another oxide with similar properties is $BiFeO_3$ $(T_N = 670$ K, $T_{CE} = 1110$ K),[11] where ferroelectricity arises from the stereochemical activity of the Bi lone pair.[12] $BiMnO_3$ is one of the few materials recently shown to be simultaneously ferromagnetic and ferroelectric $(T_C = 450$ K, $T_{CE} = 105$ K).[13] While $BiFeO_3$ and $BiMnO_3$ are proper ferroelectrics, $TbMnO_3$, which is an improper ferroelectric, shows interesting features wherein spiral magnetic ordering is the source of the ferroelectricity.[14] One of the new ideas is that charge-ordered and orbital ordered perovskites could exhibit ferroelectric magnetism due to coupling between magnetic and charge-ordering.[15] Beside single phase oxide materials, several two-phase systems

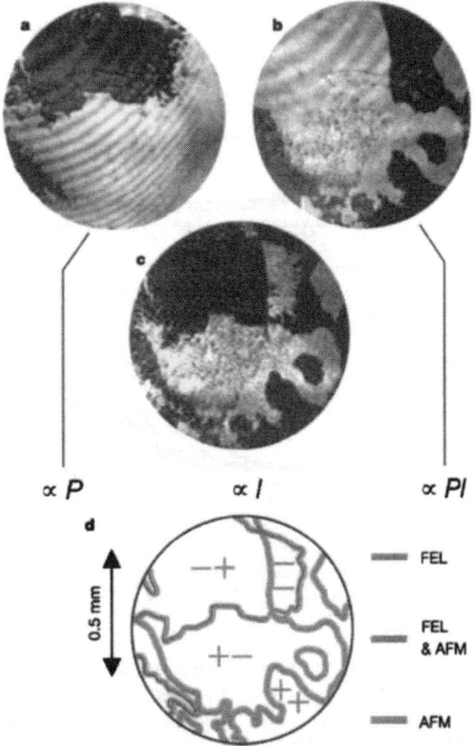

Fig. 2 Coexistance of electric and magnetic domains in a $YMnO_3$ sample at 6 K. (a) Exposed with second harmonic light from $\chi(P)$, where χ is the nonlinear optical susceptibility and P is the ferroelectric order parameter. Dark and bright areas correspond to opposite ferroelectric (FEL) domains. (b) Exposed with ferroelectromagentic (FEM) second harmonic light from $\chi(PI)$ where I is the antiferromagnetic (AFM) order parameter. Bright and dark regions are distinguished by an opposite sign of the product PI. (c) Dark and bright areas correspond to opposite AFM domains. (d) Topology of FEL (red) walls and AFM (green) walls in the sample with $\pm$ signs of the corresponding colour indicating the orientation of the FEL and AFM order parameters in selected domains (from ref. 10).

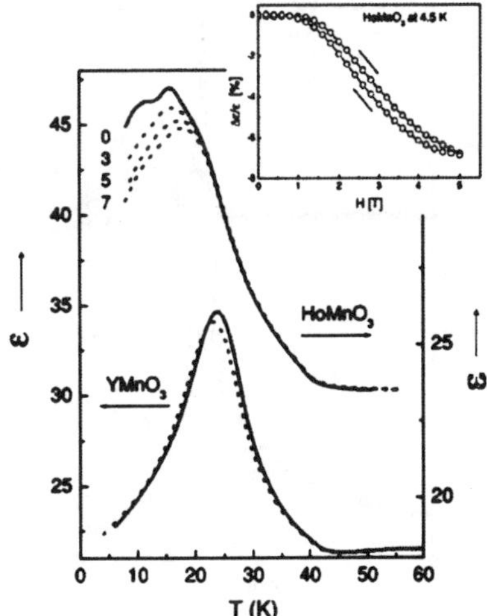

Fig. 3 Magnetodielectric effect in $YMnO_3$ (left scale, full line: $H = 0$; dotted line: $H = 7$ T) and in $HoMnO_3$ (right scale, full line: $H = 0$; dotted lines: $H = 3,5,7$ T from top to bottom). Inset: relative change of the dielectric constant as a function of the magnetic field for $HoMnO_3$ at 4.5 K (from ref. 22).

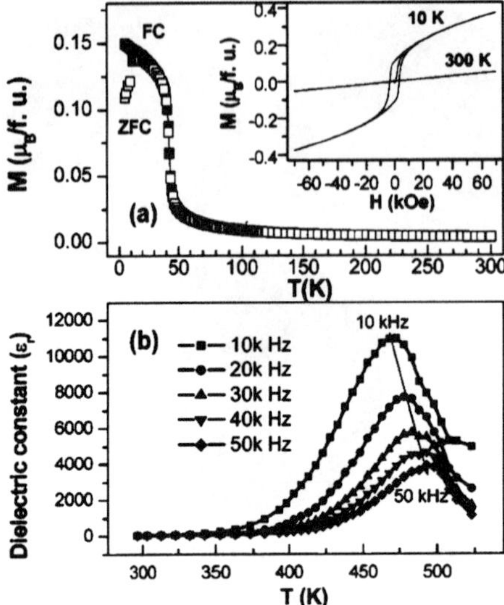

Fig. 4 (a) Temperature variation of magnetization of InMnO$_3$ at 100 Oe. Inset shows the magnetic hysteresis loops at 10 and 300 K. (b) Temperature variation of the dielectric constant for different frequencies (from ref. 24).

have been shown to exhibit magneto-electric coupling,[4] typical examples being BaTiO$_3$–CoFe$_2$O$_4$, Tb$_x$Dy$_{1-x}$Fe$_2$–Pb(Zr,Ti)O$_3$ (PZT) and La$_{0.7}$Sr$_{0.3}$MnO$_3$–PZT. In this article, we restrict our discussion to the multiferroic properties of different classes of single phase materials with special attention to their special features and newer concepts.

Rare earth manganites, LnMnO$_3$ (Ln = rare earth): polyhedral tilting and spiral magnetic ordering

Hexagonal YMnO$_3$ (space group $P6_3cm$) is a proper ferroelectric and an A-type antiferromagnet with non-collinear Mn spins oriented in a triangular arrangement. Ferroelectricity in YMnO$_3$ is driven by electrostatic and size effects unlike in other perovskite oxides, where the transition is associated with changes in the chemical bonds. Off-centring of the Mn ions is energetically unfavourable in YMnO$_3$ and ferroelectricity arises from the buckling of the MnO$_5$ polyhedra, combined with the displacement of Y ions and the layered MnO$_5$ network.[16] In Fig. 1, we show the crystal structures of the ferroelectric and paraelectric phases of YMnO$_3$. The dielectric constant and tanδ measurements near T_N (80 K) indicate coupling between the two orders and a negative magnetoresistive effect increasing with cooling reaching a value up to 15% at $\sim$230 K.[17] Coupling between electric and magnetic ordering in YMnO$_3$ is accompanied by the formation of domains and domain walls (Fig. 2). Domain wall interactions are seen in spatial maps obtained by imaging with optical second harmonic generation.[10] The antiferromagnetic domain walls interact with the lattice strain (ferroelectric walls), coupling between the two being mediated by the piezomagnetic effect.[18] This lowers the total energy of the system and leads to a piezomagnetic clamping of the electric and magnetic order parameters. Neutron diffraction experiments show that the structural parameters of YMnO$_3$

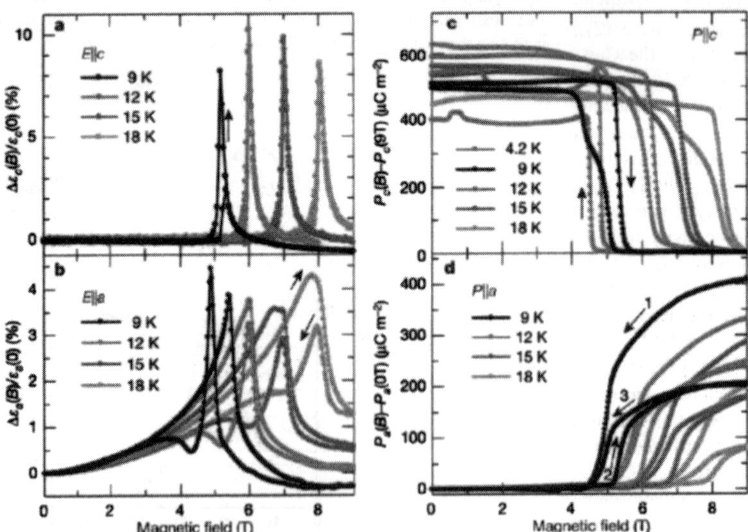

Fig. 5 Magnetocapacitance and magnetoelectric effects in TbMnO$_3$. Magnetic field-induced change in the dielectric constant (a) and (b), electric polarization along the c and a axes, respectively (c) and (d). Magnetic fields are applied along the b axis. The data for (d) were collected after magnetic field cooling. The numbers in (d) denote the order of measurements at 9 K (from ref. 14).

abruptly change at T_N indicating spin-lattice coupling.[19] Thermal conductivity undergoes a sudden increase upon magnetic ordering and enhances in magnitude below T_N.[20]

Just like YMnO$_3$, hexagonal HoMnO$_3$ also exhibits a dielectric anomaly at T_N (75 K).[21] An electrically driven magnetic phase transition has also been observed in HoMnO$_3$. Orthorhombic HoMnO$_3$ and YMnO$_3$ show a large increase in the dielectric constant at T_N due to strong magnetoelectric coupling (Fig. 3).[22] Hexagonal LuMnO$_3$ shows a dependence of the dielectric constant in the T_N region with a weak magnetoelectric coupling. Magnetic exchange coupling in LuMnO$_3$ is mainly in the ab plane of the MnO$_5$ trigonal bipyramids, while the electric dipole moment of LuO$_7$ is oriented along the hexagonal c axis.[23] InMnO$_3$, with a hexagonal structure similar to that of YMnO$_3$, has been shown to be a canted antiferromagnet below 50 K (T_N) and a ferroelectric below 500 K with a small polarization (Fig. 4).[24] InMnO$_3$ is predicted to exhibit a weak piezoelectric response to uniaxial strain.

Considerable research has been carried out recently on TbMnO$_3$ and related manganites where spin frustration causes sinusoidal antiferromagnetic ordering. The collinear sinusoidal modulated magnetic structure is observed along the b axis accompanied by spin polarization. Large magnetocapacitance and magnetoelectric effects are observed in TbMnO$_3$ due to the switching of the electric domains by the magnetic field (Fig. 5).[14] DyMnO$_3$ shows similar properties.[25] It is interesting to compare the magnetic properties of TbMnO$_3$ and DyMnO$_3$ with those of LaMnO$_3$. In LaMnO$_3$, orbital ordering between the neighboring spins gives rise to ferromagnetic exchange in the ab plane and antiferromagnetic exchange along the c axis. Replacement of La by Tb or Dy increases the structural distortion and the next-nearest neighbor antiferromagnetic exchange frustrates the nearest neighbor ferromagnetic ordering in the ab plane sinusoidal collinear along the b axis.

Bismuth-based compounds: lone pair effect

The properties of the bismuth compounds are largely determined by the

Bi 6s lone pair.[12] BiFeO$_3$ is an incommensurate antiferromagnet and a commensurate ferroelectric at room temperature.[11,26] The spins are not collinear and take a long wavelength-spiral form and the magnetoelectric effect is, therefore, not linear and occurs in the presence of a large magnetic field[27] or by appropriate chemical substitution[28] and in epitaxial thin films.[29] BiMnO$_3$ is probably the only single phase material which is truly biferroic, but the observed polarization is small (0.12 μC cm^{-2} at 87 K).[13] In Fig. 6 we show the magnetic

and dielectric properties of BiMnO$_3$ along with the magnetic field-induced changes in the dielectric constant.[13,30] Ferromagnetism in BiMnO$_3$ is due to orbital ordering[31] and ferroelectric ordering is accompanied by a structural transition.[13] It has been shown recently that BiMnO$_3$ is centrosymmetric at room temperature with the centrosymmetric space group $C2/c$.[32] Theoretical calculations also seem to suggest a centrosymmetric structure. It is possible that this material is locally non-centrosymmetric and globally centrosymmetric. We shall

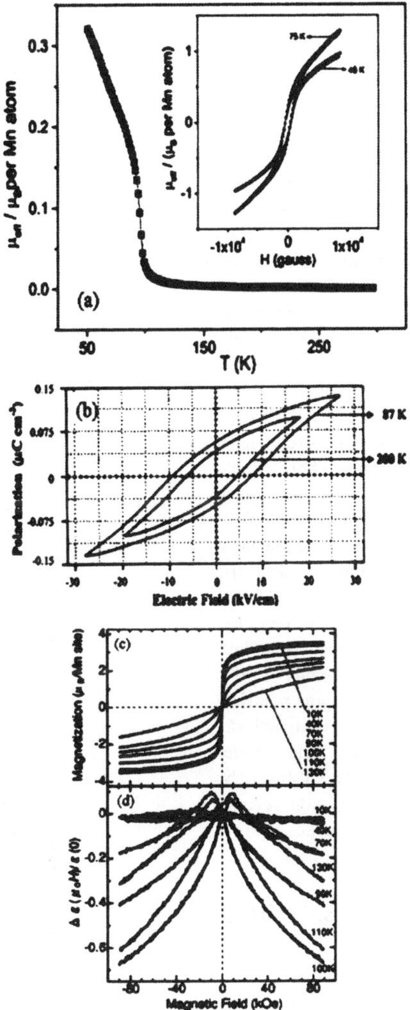

Fig. 6 (a) Temperature variation of magnetization of BiMnO$_3$ at 500 Oe. Inset shows the hysteresis loops at 75 and 45 K, (b) P–E hysteresis loops of polycrystalline BiMnO$_3$ (from ref. 13). Isothermal (c) magnetization and (d) field-induced change in dielectric constant as a function of a magnetic field at different temperatures (from ref. 30).

discuss this aspect later when we examine rare earth chromites.

BiCrO$_3$ exhibits a ferroelectric transition at 440 K and parasitic ferromagnetism below 114 K.[33] Bi$_2$Mn$_{4/3}$Ni$_{2/3}$O$_6$ is a polar oxide showing a magnetic response of a concentrated spin-glass below 35 K.[34] Bi$_2$MnNiO$_6$ shows both the ferroelectric and ferromagnetic properties.[35] Interestingly La$_2$NiMnO$_6$ which is a ferromagnetic semiconductor has recently been shown to exhibit magneto-capacitance and magnetoresistance properties at temperatures upto 280 K.[36]

LnMn$_2$O$_5$: frustrated magnetism

LnMn$_2$O$_5$ type manganites are orthorhombic solids where the Mn^{3+} and Mn^{4+} ions occupy different crystallographic sites, with the Mn^{3+} ion at the base centre of a square pyramid and octahedrally coordinated Mn^{4+} ions.[37,38] They show sequential magnetic transitions: incommensurate sinusoidal orderings of magnetic Mn spins, commensurate antiferromagnetic ordering, re-entrant transition into the incommensurate sinusoidal state and finally ordering due to rare earth spins.[39–44] These manganites exhibit electric polarization induced by collinear spin order in a frustrated magnetic system. Ferroelectricity occurs in the temperature range 25–39 K and antiferromagnetic behaviour in the range 39–45 K,[45] and the magnetic transitions are accompanied by dielectric anomalies. In a magnetic field, DyMn$_2$O$_5$ shows a 100% change in the dielectric constant at 3 K, and a similar behaviour is noticed in other compounds of this family as well. In TbMn$_2$O$_5$ the polarization is 0.04 μC cm^{-2} and it is magnetically reversed.[46] The magnetic field rotates the electric polarization of TbMn$_2$O$_5$ by 180°. Reproducible polarization reversal in TbMn$_2$O$_5$ by a magnetic field is shown in Fig. 7.

Rare earth chromites, LnCrO$_3$: role of local non-centrosymmetry

YCrO$_3$ shows canted antiferromagnetism (T_N = 140 K) and a dielectric anomaly like a ferroelectric at 473 K[47] (Fig. 8). Dielectric hysteresis in YCrO$_3$ is like in leaky dielectrics with a small polarization

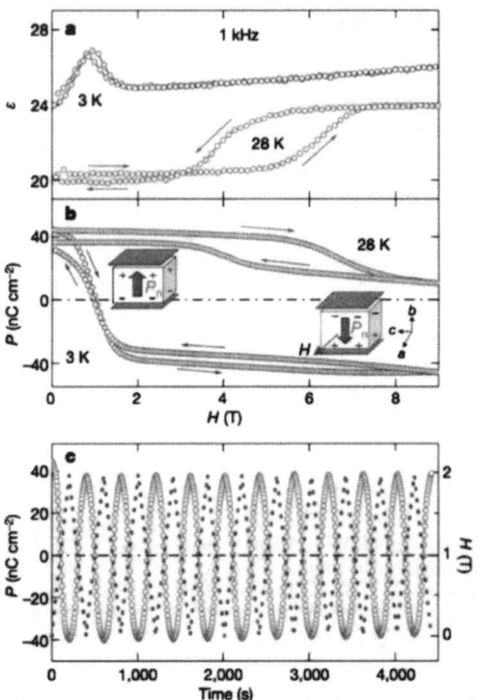

Fig. 7 Reproducible polarization reversal in TbMn$_2$O$_5$ by magnetic fields. (a) Dielectric constant *versus* applied magnetic field at 3 and 28 K. (b) Change of the total electric polarization by applied magnetic fields at 3 and 28 K. (c) Polarization flipping at 3 K by linearly varying the magnetic field from 0 to 2 T. The results clearly display highly reproducible polarization switching by magnetic fields (from ref. 46).

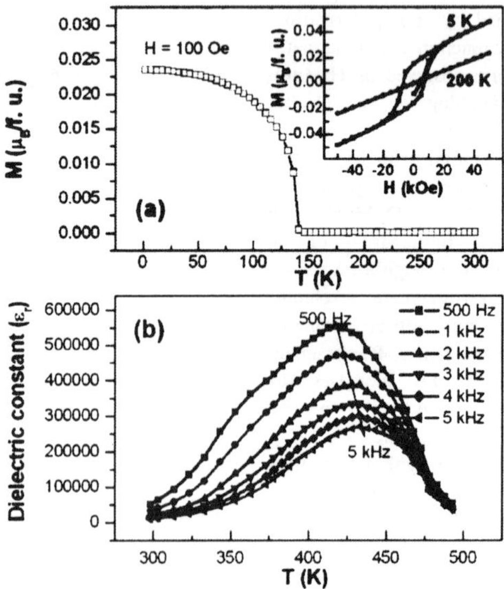

Fig. 8 (a) Temperature variation of magnetization of YCrO$_3$ at 100 Oe. The inset shows the hysteresis loops at 5 K and 200 K. (b) Temperature variation of the dielectric constant for different frequencies (from ref. 47).

value (2 μC cm^{-2}). The apparent biferroic nature of YCrO$_3$ needs explanation since it is centrosymmetric with the space group *Pnma*. Local non-centrosymmetry has been proposed as the origin of ferroelectricity and interestingly careful studies of the pair distribution function (PDF) data in the 1.6–6 Å range show that the data below 473 K (T_{CE}) are best fitted with the non-centrosymmetric space group $P2_1$. PDF refinements over the 0–20 Å range, however, give good agreement with the centrosymmetric *Pnma* space group. The near-transition PDF data shows non-centrosymmetry below T_{CE} and centrosymmetry above

T_{CE}.[48] Typical PDF data are shown in Fig. 9. Based on these studies, it is tempting to suggest that BiMnO$_3$ could be locally non-centrosymmetric with the space group *C*2. Features of YCrO$_3$ are also shown by the heavy rare earth chromites with Ln = Ho, Er, Yb, Lu. The T_{CE} increases with the decrease in the radius of the rare earth ion while the T_N decreases (Fig. 10).[49]

Magnetic ferroelectricity due to charge-ordering

Recently, Khomskii and coworkers[15] have pointed out that coupling between

magnetic and charge-ordering in charge-ordered and orbital ordered perovskites can give rise to ferroelectric magnetism. Charge-ordering in the rare earth manganites, Ln$_{1-x}$Ca$_x$MnO$_3$, (Ln = rare earth) can be site-centred (SCO) or body-centred (BCO). The SCO behaviour occurs around x = 0.5 with a CE-type antiferromagnetic state while BCO can occur around x = 0.4 with a possible perpendicular spin structure. There is a report in the literature of the occurrence of a dielectric anomaly in Pr$_{0.6}$Ca$_{0.4}$MnO$_3$ around the charge-ordering transition temperature.[50] Although magnetic fields are noted to affect the dielectric properties of manganites,[51] there has been no definitive study of the effect of magnetic fields on the dielectric properties to establish whether there is coupling between the electrical and magnetic order parameters. Dielectric properties of Pr$_{0.7}$Ca$_{0.3}$MnO$_3$, Pr$_{0.6}$Ca$_{0.4}$MnO$_3$ and Nd$_{0.5}$Ca$_{0.5}$MnO$_3$ which have comparable radii of the A-site cations, and exhibit charge-ordering in the 220–240 K (T_{CO}) region and an antiferromagnetic transition in the 130–170 K region which have been investigated recently.[52] All these manganites exhibit dielectric constant anomalies around the charge-ordering or the magnetic transition temperatures. Magnetic fields have a marked effect on the dielectric properties, indicating the presence of coupling between the magnetic and electrical order parameters. The observation of magnetoferroelectricity in these manganites is in accord with the recent theoretical predictions of Khomskii and coworkers.[15]

It is important to note some of the important characteristics of the charge-ordered manganites in order to fully understand their ferroic properties. The most important feature is that all these manganites exhibit electronic phase separation at low temperatures ($T <$ T_{CO}).[53] They exhibit a decrease in resistivity on application of large magnetic fields (>4 T).[53,54] Application of electric fields also causes a significant decrease in the resistivity of the manganites.[54,55] On the application of electric fields, the manganites show a magnetic response.[56] Such electric field-induced magnetization may also be taken as evidence for coupling between the electric and magnetic order parameters in the

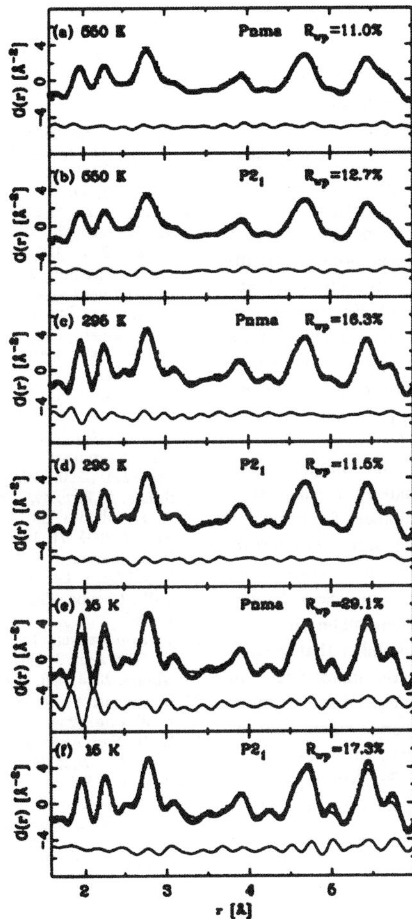

Fig. 9 PDF refinements over the r_{min} = 1.6 to r_{max} = 6 Å range for YCrO$_3$ in the paraelectric region (550 K) [(a) and (b)] and ferroelectric region (295 and 15 K) [(c)–(f)]. PDF refinements indicate that, in the paraelectric regime, the *Pnma* model is in better agreement with experimental data compared to $P2_1$ while, in the ferroelectric regime, the $P2_1$ model shows better agreement (295 and 15 K). The space group modeled and the corresponding R_{wp} value are indicated. The differences between observed and calculated data are shown below each graph (from ref. 48).

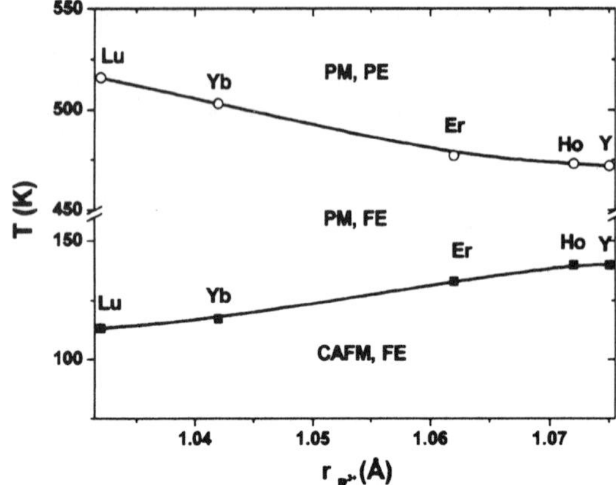

Fig. 10 Phase diagram of heavy rare earth chromites (from ref. 49).

manganites. It is likely that grain boundaries between the different electronic phases have a role in determining the dielectric behaviour. The importance of grain boundaries in giving rise to high dielectric constants has indeed been recognized.[57,58] In spite of the complexity of their electronic structure, the present study shows that the charge-ordered rare earth manganites possess multiferroic and magnetoelectric properties. Clearly, charge-ordering provides a novel route to multiferroic properties specially in the case of the manganites.

Besides the manganites, $LuFe_2O_4$ which is a charge-ordered bilayer system exhibits electric polarization at the ferrimagnetic transition temperature.[59,60] In addition to $LuFe_2O_4$ other heavy rare earth ferrites such as $ErFe_2O_4$, $YbFe_2O_4$ and YFe_2O_4 would be expected to show electric polarization at the ferrimagnetic transition temperatures beside the magnetoelectric effect.

Concluding remarks

Recent research on multiferroic systems gives an optimistic scenario wherein alternative mechanisms bring about magnetic ferroelectricity in a variety of materials. Some of the important mechanisms that have emerged relate to tilting of metal–oxygen polyhedra, spiral magnetic ordering and stereochemical activity of the Bi lone pair. In many of the multiferroic oxides, magnetic frustration appears to give rise to novel magnetic properties which then induce ferroelectricity. The role of local non-centrosymmetry and charge-ordering of mixed valent ions are two new ideas that have come to the fore. It appears that many ferroelectric materials formally possessing centrosymmetric structure may indeed have local non-centrosymmetry. In principle, non-centrosymmetry can also be achieved by making use of strain, defects and other factors. Multiferroics also offer theoretical challenges besides attractive experimental possibilities. In particular, there is considerable scope for investigations of the large family of charge-ordered manganites and related materials. It is likely that a large number of materials in this family are not only multiferroic but also magnetoelectric, considering the known effects of electric and magnetic fields on these materials.

There is every possibility that magnetoelectric effects in the manganites can be exploited technologically in memory devices, recording and other applications. The recent work of Gajek et al.[61] on thin films of $La_{0.1}Bi_{0.9}MnO_3$ promises their use in memory devices. For experimental materials scientists, the challenge of discovering materials exhibiting both magnetism and ferroelectricity with coupling between them makes the subject exciting. Some of the important aspects are related to unified theoretical development, study of spectroscopic properties and discovery of monophasic multiferroic materials showing magnetoelectric coupling.

References

1 N. A. Hill, *J. Phys. Chem. B*, 2000, **104**, 6694.
2 M. Fiebig, *J. Phys. D*, 2005, **38**, R123.
3 W. Prellier, M. P. Singh and P. Murugavel, *J. Phys.: Condens. Matter*, 2005, **17**, R803.
4 W. Eerenstein, N. D. Mathur and J. F. Scott, *Nature*, 2006, **442**, 759.
5 S. W. Cheong and M. Mostovoy, *Nat. Mater.*, 2007, **6**, 13.
6 R. Ramesh and N. A. Spaldin, *Nat. Mater.*, 2007, **6**, 21.
7 G. A. Smolensky, V. A. Isupov and A. I. Agronovskaya, *Sov. Phys. Solid State*, 1959, **1**, 150.
8 E. Ascher, H. Rieder, H. Schmid and H. Stossel, *J. Appl. Phys.*, 1966, **37**, 1404.
9 J. Hemberger, P. Lunkenheimer, R. Fichtl, H.-A. K. von Nidda, V. Tsurkan and A. Loidl, *Nature*, 2005, **434**, 364.
10 M. Fiebig, T. Lottermoser, D. Frohlich, A. V. Goltsev and R. V. Pisarev, *Nature*, 2002, **419**, 818.
11 J. R. Teage, R. Gerson and W. J. James, *Solid State Commun.*, 1970, **8**, 1073.
12 R. Seshadri and N. A. Hill, *Chem. Mater.*, 2001, **13**, 2892.
13 A. M. dos Santos, S. Parashar, A. R. Raju, Y. S. Zhao, A. K. Cheetham and C. N. R. Rao, *Solid State Commun.*, 2002, **122**, 49.
14 T. Kimura, T. Goto, H. Shintani, K. Ishizaka, T. Arima and Y. Tokura, *Nature*, 2003, **426**, 55.
15 D. V. Efremov, J. van den Brink and D. I. Khomskii, *Nat. Mater.*, 2004, **3**, 853.
16 B. B. Van Aken, T. T. M. Palstra, A. Filippetti and N. A. Spaldin, *Nat. Mater.*, 2004, **3**, 164.
17 Z. J. Huang, Y. Cao, Y. Y. Sun, Y. Y. Xue and C. W. Chu, *Phys. Rev. B*, 1997, **56**, 2623.
18 A. V. Goltsev, R. V. Pisarev, T. Lottermoser and M. Fiebig, *Phys. Rev. Lett.*, 2003, **90**, 177204.
19 S. Lee, A. Pirogov, J. H. Han, J.-G. Park, A. Hoshikawa and T. Kamiyama, *Phys. Rev. B*, 2005, **71**, 180413(R).
20 P. A. Sharma, J. S. Ahn, N. Hur, S. Park, S. B. Kim, S. Lee, J.-G. Park, S. Guha and S.-W. Cheong, *Phys. Rev. Lett.*, 2004, **93**, 177202.
21 T. Lottermoser, T. Lonkai, U. Amann, D. Hohlwein, J. Ihringer and M. Fiebig, *Nature*, 2004, **430**, 541.
22 B. Lorenz, Y. Q. Wang, Y. Y. Sun and C. W. Chu, *Phys. Rev. B*, 2004, **70**, 212412.
23 B. B. Van Aken and T. T. M. Palstra, *Phys. Rev. B*, 2004, **69**, 134113.
24 C. R. Serrao, A. K. Kundu, J. Bhattacharjee, S. B. Krupanidhi, U. V. Waghmare and C. N. R. Rao, *J. Appl. Phys.*, 2006, **100**, 076104.
25 O. Prokhnenko, R. Feyerherm, E. Dudzik, S. Landsgesell, N. Aliouane, L. C. Chapon and D. N. Argyriou, *Phys. Rev. Lett.*, 2007, **98**, 057206.
26 I. Sosnovska, T. Peterlin-Neumaier and E. Steichele, *J. Phys. C*, 1982, **15**, 4835.
27 Y. F. Popov, A. M. Kadomtseva, G. P. Vorob'ev and A. K. Zvezdin, *Ferroelectrics*, 1994, **162**, 135.
28 V. A. Murashov, D. N. Rakov, V. M. Ionov, I. S. Dubenko, Y. V. Titov

and V. I. Gorelik, *Ferroelectrics*, 1994, **162**, 11.

29 F. Bai, J. Wang, M. Wuttig, J. Li, N. Wang, A. P. Pyatakov, A. K. Zvezdin, L. E. Cross and D. Viehland, *J. Appl. Phys.*, 2005, **86**, 032511.

30 T. Kimura, S. Kawamoto, I. Yamada, M. Azuma, M. Takano and Y. Tokura, *Phys. Rev. B*, 2003, **67**, R180401.

31 A. M. dos Santos, A. K. Cheetham, T. Atou, Y. Syono, Y. Yamaguchi, K. Ohoyama, H. Chiba and C. N. R. Rao, *Phys. Rev. B*, 2002, **66**, 064425.

32 A. A. Belik, S. Iikubo, T. Yokosawa, K. Kodama, N. Igawa, S. Shamoto, M. Azuma, M. Takano, K. Kimoto, Y. Matsui and E. Takayama-Muromachi, *J. Am. Chem. Soc.*, 2007, **129**, 971.

33 S. Niitaka, M. Azuma, M. Takano, E. Nishibori, M. Takata and M. Sakata, *Solid State Ionics*, 2004, **172**, 557.

34 H. Hughes, M. M. B. Allix, C. A. Bridges, J. B. Claridge, X. Kuang, H. Niu, S. Taylor, W. Song and M. Rosseinsky, *J. Am. Chem. Soc.*, 2005, **127**, 13790.

35 M. Azuma, K. Takata, T. Saito, S. Ishiwata, Y. Shimakawa and M. Takano, *J. Am. Chem. Soc.*, 2005, **127**, 8889.

36 N. S. Rogado, J. Li, A. W. Sleight and M. A. Subramanian, *Adv. Mater.*, 2005, **17**, 2225.

37 J. A. Alonso, M. R. T. Casaos, M. J. Martinez-Lope and I. Rasines, *J. Solid State Chem.*, 1997, **129**, 105.

38 S. C. Abrahams and J. L. Bernstein, *J. Chem. Phys.*, 1967, **46**, 3776.

39 S. Quezel-Ambrunaz, F. Bertaut and G. Buisson, *Compt. Rend.*, 1964, **258**, 3025.

40 M. Schieber, A. Grill, I. Nowik, B. M. Y. Wanklyn, R. C. Sherwood and L. G. van Uitert, *J. Appl. Phys.*, 1973, **44**, 1864.

41 E. I. Golovenchits, N. V. Morozov, V. A. Sanina and L. M. Sapozhnikova, *Sov. Phys. Solid State*, 1992, **34**, 56.

42 K. Saito and K. Kohn, *J. Phys.: Condens. Matter*, 1995, **7**, 2855.

43 A. Inomata and K. Kohn, *J. Phys.: Condens. Matter*, 1996, **8**, 2673.

44 P. P. Gardner, C. Wilkinson, J. B. Forsyth and B. M. Wanklyn, *J. Phys. C: Solid State Phys.*, 1998, **21**, 5653.

45 I. Kagomiya, K. Kohn and T. Uchiyama, *Ferroelectrics*, 2002, **280**, 297.

46 N. Hur, S. Park, P. A. Sharma, J. S. Ahn, S. Guha and S.-W. Cheong, *Nature*, 2004, **429**, 392.

47 C. R. Serrao, A. K. Kundu, S. B. Krupanidhi, U. V. Waghmare and C. N. R. Rao, *Phys. Rev. B*, 2005, **72**, R220101.

48 K. Ramesha, A. Llobet, T. Profen, C. R. Serrao and C. N. R. Rao, *J. Phys.: Condens. Matter*, 2007, **19**, 102202.

49 J. R. Sahu, C. R. Serrao, N. Ray, U. V. Waghmare and C. N. R. Rao, *J. Mater. Chem.*, 2007, **17**, 42.

50 N. Biskup, A. de Andres and J. L. Martinez, *Phys. Rev. B*, 2005, **72**, 024115.

51 R. S. Freitas, J. F. Mitchell and P. Schiffer, *Phys. Rev. B*, 2005, **72**, 144429.

52 C. R. Serrao, A. Sundaresan and C. N. R. Rao, arXiv:0708.0746v1 [cond-mat.mtrl-sci].

53 V. B. Shenoy, D. D. Sarma and C. N. R. Rao, *ChemPhysChem*, 2006, **7**, 2053.

54 C. N. R. Rao, *J. Phys. Chem. B*, 2000, **104**, 5877.

55 S. Parashar, L. Sudheendra, A. R. Raju and C. N. R. Rao, *J. Appl. Phys.*, 2004, **95**, 2181.

56 A. Guha, N. Khare, A. K. Raychaudhuri and C. N. R. Rao, *Phys. Rev. B*, 2000, **62**, R11941.

57 P. Lunkenheimer, V. Bobnar, A. V. Pronin, A. I. Ritus, A. A. Volkov and A. Loidl, *Phys. Rev. B*, 2002, **66**, 052105.

58 L. Sudheendra and C. N. R. Rao, *J. Phys.: Condens. Matter*, 2003, **15**, 3029.

59 N. Ikeda, H. Ohsumi, K. Ohwada, K. Ishii, T. Inami, K. Kakurai, Y. Murakami, K. Yoshii, S. Mori, Y. Horibe and H. Kito, *Nature*, 2005, **436**, 1136.

60 M. A. Subramanian, T. He, J. Chen, N. S. Rogado, T. G. Calvarese and A. W. Sleight, *Adv. Mater.*, 2006, **18**, 1737.

61 M. Gajek, M. Bibes, S. Fusil, K. Bouzehouane, J. Fontcuberta, A. Barthelemy and A. Fert, *Nat. Mater.*, 2007, **6**, 296.

INSTITUTE OF PHYSICS PUBLISHING

JOURNAL OF PHYSICS: CONDENSED MATTER

J. Phys.: Condens. Matter **18** (2006) 4809–4818

doi:10.1088/0953-8984/18/20/005

Glassy behaviour of the ferromagnetic and the non-magnetic insulating states of the rare earth manganates $Ln_{0.7}Ba_{0.3}MnO_3$ (Ln = Nd or Gd)

Asish K Kundu[1,2], P Nordblad[2] and C N R Rao[1,3]

[1] Chemistry and Physics of Materials Unit, Jawaharlal Nehru Centre for Advanced Scientific Research, Jakkur PO, Bangalore-562064, India
[2] Department of Engineering Sciences, Uppsala University, 751 21 Uppsala, Sweden

E-mail: cnrrao@jncasr.ac.in

Received 24 March 2006
Published 2 May 2006
Online at stacks.iop.org/JPhysCM/18/4809

Abstract

While $La_{0.7}Ba_{0.3}MnO_3$ is a ferromagnetic metal (T_C = 340 K) with long-range ordering, $Nd_{0.7}Ba_{0.3}MnO_3$ shows a transition around 150 K with a small increase in magnetization, but remains an insulator at all temperatures. $Gd_{0.7}Ba_{0.3}MnO_3$ is non-magnetic and insulating at all temperatures. Low field dc magnetization and ac susceptibility measurements reveal the presence of a transition at around 150 K in $Nd_{0.7}Ba_{0.3}MnO_3$, and a complex behaviour with different ordering/freezing transitions at 62, 46 and 36 K in the case of $Gd_{0.7}Ba_{0.3}MnO_3$, the last one being more prominent. The nature of the field dependence of the magnetization, combined with the slow magnetic relaxation, ageing and memory effects, suggests that $Nd_{0.7}Ba_{0.3}MnO_3$ is a cluster glass below 150 K, a situation similar to that found for $La_{1-x}Sr_xCoO_3$. $Gd_{0.7}Ba_{0.3}MnO_3$, however, shows non-equilibrium dynamics characteristic of spin glasses, below 36 K. The difference in nature of the glassy behaviour between $Gd_{0.7}Ba_{0.3}MnO_3$ and $Nd_{0.7}Ba_{0.3}MnO_3$ probably arises because of the larger disorder arising from the mismatch between the sizes of the A-site cations in the former. Our results on $Nd_{0.7}Ba_{0.3}MnO_3$ and $Gd_{0.7}Ba_{0.3}MnO_3$ suggest that the magnetic insulating states often reported for rare earth manganates of the type $Ln_{1-x}A_xMnO_3$ (Ln = rare earth, A = alkaline earth) are likely to be associated with glassy magnetic behaviour.

1. Introduction

Among the several novel properties and phenomena exhibited by rare earth manganates of the type $Ln_{1-x}A_xMnO_3$ (Ln = rare earth, A = alkaline earth), charge ordering and electronic

[3] Author to whom any correspondence should be addressed.

phase separation are of particular interest [1–6]. Both these properties are highly sensitive to the average radius of the A-site cations, $\langle r_A \rangle$, and the size disorder arising from the mismatch between the A-site cations [4–7]. The size disorder is generally expressed in terms of the σ^2 parameter which is defined as $\sigma^2 = \sum x_i r_i^2 - \langle r_A \rangle^2$, where x_i is the fractional occupancy of A-site ions, r_i is the corresponding ionic radius and $\langle r_A \rangle$ is the weighted average radius calculated from r_i values [7]. Electronic phase separation is found to occur above a critical composition x_c in $La_{0.7-x}Ln_xCa_{0.3}MnO_3$, especially in the regime where $\langle r_A \rangle$ is close to 1.18 Å or lower, and is favoured by large size disorder [3, 6]. In this system, $\langle r_A \rangle$ decreases with increasing x, affecting the e_g bandwidth. A study of $La_{0.250}Pr_{0.375}Ca_{0.375}MnO_3$ by Deac *et al* [8] has shown two types of magnetic relaxation, one at low fields associated with the reorientation of ferromagnetic (FM) domains and another at higher fields due to the transformation between FM and non-FM phases. The presence of FM clusters and associated magnetic relaxation phenomena well below T_C has been reported for $La_{0.7-x}Y_xCa_{0.3}MnO_3$ by Freitas *et al* [9]. $Nd_{0.7}Sr_{0.3}MnO_3$ with a well defined ferromagnetic T_C exhibits ageing phenomena in the ferromagnetic phase indicating magnetic frustration and disorder [10]. Lopez *et al* [11] have provided evidence for two competing magnetic phases in $La_{0.5}Ca_{0.5}MnO_3$ based on a magnetic relaxation study. A recent investigation of the magnetic and electric properties of $La_{0.7-x}Ln_xBa_{0.3}MnO_3$ (Ln = Pr, Nd, or Gd) where the $\langle r_A \rangle$ remains relatively large over the entire range of compositions ($\geqslant 1.216$ Å) has shown that the FM or non-magnetic insulating compositions can be rendered FM and metallic by decreasing the size disorder [12]. An insulating magnetic state is found in $Nd_{0.7}Ba_{0.3}MnO_3$, but $Gd_{0.7}Ba_{0.3}MnO_3$ is insulating and non-magnetic down to low temperatures, although the carrier concentration (Mn^{3+}/Mn^{4+} ratio) is the same as in $La_{0.7}Ba_{0.3}MnO_3$ which is a genuine FM metal [12, 13]. In this paper, we focus our interest on $Nd_{0.7}Ba_{0.3}MnO_3$ and $Gd_{0.7}Ba_{0.3}MnO_3$ and have carried out a detailed study of the magnetic properties by employing measurements of the low field dc magnetization, ac susceptibility, magnetic relaxation and memory effects. The study has revealed the presence of glassy magnetic phases in both these manganates, albeit of different varieties.

2. Experimental procedure

Polycrystalline samples of $Ln_{0.7}Ba_{0.3}MnO_3$ (Ln = Nd and Gd) were prepared by the conventional solid-state reactions. Stoichiometric mixtures of the respective rare earth oxides, $BaCO_3$ and MnO_2, were weighed in the desired proportions and milled for a few hours with propanol. The mixtures were dried, and calcined in air at 1223 K; this was followed by heating at 1273 and 1373 K for 12 h each in air. The powders thus obtained were pelletized and the pellets sintered at 1673 K for 24 h in air. Composition analysis was carried out using energy dispersive x-ray (EDX) analysis using a LEICA S440I scanning electron microscope fitted with a Si–Li detector and it confirms the composition within experimental errors. The oxygen stoichiometry was determined by iodometric titrations.

The phase purity of the manganates was established by recording the x-ray diffraction patterns in the 2θ range of $10°$–$80°$ with a Seiferts 3000 TT diffractometer using Cu Kα radiation. A Quantum Design MPMSXL superconducting quantum interference device (SQUID) magnetometer and a non-commercial low field SQUID magnetometer system [14] were used to investigate the magnetic properties of the samples. The temperature dependences of the zero-field-cooled (ZFC) and field-cooled (FC) magnetizations were measured in different applied magnetic fields. Hysteresis loops were recorded at some different temperatures in the low temperature phases of the system. The dynamics of the magnetic response was studied using ac susceptibility measurements at different frequencies and measurements of the relaxation of the low field ZFC magnetization.

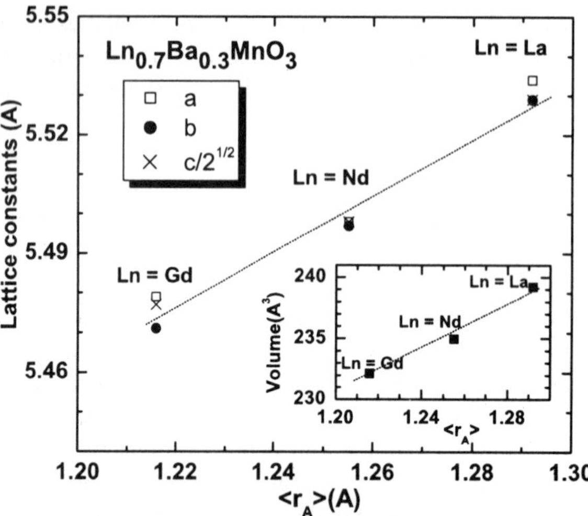

Figure 1. Variation of the lattice parameters and cell volume (inset figure) with $\langle r_A \rangle$ of Ln$_{0.7}$Ba$_{0.3}$MnO$_3$ with Ln = La, Nd and Gd.

In the measurements of the temperature dependence of the ZFC magnetization, the sample was cooled from 350 to 10 K in zero field, the field was applied at 10 K and the magnetization recorded on re-heating the sample. In the FC measurements the sample was cooled in the applied field to 10 K and the magnetization recorded on re-heating the sample, keeping the field applied. In the relaxation experiments, the sample was cooled in zero field from a reference temperature of 170 K (for Nd) and 90 K (for Gd) to a measuring temperature, T_m and kept there for a wait time, t_w. After the wait time, a small probing field was applied and the magnetization was recorded as a function of time elapsed after the field application. The electrical resistivity (ρ) measurements were carried out by a standard four-probe method with silver epoxy as electrodes in the 20–300 K temperature range.

3. Results and discussion

La$_{0.7}$Ba$_{0.3}$MnO$_3$, Nd$_{0.7}$Ba$_{0.3}$MnO$_3$ and Gd$_{0.7}$Ba$_{0.3}$MnO$_3$ possess orthorhombic structures (*Pnma* space group) and the lattice parameters decrease with the decrease in the size of the rare earth ion as expected. In figure 1, we show the variation of lattice parameters and cell volume with $\langle r_A \rangle$ to demonstrate this feature. The $\langle r_A \rangle$ values of La$_{0.7}$Ba$_{0.3}$MnO$_3$, Nd$_{0.7}$Ba$_{0.3}$MnO$_3$ and Gd$_{0.7}$Ba$_{0.3}$MnO$_3$ are 1.292, 1.255 and 1.216 Å respectively, the corresponding values of the size disorder parameter, σ^2, being 0.014, 0.020 and 0.027 Å^2 respectively. Thus, Gd$_{0.7}$Ba$_{0.3}$MnO$_3$ has the smallest $\langle r_A \rangle$ and the largest σ^2.

In figure 2(a), we show the dc magnetization behaviour of La$_{0.7}$Ba$_{0.3}$MnO$_3$, Nd$_{0.7}$Ba$_{0.3}$MnO$_3$ and Gd$_{0.7}$Ba$_{0.3}$MnO$_3$ under FC conditions ($H = 500$ Oe). La$_{0.7}$Ba$_{0.3}$MnO$_3$ shows a sharp increase in the magnetization around 340 K (T_C) corresponding to the ferromagnetic transition. There is evidence for saturation, the values of the saturation magnetization and the corresponding magnetic moment being 35 emu g^{-1} and 1.5 μ_B/f.u. Nd$_{0.7}$Ba$_{0.3}$MnO$_3$ shows an increase in the magnetization around 150 K, but the maximum magnetization value found is 18 emu g^{-1} (0.8 μ_B/f.u.) at 40 K. Gd$_{0.7}$Ba$_{0.3}$MnO$_3$ shows no evidence for a magnetic transition and the magnetization value is 5 emu g^{-1} (0.25 μ_B/f.u.) at 40 K. Clearly, the magnetic properties of the three manganates are distinctly different from

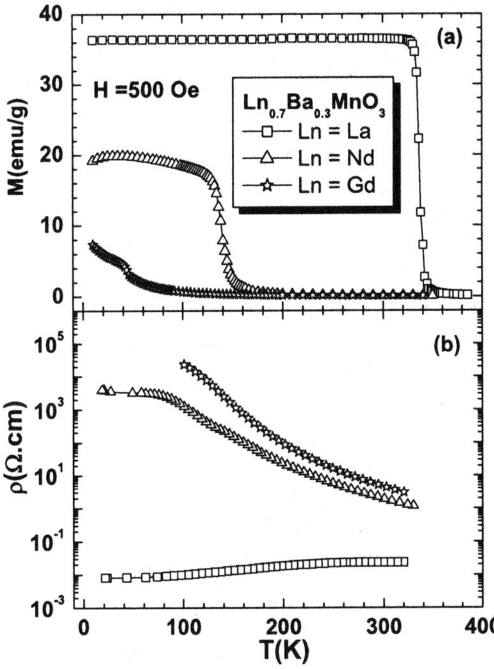

Figure 2. Temperature dependence of (a) the FC magnetization, M (at $H = 500$ Oe), and (b) the electrical resistivity, ρ, of $Ln_{0.7}Ba_{0.3}MnO_3$ with Ln = La, Nd and Gd. Note that $Nd_{0.7}Ba_{0.3}MnO_3$ is insulating at 150 K where there is a weak magnetic transition.

one another. Whereas $La_{0.7}Ba_{0.3}MnO_3$ shows metallic behaviour below T_C, $Nd_{0.7}Ba_{0.3}MnO_3$ and $Gd_{0.7}Ba_{0.3}MnO_3$ show insulating behaviour over the entire temperature range (figure 2(b)). Thus, $Nd_{0.7}Ba_{0.3}MnO_3$ is insulating at and below the 150 K transition and $Gd_{0.7}Ba_{0.3}MnO_3$ is a non-magnetic insulator at all temperatures.

The magnetization data for $La_{0.7}Ba_{0.3}MnO_3$ at low fields were similar to those obtained at higher field, showing little divergence between the ZFC and FC data. In figure 3, we present low field ZFC and FC magnetization data for $Nd_{0.7}Ba_{0.3}MnO_3$ and $Gd_{0.7}Ba_{0.3}MnO_3$. The FC magnetization of $Nd_{0.7}Ba_{0.3}MnO_3$ shows a transition around 150 K. $Gd_{0.7}Ba_{0.3}MnO_3$ exhibits a rather complex behaviour below 62 K where irreversibility between the ZFC and FC magnetization data first appears (figure 3(c)). The low temperature region is discussed later, but it is noteworthy that there are three characteristic temperatures: 62 K (onset of significant irreversibility between the ZFC and FC magnetization curves), 46 K (a maximum in the FC curve) and 36 K (a maximum in the ZFC curve), all indicating different ordering and/or freezing processes in the system.

Figure 4 shows the field variation of the magnetization at three different temperatures for $Nd_{0.7}Ba_{0.3}MnO_3$ and $Gd_{0.7}Ba_{0.3}MnO_3$. Neither of these manganates exhibits hysteresis. Below 150 K, $Nd_{0.7}Ba_{0.3}MnO_3$ shows a behaviour similar to that of a weak ferromagnet, the magnetization approaching saturation at high fields. $Gd_{0.7}Ba_{0.3}MnO_3$ does not show the $M–H$ behaviour of a ferromagnet at low temperatures, and exhibits no tendency for saturation even at high fields. The shape of the $M–H$ curve and the absence of saturation even at high fields in $Gd_{0.7}Ba_{0.3}MnO_3$ are reminiscent of magnetization curves of spin glasses [15]. The $M–H$ behaviour becomes nearly linear (paramagnetic) at 200 K in both the manganates.

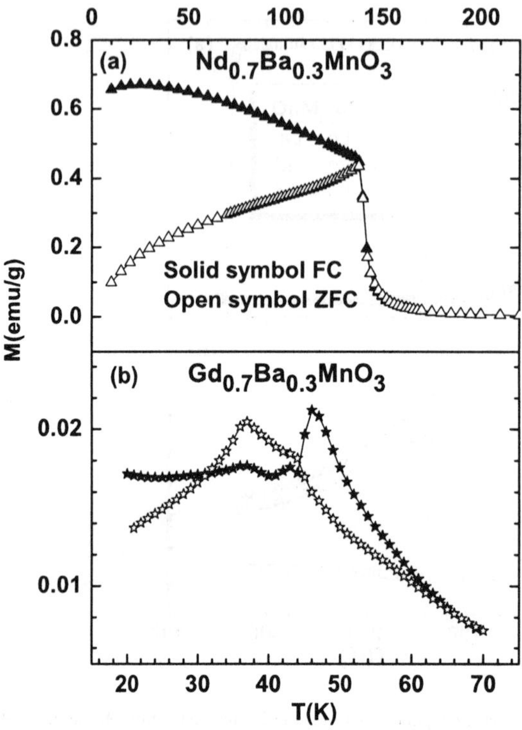

Figure 3. Temperature dependence of the ZFC (open symbols) and FC (solid symbols) magnetization, M, of (a) Nd$_{0.7}$Ba$_{0.3}$MnO$_3$ (at $H = 10\,$Oe) and (b) Gd$_{0.7}$Ba$_{0.3}$MnO$_3$ (at $H = 3\,$Oe). The features of the M–T curves remain same when the magnetic field is in the 1–10 Oe range.

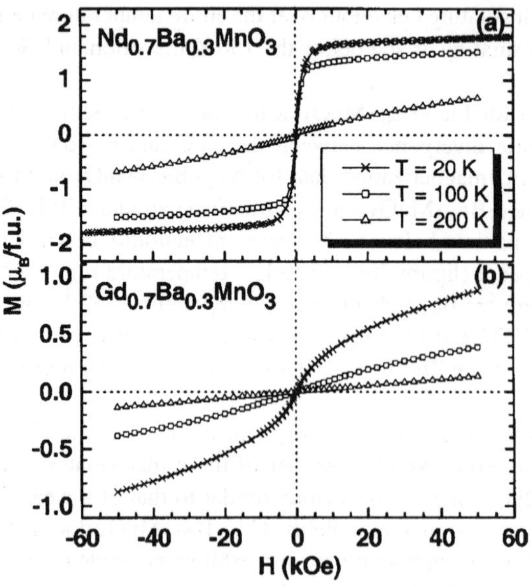

Figure 4. Typical hysteresis curves for (a) Nd$_{0.7}$Ba$_{0.3}$MnO$_3$ and (b) Gd$_{0.7}$Ba$_{0.3}$MnO$_3$ at different temperatures.

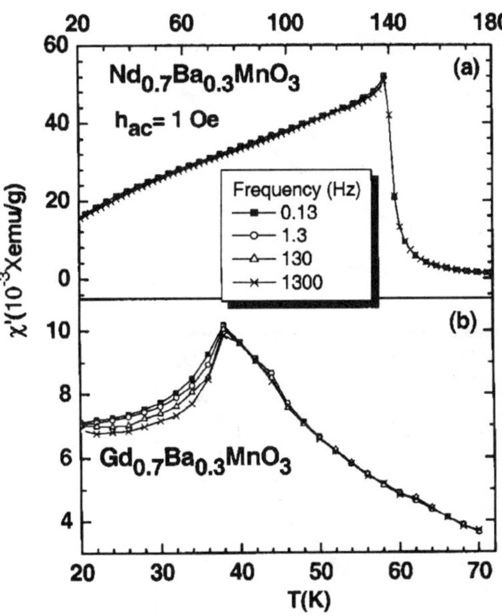

Figure 5. The temperature dependence of the in-phase ac susceptibility for (a) $Nd_{0.7}Ba_{0.3}MnO_3$ and (b) $Gd_{0.7}Ba_{0.3}MnO_3$ at different frequencies.

The temperature dependences of the ac susceptibilities of $Nd_{0.7}Ba_{0.3}MnO_3$ and $Gd_{0.7}Ba_{0.3}MnO_3$ are presented at different frequencies in figure 5. The in-phase $\chi'(T)$ component of the ac susceptibility reveals similar features to the ZFC magnetization at low fields for both the manganates. $Nd_{0.7}Ba_{0.3}MnO_3$ shows a sharp maximum below 150 K, which is frequency independent. However, there is a weak frequency dependence at temperatures below 140 K, a behaviour noted earlier for $Nd_{0.7}Sr_{0.3}MnO_3$ [10]. $Gd_{0.7}Ba_{0.3}MnO_3$ shows a shoulder around 62 K, a weak anomaly just above 46 K and a maximum at 36 K. The $\chi'(T)$ data become strongly frequency dependent below 36 K. This transition could arise from the presence of small magnetic clusters in a non-magnetic matrix just as in cobaltates of the type $La_{0.7}Ca_{0.3}CoO_3$ [16]. Other examples of oxide systems where only such short-range ferromagnetic correlations occur are known [10, 16, 17].

Time dependent ZFC magnetization measurements show that both $Nd_{0.7}Ba_{0.3}MnO_3$ and $Gd_{0.7}Ba_{0.3}MnO_3$ exhibit logarithmically slow dynamics and ageing at low temperatures. In figures 6(a) and (b), we show the time dependent ZFC magnetization, $m(t)$, measured at $T_m = 40$ K, and the corresponding relaxation rates $S(t) = (1/H)[dM_{ZFC}(T, t_w, t)/d \log_{10}(t)]$ for $Nd_{0.7}Ba_{0.3}MnO_3$. The applied field was 1 Oe and the wait times were $t_w = 100, 1000$ and 10 000 s. The results of similar measurements on $Gd_{0.7}Ba_{0.3}MnO_3$ at 30 K are presented in figures 7(a) and (b). The wait time dependence of the magnetic relaxation illustrated in figures 6 and 7 shows that both the manganates are subject to magnetic ageing at low temperatures. Relaxation experiments (not shown) at 80 K (Nd) and 40 K (Gd) reveal slow relaxation and ageing behaviour at these temperatures as well, but with a much decreased relaxation rate compared to that at low temperatures. Time dependent thermoremanent magnetization (TRM) measurements at the same temperatures yielded similar results for both systems. Magnetic ageing is a signature of spin glasses [15] and, explained within the droplet (or domain growth) model, the maximum in the relaxation rate is associated with a crossover between quasi-equilibrium and non-equilibrium dynamics [18]. The slow relaxation and ageing behaviour

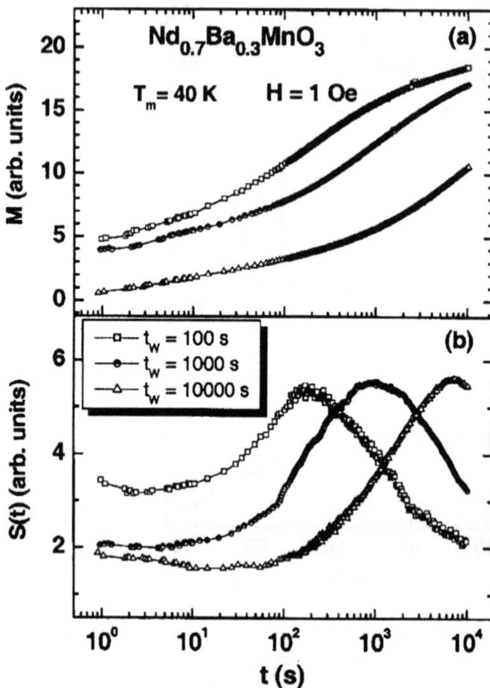

Figure 6. ZFC relaxation measurements on Nd$_{0.7}$Ba$_{0.3}$MnO$_3$ at T_m = 40 K for different waiting times, t_w = 100, 1000 and 10 000 s (H = 1 Oe).

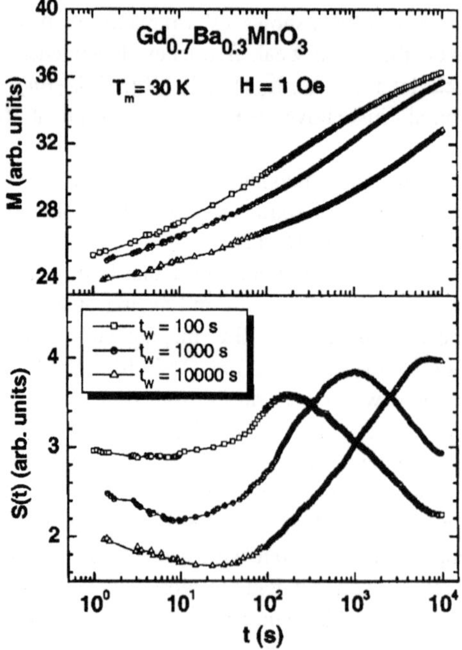

Figure 7. ZFC relaxation measurements on Gd$_{0.7}$Ba$_{0.3}$MnO$_3$ at T_m = 30 K for different waiting times, t_w = 100, 1000 and 10 000 s (H = 1 Oe).

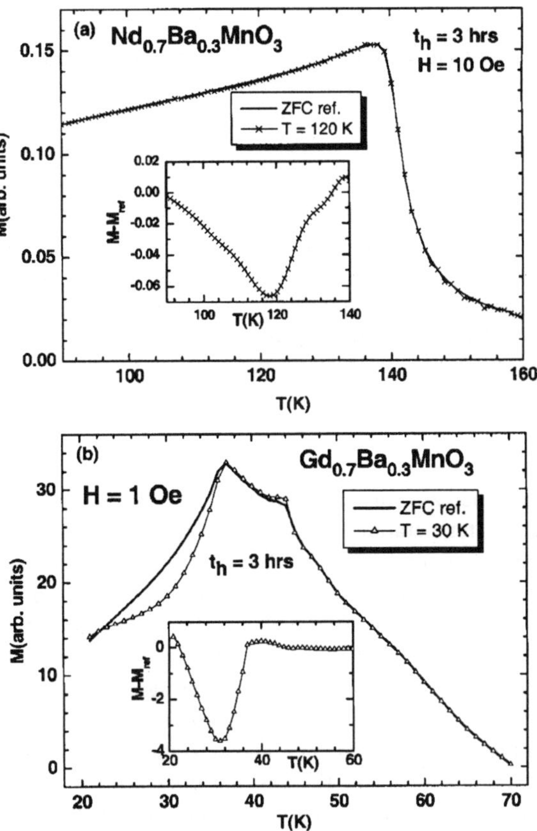

Figure 8. ZFC magnetization memory experiment on (a) $Nd_{0.7}Ba_{0.3}MnO_3$; the temperature dependence of the ZFC magnetization, M (reference curve), and on imprinting memories at temperature stops (120 K) during cooling for 3 h ($H = 10$ Oe) and the inset shows the difference ($M_{mem} - M_{ref}$) plot. (b) $Gd_{0.7}Ba_{0.3}MnO_3$; the temperature dependence of the ZFC magnetization, M (reference curve), and on imprinting memory at 30 K during cooling for 3 h ($H = 1$ Oe). The inset shows the difference ($M_{mem} - M_{ref}$) plot.

of $Nd_{0.7}Ba_{0.3}MnO_3$ and $Gd_{0.7}Ba_{0.3}MnO_3$ demonstrate that magnetic disorder and frustration occur in the low temperature phases.

Glassy dynamics in spin glasses is also manifested by a memory effect that can be demonstrated by dc magnetization or low frequency ac susceptibility experiments. We have employed zero-field-cooled magnetization versus temperature experiments [16] to investigate possible memory phenomena in the two manganates. The experiment includes a reference measurement, according to the ZFC protocol described earlier, and a similar ZFC memory experiment, the protocol of which only includes one additional feature: the cooling down of the sample is halted at a stop temperature for some hours. In a spin glass experiment, the memory curve acquires a weak dip at the temperature where the zero-field cooling was halted. To illustrate the memory effect, it is convenient to plot the difference between the reference and the memory curve. A spin glass phase (ordinary or re-entrant) has a pronounced memory behaviour, whereas a disordered and frustrated ferromagnetic phase shows little or no memory effect. In the case of $Nd_{0.7}Ba_{0.3}MnO_3$, we carried out the ZFC experiment by cooling the sample from a reference temperature of 170–90 K, where the magnetic field (10 Oe) was applied and the magnetization recorded while continuously heating the sample to 170 K. The ZFC memory curve was recorded in a similar way with the additional feature that the cooling in zero field was stopped at 120 K for 3 h. Figure 8(a) shows the two curves. A weak dip can barely be discerned in the memory curve (labelled 120 K). The difference plot $M_{mem} - M_{ref}$ versus T, shown as an inset in figure 8(a), reveals a broad but shallow memory of the stop at 120 K. In contrast, the corresponding experiment on the $Gd_{0.7}Ba_{0.3}MnO_3$ sample shows a prominent memory dip. The experiment was performed starting from 70 K and cooling the sample continuously to 20 K, with an intermediate stop at 30 K for 3 h in the memory measurement. Figure 8(b) shows the two curves. There is a significant difference between the reference and the memory curves. The difference plot shown in the inset of figure 8(b) reveals a deep, broad memory dip. The dip abruptly ceases above 36 K. The memory behaviour of $Gd_{0.7}Ba_{0.3}MnO_3$ at 36 K is clearly that of a spin glass.

4. Conclusions

$Nd_{0.7}Ba_{0.3}MnO_3$ shows an increase in magnetization at 150 K, but the value of the magnetization is small at low temperature. It is also an insulator. It shows a pronounced ageing behaviour, but a rather weak memory effect below 150 K, probably due to the presence of FM clusters in an insulating matrix. $Nd_{0.7}Ba_{0.3}MnO_3$ appears to be a cluster glass or a magnetically disordered system similar to $La_{1-x}Sr_xCoO_3$ [19]. $Gd_{0.7}Ba_{0.3}MnO_3$ appears to contain small magnetic clusters, giving rise to a spin glass state below 36 K. Low field magnetization experiments indicate that some kind of ordering/or freezing process occurs in this manganate even around 62 K, with an additional process at 46 K. The origin of these features is difficult to establish from macroscopic magnetization data. The small proportion of the clusters responsible for the weak 62 K transition does not result in a distinct glassy transition or a FM-like transition. This behaviour of $Gd_{0.7}Ba_{0.3}MnO_3$ is attributed to the large size mismatch between the A-site cations or large σ^2 value (0.028 Å^2), the mismatch being considerably smaller for $Nd_{0.7}Ba_{0.3}MnO_3$ [12, 13]. Such size mismatch favours chemical/electronic inhomogeneities. To our knowledge, this is a unique case of a perovskite manganate showing a size disorder-induced spin glass behaviour, occurring in spite of the relatively large A-site cation radius ($\langle r_A \rangle = 1.216$ Å). This behaviour is comparable to the one observed in $Nd_{0.7}Ca_{0.3}CoO_3$ [20]. It appears that the so-called FM insulating state or non-magnetic insulating state often reported in the rare earth manganates of the type $Ln_{1-x}A_xMnO_3$ arises from the glassy behaviour of the magnetic clusters in these materials, generally associated with electronic phase separation.

Acknowledgments

Financial support for this work from the Swedish agencies SIDA/SAREC and VR through the Asian–Swedish research links programme is acknowledged. The authors thank BRNS (DAE), India, for support of this research. AKK wants to thank the University Grants Commission, India, for a fellowship award and Motin for his help during sample preparation.

References

[1] Rao C N R and Raveau B (ed) 1998 *Colossal Magnetoresistance, Charge Ordering and Related Properties of Manganese Oxides* (Singapore: World Scientific)
     Tokura Y (ed) 1999 *Colossal Magnetoresistance Oxides* (London: Gordon and Breach)
[2] Dagotto E (ed) 2003 *Nanoscale Phase Separation and Colossal Magnetoresistance* (Berlin: Springer)

[3] Uehara M, Mori S, Chen C H and Cheong S W 1999 *Nature* **399** 562

[4] Rao C N R 2000 *J. Phys. Chem.* B **104** 5877

[5] Rao C N R, Kundu A K, Seikh M M and Sudheendra L 2004 *Dalton Trans.* **19** 3003
     Rao C N R and Vanitha P V 2002 *Curr. Opin. Solid State Mater. Sci.* **6** 97

[6] Sudheendra L and Rao C N R 2003 *J. Phys.: Condens. Matter* **15** 3029

[7] Rodriguez-Martinez L M and Attfield J P 2000 *Phys. Rev.* B **63** 024424

[8] Deac I G, Diaz S V, Kim B G, Cheong S W and Schiffer P 2002 *Phys. Rev.* B **65** 174426

[9] Freitas R S, Ghivelder L, Damay F, Dias F and Cohen L F 2001 *Phys. Rev.* B **64** 144404

[10] Nam D N H, Mathieu R, Nordblad P, Khiem N V and Phuc N X 2000 *Phys. Rev.* B **62** 1027

[11] Lopez J, Lisboa-Filho P N, Passos W A C, Oritz W A, Araujo-Moreira F M, de Lima O F, Schaniel D and
      Ghosh K 2001 *Phys. Rev.* B **63** 224422

[12] Kundu A K, Seikh M M, Ramesha K and Rao C N R 2005 *J. Phys.: Condens. Matter* **17** 4171 and the references
      therein

[13] Maignan A, Martin C, Hervieu M, Raveau B and Hejtmanek J 1998 *Solid State Commun.* **107** 363

[14] Magnusson J, Djurberg C, Granberg P and Nordblad P 1997 *Rev. Sci. Instrum.* **68** 3761

[15] Binder K and Young A P 1986 *Rev. Mod. Phys.* **58** 801
      Mydosh J A 1993 *Spin Glasses: An Experimental Introduction* (London: Taylor and Francis)

[16] Kundu A K, Nordblad P and Rao C N R 2005 *Phys. Rev.* B **72** 144423

[17] Mathieu R, Nordblad P, Nam D N H, Phuc N X and Khiem N V 2001 *Phys. Rev.* B **63** 174405

[18] Fisher D S and Huse D A 1988 *Phys. Rev.* B **38** 373

[19] Itoh M, Natori I, Kubota S and Matoya K 1994 *J. Phys. Soc. Japan* **63** 1486
      Nam D N H, Jonason K, Nordblad P, Khiem N V and Phuc N X 1999 *Phys. Rev.* B **59** 4189

[20] Kundu A K, Nordblad P and Rao C N R 2006 *J. Solid State Chem.* **179** 923

IV. OPEN-FRAMEWORK AND HYBRID NETWORK MATERIALS

Commentary by C.N.R. Rao

Zeolitic aluminosilicates and aluminophosphates are the most well-known examples of inorganic open-framework solids. In the last few years, a variety of metal phosphates with open architectures have been synthesized and characterized. Some of them exhibit interesting magnetic properties. These include one-, two- and three-dimensional structures. An important aspect of these materials concerns their mode of formation.[1] Are the complex 3D structures formed by making use of secondary structural building units?[2] Is there a progressive building up of structures from 1D to 3D? An attempt to answer such questions is presented in the papers of this section. The section also contains papers on open-framework metal carboxylates which constitute an important new family.[3] Interestingly, even simple metal carboxylates crystallized under appropriate conditions exhibit novel structures. Certain metal oxalates incorporate alkali halides in different forms and such hybrid structures are described.

It has been shown recently that oxyanions such as sulfate and selenite can also be used to design open architectures.[4] More importantly, novel inorganic-organic hybrid materials[5] exhibiting fascinating structures and properties are described in one of the articles. These materials can have differing dimensionalities of the inorganic and organic connectivities. There is much to be done on hybrid-network materials and there is considerable scope for materials design. Organically templated compounds with the Kagome structure possessing unusual magnetic properties are part of this large family. Two articles on Kagome compounds are also presented in this section.

References

1. C.N.R. Rao, in *Frontiers of Solid State Chemistry*, Eds. S.H. Feng and J.S. Chen, World Scientific, Singapore, 2002.
2. R. Murugavel, M.G. Walawalkar, H.W. Roesky and C.N.R. Rao, *Acc. Chem. Res.* **37**, 763 (2004).
3. C.N.R. Rao, S. Natarajan and R. Vaidhyanathan, *Angew. Chem. Int. Ed.* **43**, 1466 (2004).
4. C.N.R. Rao, J.N. Behera and M. Dan, *Chem. Soc. Rev.* **35**, 58 (2007).
5. A.K. Cheetham and C.N.R. Rao, *Science* **318**, 58 (2007).

Acc. Chem. Res. **2001**, *34*, 80–87

Aufbau Principle of Complex Open-Framework Structures of Metal Phosphates with Different Dimensionalities

C. N. R. RAO,*,† SRINIVASAN NATARAJAN,
AMITAVA CHOUDHURY,† S. NEERAJ, AND
A. A. AYI
*Chemistry and Physics of Materials Unit and
CSIR Centre of Excellence in Chemistry, Jawaharlal Nehru
Centre for Advanced Scientific Research, Jakkur P.O.,
Bangalore, 560 064, India*

Received July 25, 2000

ABSTRACT
Open-framework metal phosphates occur as one-dimensional (1D) chains or ladders, two-dimensional (2D) layers, and complex three-dimensional (3D) structures. Zero-dimensional monomers have also been isolated recently. These materials are traditionally prepared by hydrothermal means, in the presence of organic amines, but the reactions of amine phosphates with metal ions provide a facile route for the synthesis, and also throw some light on the mode of formation of these fascinating architectures. Careful studies of the transformations of monophasic zinc phosphates of well-characterized structures show that the 1D structures transform to 2D and 3D structures, while the 2D structures transform to 3D structures. The zero-dimensional monomers transform to 1D, 2D, and 3D structures. There is reason to believe that the 0D monomers, comprising four-membered rings, are the most basic structural units of the open-framework phosphates and that after an optimal precursor state, such as the ladder structure, is formed, further building may occur spontaneously. Evidence for the occurrence of self-assembly in the formation of complex structures is provided by the presence of the structural features of the one-dimensional starting material in the final products. These observations constitute the beginning of our understanding of the building-up principle of such complex structures.

1. Introduction

While supramolecular chemistry of organic compounds has developed to maturity in the past few years,[1] supramolecular inorganic chemistry is somewhat at a nascent stage. Supramolecular design provides a means to generate a variety of novel inorganic materials with complex, unusual structural features in areas such as host–guest chemistry, open-framework structures, and the like. In these compounds, one can visualize structures of different dimensionalities, the dimensionality varying from zero to three. Such complex structures would be expected to have subunits, which not only are structural motifs but also act as building blocks in the building up of the complex structures. Such building units can be considered to be synthons of complex structures. The synthons would be simple geometrical figures such as squares, cubes, or polyhedra, their corners acting as linkage points. The challenge that one faces in supramolecular inorganic materials chemistry is to establish whether such synthons exist in reality and, if so, whether one can demonstrate how they are involved in the formation of the larger structures. Müller et al.[2] have made use of the concepts of supramolecular inorganic chemistry to provide a beautiful description of large assemblies based on the chemistry of polyoxometalates. Recently, Férey[3] has described the concept of building units to understand inorganic solid-state structures, to visualize new topologies, and to conceive the formation of new solids with novel designs. On the basis of such building units, Férey also defines the notion of scale chemistry, which is concerned with the edification of the solids with building units and the consequences it has on the structure of the framework and the voids in them. Of particular interest is the fact that not only are the structures of open-framework compounds topologically interesting, but also the cavities present in them have potential applications. When one looks at the complexity as well as the beauty of the myriad of structures of both open-framework and host–guest compounds, one cannot escape the feeling that the building-up process cannot occur by conventional chemical means alone, involving simple making and breaking of chemical bonds. The formation of such organized structures would be expected to involve self-assembly or some such spontaneous process at a certain stage. In this Account, we demonstrate how the formation of complex open-framework metal phosphates with open architectures involves a building-up process from simple building units, possibly leading to an ultimate step where spontaneous self-assembly occurs.

Several classes of inorganic open-framework structures have been synthesized and characterized in the past several years. While zeolitic aluminosilicates constitute the best-known class of open-framework structures,[4] metal phosphates have been gaining considerable importance, and a variety of metal phosphates with open architectures have been reported in the past decade.[5] The open-framework phosphates are generally synthesized under hydrothermal conditions in the presence of organic

C. N. R. Rao obtained his Ph.D. degree from Purdue University and a D.Sc. degree from the Mysore University. He is Linus Pauling Research Professor at the Centre and Honorary Professor at the Indian Institute of Science. He is a member of several academies, including the Royal Society, London, the U.S. National Academy of Sciences, and the Pontifical Academy of Sciences.

Srinivasan Natarajan obtained his M.Sc. degree from the Madurai Kamaraj University and his Ph.D. from the Indian Institute of Technology, Madras. He did postdoctoral work at the Royal Institution London and at the University of California, Santa Barbara, and is a Faculty Fellow at the Centre.

Amitava Choudhury obtained his M.Sc. degree from North Bengal University and is now a Ph.D. scholar at the Indian Institute of Science.

S. Neeraj obtained his M.S. degree at the Centre and has just completed his Ph.D. work.

A. A. Ayi has an M.Sc. degree from the University of Calabar, Nigeria, where he is a Lecturer. He is on a visiting fellowship of the Third World Academy of Sciences.

* To whom correspondence should be addressed. E-mail: cnrrao@jncasr.ac.in. Fax: 91-80-8462766.
† Also at the Solid State & Structural Chemistry Unit, Indian Institute of Science, Bangalore 560012, India.

10.1021/ar000135+ CCC: $20.00© 2001 American Chemical Society
Published on Web 11/15/2000

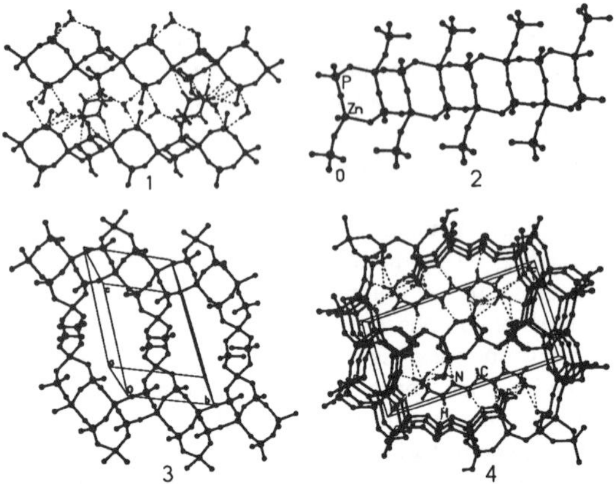

FIGURE 1. Examples of open-framework metal phosphates of different dimensionalities: **1**, one-dimensional (1D) linear chain phosphate, $[C_4N_2H_{10}][Zn(HPO_4)_2]\cdot H_2O$ with piperazine (PIP); **2**, one-dimensional (1D) ladder phosphate obtained with triethylenetetramine (TETA), $[C_6N_4H_{22}]_{0.5}[Zn(HPO_4)_2]$; **3**, two-dimensional (2D) layer phosphate $[C_6N_4H_{22}]_{0.5}[Zn_2(HPO_4)_3]$ with TETA; **4**, three-dimensional (3D) structure with 16-membered channels obtained with TETA. When counting the number of atoms forming a feature such as a ring or a channel, only metal and phosphorus atoms (T atoms) are taken into account.

amines. A noteworthy aspect of the metal phosphates is the occurrence of a hierarchy of structures with different dimensionalities. These include one-dimensional linear chains possessing corner-shared four-membered rings, one-dimensional ladders with edge-shared four-membered rings, two-dimensional layers, and three-dimensional structures with channels.[5,6] In Figure 1, we show typical examples of these structures (**1**−**4**) from the zinc phosphate family. Zero-dimensional metal phosphate monomers comprising four-membered rings have also been isolated recently.[7,8] Four-membered rings are generally the simplest units in the open-framework metal phosphates, and they seem to readily transform to six-membered, eight-membered, and higher membered rings.[9] Among the hierarchy of open-framework structures, the three-dimensional ones are most commonly observed, being associated with greater stability, and the lower dimensional architectures such as the linear chain and the ladder are somewhat rare. There has been considerable effort to understand the processes involved in the formation of open-framework structures.[10] Our knowledge of the mechanism(s) of formation of the beautiful architectures of metal phosphates with varying degrees of complexity, however, remains limited. It is difficult to readily obtain information related to the mechanism(s), partly because the materials are generally prepared under hydrothermal conditions. We know little about the nature of the species in solution or the exact role of the organic amine. The hydrothermal reaction vessel is the proverbial black box. The processes involved are kinetically controlled, and the energies associated with the different structures are likely to be comparable. Quite often, with a single organic amine, one obtains several open-framework phosphates with different structures. There

have been some suggestions with regard to the role of the amine in the formation of these structures.[11] The amine could act as a structure-directing agent or merely fill the available voids and stabilize the structure through hydrogen-bonding and other interactions. Since the amine generally gets protonated in the reaction, it also helps in charge compensation with respect to the framework. Recently, it has been suggested that phosphates of the organic amines may act as intermediates in the formation of the metal phosphates.[12]

In situ synchrotron X-ray diffraction as well as NMR studies of gallium phosphates under hydrothermal conditions have revealed the spontaneous nature of the transformation of preformed precursor units into the open-framework structure, and also the initial formation of a four-membered ring phosphate.[13] It is possible to conceive of basic building units involved in the formation of open-framework metal phosphates, since a variety of complex three-dimensional structures with varying cell sizes are often produced by the arrangement of similar building units.[3] What is of vital importance, however, is to understand the relationship between the structures of different dimensionalities and complexity. For example, it is of value to explore whether it is possible to transform one-dimensional chain or ladder structures to two-dimensional layer or three-dimensional structures under certain conditions. If such transformations do occur, one can rationalize the formation of structures of different dimensionality in terms of a "building-up" (Aufbau) principle. In the case of aluminum phosphates, it has been proposed that a one-dimensional chain, on hydrolysis and rotation of bonds, may transform to higher dimensional structures.[6] In Figure 2, we show a few such transformations schematically, but there has not been sufficient experimental evidence in support of such a hypothesis.

We have been trying to understand the formation of the hierarchy of open-framework structures in metal phosphates by several methodologies. The amine phosphate route[12] has provided some insight into the mode of formation of these materials but does not adequately reveal the nature of interconvertibility of the structures. In this context, our recent finding that a zero-dimensional monomeric zinc phosphate transforms to a layered structure on heating[7] was encouraging and prompted us to explore the transformations of zero-, one-, and two-dimensional structures to higher dimensional ones under different conditions. In this Account, we shall briefly present some of the results obtained by the study of the reactions of organic amine phosphates with metal ions, to show how they throw some light on the formation of open-framework metal phosphates. We shall then discuss the results of our investigations of the transformations of well-characterized zinc phosphates of different dimensionalities and demonstrate how these studies provide valuable insight into the building-up (Aufbau) principle of these complex architectures. In particular, we discuss the transformation of the one-dimensional (1D) ladder structures to two-dimensional (2D) layer and three-dimensional (3D) structures with channels, and the trans-

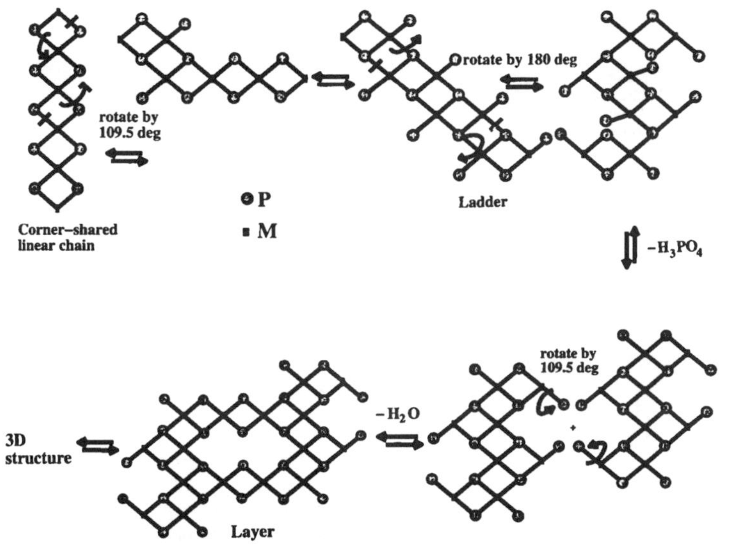

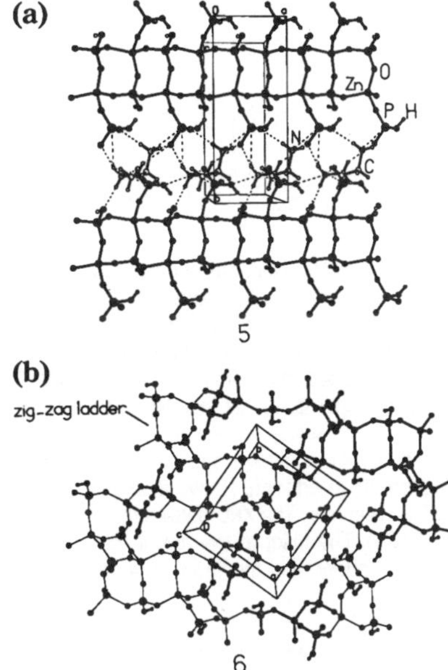

FIGURE 2. Schematic representation of possible types of transformations in open-framework phosphates.

formations of the 2D layer structures to 3D structures. We also examine the transformations of the zero-dimensional monomers to higher dimensional structures. We believe that these studies are of great significance not only to the area of open-framework structures, but also to our understanding of the nature of the processes involved in the building up of complex structures. Studies employing in situ methods and better facilities are necessary to make further progress. As pointed out by Müller et al.,[2] unraveling the Aufbau principle of complex inorganic systems and understanding such systems in terms of the topological classifications based on dimensionality constitute an important direction in the chemistry of materials.

2. Information Revealed by the Reactions of Organic Amine Phosphates

Phosphates of organic amines are often found as byproducts during the hydrothermal synthesis of open-framework metal phosphates. The role of the amine phosphates in the formation of metal phosphates was, however, not clear. Our recent studies of the reactions of well-characterized amine phosphates with metal ions have thrown some light in this regard.[12,14,15] Thus, amine phosphates react with Zn^{2+} ions to yield open-framework metal phosphates of different dimensionalities. What is noteworthy is that many of these reactions can be carried out under ambient conditions, thereby avoiding the hydrothermal route. For example, piperazine phosphate (PIPP) on reaction with Zn^{2+} ions gives the linear-chain phosphate $[C_4N_2H_{10}][Zn(HPO_4)_2]$ (1), formed by corner-shared four-membered rings (Figure 1) at temperatures as low as 85 °C. 1,3-Diaminopropane phosphate (DAPP) on reaction with Zn^{2+} ions gives a ladder phosphate, $[C_3N_2H_{12}][Zn(HPO_4)_2]$ (5), comprising edge-shared four-membered rings (Figure 3) and on prolonged reaction yields a layered structure, $[C_3N_2H_{12}][Zn_2(HPO_4)_3]$ (6). These

FIGURE 3. (a) One-dimensional ladder phosphate with 1,3-diaminopropane (DAP), $[C_3N_2H_{12}][Zn(HPO_4)_2]$ (5), and (b) the two-dimensional layer phosphate, $[C_3N_2H_{12}][Zn_2(HPO_4)_3]$ (6), obtained by the reaction of Zn^{2+} ions with 1,3-diaminopropane phosphate (DAPP). We can see the features of a zigzag ladder in 6.

reactions could be carried out in the 30–50 °C temperature range. The layers in 6 are formed from a zigzag chain of four-membered rings, constructed from two Zn and P atoms ($Zn_2P_2O_4$ units), that are connected to each other via two PO_4 units, creating a bifurcation within the layer.

Reaction of DAPP with Zn^{2+} ions in aqueous solution at 30 °C for 24 h gives a product whose XRD pattern shows lines due to ladder structure 5 ($d_{011} = 9.76$ Å), while the XRD pattern of the product obtained from the reaction

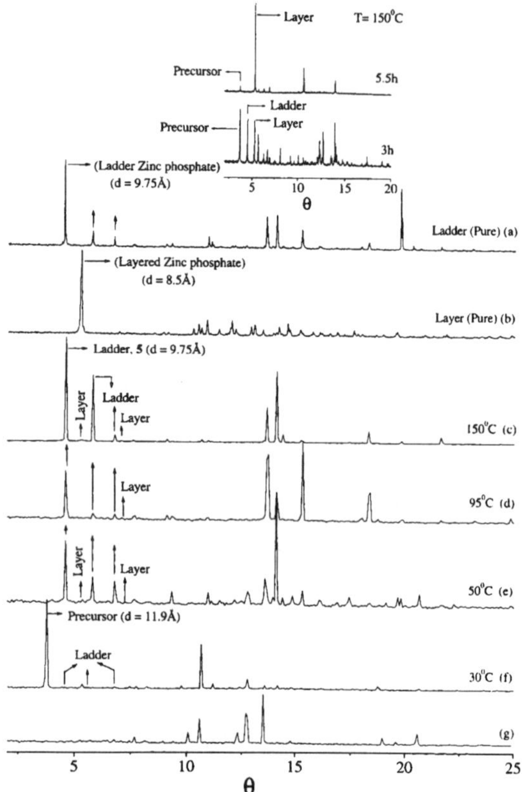

FIGURE 4. X-ray diffraction patterns of the products of the reaction of DAPP with Zn^{2+} ions. (a) XRD pattern of the monophasic zinc phosphate with a ladder structure, and (b) XRD pattern of the monophasic layered zinc phosphate. The diffraction patterns (c)−(f) are those of the products obtained from the reaction of DAPP with Zn^{2+} ions at different temperatures, as indicated (duration of reaction, ∼24 h). Notice the presence of reflections due to the ladder and the layer structures in the patterns and the time evolution of phases. The XRD pattern (f), in addition, shows a unique reflection due to an unidentified precursor. The XRD pattern (g) is that of the amine phosphate (DAPP). The inset at the top of the figure shows the formation of the ladder and the layer structures and their time evolution at 150 °C.

at 50 °C (24 h) shows a reflection due to the ladder structure, **5**, as well as the layered structure, **6** (d_{002} = 8.5 Å), as can be seen from Figure 4. Furthermore, immediately after mixing DAPP with Zn^{2+} ions at room temperature, in addition to the weak line due to the ladder phase, one finds another reflection (d = 11.9 Å) due to an uncharacterized precursor phase (Figure 4). This precursor phase subsequently transforms to the ladder, and finally to the layered structure (see Figure 4), suggesting that the transformation to the layer structure probably occurs through the ladder phase. These results are also supported by in situ ^{31}P NMR studies carried out at 85 °C, which showed the disappearance of the amine phosphate signal followed by the immediate appearance of a signal due to the precursor phase, before the ladder phase is formed. We have followed the time evolution of

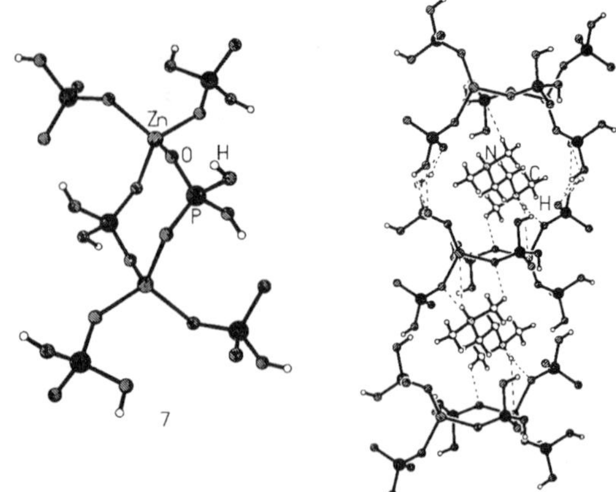

FIGURE 5. Monomeric zinc phosphate possessing a four-membered ring $[C_6N_2H_{18}][Zn(HPO_4)(H_2PO_4)_2]$ (**7**), formed by the reaction of *N,N,N′,N′*-tetramethylethyelenediamine (TMED) with Zn^{2+} ions in the presence of phosphoric acid.

the products of the reaction of DAPP with Zn^{2+} ions by powder X-ray diffraction (XRD). The reaction at 150 °C initially gives a mixture of the ladder structure, **5**, and the layered structure, **6**. After a period of 3 h (see inset Figure 4), reflections due to the various phases can be seen. After 5.5 h, we mainly see reflections due to the layered phase, **6**, suggesting that the ladder, **5**, transforms to the layer, **6**.

We have been able to prepare a four-membered-ring zinc phosphate monomer, $[C_6N_2H_{18}][Zn(HPO_4)(H_2PO_4)_2]$ (**7**) by the reaction of *N,N,N′,N′*-tetramethylethylenediamine phosphate (TMEDP) with Zn^{2+} ions under ambient conditions (Figure 5). The structure consists of four-membered rings formed by ZnO_4 and $PO_2(OH)_2$ tetrahedra, with the $PO_3(OH)$ and $PO_2(OH)_2$ moieties hanging from the Zn center. Having discovered that the amine phosphates react with Zn^{2+} ions in a facile manner, we have employed the amine phosphate route as an effective method for the synthesis of a variety of open-framework metal phosphates.[14−17] Thus, starting with PIPP, we have isolated a large number of open-framework structures with various metal ions.[14]

3. Transformations of One-Dimensional Zinc Phosphates

As mentioned earlier, one-dimensional metal phosphates can possess either linear chain or ladder structures. The linear chain zinc phosphates do not have pendant phosphate groups, and we have found them to be quite unreactive, unlike the one-dimensional ladder structures with pendant phosphate groups (see Figures 1 and 3). We have carried out transformations of the ladder structures **2** and **5** under different conditions.[18] On heating the ladder structure, $[C_6N_4H_{22}]_{0.5}[Zn(HPO_4)_2]$ (**2**), in water (2:H$_2$O = 1:100) at 150 °C for 100 h, we obtained the 3D structure $[C_6N_4H_{22}]_{0.5}[Zn_3(PO_4)_2(HPO_4)]$ (**4**). The structure of **4** is built up of ZnO_4 and PO_4 tetrahedra sharing vertexes, forming a 3D structure (Figure 1), the connectivity between the ZnO_4 and PO_4 units giving rise to 16-membered 1D

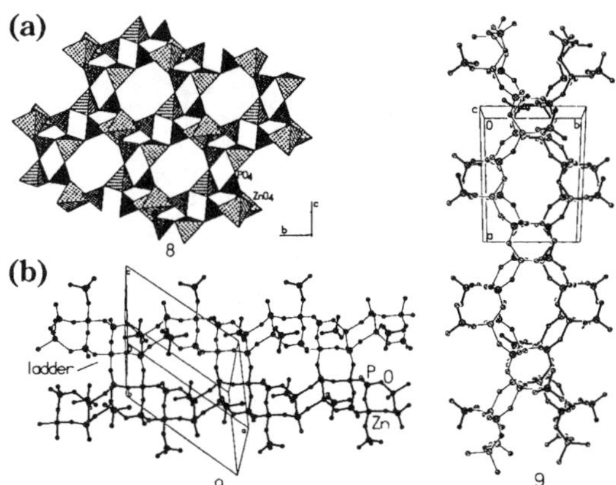

FIGURE 6. 3D zinc phosphates with eight-membered rings obtained from the transformation of the ladder compound **2**: (a) $[C_6N_4H_{22}]_{0.5}$-$[Zn_2(PO_4)_2]$ (**8**) and (b) $[C_2N_2H_{10}][Zn_2(PO_4)_2]$ (**9**). The features of the ladder structure can be clearly seen in the 3D structure of **9**. Such features are also present in **4** and **8**.

channels along the *bc* plane. However, by heating **2** in the presence of a base such as piperazine (PIP) or triethylenetetramine (TETA) at 150 °C in water (2:amine:H_2O = 1:5:100) for relatively short periods (24 h), we obtained another 3D structure, $[C_6N_4H_{22}]_{0.5}[Zn_2(PO_4)_2]$ (**8**) (Figure 6a). The structure of **8** is also built up of ZnO_4 and PO_4 tetrahedra, but the connectivity between these units creates eight-membered channels along all the crystallographic directions. The same reaction carried out at a slightly higher temperature (165 °C) gave a new 3D structure, $[C_2N_2H_{10}]$ $[Zn_2(PO_4)_2]$ (**9**) (Figure 6b). The connectivity between ZnO_4 and PO_4 tetrahedra in **9** results in four-membered rings, which are connected to each other via oxygens. The connectivity between the four-membered rings is such that they form an edge-shared four-ring ladder. The ladders are then connected together within and out of the plane, forming eight-membered channels along the *a* axis. Along the *c* axis, the connectivity between the tetrahedra gives rise to another eight-membered channel. While forming **9**, the amine (TETA) decomposed to ethylenediamine. The 3D structure **4** with 16-membered rings appears to be the more stable phase, since we could transform **8** into **4** by heating it in water at 150 °C under acidic conditions. It is noteworthy that all the three 3D structures, **4**, **8**, and **9**, obtained by the transformation of **2** contain ladder-like features of the starting material.

By heating the ladder structure **2** in the presence of PIP at 165 °C, we were able to obtain a two-dimensional layer compound, $[C_4N_2H_{10}][Zn_2(PO_4)_2]$ (**10**), shown in Figure 7a. The ladder phosphate, **5**, obtained with DAP (Figure 3a) could be transformed to the layered structure, **6** (Figure 3b), by heating it in water (5:H_2O = 1:100) at different temperatures (50–150 °C). We have followed this transformation by recording the X-ray powder diffraction patterns at different times. By heating **5** with zinc acetate in water at 85 °C (5:ZnAce:H_2O = 1:5:100), we obtained the layered phosphate, $[C_3N_2H_{12}][Zn_4(PO_4)_2(HPO_4)_2]$ (**11**), shown in Figure 7b. Both of the layer structures, **10** and

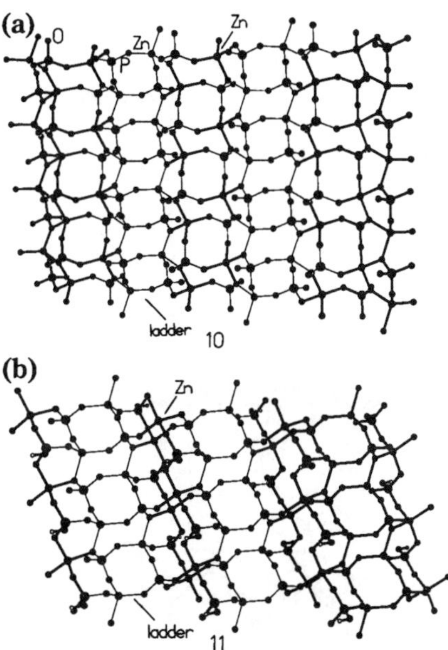

FIGURE 7. (a) 2D layer phosphate, $[C_4N_2H_{10}]$ $[Zn_2(PO_4)_2]$ (**10**), obtained from the transformation of the ladder compound **2**. (b) A 2D layer compound, $[C_3N_2H_{12}][Zn_4(PO_4)_2(HPO_4)_2]$ (**11**), obtained from the transformation of the ladder compound **5**. Both **10** and **11** exhibit features of the ladder structure, from which they are formed.

11, contain ladder-like features, as can be seen from Figure 7. The layers in **11** can be considered to be formed from the ladders in **2**, wherein the pendant groups connect the ladders, mediated by zinc ions, in an "*in-* and *out-of-plane*" fashion to result in chains with alternating three- and four-membered rings running in opposite directions. The connectivity between the chains and the parent ladder results in the formation of a tubular structure with six-membered rings capped by two or three three-membered rings on top and bottom and by two four-membered rings along the direction of the tubule. Thus, we have been able to transform the ladder structures **2** and **5** to 2D layers and 3D structures, all of which possess the structural features of the ladder, indicative of the likely occurrence of self-assembly in the formation of the complex structures.

We attempted to transform linear chain zinc phosphates of type **1** by carrying out reactions under different conditions but were generally unsuccessful. The only case of transformation of a linear chain phosphate was found with $[C_{10}N_4H_{26}][Zn(HPO_4)_2]$ (**12**), prepared with 1,4-bis(3-aminopropyl)piperazine, which on heating in the presence of phosphoric acid at 150 °C (**12**:H_3PO_4:H_2O = 1:5:100) yielded $[C_{10}N_4H_{26}]$ $[Zn_3(PO_4)_2(HPO_4)]$ (**13**) with tubular layers (Figure 8). The tubules themselves are made of "strips" (formed by fusion of two linear chains via a three-coordinated oxygen) in which the three-membered rings get capped by Zn atoms alternately on top and bottom. We have also been able to transform the ladder phosphate **2** to the linear chain phosphate **1** (Figure 1a) by heating it with PIP in water. It appears that the transformation of

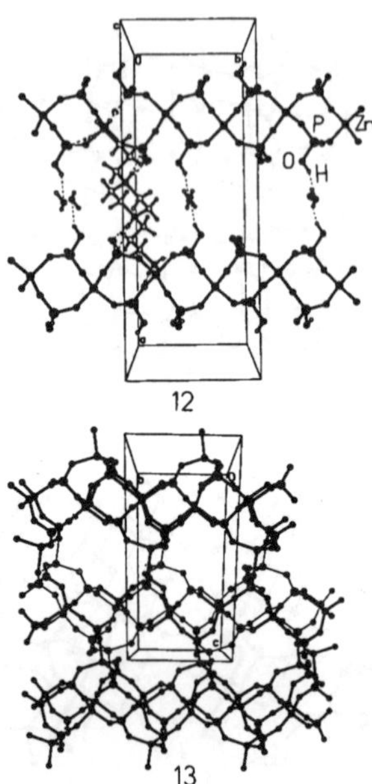

FIGURE 8. Linear chain phosphate, $[C_{10}N_4H_{26}][Zn(HPO_4)_2]$ (12), which transforms to a tubular layered phosphate, $[C_{10}N_4H_{26}]$ $[Zn_3(PO_4)_2$-$(HPO_4)]$ (13). The features of the linear chain can be delineated in the tubular structure of 13.

the ladder to other structures involves the deprotonation of the pendant HPO_4 groups when heated with an organic amine and/or water. In such a reaction, the ladder can go to a higher dimensional or to a one-dimensional linear chain structure where the pendant HPO_4 groups are absent.

4. Transformations of Two-Dimensional Zinc Phosphates

We have carried out the reactions of 2D layered zinc phosphates to see whether they transform to 3D structures. Thus, the layer structure $[C_6N_4H_{22}]_{0.5}[Zn_2(HPO_4)_3]$ (3), on heating in water at 150 °C (3:H_2O = 1:200), gave the 3D structure 4 with 16-membered channels. It must be recalled that we could obtain this 3D structure from the ladder structure, 2, as well. Heating the tubular layer phosphate obtained with TETA, $[C_6N_4H_{22}]_{0.5}[Zn_3(PO_4)_2$-$(HPO_4)]$ (14), at 150 °C in water (14:H_2O = 1:100), produced the 3D structure, 8.

5. Transformations of Zero-Dimensional Monomers

The zero-dimensional monomeric zinc phosphate, 7, in Figure 5, transforms to the layered compound, $[C_6N_2H_{18}]$-$[Zn_3(H_2O)_4(HPO_4)_4]$ (15), shown in Figure 9, on heating in water (7:H_2O = 1:100) at 50 °C. On heating 7 with piperazine in water at 60 °C (7:PIP:H_2O = 1:1:500), we obtained a 3D structure $[C_4N_2H_{10}][Zn(H_2O)Zn(HPO_4)(PO_4)]_2$, which has been described in the literature.[19] An in situ ^{31}P NMR study at 85 °C showed that the intensity of the signal due to the monomer decreases with time, accompanied by an

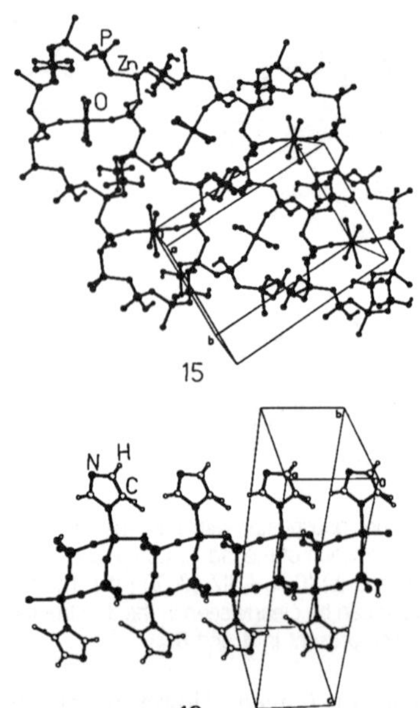

FIGURE 9. Layer phosphate, 15, and ladder phosphate, 16, obtained from the transformation of the zero-dimensional monomer, 7.

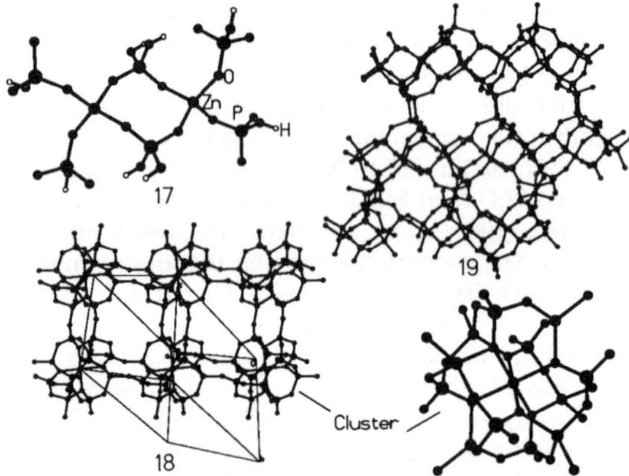

FIGURE 10. Monomeric zinc phosphate, 17, which transforms to the 3D structure, 18, and the tubular layer structure, 19.

increase in the signal due to the ladder structure. The monomer structure disappears after ~4 h, and the intensity of the signal due to the ladder starts decreasing thereafter due to subsequent transformation. Heating the monomer 7 with imidazole in water (7:imidazole:H_2O = 1:1:500) at 60 °C gave a ladder compound of the formula $[C_3N_2H_5][Zn(HPO_4)]$, 16, where the pendant HPO_4 groups are replaced by imidazole molecules (Figure 9).

The monomeric zinc phosphate obtained with tris(2-aminoethyl)amine (TREN), $[C_6N_4H_{21}][Zn(HPO_4)_2(H_2PO_4)]$ (17), on heating in water (17:H_2O = 1:100) at 180 °C, transformed to a 3D structure, $[C_6N_4H_{21}]_{1.33}[Zn_7(PO_4)_6]$ (18), shown in Figure 10. The structure of 18 is made of ZnO_6, ZnO_4, and PO_4 polyhedra, which connect in such a way as to give rise to a Zn_7 cluster. The clusters are arranged

Scheme 1. Various Types of Transformations in Open-Framework Zinc Phosphates: (a) Transformations of One-Dimensional Ladder and 2D Layer Compounds and (b) Transformations of 0D Monomer[a]

(a)

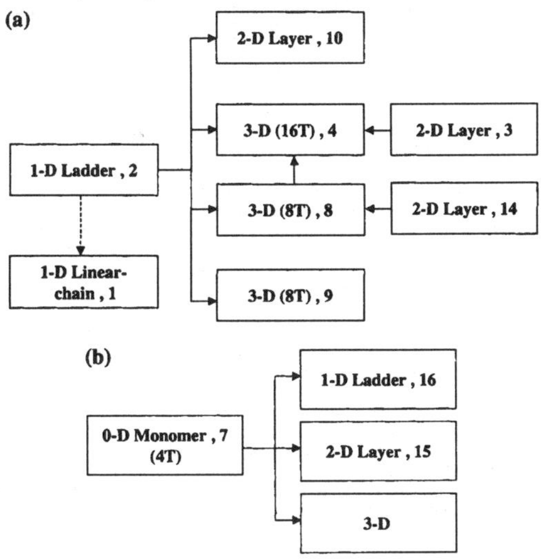

(b)

[a] T refers to tetrahedral framework atoms (Zn or P in the present case).

in such a manner that each cluster is displaced by half the length of *c*-axis from its neighbor, forming a honeycomb network. The next layer of clusters is identical to the first one but is displaced along the *a*-axis by half the unit cell, so that the honeycomb channels get capped. This type of three-dimensional AB-type stacking results in the formation of eight-membered channels along the *b*-axis between two such layers. On reaction of the monomer **17** with zinc acetate in water (**17**:ZnAce:H_2O = 1:1:100), a tubular layer structure of the composition $[C_6N_4H_{21}][NH_4][Zn_6(PO_4)_4(HPO_4)_2]\cdot H_2O$ (**19**) was obtained (Figure 10).

6. Conclusions and Outlook

The transformations of lower dimensional structures to higher dimensional ones in the open-framework zinc phosphates are truly fascinating. In Scheme 1, we summarize the important observations, to highlight the changes in dimensionality and the interconvertibility of structures. Among the zinc phosphates investigated, the one-dimensional ladder structure is the most reactive, rather than the linear chain phosphate. The exact nature of the reactive low-dimensional structure may vary from one system to another, as exemplified by the linear chain gallium fluorophosphate described recently.[20] This one-dimensional gallium fluorophosphate, formed at room temperature, transforms to a 3D structure under hydrothermal conditions. This is comparable to the transformations of the ladder structures **2** and **5**, described in Figures 2 and 6 as well as Scheme 1. The transformations of the monomeric zero-dimensional zinc phosphates comprising four-membered rings to one-, two-, and three-dimensional structures are especially noteworthy, since the four-membered ring is the common structural unit in open-framework metal phosphates. It is possible that the four-membered ring is related to the building units described

by Férey.[3] Thus, the hexameric unit of Férey has two four-membered rings.

Since the four-membered ring appears to be the first unit formed in the process of building of these complex open-framework structures,[13] it is possible that the four-membered rings initially form a one-dimensional chain or a ladder structure, which then transforms to 2D and 3D structures. In principle, we could consider the four-membered rings or/and the one-dimensional structures as synthons of the more complex structures. Self-assembly of ladder structures is suggested by the present study. The spontaneous assembly observed in in situ X-ray diffraction studies[13] could also correspond to the transformation of a 1D or 2D structure to a 3D structure by self-assembly. The formation of six-membered, eight-membered, and higher membered rings, commonly present in the open-framework phosphates, may follow that of the 0D/1D structures, the four-membered rings themselves transforming to the higher rings as suggested in the literature.[9] Although we are not yet in position to provide a definitive mechanism for the formation of complex 3D structures, it seems reasonable to state that the 3D and the 2D structures are likely to be formed through the 0D and 1D precursors. Clearly, there is a need for detailed in situ X-ray diffraction and NMR studies for fully characterizing the precursor states and the transformations. It would be of value to establish the occurrence of progressive 0D−1D−2D−3D transformations. It is therefore necessary to pursue the study of the transformations of well-characterized monophasic compounds of lower (zero,one, and two) dimensionalities.

In our studies of the open-framework zinc oxalates, we have recently isolated monomeric, dimeric, 1D linear chain, 2D layer, and 3D structures by the reaction of amine oxalates with Zn^{2+} ions,[21] suggesting thereby that the presence of a hierarchy of structures is not unique to the phosphates alone. We believe that the evidence provided by our studies for the existence of an Aufbau principle of open-framework complex structures is of considerable significance. Many other complex inorganic structures are also likely to be formed by similar building-up processes, involving basic building units and self-assembly.

References

(1) *Supramolecular Chemistry, Concepts and Perspectives*; Lehn, J.-M., Ed.; VCH: Weinheim, 1995.

(2) Müller, A.; Reuter, H.; Dillinger, S. Supramolecular inorganic chemistry: small guests in small and large hosts. *Angew. Chem., Int. Ed. Engl.* **1995**, *34*, 2328.

(3) Férey, G. Building units, design and scale chemistry. *J. Solid State Chem.* **2000**, *152*, 37.

(4) *Hydrothermal Chemistry of Zeolites*; Barrer, R. M., Ed.; Academic Press: London, 1982. *An Introduction to molecular sieves*; Dyer, A., Ed.; Wiley: Chichester, 1988.

(5) Cheetham, A. K.; Ferey, G.; Loiseau, T. Open-framework inorganic materials. *Angew. Chem., Int. Ed.* **1999**, *38*, 3268.

(6) Oliver, S.; Kuperman, A.; Ozin, G. A. A new model for aluminophosphate formation: Transformation of a linear chain aluminophosphate to chain, layer and framework structures. *Angew. Chem., Int. Ed.* **1998**, *37*, 46.

(7) Neeraj, S.; Natarajan, S.; Rao, C. N. R. Isolation of a zinc phosphate primary building unit [C$_6$N$_2$H$_{18}$]$^{2+}$[Zn(HPO$_4$)(H$_2$PO$_4$)$_2$]$^{2-}$ and its transformation to open-framework phosphate [C$_6$N$_2$H$_{18}$]$^{2+}$[Zn$_3$(H$_2$O)$_4$(HPO$_4$)$_4$]$^{2-}$. *J. Solid State Chem.* **2000**, *150*, 417.

(8) Ayyappan, S.; Cheetham, A. K.; Natarajan. S.; Rao, C. N. R. A novel monomeric Tin(II) Phosphate: [N(C $_2$H$_5$NH$_3$)$_3$]$^{3+}$[Sn(PO$_4$)(HPO$_4$)]$^{3-}$· 4H$_2$O, connected through hydrogen bonding. *J. Solid State Chem.* **1998**, *139*, 207.

(9) Chidambaram, D.; Neeraj, S.; Natarajan, S.; Rao, C. N. R. Open-framework zinc phosphates synthesized in the presence of structure-directing organic amines. *J. Solid State Chem.* **1999**, *147*, 154. Ayyappan, S.; Bu, X.; Cheetham, A. K.; Natarajan, S.; Rao, C. N. R. A simple ladder-tin phosphate and its layered relative. *Chem. Commun.* **1998**, 218.

(10) Thomas, J. M. New Microcrystalline Catalysts. *Philos. Trans. R. Soc. London A*. **1990**, *333*, 173. Davis, M. E. The quest for extra-large pore, crystalline molecular sieves *Chem. Eur. J.* **1997**, *3*, 1745. Francis, R. J.; O'Hare, D. The kinetics and mechanisms of the crystallization of microporous materials. *J. Chem. Soc., Dalton Trans.* **1998**, 3133.

(11) Davis, M. E.; Lobo, R. F. Zeolite and Molecular sieve synthesis. *Chem. Mater.* **1992**, *4*, 756. Férey, G. The new microporous compounds and their design. *C. R. Acad. Ser. Paris Ser. II* **1998**, 1.

(12) Neeraj, S.; Natarajan, S.; Rao, C. N. R. Amine phosphates as intermediates in the formation of open-framework structures. *Angew. Chem., Int. Ed.* **1999**, *38*, 3480.

(13) Francis, R. J.; O'Brien, S.; Fogg, A. M.; Halasyamani, P. S.; O'Hare, D.; Loiseau, T.; Férey, G. Time-resolved in-situ energy and angular dispersive X-ray diffraction studies of the formation of the microporous gallophosphates ULM-5 under hydrothermal conditions. *J. Am. Chem. Soc.* **1999**, *121*, 1002. Taulelle, F.; Haoqs, M.; Gerardin, C.; Estournes, C.; Loiseau, T.; Férey G. NMR of microporous compounds: From in-situ reactions to solid paving. *Colloids Interfaces* **1999**, *158*, 229.

(14) Rao, C. N. R.; Natarajan, S.; Neeraj, S. Exploration of a simple universal route to the myriad of open-framework metal phosphates. *J. Am. Chem. Soc.* **2000**, *122*, 2810.

(15) Rao, C. N. R.; Natarajan, S.; Neeraj, S. Building open-framework metal phosphates from amine phosphates and a monomeric four-membered ring phosphate. *J. Solid State Chem.* **2000**, *152*, 302.

(16) Natarajan, S.; Neeraj, S.; Choudhury, A.; Rao, C. N. R. Three-dimensional open-framework cobalt(II) phosphates by novel routes. *Inorg. Chem.* **2000**, *39*, 1426.

(17) Choudhury, A.; Natarajan, S.; Rao, C. N. R. A layered aluminum phosphate by the amine phosphate by the amine phosphate route. *Int. J. Inorg. Mater.* **2000**, *2*, 87. Natarajan, S.; Neeraj, S.; Rao, C. N. R. Three-dimensional open-framework Co^{II} and Zn^{II} phosphates synthesized via the amine phosphate route. *Solid State Sci.* **2000**, *2*, 87.

(18) The general procedure employed in these studies was to isolate a specific low-dimensional structure (monomer, ladder, or layer), solve its structure by single-crystal X-ray crystallography, and then study its transformation in an appropriate medium at a desired temperature. The medium could be just water or water with additives such as an amine, phosphoric acid, or zinc acetate. After the reaction, the product was analyzed by single-crystal X-ray crystallography.

(19) Feng, P.; Bu, X.; Stucky, G. D. Designed assemblies in open-framework materials synthesis: An interrupted sodalite and an expanded sodalite. *Angew. Chem., Int. Ed. Engl*. **1995**, *34*, 1745.

(20) Walton, R. I.; Millange, F.; Le Bail, A.; Loiseau, T.; Serre, C.; O'Hare, D.; Férey G. The room-temperature crystallization of a one-dimensional gallium fluorophosphate, a precursor to three-dimensional microporous gallium flurophosphates. *Chem. Commun*. **2000**, 203.

(21) Vaidhyanathan, R.; Natarajan, S.; Rao, C. N. R. Synthesis of a hierarchy of open-framework zinc oxalates from amine oxalates, communicated.

AR000135 +

Acc. Chem. Res. **2004**, *37*, 763–774

Transformations of Molecules and Secondary Building Units to Materials: A Bottom-Up Approach

R. MURUGAVEL,*,† M. G. WALAWALKAR,‡
MEENAKSHI DAN,§ H. W. ROESKY,*,⊥ AND
C. N. R. RAO*,§

*Department of Chemistry, IIT-Bombay, Powai,
Mumbai-400 076, India, Applied Chemistry Division,
Mumbai University Institute of Chemical Technology,
Matunga, Mumbai-400 019, India, Chemistry and Physics of
Materials Unit, Jawaharlal Nehru Center of Advanced
Scientific Research, Jakkur, Bangalore-560 064, India,
Solid State and Structural Chemistry Unit,
Indian Institute of Science, Bangalore-560 012, India, and
Institute of Inorganic Chemistry, University of Göttingen,
D-37077 Göttingen, Germany*

Received March 18, 2004

ABSTRACT

A variety of complex inorganic solids with open-framework and other fascinating architectures, involving silicate, phosphate, and other anions, have been synthesized under hydrothermal conditions. The past few years have also seen the successful synthesis and characterization of several molecular compounds that can act as precursors to form open-framework and other materials, some of them resembling secondary building units (SBUs). Transformations of rationally synthesized molecular compounds to materials constitute an important new direction in both structural inorganic chemistry and materials chemistry and enable possible pathways for the rational design of materials. In this article, we indicate the potential of such a bottom-up approach, by briefly examining the transformations of molecular silicates and phosphates. We discuss stable organosilanols and silicate secondary building units, phosphorous acids and phosphate secondary building units, di- and triesters of phosphoric acids, and molecular phosphate clusters and polymers. We also examine the transformations of metal dialkyl phosphates and molecular metal phosphates.

1. Introduction

In 1756, Cronstedt discovered a class of minerals that on heating produced steam and called them *zeolites* from Greek, meaning *boiling stone*.[1,2] This was followed by the observations of Damour who noticed in 1840 that zeolites could be reversibly dehydrated and of Grandjean (1909)

who reported that dehydrated zeolites adsorb small molecules. Weigel and Steinhoff found that the zeolite *chabazite* adsorbs methanol, ethanol, and formic acid but not acetone, ether, and benzene. MacBain interpreted this observation in 1926 in terms of molecular size differentiation at about 5 Å due to dehydration of the zeolites without the loss of the space lattice. Milton produced the first man-made zeolite in 1949.[1,2] Since then, zeolite science has emerged as one of the important areas attracting the attention of chemists, engineers, materials scientists, and physicists. Metal phosphates, on the other hand, exist in nature as minerals or as constituents of living animals and play an important role in energy conversion cycles. From a structural viewpoint, metal phosphates are similar to silicates and for this reason, studies on metal phosphate and related phosphonate materials have received considerable attention.

The search for new zeolite-like structures was initially extended to aluminophosphate-based molecular sieves,[3] and these explorations produced a variety of exotic compounds with open-framework structures, which include besides metal phosphates,[4] carboxylates,[5] sulfates,[6] selenites, and selenates.[7] Growth of this area has been rapid in the past decade necessitating new editions of the *Atlas of Zeolites*.[2] It is noteworthy that metal–organic framework (MOF) solids have been included as zeolite types in the latest edition of the *Atlas of Zeolites*.

While numerous framework structures incorporating a variety of metals and nonmetals have been assembled under laboratory conditions and structurally characterized and their properties have been explored, the present synthetic methodologies available for their preparation (viz., solvothermal and hydrothermal synthesis) do not offer much control over the final products. Typically, a metal source, a ligand, and an organic or inorganic structure-directing template are heated together in a sealed vessel for several days and cooled slowly to crystallize the framework solid. Although some generalizations can be made with regard to these reactions and to predicting the role played by the structure-directing

* Corresponding author. Fax: +91-080-23622760. E-mail: cnrrao@jncasr.ac.in (C. N. R. Rao).
† IIT-Bombay.
‡ Mumbai University Institute of Chemical Technology.
§ Jawaharlal Nehru Centre for Advanced Scientific Research and Indian Institute of Science.
⊥ University of Göttingen.

Ramaswamy Murugavel is an Associate Professor of Chemistry at the Indian Institute of Technology-Bombay and has recently received the Swarnajayanti Fellowship from Government of India for his work on synthetic inorganic chemistry applied to materials science and catalysis.

Mrinalini Walawalkar is a Reader at the University Institute of Chemical Technology, Mumbai. Previously she was a Scientific Officer at IIT-Bombay and an Alexander-von-Humboldt Fellow at Bochum and Göttingen Universities. Her current research interests are in the area of inorganic materials.

Meenakshi Dan is a student of the Integrated Ph.D. program at the Indian Institute of Science, Bangalore. She has a M.S. degree in Chemistry from the Indian Institute of Science and works on open-framework materials.

Herbert Roesky is a Professor and Director at the Institute of Inorganic Chemistry at the University of Göttingen. He is a recipient of ACS Award for Creative Work in Fluorine Chemistry, ACS Award for Inorganic Chemistry, and Wilkinson Prize for Creativity in Inorganic Chemistry. He is a foreign member of the Indian Academy of Sciences and the French Academy.

C. N. R. Rao is Linus Pauling Research Professor at the Jawaharlal Nehru Centre for Advanced Scientific Research and Honorary Professor at the Indian Institute of Science. He is a member of several academies including the Royal Society, London, the U. S. National Academy of Sciences, the French Academy, the Japan Academy, and the Pontifical Academy of Sciences. He has received several awards including the Hughes medal from the Royal Society, Einstein Gold medal of UNESCO, and Somiya award of IUMRS.

10.1021/ar040083e CCC: $27.50 © 2004 American Chemical Society
Published on Web 08/06/2004

Chart 1. Schematic Representation of the Formation of Zeolites from TO₄ Units

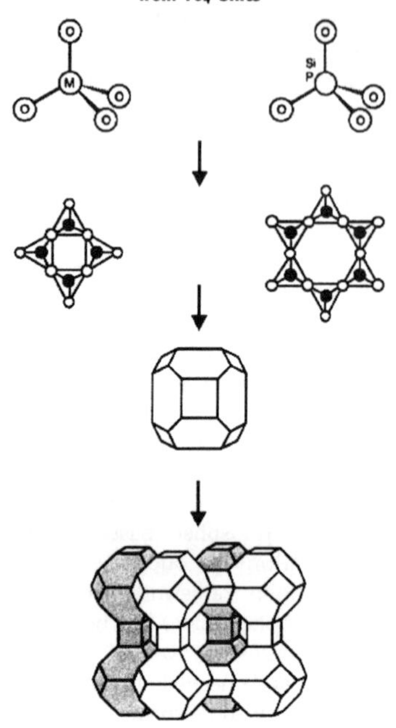

Chart 2. Tetrahedral Primary Building Units

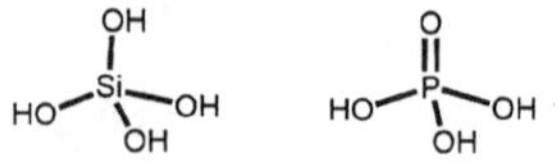

Chart 3. Silanols and Phosphorous Acids Used as Building Blocks for the Synthesis of Model Compounds

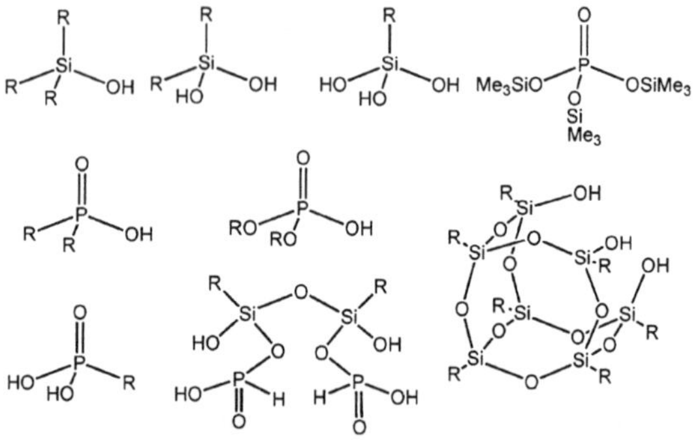

discussion to silicates and phosphates due to the intense activity on these materials both at the molecular and extended framework levels. For the sake of completeness, a brief discussion on the conversion of the molecular compounds to condensed solids and inorganic–organic hybrid materials with potential use as solid-state catalysts is also included.

2. Stable Organosilanols and Silicate SBUs

While $Si(OH)_4$ and $P(O)(OH)_3$ could serve as ideal starting materials for the incorporation of $SiO_4{}^{4-}$ and $PO_4{}^{3-}$ anions (Chart 2), their use in synthesizing molecular siloxanes and phosphates is somewhat limited since their synthesis is conducted in a homogeneous organic medium. For example, orthosilicic acid is metastable and not available in a free state. Even if $Si(OH)_4$ could be made available, its insoluble nature precludes its chemistry from being explored in solution. Similar problems are encountered in using anhydrous $P(O)(OH)_3$. Synthetic chemists have sought to use alternative starting materials for the incorporation of Si–O and P–O linkages. This has led to the use of a variety of organosilanols,[9] esters of orthosilicic acid, phosphonic acids, and esters of orthophosphoric acid as starting materials (Chart 3).

Although the chemistry of metal-containing molecular silicates (*metallosiloxanes*) is more than a century old,[10] their use as model compounds for zeolites and as precursors for the preparation of metallosilicate materials has gained importance only in the past decade.[11] While the early work on metallosiloxanes provided insights into the nature of Si–O–M linkages, its direct application to zeolites was somewhat remote since monosilanols, [R₃Si-(OH)], produce metallosiloxanes, [M(OSiR₃)ₙ], that barely resemble any of the SBUs of zeolites.[12] It is no wonder that these compounds have often been referred to as a special class of alkoxides rather than molecular silicates. During the 1970s and 1980s, the synthesis of cyclic metallosiloxanes was reported starting from silanediols of

templates to some extent, it is not yet possible to choose a given set of reactants and reaction conditions and predict *before-hand* the major product that would form in preference to other possible products. In other words, although framework solids and synthetic zeolites are essentially formed by metal ions in association with anions such as $PO_4{}^{3-}$ and $SiO_4{}^{4-}$ (Chart 1), it is not yet possible to drive these reactions toward a particular framework solid.

The lack of suitable starting materials, combined with the fact that the reactions are generally carried out in a heterogeneous medium, explains the difficulties faced in designing framework solids. In light of recent developments in the molecular chemistry of siloxanes and phosphates, it is beginning to appear that routes to *designer framework solids* may indeed become possible soon. While there have been numerous literature reports of both hydrothermal synthesis of framework solids and laboratory preparation of model compounds often soluble in organic solvents,[3–8] there have been very few attempts to convert molecular compounds into framework structures through simple chemical transformations. The objective of this Account is to highlight some of the recent developments in the area of molecular compounds that resemble several of the secondary building units (SBUs) identified in extended structures[8] and demonstrate possible strategies and some successes wherein these small molecular units are transformed into framework solids or vice versa. In doing so, it is by no means possible to cover all aspects of this burgeoning area, and we shall therefore refer to pertinent reviews describing materials and structures and limit ourselves solely to the theme relating to molecule/SBU to materials transformations. We have restricted the

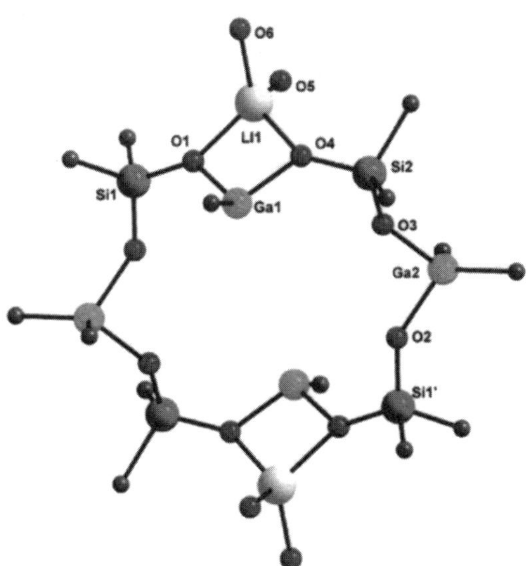

FIGURE 1. Molecular structure of the core of the lithium gallium siloxane showing the S8R.

Chart 4. Organic Soluble Molecular Silicates Derived from RSi(OH)₃

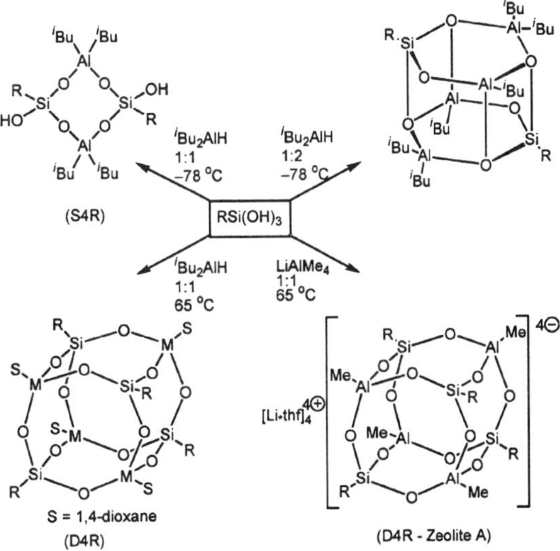

Chart 5. Interesting Metallosiloxanes Derived from RSi(OH)₃

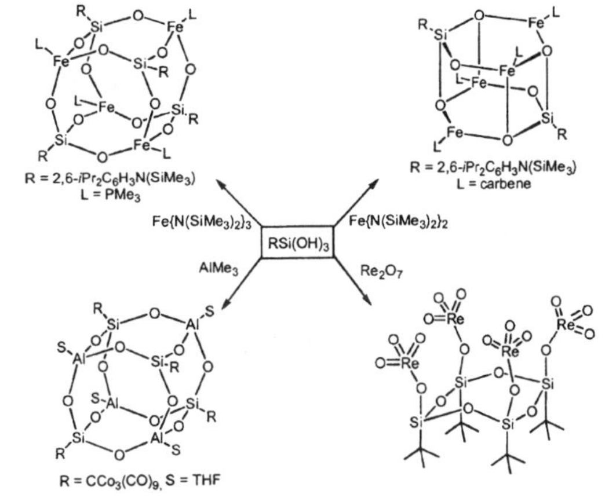

the type [R₂Si(OH)₂] and [HOSiR₂−O−SiR₂OH].[13] We recently demonstrated that even a simple silanediol such as Ph₂Si(OH)₂ could be used for the preparation of the only known soluble compound with a single-8-ring (S8R) motif from its reaction with LiGaMe₄ (Figure 1). Interestingly, the cavity formed inside the 8-ring is large enough in this case to accommodate a (LiOH)₂ unit.[14]

Modern metallosiloxane chemistry *mainly* deals with three types of silanols: silanetriols,[11] incompletely condensed silasesquioxanes,[15] and tri(*tert*-butyl) ester of orthosilicic acid.[16] Research on these compounds has contributed to the development of model compounds for zeolites, new homogeneous catalysts, and low-temperature pathways to metal−silica materials.

Silanetriols. A major break-through in silanol research occurred in the 1990s with reports appearing on the high yield synthesis of silanetriols, [RSi(OH)₃].[17] Silanetriols were traditionally considered to be highly unstable due to the well-known self-condensation reactions through which they form condensed siloxanes. However, the use of a sterically demanding and kinetically stabilizing substituent on silicon allowed the isolation of silanetriols in multigram quantities. For example, the silanetriol [(2,6-*i*Pr₂C₆H₃)(SiMe₃)NSi(OH)₃], is a unique material that is stable in air for at least 3 years![11] It is lipophilic and soluble in a variety of organic solvents. While the bulky 2,6-diisopropylphenyl group no doubt provides the kinetic impedance to Si−OH condensation, the role played by the SiMe₃ group in stabilizing this silanetriol is also crucial. This silanetriol has served as a versatile starting material for the synthesis of more than 100 metallosiloxanes, highlighting the use of these compounds as models for metal-embedded silica surfaces and single-source precursors for new catalytic materials. It is interesting that the reactions can often be fine-tuned by paying careful

attention to the nature of the metal precursor, the reaction temperature, and the polarity of the medium. For example, the reactions of RSi(OH)₃ with organoaluminum reagents proceed at temperatures between −78 and 80 °C and produce cyclic and cage-soluble aluminosiloxanes, which model zeolites (Chart 4).[18] The structural varieties that can result from silanetriol and metal precursor reactions are shown in Chart 5.[19]

The successful preparation of exceptionally stable and soluble organosilicon trihydroxides has led to the development of exciting new chemistry, and clearly there are several opportunities for further work, especially in the synthesis of framework structures starting from the SBUs shown in Charts 4 and 5.

Silsesquioxanes. At the same time as silanetriols were stabilized, another significant observation was made in the hydrolysis of silyl trihalides. Incompletely condensed silsesquioxanes, R₇Si₇O₉(OH)₃ (Chart 6) are formed along with other products during the hydrolysis of RSiCl₃ (R =

Chart 6. Silasesquioxanes Derived from (cy)$_7$Si$_7$O$_9$(OH)$_3$

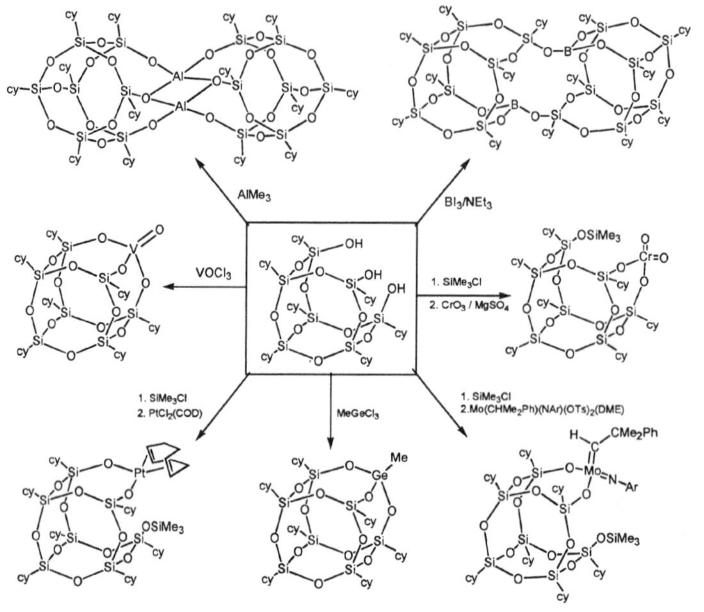

Chart 7. Thermally Unstable Metal Siloxides Derived from Tri(*tert*-butoxy)silanol

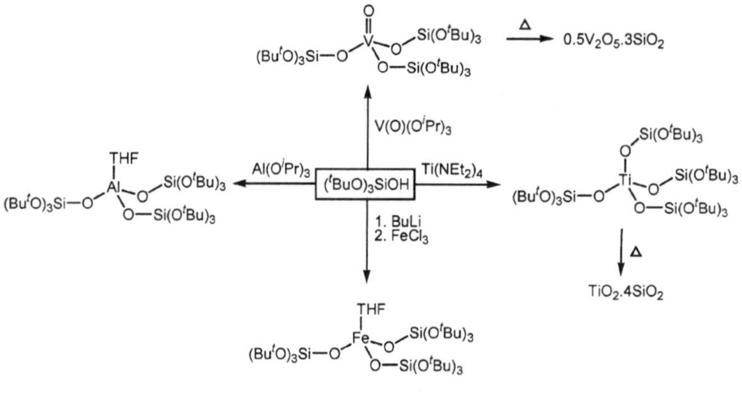

cyclohexyl) in commercial acetone.[20] Silsesquioxanes, R$_7$-Si$_7$O$_9$(OH)$_3$, by the virtue of having three Si—OH hydroxyl groups pointing toward an open corner of a cube, are very important as model compounds for silica surfaces and also as starting materials for the preparation of metallasiloxanes through corner-capping reactions (Chart 6).[15] Some of these metallasilsesquioxanes are found to be useful as catalysts in organic transformations.[21] These aspects have been discussed elsewhere[15,21] and will not be elaborated here.

Silicate and Hybrid Materials through Molecular Siloxanes. Renewed interest in monosilanols and silanediols is primarily because compounds obtained from [(*t*BuO)$_3$SiOH] and [(*t*BuO)$_2$Si(OH)$_2$] serve as better single-source precursors for ceramic materials. A number of metallasiloxanes have been synthesized starting from [(*t*BuO)$_3$SiOH], and a selection of these compounds is

shown in Chart 7. Most of these metallosiloxanes serve as excellent precursors for the preparation of condensed-phase metallosilicate materials (or metal oxide/silica composites) with precise stoichiometries.[16,22,23] For example, direct thermolysis of the titanosiloxane [Ti{(OSi(O*t*Bu)$_3$)$_4$}] results in the removal of organic material through β-hydride elimination and condensation of hydroxyl groups to result in TiO$_2$:4SiO$_2$. Using similar strategies, oxides of vanadium, chromium, molybdenum, tungsten, zinc, manganese, and several other metals have been stabilized in a silica matrix.[23] Aluminosilicates have been synthesized using this methodology starting from [Al{(OSi(O*t*Bu)$_3$)$_3$}]. Several of the metallosilicate materials prepared through this pathway are useful catalysts in organic transformations.[24]

An interesting variant of metal—tri(*tert*-butoxy)siloxide chemistry was described recently. Thus, the thermal

Chart 8. Conversion of Zr[OSi(OtBu)$_3$]$_4$ into Ceramic and Inorganic/Organic Hybrid Materials

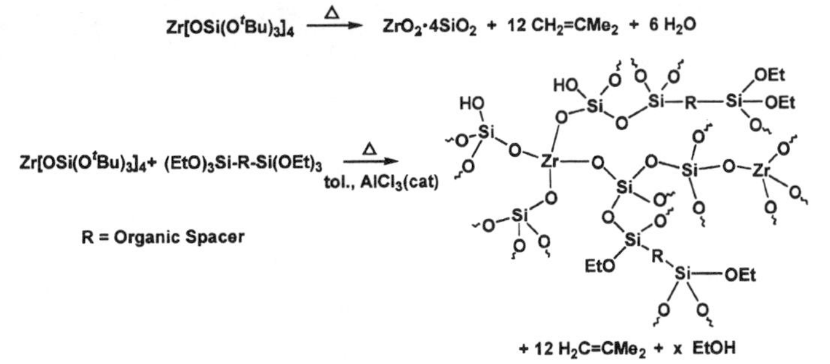

$$Zr[OSi(O^tBu)_3]_4 \xrightarrow{\Delta} ZrO_2 \cdot 4SiO_2 + 12\ CH_2{=}CMe_2 + 6\ H_2O$$

Zr[OSi(O^{t}Bu)$_3$]$_4$ + (EtO)$_3$Si-R-Si(OEt)$_3$ $\xrightarrow[\text{tol., AlCl}_3\text{(cat)}]{\Delta}$

+ 12 H$_2$C=CMe$_2$ + x EtOH

R = Organic Spacer

Chart 9. Silanetriolate Anion vs Phosphonate Anion

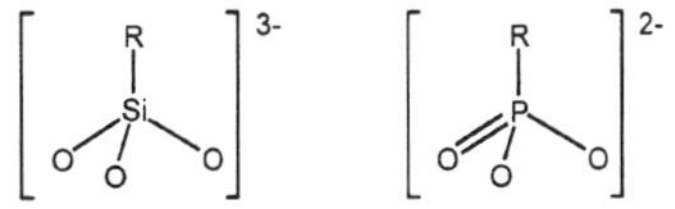

decomposition of Zr[OSi(OtBu)$_3$]$_4$ occurs in toluene or in the solid state under mild conditions (150 °C) to give (ZrO$_2$·4SiO$_2$) with all of the component elements originating from a single molecular species (Chart 8). On the other hand, the cothermolytic synthesis of the hybrid network materials can be carried out in toluene at 155 °C in the presence of AlCl$_3$ catalyst.[25] It is postulated that the condensation of Zr[OSi(OtBu)$_3$]$_4$ with (EtO)$_3$Si−R−Si(OEt)$_3$ proceeds through the initial thermal decompositon of Zr-[OSi(OtBu)$_3$]$_4$ to give Zr[OSi(OH)$_3$]$_4$. The reactive Si−OH groups then self-condense or condense with Si−OEt groups from (EtO)$_3$Si−R−Si(OEt)$_3$ to form an inorganic/ organic network material via the concomitant elimination of ethanol. Gel formation occurs rapidly within ca. 10 min and affords high-surface-area xerogels upon drying. The as-synthesized materials are highly functionalized with surface OH groups up to 4.8 sites nm^{-2}. This method offers an opportunity to synthesize a wide variety of hybrid materials that combine transition metals with diverse organic spacers in an organic medium excluding water.

3. Phosphorus Acids and Phosphate SBUs

Unlike the silanols, alkyl phosphorus acids, such as phosphinic acids [R$_2$P(O)(OH)] and phosphonic acids [RP-(O)(OH)$_2$] (Chart 2), are easier to prepare and handle under laboratory conditions.[26] Metal phosphates derived from these acids are similarly more stable than the corresponding metallasiloxanes. This has resulted in an outburst of activity in building molecular phosphinates and phosphonates that often resemble the smaller secondary building units of zeolites and other framework solids, apart from the synthesis of a large number of layered inorganic phosphonates. The chemistry of hundreds of layered metal phosphonates synthesized from phenylphosphonic acid, alkyl and aryl diphosphonic acids, and phosphonic acids containing additional carboxylic acid functional groups has been documented well in the literature in the last three decades.[26] Structures of these phosphonates do not resemble the SBUs of framework phosphates, nor are they soluble in organic solvents for further reaction chemistry. We will focus our attention here on soluble cyclic and cage-like organic phosphinates and phosphonate molecules, which resemble SBUs of framework solids.

Phosphonic Acids. Phosphonic acids, RP(O)(OH)$_2$, when deprotonated afford RPO$_3{}^{2-}$ which, in principle, should behave similarly to the silanetriols, RSi(OH)$_3$ (Note that SiO$_2$ is isoelectronic to AlPO$_4$) (Chart 9).[26] For reasons of solubility, most of the efforts to generate SBUs have been carried out with *tert*-butylphosphonic acid, although reactions with phenyl- and methylphosphonic acid have also been examined. Thus, the reaction of t-BuP(O)(OH)$_2$ with any group 13 trialkyl leads to a three-dimensional cage with a double-4-ring (D4R) core (cf. Chart 4)[27] as in the case of reactions of silanetriols with triaalkylalumnium. While in compounds derived from the phosphonic acids one alkyl group is intact on each heteroatom, the fourth coordination site is a solvent molecule around the heteroelement in the analogous siloxane cubes.

There appears to be a steric control in determining whether a cubic cage (D4R) is always formed in these. A report by Mason suggests that in the reaction of bulky t-Bu$_3$Ga with phenylphosphonic acid, it is possible to obtain both an eight-membered ring compound (single-4-ring, S4R; Chart 10, structure **A**) and a cubic cage (D4R).[28] The reaction of Me$_3$Al with *tert*-butylphosphonic acid underscores such steric control.[29] In this reaction, apart from the cubic tetrameric aluminophosphonate **B** (D4R), a hexameric drum-like product with a double-6-ring (D6R) core (Chart 10, structure **C**) is obtained in considerable yields. While in the former case the bulkiness of the *tert*-butyl group on gallium is responsible for the formation of the S4R structure, in the latter the rather small size of the methyl group on aluminum is the main cause of the ready assembly of the D6R structure. The reactions of Cp*TiMe$_3$ with alkyl and aryl phosphonic acids offer easy access to novel oxygen bicapped distorted D4R structures (Chart 10, structure **D**).[30]

Ionic phosphonates, which contain loosely bound alkali metal ions, are of interest from the point of view of modeling zeolites containing dissolved metal ions displaying conductive hyperlattices. Reactions of tBuP(O)(OH)$_2$ with alkali metal salts of tetraalkylgallates provide a convenient route to ionic phosphates (Chart 11).[31] A

Chart 10. Neutral Phosphonates Mimicking S4R, D4R, D6R, and Bicapped D4R SBUs

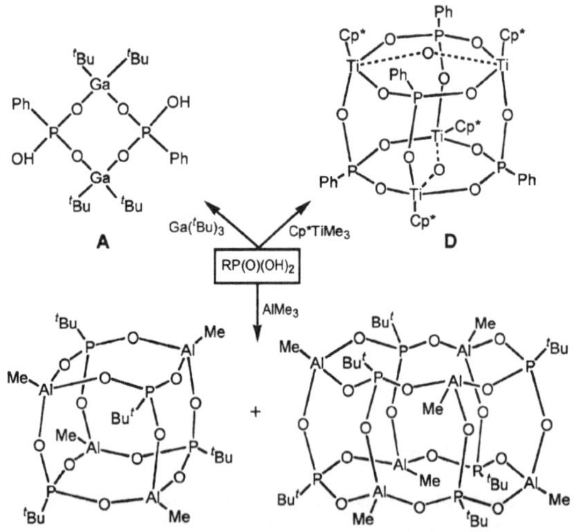

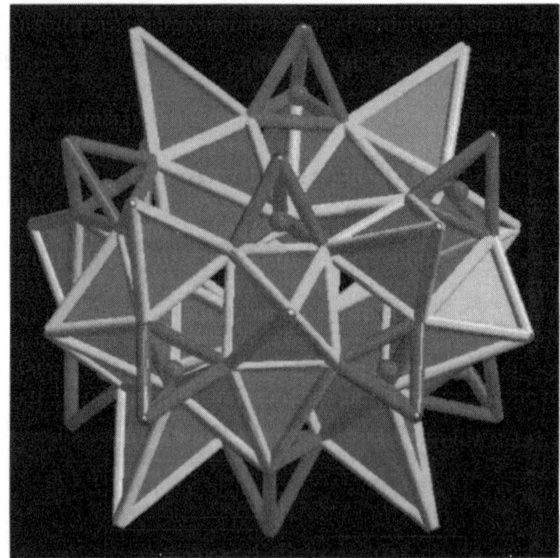

FIGURE 2. Polyhedral arrangement in $[(MeCd)_{10}((THF)Cd)_4Cd_6(\mu_4\text{-}O)_2(\mu_3\text{-}OH)_2(tBuPO_3)_{12}]$ showing cadmium (closed polyhedra with green interior and yellow edges) and phosphorus (open polyhedra in pink) atoms.

notable feature is that the size of the alkali metal ions regulates the final structure. Thus, the lithium compound is made of two 12-membered gallophosphonate bowls, which sandwich four lithium ions arranged in the form of a one-dimensional wire.[31a] The gallophosphonates derived from both $NaGaMe_4$ and $KGaMe_4$ are essentially made up of an eight-membered gallophosphonate bowl, which surrounds an ensemble of Na or K ions. Because of the larger size of Na and K ions, they can accommodate higher coordination numbers and form agglomerates rather than the linear wire as in the lithium case.[31b]

Nanosized Metal Phosphonate Aggregates. Several phosphonate clusters of a very large size have been synthesized in recent years. For example, the use of the organometallic precursor $Cs[AliBu_3F]$ has allowed the synthesis of a new supercluster $[Cs_3(THF)_3F(iBuAl)_3$-$(tBuPO_3)_4]_2[(iBuAl)_2Al_2(\mu\text{-}F)_2(tBuPO_3)_4]$,[32] which contains several structural features closely related to the motifs present in layered and three-dimensional alumino- and gallophosphates (C6R or 6≡1 found in BPH and AFS types of layered aluminophosphates). The reaction of Et_2Zn with $tBuP(O)(OH)_2$ leads to a rugby-ball shaped dodecanuclear Zn-cluster, $[Zn_4(\mu_4\text{-}O)(Zn\cdot THF)_2(ZnEt)_6(tBuPO_3)_8]$.[26] The reaction of Me_2Cd with $tBuP(O)(OH)_2$ solution yields a still larger aggregate, $[(MeCd)_{10}((THF)Cd)_4Cd_6(\mu_4\text{-}O)_2(\mu_3\text{-}OH)_2$-$(tBuPO_3)_{12}]$, which contains 68 heteroatoms arranged in the form of a hollow sphere (Figure 2).[33] This cluster contains a skeleton made up of 20 cadmium atoms and 12 phosphorus atoms held together by 36 surface oxygen atoms along with a pair of oxo and hydroxo groups in the interior of the sphere (Figure 2). The inorganic core is embedded in the organic sheath made up of 10 methyl groups, four THF, and 12 *tert*-butyl groups. Within the inorganic core, the diagonally opposite Cd−Cd and P−P distances are 10.9 and 9.8 Å, respectively. This giant molecule can encapsulate three C_{60} molecules.

It is possible to use ancillary ligands in addition to phosphonic acids in building up nanosized cluster compounds of late transition metal ions. Thus, the reaction of $CuCl_2$ with *tert*-butylphosphonic acid in the presence of 3,5-dimethylpyrazole affords a dodecanuclear copper phosphonate with an interesting cage structure.[34] Similarly, large vanadium phosphonate clusters with up to 18 vanadium atoms have been assembled from phosphonic acids.[35]

4. Phosphoric Acid Esters as Primary Building Units

Much of the work on the use of phosphate esters for building up molecular metallaphosphates has emerged only in the past decade. While di(*tert*-butyl) phosphate (dtbp-H) has been the most widely used,[36,37] studies have also been carried out with the tris(trimethylsilyl) ester, $(Me_3SiO)_3P{=}O$.[38] Compared to phosphonic acid chemistry, the chemistry of phosphate esters assumes importance, because of the presence of the reactive/thermally labile P−OR linkages relative to the stable P−C linkages in the phosphonic acids. Although phosphonic acids provide model compounds that can act as SBUs of zeolite structures and nanoscopic clusters, expansion of these SBUs and clusters to framework solids would require further functionalization of the alkyl groups on phosphorus. Conversely, the generation of P−OH groups on the surface of the molecular cluster formed by phosphorus esters is a relatively easier task and is achieved either by β-hydride elimination of P−O*t*Bu groups or by the facile hydrolysis of P−OSiMe_3 groups in solution. Several metal phosphate molecules have been synthesized starting from these two

Chart 11. Anionic Gallium Phosphonates

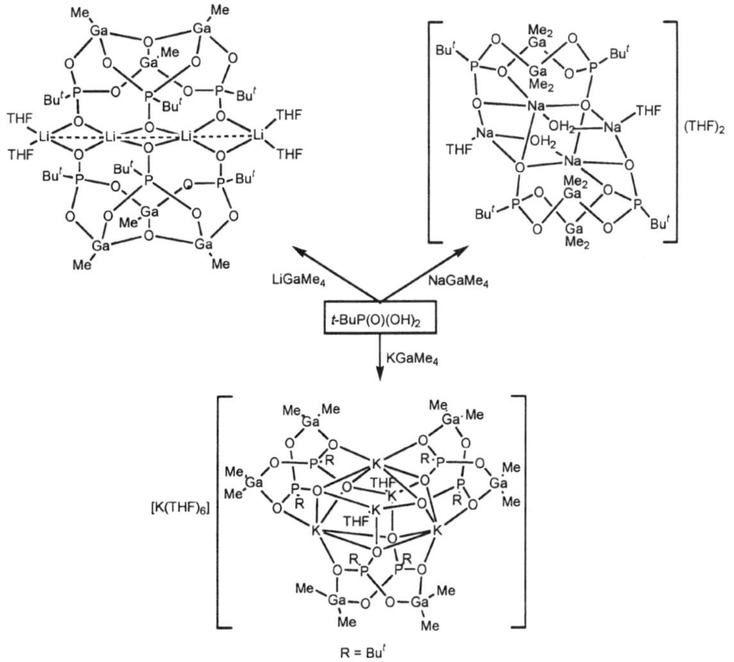

t-BuP(O)(OH)₂

KGaMe₄

Chart 12. Complexes of Di(*tert*-butyl) Phosphate with Aluminum

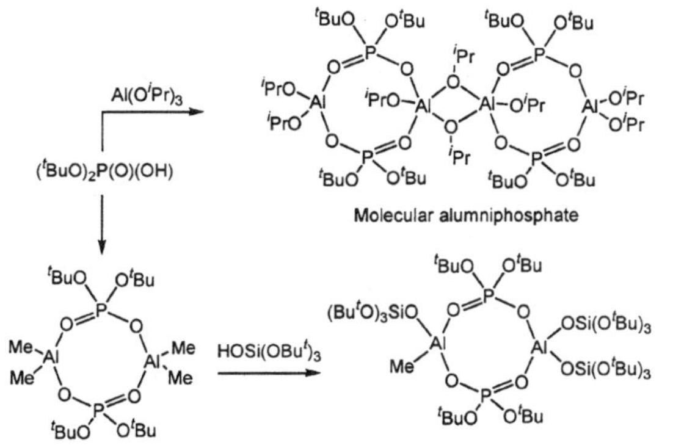

Al(O*i*Pr)₃

(*t*BuO)₂P(O)(OH)

HOSi(OBu')₃

Chart 13. Synthesis of Tetranuclear Phosphates

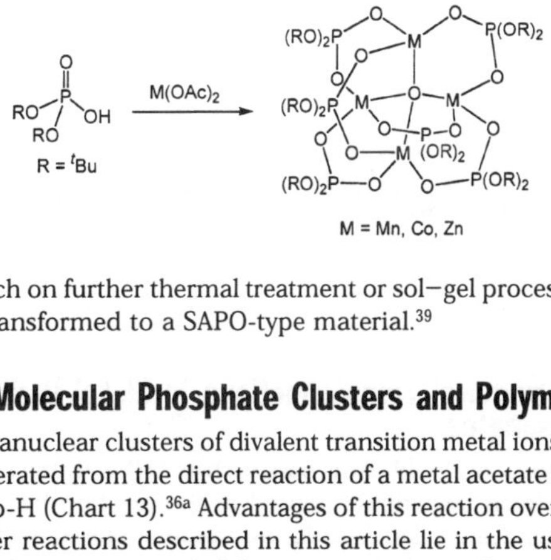

which on further thermal treatment or sol–gel processing is transformed to a SAPO-type material.[39]

5. Molecular Phosphate Clusters and Polymers

Tetranuclear clusters of divalent transition metal ions are generated from the direct reaction of a metal acetate with dtbp-H (Chart 13).[36a] Advantages of this reaction over the other reactions described in this article lie in the use of metal acetates instead of the more expensive and moisture-sensitive metal alkyls and the use of methanol as a solvent in aerobic conditions. It is to be noted, however, that organometallic routes can also be used to synthesize these tetrameric compounds.[37a] Small changes in the reaction conditions and the use of mild Lewis bases bring about changes in the structures. For example, polymeric phosphates are formed by the addition of THF or a mild base during the course of the reaction (Chart 14). These one-dimensional polymers are made up of M–O–P inorganic backbones covered with a hydrophobic sheath of *tert*-butyl groups.[36b] Thermal decomposition of the polymers in the 300–400 °C range produces the corresponding metal metaphosphate (Chart 15).[36b]

The reaction of 2 equiv of dtbp-H with Al₂Me₆ affords an eight-membered ring by the loss of 2 equiv of methane.[37c] On the other hand, the reaction with [Al-(O*i*Pr)₃]₄ yields a centrosymmetric tetramer with two eight-membered rings connected by bridging isopropoxides (Chart 12). The presence of reactive Al–Me, Al–O*i*Pr, and P–O*t*Bu linkages in these S4R products renders them useful as starting materials to build higher structures. For example, the cyclic aluminophosphate with Al–Me groups reacts readily with 3 equiv of (*t*BuO)₃SiOH to give a molecular silicoaluminophosphate (M-SAPO) (Chart 12),

phosphate esters; we shall limit the discussion here to those that can act as precursors for ceramic phosphates or have the potential to act as starting materials for building up framework structures.

Chart 14. Preparation of 1D Polymeric Phosphates

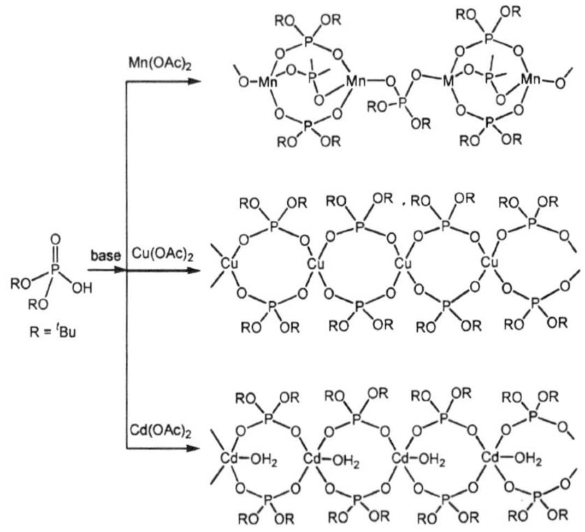

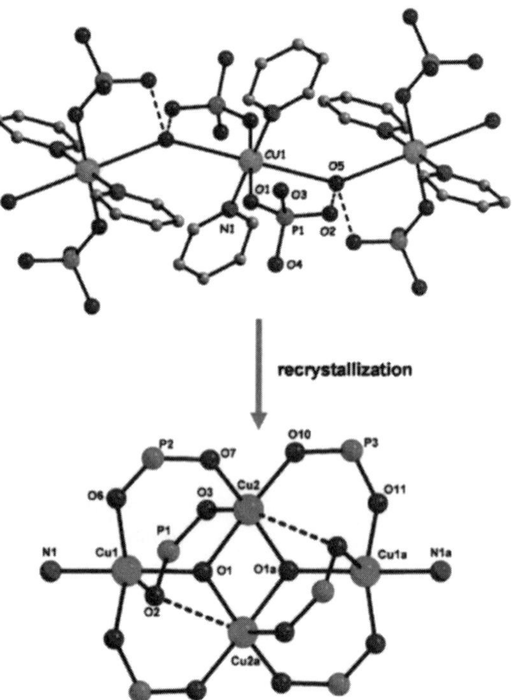

FIGURE 3. Conversion of a 1D polymer to a cage structure by simple recrystallization.

Chart 15. Preparation of Phosphate Materials from Molecular Complexes

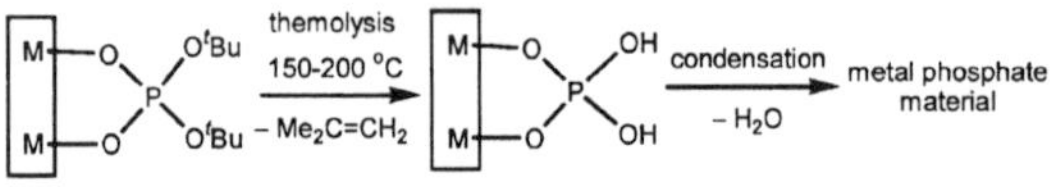

Chart 16. Linear Polymer and Tetrameric Cage Phosphates from the Same Reaction

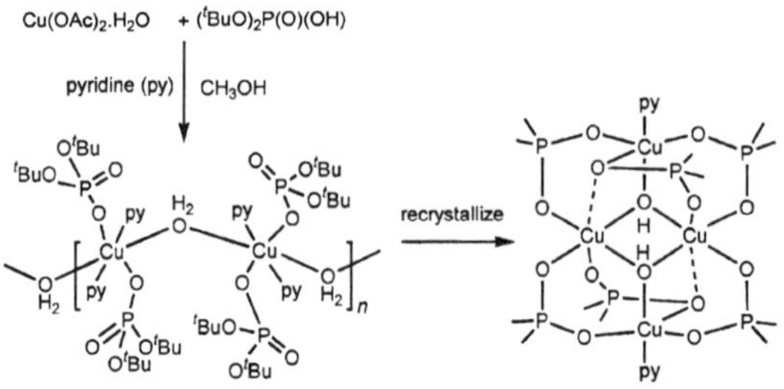

Chart 17. Linear Polymer → 2D Grid Structure Conversion at Room Temperature

P = O(O)P(O^{t}Bu)$_2$

6. Structural Transformations in Metal Dialkyl Phosphates

If instead of a mild Lewis base such as THF a strong Lewis base such as pyridine is used in the reaction between copper acetate and dtbp-H, one obtains yet another type of one-dimensional polymeric phosphate (Chart 16).[36c] Recrystallization of this water-bridged polymer in a DMSO/THF/CH$_3$OH mixture results in the transformation of the polymer to a more stable tetranuclear copper cluster [Cu$_4$-(μ_3-OH)$_2$(dtbp)$_6$(py)$_2$] in about 60% yield (Figure 3). This observation demonstrates that even simple recrystallization can bring about an interesting structural transformation at room temperature in such materials.

Encouraged by the 1D polymer → tetrameric cage transformation, other possibilities of transforming metal dialkyl phosphates have been investigated. Interesting among the results obtained so far is the transformation of the 1D cobalt-dtbp polymer into a 2D polymeric grid structure by reaction with 2 equiv of 4,4-bipyridine in

methanol at room temperature as shown in Chart 17.[40] This reaction is quantitative with no side products.

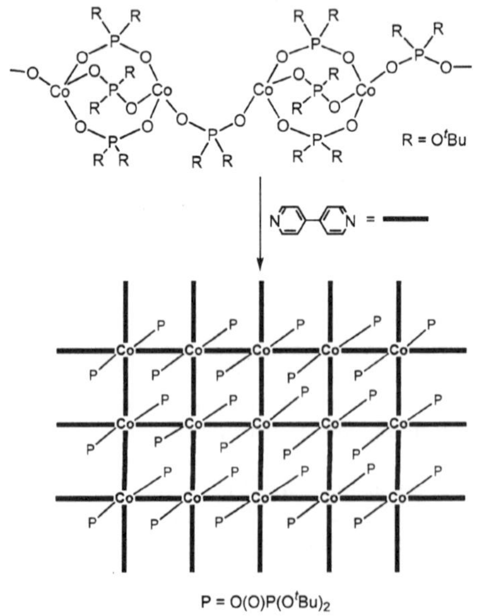

The above two examples are representative of how it is possible to either increase or decrease the dimensionality of a metal dialkyl phosphate, and clearly other possibilities exist in this area. Especially interesting would be the exploitation of the hydrolyzable P–OtBu linkages in the presence of added amines (structure directors) to build up 3D framework phosphates.

7. Molecular Metal Phosphates from P(O)(OH)₃ and Their Conversion to Higher Dimensional Structures

While all the above molecular phosphates were prepared starting from phosphonic acids and phosphate esters, there are a few examples of molecular phosphates synthesized from phosphoric acid in aqueous medium. Although under hydrothermal conditions the reactions of phosphoric acid with metal ions generally result in extended open framework structures, it has been possible to isolate molecular zero-dimensional metal phosphates.[41,42]

A zero-dimensional tin phosphate comprising a four-membered ring was isolated a few years ago.[41] Similar molecular zinc phosphates were isolated by carrying out the reaction between zinc oxide, phosphoric acid, and an amine at 80 °C.[42a] Another four-membered ring zinc phosphate was obtained by the reaction between a zinc salt, phosphoric acid, and the amine at room temperature over an extended period of time.[42b] This four-membered ring phosphate could also be obtained by the reaction of the amine phosphate with the zinc salt. What is interesting about these four-membered ring metal phosphates is their close resemblance to the S4R secondary building unit. It has been shown recently that the four-membered ring zinc phosphate transforms to one-dimensional, two-dimensional, and three-dimensional structures under relatively mild conditions (Figure 4).[43,44] Interestingly, one-dimensional phosphates with linear chains or ladder structures also transform to two-dimensional and three-dimensional structures, while the layered two-dimensional structures transform to three-dimensional structures.[45,46] These studies show that the formation of complex three-dimensional structures may involve a progressive building up process, as indeed verified by synchrotron studies.[47] Acid degradation of three-dimensional zinc phosphates with channel structures have been found to yield low-dimensional structures, showing thereby the reversibility of the 1D ↔ 2D ↔ 3D transformation process.[48] What is especially noteworthy is the recent observation of the transformation of a four-membered ring zinc phosphate to a linear chain structure at room temperature on reaction with piperazine. The linear chain then transforms to a sodalite-related structure with a 4⁶8⁸ cage at 50 °C (Figure 5).[49] A transformation of metal squarates to sodalite structures has been observed recently,[50] showing thereby that the four-membered ring motif can generally act as a SBU.

In the case of aluminum phosphates, Ozin et al.[3b] have pointed out the possibility of the transformation of a chain structure to other chains or to layered and three-dimensional open-framework structures through a hydrolysis–condensation self-assembly pathway. We have recently carried out a few interesting molecule–material

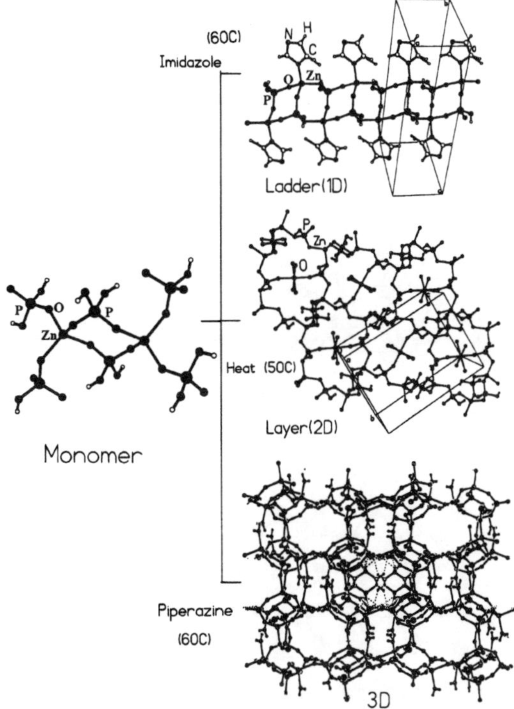

FIGURE 4. Transformations of a zero-dimensional monomeric zinc phosphate to 1D (ladder), 2D (layer), and 3D structures.

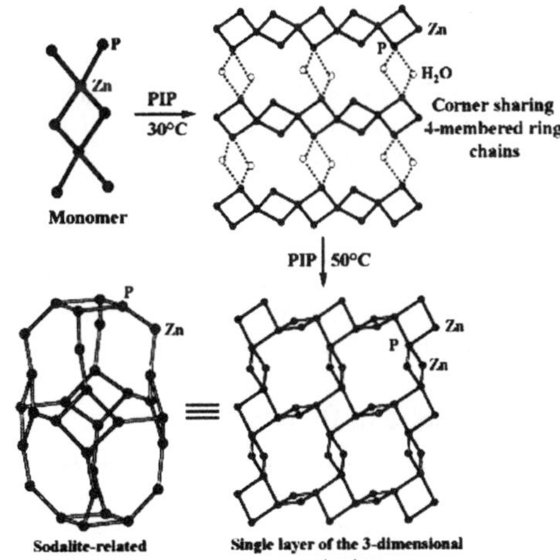

FIGURE 5. Transformation of a four-membered ring zinc phosphate (SBU) to a linear chain phosphate and the transformation of the latter to a 3D sodalite-related structure with a 4⁶8⁸ cage.

transformations in aluminophosphates. Thus, we found that the centrosymmetric, tetrameric molecular aluminophosphate (see Chart 12) transforms to a layered structure[51] of composition [NH₃(CH₂)₂NH₃]²⁺[Al₂(OH)₂(PO₄)₂(H₂O)]²⁻·H₂O on reaction with ethylenediamine in aqueous medium at 150 °C as shown in Figure 6. Interestingly, this structure can also be obtained as the transformation

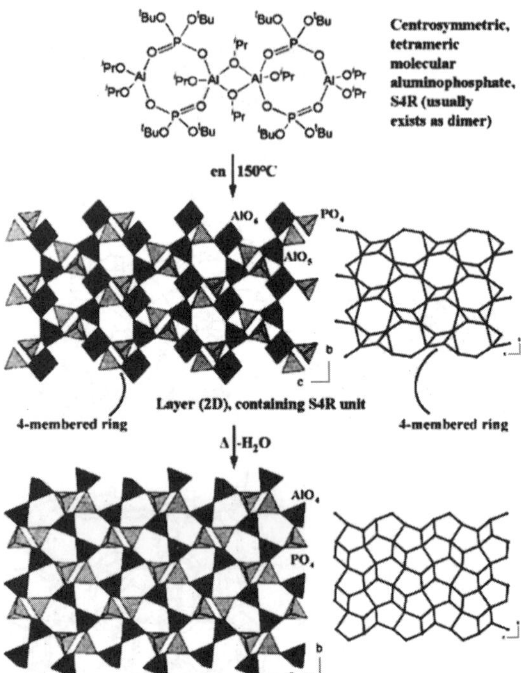

FIGURE 6. Transformation of a four-membered ring aluminophosphate (S4R) to a layered phosphate, which then undergoes transformation to another layered structure by the loss of water molecules. Note that the 2D component contains four-membered rings.

be rewarding to pursue investigations on the transformation of various molecular precursors including those involving aluminophosphate and silicate motifs.

product of a one-dimensional ladder compound of composition $[NH_3(CH_2)_2NH_3][AlP_2O_8H]$.[52] This layered structure transforms to another layered structure by loss of water molecules, followed by reorganization of the aluminophosphate layer (Figure 6). The molecular aluminum phosphate compound gives rise to a three-dimensional structure with channels on reacting with piperazine in aqueous medium at 150 °C. We are exploring reactions of this molecular aluminum phosphate precursor and similar compounds in various media.

8. Conclusions and Outlook

The discussion in the previous sections should suffice to indicate the importance of the transformations of molecular compounds to materials. Studies of such transformations are only making a beginning, and there appears to be a great future for such investigations. Thus, the recent synthesis of a sodalite-related structure from a molecular zinc phosphate is noteworthy, besides the transformation of the centrosymmetric, tetrameric, molecular aluminophosphate to open-framework structures. Preliminary investigations[53] have shown that interesting transformations also occur in metal carboxylates. Thus, molecular zinc oxalate monomers and dimers are found to transform to chain or three-dimensional structures on heating with piperazine in an aqueous medium. The chain structure transforms to a layer structure. We believe that it would

References

(1) Milton, R. M. Commercial Development of Molecular Sieve Technology. *Molecular sieves; papers read at the conference held at the School of Pharmacy, University of London, 4th–6th April, 1967;* Society of Chemical Industry: London, 1968; pp 199–203.

(2) (a) Barrer, R. M. *Hydrothermal Chemistry of Zeolites;* Academic Press: London, 1982. (b) Baerlocher, Ch.; Meier, W. M.; Olson, D. H. *Atlas of Zeolite Framework Types,* Elsevier: Amsterdam, 2001.

(3) (a) Wilson, S. T.; Lok, B. M.; Messina, C. A.; Cannan, T. R.; Flanigen, E. M. Aluminophosphate molecular sieves: A new class of microporous crystalline inorganic solids. *J. Am. Chem. Soc.* **1982,** *104,* 1146–1147. (b) Oliver, S.; Kuperman, A.; Ozin, G. A. A new model for aluminophosphate formation: Transformation of a linear chain aluminophosphate to chain, layer, and framework structures. *Angew. Chem., Int. Ed.* **1998,** *37,* 46–62.

(4) Cheetham, A. K.; Férey, G.; Loiseau, T. Open-framework inorganic materials. *Angew Chem., Int. Ed.* **1999,** *38,* 3268–3292.

(5) Rao, C. N. R.; Natarajan, S.; Vaidhyanathan, R. Metal Carboxylates with Open Architectures. *Angew. Chem., Int. Ed.* **2004,** *43,* 1466–1496.

(6) (a) Choudhury, A.; Krishnamoorthy, J.; Rao, C. N. R. An approach to the synthesis of organically templated open-framework metal sulfates by the amine-sulfate route. *Chem. Commun.* **2001,** 2610–2611. (b) Paul, G.; Choudhury, A.; Sampathkumaran, E. V.; Rao, C. N. R. Organically templated mixed-valent iron sulfates possessing Kagome and other types of layered networks. *Angew. Chem., Int. Ed.* **2002,** *41,* 4297–4300. (c) Paul, G.; Choudhury, A.; Rao, C. N. R. Organically templated linear and layered cadmium sulfates. *J. Chem. Soc., Dalton Trans.* **2002,** 3859–3867. (d) Doran, M.; Norquist, A. J.; O'Hare, D. $[NC_4H_{12}]_2[(UO_2)_6(H_2O)_2(SO_4)_7]$: the first organically templated actinide sulfate with a three-dimensional framework structure. *Chem. Commun.* **2002,** 2946–2947. (e) Behera, J. N.; Paul, G.; Choudhury, A.; Rao, C. N. R. An organically templated Co(II) sulfate with the kagome lattice. *Chem. Commun.* **2004,** 456–457.

(7) (a) Harrison, W. T. A.; Phillips, M. L. F.; Stanchfield, J.; Nenoff, T. M. $(CN_3H_6)_4[Zn_3(SeO_3)_5]$: The first organically templated selenite. *Angew Chem., Int. Ed.* **2000,** *39,* 3808–3809. (b) Choudhury, A.; Udayakumar, D.; Rao, C. N. R. Three-Dimensional Organically Templated Open-Framework Transition Metal Selenites. *Angew. Chem., Int. Ed.* **2002,** *41,* 158–161. (c) Udayakumar, D.; Rao, C. N. R. Organically templated three-dimensional open-framework metal selenites with a diamondoid network. *J. Mater. Chem.* **2003,** *13,* 1635–1638. (d) Behera, J. N.; Ayi, A. A.; Rao, C. N. R. The First Organically Templated Open-Framework Metal Selenate with a Three-Dimensional Architecture. *Chem. Commun.* **2004,** 456–457.

(8) (a) Férey, G. Building units design and scale chemistry. *J. Solid State Chem.* **2000,** *152,* 37–48. (b) Férey, G. Microporous solids: From organically templated inorganic skeletons to hybrid frameworks...ecumenism in chemistry. *Chem. Mater.* **2001,** *13,* 3084–3098.

(9) Lickiss, P. D. Polysilanols. In *The Chemistry of Organosilicon Compounds;* Rappoport, Z., Apeloig, Y. Eds.; John Wiley & Sons: 2001; Vol. 3, Chapter 12, pp 695–744.

(10) Murugavel, R.; Voigt, A.; Walawalkar, M. G.; Roesky, H. W. Hetero- and metallasiloxanes derived from silanediols, disilanols, silanetriols, and trisilanols. *Chem. Rev.* **1996,** *96,* 2205–2236.

(11) Murugavel, R.; Chandrasekhar, V.; Roesky, H. W. Discrete silanetriols: Building blocks for three-dimensional metallasiloxanes. *Acc. Chem. Res.* **1996,** *29,* 183–189.

(12) Schmidbaur, H. Advances in Heterosiloxane Chemistry. *Angew. Chem.* **1965,** *77,* 206–216.

(13) (a) Beckmann, J.; Jurkschat, K. Stannasiloxanes: from rings to polymers. *Coord. Chem. Rev.* **2001,** *215,* 267–300. (b) Lorenz, V.; Fischer, A.; Giessmann, S.; Gilje, J. W.; Gun'ko, Y.; Jacob, K.; Edelmann, F. T. Disiloxanediolates and polyhedral metallasilsesquioxanes of the early transition metals and f-elements. *Coord. Chem. Rev.* **2000,** *206–207,* 321–368. (c) King, L.; Sullivan, A. C. Main group and transition metal compounds with silanediolate $[R_2SiO_2]^{2-}$ and α,ω-siloxanediolate $[O(R_2SiO)_n]^{2-}$ ligands. *Coord. Chem. Rev.* **1999,** *189,* 19–57.

(14) Murugavel, R.; Walawalkar, M. G.; Prabusankar, G.; Davis, P. Synthesis and structure of a novel lithium gallosiloxane containing a $Ga_4Si_4O_8$ macrocycle analogous to the S8R building unit of zeolites. *Organometallics* **2001,** *20,* 2639–2642.

(15) (a) Feher, F. J.; Budzichowski, T. A. Silasesquioxanes as ligands in inorganic and organometallic chemistry. *Polyhedron*, **1995**, *14*, 3239–3253. (b) Duchateau, R. Incompletely condensed silsesquioxanes: Versatile tools in developing silica-supported olefin polymerization catalysts. *Chem. Rev.* **2002**, *102*, 3525–3542.

(16) Fujdala, K. L.; Tilley, T. D. New vanadium tris(*tert*-butoxy)siloxy complexes and their thermolytic conversions to vanadia-silica materials. *Chem. Mater.* **2002**, *14*, 1376–1384.

(17) Murugavel, R.; Chandrasekhar, V.; Voigt, A.; Roesky, H. W.; Schmidt, H.-G.; Noltemeyer, M. New lipophilic air-stable silanetriols: first example of an X-ray crystal structure of a silanetriol with Si–N bonds. *Organometallics* **1995**, *14*, 5298–5301 and references therein.

(18) (a) Montero, M. L.; Voigt, A.; Teichert, M.; Usón, I.; Roesky, H. W. Soluble aluminosilicates with frameworks of minerals. *Angew. Chem., Int. Ed. Engl.* **1995**, *34*, 2504–2506. (b) Montero, M. L.; Usón, I.; Roesky, H. W. Soluble organic derivatives of aluminosilicates with $Al_2Si_2O_4$ and $Al_4Si_2O_6$ frameworks. *Angew. Chem., Int. Ed. Engl.* **1994**, *33*, 2103–2104.

(19) Nehete, U. N.; Anantharaman, G.; Chandrasekhar, V.; Murugavel, R.; Walawalkar, M. G.; Roesky, H. W.; Vidovic, D.; Magull, J.; Samwer, K.; Sass, B. Polyhedral Ferrous and Ferric Siloxanes. *Angew. Chem., Int. Ed.* **2004**, *43*, 3832–3835.

(20) Feher, F. J.; Newman, D. A.; Walzer, J. F. Silsesquioxanes as models for silica surfaces. *J. Am. Chem. Soc.* **1989**, *111*, 1741–1748.

(21) Abbenhuis, H. C. L. Advances in homogeneous and heterogeneous catalysis with metal-containing silsesquioxanes. *Chem.—Eur. J.* **2000**, *6*, 25–32.

(22) Coles, M. P.; Lugmair, C. G.; Terry, K. W.; Tilley, T. D. Titania-silica materials from the molecular precursor Ti[OSi((OBu^t)$_3$]$_4$: Selective epoxidation catalysts. *Chem. Mater.* **2000**, *12*, 122–131.

(23) See also the cited references of Tilley et al. in ref 16.

(24) Fujdala, K. L.; Tilley, T. D. Design and synthesis of heterogeneous catalysts: the thermolytic molecular precursor approach. *J. Catal.* **2003**, *216*, 265–275.

(25) Brutchey, R. L.; Goldberger, J. E.; Koffas, T. S.; Tilley, T. D. Nonaqueous, molecular precursor route to hybrid inorganic/organic zirconia-silica materials containing covalently linked organic bridges. *Chem. Mater.* **2003**, *15*, 1040–1046.

(26) Walawalkar, M. G.; Roesky, H. W.; Murugavel, R. Molecular phosphonate cages: Model compounds and starting materials for phosphate materials. *Acc. Chem. Res.* **1999**, *32*, 117–126.

(27) Walawalkar, M. G.; Murugavel, R.; Roesky, H. W.; Schmidt, H.-G. The first molecular borophosphonate cage: Synthesis, spectroscopy, and single-crystal X-ray structure. *Organometallics* **1997**, *16*, 516–518.

(28) Mason, M. R. Molecular phosphates, phosphonates, phosphinates, and arsonates of the group 13 elements. *J. Cluster Sci.* **1998**, *9*, 1–23.

(29) Yang, Y.; Walawalkar, M. G.; Pinkas, J.; Roesky, H. W.; Schmidt, H.-G. Molecular aluminophosphonate: Model compound for the isoelectronic double-six-ring (D6R) secondary building unit of zeolites. *Angew. Chem., Int. Ed.* **1998**, *37*, 96–98.

(30) Walawalkar, M. G.; Horchler, S.; Dietrich, S.; Chakraborty, D.; Roesky, H. W.; Schaefer, M.; Schmidt, H.-G.; Sheldrick, G. M.; Murugavel, R. Novel organic-soluble molecular titanophosphonates with cage structures comparable to titanium-containing silicates. *Organometallics* **1998**, *17*, 2865–2868.

(31) (a) Walawalkar, M. G.; Murugavel, R.; Voigt, A.; Roesky, H. W.; Schmidt, H.-G. A novel molecular gallium phosphonate cage containing sandwiched lithium ions: Synthesis, structure, and reactivity. *J. Am. Chem. Soc.* **1997**, *119*, 4656–4661. (b) Walawalkar, M. G.; Murugavel, R.; Roesky, H. W.; Usón, I.; Kraetzner, R. Gallophosphonates containing alkali metal ions. 2. Synthesis and structure of gallophosphonates incorporating Na$^+$ and K$^+$ ions. *Inorg. Chem.* **1998**, *37*, 473–478.

(32) Yang, Y.; Pinkas, J.; Schaefer, M.; Roesky, H. W. Molecular model for aluminophosphates containing fluoride as a structure-directing and mineralizing agent. *Angew. Chem., Int. Ed.* **1998**, *37*, 2650–2653.

(33) Anantharaman, G.; Walawalkar, M. G.; Murugavel, R.; Gábor, B.; Herbst-Irmer, R.; Baldus, M.; Angerstein, B.; Roesky, H. W. A nanoscopic molecular cadmium phosphonate wrapped in a hydrocarbon sheath. *Angew. Chem., Int. Ed.* **2003**, *42*, 4482–4485.

(34) Chandrasekhar, V.; Kingsley, S. A dodecanuclear copper (II) cage containing phosphonate and pyrazole ligands *Angew. Chem., Int. Ed.* **2000**, *39*, 2320–2322.

(35) Salta, J.; Chen, Q.; Chang, Y. D.; Zubieta, J. The oxovanadium-organophosphonate system – Complex cluster structures [(VO)$_6$(tBuPO$_3$)$_8$Cl], [(VO)$_4$(PhPO$_2$OPO$_2$Ph)$_4$Cl]$^-$, and [V$_{18}$O$_{25}$(H$_2$O)$_2$-(PhPO$_3$)$_{20}$Cl$_4$]$^{4-}$ with encapsulated chloride anions prepared from simple precursors. *Angew. Chem., Int. Ed. Engl.* **1994**, *33*, 757–760.

(36) (a) Murugavel, R.; Sathiyendiran, M.; Walawalkar, M. G. Di-*tert*-butyl phosphate complexes of cobalt(II) and zinc(II) as precursors for ceramic M(PO$_3$)2 and M$_2$P$_2$O$_7$ materials: Synthesis, spectral characterization, structural studies, and role of auxiliary ligands. *Inorg. Chem.* **2001**, *40*, 427–434. (b) Sathiyendiran, M.; Murugavel, R. Di-*tert*-butyl phosphate as synthon for metal phosphate materials via single-source coordination polymers [M(dtbp)$_2$]$_n$ (M = Mn, Cu) and [Cd(dtbp)$_2$(H$_2$O)]$_n$ (dtbp-H = (tBuO)$_2$P(O)OH). *Inorg. Chem.* **2002**, *41*, 6404–6411. (c) Murugavel, R.; Sathiyendiran, M.; Pothiraja, R.; Walawalkar, M. G.; Mallah, T.; Riviére, E. Monomeric, tetrameric, and polymeric copper di-*tert*-butyl phosphate complexes containing pyridine ancillary ligands. *Inorg. Chem.* **2004**, *43*, 945–953. (d) Murugavel, R.; Sathiyendiran, M.; Pothiraja, R.; Butcher, R. J. O–H Bond elongation in coordinated water through intramolecular P=O···H–O bonding. 'Snapshots' in phosphate ester hydrolysis. *Chem. Commun.* **2003**, 2546–2547.

(37) (a) Lugmair, C. G.; Tilley, T. D.; Rheingold, A. L. Zinc di(*tert*-butyl)-phosphate complexes as precursors to zinc phosphates. Manipulation of zincophosphate structures. *Chem. Mater.* **1997**, *9*, 339–348. (b) Lugmair, C. G.; Tilley, T. D. Di-*tert*-butyl phosphate complexes of titanium. *Inorg. Chem.* **1998**, *37*, 1821–1826. (c) Lugmair, C. G.; Tilley, T. D.; Rheingold, A. L. Di(*tert*-butyl)-phosphate complexes of aluminum: Precursors to aluminum phosphate xerogels and thin films. *Chem. Mater.* **1999**, *11*, 1615–1620.

(38) Pinkas, J.; Chakraborty, D.; Yang, Y.; Murugavel, R.; Noltelmeyer, M.; Roesky, H. W. Reactions of trialkyl phosphates with trialkyls of aluminum and gallium: New route to alumino- and gallophosphate compounds via dealkylsilylation. *Organometallics* **1999**, *18*, 523–528.

(39) Fujdala, K. L.; Tilley, T. D. An efficient, single-source molecular precursor to silicoaluminophosphates. *J. Am. Chem. Soc.* **2001**, *123*, 10133–10134.

(40) Pothiraja, R.; Murugavel, R. Unpublished results.

(41) Ayyappan, S.; Cheethan, A. K.; Natarajan, S.; Rao, C. N. R. A Novel Monomeric Tin(II) Phosphate, [N(C$_2$H$_5$NH$_3$)$_3$]$^{3+}$[Sn(PO$_4$)(HPO$_4$)]$^{3-}$·4H$_2$O, Connected through hydrogen Bonding. *J. Solid State Chem.* **1998**, *139*, 207–210.

(42) (a) Harrison, W. T. A.; Hannooman, L. Two New Tetramethylammonium Zinc Phosphates: N(CH$_3$)$_4$·Zn(HPO$_4$)(H$_2$PO$_4$), an Open Framework Phase Built up from a Low-density 12-Ring topology, and N(CH$_3$)$_4$·Zn(H$_2$PO$_4$)$_3$, a Molecular Cluster. *J. Solid State Chem.* **1997**, *131*, 363–369. (b) Neeraj, S.; Natarajan, S.; Rao, C. N. R. Isolation of a Zinc Phosphate Primary Building Unit, [C$_6$N$_2$H$_{18}$]$^{2+}$[Zn(HPO$_4$)(H$_2$PO$_4$)$_2$]$^{2-}$, and Its Transformation to an Open-Framework Phosphate, [C$_6$N$_2$H$_{18}$]$^{2+}$[Zn$_3$(H$_2$O)$_4$(HPO$_4$)$_4$]$^{2-}$. *J. Solid State Chem.* **2000**, *150*, 417–422.

(43) Rao, C. N. R.; Natarajan, S.; Choudhury, A.; Neeraj, S.; Ayi, A. A. Aufbau Principle of Complex Open-Framework Structures of Metal Phosphates with Different Dimensionalities. *Acc. Chem. Res.* **2001**, *34*, 80–87.

(44) Ayi, A. A.; Choudhury, A.; Natarajan, S.; Neeraj, S. Rao, C. N. R. Transformations of Low-Dimensional Zinc Phosphates to complex Open-Framework Structures. Part 1: Zero-dimensional to One-, Two- and Three-Dimensional Structures. *J. Mater. Chem.* **2001**, *11*, 1181–1191.

(45) (a) Choudhury, A.; Neeraj, S.; Natarajan, S.; Rao, C. N. R. Transformations of the Low-Dimensional Zinc Phosphates to Complex Open-Framework Structures. Part 2: One-Dimensional Ladder to Two- and Three-Dimensional Structures. *J. Mater. Chem.* **2001**, *11*, 1537–1546. (b) Choudhury, A.; Neeraj, S.; Natarajan, S.; Rao, C. N. R. Transformations of Two-Dimensional Layered Zinc Phosphates to Three-Dimensional and One-Dimensional Structures. *J. Mater. Chem.* **2002**, *12*, 1044–1052.

(46) Walton, R. I.; Millange, F.; Le Bail, A.; Loiseau, T.; Serre, C.; O'Hare, D.; Férey, G. The Room-Temperature Crystallization of a One-Dimensional Gallium Fluorophosphate, Ga(HPO$_4$)$_2$F. H$_3$N(CH$_2$)$_3$-NH$_3$·2H$_2$O, a Precursor to Three-Dimensional Microporous Gallium Fluorophosphates. *Chem. Commun.* **2000**, 203–204.

(47) (a) Walton, R. I.; Norquist, A. J.; Neeraj, S.; Natarajan, S.; Rao, C. N. R.; O'Hare, D. Direct *in situ* Observation of Increasing Structural Dimensionality During the Hydrothermal Formation of Open-Framework Zinc Phosphates. *Chem. Commun.* **2001**, 1990–1991. (b) Francis, R. J.; O'Brein, S.; Fogg, A. M.; Halasyamani, P. S.; O'Hare, D.; Loiseau, T.; Ferey, G. Time-Resolved *In-Situ* Energy and Angular Dispersive X-ray Diffraction Studies of the Formation of the Microporous Gallophosphate ULM-5 under Hydrothermal Conditions. *J. Am. Chem. Soc.* **1999**, *121*, 1002–1015.

(48) Choudhury, A.; Rao, C. N. R. Understanding the Building-Up Process of Three-Dimensional Open-Framework Metal Phosphates: Acid Degradation of the 3D Structures to Lower Dimensional Structures. *Chem. Commun.* **2003**, 366–367.

(49) Dan, M.; Udayakumar, D.; Rao, C. N. R. Transformation of a 4-membered Ring Zinc Phosphate SBU to a Sodalite-Related 3-Dimensional Structure through a Linear Chain Structure. *Chem. Commun.* **2003**, 2212–2213.

(50) Neeraj, S.; Noy, M. L.; Rao, C. N. R.; Cheetham, A. K. Sodalite Networks Formed by Metal Squarates. *Solid State Sci.* **2002**, *4*, 1231–1236.

(51) Kongshaug, K. O.; Fjellvag, H.; Lillerud, K. P. Layered aluminophosphates II. Crystal Structure and Thermal Behaviour of the Layered Aluminophosphate UiO-15 and its High-Temperature Variants. *J. Mater. Chem.* **1999**, *9*, 1591–1598.

(52) Williams, I. D.; Yu, J.; Gao, Q.; Chen, J.; Xu, R. New Chain Architecture for a One-Dimensional Aluminophosphate, [H₃NCH₂CH₂NH₃][AlP₂O₈H]. *Chem. Commun.* **1997**, 1273–1274.

(53) Dan, M.; Ayi, A. A.; Rao, C. N. R. Unpublished results.

AR040083E

CRITICAL REVIEW   www.rsc.org/csr | Chemical Society Reviews

Organically-templated metal sulfates, selenites and selenates

C. N. R. Rao,†* J. N. Behera and Meenakshi Dan

Received 16th October 2005
First published as an Advance Article on the web 30th January 2006
DOI: 10.1039/b510396g

The literature on inorganic open-framework materials abounds in the synthesis and characterization of metal silicates, phosphates and carboxylates. Most of these materials have an organic amine as the template. In the last few years, it has been shown that anions such as sulfate, selenite and selenate can also be employed to obtain organically templated open-framework materials. This tutorial review provides an up-to-date survey of organically templated metal sulfates, selenites and selenates, prepared under hydrothermal conditions. The discussion includes one-, two-, and three-dimensional structures of these materials, many of which possess open architectures. The article should be useful to practitioners of inorganic and materials chemistry, besides students and teachers. The article serves to demonstrate how most oxy-anions can be used to build complex structures with metal–oxygen polyhedra.

Introduction

Inorganic open-framework compounds constitute an important class of materials that has attracted much attention as evidenced by the vast number of research papers published in the last few years. Although work in this area started with aluminium silicates[1,2] because of their important uses in sorption and catalysis, much of the recent work pertains to the structure and characterization of open-framework metal phosphates and carboxylates. Both these families of open-framework materials have been reviewed recently.[3–7] Since open-framework silicates and phosphates can essentially be considered as resulting from different arrangements of the metal–oxygen polyhedra and the anion tetrahedra, it seemed rational to examine whether open-framework compounds similar to metal phosphates are readily formed by an oxy-anions such as the sulfate. The main difference between the last two anions and the silicate and phosphate ions relates to the formal oxidation states or the charge on the central atom. In the bonds formed with metals, sulfur would have a higher charge compared to Si–O–M or P–O–M bonds. The sulfate bonds are also least covalent. Sulfates should, however, form stable structures with relatively low valent metal ions. Sulfates rarely form rings unlike silicates and the latter polymerize readily. In spite of such factors, it was not obvious why the sulfate ion should not form open-framework structures with metal–oxygen polyhedra. Recent efforts in this direction have indeed been successful in isolating several organically templated open-framework metal sulfates. A few amine-templated metal selenites and selenates have also been reported recently.

Chemistry and Physics of Materials Unit and CSIR Centre of
Excellence in Chemistry, Jawaharlal Nehru Centre for Advanced
Scientific Research, Jakkur P.O., Bangalore 560064, India.
E-mail: cnrrao@jncasr.ac.in; Fax: +91-80-2362-2760
† Also at, Solid State and Structural Chemistry Unit, Indian Institute
of Science, Bangalore 560012, India

C. N. R. Rao obtained his PhD from Purdue University and the DSc degree from the University of Mysore. He has served as a faculty member at the Indian Institute of Technology (Kanpur) and the Indian Institute of Science (Bangalore) from 1959 to 1994 and has been associated since 1989 with the Jawaharlal Nehru Center for Advanced Scientific Research (Bangalore), where he is the Linus Pauling Research Professor. His research interests are mainly in solid-state and materials chemistry. He is a Fellow of the Royal Society, a Foreign Associate of the US National Academy of Sciences and French Academy of Sciences, and a Foreign Member of several

C. N. R. Rao

academies including those of Russia and Japan. He is also a member of the Pontifical Academy of Sciences and an Honorary Fellow of the RSC. He is the recipient of several awards including the Dan-David prize from Israel, Hughes medal from the Royal Society, Einstein Gold Medal of UNESCO, Somiya award of IUMRS and the first India Science Award.

J. N. Behera

J. N. Behera obtained his MSc and BEd degrees from Utkal University and is now pursuing his PhD studies at the Indian Institute of Science, Bangalore. His work pertains to open-framework materials.

Many of these sulfates, selenites and selenates have open-framework structures.

The open-framework metal sulfates, selenites and selenates are synthesized under hydro/solvothermal conditions in the presence of organic amines. The organic amines play different roles, as true templates, as structure directors and as space fillers.[8,9] In some instances the parent amine is taken as part of the reaction mixture while in some others the amine is taken in the form of an amine sulfate which serves as the source of both the amine and the sulfate.[10] The use of amine sulfates as reaction precursors and their reactions with metal ions is in its infancy, unlike the use of amine phosphates in open-framework phosphate chemistry.[11,12] Organically templated metal sulfates, selenites and selenates possessing 1D inorganic chain structures, 2D layer structures, as well as 3D structures with channels have been synthesized and characterized. The inorganic frameworks are generally connected by hydrogen bonds to the protonated amine molecules and the extra framework water molecules when present. In this article, we briefly review the present status of the organically templated metal sulfates, selenites and selenates of different dimensionalities.

One-dimensional metal sulfates

Single-stranded chain. This simple topology is found in $[Zn(SO_4)(H_2O)_2(C_{10}N_2H_8)]$, **I**, which consists of chains formed by $[ZnO_4N_2]$ octahedra and $[SO_4]$ tetrahedra joined by common vertices.[13] The octahedral coordination around the zinc center is provided by two μ-O groups from two adjacent sulfate ligands in the chain and two nitrogens from the chelating 2,2′-bipyridine ligand. Each sulfate ligand uses two oxygen atoms to link two ZnO_4N_2 octahedra each of which, in turn, shares vertices with two SO_4 groups. This generates the –Zn–O–S–O–Zn– wire-like chain shown in Fig. 1a. Two adjacent chains interact through hydrogen bonds formed by the aqua ligand and terminal S=O bonds. A similar single stranded structure has been found in a vanadium sulfates.[14,15] The $[M(TO_4)\phi_4]$ chain backbone is the simplest possible 1D structure and is the basic motif in the mineral chalcanthite.[16]

Double-stranded chain. This topology is found in $[H_2N(CH_2)_2NH_2][FeF_3SO_4]$, **II**, comprising strictly alternating

Meenakshi Dan is a student of the Integrated PhD program at the Indian Institute of Science, Bangalore. She has obtained the MS degree in chemistry from Indian Institute of Science and is working on open-framework materials for the PhD degree.

Meenakshi Dan

FeF_3O_3 octahedra and SO_4 tetrahedral units, to form a four-member ring ladder as shown in Fig. 1b.[17] The ladder runs along the a-axis with the FeF_3O_3 octahedra sharing vertices with triply-bridging SO_4 groups. The structure can be viewed as a $[M(TO_4)\varphi_4]$ chalcanthite chain, polymerized by the corner-sharing of polyhedra from the adjacent chains to form the (O–T) ladder. The structure of the mineral banattite, $[Cu^{2+}(SO_4)(H_2O)_3]$, is based on a skewed arrangement of chalcanthite chains where it forms six membered rings as in **II**. The formation of octahedral–tetrahedral (O–T) ladder in the metal sulfate family is noteworthy as there is only one other report of an O–T ladder in the literature,[18] unlike the relative large number of T–T ladders.[19–22] It is possible that higher dimensional structures can be obtained from these ladder structures as in the case of the zinc phosphates.[23]

Kröhnkite-type chain. In $[HN(CH_2)_6NH][(VO)_2(OH)_2(SO_4)_2].H_2O$, **III**, there are five-coordinated vanadium sites [V(1) and V(2)] with the square-pyramidal geometry with the

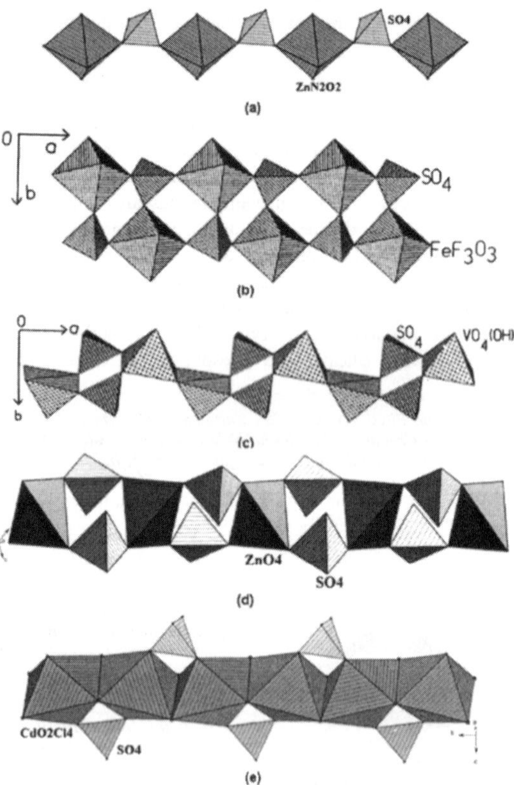

Fig. 1 (a). Strands of ZnN_2O_4 octahedra and SO_4 tetrahedra forming wire like chains in $[Zn(SO_4)(H_2O)_2(2,2′\text{-bpy})]$, **I**, from ref. 13. (b) Double stranded chains of $[FeF_3(SO_4)]_n^{2n-}$ in **II**, from ref. 17. (c) Kröhnkite-type chain motif of $[(VO)_2(OH)_2(SO_4)_2]_n^{2n-}$ in **III**, from ref. 15. (d) View of the $[Zn(SO_4)_2]_n^{2n-}$ units along the b-axis showing the chains of corner sharing '4-ring' built from ZnO_4 and SO_4 tetrahedra in **IV**, from ref.13. (e) Linerite- type chain of $[CdCl_2(SO_4)]_n^{2n-}$ in **V**, redrawn from ref. 10.

basal plane defined by the oxygens of adjacent sulfate groups, two hydroxyl donors and a terminal oxide group.[15] Two square-pyramids, $V(1)O_3(OH)_2$ and $V(2)O_3(OH)_2$, share a common edge *via* two hydroxyl oxygens, with their apices on the opposite sides of the basal plane, to form V_2O_8 dimers. These dimers are capped by the sulfate tetrahedra by sharing the corners to form four-membered rings which propagate along the *a*-axis (Fig. 1c). This structure is related to that of the mineral kröhnkite $Na_2Cu^{II}(SO_4)_2 \cdot 2H_2O$.[16] Kröhnkite-type compounds contain infinite $[M(XO_4)_2(H_2O)_2]$ chains, where M is a divalent (Mn, Fe, Co, Ni, Cu, Zn or Cd) or a trivalent (Al, Fe or In) cation and X is pentavalent (P or As) or hexavalent (S, Se, Cr or Mo). In these chains, the MO_6 octahedra are corner-linked to bridging XO_4 tetrahedra to give the 1D-chains. The kröhnkite chain topology with tetrahedral atoms has been observed in open-framework metal phosphates[24-26] as well as in some metal sulfates and selenates.[27] The common occurrence of the kröhnkite-type chain is probably because the corner-link between the MO_6 and SO_4 groups is flexible, involving little steric constraint.

Chain consisting of four-membered rings. A typical case is that of $[Zn(SO_4)_2(H_2O)_2]^{2-}$, **IV**, consisting of infinite zinc sulfate chains containing four-membered rings obtained from strictly alternating ZnO_4 and SO_4 tetrahedral units, which are linked through their corners (Fig. 1d).[13] Two oxygen atoms of a SO_4 unit bridge adjacent ZnO_4 units while the other two form terminal S=O groups. Tetrahedra-based one-dimensional metal phosphates with a chain topology formed by MO_4 (M = Al, Ga and Zn) and PO_4 tetrahedra are known.[28-32] Such a corner-shared chain structure has been considered to be a precursor to 3D open-framework metal phosphates.[33,34]

Linarite-type chain. Four linear cadmium sulfates of linarite topology have been reported recently.[10,35] In the linarite chain, $[CdX_2SO_4]^{2-}$, **V**, CdX_4O_2 (X = Cl, Br) octahedra share edges in a *trans*-fashion with the sulfate tetrahedra grafted on to the chain as a symmetrical bridge. Two Cd–μ-X– Cd linkages between adjacent Cd atoms lead to infinite linear chains of *trans* edge-sharing CdX_4O_2 octahedra. Two of the sulfate oxygens bond to adjacent cadmium sites of the edge-shared CdX_4O_2 in the common symmetrical bridging mode, forming a synthetic analogue of the linarite chain[36] (Fig. 1e). In linarite, $Pb^{2+}[Cu^{2+}(OH)_2(SO_4)]$, octahedra share two *trans* edges to form a $[M\varphi_4]$ chain. This is decorated by flanking tetrahedra that adopt a staggered arrangement on either side of the chain. The structure of wherryite, $Pb_7^{2+}[Cu^{2+}(OH)(SO_4)(SiO_4)]_2$-$(SO_4)$, is based on linarite chains that are decorated by (SiO_4) tetrahedra. The $[M(TO_4)\varphi]$ chains extend along the *c*-axis and pack along the vertices of a primitive orthogonal plane lattice. The additional SO_4 groups that do not link to the $(Cu^{2+}\varphi_6)$ octahedra are linked by [7]- and [8]- coordinated Pb^{2+} cations. Infinite chains of $M(TO_4)\Phi_2$ stoichiometry in the linarite group of minerals are rare and occur in layered metal phosphates[37-39] such as tsumcorite and bermanite.

Butlerite-type chain. Open-framework metal sulfates form the $[M(TO_4)\varphi_3]$ chain which is the basic motif of butlerite,[40] wherein tetrahedra alternate along the chain and link to the

trans vertices of the octahedra. The structure of $[H_2N(CH_2)_4NH_2][MF_3SO_4]$ (M = V, Fe), **VI**, consists of MF_4O_2 octahedra sharing vertexes with similar neighbors through the fluorine (Fig. 2a).[15] The sulfate tetrahedra are bridged onto the *trans* vertex of the metal octahedra along the chain. The *trans* orientation of the bridging F atom creates a zigzag {–F–M–F–M–} backbone to the linear chain of MF_4O_2 octahedra. In the sulfate tetrahedra, two oxygens bond to the adjacent M-sites of the vertex-shared MF_4O_2 octahedra in a symmetrical bridge, the remaining two forming terminal S=O bonds. The individual chains are held together by hydrogen bond interactions involving diprotonated amine molecules. A similar chain of *cis* corner-sharing octahedra occurs in fibroferrite, $[Fe^{3+}(OH)(H_2O)_2(SO_4)][H_2O]_3$, giving rise to a helical configuration. In butlerite $[Fe^{3+}(OH)(H_2O)_2(SO_4)]$, parabutlerite, $[Fe^{3+}(OH)(H_2O)(SO_4)]$, and fibroferite, the

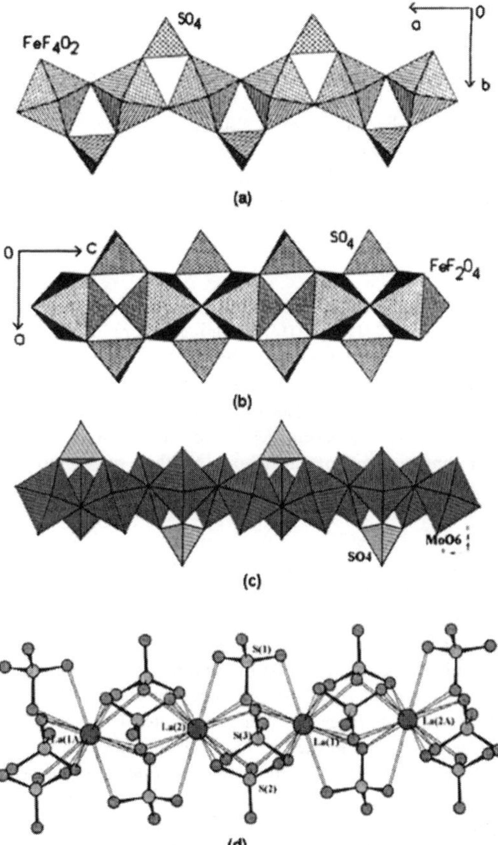

Fig. 2 (a). Butlerite-type chains with alternating up-down bridging of sulfate tetrahedron in $[MF_3SO_4]^{2-}$, **VI**, from ref. 15. (b) Tancoite-type chains in metal sulfates in **VII**, from ref. 17. (c) α-Molybdena backbone in molybdenum sulfate, $[(MoO_3)_3(SO_4)]_n^{2n-}$, **VIII**, redrawn from ref. 46, copyright (2004) with permission from the American Chemical Society. (d) Infinite chains of $[La(SO_4)_3]_n^{3n-}$ in **IX** running along the *a*-axis, redrawn from ref. 48, copyright (2003) with permission from Elsevier.

chains are linked solely by hydrogen bonds as there are no interstitial cations whereas in organically templated compounds the chains are held together by the hydrogen bond assembly of the diprotonated amine molecules located in the inter-chain space to form the 3D assembly. The mineral uklonskovite consists of topologically identical $Mg(SO_4)(OH)(H_2O)_2$ chains, linked by 8-coordinated Na.

Tancoite-type chain. The tancoite chain is commonly found in open-framework materials. Tancoite is a phosphate mineral of the formula, $LiNa_2Hal(PO_4)_2(OH)$, possessing a one-dimensional chain structure.[41] The chain has the composition, $[M(TO_4)_2L]_n$ (M and T are cations of different coordination, usually octahedral and tetrahedral; L = anionic ligand, *e.g.* O^{2-}, OH^- or F^-), and is found in vanadium and iron sulfates.[15,17] In $[FeF(SO_4)_2]^{3-}$, **VII**, the metal octahedra share vertices with similar neighbors to produce a linear zigzag chain, and the sulfate tetrahedra are grafted on to the chain in a symmetrical bridge. The *trans* orientation of the bridging OH/F group creates a {–M–(OH)/F–M–(OH)/F–M–} backbone in the linear chain and allows the sulfate moiety to bridge symmetrically to form a tancoite type topology as shown in Fig. 2b. Such a structure is also found in open-framework metal phosphates[42] as well as in several phosphate, sulfate and silicate minerals.[43] The 1-D tancoite type gallium phosphate chain has been shown to undergo facile transformation to 3-D open-framework structures under hydrothermal conditions.[44,45]

α-Molybdena sulfate. A sulfated α-molybdena, $[C_5H_{14}N_2][(MoO_3)_3(SO_4)].H_2O$, **VIII**, containing an organic amine has been reported.[46] It has three distinct Mo^{6+} centers, each of which is coordinated to six oxygens in a distorted octahedral geometry. Three of the oxide ligands in each MoO_6 octahedron bridge between adjacent Mo^{6+} centers in a μ^3 fashion, exhibiting a structure analogous to that of α-molybdena (Fig. 2c). This chain topology has also been found in $Rb_2SMo_3O_{13}$.[47]

Lanthanum and uranium sulfates. Lanthanum forms a one-dimensional compound of the composition, $[C_4N_3H_{16}][La(SO_4)_3]\cdot H_2O$,[48] **IX**, where the edge linkage of the LaO_{12} polyhedra and the SO_4 tetrahedra provides the building unit, the units being connected by the sharing of La atoms to give infinite chains of $[La(SO_4)_3]_n^{3-}$ along the *a*-axis. The LaO_{12} polyhedra share faces *via* three-coordinated oxygens and the SO_4 tetrahedra are grafted on to the chains by sharing corners and edges (Fig. 2d). Several uranium sulfates are reported to occur in the presence of structurally related organic templating agents by O'Hare and coworkers.[49–56] One-dimensional chains are found in $[C_5N_2H_{14}][UO_2(H_2O)(SO_4)_2]$,[49] where the uranium ion is seven-coordinated in the pentagonal bipyramidal geometry. The two adjacent $[UO_7]$ pentagonal bipyramids are linked by the corner sharing sulfate tetrahedra. Adjacent $Ur\phi_5$ pentagonal bipyramids have their uranyl ions oriented approximately perpendicular to the chain length. The $Ur\phi_5$ polyhedra share corners with the sulfate tetrahedra in such a manner that each $Ur\phi_5$ pentaganal bipyramid is linked to four sulfate tetrahedra, the fifth equatorial ligand of the $Ur\phi_5$ pentaganal bipyramid being shared with an aqua ligand. This chain topology is analogous to the mineral $Mn[(UO_2)(SO_4)_2(H_2O)].4H_2O$ and $[(UO_2)(H_2PO_4)_2(H_2O)].(H_2O)_2$, and also occurs in arsenate and phosphate minerals such as brondtite, $Ca_2[Mn^{2+}(AsO_4)_2(H_2O)_2$, talmessite, $Ca_2[Mg(AsO_4)_2(H_2O)_2]$ and fairfieldite, $Ca_2[Mn^{2+}(PO_4)_2(H_2O)_2]$.

Two-dimensional metal sulfates

Organically templated metal sulfates possessing two-dimensional structures are not that common. One example is that of $[H_3N(CH_2)_2NH_3)][Fe_2F_2(SO_4)_2(H_2O)_2]$, **X**, built up from FeF_2O_4, $FeF_2O_2(H_2O)_2$ octahedra and sulfate tetrahedra by sharing their vertices.[17] The FeF_2O_4 and $FeF_2O_2(H_2O)_2$ octahedra are alternately connected by their *trans*-fluorine vertexes to form an infinite {–F–Fe–F–Fe–} backbone analogous to the mineral butlerite[40] (Fig. 3a). The butlerite-type chains are fused together to form the layer architecture. This layer structure is similar to that of the iron fluorophosphates (ULM-10),[57] which consist of infinite chains of alternating corner-sharing $Fe^{3+}F_2O_4$ and $Fe^{2+}F_2O_2(H_2O)_2$ octahedra decorated by phosphate tetrahedra fused together to form the layered structure. ULM-11 and the mineral curetonite, $Ba_2[(Al, Ti)Al(PO_4)_2(OH, O_2)F]$, possess similar structures, being built up of the laucite motif.[58] Jacobson *et al.*[59] have reported two niobium phosphates and one titanium phosphate with similar structures.

In $[H_2N(CH_2)_4NH_2][Fe_2^{III}Fe_3^{II}F_{12}(SO_4)_2(H_2O)_2]$, **XI**, two iron octahedra are vertex-shared through *trans* –Fe–F–Fe– linkages to form an infinite chain along the *a*-axis.[60] These chains are connected by a trimer of edge-sharing octahedra along the <001> direction to form a layered network in the *ac*-plane. The triangular lattices formed by the trimers while connecting the chains are capped by the sulfate tetrahedra, thereby forming a 10-membered aperture within the layer (Fig. 3b). The layers are stacked one over the other along the *b*-axis in the AAA... fashion and are held together by hydrogen bonding interactions with the diprotonated amine molecules residing in the interlamellar space.

Transition metal sulfates with the kagome lattice. The most common iron sulfates with the kagome lattice are the jarosites which show magnetic frustration or low temperature antiferromagnetism.[61] The organically templated iron sulfate $[HN(CH_2)_6NH][Fe^{III}Fe_2^{II}F_6(SO_4)_2][H_3O]$, **XII**, is an unusual example of an iron compound with a perfect kagome structure.[60] The structure consists of anionic layers of vertex-sharing $Fe^{III}F_4O_2$ and $Fe^{II}F_4O_2$ octahedra and SO_4 tetrahedra units, which are fused together by $Fe^{III/II}$–F–$Fe^{III/II}$ and $Fe^{III/II}$–O–S moieties. Each $Fe^{III/II}F_4O_2$ unit shares four of its $Fe^{III/II}$–F vertices with similar neighbors with the $Fe^{III/II}$–O bonds roughly aligned in the *ab*-plane. The $Fe^{III/II}$–O bond is canted from the *ab*-plane and this $Fe^{III/II}$–O vertex effectively forces a three-ring trio of apical $Fe^{III/II}$–O bonds closer together to allow them to be capped by the SO_4 tetrahedra. The three- and six- rings of octahedra result from the in-plane connectivity and is typical of a kagome lattice with hexagonal tungsten bronze-type sheets (Fig. 3c). Simple calculations have been carried out to understand the properties of this mixed

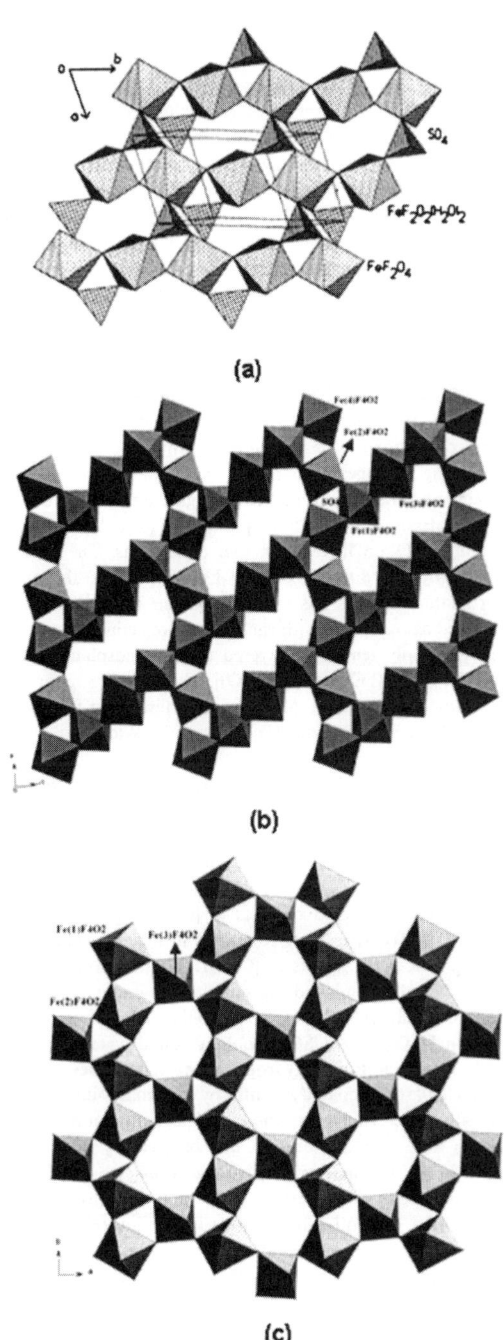

Fig. 3 (a). Fusion of butlerite-type chains to form the layered structure in $[M_2F_2(SO_4)_2(H_2O)_2]_n^{2n-}$, **X**, (M = Fe, Mn) from ref. 17. (b) Layered in $[Fe_2^{III}Fe_3^{II}F_{12}(SO_4)_2(H_2O)_2]^{4-}$, **XI**, with symmetrical capping of the sulfate tetrahedra in the triangular lattice and the 10-membered aperture within it, redrawn from ref. 60. (c) Polyhedral view of the kagome layer in **XII**, from ref. 60.

valent kagome compound.[62] A layered iron sulfate with the kagome lattice with Fe in the +2 state has been reported.[63] An organically templated iron fluorosulfate $[H_3N(CH_2)_2NH_2-(CH_2)_2NH_2(CH_2)_2NH_3][Fe_3^{II}F_6(SO_4)_2]$ with a slightly distorted kagome lattice arising from the presence of the three edge-sharing Fe^{II} octahedra instead of corner-sharing octahedra in the layered network has also been characterized.[64] Magnetic properties of the Fe(II) and the mixed valent Fe kagome compounds are distinctly different from those of the Fe(III) sulfates possessing the kagome lattice. Unlike the Fe(III) kagome compounds which show magnetic frustration or antiferromagnetism at low temperatures, the Fe(II) compound as well as the mixed-valent Fe compounds with kagome structures show ferromagnetic interactions at low temperatures. Similar features are shown by a Fe(III) compound with a distorted kagome lattice. An amine-templated cobalt(II) sulfate with the kagome lattice, is, however, found to exhibit magnetic properties comparable to those of the Fe(III) compound.[65] It appears that only those kagome compounds containing transition metal ions with integral spins (*e.g.*, Fe(II), V(III)) exhibit ferri/ferromagnetic interactions at low temperatures.

Nickel sulfate. The two-dimensional Ni sulfate, $[C_4N_2H_{12}][Ni_3F_2(SO_4)_3(H_2O)_2]$,[66] **XIII**, is built up of $Ni(1)F_2O_4$ and $Ni(2)F_2O_4$ octahedra and SO_4 tetrahedra sharing edges as M–μ-F–M bonds and vertices as Ni–O–S bonds. The $Ni(1)F_2O_4$ octahedra occur as dimers by sharing edges with two fluorines to form the $Ni_2F_2O_8$ unit. The $Ni(2)F_2O_4$ octahedra link to the anion at one end of the shared edges to form the trimeric $Ni_3F_3O_{12}$ units. The sulfate tetrahedra cap the trimers in such a way that they share three corners with three octahedra giving rise to the formation of a hexameric unit. These units are connected to one another through corner-sharing Ni–O–S linkages to form sinusoidal chains. The chains are covalently bonded to each other by the SO_4 group forming the layer structure with an eight-membered aperture in the *ab*-plane (Fig. 4a). The inorganic layers are stacked along the *c*-axis in the ABAB fashion and the interlayer space is occupied by the diprotonated amine molecules, which ensure the stability of the structure through extensive hydrogen bonding. This structure is obtained with different amines such as piperazine, 1,4-diazobicyclo-[2.2.2]octane (DABCO) and 1,3-diamonoprane (DAP). The structure of **XIII** can be compared to that of the mineral sulfoborite,[67] which is made up of complex sheets of MgO_6 and SO_4 units. In the mineral amarantite, octahedral tetramers are polymerized to form octahedral chains parallel to [100] and linked by SO_4 tetrahedra, with the linkage between the sheets being provided by $[B(OH)_4]$ tetrahedra.

Cadmium sulfate. In the two-dimensional cadmium sulfate $[H_3N(CH_2)_3[Cd_2(H_2O)_2(SO_4)_3]$,[35] **XIV**, one cadmium is octahedrally coordinated to six oxygens, of which four are from the corner-sharing sulfate ions and two are from the terminal water molecules. The second cadmium atom is coordinated to six sulfate ions of which four are from the corner-sharing sulfate ions and two are from the edge sharing sulfate ions. The inorganic layer is built up of four-membered rings formed by CdO_6 octahedra and two SO_4 tetrahedra linked through the

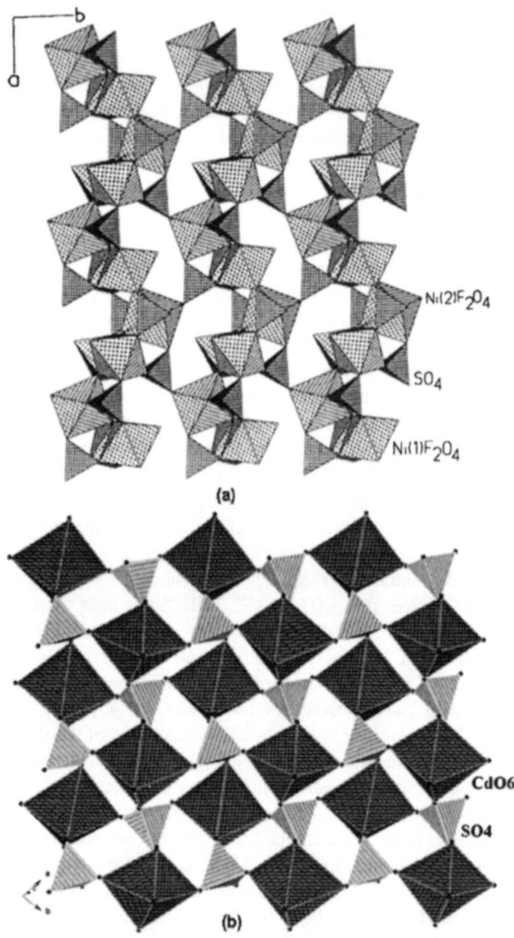

Fig. 4 (a). Two-dimensional $[Ni_3F_2(SO_4)_3(H_2O)_2]^{2-}$, **XIII**, formed by the sinusoidal chains which, in turn, consist of hexameric units from ref. 66. (b) (O–T) two-dimensional cadmium sulfate of $[Cd_2(SO_4)_3(H_2O)_2]^{2-}$ in **XIV**, redrawn from ref. 35. Note the fusion of the four-membered ring ladders to form the layered structures.

vertices. The connectivity between the four membered rings leads to the formation of a four-membered ladder with strictly alternating CdO_6 octahedra and SO_4 tetrahedra. The ladder is then fused in the *ab*-plane to form the two-dimensional framework of the inorganic layer (Fig. 4b). The four-membered rings surround both the cadmium atoms with direct linkage to four SO_4 tetrahedra.

Rare-earth sulfates. The 2D lanthanum sulfates adopt a corrugated layer structure with 8-membered ring apertures, wherein the lanthanum polyhedra get connected by the sulfate tetrahedra in bidentate or monodentate fashion. The coordination around the metal centre is then completed by either coordinating to water, as observed in $[C_6N_2H_{14}]_2[C_2N_2H_8]$-$[Ln_2(H_2O)_4(SO_4)_4][SO_4]\cdot3H_2O$, **XV**,[68] or by coordinating to other terminal sulfate tetrahedra by sharing edges or

corners.[69,70] The sulfate tetrahedra bonded to the Ln centres in the layers can get further bonded to other Ln atoms forming 4-membered ring ladders grafted onto the layers comprising 8-membered rings, as observed in $[C_2N_2H_{10}][La_2(H_2O)_4$-$(SO_4)_4]\cdot2H_2O$.[71] A layer with an eight-membered aperture formed by the joining of NdO_9 polyhedra and SO_4 tetrahedra is shown in Fig. 5a. The corrugated layer structure of the amine-templated lanthanum sulfates can be compared to that of β-$(NH_4)La(SO_4)_2$[72] or $(N_2H_5)Nd(H_2O)(SO_4)_2$,[73] except that in the latter hydrazinium groups are connected to the Nd atom. The interlamellar spacing depends on the size and orientation of the guest species, with the distance between the planes increasing with the increase in the size of the guest amine molecules.

The other structures of layered lanthanum sulfates are based on LnO_3 layers with (6, 3) nets as in $[C_3N_2H_{12}][La_2(SO_4)_4]$, **XVI**, (Fig. 5b). Here, the La centres are bridged by the sulfate tetrahedra to form 4-membered rings, which then join together by sharing of edges or corners. These chains or ladders of 4-membered rings fuse together through the three-coordinated oxygen atoms of the sulfate groups to form layers having infinite Ln–O–Ln linkages in two dimensions. The network formed by the linking of the Ln centres by the three-coordinated oxygens has the composition LnO_3 and can be described as (6, 3) nets with each Ln centre acting as a node.[68] An organically templated layered cerium phosphate sulfate, $[C_2N_2H_{10}][Ce^{III}(PO_4)(HSO_4)(H_2O)]$, has been reported.[74] Lanthanide compounds along with a few open-framework sulfates have been described in a review by Wickleder.[75]

Uranium sulfates. In contrast to the 2D-lanthanum sulfates, the uranium sulfates are made up of 4-membered uranium sulfate rings connected by the sulfate tetrahedra, giving rise to layers of different topologies. Depending on the mode of connectivity of the 4-membered rings, the layers comprise entirely of 4-membered rings as in $[C_3N_2H_5][(UO_2)_2(SO_4)_3]$, USO-8,[49] **XVII**, (Fig. 5c), contain 6-membered rings along with 4-membered rings as in $[C_3N_2H_{12}][(UO_2)_2(H_2O)(SO_4)_3]$, USO-2,[56] or 12-membered rings along with 4-membered rings as in $[C_6N_2H_{14}][UO_2(H_2O)(SO_4)_2]$, USO-5.[55] The inorganic layers in USO-2 and USO-5 contain bound water molecules which form additional hydrogen bonds. Two new layered actinide sulfates prepared recently by the amine-sulfate route, have structures similar to those of the lanthanum sulfates, wherein the uranium centres are bridged by the sulfate tetrahedra by the sharing of edges or corners to form layers with 8-membered ring apertures.[76] By employing hydrofluoric acid in the hydrothermal synthesis, six-layered uranium sulfates have been prepared.[51] The fluoride ligands join adjacent uranium centers through U–F–U linkages either by sharing an edge to form the dimer as in USFO-3 or corners to form a chain, which is then linked by the SO_4 tetrahedra to form layers as in USFO-5.

Thorium sulfate. In $[HN(CH_2)_6NH]_2[Th_2(SO_4)_6(H_2O)_2]\cdot$ $2H_2O$, **XVIII**,[13] two crystallographically distinct metal–oxygen polyhedra are capped by four sulfate ions in Q_2 connectivity to form the Th_2S_4 unit. The $Th(2)O_9$ polyhedra share edges with the $S(2)O_4$ tetrahedra to form a sinusoidal chain along the

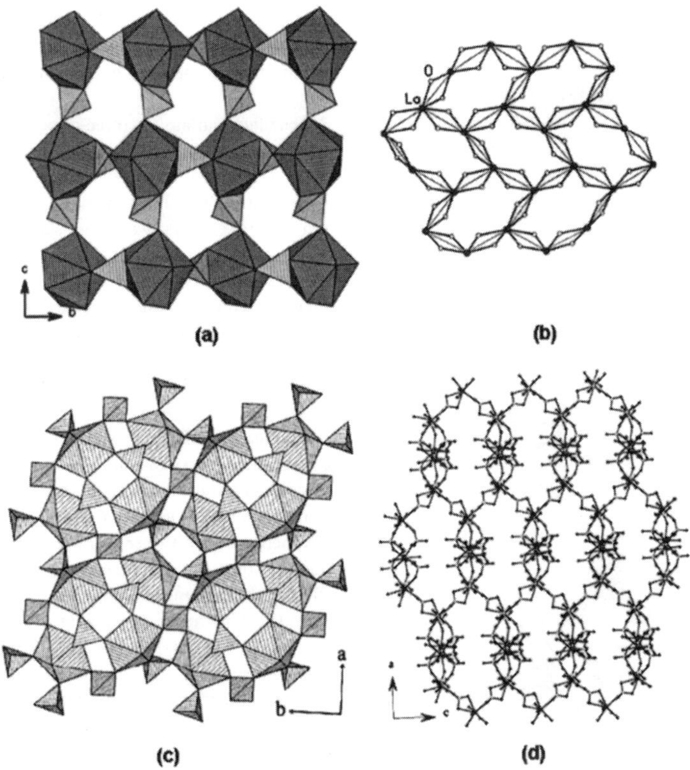

Fig. 5 (a). Polyhedral view of the inorganic layer in $[Ln_2(SO_4)_4(H_2O)_4]^{4-}$ **XV** (Ln = La, Pr or Nd), formed by joining the NdO_9 polyhedra and SO_4 tetrahedra, redrawn from ref. 68. (b) The infinite two-dimensional La–O–La linkages in $[La_2(SO_4)_4]^{2-}$ in **XVI**, from ref. 68. (c) Two-dimensional uranium sulfate in $[UO_2(SO_4)_3]^{2-}$, **XVII**, formed by $[UO_7]$ and $[SO_4]$ units from ref. 49, copyright (2003) from the Royal Society of Chemistry. (d) Two-dimensional sheet of $[HN(CH_2)_6NH]_2[Th_2(SO_4)_6(H_2O)_2].2H_2O$, **XVIII** consist of cages along the ac-plane from ref. 13.

c-axis. Two such adjacent chains arranged in the ABAB... fashion connect the $Th(1)_2S_4$ unit at both their crests and troughs to form cages. The cages are extended in the ac-plane to generate a two-dimensional sheet as shown in Fig. 5d. The cages are occupied by amine and water molecules which form extensive hydrogen bonds with the framework oxygens.

Three-dimensional metal sulfates

Nickel sulfate. The prominent feature of the three-dimensional nickel sulfate, $[C_4N_2H_{12}][Ni_2F_4(SO_4) H_2O]$,[66] **XIX**, is a dimer of edge-sharing octahedra. The Ni(1) and Ni(2) octahedral units share edges through fluorine (Ni(1)–F(1), F(2) and Ni(2)–F(3), F(4)) to form the dimer units $Ni_2(1)F_6O_2(H_2O)_2$ and $Ni_2(2)F_6O_4$. The dimers are linked by sharing octahedral edges to form two distinct infinite sinusoidal chains along the [001] direction. These chains are alternately stacked one over the other along the [100] direction and are interlinked by sharing corners with the tetrahedral corners of the SO_4 groups to form a layer with an 11-membered aperture in the [101] plane. **XIX** seems to be the first example of an 11-membered aperture in open-framework compounds. In the layer, each sulfate shares three corners with three octahedra of the adjacent chains. It is connected to the

Ni(1) octahedral chain by sharing one oxygen atom, whereas it shares two oxygen atoms with two neighboring octahedra of the Ni(2) chain. The layers are cross-linked by SO_4 tetrahedra along the b-axis through corner-sharing S–O–Ni linkages to form the three-dimensional structure. The cross inking of the layers gives rise to an elliptical channel along the c-axis as shown in Fig. 6a. The channels are filled by the amine molecules, which interact with the framework oxygen and fluorine atoms through F–H···O and C–H···O hydrogen bonds. This structure is somewhat related to the mineral phosphopherite,[77] which contains layers of sinusoidal chains made up of edge-shared octahedral trimers cross-linked by tetrahedral PO_4 units.

Zinc sulfate. $[(CN_3H_6)_2][Zn(SO_4)_2]$,[78] **XX**, is built up of alternating ZnO_4 and SO_4 tetrahedra, sharing vertices. The polyhedral connectivity of the ZnO_4 and SO_4 results in an infinite three-dimensional network. This gives a 12-ring system, built up of six ZnO_4 and six SO_4 units. The network possesses two-dimensional intersecting channels propagating along the <100> and <010> directions, with no channels apparent in the <001> direction. Well-ordered guanidinium cations occupy all the 12-ring windows (Fig. 6b) and interact with the zinc sulfate framework through N–H···O hydrogen

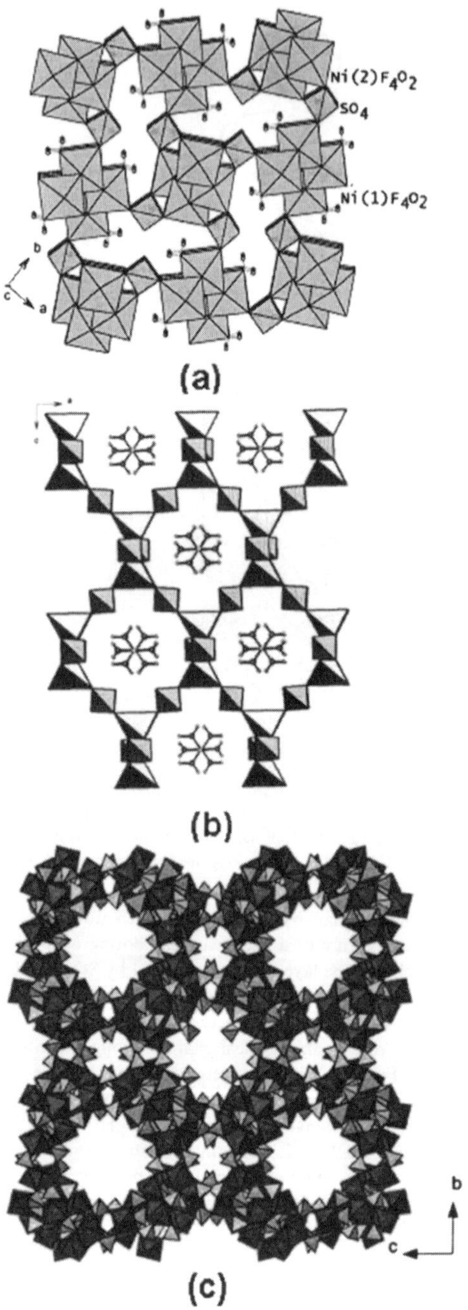

Fig. 6 (a). Three-dimensional Ni(II) sulfate in [C₄N₂H₁₂][Ni₂F₄(SO₄) H₂O], **XIX**, showing 10-membered elliptical channels from ref. 66. (b) Polyhedral view of three-dimensional zinc sulfate comprising ZnO₄ and SO₄ tetrahedra in [(CN₃H₆)₂][Zn(SO₄)₂], **XX**, redrawn from ref. 78. Note the presence of 12-membered channels. (c) Three-dimensional structure of scandium with mixed sulfate and phosphate tetrahedra in **XXI**, redrawn from ref. 81, copyright (2002) from the Royal Society of Chemistry.

bonds. The structure of **XX** resembles that of diamond and is akin to that of a three-dimensional zinc phosphite[79] and of a guanidinium-templated zinc phosphate.[80]

Scandium sulfate. A three-dimensional scandium sulfate in which secondary building units of the formula $Sc_7(S,P)_{12}O_{48}$, **XXI**, templated with azamacrocycles has been synthesized hydrothermally.[81] The structure is made up of ScO_6 octahedra that share corners with the sulfate tetrahedra. The main feature of the framework is that supercages are formed at the corners and at the centre of the cubic unit cell. Each supercage is connected to six other supercages *via* smaller cages, in which protonated cyclen molecules are present. Each small cage links two supercages in such a way that in a unit cell there are two supercages and six smaller cages (Fig. 6c).

Rare-earth sulfates. Compared to the number of 2D rare-earth sulfates, very few 3D rare-earth sulfates are known. The three-dimensional lanthanum sulfates reported hitherto contain 8-membered ring units, which are connected by the sulfate tetrahedra or by the three-coordinating oxygens of the SO_4 unit to form the 3D framework. In $[C_4N_2H_{12}]_2[La_2(H_2O)_2$-$(SO_4)_4]$, the 8-membered rings join together by sharing edges to form the 3D framework, giving rise to 8-ring apertures down the *b*-axis.[70] $[C_2N_2H_{10}]_2[Ln_2(H_2O)_2(SO_4)_5]$ possesses a α-Po structure wherein the 8-membered rings are bridged by the sulfate tetrahedra along the three directions, resulting in the formation of a network of interconnected 8-membered ring channels. Such a connectivity also results in the formation of 16-membered ring channels down the *a*-axis of the unit cell.[68] In $[C_4N_2H_{12}][Nd_2(H_2O)_2(SO_4)_4]$, (**XXII**), the 8-membered rings get connected by the sulfate tetrahedra in one direction and three-coordinated oxygen atoms in another direction, thereby resulting in the formation of channels with 12-ring apertures around the 8-ring channels[68] (Fig. 7a). In all the three 3D lanthanum sulfates, the protonated amine molecules reside in the channels and form hydrogen bonds with the framework oxygens.

Organically templated 3d–4f mixed metal sulfates of the formula $[C_2N_2H_{10}][La_2M(H_2O)_2(SO_4)_6]$ (M = Co, Ni), have been recently synthesized by Clearfield *et al.*[82] The three-dimensional anionic framework is formed by the bridging of La(III) and divalent transition metal ions by the sulfate anions. 8-Membered ring tunnels run along the *c*-axis, which are occupied by the protonated organic cations.

There is one report of a three-dimensional actinide sulfate, $[C_4NH_{12}]_2[(UO_2)_6(H_2O)_2(SO_4)_7]$, MUS-1 **XXIII**,[83] formed by the bridging of the uranium centres by the sulfate tetrahedra. Each sulfate tetrahedron binds to four different uranium atoms to form the three-dimensional network (Fig. 7b). MUS-1 is structurally similar to $Mg(UO_2)_6(MoO_4)_7(H_2O)_{15}$ and Sr-$(UO_2)_6(MoO_4)_7(H_2O)_{19}$.[84] Channels run along the *c*-axis of the unit cell and are occupied by the tetramethylammonium cations.

Open-framework metal selenites

Selenite-based frameworks are of interest due to the possible role of the lone pair of electrons as an invisible

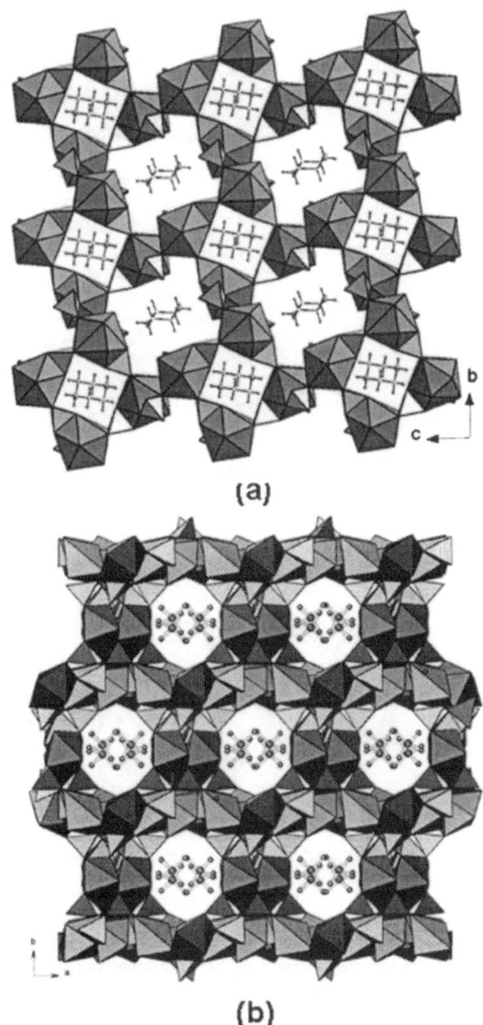

(a)

(b)

Fig. 7 (a). The three-dimensional framework of neodymium sulfate with eight and 12-membered channels in [C$_4$N$_2$H$_{12}$][Nd$_2$(H$_2$O)$_2$(SO$_4$)$_4$], **XXII**, redrawn from ref. 68. (b) Three-dimensional structure of MUS-1 in [C$_4$NH$_{12}$]$_2$[(UO$_2$)$_6$(H$_2$O)$_2$(SO$_4$)$_7$], **XXIII**, as viewed along the [001] direction, redrawn from ref. 83, copyright (2002) from the Royal Society of Chemistry.

structure-directing agent. The stereochemically active lone pair of electrons in Se(IV) leads to a pyramidal coordination for the selenite species. Due to the presence of lone pair, metal selenites tend to crystallize in non-centrosymmetric structures giving rise to interesting physical properties, such as non-linear optical second harmonic generation (SHG).[85,86] The low reduction potential of the Se^{IV}/Se^0 couple, which under hydrothermal conditions may cause reduction to metallic Se, renders the synthesis of amine-templated open-framework metal selenites rather difficult. The first organically templated metal selenite with a layered structure was reported by Harrison *et al.*[87] Rao *et al.*[88,89] reported the first

three-dimensional organically templated iron(III) and zinc selenites. Since then, there have been a few other reports of 2D and 3D metal selenites

One-dimensional zinc selenite. [C$_3$N$_2$H$_{12}$]$_4$[Zn$_4$(SeO$_3$)$_8$] is formed by the connectivity of the strictly alternating ZnO$_4$ and SeO$_3$ units which results in the formation of corner sharing 4-membered ring chains, similar to that observed in metal phosphates and sulfates.[90]

Two-dimensional selenites. Depending on the mode of connectivity between the selenite units and the different metal atoms, the two-dimensional metal selenites show a rich variety of architectures. The connectivity of the metal centres by the selenite units wherein the selenite oxygens occupy the equatorial positions around the metal atom results in the formation of layers with 8-membered ring apertures (Fig. 8a). The coordination around the metal atom is completed by bonding to halo or oxo groups, as in [C$_2$N$_2$H$_{10}$]-[CdCl$_2$(HSeO$_3$)$_2$] and [C$_2$N$_2$H$_{10}$][(VO)(SeO$_3$)$_2$], **XXIV**.[91,92] Such a layered topology is found in a copper hydrogen selenite,[93] which is not an open-framework and in a few open-framework layered phosphates such as the layered tin(II) phosphate and phosphatoantimonate.[94,95] The metal centres can also be connected by the selenite groups to form

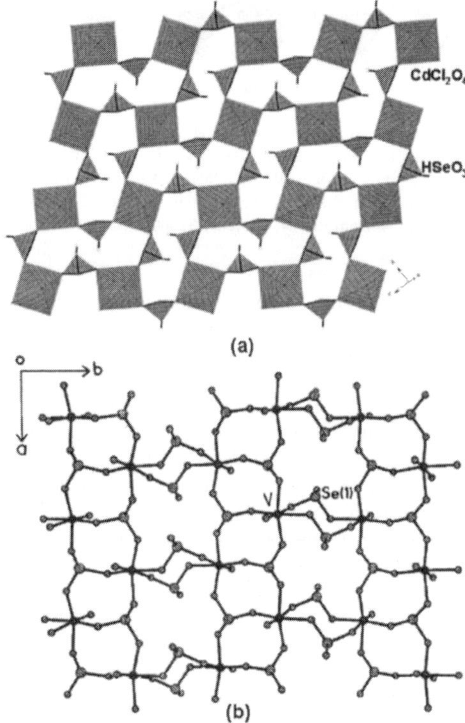

Fig. 8 (a) The layer structure of [C$_2$N$_2$H$_{10}$][CdCl$_2$(HSeO$_3$)$_2$], **XXIV**, with 8-membered ring aperture, redrawn from ref. 91. (b) Layer formed in [C$_6$N$_2$H$_{14}$][(VO)$_2$(HSeO$_3$)$_2$(SeO$_3$)$_2$].2H$_2$O, **XXV**, by the joining of the 4-membered ring ladders by HSeO$_3$ units from ref. 96.

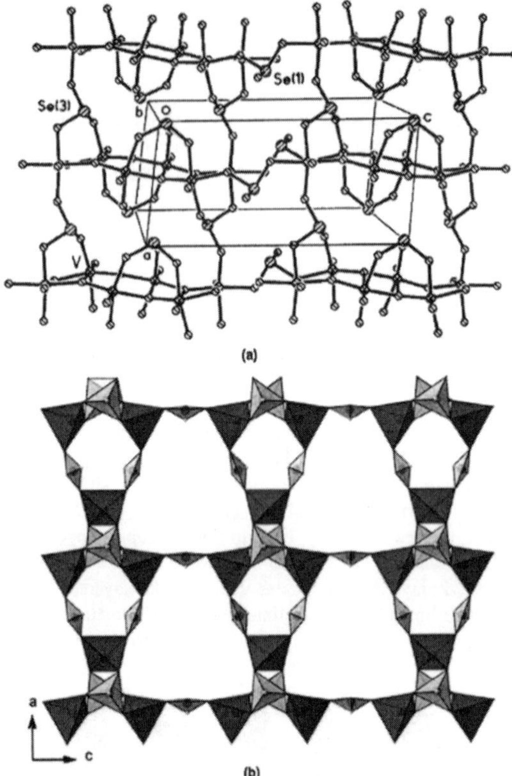

(a)

(b)

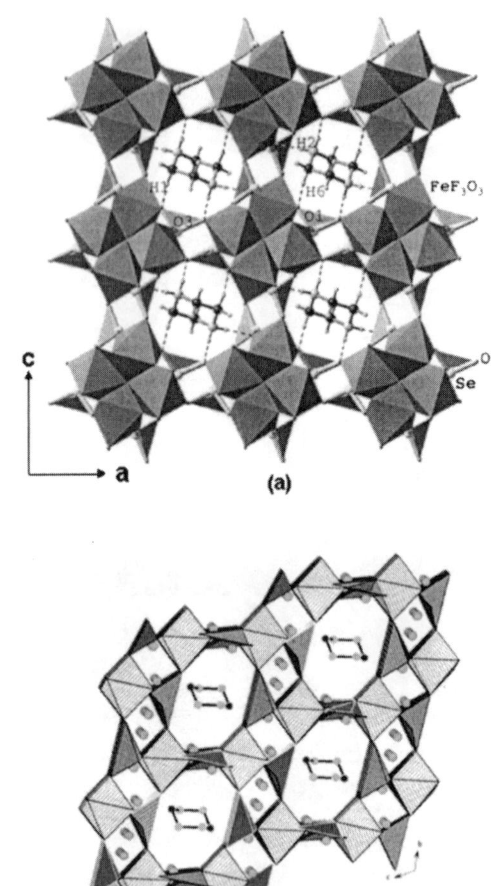

(a)

(b)

Fig. 9 (a) Layer structure formed by the joining of vanadium–oxygen metal clusters, which act as the secondary building units in $[C_2N_2H_{10}][(V^{iv}O)_2(V^vO)O_2(SeO_3)_3]\cdot1.25H_2O$, **XXVI**, from ref. 96. (b) Polyhedral view of the sheet topology of $[CN_3H_6]_4[Zn_3(SeO_3)_5]$ in **XXVII**, redrawn from ref. 87, copyright (2000) with permission from Wiley-VCH.

Fig. 10 (a) Polyhedral view of the inorganic framework of $[A][Fe_4F_6(SeO_3)_4]$, **XXVIII**, showing one-dimensional 8-ring channels from ref. 88. (b) Three-dimensional framework structure of $[C_4N_2H_{12}][M_2(HSeO_3)_2(Se_2O_5)_2]$, **XXIX**, showing 10-membered channels, redrawn from ref. 89.

4-membered ring units which get joined to form layers with 4- and 8-membered ring apertures (Fig. 8b) as in $[C_6N_2H_{14}][(VO)_2(HSeO_3)_2(SeO_3)_2]\cdot2H_2O$, **(XXV)**,[96] Thus, the structure can be described as being formed by the joining of the four-membered ring ladders by the $HSeO_3$ groups. The importance of the ladder structure as a possible precursor of the 3D metal phosphates has been pointed out.[19] The layer structure in the metal selenite is comparable to that of the indium phosphate, $[C_5NH_6][In(HPO_4)_2]$.[97] The metal atoms and the selenite groups can join together to form clusters, which then get connected to form layers, as in the vanadium selenite, $[C_2N_2H_{10}][(V^{iv}O)_2(V^vO)O_2(SeO_3)_3]\cdot1.25H_2O$, **XXVI**. In this compound, the clusters are joined by the selenite groups to form layers with 8-membered apertures[96] (Fig. 9a). Similarly, in $[CN_3H_6]_4[Zn_3(SeO_3)_5]$, **XXVII**, the layer is formed by the linking of $Zn_3Se_2O_{12}$ clusters by the selenite groups (Fig. 9b).[87]

Three-dimensional selenites. In the 3D selenites of the formula, $[A][Fe_4F_6(SeO_3)_4]$, **XXVIII**, obtained with different amines,[88] the Fe centres join together through bridging

fluorine atoms to form tetrameric $Fe_4F_6O_{12}$ clusters, which then get connected by the selenite groups to form the three-dimensional architecture. Such a connectivity results in the formation of 3-, 4-, 5- and 6-membered rings in the structure. There is also a 8-ring 1D channel along the b-axis of the unit cell (Fig. 10a). The presence of the near-tetrahedral $Fe_4F_6O_{12}$ cluster makes it an interesting geometrically frustrated magnetic material.

A three-dimensional open-framework metal selenite of the formula $[C_4N_2H_{12}][M_2(HSeO_3)_2(Se_2O_5)_2]$, **XXIX**, (M = Zn, Co or Ni), containing both the selenite and the diselenite units has been reported.[89] The structure possesses a diamondoid network and has intersecting two-dimensional 10-membered channels. The three-dimensional framework is formed by the connectivity of the zinc-diselenite layers by the selenite groups acting as pillars. This connectivity results in the formation of

4- and 6-membered rings in the structure. In both the three-dimensional metal selenite frameworks, protonated amine molecules reside in the channels and forming extensive hydrogen bonds with the framework atoms (Fig. 10b).

Three-dimensional inorganic–organic hybrid metal selenites, wherein the metal selenite layers get crosslinked by the organic moiety to form the 3D architecture have been reported. In $[C_2N_2H_8][Zn_2(SeO_3)_2]$, the zinc selenite layers get connected by the pillaring ethylene diamine molecules in an end-to-end fashion resulting in the formation of channels along the a- and c-axes.[88] An allotropic form of $[C_2N_2H_8][Zn_2(SeO_3)_2]$, reported by Ferey et $al.$,[90] differs only in the topology of the zinc selenite layers. In $[M(C_{10}N_2H_{10})(H_2O)V_2Se_2O_{10}]$, M = Co, Ni, the Co/Ni–V–SeO$_3$ layers get crosslinked by the 4,4′-bipyridyl units to form the three-dimensional architecture.[98]

Open-framework metal selenates

It is only recently that it has been possible to prepare open-framework metal selenates. The difficulty in the synthesis of these compounds lies in the instability of Se(VI) in the basic medium. It, therefore, becomes necessary to prepare them in an acidic medium.

One-dimensional selenates. The framework structures in $[Cd(H_2O)_2(SeO_4)_2]^{2-}$ [99] and $[Zn(H_2O)_2(SeO_4)_2]^{2-}$, **XXX**,[100] has the kröhnkite-type topology. In this structure, a pair of selenate tetrahedra shares corners to form the one-dimensional chain structure bridging the neighboring metal octahedra as shown in Fig. 11a. Fleck et $al.$[101] have reviewed compounds with the kröhnkite-type topology. The structure of the kröhnkite-type infinite chain is composed of a distorted MO$_6$ octahedron corner-linked with the XO$_4$ tetrahedron, the remaining two opposite corners in the $trans$ position being occupied by water molecules. Each tetrahedron connects the neighboring octahedra, forming infinite $[M(XO_4)_2(H_2O)_2]$ chains. The simplest chain of the formula [Zn(SeO$_4$)(phen)(H$_2$O)], consisting of strictly alternating zinc octahedra and selenate tetrahedra similar to that of zinc sulfate [Zn(SO$_4$)(H$_2$O)$_2$(2,2-bpy)], **I**, has been reported.[100]

Two-dimensional selenates. The structure of $[C_4N_2H_{14}]$ $[La_2(SeO_4)_4].H_2O$, **XXXI**,[102] involves LaO sheets formed by La–μ-O–La linkages forming chains running parallel to the b-axis with the alternate La atoms of the chains being bridged by La–μ-O–La linkages along the perpendicular direction which forms a hexanuclear lanthanide ring. The hexanuclear lanthanide rings are surrounded by six similar rings resulting in the formation of a (6,3) net. Along the chain, two LaO$_9$ polyhedra are capped on either side by the Se(2)O$_4$ units which are connected to two La atoms in the neighboring chains through a three-coordinate oxygen (O8). In the perpendicular direction, the bridging La-atoms are capped by two Se(1)O$_4$ units which share an edge with a LaO$_9$ polyhedron of the neighboring chain. Accordingly, the SeO$_4$ units are involved in both capping and bridging the La atoms in the LaO sheets. Such a dual connectivity of the SeO$_4$ units produces corner-shared four-membered rings propagating along the a-direction

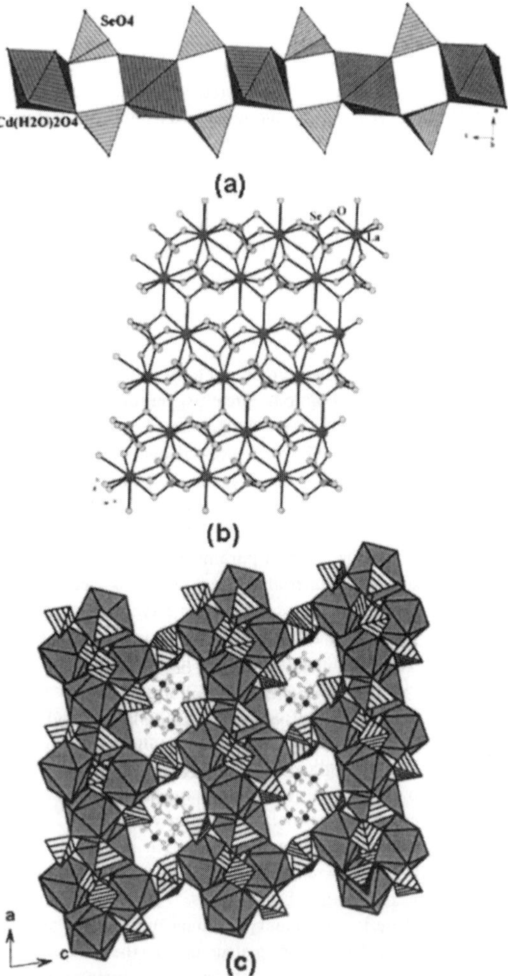

Fig. 11 (a) Polyhedral structure of the $[Cd(SeO_4)_2(H_2O)_2]^{2-}$ chain in **XXX** along the c-axis from ref. 99. (b) The inorganic layer of $[La_2(SeO_4)_4]^{2-}$, in **XXXI** with the capping and bridging of the LaO$_9$ dimers by two selenate units, resulting in corner-shares four-membered rings along the a-axis from ref. 102. (c) Polyhedral view of three-dimensional lanthanum selenate comprising of La$_2$Se$_4$ building units in $[C_2N_2H_{10}][La_2(SeO_4)_4(H_2O)_3].H_2O$, **XXXII**, from ref. 103. Note the presence of 12-membered channels.

as shown in Fig. 11b. $[C_2N_2H_{10}][Ln_2(SeO_4)_4(H_2O)_4]$, constructed by capping two edge-shared LaO$_9$ polyhedra by SeO$_4$ groups has also been described.[102]

Three-dimensional selenate. A three-dimensional selenate of the formula, $[C_2N_2H_{10}][La_2(SeO_4)_4(H_2O)_3]\cdot H_2O$, **XXXII**, comprising La$_2$Se$_4$ building units and possessing 12-membered channels, has been prepared in an acidic medium under hydrothermal conditions.[103] The complex network of LaO$_8$, LaO$_9$ and SeO$_4$ moieties gives rise to a three-dimensional structure possessing channels along all the three axes. Connectivity between two building units with the La$_2$Se$_4$

stoichiometry through La–O–Se–O–La linkages gives rise to infinite chains along the <100> direction, which are connected through a bridging Se(2)O$_4$ to form a two-dimensional layer in the *ab*-plane, with 4- and 8-membered apertures. The layers are stacked one over the other along the *c*-axis and connected by the Se(4)O$_4$ tetrahedra which share corners with the La(1)–O polyhedra from the adjacent layers forming the 3-dimensional network. The 3D structure has 12-membered·channels along the *a*-and *b*-axes, and 8-membered channels along the *c*-axis. Amine molecules reside in the 12-membered channels forming hydrogen bonds with the framework oxygens (Fig. 11c).

Conclusions

The discussion in the earlier sections should suffice to demonstrate how oxyanions such as, sulfate, selenite and selenate could be used to design novel inorganic open-framework materials. The structures of many of the sulfates and selenates are close to those of the phosphates. Although several open-framework metal sulfates have been synthesized and characterized, there is need for further work to investigate the effects of various reaction parameters and solvents on the formation of these materials. Great effort is necessary to synthesize 3D metal selenates since we have one organically templated metal selenate to-date. To our knowledge, no organically-templated metal sulfite has been reported hitherto. It would be worth exploring reactions of organic amine sulfites with metal ions. It should be possible to make use combinations of these oxyanions with other anions such as the phosphate and the silicate to produce novel architectures. There is also need for efforts to obtain open-framework metal sulfates, selenites and selenates with empty channels, in order to study sorption and other properties. An important aspect that is worthy of study relates to the possible transformation of chain sulfates to 2D and 3D structures. It would be of interest to know if there is a secondary building unit for the sulfates similar to the 4-membered ring structures in the case of open-framework phosphates. The synthesis and transformation of such a 4-membered ring sulfate should be examined.

Acknowledgements

The authors thank the DRDO (India) for support of this research.

References

1 D. W. Breck, *Zeolite Molecular Sieves*, Wiley, New York, 1974.
2 W. M. Meier, D. H. Oslen and C. Baerlocher, *Atlas of Zeolite Structure Types*, Elsevier: London, 1996.
3 A. K. Cheetam, G. Ferey and T. Loiseau, *Angew. Chem., Int. Ed.*, 1999, **38**, 3268.
4 C. N. R. Rao, S. Natarajan, A. Choudhury, S. Neeraj and A. A. Ayi, *Acc. Chem. Res.*, 2001, **34**, 80.
5 C. N. R. Rao, S. Natarajan and R. Vaidyanathan, *Angew. Chem., Int. Ed.*, 2004, **43**, 1466.
6 O. M. Yaghi, M. O'Keefe, N. W. Ockwig, H. K. Chae, M. Eddaoudi and J. Kim, *Nature*, 2003, **423**, 705.
7 O. M. Yaghi, H. L. Li, C. Davis, D. Richardson and T. L. Groy, *Acc. Chem. Res.*, 1998, **31**, 474.
8 M. E. Davis and R. F. Lobo, *Chem. Mater.*, 1992, **4**, 756 and the references therein.
9 A. Choudhury, S. Natarajan and C. N. R. Rao, *Inorg. Chem.*, 2000, **39**, 4295.
10 A. Choudhury, J. Krishnamoorthy and C. N. R. Rao, *Chem. Commun.*, 2001, 2610.
11 S. Neeraj, S. Natarajan and C. N. R. Rao, *Angew. Chem., Int. Ed.*, 1999, **38**, 3480.
12 C. N. R. Rao, S. Neeraj and S. Natarajan, *J. Am. Chem. Soc.*, 2000, **122**, 2810.
13 J. N. Behera and C. N. R. Rao, *Z. Anorg. Allg. Chem.*, 2005, **631**, 3030.
14 M. I. Khan, S. Cevik and R. J. Doedens, *Chem. Commun.*, 2001, 1930.
15 G. Paul, A. Choudhury, R. Nagarajan and C. N. R. Rao, *Inorg. Chem.*, 2003, **42**, 2004.
16 F. C. Hawthrone, S. V. Krivovichev and P. C. Burns, *Rev. Mineral. Geochem.*, 2000, **40**, 1.
17 G. Paul, A. Choudhury and C. N. R. Rao, *Chem. Mater.*, 2003, **15**, 1174.
18 S. Chakrabarti and S. Natarajan, *Angew. Chem., Int. Ed.*, 2002, **41**, 1224.
19 A. Choudhury, S. Neeraj, S. Natarajan and C. N. R. Rao, *J. Mater. Chem.*, 2001, **11**, 1537.
20 X. Bu, T. E. Gier and G. D. Stutkey, *Chem. Commun.*, 1997, 2271.
21 W. T. A. Harrison, Z. Bircsak, L. Hannooman and Z. Zhang, *J. Solid State Chem.*, 1998, **136**, 93.
22 S. Neeraj, S. Natarajan and C. N. R. Rao, *Chem. Mater.*, 1999, **11**, 1390.
23 A. Choudhury, S. Neeraj, S. Natarajan and C. N. R. Rao, *J. Mater. Chem.*, 2001, **11**, 1537.
24 A. M. Chippindale and C. Turner, *J. Solid State Chem.*, 1997, **128**, 317.
25 T. Loiseau, F. Serpaggi and G. Férey, *Chem. Commun.*, 1997, 1093.
26 A. Choudhury, S. Natarajan and C. N. R. Rao, *J. Chem. Soc., Dalton Trans.*, 2000, 2595.
27 U. Kolitsch, *Acta Crystallogr., Sect. C: Cryst. Struct. Commun.*, 2004, **60**, i3–i6.
28 Q. Gao, J. Chen, S. Li, R. Xu, J. M. Thomas, M. Light and M. B. Hursthouse, *J. Solid State Chem.*, 1996, **127**, 145.
29 W. Tieli, Y. Long and P. Wenqin, *J. Solid State Chem.*, 1990, **89**, 392.
30 R. H. Jones, J. M. Thomas, R. Xu, Y. Xu, Q. Huo, A. K. Cheetham and J. Chen, *J. Chem. Soc., Chem. Commun.*, 1990, 1170.
31 A. M. Chippindale, A. D. Bond, A. D. Law and A. R. Cowley, *J. Solid State Chem.*, 1998, **136**, 227.
32 A. A. Ayi, S. Neeraj, A. Choudhury, S. Natarajan and C. N. R. Rao, *J. Phys. Chem. Solids*, 2001, **62**, 1481.
33 S. Oliver, A. Kuperman, A. Lough and G. A. Ozin, *Chem. Mater.*, 1996, **8**, 2391.
34 S. Oliver, A. Kuperman and G. A. Ozin, *Angew. Chem., Int. Ed.*, 1998, **37**, 46.
35 G. Paul, A. Choudhury and C. N. R. Rao, *J. Chem. Soc., Dalton Trans.*, 2002, 23859.
36 M. I. Khan, S. Cevik and R. J. Doedens, *Inorg. Chim. Acta*, 1999, **292**, 112.
37 M. Cavellec, D. Riou, J. M. Greneche and G. Ferey, *Inorg. Chem.*, 1997, **36**, 2187.
38 Z. A. D. Lethbridge, P. Lightfoot, R. E. Morris, D. S. Wragg, P. A. Wright, A. Kvick and G. Vaughan, *J. Solid State Chem.*, 1999, **142**, 455.
39 R. H. Jones, J. M. Thomas, H. Qisheng, M. B. Hursthouse and J. Chen, *J. Chem. Soc., Chem. Commun.*, 1991, 1520.
40 L. Fafani, A. Nunzi and P. F. Zanazzi, *Am. Mineral.*, 1971, **56**, 751.
41 R. A. Ramik, B. D. Sturman, P. J. Dunn and A. S. Poverennukh, *Can. Mineral.*, 1980, **18**, 185.
42 N. Simon, T. Loiseau and G. Ferey, *J. Mater. Chem.*, 1999, **9**, 585.
43 F. C. Hawthrone, *Acta Crystallogr., Sect. B: Struct. Sci.*, 1994, **50**, 481 and the references therein.
44 C. Livage, F. Millange, R. I. Walton, T. Loiseau, N. Simon, D. O'Hare and G. Ferey, *Chem. Commun.*, 2001, 994.
45 R. I. Walton, F. Millange, A. L. Bail, T. Loiseau, C. Serre, D. O'Hare and G. Férey, *Chem. Commun.*, 2000, 203.
46 J. R. Gutnick, E. A. Muller, A. N. Sarjeant and A. J. Norquist, *Inorg. Chem.*, 2004, **43**, 6528.

47 J. Fuch, H.-U. Kreusler and A. Z. Forster, *Z. Naturforsch., B*, 1979, **34**, 1683.
48 Y. Xing, Y. Liu, Z. Shi, H. Meng and W. Pang, *J. Solid State Chem.*, 2003, **174**, 381.
49 A. J. Norquist, M. B. Doran, P. M. Thomas and D. O'Hare, *Dalton Trans.*, 2003, 1168.
50 A. J. Norquist, P. M. Thomas, M. B. Doran and D. O'Hare, *Chem. Mater.*, 2002, **14**, 5179.
51 M. B. Doran, B. E. Cockbain and D. O'Hare, *Dalton Trans.*, 2005, 1774.
52 A. J. Norquist, M. B. Doran and D. O'Hare, *Solid State Sci.*, 2003, **5**, 1149.
53 A. J. Norquist, M. B. Doran, P. M. Thomas and D. O'Hare, *Inorg. Chem.*, 2003, **42**, 5949.
54 M. B. Doran, A. J. Norquist and D. O'Hare, *Inorg. Chem.*, 2003, **42**, 6989.
55 M. B. Doran, B. E. Cockbain and D. O'Hare, *Dalton Trans.*, 2004, 3810.
56 P. M. Thomas, A. J. Norquist, M. B. Doran and D. O'Hare, *J. Mater. Chem.*, 2003, **13**, 88.
57 M. Cavellec, D. Riou and G. Ferey, *J. Solid State Chem.*, 1994, **112**, 441.
58 M. Cavellec, D. Roiu and G. Ferey, *Eur. J. Solid State Inorg. Chem.*, 1995, **32**, 271.
59 X. Wang, L. Liu, H. Cheng, K. Ross and A. J. Jacobson, *J. Mater. Chem.*, 2000, **10**, 1203.
60 G. Paul, A. Choudhury, E. V. Sampathakumaran and C. N. R. Rao, *Angew. Chem., Int. Ed.*, 2002, **41**, 1224.
61 J. E. Dutrizac and S. Kaiman, *Can. Mineral.*, 1976, **14**, 151.
62 C. N. R. Rao, G. Paul, A. Choudhury, E. V. Sampathkumaran, A. K. Raychaudhuri, S. Ramasesha and I. Rudra, *Phys. Rev. B: Condens. Matter*, 2003, **67**, 134425.
63 C. N. R. Rao, E. V. Sampathkumaran, R. Nagarajan, G. Paul, J. N. Behera and A. Choudhury, *Chem. Mater.*, 2004, **16**, 1441.
64 G. Paul, A. Choudhury and C. N. R. Rao, *Chem. Commun.*, 2002, 1904.
65 J. N. Behera, G. Paul, A. Choudhury and C. N. R. Rao, *Chem. Commun.*, 2004, 456.
66 J. N. Behera, K. V. Gopalkrishnan and C. N. R. Rao, *Inorg. Chem.*, 2004, **43**, 2636.
67 F. C. Hawthrone, S. V. Krivovichev and P. C. Burns, *Rev. Mineral. Geochem.*, 2000, **40**, 55.
68 M. Dan, J. N. Behera and C. N. R. Rao, *J. Mater. Chem.*, 2004, **14**, 1257.
69 L. Liu, H. Meng, G. Li, Y. Cui, X. Wang and W. Pang, *J. Solid State Chem.*, 2005, **178**, 1003.
70 T. Bataille and D. Louer, *J. Mater. Chem.*, 2002, **12**, 3487.
71 Y. Xing, Z. Shi, G. Li and W. Pang, *Dalton Trans.*, 2003, 940.
72 P. Benard-Rocherulle, H. Tronel and D. Louer, *Powder Diffr.*, 2002, **17**, 290.
73 S. Govindarajan, K. C. Patil, H. Manohar and P. E. Werner, *J. Chem. Soc., Dalton Trans.*, 1986, 119.
74 D. Wang, R. Yu, Y. Xu, S. Feng, R. Xu, N. Kumada, N. Kinomura, Y. Matsumura and M. Takano, *Chem. Lett.*, 2002, 1120.
75 M. S. Wickleder, *Chem. Rev.*, 2002, **102**, 2011.
76 A. J. Norquist, M. B. Doran and D. O'Hare, *Inorg. Chem.*, 2005, **44**, 3837.
77 D. M. C. Huminicki and F. C. Hawthorne, *Rev. Miner. Geochem.*, 2002, **48**, 193.
78 C. N. Morimoto and E. C. Lingafelter, *Acta Crystallogr., Sect. B: Struct. Crystallogr. Cryst. Chem.*, 1970, **B26**, 341.
79 W.T.A. Harrison, M. L. F. Phillips and T. M. Nenoff, *J. Chem. Soc., Dalton Trans.*, 2001, 2459.
80 W.T.A. Harrison and M. L. F. Phillips, *Chem. Mater.*, 1997, **9**, 1837.
81 I. Bull, P. S. Wheatley, P. Lightfoot, R. E. Morris, E. Sastre and P. A. Wright, *Chem. Commun.*, 2002, 1180.
82 Y.-P. Yuan, R.-Y. Wang, D.-Y. Kong, J.-G. Mao and A. Clearfield, *J. Solid State Chem.*, 2005, **178**, 2030.
83 M. Doran, A. J. Norquist and D. O'Hare, *Chem. Commun.*, 2002, 2946.
84 V. V. Tabachenko, L. M. Kovba and V. N. Serezhkin, *Khoord Khim.*, 1984, **10**, 558.
85 P. S. Halasyamani and K. R. Poeppelmier, *Chem. Mater.*, 1998, **10**, 2753.
86 Y. Porter, N. S. P. Bhuvanesh and P. S. Halasyamani, *Inorg. Chem.*, 2001, **40**, 1172.
87 W. T. A. Harrison, M. L. F. Phillips, J. Stanchfield and T. M. Nenoff, *Angew. Chem., Int. Ed.*, 2000, **39**, 3808.
88 A. Choudhury, D. Udayakumar and C. N. R. Rao, *Angew. Chem., Int. Ed.*, 2002, **41**, 158.
89 D. Udayakumar and C. N. R. Rao, *J. Mater. Chem.*, 2003, **13**, 1635.
90 F. Millange, C. Serre, T. Cabourdin, J. Marrot and G. Ferey, *Solid State Sci.*, 2004, **6**, 229.
91 I. Pasha, A. Choudhury and C. N. R. Rao, *Solid State Sci.*, 2003, **5**, 257.
92 Z. Dai, Z. Shi, G. Li, X. Chen, X. Lu, Y. Xu and S. Feng, *J. Solid State Chem.*, 2003, **172**, 205.
93 H. Effenberger, *Z. Kristallogr.*, 1985, **173**, 265.
94 S. Natarajan and A. K. Cheetham, *J. Solid State Chem.*, 1998, **140**, 435.
95 Y. Piffard, V. Verbeare, S. Oyetola, S. Courant and M. Tournoux, *Eur. J. Solid State Inorg. Chem.*, 1989, **26**, 113.
96 I. Pasha and C. N. R. Rao, *Inorg. Chem.*, 2003, **42**, 409.
97 A. M. Chippindale and S. J. Brech, *Chem. Commun.*, 1996, 2781.
98 Z. Dai, X. Chen, Z. Shi, D. Zhang, G. Li and S. Feng, *Inorg. Chem.*, 2003, **42**, 908.
99 I. Pasha, A. Choudhury and C. N. R. Rao, *J. Solid State Chem.*, 2003, **174**, 386.
100 M.-L. Feng, J.-G. Mao and J.-L. Song, *J. Solid State Chem.*, 2004, **177**, 3529.
101 M. Fleck, U. Kolitsch and B. Hertweck, *Z. Kristallogr.*, 2002, **217**, 435.
102 D. Udayakumar, M. Dan and C. N. R. Rao, *Eur. J. Inorg. Chem.*, 2004, 1733.
103 J. N. Behera, A. A. Ayi and C. N. R. Rao, *Chem. Commun.*, 2004, 968.

Inorg. Chem. 2007, 46, 2511−2518

Inorganic Chemistry

: Article

Coordination Polymers and Hybrid Networks of Different Dimensionalities Formed by Metal Sulfites

K. Prabhakara Rao and C. N. R. Rao*

Chemistry and Physics of Materials Unit and CSIR Centre of Excellence in Chemistry, Jawaharlal Nehru Centre for Advanced Scientific Research, Jakkur P.O., Bangalore 560064, India

Received October 17, 2006

In our effort to explore the use of the sulfite ion to design hybrid and open-framework materials, we have been able to prepare, under hydrothermal conditions, zero-dimensional $[Zn(C_{12}H_8N_2)(SO_3)] \cdot 2H_2O$, I ($a = 7.5737(5)$ Å, $b = 10.3969(6)$ Å, $c = 10.3986(6)$ Å, $\alpha = 64.172(1)°$, $\beta = 69.395(1)°$, $\gamma = 79.333(1)°$, $Z = 2$, and space group $P\bar{1}$), one-dimensional $[Zn_2(C_{12}H_8N_2)(SO_3)_2(H_2O)]$, II ($a = 8.0247(3)$ Å, $b = 9.4962(3)$ Å, $c = 10.2740(2)$ Å, $\alpha = 81.070(1)°$, $\beta = 80.438(1)°$, $\gamma = 75.66(5)°$, $Z = 2$, and space group $P\bar{1}$), two-dimensional $[Zn_2(C_{10}H_8N_2)(SO_3)_2] \cdot H_2O$, III ($a = 16.6062(1)$ Å, $b = 4.7935(1)$ Å, $c = 19.2721(5)$ Å, $\beta = 100.674(2)°$, $Z = 4$, and space group $C2/c$), and three-dimensional $[Zn_4(C_6H_{12}N_2)(SO_3)_4(H_2O)_4]$, IV ($a = 11.0793(3)$ Å, $c = 8.8246(3)$ Å, $Z = 2$, and space group $P42nm$), of which the last three are coordination polymers. A hybrid open-framework sulfite−sulfate of the composition $[C_2H_{10}N_2][Nd(SO_3)(SO_4)(H_2O)]_2$, V ($a = 9.0880(3)$ Å, $b = 6.9429(2)$ Å, $c = 13.0805(5)$ Å, $\beta = 91.551(2)°$, $Z = 2$, and space group $P2_1/c$), with a layered structure containing metal−oxygen−metal bonds has also been described.

Introduction

Coordination polymers, open-framework materials, and hybrid compounds built up with various anions have been described in the literature. The most common anions employed in open-framework structures are silicates and phosphates.[1−3] Metal carboxylates with a variety of structures and dimensionalities have also been described in the recent literature.[4,5] In recent years, other oxyanions such as sulfate, selenate, selenite, and tellurite have also been employed to design these structures.[6,7] Surprisingly, coordination polymers and open-framework structures employing the sulfite ion do not appear to have been synthesized and characterized hitherto. This is not entirely surprising since the sulfur atom in the sulfite ion is in the 4+ oxidation state, which is relatively unstable under hydrothermal and acidic conditions. The sulfite ion gets readily oxidized to the sulfate ion. It is possible that in the sulfite-based materials, the lone pair of electrons may act as an invisible structure-directing agent. Due to the lone pair, many of the compounds analogous to the sulfites such as selenites and tellurites crystallize in noncentrosymmetric structures, with interesting nonlinear physical properties, such as second harmonic generation.[8]

In this article, we describe the synthesis and structures of the first examples of metal sulfite-based coordination polymers and hybrid structures with different dimensionalities between one and three as well as a zero-dimensional coordination compound. These compounds have the compositions $[Zn(C_{12}H_8N_2)(SO_3)] \cdot 2H_2O$, I, $[Zn_2(C_{12}H_8N_2)(SO_3)_2(H_2O)]$, II, $[Zn_2(C_{10}H_8N_2)(SO_3)_2] \cdot H_2O$, III, and $[Zn_4(C_6H_{12}N_2)(SO_3)_4(H_2O)_4]$, IV. In addition, we have obtained a layered sulfite−sulfate, $[C_2H_{10}N_2][Nd(SO_3)(SO_4)(H_2O)]_2$, V, containing metal−oxygen−metal bonds. The present results suggest

* To whom correspondence should be addressed. E-mail: cnrrao@jncasr.ac.in. Fax: +91-80-22082760, 22082766.

(1) Breck, D. W. *Zeolite Molecular Sieves*; Wiley: New York, 1974.
(2) Meier, W. M.; Oslen, D. H.; Baerlocher, C. *Atlas of Zeolite Structure Types*; Elsevier: London, 1996.
(3) Cheetham, A. K.; Férey, G.; Loiseau, T. *Angew. Chem., Int. Ed.* **1999**, *38*, 3268.
(4) (a) Rao, C. N. R.; Natarajan, S.; Vaidhyanathan, R.; *Angew. Chem., Int. Ed.* **2004**, *43*, 1466. (b) Ocwig, N. C.; Delgado-Friedrichs, O.; O'Keefe, M.; Yaghi, O. M. *Acc. Chem. Res.* **2005**, *38*, 176. (c) Forster, P. M.; Cheetham, A. K. *Top. Catal.* **2003**, *24*, 79. (d) Férey, G.; M-Draznieks, C.; Serre, C.; Millange, F.; Dutour, J.; Surblé, S.; Margiolaki, I. *Science* **2005**, *309*, 2040. (e) Kitaura, R.; Fujimoto, K.; Noro, S.; Kono, M.; Kitagawa, S. *Angew. Chem., Int. Ed.* **2002**, *41*, 133. (f) Zhao, X.; Xiao, B.; Fletcher, A. J.; Thomas, K. M.; Bradshaw, D.; Rosseinsky, M. J. *Science* **2004**, *306*, 1012.
(5) Cheetham, A. K.; Rao, C. N. R. *Chem. Commun.* **2006**, 4780.
(6) Rao, C. N. R.; Behra, J. N; Dan, M. *Chem. Soc. Rev.* **2006**, *35*, 375.
(7) Xie, J-Y; Mao, J-G. *Inorg. Chem. Commun.* **2005**, *8*, 375.

(8) Halasyamani, P. S.; Poeppelmier, K. R. *Chem. Mater.* **1998**, *10*, 2753.

Table 1. Crystal Data and Structure Refinement Parameters for I−V

structure params	I	II	III	IV	V
empirical formula	$ZnSO_5N_2C_{12}H_{12}$	$Zn_2S_2O_7N_2C_{12}H_{10}$	$Zn_2S_2O_7N_2C_{10}H_{10}$	$Zn_4S_4O_{16}N_2C_6H_{20}$	$Nd_2S_4O_{16}N_2C_2H_{14}$
fw	361.67	489.08	465.06	765.96	738.87
crystal syst	triclinic	triclinic	monoclinic	tetragonal	monoclinic
space group	$P\bar{1}$ (No. 2)	$P\bar{1}$ (No. 2)	$C2/c$ (No.15)	$P42nm$ (No. 102)	$P2_1/c$ (No. 14)
a /Å	7.5737(5)	8.0247(3)	16.6062(1)	11.0793(3)	9.0880(3)
b /Å	10.3969(6)	9.4962(3)	4.7935(1)	11.0793(3)	6.9429(2)
c /Å	10.3986(6)	10.2740(2)	19.2721(5)	8.8246(3)	13.0805(5)
α /deg	64.172(1)	81.070(1)	90	90	90
β /deg	69.395(1)	80.438(1)	100.674(2)	90	91.551(2)
γ /deg	79.333(1)	75.66(5)	90	90	90
V /Å^3	689.38(7)	742.63(4)	1507.55(5)	1083.23(6)	825.04(5)
Z	2	2	4	2	2
D (calc)/g cm^{-3}	1.742	2.187	2.049	2.348	2.974
μ /mm^{-1}	1.955	3.554	3.495	4.839	6.816
total data collected	2822	3095	2897	4268	13359
unique data	1931	2115	1084	763	1514
R indices R_1,[a] R_2[b] $[I > 2\sigma(I)]$	0.0387, 0.0978	0.0369, 0.0946	0.0248, 0.0642	0.0272, 0.0706	0.0175, 0.0472
R indices R_1,[a] R_2[b] (all data)	0.0417, 0.0993	0.0399, 0.0957	0.0267, 0.0650	0.0293, 0.0726	0.0176, 0.0473

[a] $R_1 = \Sigma||F_o| - |F_c||/\Sigma|F_o|$. [b] $R_2 = \{\Sigma[w(F_o^2 - F_c^2)^2]/\Sigma[w(F_o^2)^2]\}^{1/2}$, $w = 1/[\sigma^2(F_o)^2 + (aP)^2 + bP]$, $P = [\max(F_o^2,0) + 2(F_c)^2]/3$, where $a = 0.0402$ and $b = 0.0$ for **I**, $a = 0.0567$ and $b = 0.0$ for **II**, $a = 0.0369$ and $b = 0.7579$ for **III**, $a = 0.0396$ and $b = 0.1992$ for **IV**, and $a = 0.0225$ and $b = 1.710$ for **V**.

that it should be possible to utilize the sulfite ion to build up hybrid network materials.

Experimental Section

Synthesis. Ammonium sulfite monohydrate, 4,4′-bipyridyl, 1,4-diazabicyclo[2.2.2]octane (DABCO), neodymium(III) nitrate hexahydrate (Aldrich), 1,10-phenanthroline, ethylenediamine (SD Fine, India), and zinc acetate dihydrate (Qualigens, India) of high purity were used for the synthesis.

Compounds **I**−**V** were synthesized by the hydrothermal method by heating the homogenized reaction mixture in a 23 mL (for **IV**, 7 mL) PTFE-lined bomb at 125 °C (for **V**, 150 °C) over a period of 72 h (for **IV**, 120 h) under autogenous pressure. The pH of the starting reaction mixture was 7, and the pH after the reaction showed little change. The products of the hydrothermal reactions were vacuum-filtered and dried under ambient conditions. The products, containing light yellow block-shaped crystals in the case of **I** and **II**, colorless flakes in **III** and **IV**, and needle-shaped pink crystals of **V**, were isolated in 50−60% yield. The compositions of the starting mixtures for **I**−**V** were as follows: **I**, Zn(OAc)$_2$·2H$_2$O (0.2194 g, 1 mmol), (NH$_4$)$_2$SO$_3$·H$_2$O (0.2683 g, 2 mmol), ethylenediamine (0.07 mL, 1 mmol), 1,10-phenanthroline (0.1982 g, 1 mmol), and H$_2$O (5 mL, 278 mmol); **II**, Zn(OAc)$_2$·2H$_2$O (0.2194 g, 1 mmol), (NH$_4$)$_2$SO$_3$·H$_2$O (0.2683 g, 2 mmol), ethylenediamine (0.07 mL, 1 mmol), 1,10-phenanthroline (0.0660 g, 0.333 mmol), and H$_2$O (5 mL, 278 mmol); **III**, Zn(OAc)$_2$·2H$_2$O (0.1097 g, 0.5 mmol), (NH$_4$)$_2$SO$_3$·H$_2$O (0.1342 g, 1 mmol), ethylenediamine (0.034 mL, 0.5 mmol), 4,4′-bipyridyl (0.0195 g, 0.125 mmol), and H$_2$O (5 mL, 278 mmol); **IV**, Zn(OAc)$_2$·2H$_2$O (0.0768 g, 0.35 mmol), (NH$_4$)$_2$SO$_3$·H$_2$O (0.0536 g, 0.4 mmol), DABCO (0.0112 g, 0.1 mmol), and H$_2$O (2 mL, 111 mmol); and **V**, Nd(NO$_3$)$_3$·6H$_2$O (0.2192 g, 0.5 mmol), (NH$_4$)$_2$SO$_3$·H$_2$O (0.1342 g, 1 mmol), ethylenediamine (0.02 mL, 0.25 mmol), and H$_2$O (5 mL, 278 mmol).

Characterization. CHNS analysis was carried using the Thermo Finnigan FLASH EA 1112 CHNS analyzer. Energy dispersive analysis of X-rays (EDAX) was carried using the OXFORD EDAX system. Infrared spectroscopic studies of KBr pellets were recorded in the mid-IR region (Bruker IFS-66v). Thermogravimetric analysis was carried out (Metler-Toledo) in nitrogen atmosphere (flow rate

= 50 mL/min) in the temperature range of 25−900 °C (heating rate = 10 °C/min).

The structural compositions of **I**−**V** are consistent with the elemental analysis. For **I**, ZnSO$_5$N$_2$C$_{12}$H$_{12}$, Anal. Calcd: C, 39.85; H, 3.34; N, 7.74; S, 8.87. Found: C, 39.72; H, 3.28; N, 7.72; S, 8.81. For **II**, Zn$_2$S$_2$O$_7$N$_2$C$_{12}$H$_{10}$, Anal. Calcd: C, 29.47; H, 2.06; N, 5.73; S, 13.11. Found: C, 29.55; H, 2.11; N, 5.69; S, 13.15. For **III**, Zn$_2$S$_2$O$_7$N$_2$C$_{10}$H$_{10}$, Anal. Calcd: C, 25.83; H, 2.17; N, 6.02; S, 13.79. Found: C, 25.75; H, 2.08; N, 6.15; S, 13.84. For **IV**, Zn$_4$S$_4$O$_{16}$N$_2$C$_6$H$_{20}$, Anal. Calcd: C, 9.40; H, 2.63; N, 3.66; S, 16.75. Found: C, 9.27; H, 2.78; N, 3.65; S, 13.78. For **V**, Nd$_2$S$_4$O$_{16}$N$_2$C$_2$H$_{14}$, Anal. Calcd: C, 3.25; H, 1.91; N, 3.79; S, 17.36. Found: C, 3.19; H, 1.87; N, 3.66; S, 17.28. EDAX gave the expected metal/sulfur ratios of 1:1 for **I**−**IV** and of 1:2 for **V**.

Infrared spectra of **I**−**IV** showed the characteristic bands of the SO$_3^{2-}$ ion[9] around 970(ν_1), 620(ν_2), 930(ν_3), and 470(ν_4) cm^{-1}. The spectra of **I**−**III** showed the C−H stretching bands of the aromatic ring[10,11] in the region of 3090−3000 cm^{-1}. In **IV**, the C−H stretching vibrations of the methylene group of DABCO molecule were observed[10] around 2870 and 2850 cm^{-1}, in addition to the bands, due to the bending modes. In **V**, in addition to the SO$_3^{2-}$ bands, bands in the regions of 1015−850 and 640−580 cm^{-1} due to the SO$_4^{2-}$ ion were observed.[11]

X-ray Diffraction and Crystal Structures. Powder XRD patterns of the products were recorded using Cu Kα radiation (Rich-Seifert, 3000TT). The patterns agreed with those calculated by single-crystal structure determination. A suitable single crystal of each compound was carefully selected under a polarizing microscope and glued to a thin glass fiber. Crystal structure determination by X-ray diffraction was performed on a Siemens Smart-CCD diffractometer equipped with a normal focus, 2.4 kW sealed tube X-ray source (Mo Kα radiation, $\lambda = 0.71073$ Å) operating at 40

(9) Barbara, S. *Infrared Spectroscopy: Fundamentals and Applications*; Wiley: New York, 2004.

(10) Silverstein, R. M.; Bassler, G. C.; Morrill, T. C. *Spectrometric Identification of Organic Compounds*; John Wiley & Sons: New York, 1963.

(11) (a) Nakamoto, K. *Infrared and Raman Spectra of Inorganic and Coordination Compounds*; Wiley: New York, 1978. (b) Rao, C. N. R.; Sampathkumaran, V.; Nagarajan, R.; Paul, G.; Behra, J. N; Choudhury, A. *Chem. Mater.* **2004**, *16*, 1441.

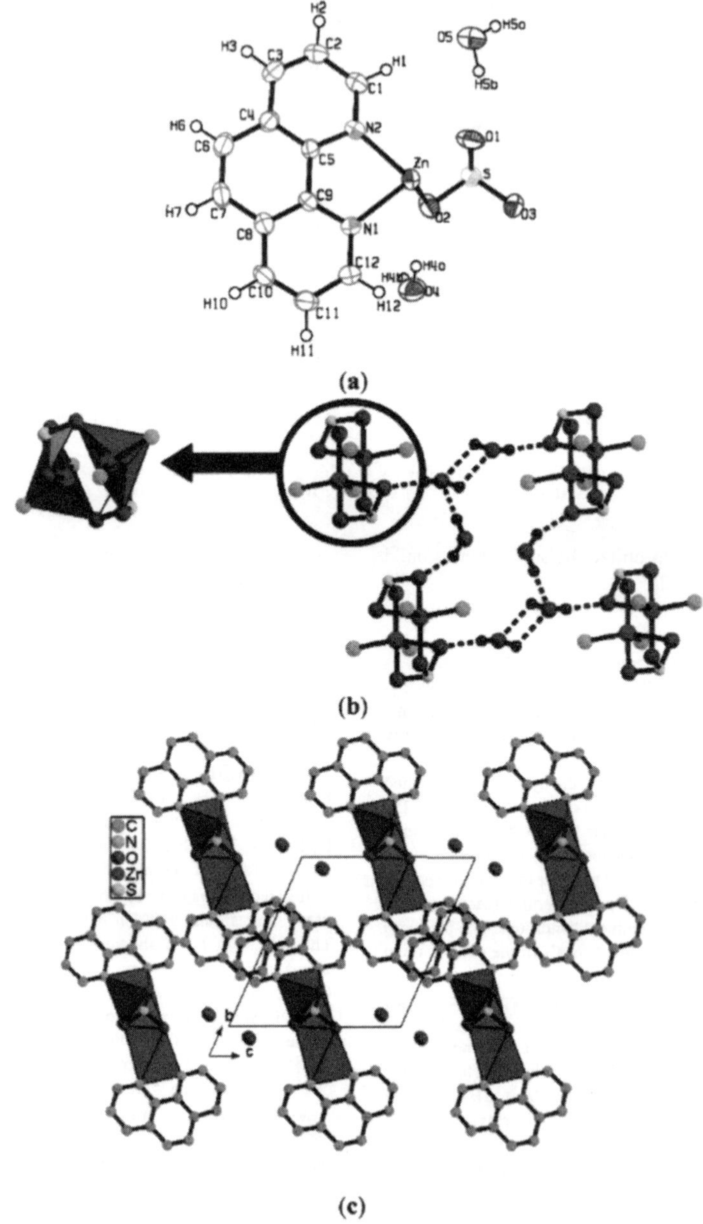

Figure 1. (a) ORTEP diagram (at 50% probability) of **I** showing the atom-labeling scheme. (b) $Zn_2S_2O_6N_4$ polyhedra and the hydrogen-bonded network. (c) Packing diagram, viewed along the *a*-axis of **I**.

kV and 40 mA. An empirical absorption correction based on symmetry equivalent reflections was applied using the SADABS program.[12] The structure was solved and refined using the WinGX suite of programs.[13] The graphic programs[14,15] DIAMOND and ORTEP were used to draw the structures. The final refinement included atomic positions for all the atoms, anisotropic thermal parameters for all the non-hydrogen atoms, and isotropic thermal parameters for the hydrogen atoms. The aromatic hydrogen atoms for **I**−**III**, methylene hydrogen atoms of the DABCO unit (for **IV**), and ethylenediammonium ion (for **V**) were introduced in the

(12) Sheldrick, G. M. *SADABS: Siemens Area Detector Absorption Correction Program;* University of Gottingen: Gottingen, Germany, 1994.

(13) Farrugia, L. J. *J. Appl. Cryst.* **1999**, *32*, 837.

(14) Pennington, W. T. *DIAMOND Visual Crystal Structure Information System. J. Appl. Cryst.* **1999**, *32*, 1029.

(15) (a) Johnson, C. K. *ORTEP*; Oak Ridge National Laboratory: Oak Ridge, TN, 1976. (b) Farrugia, L. J. *ORTEP. J. Appl. Cryst.* **1997**, *30*, 565.

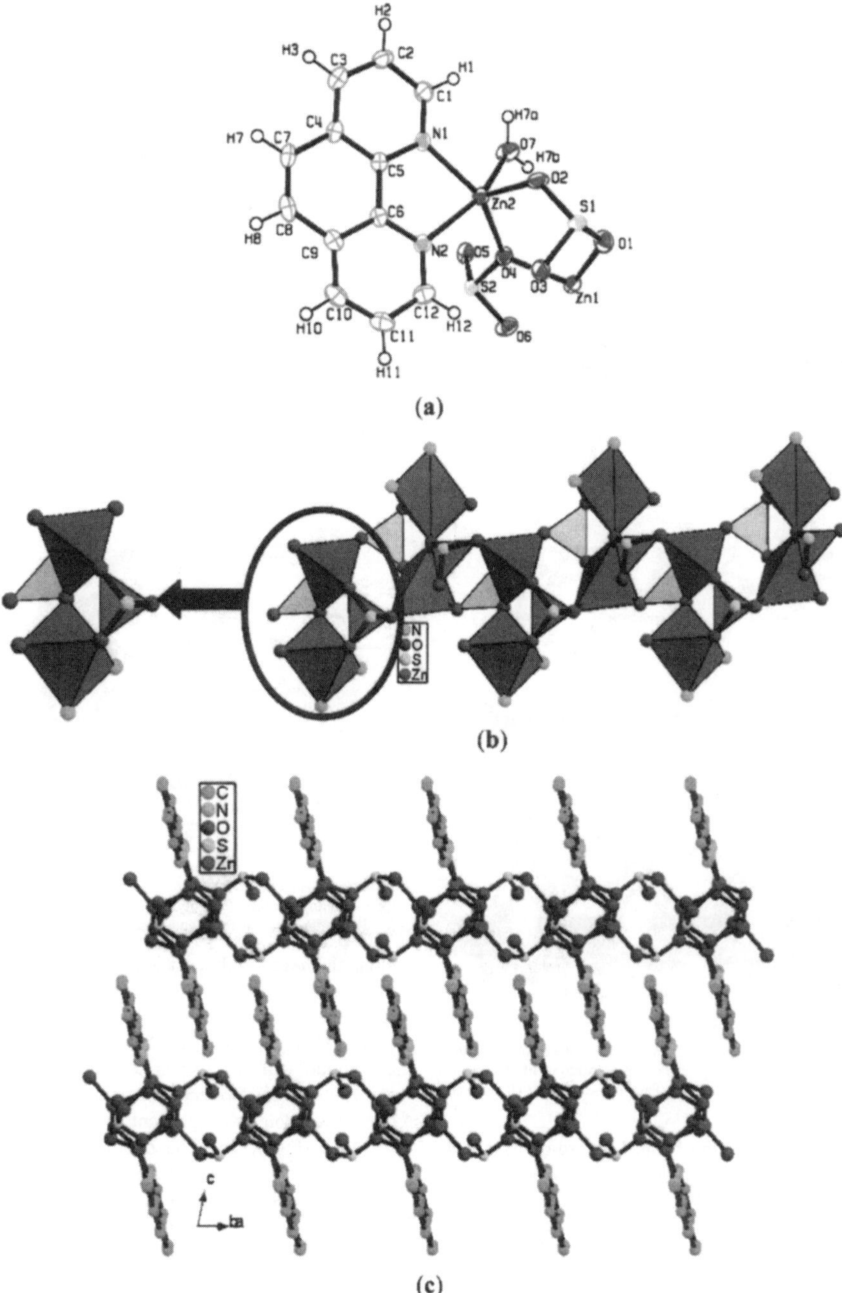

Figure 2. (a) ORTEP diagram (at 50% probability) of **II** showing the atom-labeling scheme. (b) Polyhedral representation of $Zn_2S_2O_9N_2$ unit and the one-dimensional chain. (c) Packing diagram, viewed along $\sim b$-axis of **II**.

calculated positions and refined isotropically. Hydrogens of the water molecules were located by difference-Fourier maps and included in the final refinement. The O–H bond lengths of all the water molecules were constrained to 0.950 Å. Details of the structure solution and final refinements for compounds **I–V** are given in Table 1.

Results and Discussion

We have obtained three zinc sulfite-based coordination polymers involving different amines. The structure is one-dimensional with phenanthroline, two-dimensional with 4,4′-bipyridine, and three-dimensional with DABCO. With

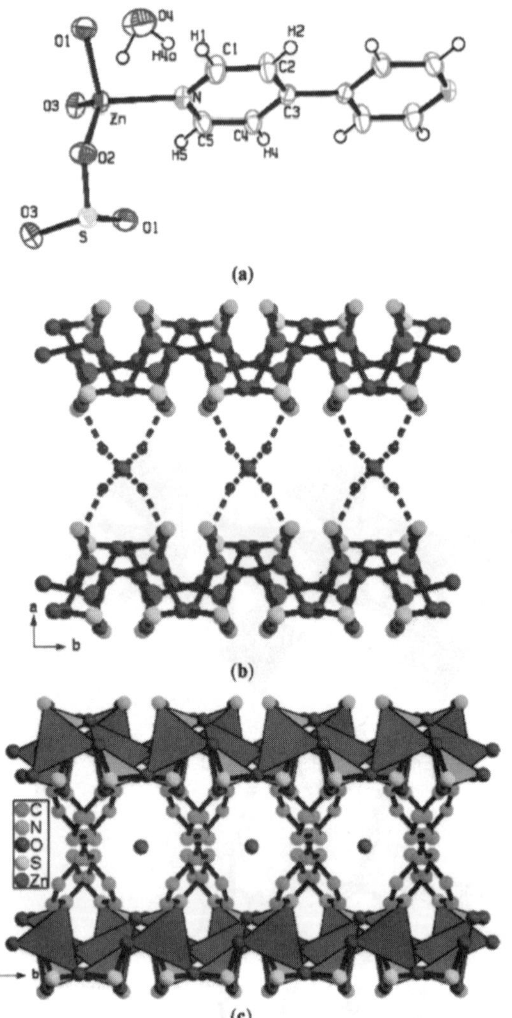

Figure 3. (a) ORTEP diagram (at 50% probability) of **III** showing the atom-labeling scheme. (b) Ball-and-stick representation of the two-dimensional hydrogen-bonded network. (c) Packing diagram of **III**, viewed along the *c*-axis.

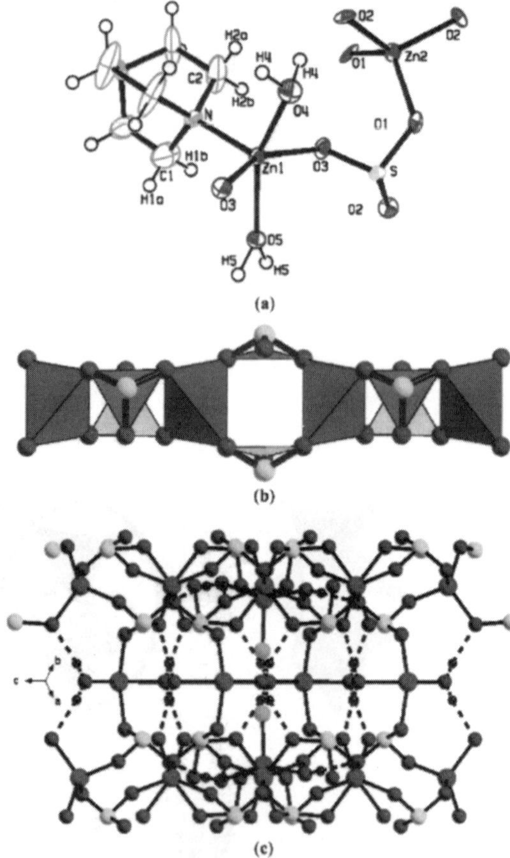

Figure 4. (a) ORTEP diagram (at 50% probability) of **IV** showing the atom-labeling scheme. (b) Polyhedral representation of the twisted chain. (c) Three-dimensional hydrogen-bonded network of **IV**.

phenanthroline, we have also obtained a zero-dimensional coordination compound. The zero-dimensional compound has the formula $[Zn(C_{12}H_8N_2)(SO_3)]\cdot2H_2O$ (**I**), with an asymmetric unit containing 21 non-hydrogen atoms wherein one crystallographically distinct Zn^{2+} ion, one SO_3^{2-} ion, one phenanthroline group (C1–C12, N1 and N2), and two water molecules are present (Figure 1a). The zinc ion is five-coordinated to three oxygens of the sulfite ion and two nitrogens of phenanthroline. The Zn–O and Zn–N bond distances are in the range of 2.028(3)–2.265(3) Å. The ZnO_3N_2 polyhedron shares an edge and a corner of two different sulfite ions forming $Zn_2S_2O_6N_4$ units (Figure 1b), which are hydrogen bonded to two water molecules along the *b*- and *c*-axes with O–O distances in the range of 2.765(6)–2.867(7) Å. The molecular units are packed

along the *a*-axis with $\pi–\pi$ interactions between the phenanthroline rings (Figure 1c), the shortest C–C distance being 3.433(3) Å.

$[Zn_2(C_{12}H_8N_2)(SO_3)_2(H_2O)]$, **II**, is a one-dimensional compound with an asymmetric unit of 25 non-hydrogen atoms containing two crystallographically distinct Zn^{2+} ions, two SO_3^{2-} ions, one phenanthroline unit, and one water molecule (Figure 2a). The Zn1 ion is tetrahedrally coordinated to the oxygens of the sulfite ions, whereas the Zn2 ion is five-coordinated to two oxygens of the sulfite, two nitrogens of the phenanthroline, and a water molecule. The Zn–O and Zn–N bond distances are in the range of 1.953(3)–2.156(3) Å. The $Zn1O_4$ tetrahedron and the $Zn2O_3N_2$ polyhedron share an oxygen of one sulfite ion and are connected to two of the sulfite ions forming $Zn_2S_2O_9N_2$ units (Figure 2b), which share corners along the *a*-axis to form a one-dimensional chain (Figure 2b). The chains are hydrogen bonded to water molecules along the *b*-axis forming a two-dimensional layer along the *a*-axis allowing $\pi–\pi$ interaction (with the shortest C–C distance of 3.468(3) Å) between the phenanthroline units (Figure 2c). There is O···H···O hydrogen bonding between water and sulfite oxygens within the

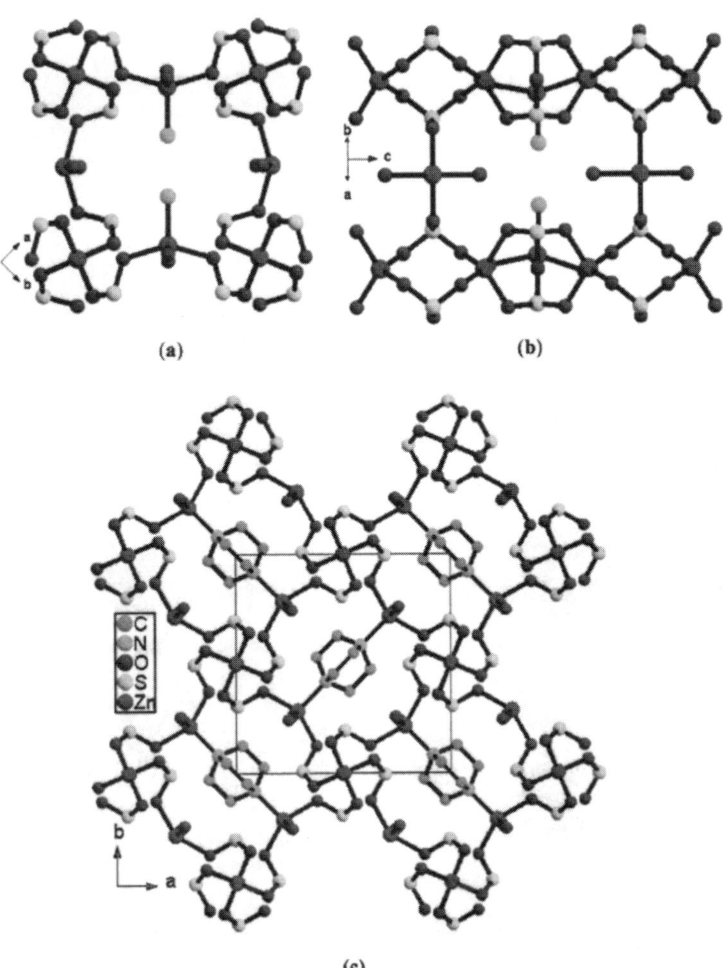

Figure 5. (a) Two-dimensional micropore, viewed along the *c*-axis. (b) Two-dimensional micropore, viewed along the *ab*-plane. (c) Three-dimensional packing diagram, viewed along the *c*-axis of **IV**.

same chain or in the adjacent chain. The O—O distances are in the range of 2.750(4)—2.881(4) Å. The chain compound **II** has a lower water content (or lower H_2O/Zn ratio) component compared with the molecular compound **I**.

$[Zn_2(C_{10}H_8N_2)(SO_3)_2]\cdot H_2O$, **III**, is a two-dimensional compound with an asymmetric unit of 11.5 non-hydrogen atoms containing one crystallographically distinct Zn^{2+} ion, one SO_3^{2-} ion, half of the 4,4′-bipyridine unit, and half of a water molecule (Figure 3a). The Zn atom at the 8*f* crystallographic position is tetrahedrally coordinated to three oxygens of the sulfite ion and a nitrogen atom. The Zn—O and Zn—N bond distances are in the 1.940(2)—2.037(2) Å range. Each ZnO_3N tetrahedron is connected to three different sulfite ions, and each sulfite ion is coordinated to three different Zn ions, forming a one-dimensional ladder along the *b*-axis. The ladders are hydrogen bonded to the water molecules with a O—O distance of 2.956(4) Å along the *a*-axis forming a two-dimensional hydrogen-bonded network (Figure 3b). The hydrogen-bonded layers are diagonally

connected by the 4,4′-bipyridine units to the *ac*-plane forming the two-dimensional layered structure (Figure 3c).

$[Zn_4(C_6H_{12}N_2)(SO_3)_4(H_2O)_4]$, **IV**, is a three-dimensional compound, crystallizing in the noncentrosymmetric space group $P4_2nm$ (No. 103). The asymmetric unit of **IV** contains eight non-hydrogen atoms with two crystallographically distinct Zn^{2+} ions with half occupancy, one SO_3^{2-} ion, one-quarter of the DABCO molecule, and one water molecule (Figure 4a). The Zn1 atom at the 4*c* crystallographic position is five-coordinated to an oxygen of the sulfite ion, two oxygens of the water molecules, and a nitrogen atom. The Zn2 atom at the 4*b* crystallographic position is tetrahedrally coordinated to two oxygens of the sulfite ion. The Zn—O and Zn—N bond distances are in the range of 1.969(7)—2.11(1) Å. Each $Zn2O_4$ tetrahedron shares its corners with four different sulfites, forming a twisted one-dimensional ladder along the *c*-axis (Figure 4b). The ladders are connected through $Zn1O_4N$ polyhedra by the sharing of two corners along the *ab*-plane as well as along the diagonal to the *ab*-

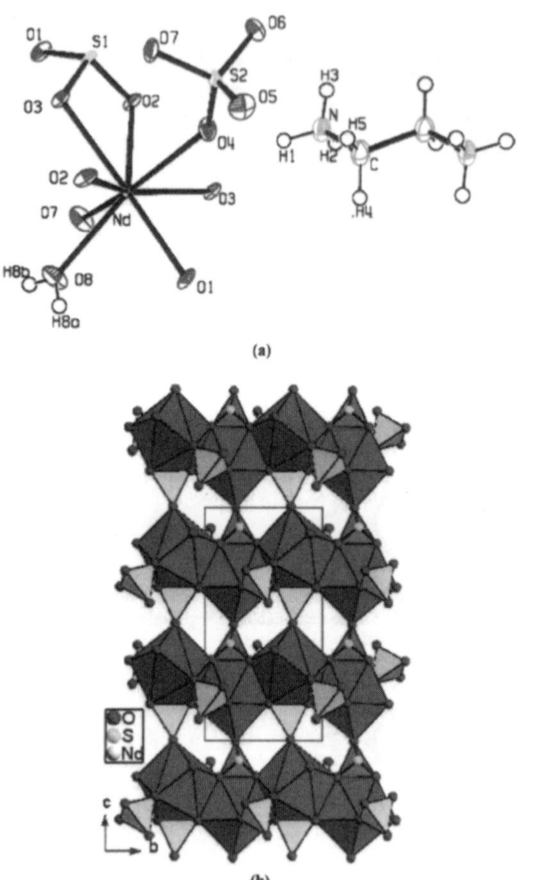

Figure 6. (a) ORTEP diagram (at 50% probability) of **V** showing the atom-labeling scheme. (b) Polyhedral representation of the anionic layer, [Nd(SO₃)(SO₄)(H₂O)]⁻, viewed along the *c*-axis of **V**.

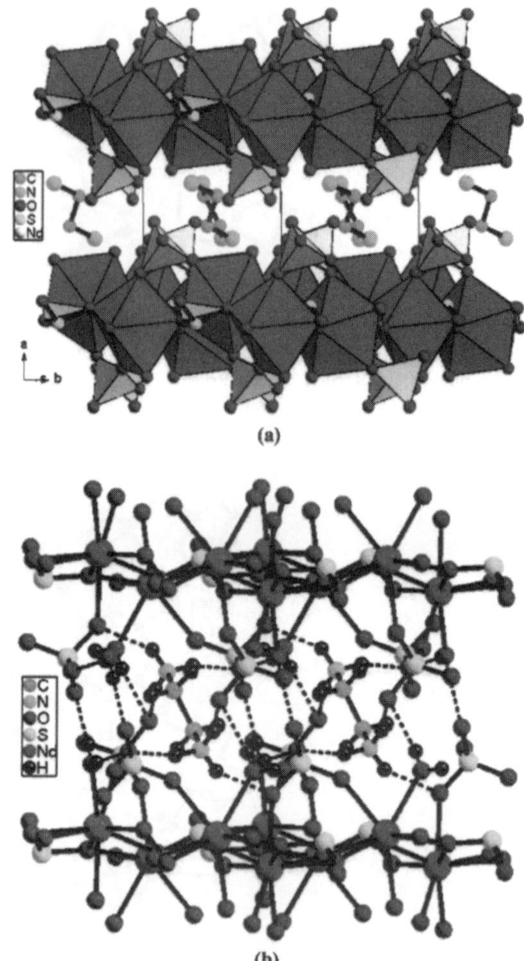

Figure 7. (a) Packing diagram of **V** viewed along the *c*-axis. (b) Three-dimensional hydrogen-bonded network of **V**.

plane, forming a three- dimensional framework. The three-dimensional framework is stabilized by hydrogen bonding between the coordinated water molecules and the oxygens of the sulfite ion with O–O distances in the range of 2.679(9)–2.715(8) Å in all three crystallographic directions (Figure 4c). This arrangement leads to two-dimensional micropores along the *c*-axis and along the *ab*-plane as shown in parts a and b of Figure 5, respectively. The DABCO molecule plays a dual role by satisfying the coordination of Zn1 and also directing the structure of the three-dimensional framework as shown in Figure 5c.

[C₂H₁₀N₂][Nd(SO₃)(SO₄)(H₂O)]₂, **V**, is an open-framework two-dimensional compound, crystallizing in the space group *P*2₁/*c* (No. 14). The asymmetric unit contains 13 non-hydrogen atoms with one crystallographically distinct Nd^{3+} ion, one SO_3^{2-} ion, one SO_4^{2-} ion, half of the ethylenediammonium ion, and a water molecule (Figure 6a). The Nd^{3+} ion at the 4*e* crystallographic position is eight-coordinated to three oxygens of the sulfite ion, two oxygens of the sulfate ion, and a water molecule. The Nd–O bond distances are in the range of 2.367(3)–2.544(2) Å. Each NdO₈ polyhedron

shares an edge with an adjacent NdO₈ polyhedron through two different sulfites (one corner of each), forming helical one-dimensional Nd–O–Nd chains along the *b*-axis (Figure 6b). The {NdO₈}∞ helical chains are connected through corners of the sulfite ions forming two-dimensional layers along the *c*-axis. The sulfate ion bridges two adjacent NdO₈ polyhedra through an O4–O7 edge on either side of the {NdO₈}∞ helical chain. The O5 and O6 oxygens of the sulfate ions are pointed perpendicular to the layer. This arrangement leads to two-dimensional [Nd(SO₃)(SO₄)(H₂O)]⁻ anionic layers (Figure 6b), which are stacked one over the other along the *a*-axis, separated by the charge-compensating protonated ethylenediamine units as shown in Figure 7a. The ethylenediammonium ion forms hydrogen bonds (N–H⋯O) with O4–O6 of the sulfate ion, (d(N⋯O4) 2.940(4), (d(H1⋯O4) 2.13(5) Å, θ(N–H1⋯O4) 149(4)°; (d(N⋯O5) 2.871(4), (d(H2⋯O5) 1.84(5) Å, θ(N–H2⋯O5) 165(4)°; (d(N⋯O6) 2.853(4), (d(H3⋯O6) 2.05(6) Å, θ(N–H3⋯O6) 152(5)°. The coordinated water molecule forms hydrogen

Table 2. S−O Bond Distances in Compounds I−V

compound	S−O	distance (Å)	compound	S−O	distance (Å)
I	S−O1	1.537(3)	IV	S−O1	1.550(8)
	S−O2	1.531(3)		S−O2	1.515(7)
	S−O3	1.534(3)		S−O3	1.518(3)
II	S1−O1	1.521(3)	V	S1−O1	1.496(3)
	S1−O2	1.547(3)		S1−O2	1.531(3)
	S1−O3	1.536(3)		S1−O3	1.561(3)
	S2−O4	1.588(3)		S2−O4	1.480(3)
	S2−O5	1.519(3)		S2−O5	1.454(3)
	S2−O6	1.489(3)		S2−O6	1.462(3)
III	S−O1	1.529(2)		S2−O7	1.475(3)
	S−O2	1.519(2)			
	S−O3	1.523(2)			

bonds (O8−H···O) with O5 and O6 of the sulfate ion within the chain, (d(O8···O5) 2.732(4), (d(H8A···O5) 1.98(5) Å, θ(O8−H8A···O5) 155(5)°; (d(O8···O6) 2.811(4), (d(H2···O5) 2.11(6) Å, and θ(O8−H8A···O6) 168(6)°) as can be seen from Figure 7b.

In I−V, the sulfite ion has pyramidal geometry with the lone pair of electrons directed toward the fourth tetrahedral vertex. The S1−O bond distances are in the range of 1.489(3)−1.588(3) Å. In V, the sulfate ion has the tetrahedral geometry, with S2−O bond distances in the 1.454(3)−1.480(3) Å range (Table 2).

Thermal Analysis. Thermogravimetric analysis of I−V reveals that they are stable up to temperatures of 50, 150, 150, 280, and 350 °C, respectively, after which time they undergo weight losses of 77, 66, 64, 57, and 22%, respectively, in the temperature range of 30−700 °C. Compounds I−III and V undergo weight loss in two steps, the second step being that of major weight loss. Compound IV undergoes weight loss in a single step. In compound V, the first step is that of major weight loss. Powder X-ray diffraction patterns of the decomposed products of I−IV at 700 °C correspond to ZnO (JCPDS file card 1-1136). The powder X-ray diffraction pattern of the decomposed product of V at 700 °C corresponds to $Nd_2(SO_4)_3$ (JCPDS file card 23-1265).

Conclusions

In conclusion, zinc sulfite-based coordination polymers and hybrid networks of different dimensionalities have been synthesized and characterized. An open-framework hybrid sulfite−sulfate, where both SO_3^{2-} and SO_4^{2-} are part of a two-dimensional layer, has also been synthesized and characterized. On the basis of the present studies, it appears that many more metal sulfite-based open-framework and hybrid materials can be synthesized.

Supporting Information Available: A crystallographic information file (CIF) for the compounds I−V is available free of charge via the Internet at http://pubs.acs.org.

IC061988M

Inorg. Chem. 2008, 47, 823–831

Inorganic Chemistry
: Article

Hybrid Structures Formed by Lead 1,3-Cyclohexanedicarboxylates

A. Thirumurugan, R. A. Sanguramath, and C. N. R. Rao*

Chemistry and Physics of Materials Unit, Jawaharlal Nehru Centre for Advanced Scientific Research, Jakkur P. O., Bangalore 560064, India

Received July 4, 2007

By the employment of hydrothermal methods, four lead 1,3-cyclohexanedicarboxylates with the compositions $Pb(1,3\text{-CHDC})(H_2O)$, **I**, $[(OPb_4)_2(OH)_2(C_2O_4)(1,3\text{-CHDC})_4]\cdot H_2O$, **II**, $Pb_2(1,3\text{-CHDC})_2(H_2O)$, **III**, and $(OPb_3)(1,3\text{-CHDC})_2$, **IV**, have been prepared and characterized. Of these, **I** and **II** have layered structures while **III** and **IV** have three-dimensional structures. **I–III** are hybrid structures possessing extended inorganic connectivity in one or two-dimensions (I^n, $n = 1$ or 2) involving infinite Pb–O–Pb linkages along with zero or one-dimensional organic connectivity (O^m, $m = 0$ or 1). **I** contains two types of layers with different connectivities (I^1O^1 and I^2O^0). **III** is a truly 3-D hybrid compound with I^3O^1 type connectivity. **IV** has three-dimensional organic connectivity (O^3) but no inorganic connectivity (I^0). The conformation of the CHDC anion is e,e in **I–III** and a,e in **IV**. In all these compounds, the lead atom has hemi- or holodirected coordination geometry.

Introduction

Besides aluminosilicates and phosphates, metal carboxylates form a wide variety of open framework structures.[1–3] Metal carboxylates have also enabled the design and synthesis of novel architectures, some of which exhibit useful properties such as sorption, catalysis, luminescence, and magnetism. Thus, Yaghi and co-workers[4] have used the zinc dicarboxylates to synthesize a MOF-5-based *reticular* family of materials with low density and high capacity for hydrogen storage. Férey and co-workers[5] have reported trinuclear chromium cluster-based benzene di- and tricarboxyaltes with unit cell volumes going up to 700 000 Å³. Of particular interest are the inorganic–organic hybrid compounds formed by metal carboxylates.[6–11] These hybrid compounds possess inorganic and organic connectivities of different dimension-

alities. A classification has been provided recently on the basis of the dimensionality (n,m) of the inorganic (I) and the organic (O) connectivities (I^nO^m).[11] According to this classification, simple molecular coordination compounds would be of the type I^0O^0. Hybrid compounds with inorganic connectivity in one or more dimensions ($n > 0$) containing extended M–X–M (X = O, N, Cl, S) connectivity are known. The literature on metal benzenedicarboxylates shows that the 1,4-benzenedicarboxylates generally form three-dimensional hybrid structures while the 1,2-dicarboxylates favor two-dimensional hybrid structures, both with extended M–O–M linkages.[12–14] Recent studies of cyclohexanedicarboxylates (CHDCs) suggests that both 1,2- and 1,4-CHDCs form hybrid compounds with extended M–O–M connectivity, with the 1,2-CHDCs generally possessing layered structures.[15–23] To our knowledge there are not many

---

* To whom correspondence should be addressed. E-mail: cnrrao@jncasr.ac.in. Fax: +91-80-22082760, 22082766.

(1) Cheetham, A. K.; Férey, G.; Loiseau, T. *Angew. Chem., Int. Ed.* **1999**, *38*, 3268.
(2) Rao, C. N. R.; Natarajan, S.; Vaidhyanathan, R. *Angew. Chem., Int. Ed.* **2004**, *43*, 1466.
(3) Maspoch, D.; Ruiz-Molina, D.; Veciana, J. *Chem. Soc. Rev.* **2007**, *36*, 770.
(4) Rosi, N. L.; Eukert, J.; Eddaoudi, M.; Vodak, D. T.; Kim, J.; O'Keeffe, M.; Yaghi, O. M. *Science* **2003**, *300*, 1127.
(5) Férey, G.; Mellot-Draznieks, C.; Serre, C.; Millange, F.; Dutour, J.; Surblé, S.; Margiolaki, I. *Science* **2005**, *309*, 2040.
(6) Guillou, N.; Livage, C.; van Beek, W.; Nogues, M.; Férey, G. *Angew. Chem., Int. Ed.* **2003**, *42*, 644.
(7) Forster, P. M.; Cheetham, A. K. *Angew. Chem., Int. Ed.* **2003**, *42*, 457.
(8) Vaidhayanathan, R.; Natarajan S.; Rao, C. N. R. *Dalton Trans.* **2003**, 1459.

(9) Forster, P. M.; Burbank, A. R.; Livage, C.; Férey, G.; Cheetham, A. K. *Chem. Commun.* **2004**, 368.
(10) Thirumurugan, A.; Natarajan, S. *J. Mater. Chem.* **2005**, *15*, 4588.
(11) Cheetham, A. K.; Rao, C. N. R.; Feller, R. K. *Chem. Commun.* **2006**, 4780.
(12) Miller, S. R.; Wright, P. A.; Serre, C.; Loiseau, T.; Marrot, J.; Férey, G. *Chem. Commun.* **2005**, 3850.
(13) Dale, S. K.; Elsegood, M. R. J.; Kainth, S. *Acta Crystallogr.* **2004**, *C60*, m76.
(14) Thirumurugan, A.; Rao, C. N. R. *J. Mater. Chem.* **2005**, *15*, 3852.
(15) Kim, Y. J.; Jung, D. Y. *Chem. Commun.* **2002**, 908.
(16) Qi, Y. J.; Wang, Y. H.; Hu, C. W.; Cao, M. H.; Mao, L.; Wang, E. B. *Inorg. Chem.* **2003**, *42*, 73.
(17) Bi, W.; Cao, R.; Sun, D.; Yaun, D.; Li, X.; Wang, Y.; Li, X.; Hong, M. *Chem. Commun.* **2004**, 2104.
(18) Gong, Y.; Hu, C. W.; Li, H.; Huang, K. L; Tang, W. *J. Solid State Chem.* **2005**, *178*, 3153.

Table 1. Synthetic Conditions for the 1,3-CHDCs, **I**–**IV**

compd no.	formula	composition (mol ratio)				temp, °C	yield based on Pb, %
		$Pb(NO_3)_2$	$1,3-H_2CHDC$	NaOH (5 M soln)	water		
I	$Pb(1,3-C_8H_{10}O_4)(H_2O)$	0.330 g, 1 mM (2)	0.088 g, 0.5 mM (1)	1 mM, 0.2 mL (2)	278 mM, 5 mL	150	38
II	$[(OPb_4)_2(1,3-C_8H_{10}O_4)_4(C_2O_4)(OH)_2]\cdot H_2O$	0.330 g, 1 mM (2)	0.088 g, 0.5 mM (1)	1.5 mM, 0.3 mL (3)	278 mM, 5 mL	200	31
III	$Pb_2(1,3-C_8H_{10}O_4)_2(H_2O)$	0.330 g, 1 mM (2)	0.088 g, 0.5 mM (1)	0.5 mM, 0.1 mL (1)	278 mM, 5 mL	200	36
IV	$OPb_3(1,3-C_8H_{10}O_4)_2$	0.330 g, 1 mM (2)	0.088 g, 0.5 mM (1)	1.5 mM, 0.3 mL (3)	278 mM, 5 mL	180	39

investigations of the 1,3-CHDC compounds.[22–24] We have, therefore, carried out an investigation on the synthesis and characterization of 1,3-CHDC compounds and indeed found that they form hybrid organic–inorganic compounds with extended inorganic connectivity of varied dimensions. We also find that the CHDC anion in these compounds can be in the equatorial–equatorial (*e,e*) as well as the axial–equatorial (*a,e*) conformations.[25] The lead atoms in the 1,3-CHDC compounds exhibit hemi- as well as holodirected coordination geometry.

Experimental Section

Materials. $Pb(NO_3)_2$ (Qualigens, Mumbai, India, 99%), 1,3-cyclohexanedicarboxylicacid, $1,3-H_2CHDC$ [Aldrich, 98%, mixture of cis and trans (anti) compounds], and NaOH (Merck, Mumbai, India, 99%) of high purity and double distilled water were used for the synthesis.

Synthesis and Characterization. The Pb 1,3-CHDCs were synthesized under hydrothermal conditions by heating homogenized reaction mixtures in a 23 mL PTFE-lined bomb in the temperature range 150–200 °C for 72 h under autogenous pressure. The pH of the starting reaction mixture was generally in the range 5–6. The pH after the reaction did not show appreciable change. The products of the hydrothermal reactions were vacuum-filtered and dried under ambient conditions. The starting compositions and synthetic conditions for the different compounds synthesized by us are given in Table 1. All the compounds were obtained as single phase, except **II**, whose crystals were obtained admixed with small quantities of polycrystalline **I** and **IV** powder. The crystals were separated under a polarizing microscope and used for all the characterization. The yield based on the lead was generally grater than 30%.

Elemental analyses of **I**–**IV** were satisfactory. Anal. Calcd for **I** ($C_{16}H_{24}Pb_2O_{10}$): C, 24.29; H, 3.04. Found: C, 24.22; H, 3.12. Calcd for **II** ($C_{34}H_{44}Pb_8O_{25}$): C, 16.26; H, 1.75. Found: C, 16.33; H, 1.69. Calcd for **III** ($C_{16}H_{22}Pb_2O_9$): C, 24.86; H, 2.85. Found: C, 24.80; H, 2.91. Calcd for **IV** ($C_{16}H_{20}Pb_3O_9$): C, 19.64; H, 2.05. Found: C, 19.71; H, 2.01.

Powder XRD patterns of the products were recorded using Cu Kα radiation (Rich-Seifert, 3000TT). The patterns agreed with those calculated for single-crystal structure determination. Thermogravimetric analysis (TGA) was carried out (Metler-Toledo) in an oxygen atmosphere (flow rate = 50 mL/min) in the temperature range 25–800 °C (heating rate = 5 °C/min).

Infrared (IR) spectra of KBr pellets of **I**–**IV** were recorded in the mid-IR region (Bruker IFS-66v). Compounds **I**–**IV** show characteristic bands for the functional groups.[26–28] The bands around 1550 and 1400 cm^{-1} are assigned to carboxylate $\nu_{asC=O}$ and $\nu_{sC=O}$ stretching, and the absence of a band at 1700 cm^{-1} confirms the binding of carboxylate group to the lead cation. The bands at 3558 (ν_{asO-H}), 3470 (ν_{sO-H}), 1245 (δ_{O-H} in-plane), and 616 cm^{-1} (δ_{O-H} out-of-plane) indicate the presence of hydroxyl groups and their ligation to the lead cation in **II**.

Thermogravimetric analyses for **I**–**IV** are as follows. For **I**, the first weight loss of 4.96% (calcd 4.55%) occurred around 180 °C and the second weight loss of 43.59% (calcd 43.02%) was in the 280–470 °C range. For **II**, the total weight loss of 33.2% (calcd 32.69%) occurred in the 130–450 °C range. For **III**, the first weight loss of 2.46% (calcd 2.33%) occurred around 150 °C and the second weight loss of 31.63% (calcd 31.07%) was in the 300–450 °C range. For **IV**, the total weight loss of 35.0% (calcd 34.78%) occurred in the 250–425 °C range. The total weight loss matches very well with the loss of CO_2, with (**I**–**III**) or without (**IV**) H_2O, and the formation of PbO (PDF No. 00-004-0561) in all the cases.

A suitable single crystal of each compound was carefully selected under a polarizing microscope and glued to a thin glass fiber. Crystal structure determination by X-ray diffraction was performed on a Bruker-Nonius diffractometer with Kappa geometry attached with an APEX-II CCD detector and a graphite monochromator for the X-ray source (Mo Kα radiation, $\lambda = 0.710\,73$ Å) operating at 50 kV and 30 mA. An empirical absorption correction based on symmetry-equivalent reflections was applied using the SADABS program.[29] The structure was solved and refined using the SHELXTL-PLUS suite of programs.[30] For the final refinement the hydrogen atoms on the cyclohexane rings were placed geometrically and held in the riding mode. The hydrogen atoms on the water molecules were located in the difference Fourier map, and the O–H distance was constrained to 0.85 Å. Final refinement included atomic positions for all the atoms, anisotropic thermal parameters for all the non-hydrogen atoms, and isotropic thermal parameters for the hydrogen atoms. All the hydrogen atoms were included in the final refinement, except the hydrogen atoms on O(100) in **III**; the hydrogen atoms of this water molecule could not be located in the difference Fourier map. Some of the carbon atoms of the CHDC anion (C14–C17) were disordered, so the C–C distances involving

(19) Inoue, M.; Atake, T.; Kawaji, H.; Tojo, T. *Solid State Commun.* **2005**, *134*, 303.

(20) Chen, J.; Ohba, M.; Zhao, D.; Kaneko, W.; Kitagawa, S. *Cryst. Growth Des.* **2006**, *6*, 664.

(21) Yang, J.; Li, G.-D.; Cao, J.-J.; Yue, Q.; Guang-Hua Li, G.-H.; Chen, J.-S. *Chem.—Eur. J.* **2007**, *13*, 3218.

(22) Thirumurugan, A.; Avinash, M. B.; Rao, C. N. R. *Dalton. Trans.* **2006**, 221.

(23) Rao, K. P.; Thirumurugan, A.; Rao, C. N. R. *Chem.—Eur. J.* **2007**, *13*, 3193.

(24) Surblé, S.; Serre, C.; Millange, F.; Pellé, F.; Férey, G. *Solid State Sci.* **2007**, *9*, 131.

(25) Nasipuri, D. *Stereochemistry of Organic Compounds: Principles and Applications*; Wiley: New York, 1991.

(26) Barbara, S. *Infrared Spectroscopy: Fundamentals and Applications*; Wiley: New York, 2004.

(27) Silverstein, R. M.; Bassler, G. C.; Morrill, T. C. *Spectrometric Identification of Organic Compounds*; John Wiley & Sons: New York, 1963.

(28) Nakamoto, K. *Infrared and Raman Spectra of Inorganic and Coordination Compounds*; Wiley: New York, 1978.

(29) Sheldrick, G. M. *SADABS Siemens Area Detector Absorption Correction Program*; University of Göttingen: Gottingen, Germany, 1994.

(30) Sheldrick, G. M. *SHELXTL-PLUS Program for Crystal Structure Solution and Refinement*; University of Göttingen: Gottingen, Germany, 1997.

Lead 1,3-Cyclohexanedicarboxylates

Table 2. Crystal Data and Structure Refinement Parameters for the 1,3-CHDCs, **I**–**IV**

param	I	II	III	IV
empirical formula	$C_{16}H_{24}Pb_2O_{10}$	$C_{34}H_{44}Pb_8O_{25}$	$C_{16}H_{22}Pb_2O_9$	$C_{16}H_{20}Pb_3O_9$
fw	790.73	2510.20	772.74	977.89
cryst system	orthorhombic	monoclinic	triclinic	monoclinic
space group	$Pbca$ (No. 61)	$C2/m$ (No. 12)	$P\bar{1}$ (No. 2)	$C2/c$ (No. 12)
$a/\text{Å}$	14.0078(3)	8.1086(2)	9.1442(5)	16.3279(6)
$b/\text{Å}$	8.8268(2)	29.9463(8)	9.7893(5)	13.6071(5)
$c/\text{Å}$	31.6165(7)	11.5853(3)	11.0738(6)	18.8434(6)
α/deg	90	90	92.922(3)	90
β/deg	90	103.0100(10)	97.147(2)	112.182(2)
γ/deg	90	90	107.695(2)	90
$V/\text{Å}^3$	3909.19(15)	2740.96(12)	932.89(9)	3876.7(2)
Z	8	2	2	8
D(calc)/g cm^{-3}	2.687	3.041	2.751	3.351
μ/mm^{-1}	17.257	24.548	18.071	26.041
tot. data collcd	50 632	19 662	17 716	31 818
unique data	3339	2407	3343	3347
obsd data [$I > 2\sigma(I)$]	3032	2201	3035	2752
R_{merg}	0.0201	0.0214	0.0303	0.0308
goodness of fit	1.308	1.218	1.149	1.071
R indexes [$I > 2\sigma(I)$]	$R_1 = 0.0282^a$	$R_1 = 0.0344^a$	$R_1 = 0.0300^a$	$R_1 = 0.0341^a$
	$wR_2 = 0.0688^b$	$wR_2 = 0.1042^b$	$wR_2 = 0.0821^b$	$wR_2 = 0.0683^b$
R indexes (all data)	$R_1 = 0.0321^a$	$R_1 = 0.0418^a$	$R_1 = 0.0344^a$	$R_1 = 0.0496^a$
	$wR_2 = 0.0701^b$	$wR_2 = 0.1232^b$	$wR_2 = 0.0930^b$	$wR_2 = 0.0748^b$

^{a} $R_1 = \Sigma||F_o| - |F_c||/\Sigma|F_o|$. ^{b} $wR_2 = \{\Sigma[w(F_o^2 - F_c^2)^2]/\Sigma[w(F_o^2)^2]\}^{1/2}$; $w = 1/[\sigma^2(F_o)^2 + (aP)^2 + bP]$, and $P = [\max(F_o^2,0) + 2(F_c)^2]/3$, where $a = 0.0182$, $b = 38.2775$ for **I**, $a = 0.0658$, $b = 72.9605$ for **II**, $a = 0.0467$, $b = 6.7206$ for **III**, and $a = 0.0658$, $b = 72.9605$ for **IV**.

these atoms were constrained to 1.550 Å. Details of the structure solution and final refinements for the compounds **I**–**IV** are given in Table 2.

Results and Discussion

Structures of Lead 1,3-Cyclohexanedicarboxylates. Four different lead 1,3-cyclohexanedicarboxylates, **I**–**IV**, have been synthesized and characterized. Of these, **I** and **II** possess layered structures while **III** and **IV** have three-dimensional structures. The 1,3-CHDC anions in these compounds exhibit seven different coordinational modes with different connectivities and conformations as shown in Figure 1. The lead cations are in either hemi- or holodirected geometry with the coordination numbers in the range 5–8 (Figure 2). We shall now discuss the structures of these different compounds.

Two-Dimensional Pb(1,3-CHDC)(H₂O). The 1,3-cyclohexanedicarboxylate Pb(1,3-CHDC)(H₂O), **I**, has a two-dimensional structure with an asymmetric unit of 28 non-hydrogen atoms (Figure 3a). There are two crystallographically distinct Pb^{2+} ions Pb(1) and Pb(2), two CHDC anions, and two terminal water molecules in the asymmetric unit. The anions exhibit two types of coordination modes with acid-1 having (1121) connectivity (Figure 1a) and acid-2 having (1212) connectivity[31] (Figure 1b). Both of them are in the anti, *e,e* conformation with torsional angles in the 172.33–172.66° range. Pb(1) is hemidirected and five-coordinated by oxygen atoms (PbO₅) from four different CHDC anions (acid-1) (Figure 2a). Two of the oxygens have μ_3 connections linking Pb(1) with two other Pb(1) atoms. Thus, Pb(1)O₅ polyhedra share corners with each other forming an infinite one-dimensional Pb–O–Pb chain. The chains are further

connected by CHDC anions (acid-1), each anion binding to four different Pb^{2+} cations of adjacent Pb–O–Pb chains to form an infinite two-dimensional layer structure. This gives rise to a I^1O^1 type hybrid structure as per the recent classification[11] (Figure 3b). The active lone pair of the hemidirected Pb(1)O₅ polyhedra projects on both sides of the layer (layer-1) as in Figure 4. The Pb–O bond lengths are in the 2.388–2.805 Å range.

Pb(2) is eight-coordinated by oxygen atoms (PbO₈) and has a holodirected geometry (Figure 2a). Among the eight oxygens of (PbO₈), two are from two different terminal water molecules and the other six from four different CHDC anions (acid-2). Four among these six oxygens have μ_3 connections linking Pb(2) with three other Pb(2)'s. Thus, each Pb(2)O₈ polyhedron shares an edge with another Pb(2)O₈, forming a dimer of Pb₂O₁₄. The dimers are further connected to each other by sharing a corner each with four dimers, forming an infinite two-dimensional Pb–O–Pb layer with a (4,4) square lattice and a hybrid connectivity of the type I^2O^0 (Figure 3c). The CHDC anions (acid-4) decorate both sides of the layer (layer-2) as in Figure 4. The Pb–O bond lengths are in the 2.453–2.834 Å range. There is no apparent hydrogen bonding in **I**, and van der Waals interactions between layers appear to be the main stabilizing factor (Figure 4).

Two-Dimensional [(OPb₄)₂(OH)₂(C₂O₄)(1,3-CHDC)₄]· H₂O. We have been able to obtain a lead oxalate–cyclohexanedicarboxylate, [(OPb₄)₂(OH)₂(C₂O₄)(1,3-CHDC)₄]· H₂O, **II**, where the oxalate moiety was generated in situ from the 1,3-CHDC under the hydrothermal synthesis. Similar cases, where the ligands are formed in situ during the synthesis, have been reported in the literature.[32-34] The formation of the oxalate in the synthesis of **II** is rather

---

(31) For description of connectivity, see the following: Massiot, D.; Drumel, S.; Janvier, P.; B-Doeuff, M.; Buujoli, B. *Chem. Mater.* **1997**, *9*, 6.

(32) Sun, D.; Cao, R.; Liang, Y.; Shi, Q.; Su, W.; Hong, M. *J. Chem. Soc., Dalton Trans.* **2001**, 2335.

(33) Thirumurugan, A.; Natarajan, S. *Eur. J. Inorg. Chem.* **2004**, *4*, 762.

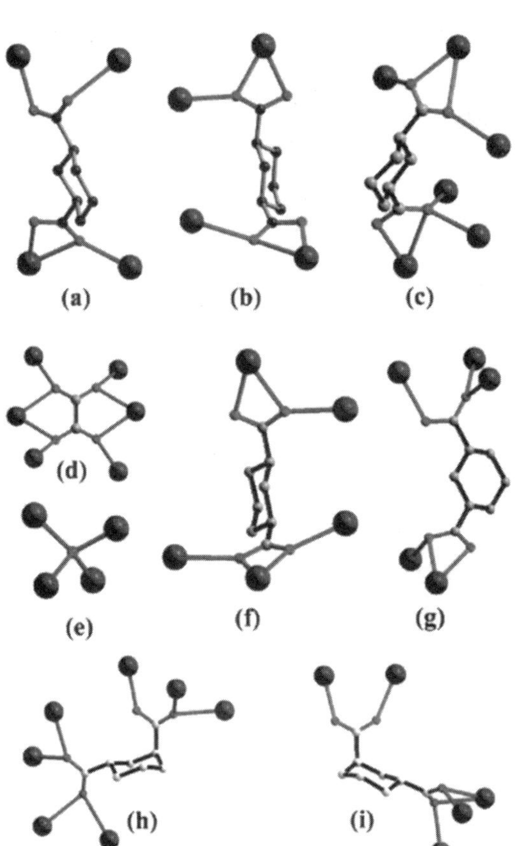

Figure 1. Coordination modes of the 1,3-CHeDC, oxalate, and oxo anions in **I**–**IV**.

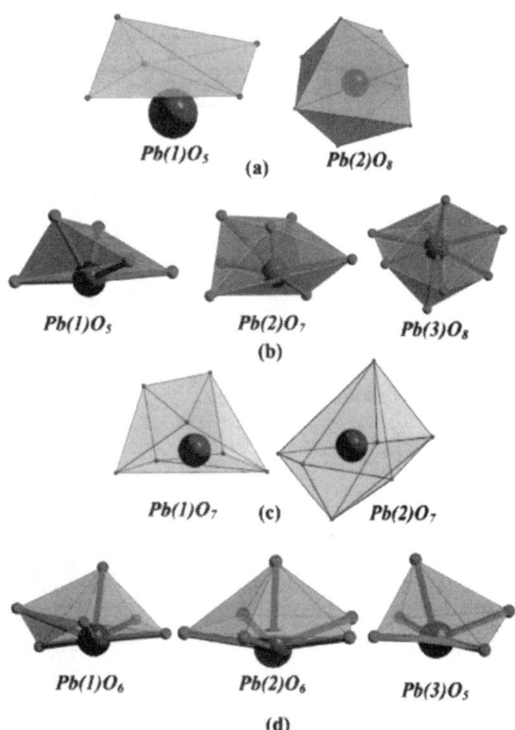

Figure 2. PbO$_n$ ($n = 5-8$) polyhedra showing the coordination geometry of lead atoms.

unusual. This can happen if 1,3-H₂CHDC undergoes an oxidative decarboxylation to produce cyclohexene or cyclohexadiene. Cyclic olefins in the gas phase are known to produce oxalic acid on oxidation at high temperatures.[35–37]

II has a two-dimensional structure with an asymmetric unit of 16.75 non-hydrogen atoms (Figure 5a). The Pb²⁺ cations are in three crystallographically distinct positions with Pb(1) and Pb(3) landing with 0.5 occupancies in 4*f* and 4*h* special positions, respectively, and Pb(2) with a full occupancy. One CHDC anion, one-quarter of the oxalate anion (with C at 4*h*), one hydroxyl anion (with the O at 4*f*), one independent oxo dianion (at 4*f*), and one-quarter of a lattice water molecule (at 2*a*) are also in the asymmetric unit. Three of the four anions are shown in Figure 1c–e. The CHDC anion in the anti, *e,e* conformation with a torsional angle of 176.74(2)° has (2223) connectivity and binds to six Pb²⁺ cations [three Pb(2) and three Pb(3)]. The oxalate anion has (2222) connectivity and binds to six Pb²⁺ cations [two Pb-

(1) and four Pb(2)]. The independent oxo dianion binds to four Pb²⁺ cations [two Pb(1) and two Pb(2)] to form an OPb₄ tetrahedron. The hydroxyl anion binds to three Pb²⁺ cations [one Pb(1) and two Pb(2)].

Pb(1) is hemidirected and coordinated by five oxygen atoms (PbO₅) (Figure 2b). Among the five oxygens of Pb-(1)O₅, two with μ_3 connectivity are from a single oxalate anion, another two with μ_4 connectivity are from two different oxo dianions, and the fifth one with a μ_4 connectivity is from the hydroxyl anion. Thus, the oxygens with μ_3 and μ_4 connections link Pb(1) with two other Pb(1) atoms by sharing edges and two different Pb(2) atoms by sharing edges. The Pb–O bond lengths are in the 2.314–2.635 Å range.

Pb(2) is slightly hemidirected and coordinated by seven oxygen atoms (PbO₇) (Figure 2b). Among the seven oxygens of Pb(2)O₇, the oxo and the hydroxyl anions have μ_4 connectivity; the other one with μ_3 connectivity is from the oxalate anion. The remaining four (one with monodentate, one with μ_3, and two with μ_4 connectivity) are from three different CHDC anions. These oxygens link Pb(2) with two Pb(2) and two Pb(1) atoms, by sharing edges. They also connect Pb(2) with two Pb(3) atoms, by sharing a face each. The Pb–O bond lengths are in the 2.459–2.919 Å range.

Pb(3) is holodirected and coordinated by eight oxygen atoms (PbO₈) (Figure 2b). All the eight oxygens are from

(34) Gong, Y.; Li, H.; Li, Y. G.; Wang, Y. H.; Tang, W.; Hu, C. W. *J. Coord. Chem.* **2007**, *60*, 61.
(35) Serguchev; Beletskaya. *Russ. Chem. Rev.* **1980**, *49*, 1119.
(36) Badanyan; Minasyan; Vardapetyan. *Russ. Chem. Rev.* **1987**, *56*, 740.
(37) Kawamura, K.; Ikushima, K. *Environ. Sci. Technol.* **1993**, *27*, 2227.

Lead 1,3-Cyclohexanedicarboxylates

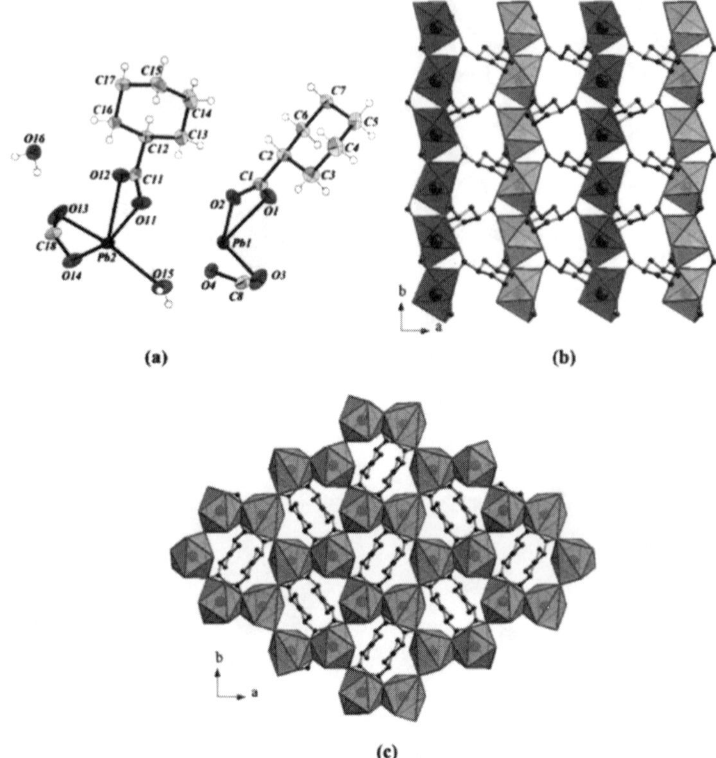

Figure 3. (a) ORTEP plot of **I**. Thermal ellipsoids are shown at 50% probability. Views are shown of the layers and the infinite Pb−O−Pb linkages in (b) layer-1 and (c) layer-2 of Pb(1,3-CHDC)(H₂O), **I**.

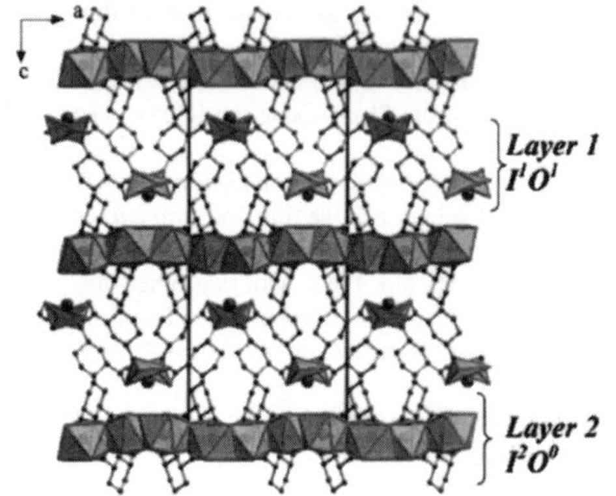

Figure 4. Packing arrangement in Pb(1,3-CHDC)(H₂O), **I** (viewed along the *b*-axis).

six different CHDC anions. Among the eight oxygens, two have μ_3 connectivity and the other two μ_4 connectivity. These oxygens link Pb(3) with two other Pb(3) atoms, by sharing edges, and with two different Pb(2) atoms, by sharing faces. The Pb−O bond lengths are in the 2.583−2.905 Å range. These connectivities lead to the formation of a secondary building unit (SBU) of the composition Pb₈(OH)₂O₂₆, which

includes two edge-shared OPb₄ tetrahedra and two hydroxyl groups (Figure 5b). The SBU is made up of two Pb(1)O₅, four Pb(2)O₇, and two Pb(3)O₈ polyhedra. The oxalate anion connects the adjacent SBUs to form an infinite 1-D chain. Each SBU is connected to four other SBUs, leading to the formation of a (4,4) square lattice with infinite two-dimensional Pb−O−Pb connectivity (I^2O^0) (Figure 5c). The CHDC anions (acid-4) decorate both sides of the layer, and the lattice water molecule resides in the 1-D channel (7.1 Å × 3.0 Å) between the layers (Figure 6). The water molecules are H-bonded to the hydroxyl groups projecting in the channel. Platon-Solv analysis showed 17.2% of the unit cell volume to be solvent accessible.[38,39]

Three-Dimensional Pb₂(1,3-CHDC)₂(H₂O). The 1,3-cyclohexanedicarboxylate Pb₂(1,3-CHDC)₂(H₂O), **III**, has a three-dimensional structure with an asymmetric unit of 27 non-hydrogen atoms (Figure 7a). There are two crystallographically distinct Pb²⁺ ions, two CHDC anions, and one terminal water molecule in the asymmetric unit. On the basis of the coordination modes, the anions can be classified into two types: (a) acid-1 with (1222) connectivity (Figure 1f); (b) acid-2 with (1221) connectivity (Figure 1g). Both of them

(38) Spek, A. L. *J. Appl. Crystallogr.* **2003**, *36*, 7.
(39) Spek, A. L. *PLATON, A Multipurpose Crystallographic Tool*; Utrecht University: Utrecht, The Netherlands, 2005.

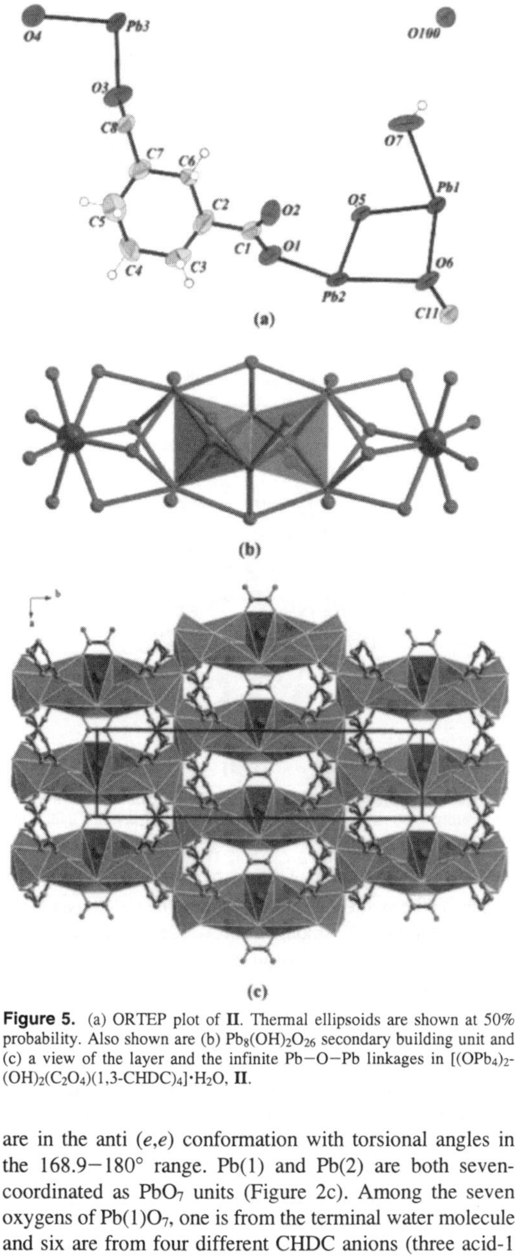

Thirumurugan et al.

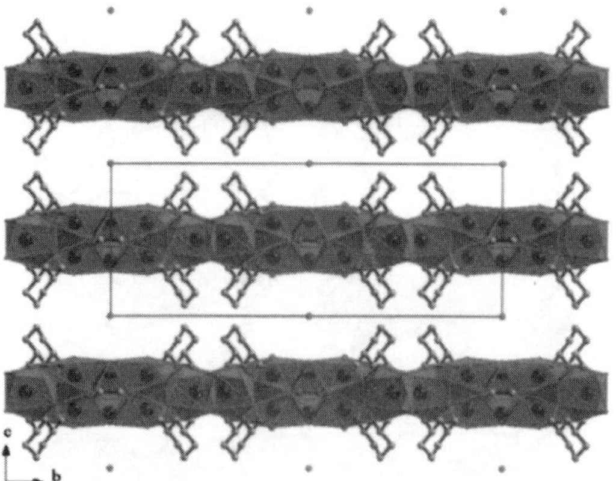

Figure 6. (a) Packing arrangement in $[(OPb_4)_2(OH)_2(C_2O_4)(1,3\text{-CHDC})_4]\cdot H_2O$, **II** (viewed along the a-axis), and (b) view of the pores (lattice water residing in the pore).

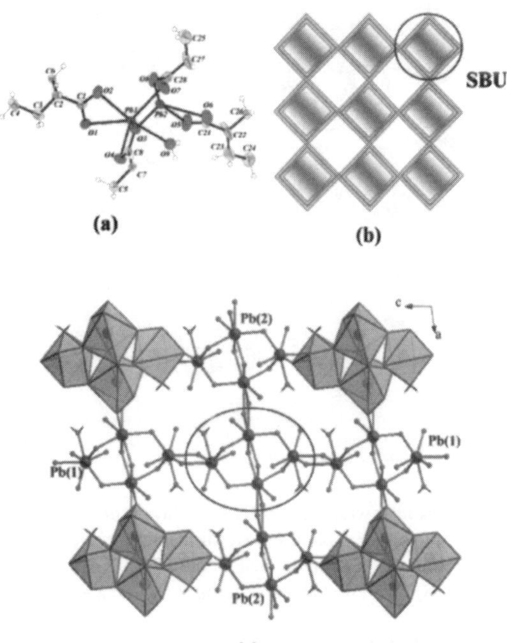

Figure 7. (a) ORTEP plot of **III**. Thermal ellipsoids are shown at 50% probability. Also shown are (b) a schematic of the (4,4) square lattice with the SBUs and (c) a view of the layer with infinite Pb–O–Pb linkages in $[(OPb_4)_2(OH)_2(C_2O_4)(1,3\text{-CHDC})_4]\cdot H_2O$, **II**.

Figure 5. (a) ORTEP plot of **II**. Thermal ellipsoids are shown at 50% probability. Also shown are (b) $Pb_8(OH)_2O_{26}$ secondary building unit and (c) a view of the layer and the infinite Pb–O–Pb linkages in $[(OPb_4)_2(OH)_2(C_2O_4)(1,3\text{-CHDC})_4]\cdot H_2O$, **II**.

are in the anti (e,e) conformation with torsional angles in the 168.9–180° range. Pb(1) and Pb(2) are both seven-coordinated as PbO_7 units (Figure 2c). Among the seven oxygens of $Pb(1)O_7$, one is from the terminal water molecule and six are from four different CHDC anions (three acid-1 and one acid-2). Four of these oxygens have μ_3 connections linking one Pb(1) with another Pb(1) through two oxygen atoms and with two different Pb(2) atoms through two oxygens. Thus, the $Pb(1)O_7$ polyhedra share edges with each other forming a dimeric $Pb(1)_2O_{13}$ unit. The Pb–O bond lengths are in the 2.418–2.899 Å range.

The seven oxygens of $Pb(2)O_7$ come from six different CHDC anions (two acid-1 and four acid-2). Six of these oxygens have μ_3 connections linking Pb(2) with two different Pb(1) through two oxygen atoms and with two other Pb(2)

through four oxygen atoms. Thus, each $Pb(2)O_7$ polyhedron shares an edge with another forming a dimeric $Pb(2)_2O_{13}$ unit. The dimers are connected to each other through two μ_3 oxygens, forming an infinite one-dimensional Pb–O–Pb chain. The chains are connected by the $Pb(1)_2O_{13}$ dimer into a two-dimensional Pb–O–Pb layer (Figure 7c). The layers can also be viewed as a (4,4) square lattice consisting of tetranuclear secondary building units (SBU) of Pb_4O_{18} with two Pb(1) and Pb(2) each (Figure 7b). Acid-1 which connects

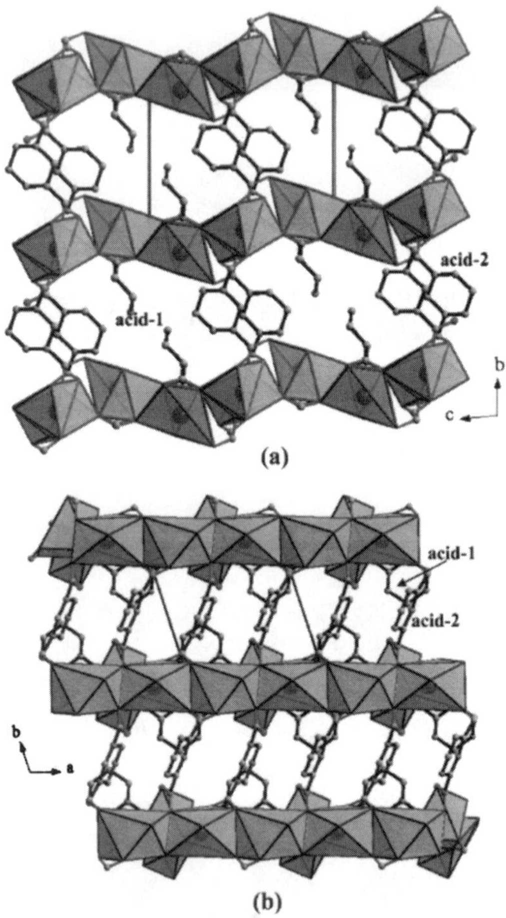

(a)

(b)

Figure 8. 3D-structure of $Pb_2(1,3\text{-CHDC})_2(H_2O)$, **III**, (a) viewed along the a-axis and (b) viewed along the c-axis.

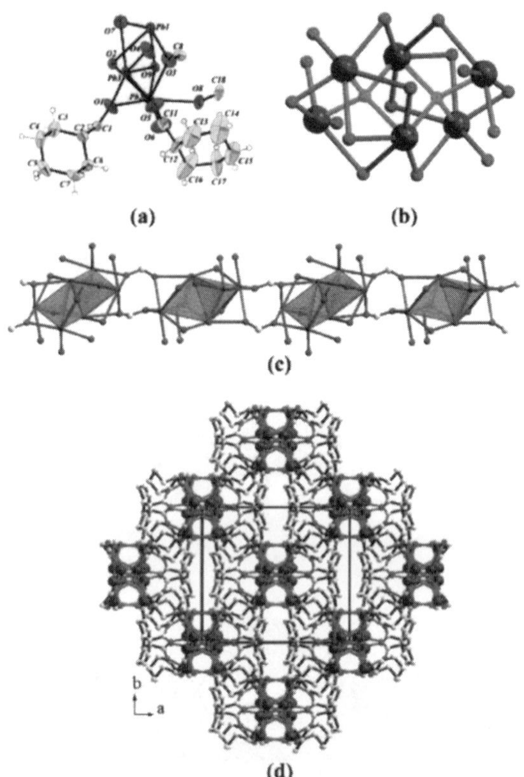

(a)

(b)

(c)

(d)

Figure 9. (a) ORTEP plot of **IV**. Thermal ellipsoids are shown at 50% probability. Also shown are (b) the Pb_6O_{18} secondary building unit, (c) the chain formed by the connectivity of SBUs through the carboxylate, and (d) the 3D-structure of $(OPb_3)(1,3\text{-CHDC})_2$, **IV**, viewed along the c-axis.

the Pb cations within the layer (intralayer) decorates both sides of the layer. The layers are further connected by acid-2, forming a three-dimensional structure with the I^2O^1 type (Figure 8). The Pb—O bond lengths are in the 2.450–2.685 Å range.

Three-Dimensional $(OPb_3)(1,3\text{-CHDC})_2$. $(OPb_3)(1,3\text{-CHDC})_2$, **IV**, has a three-dimensional structure without any inorganic connectivity (I^0O^3). It has an asymmetric unit with 28 non-hydrogen atoms (Figure 9a). There are three Pb^{2+} cations in three crystallographically distinct positions, two CHDC anions, and one independent oxo dianion in the asymmetric unit. Both the CHDC anions (acid-1 and acid-2, Figure 1h,i) are in the cis (a,e) conformation with the torsional angle of 120.95(1) and 99.31(1)°. They have (1222) and (1122) connectivities, respectively. Acid-1 binds to seven Pb^{2+} cations [two Pb(1), three Pb(2), and two Pb(3)], and acid-2 binds to five Pb^{2+} cations [one Pb(1), two Pb(2), and two Pb(3)]. The independent oxo dianion binds to four Pb^{2+} cations [two Pb(1), one Pb(2), and one Pb(3)] to form a OPb_4 tetrahedron.

Pb(1) is hemidirected and coordinated by six oxygen atoms (PbO_6) (Figure 2d). Among the six oxygens of $Pb(1)O_6$, four with μ_3 connectivity are from three different CHDC anions and the remaining two with μ_4 connectivity are from two different oxo dianions. These oxygen atoms link Pb(1) with an another Pb(1) by sharing an edge, with two different Pb-(2) atoms by sharing an edge with one and by sharing a face with another, and with two different Pb(3) by sharing a corner to one and by sharing an edge to another. The Pb—O bond lengths are in the 2.286–2.871 Å range.

Pb(2) is also hemidirected and coordinated by six oxygen atoms (PbO_6) (Figure 2d). Among the six oxygens of Pb-$(2)O_6$, one oxo anion has μ_4 connectivity. The remaining oxygens are from five different CHDC anions with one oxygen having μ_2 connectivity and other four with μ_3 connectivity. These oxygens link Pb(2) with two Pb(1) and also connect Pb(2) with one Pb(3) by edge sharing. The Pb—O bond lengths are in the 2.245–2.764 Å range.

Pb(3) is also hemidirected but coordinated by five oxygen atoms (PbO_5) (Figure 2d). Among the six oxygens of Pb-$(2)O_6$, one oxo anion is having μ_4 connectivity. The remaining four oxygens are from four different CHDC anions; two have μ_3 connectivity, and the other two have μ_2 connectivity. These oxygens link Pb(3) with two other Pb(1) and one Pb-

(2), by sharing an edge. The Pb—O bond lengths are in the 2.305—2.700 Å range. These connectivities lead to the formation of a secondary building unit (SBU) of Pb_6O_{18}, which includes two edge-shared OPb_4 tetrahedra (Figure 9b). The SBU is constructed by sharing $Pb(1)O_6$, $Pb(2)O_6$, and $Pb(3)O_5$ polyhedra. A carboxylate of acid-2 connects the adjacent SBUs to form an infinite 1-D chain (Figure 9c), and the chains get connected to four other adjacent chains by CHDC anions to form the infinite three-dimensional structure (I^0O^3) (Figure 9d).

Coordination Geometry of Lead. The coordination geometry of the PbO_n polyhedra in Pb(II) compounds is hemidirected for low coordination numbers (2—5) and holodirected for high coordination numbers (9, 10). For intermediate coordination numbers (6—8), either type of stereochemistry is found.[40,41] Ab initio molecular orbital studies of gas-phase Pb(II) complexes show that a hemidirected geometry occurs if the ligand coordination number is low, the ligands are hard, and there are attractive interactions between the ligands.[42,43] In such cases, the lone pair orbital has p character and fewer electrons are transferred from the ligands to the bonding orbitals of Pb(II), giving rise to more ionic bonds. Holodirected geometry occurs when the coordination number is high and the ligands are soft and bulky or show strong interligand repulsion. The lone pair orbital has negligible p character when the geometry is holodirected, and the bonds are more covalent than in the hemidirected structures. The Pb(II) cations in **I** exhibit both hemi- and holodirected geometry with coordination numbers 5 and 8, respectively. The Pb(II) cations in **II** also exhibit both types of geometry. $Pb(1)O_5$ and $Pb(2)O_7$ are hemidirected with coordination numbers 5 and 7, respectively, whereas $Pb(3)$-O_8 is holodirected with coordination number 8. In **III**, seven-coordinated Pb(1) and Pb(2) show hemi- and holodirected geometry. The hemidirectness of $Pb(1)O_7$, though not very prominent, nevertheless may be due to the presence of the terminal water molecule and six carboxylate oxygens. The holodirected $Pb(2)O_7$ has only carboxylate oxygens. All the Pb(II) cations in **IV** exhibit only hemidirected geometry with coordination numbers 6 and 5.

An examination of the inorganic and organic connectivities in these hybrid compounds would be in order. In Table 3, we have shown a matrix with organic and inorganic connectivities.[2,11] We show the range of possibilities in terms of the dimensionality of the organic and the inorganic connectivities. The overall dimensionality of the structure is then represented with the notation I^nO^m, the sum of the exponents, $m + n$, giving the overall dimensionality of the structure. The entire family of molecular coordination compounds are contained within a single box (I^0O^0) in Table 3 {i.e., both M—L—M (m) and inorganic connectivity (n) = 0}. The remaining three boxes in the first column represent the coordination polymers with overall dimensionality 1—3.

Table 3. Description of Hybrid Carboxylate Inorganic Connectivity, I^n

Inorganic Connectivity, I^n

Organic Connectivity, O^m	n \ m	0	1	2	3
	0	Molecular coordination compounds	Nd(DPA)[44]	Cd(1,2-C)[22] Pb(1,2-C)[23] Pb(1,3-C),I Pb(1,3-C),II	Ni(Succ)[45] Ni(Glut)[46] Cd(Malon)[47]
	1	Cd(1,4-C)[21] Mn(1,3-C)Ph[22]	Cd(1,2-Ce)[23] Pb(1,3-C),I	Pb(1,3-+1,4-C)[23] Pb(1,3-C),III	—
	2	Cd(1,3-C)[22]	La(1,4-C)[23]	—	—
	3	Pb(1,3-C),IV Many Known	—	—	—

(a) B = Benzenedicarboxylate, C = Cyclohexanedicarboxylate Ce = Cyclohexenedicarboxylate, Ph = 1,10-phenanthroline, DPA = Diphenate , Succ = Succinate, Glut = Glutarate and Malon = malonate

(b) The only example of 3-D inorganic connectivity (with zero organic connectivity) are aliphatic dicarboxylates such as propionate, oxalte, malonate, succinate and glutarate.

The three boxes in the second column represent hybrid compounds with one-dimensional inorganic connectivity (I^1) and with an overall dimensionality between 1 and 3. The two boxes in the third column represent hybrid compounds with two-dimensional inorganic connectivity (I^2) and with the overall dimensionality of 2 or 3. The first box in the fourth column represents a rare class of hybrid compound (I^3O^0) with a three-dimensional inorganic connectivity (I^3) and with the overall dimensionality of 3. There are examples of all of these classes of hybrid materials. Currently there are no examples for the empty boxes in the bottom right part of the Table 3. As seen from the table, there are many dicarboxylates with 1-, 2-, and 3-D organic connectivity and with zero and 1-D inorganic connectivity.[22,23] There are very few compounds with 2- or 3-D inorganic connectivity.[45−47] Another notable observation is that all the three isomers (1,2-, 1,3-, and 1,4-) of cyclohehanedicarboxylic acids favor the formation of hybrid compounds, but the presence of secondary ligands and chelating amines such as 1,10-phenanthroline and 2,2-bipyridine decreases the overall dimensionality of the hybrid compounds.[21−23] In 1,3- and 1,4-CHDC compounds,[21] the cis (a,e) conformation also does not favor extended inorganic connectivity.

(40) Ayyappan, S.; Diaz de Delgado, G.; Cheetham, A. K.; Férey, G.; Rao, C. N. R. *J. Chem. Soc., Dalton Trans.* **1999**, 2905.

(41) Glowiak, T.; Kozlowski, H.; Erre, L. S.; Micera, G.; Gulinati, B. *Inorg. Chim. Acta* **1992**, *202*, 43.

(42) S-Livny, L.; Glusker, J. P.; Bock, C. W. *Inorg. Chem.* **1998**, *37*, 1853.

(43) Watson, C. W.; Parker, S. C. *J. Phys. Chem. B* **1999**, *103*, 1258.

(44) Thirumurugan, A.; Pati, S. K.; Green, M. A.; Natarajan, S. *J. Mater. Chem.* **2003**, *13*, 2937.

(45) Guillou, N.; Livage, C.; Drillon, M.; Férey, G. *Angew. Chem., Int. Ed.* **2003**, *42*, 5314.

(46) Forster, P. M.; Cheetham, A. K. *Angew. Chem., Int. Ed.* **2002**, *41*, 457.

(47) Vaidhyanathan, R.; Natarajan, S.; Rao, C. N. R. *Dalton Trans.* **2003**, 1459.

Lead 1,3-Cyclohexanedicarboxylates

Concluding Remarks. We have successfully synthesized and characterized four 1,3-CHDCs of lead with layered and 3-dimensional structures. Compounds **I**–**III** contain 1,3-CHDC anions in the anti (e,e) conformation, exhibiting infinite Pb–O–Pb linkages, but compound **IV** with 1,3-CHDC anions in the cis (a,e) conformation does not have infinite Pb–O–Pb linkages. The lead atom is in either hemi- or holodirected in **I**–**IV**. The hybrid 1,3-CHDC compounds **I**–**III** show 1- or 2-D inorganic connectivity along with zero or 1-D organic connectivity. Compound **IV**, however, shows 3-D organic connectivity with no inorganic connectivity.

Acknowledgment. A.T. thanks the Council of Scientific and Industrial Research (CSIR), Government of India, for the award of the Senior Research Fellowship.

Supporting Information Available: A crystallographic information file (CIF) for compounds **I**–**IV**. This material is available free of charge via the Internet at http://pubs.acs.org.

IC701323Q

Inorg. Chem. 2006, 45, 9475-9479

Inorganic Chemistry

: Article

Synthesis and Magnetic Properties of an Amine-Templated Fe^{2+} ($S = 2$) Sulfate with a Distorted Kagome Structure

J. N. Behera and C. N. R. Rao*

Chemistry and Physics of Materials Unit, Jawaharlal Nehru Centre for Advanced Scientific Research, Jakkur P. O., Bangalore 560064, India, and Solid State and Structural Chemistry Unit, Indian Institute of Science, Bangalore 560012, India

Received August 3, 2006

An organically templated iron(II) sulfate of the composition $[H_3N(CH_2)_2NH_2(CH_2)_2(NH_3)]_4[Fe^{II}_9F_{18}(SO_4)_6]\cdot9H_2O$ with a distorted Kagome structure has been synthesized under solvothermal conditions in the presence of diethylenetriamine. The distortion of the hexagonal bronze structure comes from the presence of two different types of connectivity between the FeF_4O_2 octahedra and the sulfate tetrahedra. This compound exhibits magnetic properties different from those of an Fe(II) compound with a perfect Kagome structure and is a canted antiferromagnet at low temperatures.

Introduction

Transition metal compounds with the Kagome structure have been investigated extensively because of their interesting magnetic properties. The literature abounds with studies of Kagome compounds of Fe^{3+} ($S = 5/2$), such as the family of jarosites, all of which exhibit magnetic frustration or low-temperature antiferromagnetism.[1] Most of the Fe^{3+} jarosites investigated are not pure because of the presence of site defects. It is only recently that Nocera et al.[2,3] prepared a pure Fe^{3+} jarosite by redox-based hydrothermal methods. In this compound, the Fe^{3+} Kagome layer shows an antiferromagnetic transition at 61.4 K. A pure $S = 1/2$ copper Kagome compound has been found to show spin frustration,[4] while a Co^{2+} ($S = 3/2$) compound shows properties similar to those of the Fe^{3+} jarosites.[5] On the other hand, V^{3+} ($S = 1$) Kagome compounds exhibit ferromagnetic coupling within the triangles of Kagome layer.[6] Ferromagnetic interactions

are also found in Fe^{2+} ($S = 2$) Kagome compounds.[7] A recent study of a Ni^{2+} ($S = 1$) Kagome compound has shown it to be a canted antiferromagnet.[8] The Fe^{2+} Kagome compounds investigated hitherto are $[H_3N(CH_2)_6NH_3][Fe^{II}_{1.5}F_3(SO_4)]\cdot0.5H_2O$, **I**,[7a] and $[H_3N(CH_2)_2NH_2(CH_2)_2NH_2(CH_2)_2NH_3]$-$[Fe^{II}_3F_6(SO_4)_2]$, **II**.[7b] **I** has a perfect Kagome structure with uniform hexagonal bronze layers, and **II** has a slightly distorted structure. A significant question in this context concerns the relationship between the distortion of the hexagonal layers and the magnetic properties of the Kagome compounds. We have now prepared a compound of the formula $[H_3N(CH_2)_2NH_2(CH_2)_2NH_3]_4[Fe^{II}_9F_{18}(SO_4)_6]\cdot9H_2O$, **III**, which is highly distorted compared to **I** and **II** and shows different magnetic properties. We describe the structure and properties of **III** in this article.

Experimental Section

Synthesis and Characterization. In a typical synthesis of $[H_3N(CH_2)_2NH_2(CH_2)_2NH_3]_4[Fe^{II}_9F_{18}(SO_4)_6]\cdot9H_2O$, **III**, 0.263 g of iron(III) citrate was dispersed in an ethanol (EtOH)−water mixture (5.8 and 1.8 mL, respectively) under constant stirring. To this mixture, 0.22 mL of sulfuric acid (98%) and 0.38 mL of diethyl-

---

* To whom correspondence should be addressed. E-mail: cnrrao@jncasr.ac.in. Fax: +91-80-2208-2760.

(1) (a) Ramirez, A. P. *Annu. Rev. Mater. Sci.* **1994**, *24*, 453. (b) Greedan, J. E. *J. Mater. Chem.* **2001**, *11*, 37. (c) Nocera, D. G.; Bartlett, B. M.; Grohol, D.; Papoutsakis, D.; Shores, M. P. *Chem.—Eur. J.* **2004**, *10*, 3850. (d) Wills, A. S.; Harrison, A.; Ritter, C.; Smith, R. I. *Phys. Rev. B* **2000**, *61*, 6156. (e) Wills, A. S.; Harrison, A. *J. Chem. Soc., Faraday Trans.* **1996**, *92*, 2161.
(2) Bartlett, B. M.; Nocera, D. G. *J. Am. Chem. Soc.* **2005**, *127*, 8985.
(3) Grohol, D.; Nocera, D. G.; Papoutsakis, D. *Phys. Rev. B* **2003**, *67*, 064401.
(4) Shores, M. P.; Nytko, E. A.; Bartlett, B. M.; Nocera, D. G. *J. Am. Chem. Soc.* **2005**, *127*, 13462.
(5) Behera, J. N.; Paul, G.; Choudhury, A.; Rao, C. N. R. *Chem. Commun.* **2004**, 456.

(6) (a) Grohol, D.; Papoutsakis, D.; Nocera, D. G. *Angew. Chem., Int. Ed.* **2001**, *40*, 1519. (b) Grohol, D.; Huang, Q.; Toby, B. H.; Lynn, J. W.; Lee, Y. S.; Nocera, D. G. *Phys. Rev. B* **2003**, *68*, 094404. (c) Papoutsakis, D.; Grohol, D.; Nocera, D. G. *J. Am. Chem. Soc.* **2002**, *124*, 2647.
(7) (a) Rao, C. N. R.; Sampathkumaran, E. V.; Nagarajan, R.; Paul, G.; Behera, J. N.; Choudhury, A. *Chem. Mater.* **2004**, *16*, 1441. (b) Paul, G.; Choudhury, A.; Rao, C. N. R. *Chem. Commun.* **2002**, 1904.
(8) Behera, J. N.; Rao, C. N. R. *J. Am. Chem. Soc.* **2006**, *128*, 9334.

10.1021/ic061457y CCC: $33.50 © 2006 American Chemical Society
Published on Web 10/13/2006

Behera and Rao

enetriamine (DETA, 99.98%) were added, followed by the addition of 0.36 mL of HF (40%). The resultant mixture with the molar composition of iron(III) citrate/$4H_2SO_4$/3.5DETA/100EtOH/$100H_2O$/8HF had an initial pH of 4 after it had been stirred for 2 h. The mixture was taken in a 23 mL PTFE-lined acid-digestion bomb and heated at 180 °C for 4 days. After it was cooled to room temperature, the product containing thin plate-shaped crystals of **III** (yield 30% with respect to Fe) was filtered and washed with water and then with ethanol.

The initial characterization of **III** was carried out by powder X-ray diffraction (PXRD), energy-dispersive analysis of X-rays (EDAX), thermogravimetric analysis (TGA), and IR spectroscopy. Magnetic measurements on powdered samples were performed at temperatures between 2 and 300 K, in a vibrating sample magnetometer using a physical property measurement system (quantum design). PXRD patterns indicated the products to be new materials and monophasic, the patterns being consistent with those generated from single crystal X-ray diffraction. EDAX gave the expected metal/sulfate ratio of 3:2. The fluoride test was performed qualitatively, and quantitative analysis was performed by field emission scanning electron microscopy (FE-SEM). Bond valence sum calculations and the absence of electron density near fluorine in the difference Fourier map also provide evidence for the presence of fluorine. The water content of **III** was established by thermogravimetric analysis (TGA) to be close to the value given by the formula.

The infrared spectrum of **III** showed characteristic bands in the $980-1010$ cm^{-1} region from ν_1 and in the $1090-1140$ cm^{-1} region from ν_3 of $SO_4{}^{2-}$. The bending mode of $SO_4{}^{2-}$ was in the $450-600$ cm^{-1} region. The stretching and bending modes of the NH_2/$NH_3{}^+$ groups and H_2O were also in the expected ranges.[9]

Single-Crystal Structure Determination. A suitable single crystal of compound **III** was carefully selected under a polarizing microscope and mounted at the tip of the thin glass fiber using cyanoacrylate adhesive. The single-crystal structure determination by X-ray diffraction was performed on a Siemens SMART-CCD diffractometer equipped with a normal focus, 2.4 kW sealed-tube X-ray source (Mo Kα radiation, $\lambda = 0.71073$ Å) operating at 40 kV and 40 mA. The structure was solved by direct methods using SHELXS-97,[10] which readily revealed all the heavy-atom positions (Fe and S) and allowed us to locate the other non-hydrogen (C, N, O, and F) positions from the difference Fourier maps. An empirical absorption correction based on symmetry-equivalent reflections was applied using SADABS.[11] All the hydrogen positions were found in the difference Fourier maps. For the final refinement, the hydrogen atoms of the amine were placed geometrically and held in the riding mode. The last cycles of refinement included atomic positions, anisotropic thermal parameters for all the non-hydrogen atoms, and isotropic thermal parameters for hydrogen atoms of the amine. The hydrogen positions for water molecules were excluded from the final refinement. Full -matrix least-squares structure refinement against $|F^2|$ was carried out using the SHELXL-97[12] package of programs. Details of the structure determination and

(9) Nakamoto, K. *Infrared and Raman Spectra of Inorganic and Coordination Compounds*; Wiley-Interscience, New York, 1978.

(10) Sheldrick, G. M. *SHELXS-97, Program for Crystal Structure Determination*; University of Göttingen: Göttingen, Germany, 1997. (b) Sheldrick, G. M. *Acta Crystallogr., Sect. A* **1990**, *46*, 467.

(11) Sheldrick, G. M. *SADABS: Siemens Area Detector Absorption Correction Program*; University of Göttingen: Göttingen, Germany, 1994.

(12) Sheldrick, G. M. *SHELXTL-PLUS, Program for Crystal Structure Solution and Refinement*; University of Göttingen: Göttingen, Germany, 1997.

Table 1. Crystal Data and Structure Refinement Parameters for **III**

empirical formula	$C_{16}H_{82}F_{18}N_{12}O_{33}S_6FeII_9$
formula mass	2008.01
cryst syst	triclinic
space group	$P\bar{1}(2)$
a (Å)	11.0438(2)
b (Å)	15.8352(2)
c (Å)	19.5662(4)
α (deg)	77.4810(10)
β (deg)	74.2310(10)
γ (deg)	71.1710(10)
vol (Å^3)	3085.31(10)
Z	2
T (°C)	20
ρ_{calcd} (g cm^{-3})	2.161
λ(Mo Kα) (Å)	0.71073
μ (mm^{-1})	2.403
θ range (deg)	$1.09-23.24$
total data collected	8682
R_{int}	0.0475
R [$I > 2\sigma(I)$]	R1 = 0.0642^a, wR2 = 0.1766^b
R (all data)	R1 = 0.112, wR2 = 0.2129
GOF (S)	0.906

^{a} R1 $= \Sigma|F_0| - |F_c|/\Sigma|F_0|$. ^{b} wR2 $= \{[w(F_0{}^2 - F_c{}^2)^2] /[w(F_0{}^2)^2]\}^{1/2}$, $w = 1/[\sigma^2(F_0)^2 + (aP)^2 + bP]$, $P = [F_0{}^2 + 2F_c{}^2]/3$, where $a = 0.1302$ and $b = 0$.

final refinements for **III** are listed in Table 1. The positions of the fluorine atoms in **III** were located primarily by examination of their thermal parameters. Their assignment as oxygen instead of fluorine invariably leads to nonpositive definite values when they were refined with anisotropic displacement parameters. The powder X-ray diffraction pattern of **III** was in good agreement with the simulated pattern based on the single-crystal data, indicative of phase purity.

Results and Discussion

$[H_3N(CH_2)_2NH_2(CH_2)_2NH_3]_4[Fe^{II}_9F_{18}(SO_4)_6]\cdot 9H_2O$, **III**, has an asymmetric unit with 85 non-hydrogen atoms, of which 58 belong to the inorganic framework and 37 belong to the extraframework including nine water molecules (Figure 1a). There are 10 crystallographically distinct Fe atoms and six S atoms with all the Fe atoms in octahedral geometry. The Fe atoms have fluorine and oxygen neighbors to form FeF_4O_2 octahedra. There are two types of octahedral arrangements around the metal ion. In one, four F atoms are in equatorial positions, and two O are in the axial position as in jarosites; the other type has three F and one O in equatorial positions and one F and one O in axial positions (Figure 1b). Anionic layers of vertex-sharing $Fe^{II}F_4O_2$ octahedra and SO_4 tetrahedra are linked by Fe−F−Fe and Fe−O−S bonds. Note that in the perfect Kagome lattice of **I**, the six coordination of the metal ion is satisfied by the presence of four F atoms in equatorial positions, with the two axial positions occupied by the oxygen atoms of the sulfate (Figure 1c). As a result of the two types of Fe octahedra, the hexagonal structure of **III** gets distorted from that of a perfect Kagome lattice. Because 33% of the sulfate tetrahedra share equitorial oxygen with the iron octahedra, they do not lie perfectly on the triangular lattice and, instead, get tilted. The bridging fluorines connect Fe(II) ions with an Fe−F−Fe angle of $124.6-132.7°$ to form a triangular μ-fluoro trimer, which is capped by the sulfate anion. Because of the presence of F and O in both axial and equatorial positions of the octahedra, two types of triangular lattices are created, and the structure

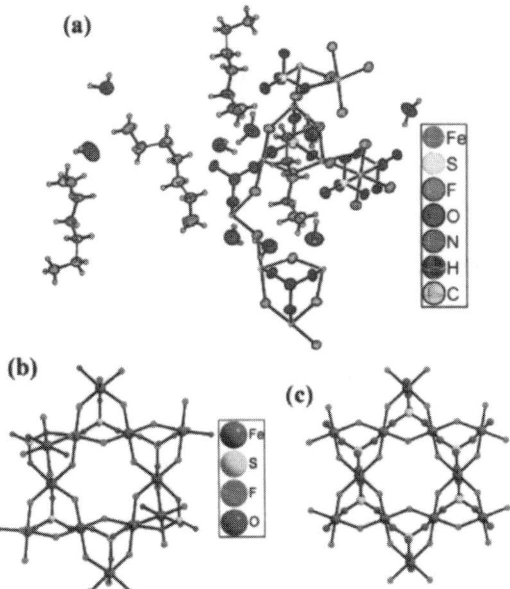

Figure 1. (a) ORTEP plot of $[C_{16}N_{12}H_{64}][Fe^{II}_9F_{18}(SO_4)_6]0\cdot9H_2O$, **III**. Thermal ellipsoids are given at 50% probability. The connectivity of the iron octahedra and the sulfate tetrahedra in **III** and in the perfect Kagome structure of **I** are shown in panels b and c, respectively.

gets distorted from the normal Kagome lattice. One type of three-ring trio is formed by the vertex-sharing of the FeF_4O_2 octahedra creating a perfect three-membered ring which is capped by the sulfate tetrahedron as in jarosites. On the other hand, the three-ring trio formed by the sharing of an equatorial corner between two octahedra and the axial corners shared with two equatorial positions a third octahedron gives rise to a distorted triangular unit. The presence of both perfect and distorted triangular units gives rise to symmetric and unsymmetric hexagons in **III**.

The Fe−O bond distances in **III** are in the range of 2.127(5)−2.176(5) Å with $(Fe-O)_{av}$ = 2.158 Å. The Fe−F bond distances are in the range of 2.008(4)−2.119(4) Å with $(Fe-F)_{av}$ = 2.068 Å. Bond valence calculations[13] using r_0(Fe−F) = 1.65 Å and r_0(Fe−O) = 1.734 Å (Fe(1) = 1.96, Fe(2) = 1.97, Fe(3) = 1.98, Fe(4) = 1.89, Fe(5) = 1.98, Fe(6) = 1.90, Fe(7) = 1.85, Fe(8) = 1.96, Fe(9) = 1.91, and Fe(10) = 1.93) and the average bond distance value indicate that the valence state of all the iron atoms is +2. The bond lengths and angles are in good agreement with those reported for the Fe^{2+} in fluorine and oxygen environments.[7] The Mössbauer spectrum shows the presence of the characteristic signal from Fe^{2+}. The position of fluorine atoms is supported by the calculated bond valence sums which lie in the 0.64−0.72 range. The framework stoichiometry of $[Fe^{II}_9F_{18}(SO_4)_6]$ with a −12 charge is balanced by the presence of triprotonated DETA molecules residing in the crest and trough regions of the corrugated layers and in the interlayer space where water molecules are also present

Figure 2. View down the c axis showing the stacking of the layers in **III**. DETA molecules are not shown in the interlayer space for purpose of clarity.

(Figure 2). The layers in **III** are stacked along the b axis in an ABAB fashion and are held together by hydrogen bonding interactions with the triprotonated amine and water molecules.

In Figure 3, we compare the structure of **III** with that of the perfect Kagome structure of **I** and the slightly distorted structure of **II**. Note that **I** contains only vertex-shared octahedra and, hence, symmetrical hexagons characteristic of the tungsten bronze layer (Figure 3a).[14] In **II**, the distorted structure results from the edge sharing of octahedra with two types of triangular units (Figure 3b). We realized that the distortion in **III** (Figure 3c) is far greater than that in **II**.

The TGA curve of **III** (N_2 atmosphere, range of 30−900 °C, heating rate 5 °C/min, given as Supporting Information) showed weight losses in three steps. The first weight loss corresponds to the loss of guest water molecules in the 60−200 °C range [obsd = 7.8%, calcd = 8%]; the major weight loss in the 200−400 °C range corresponds to the loss of amine and HF [obsd = 30.6%, calcd = 31.8%]. The third weight loss in the region of 400−800 °C corresponds to the removal of fluorine and the decompositions of sulfate [obsd = 27%, calcd = 25%]. The sample heated at 900 °C diffracts weakly, and the PXRD corresponds mixture of Fe_2O_3 and FeO (JCPDS files 00-003-0800 and 00-002-1186, respectively).

The magnetic properties of the jarosites are known to be sample dependent.[1d,1e] It is expected that any small perturbation would have a strong effect on the ground-state manifold. In Figure 4, we present the magnetic data on **III**. From the high-temperature inverse-susceptibility data of **III**, recorded at 1000 Oe (see inset a of Figure 4), we estimate a Weiss temperature of −105 K (θ_p), suggesting dominant antiferromagnetic exchange interactions. The effective magnetic moment per iron atom is 5.02 μ_B, which is close to the spin-only (S = 2) value of 4.9 μ_B for Fe^{2+} and is comparable to the values reported for similar compounds.[15] Furthermore,

(13) Brown, I. D.; Altermatt, D. *Acta Crystallogr., Sect. B* **1985**, *47*, 244.

(14) Magneli, A. *Acta Chem. Scand.* **1953**, *7*, 315.
(15) (a) Paul, G.; Choudhury, A.; Rao, C. N. R. *Chem. Mater.* **2003**, *15*, 1174. (b) Fu, A.; Huang, X.; Li, J.; Yuen, T.; Lin, C. L. *Chem.−Eur. J.* **2002**, *8*, 2239.

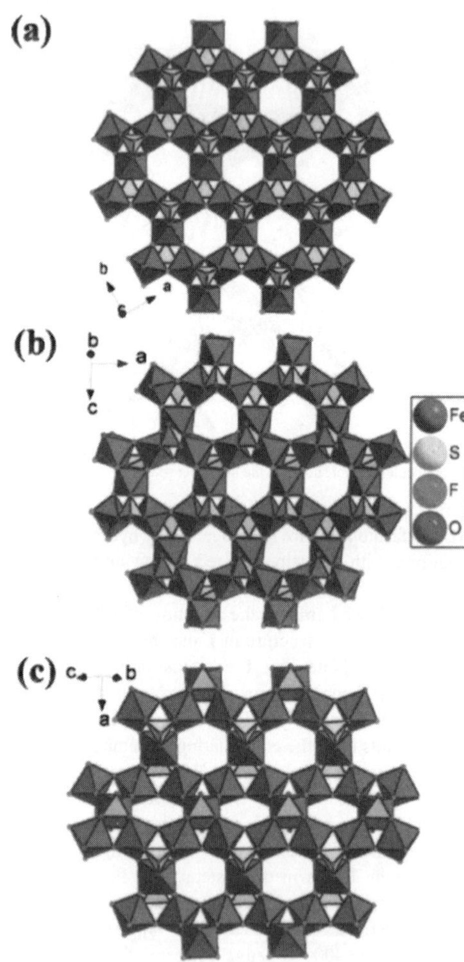

Figure 3. Polyhedral views of the hexagonal Kagome layers in **I**, **II**, and **III** are shown in panels a, b, and c, respectively. Notice how c has the most distorted structure and a has the perfect structure.

III shows magnetic hysteresis at low temperatures (inset b of Figure 4). The magnetization increases with the field without any evidence for saturation, implying an antiferromagnetic component as well, in agreement with the negative θ_p. The magnetic behavior of **III** is quite different from that of **I** or **II**. In Figure 5a and b, we compare the magnetic susceptibilities of **I**, **II**, and **III** measured at 100 Oe under field-cooled (FC) and zero-field-cooled (ZFC) conditions. Particularly important is the fact that magnetic susceptibility of this distorted compound (**III**) is very different from the previous compounds (**I** and **II**), mainly in the intermediate temperature regime, down to very low temperature. The broad nature of the ZFC susceptibility around $T \approx 10$ K suggests magnetic polarizations caused by some interactions, the strength of which is in the intermediate energy scale (kT ≈ 10 K). Since this appears in a distorted system, the magnetic polarizations must have a geometrical origin. It is by now quite well-known that distortions in a corner shared

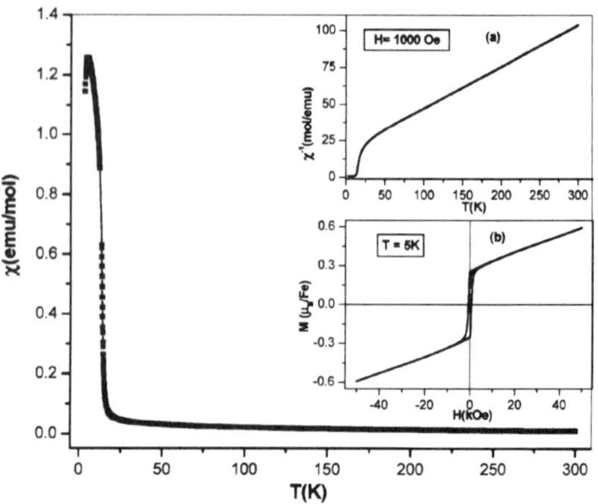

Figure 4. Temperature dependence of the magnetic susceptibility of **III** at 1000 Oe. Inset a shows the temperature variation of the inverse susceptibility at 1000 Oe. Inset b shows magnetic hysteresis at 5 K.

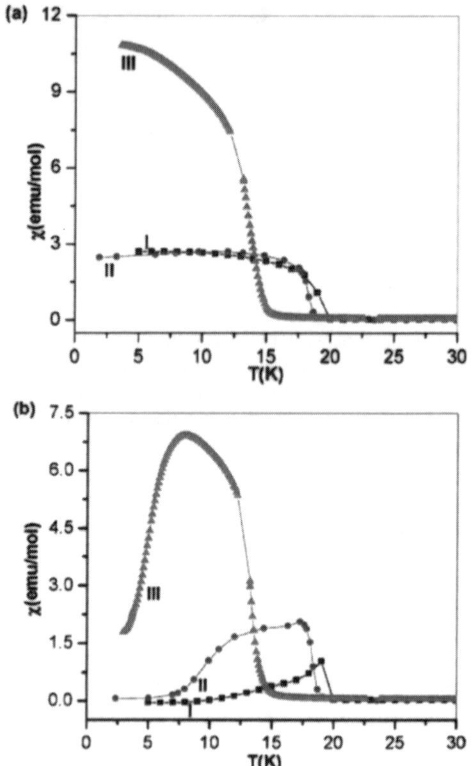

Figure 5. Comparison of the magnetic susceptibility data of **III** at 100 Oe with **I** and **II** under (a) field-cooled (FC) and (b) zero-field-cooled (ZFC) conditions.

Kagome lattice leads to Dzyaloshinsky−Moriya (DM) interactions, the in-plane components of which can give rise to magnetic polarizations in the out-of-plane directions with canted vectors of magnetic moments.[16] It is interesting to

(16) Elhajal, M.; Canals, B.; Lacroix, C. *Phys. Rev. B* **2002**, *66*, 014422.

note that the degree of this DM interactions can be tuned by controlled synthetic methods, as evident from Figure 5. The fact that the DM interactions can be controlled has a potential in designing frustrated systems for various magnetic applications. It is noteworthy that the different degrees of distortion lead to significant differences in the magnetic properties of the Fe(II) Kagome compounds, a feature different from those of the Fe(III) jarosites with a d^5 ion. If such interactions with geometrical origin can be controlled in both integer spin and half-odd-integer spin Kagome system[17] requires further studies.

Conclusions

An organically templated iron(II) sulfate with a highly distorted Kagome lattice has been synthesized and character-

ized. The magnetic properties of this compound are distinctly different from those of other Fe(II) Kagome compounds possessing perfect or slightly distorted Kagome structures; the distorted Kagome compound prepared by us shows canted antiferromagnetism at low temperatures.

Acknowledgment. J.N.B. thanks CSIR, India, for a research fellowship. The authors thank Dr. S. K. Pati for helpful discussions.

Supporting Information Available: TGA curve, atomic coordinates, important bond distances and angles, hydrogen bonding interactions, and X-ray crystallographic information in CIF format for the structure determination of **III**. This material is available free of charge via the Internet at http://pubs.acs.org.

(17) Pati, S. K.; Rao, C. N. R. *J. Chem. Phys.* **2005**, *123*, 234703.

IC061457Y

PAPER   www.rsc.org/dalton | Dalton Transactions

Synthesis, structure and magnetic properties of an amine-templated Mn^{2+} ($S = 5/2$) sulfate with the Kagome structure

J. N. Behera[a,b] and C. N. R. Rao*[a,b]

Received 21st September 2006, Accepted 4th January 2007
First published as an Advance Article on the web 15th January 2007
DOI: 10.1039/b613732f

In pursuit of a compound with the Kagome structure, formed by a non-Fe^{3+} transitional metal ion with a spin of 5/2, we have synthesized an amine-templated Mn^{2+} sulfate under solvothermal conditions. This compound with a perfect Kagome structure shows evidence for antiferromagnetic interactions with no long-range order.

Introduction

Transition metal compounds with the Kagome structure are not only interesting structurally but also because of their magnetic properties. There have been several studies of Kagome compounds of Fe^{3+} ($S = 5/2$) such as the jarosites, in the past few years, all of which exhibit magnetic frustration or low-temperature antiferromagnetism.[1] The Fe^{3+} jarosites are generally not pure owing to the presence of site defects. Nocera *et al.*[2] have recently prepared a pure Fe^{3+} jarosite by redox-based hydrothermal methods and found the Fe^{3+} Kagome layer in this compound shows an antiferromagnetic transition at 61.4 K. On the other hand, V^{3+} ($S = 1$) and Ni^{2+} ($S = 1$) Kagome compounds appear to show the presence of ferromagnetic interaction and less evidence of frustration.[3,4] A theoretical study employing quantum many-body Heisenberg models has shown that Kagome compounds where transition metal ions with half-odd integer spins would exhibit magnetic-frustration while those with transition metal ions with integral spins may show evidence for ferromagnetic interactions and less of frustration.[5] In view of the above observation, we considered it as important to prepare a Kagome compound with a transition metal ion other than Fe^{3+}, with $S = 5/2$. We have now been able to synthesize an amine-templated manganese(II) sulfate of the composition, $[C_4N_2H_{12}][NH_4]_2[Mn^{II}_3F_6(SO_4)_2]$, **I**, with the Kagome structure, under solvothermal conditions. We describe the structure and magnetic properties of this compound in this article. Interestingly, we find that the magnetic properties of this compound differ from those of Fe^{3+} jarosites[2] as well as Kagome compounds where the transition metal ions have integral spins.[3,4]

Results and discussion

The asymmetric unit (Fig. 1a) of **I** consists of 27 non-hydrogen atoms, out of which 19 belong to the inorganic-framework and 8 to the extra-framework guest molecules, including two nitrogens of the two ammonium ions. It consists of vertex-sharing $Mn^{II}F_4O_2$ octahedra and SO_4 tetrahedral units, which are fused together by Mn–F–Mn and Mn–O–S bonds. The distorted octahedron around

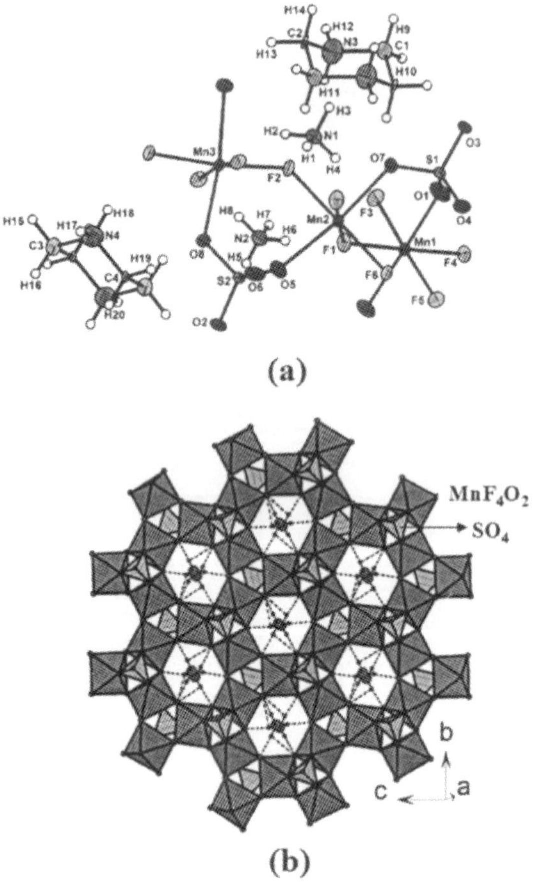

[a]*Chemistry and Physics of Materials Unit, Jawaharlal Nehru Centre for Advanced Scientific Research, Jakkur P. O., Bangalore, 560064, India*
[b]*Solid State and Structural Chemistry Unit, Indian Institute of Science, Bangalore, 560012, India. E-mail: cnrrao@jncasr.ac.in; Fax: +91-80-2208-2760*

Fig. 1 (a) Labeled asymmetric unit of $[C_4N_2H_{12}][NH_4]_2[Mn^{II}_3F_6(SO_4)_2]$, **I**. Thermal ellipsoids are given at 50% probability. (b) Polyhedral view of the hexagonal Kagome layer in **I**. Note the presence of the ammonium ion in the hexagonal channel.

Mn^{2+} has the oxygen atoms from the sulfate along the z-axis and of the F atoms along the xy-plane. Each MnF_4O_2 unit shares four of its Mn–F vertices with similar neighbors, with the Mn–F–Mn bonds roughly aligned in the bc-plane. The Mn–O bond is canted from the bc-plane and the Mn–O vertex forces a three-ring trio of apical Mn–O bonds close together to be capped by the SO_4 tetrahedra. The sulfate groups are positioned alternatively up and down about the hexagonal layer. The three- and six-rings of the octahedra forming the in-plane connectivity can be seen in Fig. 1(b). Such a layer consisting of a hexagonal tungsten bronze layer is characteristic of the Kagome structure.[6]

The Mn–O bond distances in **I** are in the range 2.179(3)–2.246(3) Å, [(Mn(1)–O)av = 2.210(3), (Mn(2)–O)av = 2.194(3) and (Mn(3)–O)av = 2.236(3) Å]. The Mn–F bond distances are in the range 2.097(2)–2.126(2) Å [(Mn(1)–F)av = 2.110(2), (Mn(2)–F)av = 2.116(2) and (Mn(3)–F)av = 2.108(2) Å]. Selected values of the bond distances and bond angles in **I** are listed in Table 1. The values of the bond angles and distances indicate a distorted octahedral coordination of Mn and near-perfect tetrahedral coordination of sulfur. Bond valence sum calculations[7] using r_0 (Mn–F) of 1.698 and r_0 (Mn–O) of 1.790 Å [Mn(1) = 1.95, Mn(2) = 1.96 and Mn(3) = 1.92] and the values of the bond distances indicate the valence state of the Mn atoms to be +2. The bond valence sums are also consistent with the presence of bridging fluorine

Table 1 Bond lengths [Å] and angles [°] for **I**[a]

Mn(1)–F(1)	2.106(2)	Mn(3)–F(6)	2.108(2)
Mn(1)–F(2)	2.111(2)	Mn(3)–F(2)#2	2.108(2)
Mn(1)–F(3)	2.111(2)	Mn(3)–F(5)#3	2.122(2)
Mn(1)–F(4)	2.115(2)	Mn(3)–O(5)#3	2.227(3)
Mn(1)–O(1)	2.182(3)	Mn(3)–O(6)	2.246(2)
Mn(1)–O(2)#1	2.238(3)	S(1)–O(7)	1.461(3)
Mn(2)–F(3)#2	2.103(2)	S(1)–O(3)	1.464(3)
Mn(2)–F(1)	2.115(2)	S(1)–O(6)	1.480(3)
Mn(2)–F(5)	2.120(2)	S(1)–O(2)	1.485(3)
Mn(2)–F(6)	2.126(2)	S(2)–O(1)	1.458(3)
Mn(2)–O(3)	2.179(3)	S(2)–O(8)	1.467(3)
Mn(2)–O(4)	2.210(3)	S(2)–O(5)	1.477(3)
Mn(3)–F(4)#3	2.097(2)	S(2)–O(4)	1.478(3)
C(1)–C(2)#5	1.487(5)	C(3)–N(4)	1.513(5)
C(1)–N(3)	1.509(6)	N(4)–C(4)	1.479(5)
N(3)–C(2)	1.493(5)	C(4)–C(3)#6	1.485(5)
F(1)–Mn(1)–F(2)	178.76(9)	F(6)–Mn(3)–F(5)#3	172.67(8)
F(1)–Mn(1)–F(3)	87.89(9)	F(2)#2–Mn(3)–F(5)#3	88.14(8)
F(2)–Mn(1)–F(3)	90.93(8)	F(4)#3–Mn(3)–O(5)#3	104.36(9)
F(1)–Mn(1)–F(4)	94.06(8)	F(6)–Mn(3)–O(5)#3	89.45(10)
F(2)–Mn(1)–F(4)	87.12(8)	F(5)#3–Mn(3)–O(5)#3	83.73(9)
F(3)–Mn(1)–F(4)	177.59(9)	F(4)#3–Mn(3)–O(6)	89.08(9)
F(1)–Mn(1)–O(1)	92.49(10)	F(6)–Mn(3)–O(6)	103.02(9)
F(2)–Mn(1)–O(1)	87.12(9)	F(2)#2–Mn(3)–O(6)	82.08(9)
F(3)–Mn(1)–O(1)	88.34(10)	F(5)#3–Mn(3)–O(6)	84.23(9)
F(4)–Mn(1)–O(1)	90.17(10)	O(5)#3–Mn(3)–O(6)	162.17(11)
F(1)–Mn(1)–O(2)#1	85.41(9)	O(7)–S(1)–O(3)	110.76(16)
F(2)–Mn(1)–O(2)#1	94.99(9)	O(7)–S(1)–O(6)	109.61(15)
F(3)–Mn(1)–O(2)#1	92.26(9)	O(3)–S(1)–O(6)	108.77(16)
F(4)–Mn(1)–O(2)#1	89.30(9)	O(7)–S(1)–O(2)	109.51(15)
O(1)–Mn(1)–O(2)#1	177.79(10)	O(3)–S(1)–O(2)	109.57(16)
F(3)#2–Mn(2)–F(1)	176.59(8)	O(6)–S(1)–O(2)	108.59(16)
F(3)#2–Mn(2)–F(5)	85.59(8)	O(1)–S(2)–O(8)	111.17(16)
F(1)–Mn(2)–F(5)	91.80(9)	O(1)–S(2)–O(5)	109.51(17)
F(3)#2–Mn(2)–F(6)	91.99(9)	O(8)–S(2)–O(5)	108.58(15)
F(1)–Mn(2)–F(6)	90.62(9)	O(1)–S(2)–O(4)	109.57(17)
F(5)–Mn(2)–F(6)	177.57(8)	O(8)–S(2)–O(4)	109.28(15)
F(3)#2–Mn(2)–O(3)	99.11(9)	O(5)–S(2)–O(4)	108.69(16)
F(1)–Mn(2)–O(3)	83.27(9)	Mn(1)–F(1)–Mn(2)	129.15(10)
F(5)–Mn(2)–O(3)	94.36(10)	Mn(3)#1–F(2)–Mn(1)	130.99(10)
F(6)–Mn(2)–O(3)	85.88(10)	Mn(2)#1–F(3)–Mn(1)	126.04(10)
F(3)#2–Mn(2)–O(4)	88.32(10)	Mn(3)#4–F(4)–Mn(1)	123.99(10)
F(1)–Mn(2)–O(4)	89.73(9)	S(2)–O(1)–Mn(1)	136.31(17)
F(5)–Mn(2)–O(4)	95.35(9)	S(1)–O(2)–Mn(1)#2	125.93(15)
F(6)–Mn(2)–O(4)	84.71(9)	Mn(2)–F(5)–Mn(3)#4	130.52(10)
O(3)–Mn(2)–O(4)	168.20(11)	Mn(3)–F(6)–Mn(2)	124.83(10)
F(4)#3–Mn(3)–F(6)	87.42(9)	S(1)–O(3)–Mn(2)	134.42(16)
F(4)#3–Mn(3)–F(2)#2	171.13(8)	S(2)–O(4)–Mn(2)	126.28(15)
F(6)–Mn(3)–F(2)#2	93.88(9)	S(2)–O(5)–Mn(3)#4	126.19(16)
F(4)#3–Mn(3)–F(5)#3	91.66(9)	S(1)–O(6)–Mn(3)	124.99(15)
C(2)#5–C(1)–N(3)	108.9(3)	C(4)#6–C(3)–N(4)	110.4(3)
C(2)–N(3)–C(1)	110.1(3)	C(4)–N(4)–C(3)	109.8(3)
C(1)#5–C(2)–N(3)	112.0(3)	N(4)–C(4)–C(3)#6	111.5(3)

[a] Symmetry transformations used to generate equivalent atoms: #1 x, $-y + 1/2$, $z - 1/2$; #2 x, $-y + 1/2$, $z + 1/2$; #3 x, $y - 1$, z; #4 x, $y + 1$, z; #5 $-x$, $-y$, $-z + 2$; #6 $-x + 1$, $-y - 1$, $-z + 2$.

atoms, the value being in the range 0.64–0.66. Thus, the framework stoichiometry of $[Mn^{II}_3F_6(SO_4)_2]$ with a 4– charge requires the amine to be doubly protonated alongside the presence of two ammonium ions. The anionic Kagome sheets of $[Mn^{II}_3F_6(SO_4)_2]^{4-}$ are stacked one over the other in an ABAB fashion and are held strongly by the hydrogen bond interaction of the amine and ammonium ions residing in the inter-layer space (Fig. 2). The amine and the ammonium ions form N–H $\cdots$ O ($\theta_{[N-H\cdots O]}$ = 104–173(4)°, d(N $\cdots$ O) = 2.936(4)–2.940(5) Å), N–H $\cdots$ F ($\theta_{[N-H\cdots F]}$ = 126–177(3)°, d(N $\cdots$ F) = 2.770(4)–3.128(4) Å), C–H $\cdots$ O ($\theta_{[C-H\cdots O]}$ = 102–164°, d(C $\cdots$ O) = 2.770(4)–3.210(4) Å) and C–H $\cdots$ F hydrogen bonds ($\theta_{[C-H\cdots F]}$ = 138–152°, d(C $\cdots$ F) = 3.215(4)–3.402(4) Å) with the framework oxygens and fluorines.

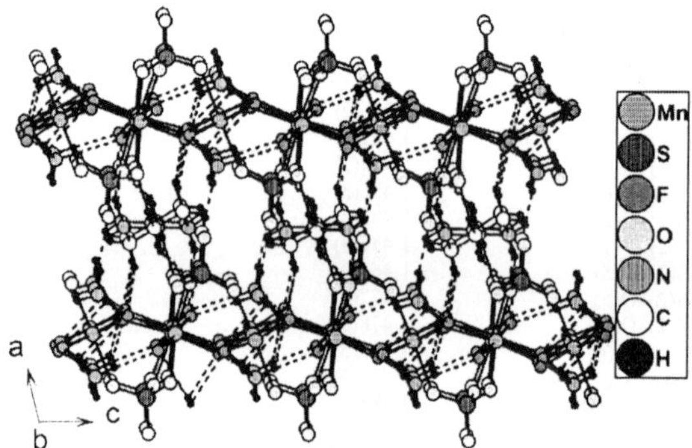

Fig. 2 Structure of $[C_4N_2H_{12}][NH_4]_2[Mn^{II}_3F_6(SO_4)_2]$, **I**, showing the stacking of layers and the presence of piperazine in the interlayer space.

Thermogravimetric analysis of **I** showed a two-step weight loss, the first loss corresponding to the loss of ammonia, amine and HF in the range 130–400 °C (obs. = 32.6%, calc. = 33.6%). The second weight loss occurs in the 460–850 °C range (obs. = 31%, calc. = 28%). The end product was found to be Mn_3O_4 (JCPDS file card PDF # 00-001-1127).

Magnetic properties

We show the variable temperature magnetic susceptibility (χ) data of **I** recorded at 100 and 1000 Oe in Fig. 3. In the paramagnetic region, the susceptibility follows the Curie–Weiss law with a negative Weiss temperature of 30 K as obtained from the fit of the χ_M^{-1} data in the 100–300 K range (inset of Fig. 3a). The negative θp value suggests that the exchange interaction is antiferromagnetic. This θp value is lower than that of the Fe^{3+} Kagome compounds (>600 K). The effective magnetic moment of Mn is 5.96 μ_B, almost equal to the spin only S = 5/2 value of 5.92 μ_B since there is negligible orbital contribution to the moment and it is comparable to that of manganese(II) compounds reported in the literature.[8] The field-cooled (FC) and the zero-field cooled (ZFC) magnetic susceptibility data measured at 100 Oe show little difference as can be seen from Fig. 3(b). The susceptibility data reveal a shoulder around 10 K which becomes broader as the magnetic field is increased to 1000 Oe. The susceptibility data do not change significantly with the field and the compound does not show any magnetic polarization at higher magnetic fields. This behavior suggests that the material is essentially paramagnetic with antiferromagnetic (AFM) interaction exhibiting only short-range order. Long-range AFM

order would have given a sharp transition as in the Fe^{3+} jarosites. The linear field dependence of magnetization at low temperatures (see inset of Fig. 3b) also supports the above observations.

Conclusions

An amine-templated Mn^{2+} sulfate with the Kagome structure, which is the first example of a S = 5/2 Kagome compound analogous to the well-known Fe^{3+} jarosites has been synthesized under solvothermal conditions. In the Mn^{2+} Kagome compound, **I**, studied by us, we do not observe any magnetic frustration or long-range AFM ordering as in the other S = 5/2 and other integer spins Kagome compounds. The absence of any ferromagnetic interaction in **I** is in accordance with theoretical predictions, which require integer spins to metal ions. It is noteworthy that the Co^{2+} (S = 3/2) and Cr^{3+} (S = 3/2) compounds with the Kagome structure have properties comparable to those of the analogous Fe^{3+} jarosites.[9,10] A pure S = 1/2 copper Kagome compound is reported to exhibit spin frustration.[11]

Experimental

Synthesis and characterization

In a typical synthesis of $[C_4N_2H_{12}][NH_4]_2[Mn^{II}_3F_6(SO_4)_2]$, **I**, 0.502 g of $Mn(NO_3)_2\cdot4H_2O$ was dissolved in 4.6 ml of ethylene glycol (EG) under constant stirring. To this mixture, 0.22 ml of sulfuric acid (H_2SO_4, 98%) and 0.344 g of piperazine (PIP) were added, followed by the addition of 0.35 ml HF (40%). The resultant mixture with a molar composition of $2Mn(NO_3)_2\cdot4H_2O : 4H_2SO_4 : 4PIP : 80EG : 8HF$ had an initial pH < 2 after stirring for 2 h. The mixture was taken in a 23 ml PTFE-lined acid digestion bomb and heated at 150 °C for 3 d. After cooling to room temperature, the product containing colorless hexagonal crystals (yield: 50% with respect to Mn) was filtered off and washed with water and then with ethanol.

I was characterized by powder X-ray diffraction (PXRD), energy dispersive analysis of X-rays (EDAX), chemical analysis, thermogravimetric analysis (TGA) and IR spectroscopy. EDAX analysis indicated the ratio of Mn : S to be 3 : 2. The presence of fluorine was confirmed by analysis and the percentage of fluorine estimated by EDAX in a field emission scanning electron microscope was also satisfactory. Thermogravimetric analysis also confirms the stoichiometry of the compound. Bond valence sum calculations[6] and the absence of electron density near fluorine in the difference Fourier map also provide evidence for the presence of fluorine. The sulfate content was found to be 30.8% compared to the expected 32% on the basis of the formula.

The IR spectrum of **I** showed characteristic bands for the amine as well as the ammonium moieties. The ammonium ions result from the decomposition of the piperazine molecules used in the starting synthesis mixture. The stretching mode of the –N–H bond (of the amine) is observed around 3004 cm^{-1} (v_1). The N–H bending modes of the amine and NH_4^+ are observed in the range 1440–1586 cm^{-1}. Strong bands in the region 850–1015 cm^{-1} correspond to v_1 and v_3 while bands in the region 583–644 cm^{-1} can be assigned to v_2 and v_4 fundamental modes of the sulfate ion.[12]

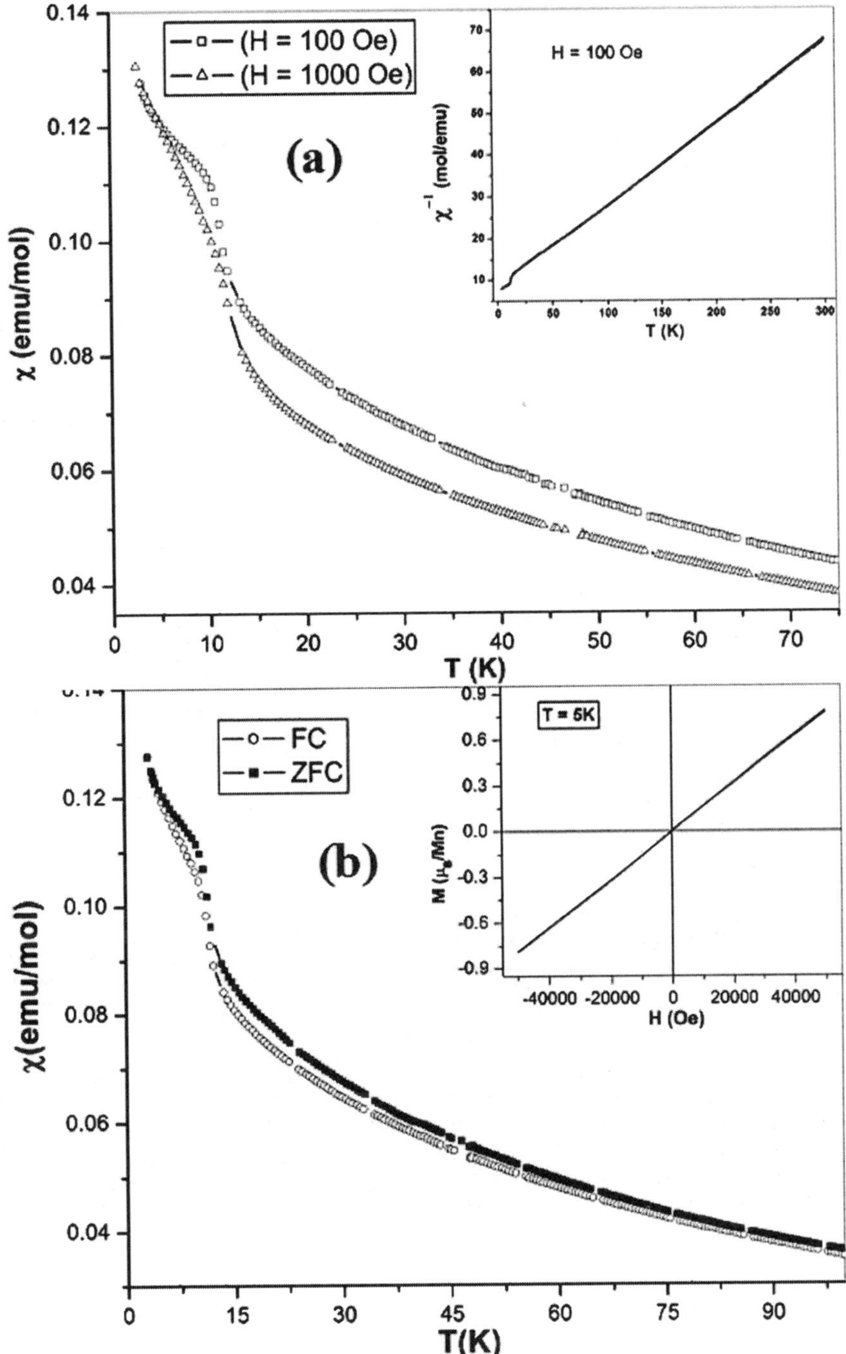

Fig. 3 (a) Temperature variation of χ of $[C_4N_2H_{12}][NH_4]_2[Mn^{II}_3F_6(SO_4)_2]$, **I** under ZFC conditions ($H = 100$ and 1000 Oe). Inset shows the variation of inverse susceptibility. (b) Temperature dependence of the magnetic susceptibility of **I** under field-cooled (FC) and zero-field-cooled (ZFC) conditions (100 Oe). Inset shows the $M–H$ curve at 5 K.

Table 2 Crystal data and structure refinement parameters for **I**

Empirical formula	$C_4H_{20}F_6Mn_3N_4O_8S_2$
Crystal system	Monoclinic
Space group	$P2(1)/c$
Crystal size/mm	$0.2 \times 0.16 \times 0.08$
a/Å	17.6128(2)
b/Å	7.60040(10)
c/Å	13.05010(10)
β/°	104.8680(10)
V/Å^3	1688.45(3)
Z	4
Formula mass	595.20
ρ_{calc}/g cm^{-3}	2.342
λ(MoKα)/Å	0.71073
μ/mm^{-1}	2.566
θ range/°	1.20 to 25.68
Total data collected	18444
Limiting indices	$-21 \leq h \leq 21, -9 \leq k \leq 7, -15 \leq l \leq 15$
Unique data	3197
Refinement method	Full-matrix least squares on $|F^2|$
R_{int}	0.0377
$R[I > 2\sigma(I)]$	$R_1{}^a = 0.0324, wR_2{}^b = 0.0862$
R (all data)	$R_1 = 0.0417, wR_2 = 0.0924$
Goodness of fit (S)	1.044
Largest difference map peak and hole/e Å	0.739 and -0.786

$^a R_1 = \sum |F_o| - |F_c|/|F_o|.$ $^b wR_2 = \{[w(F_o{}^2 - F_c{}^2)^2]/[w(F_o{}^2)^2]\}^{1/2}, w = 1/[\sigma^2(F_o)^2 + (aP)^2 + bP]; P = F_o{}^2 + 2F_c{}^2]/3;$ where $a = 0.0367$ and $b = 6.0435$.

Magnetic measurements were performed at temperatures between 2 and 300 K, in a vibrating sample magnetometer using a physical property measurement system (quantum design).

Single-crystal structure determination

Single crystal structure determination by X-ray diffraction was performed at 298 K on a Bruker AXS-CCD diffractometer equipped with a normal focus, 2.4 kW sealed-tube X-ray source (Mo-Kα radiation, $\lambda = 0.71073$ Å) operating at 40 kV and 40 mA. The structure was solved by direct methods using SHELXS-97.[13] An empirical absorption correction based on symmetry equivalent reflections was applied using SADABS.[14] All the hydrogen positions of the ammonium ions were found in the difference Fourier maps. For the final refinement, hydrogen atoms of the amine were placed geometrically and held in the riding mode. Full-matrix least-squares structure refinement against $|F^2|$ was carried out using the SHELXL-97 package of programs.[15] Details of the structure determination and final refinements for **I** are listed in Table 2. The positions of the fluorine atoms in **I** were located primarily by looking at their thermal parameters. Assigning them as oxygen instead of fluorine invariably leads to nonpositive definite values when they were refined with anisotropic displacement parameters. The powder X-ray diffraction pattern of **I** was in good agreement with the simulated pattern based on the single-crystal data, indicative of phase purity.

CCDC reference number 621801.

For crystallographic data in CIF or other electronic format see DOI: 10.1039/b613732f

Acknowledgements

JNB thanks CSIR, India for a senior research fellowship.

References

1 (a) A. P. Ramirez, *Annu. Rev. Mater. Sci.*, 1994, **24**, 453; (b) J. E. Greedan, *J. Mater. Chem.*, 2001, **11**, 37; (c) D. G. Nocera, B. M. Bartlett, D. Grohol, D. Papoutsakis and M. P. Shores, *Chem.–Eur. J.*, 2004, **10**, 3850; (d) A. S. Wills, A. Harrison, C. Ritter and R. I. Smith, *Phys. Rev. B: Condens. Matter*, 2000, **61**, 6156; (e) A. S. Wills and A. Harrison, *J. Chem. Soc., Faraday Trans.*, 1996, **92**, 2161.
2 (a) B. M. Bartlett and D. G. Nocera, *J. Am. Chem. Soc.*, 2005, **127**, 8985; (b) D. Grohol, D. G. Nocera and D. Papoutsakis, *Phys. Rev. B: Condens. Matter*, 2003, **67**, 064401.
3 (a) D. Papoutsakis, D. Grohol and D. G. Nocera, *J. Am. Chem. Soc.*, 2002, **124**, 2647; (b) D. Grohol, D. Papoutsakis and D. G. Nocera, *Angew. Chem., Int. Ed.*, 2001, **40**, 1519.
4 J. N. Behera and C. N. R. Rao, *J. Am. Chem. Soc.*, 2006, **128**, 9334.
5 S. K. Pati and C. N. R. Rao, *J. Chem. Phys.*, 2005, **123**, 234703.
6 B. Gerand, G. Nowogrocki, J. Guenot and M. Figlarz, *J. Solid State Chem.*, 1979, **29**, 429.
7 I. D. Brown and D. Altermatt, *Acta Crystallogr., Sect. B*, 1985, **41**, 244.
8 (a) Z. A. D. Lethbridge, M. J. Smith, S. K. Tiwary, A. Harrison and P. Lightfoot, *Inorg. Chem.*, 2004, **43**, 11; (b) Z. A. D. Lethbridge, A. F. Congreve, E. Esslemont, A. M. Z. Slawin and P. Lightfoot, *J. Solid State Chem.*, 2003, **172**, 212.
9 J. N. Behera, G. Paul, A. Choudhury and C. N. R. Rao, *Chem. Commun.*, 2004, 456.
10 (a) S.-H. Lee, C. Broholm, M. F. Collins, L. Heller, A. P. Ramirez, Ch. Kloc, E. Bucher, R. W. Erwin and N. Lacevic, *Phys. Rev. B: Condens. Matter*, 1997, **56**, 8091; (b) A. Keren, K. Kojima, L. P. Le, G. M. Luke, W. D. Wu, Y. J. Uemura, M. Takano, H. Dabkowska and M. J. P. Gingras, *Phys. Rev. B: Condens. Matter*, 1996, **53**, 6451.
11 M. P. Shores, E. A. Nytko, B. M. Bartlett and D. G. Nocera, *J. Am. Chem. Soc.*, 2005, **127**, 13462.
12 K. Nakamoto, *Infrared and Raman Spectra of Inorganic and Coordination Compounds*, Wiley, New York, 1978.
13 G. M. Sheldrick, *SADABS Siemens Area Detector Absorption Correction Program*, University of Göttingen, Göttingen, Germany, 1994.
14 (a) G. M. Sheldrick, *SHELXS-97 Program for crystal structure determination*, University of Göttingen, Göttingen, Germany, 1997; (b) G. M. Sheldrick, *Acta Crystallogr.*, 1990, **35**, 467.
15 G. M. Sheldrick, *SHELXTL-97 Program for Crystal Structure Solution and Refinement*, University of Göttingen, Göttingen, Germany, 1997.

FEATURE ARTICLE   www.rsc.org/chemcomm | ChemComm

Structural diversity and chemical trends in hybrid inorganic–organic framework materials

Anthony K. Cheetham,*[a] C. N. R. Rao*[b] and Russell K. Feller[a]

Received (in Cambridge, UK) 18th July 2006, Accepted 20th October 2006
First published as an Advance Article on the web 7th November 2006
DOI: 10.1039/b610264f

Hybrid framework compounds, including both metal–organic coordination polymers and systems that contain extended inorganic connectivity (extended inorganic hybrids), have recently developed into an important new class of solid-state materials. We examine the diversity of this complex class of materials, propose a simple but systematic classification, and explore the chemical and geometrical factors that influence their formation. We also discuss the growing evidence that many hybrid frameworks tend to form under thermodynamic rather than kinetic control when the synthesis is carried out under hydrothermal conditions. Finally, we explore the potential applications of hybrid frameworks in areas such as gas separations and storage, heterogeneous catalysis, and photoluminescence.

1 Introduction

The purpose of this feature article is to give an overview of developments in the field of hybrid inorganic–organic framework structures over recent years, especially during the last decade. We have not attempted to be comprehensive because of the huge amount of activity in the area, but instead we have focused on placing these developments in a broader context. We shall illustrate the enormous chemical and structural diversity of these materials and discuss some of the systematic trends that are starting to appear in synthetic routes for

[a]Materials Research Laboratory, University of California, Santa Barbara, CA, 93106-5121, USA. E-mail: cheetham@mrl.ucsb.edu
[b]Jawaharlal Nehru Centre for Advanced Scientific Research, Jakkur P.O., Bangalore, 560 064, India. E-mail: cnrrao@jncasr.ac.in

Tony Cheetham was a member of the Chemistry faculty at Oxford, 1974–1991, and has been at the University of California at Santa Barbara since 1991. He is Professor in both the Materials and Chemistry Departments at UCSB, and since 2004 has been the Director of the new International Center for Materials Research (ICMR). Cheetham's research interests lie in the area of functional inorganic materials and currently include hybrid framework materials, phosphors for solid state lighting, and inorganic nanoparticles.

C. N. R. Rao is the National Research Professor of India, Linus Pauling Research Professor at the Jawaharlal Nehru Centre for Advanced Scientific Research, and Honorary Professor at the Indian Institute of Science. His research interests are in the chemistry of materials. He has authored nearly 1000 research papers and edited or written 30 books in materials chemistry.

Russell K. Feller graduated in Chemistry from the University of California at Los Angeles in 2003 and is currently a graduate student in the Chemistry Department at UCSB. His PhD research project focuses on hybrid framework materials of transition metals.

hybrids. We shall also examine some of the emerging application for materials in this exciting area.

There is an extensive class of purely inorganic framework materials based upon extended arrays such as chains, sheets or 3-D networks. The silicate and aluminosilicate minerals, which were classified by Pauling almost 70 years ago, constitute the most versatile group. Indeed, their dimensionalities can range from 0, as in simple silicates such as zircon that are based upon isolated orthosilicate SiO_4^{4-} units, through 1-D silicate chains (*e.g.* pyroxenes), 2-D sheets (*e.g.* micas and clays) to 3-D arrays (*e.g.* quartz). Zeolites represent a particularly interesting sub-class of these aluminosilicate frameworks, since their architectures display nanoporosity that can be harnessed for applications in separations, catalysis and so on.[1] More recently it has been shown that a wide range of other inorganic families, especially phosphates, can form framework structures with varying dimensionalities. This is true, for example, of aluminium phosphates, tin(II) phosphates, zinc phosphates and so on.[2] Fig. 1 shows examples from the case of tin phosphates.

In the world of organic solids, by contrast, such structural diversity is less well represented. Molecular organics (*i.e.* 0-D) are ubiquitous, of course, but extended arrays are largely limited to 1-D chains, such as those found in polymer systems ranging from polyolefins to block copolymers and proteins. With the exception of covalent organic frameworks (COFs) that contain borate,[3] extended 2-D and 3-D organic arrays are essentially unknown, aside from cross-linked polymers and examples based upon molecular units that assemble into networks *via* hydrogen bonding rather than covalent bonding.[4] In the light of this basic distinction between inorganic and organic networks, it is interesting to examine the structural diversity of hybrid inorganic–organic frameworks.

We define hybrid inorganic–organic framework materials as compounds that contain both inorganic and organic moieties as integral parts of a network with infinite bonding connectivity in at least one dimension. This definition excludes systems that are molecular or oligomeric, such as the

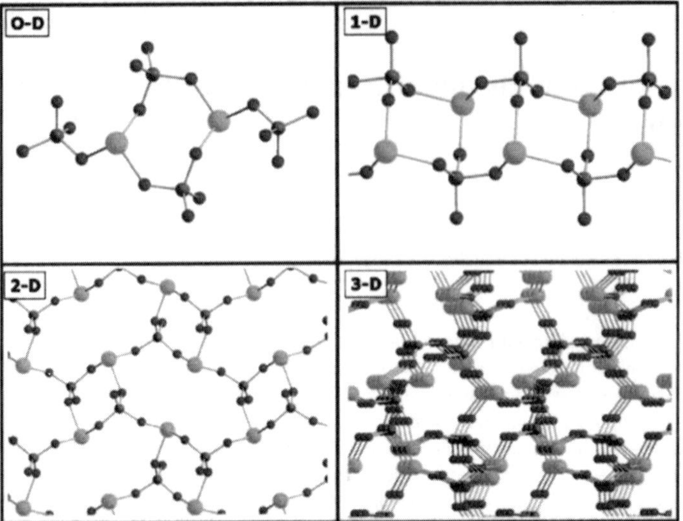

Fig. 1 Four tin phosphates, each containing the same $Sn_2P_2O_4$ motif, with monomeric, 1-D, 2-D, and 3-D structures. Blue spheres denote tin, green phosphorus, and red oxygen.

supramolecular assemblies described by Lehn, Hosseini, Stang, Fujita and many others. It also excludes systems in which the organic is merely a guest inside an inorganic cavity, as is often observed in zeolites and mesoporous materials, and hybrid composites, in which the inorganic and organic components are present as separate phases. Most of the known hybrid frameworks may conveniently be divided into two categories.[5] The *coordination polymers*, or metal organic frameworks (MOFs) as they are also known, can be defined as extended arrays composed of isolated metal atoms or clusters that are linked by polyfunctional organic ligands, L; these are based upon M–L–M connectivity. Second, there are systems that contain extended arrays of inorganic connectivity, which we shall refer to for convenience as *extended inorganic hybrids*. At present, the vast majority of known materials in this area are based upon oxygen bridges. These *hybrid metal oxides*, which often contain infinite metal–oxygen–metal (M–O–M) arrays as a part of their structures, represent a sub-group of a larger class in which there is extended M–X–M bonding *via* other atoms such as Cl, N or S, or *via* inorganic groups such as phosphate.

Examples of 1-D coordination polymers are relatively common in the early literature, even though they were not seen at the time as part of a vast and remarkable family of materials. Examples include porphyrin coordination polymers (Fig. 2) with interesting magnetic properties that were first discovered by Basolo and co-workers in the 1970s and characterized by X-ray diffraction at a later date.[6] Early examples in the 3-D coordination polymer area can be found in the work of Gravereau, Garnier and Hardy in the late 1970s, in which zeolitic materials with ion-exchange properties were made by linking hexacyanoferrate units with tetrahedrally coordinated Zn^{2+} cations.[7] There were also early examples of hybrid materials with extended inorganic connectivity, the

most notable being the layered zirconium phosphonates such as the one shown in Fig. 3.[8]

Interest in the hybrid area began to accelerate in the 1990s, when several groups, particularly those of Robson, Hoffman and Yaghi, recognized that rigid, polyfunctional organic molecules could be used to bridge metal cations or clusters into extended arrays. Robson published a landmark paper in 1990,[9] laying the groundwork for an important part of the field of crystal engineering – the science of predicting basic

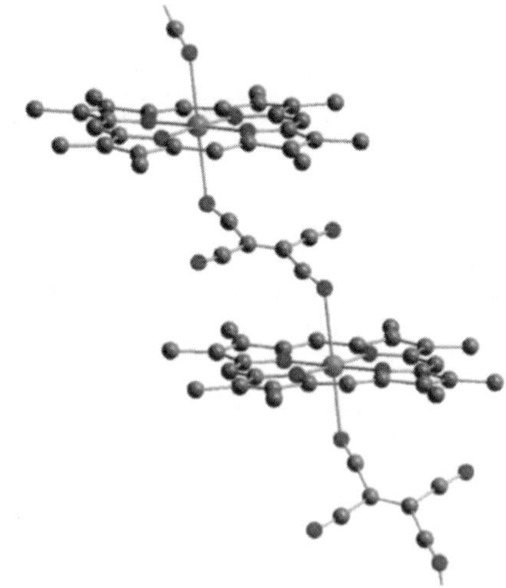

Fig. 2 The 1-D coordination polymer octaethylporphyrinato-manganese(III) tetracyanoethenide (ethyl groups omitted for clarity).[6] The two tetracyanoethenide anions are not crystallographically equivalent, but slightly tilted with respect to each other. Pink spheres denote manganese, blue nitrogen, and gray carbon.

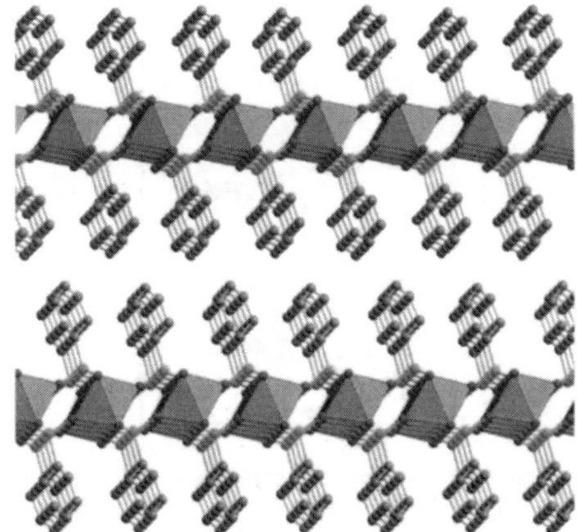

Fig. 3 A side view of the layered structure of zirconium bis(phenyl-phosphonate), with edge-to-edge interactions between phenyl rings on neighboring sheets.[8] Green spheres denote phosphorus, red oxygen, and gray carbon, with ZrO_6 octahedra in blue.

networks with potentially useful characteristics and then using appropriate molecular building blocks to synthesize them.[10] For the synthesis of *porous* materials, networks are often envisioned where rigid organic molecules and metal atoms or clusters replace bonds and atoms in classical inorganic structures.[11]

The purpose of our feature article is to discuss some of the emerging trends in this rapidly developing field rather than to provide a comprehensive review of hybrids. This is a very complex and diverse field, so we are suggesting a systematic classification of hybrid frameworks that places new and existing materials in a simple, rational context. We have also attempted to identify the structural and chemical trends that are beginning to emerge from the literature and to pinpoint some of the areas where there are important gaps and opportunities. Recent reviews of various aspects of hybrid materials include those of Clearfield, Rowsell and Yaghi, Rosseinsky, and Rao *et al.*[12]

2 Chemical and structural diversity

2.1 Chemical diversity

Hybrid frameworks are found for a wide range of metals and involve a diverse range of organic ligands. Most of the published work involves transition metals, including zinc, but there is a growing body of literature around rare-earth based systems, which are of interest for their optical properties. In addition, there has been a certain amount of effort with p-block elements, especially aluminium, gallium and tin, plus a recent growth of interest in magnesium, driven by the search for lightweight materials for hydrogen storage.

In terms of organic ligands, much of the recent focus has been on connectivity through oxygen atoms of carboxylic acid groups, and there has been a recent review of this field by Rao and colleagues.[12] Rigid dicarboxylic acids, such as benzene-1,4-dicarboxylic acid, have proved very versatile, as have the simple but more flexible aliphatic systems, such as succinic and

glutaric acids. The simplest member of this family, oxalic acid, has been used extensively. As will be discussed later, monocarboxylic acids can also form hybrids, and there has been some recent effort with formic and acetic acids. Nor is the field limited to carboxylic acids, since phosphonic acids and phenolic acids can also form hybrid frameworks. Beyond network formation involving M–O linkages, there has been a reasonable amount of work with other types of ligands, such as pyridyls and imidazoles, as well as mixed ligands that offer the possibility of more than one type of connection, *e.g.* M–O plus M–N or M–S. Much remains to be explored in the area of these more complex linkages.

We shall aim to illustrate the diversity of chemical types in the choice of examples that will be given in subsequent sections. Clearly the structures are strongly influenced by the coordination preferences of the metals as well as by the variety of ways in which different ligands can coordinate to metals.

2.2 Coordination polymers

The term coordination polymer owes its origins to the analogy with coordination compounds, in which ligands, organic or otherwise, are coordinated to monomeric metal centers. Coordination compounds can be thought of loosely as the monomers of coordination polymers. Given the enormous volume of work on coordination compounds over the last century, starting with the pioneering work of Werner for which he was awarded the Nobel Prize in Chemistry in 1913, it is hardly surprising that the field of coordination polymers is turning out to be so rich and varied.

A vast range of coordination polymers or supramolecular architectures with different dimensionalities – 1-D, 2-D and 3-D – have been discovered in recent years. Fig. 4 illustrates an example of a 1-D chain system involving Ag–N bonding through a linear 4,4'-bipyridyl group.[13] Since the linkage through the silver is also linear, the chain, itself, is too. By

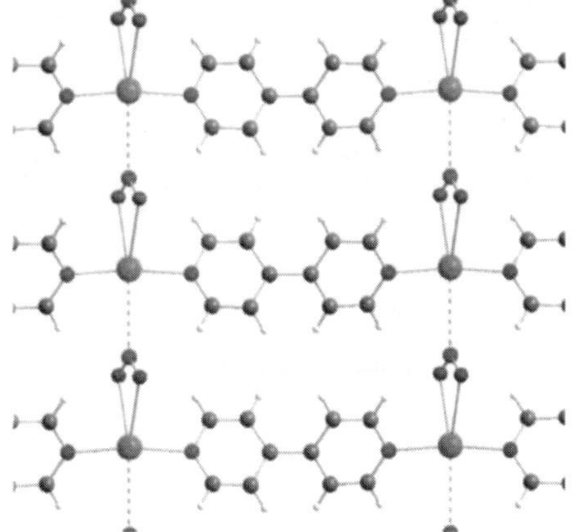

Fig. 4 A linear coordination polymer of silver with 4,4'-bipyridyl and bidentate nitrite anions,[13] There is a 3.0 Å contact between the silver and the nitrogen of the nitrite ion, shown as a dotted line. If this was a full covalent bond, the system would be two-dimensional. Gray spheres denote carbon, white hydrogen, dark blue nitrogen, red oxygen, and green silver.

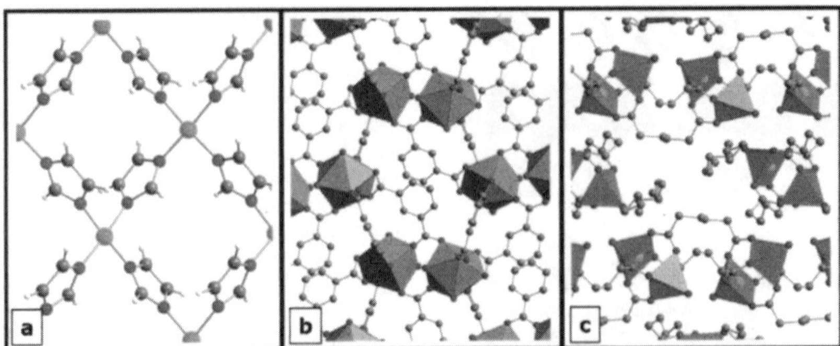

Fig. 5 Some 2-D coordination polymers: (a) nickel bis(imidazolate), containing neutral sheets of square planar metal centers,[16] (b) a plan view of the europium isophthalate structure,[17] showing nine-coordinate Eu^{3+}, and (c) copper adipate,[18] which consists of both chains and sheets, viewed along the chain axis. Gray spheres denote carbon, white hydrogen, blue nitrogen, red oxygen, and green nickel, with EuO_9 polyhedra in orange and CuO_5 square pyramids in blue.

contrast, when tetrahedrally coordinated zinc ions are connected through 4,4′-bipyridyl, this gives rise to a zigzag chain.[14] The metal center does not have to be mononuclear. For example, the common copper acetate dimer, with its paddle wheel geometry, can form linear chains through simple bidentate ligands.[15] The scope of these 1-D coordination polymers is enormous, since they may be neutral or charged (in which case they require compensating cations or anions) and they often contain solvent molecules in voids or channels.

Turning to 2-D systems, Fig. 5(a) shows a very simple example in which nickel ions are connected *via* square planar coordination by imidazolate anions to form a very simple neutral layered structure.[16] A more complex case based upon a combination of rare-earth ions with isophthalic acid is shown in Fig. 5(b); the thiophene derivatized version of the same ligand combined with Tb^{3+} yields a product with enhanced

green luminescence.[17] As a final example to illustrate the versatility of this area, we show in Fig. 5(c) a 2-D coordination polymer based on the copper carboxylate dimer, linked through adipic acid, in which a 2-D layer alternates with 1-D chains that contain the same basic building blocks.[18] Here we note that the paddle wheels are linked *via* the dicarboxylic acid rather than by ligands in the axial position of the Jahn–Teller distorted copper coordination sphere.

Robson's early work yielded some elegant examples of crystal engineering in which 3-D networks of simple, known structure types, such as the diamondoid and ReO_3 structures, were built from suitable combinations of metal ions and rigid linkers. Fig. 6 shows a simple example based upon the PtS structure.[19] Others have exploited the same concept, including Carlucci *et al.*,[20] who reported some remarkable open-framework structures based upon silver in combination with ligands such as pyrazine (Fig. 7). Some of the most striking examples

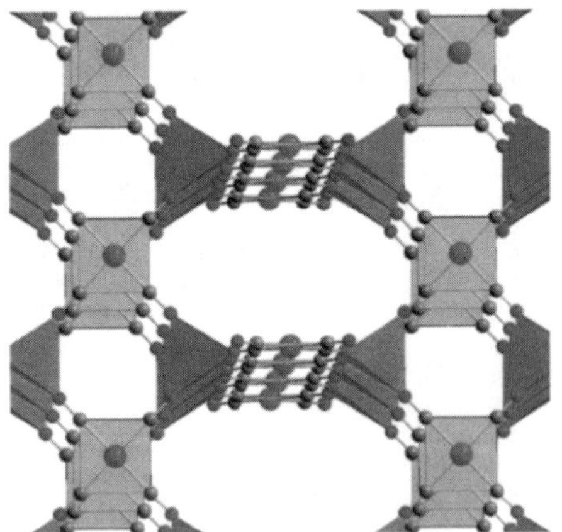

Fig. 6 A mixed copper–platinum tetracyanide, structurally analogous to PtS, with sulphur replaced by a bridging copper tetracyano complex.[19] Pink spheres denote platinum, gray carbon, and blue nitrogen, with PtC_4 square planes in pink and CuN_4 tetrahedra in blue. Reproduced with permission. Copyright 1990, Royal Society of Chemistry.

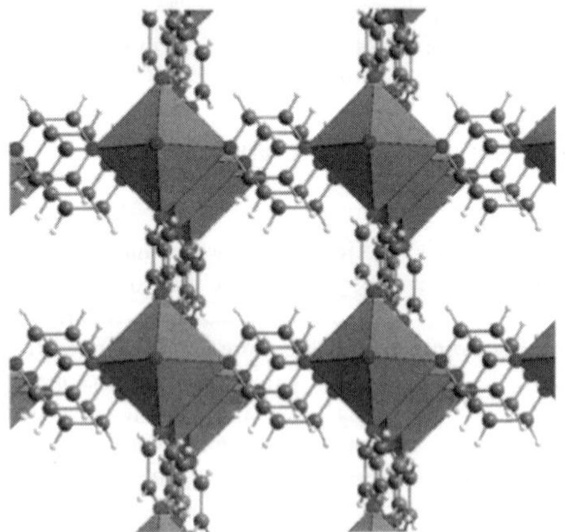

Fig. 7 Silver pyrazine hexafluoroantimonate,[20] a porous framework with an ReO_3-like structure (SbF_6^- groups omitted for clarity). Gray spheres denote carbon, blue nitrogen, and white hydrogen, with AgN_6 octahedra in green. Reproduced with permission. Copyright 1995, Wiley–VCH.

of porous 3-D coordination polymers can found in the work of Yaghi, O'Keeffe and co-workers, in which they have exploited bridging of simple Zn_4O groups *via* rigid aromatic dicarboxylates such as benzene-1,4-dicarboxylic acid to build networks with remarkably low densities and high porosity, such as MOF-5 (Fig. 8).[21] They have shown that large families based upon the same architecture can be created by altering the length or other chemical details of the organic linker.[22] In the case of the *reticular* family based upon MOF-5, for example, they have made as many as 16 derivatives with the same basic architecture. It should also be noted that in addition to architectures based upon the topologies of simple inorganic structures, there has also been success in building porous hybrids based upon known zeolite structures. These include zinc, cadmium and indium coordination polymers that adopt the ABW, BCT, MTN, RHO and SOD topologies.[23] Imidazole-based ligands are particularly effective for this purpose since they can mimic the Si–O–Si angles that are found in typical zeolites.

One of the complications that can arise when the structures are very open is that the networks can interpenetrate or interweave, thereby reducing or eliminating the porosity. There are some remarkable examples of such behavior, including cases where multiple interpenetration is observed.[24] Paradoxically, Yaghi has shown that interpenetration can be attractive for certain gas storage applications, since it can increase the available surface area per unit volume (see section 5.1).

A very recent and exciting example of a 3-D coordination polymers can found in the work of Férey *et al.*[25] on the use of trinuclear chromium clusters in combination with benzene-1,4-dicarboxylic acid. This reaction yields porous structures with unit cells volumes of up to 700 000 $Å^3$, *i.e.* similar to that of a small protein! In the absence of single crystals for structure determination, the structures were solved by the ingenious use of Monte Carlo simulations with simulated annealing.

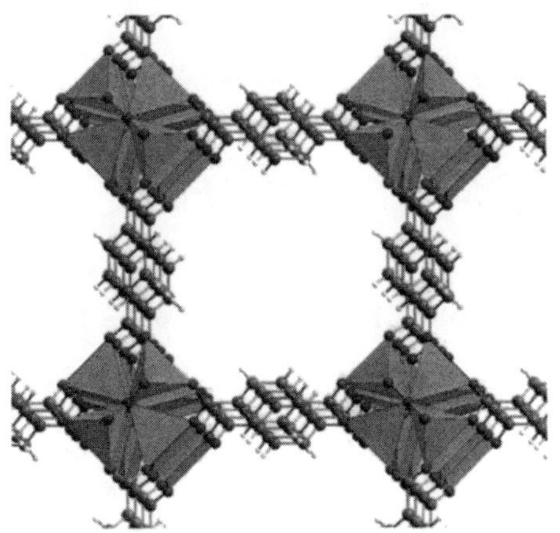

Fig. 8 MOF-5,[21] a porous cubic zinc terephthalate which is topologically analogous to ReO_3. Gray spheres denote carbon, red oxygen, and white hydrogen, with ZnO_4 tetrahedra in blue. Reproduced with permission. Copyright 1999, Macmillan Publishers Ltd.

Finally in this brief section on the exploding field of coordination polymers, we should mention that there growing body of work on systems that contain more than one metal or more than one ligand type, or both. For example, some beautiful open architectures can form when transition metals are combined with rare-earths in the presence of glycolate and water as ligands.[26] Assemblies that contain more than one metal can often be facilitated by the use of ligands with multiple N and O donor atoms, taking advantage of the different ligand affinities of transition metals and rare earths. Equally, the use of more than one ligand with a single metal can yield interesting results, as in the work mixed oxalate–diphosphonates and oxalate–dicarboxylates.[27] A variation on this theme involves a single metal with a carboxylate ligand such as oxalate in combination with phosphate as a second anion.[28] Materials of this type were initially made serendipitously when oxalate salts were being used as precursors in the synthesis of metal phosphate frameworks. These examples illustrate the huge scope of this burgeoning area, with its unlimited permutations of metals and ligands.

2.3 Extended inorganic hybrids

The area of coordination polymers is only one sub-field of the broader domain of hybrid framework materials, since far more structural permutations become accessible if we allow for the possibility of extended inorganic connectivity. This idea is illustrated schematically in Fig. 9, which compares 1-D and 2-D coordination polymers with a system that has inorganic connectivity in two dimensions and is connected in the third dimension by organic linkers. Such extended inorganic hybrid materials not only open up a vast area of new chemical and structural permutations, but they also provide a basis for creating materials with properties that are traditionally found in metal oxides. Thus we have the tantalizing possibility of making hybrid materials that are metallic, superconducting, or high temperature ferromagnets. In this section of the paper, we describe some of the progress that has already been made in this fascinating area.

Early examples of 1-D hybrid metal halides include the famous Wolfram's red salt, which contains {Pt(EtNH$_2$)$_4$} units linked into infinite chains by Pt–Cl–Pt bridges (Fig. 10(a)).[29] In this, and other more recent examples such as the zinc phosphonate chain of Stucky and co-workers (Fig. 10(b)),[30] the organic ligands simply decorate the inorganic chains, rather than cross-linking them. However, 1-D inorganic chains can also be cross-linked to make a layered structure rather than a 1-D network, as in the case of the metal succinate $Ni_7(OH)_6(H_2O)_3(C_4H_4O_4)_4 \cdot 7H_2O$ (Fig. 11).[31] A beautiful example of a cross-linked inorganic chain is found in transition-metal gallates, which comprises chains of *trans*-corner-sharing MO_6 octahedra cross-linked into a 3-D network by the gallate ions (Fig. 12).[32] Note that all the oxygen atoms of the MO_6 octahedra are supplied by the gallate ligands; the resulting topology of the inorganic network looks like the chain found in Rb_2FeF_5,[33] thereby revealing the striking resemblance between this area and classical solid-state chemistry. Note, too, that the array contains channels that accommodate zeolitic water molecules.

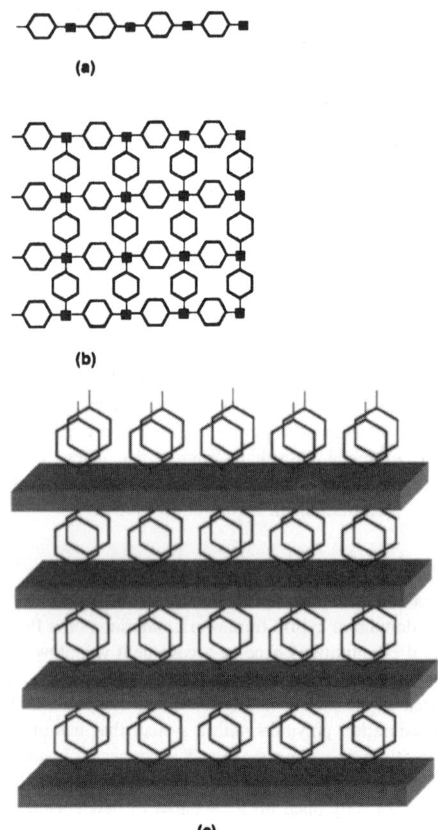

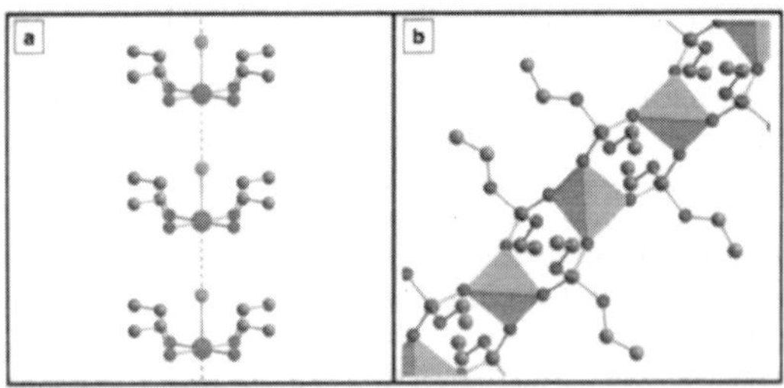

Fig. 9 Schematic representation of coordination polymers and extended inorganic hybrids; (a) and (b) show 1-D and 2-D coordination polymers, respectively, while (c) shows a system that has inorganic connectivity in two dimensions and is connected in the third dimension by organic linkers.

Fig. 10 Some 1-D extended inorganic hybrid materials: (a) Wolfram's Red salt, a 1-D platinum chloride polymer.[29] Controversy continues as to whether the coordination geometry is square pyramidal as shown here, or alternating octahedral and square planar; and (b) zinc diethylphosphate,[30] an inorganic chain decorated by organic groups. Light green spheres denote chlorine, gray carbon, dark blue nitrogen, pink platinum, red oxygen, and dark green phosphorus, with ZnO_4 tetrahedra in light blue.

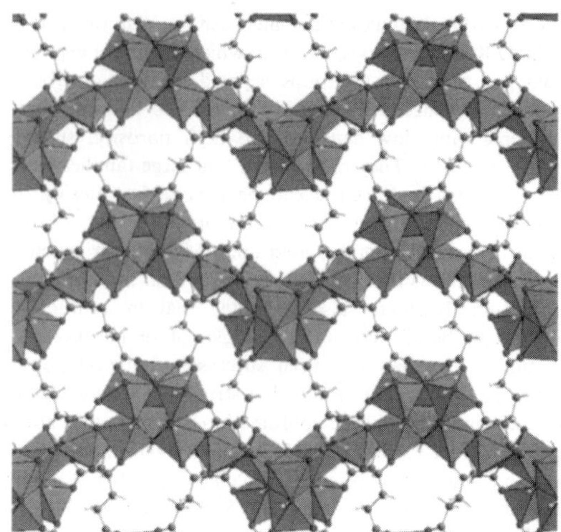

Fig. 11 A nickel succinate containing inorganic chains, bridged by organic groups along a second dimension to form sheets.[31] Gray spheres denote carbon, red oxygen, and white hydrogen, with NiO_6 octahedra in green. Reproduced with permission. Copyright 2003, Wiley–VCH.

Many nice examples of 2-D hybrid oxides can be derived from Clearfield's α-zirconium phosphate structure, which contains 2-D sheets of ZrO_6 octahedra sandwiched between phosphate layers.[34] Alberti *et al.*[8] was able to increase the interlayer spacing by taking advantage of monophosphonates, which acted as spacers between the layers, albeit with no bonding connection between them (Fig. 3). As with some of the examples in the previous paragraph, the organic groups are decorating the layers in the Alberti structures. In order to build three-dimensional frameworks, Dines *et al.* demonstrated that

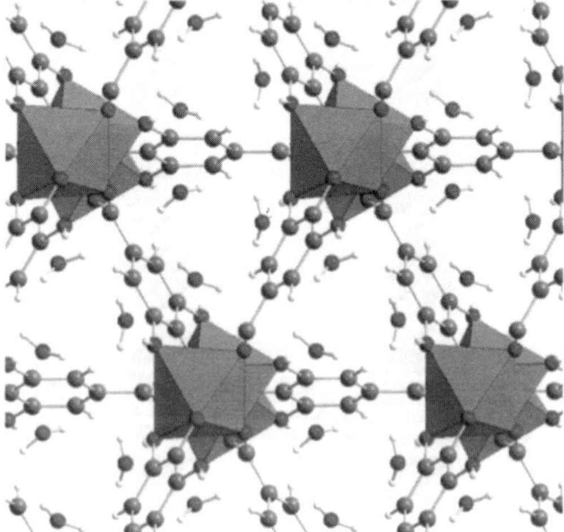

Fig. 12 Nickel gallate,[32] a framework containing inorganic chains bridged by organic groups into a 3-D network. Gray spheres denote carbon, red oxygen, and white hydrogen, with NiO_6 octahedra in green. Reproduced with permission. Copyright 2006, Elsevier Ltd.

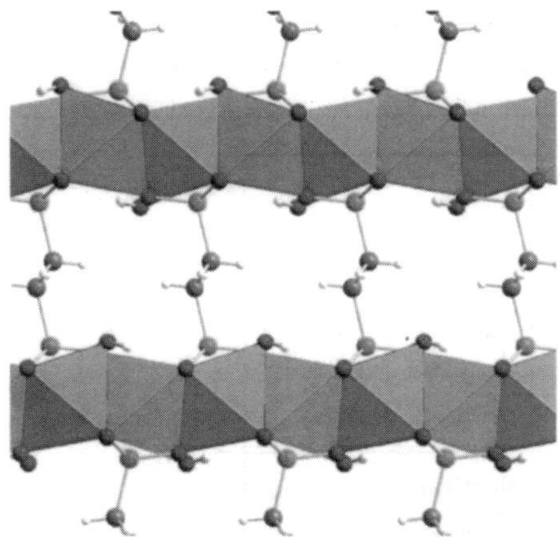

Fig. 13 A pillared cobalt(II) ethanediphosphonate with inorganic sheets connected in the third dimension by organic groups.[35] Gray spheres denote carbon, red oxygen, white hydrogen, and green phosphorus, with CoO_6 octahedra in blue. Reproduced with permission. Copyright 2005, American Chemical Society.

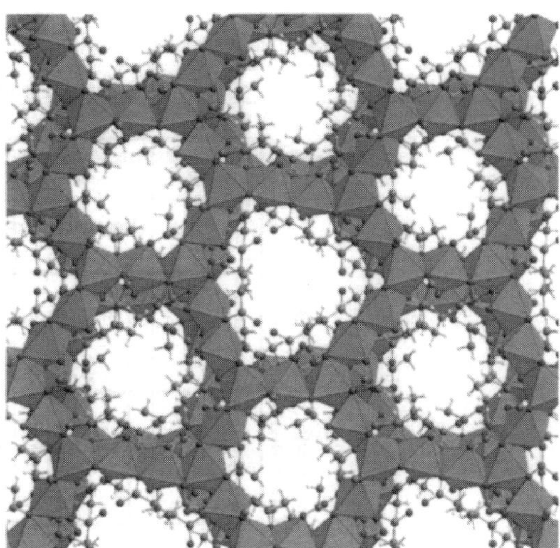

Fig. 14 A nickel succinate containing a channelled 3D nickel–oxygen network decorated by organic groups.[38] Gray spheres denote carbon, red oxygen, and white hydrogen, with NiO_6 octahedra in green. Reproduced with permission. Copyright 2002, Wiley–VCH.

is possible to create connections between the inorganic layers by using diphosphonates (Fig. 13).[36] These materials are not porous, but porosity can be created by using a combination of diphosphonate linkers and shorter monophosphonate groups.[37] There has also been some work to add functionality to the zirconium-based materials through careful choice of the diphosphonate used in the synthesis. For example, Vermuelen and Thompson used viologens, rigid organic molecules of the formula $[H_2O_3PCH_2CH_2–bipyridinium–CH_2CH_2PO_3H_2]^{2+}$, to bridge zirconium phosphonate layers.[38]

In view of the existence of 1-D and 2-D inorganic connectivity in hybrid materials, there has been considerable interest in the possible existence of 3-D systems. A striking example was found in 2002 with the discovery of a nickel succinate with a 3-D network of corner- and edge-sharing NiO_6 octahedra (Fig. 14).[39] Additional examples have since been found in other systems, including cadmium malonate[40] and nickel glutarate.[41] In all of these systems, the frameworks are open with the organic groups lining the pores, as in Fig. 14, and the structures show enhanced thermal stability on account of the inertness of the inorganic skeleton. Nickel succinate, for example, is stable to 400 °C in air.

2.3 Classification of hybrid framework structures

The examples of hybrid materials presented in the previous two sections underline the enormous structural diversity of this exciting class of materials. For example, while we have focused on the dimensionality with respect to either M–ligand–M or extended inorganic connectivity, it is clear that many systems exhibit both types of linkages. For example, the cobalt diphosphonate shown in Fig. 13[35] is 2-D with respect to inorganic connectivity and 1-D with respect to M–ligand–M connectivity. Overall, therefore, the dimensionality of the network is 3-D. Table 1 shows the full range of possibilities in terms of M–ligand–M or extended inorganic dimensionalities.

It is interesting to reflect that virtually the whole of organometallic chemistry and much of classical coordination chemistry is contained within a single box in this table (*i.e.* both M–L–M and inorganic connectivity = 0). For conciseness, we refer to this type as I^0O^0 (I = inorganic and O = organic); note that the sum of the exponents gives the overall dimensionality of the structure. The 3-D nickel succinate (Fig. 14) falls in the box with M–L–M = 0 and inorganic connectivity = 3 (I^3O^0) while the MOF-5 structure (Fig. 8) has M–L–M = 3 and inorganic connectivity = 0 (I^0O^3). There are known examples of all of these classes of hybrid material. Wolfram's red salt (Fig. 10(a)), for example, would be I^1O^0, since there is no bonding between the chains, whereas the 1-D coordination polymers of Basolo (Fig. 2) would be I^0O^1. However, some classes, such as I^0O^3, are quite common, while other, such as I^1O^1, are relatively rare.

The question arises as to whether there are materials that can be classified within the empty boxes in the bottom right part of the table. For example, can we have materials in which the connectivity is 2-D with respect to both M–L–M and M–X–M. We currently know of no such materials, but the possibility exists that these may be found in materials that have not yet been discovered.

2.4 Dense and open frameworks

In the previous three sections, we have not attempted to differentiate between dense and open hybrid framework structures. As with inorganic silicates and aluminosilicates, there is no fundamental chemical difference between the dense and open hybrid structures, though their properties and applications are often quite distinct. There are also differences in the synthetic strategies that are needed to synthesize them, and in section 4 we shall discuss some of the reaction

Table 1 Proposed classification of hybrid materials, showing the dimensionality of different structures with respect to both organic connectivity between metal centers (O^n) and extended inorganic connectivity (I^n) (see text for explanation)

Dimensionality of inorganic connectivity, I^n ($n = 0$–3)

Metal–organic–metal connectivity, O^n ($n = 0$–3)		0	1	2	3
	0	Molecular complexes I^0O^0	Hybrid inorg. chains I^1O^0	Hybrid inorg. layers I^2O^0	3-D Inorg. hybrids I^3O^0
	1	Chain coordination polymers I^0O^1	Mixed inorg.– organic layers I^1O^1	Mixed inorg.– organic 3-D framework I^2O^1	—
	2	Layered coordination polymers I^0O^2	Mixed inorg.– organic 3-D framework I^1O^2	—	—
	3	3-D Coordination polymers I^0O^3	—	—	—

conditions that are likely to favor the formation of open structures. One strategy that should be mentioned here concerns the use of organic template molecules in the creation of open hybrid structures. Template molecules or structure-directing agents, especially quaternary amines, have been extensively used in zeolite synthesis, but the strategy has not been widely adopted in the hybrid area. Exceptions include the use of amines and other templates in the synthesis of transition-metal diphosphonates, aluminium monophosphonates, and cobalt squarates.[42] In the latter instance, Dan *et al.* prepared $[C_6N_2H_{14}]_2[Co_2(C_4O_4)_3(H_2O)_4]$ and $[C_3N_2H_5]_2$-$[Co_2(C_4O_4)_3(H_2O)_4]$ under hydrothermal conditions in the presence of quaternary amines. Both compounds contain chains formed by cobalt dimers linked by the squarate units, the chains being connected through hydrogen bonding interactions *via* the amines. These materials would be classified as I^0O^1 in the classification shown in Table 1, since we do not include organic connectivity through hydrogen bonding. Finally we should mention a rather unusual example of *inorganic* templating, in which Rao and co-workers prepared an open-framework cadmium oxalate that formed around an alkali halide assembly.[43]

3 Chemical trends

We now turn to the intriguing question of what chemical factors influence whether a particular system will form coordination polymers rather than frameworks with extended inorganic connectivity, or low dimensional rather than high dimensional networks. The findings so far in this area are relatively sparse, but a few systematic trends are beginning to emerge and will be discussed in the following sub-sections.

3.1 Effects of ligand geometry and flexibility on dimensionality

A growing number of materials have been made recently that involve the use of 1,2- 1,3- or 1,4-cyclohexanedicarboxylates (CHDCs) or cyclohexenedicarboxylates.[44] In the case of the CHDCs, the structural trends for the hybrids formed by the three different isomers have been examined systematically with cadmium- and manganese-containing systems;[45] each of the organics can be found as both a *cis*- and a *trans*-isomer. Two-dimensional layered structures of all three of the 1,2-, 1,3- and 1,4-cyclohexanedicarboxylates were made, but infinite metal-oxygen-metal linkages were observed only in the case of the 1,2-dicarboxylate (Fig. 15), the remaining phases being coordination polymers. Only with the close proximity of the carboxylate groups that is found in the 1,2 compound can the metals be sufficiently close to sustain infinite inorganic connectivity, while the 1,3 and 1,4 ligands all provide excellent linkages for coordination polymers with varying dimensionalities. This conclusion is further corroborated by work on cobalt and manganese 4-cyclohexene-1,2-dicarboxylates.[46] We note that the geometry of the 1,2 compound is similar to that of succinic acid, which readily forms extended inorganic connectivity (Fig. 14).

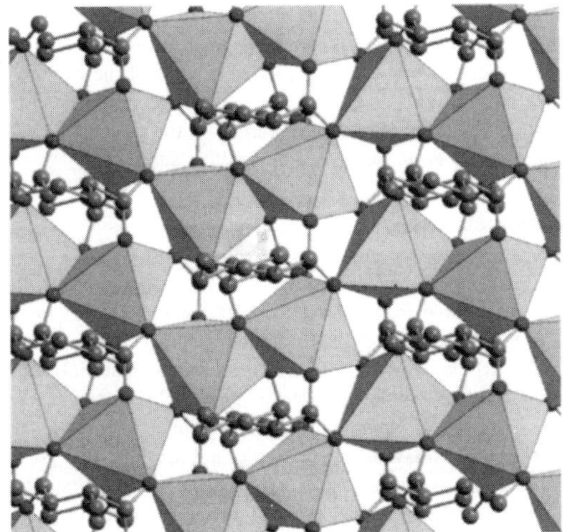

Fig. 15 A plan view of the inorganic sheet structure of cadmium cyclohexane-1,2-dicarboxylate.[45] Gray spheres denote carbon and red oxygen, with CdO_6 octahedra in yellow. Reproduced with permission. Copyright 2006, Royal Society of Chemistry.

Chirality offers another facet of ligand geometry that is very interesting. Many organic ligands, such as tartaric acid, can be obtained in an enantiomerically pure form (many occur naturally, of course), and there is much interest in the formation of chiral open frameworks that might be used as enantiomerically-selective catalysts (see section 5.2). One concern has been whether the chiral ligands are sufficiently robust to survive the reaction conditions that are required for hybrid framework formation without racemization, but this does not appear to be a general problem.[47] A considerable amount of work has been done with naturally occurring amino acids.[48] A nice example of the subtlety that can arise in this area is illustrated by a recent study with nickel aspartate.[49] The structure obtained with the L-aspartate contains a chiral helix based upon a chain of edge-sharing NiO_6 octahedra (space group $P2_12_12_1$), whereas the structure of the racemic analogue contains both left-handed and right-handed forms of the same chain (space group $I4_1/a$). In the more general case, it seems possible that the topologies of chiral structures might in some cases be quite different from those of their racemic analogues.

In addition to the geometry of the ligand, the degree of flexibility is also important in determining the type of structure that can be formed. In particular, ligands with greater flexibility are more likely to be able to adapt to the geometries found in extended inorganic hybrids, such as metal–oxygen–metal linkages. This has been seen in transition-metal diglycolates and iminodiacetates, where a range of hybrid metal oxides has been reported.[50] Furthermore, the use of monocarboxylates, which are clearly not ideal as linkers for coordination polymers, entirely eliminates the need for ligand flexibility, thereby facilitating the formation of frameworks with infinite inorganic connectivity. This is found, for example, in the case of transition-metal cyclopropane monocarboxylates[51] as well as in formates.[101] Lack of ligand flexibility is not only a disadvantage for forming extended inorganic hybrids, but there is also evidence that it may restrict the range of structures that may form in any particular system. Our recent work on transition-metal gallates illustrates this point, because in spite of many attempts to synthesize alternative hybrid frameworks, we have never succeeded in making anything other than the one shown in Fig. 12.[32] Other systems with more flexible ligands, such as the cobalt succinates, form large numbers of different structures (see section 4.1). We ascribe this difference to the limitations on the bonding options, which are very severe for the gallate ion compared with flexible ligands such as succinate.

3.3 Influence of metal ion properties

Turning to the role played by the metal ions in determining the types of structures that can form, a number of important factors are apparent. Most obviously, the preferred coordination number and geometry of the metal ion is a key issue, just as it is in classical coordination chemistry. Divalent and trivalent first-row transition metals, *e.g.* Mn^{2+}, Fe^{2+}, Co^{2+}, Ni^{2+}, Mn^{3+}, Fe^{3+} and so on, all have well known coordination preferences that often depend upon the identity of the ligand environment. Mn^{2+} is typically octahedrally coordinated by oxygen, while Co^{2+} is more versatile and ranges from tetrahedral through pentacoordinated to octahedral. The optical properties change accordingly as the geometry influences the ligand field splitting. Cr^{3+} is always octahedral, while Ni^{2+} is usually octahedral when surrounded by oxygen, but may be square planar in nitrogen environments (this is apparent in the imidazole network shown in Fig. 5(a)). Zn^{2+}, which has been widely used in studies on hybrid frameworks, is very versatile and behaves somewhat like Co^{2+}. Isolated Cu^{2+}, by contrast, is relatively inflexible and is constrained by its need to accommodate the Jahn–Teller distortion that is characteristic of d^9 ions. The rare-earths ions, however, are entirely different from the transition metals, preferring coordination numbers greater than 6 and often 7, 8 or 9 with a wide variety of geometries (see Fig. 16 for a typical example[52]).

Another factor that is frequently apparent is that certain metal ions form well-defined and robust clusters that recur in many hybrid materials. Cu^{2+}, for example, is well known for forming a large number of molecular carboxylate clusters that contain the characteristic paddle wheel dimer shown for the case of copper acetate in Fig. 17(a), so it is not a surprise that this unit is ubiquitous in copper coordination polymers such as the highly porous 3-D Cu trimesate system described by Williams and co-workers[53] and the mixed 1-D and 2-D adipate illustrated in Fig. 5(c). Similarly, Yaghi and O'Keeffe have utilized the Zn_4O cluster (Fig. 17(b)) that is a found in basic zinc acetate, $Zn_4O(OCOCH_3)_6$, using this unit as the primary inorganic node in a huge range of zinc dicarboxylate coordination polymers,[21,22] of which MOF-5 is a prototypic example (Fig. 8). A third example of such a cluster is the trimeric Cr^{3+} cluster, Cr_3O (Fig. 17(c)), which was used in combination with terephthalic acid linkers to create the huge unit cells that are found in MIL-101 and related materials.[25] Silica clusters based upon silsesquioxanes have also been

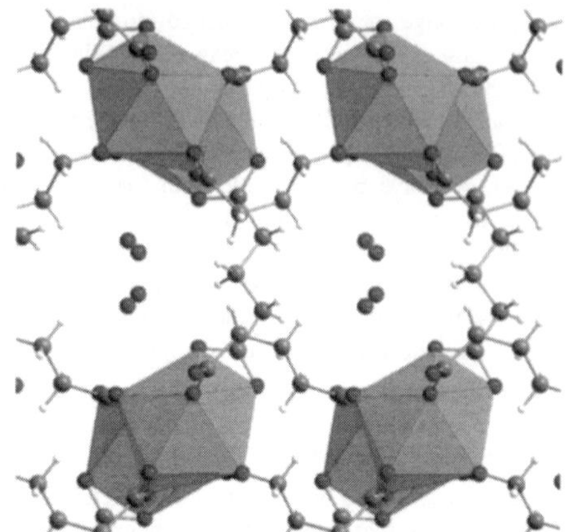

Fig. 16 Neodymium glutarate, a coordination polymer containing chains of edge-sharing NdO_9 polyhedra.[52] Gray spheres denote carbon, red oxygen, and white hydrogen, with NdO_9 polyhedra in blue. Reproduced with permission. Copyright 1998, Royal Society of Chemistry.

exploited, though many of the resulting hybrid frameworks are non-crystalline.[54]

A third point connected with the identity of the metal ions concerns their kinetic stability, or lack thereof. Most M^{2+} and M^{3+} aquo-coordination complexes show rapid ligand exchange with rate constants in the range 10^3 to 10^8 s^{-1}, but a small number of very inert ions have extremely slow ligand exchange rates. Cr^{3+} is the most common and striking example, with a rate constant of $\sim 10^{-5}$ s^{-1}, *i.e.* about eight orders of magnitude slower than Fe^{3+}! The origin of this huge difference lies in the large ligand field stabilization energy of the d^3 Cr^{3+} ion that is also responsible for its strong octahedral coordination preference. Clearly the Cr^{3+} ion provides an excellent basis for creating very stable hybrid frameworks, as has been demonstrated by the work on Férey on the MIL-101, though it can also pose synthetic challenges due to the inertness of the starting materials.

A final point concerning the nature of the metal ion relates to those cations, mainly d^{10}, that readily form linear complexes, *e.g.* Ag^+. As with certain types of ligands,

these ions are only able to support the formation of 1-D coordination polymers when they have linear coordination (Fig. 4). This behavior was exploited in some of the very early work by Robson,[9] and continues to be a useful approach for creating low-dimensional structures, as the recent work of Abu-Youssef *et al.* on silver quinoxalines shows.[55]

4 Synthetic trends

4.1 Effect of reaction temperature and pH

It is reasonable to ask how over 100 years of effort in the field of coordination chemistry failed to uncover the existence of the whole world of hybrid framework materials until very recently. With the knowledge of hindsight, we can see that this extraordinary omission arose primarily because the classical coordination chemists did not explore the use of temperature as a variable during synthesis. The first clear insight into the influence of reaction temperature on hybrid formation arose from a series of experiments in which cobalt(II) hydroxide was reacted with succinic acid in a 1 : 1 molar ratio at five temperatures between 60 and 250 °C.[56] Remarkably, and perhaps serendipitously, this yielded a series of five different phases with clear trends in the structures and compositions. In particular, the phases became more dense and less hydrated with increasing temperature, transitioning from a hydrated 1-D coordination polymer at the lowest temperature to an anhydrous, 2-D hybrid metal oxide at the highest (Fig. 18). This showed for the first time that hybrid framework formation is strongly influenced by classical thermodynamic factors, such as condensation due to entropy-driven dehydration reactions at higher temperatures.

In a more comprehensive study of the cobalt succinate system by high throughput experimentation, the trends as a function of pH and time were also examined.[57] It was found that extended inorganic hybrid structures are also favored at high pH, where the formation of M-O-M linkages arise due to the elimination of water or hydroxide groups by condensation reactions. The evolution of reaction products as a function of time, however, showed relatively few changes, further supporting the idea that thermodynamic factors can be very important in hybrid synthesis. The work also lead to the discovery of new

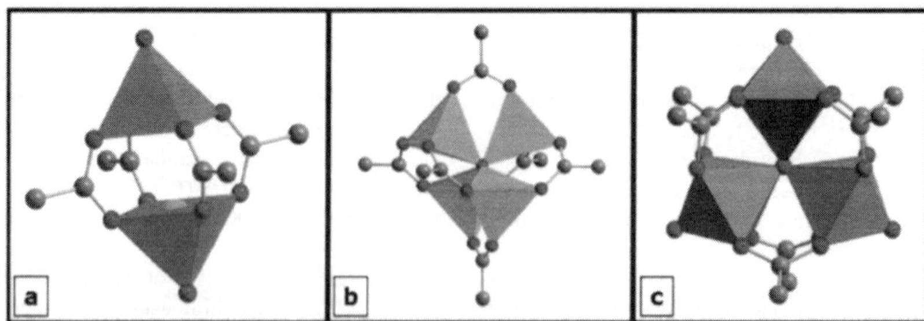

Fig. 17 Commonly recurring structural motifs: (a) the "paddlewheel" dimer of CuO_5 square pyramids, (b) the tetrahedron of ZnO_4 tetrahedra sharing a central oxygen, and (c) the trigonal planar trimer of CrO_6 octahedra sharing a central oxygen. Gray spheres denote carbon and red oxygen; CuO_5, ZnO_4 and CrO_6 polyhedra are shown in dark blue, pale blue and green, respectively.

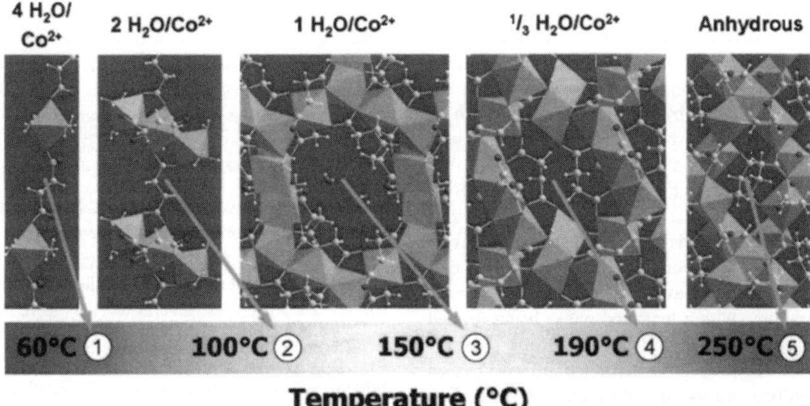

Fig. 18 Formation temperatures of five cobalt succinates, showing the trend toward greater inorganic connectivity and less hydration at higher temperatures.[56] Gray spheres denote carbon, white hydrogen, and red oxygen, with CoO_6 octahedra in pink.

phases,[58] underlining the power of high throughput methods in materials discovery.

Another interesting example of the influence of pH is apparent in the work of Stock and Bein on manganese with the phosphonocarboxylic acid, p-$H_2O_3PCH_2C_6H_4CO_2H$.[59] At low pH values, only one oxygen of the phosphonate group is deprotonated and the system is limited to forming a 1-D coordination polymer (Fig. 19). As the pH is raised, the second proton of the phosphonate group is released and the dimensionality increases to form a 2-D coordination polymer, while at the highest pH, the carboxylate becomes deprotonated and the dimensionality can increase to 3-D.

It is now becoming clear that the trends observed in the cobalt succinate and zinc phosphonocarboxylate systems are quite typical of hybrid materials synthesized under hydrothermal conditions. For example, looking back at work on the nickel(II) diphosphonates, $Ni_4(O_3PCH_2PO_3)_2 \cdot nH_2O$ (n = 3, 2, 0), we see both an increase of dimensionality and multiple coordination changes during a temperature-driven quasi-topotactic dehydration reaction *in the solid state*.[60] With the knowledge of hindsight, we can also ascribe the increase in dimensionality with temperature in the system cobalt pyridine-3,4-dicarboxylate to the same effect.[61] However, we do not wish to imply that the synthesis of hybrid frameworks always proceeds under thermodynamic control; we shall return to this point in section 4.3.

4.2 Influence of solvent

The nature of the solvent is an important parameter in hybrid synthesis, especially as it can sometimes be problematic to identify a solvent that is suitable for both the inorganic and the organic reactants. Obviously, the use of non-aqueous or mixed solvents has been widely adopted, just as it has in the case of purely inorganic frameworks, and the use of immiscible biphasic solvents,[18] whereby the products form at the solvent interface, is an interesting strategy. Ionic liquids, which have been shown to be effective in the synthesis of inorganic framework materials,[62] are just starting to be applied to hybrid frameworks.[63]

4.3 Kinetic *vs.* thermodynamic factors

The emergence of clear trends in the synthesis of hybrid framework materials, as discussed in section 4.1, points to the likelihood that thermodynamic factors are more dominant than is found to be the case in zeolites and other inorganic framework materials. In the synthesis of aluminosilicate zeolites, it is well-known that hydrothermal crystallization often proceeds under kinetic control, with successive crystallization of increasingly stable phases as a function of time, according to the Ostwald step rule.[64] The high throughput work on cobalt succinates[57] reveals very little change as a function of time, aside from reactions that start with a large percentage of solid cobalt(II) hydroxide, in which case the

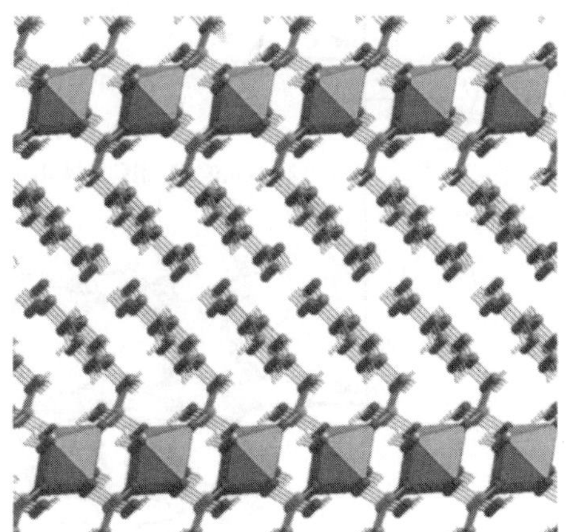

Fig. 19 The low-pH chain structure of manganese p-(phosphonomethyl)benzoate, with all carboxylate oxygens and half the phosphonate oxygens protonated.[59] Gray spheres denote carbon, red oxygen, white hydrogen, and green phosphorus, with MnO_6 octahedra in blue. Reproduced with permission. Copyright 2006, Royal Society of Chemistry.

kinetics of dissolution appear to be important. Other recent work also points towards the importance of thermodynamic control. A combined experimental and computational study of layered aluminium alkyldiphosphonates has shown that the stacking sequence is controlled by packing considerations that depend upon the number of carbon atoms in the alkyl chain; with even numbers the inorganic layers are more stable if they stack in an AAAA sequence, and with odd numbers the ABAB packing is more stable.[65] In a very recent study of zinc 4-cyclohexene-*cis*-1,2-dicarboxylates,[66] a similar combination of experimental and computational methods has shown that the temperature dependent behavior is controlled by thermodynamic considerations, with a hydrated phase being formed at temperatures below 100 °C and an anhydrous phase at higher temperatures. However, time-dependent effects are seen due to competition between the rate of product crystallization and the rate of isomerization of the *cis* ligand to the more stable *trans* form (Fig. 20).

The likelihood that thermodynamics play a strong role in hybrid framework formation is not very surprising with the knowledge of hindsight. In the case of aluminosilicate zeolites, crystallizations and transformations involve the making and breaking of very strong Si–O or Al–O bonds, whereas transition metal-ligand bonds, *e.g.* M–O, M–N *etc.* in hybrids are relatively weak. For example, $M-OH_2$ bonds for most divalent and trivalent transition metals give rates of water exchange that are in the range 10^3–10^8 s^{-1}, as mentioned in section 3.3, though a small number of very inert ions such as Cr^{3+} have extremely slow ligand exchange rates. We might therefore expect to see stronger kinetic control in systems containing these kinetically inert ions. Furthermore, it seems probable that very open frameworks such as MOF-5 and MIL-101, which are synthesized at relatively low temperatures, form under kinetic control, as is found with the zeolitic aluminosilicates. Further research is needed to clarify this question.

The use of microwave radiation in hybrid synthesis is a closely related issue that is beginning to attract attention, and in some cases the enhancement of the reaction rates is two–three orders of magnitude.[67] The reasons for this strong enhancement are not yet clear, and nor is it apparent whether the use of microwaves affects the overall outcome of the reaction in terms of which products are formed. Further work in this area should be very illuminating.

4.4 Reaction pathways and building-up processes

Although a wide variety of hybrid framework materials with different dimensionalities have been synthesized and characterized in recent years, we do not yet know the mode of formation of these materials. In the case of inorganic open-framework structures, several workers have identified discrete secondary building units (SBUs), such as SBU-4 (formed by two metal oxygen polyhedra and two anionic polyhedra), which are believed to be involved in the building up process.[68–70] In the case of metal phosphates, for example, zero-dimensional units comprising four-membered rings can be transformed to chains, sheets and three-dimensional structures under relatively mild conditions. A recent study of zinc oxalates has shown that zero-dimensional dimeric units undergo transformations of this type as a function of temperature and time (Fig. 21).[71] Further evidence supporting such a mechanism has been found in the tin phosphonates, where the dimensionality can be controlled by blocking certain reaction pathways by means of using unreactive substituent

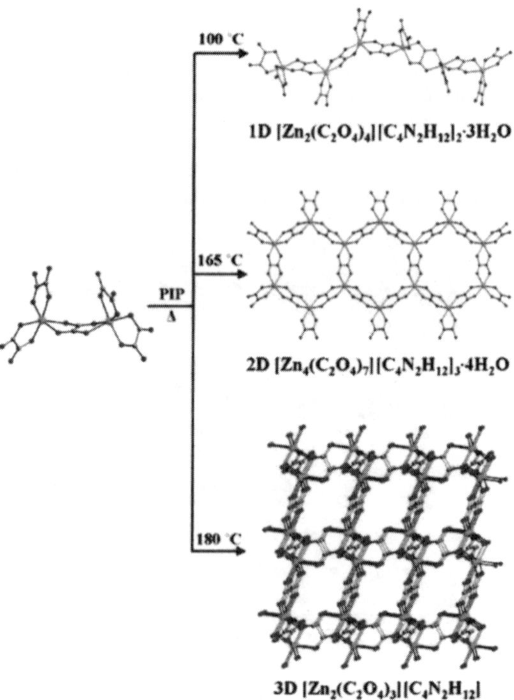

Fig. 21 Zero-dimensional dimeric zinc oxalate species transform progressively to chain, layered and 3-D structures with increasing temperature.[71] Reproduced with permission. Copyright 2005, Wiley-VCH.

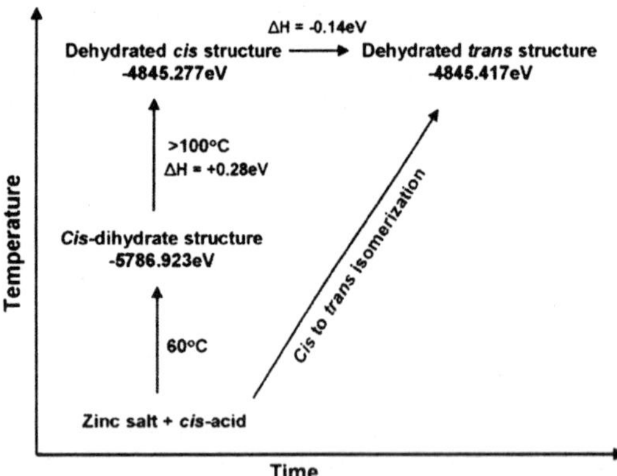

Fig. 20 The role of thermodynamics and kinetics in the zinc cyclohexene-1,2-dicarboxylate system as a function of temperature and time.[66]

groups.[72] Specially noteworthy is the transformation found in metal squarates, where it has been possible to form a sodalite structure from the assembly of six squarate units with divalent transition metals.[73] There is currently no *in situ* evidence for a building up process of the type that has been found in the purely inorganic systems, but a recent EXAFS study indicates that a trimeric iron oxide SBU remains intact during the crystallization of MIL-89.[74]

5 Properties and potential applications

5.1 Adsorption and separation processes

In comparison to the enormous number of commercial applications of the aluminosilicate zeolites and related inorganic materials in the fields of catalysis, separations, ion-exchange, and so on, the potential uses of hybrid frameworks are only gradually beginning to emerge. The most obvious possibilities concern highly porous hybrids, such as the MOF-5 and MIL-101 types of structures. In the case of MOF-5, which is a very low-density material, there is a good deal of data on gas storage capacities, including hydrogen and methane.[75] The locations of the hydrogen adsorption sites in MOF-5 have been determined at 4 K by single crystal neutron diffraction and inelastic neutron scattering.[76] The total hydrogen adsorption capacity of the MOF materials is not as high as was originally expected, no doubt because the pores are too large and molecules prefer to adsorb at surfaces. However, this problem can be alleviated and the capacity increased by using interpenetrating networks[77] or by using lightweight hybrid materials with smaller cavities, such as magnesium formate.[78] Another attractive strategy is to use hybrids that contain coordinatively unsaturated metal sites that are able to bind dihydrogen through a weak chemisorptive interaction.[79] Systems with smaller cavities are also of interest for adsorbing other molecules; for example, N_2 and CO_2 have been studied both experimentally and computationally in aluminium methylphosphonates.[80]

The system MIL-53, based upon trivalent ions, *e.g.* Cr(III), Fe(III), Sc(III) *etc*, in combination with benzene-1,4-dicarboxylate, is another highly porous materials with quite different properties.[81] The structure comprises parallel 1-D chains of corner-sharing MO_6 octahedra that are cross-linked to form a 3-D network by the benzene-1,4-dicarboxylate groups (Fig. 22). In terms of the classification in section 2.3, MIL-53 would be a (I^1O^2) system. The surface area of the dehydrated form of the first member of the MIL-53 family, the Cr(III) phase, is ~1500 m^2 g^{-1} and it is stable to 500 °C. What is remarkable about the MIL-53 architecture is that it is sufficiently flexible that it can adapt its structure to accommodate sorbates of different sizes by means of a so-called "breathing effect".[81] The dehydrated structure is very open and actually contracts when water is adsorbed due to hydrogen bonding between the water molecules and the oxygen atoms of the benzene-1,4-dicarboxylate groups (Fig. 22). The case of MIL-88, which is an iron(III) fumarate structure, is similar in the sense that it can contract and expand with a considerable change in volume.[82] However, unlike MIL-53, it expands with the addition of solvent due to the unusual flexibility of the framework. For example, the cell volume of the anhydrous form is 1135 $Å^3$, while that of the fully hydrated form is 2110 $Å^3$.

As discussed in section 3.2, hybrid frameworks provide a unique opportunity to create interesting enantiomerically pure (homochiral), porous networks. One of the motivations for so doing is the possible applications of such networks in the area of chiral separations, which was first demonstrated in 2000 by Rosseinsky and co-workers.[83] In more recent work, a homochiral network based upon nickel benzene-1,3,5-tricarboxylate showed a modest enantiomeric excess (ee) of ~8% for the adsorption of a simple naphthol derivative.[84] In general, it is found that the enantiomeric discrimination depends upon the relative sizes of the cavities and the sorbate molecules, with better selectivity being found when the size match is close. It also appears that ee values are higher for catalytic applications than chiral separations, as described in the following section.

5.2 Catalytic applications

It might seem likely that porous hybrid frameworks would be generally inferior to conventional zeolitic materials for applications in heterogeneous catalysis, given their relative instability, their lack of strong acidity, and their relatively costly synthesis. However, hybrids offer certain advantages

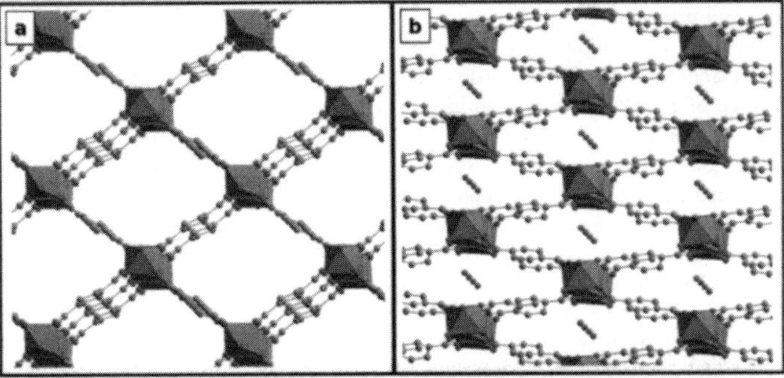

Fig. 22 Expanded (anhydrous) and contracted (hydrated) forms of the chromium terephthalate MIL-53, drawn to the same scale.[81] Gray spheres denote carbon and red oxygen, with CrO_6 octahedra in green. Reproduced with permission from Serre *et al.* Copyright 2002, American Chemical Society.

because of (i) the relative ease with which they can be functionalized, *post-synthesis*, (ii) the simplicity with which the pore size can be tuned over a wide range of sizes, and (iii) the manner in which *enatiomerically-pure* chiral frameworks can be created. The examples shown below will illustrate that many of these advantages are now beginning to be realized; further details are given in a recent review of the area.[85]

The first study of catalysis by a nanoporous coordination polymer used a simple lamellar coordination polymer of Cd and 4,4'-bipyridine in which several aldehydes were tested for cyanosilation with cyanotrimethylsilane.[86] Good yields were found for smaller molecules, with progressively poorer yields for larger ones, while control experiments using $Cd(NO_3)_2$ and 4,4'-bipyridine as catalysts resulted in no reaction, confirming the role of the coordination polymer's porosity in catalyzing the reaction.

Catalysis with organically pillared zirconium phosphate/ phosphonate catalysts has recently been reviewed by Clearfield and Wang.[87] In one of the earlier demonstrations, a highly acidic catalyst was created by post-synthesis sulfonation of aromatic rings in zirconium phosphate-based systems,[88] though the acid groups are only stable up to ~130 °C, thereby limiting the utility of these systems. However, a sample saturated by acetone at room temperature produced the polymerization products mesityl oxide, isophorone, and 1,3,5-trimethoxybenzene as verified by *in situ* NMR,[89] confirming that the samples are indeed catalytic. Similarly, Pt- or Pd-containing viologen-bridged phosphonates are active catalysts for the photochemical production of H_2 gas from water[90] and for the production of H_2O_2 from streams of H_2 and O_2,[91] though there is uncertainty as to whether the reactions take place in the micropores of the catalysts.

One of the major limitations of aluminosilicate zeolites is their current inability to impart shape selectivity based on molecular handedness, primarily because it is extremely difficult to make enantiomerically pure materials. Because of the relative ease of synthesizing accessible chiral channels in hybrid materials, a significant amount of attention is now being devoted to developing materials with chiral pores and studying their catalytic activity. The area has recently been reviewed by Lin.[92] One potential problem is that chiral organics may not survive hydrothermal reaction conditions enantiomerically intact, but Williams and co-workers[47] have shown that this is not a general problem.

In the first demonstration of enantiomerically selective catalysis with hybrids, Kim and coworkers[93] demonstrated that a tartaric acid derivative of a well known oxo-bridged trinuclear zinc carboxylate cluster forms a chiral, layered phase, POST-1, which shows good activity toward transesterification. Tests on a racemic mixture of reactants resulted in a modest 8% enantiomeric excess of either *S* or *R* enantiomers depending on the chirality of the framework. Although a chiral catalyst is not always sufficient to promote a preference for a particular handedness in products, some of the recent results from Lin and co-workers are very impressive. For example, a cadmium-containing hybrid with large channels and over 50% void space gave 93% ee for the addition of diethyl zinc to 1-naphthaldehyde, rivaling the results obtained from homogeneous analogs.[94]

In an unexpected finding, it has been shown that the size of products is not necessarily limited by the size of molecules that are able to escape from the pores (in stark contract to the shape-selectivity found in zeolites). Li and coworkers generated a 3D porous structure from a known 1D cobalt biphenyldicarboxylate chain by replacing ligand water molecules with 4,4'-bipyridine under solvothermal conditions.[95] This 3D structure is very stable and porous, but decomposes back to the 1D hydrated chain when immersed in water. Treatment of the 1D chains with amine solvothermally regenerates the 3D structure. The ability to break and rebuild this phase reversibly leads to the name RPM-1, for Rutger's Recyclable Porous Material #1. Recognizing that the recyclable nature of the material might allow for the formation of larger molecules than the cavities would normally permit, Li and coworkers tested the material for photochemical catalysis of dibenzyl ketone. The reaction resulted in a significantly different mixture of products than is seen in zeolite hosts.

5.3 Other applications

Many applications beyond gas storage and catalysis have been proposed for hybrid systems. For example, several groups, particularly those of Ward[96] and Lin,[97] have developed strategies to engineer non-centrosymmetric frameworks for use as nonlinear optical materials. There has also been considerable interest in using hybrid systems to create porous magnets.[98] Although no hybrids currently display long-range ordering temperatures above 100 K,[99] several promising strategies have been proposed, such as using radical organic ligands to couple metal centers.[100] Other interesting observations include the modulation of the magnetic ordering temperature by guest molecules in porous manganese formate,[101] ferrimagnetic-ferromagnetic transformations in nickel cyclohexane-1,4-dicarboxylates,[102] and complex exchange interactions in trimetallic 4f–3d coordination polymers based upon Cu–Gd–Fe.[103]

There is also considerable interest in the optical properties of hybrid frameworks on account of their tunability and their capacity to incorporate a wide range of metal ion and organic ligand chromophors. There are a number of examples of photoluminescent behavior in rare-earth containing hybrids that could be of interest for applications as phosphors or fluorescent probes.[17,104] In the case of porous materials, the photoluminescent lifetimes of Eu^{3+}-doped gadolinium glutarates have been shown to depend on the degree of dehydration because the coordinating water molecules act as relaxation agents.[105] Other systems show fluorescent emissions due to charge transfer between metal ions and ligands or π to π^* transitions within the ligands.[106] For commercial applications, of course, it will be necessary for the materials to have sufficient chemical and optical stability.

Other interesting avenues are also being explored. For example, soluble 1-D coordination polymers based upon dendrimers in combination with palladium have been made;[107] such soluble, low-dimensional polymers are of interest for liquid crystalline behavior and use in nanocomposites. One- and two-dimensional coordination polymers have also been used as templates for the formation of zinc oxide nanorods and radial nanoneedles,[108] while silver-containing coordination polymers are being explored for their antimicrobial activity.[55] Finally, we would like to note that thin films of layered metal diphosphonates have been used as intercalation sensors for small molecules.[109]

6 Future prospects

The purpose of this brief overview has been to illustrate the progress that has been made in many aspects of the hybrid frameworks area in the last decade. The diversity of chemical and structural types is enormous and grows by the day, and a better understanding of the factors that influence hybrid formation is beginning to emerge. As this understanding improves, our ability to design new materials for specific uses will also improve, and this will be reflected in a greater range of applications than we see at present. This is most certainly an extraordinarily rich area that will be seen in the future as one of the most important developments in the history of materials chemistry. Many avenues still remain to be explored, including some that are mentioned in the above discussion, and we encourage the community to put further effort into this exciting field.

Acknowledgements

This work was supported by the National Science Foundation under Award No. DMR05-20415 to the MRSEC center at UCSB and Award No. DMR04-09848 to the International Center of Materials Research at UCSB.

References

1 A. Corma, *J. Catal.*, 2003, **216**, 298; M. E. Davis, *Nature*, 2002, **417**, 813; J. M. Thomas, *Sci. Am.*, 1992, **266**, 112.

2 A. K. Cheetham, G. Férey and T. Loiseau, *Angew. Chem., Int. Ed.*, 1999, **38**, 3268.

3 A. P. Côté, A. I. Benin, N. W. Ockwig, M. O'Keeffe, A. J. Matzger and O. M. Yaghi, *Science*, 2005, **310**, 1166.

4 G. R. Desiraju, *Angew. Chem., Int. Ed. Engl.*, 1995, **34**, 2311.

5 A. K. Cheetham and P. M. Forster, in *Chemistry of Nanomaterials*, ed. C. N. R. Rao, A. Müller and A. K. Cheetham, Wiley–VCH, Weiheim, 2003, pp. 589–619.

6 D. A. Summerville, T. W. Cape, E. D. Johnson and F. Basolo, *Inorg. Chem.*, 1978, **17**, 3297; J. S. Miller, C. Vazquez, N. L. Jones, R. S. McLean and A. J. Epstein, *J. Mater. Chem.*, 1995, **5**, 707.

7 P. Gravereau, E. Garnier and A. Hardy, *Acta Crystallogr., Sect. B: Struct. Crystallogr. Cryst. Chem.*, 1979, **35**, 2843.

8 G. Alberti, U. Constantino, S. Allulli and N. Tomassini, *Inorg. Nucl. Chem.*, 1978, **40**, 1113; M. D. Poojary, H.-L. Hu, F. Campbell, III and A. Clearfield, *Acta Crystallogr., Sect. B: Struct. Sci.*, 1993, **49**, 996.

9 B. F. Hoskins and R. Robson, *J. Am. Chem. Soc.*, 1990, **112**, 1546.

10 R. Robson, *J. Chem. Soc., Dalton Trans.*, 2000, 3735; B. Moulton and M. J. Zaworotko, *Chem. Rev.*, 2001, **101**, 1629; B. Moulton and M. J. Zaworotko, *Curr. Opin. Solid State Mater. Sci.*, 2002, **6**, 117.

11 M. O'Keeffe, M. Eddaoudi, H. Li, T. Reineke and O. M. Yaghi, *J. Solid State Chem.*, 2000, **152**, 3.

12 A. Clearfield, *Curr. Opin. Solid State Mater. Sci.*, 2002, **6**, 495; J. L. C. Rowsell and O. M. Yaghi, *Microporous Mesoporous Mater.*, 2004, **73**, 3; M. J. Rosseinsky, *Microporous Mesoporous Mater.*, 2004, **73**, 15; C. N. R. Rao, S. Natarajan and R. Vaidhayanathan, *Angew. Chem., Int. Ed.*, 2004, **43**, 1466.

13 A. J. Blake, N. R. Champness, M. Crew and S. Parsons, *New J. Chem.*, 1999, 13.

14 J. T. Sampanthar and J. J. Vittal, *J. Chem. Soc., Dalton Trans.*, 1999, 1993.

15 W. J. Belcher, C. A. Longstaff, M. R. Neckenig and J. W. Steed, *Chem. Commun.*, 2002, 1602.

16 N. Masciocchi, F. Castelli, P. M. Forster, M. M. Tafoya and A. K. Cheetham, *Inorg. Chem.*, 2003, **42**, 6147.

17 A. de Bettencourt-Dias, *Inorg. Chem.*, 2005, **44**, 2734.

18 P. M. Forster, P. Thomas and A. K. Cheetham, *Chem. Mater.*, 2002, **14**, 17.

19 R. W. Gable, B. F. Hoskins and R. Robson, *J. Chem. Soc., Chem. Commun.*, 1990, 762.

20 L. Carlucci, G. Ciani, D. M. Proserpio and A. Sironi, *Angew. Chem., Int. Ed. Engl.*, 1995, **34**, 1895.

21 H. Li, M. Eddaoudi, M. O'Keeffe and O. M. Yaghi, *Nature*, 1999, **402**, 276.

22 N. L. Rosi, M. Eddaoudi, J. Kim, M. O'Keeffe and O. M. Yaghi, *CrystEngComm*, 2002, **4**, 401.

23 Q. R. Fang, G. S. Zhu, J. Y. Sun, Y. Wei, S. L. Qiu and R. R. Xu, *Angew. Chem., Int. Ed.*, 2005, **44**, 3845; Q. R. Fang, G. S. Zhu, M. Xue, J. Y. Sun and S. L. Qiu, *Dalton Trans.*, 2006, 2399; Y. L. Liu, V. C. Kravtsov, R. Larsen and M. Eddaoudi, *Chem. Commun.*, 2006, 1488; K. S. Park, Z. Ni, A. P. Cote, J. Y. Choi, R. D. Huang, F. J. Uribe-Romo, H. K. Chae, M. O'Keeffe and O. M. Yaghi, *Proc. Natl. Acad. Sci. U. S. A.*, 2006, **103**, 10186.

24 L. Carlucci, G. Ciani, P. Macchi and D. M. Proserpio, *Chem. Commun.*, 1998, 1837.

25 G. Férey, C. Mellot-Draznieks, C. Serre, F. Millange, J. Dutour, S. Surblé and I. Margiolake, *Science*, 2005, **309**, 2040; C. Mellot-Draznieks, J. Dutour and G. Férey, *Angew. Chem., Int. Ed.*, 2004, **43**, 6290.

26 A. C. Rizzi, R. Calvo, R. Baggio, M. T. Garland, O. Pena and M. Perec, *Inorg. Chem.*, 2002, **41**, 5609.

27 B. Adair, S. Natarajan and A. K. Cheetham, *J. Mater. Chem.*, 1998, **8**, 1477; N. Stock, G. D. Stucky and A. K. Cheetham, *Chem. Commun.*, 2000, 2277; M. Dan, G. Cotterau and C. N. R. Rao, *Solid State Sci.*, 2005, **7**, 437.

28 Y. F. Huang and K. H. Lii, *J. Chem. Soc., Dalton Trans.*, 1998, 4085; Z. A. D. Lethbridge and P. Lightfoot, *J. Solid State Chem.*, 1999, **143**, 58; A. Chowdhury, S. Natarajan and C. N. R. Rao, *Chem. Mater.*, 1999, **11**, 2316; J. Do, R. P. Bontchev and A. J. Jacobson, *Inorg. Chem.*, 2000, **39**, 3230.

29 B. M. Craven and D. Hall, *Acta Crystallogr.*, 1961, **14**, 475; S. Sato and K. Kobayashi, *Acta Crystallogr., Sect. C: Cryst. Struct. Commun.*, 1987, **43**, 1863.

30 W. T. A. Harrison, T. M. Nenoff, T. E. Gier and G. D. Stucky, *Inorg. Chem.*, 1992, **31**, 5395.

31 N. Guillou, C. Livage, W. van Beek, M. Nogues and G. Férey, *Angew. Chem., Int. Ed.*, 2003, **42**, 644.

32 R. K. Feller and A. K. Cheetham, *Solid State Sci.*, 2006, **8**, 1121; C. H. Wunderlich, R. Weber and G. Bergerhoff, *Z. Anorg. Allg. Chem.*, 1991, **598**, 371.

33 A. Tressaud, J. L. Soubeyroux, J. M. Dance, R. Sabatier and P. Hagenmuller, *Solid State Commun.*, 1981, **37**, 479.

34 A. Clearfield and J. A. Stynes, *J. Inorg. Nucl. Chem.*, 1964, **26**, 117.

35 C. A. Merrill and A. K. Cheetham, *Inorg. Chem.*, 2005, **44**, 5273.

36 M. B. Dines, P. M. DiGiacomo, K. P. Callahan, P. C. Griffith, R. H. Lane and R. E. Cooksey, in *Chemically Modified Surfaces in Catalysis and Electrocatalysis*, ed. J. S. Miller, ACS Symposium Series 192, American Chemical Society, Washington, D. C., 1982, p. 223; G. Alberti, R. Viviani and S. Murcia Mascarós, *J. Mol. Struct.*, 1998, **470**, 81.

37 A. Clearfield, *Chem. Mater.*, 1998, **10**, 2801.

38 L. A. Vermeulen and M. E. Thompson, *Nature*, 1992, **358**, 656.

39 P. M. Forster and A. K. Cheetham, *Angew. Chem., Int. Ed.*, 2002, **41**, 457.

40 R. Vaidhayanathan, S. Natarajan and C. N. R. Rao, *Dalton Trans.*, 2003, 1459.

41 N. Guillou, C. Livage, M. Drillon and G. Férey, *Angew. Chem., Int. Ed.*, 2003, **42**, 5314.

42 V. Soghomonian, Q. Chen, R. C. Haushalter and J. Zubieta, *Angew. Chem., Int. Ed. Engl.*, 1995, **34**, 223; V. J. Carter, P. A. Wright, J. D. Gale, R. E. Morris, E. Sastre and

J. Perez Pariente, *J. Mater. Chem.*, 1997, **7**, 2287; L. M. Zheng, H. H. Song, C. H. Lin, S. L. Wang, Z. Hu, Z. Yu and X. Q. Xin, *Inorg. Chem.*, 1999, **38**, 4618; L. M. Zheng, H. H. Song, C. Y. Duan and X. Q. Xin, *Inorg. Chem.*, 1999, **38**, 5061; H. H. Song, L. M. Zheng, Z. M. Wang, C. H. Yan and X. Q. Xin, *Inorg. Chem.*, 2001, **40**, 5025; M. Dan, K. Sivashanker, A. K. Cheetham and C. N. R. Rao, *J. Solid State Chem.*, 2003, **174**, 60.

43 R. Vaidhyanathan, S. Natarajan and C. N. R. Rao, *Chem. Mater.*, 2001, **13**, 3524.

44 Y. Kim and D. Y. Jung, *Chem. Commun.*, 2002, 908; M. Kurmoo, H. Kumagai, M. Akita-Tanaka, K. Inoue and S. Tagaki, *Inorg. Chem.*, 2006, **45**, 1627; J. Chen, M. Ohba, D. Zhao, W. Kaneko and S. Kitagawa, *Cryst. Growth Des.*, 2006, **6**, 664.

45 A. Thirumurugan, M. B. Avinash and C. N. R. Rao, *Dalton Trans.*, 2006, 221.

46 D. S. Kim, P. M. Forster, G. Diaz de Delgardo, S.-E. Park and A. K. Cheetham, *Dalton Trans.*, 2004, 3365.

47 S. Thushari, J. A. K. Cha, H. H.-Y. Sung, S. S.-Y. Chui, A. L.-F. Leung, Y.-F. Yen and I. D. Williams, *Chem. Commun.*, 2005, 5515.

48 J. B. Weng, M. C. Hong, Q. Shi, R. Cao and A. C. S. Chan, *Eur. J. Inorg. Chem.*, 2002, 2553.

49 E. V. Anohkhina and A. J. Jacobson, *J. Am. Chem. Soc.*, 2004, **126**, 3044.

50 P. M. Forster and A. K. Cheetham, *Microporous Mesoporous Mater.*, 2004, **73**, 57.

51 P. M. Forster and A. K. Cheetham, *Solid State Sci.*, 2003, **5**, 635.

52 F. Serpaggi and G. Férey, *J. Mater. Chem.*, 1998, **8**, 2737.

53 S. S. Y. Chui, S. M. F. Lo, J. P. H. Y. Charmant, A. G. Orpen and I. D. Williams, *Science*, 1999, **283**, 1148.

54 J. J. Morrison, C. J. Love, B. W. Manson, I. J. Shannon and R. E. Morris, *J. Mater. Chem.*, 2002, **12**, 3208; R. E. Morris, *J. Mater. Chem.*, 2005, **15**, 931.

55 M. A. M. Abu-Youssef, V. Langer and L. Öhrström, *Dalton Trans.*, 2006, 2542.

56 P. M. Forster, A. R. Burbank, C. Livage, G. Férey and A. K. Cheetham, *Chem. Commun.*, 2004, 368.

57 P. M. Forster, N. Stock and A. K. Cheetham, *Angew. Chem., Int. Ed.*, 2005, **44**, 7608.

58 P. M. Forster, A. R. Burbank, M. C. O'Sullivan, N. Guillou, C. Livage, G. Férey, N. Stock and A. K. Cheetham, *Solid State Sci.*, 2005, **7**, 1549.

59 N. Stock and T. Bein, *J. Mater. Chem.*, 2005, **15**, 1384.

60 Q. M. Gao, N. Guillou, M. Nogues, G. Férey and A. K. Cheetham, *Chem. Mater.*, 1999, **11**, 2937.

61 M.-L. Tong, S. Kitagawa, H.-C. Chang and M. Ohba, *Chem. Commun.*, 2004, 418.

62 E. R. Cooper, C. D. Andrews, P. S. Wheatley, P. B. Webb, P. Wormald and R. E. Morris, *Nature*, 2004, **430**, 1012.

63 J.-H. Liao, P.-C. Wu and Y.-H. Bai, *Inorg. Chem. Commun.*, 2005, **8**, 390; C.-Y. Sheu, S.-F. Lee and K.-H. Lii, *Inorg. Chem.*, 2006, **45**, 1891; Z. J. Lin, D. S. Wragg and R. E. Morris, *Chem. Commun.*, 2006, 2021.

64 R. A. van Santen, *J. Phys. Chem.*, 1984, **88**, 5768.

65 H. G. Harvey, B. Slater and M. P. Attfield, *Chem.-Eur. J.*, 2004, **10**, 3270.

66 C. Lee, C. Mellot-Draznieks, B. Slater, G. Wu, W. T. A. Harrison, C. N. R. Rao and A. K. Cheetham, *Chem. Commun.*, 2006, 2687.

67 S. H. Jhung, J.-H. Lee, P. M. Forster, G. Férey, A. K. Cheetham and J.-S. Chang, *Chem.-Eur. J.*, 2006, **12**, 7699.

68 G. Férey, *J. Solid State Chem.*, 2000, **152**, 37.

69 S. Oliver, A. Kuperman and G. A. Ozin, *Angew. Chem., Int. Ed.*, 1998, **37**, 46; C. N. R. Rao, S. Natarajan, A. Choudhury, S. Neeraj and A. A. Ayi, *Acc. Chem. Res.*, 2003, **34**, 80; A. J. Norquist and D. O'Hare, *J. Am. Chem. Soc.*, 2004, **126**, 6673.

70 R. Murugavel, M. G. Walawalkar, M. Dan, H. W. Roesky and C. N. R. Rao, *Acc. Chem. Res.*, 2004, **37**, 763.

71 M. Dan and C. N. R. Rao, *Angew. Chem., Int. Ed.*, 2005, **45**, 281.

72 B. A. Adair, S. Neeraj and A. K. Cheetham, *Chem. Mater.*, 2003, **15**, 1518.

73 S. Neeraj, M. L. Noy, C. N. R. Rao and A. K. Cheetham, *Solid State Sci.*, 2002, **4**, 1231.

74 S. Surble, F. Millange, C. Serre, G. Férey and R. I. Walton, *Chem. Commun.*, 2006, 1518.

75 J. L. C. Rowsell, A. R. Millward, K. S. Park and O. M. Yaghi, *J. Am. Chem. Soc.*, 2004, **126**, 5666; T. Duren, L. Sarkisov, O. M. Yaghi and R. Q. Snurr, *Langmuir*, 2004, **20**, 2683.

76 E. C. Spencer, J. A. K. Howard, G. J. McIntyre, J. L. C. Rowsell and O. M. Yaghi, *Chem. Commun.*, 2006, 278; J. L. C. Rowsell, E. C. Spencer, J. Eckert, J. A. K. Howard and O. M. Yaghi, *Science*, 2005, **309**, 1350.

77 J. L. C. Rowsell and O. M. Yaghi, *Angew. Chem., Int. Ed.*, 2005, **44**, 4670.

78 J. A. Rood, B. C. Noll and K. W. Henderson, *Inorg. Chem.*, 2006, **45**, 5521.

79 P. M. Forster, J. Eckert, B. Heiken, J. B. Parise, J. W. Yoon, S. H. Jhung, J.-S. Chang and A. K. Cheetham, *J. Am. Chem. Soc.*, in press.

80 C. Schumacher, J. Gonzalez, P. A. Wright and N. A. Seaton, *Phys. Chem. Chem. Phys.*, 2005, **7**, 2351.

81 C. Serre, F. Millange, C. Thouvenot, M. Nogues, G. Marsolier, D. Louer and G. Férey, *J. Am. Chem. Soc.*, 2002, **124**, 13519; T. Loiseau, C. Serre, C. Huguenard, G. Fink, F. Taulelle, M. Henry, T. Bataille and G. Férey, *Chem.-Eur. J.*, 2004, **10**, 1373; T. R. Whitfield, X. Q. Wang, L. M. Liu and A. J. Jacobsen, *Solid State Sci.*, 2005, **7**, 1096.

82 C. Mellot-Draznieks, C. Serre, S. Surble, N. Auderbrand and G. Férey, *J. Am. Chem. Soc.*, 2005, **127**, 16273.

83 C. J. Kepert, T. J. Prior and M. J. Rosseinsky, *J. Am. Chem. Soc.*, 2000, **122**, 5158.

84 D. Bradshaw, T. J. Prior, E. J. Cussen, J. B. Claridge and M. J. Rosseinsky, *J. Am. Chem. Soc.*, 2004, **126**, 6106.

85 P. M. Forster and A. K. Cheetham, *Top. Catal.*, 2003, **24**, 79.

86 M. Fujita, Y. J. Kwon, S. Washizu and K. Ogura, *J. Am. Chem. Soc.*, 1994, **116**, 1151.

87 A. Clearfield and Z. K. Wang, *J. Chem. Soc., Dalton Trans.*, 2002, 2937.

88 C.-Y. Yang and A. Clearfield, *React. Polym.*, 1987, **5**, 13.

89 A. Clearfield, Z. Wang and P. Bellinghausen, *J. Solid State Chem.*, 2002, **167**, 376.

90 H. Byrd, A. Clearfield, D. Poojary, K. P. Reis and M. E. Thompson, *Chem. Mater.*, 1996, **8**, 2239.

91 A. Dokoutchaev, V. V. Krishnan, M. E. Thompson and M. Balasubramanian, *J. Mol. Struct.*, 1998, **470**, 191; K. P. Reis, V. K. Joshi and M. E. Thompson, *J. Catal.*, 1996, **161**, 62.

92 W. Lin, *J. Solid State Chem.*, 2005, **178**, 2486.

93 J. Soo Seo, D. Whang, H. Lee, S. I. Jun, J. Oh, Y. J. Jeon and K. Kim, *Nature*, 2000, **404**, 982.

94 C.-D. Wu, A. Hu, L. Zhang and W. Lin, *J. Am. Chem. Soc.*, 2005, **127**, 8940.

95 L. Pan, N. Ching, X.-Y. Huang and J. Li, *Inorg. Chem.*, 2000, **39**, 5333.

96 K. T. Holman, A. M. Pivovar, J. A. Swift and M. D. Ward, *Acc. Chem. Res.*, 2001, **34**, 107.

97 O. R. Evans and W. B. Lin, *Acc. Chem. Res.*, 2002, **35**, 511.

98 G. Férey, *Chem. Mater.*, 2001, **13**, 3084.

99 K. Barthelet, J. Marrot, D. Riou and G. Férey, *Angew. Chem., Int. Ed.*, 2002, **41**, 281.

100 D. Maspoch, D. Ruiz-Molina, K. Wurst, N. Domingo, M. Cavallini, F. Biscarini, J. Tejada, C. Rovira and J. Veciana, *Nat. Mater.*, 2003, **2**, 190.

101 Z. M. Wang, B. Zhang, H. Fujiwara, H. Kobayashi and M. Kurmoo, *Chem. Commun.*, 2004, 416.

102 M. Kurmoo, H. Kumagai, M. Akita-Tanake, K. Inoue and S. Takagi, *Inorg. Chem.*, 2006, **45**, 1627.

103 H. Z. Kou, B. C. Zhou and R. J. Wang, *Inorg. Chem.*, 2003, **42**, 7658.

104 X. D. Gou, G. S. Zhu, Q. R. Fang, M. Xue, G. Tian, J. Y. Sun, X. T. Li and S. L. Qiu, *Inorg. Chem.*, 2005, **44**, 3850.

105 F. Serpaggi, T. Luxbacher, G. Férey and A. K. Cheetham, *J. Solid State Chem.*, 1999, **145**, 580.

106 Q. R. Fang, G. S. Zhu, M. Xue, J. Y. Sun, F. X. Sun and S. L. Qiu, *Inorg. Chem.*, 2006, **45**, 3582; J. Lu, K. Zhao, Q. R. Fang, J. Q. Xu, J. H. Yu, X. Zhang, H. Y. Bie and T. G. Wang, *Cryst. Growth Des.*, 2005, **5**, 1091.

107 H. Tokuhisa and M. Kanesato, *Langmuir*, 2005, **21**, 9728.

108 Z. Li, Y. Xiong and Y. Xie, *Nanotechnology*, 2005, **16**, 2303.

109 T. E. Mallouk and J. A. Gavin, *Acc. Chem. Res.*, 1998, **31**, 209.

JOURNAL OF
Materials
CHEMISTRY

An organic channel structure formed by the supramolecular assembly of trithiocyanuric acid and 4,4′-bipyridyl

Anupama Ranganathan, V. R. Pedireddi, Swati Chatterjee and C. N. R. Rao*

Chemistry & Physics of Materials Unit, Jawaharlal Nehru Centre for Advanced Scientific Research, Jakkur P.O., Bangalore 560 064, India

Received 18th February 1999, Accepted 5th July 1999

Trithiocyanuric acid (TCA) and 4,4′-bipyridyl (BP) form hydrogen-bonded co-crystals with aromatic compounds such as benzene, toluene, p-xylene and anthracene. The TCA–BP co-crystal is composed of cavities formed by the N–H···N hydrogen bonds between the two molecules, and the three-dimensional structure contains channels of approximately 10 Å where aromatic molecules are accommodated. The molar ratios of TCA, BP and the aromatic compound in the co-crystals are 2:1:1 or 2:1:0.5. Benzene, toluene and p-xylene are removed from the channels around 190, 183 and 170 °C respectively, and these aromatic guests can be reintroduced into the empty channels of the apo-hosts. The apo-hosts with empty channels have reasonable thermal stability and exhibit shape selectivity in that the empty channels accommodate p-xylene but not m- or o-xylene or mesitylene.

1 Introduction

A variety of inorganic microporous solids have been synthesized in recent years because of their use in catalysis and separation technology.[1–3] These structures are generally prepared by using structure-directing template molecules such as organic amines. Organic amines are also employed extensively to synthesize open-framework metal phosphates and carboxylates.[3] A few organic porous structures containing transition metal ions have been synthesized.[4,5] The supramolecular assembly formed by trimesic acid (benzene-1,3,5-tricarboxylic acid) with cobalt nitrate and pyridine reported by Yaghi *et al.*[4] is a good example of such a metal–organic microporous material. The recently reported nanoporous material formed between trimesic acid and cupric nitrate reported by Chui *et al.*[6] highlights the role of metal–organic interactions in supramolecular assemblies. There is also considerable interest in designing organic microporous solids by making use of supramolecular assemblies. Thus, hexagonal channel structures formed by trimesic acid and its anion have been described.[7] There are also other organic channel structures comprising large pores in three-dimensional structures reported in the literature.[8,9] The nanoporous molecular sandwiches formed by guanidium ions and disulfonate ions serve as representative examples.[8] It has been shown recently that the hexagonal rosette structure formed through hydrogen bonding between cyanuric acid and melamine gives rise to channels in the three-dimensional structure.[10] Self-assembly of three-dimensional networks with large chambers has been reported by using a pyridone as the tecton.[11] Inclusion compounds of urea and thiourea as well as the quinol clathrates also belong to this class of organic solids, but they are not engineered supramolecular designs. We have been exploring several possible designs of organic porous solids for some time and have found that the hydrogen bonded co-crystal between trithiocyanuric acid (TCA) and 4,4′-bipyridyl (BP) has channels in the three-dimensional structure that can accommodate benzene.[12] We have carried out detailed investigations to explore whether other aromatic compounds can be introduced in these channels and if so, whether there is shape selectivity. The study has shown that the supramolecular hydrogen-bonded structure formed by TCA and BP accommodates several aromatic molecules, with some shape selectivity and thermal stability.

2 Experimental

The 2:1 co-crystal of TCA with BP was first prepared by the co-crystallization of the two compounds in methanol solution, but the crystals were not very stable, because of the loss of methanol. After the loss of methanol, the crystallinity was not sufficiently good to carry out a single crystal study. In the presence of benzene, toluene, o-xylene, m-xylene, p-xylene or anthracene, however, the crystals of the 2:1 co-crystal of TCA and BP were stable. In a typical preparation, 106 mg of TCA and 47 mg of BP were taken up in 15 mL of methanol in the presence of an aromatic compound (5 mL). In the case of anthracene, the same quantities of TCA and BP as above were taken up in 15 mL of methanol along with 55 mg of anthracene. After slow evaporation, crystals containing the 2:1 co-crystal of TCA with BP, containing different proportions of the aromatic compounds were obtained. We designate the co-crystals containing benzene, toluene, o-xylene, m-xylene, p-xylene, and anthracene by **1**, **2**, **3**, **4**, **5** and **6** respectively. We could not prepare the 2:1 TCA–BP cocrystal with mesitylene. The crystals were generally needle-shaped and yellow in colour. The as-prepared crystals were dried at room temperature over a period of two days prior to the characterization studies.

Crystals of **1–6** prepared as above were employed for the determination of molecular structure by single crystal X-ray diffraction. The relevant details of the crystal structures of the co-crystals are listed in Table 1.† The intensity data were collected on a SMART system, Siemens, equipped with a CCD area detector,[13] using Mo-Kα radiation. The structures were solved and refined using SHELXTL[14] software. The refinements were uncomplicated and converged to good R-factors as

---

† CCDC reference number 1145/171. See http://www.rsc.org/suppdata/jm/1999/2407 for crystallographic files in .cif format.

Table 1 Crystal data for the complexes, 1–6, formed between trithiocyanuric acid and 4,4′-bipyridyl along with various guest molecules

	1	2	3	4	5	6
Formulae	$2(C_3H_3N_3S_3)$: $C_{10}H_8N_2$:C_6H_6	$2(C_3H_3N_3S_3)$: $C_{10}H_8N_2$:C_7H_8	$2(C_3H_3N_3S_3)$: $C_{10}H_8N_2$:C_8H_{10}	$2(C_3H_3N_3S_3)$: $C_{10}H_8N_2$:C_8H_{10}	$2(C_3H_3N_3S_3)$: $(C_{10}H_8N_2)$: $0.5(C_8H_{10})$	$2(C_3H_3N_3S_3)$: $C_{10}H_8N_2$: $0.5(C_{14}H_{10})$
Mol. wt.	588.82	602.85	616.87	616.87	872.23	599.82
Crystal system	triclinic	triclinic	triclinic	triclinic	triclinic	triclinic
Space group	$P\bar{1}$	$P\bar{1}$	$P\bar{1}$	$P\bar{1}$	$P\bar{1}$	$P\bar{1}$
a/Å	7.018(1)	7.112(1)	7.282(1)	10.324(1)	10.359(1)	10.472(1)
b/Å	10.344(1)	10.489(1)	10.621(1)	11.615(1)	11.467(1)	11.419(1)
c/Å	10.757(1)	10.681(1)	10.648(1)	12.319(1)	18.412(1)	12.932(1)
α/°	63.75(1)	65.40(1)	66.30(1)	102.16(1)	83.65(1)	100.46(1)
β/°	75.78(1)	76.58(1)	76.53(1)	93.46(1)	75.00(1)	107.69(2)
γ/°	73.82(1)	73.05(1)	70.42(1)	90.68(1)	67.94(1)	111.28(2)
Cell vol./Å^3	665.8(1)	687.4(1)	705.9(1)	1441.0(2)	1957.7(3)	1296.5(2)
Z	1	1	1	2	1	2
μ/mm^{-1}	0.54	0.53	0.52	0.51	0.55	0.56
T/K	293	293	293	293	293	293
Total reflectn.	2581	2887	2781	6134	7548	4996
Non-zero reflectn.	1838	1928	1970	4109	5489	3556
R	0.032	0.063	0.067	0.082	0.058	0.049
R_w	0.077	0.183	0.184	0.219	0.132	0.098

shown in Table 1. Except for the atoms of the guest molecules (toluene, *o*-xylene and *m*-xylene) in the complexes **2**, **3** and **4**, all the other atoms are oriented in well resolved positions. Non-hydrogen atoms in all the structures except guest molecules in **2**, **3** and **4** were refined anisotropically while the hydrogen atoms were refined isotropically, except in the structures where the hydrogen atoms were placed in the calculated positions using AFIX routines of SHELXTL. The guest molecules in **2** and **5** were refined isotropically only owing to the disorder. The crystal structures reveal that the ratio of TCA, BP and the aromatic molecules in the co-crystals **1**–**4** was 2:1:1. The ratio was 2:1:0.5 in **5** and **6**. We see from Table 1 that the unit cells of **1**–**3** are similar and the unit cells of **4**–**6** form another set. As an illustrative example, we list the structural parameters of **2** in Table 2. These data typify the structures of the TCA–BP co-crystals examined here. We have not given the data for the other co-crystals for the purpose of brevity. Intermolecular interactions were calculated using the PLATON[15] programme.

Thermogravimetric analysis (TGA) of the co-crystals was carried out over the temperature range 25 °C to 200 °C by employing a Mettler Toledo instrument. These curves showed mass loss due to the removal of the aromatic guest molecules. The apo-hosts thus obtained did not dissolve in benzene and other aromatic solvents. The apo-hosts were immersed in the respective aromatic liquid for several hours. The crystals were then taken out, and the TGA repeated. This procedure was repeated more than once to find out whether the inclusion of the guest molecule was reversible and also whether there was any change in the temperature of decomposition or the proportion of the aromatic compound in the co-crystal, with such cycling.

3 Results and discussion

Co-crystallization of TCA with BP from methanol solution gives a 2:1 hydrogen bonded co-crystal containing the solvent of crystallization. The crystal structure of the co-crystal reveals the presence of intermolecular N–H$\cdots$N hydrogen bonds between TCA and BP as shown in Fig. 1. This hydrogen bonded structure has a cavity of 10 Å (calculated using CERIUS, version 3.0), occupied by methanol molecules. These crystals were unstable under ambient conditions because of the evaporation of methanol. However the 2:1:1 co-crystals of TCA with BP incorporating benzene, **1**, or toluene, **2**, were highly stable. These crystals gave the same two-dimensional hydrogen bonded structure with a cavity as in Fig. 1. Besides

Table 2 Bond lengths and angles of **2**

Distances/Å		Angles/°	
S(23)–C(23)	1.654(6)	C(21)–N(23)–C(23)	123.3(5)
S(22)–C(22)	1.651(6)	C(16)–C(11)–C(12)	116.5(6)
S(21)–C(21)	1.662(6)	N(14)–C(15)–C(16)	122.4(7)
N(23)–C(21)	1.357(8)	C(22)–N(22)–C(23)	125.4(5)
N(23)–C(23)	1.369(8)	C(13)–C(12)–C(11)	119.6(6)
C(11)–C(16)	1.387(9)	N(23)–C(23)–N(22)	115.8(5)
C(11)–C(12)	1.393(9)	N(23)–C(23)–S(23)	123.4(4)
C(15)–N(14)	1.339(8)	N(22)–C(23)–S(23)	120.9(5)
C(15)–C(16)	1.393(9)	C(11)–C(16)–C(15)	120.5(6)
N(22)–C(22)	1.363(7)	C(22)–N(21)–C(21)	126.1(5)
N(22)–C(23)	1.378(7)	N(21)–C(22)–N(22)	113.4(5)
C(12)–C(13)	1.392(9)	N(21)–C(22)–S(22)	122.8(5)
N(21)–C(22)	1.361(8)	N(22)–C(22)–S(22)	123.7(5)
N(21)–C(21)	1.369(7)	C(13)–N(14)–C(15)	117.5(6)
N(14)–C(13)	1.327(9)	N(23)–C(21)–N(21)	115.9(6)
C(4)–C(7)	1.26(3)	N(23)–C(21)–S(21)	122.6(4)
C(4)–C(5)	1.47(3)	N(21)–C(21)–S(21)	121.5(5)
C(4)–C(6)	1.78(3)	N(14)–C(13)–C(12)	123.5(7)
C(3)–C(5)	1.35(4)	C(7)–C(4)–C(5)	71(2)
C(3)–C(6)	1.53(4)	C(7)–C(4)–C(6)	115(2)
C(3)–C(2)	1.97(5)	C(5)–C(3)–C(6)	50(2)
C(5)–C(6)	1.23(3)	C(5)–C(3)–C(2)	88(3)
C(5)–C(7)	1.60(3)	C(6)–C(3)–C(2)	38(2)
C(2)–C(6)	1.20(3)	C(6)–C(5)–C(3)	72(3)
		C(6)–C(5)–C(4)	82(2)
		C(3)–C(5)–C(4)	154(4)
		C(6)–C(5)–C(7)	130(3)
		C(3)–C(5)–C(7)	158(4)
		C(6)–C(2)–C(3)	51(2)
		C(2)–C(6)–C(5)	148(3)
		C(2)–C(6)–C(3)	91(3)
		C(5)–C(6)–C(3)	57(2)
		C(2)–C(6)–C(4)	157(3)
		C(5)–C(6)–C(4)	55(2)
		C(3)–C(6)–C(4)	112(3)
		C(4)–C(7)–C(5)	61(2)

the strong N–H$\cdots$N bonds (H$\cdots$N, 1.82 Å) responsible for the formation of the cavity, there are weak C–H$\cdots$S bonds (H$\cdots$S, 2.91, 2.93 Å) between the TCA and BP molecules. In addition, there are N–H$\cdots$S hydrogen bonds between TCA molecules (H$\cdots$S, 2.5 Å). The three-dimensional structure of **2** is shown in Fig. 2. The structure clearly reveals the presence of channels formed by the stacking of the layers with cavities. The channels accommodate aromatic molecules as can be seen in Fig. 2(a). Fig. 2(b) shows another projection of **2** where we see layers of TCA·BP with toluene molecules residing in the regions of the

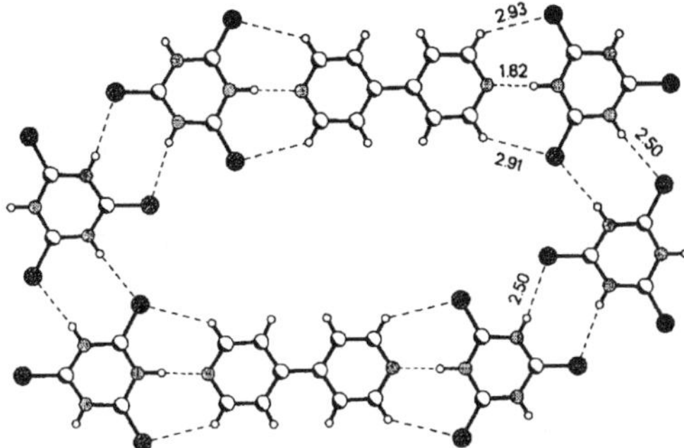

Fig. 1 Self assembly of TCA and BP forming a layered network with a cavity.

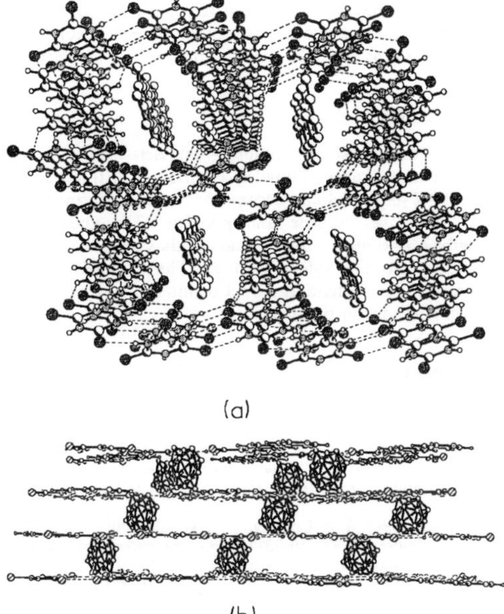

(a)

(b)

Fig. 2 (a) Three-dimensional structure of the 2:1 TCA–BP co-crystal containing toluene, 2, in the channels. (b) Stacking arrangement of molecular sheets perpendicular to the channels.

cavity. The crystals were stable and heating them up to 180 °C or slightly higher, did not destroy them.

The hydrogen-bonding pattern observed in 1 is interesting in terms of the multi-point recognition.[16] The hydrogen-bonding pattern with one N–H···N and two C–H···S bonds corresponds to the highly robust three-point recognition pattern. Different possible three-point recognition patterns of the type are shown below. While the patterns represented in (a) and (b) correspond to the motifs comprising strong hydrogen bonds, pattern (c) is a result of weak C–H···O hydrogen bonds.[17] A search of the Cambridge Structural Database[18] (CSD) suggests that the pattern present in 1 corresponding to (d) is indeed unique. No other such system appears to be reported in the literature.

Crystal structures of 3–6 show that the cavities in these co-crystals are similar to those in 2, a minor exception being 4 containing *m*-xylene, where the size and shape of the cavity are somewhat different. In Fig. 3, we show the two-dimensional

structure of the TCA–BP in 4 to illustrate the small differences between the cavities in this co-crystal and 1. Here, the intermolecular N–H···N as well as the C–H···S bonds are slightly longer but the N–H···S bonds between the TCA molecules are similar to those in 2 (Fig. 1). A noticeable difference is the slight asymmetry in the cavity arising from the different intermolecular hydrogen bond distances on the two sides. In Fig. 4a, we show the three-dimensional structure of 6 where the guest molecule is anthracene. In Fig. 4b, we show another projection of 6 showing TCA·BP layers wherein anthracene molecules protrude through the cavity. Fig. 2 and 4 serve to demonstrate the similarity of the three-dimensional channels in the various co-crystals.

In Fig. 5, we show the TGA curve of the benzene co-crystal 1, recorded at a heating rate of 2 °C min^{-1}. All the benzene is removed (mass loss, 13.9%) around 190 °C which is well above the boiling point of benzene (see curve 1). After the removal of benzene, the apo-host crystals were soaked in benzene for 36 hours and again subjected to TGA. (These crystals are insoluble in benzene as evidenced by the retention of the crystal morphology and absence of the relevant spots on the TLC plates.) The crystals showed the loss of benzene at a slightly lower temperature (158 °C), but the quantity of benzene in the channels is less by about 4% (see curve 2 in Fig. 5), compared to the as-prepared co-crystal. A repetition of this procedure showed that the removal of benzene continued to occur around 158 °C with the same mass loss (curve 3 in Fig. 5). It appears that after the benzene in the channels of the initial co-crystal is removed, there are some definitive changes in the channel structure, with the empty channels accommodating benzene to a smaller extent. The lower temperature at which benzene is removed also suggests that the guest molecules do not interact as strongly with the host channels. One difference that we notice is that the weak C–H···S interactions present in the as-formed co-crystal is absent after the removal of the guest molecules. The crystal structure of the TCA–BP co-crystal 1 heated to 200 °C (to remove all the benzene) gave a single crystal X-ray diffraction pattern, but the data were not sufficiently good to obtain the structure. We could not determine the structure after reintroducing benzene. Powder

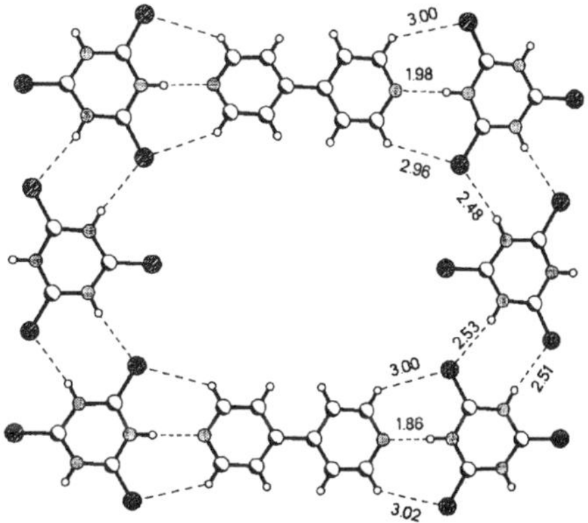

Fig. 3 The two-dimensional layered network in the TCA–BP co-crystal containing *m*-xylene, 4.

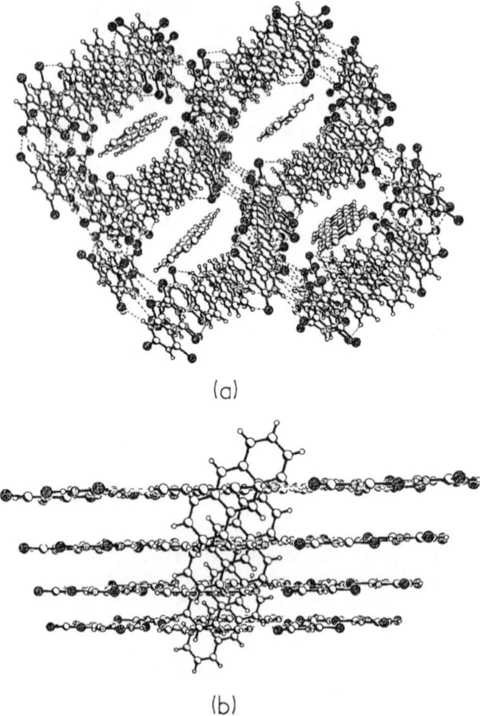

(a)

(b)

Fig. 4 (a) Three-dimensional structure of the 2:1 TCA–BP co-crystal containing anthracene, **6**. (b) Stacking arrangement of planar molecular sheets perpendicular to the channels.

XRD patterns were identical after two cycles of guest removal, but the patterns of the apo-host were generally poorer in quality.

TGA of the toluene co-crystal **2** showed a mass loss of 14.8% at 183 °C. The host crystals with the empty channels were soaked in toluene for 36 hours and then subjected to TGA. These crystals showed a mass loss of 9.2% at 166 °C, a behaviour similar to that of the benzene co-crystal, **1**. TGA of the p-xylene cocrystal, **5**, showed a mass loss (12.2%) due to the removal of the aromatic guest around 167 °C. After reintroduction of p-xylene in the channels, the mass loss occurred around 139 °C, but the magnitude of mass loss was smaller

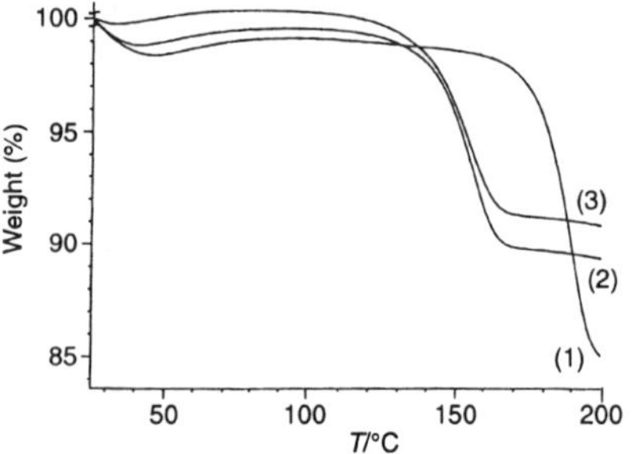

Fig. 5 Thermogravimetic analysis (heating rate 2 °C min^{-1}) of the 2:1 TCA–BP co-crystal containing benzene. Curve 1 is that of the as-prepared sample. Curve 2 was obtained with a sample obtained by immersing the host crystal with empty channels in the aromatic liquid for several hours. Curve 3 was obtained after repeating this procedure for a second time.

(8%). The temperature of decomposition and the mass loss did not change appreciably after further cycling in these cases as well. The co-crystals of o-, m-xylene, **3** and **4**, however, behaved entirely differently. Once the guest molecules were removed, they could not be reintroduced in the channels even after soaking the host crystals with empty channels for long periods. This observation suggests that there is some selectivity in the channels of the TCA–BP co-crystal. This can be understood from the differences in the shapes of p-, o- and m-xylenes. In p-xylene, the two methyl groups are symmetrically positioned and favor the ready re-incorporation into the channels of the host crystal. This is not the case in the other two isomers. Accordingly, the empty channel of the TCA–BP co-crystal only takes in p-xylene from a mixture of the three xylene isomers. It should be recalled that the channel does not accommodate mesitylene as well. In the apo-host obtained by removing benzene from co-crystal **1**, toluene could be introduced readily, but not p-xylene. We have attempted to understand the selectivity of the apo-host channels by carrying out powder XRD measurements before and after the removal of the guest molecules. Although the apo-hosts are still crystalline, we have not yet been able to come to any meaningful conclusions from the XRD patterns since the data are not entirely sufficient.

4 Conclusions

The stable organic solid containing channels formed by the supramolecular hydrogen-bonded assembly of trithiocyanuric acid and 4,4′-bipyridyl can accommodate aromatic molecules such as benzene, toluene and p-xylene. The apo-host is thermally stable up to 200 °C and exhibits shape selectivity with respect to the xylene isomers. The channels do not accommodate mesitylene. It would be interesting to carry out chemical reactions in these channels.

5 References

1 (a) G. Ozin, *Adv. Mater.*, 1992, **4**, 612; (b) J. M. Thomas, *Angew. Chem., Int. Ed. Engl.*, 1994, **33**, 913.
2 (a) S. Zone and M. E. Davis, *Curr. Opin. Solid State Mater. Sci.*, 1996, **1**, 107; (b) A. K. Cheetham, *Science*, 1994, **264**, 794; (c) X. Bu, P. Feng and G. D. Stucky, *J. Chem. Soc., Chem. Commun.*, 1995, 1337.
3 (a) S. Ayyappan, A. K. Cheetham, S. Natarajan and C. N. R. Rao, *Chem. Mater.*, 1998, **10**, 3746 and references cited therein; (b) S. Natarajan, S. Ayyappan, A. K. Cheetham and C. N. R. Rao, *Chem. Mater.*, 1998, **10**, 1627 and references cited therein.
4 (a) O. M. Yaghi, G. Li and H. Li, *Nature*, 1995, **378**, 703; (b) O. M. Yaghi, H. Li and T. L. Groy, *J. Am. Chem. Soc.*, 1996, **118**, 9096; (c) O. M. Yaghi and H. Li, *J. Am. Chem. Soc.*, 1995, **117**, 10401; (d) H. Li, M. Eddaoudi, T. L Groy and O. M. Yaghi, *J. Am. Chem. Soc.*, 1998, **120**, 8571.
5 (a) D. Venkataraman, G. B. Gardner, S. Lee and J. S. Moore, *J. Am. Chem. Soc.*, 1995, **117**, 11600; (b) B. F. Abrahams, J. Coleiro, B. Hoskins and R. Robson, *Chem. Commun.*, 1996, 603.
6 S. S. Y. Chui, S. M. F. Lo, J. P. H. Charmant, A. G. Orpen and I. D. Williams, *Science*, 1999, **283**, 1148.
7 (a) S. V. Kolotuchin, E. E. Fenlon, S. R. Wilson, C. J. Loweth and S. C. Zimmerman, *Angew. Chem., Int. Ed. Engl.*, 1995, **34**, 2654; (b) R. Melendez, C. V. K. Sharma, M. J. Zaworotko, C. Baner and R. D. Rogers, *Angew. Chem., Int. Ed. Engl.*, 1996, **35**, 2213; (c) S. C. Zimmerman, *Science*, 1997, **276**, 543.
8 (a) V. A. Russel, C. C. Evans, W. Li and M. D. Ward, *Science*, 1997, **276**, 575; (b) V. A. Russel and M. D. Ward, *Chem. Mater.*, 1996, **8**, 1654.
9 J. S. Moore and S. Lee, *Nature*, 1995, **374**, 792.
10 A. Ranganathan, V. R. Pedireddi and C. N. R. Rao, *J. Am. Chem. Soc.*, 1999, **121**, 1752.
11 M. Simard, D. Su and J. D. Wuest, *J. Am. Chem. Soc.*, 1991, **113**, 4696.
12 V. R. Pedireddi, S. Chatterjee, A. Ranganathan and C. N. R. Rao, *J. Am. Chem. Soc.*, 1997, **119**, 10867.
13 Siemens, *SMART System*, Siemens Analytical X-ray Instruments Inc., Madison, WI, U.S.A., 1995.

14 G. M. Sheldrick, SHELXTL, *Users Manual*, Siemens Analytical X-ray Instruments Inc., Madison, WI, U.S.A., 1993.
15 A. L. Spek, PLATON, *Molecular Geometry Program*, University of Utrecht, The Netherlands, 1995.
16 (*a*) J. C. MacDonald and G. M. Whitesides, *Chem. Rev.*, 1994, **94**, 2383; (*b*) G. M. Whitesides, E. E. Simanek, J. P. Mathias, C. T. Seto, D. N. Chin, M. Mammen and D. M. Gordon, *Acc. Chem. Res.*, 1995, **28**, 37.
17 (*a*) G. R. Desiraju, *Angew. Chem., Int. Ed. Engl.*, 1995, **34**, 2311; (*b*) G. R. Desiraju, *Crystal Engineering: The Design of Organic Solids*, Elsevier, New York, 1989; (*c*) G. R. Desiraju, *Science*, 1997, **278**, 404; (*d*) G. R. Desiraju, *Curr. Opin. Solid State Mater. Sci.*, 1997, 451.
18 F. H. Allen and O. Kennard, *Chemical Design Automation News*, 1993, **8(1)**, 31.

Paper 9/05344A

1752   *J. Am. Chem. Soc.* **1999**, *121*, 1752−1753

Hydrothermal Synthesis of Organic Channel Structures: 1:1 Hydrogen-Bonded Adducts of Melamine with Cyanuric and Trithiocyanuric Acids

Anupama Ranganathan, V. R. Pedireddi, and C. N. R. Rao*

*Chemistry & Physics of Materials Unit
Jawaharlal Nehru Centre for Advanced Scientific Research
Jakkur P.O. Box 6436, Bangalore 560 064, India*

Received November 12, 1998

Whitesides and co-workers[1] have extensively cited the 1:1 hydrogen-bonded adduct between cyanuric acid (**CA**) and melamine (**M**)-forming rosettes, as an outstanding case of noncovalent synthesis. The hydrogen-bonded **CA·M** lattice is the template on which these workers based the design of a family of self-assembled hydrogen-bonded aggregates, stable in solution.[2,3] The **CA·M** adduct is expected to form a hexagonal network through the formation of hydrogen bonds between the two molecules, but the structure has not been established by single-crystal X-ray crystallography.[3] The crystal structure of **CA·M·3HCl** reported by Wang et al.[4] reveals one-dimensional tapes rather than the rosette structure. The difficulty in growing crystals of the **CA·M** adduct is partly because **CA** and **M** are both highly hydrogen-bonded solids melting at high temperatures, with limited solubility in most organic solvents. Mixtures of **CA** and **M** in water or alcohol solution give precipitates containing the adduct, but do not yield crystals suitable for X-ray crystallography.[1,3] We felt that to obtain single crystals of the **CA·M** adduct, it may be necessary to employ more drastic conditions than normally employed. We have, therefore, made use of hydrothermal conditions,[5] commonly employed in the synthesis of quartz, zeolites, and inorganic open-frame structures, to obtain the crystals of the 1:1 adduct. We have indeed been able to obtain good crystals of the 1:1 **CA·M** adduct by this means. Crystals of the 1:1 adduct of trithiocyanuric acid (**TCA**) and **M** could also be

prepared hydrothermally. We report the structures of both of these fascinating adducts of **M** in this paper.

The procedure for preparing the crystals of the **CA·M** adduct was as follows. Aqueous solutions of cyanuric acid (0.1 mmol) and melamine, (0.1 mmol) were mixed in a Teflon flask, and the mixture (15 mL) was kept in a stainless steel bomb. The bomb was sealed and maintained in a furnace at 180 °C. Rectangular platelike crystals of good quality separated from the solution, upon cooling the bomb to room temperature over a period of 4 h. These crystals were used for collecting intensity data on the single-crystal

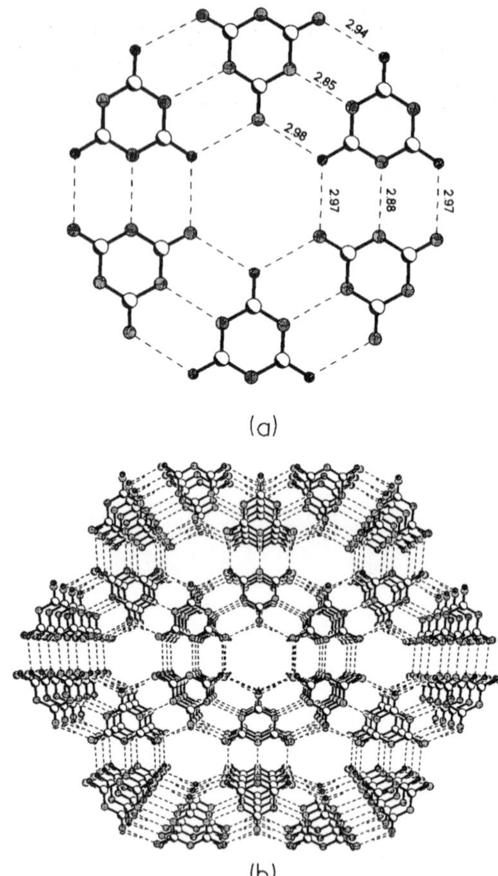

Figure 1. (a) A rosette formed between cyanuric acid (**CA**) and Melamine (**M**) in the adduct of **CA·M** with a cavity diameter of approximately 4 Å. Dashed lines represent intermolecular hydrogen bonds. (b) Three-dimensional arrangement of the **CA·M** adduct forming channels along the crystallographic *c*-axis.

X-ray diffractometer. Crystals of the 1:1 **TCA·M** adduct were obtained by a similar procedure except that the temperature was maintained at 100 °C.[6]

The 1:1 **CA·M** adduct[7] has an asymmetric unit consisting of superimposed **CA** and **M** molecules.[8] Packing analysis shows that

(1) Whitesides, G. M.; Simanek, E. E.; Mathias, J. P.; Seto, C. T.; Chin, D. N.; Mammen, M.; Gordon, D. M. *Acc. Chem. Res.* **1995**, *28*, 37. Whitesides, G. M.; Mathias J. P.; Seto, C. T. *Science* **1991**, *254*, 1312.

(2) Seto C. T.; Whitesides, G. M. *J. Am. Chem. Soc.* **1993**, *115*, 905.

(3) Mathias, J. P.; Simanek, E. E.; Zerkowski, J. A.; Seto, C. T.; Whitesides, G. M. *J. Am. Chem. Soc.* **1994**, *116*, 4316. (b) Zerkowski, J. A.; Seto, C. T.; Wierda, D. A.; Whitesides, G. M. *J. Am. Chem. Soc.* **1990**, *112*, 9025.

(4) Wang, Y.; Wei, B.; Wang, Q. *J. Crystallogr. Spectrosc. Res.* **1990**, *20*, 79.

(5) Rao, C. N. R. *Chemical Approaches to the Synthesis of Inorganic Materials*; John Wiley: New York, 1993.

(6) **TCA** decomposed when the temperature was maintained at 180 °C as in the case of **CA**.

(7) Crystal data for the **CA·M** adduct: $(C_3H_3N_3O_3):(C_3H_6N_6)$, $M = 255.22$, crystal dimensions, $0.3 \times 0.2 \times 0.2$ mm, monoclinic, space group, $C2/m$, $a = 14.853(2)$ Å, $b = 9.641(2)$ Å, $c = 3.581(1)$ Å, $\beta = 92.26(1)°$, $V = 512.4$-(2) Å^3, $Z = 2$, $\rho_{calcd} = 1.654$, μ(Mo Kα) = 0.136 mm^{-1}, $F(000) = 264$, $\lambda = 0.710\ 73$ Å, Smart CCD area detector, Siemens, $\omega-2\theta$ scan, $2° < \theta < 24°$ $(-16 \leq h \leq 14, -8 \leq k \leq 10, -3 \leq l \leq 3)$, 1095 total reflections, 395 independent reflections which were used for refinement. The structure was solved by direct methods (SHELXTL-PLUS) and refined by full-matrix least squares on F^2 (SHELX-93; G. M. Sheldrick, Gottingen, 1993) to R1 = 0.056 and wR2 = 0.141. Residual density, min/max $-0.495/0.393$ e·Å^{-3}. Hydrogen atoms were not included in the refinement, but including the hydrogen atoms in the calculated positions improves the *R*-factor by 1%.

(8) The best stacking arrangement was one with alternating **CA** and **M** molecules (with 50% occupation each), as established by the low *R*-factor and also by energy calculations using the MOPAC program. The least stable arrangement is the one with **CA** on top of another **CA**, as one would expect.

CA and **M** are held together by N–H···O and N–H···N hydrogen bonds yielding the hexamer (rosette) shown in Figure 1a. The interatomic N···O and N···N distances corresponding to N–H···O and N–H···O bonds are in the range of 2.94–2.98 and 2.85–2.88 Å, respectively. The hexamers are arranged in two dimensions to form planar sheets, the structure being exactly as predicted by Whitesides.[1] A significant feature is that the planar sheets are stacked in three dimensions[8] to give channels with a diameter of 4 Å (Figure 1b). These channels are comparable to the cavities in cryptands.[9]

Crystal structure determination of the **TCA·M** adduct[10] reveals that it has features similar to those of the **CA·M** adduct, except that the N–H···O hydrogen bonds are replaced by the N–H···S hydrogen bonds as shown in Figure 2a. The N···S and N···N distances corresponding to N–H···S and N–H···N bonds are in the range of 2.96–2.99 and 2.86–2.88 Å, respectively. The sheets are stacked in a three-dimensional channel arrangement as depicted in Figure 2b. The diameter of the channel is approximately 4 Å here as well.

The present study illustrates the efficacy of the hydrothermal method to prepare crystals of organic materials in certain special situations, as in the present case where the compounds are highly hydrogen-bonded (with limited solubility in common solvents) and give precipitates when mixed in water.[11]

It is noteworthy that the adducts of **M** with **CA** and **TCA** constitute members of a rare class of organic solids containing channels.[12] Trimesic acid is a good example of a system forming hexagonal channels.[13] The other example of such a solid is the adduct of **TCA** with 4,4′-bipyridyl.[14] Experiments to explore whether the channels in the **CA·M** adduct can be filled by cations and other species are in progress.

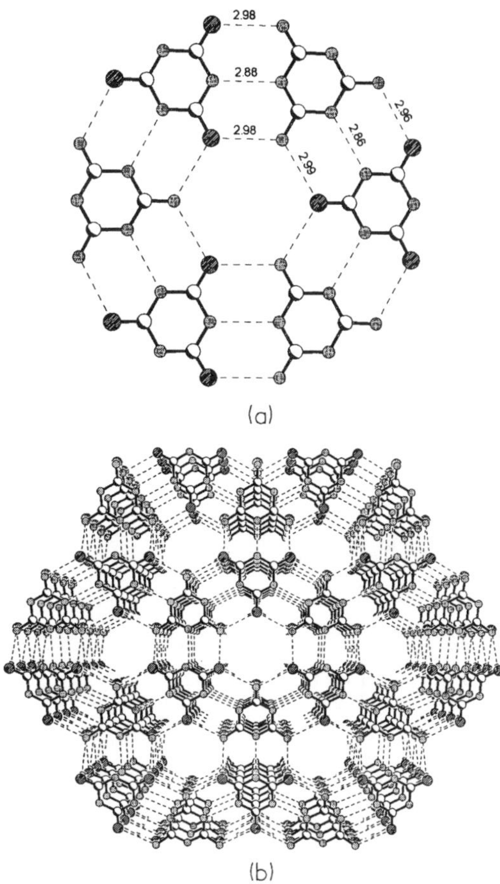

Figure 2. (a) Hexagonal network (rosette) formed between trithiocyanuric acid (**TCA**) and Melamine (**M**) in the adduct of **TCA·M** with a cavity diameter of approximately 4 Å. (b) Three-dimensional arrangement of the **TCA·M** adduct forming channels along crystallographic *c*-axis.

Supporting Information Available: Two figures (ORTEP drawings of the adducts **CA·M** and **TCA·M** with labeling scheme) and 6 tables containing general crystallographic information, fractional coordinates of non-hydrogen atoms, list of bond lengths and angles, and anisotropic displacement parameters (PDF). This material is available free of charge via the Internet at http://pubs.acs.org.

JA9839280

CA on top of **M** is more stable by 37 kcal mol⁻¹ relative to **CA** on **CA**. The packing coefficient of the **CA·M** adduct calculated by using Cerius is 67.5%.

(9) Vogtle, F. *Supramolecular Chemistry*; Wiley: Chichester, 1971.

(10) Crystal data for the **TCA·M** adduct: $(C_3H_3N_3S_3):(C_3H_6N_6)$ $M = 303.39$, crystal dimensions, $0.3 \times 0.25 \times 0.35$ mm, monoclinic, space group, $C2/m$, $a = 14.862(3)$ Å, $b = 9.650(2)$ Å, $c = 3.588(1)$ Å, $\beta = 92.32(1)°$, $V = 514.2(2)$ Å³, $Z = 2$, $\rho_{calcd} = 1.901$, μ(Mo Kα) = 0.716 mm⁻¹, $F(000) = 294$, $\lambda = 0.710\,73$ Å, Smart CCD area detector, Siemens, $\omega{-}2\theta$ scan, $2° < \theta < 24°$ ($-13 \leq h \leq 16$, $-10 \leq k \leq 10$, $-3 \leq l \leq 3$), 1091 total reflections, 397 independent reflections which were used for refinement. The structure was solved by direct methods (SHELXTL-PLUS) and refined by full-matrix least squares on F^2 (SHELX-93; G. M. Sheldrick, Gottingen, 1993) to R1 = 0.062 and wR2 = 0.134. Residual density, min/max −0.396/0.487e·Å⁻³. Hydrogen atoms were not included in the refinement, but including the hydrogen atoms in the calculated positions improves the *R*-factor by 1%. The best stacking arrangement is **TCA** on top of **M** (see (8) for details). The packing coefficient of the adduct is 74.6% as calculated from Cerius.

(11) The powder X-ray diffraction pattern of the **CA·M** adduct simulated on the basis of structure found by us agrees with the experimental powder pattern of the precipitate or of the polycrystalline sample.

(12) Zimmerman, S. C. *Science* **1997**, *276*, 544.

(13) Kolotuchin, S. V.; Fenlon, E. E.; Wilson, S. R.; Loweth, C. J.; Zimmerman, S. C. *Angew. Chem., Int. Ed. Engl.* **1995**, *34*, 2654.

(14) Pedireddi, V. R.; Chatterjee, S.; Ranganathan, A.; Rao, C. N. R. *J. Am. Chem. Soc.* **1997**, *119*, 10867.

V. NANOMATERIALS

Commentary by C.N.R. Rao

Chemistry plays a major role in the study of nanomaterials[1,2]. Nanoparticles which constitute an important class of nanomaterials are not new to chemistry.[3] One of the earliest experiments on nanoparticles was carried out by Michael Faraday who prepared sols of gold and other metals. Nanoparticles of a variety of metals, semiconductors and ceramic materials have been prepared and characterized in recent years.[4] An aspect of great interest concerns the size-dependent properties of nanoparticles.[3] Quantum confinement in metal and semiconductor nanoparticles has been investigated widely. Thus, very small metal nanoparticles cease to be metallic and show different reactivities compared to the bulk. Semiconductor nanoparticles exhibit size-dependent energy gaps and the resulting shifts in electronic absorption and emission spectra. Metal quantum dots exhibit Coulomb blockade. The articles in this section discuss the synthesis, physical properties, electronic structure, assembly and chemical reactivity of nanoparticles. Certain unusual properties such as ferromagnetism exhibited universally by nanoparticles of the otherwise nonmagnetic oxides are presented.

Soon after the laboratory synthesis of fullerenes in 1990, carbon nanotubes were discovered by Iijima. The discovery of the carbon nanotubes has given great impetus to investigate various one-dimensional materials.[5] The author initiated investigations of carbon nanotubes (CNTs) in the early 1990s. These include synthesis, opening, filling and functionalization of nanotubes. An important discovery was the precursor synthesis of carbon nanotubes by employing organometallics. This method also yielded aligned CNT bundles and junction nanotubes. Doping CNTs with nitrogen was accomplished for the first time in 1997, followed by boron doping. Properties of CNTs including those of single-walled and double-walled nanotubes as well as Y-junction nanotubes have been investigated. In this section, representative articles covering some of the above investigations of CNTs are included.

In the last few years, nanotubes of several inorganic layered materials such MoS_2, NbS_2 and BN have been prepared and characterized.[5] The area of inorganic nanotubes has been reviewed in one of the articles and nanotubes of BC_4N have been described in another.

Nanowires constitute a major family of one-dimensional materials. Besides confinement, they exhibit several interesting properties. Inorganic nanowires can be prepared by several methods. A few articles dealing with the synthesis, characterization and applications of inorganic nanowires are presented. Growth mechanism of nanorods has been examined in one of the article.

Functionalization and solubilization of nanomaterials are important components of nanomaterials chemistry. A few of the articles in this section cover these aspects. Assembling nanomaterials is the subject matter of a few of the articles, while others deal with new techniques of preparing nanomaterials (e.g. hydrogel method). The use of the liquid-liquid interface to generate nanofilms of inorganic materials in a novel discovery. This method is described in detail in an article. Graphene which is a current hot topic in condensed matter science is discussed in the last article.

References

1. C.N.R. Rao, A. Müller and A.K. Cheetham (Eds.), *Chemistry of Nanomaterials*, Vols. 1 & 2, Wiley-VCH, (2004).
2. C.N.R. Rao, A. Müller and A.K. Cheetham (Eds.), *Nanomaterials Chemistry, Recent Developments and New Directions*, Wiley-VCH, 2007.
3. J. Jortner and C.N.R. Rao, *Pure Appl. Chem.* **74**, 1491 (2002).
4. C.N.R. Rao, P.J. Thomas and G.U. Kulkarni, *Nanocrystals: Synthesis, Properties and Applications*, Springer-Verlag, 2007.
5. C.N.R. Rao and A. Govindaraj, *Nanotubes and Nanowires*, Royal Society of Chemistry, London, 2005.

Size-Dependent Chemistry: Properties of Nanocrystals

C. N. R. Rao,*[a] G. U. Kulkarni,[a] P. John Thomas,[a] and Peter P. Edwards*[b]

Abstract: Properties of materials determined by their size are indeed fascinating and form the basis of the emerging area of nanoscience. In this article, we examine the size dependent electronic structure and properties of nanocrystals of semiconductors and metals to illustrate this aspect. We then discuss the chemical reactivity of metal nanocrystals which is strongly dependent on the size not only because of the large surface area but also a result of the significantly different electronic structure of the small nanocrystals. Nanoscale catalysis of gold exemplifies this feature. Size also plays a role in the assembly of nanocrystals into crystalline arrays. While we owe the beginnings of size-dependent chemistry to the early studies of colloids, recent findings have added a new dimension to the subject.

Keywords: colloids · nanostructures · self-assembly · semiconductors

Introduction

Steric effects—arising from bulky groups—are well known as key factors in determining the reactivity of organic molecules. Otherwise, we do not ordinarily concern ourselves with the physical dimension of a system as a factor in determining its intrinsic properties except in intercalation chemistry or some such situation where the pore or cavity in a host lattice molecule can accommodate guest species of a particular size. Size, however, becomes *the* sole controlling factor while dealing with the science and application of the so-called

nanoparticles, that cover a size range between 1–100 nm. Nanoparticles within this size domain are thus intermediate between the atomic and molecular size regimes on one hand, and the macroscopic, bulk on the other. In this regime, size-dependent properties manifest themselves when the size of an individual particle is sufficiently small. A key aspect of the rapidly developing area of nanoscience and nanotechnology concerns itself with the size dependence of intrinsic properties of materials. It is, therefore, instructive to look at typical properties of metals and semiconductors in the size-dependent regime of nanoparticles. We shall discuss three important aspect of the nanocrystals of these materials, electronic structure and properties, chemical reactivity and self-assembly. Size dependent structural and thermodynamic properties of nanoparticles such as bond lengths, melting point and specific heat have already been reviewed[1] in the literature and will not be discussed in any detail in the present article.

Two centuries ago the study of nanoscale solid particles, dispersed within a liquid host, played a pivotal role in establishing colloid science. During the final decades of the last century, colloid science was, perhaps, something of an intellectual backwater—with one or two notable exceptions. However, significant advances in both experimental and theoretical aspects of the subject, and, of course, the emergence of, and explosion of interest in, the broad area of nanoscience and nanotechnology, have now set the scene for a renaissance in colloid science. Interestingly, it does appear that many scientists in the modern field of nanoscience may not even recognize its colloidal source. This is unfortunate, since there is a wealth of information and expertise in that old and venerable science.

An important example of the parentage of the modern subject of metal nanoparticles derives from the work of Faraday in the 1850s. During that period, Faraday carried out groundbreaking studies of nanoscale gold particles in aqueous solution. He established the first quantitative basis for the area, noting that these colloidal metal sols ("pseudosolutions") are thermodynamically unstable, and that the individual gold nanoparticles must be stabilized kinetically against aggregation. Once the nanoparticles coagulate, the process cannot be reversed. Remarkably, Faraday also identified the very essence of the nature of colloidal, nanoscale particles of metals; specifically, for the case of gold, he concluded (in

[a] Prof. Dr. C. N. R. Rao, Prof. Dr. G. U. Kulkarni, P. J. Thomas
Chemistry and Physics of Materials Unit
Jawaharlal Nehru Center for Advanced Scientific Research
Jakkur P.O., Bangalore, 560 064 (India)
E-mail: cnrrao@jncasr.ac.in

[b] Prof. Dr. P. P. Edwards
School of Chemistry
University of Birmingham, Edgbaston,
Birmingham, B15 2TT (UK)
E-mail: p.p.edwards@bham.ac.uk

1857!) ... *"the gold is reduced in exceedingly fine particles which becoming diffused, produce a beautiful fluid ... the various preparations of gold whether ruby, green, violet or blue ... consist of that substance in a metallic divided state"*.

During that century, colloidal phenomena played a pivotal role in the genesis of physical chemistry by establishing a connection between descriptive chemistry and theoretical physics. For example, Einstein provided the relationship between Brownian motion and diffusion coefficient of colloidal particles.

Today, the overwhelming importance now associated with the nanoscale in both science and technology means that the scene is once again set for this key subject to impact upon the development of not only chemistry, but also physics and materials science. In the following sections we attempt to highlight a few of the key issues relating to nanparticles where size determines their properties.

Electronic structure and properties: The electronic structure of a nanocrystal critically depends on its very size. For small particles, the electronic energy levels are not continuous as in bulk materials, but discrete, due to the confinement of the electron wavefunction because of the physical dimensions of the particles (see Figure 1). The average electronic energy level spacing of successive quantum levels, δ, known as the so-called Kubo gap, is given by, $\delta = 4E_f/3n$, where E_f is the Fermi energy of the bulk material and n is total number of valence electrons in the nanocrystal. Thus, for an individual silver nanoparticle of 3 nm diameter containing approximately one

thousand silver atoms, the value of δ would be 5–10 meV. Since the thermal energy at room temperature, $kT \cong 25$ meV, a 3 nm particle would be metallic ($kT > \delta$). At low temperatures, however, the level spacings specially in small particles, may become comparable to kT, rendering them nonmetallic.[2] Because of the presence of the Kubo gap in individual nanoparticles, properties such as electrical conductivity and magnetic susceptibility exhibit quantum size effects.[3] The resultant discreteness of energy levels also brings about fundamental changes in the characteristic spectral features of the nanoparticles, especially those related to the valence band.

Extensive investigations of metal nanocrystals of various sizes obtained, for example by the deposition of metals on amorphised graphite and other substrates, by X-ray photoelectron spectroscopy and related techniques[4, 5] have yielded valuable information on their electronic structure. An important result from these experiments is that as the metal particle size decreases, the core-level binding energy of metals such as Au, Ag, Pd, Ni and Cu increases sharply. This is shown in the case of Pd in Figure 2, where the binding energy

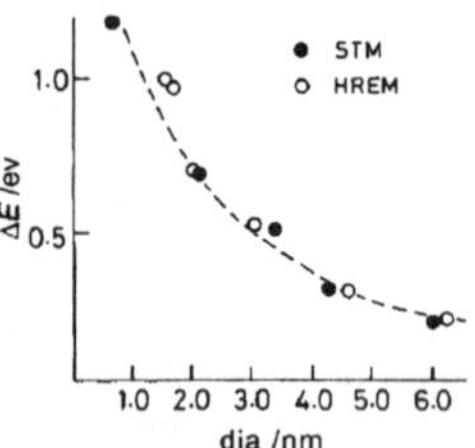

Figure 2. Variation of the shift, ΔE, in the core-level binding energy (relative to the bulk metal value) of Pd with the nanoparticle diameter. The diameters were obtained from HREM and STM images (reproduced with permission from ref. [3]).

increases by over 1 eV at small size. The variation in the binding energy is negligible at large coverages or particle size, since the binding energies are close to those of the bulk, macroscopic metals. The increase in the core-level binding energy in small particles occurs due to the poor screening of the core-hole and is a manifestation of the *size-induced metal-nonmetal transition* in nanocrystals. Further evidence for the occurrence of such a metal-nonmetal transition driven by the size of the individual particle is provided by other electron spectroscopic techniques such as UPS, BIS.[5–7] All these measurements indicate that an electronic gap manifests itself for a nanoparticle having diameters of 1–2 nm possessing 300 ± 100 atoms.

Photoelectron spectroscopic measurements[6] on mass-selected Hg_n nanoparticles ($n = 3$ to 250) in the gas phase reveal that the characteristic HOMO–LUMO (s–p) energy gap decreases gradually from ~ 3.5 eV for $n = 3$ to ~ 0.2 eV for $n = 250$, as shown in Figure 3. The band gap closure is predicted at $n \sim 400$. The metal–nonmetal transition in gaseous Hg nanoparticles was examined by Rademann and co-workers[7] by measuring the ionization energies (IE). For $n < 13$, the dependence of IE on n suggested a different type of bonding. A small Hg particle with atoms in the $6s^2 6p^0$ configuration

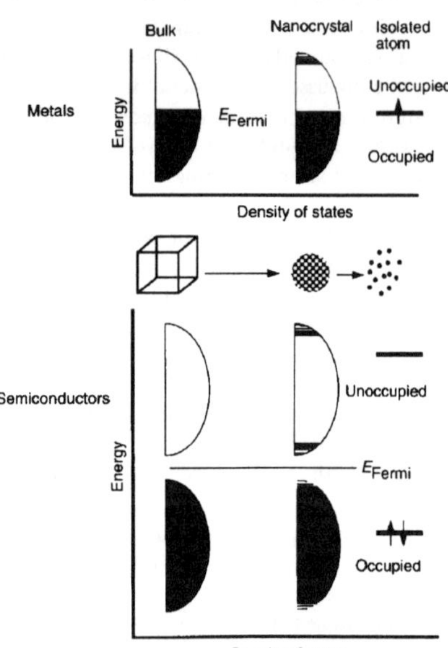

Figure 1. Density of states for metal (a) and semiconductor (b) nanocrystals. In each case, the density of states is discrete at the band edges. The Fermi level is in the center of a band in a metal, and so kT may exceed the electronic energy level spacing even at room temperatures and small sizes. In contrast, in semiconductors, the Fermi level lies between two bands, so that the relevant level spacing remains large even at small sizes. The HOMO–LUMO gap increases in semiconductor nanocrystals of smaller sizes.

 © WILEY-VCH Verlag GmbH, 69451 Weinheim, Germany, 2002 0947-6539/02/0801-0030 $ 17.50+.50/0

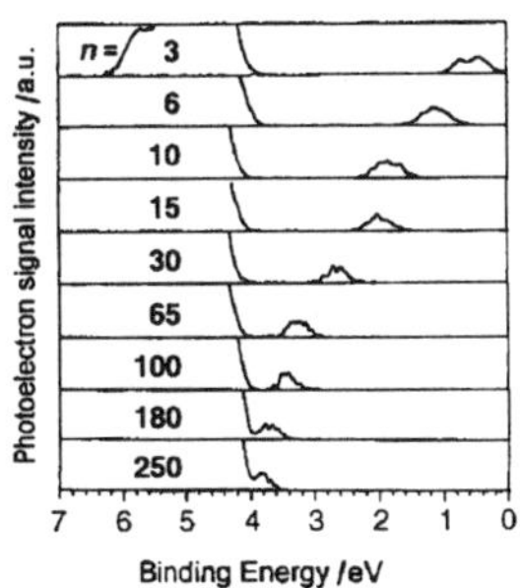

Figure 3. Photoelectron spectra of Hg clusters of varying nuclearity. The 6p feature moves gradually towards the Fermi level, emphasizing that the band gap shrinks with increase in cluster size (reproduced with permission from ref. [6]).

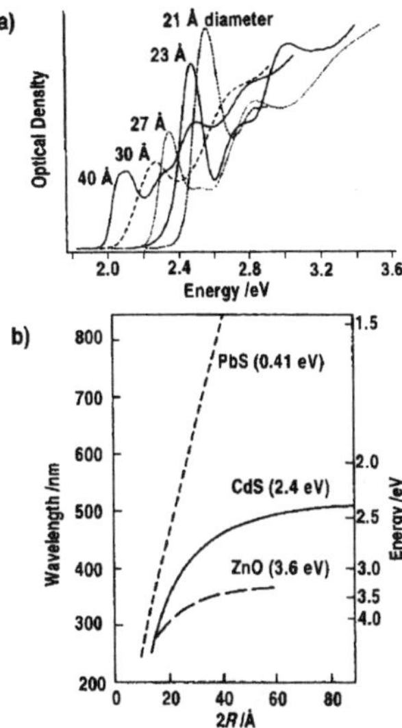

Figure 4. a) Absorption spectra of CdSe nanocrystals (at 10 K) of various diameters (reproduced with permission from ref. [12a]). b) Wavelength of the absorption threshold and band gap as a function of the particle diameter for various semiconductors. The corresponding energy gap in the bulk state is given in parenthesis (reproduced with permission from ref. [12b]).

held together by relatively weak van der Waals forces, is essentially nonmetallic. As the nanoparticle grows in size, the atomic 6s and 6p levels broaden into bands and an insulator – metal transition appears to occur—driven by the physical dimensions of the individual particle. Note that this is the same element, Hg, behaving as either a metal or a nonmetal, depending upon its physical size!

The electronic absorption spectrum of metal nanocrystals in the visible region is dominated by the plasmon band. This absorption is due to the collective excitation of the itinerant electron gas on the particle surface and is characteristic of a metal nanocrystal of the given size. In colloids, surface plasmon excitations impart characteristic colours to the metal sols, the beautiful wine-red colour of gold sols being well-known. Gold nanocrystals of varying diameters between 2 and 4 nm exhibit distinct bands around ∼525 nm, the intensity of which increases with size.[9, 10] The intensity of this feature becomes rather small in the case of 1 nm diameter particles basically due to a reduced less number of "itinerant" electrons in the electron cloud. With a change in temperature, the intensity of the plasmon band decreases as seen in the case of Au nanocrystals.[10] In contrast to the situation for metals, exciton peaks dominate the absorption of semiconductor nanocrystals in the visible region.[11, 12a] Thus CdS, a yellow solid, exhibits an exciton peak around 600 nm, which gradually shifts into the UV region as the nanocrystal diameters are varied below 10 nm (see Figure 4).[12b] Similar effects have been observed in the case of PbS and ZnO.[12b]

Direct information on the gap states in nanocrystals of metals and semiconductors is obtained by scanning tunneling spectroscopy (STS). This technique provides the desired sensitivity and spatial resolution making it possible to carry out tunneling spectroscopic measurements on individual particles. A systematic STS study of Pd, Ag, Cd and Au nanoparticles of varying sizes deposited on a graphite substrate has been carried out under ultrahigh vacuum conditions, after having characterized the nanoparticles by XPS and STM.[13] The I-V spectra of bigger particles were featureless while those of the small particles (< 1 nm) showed well-defined peaks on either side of zero-bias due to the presence of a gap. Ignoring gap values below 25 meV ($\sim kT$), it is seen that small particles of ≤1 nm diameter are in fact

nonmetallic! (Figure 5a) From the various studies discussed hitherto, it appears that the size-induced metal – insulator transition in metal nanocrystals (Figure 1) occurs in the range of 1 – 2 nm diameter or 300 ± 100 atoms. The band gap of CdS nanocrystals estimated by the above method yielded a value of 2.9 eV for a 3.1 nm diameter nanocrystal[14] and the gap increases with the decrease in size (Figure 5b).

Theoretical calculations of the electronic structure of metal and semiconductor nanocrystals throw light on the size-

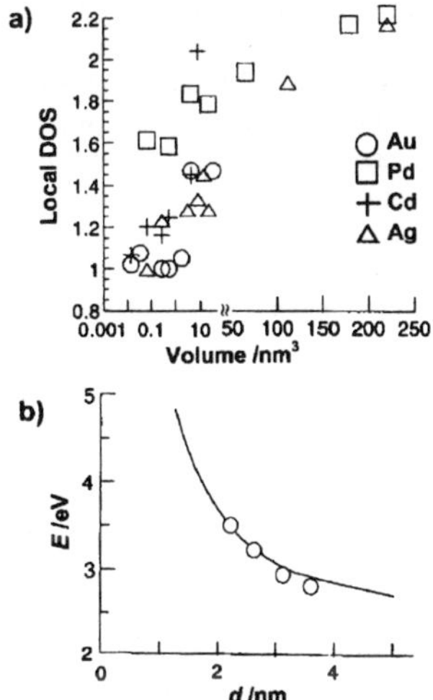

Figure 5. a) Variation of the nonmetallic band gap with nanocrystal size, b) in CdS nanocrystals (reproduced with permission from refs. [13, 14]).

　　　　　　　　　　　　　　　　C. N. R. Rao, P. P. Edwards et al.

induced changes in the electronic structure. Rosenblit and Jortner[15] calculated the electronic structure of a model metal cluster and predicted electron localization to occur in a cluster of diameter ~ 0.6 nm. A molecular orbital calculation on Au_{13} cluster[16] in icosahedral and cuboctahedral structures shows that the icosahedral structure undergoes Jahn–Teller distortion while the cuboctahedral structure does not distort. The onset of the metallic state is barely discernible in the Au_{13} cluster. Relativistic density functional calculations of gold clusters,[17] with $n = 6$ to 147 show that the average interatomic distance increases with the nuclearity of the cluster. The HOMO–LUMO electronic gap decreases with particle size from 1.8 eV for Au_6 (~ 0.5 nm diameter) to 0.3 eV for Au_{147} (~ 2 nm diameter). Ab initio molecular dynamics simulations of aluminum clusters,[18] with $n = 2–6$, 12, 13, 55 and 147 reveal that the minimum energy structures of Al_{13} and Al_{55} to be distorted icosahedra whereas Al_{147} is a near cuboctahedron. The HOMO–LUMO gap increases from ~ 0.5 eV for Al_2 to ~ 2 eV for Al_{13}; the gap is around 0.25 eV for Au_{55} and decreases to ~ 0.1 eV for Au_{147}. In the case of semiconductor nanocrystals, it is shown using tight binding approximation that the band gaps nearly reach the bulk values at sizes of around 5 nm.[19] The convergence of the cluster properties towards those of the corresponding bulk materials with increase in size is noteworthy.

In a bulk metal, the energy required to add or remove an electron is its work function. In a molecule, the corresponding energies, electron affinity and ionization potential, respectively, are, however, nonequivalent. Nanocrystals being intermediary, the two energies differ only to a small extent,[20] the difference being the charging energy, U. This is a Coulombic energy and is different from electronic energy gap. Further, Coulombic states can be similar for both semiconductor and metallic nanocrystals unlike the electronic states. A manifestation of single electron charging is the Coulomb staircase behaviour observed in the tunneling spectra,[21] when a nanocrystal, covered with an insulating ligand shell is held between two tunnel junctions. A typical staircase along with its theoretical fit is shown in Figure 6a. Such measurements have also been carried out on Pd and Au nanocrystals in the size range, 1.5–6.5 nm.[22] The charging energies follow a scaling law[23] of the form, $U = A + B/d$, where A and B are constants, characteristic of the metal and d is the particle diameter(see Figure 6b).

Magnetic properties of nanoparticles of transition metals such as Co, Ni show marked variations with size. It is well known that in the nanometric domain, the coercivity of the particles tends to zero.[23] Thus, the nanocrystals behave as superparamagnets with no associated coercivity or retentivity. The blocking temperature which marks the onset of this superparamagnetism also increases with the nanocrystal size. Further, the magnetic moment per atom is seen to increase as the size of a particle decreases[25] (see Figure 7).

Chemical reactivity: The surface area of nanocrystals increases markedly with the decrease in size. Thus, a small metal nanocrystal of 1 nm diameter will have $\sim 100\%$ of its atoms on the surface. A nanocrystal of 10 nm diameter on the other hand, would have only 15% of its atoms on the surface. A

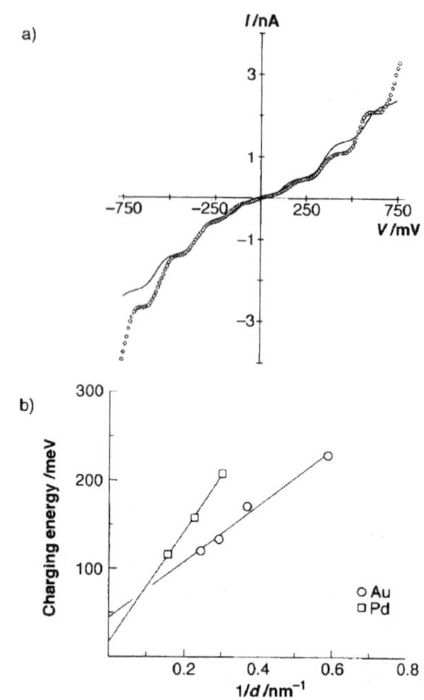

Figure 6. a) I-V characteristics of an isolated 3.3 nm Pd nanocrystal (dotted line) and the theoretical fit (solid line) obtained at 300 K using a semiclassical model according to which, the observed capacitance (C) may be resolved into two components $C1$ and $C2$ and the resistance (R) into $R1$ and $R2$, such that $C = C1 + C2$, $R = R1 + R2$. For $C1 \ll C2$ and $R1 \ll R2$, the model predicts steps in the measured current to occur at critical voltages, $V_c = n_c e/(C + (1/C)q_0 + e/2)$, where q_0 is the residual charge. b) Variation of the charging energies of Pd and Au nanocrystals with inverse diameters (d) (reproduced with permission from ref. [22]).

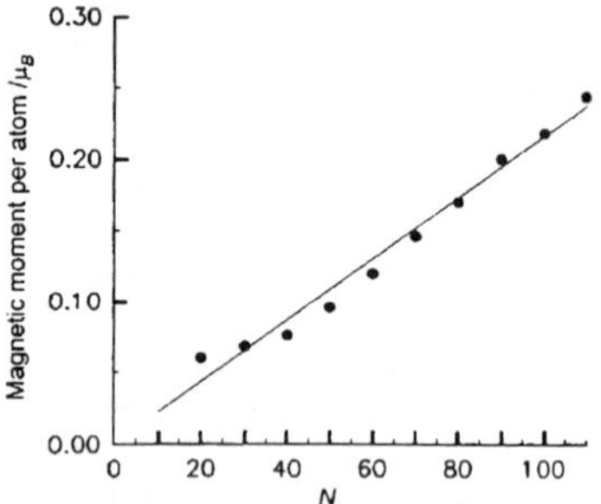

Figure 7. Variation of the magnetic moment per atom as a function of the number of atoms in Co clusters at $T = 170$ K and $H = 0.53$ T. The continuous curve is the theoretical value and the solid circles are the experimental results (reproduced with permission from ref. [25b]).

small nanocrystal with a higher surface area would be expected to be more reactive. Furthermore, the qualitative change in the electronic structure arising due to quantum confinement in small nanocrystals will also bestow unusual catalytic properties on these particles, totally different from those of the bulk metal. We illustrate these important aspects with a few examples from the recent literature.

A low temperature study[26] of the interaction of elemental O_2 with Ag nanocrystals of various sizes (Figure 8) has

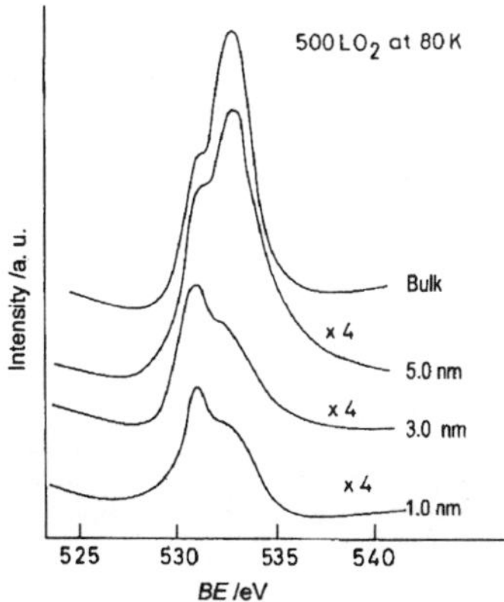

Figure 8. Change in the O(1s) spectra of Ag clusters exposed to 500 L O₂ at 80 K. The diameters of the clusters have been estimated from metal coverage. The lower binding energy peak at 531 eV corresponds to O⁻ while that at 533 eV arises due to molecular oxygen (reproduced with permission from ref. [26]).

revealed the capability of smaller nanocrystals to dissociate dioxygen to atomic oxygen species. On bulk Ag, the adsorbed oxygen species at 80 K is predominantly O_2^-. This interaction of O_2 with Ag—*dependent on its particle size*—is remarkable. Another important example is the reaction of H_2S with Ni nanocrystals giving rise to S^{2-} species, with nanocrystals of different sizes exhibiting different temperature profiles (see Figure 9). Unlike bulk nickel, small nanocrystals show less dependence in their catalytic activity on ambient temperature.

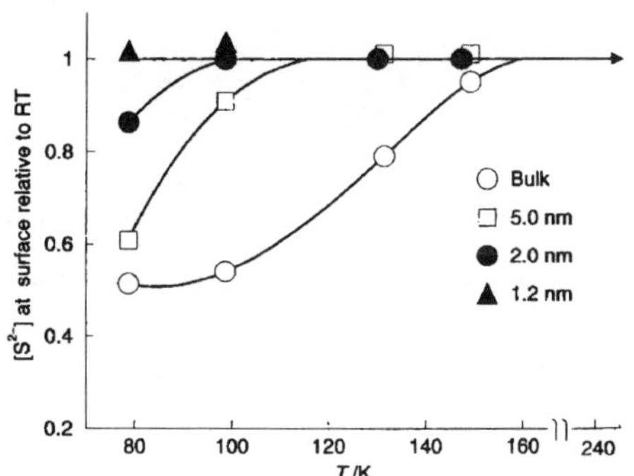

Figure 9. Variation of the normalized areas of the signal for the core-level transitions of S^{2-} with temperature for different sizes of Ni clusters deposited on graphite (reproduced with permission from ref. [26]).

The ability of Cu, Pd and Ni nanoparticles to absorb CO has been thoroughly investigated. Carbon monoxide from a bulk Cu surface desorbs above 250 K. Small Cu particles, however, retain CO up to much higher temperatures (Figure 10).[27] A similar observation has been made in case of Pd particles.[28] The results obtained with Ni particles are more even interesting. In addition to showing a trend similar to the above, small Ni particles are also capable of dissociating CO to form carbidic species on the particle surface (Figure 11).[29]

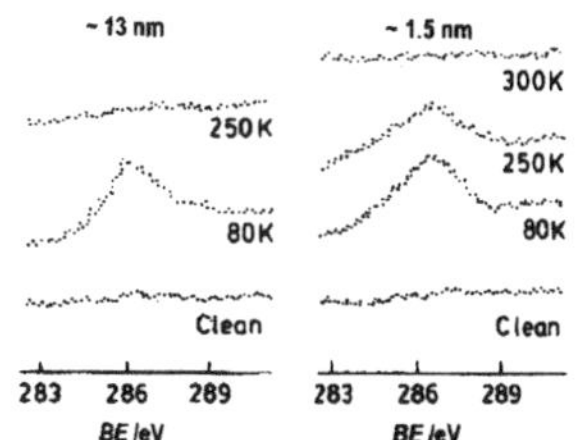

Figure 10. Change in the C(1s) spectra of CO adsorbed on Cu. The feature at 286 eV corresponds to molecularly adsorbed CO (reproduced with permission from ref. [27]).

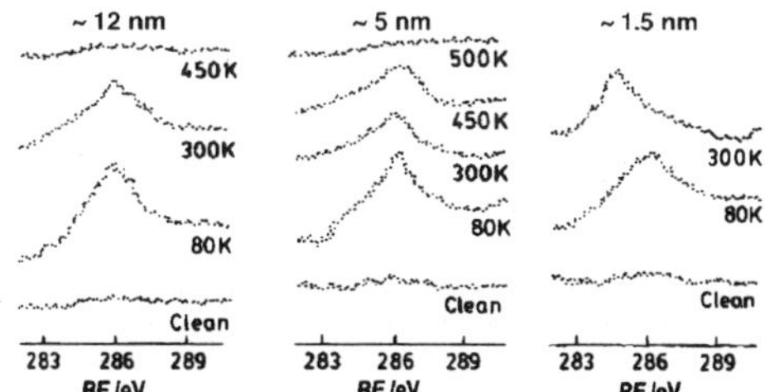

Figure 11. Change in the C(1s) spectra of CO adsorbed on Ni clusters with temperature. The feature at 286 eV corresponds to molecularly adsorbed CO while that at 284 eV arises due to the formation of carbidic species (reproduced with permission from ref. [27]).

This could be due to the Ni(3d) level in small clusters coming close to the anti-bonding energy level of CO(2π*).

Bulk Au is a noble metal. Goodman and co-workers,[30] however, found that Au nanocrystals supported on a titania surface show a marked size-effect in their catalytic ability for CO oxidation reaction, with Au nanoparticles in the range of 3.5 nm exhibiting the maximum chemical reactivity (Figure 12a). A metal to non-metal transition as observed in the I-V spectra (Figure 12b), as the cluster size is decreased below 3.5 nm³ (consisting of ca. 300 atoms). This result is quite similar to that obtained with Pd particles supported on oxide substrate.[31] In another study of Au particles supported on a zinc oxide surface, smaller particles (<5 nm) exhibited a marked tendency to adsorb CO while those with diameters above 10 nm did not significantly adsorb CO (Figure 13).[32] The increased activity of these metal particles is attributed to the charge transfer between the oxide support and the particle surface.

Self-assembly of nanocrystals: Just as individual atoms aggregate to form crystals, nanocrystals themselves act as building units to form particle superlattices. Thus, monodispersed nanocrystals suitably covered by ligands such as alkane thiols, when transferred to a flat substrate, spontaneously assemble into two-dimensional lattices.[33–37] In Figure 14, we show typical arrays of 2.5 and 3.2 nm Pd nanocrystals coated with octane thiol. The diameter of the nanocrystal, *d*, and the length of the protecting ligand, *l*, play an important role in determining the very nature of the assembly.[38, 39] A study of the two-dimensional arrays formed by Pd nanocrystals of varying diameters covered with alkane thiols of different chain lengths has enabled to obtain an experimental stability

____

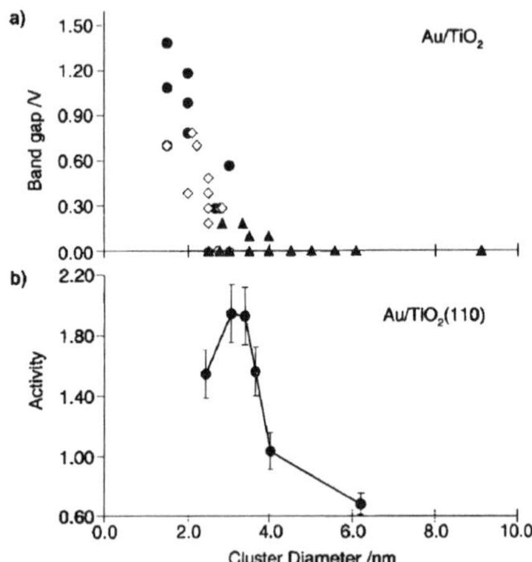

Figure 12. a) Cluster band gap measured by STS as a function of Au cluster size. The band gaps were obtained while the corresponding topographic scan was acquired on various Au coverages ranging from 0.2 to 4.0 ML. two-dimensional clusters ○; three-dimensional clusters, two atom layers in height □; three-dimensional clusters with three atom layers or greater in height △. b) The activity (*CO atoms/total Au atoms*) for CO oxidation at 350 K as a function of the Au cluster diameter supported on a TiO₂ surface. The CO/O₂ mixture was 1:5 at a total pressure of 40 Torr (reproduced with permission from ref. [30]).

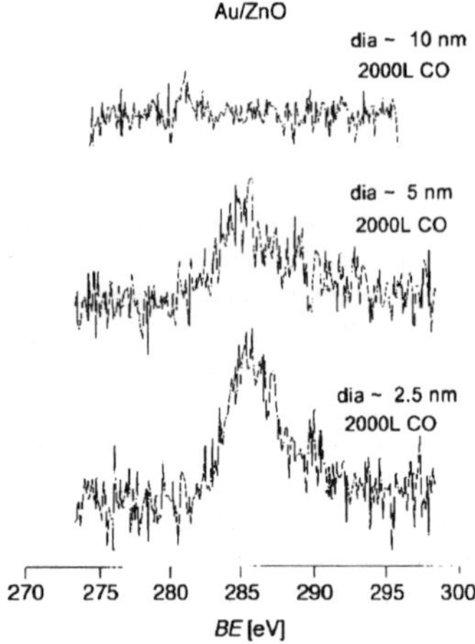

Figure 13. C(1s) core-level spectra of CO adsorbed on Au particles supported on a ZnO substrate. The feature at ∼285 eV corresponds to molecularly absorbed CO. The diameters have been obtained from the metal coverages.

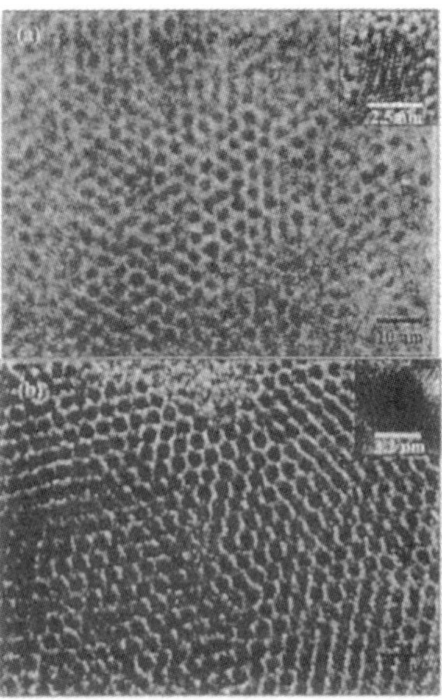

Figure 14. TEM micrograph showing hexagonal arrays of thiolized Pd nanocrystals: a) 2.5 nm, octane thiol, b) 3.2 nm, octane thiol (reproduced with permission from ref. [40]).

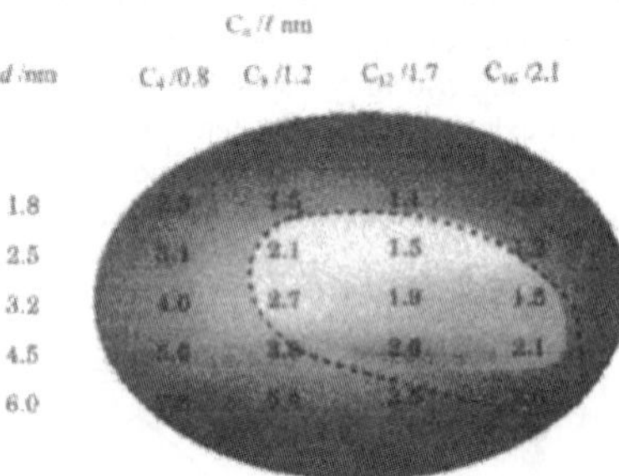

Figure 15. The $d-l$ phase diagram for Pd nanocrystals thiolized with different alkane thiols. The mean diameter, d, was obtained from the TEM measurements on as-prepared sols. The length of the thiol, l, is estimated by assuming an all-*trans* conformation of the alkane chain. The thiol is indicated by the number of carbon atoms, C_n. The bright area in the middle encompasses systems which form close-paced organizations of nanocrystals. The surrounding darker area includes disordered or low-order arrangements of nanocrystals. The area enclosed by the dashed line is derived from calculations from the soft sphere model (reproduced with permission from ref. [40]).

diagram (Figure 15) of the superlattices in terms of d and l.[40] In this Figure, the bright area in the middle is the most favourable d/l regime, corresponding to extended close packed organization, such as those illustrated in Figure 14. The d/l values for the most favorable regions are in the range 1.5–3.8. The area shaded dark in Figure 15 includes d/l regime giving rise to various short-range organizations; these are formed when the particles are small and the chain length is large or vice versa. The experimental results have been compared with empirical calculations based on a soft sphere

model,[41] involving an attractive van der Waals term and a repulsive steric term (see Figure 15).

The ability to synthesize lattices of nanocrystals have led to explorations of their collective physical properties. Thus, it is observed in the case of Co nanocrystals (5.8 nm) that, accompanying lattice formation, the blocking temperature increases.[42] FePt alloy nanocrystals yield ferromagnetic assemblies for which the coercivity is tunable by controlling the parameters such as Fe:Pt ratio and the particle size.[43] The evolution of collective electronic states in CdSe nanocrystals have been followed by optical spectroscopic methods. Compared with isolated nanocrystals, those in the lattice exhibited broader bands.[44] Various investigations have been carried out on the electrical transport properties of nanocrystalline lattices.[45, 46] Heath and co-workers have successfully demonstrated a reversible Mott–Hubbard metal–nonmetal transition with Ag nanocrystals (3 nm) capped with octane thiol.[47] Since then, this has been a subject of intense theoretical study.[48]

Conclusion

We have hitherto discussed in the earlier sections, electronic structure and properties, chemical reactivity and self-assembly of nanocrystals, particularly those of metals. The discussion should suffice to illustrate how size if a crucial factor in deciding the chemistry in the nano regime. These size dependent properties form the basis of nanoscience, where the properties are exploited for possible applications.

[1] a) J. S. Vermaak, L. W. Mays, D. Kuhlmann-Wilsdorf, *Surf. Sci.* **1968**, *12*, 128; b) G. Apai, J. F. Hamilton, J. Stohor, A. Thompson, *Phys. Rev. Lett.* **1979**, *40*, 165; c) A. A. Montano, G. K. Shenoy, E. E. Apl, W. Schulze, J. Urban, *Phys. Rev. Lett.* **1986**, *56*, 2076; d) H. J. Wasswemann, J. S. Vermaak, *Surf. Sci.* **1970**, *22*, 164; e) P. A. Buffat, J. P. Borel, *Phys. Rev. A* **1976**, *13*, 2287; f) W. P. Halperin, *Rev. Mod. Phys.* **1986**, *58*, 533.

[2] a) P. P. Edwards, R. L. Johnston, C. N. R. Rao in *Metal Clusters in Chemistry* (Eds.: P. Braunstein, G. Oro, P. R. Raithby), Wiley-VCH, **1998**; b) A. I. Kirkland, D. A. Jefferson, D. G. Duff, *Annual Reports C*, Royal Society of Chemistry, **1993**, p. 247.

[3] H. N. Aiyer, V. Vijayakrishnan, G. N. Subanna, C. N. R. Rao, *Surf. Sci.* **1994**, *313*, 392.

[4] a) D. C. Johnson, R. E. Benfield, P. P. Edwards, W. J. H. Nelson, M. D. Vargas, *Nature* **1985**, *314*, 231; b) Y. Volokitin, J. Sinzig, L. J. de Jongh, G. Schmid, M. N. Vargaftik, I. I. Moiseev, *Nature* **1996**, *384*, 624.

[5] V. Vijayakrishnan, A. Chainani, D. D. Sarma, C. N. R. Rao, *J. Phys. Chem.* **1992**, *96*, 8679.

[6] R. Busani, M. Folker, O. Chesnovsky, *Phys. Rev. Lett.* **1998**, *81*, 3836.

[7] K. Rademann, O. D. Rademann, M. Schlauf, V. Even, F. Hensel, *Phys. Rev. Lett.* **1992**, *69*, 3208.

[8] H. Haberland, B. von Issendrof, Y. Yufeng, J. Kolar, G. Thanner, *Z. Phys. D: At. Mol. Clusters* **1993**, *26*, 8.

[9] K. V. Sarathy, G. Raina, R. T. Yadav, G. U. Kulkarni, C. N. R. Rao, *J. Phys. Chem. B* **1997**, *101*, 9876.

[10] S. Link, M. A. El-Sayed, *J. Phys. Chem. B* **1999**, *103*, 4212.

[11] A. P. Alivisatos, *J. Phys. Chem.* **1996**, *100*, 13226.

[12] a) D. M. Mittleman, R. W. Schoenlein, J. J. Shiang, V. L. Colvin, A. P. Alivisatos, C. V. Shank, *Phys. Rev. B* **1994**, *49*, 14435; b) A. Henglein, *Ber. Bunsenges. Phys. Chem.* **1995**, *99*, 903.

[13] C. P. Vinod, G. U. Kulkarni, C. N. R. Rao, *Chem. Phys. Lett.* **1998**, *289*, 329.

[14] M. Miyake, T. Torimoto, T. Sakata, H. Mori, S. Kuwabata, H. Yoneyama, *Langmuir* **1997**, *13*, 742.

[15] M. Rosenblit, J. Jortner, *J. Phys. Chem.* **1994**, *98*, 9365.

[16] R. A. Perez, A. F. Ramos, G. L. Malli, *Phys. Rev. B* **1989**, *39*, 3005.

[17] O. D. Haberlen, S. C. Chung, M. Stener, N. J. Rosch, *J. Chem. Phys.* **1997**, *106*, 5189.

[18] S. H. Yang, D. A. Drabold, J. B. Adams, A. Sachdev, *Phys. Rev. B* **1993**, *47*, 1567.

[19] P. E. Lippens, M. Lannoo, *Phys. Rev. B* **1989**, *39*, 10935.

[20] C. P. Collier, T. Vossmeyer, J. R. Heath, *Annu. Rev. Phys. Chem.* **1998**, *49*, 371.

[21] Single Charge Tunneling, Coulomb Blockade Phenomena in Nanostructures (Eds.: H. Grabert, M. H. Devoret), *NATO-ASI Ser. B* **1992**, *294*.

[22] P. J. Thomas, G. U. Kulkarni, C. N. R. Rao, *Chem. Phys. Lett.* **2000**, *321*, 163.

[23] J. Jortner, *Z. Phys. D: At. Mol. Clusters* **1992**, *24*, 247.

[24] C. P. Bean, J. D. Livingston, *J. Appl. Phys.* **1959**, *30*, 1208.

[25] a) Van de Heer, P. Milani, A. Chatelain, *Z. Phys. D: At. Mol. Clusters* **1991**, *19*, 241; b) S. N. Khanna, S. Linderoth, *Phys. Rev. Lett.* **1991**, *67*, 742.

[26] C. N. R. Rao, V. Vijayakrishnan, A. K. Santra, M. W. J. Prins, *Angew. Chem.* **1992**, *104*, 1110; *Angew. Chem. Int. Ed. Engl.* **1992**, *31*, 1062.

[27] A. K. Santra, S. Ghosh, C. N. R. Rao, *Langmuir* **1994**, *10*, 3937.

[28] E. Gillet, S. Channakhone, V. Matolin, M. Gillet, *Surf. Sci.* **1986**, *152/153*, 603.

[29] a) D. L. Doering, J. T. Dickinson, H. Poppa, *J. Catal.* **1982**, *73*, 91; b) D. L. Doering, H. Poppa, J. T. Dickinson, *J. Catal.* **1982**, *73*, 104.

[30] M. Valden, X. Lai, D. W. Goodman, *Science* **1998**, *281*, 1647.

[31] C. Xu, X. Lai, G. W. Zajac, D. W. Goodman, *Phys. Rev. B* **1997**, *56*, 13464.

[32] Unpublished results from our laboratory.

[33] C. N. R. Rao, G. U. Kulkarni, P. J. Thomas, P. P. Edwards, *Chem. Soc. Rev.* **2000**, *29*, 27.

[34] A. N. Shipway, E. Katz, I. Willner, *Chem. Phys. Chem.* **2000**, *1*, 18.

[35] M. P. Pileni, *New J. Chem.* **1998**, 693.

[36] G. Schmid, L. F. Chi, *Adv. Mater.* **1998**, *10*, 515.

[37] W. P. Wuelfing, F. P. Zamborini, A. C. Templeton, X. Wen, H. Yoon, R. W. Murray, *Chem. Mater.* **2001**, *13*, 87.

[38] R. L. Whetten, M. M. Shafigullin, J. T. Khoury, T. G. Schaaf, I. Vezmar, M. M. Alvarez, A. Wilkinson, *Acc. Chem. Res.* **1999**, *32*, 397.

[39] P. C. Ohara, D. V. Leff, J. R. Heath, W. M. Gelbart, *Phys. Rev. Lett.* **1995**, *75*, 3466.

[40] P. J. Thomas, G. U. Kulkarni, C. N. R. Rao, *J. Phys. Chem. B* **2000**, *104*, 8138.

[41] B. A. Korgel, S. Gullam, S. Conolly, D. Fitzmaurice, *J. Phys. Chem. B* **1998**, *102*, 8379.

[42] a) V. Russier, C. Petit, J. Legrand, M. P. Pileni, *Phys. Rev. B* **2000**, *62*, 3910; b) C. Petit, M. P. Pileni, *Appl. Surf. Sci.* **2000**, *162–163*, 519; c) C. Petit, A. Taleb, M. P. Pileni, *Adv. Mater.* **1998**, *10*, 259.

[43] S. Sun, C. B. Murray, D. Weller, L. Folks, A. Maser, *Science* **2000**, *287*, 1989.

[44] a) M. V. Artemyev, A. I. Bibik, L. I. Gurinovich, S. V. Gopenko, U. Woggon, *Phys. Rev. B* **1999**, *60*, 1504; b) C. R. Kagan, C. B. Murray, M. G. Bawendi, *Phys. Rev. B* **1996**, *54*, 8633; c) M. V. Artemyev, U. Woggon, H. Jaschinski, L. I. Gurinovich, S. V. Gopenko, *J. Phys. Chem. B* **2000**, *104*, 11617.

[45] M. Brust, D. Bethell, D. J. Schiffrin, C. J. Kiely, *Adv. Mater.* **1995**, *7*, 795.

[46] a) D. B. Janes, V. R. Kolagunta, R. G. Osifchin, J. D. Bielefeld, R. P. Andres, J. I. Henderson, C. P. Kubiak, *Superlattices Microstruct.* **1995**, *18*, 275; b) A. W. Snow, H. Wohltjen, *Chem. Mater.* **1998**, *10*, 947; c) Y. Liu, Y. Wang, R. O. Claus, *Chem. Phys. Lett.* **1998**, *298*, 315; d) H. M. Jaeger, personal communication, **2001**.

[47] a) G. Markovich, C. P. Collier, S. E. Hendricks, F. Remacle, R. D. Levine, J. R. Heath, *Acc. Chem. Res.* **1999**, *32*, 415; b) C. Medeiros-Riberio, D. A. A. Phlberg, R. S. Williams, J. R. Heath, *Phys. Rev. B* **1999**, *59*, 1633.

[48] F. Remacle, R. D. Levine, *Chem. Phys. Chem.* **2001**, *2*, 20.

J. Phys. Chem. B **2001**, *105*, 2515−2517

Magic Nuclearity Giant Clusters of Metal Nanocrystals Formed by Mesoscale Self-Assembly

P. John Thomas, G. U. Kulkarni, and C. N. R. Rao*

Chemistry and Physics of Materials Unit, Jawaharlal Nehru Center for Advanced Scientific Research, Jakkur, Bangalore-560 064, India

Received: June 16, 2000; In Final Form: December 5, 2000

Magic nuclearity giant clusters formed by the mesoscale self-assembly of Pd nanocrystals of 2.5 nm diameter (nuclearity, ∼561) have been identified by transmission electron microscopy. The clusters have discrete diameters, corresponding to those expected for magic nuclearity of 13, 55, 147, 309, 561, and 1415 and corresponding to closed shells of 1, 2, 3, 4, 5, and 7, respectively. Imaging at different tilt angles has provided confirmation of the spherical nature of these giant clusters. Giant clusters of magic nuclearity have also been found with Pd nanocrystals of ∼3.2 nm diameter (nuclearity, ∼1415).

Introduction

Mesoscale self-assembly of polyhedral objects is a topic of great current interest.[1] The occurrence of self-assembly in nanometric dimensions through weak forces has been well documented. Thus, cooperative assemblies of ligated metal and semiconductor nanocrystals as well as of colloidal polymer spheres appear to form through the mediation of electrostatic and capillary forces. Typical examples of mesoscale assemblies are provided by the ordering of nanocrystals of Au[2,3] and CdSe[4] and of polystyrene spheres.[5] Besides ordered two-dimensional arrays, such forces have also been exploited to obtain giant nanocrystal aggregates measuring tens of nanometers,[6] 2D rings of micrometer diameters,[7] and dendrimeric structures of Au nanocrystals.[8] The ability to engineer such assemblies may extend the reach of current lithographic techniques. In this context, the synthesis and programmed assembly of well-defined metal nanocrystals assumes significance.[9] Metal nanocrystals with magic number of atoms 13, 55, 309, 561, and 1415 corresponding to 1, 2, 4, 5, and 7 closed shells, respectively, have been prepared by chemical means.[10−13] We have obtained two-dimensional arrays of Pd_{561} and Pd_{1415} nanocrystals by employing alkanethiol spacers.[14] Schmid et al.[15,16] have arranged Au_{55} clusters on a polymer film and have isolated microcrystals of the same.

The assembly of magic nuclearity nanocrystals into giant clusters containing a magic number of initial nanocrystals has been a subject of fascination.[17] Initial electrophoresis experiments of Schmidt[17] with Au nanocrystals indicated the formation of $(Au_{13})_{13}$ types of superclusters. Further, a high mass secondary ion peak observed also seemed to support this contention.[18,19] Theoretical calculations based on the embedded atom method have indicated that such a growth of metal nanocrystals is indeed a possibility.[20] In Figure 1, is shown a schematic illustration of a $(M_{55})_{55}$ giant cluster.

In our experiments with monodisperse Pd nanocrystals in a ethanol−water mixture,[11] we observed some giant aggregates as a result of mesoscalar self-assembly. In this article, we report electron microscopic investigation of the aggregates of Pd_{561} and Pd_{1415} nanocrystals. The aggregates have diameters that could be assigned to values expected of giant clusters containing

Figure 1. Schematic illustration of a $(M_{55})_{55}$ giant cluster. The structure of one of the nanocrystals is shown.

a magic number of the initial nanocrystals. Thus, it has been possible to obtain giant clusters containing approximately 561 nanocrystals, each of which comprises around 561 atoms.

Experimental Section

Monodisperse Pd nanocrystals were prepared in a ethanol−water mixture in the presence of poly(vinylpyrrolidone) (PVP) following the procedure of Teranishi et al.[12] Typically, a 15 mL of a 2.0 mM aqaueous solution of H_2PdCl_4 was reduced by refluxing it with a mixture of 15 mL of absolute ethanol and 25 mL of water containing 33.3 mg of PVP (M_w ∼ 40 000 g mol^{-1}) for ∼3 h. The obtained sols were examined with a JEOL-3010 transmission electron microscope (TEM), operating at 300 kV. Samples for TEM were prepared by depositing a drop of the sol on a holey carbon grid and allowing it to dry slowly in air and then in a desiccator overnight.

Results and Discussion

A scan of the grid revealed the presence of a large number of uniform spherical particles. Figure 2 shows a typical TEM image of the dispersion. In addition to isolated nanocrystals, we observed pairs of particles separated by a nearly uniform distance. A few were seen as sitting on top of each other. The histogram in Figure 2 shows that the metal cores are of a uniform diameter of 2.5 nm (σ = 6%). Furthermore, high-

* Correponding author. E-mail: cnrrao@jncasr.ac.in.

10.1021/jp0021998 CCC: $20.00   © 2001 American Chemical Society
Published on Web 03/08/2001

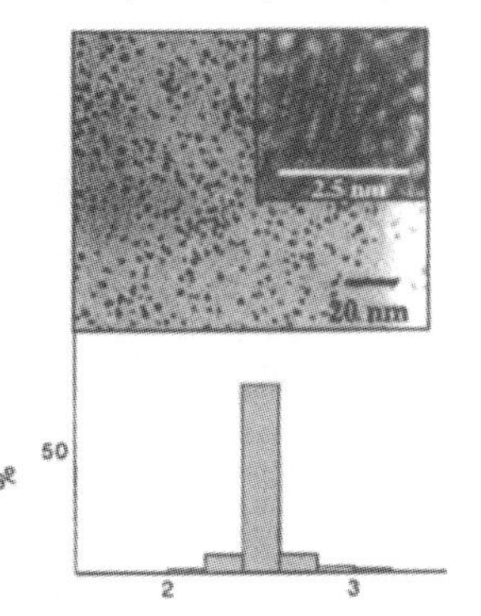

Figure 2. (a) TEM micrograph showing Pd$_{561}$ nanocrystals. The inset shows a high-resolution image of an individual nanocrystal. (b) Histogram showing the size distribution (in percentage).

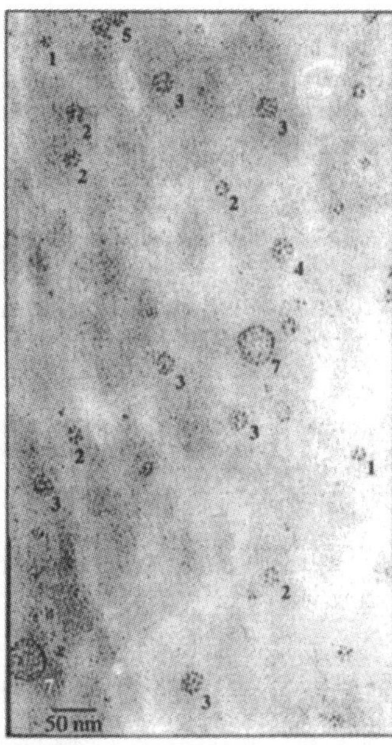

Figure 3. TEM image of Pd$_{561}$ nanocrystals forming giant clusters. The numbers correspond to the proposed number of nanocrystal shells, n.

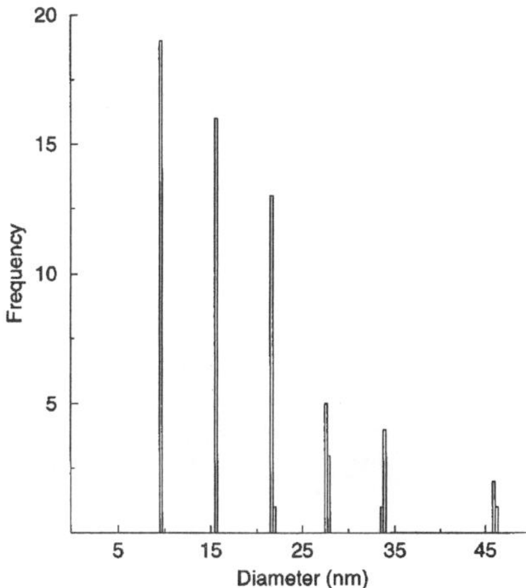

Figure 4. Histogram showing the distribution in the diameters of giant clusters.

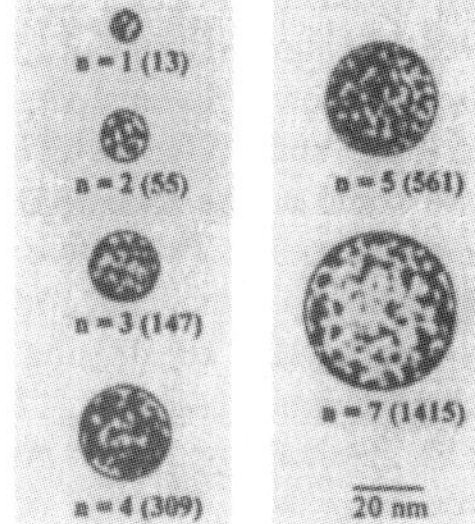

Figure 5. Giant clusters of different magic nuclearities, (Pd$_{561}$)$_n$, the circles corresponding to the diameters of the clusters calculated on the basis of the effective volume of an individual nanocrystal.

TABLE 1: Giant Clusters of Pd$_{561}$ Nanocrystals

diameters measd (nm)	no. of nanocrystals	diameters measd (nm)	no. of nanocrystals
9.6 ± 0.1	13 ± 1	27.7 ± 0.1	309 ± 34
15.6 ± 0.1	55 ± 6	33.8 ± 0.1	561 ± 61
21.6 ± 0.1	147 ± 16	46.0 ± 0.15	1415 ± 156

resolution TEM images (see inset of Figure 2) revealed 11 [111] lattice fringes with each nanocrystal,[13] indicating the atomic nuclearity to be close to 561.

Besides the 2.5 nm nanocrystals, the grid also contained bigger features resembling aggregates of nanocrystals. Such aggregates were spherical and seemed to exhibit an unusual preference for a few definite sizes. In Figure 3, a typical low magnification ($\times 80\,000$) TEM image revealing the presence of giant aggregates of nanocrystals is shown. A histogram showing the frequency distribution of the observed diameters is given in Figure 4. Evidently, the giant clusters exhibit preference for specific diameters, 9.6, 15.6, 21.6, 27.7, 33.8, and 46.0 nm, with the smaller giant clusters being more abundant. To estimate the number of nanocrystals present in these aggregates, we calculated the effective volume of a nanocrystal on the basis of the distance between adjacent nanocrystals in TEM images (Figure 2). The pairs of particles were chosen carefully such that the

two metal cores were distinct. The mean value obtained from hundred such measurements was 4.1 ± 0.1 nm. This value is more suitable than the diameter of an isolated nanocrystal, the latter being 3.3 nm as estimated by STM measurements.[21] We assume that the effective volume thus estimated takes into account the free volume in the nanocrystal aggregate. In other words, a particle pair is taken to adequately describe interactions in the bigger aggregates. In Table 1, we list the observed diameters of the giant clusters along with the estimates of the nuclearities. To our surprise, the estimated values compare closely with the magic nuclearities of 13, 55, 147, 309, 561, and 1415 corresponding to closed shells 1, 2, 3, 4, 5, and 7, respectively. In Figure 5, we have projected the calculated perimeters enclosing the different giant clusters. The close agreement between the observed and calculated diameters is indeed gratifying.

The giant aggregates obtained are distinct from the two-dimensional rings obtained by other groups.[7,22] To establish the spherical nature of the giant clusters observed by us, they were imaged at different tilts ($\pm 17.5°$) along two perpendicular axes in the focal plane. In Figure 6, we show such images obtained

J. Phys. Chem. B, Vol. 105, No. 13, 2001 **2517**

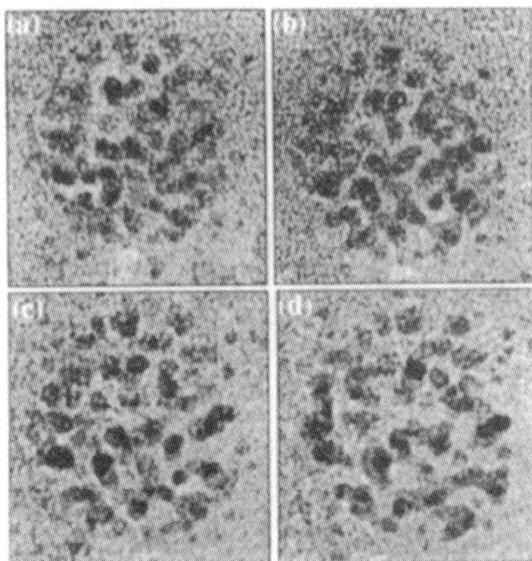

Figure 6. A $(Pd_{561})_{561}$ giant cluster imaged at various tilt angles along two perpendicular axes in the focal plane: (a) $-18.7°$, $1.4°$; (b) $-7.3°$, $-1.4°$; (c) $0°$, $1.4°$; (d) $12.4°$, $-1.4°$.

with a giant cluster of nuclearity 561. The projected image appears circular over the entire tilt range of 35°, along both the directions. We observe a few individual nanocrystals to shift their positions in the image under different tilt conditions, though the exact trajectories could not be followed. A series of such tilt experiments confirmed the spherical nature of the aggregates. Furthermore, we were able to obtain lattice-resolved images of the individual Pd_{561} nanocrystals constituting the giant cluster. Such images revealed the characteristic 11 [111] lattice fringes. Attempts to image the internal structure of giant clusters were, however, not successful probably because of the interference from the polymeric ligand shell of the nanocrystals.

We have previously obtained two-dimensional arrays of metal nanocrystals by using monothiols[2,9,14,23] and three-dimensional layered assemblies with metal and semiconductor nanocrystals by employing dithiols.[24] Brushlike ligands in these cases direct the assembly process along definite directions. We believe that the giant clusters discussed in this paper are a result of self-assembly in three-dimensions facilitated by the relatively passive and isotropic PVP coating around the nanocrystals. Unlike in a two-dimensional assembly, where the solvent plays an important role,[24] the formation of giant clusters does not seem to be influenced by the choice of the solvent. We are able to obtain the giant clusters from a wide variety of solvents such as water, ethanol, ethanol water mixtures, or other types of solvents such as ethylene glycol. In all cases, there were no significant changes in the size distribution of the obtained giant nanocrystals. These giant clusters could be reproducibly obtained over several trials starting with sols of widely differing concentrations and employing PVP of molecular mass $\sim 160,000$ g mol^{-1} as well.

We have made attempts to investigate the process of self-assembly of nanocrystals of a different magic nuclearity. Experiments with Pd nanocrystals of 3.2 nm diameter, corresponding to a nuclearity of 1415, have revealed the formation of giant clusters with possible nuclearities of 147 and 309, as shown in Figure 7. However, the slightly wider distribution in the pristine nanocrystal diameter (possibly due to a mixture of the 7 and 8 shell clusters[13]) seems to hinder the facile formation of uniform giant clusters. A majority of these clusters ($\sim$60%) display elongation along one axis. A narrow size distribution in the initial nanocrystal sol appears to be essential for the formation of the magic nuclearity giant clusters.

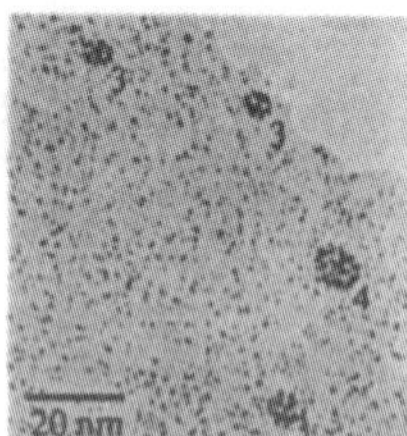

Figure 7. TEM image of giant clusters of Pd_{1415} nanocrystals. The numbers correspond to the proposed number of nanocrystal shells, n.

Conclusion

Giant clusters of Pd nanocrystals are obtained by the mesoscopic assembly of nanocrystals of uniform size. This observation provides an illustration of the principle of self-similarity. These clusters of metal nanocrystals are perhaps bound by the same laws that endow special stability to magic numbered nanocrystals.[15] It would therefore appear that these laws are invariant under scaling. The giant clusters of Pd_{561} nanocrystals obtained by us are different from other types of ligand directed nanocrystal aggregates which are formed mostly due to the directional nature of the ligand shell.[16,6]

References and Notes

(1) Terfort, A.; Bowden, N.; Whitesides, G. M. *Nature* **1997**, *386*, 162.

(2) Vijayasarathy, K.; Raina, G.; Yadav, R. T.; Kulkarni, G. U.; Rao, C. N. R. *J. Phys. Chem.* **1998**, *B101*, 9876.

(3) Whetten, R. L.; Khoury, J. T.; Alvarez, M. M.; Murthy, S.; Vezmar, I.; Wang, Z.; Stephens, P. W.; Clevend, L. Ch.; Luedtke, W. D.; Landman, U. *Adv. Mater.* **1996**, *8*, 428.

(4) Murray, C. B.; Kagan, C. R.; Bawendi, M. G. *Science* **1995**, *270*, 1335.

(5) Trau, M.; Saville, D. A.; Aksay, I. A. *Science* **1996**, *272*, 706.

(6) Boal, A. K.; Ilhan, F.; Derouchey, J. E.; Thurn-Albrecht, T.; Russull, T. P.; Rotello, V. M. *Nature* **2000**, *404*, 746.

(7) Ohara, P. C.; Heath, J. R.; Gelbart, W. M. *Angew. Chem., Int. Ed. Engl.* **1997**, *36*, 1078.

(8) Seshadri, R.; Subbanna, G. N.; Vijayakrishnan, V.; Kulkarni, G. U.; Ananthakrishna, G.; Rao, C. N. R. *J. Phys. Chem.* **1995**, *99*, 5639.

(9) Rao, C. N. R.; Kulkarni, G. U.; Thomas, P. J.; Edwards, P. P. *Chem. Soc. Rev.* **2000**, *29*, 27.

(10) Rao, C. N. R. *Chemical approaches to the Synthesis of Inorganic Materials*; Wiley Eastern: New Delhi, 1994.

(11) Vargaftik, M. N.; Moiseev, I. I.; Kochubey, D. I.; Zamareev, K. I. *Faraday Discuss.* **1991**, *92*, 13.

(12) Teranishi, T.; Hori, H.; Miyake, M. *J. Phys. Chem.* **1997**, *B101*, 5774.

(13) Schmid, G.; Harms, M.; Malm, J. O.; Bovin, J. O.; Ruitenbeck, J. V.; Zandbergen, H. W.; Fu, W. T. *J. Am. Chem. Soc.* **1993**, *113*, 2046.

(14) Thomas, P. J.; Kulkarni, G. U.; Rao, C. N. R. *J. Phys. Chem.* **2000**, *B104*, 8138.

(15) Schmid, G.; Bäumle, M.; Beyer, N. *Angew. Chem., Int. Ed. Engl.* **2000**, *39*, 181.

(16) Schmid, G.; Zaika, W. M.; Pugin, R.; Sawitowski, T.; Majoral, J.-P.; Caminade, A.-M.; Turrin, C.-O. *Chem. Eur. J.* **2000**, *6*, 1693.

(17) Schmid, G. *Polyhedron.* **1988**, *7*, 2321. (b) Schmid, G.; Klein, N. *Angew. Chem., Int. Ed. Engl.* **1986**, *25*, 922.

(18) Feld, H.; Leute, A.; Rading, D.; Benninghoven, A.; Schmid, G. *J. Am. Chem. Soc.* **1990**, *112*, 8166.

(19) McNeal, C. J.; Winpenny, R. E. P.; Hughes, J. M.; Macfarlane, R. D.; Pignolet, L. H.; Nelson, L. T. J.; Gardner, T. G.; Irgens, I. H.; Vigh, G.; Fackler, J. P. *Inorg. Chem.* **1993**, *32*, 5582.

(20) Fristche, H.-G.; Muller, H.; Fehrensen, B. *Z. Phys. Chem.* **1997**, *199*, 87.

(21) Thomas, P. J.; Kulkarni, G. U.; Rao, C. N. R. *Chem. Phys. Lett.* **2000**, *321*, 163.

(22) Shafi, K. V. P. M.; Felner, I.; Mastai, Y.; Gedanken, A. *J. Phys. Chem.* **1999**, *B103*, 3358.

(23) Vijayasarathy, K.; Kulkarni, G. U.; Rao, C. N. R. *Chem. Commun.* **1997**, 573.

(24) Vijayasarathy, K.; Thomas, P. J.; Kulkarni, G. U.; Rao. C. N. R. *J. Phys. Chem.* **1999**, *B103*, 399.

(25) Korgel, B. A.; Fitzmaurice, D. *Phys. Rev. Lett.* **1998**, *80*, 3531.

Acc. Chem. Res. **2002**, *35*, 998–1007

ARTICLES

Carbon Nanotubes from Organometallic Precursors

C. N. R. RAO* AND A. GOVINDARAJ

Chemistry and Physics of Materials Unit and CSIR Centre of Excellence in Chemistry, Jawaharlal Nehru Centre for Advanced Scientific Research, Jakkur P. O., Bangalore 560 064, India

Received January 7, 2002

ABSTRACT
Multiwalled as well as single-walled carbon nanotubes are conveniently prepared by the pyrolysis of organometallic precursors such as metallocenes and phthalocyanines in a reducing atmosphere. More importantly, pyrolysis of organometallics alone or in mixture with hydrocarbons yields aligned nanotube bundles with useful field emission and hydrogen storage properties. By pyrolysis of organometallics in the presence of thiophene, Y-junction nanotubes are obtained in large quantities. The Y-junction tubes have a good potential in nanoelectronics. Carbon nanotubes prepared from organometallics are useful to prepare nanowires and nanotubes of other materials such as BN, GaN, SiC, and Si_3N_4.

Introduction

Carbon nanotubes were discovered as a microscopic miracle in the cathode deposits obtained in the arc evaporation of graphite.[1] The arc method has since been improved and modified to obtain good yields of both multiwalled and single-walled nanotubes. Carbon nanotubes are conveniently obtained by carrying out the pyrolysis of hydrocarbons such as ethylene and acetylene over nanoparticles of iron, cobalt, and other metals dispersed over a solid substrate.[2-4] The presence of nanoparticles is essential not only to form nanotubes but also to control the diameter of the nanotube to some extent.[5] Since a carbon source as well as metal nanoparticles is necessary for producing carbon nanotubes by the pyrolysis of hydrocarbons, the strategy of employing an appropriate organometallic precursor which can serve as a dual source of both the carbon and the metal nanoparticles was explored by us. The very first experiments carried out on the pyrolysis of organometallic precursors

such as metallocenes and iron pentacarbonyl were successful in producing multiwalled nanotubes.[6,7] We have employed this method not only to produce multiwalled and single-walled nanotubes but also to make aligned nanotube bundles and Y-junction nanotubes. In this Account, we present the salient features of the various types of nanotubes obtained by us in Bangalore by employing the organometallic route and also examine some of the properties of the nanotubes so produced. Aligned nanotube bundles are expected to have applications in electronic displays and hydrogen storage while Y-junction nanotubes could be useful as building blocks in nanoelectronics. The nanotubes produced from organometallics are also usefully employed as starting materials to prepare other types of nanostructures.

Multiwalled and Single-Walled Nanotubes

Sen et al.[6] prepared multiwalled carbon nanotubes (MWNTs) and metal-filled onionlike structures by the pyrolysis of metallocenes such as ferrocene, cobaltocene, and nickelocene under reducing conditions, wherein the precursor acts as the source of the metal as well as carbon. The pyrolysis setup consists of stainless steel gas flow lines and a two-stage furnace system fitted with a quartz tube (25 mm i.d.) as shown in Figure 1a, the flow rate of the gases being controlled by the use of mass flow controllers. In a typical preparation, a known quantity (100 mg) of the metallocene (presublimed 99.99% purity) is taken in a quartz boat and placed at the center of the first furnace, and a mixture of Ar and H_2 of the desired composition is passed through the quartz tube. The metallocene is sublimed by raising the temperature of the first furnace to 200 °C at a controlled heating rate (20 °C/min). The metallocene vapor so generated is carried by the Ar–H_2 gas stream into the second furnace, maintained at 900 °C, where the pyrolysis occurs. The main variables in the experiment are the heating rate of ferrocene, the flow rate of Ar gas, and the pyrolysis temperature. Ferrocene vapor carried by a 75% Ar + 25% H_2 mixture at a flow rate of 900 sccm (sccm = standard cubic centimeter per minute) into the furnace yields large quantities of carbon deposits, mainly containing carbon nanotubes. These deposits are examined by a scanning electron microscope (SEM) and transmission electron microscope (TEM). In Figure 2a, we show a TEM image of nanotubes produced by the pyrolysis of ferrocene. To increase the yield of the multiwalled carbon nanotubes with metallocene, vapors of an additional hydrocarbon source were mixed along with the metallocene vapor in the first furnace. Thus, pyrolysis of benzene in the presence of ferrocene or $Fe(CO)_5$ gives high yields of multiwalled nanotubes, the wall thickness of the nanotubes depending on the proportion of the carbon

C. N. R. Rao obtained his Ph.D. degree from Purdue University and D.Sc. degree from the University of Mysore. He is Linus Pauling Research Professor at the JNCASR and Honorary Professor at the Indian Institute of Science, Bangalore, India. He is a member of several academies including the Royal Society (London), U.S. National Academy of Sciences, French Academy of Sciences, and Japan Academy. His main research interests are in solid state and materials chemistry and nanomaterials. The most recent award received by him is the Hughes medal for physical sciences from the Royal Society.

A. Govindaraj obtained his Ph.D. degree from the University of Mysore and is a Scientific Officer at the Indian Institute of Science, Bangalore, India, and Honorary Senior Research Officer at the JNCASR. His main research interests are nanotubes and fullerenes. He is a recipient of the MRS (India) medal.

* To whom correspondence should be addressed. Fax: 91-80-8462760. E-mail: cnrrao@jncasr.ac.in.

10.1021/ar0101584 CCC: $22.00

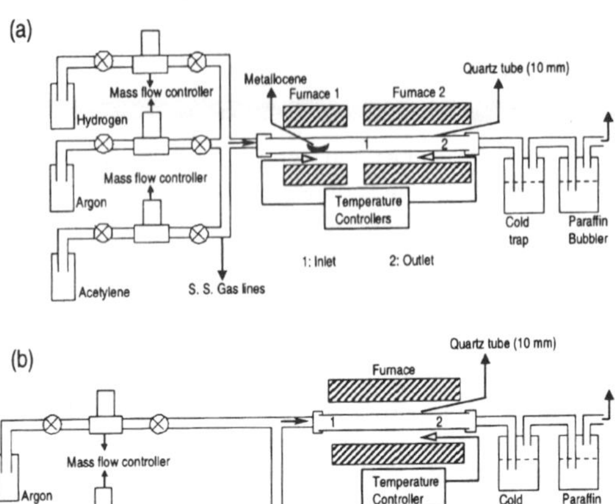

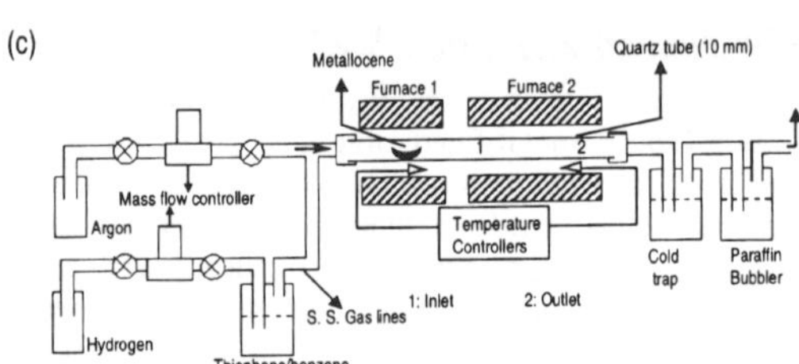

FIGURE 1. Pyrolysis apparatus employed for the synthesis of carbon nanotubes by pyrolysis of mixtures of (a) metallocene + C_2H_2, (b) $Fe(CO)_5$ + C_2H_2, and (c) metallocene + benzene or thiophene. The numbers 1 and 2 indicated in the figure represents inlet and outlet, respectively.[6]

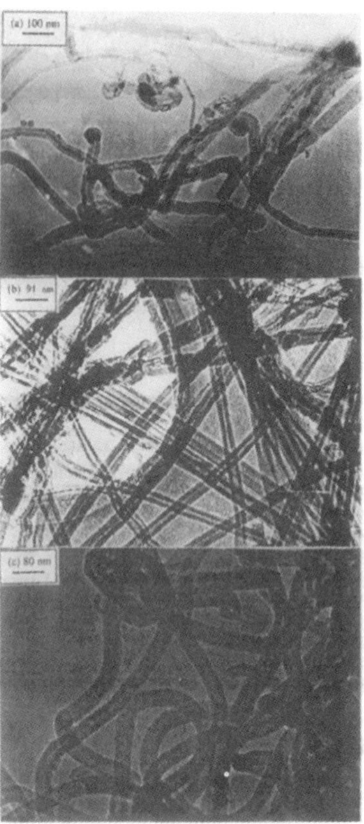

FIGURE 2. TEM image of a multiwalled carbon nanotube obtained by pyrolysis of (a) ferrocene at 900 °C in a mixture of 75% Ar/25% H_2 at a flow rate of 900 sccm, (b) a mixture of C_2H_2 (25 sccm) and ferrocene at 1100 °C at 1000 sccm Ar flow, and (c) a mixture of nickelocene and benzene at 900 °C in 85% Ar and 15% H_2 mixture at a flow rate of 1000 sccm.[6,7]

source and the metal precursor.[7] In Figure 2b, we show a TEM image of nanotubes obtained by the pyrolysis of a mixture of C_2H_2 (25 sccm) and ferrocene at 1100 °C in a Ar flow rate of 1000 sccm. The image clearly reveals that the addition of hydrocarbon not only increases the yield of hollow MWNTs but also reduces the amount of carbon-coated metal nanoparticles. In Figure 2c we show a TEM image of MWNTs obtained by the pyrolysis of a mixture of nickelocene and benzene at 900 °C in 85% Ar and 15% H_2 mixture at a flow rate of 1000 sccm. Besides metallocenes, one can also employ metal phthalocyanines as precursors to prepare MWNTs.[8,9]

To prepare single-walled nanotubes (SWNTs), alternate synthetic strategies have been explored. Under controlled conditions of pyrolysis, dilute hydrocarbon–organometallic mixtures yield SWNTs.[10,11] High-resolution TEM images of SWNTs, obtained by the pyrolysis of a nickelocene–acetylene mixture at 1100 °C,[11] are shown in the Figure 3a,b. The diameter of the SWNT in Figure 3a is 1.4 nm. It may be recalled that the pyrolysis of nickelocene in admixture with benzene under similar conditions primarily yields MWNTs. Acetylene appears to be a better carbon source for the preparation of SWNTs, since it contains a smaller number of carbon atoms. The bottom portion of the SWNT in Figure 3a shows an amorphous

carbon coating around the tube, commonly observed in many of the preparations. This can be avoided by reducing the proportion of the hydrocarbon and mixing hydrogen in the Ar stream. Pyrolysis of cobaltocene and acetylene under similar conditions gives rise to isolated SWNTs. In Figure 3c,d, we show the high-resolution electron microscope (HREM) images of the SWNTs obtained by the pyrolysis of ferrocene with methane at 1100 °C. Surprisingly, the pyrolysis of nickelocene or cobaltocene in admixture with methane under similar conditions did not give SWNTs in good yield. Pyrolysis of binary mixtures (1:1 by weight) of the metallocenes along with acetylene gives good yields of SWNTs, due to the beneficial effect of binary alloys.[12]

SWNTs are obtained in good yields by the pyrolysis of acetylene in mixture with $Fe(CO)_5$ at 1100 °C employing a setup of the type shown in Figure 1b. We show TEM images of the SWNTs so obtained in Figure 4. Pyrolysis of ferrocene–thiophene mixtures yield SWNTs, but the yield is somewhat low. Pyrolysis of a mixture of benzene and thiophene along with ferrocene (Figure. 1c), however, gives a high yield of SWNTs.[13]

FIGURE 3. (a, b) HREM images of SWNTs obtained by the pyrolysis of nickelocene and C_2H_2 at 1100 °C in a flow of Ar (1000 sccm) with C_2H_2 flow rate of 50 sccm. (c, d) HREM images of SWNTs obtained by the pyrolysis of ferrocene and CH_4 at 1100 °C in a flow of Ar (990 sccm) with CH_4 flow rate of 10 sccm.[11]

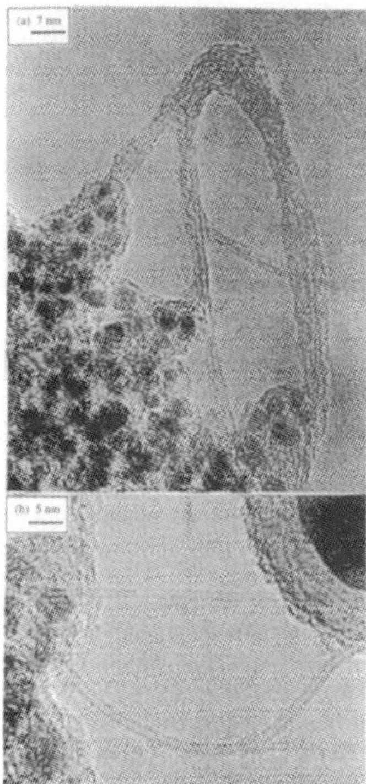

FIGURE 4. (a, b) TEM images of SWNTs obtained by the pyrolysis of $Fe(CO)_5$ and C_2H_2 (flow rate = 50 sccm) at 1100 °C in Ar (flow rate = 1000 sccm) flow.[10,11]

TEM examination of the various carbonaceous products obtained from the pyrolysis of hydrocarbons and organometallic precursors indicates that the size of the catalyst particle plays an important role with regard to the nature of the product. In the case of organometallic precursors, it seems that metal nanoclusters of ~1 nm diameter are produced under controlled conditions. When the concentration of the organometallic precursor is high, MWNTs are formed around the metal particles of 5–20 nm diameter. This is true of carbon nanotubes obtained by the metallocene route.[6,7] In the higher size range of ≥50 nm, graphite-covered metal particles are formed predominantly.[15]

Aligned Carbon Nanotube Bundles

Since the pyrolysis of mixtures of organometallic precursors and hydrocarbons yields good quantities of multi-walled nanotubes,[6,7] we considered the possibility of obtaining aligned nanotube bundles under appropriate conditions. For this purpose, we carried out pyrolysis of metallocenes along with other hydrocarbon sources, in the apparatus shown in Figure 1a.[10,15,16] To obtain aligned nanotube bundles, a typical heating rate 50 °C/min of the first furnace and an Ar flow rate of 1000 sccm have been employed. Compact aligned nanotube bundles could be obtained by introducing C_2H_2 (50–100 sccm) during the sublimation of ferrocene. Scanning electron microscope images of aligned nanotubes obtained by the pyrolysis of ferrocene are shown in Figure 5. The image in Figure 5a shows the top view of the aligned nanotubes, wherein the nanotube tips are seen. The image in Figure 5b shows the side view. Aligned nanotubes obtained by the pyrolysis of ferrocene with different alkanes are shown in the SEM images in Figure 6. The average length of the nanotubes is generally around 60 μm with methane and acetylene. In the case of methane, the nanotubes are aligned, but the packing density is not high. The nanotube bundles obtained with ferrocene + acetylene mixtures appear to have a packing density greater than that obtained with Fe/silica catalysts.[17] SEM images of large bundles of the aligned nanotubes are shown in Figure 7. A small proportion of graphite-covered metal nanoparticles is often present along with the nanotubes. In Table 1, we sum-

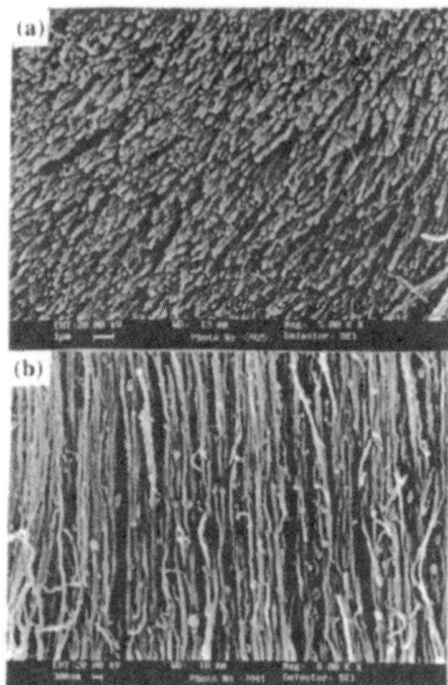

FIGURE 5. SEM images of aligned carbon nanotubes obtained by the pyrolysis of ferrocene. (a) and (b) show views of the aligned nanotubes along and perpendicular to the axis of the nanotubes, respectively.[16]

marize the products obtained by the pyrolysis of hydrocarbon + organometallic mixtures at 1100 °C in a stream of Ar + H_2.

Andrews et al.[18] have carried out the pyrolysis of ferrocene−xylene mixtures to obtain aligned carbon nanotubes. Pyrolysis of Fe(II) phthalocyanine also yields aligned nanotubes.[19] Chen et al.[20] have grown three-dimensional micropatterns of well-aligned carbon nanotubes on photolithographically prepatterned substrates, by the pyrolysis of iron(II) phthalocyanine in an Ar/H_2 atmosphere around 950 °C. They achieved the photopatterning by photolithographic cross-linking of a chemically amplified photoresist layer spin-cast on a quartz plate or a silicon wafer coupled with solution development. Owing to the appropriate surface characteristics, the patterned photoresist layer supports aligned nanotube growth. The difference in the chemical nature between the surfaces covered and uncovered by the photoresist film causes a region-specific growth of nanotubes with different tubular lengths and packing densities leading to the formation of three-dimensional aligned nanotube patterns.

Considering all aspects, we feel that the pyrolysis of organometallic precursors is the most convenient means of preparing aligned MWNTs. The other methods reported

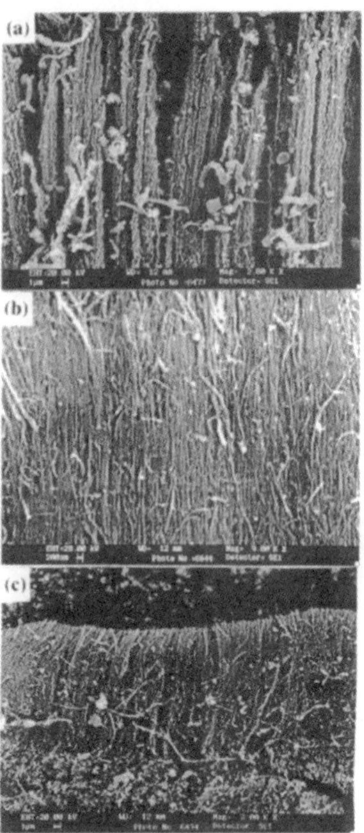

FIGURE 6. SEM images showing the bundles of aligned nanotubes obtained by the pyrolysis of ferrocene along with (a) methane, (b) acetylene, and (c) butane (hydrocarbon at 50 sccm) at 1100 °C in a Ar flow of 950 sccm.[15]

in the literature, such as hydrocarbon pyrolysis on patterned metal films,[21] are more difficult and are not amenable for large-scale synthesis. The advantage of the precursor method is that the aligned bundles are produced in one step, at a relatively low cost, without prior preparation of substrates. It is possible that the self-assembly of carbon nanotubes from precursor pyrolysis is influenced by the transition metal particles which take part in the nucleation and growth of the nanotubes. The metal often gets encapsulated to form nanorods or nanoparticles inside the carbon nanotubes. Ferromagnetism of these metal particles is also a property of significance. The iron nanorods encapsulated in the nanotubes exhibit a complex behavior with respect to magnetization reversal and could be useful as probes in magnetic force micros-

Table 1. Products Obtained by the Pyrolysis of Hydrocarbon−Organometallic Mixtures at 1100 °C

organometallic precursor	hydrocarbon	inlet	outlet
ferrocene	acetylene	aligned MWNTs, nanorods	metal nanoparticles*
nickelocene	acetylene	MWNTs	SWNTs
cobaltocene	acetylene	MWNTs	SWNTs
ferrocene + nickelocene	acetylene	aligned MWNTs	SWNTs
ferrocene + cobaltocene	acetylene	aligned MWNTs	SWNTs
nickelocene + cobaltocene	acetylene	MWNTs	SWNTs
Fe(CO)$_5$	acetylene	MWNTs	SWNTs (bundles)

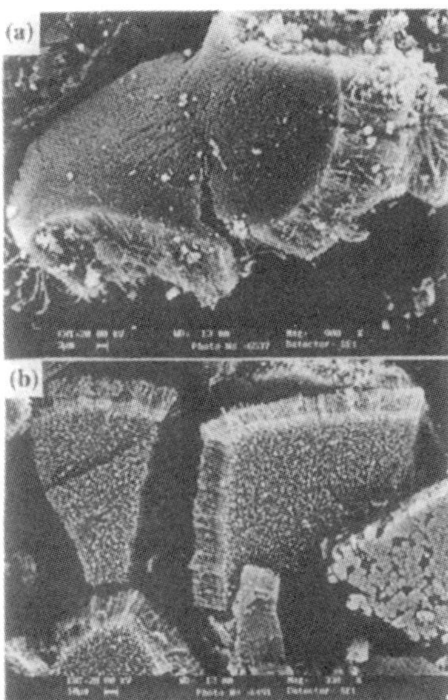

FIGURE 7. SEM images showing the bundles of aligned nanotubes obtained by pyrolyzing ferrocene along with (a) acetylene and (b) butane.[15]

copy. Aligned carbon nanotubes are potential candidates for use as field emitters,[14] and the easy synthesis from organometallic precursors is therefore of importance.

Y-Junction Carbon Nanotubes

Device miniaturization in semiconductor technology is expected to reach its limits due to the inherent quantum effects as one goes toward smaller size. In such a scenario, an alternative would be nanoelectronics based on molecules. The possible use of carbon nanotubes in nanoelectronics has aroused considerable interest. For such applications it is important to be able to connect the nanotubes of different diameters and chirality.[22–24] Complex three-point nanotube junctions have been proposed as the building blocks of nanoelectronics, and in this regard Y- and T-junctions have been considered as prototypes.[25,26] While we would expect an equal number of five- and seven-membered rings to create nanotube junctions, it appears that they can be created with an equal number of five- and eight-membered rings as well.[26] To date, there have been no practical devices made of real three-point nanotube junctions. However, junctions consisting of crossed nanotubes have been fabricated to study their transport characteristics.[27] Y-junction nanotubes have been produced by using Y-shaped nanochannel alumina as templates.[28] We have prepared Y-junction nanotubes in large quanties by carrying out the pyrolysis of a mixture of a metallocene with thiophene.[29,30]

The experimental setup employed by us for the synthesis of the Y-junction nanotubes employed a two-stage furnace system similar to that described earlier[29] (Figure 1c). A known quantity of metallocene was sublimed in the first furnace and carried along with a flow of argon (Ar)

gas to the pyrolysis zone in the second furnace. Simultaneously hydrogen was bubbled through thiophene and was mixed with the argon–metallocene vapors at the inlet of the furnace and carried to the pyrolysis zone. Pyrolyzing the mixed vapors at 1000 °C yielded Y-junction nanotubes in plenty. Pyrolysis of Ni/Fe phthalocyanine in mixture with thiophene was carried out in a similar manner taking phthalocyanine in place of the metallocenes to obtain good Y-junction nanotubes. Pyrolysis of various organometallics with sulfur-containing compounds has shown that pyrolysis of thiophene with nickelocene, ferrocene, and cobaltocene yields excellent Y-junction nanotubes. A TEM image of a Y-junction nanotube obtained by the pyrolysis of nickelocene/thiophene mixture is shown in Figure 8a. A TEM image revealing the presence of several Y-junction carbon nanotubes is shown in Figure 8b. Many of the nanotubes show multiple Y-junctions.

Pyrolysis of thiophene with Fe or Ni phthalocyanine or iron pentacarbonyl also yields Y-junction nanotubes. In Figure 9a, we show few Y-junctions, where as in Figure 9b we show the multiple junctions formed continuously by the pyrolysis of thiophene with Ni phthalocyanine. While the pyrolysis of nickelocene with CS_2 yields similar junctions, the yield and quality of the nanotubes is not satisfactory. At higher flow rates of CS_2, with ferrocene it pyrolyses and gives Y-junction carbon fibers. In Figure 9c we show the Y-junctions obtained by the pyrolysis of thiophene with Fe phthalocyanine. Pyrolysis of thiophene with $Fe(CO)_5$ carried out by bubbling H_2 (50–100 sccm) through the pentacarbonyl along with Ar (150–200 sccm) bubbled through thiophene showed the presence of interesting junction structures as revealed in Figure 9d. The pyrolysis of thiophene over a $Ni(Fe)/SiO_2$ catalyst provides an alternative procedure to the metallocene route, but the yield of Y-junction tubes is rather low. The availability of large quantities of Y-junctions should render them useful for exploitation in nanoelectronics. Compositional analysis at the Y-junction has shown absence of sulfur indicating that the junction is formed by the curvature caused by different carbon rings. The metal nanoparticles abstract the sulfur from thiophene forming a sulfide, the remaining carbon fragment probably being involved in ring formation.

HREM images show that the graphitic layers bend parallely with respect to the junction in many of these nanotubes. Our studies in collaboration with Torsteen Seeger and Manfred Ruhle have shown that the Y-juction tubes are entirely composed of carbon with no sulfur impurity. These observations suggest the presence of equal numbers of five- and seven-/eight-membered rings at the junctions. The metal particles formed in the pyrolysis contain both sulfur and carbon.

Scanning tunneling spectroscopic studies of Y-junction carbon nanotubes show interesting diodelike device characteristics at the junctions. A typical $I–V$ curve obtained from positioning the tip atop a Y-junction (the point of contact between the three arms) as well as on the individual arms of the Y-junction is shown in Figure 10. The $I–V$ plot at the junction is asymmetric (Figure 10b)

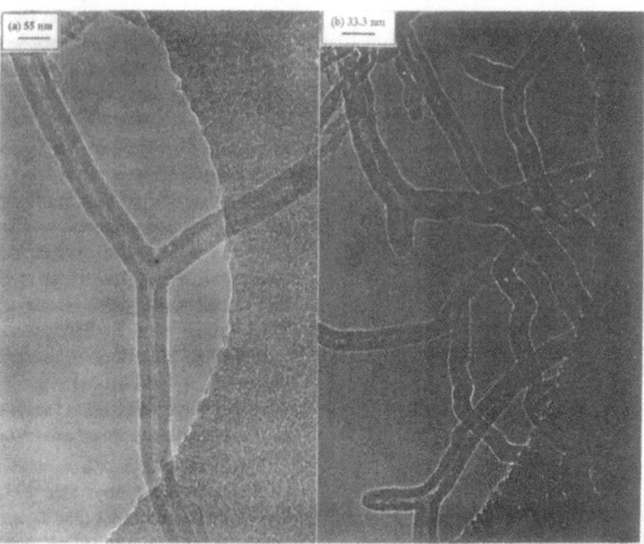

FIGURE 8. (a, b) TEM images of Y-junction carbon nanotubes obtained by the pyrolysis of nickelocene and thiophene at 1000 °C.[29]

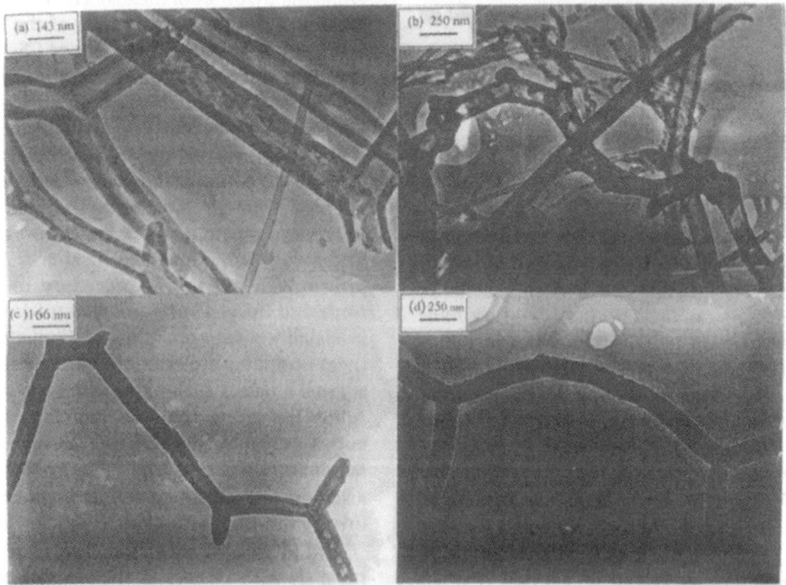

FIGURE 9. TEM images of Y-junction nanotubes: (a, b) obtained by the pyrolysis of thiophene along with Ni phthalocyanine at 1000 °C; (c) obtained by the pyrolysis of thiophene along with Fe phthalocyanine at 1000 °C; (d) obtained by the pyrolysis of an Fe(CO)$_5$–thiophene mixture at 1000 °C.[30]

with respect to bias polarity, unlike that along the arm (Figure 10a). The insets in Figure 10 gives a plot of differential conductance vs bias with respect to zero bias which is symmetric in Figure 10a and asymmetric in Figure 10b. Such asymmetry is characteristic of a junction diode, and this in turn indicates the existence of intramolecular junctions in the carbon nanotubes. The findings discussed above open up the possibility of assembling carbon nanotubes possessing novel devicelike properties[29,31–33] into multifunctional circuits and ultimately toward the realization of a carbon nanotube based

computer chip. Rueckes et al.[34] have described the concept of carbon-nanotube-based nonvolatile random access memory for molecular computing. The viability of the concept has been demonstrated.

Nanorods, Nanowires, and Nanotubes of Other Materials

Preparation of metal nanorods covered by carbon has been reported in the recent literature.[10,35–37] The organometallic precursor route provides a means of preparing

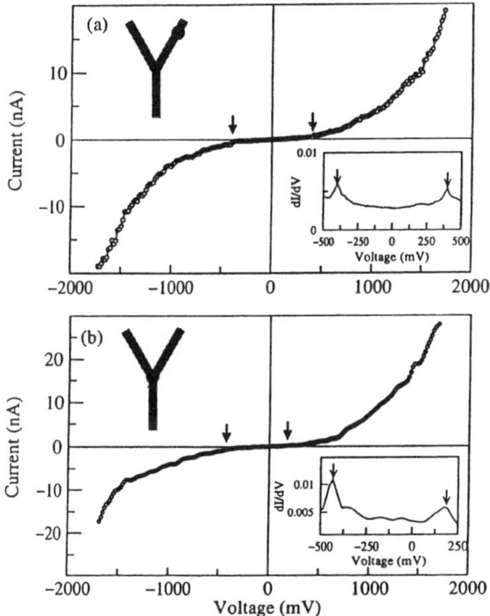

FIGURE 10. $I-V$ curves collected from different points in a Y-junction carbon nanotube (a) from one of the arms away from the junction and (b) from the junction of the three arms. Inset show plots of dI/dV vs bias voltage. The observed gaps are indicated by arrows.[29]

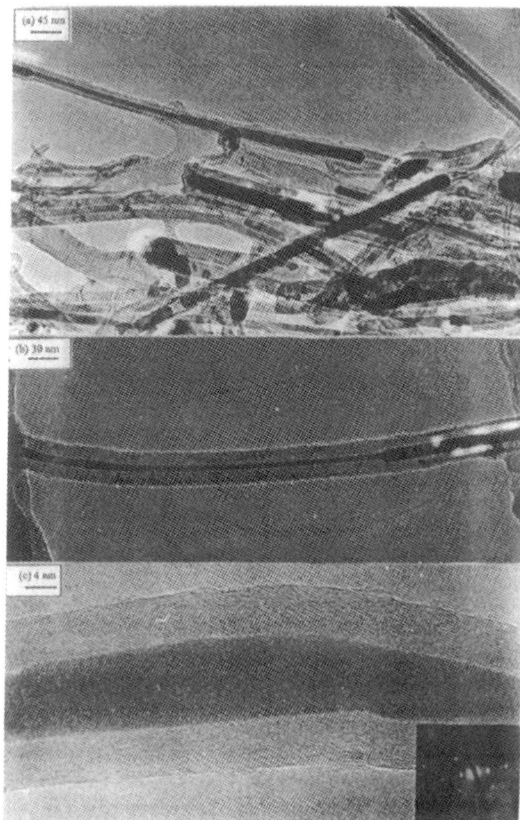

FIGURE 11. (a, b) TEM images of the iron nanorods encapsulated inside the carbon nanotubes from aligned nanotube bundles. (c) HREM image of a single-crystal iron nanorod encapsulated inside a carbon nanotube. The inset in (c) represents the selected area electron diffraction (SAED) pattern of a iron nanorod.[16,35]

metal nanorods. The pyrolysis of ferrocene + hydrocarbon mixtures or of ferrocene alone yields iron nanorods encapsulated inside the carbon nanotubes as evidenced from TEM, the proportion of the nanorods depending on the proportion of ferrocene. Typical TEM images of such nanorods are shown in Figure 11a,b. The selected area electron diffraction (SAED) pattern of the nanorods (inset in Figure 11c) shows spots due to (010) and (011) planes of α-Fe. The HREM image of the iron nanorod in Figure 11c shows well-resolved (011) planes of α-Fe in single-crystalline form. X-ray diffraction patterns show the presence of α-Fe with a small portion of Fe_3C as the minor phase. In addition to the nanorods, iron nanoparticles (20−40 nm diameter) encapsulated inside the graphite layers are also obtained in many of the preparations. The iron nanorods and nanoparticles are well protected against oxidation by the graphitic layers. Ni/Fe phthalocyanines can be used in the place of metallocenes to prepare metal nanowires. However, the yields are not as good as in the case of metallocenes.

There are several reports on the preparation of SiC nanowires in the literature but fewer on the preparation of Si_3N_4 nanowires.[38,39] The methods employed for the synthesis of SiC nanowires have been varied. Since both SiC and Si_3N_4 are products of the carbothermal reduction of SiO_2, it should be possible to establish conditions wherein one set of specific conditions favor one over the other. We have been able to prepare Si_3N_4 nanowires,[40] by reacting multiwalled carbon nanotubes produced by ferrocene pyrolysis with ammonia and silica gel at 1360

°C. The MWNTs were used as a carbon source in the carbothermal reduction because they have higher thermal stability compared to activated carbon. The reaction of MWNTs with silica gel and NH_3 at 1360 °C yields a mixture of α- and β-Si_3N_4. SEM images of the product obtained by this reaction showed the nanowires to have large diameters (5−7 μm), with lengths of the order of hundreds of micrometers (Figure 12a). By addition of catalytic iron particles (0.5 at. %), we obtain β-SiC nanowires under the same conditions[40] (Figure 12b).

Several chemical methods of preparing boron nitride nanotubes and nanowires have been investigated.[41] The general methods involve reacting boric acid with ammonia in the presence of MWNTs. Good yields of clean BN nanotubes are obtained with MWNTs. Aligned BN nanotubes are obtained by heating aligned bundles of MWNTs with boric acid in the presence of ammonia at the 1000°− 1300 °C range. The SEM images of aligned BN nanotube bundles shown in Figure 13 a,b clearly reveal BN nanotubes aligned in two different orientations. The carbon MWNTs appear to not only take part in the reaction but also serve as templates. The average outer diameter of the

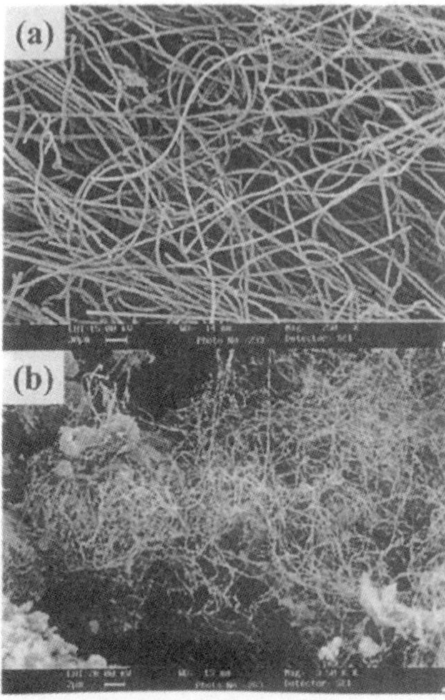

FIGURE 12. (a) SEM image of Si_3N_4 nanowires obtained by the reaction of aligned multiwalled nanotubes (produced by metallocene route) with silica gel at 1360 °C. (b) SiC nanowires.[40]

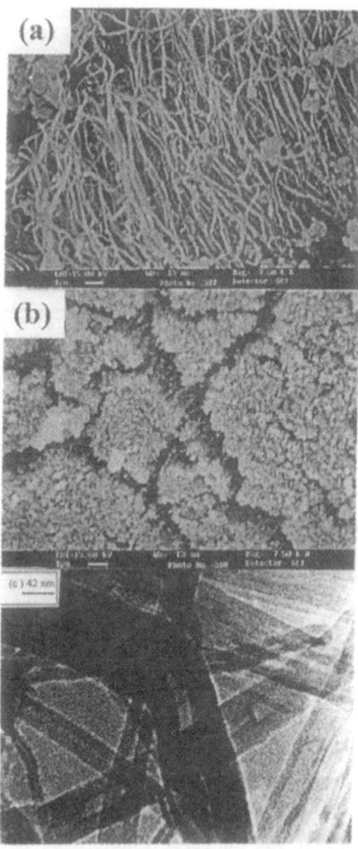

FIGURE 13. (a, b) SEM images of BN nanotubes obtained by heating aligned bundles of MWNTs with boric acid in the presence of ammonia at 1300 °C. (c) TEM image of the BN nanotubes obtained by the same procedure as in (a) and (b).[41]

aligned BN nanotubes varies from 15 to 40 nm as revealed by the TEM image in Figure 13c. This suggests that, during the formation of BN nanotubes, the carbon MWNTs not only take part in the reaction but also serve as templates. It is noteworthy that the BN nanotubes can be produced at a temperature as low as 1000 °C by this procedure. On the basis of elemental analysis and X-ray diffraction, it is found that the carbon content of the BN nanotubes becomes marginal if the initial proportion of carbon nanotubes is kept low.

Gallium nitride nanowires have been prepared by us by employing several procedures involving the use of carbon nanotube templates.[42] We have employed gallium acetylacetonate as the precursor for the in-situ production of small particles of the oxide (GaO_x) which then react with NH_3 vapor around 900 °C in the presence of nanotubes. When multiwalled carbon nanotubes prepared by arc discharge are used as templates, the yield of the GaN nanowires is excellent and the diameter of the majority of the nanowires is in the 35–100 nm range. The length of the nanowires extends to a few micrometers. The linear nanowires are generally single crystalline, showing a layer spacing of 0.276 nm corresponding to the [100] planes. This variation in the diameter of the nanowires occurs because of the nonuniformity in the diameter of the nanotubes. The diameter of the nanowires could be reduced to 20 nm by using single-walled nanotubes in place of multiwalled nanotubes. The growth direction of the nanowires is nearly perpendicular to the [100] planes. The nanowires show satisfactory photoluminescence characteristics similar to those of bulk GaN. The non-carbon

nanowires and nanotubes have potential applications. For example, GaN nanowires, suitably doped, can have uses in nanoelectronics and in optical devices. BN nanotubes and Si_3N_4 nanowires are useful ceramic materials.

Properties

Field emission properties of carbon nanotubes have direct applications in vacuum microelectronic devices.[43–46] We have found that carbon nanotubes produced by the pyrolysis of ferrocene on a pointed tungsten tip exhibit high emission current densities with good performance characteristics.[47] In Figure 14a we show a typical $I–V$ plot for the carbon nanotube covered tungsten tip for currents ranging from 0.1 nA to 1 mA. The applied voltage was 4.3 kV for a total current of 1 μA and 16.5 kV for 1000 μA. The Fowler–Nordheim (F–N) plot shown in Figure 14b has two distinct regions. The behavior is metal-like in the low-field region, while it saturates at higher fields as the voltage is increased. We have obtained a field emission current density of 1.5 A cm^{-2} at a field of 290 V/mm, a value considerably higher than that found with planar cathodes.

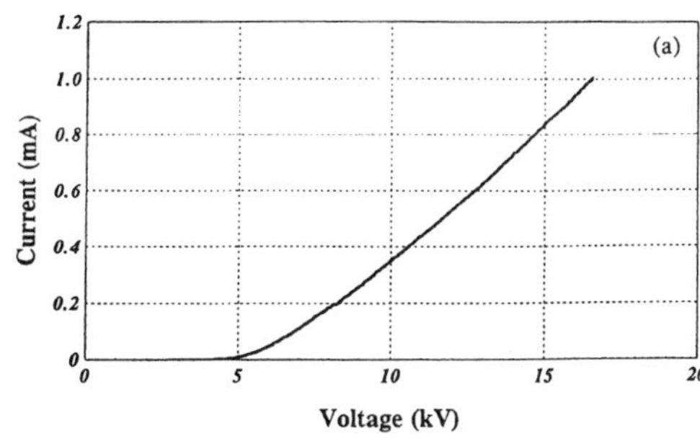

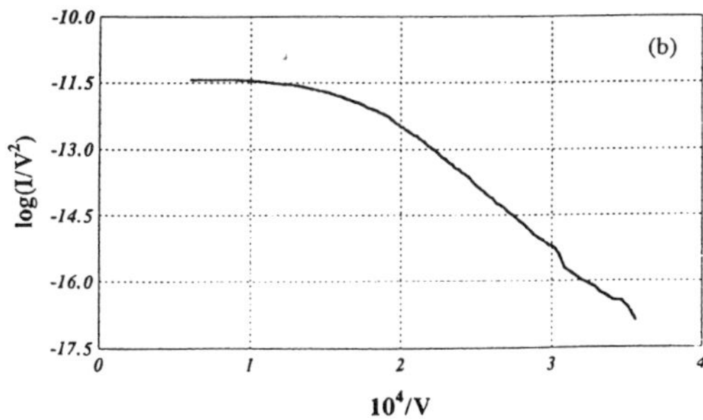

FIGURE 14. *I−V* characteristics showing field emission currents in the range of 0.1 nA to 1 mA. (b) Fowler−Nordheim plot corresponding to the data in (a).[47]

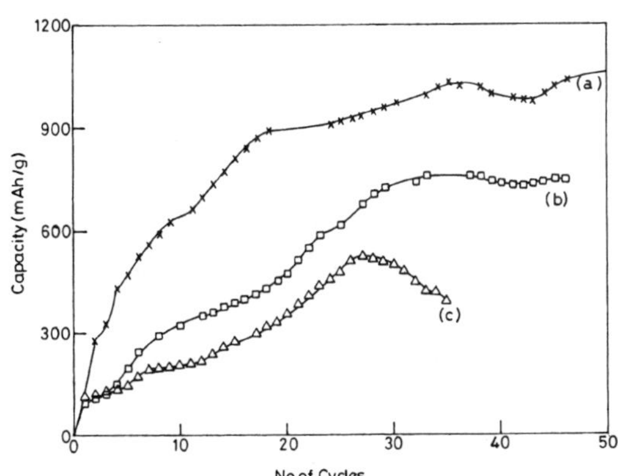

FIGURE 15. Comparison of electrochemical charging capacity of hydrogen of (a) aligned MWNT bundles, (b) SWNTs, and (c) MWNTs (arc-generated).[49]

Accordingly, the field enhancement factor calculated from the slope of the F−N plot in the low-field region is also large. The field emission micrographs reveal the lobe structure symmetries typical of carbon nanotube bundles. The emission current is remarkably stable over an operating period of more than 3 h for various current values in the 10−500 mA range. The relative fluctuations decrease with increasing current level, and the emitter can be operated continuously at the high current levels for at least 3 h without any degradation in the current. It appears that aligned carbon nanotubes from ferrocene pyrolysis hold promise as good field electron emission sources.

Carbon nanotubes are considered to be good hosts for hydrogen storage, although there is some controversy about the magnitude of the hydrogen uptake.[35,48] High-pressure adsorption experiments carried out by us with G. Gundiah show that the storage capacity of compact aligned nanotubes prepared by the pyrolysis of ferrocene−hydrocarbon mixtures is between 3 and 4 wt % (143 bar, 27 °C). Electrochemical hydrogen storage in these nanotubes is also substantial. In Figure 15 a−c we show (for comparison) the plots of electrochemical charging capacity of aligned nanotubes, SWNTs, and MWNTs (arc-generated), respectively.[49] Electrodes made out of aligned MWNTs clearly demonstrate higher electrochemical charging capacities up to 1100 mA h g^{-1} which correspond to a hydrogen storage capacity of 3.75 wt %. SWNTs and MWNTs (arc-generated), however, show capacity in the range of 2−3 wt %.

Concluding Remarks

Although arc evaporation of graphite has traditionally been found to yield both single-walled and multiwalled carbon nanotubes, the pyrolysis of organometallic precursors with or with out the presence of additional carbon sources seems to provide a direct and effective method of producing nanotubes of various kinds. A particularly important finding is the one-step synthesis of aligned carbon nanotubes and Y-junction nanotubes which cannot be made by arc evaporation or other methods easily. It is also noteworthy that nanotubes produced by organometallic precursors may also find applications in field emission and hydrogen storage. The successful synthesis of nanowires of gallium nitride and silicon nitride and of boron nitride nanotubes by using carbon nanotubes produced from organometallic precursors is also of interest.

The authors thank the Department of Science and Technology, Government of India, and the DRDO (India) for supporting this research. They acknowledge productive collaboration with Dr. B. C. Satishkumar, Dr. R. Sen, Mr. F. L. Deepak, and Mr. G. Gundiah.

References

(1) Iijima, S. Helical microtubules of graphitic carbon. *Nature* **1991**, *354*, 56.
(2) Jose-Yacaman, M.; Miki-Yoshida, M.; Rendon, L.; Santiesteban, T. G. Catalytic growth of carbon microtubles with fullerene structure. *Appl. Phys. Lett.* **1993** *62*, 202.
(3) Ivanov, V.; Nagy, J. B.; Lambin, Ph.; Lucas, A.; Zhang, X. B.; Zhang, X. F.; Bernaerts, D.; Van Tendeloo, G.; Amelinckx, S.; Van Landuyt, J. The study of carbon nanotubules produced by catalytic method. *Chem. Phys. Lett.* **1994** *223*, 329.
(4) Hernadi, K.; Fonseca, A.; Nagy, J. B.; Bernaerts, D.; Riga, J.; Lucas, A. Catalytic synthesis and purification of carbon nanotubes. *Synth. Met.* **1996** *77*, 31.
(5) Rodriguez, N. M. A review of catalytically grown carbon nanofibers. *J. Mater. Res.* **1993** *8*, 3233.
(6) Sen, R.; Govindaraj, A.; Rao, C. N. R. Carbon nanotubes by metallocene route. *Chem. Phys. Lett.* **1997**, *267*, 276.
(7) Sen, R.; Govindaraj, A.; Rao, C. N. R. Metal-filled and hollow carbon nanotubes obtained by the decomposition of metal containing free precursor molecules. *Chem. Mater.* **1997**, *9*, 2078.
(8) Yudasaka, M.; Kikuchi, R.; Ohki, Y.; Yoshimura, S. Nitrogen-containing carbon nanotube growth from Ni-phthalocyanine by chemical vapor deposition *Carbon* **1997**, *35*, 195.
(9) Fan, S.; Chapline, M. C.; Franklin, N. R.; Tombler, T. W.; Cassel, A. M.; Dai, H. Self-oriented regular arrays of carbon nanotubes and their field emission devices. *Science* **1999** *283*, 512.

(10) Rao, C. N. R.; Govindaraj, A.; Sen, R.; Satishkumar, B. C. Synthesis of multiwalled and single-walled nanotubes, aligned bundles and nanorods by employing organometallic precursors. *Mater. Res. Innovations* **1998** *2*, 128.

(11) Satishkumar, B. C.; Govindaraj, A.; Sen, R.; Rao, C. N. R. Single-walled nanotubes by the pyrolysis of acetylene-organometallic precursors. *Chem. Phys. Lett.* **1998** *293*, 47.

(12) Seraphin, S.; Zhou, D. Single-walled carbon nanotubes produced at high yield by mixed catalysts. *Appl. Phys. Lett.* **1994** *64*, 2087.

(13) Cheng, H. M.; Li, F.; Su, G.; Pan, H. Y.; He, L. L.; Sun, X.; Dresselhaus, M. S. Large-scale and low-cost synthesis of single-walled carbon nanotubes by the catalytic pyrolysis of hydrocarbons. *Appl. Phys. Lett.* **1998** *72*, 3282.

(14) de Heer, W. A.; Bonard, J. M.; Fauth, K.; Chatelain, A.; Forro, L.; Ugarte, D. Electron field emitters based on carbon nanotube films. *Adv. Mater.* **1997**, *9*, 87.

(15) Satishkumar, B. C.; Govindaraj, A.; Rao, C. N. R. Bundles of aligned carbon nanotubes obtained by the pyrolysis of ferrocene-hydrocarbon mixtures: role of the metal nanoparticles produced in situ. *Chem. Phys. Lett.* **1999** *307*, 158.

(16) Rao, C. N. R.; Sen, R.; Satishkumar, B. C.; Govindaraj, A. Aligned nanotube bundles from ferrocene pyrolysis. *Chem. Commun.* **1998** 1525.

(17) Pan, Z. W.; Xie, S. S.; Chang, B. H.; Sun, L. F.; Zhou, W. Y.; Wang, G. Direct growth of aligned open carbon nanotubes by chemical vapor deposition. *Chem. Phys. Lett.* **1999** *299*, 97.

(18) Andrews, R.; Jacques, D.; Rao, A. M.; Derbyshire, F.; Qian, D.; Fan, X.; Dickey, E. C.; Chen, J. Continuous production of aligned carbon nanotubes: a step closer to commercial realization. *Chem. Phys. Lett.* **1999** *303*, 467.

(19) Huang, S.; Mau, A. W. H.; Turney, T. W.; White, P. A.; Dai, L. Patterned growth of well-aligned carbon nanotubes: A soft-lithographic approach *J. Phys. Chem. B* **2000** *104*, 2193.

(20) Chen, Q.; Dai, L. Three-dimensional micropatterns of well-aligned carbon nanotubes produced by photolithography. *J. Nanosci. Nanotechnol.* **2001** *1*, 43.

(21) Terrones, M.; Grobert, N.; Zhang, J. P.; Terrones, H.; Olivares, J.; Hsu, H. K.; Hare, J. P.; Cheetham, A. K.; Kroto, H. W.; Walton, D. R. M. Preparation of aligned carbon nanotubes catalysed by laser-etched cobalt thin films. *Chem. Phys. Lett.* **1998** *285*, 299.

(22) Chico, L.; Crespi, V. H.; Benedict, L. X.; Louie, S. G.; Cohen, M. L. Pure carbon nanoscale devices: Nanotube heterojunctions. *Phys. Rev. Lett.* **1996** *76*, 971.

(23) Kouwenhoven, L. Single-molecule transistors. *Science* **1997**, *275*, 1896.

(24) McEuen, P. L. Nanotechnology: Carbon-based electronics. *Nature (London)* **1998** *393*, 15.

(25) Menon, M.; Srivastava, D. Carbon nanotube "T-junctions": Nanoscale metal-semiconductor-metal contact devices. *Phys. Rev. Lett.* **1997**, *79*, 4453.

(26) Menon, M.; Srivastava, D. Carbon nanotube based molecular electronic devices. *J. Mater. Res.* **1998** *13*, 2357.

(27) Fuhrer, M. S.; Nygard, J.; Shih, L.; Forero, M.; Yoon, Y. G.; Mazzoni, M. S. C.; Choi, H. J.; Ihm, J.; Louie, S. G.; Zettl, A.; McEuen, P. L. Crossed nanotube junctions. *Science* **2000** *288*, 494.

(28) Li, J.; Papadopoulos, C.; Xu, J. Nanoelectronics: Growing Y-junction carbon nanotubes. *Nature (London)* **1999** *402*, 253.

(29) Satishkumar, B. C.; Thomas, P. J.; Govindaraj, A.; Rao, C. N. R. Y-junction carbon nanotubes. *Appl. Phys. Lett.* **2000** *77*, 2530.

(30) Deepak, F. L.; Govindaraj, A.; Rao, C. N. R. Synthetic strategies for Y-junction carbon nanotubes. *Chem. Phys. Lett.* **2001** *345*, 5.

(31) Tans, S. J.; Verschueren, A. R. M.; Dekker, C. Room-temperature transistor based on a single carbon nanotube *Nature* **1998** *393*, 49.

(32) Martel, R.; Schmidt, T.; Shea, H. R.; Hertel, T.; Avouris, Ph. Single- and multiwall carbon nanotube field-effect transistors. *Appl. Phys. Lett.* **1998** *73*, 2447.

(33) Yao, Z.; Postma, H. W. Ch.; Balents, L.; Dekker, C. Carbon nanotube intramolecular junctions. *Nature* **1999** *402*, 273.

(34) Rueckes, T.; Kim, K.; Joseluich, E.; Tsang, G. Y.; Cheung, C. L.; Leiber, C. M. Carbon nanotube-based nonvolatile random access memory for molecular computing. *Science* **2000** *289*, 94.

(35) Rao, C. N. R.; Satishkumar, B. C.; Govindaraj, A.; Nath, M. Nanotubes. *Chem. Phys. Chem.* **2001** *2*, 78.

(36) Morales, A. M.; Leiber, C. M. A laser ablation method for the synthesis of crystalline semiconductor nanowires. *Science* **1998** *279*, 208.

(37) Zhang, Y.; Suenaga, K.; Colliex, C.; Iijima, S. Coaxial nanocable: Silicon carbide and silicon oxide sheathed with boron nitride and carbon. *Science* **1998** *281*, 973.

(38) Han, W.; Fan, S.; Li, Q.; Gu, B.; Zhang, X.; Yu, D. Synthesis of silicon nitride nanorods using carbon nanotube as a template. *Appl. Phys. Lett.* **1997**, *71*, 2271.

(39) Wang, M. J.; Wada, H. Synthesis and characterization of silicon nitride whiskers. *J. Mater. Sci.* **1990** *25*, 1690. Wu, X. C.; Song, W. H.; Zhao, B.; Huang, W. D.; Pu, M. H.; Sun, Y. P.; Du, J. J. Synthesis of coaxial nanowires of silicon nitride sheathed with silicon and silicon oxide. *Solid State Commun.* **2000** *115*, 683.

(40) Gundiah, G.; Madhav, G. V.; Govindaraj, A.; Rao, C. N. R. Synthesis and characterization of silicon carbide, silicon oxynitride and silicon nitride nanowires. *J. Mater. Chem.* **2002** *12*, 1606–1611.

(41) Deepak, F. L.; Mukhopadhyay, K.; Vinod, C. P.; Govindaraj, A.; Rao, C. N. R. Boron nitride nanotubes and nanowires. *Chem. Phys. Lett.* **2002** *353*, 345.

(42) Deepak, F. L.; Govindaraj, A.; Rao, C. N. R. Single-crystal GaN nanowires. *J. Nanosci. Nanotechnol.* **2001** *1*, 303.

(43) de Heer, W. A.; Chatelain, A.; Ugarte, D. A carbon nanotube field-emission electron source. *Science* **1995** *270*, 1179.

(44) Wang, Q. H.; Corrigan, T. D.; Dai, J. Y.; Chang, R. P. H.; Krauss, A. R. Field emission from nanotube bundle emitters at low fields. *Appl. Phys. Lett.* **1997**, *70*, 3308.

(45) Bonard, J.-M.; Maier, F.; Stockli, T.; Chatelain, A.; de Heer, W. A.; Salvetat, J.-P.; Ferro, L. Field emission properties of multiwalled carbon nanotubes. *Ultramicroscopy* **1998** *73*, 7.

(46) Saito, Y.; Hamaguchi, K.; Hata, K.; Tohji, K.; Kasuya, A.; Nishina, Y.; Uchida, K.; Tasaka, Y.; Ikazaki, F.; Yumura, M. Field emission from carbon nanotubes; Purified single-walled and multiwalled tubes. *Ultramicroscopy* **1998** *73*, 1.

(47) Sharma, R. B.; Tondare, V. N.; Joag, D. S.; Govindaraj, A.; Rao, C. N. R. Field emission from carbon nanotubes grown on a tungsten tip. *Chem. Phys. Lett.* **2001** *344*, 283.

(48) Dresselhaus, M. S.; Williams, K. A.; Eklund, P. C. Hydrogen adsorption in carbon materials. *MRS Bull.* **1999** *24*, 45.

(49) Rajalakshmi, N.; Dhathathreyan, K. S.; Govindaraj, A.; Rao, C. N. R. To be submitted for publication.

AR0101584

Inorganic nanotubes †

C. N. R. Rao *[a,b] and Manashi Nath [a,b]

[a] *CSIR Centre for Excellence in Chemistry and Chemistry and Physics of Materials Unit, Jawaharlal Nehru Centre for Advanced Scientific Research, Jakkur P.O., Bangalore, 560 064, India. E-mail: cnrrao@jncasr.ac.in*
[b] *Solid State and Structural Chemistry Unit, Indian Institute of Science, Bangalore, 560 012, India*

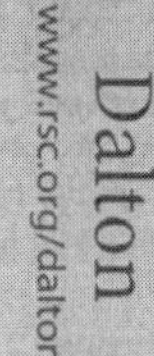

Received 13th September 2002, Accepted 14th November 2002
First published as an Advance Article on the web 2nd December 2002

Carbon nanotubes were discovered in 1991. It was soon recognized that layered metal dichalcogenides such as MoS_2 could also form fullerene and nanotube type structures, and the first synthesis was reported in 1992. Since then, a large number of layered chalcogenides and other materials have been shown to form nanotubes and their structures investigated by electron microscopy. Inorganic nanotubes constitute an important family of nanostructures with interesting properties and potential applications. In this article, we discuss the progress made in this novel class of inorganic nanomaterials.

1. Introduction

In 1991, Iijima[1] observed some unusual structures of carbon under the transmission electron microscope wherein the graphene sheets had rolled and folded onto themselves to form hollow structures. Iijima called them *nanotubes* of carbon which consisted of several concentric cylinders of graphene sheets. Graphene sheets are hexagonal networks of carbon and these layers get stacked one above the other in the *c*-direction to form bulk graphite. Following the initial discovery, intense research has been carried out on carbon nanotubes (CNTs).[2] The nanotubes can be open-ended or closed by caps containing five-membered rings. They can be multi- (MWNTs) or single-walled (SWNTs). We show a typical high-resolution electron

---

† The illustration of John Dalton (reproduced courtesy of the Library and Information Centre, Royal Society of Chemistry) marks the 200th anniversary of his investigations which led to the determination of atomic weights for hydrogen, nitrogen, carbon, oxygen, phosphorus and sulfur.

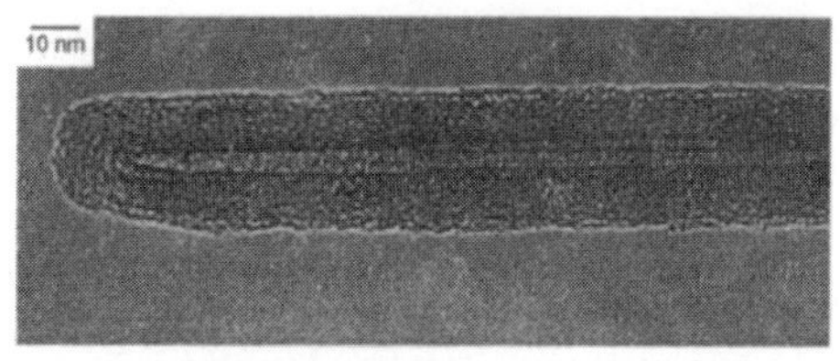

Fig. 1 A typical TEM image of a closed, multi-walled carbon nanotube. The separation between the graphite layers is 0.34 nm.

microscope (HREM) image of a multi-walled nanotube in Fig. 1. Depending on the way the graphene sheets fold, nanotubes are classified as armchair, zigzag or chiral as shown in Fig. 2. The electrical conductivity of the nanotubes depends on the nature of folding.

Several layered inorganic compounds possess structures comparable to the structure of graphite, the metal dichalcogenides being important examples. The metal dichalcogenides, MX_2 (M = Mo, W, Nb, Hf; X = S, Se) contain a metal layer sandwiched between two chalcogen layers with the metal in a trigonal pyramidal or octahedral coordination mode.[3] The MX_2 layers are stacked along the *c*-direction in ABAB fashion. The MX_2 layers are analogous to the single graphene sheets in the graphite structure (Fig. 3). When viewed parallel to the *c*-axis, the layers show the presence of dangling bonds due to the absence of an X or M atom at the edges. Such unsaturated bonds at the edges of the layers also occur in graphite. The dichalcogenide layers are unstable towards bending and have a high propensity to roll into curved structures. Folding in the

---

C. N. R. Rao was born in Bangalore, India in 1934. He received his MSc degree from the Banaras Hindu University, PhD from Purdue University and DSc from the University of Mysore. Following postdoctoral work at the University of California, Berkeley, he joined the faculty at the Indian Institute of Science, Bangalore. He later moved to the Indian Institute of Technology, Kanpur where he was Professor and Head of the Department of Chemistry. He returned to the Indian Institute of Science, Bangalore in 1976 to start a new

unit devoted to Solid State and Structural Chemistry. He also founded the Jawaharlal Nehru Centre for Advanced Scientific Research, Bangalore in 1989. His main research interests are in solid state and materials chemistry. He is a Fellow of the Royal Society, London, and is a member of a several other academies. He is an honorary fellow of the Royal Society of Chemistry and is the recipient of the Hughes Medal of the Royal Society.

C. N. R. Rao

Manashi Nath was born in Calcutta in 1976 and received her BSc degree from the Presidency College, Calcutta in 1997. She is a student of the integrated PhD programme of the Indian Institute of Science, Bangalore and received her MSc degree in 2000. She has worked mainly on inorganic nanotubes.

Manashi Nath

DOI: 10.1039/b208990b

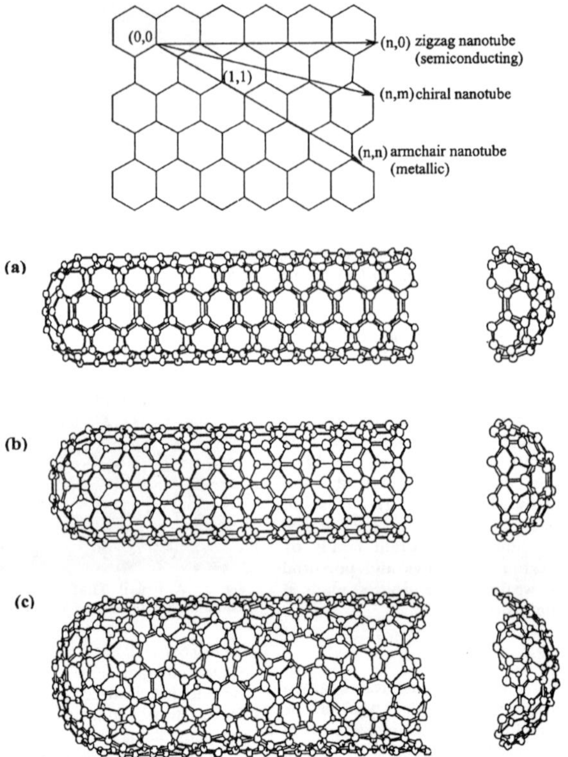

Fig. 2 Schematic representation of the folding of a graphene sheet into (a) zigzag, (b) armchair and (c) chiral nanotubes.

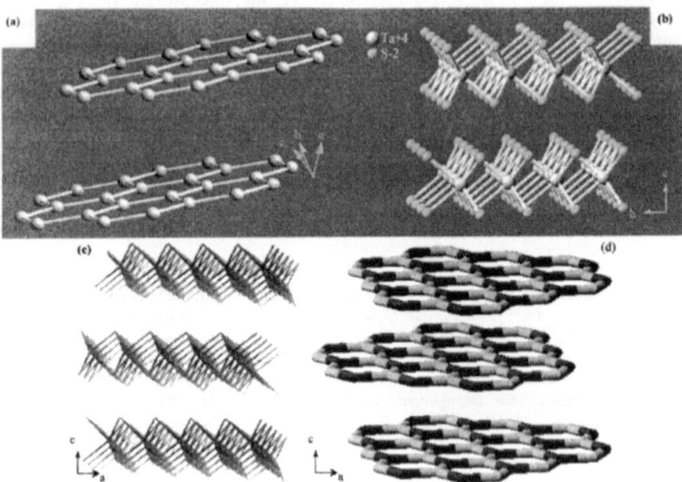

Fig. 3 Comparison of the structures of (a) graphite and inorganic layered compounds such as (b) NbS_2/TaS_2; (c) MoS_2; (d) BN. In the layered dichalcogenides, the metal is in trigonal prismatic (TaS_2) or octahedral coordination (MoS_2).

layered transition metal chalcogenides (LTMCs) was recognized as early as 1979, well before the discovery of the carbon nanotubes. Rag-like and tubular structures of MoS_2 were reported by Chianelli[4] who studied their usefulness in catalysis.

The folded sheets appear as crystalline needles in low magnification transmission electron microscope (TEM) images, and were described as layers that fold onto themselves (Fig. 4). These structures indeed represent those of nanotubes. Tenne *et al.*[5] first demonstrated that Mo and W dichalcogenides are capable of forming nanotubes (Fig. 5a). Closed fullerene-type structures (*inorganic fullerenes*) also formed along with the nanotubes (Fig. 5b). The dichalcogenide structures contain concentrically nested fullerene cylinders, with a less regular structure than in the carbon nanotubes. Accordingly, MX_2 nanotubes have

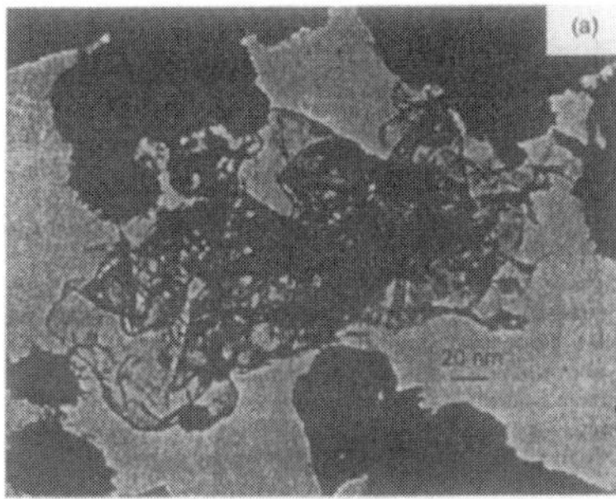

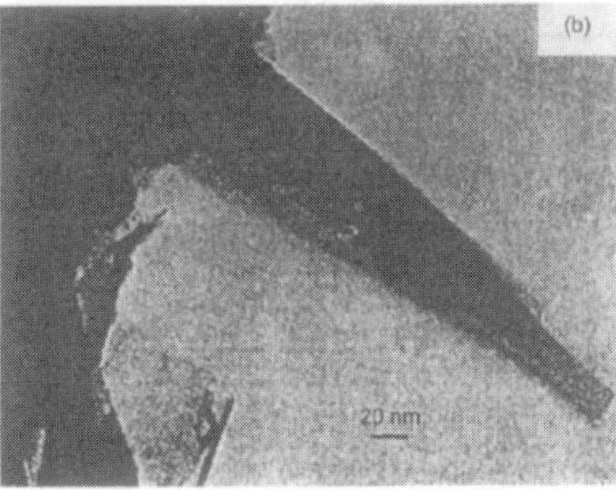

Fig. 4 Low-magnification TEM images of (a) highly folded MoS_2 needles and (b) a rolled sheet of MoS_2 folded back on itself. (Reproduced with permission from ref. 4).

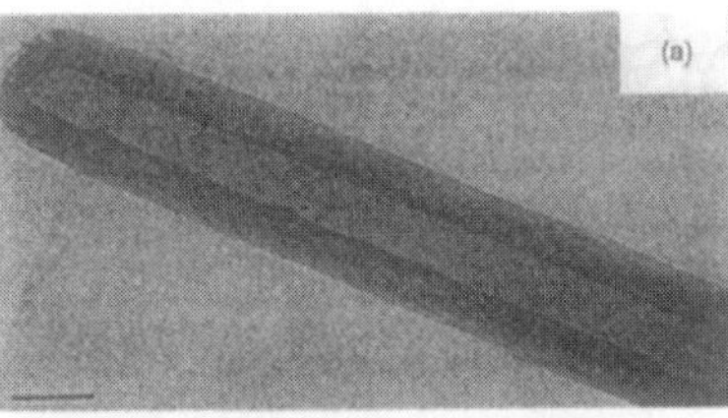

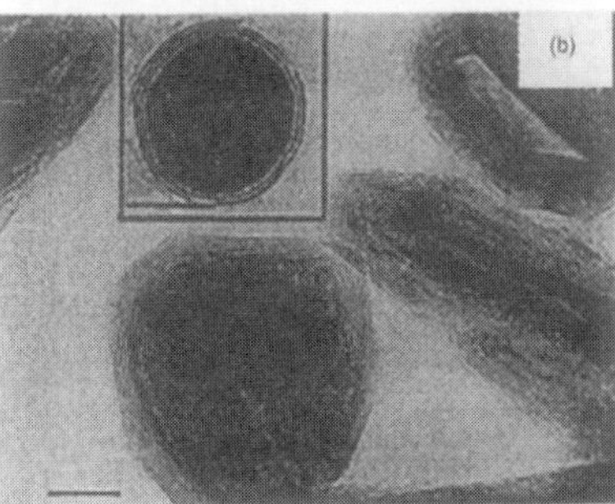

Fig. 5 TEM images of (a) a multi-walled nanotube of WS_2 and (b) hollow particles (inorganic fullerenes) of WS_2. (Reproduced with permission from refs. 5a and b, respectively).

varying wall thickness and contain some amorphous material on the exterior of the tubes. Nearly defect-free MX_2 nanotubes are rigid as a consequence of their structure and do not permit plastic deformation. The folding of a MS_2 layer in the process of forming a nanotube is shown in the schematic in Fig. 6. Considerable progress has been made in the synthesis of the nanotubes of Mo and W dichalcogenides in the last few years (Table 1). There has been some speculation on the cause of folding and curvature in the LTMCs. Stoichiometric LTMC chains and layers such as those of TiS_2 possess an inherent ability to bend and fold, as observed in intercalation reactions.

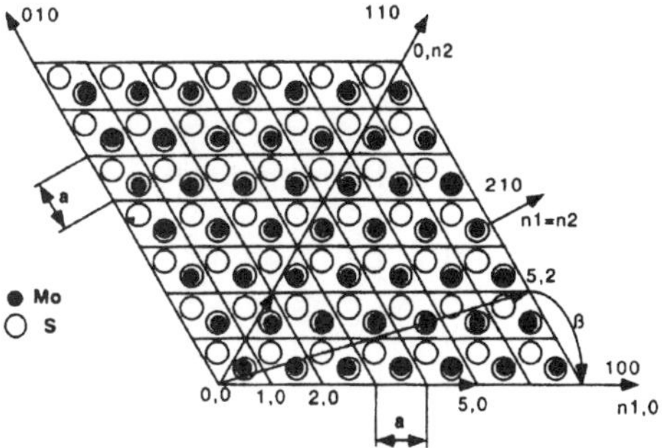

Fig. 6 Schematic illustration of the bending of a MoS_2 layer. (Reproduced with permission from ref. 19).

The existence of alternate coordination and therefore of stoichiometry in the LTMCs may also cause folding. Lastly, a change in the stoichiometry within the material would give rise to closed rings.

Transition metal chalcogenides possess a wide range of interesting physical properties. They are widely used in catalysis and as lubricants. They have both semiconducting and superconducting properties. With the synthesis and characterization of the fullerenes and nanotubes of MoS_2 and WS_2, a wide field of research has opened up enabling the successful synthesis of nanotubes of other metal chalcogenides. It may be recalled that the dichalcogenides of many of the Group 4 and 5 metals have layered structures suitable for forming nanotubes.

Curved structures are not only limited to carbon and the dichalcogenides of Mo and W. Perhaps the most well-known example of a tube-like structure with diameters in the nm range is formed by the asbestos mineral (*chrysotil*) whose fibrous characteristics are determined by the tubular structure of the fused tetrahedral and octahedral layers. The synthesis of mesoporous silica with well-defined pores in the 2–20 nm range was reported by Beck and Kresge.[6] The synthetic strategy involved the self-assembly of liquid crystalline templates. The pore size in zeolitic and other inorganic porous solids is varied by a suitable choice of the template. However, in contrast to the synthesis of porous compounds, the synthesis of nanotubes is somewhat more difficult.

Nanotubes of oxides of several transition metals as well as of other metals have been synthesized employing different methodologies.[7–12] Silica nanotubes were first produced as a spin-off product during the synthesis of spherical silica particles by the hydrolysis of tetraethylorthosilicate (TEOS) in a mixture of water, ammonia, ethanol and D,L-tartaric acid.[8] Since self-assembly reactions are not straightforward with respect to the desired product, particularly its morphology, templated reactions have been employed using carbon nanotubes to obtain nanotube structures of metal oxides.[9,10] Oxides such as V_2O_5 have good catalytic activity in the bulk phase. Redox catalytic activity is also retained in the nanotubular structure. There have been efforts to prepare V_2O_5 nanotubes by chemical methods as well.[11]

Boron nitride (BN) crystallizes in a graphite-like structure and can be simply viewed as replacing a C–C pair in the graphene sheet with the iso-electronic B–N pair. It can, therefore, be considered as an ideal precursor for the formation of BN nanotubes. Replacement of the C–C pairs partly or entirely by the B–N pairs in the hexagonal network of graphite leads to the formation of a wide array of two-dimensional phases that can form hollow-cage structures and nanotubes. The possibility of replacing C–C pairs by B–N pairs in the hollow cage structure of C_{60} was predicted[13] and verified experimentally.[14] BN-doped carbon nanotubes have been prepared.[15] Pure BN nanotubes have been generated by employing several

Table 1 Synthetic strategies for various chalcogenide nanotubes

Chalcogenide	Synthetic strategy	Ref.	Chalcogenide	Synthetic strategy	Ref.
WS_2	(i) Heating MoO_3 in the presence of forming gas followed by heating in H_2S	5a	$Mo_xW_yC_zS_2$	Pyrolysis of H_2S over carbon-containing W and Mo-oxide complexes	43
	(ii) WS_3 and $(NH_4)_2WS_4$ decomposition in H_2	21	$Nb–W–S$	Heating Nb_2O_5 coated $W_{18}O_{49}$ nanorods in H_2S at 1100 °C	41
	(iii) Pyrolysis of H_2S/N_2 over WO_3-coated MWNTs	45a,b			
	(iv) Laser ablation of WS_2 target	64			
MoS_2	(i) Heating MoO_3 in the presence of forming gas followed by heating in H_2S	5b,c	$Mo–Ti–S$	Pyrolysis of H_2S/N_2 mixture over Ti–Mo alloy at elevated temperatures	42
	(ii) Decomposition of $(NH_4)_2MoS_4$ in the pores of anodic alumina	31	NbS_2	(i) NbS_3 decomposition	23
	(iii) Decomposition of $(NH_4)_2MoS_4$ MoS_3 in H_2	21		(ii) CNT templated reaction	46
	(iv) Hydrothermal treatment of ammonium thiomolybdate with ethylenediamine	62	TaS_2	TaS_3 decomposition	23
	(v) MoS_2 powder covered with Mo foil, heated to 1300 °C in H_2S	61			
	(vi) Laser ablation of MoS_2 target	64	HfS_2	Decomposition of HfS_3	24
			ZrS_2	Decomposition of ZrS_3	24
MoS_2I_x	MoS_2 with I_2 and C_{60} as carrier	63	$NbSe_2$	(i) $NbSe_3$ decomposition	70
				(ii) Electron irradiation of $NbSe_2$	69
$MoSe_2$	(i) MoO_3 + H_2Se	20a			
	(ii) $MoSe_3$ and $(NH_4)_2MoSe_4$ decomposition in H_2	22	CdS, CdSe	Surfactant-assisted synthesis	28,75
			ZnS	Sulfidization of ZnO columns by H_2S at 400 followed by etching the core	77
WSe_2	(i) WO_3 + Se vapors at 650–850 °C	20a	NiS	Treatment of $Ni(NH_3)_4^{2+}$ complex with CS_2 in aqueous ammonia	78
	(ii) Electron irradiation of WSe_2	20b			
	(iii) WSe_3 and $(NH_4)_2WSe_4$ decomposition in H_2	22	$Cu_{5.5}FeS_{6.5}$	Hydrothermal reaction between Cu and S in presence of $LiOH.H_2O$ and trace amount of Fe	79

procedures, yielding nanotubes with varying wall thickness and morphology.[16,17] It is therefore quite possible that nanotube structures of other layered materials can be prepared as well. For example, many metal halides (*e.g.*, NiCl$_2$), oxides (GeO$_2$) and nitrides (GaN) crystallize in layered structures. There is considerable interest at present to prepare exotic nanotubes and to study their properties.

In this article, we discuss the synthesis and characterization of nanotubes of chalcogenides of Mo, W and other metals, metal oxides, BN and other materials and present the current status of the subject. We briefly examine some of the important properties of the inorganic nanotubes and indicate · possible future directions.

2. General synthetic strategies

Several strategies have been employed for the synthesis of carbon nanotubes.[2] They are generally made by the arc evaporation of graphite or by the pyrolysis of hydrocarbons such as acetylene or benzene over metal nanoparticles in a reducing atmosphere. Pyrolysis of organometallic precursors provides a one-step synthetic method of making carbon nanotubes.[18] In addition to the above methods, carbon nanotubes have been prepared by laser ablation of graphite or electron-beam evaporation. Electrochemical synthesis of nanotubes as well as growth inside the pores of alumina membranes have also been reported. The above methods broadly fall under two categories. Methods such as the arc evaporation of graphite employ processes which are far from equilibrium. The chemical routes are generally closer to equilibrium conditions. Nanotubes of metal chalcogenides and boron nitride are also prepared by employing techniques similar to those of carbon nanotubes, although there is an inherent difference in that the nanotubes of inorganic materials such as MoS$_2$ or BN would require reactions involving the component elements or compounds containing the elements. Decomposition of precursor compounds containing the elements is another possible route.

Nanotubes of dichalcogenides such as MoS$_2$, MoSe$_2$ and WS$_2$ are also obtained by employing processes far from equilibrium such as arc discharge and laser ablation.[19] By far the most successful routes employ appropriate chemical reactions. Thus, MoS$_2$ and WS$_2$ nanotubes are conveniently prepared starting with the stable oxides, MoO$_3$ and WO$_3$.[5] The oxides are first heated at high temperatures in a reducing atmosphere and then reacted with H$_2$S. Reaction with H$_2$Se is used to obtain the selenides.[20] Recognizing that the trisulfides MoS$_3$ and WS$_3$ are likely to be the intermediates in the formation of the disulfide nanotubes, the trisulfides have been directly decomposed to obtain the disulfide nanotubes.[21] Diselenide nanotubes have been obtained from the metal triselenides.[22] The trisulfide route is indeed found to provide a general route for the synthesis of the nanotubes of many metal disulfides such as NbS$_2$[23] and HfS$_2$.[24] In the case of Mo and W dichalcogenides, it is possible to use the decomposition of the precursor ammonium salt, such as (NH$_4$)$_2$MX$_4$ (X = S, Se; M = Mo, W) as a means of preparing the nanotubes.[21] Other methods employed for the synthesis of dichalcogenide nanotubes include hydrothermal methods where the organic amine is taken as one of the components in the reaction mixture (Table 1).

The hydrothermal route has been used for synthesizing nanotubes and related structures of a variety of other inorganic materials as well. Thus, nanotubes of several metal oxides (*e.g.*, SiO$_2$,[25] V$_2$O$_5$,[11] ZnO[26]) have been produced hydrothermally. Nanotubes of oxides such as V$_2$O$_5$ are also conveniently prepared from a suitable metal oxide precursor in the presence of an organic amine or a surfactant.[27] Surfactant-assisted synthesis of CdSe and CdS nanotubes has been reported. Here the metal oxide reacts with the sulfidizing/selenidizing agent in the presence of a surfactant such as TritonX.[28]

Sol–gel chemistry is widely used in the synthesis of metal oxide nanotubes, a good example being that of silica[8] and TiO$_2$.[29] Oxide gels in the presence of surfactants or suitable templates form nanotubes. For example, by coating carbon nanotubes (CNTs) with oxide gels and then burning off the carbon, one obtains nanotubes and nanowires of a variety of metal oxides including ZrO$_2$, SiO$_2$ and MoO$_3$.[10,30] Sol–gel synthesis of oxide nanotubes is also possible in the pores of alumina membranes. It should be noted that MoS$_2$ nanotubes are also prepared by the decomposition of a precursor in the pores of an alumina membrane.[31]

Boron nitride nanotubes have been obtained by striking an electric arc between HfB$_2$ electrodes in a N$_2$ atmosphere.[32] BCN and BC nanotubes are obtained by arcing between B/C electrodes in an appropriate atmosphere. A greater effort has gone into the synthesis of BN nanotubes starting with different precursor molecules containing B and N. Decomposition of borazine in the presence of transition metal nanoparticles and the decomposition of the 1 : 2 melamine–boric acid addition compound yield BN nanotubes.[16] Reaction of boric acid or B$_2$O$_3$ with N$_2$ or NH$_3$ at high temperature in the presence of activated carbon, carbon nanotubes or catalytic metal particles has been employed to synthesize BN nanotubes.[17]

3. Nanotubes of Mo and W dichalcogenides

Nanotubes of the disulfides and diselenides of Mo and W have been prepared by employing several strategies. The first synthesis of the MoS$_2$ and WS$_2$ nanotubes was carried out by Tenne *et al.*[5] by treating the metal oxides in an atmosphere of forming gas (95% N$_2$ + 5% H$_2$), followed by heating in a stream of H$_2$S at elevated temperatures. Initially, the oxides are reduced to the suboxides which are then converted to the sulfides. Onion-like structures analogous to nested fullerenes were also obtained in considerable yields (Fig. 5). The heating arrangement with a tubular furnace employed for the preparation of MoS$_2$ nanotubes is shown in Fig. 7. MoO$_3$ being sublimable, the

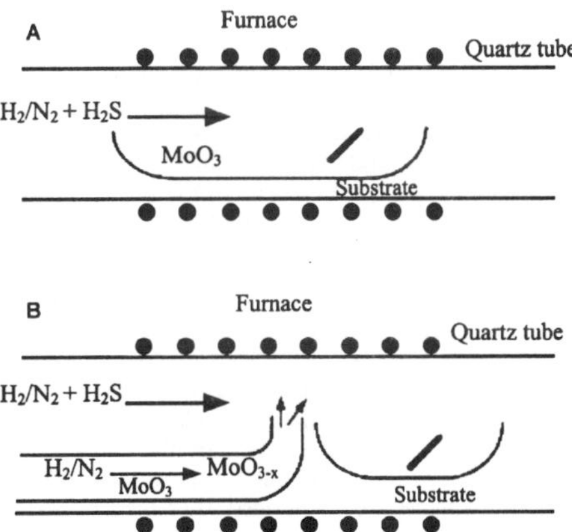

Fig. 7 (A) Schematic of the gas-phase reactor used to produce MoS$_2$ nanotubes; (B) a modification of the above where molybdenum suboxide flow was introduced through a separate tube and the collecting substrate was placed behind a high wall. (Reproduced with permission from ref. 5c).

growth of MoS$_2$ occurred from the vapor phase. On the other hand WO$_{3-x}$, not being volatile, the growth of WS$_2$ took place at the vapor–solid interface. Following the initial synthesis, various modifications have been adopted. Instead of starting with oxide particles, preformed morphologies like needles and whiskers containing the desired shapes of the nanotube products were also used as precursors. Thus, thermal treatment of the oxide needles in H$_2$S gave WS$_2$ nanotubes.[19] In these processes, the metal trisulfide is first formed on heating the oxide in an excess of H_2S. The trisulfide loses sulfur on annealing and crystallizes in the form of a disulfide nanoparticle or nanotube. MoS_2 nanotubes have also been prepared from the MoS_3 by sending a pulse through a STM tip.[33] The tip induced the crystallization of amorphous MoS_3 films deposited on a gold substrate producing inorganic fullerenes, some of them containing MoS_3 in the core, while the rest are hollow nanoparticles (Fig. 8).

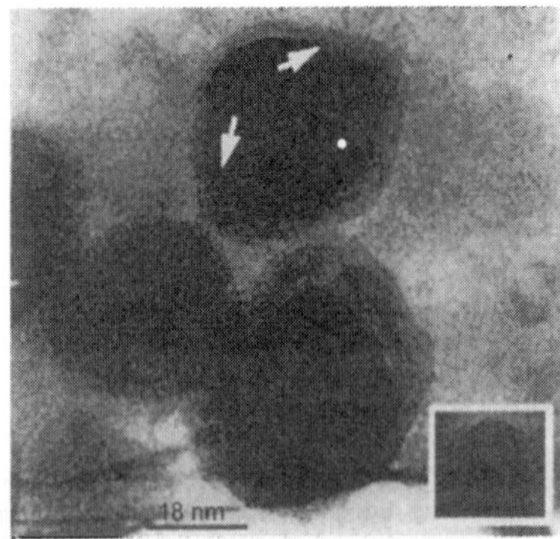

Fig. 8 Crystalline IF–MoS_2 nanoparticles with a MoS_3 core. Inset shows a fullerene-like nanoparticle of ~15 nm diameter. (Reproduced with permission from ref. 33).

On heating the nanorods or needles of the precursor oxide particles in an atmosphere of flowing H_2S or H_2Se, the outermost oxide layers are first converted to the chalcogenide layers. Further agglomeration of the oxidic particles is inhibited by the inert layer coating. Fast diffusion of H_2 into the precursor nanoparticle leads to the reduction of the core to the suboxide form. The second stage of the reaction involves the slow diffusion of the sulfur vapor into the core and the conversion of the oxide core to the sulfide. This mechanism has been substantiated by the presence of an incompletely sulfidized oxide core or even the metal in some of the WS_2 nanotubes.[34,35] The growth mechanism of the disulfide nanotubes from the suboxide particles comprises several steps. In the first step, the oxide particles react with H_2S to form one or two layers of the sulfide. This prevents further aggregation of the precursor particles and the formation of larger particles. Diffusion of H_2 into the particle and the out-diffusion of O_2 leads to the reduction of the oxide particles and the creation of crystallographic shear planes. In the next step of the reaction, sulfur vapor diffuses slowly into the precursor particles converting the suboxide core to the sulfide, which becomes hollow at the end of the process. The growth front is near the core of the precursor particle.

The knowledge that MoS_3 and WS_3 are the stable phases at low temperatures in the excess sulfur regime,[36] and that these amorphous trisulfides are first formed when the metal trioxide reacts with H_2S,[37] underscores the role of the trisulfide as a likely intermediate in the formation of the disulfide. A study of the binary metal–sulfur (Mo(W)–S) phase diagram reveals that the inorganic fullerene phase is obtained at the phase boundary between the amorphous MS_3 and the crystalline MS_2 phases. The loss of sulfur from the MS_3 phase triggers the nucleation of the MS_2 nanoparticles. By starting with a trisulfide, it should therefore be possible to obtain the disulfide nanostructure by thermal decomposition or by reacting with hydrogen.[38,39]

$$MS_3 \rightarrow MS_2 + S \quad (M = W \text{ or } Mo)$$
$$MS_3 + H_2 \rightarrow MS_2 + H_2S$$

Thermal decomposition of amorphous Mo and W trisulfides has been investigated under a steady gas flow, to explore the

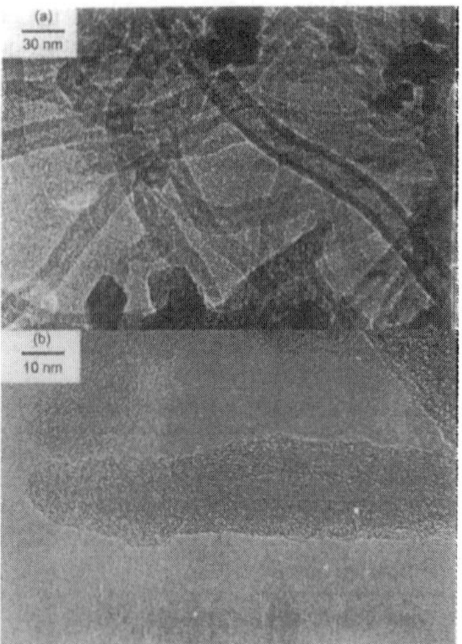

Fig. 9 TEM images of MoS_2 nanotubes grown by the decomposition of MoS_3. (Reproduced with permission from ref. 21).

formation of the disulfide nanotubes under a steady flow of hydrogen (100 sccm).[21] The decomposition products consisted of a high proportion of disulfide nanotubes (Fig. 9).

Mo and W trisulfides themselves are prepared by thermal decomposition of the ammonium thiometallates, $(NH_4)_2MS_4$ (M = Mo, W).[38] Accordingly, direct decomposition of the ammonium thiometallates in H_2 yields the MoS_2 and WS_2 nanotubes (Fig. 10).[21] In addition to providing a direct method

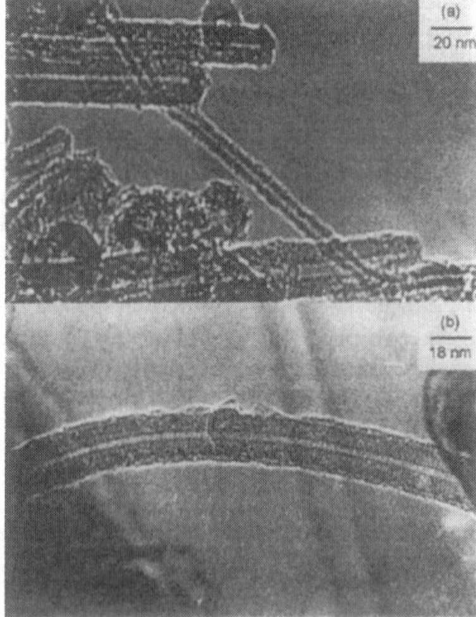

Fig. 10 (a) Low-resolution TEM images of MoS_2 nanotubes grown by the decomposition of ammonium thiomolybdate; (b) HREM image of the MoS_2 nanotube. (Reproduced with permission from ref. 21).

for the preparation of dichalcogenide nanotubes,[21] the trichalcogenide or the ammonium chalcometallate route enables the easy synthesis of nanotubes of the other layered dichalcogenides as well.[22–24]

$$(NH_4)_2MX_4 + H_2S \rightarrow MX_2 + H_2X + NH_3$$
$$(X = S, Se; M = Mo, W)$$
$$MX_3 + H_2 \rightarrow MX_2 + H_2X$$
$$MX_3 \rightarrow MX_2 + X$$

Thus, nanotubes of Mo and W diselenides have been prepared by the decomposition of the triselenide or the ammonium selenometallate at elevated temperatures under a flow of H_2.[22] Apart from the nanotubes, nanorods of WSe_2 were also obtained. Some of the nanorods were attached to an amorphous particle at the tip (Fig. 11a). Several single-walled WSe_2

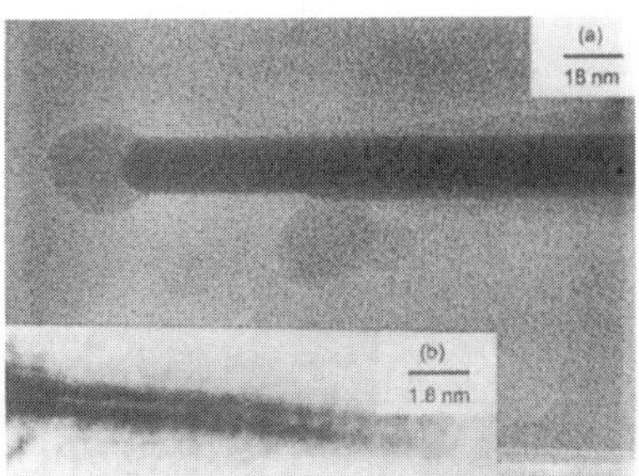

Fig. 11 (a) TEM image of a WSe_2 nanorod with a particle attached at the tip, grown by the decomposition of WSe_3; (b) single-walled WSe_2 nanotube. (Reproduced with permission from ref. 22).

nanotubes were also observed (Fig. 11b). This strategy has been employed to obtain W doped MoS_2 nanotubes.[40] Thus, solid solutions of the ammonium thiometallates, $(NH_4)_2Mo_{1-x}W_xS_4$ with varying ratios of Mo : W were used as precursors, to yield $Mo_{1-x}W_xS_2$ nanotubes (x = 0.15–0.5) on thermal decomposition under a gas flow (H_2, He or Ar) at ~770 °C (Fig. 12).

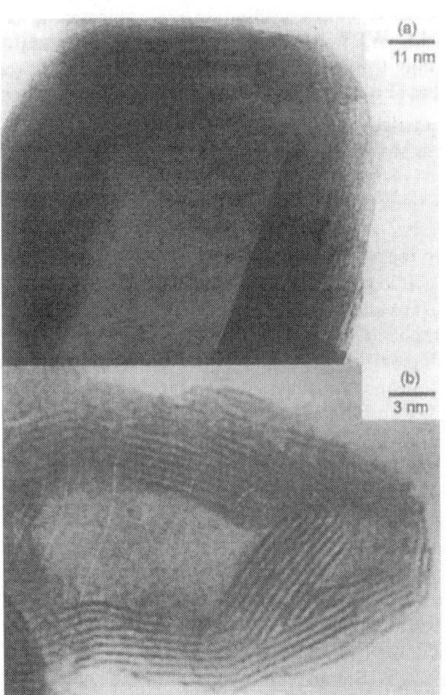

Fig. 12 HREM image of $Mo_{1-x}W_xS_2$ nanotubes showing a rectangular tip in (a) and (b) bamboo-like stacking. The layer separation in the walls is ~0.62 nm. (Reproduced with permission from ref. 40).

SWNTs of the doped sulfide structures were also observed. In general, the yield of nanotubes decreased with the increase in W content in the host MoS_2 lattice. Increasing the W content in MoS_2 also increases the layer mismatch in the tube walls and defects.

Other metals have been doped in the host disulfide layers of the nanotubes by using methodologies similar to those employed for the parent disulfides. Thus, composite nanotubes

of $W–Nb–S$,[41] $Ti–Mo–S$[42] and $Mo–W–C–S$[43] have been prepared where the precursor was a mixed oxide of the two metals. In the case of Ti–Mo–S, the Ti–Mo alloy heated in oxygen to yield the mixed oxide, which was further heated in forming gas followed by H_2S to produce Ti doped MoS_2 nanotubes (Fig. 13).

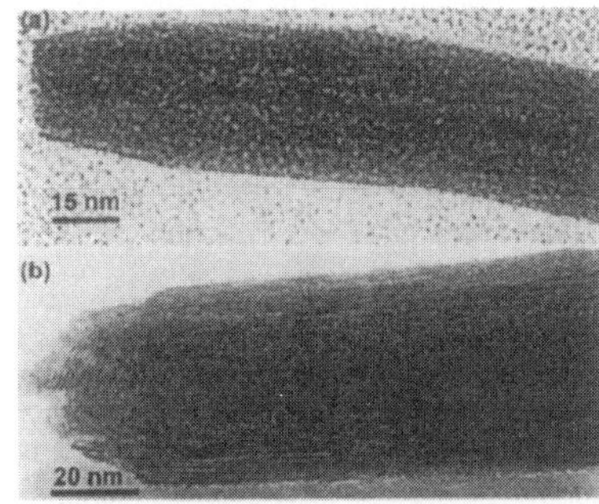

Fig. 13 TEM images of hollow Ti–Mo–S nanotubes. (Reproduced with permission from ref. 42).

The W–Nb mixed metal oxide was prepared by sonicating the solution containing the precursors. Treatment of the mixed metal oxide precursors in H_2S produced the Nb doped WS_2 nanotubes (Fig. 14). The mechanism of formation of the com-

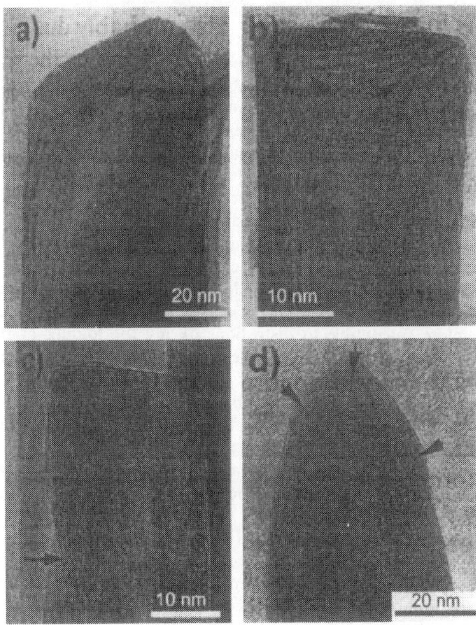

Fig. 14 HREM images of Nb–W–S nanotubes showing various tube closures: (a) an irregular tube closure; (b) a 90° wall–tip junction; (c) another near 90° closure (the arrow points to the buckling defect in the tube wall); (d) an irregular closure with severe bending defects. (Reproduced with permission from ref. 41).

posite nanotubes is similar to that of the binary nanotubes of MoS_2 and WS_2 and involves a layer by layer conversion of the mixed metal oxide to the sulfide. Some of the layers in the nanotube walls terminate abruptly, probably because of the exhaustion of the growing materials at that edge. The metal particles present at the terminated edge also inhibit the growth,[44] as in the case of metal oxide promoted growth of the BN nanotubes.

The methods of preparation discussed above do not involve any template, and the nanoparticles of the oxide or the tri-sulfide act as nucleation centers for tube growth. Recently, CNTs have been used as templates to grow MoS_2, WS_2 and NbS_2 coated carbon nanotubes, some of which contain 1–2 layers of the chalcogenide at the exterior.[45,46] The CNTs were coated with the metal oxide or its precursor and treated in a $H_2S/H_2/N_2$ atmosphere at elevated temperatures to convert the oxide to the sulfide. However, the CNT core was not removed in the nanostructures (Fig. 15).

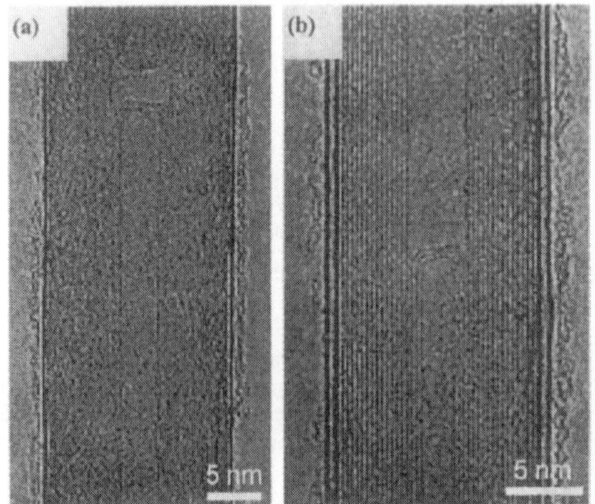

Fig. 15 HREM image of a CNT coated with (a) a single layer and (b) a double layer of WS_2. (Reproduced with permission from ref. 45*a*).

Confined reactions have been used to control the size of MoS_2 nanotubes. For example, MoS_2 nanotubes have been grown in the voids of an anodic alumina membrane, by decomposing $(NH_4)_2MoS_4$ inside the pores of the membrane.[31] The membrane is dissolved by treatment with alkali to yield free tubes. This method yields large diameter MoS_2 nanotubes.

As mentioned earlier, the structure of MoS_2 consists of the disulfide layers stacked along the *c*-direction.[3] This implies that the S–S interaction between the MoS_2 slabs is weaker compared to the intralayer interactions. The S–S interlayer distances are therefore susceptible to distortions during the folding of the layers. This is exemplified by the slight expansion of the *c*-axis (2%) in the MoS_2 nanotubes.[19] High resolution (HREM) images of the disulfide nanotubes show stacking of the (002) planes parallel to the tube axis. The distance between the layer fringes corresponds to the *d*(002) spacing.

Nanotubes of the disulfides are open-ended or capped. However, the closed end of the nanotube is not exactly spherical. Polygonal caps and rectangular tips are frequently observed in the disulfide nanotubes[47] (Fig. 16). Various types of open ends of nanotubes have been observed. They include flat open ends, conical open ends and also open-ended tubes where the layers at the tip arrange in peculiar ways giving the appearance of a pseudo closed tip.[47,48] The outer layer of the disulfide nanotubes and inorganic fullerenes are almost always complete, but the inner layers show defects, dislocations and terminated growth. This results in differences in the wall thickness on the two sides and in the varying diameter of the inner core along the length of the tube wall. This type of terminated growth is not observed in the CNTs, but has been found in metal-filled CNTs.[49] The terminated layer may be a manifestation of the defective structure of the starting precursor, or can be due to the absence of growing material at the edges. Layer defects have been observed in WS_2 nanotubes. They can be traced back to the crystallographic shear in the starting precursor phase.[50] Rémskar *et al.*[51] showed that each WS_2 layer has to satisfy the stacking order and orientation relationship with respect to the previous layer

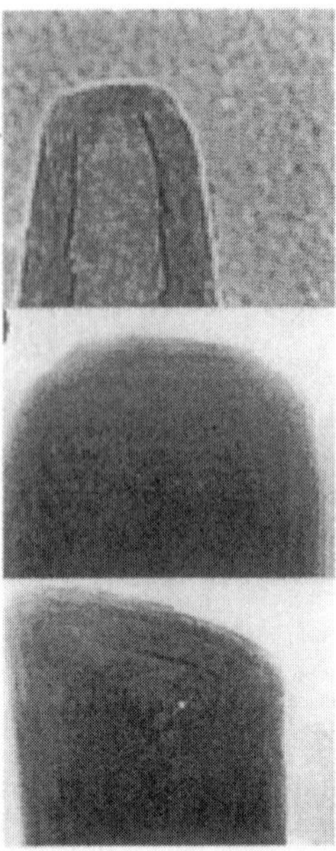

Fig. 16 Various closed tips observed in WS_2 tubes possibly containing square or octagonal defects. (Reproduced with permission from ref. 47).

and the strain involved can be relaxed by the formation of edge dislocations. The outer cylinders are subjected to less stress than the inner cylinders, and this internal stress stabilizes the inner shells, while the outer layers become easy to deform resulting in a greater number of defects near the edges.

The dichalcogenide nanotubes exhibit positive and negative curvatures similar to those in the CNTs.[52] In the CNTs, the curvature is believed to be due to the presence of a pentagonal ring (positive curvature) and a heptagonal ring (negative curvature) in the otherwise hexagonal network of the graphene sheet.[53] In the MX_2 layers, the absence of a central M or X atom can give rise to triangular or rhombohedral point defects, which can cause curvature in the MX_2 layers similar to those obtained in the graphene sheets (Fig. 17).[5,19,54] These point defects lead to topological defects and a combination of topological defects can cause tube closure.[55] Typical topological defects are square-like and octagonal-like defects. Based on the TEM images and electron diffraction (ED) patterns, it has been shown that a combination of these can close the armchair and zigzag tubes.[55] The presence of a rhombohedral point defect gives sharp edges at the corners of a rectangular inorganic fullerene. Similar point defects provide nearly 90° bends at the corners of the disulfide nanotubes[47] as well as the doped disulfide nanotubes (Fig. 18).[41] Analogous to the CNTs, the MX_2 tubes exhibit the energy minimized zigzag and armchair type tube morphologies.[47,56,57]

In $Mo_{1-x}W_xS_2$ nanotubes, defects in the layers, such as edge mismatch, growth termination of some of the layers and disordered layer stacking, increase with increasing tungsten doping.[40] Layer mismatch at the edges has also been observed in composite nanotubes such as $Nb–WS_2$[41] and $Ti–MoS_2$,[42] due to the simultaneous growth of the disulfide layers from different points on the precursor particle. In the $Mo_{1-x}W_xS_2$ nanotubes, W occurs in the MoS_2 layers, rather than having a structure with alternate layers of MoS_2 and WS_2.[58] The intra-layer doping of W in the MoS_2 layer does not cause a major structural deformation as the lattice constants of the MoS_2 and WS_2 are comparable. However, a slight expansion in the *c*-axis occurs in composite nanotubes similar to that in the MoS_2 nanotubes, due to the increased strain in curving the doped layers. With the

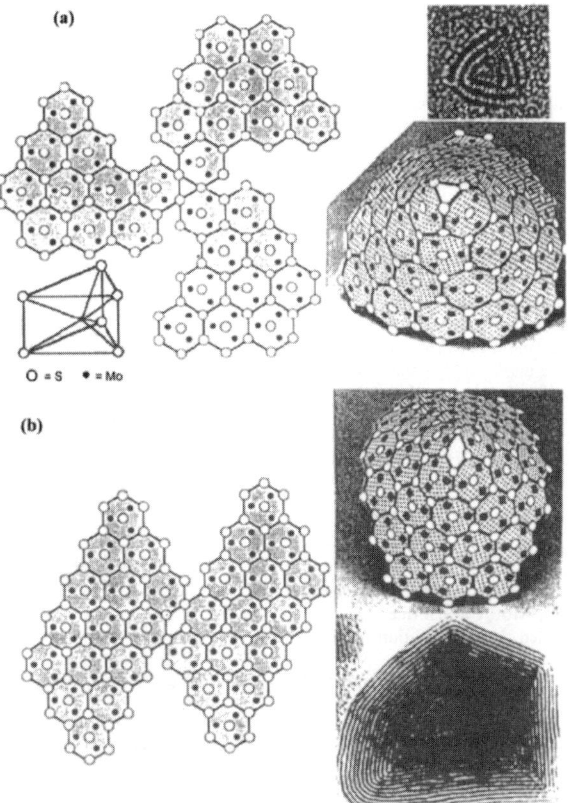

Fig. 17 Illustration of various point defects which exists in the vertices of IF MoS_2: (a) a triangular point defect; (b) a rhombohedral point defect. Insets show IF structures that are likely to contain such point defects. (Reproduced with permission from ref. 19).

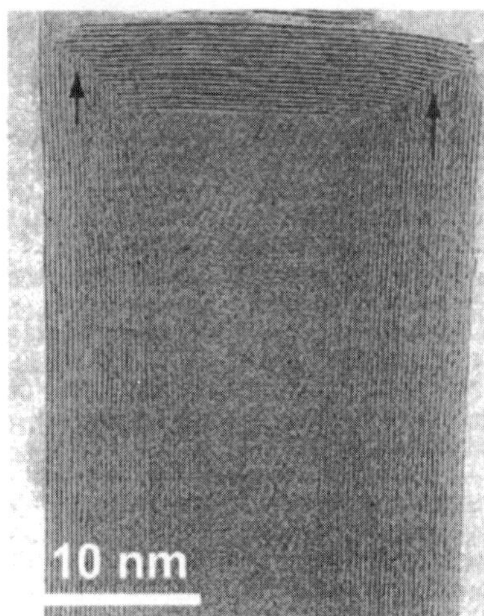

Fig. 18 TEM image showing the sharp 90° bend observed in the W–Nb–S nanotubes. (Reproduced with permission from ref. 41a).

increasing W doping, the yield of the nanotubes decreases considerably, while other nanostructures like nanorods and nanowires are obtained in moderate yield.

Disulfide nanotubes formed from the trisulfide precursors are structurally similar to those obtained by other means. Some of the nanotubes obtained by this method are open-ended, while others are closed. The nanotube tips are mostly non-spherical, polygonal with faceted edges. In the TEM images of MoS_2, WS_2, and $Mo_xW_{1-x}S_2$ nanotubes, the resolution at the tip of some of the nanotubes is lost. The ED patterns of the MoS_2 and WS_2 nanotubes prepared by this method match those reported for the armchair and zigzag tubes. It has been noticed that placing a small amount of sulfur powder near the inlet of the reaction zone, increases the yield of the nanotubes in some of the reactions. This observation suggests that the diffusion of vapors inside the particle may play a role in the formation of the nanostructures. Tenne et al.[59] have shown that the yield and diameter of the nanostructures depend on the diffusion length of the reactant vapors. In this study, MoO_3 was reduced to the suboxide under N_2 which was then carried into another reaction zone where it was converted to the sulfide. A quartz rod was used to precipitate the reaction products in order to quantify and hence study the influences of temperature, gas flow rates and diffusion rates of the reactant gases on the morphology and formation of the nanostructures.

The growth of the disulfide nanotubes from the trisulfide precursors starts with the reduction of the trisulfide precursors to a nanoparticulate form on heating in a gas flow. As the temperature reaches the decomposition temperature of the trisulfide, a few layers of the disulfide are formed at the periphery of the trisulfide particle. The growth then proceeds from "inside outwards" as the trisulfide at the core decomposes to form the disulfide layers that starts projecting outwards. This mechanism is supported by the presence of the precursor particle at the tip of some of the nanostructures, the diameter of such particles being greater than the outer diameter of the nanostructures (Fig. 11a).[22] There is a simultaneous growth of the disulfide layers from different points in the precursor particle, thus resulting in a layer mismatch at the edges where the growing layers meet. In the case of WSe$_2$, while the nanostructures show lattice fringes with the layer separation matching that reported for the bulk, the particle at the tip appears to be due to the amorphous triselenide.

Metal-catalyzed CNT growth is diffusion-controlled, and the carbon vapor adsorbed onto the metal catalyst particles, diffuses to the rear end, from where the graphite layer grows thus encapsulating the particle inside the growing nanotubes in some cases.[60] The dichalcogenide tube growth is also diffusion-controlled and involves the diffusion of H$_2$S into the inner core of the precursor particle, which is slowly converted to the disulfide. However, in the nucleation-mediated growth of the dichalcogenide nanotubes where the trisulfide precursor acts as the nucleation center, the disulfide layer grows from the inner core of the particle and proceeds outwards.

MoS$_2$ nanotubes are also obtained by heating MoS$_2$ powder covered with a Mo foil in the presence of H$_2$S.[61] MoS$_2$ evaporates on heating and is deposited on the Mo foil. The hollow tubes formed exhibit a zigzag arrangement of the layers in the tube walls. In a separate hydrothermal synthesis, molydenum polysulfide microtubules and hollow fibers were grown at room temperature from a solution containing condensed ammonium thiomolybdate and ethylene diamine. Initially, a condensed phase of the formula $(NH_3OH)_{3.9}MoS_{4.8}$ was formed. Thermal decomposition of these phases led to the formation of highly dispersed tubular MoS$_2$.[62] The amine plays an important role in nanotube formation, the absence of which yields bulk MoS$_2$. Ethylene diamine dihydrate appears to form chain-like structures, providing a support for formation of the nanotube. The amine is washed off to separate the hollow tubes. Bundles containing MoS$_2$I$_x$ single-walled nanotubes have been prepared in the presence of C$_{60}$ as a carrier.[63] The presence of C$_{60}$ was crucial in the growth process as the nanotubes failed to grow in their absence. The single-walled tubes are hexagonally packed in bundles with a cell constant of 4.0 Å along the bundle axis and 9.6 Å perpendicular to the axis.

Electron beam irradiation and laser ablation has been successful in producing the metal dichalcogenide nanotubes. Laser ablation of MoS$_2$ and WS$_2$ targets produces substantial amounts of inorganic fullerenes and nanotubes.[64] Electron beam irradiation of bulk WS$_2$ powder yields various nanostructures of WS$_2$ including inorganic fullerenes, nanotubes and nanorods.[65]

The disulfide nanotubes have been characterized by Raman spectroscopy. The Raman bands for the MoS$_2$ nanotubes are similar to those of the bulk 2H-phase, except that the bands of the nanostructures show slight broadening. The composite nanotubes like those of Ti–Mo–S and Nb–W–S were also characterized by Raman spectroscopy.[41,42] The bands were similar to those of the undoped nanotubes, except for the appearance of a new band at 313 cm^{-1} which has been assigned to the increase in disorder owing to the presence of Nb.[41]

4. Nanotubes of dichalcogenides of Group 4 and 5 metals

With the success in synthesising nanotubes of Mo and W dichalcogenides starting from the trichalcogenides,[21,22] it was realised that this methodology could prove useful to prepare the nanostructures of other metal dichalcogenides, even though the trichalcogenides may be crystalline (rather than amorphous as in the case of Mo and W), since the dichalcogenide phase appears at the boundary of the trichalcogenide phase, and the thermal decomposition of the bulk samples of the latter is known to yield dichalcogenides.[39]

Nath and Rao[23,24] have investigated the thermal decomposition of the trisulfides of Group 4 and 5 metals prepared by the conventional solid state synthesis route. The decomposition of the trisulfides of Zr, Hf, Nb and Ta in a reducing atmosphere at elevated temperatures has indeed produced good yields of nanostructures, including nanotubes and nanorods.

4.1. HfS$_2$ nanotubes

Nanotubes of HfS$_2$ are obtained by the decomposition of HfS$_3$ in an atmosphere of H$_2$ + Ar (1 : 9), at 1170 K.[24] The product contained a good yield of the nanostructures as can be seen from the SEM image in Fig. 19a. EDX analysis revealed the

Fig. 19 (a) SEM image of the HfS$_2$ nanostructures; (b) and (c) low-resolution TEM images showing hollow nanotubes. The tube in (c) has a flat tip.

chemical composition to be HfS$_2$, while the XRD pattern showed it to be hexagonal. There is a slight expansion of the lattice in the c-direction (~1%) of the nanotubes as compared to bulk HfS$_2$. The lattice expansion is less than that observed with the MoS$_2$ and WS$_2$ nanotubes (2–3%).[19] This can be attributed to the fact that the mean compressibility factor of the c-axis in HfS$_2$ is higher than that observed for MoS$_2$.[66] The nanostructures as can be seen from the SEM image on Fig. 19a are quite lengthy, some being more than a micron long. Interestingly, a large proportion of these nanostructures is nanotubes. The low-resolution TEM image in Fig. 19b shows several nanotubes, some of which are closed with a non-spherical, nearly rectangular tip. The image in Fig. 19c shows a single nanotube with a rectangular tip, showing ripple-like undulations near the bend, which can arise from the strain involved in bending the layers. On close inspection, layer fringes are visible along the tube walls. Interrupted layer growth is observed in the inner edge of the tube wall, causing terminated layers and has non-uniformity in the wall thickness. This type of rectangular tip and terminated layer growth has also been observed in MoS$_2$, WS$_2$ and Mo$_{1-x}$W$_x$S$_2$ nanotubes.[40,47]

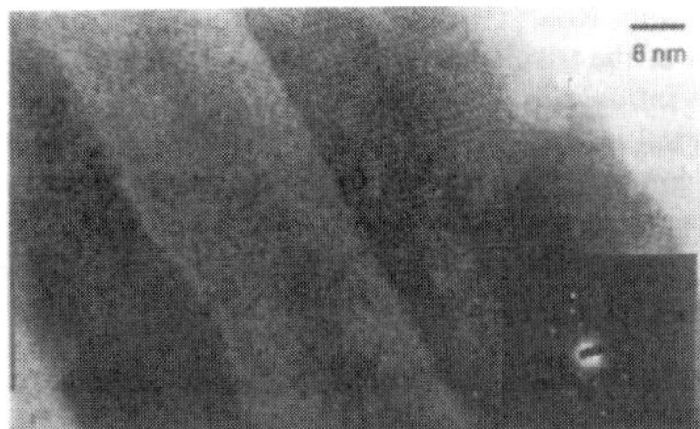

Fig. 20 HREM image of the HfS$_2$ nanotubes, showing a layer separation of ~0.6 nm in the walls. Inset shows a typical ED pattern.

A high-resolution image of a HfS$_2$ nanotube is shown in Fig. 20. The layers are separated by ~5.8 Å corresponding to the spacing of the (001) planes. Several terminated layers are observed at the outer edge of the tube wall possibly due to the absence of the growing material at these edges. A considerable number of defects and edge dislocations are also present along the length of the tube wall. The inset in Fig. 20 shows a ED pattern of the nanotube, characteristic of the hexagonal arrangement of the layers. Bragg spots corresponding to d(002) plane (2.923 Å) are seen. The ED pattern together with the high-resolution image indicate that the growth axis of the nanotube is perpendicular to the c-direction.

Bulk HfS$_2$ is an indirect band gap semiconductor with an indirect band gap energy of ~2.1 eV.[67] The reflectance spectrum of the nanotubes shows a small blue shift compared with the bulk. The photoluminescence spectrum of the nanotubes shows a band at 676 nm due to trapped states, and the band is blue-shifted with respect to that of bulk HfS$_2$ powder (see Fig. 21a). The Raman spectrum of the HfS$_2$ nanotubes is shown in Fig. 21b. It shows a band due to the A$_{1g}$ mode, corresponding to the S atom vibration along the c-axis perpendicular to the basal plane, and another due to the E$_g$ mode due to the movement of the S and Hf atoms in the basal plane.[68] The full-width at half maximum (FWHM) of the A$_{1g}$ band is 11 cm^{-1} in the nanotubes compared to 8 cm^{-1} for the bulk sample. Such broadening of the Raman band has been noted with MoS$_2$ and WS$_2$ nanotubes.[19]

4.2. ZrS$_2$ nanotubes

ZrS$_2$ nanotubes admixed with nanorods have been prepared by the thermal decomposition of ZrS$_3$ under H$_2$ + Ar at 1170 K.[24] Many of the nanotubes exhibit rectangular tips. The inner wall of some of the ZrS$_2$ nanotubes show non-uniformity near the tip resulting from the discontinuous growth of the ZrS$_2$ layers.

4.3. NbS$_2$ nanotubes

NbS$_3$ when heated in a stream of H$_2$ (100 sccm) at 1000 °C for 30–60 min, produces good yield of nanostructures.[23] A c-axis expansion of ~3% is observed in the nanotubes. The SEM image in Fig. 22a reveals a high yield of the nanostructures from the thermal decomposition of the trisulfide. These nanostructures contained a considerable amount of nanotubes as seen from the TEM images. The TEM image in Fig. 22b shows a couple of nanotubes, having closed rectangular tips. While some of the nanotubes are closed with flat non-spherical polygonal tips, most of the tubes are open at one or both ends. The image in Fig. 22c shows the high-resolution image of an open nanotube. The layer separation in the walls is ~6 Å, corresponding to the (002) plane of bulk NbS$_2$. The ED pattern in the inset shows the nanotube to be single-crystalline with Bragg spots corresponding to the known d values. Some of the diffraction spots show diffuse scattering or streaking due to bent layers or disorder. The NbS$_2$ nanostructures were not superconducting.

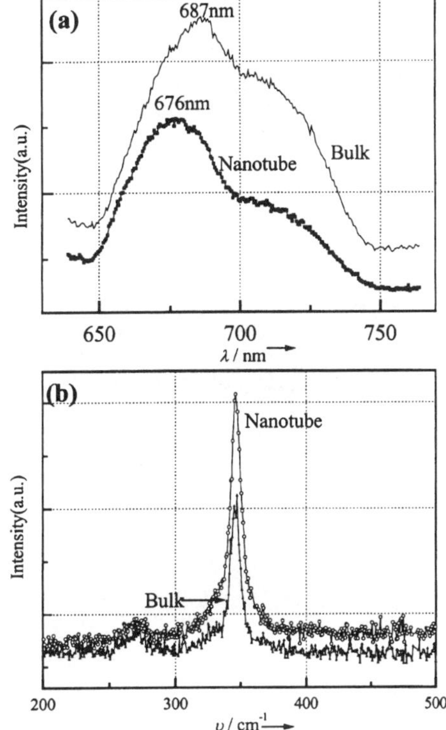

Fig. 21 (a) Photoluminescence spectra of bulk HfS$_2$ and HfS$_2$ nanotubes. (b) Raman spectrum of bulk HfS$_2$ and HfS$_2$ nanotubes.

4.4. TaS$_2$ nanotubes

TaS$_2$ nanotubes could be prepared by decomposing TaS$_3$ at 1270 K under H$_2$ (100 sccm). TEM observation of the product showed the presence of hollow core nanotubes in the product. The inner wall of some of the TaS$_2$ nanotubes showed non-uniformity along the length of the tube. Some of the tubes are closed with flat rectangular tips. In both NbS$_2$ and TaS$_2$ nanotubes, most of the tube tips are nearly flat or rectangular.[23] This is in sharp contrast to the CNTs.

4.5. NbSe$_2$ nanotubes

Galván *et al.*[69] have reported that NbSe$_2$ nanotubes can be prepared by the use of intense electron irradiation. The nanotubes and nanorods of NbSe$_2$ have been prepared by the decomposition of the triselenide, at ~970 K under a gas flow of Ar.[70] The product contained a mixture of nanotubes and nanorods. The powder X-ray diffraction was characteristic of 6aH–NbSe$_2$, with no appreciable shift in the d(002) line position relative to the bulk sample. Most of the NbSe$_2$ nanotubes are open-tipped. The NbSe$_2$ nanorods generally possess smaller diameters as well as shorter lengths compared to the nanotubes. Under the most favourable conditions, the ratio of the nanotubes to nanorods was around 2 : 1. The rate of argon flow appears to have an effect on the ratio of nanotubes and nanorods in the final product. At a smaller flow rate, there was a higher percentage of nanotubes in the final product, while higher flow rates gave a greater percentage of nanorods. High argon flow also gave rise to platelets, which also occurred, on increasing the temperature.

Fig. 23a shows the HREM image of a NbSe$_2$ nanotube in the tip region. The structure of the tip deviates from the hemi-

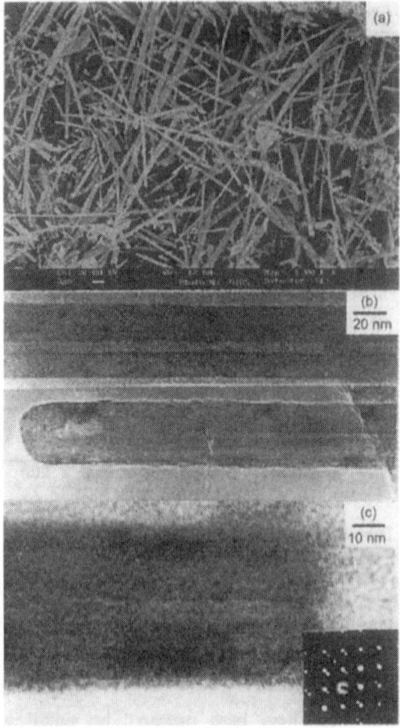

Fig. 22 (a) SEM image of NbS$_2$ nanotubes; (b) low-resolution TEM image and (c) HREM image of NbS$_2$ nanotubes. Inset shows the typical ED pattern. The layer separation is 0.62 nm in the walls. (Reproduced with permission from ref. 23).

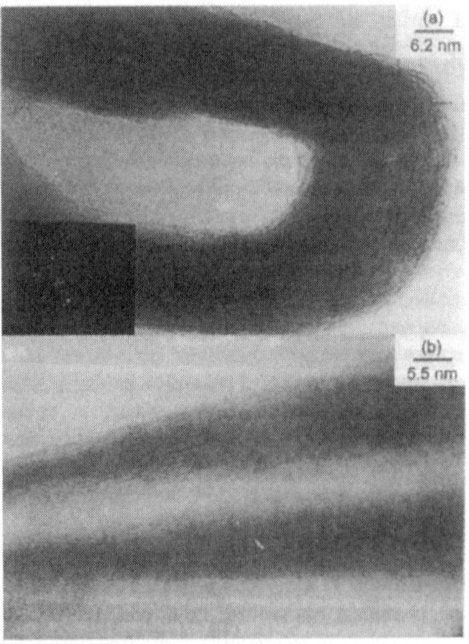

Fig. 23 HREM images of the NbSe$_2$ nanotubes. The tube in (a) has a closed tip with a 90° bend in one corner, while the tube in (b) is near-conical in shape due to several terminated layers in the walls. The inset in (a) shows a typical ED pattern. (Reproduced with permission from ref. 70).

spherical nature, and has a 90° bend at one corner while the other corner is comparatively smooth. The outer diameter of the tube is ~57 nm and the wall contains ~16 layers. The inner wall of the tube shows evidence for terminated layer growth. Sharp facets of the type reported by Tenne *et al.*[71] in the fuller-

ene structures of NbS$_2$ are however not present in these nanotubes. The layers near the tip are somewhat less regularly ordered, but the number of defects, dislocations and stacking faults in the NbSe$_2$ nanotubes is generally much smaller than that in the other chalcogenide nanotubes.[23,24] Fig. 23b shows a HREM image of a nanotube with a near-conical closed tip showing an interlayer separation of ~6.2 Å. This tube has a tapering shape due to the disrupted growth of the layers as it proceeds towards the tip. As a result, the tube walls near the tip are thinner than that along the body. There are approximately 10 layers near the tip and ~14 layers along the body of the tube, ~77 nm away from the tip.

It is interesting that most of the NbSe$_2$ nanotubes contain more than 10 layers with one or two containing a smaller number of layers. Some of the nanotubes exhibit a different type of stacking due to the presence of different polytypes just as in BN nanotubes where the local rhombohedral stacking occurs within the hexagonal phase.[44]

The Raman spectrum of NbSe$_2$ single crystals exhibit three first order lines at 29.6, 230.9 and 238.3 cm^{-1}.[72] The low frequency line at 29.6 cm^{-1} is due to the rigid layer vibration mode (E_g^2) which accounts for the weak interlayer bonding. The high-frequency lines at 230.9 and 238.3 cm^{-1} are due to the A_{1g} and E_{2g}^1 modes, respectively. The high frequency Raman modes in the NbSe$_2$ nanostructures were found to be identical to those of the bulk crystal. Bulk NbSe$_2$ shows a photoluminescence band of very weak intensity band at around 825 nm possibly due to trapped states. The band is shifted to 820 nm in the nanostructures.

NbSe$_2$ is a metallic conductor, becoming superconducting at low temperatures.[36] The NbSe$_2$ nanostructures show metallic conductivity from 300 K down to lower temperatures (Fig. 24),

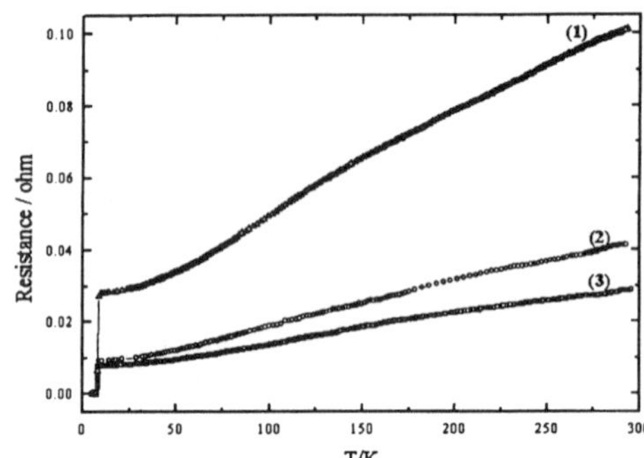

Fig. 24 Resistance *vs.* temperature curves of the NbSe$_2$ nanostructures, (2) and (3), are compared with the behavior of the bulk sample in (1).

in agreement with theoretical predictions.[57] The resistance–temperature curve for the bulk sample has a much higher slope than for the nanostructures, possibly because of the dominance of temperature-independent scattering in the latter. The NbSe$_2$ nanostructures become superconducting at 8.3 K (Fig. 24). The superconducting transition temperature of the bulk sample was found at 8.6 K, indicating a very small decrease in T_C, if at all, in the nanostructures. It has been suggested that the superconducting T_C of NbSe$_2$ is dependent on factors such as stoichiometry and the number of layers of NbSe$_2$ when the number of layers goes below 6, the T_C is expected to shift to 3 K.[73] The NbSe$_2$ nanostructures studied had diameters in excess of 30 nm, with the number of layers greater than 10. It would be worthwhile to prepare NbSe$_2$ nanostructures containing a smaller number of layers to study whether the T_C is substantially affected compared to the bulk material. In this context, it must be noted that MgB$_2$ nanowires of 50–200 nm diameter have been shown recently to have a superconducting T_C identical to that of the bulk material.[74]

5. Nanotubes of other metal chalcogenides

Nanotubes and nanowires of II–VI semiconductor compounds such as CdS and CdSe have been obtained by a soft chemical route involving surfactant-assisted synthesis.[28,75] For CdSe nanotubes, the metal oxide was reacted with the selenidizing reagent in the presence of a surfactant such as Triton 100X. Substantial amounts of nanotubes were obtained by this method (Fig. 25a and b). Annealing of the as-prepared nano-

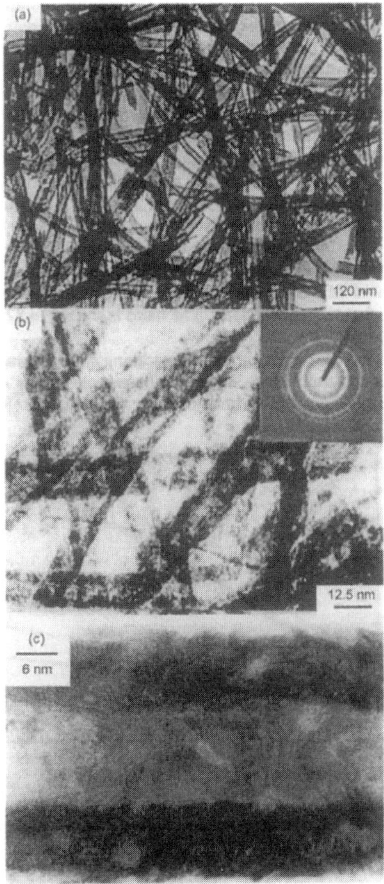

Fig. 25 (a) and (b) low-resolution TEM images of CdSe nanotubes. Inset shows a typical ED pattern; (c) HREM image of the CdSe nanotube showing walls containing several nano-crystallites. (Reproduced with permission from ref. 28).

tubes appears to improve the crystallinity of the nanotubes (Fig. 25c). To obtain nanotubes of CdS, a similar procedure was followed, except that thioacetamide was used as the sulfidizing agent in place of NaHSe. The formation of the nanotubes and nanowires is controlled by varying the surfactant concentration which in turn changes the morphology and shape of the micellar cavity. Thus, in both CdS and CdSe, a higher concentration of the surfactants produce nanotubes, while a lower concentration of the surfactants preferentially produce nanowires. The surfactant is readily removed from the nanotubes by washing with a solvent. Both the CdSe and CdS nanotubes seem to be polycrystalline formed by aggregates of

nanoparticles.[76] The nanotubes of CdSe, though extended in one direction show quantum confinement and the absorption band is blue-shifted to 550 nm from 650 nm of the bulk sample. A quantum confinement effect is also observed in the PL spectrum of the nanotubes, which shows a band at 560 nm compared to 750 nm for the bulk sample (Fig. 26). The CdS nano-

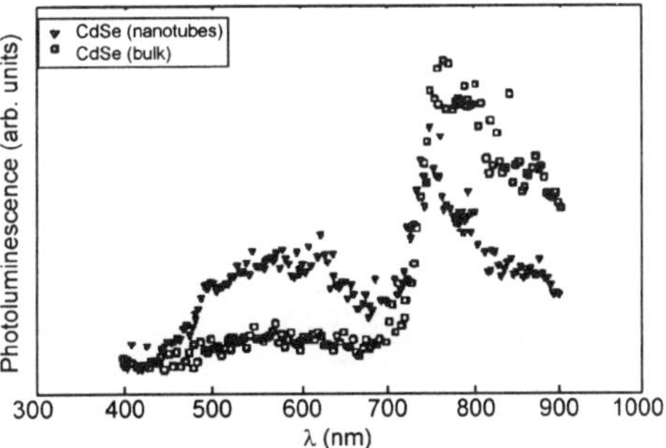

Fig. 26 Photoluminescence spectra of nanotubes of CdSe and bulk CdSe. (Reproduced with permission from ref. 28).

tubes also show a blue-shift of the absorption maximum due to confinement effects. The Raman spectrum of CdSe nanotubes show modes at 208 and 198 cm^{-1}, the former arising from the longitudinal optic phonon, red-shifted due to phonon confinement.[76]

ZnS nanotubes have been prepared by sulfidizing ZnO templates obtained in columnar form by electrochemical deposition.[77] Heating the ZnO column in H$_2$S above 400 °C gave the ZnS coated ZnO columns. The ZnO cores were then etched out giving hollow ZnS tubes.

Hydrothermal methods have been successfully used to synthesize a variety of nanotubes and nanorods. Layer-rolled structures of nickel sulfide have been obtained by sulfidizing the Ni(NH$_3$)$_4$$^{2+}$ complex by CS$_2$ in the presence of aqueous ammonia[78] under hydrothermal conditions. Nanotubes of Cu$_{5.5}$FeS$_{6.5}$[79] have been prepared hydrothermally. These nanotubes are large in diameter, have thick walls and are mostly open-ended. The crystallinity of the thin-walled nanotubes is not generally satisfactory. Cu$_{5.5}$FeS$_{6.5}$ has a layered structure characterized by S–S pairs and MS$_4$ tetrahedra. The nanotubes are thought to be formed by rolling the basal planes. The ED patterns collected from the nanotubes indicates that they are of the armchair and zigzag types. In the hydrothermal synthesis of the nanotubes and related structures, organic amines seem to act as the templating agents. The amine probably intercalates into the lattice of the precursor which is subsequently decomposes to yield the chalcogenide nantoubes around the templating amine.

6. Metal oxide nanotubes

Metal oxide nanotubes have been prepared by using a variety of techniques including templating reactions, sol–gel chemistry and hydrothermal methods.[7] Besides CNTs, organic surfactants have been employed to grow oxidic nanotubes. Pores of anodic alumina have been used as confining templates for the growth of TiO$_2$ and Al$_2$O$_3$ nanotubes, as well as coaxial nanotubes of TiO$_2$ sheathed SiO$_2$.[80,92]

6.1. VO$_x$ nanotubes

Vanadium oxides deserve special attention owing to their structural flexibility and interesting catalytic, electrochemical and other properties. V$_2$O$_5$ can be regarded as a layer structure in which the VO$_5$ square pyramids are connected by sharing corners and edges and thereby forms the layer.[81] The interlayer interaction is weak as shown by the long V–O distances.[82] This structural characteristic permits the intercalation of various cationic species in the interlamellar space.[83] The cations include alkylammonium ions which are readily intercalated in between the layers under hydrothermal conditions.[84] Thus, this system is analogous to graphite and the layered dichalcogenides. In 1998 Nesper and co-workers[11,85] synthesized nanotubules of alkyl-ammonium intercalated VO_x by hydrothermal means. The vanadium alkoxide precursor was hydrolyzed in the presence of hexadecylamine and the hydrolysis product (lamellar structured composite of the surfactant and the vanadium oxide) yielded VO_x nanotubes along with the intercalated amine under hydro-thermal conditions (Fig. 27). The interesting feature of this

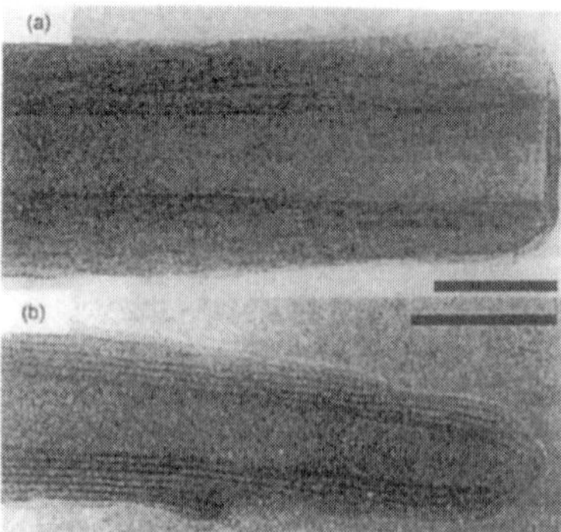

Fig. 27 TEM images of VO_x nanotubes with intercalated amine having varying chain lengths; (a) C_4VO_x–NT; (b) C_{16}–VO_x–NT. The length of the bar is 50 nm. (Reproduced with permission from ref. 85).

vanadium oxide nanotube is the presence of vanadium in the mixed valent state, thereby rendering it redox-active. The template could not be removed by calcination as the structural stability was lost above 523 K. Nevertheless, it was possible to partially extract the surfactant under mildly acidic conditions. These workers have later shown that the alkylamine intercalated in the intertubular space could be exchanged with other alkylamines of varying chain lengths as well as α,ω-diamines.[85] The distance between the layers in the VO_x nanotubes can be controlled by the length of the –CH_2– chain in the amine template. The tubes obtained with the diamine had thicker walls compared to those obtained with the monoamines. A change in the interlayer distance observed in the XRD pattern as a function of the alkyl chain length is rendered to be due to the mono-layer arrangement of the amine molecules in between the layers. In the VO_x nanotubes, the amine may be acting as a structure-directing agent and is incorporated in the tube walls.

Most of the VO_x nanotubes obtained by the hydrothermal method are open-ended. Very few closed tubes had flat or pointed conical tips. Cross-sectional TEM images of the nano-tubular phases show that instead of concentric cylinders, (*i.e.* layers that fold and close within themselves), the tubes are made up of single or double layer scrolls providing a serpentine-like morphology.[85,86] The scrolls are seen as circles that do not close in the images (Fig. 28). Non-symmetric fringe patterns in the tube walls exemplify that most of the nanotubes are not rota-tionally symmetric and carry depressions and holes in the walls. Diamine-intercalated VO_x nanotubes are multilayer scrolls with narrow cores and thick walls, composed of packs of several vanadium oxide layers (Fig. 29). The diamine-containing VO_x nanotubes also show a smaller number of holes in the wall structure and the tubes are well ordered with uniform distances throughout the tube length.[85] The scroll-like structure of the nanotubes may be the most obvious cause for their high struc-tural flexibility. This is also why facile exchange reactions occur.

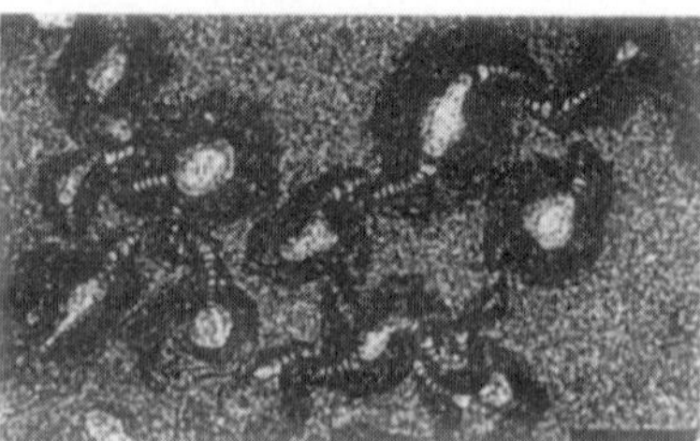

Fig. 28 Cross-sectional TEM images of monoamine-intercalated VO_x nanotubes showing serpentine-like scrolls.

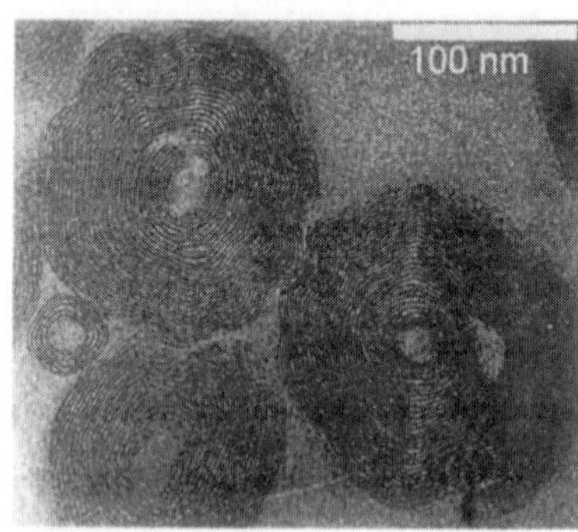

Fig. 29 Cross-sectional TEM images of diamine-intercalated VO_x nanotubes showing a larger thickness of the tube walls and a smaller inner core. (Reproduced with permission from ref. 85).

It must be pointed out that the VO_x scroll-like structures are not genuine nanotubes of the type formed by carbon or metal dichalcogenides.

The alkylammonium intercalated VO_x nanotubes are para-magnetic and show semimetallic conductivity probably due to the mixed valent V centers. The percentage content of V(IV) calculated on the basis of the effective magnetic moment is ~45–50%.[85] The structure contains VO_4 tetrahedra and VO_5 square-pyramids simultaneously.

The as-synthesized amine intercalated VO_x nanotubes could be aligned on glass substrates by using micromolding in capillaries (MIMIC).[86] An elastomeric polydimethylsiloxane (PDMS) stamp where parallel capillaries of 5 μm were pat-terned was used as the mould.[86] The capillaries were filled with a VO_x nanotube suspension in octanol. After evaporation of the solvent and removal of the mould, long lines of assemblies of well aligned nanotubes were obtained.

Vanadium oxide nanotubes containing primary monoamines with long alkyl chains have been prepared by employing non-alkoxide vanadium precursors such as $VOCl_3$ and V_2O_5. The amine complexes of the vanadium precursors are then hydro-lyzed. Hydrothermal treatment of the precursors gives good yields of VO_x nanotubes incorporating the amines.[27] The dis-tance between the layers is proportional to the length of the alkyl amine chain which acts as the structure-directing tem-plate. Cross-sectional TEM images demonstrate the predomin-ance of serpentine-like scrolls rather than of concentric tubes.

Mn–V-oxides have been prepared starting from vanadium oxide–dodecylamine composite nanotubes.[87] The composite nanotubes were prepared by mixing V_2O_5 with dodecylamine in the presence of ethanol and water. The amine templates can be easily substituted or even ion-exchanged with ions like Mn^{2+} in an aqueous alcohol solution to obtain the Mn–V–O nanotubes. Most of the nanotubes had open ends, while some of them had closed ends, with the side of the tubes wrapped around the end to close it. The Mn^{2+} ions replace the organic cations in the structures and hence are intercalated in between the layers.

6.2. Use of carbon nanotubes as templates for the growth of oxide nanotubes

Carbon nanotubes have been successfully used as removable templates for the synthesis of a variety of oxide nanotubes. Ajayan *et al.*[9] reported the preparation of V_2O_5 nanotubes by using partially oxidized carbon nanotubes as templates. Apart from coating of CNTs by the oxide phase, metal oxide fillings in the internal cavities and thin oxide layers between the concentric shells of the tubes were also obtained (Fig. 30). A mix-

Fig. 30 TEM images of the VO_x coated carbon nanotubes. (a) Low-magnification image showing partial coating; (b) image at higher magnification showing the 0.34 nm fringes of the CNT and oxide coatings of uniform thickness on the upper and lower parts of the tubes. (Reproduced with permission from ref. 9).

ture of partially oxidized CNTs and V_2O_5 powder was annealed in air to produce a nanotube–oxide composite. The external coating of the tubes consists of crystalline V_2O_5 layers, which grow with the *c*-axis parallel to that of the nanotube layers. Intercalation of the oxide occurs where there are missing shells of the nanotubes. The coating on the isolated nanotubes is of uniform thickness, but does not always cover the entire length of the nanotube surface. Continuous cylindrical oxidic films on the nanotube surface are generally formed over a length of a few hundred nanometers. Attempts have been made to remove the carbon nanotube templates by oxidation at 650 °C. These samples show a dramatic increase in the ratio of the coated to the uncoated tubes. On observing the tips of the remaining coated tubes, oxidation was found to have occurred from the inner wall of the tube and progressed outwards leaving few outer layers and the outer coating of the oxide. This is in contrast to the oxidation of pure nanotubes where the etching of the layers starts from outside and proceeds inwards leaving chisel-shaped tips. The complete removal of the nanotube template resulted in larger hollow oxidic structures with thin skins.

Rao *et al.*[10] have prepared a variety of oxide nanotubes including SiO_2, Al_2O_3, V_2O_5, MoO_3 and RuO_2 employing carbon nanotube templates. By using multi-walled carbon nanotubes as the template they also made nanotubes of zirconia and yttrium-stabilized zirconia.[30] In these preparations, acid-treated MWNT bundles were coated with a suitable precursor of the metal oxide, and the coated composites heated to high temperatures to remove the carbon template (Fig. 31). TEM images show that the carbon nanotubes get fully coated with the oxidic material on reaction with the organometallic compound followed by calcination at 500 °C. The nanotube features are essentially retained in the oxide–nanotube composites even after calcination at 500 °C (Fig. 31). Acid-treated tubes give a better coating than the pristine ones. The nanotube templates could be removed successfully by heating at 750 °C for several hours. The oxidation proceeded from the hollow regions inside the tubes. Silica nanotubes prepared by this method could be modified by doping with transition metal ions for possible use in catalysis. Thus Cu, Cr or Ni containing SiO_2 tubes have been prepared by this method by taking the relevant transition metal ions in the starting reaction mixture.[10] The transition metals are present as oxides in the final structure, and the oxide appeared as islands on the surface of the hollow silica tubes in the case of Cu.

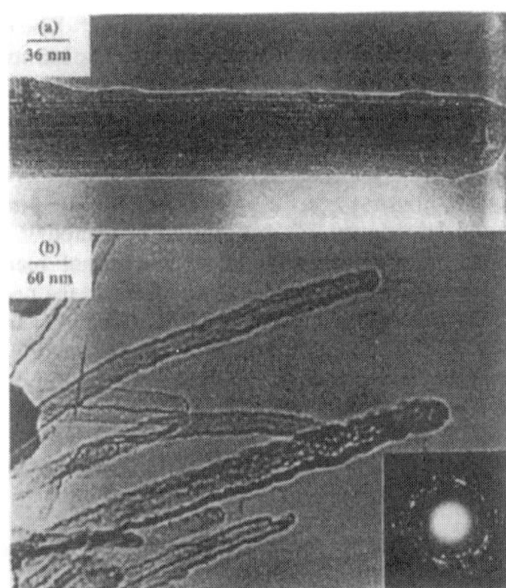

Fig. 31 (a) TEM image of a ZrO_2 coated CNT. (b) Hollow ZrO_2 nanotubes after removal of the CNT template. (Reproduced with permission from ref. 30).

6.3. TiO₂

Nanotubes of transition metal oxides that do not possess layered structures, (*e.g.*, TiO_2, ZrO_2) have been prepared by employing the sol–gel technique, membrane-confined growth and templated growth. Titania tubes were first prepared electrochemically by Hoyer.[88] A polymer mould with a negative (replicated) structure of an anodic alumina porous membrane was used for electrodeposition of the titania nanotubes. The titania phase obtained was amorphous and calcination to induce crystallization led to the deformation of the tube structure. The diameters of the tubes were in the ~70–100 nm range, being controlled by the pore size of the membrane. Needle-shaped anatase nanotubes could be precipitated from a gel containing a mixture of SiO_2 and TiO_2.[29] A mixture of titanium isopropoxide and tetraethylorthosilicate (TEOS) was hydrolyzed and gelled in an incubator, and the gel further heated to 870 K resulting in the precipitation of fine TiO_2 (anatase) crystals. This was further treated with NaOH at 380 K for 20 h to yield the TiO_2 nanotubular phase. An amorphous SiO_2-related phase present in the product could be removed by chemical treatment. The nanotubes formed by this method had a diameter of ~8 nm and lengths upto 100 nm (Fig. 32a).

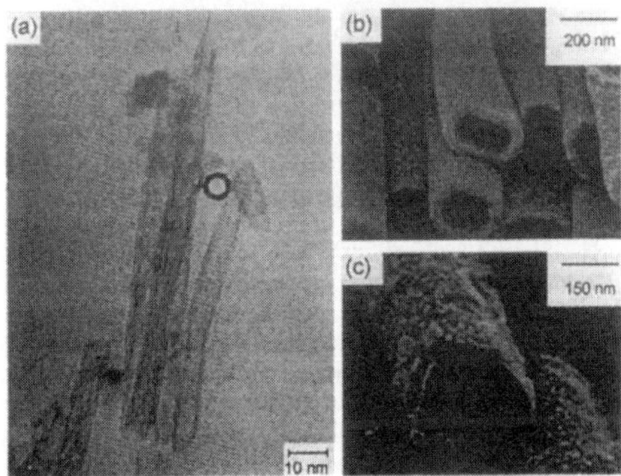

Fig. 32 (a) TEM image of TiO_2 nanotubes (Reproduced with permission from ref. 29); (b) and (c) FE-SEM images of TiO_2 nanotubes deposited from a TiF_4 solution. (Reproduced with permission from ref. 80).

Titania nanotubes are obtained by the direct deposition of titanium tetraflouride (TiF$_4$) in the pores of an alumina membrane.[80] The porous alumina membrane prepared electrochemically was immersed in an aqueous solution of TiF$_4$ and ammonia, and maintained at 60 °C. The walls of the nanotubes contained small nanoparticles of anatase (Fig. 32b). Titania rods were obtained at longer deposition times. The initial solution supersaturated with titania produces nuclei of anatase which deposit on the inner walls of the porous membranes, and give rise to the nanotubes. Photocatalytic activity of the as-deposited anatase nanotubes shows a performance similar to that of the conventionally prepared anatase powder.[80] Titania nanotubes are also prepared by using electrospun polymer fibers as templates.[89] The polymer fibers are coated with titanium oxide, by dipping in a titanium isopropoxide solution followed by dipping in ethanol–water for hydrolysis and condensation. Thermal treatment of the metal oxide coated polymer fibers results in the loss of the polymeric core, producing hollow nanotubes of titania. HREM images show that the individual particles that make the nanotube walls are crystalline (anatase phase). The sol–gel coating was able to mimic the finer details of the polymeric fiber, thereby forming nodules in the inner walls of the tubes. Well aligned, uniform arrays of titania nanotubes could be fabricated by the anodic oxidation of a pure Ti sheet in an aqueous solution containing 0.5 to 3.5 wt% of HF.[90] The upper ends of these tubes were mostly open. Recently, single crystalline nanotubes of TiO$_2$ (anatase) have been prepared by an unconstrained solution growth, by hydrolyzing TiF$_4$ under acidic conditions at 60 °C (Fig. 33).[91] A majority of these TiO$_2$ nanotubes were closed, with spherical or polyhedral caps.

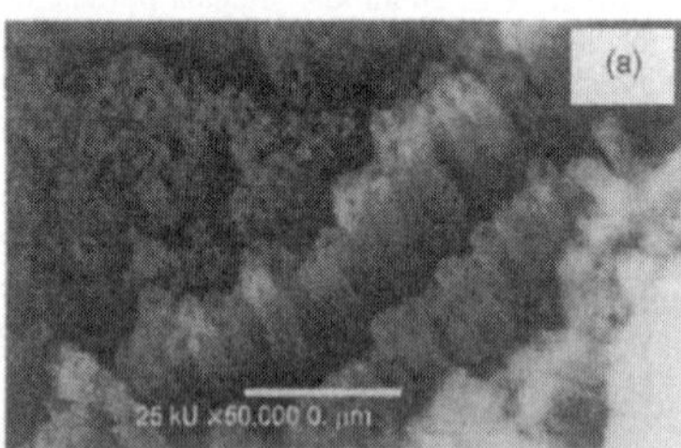

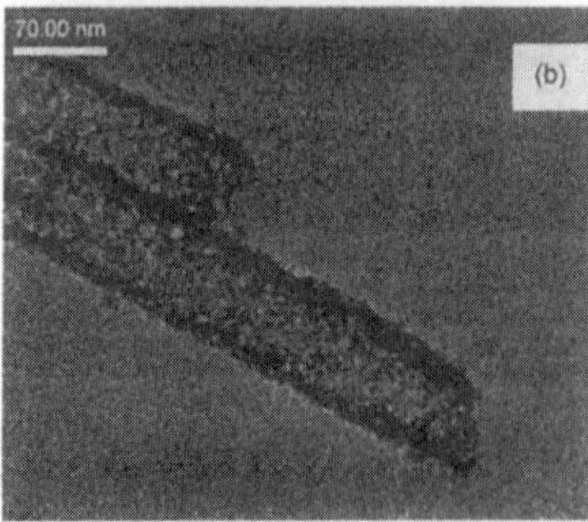

Fig. 33 (a) SEM image of unidirectionally aligned TiO$_2$ nanotubes; (b) TEM image of a TiO$_2$ nanotube formed with nanocrystalline TiO$_2$ particles. (Reproduced with permission from ref. 91).

Pores of anodic alumina have been used for the growth of coaxial nanotubes of TiO$_2$ sheathed SiO$_2$.[92] Electrochemical deposition in the pores of a polymeric alumina membrane generates arrays of TiO$_2$ nanotubes. For the coaxial nanotubes, the SiO$_2$ nanotubes are first grown inside the pores of the anodic alumina membrane. The TiO$_2$ nanotubes are then grown inside the SiO$_2$ nanotubes. The existence of Ti–O–Si bonds in the amorphous sheaths is believed to play a role in the formation of these composite nanotubes.

6.4. SiO$_2$ and Al$_2$O$_3$ nanotubes

6.4.1. SiO$_2$.

Sol–gel techniques have been extensively used to form silica gel nanotubes[93] as well as mesoporous silica nanotubes.[25] Nakamura and Matsui[93] prepared silica nanotubes as a spin-off product of sol–gel synthesis wherein, tetraethylorthosilicate (TEOS) was hydrolyzed in the presence of ammonia and D,L-tartaric acid. Mesoporous silica nanotubes were also prepared by a special post-synthesis-ammonia-hydrothermal treatment of the mesoporous silica material.[25] It was possible to simultaneously restructure the pore size, nanochannel regularity and morphology of the mesoporous material by this method. The restructured products were highly ordered and exhibited nanotubular forms of silica. This methodology gained momentum with the successful synthesis of individual nanotubes or small bundles containing a few nanotubes. Adachi et al.[94] were the first to report the synthesis of very long silica nanotubes by employing surfactant-assisted growth. Laurylamine hydrochloride was used as the surfactant template around which TEOS was hydrolyzed. Tube formation was followed by trisilylation treatment. Trimethylsilylation inactivated the silanol groups on the surface of the tube, thus inhibiting the condensation of silanol groups between the different bundles, and yielding long individual silica tubes. Another advantage of the trisilylation treatment was that the surfactant was removed without calcination. Individual silica nanotubes have also been prepared by the sol–gel method in the presence of citric acid as the structure modifier.[95] The silica nanotubes obtained were, however, mostly amorphous as determined by XRD. Electron microscopy revealed that the individual silica nanotubes were long (~0.5–20 µm), with wide diameters.

The sol–gel technique has been used to dope silica in TiO$_2$ nanotubes.[96] Different loadings of Si in the TiO$_2$ nanotubes was achieved by varying the amount of TEOS in the starting solution containing tetrapropylorthotitanate in butanediol. After several days of aging the dry gel was calcined to induce the crystallization and formation of the nanotubes. The pore size of the nanotubes decreased with increasing Si content. EDAX mapping images revealed that the Ti and Si were homogeneously distributed in the nanotubes. It is believed that Si doping influences the sintering process and suppresses the grain growth of titania nanoparticles, thus increasing the surface area of the doped tube.[96]

Mesoporous silica materials with hierarchical tubule-within-tubule structures have been prepared and characterized by spectroscopic methods.[97] Strong photoluminescence of these materials is explained as due to the presence of Si–OH complexes located on the nanotube surface, which also explains the persistence of the signal for some time after the pumping laser is turned off.

6.4.2. Al$_2$O$_3$.

Alumina nanotubes have been prepared by electrochemical means.[12] Two different preparation methods designated as normal stepwise anodization (NSA) and lateral stepwise anodization (LSA) have been employed for the purpose. The major difference between the two methods is the way the potential was applied. An aluminium film deposited on a p-type Si substrate was anodized in dilute H$_2$SO$_4$. In NSA, the potential was applied to the bottom surface of the Si substrate, while in LSA, it was applied to the top surface of the alumina film. The nanotubes formed were attached to the anodic porous alumina film (Fig. 34a). The NSA tubes were smaller than the LSA tubes. In addition to straight alumina nanotubes, branched alumina nanotubes were also obtained in the same synthesis (Fig. 34b).[12,98]

6.5. Other oxide nanotubes

Recently, ZnO nanotubes have been prepared hydrothermally.[99] In this procedure, ethanol was added to an aqueous solution containing the Zn(NH$_3$)$_4$$^{2+}$ complex, and the mixture heated to

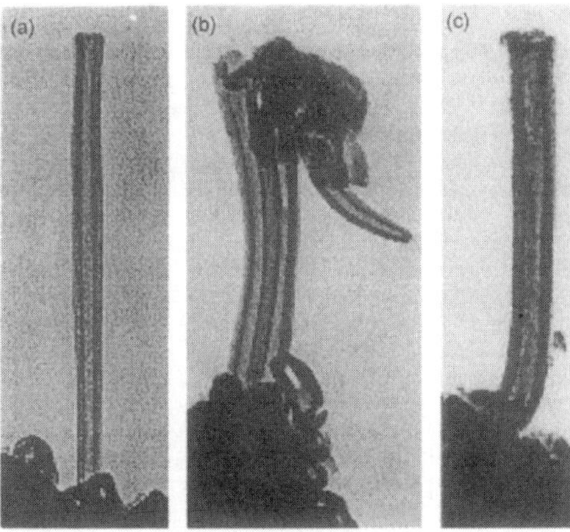

Fig. 34 TEM images of alumina nanotubes: (a and b) NSA tubes; (c) LSA tubes; (d) side-view of the branched alumina tubes attached to the APA film, (Y indicates Y-shaped branched cells, S indicates the stem of such branched cells). (Reproduced with permission from refs. 12 and 98).

450 K for 13 h in a Teflon-lined autoclave. There was, however, deviation in the peak intensities compared to the standard diffraction pattern, indicating the preferred orientation of the tubular ZnO (wurtzite). TEM images showed that the walls of the ZnO tubes were not very smooth due to the build-up by polycrystalline nanoparticles. The $Zn(NH_3)_4^{2+}$ complex is hydrolyzed in the presence of ammonia and nanoparticles of ZnO are produced along with the nanotubes as seen in the TEM images. As the temperature and duration of the reaction are increased, there is increased evolution of ammonia and the ZnO nanoparticles assemble along certain orientations and aggregate to the hollow tubular structures.

Oxides of Er, Tm, Yb and Lu have been prepared in the nanotubular form employing template-mediated reactions using dodecylsulfate assemblies.[100] These oxidic nanotubes were synthesized by the homogeneous precipitation method using urea. The oxidic phase was precipitated from the reaction mixture containing the salt of the rare earth element, sodium dodecylsulfate and water, the pH of the medium being varied by the progressive addition of urea. Hydrolysis of the rare earth salt occurs on heating to 60 °C and the oxidic phase forms a precipitate. The rare earth oxide nanotubes so obtained have small inner diameters and thin walls. Only Yb- and Lu-oxide nanotubes have been obtained reproducibly by this method.

Nanotubes of In_2O_3 and Ga_2O_3 have been synthesized by employing sol–gel chemistry and porous alumina templates.[101] Aqueous ammonia is added to In (or Ga) containing sols to

obtain precipitates which are then peptized with nitric acid to produce stable sols. The alumina membrane is immersed in the sol and then air-dried followed by annealing in air at elevated temperatures for 12 h to obtain the oxides. The alumina template is dissolved in alkali solution to give the free tubes. The hollow nanotubes so obtained have lengths of upto 10 μm. The inner diameters could be varied by selecting the template dimensions and the immersion time. The positively charged sol particles adhere to the negatively charged pore walls of the templating membrane leading to the formation of alumina–semiconductor composite nanotubes after annealing.

Nanotubes of perovskite oxides have been prepared by employing a template-mediated growth mechanism.[102] Nanotubes of the ceramic materials such as $BaTiO_3$ and $PbTiO_3$ are obtained by heating the metal acetate sol with Ti-isopropoxide in ethanol. Masked Whatman anodic membranes (200 nm pores) were used as the templates and were immersed in the mixed sol, and then air-dried. The templates could be removed after calcination followed by treatment with 6 M NaOH. TEM images revealed 50 μm long $BaTiO_3$ tubes bundled together after removal of the template. Most of the tubes obtained were open-ended (Fig. 35). Nanotubes of both $PbTiO_3$ and $BaTiO_3$

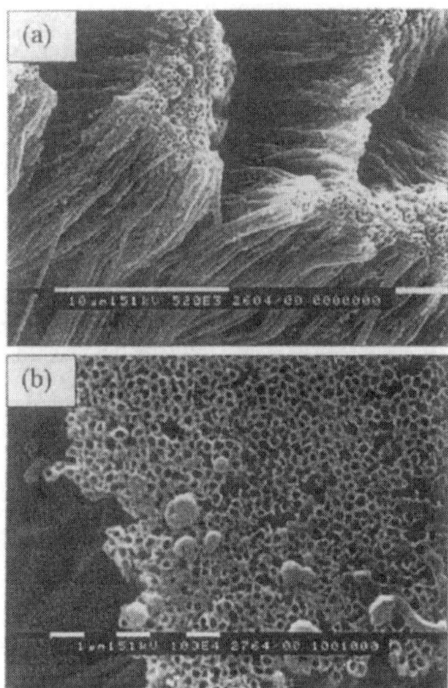

Fig. 35 SEM images of the perovskite nanotubes: (a) side-view showing the bundle formation after removal of the template; (b) top-view of $PbTiO_3$ bundles showing the open ends. (Reproduced with permission from ref. 102).

were found to be hollow throughout their length. XRD, Raman and electron diffraction measurements indicated that $BaTiO_3$ was in the cubic (paraelectric) phase, while $PbTiO_3$ was in a tetragonal ferroelectric phase after calcination. TEM images also showed that the tubes comprise small polycrystalline grains.

7. Nanotubes of BN and other nitrides

Based on theoretical calculations, the existence of nanotube structures of BN was predicted in 1994[13,103] which was soon verified by the first synthesis of BN nanotubes in 1995.[104] Based on theoretical calculations, it was predicted that unlike the carbon nanotubes whose electronic properties (metallic or semiconducting nature) are controlled by the tube diameter, wrapping, twisting and topological defects, the electronic properties of the BN nanotubes were independent of the tube diameter and chirality. In contrast to the CNTs, the BN nanotubes were predicted to be constant band-gap materials, with a large band gap of ~5.5 eV.[13]

Thin BN tubes of less than 200 nm diameter were obtained by arc discharge with hollow tungsten electrodes filled with h-BN powder. Following this initial report, a variety of methods have been employed to prepare BN nanotubes. In some cases, BN nanocages and fullerenes were also obtained.[105] The methods of synthesis of BN nanotubes include those which are far from the equilibrium, such as the electrical arc method,[32,104] arcing between h-BN and Ta rods in a N_2 atmosphere,[106] laser ablation of h-BN,[107] and continuous laser heating of h-BN surfaces.[108] The last method is particularly useful in providing long ropes of BN nanotubes with thin walls. Faceted BN polyhedra filled with metallic boron are found in a few cases. Single-walled BN nanotubes are also observed in some cases.[32,109] Recently, single-walled nanotubes of BN deposited on polycrystalline W substrates were obtained by using electron-cyclotron resonance nitrogen and electron beam boron sources.[110] Fig. 36 shows a

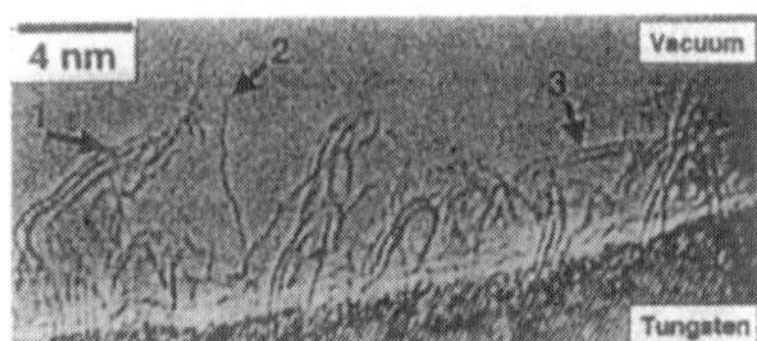

Fig. 36 HREM image of single-walled BN nanotubes adhering on a W substrate. (Reproduced with permission from ref. 110).

HREM image of the single-walled BN nanostructures adhering to the W substrate. *In-situ* HREM images show that the single-walled BN nanotubes contain both four- and eight-fold rings and the tubes are closed with fullerene-like structures. This is in contrast to the rectangular terminations observed earlier[109] which were interpreted in terms of the closure *via* four-fold rings and by nitrogen-rich pentagons.

Boron nitride nanotubes co-existing with a web and an amorphous phase were obtained by a plasma jet method wherein a sintered BN disk was subjected to a direct current arc plasma jet with Ar–H_2 as the plasma gas.[111] As revealed by the HREM image (Fig. 37), the tubes grow from the amorphous phase present at the root of the tubes. The tubes were closed by parallel bases of co-axial cylindrical shape. Solid state processes have also been employed to obtain BN nanotubes. Hexagonal BN powder was first ball-milled to generate highly disordered and amorphous nanostructures which were then annealed under N_2 to 1570 K for about 10 h.[112] Several tubes with bamboo-like morphology were observed some of which contained metal particles at the tip as a contaminant from the stainless steel reaction chamber. The nanostructure of the ball-milled powder which acts as the catalyst was crucial for the tube growth.

Apart from the above methods, some of which employ drastic conditions, processes close to equilibrium conditions such as pyrolysis and chemical vapor deposition (CVD) have also been employed to prepare BN nanotubes. The CVD growth of hollow, crystalline BN nanotubules by the pyrolysis of borazine on nickel boride catalyst particles maintained at 1270–1370 K, produced nanotubes with bulbous or flag-like caps (Fig. 38).[16b] The reaction is given by,

$$B_3N_3H_6 + Ni\ boride \longrightarrow 3BN + 3H_2$$

A root-growth mechanism has been proposed for the growth of BN nanotubes, wherein the nanotubes nucleate on the nickel boride catalyst particle often with irregular initiation caps and

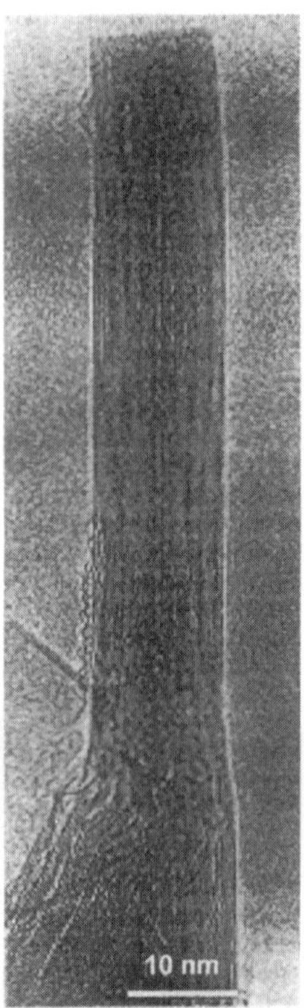

Fig. 37 TEM image of a BN nanotube having a phase boundary between the amorphous phase and the nanotube structure at the roots. (Reproduced with permission from ref. 111).

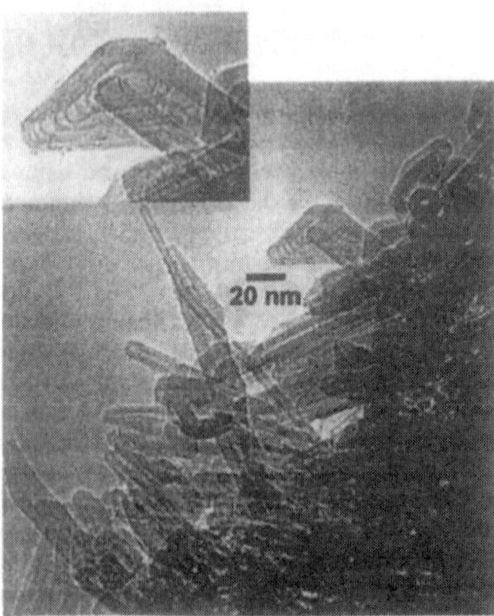

Fig. 38 TEM image showing different BN nanotube tip closures. The club-shaped tip is magnified in the inset. (Reproduced with permission from ref. 16b).

grow out by the incorporation of additional BN at the catalyst–nanotube junction. An efficient CVD method involving the thermal decomposition of the 1 : 2 adduct of melamine and boric acid in a N_2 atmosphere at 1970 K was developed for the synthesis of BN nanotubes. This method does not use any metal catalyst.[16c] The nanotubes have the stoichiometric composition and the HREM images (Fig. 39a) reveal the regular

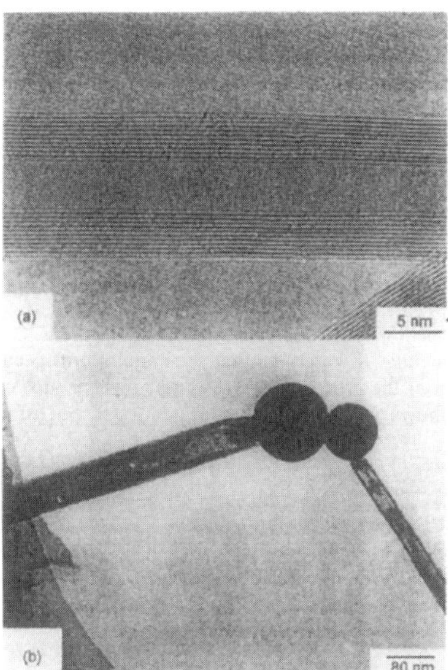

Fig. 39   (a) HREM image of a multi-walled BN tube (layer separation = 5.2 nm); (b) low-magnification image of the tubes showing that they may grow out from the bulbous tips. (Reproduced with permission from ref. 113).

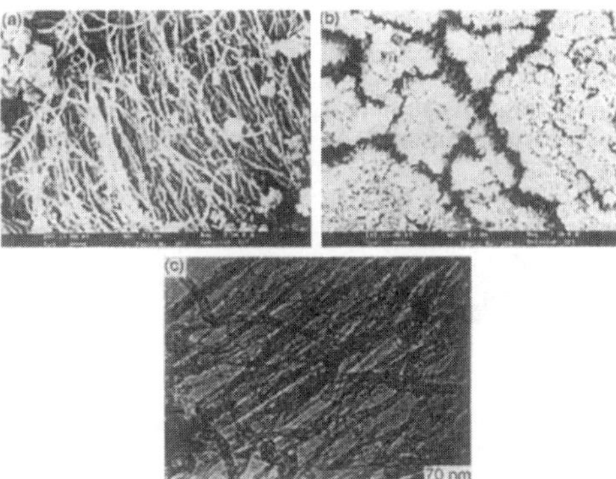

Fig. 40   SEM and TEM images of aligned BN nanotubes: (a) and (b) give side and top view SEM images, respectively; (c) TEM image of pure BN nanotube. (Reproduced with permission from ref. 17).

spacing of the BN layers (d = 5.2 nm). The nanotubes had bulbous tips filled with amorphous material (Fig. 39b) and the tip-growth mechanism seems to be valid. The amorphous material contains mainly $B_4N_3O_2H$.[113] It is believed that at the reaction temperature (1700 °C), some species such as B_2O_3 form along with the amorphous B–N–O cluster originating from decomposition of the precursor. The BN nanotube growth from the oxide phase follows the reaction

$$B_2O_3 + 4C + H_2O + N_2 \longrightarrow 2BN + 4CO + H_2$$

Boron powder has also been used in the CVD method for growing BN nanotubes. By heating a mixture of B and iron oxide in flowing ammonia gas, nanotubes and nano-bamboo structures of BN were obtained.[114] The *in-situ* generated Fe particles act as a catalyst and the growth of the BN nanotubes and nano-bamboo structures is ascribed to the vapor–liquid–solid (VLS) catalytic growth mechanism. Recently, B_2O_2 (obtained by heating B and MgO at 1300 °C) was heated in the presence of NH_3 in a BN-made reaction tube, to obtain BN nanotubes as the product with ~40% yield.[115] Most of the nanotubes obtained were open-tipped showing irregular fracture surfaces.

An exhaustive study has been carried out recently on the synthesis of BN nanotubes and nanowires by various CVD techniques.[17] The methods examined include heating boric acid with activated carbon, multi-walled carbon nanotubes, catalytic iron particles or a mixture of activated carbon and iron particles, in the presence of ammonia. With activated carbon, BN nanowires are obtained as the primary product. However, with multi-walled carbon tubes, high yields of pure BN nanotubes are obtained as the major product. BN nanotubes with different structures were obtained on heating boric acid and iron particles in the presence of NH_3. Aligned BN nanotubes are obtained when aligned multi-walled nanotubes are used as the templates (Fig. 40). Prior to this report, alignment of BN nanotubes was achieved by the synthesis of the BN nanotubule composites in the pores of the anodic alumina oxide, by the decomposition of 2,4,6-trichloroborazine at 750 °C.[116] Attempts had been made earlier to align BN nanotubes by depositing them on the surfaces of carbon fibers.[113] The tubes grow almost vertically from the surface and are nearly aligned.

Carbon nanotubes have been used successfully as templates for the growth of BN nanotubular structures.[15,17,117] MWNTs were heated to 1773 K with B_2O_3 in a N_2 atmosphere. Several metal oxides like MoO_3, V_2O_5, Ag_2O or PbO[118] were used as growth promoters, and added separately into the starting powder mixtures. Pyrolysis of borazine in the presence of acetylene over metal catalyst particles yielded B–C–N nanotubes.[15a] Previously $B_xC_yN_z$ nanotubules were prepared by the arc-discharge method.[14a] B–C–N composite rods were used as the anode and arced against pure graphite cathodes in a He gas environment to yield the nanotubules.

Boron nitride nanotubes have been used in a wide variety of ways to generate nanocables where the BN nanotubes play host to 1D nanowires or nanoclusters occupying the hollow cavity. Zhang and co-workers[119] have reported the synthesis of $(BN)_xC_y$ nanotubes filled with a SiC and SiO_2 core by laser ablation. Carbon nanotube-confined reactions involving substitution reactions are employed to synthesize SiC nanowires encapsulated in BN nanotubes.[120] The CNTs react with boron oxide vapor in the presence of N_2 to yield BN nanotubes. The SiO vapor then penetrates into the cavity of the nanotubes and reacts with the internal wall of the CNTs to give SiC nanowires. In some cases, the filling occurs through the entire length of the nanotube. In a slightly modified method, BN and $(BN)_xC_y$ nanotubes filled with boron carbide nanowires were prepared using CNTs as templates.[121] The $(BN)_xC_y$ nanotubes are formed by capillary filling of boron oxide vapor in the inner cavity of the CNTs, followed by the substitution of the inner layers of the CNTs with B_2O_3 in the presence of N_2 gas. Inside the CNTs, boron oxide reacts simultaneously with the gaseous carbon monoxide or the interior layers of the CNTs to produce the boron carbide filling. The final product contained boron carbide filled $(BN)_xC_y$ nanotubes with an outer layer of pure C and inner layers of pure BN. Fig. 41 shows an illustration of the boron carbide nanowires inside the B_xC_yN composite tubes. Pure BN tubes filled with boron carbide were also formed in the product.

Nanocables of BN nanotubes filled with Mo clusters are reported by Golberg *et al.*[122] The Mo cluster-filled BN nanotubes are prepared by the treatment of CVD grown CNTs with B_2O_3, CuO and MoO_3 in a N_2 atmosphere. It has been proposed that the filling of CNTs with MoO_3 precedes the formation of BN tubes on the CNT template. The filling of MoO_3 is then further reduced to metallic Mo by the carbon of the CNTs. Continuous filling of Mo could not be obtained by this method.

BN nanotubes have been used as hosts to oxide materials such as α-Al_2O_3 nanorods.[123] The $B_4N_3O_2H$ precursor (obtained

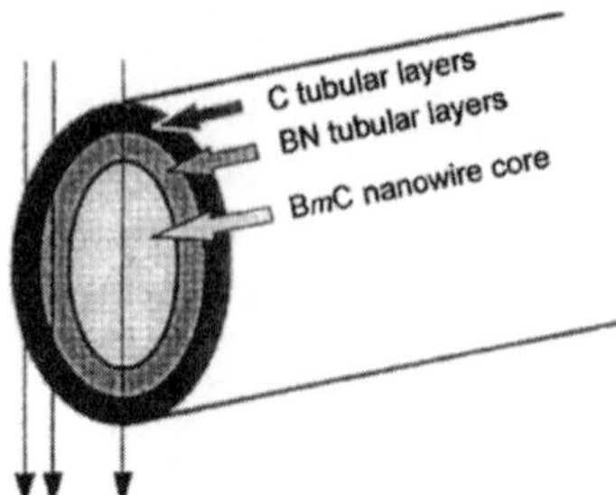

Fig. 41 Schematic illustration of a multi-phase filled B_xC_yN nanotube. (Reproduced with permission from ref. 121).

by the thermal decomposition of melamine diborate) when pyrolyzed on Si–SiO$_2$–Al$_2$O$_3$ substrate, yielded α-Al$_2$O$_3$ nanorod-filled BN nanocables in abundance.

According to Menon and Srivastava,[124] the chirality of a nanotube directly depends on the tip-end morphology, *i.e.*, for a flat tip end, a zigzag arrangement of the layers in the tube walls is energetically favorable. Rhombohedral stacking in relatively thick BN tubular fibers has been observed,[125] in contrast to the belief that the BN nanotubes are exact analogues of CNTs which usually show random stacking between the layers and no preferential tube helicity. HREM studies of the BN nanotubes reveal layered structures somewhat similar to those in the CNTs (Fig. 42a). However, there are some distinct characteristics. A

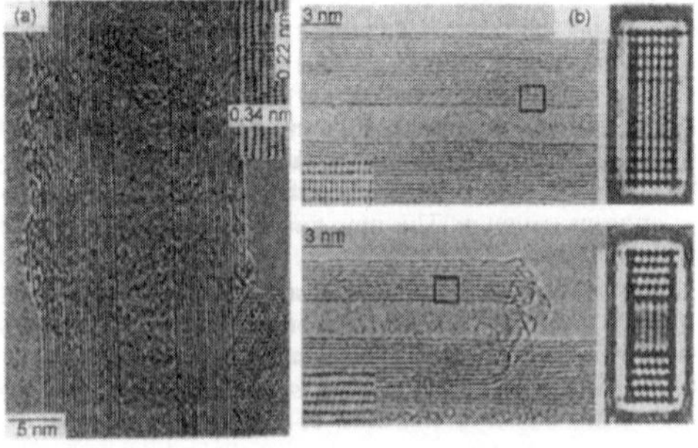

Fig. 42 (a) A BN nanotube with a uniform layer separation of 0.34 nm. The atomic columns in the wall fragments exhibit layer fringes separated by 0.22 nm and makes an angle of 12.5° with respect to the tube axis, exhibiting *r*-BN stacking; (b) hexagonal stacking and *r*-BN stacking confirmed by the computer-simulated HREM images (right hand side panels). (Reproduced with permission from ref. 120).

field emission high-resolution analytical TEM study has revealed the following characteristic features of the multi-walled BN nanotubes: (i) Hexagonal and rhombohedral (3R) stackings co-exist in nanotube shell assembly, a feature readily seen in HREM images where the layers due to 3R stacking occur in the walls and sometimes in the core besides the regular hexagonal stacking (Fig. 42a). In Fig. 42b the observed HREM images of BN nanotubes are compared with the computer simulated images for the *r*-BN and *h*-BN stacking in the walls. (ii) Flattening of the nanotube cross-section makes clear atomic resolution of the pore structure possible in a three-shelled nanotube. (iii) There is a change in chirality of tubular layers from armchair to zigzag arrangement in a 30° double-walled nanotube kink. BN nanotubes with open tips exhibit local *r*-BN stacking in the walls, while those with close tips exhibit *h*-BN stacking[120] (Fig. 42b).

BN nanotubes obtained by laser ablation and arc-discharge seem to have very few layers and the laser grown tubes self-assemble into long ropes. It is proposed that the growth of BN

nanotubes is due to surface diffusion along the external surface, which also ensures morphological stability of the open end during growth. The nanotube heights are limited by the corresponding diffusion lengths. Recombination of B and N, both in plasma and on the surface, may serve as a possible nucleation center. In the pyrolytically grown BN nanotubes, it is commonly observed that the nanotubes have bulbous tips.[113] The amorphous clusters present in the tip region may play a catalytic role in the nanotube tip-growth process similar to the metal catalyst in the CVD process of CNT growth. Formation of open/flat-tip ends are also observed in some of the BN tubes. There appears to be a preference for the growth of open BN nanotubes in metal oxide-promoted CVD synthesis. One of the reasons behind this may be that the metal atoms occurring at the edges of the growing nanotube may prevent tube closure.[44] Fig. 43 shows the presence of a dark contrast spot on the inner

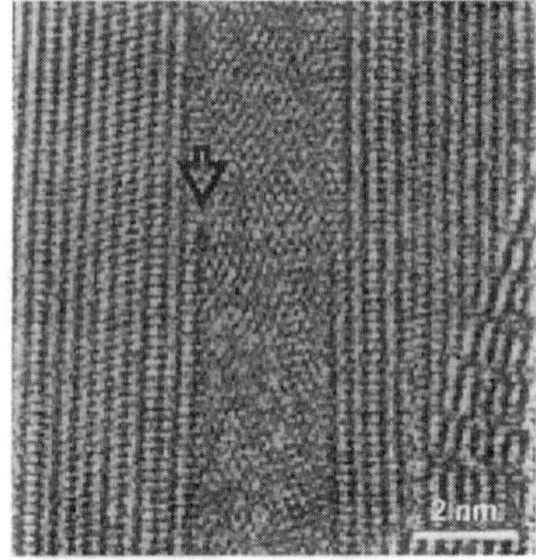

Fig. 43 The innermost terminated layer of the BN nanotube shows a spot assigned to the Pb atom (or cluster), which may prevent tube closure. (Reproduced with permission from ref. 44).

layer of the tube at the terminated edge. The spot is likely to be related to the presence of the metal atom (Pb) or the metal cluster at that edge, which prevents closure of the layer.

Recently, in their effort to prepare Si$_3$N$_4$ nanowires, Gundiah *et al.*[126] have found occasional nanotubes in the TEM images. Similarly, in the preparations of GaN nanowires, GaN nanotubes have been observed by Deepak *et al.*[127]

8. Nanotubes of other materials

Transition metal halides such as NiCl$_2$ crystallize in the CdCl$_2$ structure, with the metal halide layers held together by weak van der Waals forces. NiCl$_2$ has been shown to form closed cage structures and nanotubes.[128] These were prepared by heating NiCl$_2$·6H$_2$O initially in air to lose the water of crystallization, and then heated further at 450 °C under N$_2$ (Fig. 44).

Very few metallic nanotubes have been synthesized to date. Martin and co-workers[129] have prepared Au nanotubules with lengths upto 6 μm and inner diameters of 1 nm by using a pore-wall modified alumina membrane. Co and Fe nanotubules have been synthesized using polycarbonate membranes as templates.[130] Cu and Ni microtubules have also been prepared by the pyrolysis of composite fibers consisting of a poly(ethylenetetraphthalate) (PET) core fiber and electroless-plated metal skin at the exterior.[131] While Ni microtubules prepared by this method were single-crystalline, the Cu microtubules were polycrystalline. Ordered arrays of Ni nanotubules have been prepared by electrodeposition in the pores of an alumina membrane, the pore walls being modified with an organic amine.[132] Nickel when electrodeposited in the pores binds preferentially to the pore walls because of its strong affinity towards the

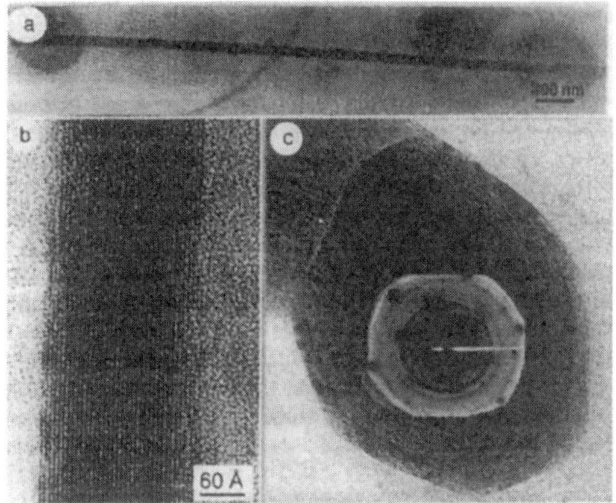

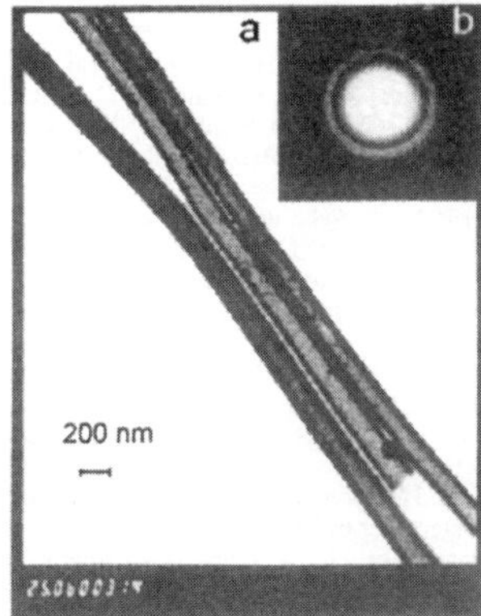

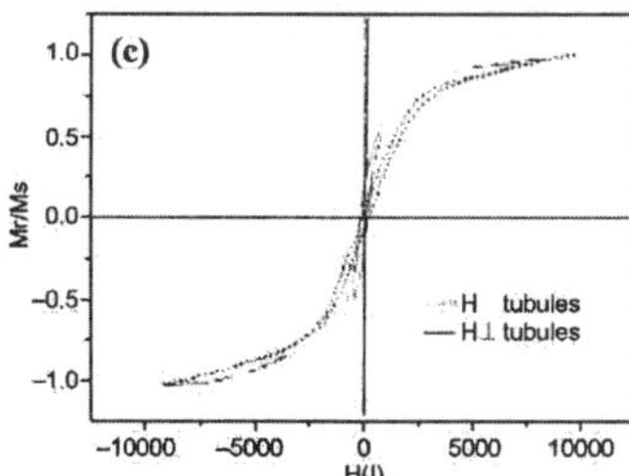

Fig. 44 (a) Low-magnification view of the NiCl$_2$ nanotube; (b) HREM image of the nanotube wall; (c) many-layered cage structure of NiCl$_2$, with the hexagonal ED pattern superimposed. (Reproduced with permission from ref. 128).

amine. In the absence of the amine in the pore walls, solid Ni nanowires were obtained. The alumina membrane could be removed by treatment with NaOH giving highly ordered arrays of Ni nanotubules (Fig. 45a). The Ni nanotubules were ferromagnetic with enhanced coercivity compared to bulk Ni (Fig. 45b).

Fig. 45 (a) TEM image showing the Ni nanotubules; (b) ED pattern of the Ni nanotubules; (c) magnetization *vs.* applied field (*M–H*) curve showing hysteresis. (Reproduced with permission from ref. 132).

Tellurium nanotubes have been prepared using the polyol method. Orthotelluric acid in ethylene glycol was added to a refluxing solution of ethylene glycol.[133] TEM images taken after stopping the reaction at different stages showed the formation of cylindrical seeds and the subsequent growth of nanotubules along the peripheral edge of the seeds. The walls of the nano-

tubes have a uniform thickness, but appear to be fragile compared to CNTs. Decreasing the amount of the Te precursor in the initial reaction mixture led to the preferential formation of solid nanorods rather than hollow tubes.

9. Useful properties of inorganic nanotubes

Various properties of carbon nanotubes of potential technological value are known.[2,134] The properties and applications of the inorganic nanotubes, however, have not been investigated as extensively as would be desirable. The electronic structures of MoS$_2$ and WS$_2$ have been examined briefly and the semiconducting nature of the nanotubes confirmed.[55,57a] It is necessary to investigate the optical, electrical and other properties of the various chalcogenide nanotubes. This is especially true of nanotubes of NbS$_2$ and such materials which are predicted to be metallic.[57b] NbSe$_2$ nanotubes have been found to be metallic at ordinary temperatures, becoming superconducting at lower temperatures.[70] Electronic and optical properties of the BN nanotubes have not yet been investigated in detail. Theoretical calculations suggest BN tubes to be insulating with a wide band gap of 5.5 eV.[13]

Carbon nanotubes have been investigated for H$_2$ storage properties.[135] It would be worthwhile to look into the H$_2$ storage ability of some of the inorganic nanotubes. The chalcogenide nanotubes with an ~6 Å van der Waals gap between the layers, are potential candidates for showing storage capacity. It has been shown recently that BN nanotubes can store a reasonable quantity of H$_2$.[136] Multi-walled BN nanotubes have been shown to possess a capacity of 1.8–2.6 wt% of H$_2$ uptake under ~10 MPa at room temperature. This value, though smaller than that reported for CNTs, nevertheless suggests the possible use of BN nanotubes as a hydrogen storage system. MoS$_2$ nanotubes could be electrochemically charged and discharged with a capacity of 260 mA h g^{-1} at 20 °C, corresponding to a formula of H$_{1.24}$MoS$_2$.[137] The high storage capacity is believed to be due to the enhanced electrochemical-catalytic activity of the highly nanoporous structure. This may find wide applications in high energy batteries.

Single-walled carbon nanotubes are known to possess extraordinary strength.[138] Mechanical properties of BN nanotubes would be worthy of exploration. Unlike carbon nanotubes, BN nanotubes are predicted to have stable insulating properties independent of their structure and morphology. Thus, BN tubes can be used as nano-insulating devices for encapsulating conducting materials like metallic wires. Filled BN nanotubes are expected to be useful in nanoscale electronic devices and for the preparation of nano-structured ceramics.

Electrochemical studies have been performed with the alkylammonium intercalated VO$_x$ nanotubes[139] as well as Mn intercalated VO$_x$ nanotubes.[87] Cyclic voltammetry studies of alkylammonium-VO$_x$ nanotubes showed a single reduction peak, which broadened on replacing the amine with Na with an additional peak. Li ion reactivity has also been tested with Mn–VO$_x$ nanotubes by reacting with *n*-butyllithium, and found that ~2 lithiums per V ion are consumed. Electrochemical Li intercalation of Mn–VO$_x$ nanotubes show that 0.5 Li ions per V atom were intercalated above 2 V.[87] This observation may be relevant to battery applications.

Conventional microfabricated AFM tips are of limited use for investigating high aspect ratio features (*i.e.* deep and narrow features), mainly because, without special treatment typical aspect ratios of such tips would be around 3 : 1 or lower. Thus, the width of such a tip at a certain height from the apex is much larger than that of the nanotubes of uniform thickness adhered to the tip. The nanotubes therefore would be more suitable for the analysis of deep and narrow structures than the commonly available tips. CNTs have been used as AFM tips and there appears to be every likelihood that extremely narrow structures can be probed.[140] WS$_2$ could be mounted on the ultrasharp Si tip following a similar methodology. These tips were tested in an AFM microscope by imaging a replica of high aspect ratio, and it was observed that these WS_2 nanotube tips provide a considerable improvement in the image quality compared to the conventional ultrasharp Si tips.[141]

The most likely application of the chalcogenide nanotubes is as solid lubricants. Mo and W chalcogenides are widely used as solid lubricants. It has been observed that the hollow nanoparticles of WS_2 show better tribological properties and act as a better lubricant compared to the bulk phase in every respect (friction, wear and life-time of the lubricant).[142] Tribological properties of $2H–MoS_2$ and WS_2 powder can be·attributed to the weak van der Waals forces between the layers which allow easy shear of the films with respect to each other. The mechanism in the WS_2 nanostructures is somewhat different and the better tribological properties may arise from the rolling friction allowed by the round shape of the nanostructures.

Recently, open-tipped MoS_2 nanotubes were prepared by the decomposition of ball-milled ammonium thiomolybdate powder under a H_2 + thiophene atmosphere, and used as a catalyst for the methanation of CO with H_2.[143] The conversion of CO to CH_4 was achieved at a much lower temperature compared to polycrystalline MoS_2 particles, and there was no deterioration even after 50 h of consecutive catalyzing cycles. This observation is of importance in the context of energy conversion of global CO_2.

10. Concluding remarks

Inorganic nanotubes have emerged to become a group of novel materials. Although this area of research started with the layered metal chalcogenides, recent results suggest that other inorganic materials can also be prepared in the form of nanotubes, as typified by the metal oxides. It is likely that many new types of inorganic nanotubes will be made in the near future. These would include metal nanotubes as well as nanotubes of inorganic compounds such as Mg_3B_2, GeO_2 and GaSe. Theoretical calculations indeed predict a stable nanotubular structure for GaSe.[144] Various layered materials could be explored for this purpose. It is noteworthy that the nanotubes of metal chalcogenides have been made by employing several methods ranging from soft chemical routes to techniques such as arc evaporation and laser ablation (Table 1).

Nanotubes of MoS_2, WS_2 and a few other layered materials are single-crystalline in the sense that the layers run through the entire structure. Some of the chalcogenide nanotubes are, however, polycrystalline, the nanotubular form being produced by an aggregation of nanoparticles, just as in some of the metal oxide nanotubes.

Properties of inorganic nanotubes such as those of MoS_2 have been investigated to some extent. However, by and large, there is much to be studied with respect to the electronic, optical and other properties of most of the inorganic nanotubes. Properties such as sorption, hydrogen storage and catalytic activity are worthy of exploration. Mechanical properties of BN, B–N–C and related nanotubes are also worthy of study.

Acknowledgements

The authors thank the Department of Science & Technology and DRDO (India) for support of this research.

References

1 S. Iijima, *Nature*, 1991, **354**, 56.
2 C. N. R. Rao, B. C. Satishkumar, A. Govindaraj and M. Nath, *Chem. Phys. Chem.*, 2001, **2**, 78.
3 (*a*) P. Ratnasamy, L. Rodrigues and A. J. Leonard, *J. Phys. Chem.*, 1973, **77**, 2242; (*b*) J. Wilson and A. D. Yoffe, *Adv. Phys.*, 1969, **269**, 193.
4 R. R. Chianelli, E. Prestridge, T. Pecorano and J. P. DeNeufville, *Science*, 1979, **203**, 1105.
5 (*a*) R. Tenne, L. Margulis, M. Genut and G. Hodes, *Nature*, 1992, **360**, 444; (*b*) L. Margulis, G. Salitra and R. Tenne, *Nature*, 1993, **365**, 113; (*c*) Y. Feldman, E. Wasserman, D. J. Srolovitch and R. Tenne, *Science*, 1995, **267**, 222.
6 (*a*) C. T. Kresge, M. E. Leonowicz, W. J. Roth, J. C. Vartulli and J. S. Beck, *Nature*, 1992, **259**, 710; (*b*) J. S. Beck, J. C. Vartulli, W. J. Roth, M. E. Leonowicz, C. T. Kressge, K. D. Schmitt, C. T. W. Chu, D. H. Olson, E. W. Sheppard, S. B. McCullen, J. B. Higgins and J. C. Scwenker, *J. Am. Chem. Soc.*, 1992, **114**, 10834.
7 G. R. Pratzke, F. Krumeich and R. Nesper, *Angew. Chem., Int. Ed.*, 2002, **41**, 2446.
8 (*a*) W. Stöber, A. Fink and E. Bohn, *J. Colloid Interface Sci.*, 1968, **26**, 62; (*b*) M. Nakamura and Y. Matsui, *J. Am. Chem. Soc.*, 1995, **117**, 2651.
9 P. M. Ajayan, O. Stephane, Ph. Redlich and C. Colliex, *Nature*, 1995, **375**, 564.
10 (*a*) B. C. Satishkumar, A. G. Govindaraj, E. M. Vogl, L. Basumallick and C. N. R. Rao, *J. Mater. Res.*, 1997, **12**, 604; (*b*) B. C. Satishkumar, A. Govindaraj, M. Nath and C. N. R. Rao, *J. Mater. Chem.*, 2000, **10**, 2115.
11 M. E. Spahr, P. Bitterli, R. Nesper, M. Müller, F. Krumeich and H. U. Nissen, *Angew. Chem., Int. Ed.*, 1998, **37**, 1263 (*Angew. Chem.*, 1998, **110**, 1339).
12 L. Pu, X. Bao, J. Zou and D. Feng, *Angew. Chem., Int. Ed.*, 2001, **40**, 1490.
13 (*a*) M. L. Cohen, *Solid State Commun.*, 1994, **92**, 45; (*b*) J. L. Corkill and M. L. Cohen, *Phys. Rev. B*, 1994, **49**, 5081; (*c*) Y. Miyamoto, A. Rubio, S. G. Louie and M. L. Cohen, *Phys. Rev. B*, 1994, **50**, 18360.
14 (*a*) K. Kobayashi and N. Kurita, *Phys. Rev. Lett.*, 1993, **70**, 3542; (*b*) Z. W. -Sieh, K. Cherrey, N. G. Chopra, X. Blasé, Y. Miyamoto, A. Rubio, M. L. Cohen, S. G. Louie, A. Zettl and P. Gronsky, *Phys. Rev. B*, 1994, **51**, 11229.
15 (*a*) R. Sen, B. C. Satishkumar, A. Govindaraj, K. R. Harikumar, G. Raina, J. P. Zhang, A. K. Cheetham and C. N. R. Rao, *Chem. Phys. Lett.*, 1998, **287**, 671; (*b*) O. Stephan, P. M. Ajayan, C. Colliex, Ph. Redlich, J. M. Lambert, P. Bernier and P. Lefing, *Science*, 1994, **266**, 1683.
16 (*a*) P. Gleize, M. C. Schouler, P. Gadelle and M. Caillet, *J. Mater. Sci.*, 1994, **29**, 1575; (*b*) O. R. Lourie, C. R. Jones, B. M. Bertlett, P. C. Gibbons, R. S. Ruoff and W. E. Buhro, *Chem. Mater.*, 2000, **12**, 1808; (*c*) R. Ma, Y. Bando and T. Sato, *Chem. Phys. Lett.*, 2001, **337**, 61.
17 F. L. Deepak, C. P. Vinod, K. Mukhopadhyay, A. Govindaraj and C. N. R. Rao, *Chem. Phys. Lett.*, 2002, **353**, 345.
18 C. N. R. Rao and A. G. Govindaraj, *Acc. Chem. Res.*, 2002 and references therein.
19 R. Tenne, M. Homyonfer and Y. Feldman, *Chem. Mater.*, 1998, **10**, 3225 and references therein.
20 (*a*) M. Hershfinkel, L. A. Gheber, V. Volterra, J. L. Hutchison, L. Margulis and R. Tenne, *J. Am. Chem. Soc.*, 1994, **116**, 1914; (*b*) T. Tsirlina, Y. Feldman, M. Homyonfer, J. Sloan, J. L. Hutchison and R. Tenne, *Fullerene Sci. Technol.*, 1998, **6**, 157.
21 M. Nath, A. Govindaraj and C. N. R. Rao, *Adv. Mater.*, 2001, **13**, 283.
22 M. Nath and C. N. R. Rao, *Chem. Commun.*, 2001, 2236.
23 M. Nath and C. N. R. Rao, *J. Am. Chem. Soc.*, 2001, **123**, 4841.
24 M. Nath and C. N. R. Rao, *Angew. Chem., Int. Ed.*, 2002, **41**, 3451.
25 A. P. Lin, C. Y. Mou and S. D. Liu, *Adv. Mater.*, 2000, **12**, 103.
26 J. Zhang, L. Sun, C. Liao and C. Yan, *Chem. Commun.*, 2002, 262.
27 M. Niederberger, H.-J. Muhr, F. Krumeich, F. Bieri, D. Günther and R. Nesper, *Chem. Mater.*, 2000, **12**, 1995.
28 C. N. R. Rao, A. G. Govindaraj, F. L. Deepak, N. A. Gunari and M. Nath, *Appl. Phys. Lett.*, 2001, **78**, 1853.
29 T. Kasuga, M. Hiramatsu, A. Hason, T. Sekino and K. Niihara, *Langmuir*, 1998, **14**, 3160.
30 C. N. R. Rao, B. C. Satishkumar and A. Govindaraj, *Chem. Commun.*, 1997, 1581.
31 C. M. Zelenski and P. K. Dorhout, *J. Am. Chem. Soc.*, 1998, **120**, 734.
32 A. Loiseau, F. Williame, N. Demonecy, G. Hug and H. Pascard, *Phys. Rev. Lett.*, 1996, **76**, 4737.
33 M. Homyonfer, Y. Mastai, M. Hershfinkel, V. Volterra, J. L. Hutchison and R. Tenne, *J. Am. Chem. Soc.*, 1996, **118**, 7804.
34 Y. Feldman, G. L. Frey, M. Homyonfer, V. Lyakhovitskaya, L. Margulis, H. Cohen, G. Hodes, J. L. Hutchison and R. Tenne, *J. Am. Chem. Soc.*, 1996, **118**, 5362.

35 M. Homyonfer, B. Alperson, Y. Rosenberg, L. Sapir, S. R. Cohen, G. Hodes and R. Tenne, *J. Am. Chem. Soc.*, 1997, **119**, 2693.
36 S. P. Cramer, K. S. Liang, A. J Jacobson, C. H. Chang and R. R. Chianelli, *Inorg. Chem.*, 1984, **23**, 1215.
37 G. U. Kulkarni and C. N. R. Rao, *Catal. Lett.*, 1991, **11**, 63.
38 (a) E. Diemann and A. Müller, *Coord. Chem. Rev.*, 1973, **10**, 79; (b) R. R. Chianelli and M. B. Dines, *Inorg. Chem.*, 1978, **17**, 2758; (c) A. Wildervanck and F. Jellinek, *Z. Anorg. Allg. Chem*, 1964, **328**, 309.
39 (a) C. N. R. Rao and K. P. R. Pisharody, *Prog. Solid State Chem.*, 1976, **10**, 207; (b) N. Allali, V. Gaborit, E. Prouzet, C. Geantet, M. Danot and A. Nadiri, *J. Phys. IV (France)*, 1997, **7**, C2–927; (c) E. Bjerkelund and A. Kjekhus, *Z. Anorg. Allg. Chem.*, 1964, **328**, 235; (d) F. K. McTaggart and A. D. Wadsley, *Aust. J. Chem*, 1958, **11**, 445.
40 M. Nath, K. Mukhopadhyay and C. N. R. Rao, *Chem. Phys. Lett.*, 2002, **352**, 163.
41 (a) Y. Q. Zhu, W. K. Hsu, M. Terrones, S. Firth, N. Grobert, R. J. H. Clark, H. W. Kroto and D. R. M. Walton, *Chem. Commun.*, 2001, 121; (b) Y. Q. Zhu, W. K. Hsu, M. Terrones, S. Firth, N. Grobert, R. J. H. Clark, H. W. Kroto and D. R. M. Walton, *Chem. Phys. Lett.*, 2001, **342**, 15.
42 W. K. Hsu, Y. Q. Zhu, N. Yao, S. Firth, R. J. H. Clark, H. W. Kroto and D. R. M. Walton, *Adv. Funct. Mater.*, 2001, **11**, 69.
43 W. K. Hsu, Y. Q. Zhu, C. B. Bothroyd, I. Kinlöch, S. Trasobares, H. Terrones, N. Grobert, M. Terrones, R. Escudero, G. Z. Chen, C. Colliex, A. H. Windle, D. H. Fray, H. W. Kroto and D. R. M. Walton, *Chem. Mater.*, 2000, **12**, 3541.
44 D. Golberg and Y. Bando, *Appl. Phys. Lett.*, 2001, **79**, 415.
45 (a) R. L. D. Whitby, W. K. Hsu, C. B. Bothroyd, P. K. Fearon, H. W. Kroto and D. R. M. Walton, *Chem. Phys. Chem.*, 2001, **10**, 620; (b) R. L. D. Whitby, W. K. Hsu, P. K. Fearon, N. C. Billingham, I. Maurin, H. W. Kroto, D. R. M. Walton, C. B. Bothroyd, S. Firth, R. J. H. Clark and D. Collison, *Chem. Mater.*, 2002, **14**, 2209; (c) R. L. D. Whitby, W. K. Hsu, P. C. P. Watts, H. W. Kroto and D. R. M. Walton, *Appl. Phys. Lett.*, 2001, **79**, 4574; (d) W. K. Hsu, Y. Q. Zhu, H. W. Kroto, D. R. M. Walton, R. Kamalakaran and M. Terrones, *Appl. Phys. Lett.*, 2000, **77**, 4130.
46 Y. Q. Zhu, W. K. Hsu, H. W. Kroto and D. R. M. Walton, *Chem. Commun*, 2001, 2184.
47 Y. Q. Zhu, W. K. Hsu, H. Terrones, N. Grobert, B. H. Chang, M. Terrones, B. Q. Wei, H. W. Kroto, D. R. M. Walton, C. B. Bothroyd, I. Kinlöch, G. Z. Chen, A. H. Windle and D. J. Fray, *J. Mater. Chem.*, 2000, **10**, 2570.
48 Y. Q. Zhu, W. K. Hsu, N. Grobert, B. H. Chang, M. Terrones, H. Terrones, H. W. Kroto and D. R. M. Walton, *Chem. Mater.*, 2000, **12**, 1190.
49 N. Grobert, M. Terrones, A. J. Osborne, H. Terrones, W. K. Hsu, S. Trasobares, Y. Q. Zhu, J. P. Hare, H. W. Kroto and D. R. M. Walton, *Appl. Phys. A*, 1998, **69**, 595.
50 J. Sloan, J. L. Hutchison, R. Tenne, Y. Feldman, T. Srilina and M. Homyonfer, *J. Solid State Chem.*, 1999, **144**, 100.
51 M. Rémskar, Z. Skraba, C. Ballif, R. Sanjines and F. Levy, *Surf. Sci.*, 1999, **435**, 637.
52 M. Homyonfer, Y. Feldman, L. Margulis and R. Tenne, *Fullerene Sci. Technol.*, 1998, **6**, 59.
53 (a) S. Iijima, S. Ichihashi and Y. Ando, *Nature*, 1992, **356**, 776; (b) D. E. H. Jones, *Nature*, 1991, **351**, 526; (c) S. Iijima, P. M. Ajayan and T. Ichihashi, *Phys. Rev. Lett.*, 1991, **69**, 3900.
54 R. Tenne, *Adv. Mater.*, 1995, **7**, 965.
55 G. Seifert, H. Terrones, M. Terrones, G. Jungnickel and T. Frauenheim, *Phys. Rev. Lett.*, 2000, **85**, 146.
56 G. Seifert, T. Köhler and R. Tenne, *J. Phys. Chem. B*, 2002, **106**, 2497.
57 (a) G. Seifert, H. Terrones, M. Terrones, G. Jungnickel and T. Frauenheim, *Solid State Commun.*, 2000, **114**, 245; (b) G. Seifert, H. Terrones, M. Terrones, G. Jungnickel and T. Frauenheim, *Solid State Commun.*, 2000, **115**, 635.
58 C. Thomazeau, C. Geantet, M. Lacroix, V. Harlé, S. Benazeth, C. Marhic and M. Danot, *J. Solid State Chem.*, 2001, **160**, 147.
59 A. Zak, Y. Feldman, V. Alperovich, R. Rosentsveig and R. Tenne, *J. Am. Chem. Soc.*, 2000, **122**, 11108.
60 (a) P. J. F. Harris, in *Carbon Nanotube and Related Structures*, Cambridge University Press, UK, 1999 pp. 62–107; (b) S. Amelinckx, D. Bernaerts, X. B. Zhang, G. Van Tendeloo and J. Van. Landuyt, *Science*, 1995, **267**, 1334; (c) A. Oberlin, M. Endo and T. Kayama, *J. Cryst. Growth*, 1992, **32**, 6941; (d) H. Dai, A. Z. Rinzler, P. Nikolaev, A. Thess, D. T. Colbert and R. E. Smalley, *Chem. Phys. Lett.*, 1996, **260**, 471.
61 W. K. Hsu, B. H. Chang, Y. Q. Zhu, W. Q. Han, H. Terrones, M. Terrones, N. Grobert, A. K. Cheetham, H. W. Kroto and D. R. M. Walton, *J. Am. Chem. Soc.*, 2000, **122**, 10155.
62 P. Afanasiev, C. Geantet, C. Thomazeau and B. Jouget, *Chem. Commun.*, 2000, 1001.
63 M. Remškar, A. Mrzel, Z. Skraba, A. Jesih, M. Ceh, J. Demšar, P. Stadelmann, F. Lévy and D. Mihailovic, *Science*, 2001, **292**, 479.
64 R. Sen, A. Govindaraj, K. Suenaga, S. Suzuki, H. Kataura, S. Iijima and Y. Achiba, *Chem. Phys. Lett.*, 2001, **340**, 242.
65 D. H. Galván, R. Rangel and G. Alonso, *Fullerene Sci. Technol.*, 1998, **6**, 1025.
66 H. D. Flack, *J. Appl. Crystallogr.*, 1972, **5**, 138.
67 D. L. Greenaway and R. Nitsche, *J. Phys. Chem. Solids*, 1965, **26**, 1445.
68 M. I. Nathan, M. W. Shafer and J. E. Smith, *Bull. Am. Phys. Soc.*, 1972, **17**, 336.
69 D. H. Galván, J.-H. Kim, M. B. Maple, M. Avalos-Berja and E. Adem, *Fullerene Sci. Technol.*, 2000, **8**, 143.
70 M. Nath S. Kar A. K. Raychaudhuri C. N. R. Rao, *Chem. Phys. Lett.*, in press.
71 C. Schuffenhauer, R. P-Biro and R. Tenne, *J. Mater. Chem.*, 2002, **12**, 1587.
72 (a) C. S. Wang and J. M. Chen, *Solid State Commun.*, 1974, **14**, 1145; (b) A. LeBlanc-Soreau, P. Molinié and E. Faulques, *Physica C*, 1997, **282–287**, 1937.
73 R. F. Frindt, *Phys. Rev. Lett.*, 1972, **28**, 299.
74 Y. Wu, B. Messer and P. Yang, *Adv. Mater.*, 2001, **13**, 1487.
75 A. Govindaraj, F. L. Deepak, N. A. Gunari and C. N. R. Rao, *Israel J. Chem.*, 2001, **41**, 23.
76 P. V. Teredesai, F. L. Deepak, A. Govindaraj, C. N. R. Rao and A. K. Sood, *J. Nanosci. Nanotechnol.*, 2002, **2**, 495.
77 L. Dloczik, R. Engelhardt, K. Ernst, S. Fiechter, I. Seiber and R. Könenkamp, *Appl. Phys. Lett.*, 2001, **78**, 3687.
78 X. Ziang, Y. Xie, L. Zhu, W. He and Y. Qian, *Adv. Mater.*, 2001, **13**, 1278.
79 Y. Peng, Z. Meng, C. Zhong, J. Lu, L. Xu, S. Zhang and Y. Qian, *New J. Chem.*, 2001, **25**, 1359.
80 H. Imai, Y. Takei, K. Shimizu, M. Matsuda and H. Hirashima, *J. Mater. Chem.*, 1999, **9**, 2971.
81 H. G. Bachmann, F. R. Ahmed and W. H. Barnes, *Z. Kristallogr.*, 1961, **115**, 110.
82 R. Enjalbert and J. Galy, *Acta Crystallogr., Sect. C*, 1986, **42**, 1467.
83 J. Galy, *J. Solid State Chem.*, 1992, **100**, 229.
84 M. S. Whittingham, J. Guo, R. Chen, T. Chirayil, G. Janauer and P.Y. Zavalji, *Solid State Ionics*, 1995, **75**, 257.
85 F. Krumeich, H.-J. Muhr, M. Niederberger, F. Bieri, B. Schnyder and R. Nesper, *J. Am. Chem. Soc.*, 1999, **121**, 8324.
86 H.-J. Muhr, F. Krumeich, U. P. Schönholzer, F. Bieri, M. Niederberger, L. J. Gauckler and R. Nesper, *Adv. Mater.*, 2000, **12**, 231.
87 A. Dobley, K. Ngala, S. Yang, P. Y. Zavalji and M. S. Whittingham, *Chem. Mater.*, 2001, **13**, 4382.
88 P. Hoyer, *Langmuir*, 1996, **12**, 1411.
89 R. A. Caruso, J. H. Schattka and A. Greiner, *Adv. Mater.*, 2001, **13**, 1577.
90 D. Gong, C. A. Grimes, O. K. Varghese, W. Hu, R. S. Singh, Z. Chen and E. C. Dickey, *J. Mater. Res.*, 2001, **16**, 3331.
91 S. M. Liu, L. M. Gan, L. H. Liu, W. D. Zhang and H. C. Zeng, *Chem. Mater.*, 2002, **14**, 1391.
92 M. Zhang, Y. Bando and K. Wada, *J. Mater. Res.*, 2001, **16**, 1408.
93 (a) H. Nakamura and Y. Matsui, *J. Am. Chem. Soc.*, 1995, **117**, 2651; (b) H. Nakamura and Y. Matsui, *Adv. Mater.*, 1995, **7**, 871.
94 (a) M. Harada and M. Adachi, *Adv. Mater.*, 2000, **12**, 839; (b) M. Adachi, T. Harada and M. Harada, *Langmuir*, 1999, **15**, 7079.
95 L. Wang, S. Tomura, F. Ohashi, M. Maeda, M. Suzuki and K. Inukai, *J. Mater. Chem.*, 2001, **11**, 1465.
96 Y. Zhang and A. Reller, *Chem. Commun.*, 2002, 606.
97 H. J. Chang, Y. F. Chen, H. P. Lin and C. Y. Mou, *Appl. Phys. Lett.*, 2001, **78**, 3791.
98 J. Zou, L. Pu, X. Bao and D. Feng, *Appl. Phys. Lett.*, 2002, **80**, 1079.
99 J. Zhang, L. Sun, C. Liao and C. Yan, *Chem. Commun*, 2002, 262.
100 M. Yada, M. Mihara, S. Mouri, M. Kuroki and T. Kijima, *Adv. Mater.*, 2002, **14**, 309.
101 B. Cheng and E. T. Samulski, *J. Mater. Chem.*, 2001, **11**, 2901.
102 B. A. Hernandez, K.-S. Chang, E. R. Fisher and P. K. Dorhout, *Chem. Mater.*, 2002, **14**, 480.
103 X. Blase, A. Rubio, S. G. Louie and M. L. Cohen, *Europhys. Lett.*, 1994, **28**, 335.
104 N. G. Chopra, R. G. Luyken, K. Cherrey, V. H. Crespi, M. L. Cohen, S. G. Louie and A. Zettl, *Science*, 1995, **269**, 966.
105 O. Stephan, Y. Bando, A. Loiseau, F. Williame, N. Schramechenko, T. Tamiya and T. Sato, *Appl. Phys. A: Mater. Sci. Process.*, 1998, **67**, 107.

106 M. Terrones, W. K. Hsu, H. Terrones, J. P. Zhang, S. Ramos, J. P. Hare, R. Castillo, K. Prassides, A. K. Cheetham. H. W. Kroto and D. R. M. Walton, *Chem. Phys. Lett.*, 1996, **259**, 568.
107 (*a*) D. P. Yu, X. S. Sun, C. S. Lee, I. Bello, S. T. Lee, H. D. Gu, K. M. Leung, G. W. Zhou, Z. F. Dong and Z. Zhang, *Appl. Phys. Lett*, 1998, **72**, 1966; (*b*) G. W. Zhou, Z. Zhang, Z. G. Bai and D. P. Yu, *Solid State Commun.*, 1999, **109**, 555; (*c*) M. Terauchi, M. Tanaka, H. Matsuda, M. Takeda and K. Kimura, *J. Electron Microsc.*, 1997, **1**, 75.
108 T. Laude, Y. Matsui, A. Marraud and B. Jouffrey, *Appl. Phys. Lett.*, 2000, **76**, 3239.
109 (*a*) D. Golberg, Y. Bando, W. Han, K. Kurashima and T. Sato, *Chem. Phys. Lett.*, 1999, **308**, 337; (*b*) P. W. Fowler, K. M. Rogers, G. Seifert, M. Terrones and H. Terrones, *Chem. Phys. Lett.*, 1999, **299**, 359.
110 E. Bengu and L. D. Marks, *Phys. Rev. Lett.*, 2001, **86**, 2385.
111 Y. Shimizu, Y. Moriyoshi, H. Tanaka and S. Komatsu, *Appl. Phys. Lett.*, 1999, **75**, 929.
112 Y. Chen, L. T. Chadderton, J. F. Gerals and J. S. Williams, *Appl. Phys. Lett.*, 1999, **74**, 2960.
113 R. Ma, Y. Bando, T. Sato and K. Kurashima, *Chem. Mater.*, 2001, **13**, 2965.
114 C. C. Tang, M. L. de la Chapelle, P. Li, Y. M. Liu, H. Y. Dang and S. S. Fan, *Chem. Phys. Lett.*, 2001, **342**, 492.
115 C. C. Tang, Y. Bando, T. Sato and K. Kurashima, *Chem. Commun.*, 2002, 1290.
116 K. B. Shelimov and M. Moskovits, *Chem. Mater.*, 2000, **12**, 250.
117 (*a*) W. Han, Y. Bando, K. Kurashima and T. Sato, *Appl. Phys. Lett*, 1998, **73**, 3085; (*b*) D. Golberg, W. Han, Y. Bando, L. Burgeois, K. Kurashima and T. Sato, *J. Appl. Phys.*, 1999, **86**, 2364; (*c*) D. Golberg, Y. Bando, K. Kurashima and T. Sato, *Chem. Phys. Lett.*, 2000, **323**, 185.
118 D. Golberg, Y. Bando, K. Kurashima and T. Sato, *Solid State Commun.*, 2000, **116**, 1.
119 Y. Zhang, K. Suenaga, C. Colliex and S. Iijima, *Science*, 1998, **281**, 973.
120 W. Han, Ph. K.- Redlich, F. Ernst and M. Rühle, *Appl. Phys. Lett.*, 1999, **75**, 1875.
121 W. Han, P. K. -Redlich, F. Ernst and M. Rühle, *Chem. Mater.*, 1999, **11**, 3620.
122 D. Golberg, Y. Bando, K. Kurashima and T. Sato, *J. Nanosci. Nanotechnol.*, 2000, **1**, 49.
123 R. Ma, Y. Bando and T. Sato, *Adv. Mater.*, 2002, **14**, 366.
124 M. Menon and D. Srivastava, *Chem. Phys. Lett.*, 1999, **307**, 407.
125 L. Burgeois, Y. Bando and T. Sato, *J. Phys. D: Appl. Phys.*, 2000, **33**, 1902.
126 G. Gundiah, G. V. Madhav, A. Govindaraj, Md. Motin Seikh and C. N. R. Rao, *J. Mater. Chem.*, 2002, **12**, 1606.
127 F. L. Deepak, A. Govindaraj and C. N. R. Rao, *J. Nanosci. Nanotechnol.*, 2001, **1**, 303.
128 Y. R. Hacohen, E. Grunbaum, R. Tenne, J. Sloan and J. L. Hutchison, *Nature*, 1998, **395**, 337.
129 (*a*) C. R. Martin, M. Nishizawa, K. Jirage and M. Kang, *J. Phys. Chem. B*, 2001, **105**, 1925; (*b*) J. C. Hutleen, K. B. Jirage and C. R. Martin, *J. Am. Chem. Soc.*, 2000, **120**, 6603; (*c*) C. J. Brumlik and C. R. Martin, *J. Am. Chem. Soc*, 1991, **113**, 3174.
130 G. Tourillon, L. Pontonnier, J. P. Levy and V. Langlais, *Electrochem. Solid-State Lett.*, 2000, **3**, 20.
131 C. C. Han, M. Y. Bai and J. T. Lee, *Chem. Mater.*, 2001, **13**, 4260.
132 J. Bao, C. Tie, Z. Xu, Q. Zhou, D. Shen and Q. Ma, *Adv. Mater.*, 2001, **13**, 1631.
133 B. Mayers and Y. Xia, *Adv. Mater.*, 2002, **14**, 279.
134 (*a*) S. Saito, S. I. Sawada, M. Hawada and A. Oshima, *Mater. Sci. Eng.*, 1993, **B19**, 105; (*b*) R. Saito, M. Fujita, G. Dresslhaus and M. S. Dresslhaus, *Mater. Sci. Eng.*, 1993, **B19**, 185; (*c*) R. Dagani, *Chem. Eng. News*, Jan 11 1999, 31.
135 (*a*) H. M. Cheng, Q. H. Yang and C. Liu, *Carbon*, 2001, **39**, 1447; (*b*) P. Hou, Q. Yang, S. Bai, S. Xu, M. Liu and H. Cheng, *J. Phys. Chem. B*, 2002, **106**, 963 and references therein.
136 R. Ma, Y. Bando, H. Zhu, T. Sato, C. Xu and D. Wu, *J. Am. Chem. Soc.*, 2002, **124**, 7672.
137 J. Chen, N. Kuriyama, H. Yuan, H. T. Takeshita and T. Sakai, *J. Am. Chem. Soc.*, 2001, **123**, 11813.
138 (*a*) M. Treacy and T. Ebbesen, *Nature*, 1996, **381**, 678; (*b*) C. F. Cornwell and L. T. Wille, *Solid State Commun.*, 1997, **101**, 505.
139 M. E. Spahr, P. S. Bitterli, R. Nesper, O. Haas and P. Novak, *J. Electrochem. Soc.*, 1999, **146**, 2780.
140 H. Dai, J. Hafner, A. G. Rinzler, D. T. Colbert and R. E. Smalley, *Nature*, 1996, **384**, 147.
141 A. Rothschild, S. R. Cohen and R. Tenne, *Appl. Phys. Lett.*, 1999, **75**, 4025.
142 L. Rapoport, Y. Bilik, Y. Feldman, M. Homyonfer, S. R. Cohen and R. Tenne, *Nature*, 1997, **387**, 791.
143 J. Chen, S. L. Li, Q. Xu and K. Tanaka, *Chem. Commun.*, 2002, 1722.
144 M. Cote and M. L. Cohen, *Phys. Rev. B*, 1998, **58**, R2477.

www.rsc.org/dalton | Dalton Transactions

Synthesis of inorganic nanomaterials

C. N. R. Rao,*[a,b] S. R. C. Vivekchand,[a] Kanishka Biswas[a,b] and A. Govindaraj[a,b]

Received 1st June 2007, Accepted 9th July 2007
First published as an Advance Article on the web 6th August 2007
DOI: 10.1039/b708342d

Synthesis forms a vital aspect of the science of nanomaterials. In this context, chemical methods have proved to be more effective and versatile than physical methods and have therefore, been employed widely to synthesize a variety of nanomaterials, including zero-dimensional nanocrystals, one-dimensional nanowires and nanotubes as well as two-dimensional nanofilms and nanowalls. Chemical synthesis of inorganic nanomaterials has been pursued vigorously in the last few years and in this article we provide a perspective on the present status of the subject. The article includes a discussion of nanocrystals and nanowires of metals, oxides, chalcogenides and pnictides. In addition, inorganic nanotubes and nanowalls have been reviewed. Some aspects of core–shell particles, oriented attachment and the use of liquid–liquid interfaces are also presented.

1. Introduction

Nanoscience involves a study of materials where at least one of the dimensions is in the 1–100 nm range. Properties of such materials are strongly dependant on their size and shape. Nano-materials include zero-dimensional nanocrystals, one-dimensional nanowires and nanotubes and two-dimensional nanofilms and nanowalls. Synthesis forms an essential component of nanoscience and nanotechnology. While nanomaterials have been generated by physical methods such as laser ablation, arc-discharge and evaporation, chemical methods have proved to be more effective, as they provide better control as well as enable different sizes, shapes and functionalization. Chemical synthesis of nanomaterials has been reviewed by a few authors,[1–6] but innumerable improvements and better methods are being reported continually in the last few years. In accomplishing the synthesis and manipulation of the nanomaterials, a variety of reagents and strategies have been employed besides a wide spectrum of reaction conditions. In view of the intense research activity related to nanomaterials synthesis, we have prepared this perspective to present recent developments and new directions in this area. In doing so, we have dealt with all classes of inorganic nanomaterials. In writing such an article, it has been difficult to do justice to the vast number of valuable contributions which have appeared in the literature in last the two to three years. We had to be necessarily succinct and restrict ourselves mainly to highlighting recent results.

2. Nanocrystals

Nanocrystals are zero-dimensional particles and can be prepared by several chemical methods, typical of them being reduction of salts, solvothermal synthesis and the decomposition of molecular precursors, of which the first is the most common method used

in the case of metal nanocrystals. Metal oxide nanocrystals are generally prepared by the decomposition of precursor compounds such as metal acetates, acetylacetonates and cupferronates in appropriate solvents, often under solvothermal conditions. Metal chalcogenide or pnictide nanocrystals are obtained by the reaction of metal salts with a chalcogen or pnicogen source or the decomposition of single source precursors under solvothermal or thermolysis conditions. Addition of suitable capping agents such as long-chain alkane thiols, alkyl amines and trioctylphosphine oxide (TOPO) during the synthesis of nanocrystals enables the control of size and shape. Monodisperse nanocrystals are obtained by post-synthesis size-selective precipitation.

2.1 Metals

Reduction of metal salts in the presence of suitable capping agents such as polyvinylpyrrolidone (PVP) is the common method to generate metal nanocrystals. Solvothermal and other reaction conditions are employed for the synthesis, to exercise control over their size and shape of the nanocrystals.[1,2,7] Furthermore, the sealed reaction conditions and presence of organic reagents reduce the possibility of atmospheric oxidation of the nanocrystals. The popular citrate route to colloidal Au, first described by Hauser and Lynn,[8] involves the addition of chloroauric acid to a boiling solution of sodium citrate.[9] The average diameter of the nanocrystals can be varied over the 10–100 nm range by varying the concentration of reactants. Au nanocrystals with diameters between 1 and 2 nm are obtained by the reduction of $HAuCl_4$ with tetrakis(hydroxymethyl)phosphonium chloride (THPC) which also acts as a capping agent.[10] Following the early work of Brust and co-workers,[11] the general practice employed to obtain organic-capped metal nanocrystals is to use a bi-phasic mixture of an organic solvent and the aqueous solution of the metal salt in the presence of a phase-transfer reagent. The metal ion is transferred across the organic–water interface by the phase transfer reagent and subsequently reduced to yield sols of metal nanocrystals. Metal nanocrystals in the aqueous phase can also be transferred to a nonaqueous medium by using alkane thiols to obtain organosols.[12,13] This method has been used to thiolize

[a]Chemistry and Physics of Materials Unit, DST unit on nanoscience and CSIR Centre of Excellence in Chemistry, Jawaharlal Nehru Centre for Advanced Scientific Research, Jakkur P. O., Bangalore, 560064, India. E-mail: cnrrao@jncasr.ac.in; Fax: +91 80 22082760
[b]Solid State and Structural Chemistry Unit, Indian Institute of Science, Bangalore, 560012, India

C. N. R. Rao obtained his PhD degree from Purdue University and DSc degree from the University of Mysore. He is the Linus Pauling Research Professor at the Jawaharlal Nehru Centre for Advanced Scientific Research and Honorary Professor at the Indian Institute of Science (both at Bangalore). His research interests are in the chemistry of materials. He has authored nearly 1000 research papers and edited or written 30 books in materials chemistry. A member of several academies including the Royal Society and the US National Academy of Sciences, he is the recipient of the Einstein Gold Medal of UNESCO, Hughes Medal of the Royal Society, and the Somiya Award of the International Union of Materials Research Societies (IUMRS). In 2005, he received the Dan David Prize for materials research from Israel and the first India Science Prize.

S. R. C. Vivekchand received his BSc degree from The American College, Madurai in 2001. He is a student of the integrated PhD programme of Jawaharlal Nehru Centre for Advanced Scientific Research, Bangalore and received his MS degree in 2004. He has worked primarily on material chemistry aspects of one-dimensional nanomaterials.

Kanishka Biswas received his BSc degree from Jadavpur University, Kolkata in 2003. He is a student of the integrated PhD programme of Indian Institute of Science, Bangalore and received his MS degree in 2006. He has worked primarily on the synthesis and characterization of inorganic nanomaterials.

A. Govindaraj obtained his PhD degree from University of Mysore and is a Senior Scientific Officer at the Indian Institute of Science, and Honorary Faculty Fellow at the Jawaharlal Nehru Centre for Advanced Scientific Research. He works on different types of nanomaterials. He has authored more than 100 research papers and co-authored a book on nanotubes and nanowires.

C. N. R. Rao

S. R. C. Vivekchand

Kanishka Biswas

A. Govindaraj

Pd nanocrystals with magic numbers of atoms.[14,15] A method to produce gold nanocrystals free from surfactants would be to reduce $HAuCl_4$ by sodium napthalenide in diglyme.[16]

Liz-Marzan and co-workers[17] have prepared nanoscale Ag nanocrystals by using dimethylformamide as both a stabilizing agent and a capping agent. By using tetrabutylammonium borohydride or its mixture with hydrazine, Jana and Peng[18] obtained monodisperse nanocrystals of Au, Cu, Ag, and Pt. In this method, $AuCl_3$, $Ag(CH_3COO)$, $Cu(CH_3COO)_2$, and $PtCl_4$ were dispersed in toluene with the aid of long-chain quaternary ammonium salts and reduced with tetrabutylammonium borohydride which is toluene-soluble. The reaction can be scaled up to produce gram quantities of nanocrystals. Mirkin and co-workers[19–21] have devised two synthetic routes for nanoprisms of Ag. In the first method, Ag nanoprisms (Fig. 1a) are produced by irradiating a mixture of sodium citrate and bis(*p*-sulfonatophenyl) phenylphosphine dihydrate dipotassium capped Ag nanocrystals with a fluorescent lamp. In the second method, $AgNO_3$ is reduced with a mixture of borohydride and hydrogen peroxide.[21] The latter method has been extended to synthesize branched nanocrystals of Au of the type shown in Fig. 1b and 1c.[22,23]

The shape and colour of Au nanoparticles can be altered by NAD(P)H-mediated growth in the presence of ascorbic acid.[24]

The method yields dipods, tripods and tetrapods (nanocrystals with 2, 3 and 4 arms respectively, the one with 4 arms generally being tetrahedral). Icosahedral Au nanocrystals are obtained by the reaction of $HAuCl_4$ with PVP in aqueous media.[25] Right-bipyramid (75–150 nm in edge length) nanocrystals of Ag (Fig. 2) have been prepared by the addition of NaBr during the polyol reduction of $AgNO_3$ in the presence of PVP.[26] Cu nanoparticles of pyramidal shape have been made by an electrochemical procedure.[27]

Nanoparticles of Rh and Ir have been prepared by the reduction of the appropriate compounds in the ionic liquid, 1-*n*-butyl-3-methylimidazoliumhexafluorophosphate, in the presence of hydrogen.[28] Synthesis and functionalization of gold nanoparticles in ionic liquids is also reported, wherein the colour of the gold nanoparticles can be tuned by changing the anion of ionic liquid.[29] Ru nanoparticles, stabilized by oligoethyleneoxythiol, are found to be soluble in both aqueous and organic media.[30] While Rh multipods are obtained through the seeded-growth mechanism on reducing $RhCl_3$ in ethylene glycol in the presence of PVP,[31] Ir nanocrystals have been prepared by the reduction of an organometallic precursor in the presence of hexadecanediol and different capping agents.[32] Ru, Rh and Ir nanocrystals and other nanostructures are prepared by carrying out the decomposition of

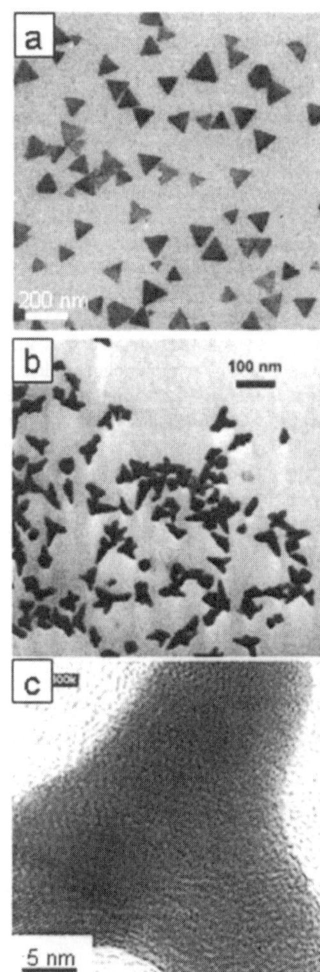

Fig. 1 (a) Ag nanoprisms obtained by controlled irradiation of bis(*p*-sulfonatophenyl) phenylphosphine dihydrate dipotassium capped Ag nanocrystals. (b) Low and (c) high-magnification (scale bar = 5 nm) TEM images of branched Au nanocrystals. Fig. 1a reprinted by permission from Macmillan Publishers Ltd.: *Nature*, 2003, **425**, 487, © 2003. Fig. 1b reprinted with permission from E. Hao, R. C. Bailey and G. C. Schatz, *Nano Lett.*, 2004, **4**, 327. © 2004 American Chemical Society.

the respective metal acetylacetonates in a hydrocarbon (decalin or toluene) or an amine (*n*-octylamine or oleylamine) around 573 K.[33] Cobalt nanoparticles of ~3 nm diameter have been synthesized by the reaction of di-isobutyl aluminium hydride with Co-$(\eta^3C_8H_{13})(\eta^4C_8H_{12})$ or Co[N(SiMe$_3$)$_2$]$_2$.[34] Monodisperse Pt nanocrystals with cubic, cuboctahedral and octahedral shapes with diameters of ~9 nm have been obtained by the polyol process.[35] The polyol process has also been employed to obtain PtBi nanoparticles.[36] AuPt nanoparticles have been successfully incorporated in SiO$_2$ films.[37] FePt nanocubes with ~7 nm diameter (Fig. 3) have been synthesized by the reaction of the oleic acid and Fe(CO)$_5$ with benzyl ether/octadecene solution of Pt(acac)$_2$.[38]

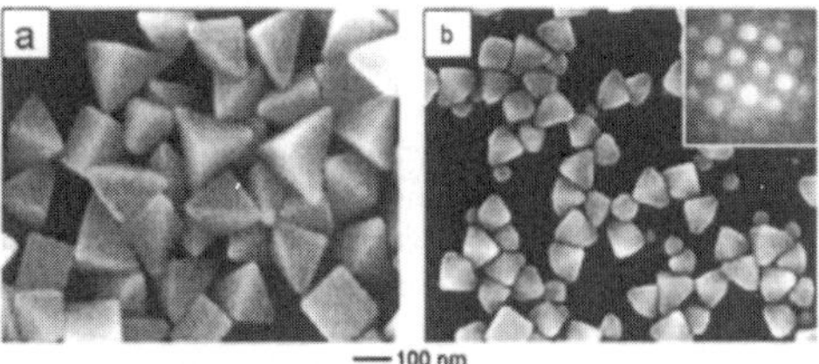

Fig. 2 SEM images of right-bipyramids approximately (a) 150 nm and (b) 75 nm in edge length. The inset in (b) shows the electron diffraction pattern obtained from a single right-bipyramid, indicating that it is bounded by (100) facets. Reprinted with permission from B. J. Wiley, Y. Xiong, Z.-Y. Li, Y. Yin and Younan Xia, *Nano Lett.*, 2006, **6**, 765. © 2006 American Chemical Society.

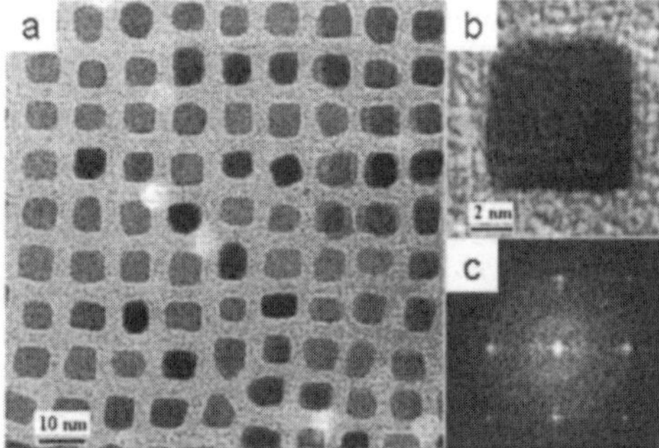

Fig. 3 TEM bright field images of (a) 6.9 nm Fe$_{50}$Pt$_{50}$ nanocubes; (b) HREM image of a single FePt nanocube; (c) Fast-Fourier transform (FFT) of the HREM in (b). Reprinted with permission from M. Chen, J. Kim, J. P. Liu, H. Fan and S. Sun, *J. Am. Chem. Soc.*, 2006, **128**, 7132. © 2006 American Chemical Society.

2.2 Metal oxides

Metal oxide nanocrystals are mainly prepared by the solvothermal decomposition of organometallic precursors. Solvothermal conditions afford high autogenous pressures inside the sealed autoclave that enable low-boiling solvents to be heated to temperatures well above their boiling points. Thus, reactions can be carried out at elevated temperatures and the products obtained are generally crystalline compared to those from other solution-based reactions.

Rockenberger *et al.*[39] described the use of cupferron complexes as precursors to prepare γ-Fe$_2$O$_3$, Cu$_2$O and Mn$_3$O$_4$ nanocrystals. CoO nanocrystals with diameters in 4.5–18 nm range have been prepared by the decomposition of cobalt cupferronate in decalin at 543 K under solvothermal conditions.[40] Magnetic measurements indicate the presence of ferromagnetic interaction in the small CoO nanocrystals. Nanocrystals of MnO and NiO are obtained from cupferronate precursors under solvothermal conditions.[41] The nanocrystals exhibit superparamagnetism accompanied by magnetic hysteresis below a blocking temperature. Nanocrystals of CdO and CuO are prepared by the solvothermal decomposition of metal-cupferronate in presence of trioctylphosphine oxide (TOPO) in toluene.[42] ZnO nanocrystals have been synthesized from the cupferron complex by a solvothermal route in toluene solution.[43] γ-Fe$_2$O$_3$ and CoFe$_2$O$_4$ nanocrystals can also be produced by the decomposition of the cupferron complexes.[44]

Metallic ReO$_3$ nanocrystals with diameters in the 8.5–32.5 nm range are obtained by the solvothermal decomposition of the Re$_2$O$_7$-dioxane complex under solvothermal conditions.[45] Fig. 4a shows a TEM image of ReO$_3$ nanocrystals of 17 nm average diameter with the size distribution histogram as an upper inset. The lower inset shows a HREM image of 8.5 nm nanocrystal. The nanocrystals exhibit a surface plasmon band around 520 nm which undergoes blue-shifts with decrease in size (Fig. 4b). Such blue-shifts in the λ_{max} with decreasing particle size is well-known in the case of metal nanocrystals.[46] Inset in Fig. 4b shows the photograph of four different sizes of ReO$_3$ nanocrystals soluble in CCl$_4$. Surface-enhanced Raman scattering of pyridine, pyrazine and pyrimidine adsorbed on ReO$_3$ nanocrystals has been observed.[47] Magnetic hysteresis is observed at low temperatures in the case of the 8.5 nm particles suggesting a superparamagnetic behaviour.

Apart from solvothermal methods, thermolysis of precursors in high boiling solvents, the sol–gel method, hydrolysis and use of micelles have been employed to synthesize the metal oxide nanocrystals. Thus, Park *et al.*[48] have used metal-oleates as precursors for the preparation of monodisperse Fe$_3$O$_4$, MnO and CoO nanocrystals. 1-Octadecene, octyl ether and trioctylamine have been used as solvents. Hexagonal and cubic CoO nanocrystals can be prepared by the decomposition of cobalt acetylaceto-nate in oleylamine under kinetic and thermodynamic conditions respectively.[49] Hexagonal pyramid-shaped ZnO nanocrystals have been obtained by the thermolysis of the Zn-oleate complex.[50] ZnO nanocrystals have been prepared from zinc acetate in 2-propanol by the reaction with water.[51] ZnO nanocrystals with cone (Fig. 5), hexagonal cone and rod shapes have been obtained by the non-hydrolytic ester elimination sol–gel reactions.[52] In this reaction, ZnO nanocrystals with various shapes were obtained by the reaction of zinc acetate with 1,12-dodecanediol in the presence of different surfactants. In this laboratory, it has been found that reactions of alcohols such as ethanol and *t*-butanol with Zn powder readily yield ZnO nanocrystals.[53]

Nanocrystals of BaTiO$_3$ are obtained by the thermal decomposition of MOCVD reagents (alkoxides such as BaTi(O$_2$CC$_7$H$_{15}$)[OCH(CH$_3$)$_2$]$_5$) in diphenyl ether containing oleic acid, followed by the oxidation of the product with H$_2$O$_2$.[54] Thermal decomposition of uranyl acetylacetonate in a mixture solution of oleic acid, oleylamine, and octadecene at 423 K gives uranium oxide nanocrystals.[55] Treatment of metal acetylacetonates under solvothermal conditions produces nanocrystals of metal oxides such as Ga$_2$O$_3$, ZnO and cubic In$_2$O$_3$.[56] Nearly monodisperse In$_2$O$_3$ nanocrystals have been obtained starting with indium acetate, oleylamine and oleic acid.[57] TiO$_2$ nanocrystals can be prepared

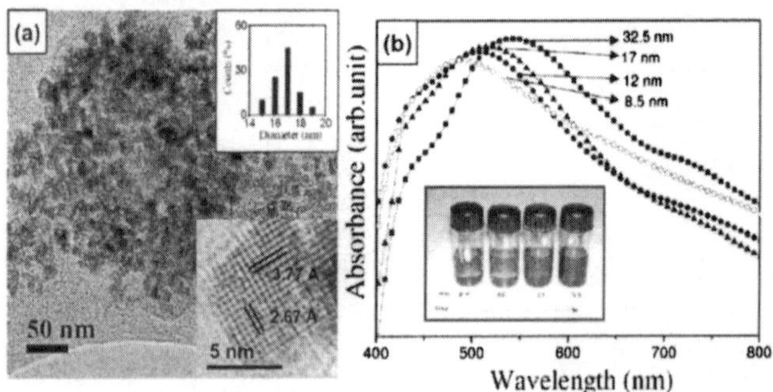

Fig. 4 (a) TEM image of ReO$_3$ nanocrystals of average diameter 17 nm. Upper inset shows the size distribution histogram. Lower inset shows the HREM image of a single 8.5 nm nanocrystal. (b) Optical absorption spectra of ReO$_3$ nanocrystals with average diameters of 8.5, 12, 17 and 32.5 nm. Inset in (b) shows a picture of four different sizes of ReO$_3$ nanocrystals dissolved in CCl$_4$. Reprinted with permission from K. Biswas and C. N. R. Rao, *J. Phys. Chem. B*, 2006, **110**, 842. © 2006 American Chemical Society.

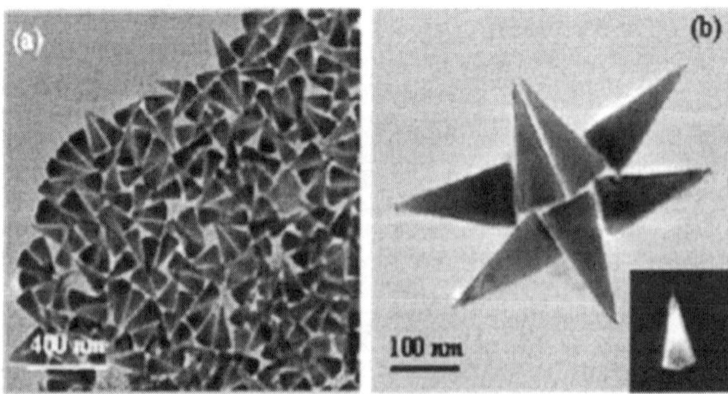

Fig. 5 (a) and (b) TEM images of cone-shaped ZnO nanocrystal. Inset in (b) shows a dark field image of a single cone-shaped nanocrystal. Reproduced with permission from J. Joo, S. G. Kwon, J. H. Yu and T. Hyeon, *Adv. Mater.*, 2005, **17**, 1873. © 2005 Wiley-VCH Verlag GmbH & Co. KGaA.

by the low-temperature reaction of low-valent organometallic precursors.[58] Pure anatase TiO_2 nanocrystals have been prepared by the hydrolysis of $TiCl_4$ with ethanol at 273 K followed by calcination at 360 K for 3 days.[59] The growth kinetics and the surface hydration chemistry have also been investigated. Pileni and co-workers[60,61] have pioneered the use of oil in water micelles to prepare particles of $CoFe_2O_4$, γ-Fe_2O_3, and Fe_3O_4. The basic reaction involving hydrolysis is now templated by a micellar droplet. The reactants are introduced in the form of a salt of a surfactant such as sodium dodecylsulfate (SDS). Thus, by adding CH_3NH_3OH to a micelle made of calculated quantities of $Fe(SDS)_2$ and $Co(SDS)_2$, nanocrystals of $CoFe_2O_4$ are obtained.

2.3 Metal chalcogenides

Nearly monodisperse Cd-chalcogenide nanocrystals (CdE; E = S, Se, Te) have been synthesized by the injection of organometallic reagents such as alkylcadmium into a hot coordinating solvent in the presence of silylchalcogenides/phosphinechalcogenides.[62] Alivisatos and coworkers[63] have produced Cd-chalcogenide nanocrystals by employing tri-butylphosphine at higher temperatures. Nanocrystals of metal chalcogenides are generally prepared by the reaction of metal salts with an appropriate sulfiding or seleniding agent under solvothermal or thermolysis conditions. Thus, toluene-soluble CdSe nanocrystals with a diameter of 3 nm have been prepared solvothermally by reacting cadmium stearate with elemental Se in toluene in the presence of tetralin.[64] The key step in the reaction scheme is the aromatization of tetralin to naphthalene in the presence of Se, producing H_2Se. Organic-soluble CdS nanocrystals are similarly prepared by the reaction of a cadmium salt with S in toluene in the presence of tetralin.[65] Fig. 6 shows the TEM images and electron diffraction patterns of TOPO-capped CdS nanocrystals prepared by this method. PbS and PbSe crystallites and nanorods can also be prepared by this method.[66] Nanocrystals of the transition metal dichalcogenides

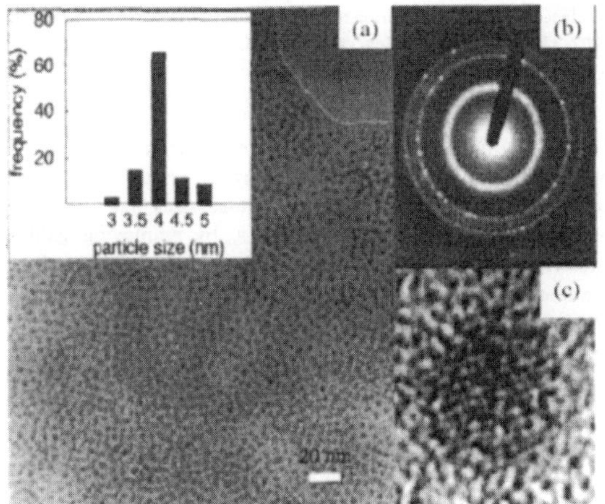

Fig. 6 (a) TEM image of TOPO-capped CdS particles. Inset shows the size distribution of the nanocrystals. The electron diffraction pattern of the nanocrystals is shown in (b) and a HREM image of a nanocrystal is shown in (c). Reprinted from U. K. Gautam, R. Seshadri and C. N. R. Rao, *Chem. Phys. Lett.*, 2003, **375**, 560, © 2003 with permission from Elsevier. http://www.sciencedirect.com/science/journal/00092614

(ME_2; M = Fe, Co, Ni, Mo; E = S or Se) with diameters in the range 4–200 nm have been prepared by a hydrothermal route.[67]

Peng *et al.*[68–71] have proposed the use of greener Cd sources such as cadmium oxide, carbonate and acetate instead of the dimethylcadmium. These workers suggest that the size distribution of the nanocrystals is improved by the use of hexadecylamine, a long-chain phosphonic acid or a carboxylic acid. The method can be extended to prepare CdS nanoparticles by the use of tri-*n*-octylphosphine sulfide (TOP-S) and hexyl or tetradecyl phosphonic acid in mixture with TOPO–TOP. Hyeon and co-workers[72] have prepared nanocrystals of several metal sulfides such as CdS, ZnS, PbS, and MnS with different shapes and sizes by the thermolysis of metal–oleylamine complexes in the presence of S and oleylamine (Fig. 7).

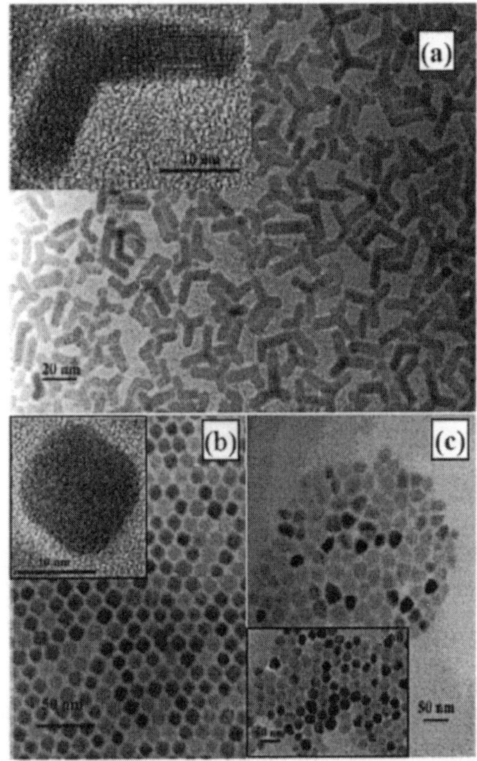

Fig. 7 (a) TEM images of a mixture of rods, bipods, and tripods of CdS nanocrystals with an average size of 5.4 nm (thickness) 20 nm (length). Inset is a HREM image of a single CdS bipod-shaped nanocrystal. (b) Low-magnification TEM image of 13 nm PbS nanocrystals. Inset shows a HREM image of a single 13 nm PbS nanocrystal. (c) Short bullet-shaped MnS nanocrystals. Inset shows hexagon-shaped MnS nanocrystals. Reprinted with permission from J. Joo, H. B. Na, T. Yu, J. H. Yu, Y. W. Kim, F. Wu, J. Z. Zhang and T. Hyeon, *J. Am. Chem. Soc.*, 2003, **125**, 11100. © 2003 American Chemical Society.

CdSe and CdTe nanocrystals can be prepared without precursor injection.[73] The method involves refluxing the cadmium precursor with Se or Te in octadecene. CdSe nanocrystals have also been synthesized using elemental selenium dispersed in octadecene without the use of trioctylphosphine.[74] ZnSe nanocrystals have been prepared in a hot mixture of a long-chain alkylamine and alkylphosphines.[75] Highly monodisperse cubic-shaped PbTe nanocrystals (Fig. 8) have been prepared with size distributions less than 7% by a rapid injection technique.[76] PbS nanocrystals are obtained by reacting the PbCl$_2$–oleylamine complex with the S–oleylamine complex without the use of any solvent.[77] Homogeneously alloyed CdS$_x$Se$_{1-x}$ (x = 0–1.) nanocrystals are prepared by the thermolysis of metal–oleylamine complexes in the presence of S and Se.[78] The band gap of these nanocrystals can be tuned by varying the composition. Thermolysis of a mixture composed of Cu and In oleates in alkanethiol yields copper indium sulfide nanocrystals with acorn, bottle, and larva shapes.[79] Nanocrystals of Ni$_3$S$_4$ and Cu$_{1-x}$S have been prepared by adding elemental sulfur to metal precursors dissolved in dichlorobenzene or oleylamine at relatively high temperatures.[80]

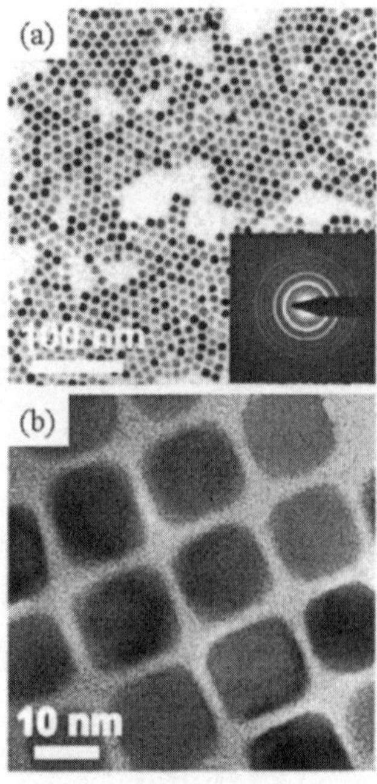

Fig. 8 (a) TEM image of as-prepared cube-like PbTe nanocrystals. Inset shows the SAED pattern. (d) Ordered array consisting of 15 nm cubic-shaped PbSe nanocrystals after size selective precipitation. Reprinted with permission from J. E. Murphy, M. C. Beard, A. G. Norman, S. P. Ahrenkiel, J. C. Johnson, P. Yu, O. I. Mićić, R. J. Ellingson and A. J. Nozik, *J. Am. Chem. Soc.*, 2006, **128**, 3241. © 2006 American Chemical Society.

Decomposition of single molecular precursors provides convenient and effective routes for the synthesis of metal chalcogenide nanocrystals. In this method, a molecular complex consisting of both the metal and the chalcogen is thermally decomposed in a coordinating solvent. For example dithiocarbamates and diselenocarbamates have been found to be good air stable precursors for sulfides and selenides of Cd, Zn and Pb.[81] Nanocrystals of Cd, Hg, Mn, Pb, Cu, and Zn sulfides have been obtained by thermal decomposition of metal hexadecylxanthates in hexadecylamine

and other solvents at relatively low temperatures (323–423 K) under ambient conditions.[82]

Hexagonal CdS nanocrystals have been obtained by the reaction of cadmium acetate dihydrate with thioacetamide in imidazolium[BMIM]-based ionic liquids.[83] Fig. 9(a) shows the TEM image along with HREM image as a top inset of CdS nanocrystals prepared in [BMIM][MeSO$_4$]. Particle size of the CdS nanoparticles varies between 3 and 13 nm with the anion of imidazolium based ionic liquid under the same reaction conditions. Addition of TOPO to the reaction mixture causes greater monodispersity as well as smaller particle size. Hexagonal ZnS and cubic PbS nanoparticles with average diameters of 3 and 10 nm respectively have been prepared by the reaction of the metal acetates with thioacetamide in [BMIM][BF$_4$]. Hexagonal CdSe nanocrystals with an average diameter of 12 nm were obtained by the reaction of cadmium acetate dihydrate with dimethylselenourea in [BMIM][BF$_4$]. CdSe nanocrystals have also been prepared using the phosphonium ionic liquid trihexyl(tetradecyl)phosphoniumbis(2,4,4 trimethylpentylphosphinate) as a solvent and capping agent.[84]

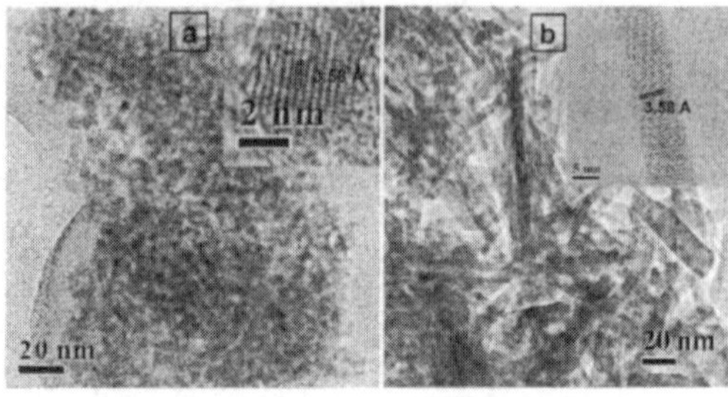

Fig. 9 (a) TEM images of 4 nm CdS nanoparticles prepared in [BMIM][MeSO$_4$]. Insets show a HREM image of the 4 nm CdS nanoparticle. (b) TEM image of CdS nanorods prepared in a [BMIM][BF$_4$] and ethylenediamine mixture. Inset shows a HREM image of a nanorod. Reproduced with permission from K. Biswas and C. N. R. Rao, *Chem.-Eur. J.*, 2007, **13**, 6123. © 2007 Wiley-VCH Verlag GmbH & Co. KGaA.

2.4 Metal pnictides

Large GaN nanocrystals (32 nm) were prepared by Xie *et al.*,[85] by treating GaCl$_3$ with Li$_3$N in benzene under solvothermal conditions. GaN nanocrystals of various sizes have been prepared under solvothermal conditions, by employing gallium cupferronate or chloride as the gallium source and 1,1,1,3,3,3-hexamethyldisilazane (HMDS) as the nitriding agent and toluene as solvent.[86] By employing surfactants such as cetyltrimethylammonium bromide (CTAB), the size of the nanocrystals could be controlled (Fig. 10a). In the case of cupferronate, the formation of the nanocrystals is likely to involve nitridation of the nascent gallium oxide nanoparticles formed by the decomposition of the cupferronate.

$$GaO_{1.5} + (CH_3)_3SiNHSi(CH_3)_3 \rightarrow GaN + (CH_3)_3SiOSi(CH_3)_3 + 1/2H_2O$$

With GaCl$_3$ as precursor, the reaction is,

$$GaCl_3 + (CH_3)_3SiNHSi(CH_3)_3 \rightarrow GaN + Si(CH_3)_3Cl + HCl$$

This method has been applied for the synthesis of AlN and InN nanocrystals (Fig. 10b).[87] The procedure yields nanocrystals with an average diameter of 10 nm for AlN, 15 nm for InN and as low

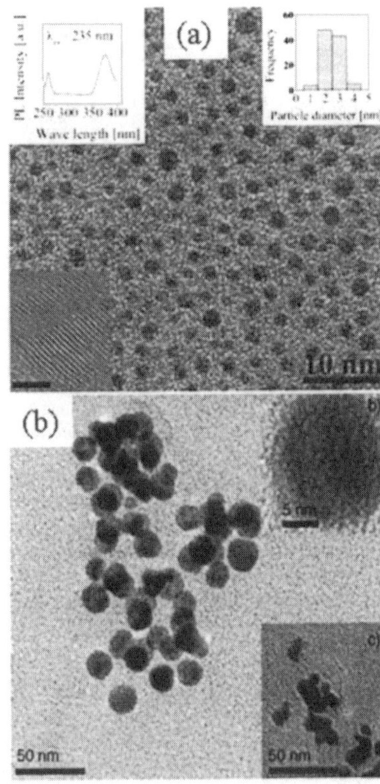

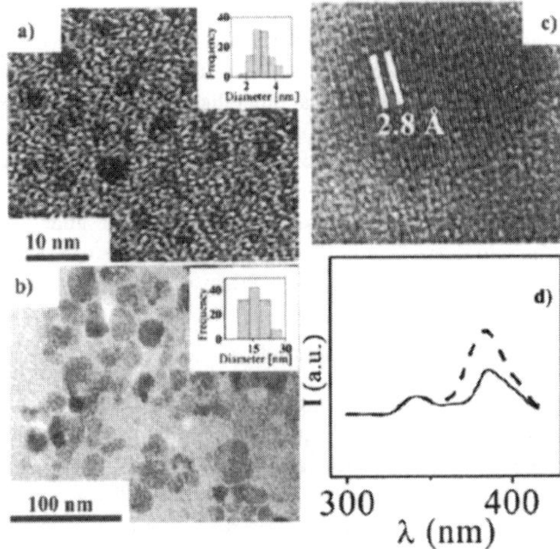

Fig. 11 (a), (b) TEM images of 3 and 15 nm GaN nanocrystals respectively prepared by the urea route. Insets at the top show particle size distributions (c) HREM image of a nanocrystals; (d) PL spectrum of 3 nm size particles at two different excitation wavelengths (solid curve 260 nm and broken curve 250 nm) (From Ref. [94]).

Fig. 10 (a) TEM image of CTAB-capped 2.5 nm GaN nanocrystals prepared starting with Ga cupferron. The upper right inset shows the size distribution. The lower inset shows a HREM image (scale bar is 2 nm) and upper inset also shows the PL spectrum of CTAB capped 2.5 nm GaN nanocrystals; (b) TEM image of InN nanocrystals of 15 nm average diameter prepared starting with In cuperron. The upper inset shows a HREM image of a single nanocrystal. The lower inset shows a TEM image of InN nanocrystals prepared starting with InCl₃. Reproduced with permission from K. Sardar and C. N. R. Rao, *Adv. Mater.*, 2004, **16**, 425 and K. Sardar, F. L. Deepak, A. Govindaraj, M. M. Seikh and C. N. R. Rao, *Small*, 2005, **1**, 91. © 2004 and 2005 Wiley-VCH Verlag GmbH & Co. KGaA.

as 4 nm for GaN. Indium-doped GaN nanocrystals with 5% and 10% In as well as 3% and 5% Mn-doped GaN nanocrystals have been prepared by this method.[88,89]

GaN nanocrystals have also been obtained by the thermal decomposition of precursor compounds such as $(C_2H_5)N \cdot Ga(N_3)_3$ and polymeric $[Ga(NH)_{3/2}]_n$.[90–92] Thus, by the thermolysis of azido compounds, Manz *et al.*[93] have obtained nanocrystals of hexagonal GaN of varying diameters. Group 13 metal nitrides (GaN, AlN, InN) have been prepared by a single source precursor route.[94] The precursors are the adducts of the metal chlorides and urea. Hexagonal nanocrystals of GaN, AlN, and InN were obtained by refluxing the precursors in tri-*n*-octylamine. Fig. 11 shows the TEM images and the PL spectrum of GaN nanocrystals prepared by the urea route. This method has also been extended for the synthesis of BN, TiN and NbN nanoparticles.[95] Solvothermal synthesis involving the reaction of $GaCl_3$ and NaN_3 yields a

poorly crystalline initial product that crystallizes as relatively large nanocrystals (∼50 nm in diameter) on annealing.[96]

Nanocrystalline InN powders are obtained by the metathesis reaction of $InBr_3$ and NaN_3 in superheated toluene and refluxing hexadecane solvents near 553 K.[97] This method has been extended to prepare $Ga_{1-x}In_xN$ (x = 0.5 and 0.75) nanocrystals. A benzene-thermal route has been developed to prepare nanocrystalline InN at 453–473K by choosing $NaNH_2$ and In_2S_3 as novel nitrogen and indium sources.[98] This route has been extended to synthesize other group III nitrides. AlN nanocrystals have been prepared by the benzene-thermal reaction between $AlCl_3$ and Li_3N.[99]

An early procedure for the preparation of phosphides and arsenides of gallium, indium and aluminium involved the de-hydrosilylation reaction. Alivisatos and co-workers[100] adapted this method to prepare GaAs nanoparticles using $GaCl_3$ and $As(SiMe_3)_3$ in quinoline. Using a similar scheme, GeSb, InSb, InAs, and InP nanocrystals were obtained.[101] This method has been modified to prepare InP, InAs, GaP, and $GaInP_2$ nanocrystals as well.[102–104] In a typical reaction, $InCl_3$ is complexed with TOPO/TOP and is reacted with a silylated pnictide such as $E(SiMe_3)_3$ (E = As, P) at 536 K, followed by growth at elevated temperatures for several days. Phase-pure FeP nanocrystals have been synthesized by the reaction of iron(III) acetylacetonate with tris(trimethylsilyl)phosphine at temperatures of 513–593 K using trioctylphosphine oxide as a solvent and dodecylamine, myristic acid, or hexylphosphonic acid as additional capping groups (ligands).[105] The treatment of $Mn_2(CO)_{10}$ with $P(SiMe_3)_3$ in (TOPO)/myristic acid at elevated temperatures produced MnP as discrete nanocrystals.[106] In the presence of a surfactant, potassium stearate, quantum-confined InP nanocrystals were hydrothermally synthesized in aqueous ammonia.[107] High quality InP nanocrystals are obtained by the reaction of $(Me)_3In$ and $P(SiMe_3)_3$ in a coordinating ester solvent.[108] O'Brien and co-workers developed a single molecular precursor route to synthesize InP and GaP nanocrystals using diorganophosphides-M(PBu*t*$_2$)$_3$ (M=Ga, In). This method has also been adapted to synthesize Cd$_3$P$_2$ using [MeCdP(Bu*t*)$_2$]$_3$. The dimer [t-Bu$_2$AsInEt$_2$]$_2$ has been synthesized and used as a single-source organometallic precursor to grow InAs nanocrystals.[109] Reduction of transition metal pnictates yields metal pnictides. Using this method nanoparticle of FeP, Fe$_2$P[110] and NiAs[111] has been prepared. Monodisperse iron, cobalt and nickel monoarsenide nanocrystals were obtained under high-intensity ultrasonic irradiation from the reaction of transition metal chlorides, arsenic and zinc in ethanol.[112]

3. Core@shell nanoparticles

Core@shell particles involving metal, semiconductor or oxide nanocrystals in the core, with shells composed of different materials have been investigated widely.[113] The method of Murray *et al.*[62] involving the decomposition of dimethyl cadmium has been adapted to synthesize nanocrystals of the type CdSe@ZnS, CdSe@ZnSe, and CdSe@CdS.[114] Core@shell growth is achieved by injecting the precursors forming the shell materials into a dispersion containing the core nanocrystals. The injection is carried out at a slightly lower temperature to force shell growth, avoiding independent nucleation. Thus, a mixture of diethylzinc and bis(trimethylsilyl)sulfide is injected into a hot solution containing the core CdSe nanocrystals to encase them with a ZnS layer.[115] O'Brien and co-workers[116] have used single-source methods to prepare core@shell nanocrystals. By successive thermolysis of unsymmetrical diseleno and dithiocarbamates, core@shell nanocrystals of the type CdSe@ZnS and CdSe@ZnSe have been prepared.[116] CdSe@CdS core@shell nanoparticles with a core diameter of ~1.5 nm have been prepared at the liquid–liquid interface starting with cadmium myristate and oleic acid in toluene and selenourea/thiourea in the aqueous medium.[117] Luminescent multi-shell nanocrystals of the composition CdSe-core CdS/Zn$_{0.5}$Cd$_{0.5}$S/ZnS-shell have been prepared by successive ion layer adhesion and reaction technique,[118] Where in the growth of the shell is carried out one monolayer at a time, by alternately injecting cationic and anionic precursors into the reaction mixture with core nanocrystals. Water-soluble CdSe@CdS core@shell nanocrystals with dendron carbohydrate anchoring groups have been prepared.[119] Cao and Banin[120] have successfully coated InAs nanocrystals with shells of InP, GaAs, CdSe, ZnSe, and ZnS. In Fig. 12, we show the TEM and HREM images of InAs@InP and InAs@CdSe core@shell nanocrystals. Using the shell layers, the bandgap of InAs can be tuned in the near-IR region.

Metal on metal core@shell structures provide a means for generating metal nanocrystals with varied optical properties. Morriss and Collins[121] prepared Au@Ag nanocrystals by reducing Au with P by the Faraday's method and Ag with hydroxylamine hydrochloride. They observed a progressive blue shift of the Au plasmon band with incorporation of Ag, accompanied by a slight broadening. For sufficiently thick shells, the plasmon band resembled that of pure Ag particles. Large Au nanoparticles prepared by the citrate method have been used as seeds for the reduction of Ag nanocrystals using ascorbic acid, with CTAB as the capping agent.[122] Au@Ag as well as Ag@Au nanocrystals were prepared by the sequential reduction using sodium citrate.[123] Au@Ag as well as Ag@Au nanoparticles are also prepared by a UV-photoactivation technique.[124] Mirkin and co-workers[125] have

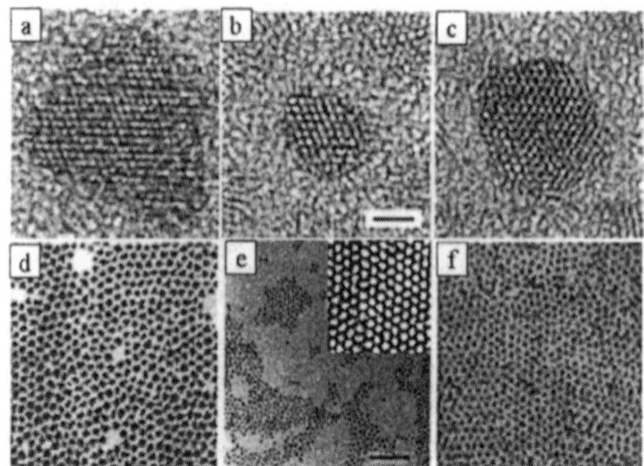

Fig. 12 HREM images of InAs@InP core@shell (frame (a), core radius 1.7 nm, shell thickness 2.5 nm), InAs core ((b), core radius 1.7 nm), and InAs@CdSe core@shell ((c), core radius 1.7 nm, shell thickness 1.5 nm). The scale bar is 2 nm. (d), (e) and (f) are low magnification TEM images of InAs@InP, InAs core and InAs@CdSe core/shell nanoparticles respectively. The scale bar is 50 nm. The inset in (e) (70 × 70 nm), displays a portion of a superlattice structure formed from the InAs cores. Reprinted with permission from Y.-W. Cao and U. Banin, *J. Am. Chem. Soc.*, 2000, **122**, 9692. © 2000 American Chemical Society.

coated Ag nanocrystals with a thin Au shell to provide stability against precipitation under physiological conditions. A thin shell has little effect on the optical properties. Au@Pd nanoparticles with controllable size from 35 to 100 nm were prepared by the chemical deposition of Pd over pre-formed 12 nm Au seeds.[126] Reactive magnetic nanocrystals are rendered passive and made easy to handle by coating them with a layer of noble metals. For example, a layer of Ag was grown *in situ* on Fe and Co nanocrystals synthesized using reverse micelles.[127] A similar procedure has been used to coat Au as well.[128] Fe@Au nanocrystals are also prepared by sequential citrate reduction followed by magnetic separation.[129] Fe@Au nanocrystals were synthesized by a wet chemical procedure involving laser irradiation of Fe nanoparticles and Au powder in a liquid medium.[130] The nanoparticles were superparamagnetic with a blocking temperature of 170 K. Fe$_3$O$_4$ of selected size were used as seeding materials for the reduction of Au precursors to produce monodisperse Fe$_3$O$_4$@Au nanoparticles.[131]

A dielectric oxide layer (*e.g.* silica) is useful as a shell material because of the stability it lends to the core and its optical transparency. The classic method of Stober for solution deposition of silica are adaptable for coating of nanocrystals with silica shells.[132] This method relies on the pH and the concentration of the solution to control the rate of deposition. The natural affinity of silica to oxidic layers has been exploited to obtain silica coating on a family of iron oxide nanoparticles including hematite and magnetite.[133] Such a deposition process is not readily extendable to grow shell layers on metals. The most successful method for silica encapsulation of metal nanoparticles is that due to Mulvaney and co-workers.[134] In this method, the surface of the nanoparticles is functionalized with aminopropyltrimethylsilane, a bifunctional molecule with a pendant silane group which is available for condensation of silica. The next step involves the slow deposition of silica in water followed by the fast deposition of silica in ethanol. Fig. 13 shows the TEM images

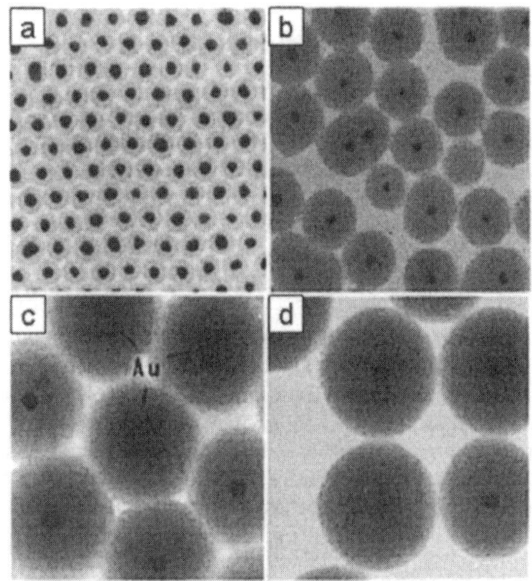

Fig. 13 TEM images of Au@SiO$_2$ particles produced during the growth of the silica shell around 15 nm Au particles. The shell thicknesses are (a) 10 nm, (b) 23 nm, (c) 58 nm, and (d) 83 nm. Reprinted with permission from L. M. Liz-Marzán, M. Giersig and P. Mulvaney, *Langmuir*, 1996, 12, 4329. © 1996 American Chemical Society.

of Au@SiO$_2$ nanocrystals with various shell thickness reported by Mulvaney's group.[135] Au@TiO$_2$ core@shell nanocrystals have been prepared by complexation of a negatively charged titanium precursor, titanium(IV) bis(ammonium lactato)dihydroxide, with poly(dimethyldiallylammonium chloride).[136] Silver nanoparticles coated with a uniform, thin shell of titanium dioxide have been synthesized by a one-pot route, where the reduction of Ag$^+$ to Ag^0 and the controlled polymerization of TiO$_2$ on the surface of silver crystallites occurs simultaneously.[137] Pradeep and co-workers[138,139] have coated Au and Ag nanoparticles with TiO$_2$ and ZrO$_2$ in a single-step process.

ReO$_3$@Au (Fig. 14a) and ReO$_3$@Ag were prepared by the reduction of metal salts over ReO$_3$ nanoparticle seeds.[140] ReO$_3$@SiO$_2$ and ReO$_3$@TiO$_2$ (Fig. 14b) core–shell nanocrystals were prepared by hydrolysis of the organometallic precursors over the ReO$_3$ nanoparticles. ReO$_3$@Au and ReO$_3$@Ag core–shell nanoparticles show composite plasmon absorption bands comprising contributions from both ReO$_3$ and Au (Ag) (Fig. 14c) whereas ReO$_3$@SiO$_2$ and ReO$_3$@TiO$_2$ show shifts in the plasmon bands depending on the refractive index of the shell material (Fig. 14d). Co@SiO$_2$ nanocrystals were prepared combining the sodium borohydride reduction in aqueous solution, the Stober method, and/or the layer-by-layer self-assembly technique.[141] Nanocrystals with the ferrimagnetic CoFe$_2$O$_4$ core and the antiferromagnetic MnO shell have been obtained by a high-temperature decomposition route with seed-mediated growth.[142]

4. Nanowires

There has been considerable interest in the synthesis, characterization and properties of nanowires of various inorganic materials.[4,143,144] Nanowires have been prepared using vapour phase

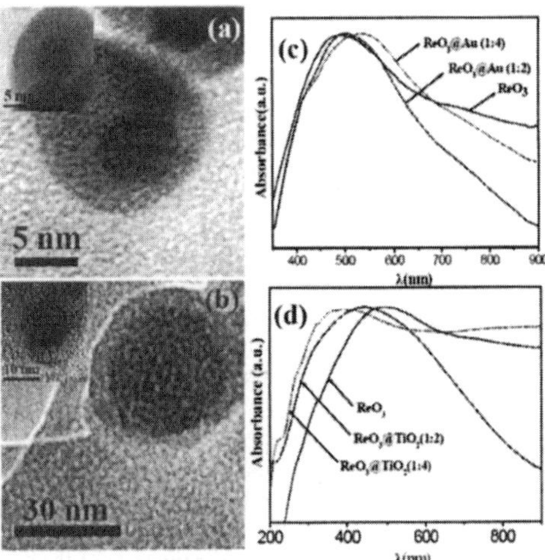

Fig. 14 TEM images of core–shell nanoparticles of (a) ReO$_3$@Au formed with a 5 nm ReO$_3$ particle. Inset shows ReO$_3$@Au formed over an 8 nm ReO$_3$ particle. (b) ReO$_3$@TiO$_2$ core–shell nanoparticle formed over a 32 nm ReO$_3$ particle with the inset showing a core–shell nanoparticle formed over a 12 nm ReO$_3$ nanoparticle. UV-visible absorption spectra of (c) ReO$_3$@Au core–shell nanoparticles (1 : 2 and 1 : 4). and (d) ReO$_3$@TiO$_2$ core–shell nanoparticles (1 : 2 and 1 : 4) with a 12 nm ReO$_3$ particle (From Ref. [140]).

methods such as vapour–liquid–solid (VLS) growth, vapour–solid (*VS*) growth, oxide-assisted growth and the use of carbothermal reactions. A variety of solution methods such as seed-assisted growth, polyol method, and oriented attachment have also been developed for the synthesis of one-dimensional nanostructures.

4.1 Metals

Metal nanowires are commonly prepared using templates such as anodic alumina or polycarbonate membranes, carbon nanotubes and mesoporous carbon.[145–148] The nanoscale channels are first impregnated with metal salts and the nanowires obtained by reduction, followed by the dissolution of the template. Nanowires of metals and semiconductors have also been grown electrochemically. This method has been employed to prepare linear Au–Ag nanoparticle chains.[149] Here, sacrificial Ni segments are placed between segments of noble metals (Au, Ag). The template pore diameter fixes the nanowire width, and the length of each metal segment is independently controlled by the amount of current passed before switching to the next plating solution for deposition of the subsequent segments. Nanowires are released by the dissolution of the template, and subsequently coated with the SiO$_2$.

Au nanorods and nanowires have been alternatively prepared by a simple solution based reduction method making use of nanoparticle seeds.[150] Au nanoparticles with ~4 nm diameter react with the metal salt along with the weak reducing agent such as ascorbic acid in the presence of a directing surfactant yielding Au nanorods. This method was extended to prepare dog-bone like nanostructures.[151] The reaction is carried out in two-steps, wherein the first step involves the addition of an insufficient amount of ascorbic acid to the growth solution, leaving some unreacted metal salt after the reaction, which is later deposited on the Au nanorods by the second addition of ascorbic acid. Addition of nitric acid enhances the proportion of Au nanorods with high aspect ratios (~20) in seed-mediated synthesis.[152] The growth of Au nanorods by the seed-assisted method does not appear to follow any reaction-limited or diffusion limited growth mechanism.[153]

A layer-by-layer deposition approach has been employed to produce polyelectrolyte-coated gold nanorods.[154] Au-nanoparticle-modified enzymes act as biocatalytic inks for growing Au or Ag nanowires on Si surfaces by using a patterning technique such as dip-pen-nanolithography.[155] Single-crystalline Au nanorods shortened selectively by mild oxidation using 1 M HCl at 343 K.[156] Aligned Au nanorods can be grown on a silicon substrate by employing a simple chemical amidation reaction on NH_2-functionalized Si substrates.[157] A seed-mediated surfactant method using a cationic surfactant has been developed to obtain pentagonal silver nanorods.[158]

A popular method for the synthesis of metal nanowires is the use of the polyol process,[159,160] wherein the metal salt is reduced in the presence of PVP to yield nanowires of the desired metal. For example, Ag nanowires have been rapidly synthesized using a microwave-assisted polyol method.[161] CoNi nanowires are obtained by heterogeneous nucleation in liquid polyol.[162] While Bi nanowires have been prepared employing $NaBiO_3$ as the bismuth source.[163] Pd nanobars are synthesized by varying the type and concentration of reducing agent as well as reaction temperature.[164]

Metal nanowires are obtained in good yields by the nebulized spray pyrolysis of a methanolic solution of metal acetates.[165] This method has been employed for the synthesis of single-crystalline nanowires of zinc, cadmium and lead (see Fig. 15). The nanowires seem to grow by the vapour–solid mechanism. ZnO nanotubes

shown in Fig. 15d can be obtained by the oxidation of zinc nanowires in air at 723 K.

4.2 Elemental semiconductors

Silicon nanowires with diameters in the 5–20 nm range have been prepared along with nanoparticles of 4 nm diameter by arc-discharge in water.[166] The Si nanowires shown in Fig. 16 were prepared in solution by using Au nanocrystals as seeds and silanes as precursors by the VLS mechanism.[167] Aligned Si nanowires are obtained by chemical vapour deposition (CVD) of $SiCl_4$ on a gold colloid deposited Si(111) substrate.[168] Gold colloids have been used for nanowire synthesis by the VLS growth mechanism. Using anodic alumina membranes as templates, Si nanowires have been synthesized on Si substrates.[169] In this method, porous anodic alumina is grown on the Si substrate followed by the electrodeposition of the gold catalyst. Epitaxial Si nanowires are then obtained subsequently by VLS growth. Presence of oxygen is important for the growth of long untappered Si nanowires by the VLS mechanism.[170]

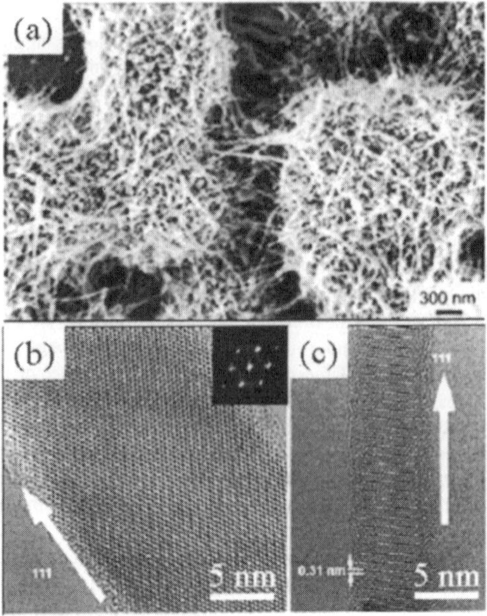

Fig. 16   (a) SEM image of Si nanowires produced from Au nanocrystals and diphenylsilane at 723 K. (b), (c) HREM images of Si showing predominantly <111> orientation. Inset in (b) shows a FFT of the image. Reproduced with permission from D. C. Lee, T. Hanrath and B. A. Korgel, *Angew. Chem., Int. Ed.*, 2005, **44**, 3573. © 2005 Wiley-VCH Verlag GmbH & Co. KGaA

Solution–liquid–solid synthesis of germanium nanowires gives high yields.[171] In this work, Bi nanocrystals were used as seeds for promoting nanowire growth in TOP, by the decomposition of GeI_2 at 623 K. A solid-phase seeded growth with nickel nanocrystals yields Ge nanowires by the thermal decomposition of diphenylgermane in supercritical toluene.[172] A patterned growth of freestanding single-crystalline Ge nanowires with uniform distribution and vertical projection has been accomplished.[173]

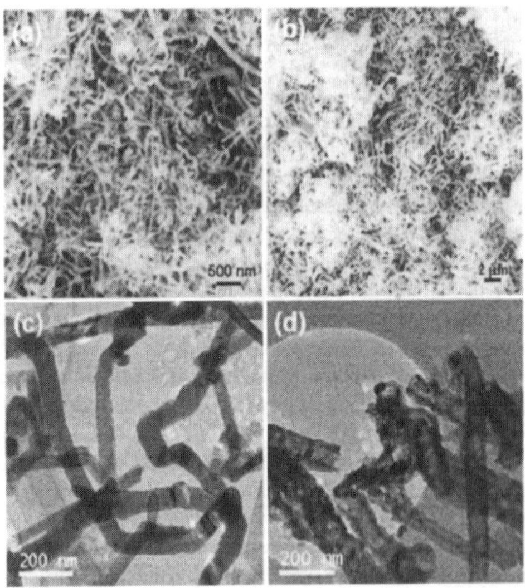

Fig. 15   (a) and (b) SEM images of zinc and cadmium nanowires obtained by the pyrolysis of the corresponding metal acetates at 1173 K. (c) TEM image of zinc nanowires and (d) TEM image of ZnO nanotubes obtained by the oxidation of Zn nanowires at 723 K. Reproduced with permission from S. R. C. Vivekchand, G. Gundiah, A. Govindaraj and C. N. R. Rao, *Adv. Mater.*, 2004, **16**, 1842. © 2004 Wiley-VCH Verlag GmbH & Co. KGaA

High yields of Ge nanowires and nanowire arrays have been obtained by low-temperature CVD by using patterned gold nanoseeds.[174] Ge nanowires have been synthesized starting from the alkoxide, by using a solution procedure involving the injection of a germanium 2,6-dibutylphenoxide solution in oleylamine into a 1-octadecene solution heated to 573 K under an argon atmosphere.[175]

A simple solution-based procedure has been discovered for the synthesis of nanowires of t-Se.[176] In this method, selenium powder is first reacted with $NaBH_4$ in water to yield NaHSe which being unstable decomposes to give amorphous selenium. The nascent selenium imparts a wine red colour to the aqueous solution. On standing for a few hours, the solution transforms into amorphous Se in colloidal form. A small portion of the dissolved selenium precipitates as t-Se nanoparticles which act as nuclei to form the one-dimensional nanowires as seen in Fig. 17. The same reaction carried out under hydrothermal conditions yields nanotubes as shown in Fig. 17e. Extending the above strategy, Te nanorods, nanowires, nanobelts and junction nanostructures have been obtained.[177] Micellar solutions of nonionic surfactants can be employed to prepare nanowires and nanobelts of t-Se.[178] Se nanowires have been prepared at room temperature by using ascorbic acid as a reducing agent in the presence of β-cyclodextrin[179] and also generated at liquid–liquid interface between water and n-butyl alcohol.[180]

single crystalline, 1-D nanostructures (nanowires and nanotubes) can be prepared in alcohol/water solutions by reacting Zn^{2+} precursor with an organic base, tetraammonium hydroxide.[183] It has been found recently that reaction of water with zinc metal powder or foils at room temperature gives ZnO nanowires.[184] A multi-component precursor has been used to produce nanoparticle nanoribbons of ZnO.[185] Porous ZnO nanoribbons are produced by the self-assembly of textured ZnO nanoparticles. Nanobelts of ZnO can be converted into superlattice-structured nanohelices by a rigid lattice rotation or twisting as seen in Fig. 18.[186] Well-aligned crystalline ZnO nanorods along with nanotubes can be grown from aqueous solutions on Si wafers, poly(ethylene terephthlate) and sapphire.[187] Atomic layer deposition was first used to grow a uniform ZnO film on the substrate of choice and to serve as a templating seed layer for the subsequent growth of nanorods and nanotubes. On this ZnO layer, highly oriented two-dimensional (2-D) ZnO nanorod arrays were obtained by solution growth using $Zn(NO_3)_2$ and hexamethylenetetramine in aqueous solution. Controlled growth of aligned ZnO nanorod arrays has been accomplished by an aqueous ammonia solution method.[188] In this method, an aqueous ammonia solution of $Zn(NO_3)_2$ is allowed to react with a zinc-coated silicon substrate at a growth temperature of 333–363 K. 3-D interconnected networks of ZnO nanowires

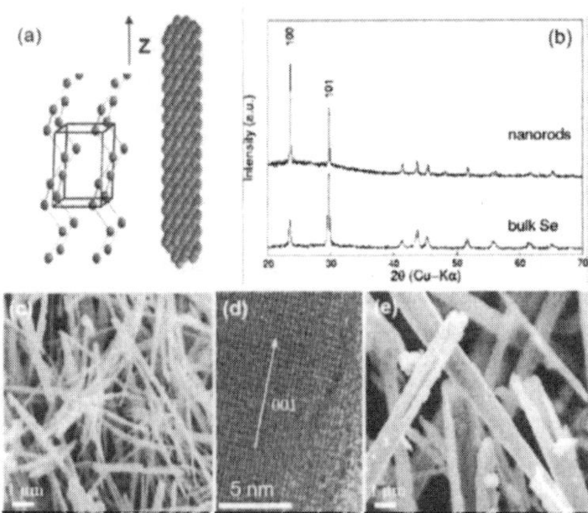

Fig. 17 (a) Crystal structure of t-Se showing a unit cell with helical chains of covalently bonded Se atoms extended along the c-axis. The growth direction of the 1D nanostructures is shown along with an atomic model of a rod. (b) XRD patterns of the t-Se nanorods and bulk selenium powder used as the starting reagent. (c) SEM image of the Se nanorods obtained after 4 days by reacting 0.025 g of Se with 0.03 g of $NaBH_4$ in 20 ml water. (d) HREM image of a nanorods (arrow indicates the growth direction of the nanorods). (e) SEM image of t-Se scrolls obtained under hydrothermal conditions (From Ref. [176]).

4.3 Metal oxides

A seed-assisted chemical reaction at 368 K is found to yield uniform, straight, thin single-crystalline ZnO nanorods on a hectogram scale.[181] Zinc oxide nanowires have been synthesized in large quantities using plasma synthesis.[182] Variable-aspect-ratio,

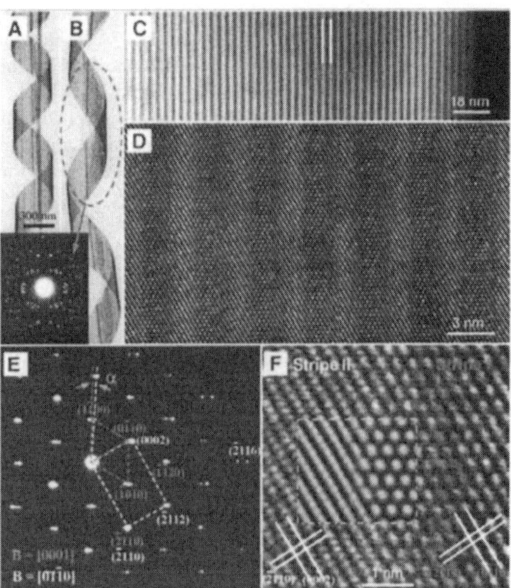

Fig. 18 (A) Typical low-magnification TEM image of a ZnO nanohelix, showing its structural uniformity. (B) Low-magnification TEM image of a ZnO nanohelix with a larger pitch to diameter ratio. The selected-area ED pattern (SAED, inset) is from a full turn of the helix. (C) Dark-field TEM image from a segment of a nanohelix. The edge at the right and side is the edge of the nanobelt. (D and E) High-magnification TEM image and the corresponding SAED pattern of a ZnO nanohelix with the incident beam perpendicular to the surface of the nanobelt, respectively, showing the lattice structure of the two alternating stripes. (F) Enlarged high-resolution TEM image showing the interface between the two adjacent stripes. From P. X. Gao, Y. Ding, W. Mai, W. L. Hughes, C. Lao and Z. L. Wang, *Science*, 2005, **309**, 1700. Reprinted with permission from AAAS. http://www.sciencemag.org

and nanorods have been synthesized by a high temperature solid–vapour deposition process.[189] Templated electrosynthesis of ZnO nanorods wherein, electroreduction of hydrogen peroxide or nitrate ions is carried out to alter the local pH within the pores of the membrane, with the subsequent precipitation of the metal oxide within the pores.[190]

1-D ZnO nanostructures have been synthesized by oxygen assisted thermal evaporation of zinc on a quartz surface over a large area.[191] Pattern- and feature-designed growth of ZnO nanowire arrays for vertical devices has been accomplished by following a pre-designed pattern and feature with controlled site, shape, distribution and orientation.[192]

The ionic liquid 1-n-butyl-3-methylimidazolium tetrafluoroborate has been used to synthesize nanoneedles and nanorods of manganese dioxide (MnO_2).[193] Crystalline silica nanowires were prepared by Deepak et al.[194] by a carbothermal procedure. Crystalline SiO_x nanowires have also been prepared by a low-temperature iron assisted hydrothermal procedure.[195]

IrO_2 nanorods can be grown by metal–organic chemical vapour deposition on sapphire substrates consisting of patterned SiO_2 as the non-growth surface.[196] By employing the hydrothermal route, uniform single-crystalline $KNbO_3$ nanowires have been obtained.[197]

MgO nanowires and related nanostructures have been produced by carbothermal synthesis, starting with polycrystalline MgO or Mg with or without the use of metal catalysts.[198] This study has been carried out with different sources of carbon, all of them yielding interesting nanostructures such as nanosheets, nanobelts, nanotrees and aligned nanowires. Orthogonally branched single-crystalline MgO nanostructures have been obtained through a simple chemical vapour transport and condensation process in a flowing Ar/O_2 atmosphere.[199]

Ga_2O_3 powder reacts with activated charcoal, carbon nanotubes or activated carbon around 1373 K in flowing Ar to give nanowires, nanorods and other novel nanostructures of Ga_2O_3 such as nanobelts and nanosheets.[200] Catalyst-assisted VLS growth of single-crystal Ga_2O_3 nanobelts has been accomplished by graphite-assisted thermal reduction of a mixture of Ga_2O_3 and SnO_2 powders under controlled conditions.[201] Zig-zag and helical one-dimensional nanostructures of α-Ga_2O_3 have been produced by the thermal evaporation of Ga_2O_3 in the presence of GaN.[202] Large scale synthesis of TiO_2 nanorods has been achieved by the non-hydrolytic sol–gel ester elimination reaction, wherein the reaction is carried out between titanium(IV) isopropoxide and oleic acid.[203] Single-crystalline and well facetted VO_2 nanowires with rectangular cross sections have been prepared by the vapour transport method, starting with bulk VO_2 powder.[204] Copious quantities of single-crystalline and optically transparent Sn-doped In_2O_3 (ITO) nanowires have been grown on gold-sputtered Si substrates by carbon-assisted synthesis, starting with a powdered mixture of the metal nitrates or with a citric acid gel formed by the metal nitrates.[205] Vertically aligned and branched ITO nanowire arrays which are single-crystalline have been grown on yttrium-stabilized zirconia substrates containing thin gold films of 10 nm thickness.[206]

Bicrystalline nanowires of hematite (α-Fe_2O_3) have been synthesized by the oxidation of pure Fe.[207] Single-crystalline hexagonal α-Fe_2O_3 nanorods and nanobelts can be prepared by a simple iron–water reaction at 673 K.[208] Mesoporous quasi-single crystalline nanowire arrays of Co_3O_4 have been grown by immersing Si or fluorine doped SnO_2 substrates in a solution of $Co(NO_3)_2$ and concentrated aqueous ammonia.[209] Networks of WO_{3-x} nanowires are obtained by the thermal evaporation of W powder in the presence of oxygen.[210] The growth mechanism involves ordered oxygen vacancies (100) and (001) planes which are parallel to the (010) growth direction. A general and highly effective one-pot synthetic protocol for producing one-dimensional nanostructures of transition metal oxides such as $W_{18}O_{49}$, TiO_2, Mn_3O_4 and V_2O_5 through a thermally induced crystal growth process starting from mixtures of metal chlorides and surfactants, has been described.[211] Self-coiling nanobelts of $Ag_2V_4O_{11}$ have been obtained by the hydrothermal reaction between $AgNO_3$ and V_2O_5.[212] Polymer-assisted hydrothermal synthesis of single crystalline tetragonal perovskite PZT ($PbZr_{0.52}Ti_{0.48}O_3$) nanowires has been carried out.[213] Nanowires of the type II superconductor $YBa_2Cu_4O_8$ have been synthesized by a biomimetic procedure.[214] The nanowires produced by the calcination of a gel containing the biopolymer chitosan and Y, Ba and Cu salts have mean diameters of 50 ± 5 nm with lengths up to 1 μm. Nanorods of V_2O_5 prepared by the polyol process self-assemble into microspheres.[215]

4.4 Metal chalcogenides

ZnS nanowires and nanoribbons with wurtzite structure can be prepared by the thermal evaporation of ZnS powder onto silicon substrates, sputter-coated with a thin ($\sim$25 Å) layer of Au film.[216] Thermal evaporation of a mixture of ZnSe and activated carbon powders in the presence of a tin-oxide based catalyst yields tetrapod-branched ZnSe nanorod architectures.[217] 1-D nanostructures of CdS have been formed on Si substrates by a thermal evaporation route.[218] The shapes of the 1-D CdS nanoforms were controlled by varying the experimental parameters such as temperature and position of the substrates. Nanorods of luminescent cubic CdS are obtained by injecting solutions of anhydrous cadmium acetate and sulfur in octylamine into hexadecylamine.[219] CdSe nanowires have been produced by the cation-exchange route.[220] By employing the cation-exchange reaction between Ag^{2+} and Cd^{2+}, Ag_2Se nanowires are transformed into single-crystal CdSe nanowires. A single-source molecular precursor has been used to obtain blue-emitting, cubic CdSe nanorods ($\sim$2.5 nm diameter and 12 nm length) at low temperatures.[221] Thin aligned nanorods and nanowires of ZnS, ZnSe, CdS and CdSe can be produced by using microwave-assisted methodology by starting from appropriate precursors.[222] An organometallic preparation of CdTe nanowires with high aspect ratios in the wurtzite structure has been described.[223] Thermal decomposition of copper-diethyldithiocarbamate (CuS_2CNEt_2) in a mixed binary surfactant solvent of dodecanethiol and oleic acid at 433 K gives rise to single-crystalline high aspect ratio ultrathin nanowires of hexagonal Cu_2S.[224] CdX(X = S, Se) nanorods have been synthesized in ionic liquids in the presence of ethylenediamine (Fig. 9b).[87] ZnSe nanorods can also be prepared by the decomposition of zinc acetate in the presence of dimethylselenourea in [BMIM][MeSO_4] ionic liquid.

Atmospheric pressure CVD can be employed to prepare arrays and networks of PbS nanowires.[225] Monodisperse PbTe nanorods of sub-10 nm diameter are obtained by sonoelectrochemical means by starting with a lead salt and TeO_2 along with nitrilotriacetic acid.[226] Taking bismuth citrate and thiourea in DMF, well-segregated, crystalline Bi_2S_3 nanorods have been synthesized by a reflux process.[227] Single-crystalline Bi_2S_3 nanowires have also been obtained by using lysozyme which controls the morphology and directs the formation of the 1D inorganic material.[228] In this method, $Bi(NO_3)_3 \cdot 5H_2O$, thiourea and lysozyme are reacted together at 433 K under hydrothermal conditions. A solvent-less synthesis of orthorhombic Bi_2S_3 nanorods and nanowires with high aspect ratios (>100) has been accomplished by the thermal decomposition of bismuth alkylthiolates in air around 500 K in the presence of the capping agent, octanoate.[229] Single-crystalline Bi_2Te_3 nanorods have been synthesized by a template free method at 373 K by the addition of thioglycolic acid or L-cysteine to a bismuth chloride solution.[230] $GeSe_2$ nanowires have been obtained by the decomposition of organoammonium selenide (See Fig. 19).[231] GeTe nanowires are obtained by a VLS process starting with GeTe powder using a Au nanoparticle catalyst.[232] Nanowires of copper indium diselenide have been prepared by the reaction of Se powder with In_2Se_3 and anhydrous $CuCl_2$ under solvothermal conditions.[233]

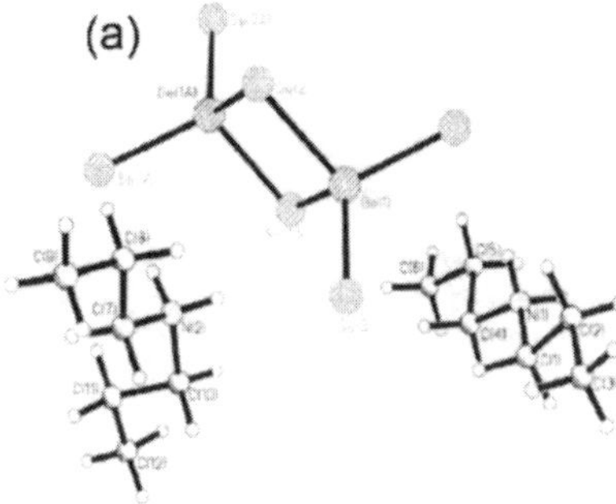

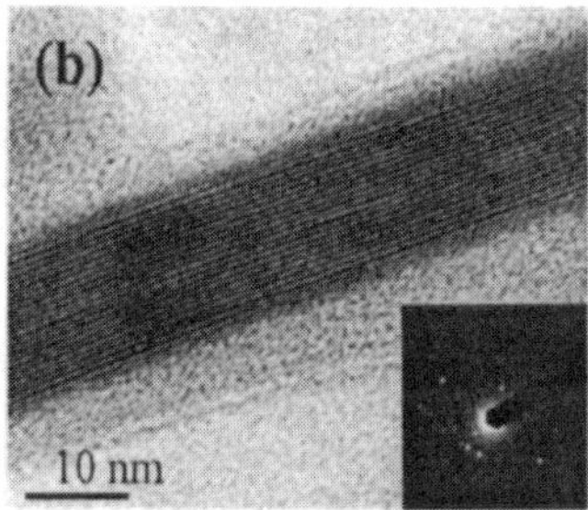

Fig. 19 (a) Crystal structure of the organoammonium germanium selenide precursor and (b) HREM image of a $GeSe_2$ nanowire. The inset shows the typical SAED pattern. (From Ref. [231]).

4.5 Metal pnictides and other nanowires

Single crystalline AlN, GaN and InN nanowires illustrated in Fig. 20 can be deposited on Si substrates covered with Au islands by using urea complexes formed with the trichlorides of Al, Ga and In as the precursors.[94] Single crystalline GaN nanowires are also obtained by the thermal evaporation/decomposition of Ga_2O_3 powders with ammonia at 1423 K directly onto a Si substrate coated with a Au film.[234] Direction-dependent homoepitaxial

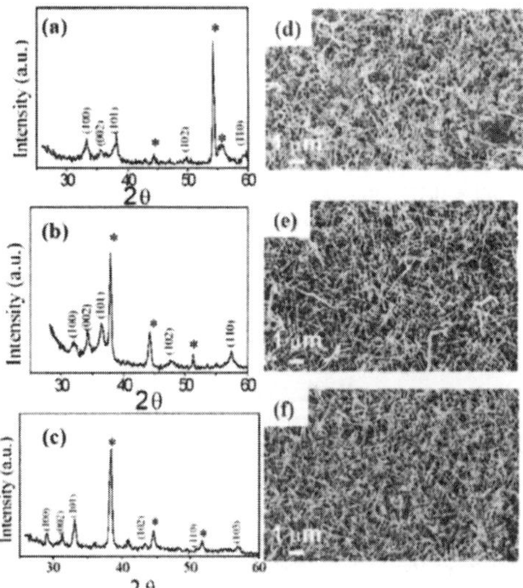

Fig. 20 XRD patterns of (a) AlN, (b) GaN and (c) InN nanowires respectively (* indicates peaks arising due to substrate or gold). SEM images of (d) AlN, (e) GaN, (f) InN nanowires. (From Ref. [94]).

growth of GaN nanowires has been achieved by controlling the Ga flux during direct nitridation in dissociated ammonia (Fig. 21).[235] The nitridation of Ga droplets at a high flux leads

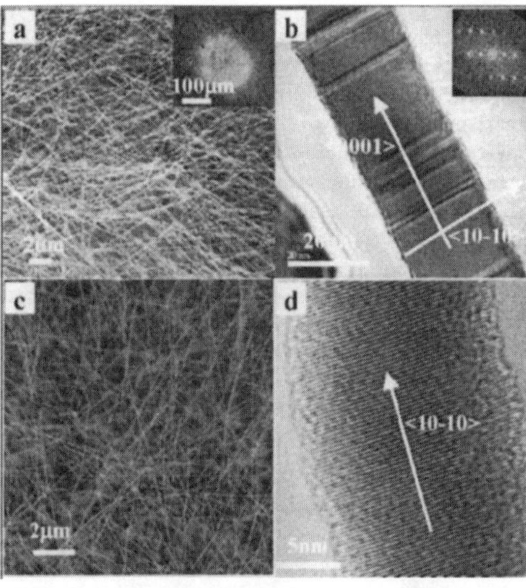

Fig. 21 (a) SEM images of GaN nanowires. The inset shows the spontaneous nucleation and growth of multiple nanowires directly from a larger Ga droplet. (b) HREM image of a GaN nanowire. The inset shows FFT of the HREM image, (c) SEM image showing GaN nanowires with diameters less than 30 nm resulting from the reactive-vapour transport of a controlled Ga flux in a NH_3 atmosphere. (d) HREM image of a GaN nanowire from the sample shown in (c) indicating that the growth direction is <10-10>. Reproduced with permission from H. Li, A. H. Chin and M. K. Sunkara, *Adv. Mater.*, 2006, **18**, 216. © 2006 Wiley-VCH Verlag GmbH & Co. KGaA

to GaN nanowire growth in the *c*-direction (<1000>) as seen in Fig. 21b, while the nitridation with a low Ga flux leads to growth in the *a*-direction (<10-10>) (Fig. 21d). InN nanowires with uniform diameters have been obtained in large quantities by the reaction of In_2O_3 powders in ammonia.[236] A general method for the synthesis of Mn-doped nanowires of CdS, ZnS and GaN based on metal nanocluster-catalyzed chemical vapour deposition has been described.[237] Vertically aligned, catalyst-free InP nanowires have been grown on InP(111)B substrates by CVD of trimethylindium and phosphine at 623–723 K.[238]

Nanowires of $InAs_{1-x}P_x$ and $InAs_{1-x}P_x$ heterostructure segments in InAs nanowires, with the P concentration varying from 22% to 100%, have been grown by VLS method.[239] Single-crystalline nanowires of LaB_6, CeB_6 and GdB_6 have been deposited on a Si substrate by the reaction of the rare-earth chlorides with BCl_3 in the presence of hydrogen.[240] Starting from BiI_3 and FeI_2, Fe_3B nanowires were synthesized on Pt and Pd (Pt/Pd) coated sapphire substrates by CVD at 1073 K.[241] The morphology of the Fe_3B nanowires can be controlled by manipulating the Pt/Pd film thickness and growth time, the typical diameter is in the 5–50 nm range and the length in the 2–30 μm range. Nanowires and nanoribbons of $NbSe_3$ have been obtained by the direct reaction of Nb and Se powders.[242] A one-pot metal–organic synthesis of single-crystalline CoP nanowires with uniform diameters has been reported.[243] The method involves the thermal decomposition of cobalt(II) acetylacetonate and tetradecylphosphonic acid in a mixture of TOPO and hexadecylamine.

4.6 Oriented attachment

Oriented attachment of nanocrystals is employed to fabricate one-dimensional as well as complex nanostructures. Thus, nanotubes and nanowires of II–VI semiconductors have been synthesized using surfactants.[244] In Fig. 22, we show TEM image of CdS nanowires and CdSe nanotubes obtained using Triton 100-X as surfactant. Oriented attachment-like growth has been observed in CdS, ZnS and CuS nanorods prepared by using hydrogels as templates.[245] The nanorods or nanotubes of CdS and other materials produced in this manner actually consists of nanocrystals. The synthesis of SnO_2 nanowires from nanoparticles has been investigated.[246] CdSe nanorods can be formed by redox-assisted asymmetric Ostwald ripening of CdSe dots to rods.[247]

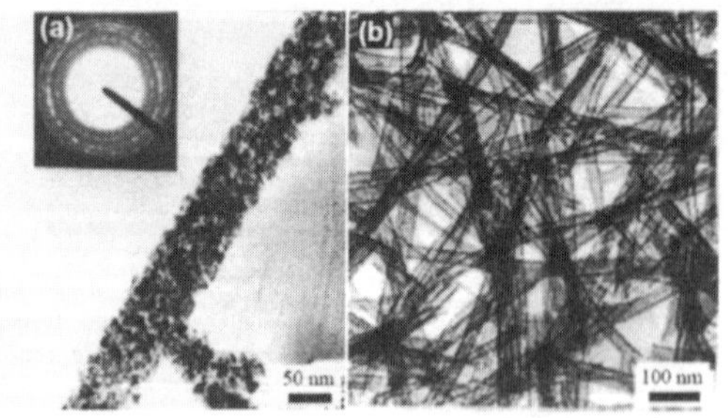

Fig. 22 TEM images of (a) CdS nanowires (b) CdSe nanotubes obtained by using Triton 100-X as the surfactant. Inset in (a) shows the SAED pattern of the CdS nanowires. Reprinted with permission from C. N. R. Rao, A. Govindaraj, F. L. Deepak, N. A. Gunari and M. Nath, *Appl. Phys. Lett.*, 2001, **78**, 1853. © 2001, American Institute of Physics.

PbSe and cubic ZnS nanowires as well as complex one-dimensional nanostructures can be obtained in solution through oriented attachment of nanocrystals.[248,249] In Fig. 23, star-shaped PbSe nanocrystals and branched nanowires are shown.

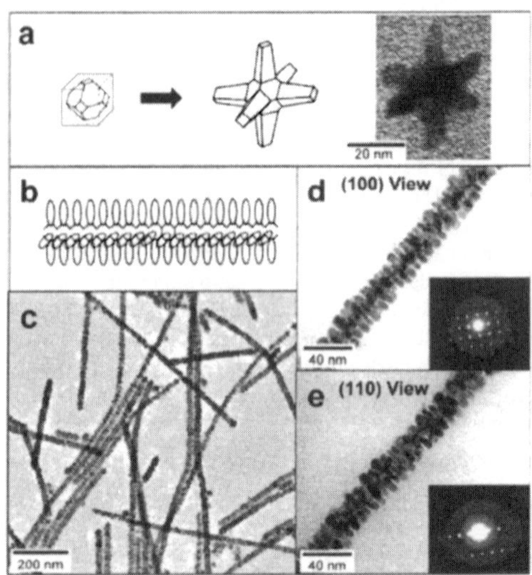

Fig. 23 (a) Star-shape PbSe nanocrystals and (b–e) radially branched nanowires. (d) TEM image of the (100) view of the branched nanowire and the corresponding selected area electron diffraction pattern. (e) TEM image of the (110) view of the branched nanowire and the corresponding selected area electron diffraction pattern. Reprinted with permission from K.-S. Cho, D. V. Talapin, W. Gaschler and C. B. Murray, *J. Am. Chem. Soc.*, 2005, **127**, 7140. © 2005 American Chemical Society.

4.7 Coaxial cables and other hybrid nanostructures

A general procedure has been proposed for producing chemically bonded ceramic oxide coatings on carbon nanotubes and inorganic nanowires wherein reactive metal chlorides are reacted with acid-treated carbon nanotubes or metal oxide nanowires, followed by hydrolysis with water.[250] On repeating the above process several times followed by calcination, oxide coatings of the desired thickness are obtained. Core-sheath CdS and polyaniline (PANI) coaxial nanocables with enhanced photoluminescence have been fabricated by an electrochemical method using a porous anodic alumina membrane as the template.[251] SiC nanowires can be coated with Ni and Pt nanoparticles (~3 nm) by plasma-enhanced CVD.[252] Single and double-shelled coaxial core-shell nanocables of GaP with SiO_x and carbon (GaP/SiO_x, GaP/C, GaP/SiO_x/C), with selective morphology and structure, have been synthesized by thermal CVD.[253] Silica-sheathed $3C-Fe_7S_8$ has been prepared on silicon substrates with $FeCl_2$ and sulfur precursors at 873–1073 K.[254]

Nanowires containing multiple GaP–GaAs junctions are grown by the use of metal–organic vapour phase epitaxy on SiO_2.[255] Silica-coated PbS nanowires have been deposited by CVD using $PbCl_2$ and S on silicon substrates at temperatures between 650 and 973 K.[256] A novel silica-coating procedure has been devised for CTAB-stabilized gold nanorods and for the hydrophobation of the silica shell with octadecyltrimethoxysilane (OTMS).[257] A combination of the polyelectrolyte layer-by-layer technique and the hydrolysis followed by condensation of tetraethoxylorthosilicate in a 2-propanol–water mixture leads to homogeneous coatings with control on shell thickness. On the other hand, the strong binding of CTAB molecules to the gold surface makes surface hydrophobation difficult but the functionalization with OTMS, which contains a long hydrophobic hydrocarbon chain, allows the particles to be transferred into nonpolar organic solvents such as chloroform.

Fabrication of InP/InAs/InP core-multishell heterostructure nanowire arrays shown in Fig. 24 has been achieved by selective area metal–organic vapour phase epitaxy.[258] These core–multishell nanowires were designed to accommodate a strained InAs quantum well layer in a higher band gap InP nanowire. Precise control over the nanowire growth direction and the heterojunction formation enabled the successful fabrication of the nanostructure in which all the three layers were epitaxially grown without the assistance of a catalyst.

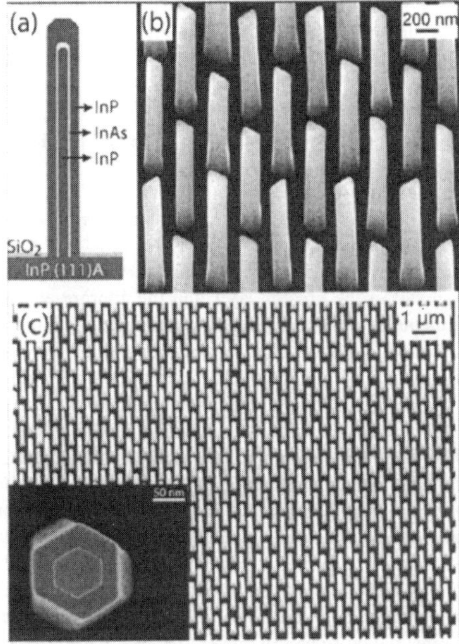

Fig. 24 (a) Schematic cross-sectional image of InP/InAs/InP core-multishell nanowire. (b) SEM image of periodically aligned InP/InAs/InP core–multishell nanowire array. (c) SEM image showing highly dense ordered arrays of core–multishell nanowires. Schematic illustration and high resolution SEM cross-sectional image of a typical core–multishell nanowire observed after anisotropic dry etching and stain etching. Inset shows the top view of a single nanowire. Reprinted with permission from P. Mohan, J. Motohisa and T. Fukui, *Appl. Phys. Lett.*, 2006, **88**, 133105. © 2006, American Institute of Physics.

5. Inorganic nanotubes

There has been considerable interest in the synthesis of fullerenes and nanotubes of inorganic materials.[4,259] These efforts have primarily been focused on layered inorganic compounds such as metal dichalcogenides (sulfides, selenides, and tellurides), halides (chlorides, bromides, and iodides), oxides and boron nitride, which possess structures comparable to the structure of graphite. Tenne and coworkers[260–261] first recognized that nanosheets of Mo and W dichalcogenides are unstable against folding and closure and that they can form fullerene-like nanoparticles and nanotubes. Nanotubes and fullerene-like nanoparticles of dichalcogenides such as MoS_2, $MoSe_2$ and WS_2 have been prepared by processes such as arc-discharge and laser ablation.[262–264] Chemical routes for the synthesis of fullerenes and nanotubes of metal chalcogenides are more versatile and popular. Nanotubes of MoS_2 and WS_2 can be obtained by heating MoO_3/WO_3 in a reducing atmosphere and then reacted with H_2S.[265] In the case of metal selenide nanotubes, H_2Se is used instead of H_2S.[266] Recognizing that MoS_3 and WS_3 are likely intermediates in the formation of the disulfides, the trisulfides have been directly decomposed in a H_2 atmosphere to obtain the disulfide nanotubes.[267] Similarly, diselenide nanotubes have been obtained by the decomposition of metal triselenides.[268] The trisulfide route provides a general route for the synthesis of the nanotubes of many metal disulfides such as NbS_2 and HfS_2.[269,270] The decomposition of precursor ammonium salts $(NH_4)_2MX_4$ (X = S, Se; M = Mo, W) also yields nanotubes.[267] The trichalcogenides are intermediates in the decomposition of the ammonium salts.

Bando and co-workers[271] have prepared BN nanotubes by the reaction of MgO, FeO and B in the presence of NH_3 at 1400 °C. Reaction of boric acid or B_2O_3 with N_2 or NH_3 at high temperatures in the presence of carbon or catalytic metal particles has been employed in the preparation of BN nanotubes.[272] Boron nitride nanotubes can be grown directly on substrates at 873 K by a plasma-enhanced laser-deposition technique.[273] Recently, GaN nanotube brushes have been prepared using amorphous carbon nanotubes templates obtained using AAO membranes.[274]

Large-scale synthesis of Se nanotubes has been carried out in the presence of CTAB.[275] Nanotubes and onions of GaS and GaSe have been generated through laser and thermally induced exfoliation of the bulk powders (see Fig. 25).[276] Single-wall nanotubes of $SbPS_{4-x}Se_x$ $(0 < x < 3)$ with tuneable bandgap have been synthesized.[277] GaP nanotubes with zinc blende structure have been obtained by the VLS growth.[278] Open-ended gold nanotube arrays have been obtained by the electrochemical deposition of Au on to an array of nickel nanorod templates followed by selective removal of the templates.[279] Free-standing, electro-conductive nanotubular sheets of indium tin oxide with different In/Sn ratios have been fabricated by using cellulose as the template.[280] A low-temperature route for synthesizing highly oriented ZnO nanotubes/nanorod arrays has been reported.[281] In this work, a radio frequency magnetron-sputtering technique was used to prepare ZnO-film-coated substrates for subsequent growth of the oriented nanostructures. Controllable syntheses of SiO_2 nanotubes with dome shaped interiors have been prepared by pyrolysis of silanes over Au catalysts.[282] High aspect-ratio, self-organized nanotubes of TiO_2 are obtained by anodization of titanium.[283] These self-organized porous structures consist of pore arrays with a uniform pore diameter of ~100 nm and an average spacing of 150 nm. The pore mouths are open on the top of the layer while on the bottom of the structure the tubes are closed by the presence of an about 50-nm thick barrier of TiO_2. Electrochemical etching of titanium under potentiostatic

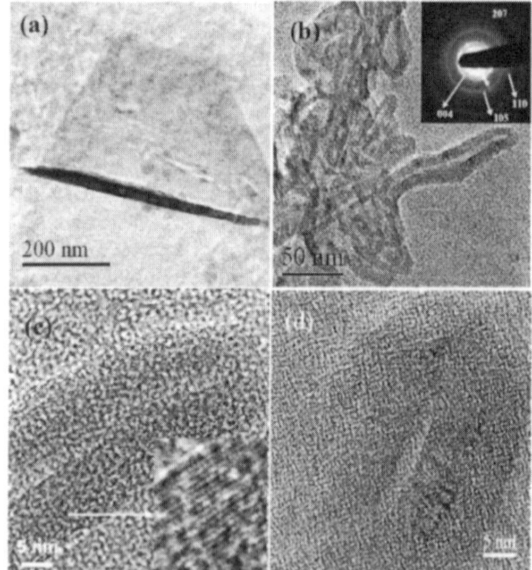

Fig. 25 TEM images revealing (a) rolling of an exfoliated GaS layer and (b) GaS nanotubes obtained by solvent irradiated laser irradiation. (c) HREM image of a nanotube ($d = 3.15$ Å). Insets shows GaS layer (d) TEM image of a GaSe onion obtained by laser irradiation in *n*-octylamine. Reprinted with permission from U. K. Gautam, S. R. C. Vivekchand, A. Govindaraj, G. U. Kulkarni, N. R. Selvi and C. N. R. Rao, *J. Am. Chem. Soc.*, 2005, **127**, 3658. © 2005, American Chemical Society.

conditions in fluorinated dimethyl sulfoxide and ethanol (1 : 1) under a range of anodizing conditions gives rise to ordered TiO_2 nanotube arrays.[284] TiO_2–B nanotubes can be prepared by the hydrothermal method.[285] Lithium is readily intercalated into the TiO_2–B nanotubes up to a composition of $Li_{0.98}TiO_2$ compared with $Li_{0.91}TiO_2$ for the corresponding nanowires. Intercalation of alkali metals into titanate nanotubes has also been investigated.[286] Highly crystalline TiO_2 nanotubes have been synthesized by hydrogen peroxide treatment of low crystalline TiO_2 nanotubes prepared by hydrothermal methods.[287] TiO_2 nanotubes with rutile structure have been prepared by using carbon nanotubes as templates.[288] Anatase nanotubes can be nitrogen doped by ion-beam implantation.[289]

RuO_2 nanotubes have been synthesized by the thermal decomposition of $Ru_3(CO)_{12}$ inside anodic alumina membranes.[290] Transition metal oxide nanotubes have been prepared in water using iced lipid nanotubes as the template.[291] Self-assembled cholesterol derivatives act as a template as well as a catalyst for the sol–gel polymerization of inorganic precursors to give rise to double-walled tubular structures of transition metal oxides.[292] Hydrothermal synthesis of single-crystalline γ-Fe_2O_3 nanotubes has been accomplished.[293] Nanotubes of single crystalline Fe_3O_4 have been prepared by wet-etching of the MgO inner cores of MgO/Fe_3O_4 core-shell nanowires.[294] Cerium oxide nanotubes are prepared by the controlled annealing of the as-formed $Ce(OH)_3$ nanotubes.[295]

Long hollow inorganic nanoparticle nanotubes with a nanoscale brick wall structure of clay mineral platelets have been synthesized by templating of block copolymer electrospun fibers with clay mineral platelets followed by interlinking of the platelets using condensation reactions.[296] Similarly, the construction of hollow inorganic nanospheres and nanotubes (inorganic nanostructures) with tunable wall thicknesses (with hollow interiors) is demonstrated by coating on self-assembled polymeric templates (nano-objects) with a thin Al_2O_3 layer by ALD, followed by removal of the polymer template upon heating.[297] The morphology of the nano-object (*i.e.*, spherical or cylindrical) is controlled by the block lengths of the copolymer. The thickness of the Al_2O_3 wall is controlled by the number of ALD cycles. Formation of ultra-long single-crystalline $ZnAl_2O_4$ spinel nanotubes, through a spinel-forming interfacial solid-state reaction of core-shell ZnO–Al_2O_3 nanowires involving the Kirkendall effect, has been reported.[298] Polycrystalline lead titanate nano- and microtubes with diameters ranging from a few tens of nanometers up to one micron were fabricated by wetting ordered porous alumina and macroporous silicon with precursor oligomers coupled with templated thermolysis.[299] WC nanotubes can be synthesized by the thermal decomposition of $W(CO)_6$ in the presence of Mg powder at 1173 K under the autogenic pressure of the precursors in a closed Swagelok reactor.[300]

6. Nanocrystaline films generated at the liquid–liquid interface

The liquid–liquid interface has been exploited recently for obtaining ultra-thin nanocrystalline films of a variety of materials. The method involves dissolving an organic precursor of the relevant metal in the organic layer and the appropriate reagent in the aqueous layer. The product formed by the reaction at the interface contains ultra-thin nanocrystalline films of the relevant material formed by closely packed nanocrystals. This simple technique has been shown to yield nanocrystals of metals such as Au, Ag, Pd and Cu, chalcogenides such as CdS, CdSe, ZnS, CoS, NiS, CuS and PbS and oxides such as γ-Fe_2O_3, ZnO and CuO.[301–310] In a typical preparation of Au nanocrystalline films, a solution of $Au(PPh_3)Cl$ in toluene was allowed to stand in contact with aqueous alkali in a beaker at 300 K. Once the two liquid layers were stable, tetrakis(hydroxymethyl)phosphonium chloride (THPC) was injected into the aqueous layer using a syringe with minimal disturbance to the toluene layer.[304] The interface first appears pink, finally growing a robust Au film at the interface. This film could be converted either to a gold organosol or a hydrosol by using appropriate capping agents in the organic and aqueous layers. The thickness and the particle size of the nanocrystals (shown in Fig. 26) is found to be dependent on the reaction conditions employed, such as reactant concentrations, reaction time and temperature.[305] The mean diameters of the nanocrystals formed at 30, 45, 60 and 75 °C are 7, 10, 12 and 15 nm respectively. The liquid–liquid interface has also been employed to prepare nanocrystalline films of binary alloys of Au–Ag and Au–Cu, and also ternary Au–Ag–Cu alloys,[306] by starting with an appropriate mixture of the corresponding metal precursors. Nanocrystalline films of metal chalcogenides such as CdS have been prepared at the toluene–water interface. In a typical preparation of a CdS nanocrystals, Na_2S was dissolved in water in a beaker and cadmium cupferronate [$Cd(cup)_2$], was dissolved in toluene by

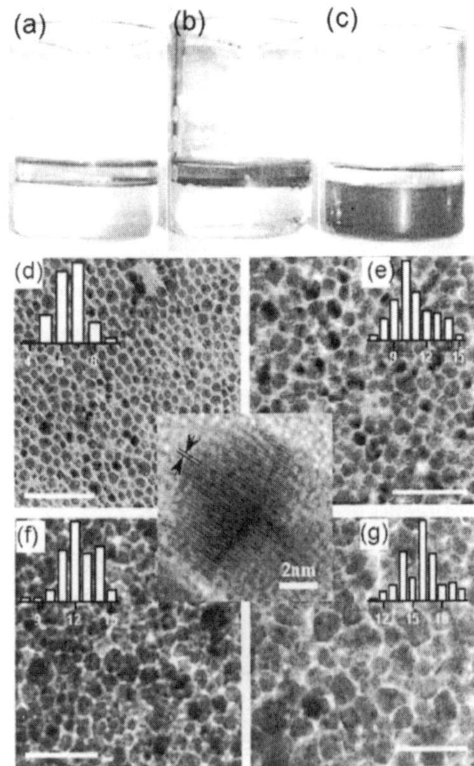

Fig. 26 (a) Nanocrystalline film of Au formed at the toluene–water interface, (b) When dodecanethiol is added to the toluene layer, the film breaks up, forming an organosol of Au, (c) Au hydrosol obtained when mercaptoundecanoic acid is added to water. TEM images of the ultra-thin nanocrystalline Au films obtained at the liquid–liquid interface after 24 h: (d) 30 °C, (e) 45 °C, (f) 60 °C and (g) 75 °C. The histograms of particle size distribution are shown. The scale bars correspond to 50 nm. A high-resolution image of an individual particle is shown at the center. Reprinted with permission from V. V. Agrawal, G. U. Kulkarni and C. N. R. Rao, *J. Phys. Chem. B*, 2005, **109**, 7300. © 2005, American Chemical Society.

ultrasonication.[307] A few drops of *n*-octylamine were added to the Cd(cup)$_2$ solution in order to make it completely soluble. The toluene solution was slowly added to a beaker containing the aqueous Na$_2$S solution. The interface started appearing yellow within a few minutes and a distinct film was formed after 10 h. γ-Fe$_2$O$_3$ nanocrystals have also been prepared at the toluene–water interface.

While many metals and metal sulfides yield films of nanocrystal assemblies at the liquid–aqueous interface, extended ultra-thin single-crystalline films are obtained in the case of CuS.[308] Single-crystalline films of CuSe, ZnS, PbS, Cu(OH)$_2$ and ZnO have also been prepared similarly. To prepare a CuS film, 75 mL of 0.12 mM Cu(cup)$_2$ solution in toluene was slowly added to an aqueous solution of 0.5 mM Na$_2$S (75 ml) taken in a crystallization dish (10 cm diameter). An excess of Na$_2$S was required in order to prevent the formation of Cu$_2$S. The interface gradually turns green, and the CuS film formed at the interface after 12 h, while the two

liquid phases remained colourless. The film grows slowly with time, first appearing as green-islands at the interface (in the initial 1–2 h) and slowly covering the entire interface. The film is fairly continuous and extends over a wide area as can be seen in Fig. 27. Small fragments on the edges of the film are also seen as the film breaks while lifting from the interface (Fig. 27a). The single-crystalline and essentially defect-free nature of the film can be inferred from the HREM image and the SAED pattern (Fig. 27b). The lattice spacing of 2.7 Å in the HREM image corresponds to the separation between (006) planes of the hexagonal CuS phase. The diffraction spots could be indexed on the basis of the hexagonal structure (*P63/mmc*, *a* = 3.792 Å and *c* = 16.34 Å). The thickness of the film was estimated to be ~50 nm from AFM and ellipsometric studies. Thicker films could be formed using higher concentration of reactants. The films prepared at higher temperatures are, however, less continuous and form flakes and rods. Nanorods and nanocrystals of various sizes and shapes as seen in Fig. 27d) and Fig. 27e were obtained upon sonication of the CuS films.

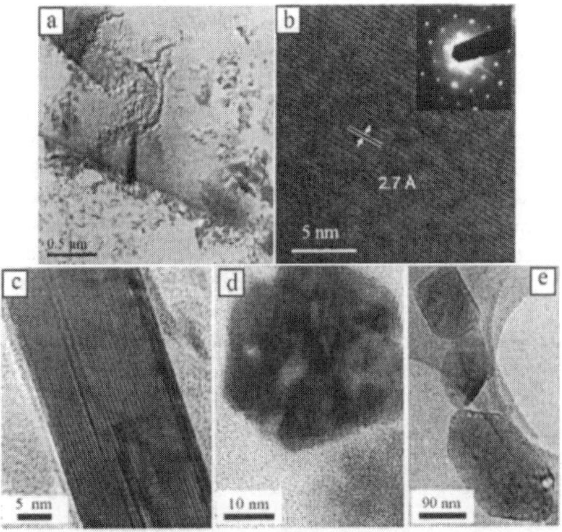

Fig. 27 (a) TEM images of an ultra-thin 1 μm of CuS obtained at the toluene–water interface. (b) HREM image of the film. Inset shows the corresponding SAED pattern. (c) HREM image of a rod like fragment obtained at 343 K. (d) Hexagonal CuS nanocrystals obtained by sonication of the film obtained with 2 mg of Cu(cup)$_2$ and (e) the nanocrystals obtained by using 1 mg Cu(cup)$_2$. Reprinted with permission from U. K. Gautam, M. Ghosh and C. N. R. Rao, *Langmuir*, 2004, **20**, 10775. © 2004, American Chemical Society.

Extended single-crystalline films of ZnS are obtained by this method starting with Zn(cup)$_2$ and Na$_2$S.[309] The UV-visible absorption spectrum of a ZnS film obtained by reacting 5 mg of Zn(cup)$_2$ in 50 ml of toluene at ~30 °C for 12 h shows an absorption band at ~320 nm, close to that of bulk ZnS. The PL spectrum of the ZnS film displays a broad emission band centered at ~425 nm due to the presence of sulfur vacancies in the ZnS lattice. A blue-shift of the absorption band to 285 nm was observed when the reaction time was restricted to 1 h, suggesting a thin, incompletely formed film. Unlike chemical methods such as the LB technique, where non-single-crystalline films are obtained by assembling nanocrystals or CVD and related techniques where stringent conditions as well as substrates are required, the interface method is simple and can be extended to a variety of materials. Using a similar approach CdSe nanofilms have also been prepared.[310]

7. Nanowalls

Thermal exfoliation of GaS and GaSe gives 'rise to nanowalls (Fig. 28). Nanowalls of carbon, which are actually interconnected 2D nanosheets of carbon vertically standing on a substrate has been described by Wu and co-workers.[311] There are reports of ZnO nanowalls and ZnO nanorods grow from the nodes of nanowalls.[312] As mentioned in the previous section, thermal exfoliation of GaSe gave rise to solid deposits in the cooler end of the sealed tube.[313] Deposits with similar morphology were obtained in the case of GaS as well. The deposits contained wall structures with smooth curved surfaces as revealed by the SEM images. The XRD pattern of the sample could be indexed on the hexagonal phase of GaSe. The EDAX spectrum recorded at various locations of the sample confirmed the Ga : Se ratio to be 1 : 1. TEM images revealed that the walls are transparent, especially at the edges, indicating a thickness of around a few nanometers. The nanowalls are single-crystalline, as established by the HREM images as well as the SAED patterns. The lattice spacing observed in the HREM image of 3.229 Å corresponds to the separation between the [100] planes of GaSe in the space group $P6_3/mmc$. In the initial stage, the exfoliated sheets deposited in the cooler end of the tube melt forming droplets. Flower-like nanostructures form around these droplets and grow with time forming extended network structures. The Ga_2O_3 nanowalls were obtained by heating GaS and GaSe nanowalls in air at 823 K. On the other hand, heating the GaS and GaSe nanowalls in NH_3, GaN nanowalls were obtained.

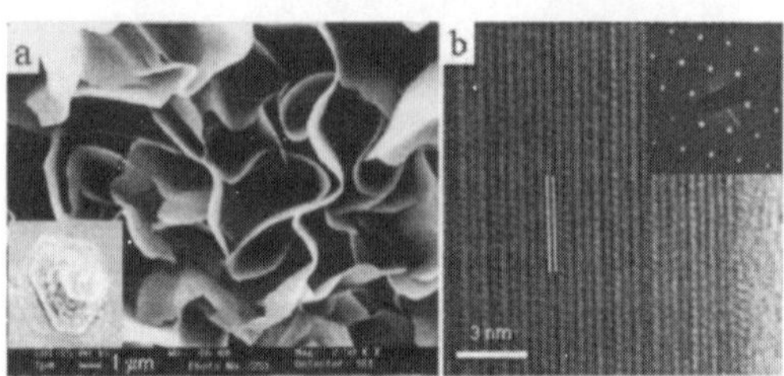

Fig. 28 (a) SEM image of GaSe nanowalls deposited at 673 K. (b) HREM images of a nanowall. The separation between the lattice planes (3.229 Å ≈ d_{100}) is shown in the image. Inset shows the SAED pattern. (From Ref. [313]).

8. Conclusions

The preceding presentation reflects the tremendous contribution of chemists to the synthesis of inorganic nanomaterials. Beside the use of known methods, the synthesis of nanomaterials has required many newer strategies geared to prepare objects of different dimensionalities. Clearly, there will be continued efforts by a large body of chemists in preparing inorganic nanomaterials, both known and new, by employing novel and improved methods. In addition to synthesis, functionalization, solubilization and assembly of nanomaterials are aspects of great relevance where

again chemistry plays a crucial role. While we have not touched on these aspects in this article, it is necessary that we recognise their importance in the manipulation and use of nanomaterials.

References

1 G. Schmid, Clusters and Colloids, From *Theory to Applications*, VCH, Weinheim, 1994.
2 *The Chemistry of Nanomaterials*, Ed. C. N. R. Rao, A. Muller and A. K. Cheetham, 2004, Wiley-VCH Verlag,Weinheim, vol. 1 & 2.
3 *Nanomaterials Chemistry: Recent Developments*, Ed. C. N. R. Rao, A. Muller and A. K. Cheetham, 2007, Wiley-VCH-Verlag, Weinheim.
4 C. N. R. Rao and A. Govindaraj, *Nanotubes and Nanowires*, RSC series on Nanoscience, London, 2005.
5 C. N. R. Rao, A. Govindaraj and S. R. C. Vivekchand, *Ann. Rep. Prog. Chem.*, Royal Society of Chemistry: London, 2006, **102**, 20.
6 C. N. R. Rao, P. J. Thomas and G. U. Kulkarni, *Nanocrystals: Synthesis, Properties and Applications*, Springer series on material science: 95, 2007.
7 (*a*) C. Burda, X. Chen, R. Narayanan and M. A. El-Sayed, *Chem. Rev.*, 2005, **105**, 1025; (*b*) M. Rajamathi and R. Seshadri, *Curr. Opin. Solid State Mater. Sci.*, 2002, **6**, 337.
8 E. A. Hauser and J. E. Lynn, *Experiments in Colloid Chemistry*, p. 18 (McGraw-Hill: New York 1940).
9 (*a*) J. Turkevich, P. C. Stevenson and J. Hillier, *Spec. Discuss. Faraday Soc.*, 1951, **11**, 55; (*b*) J. Turkevich, *Gold Bull.*, 1985, **18**, 86.
10 D. G. Duff, A. Baiker and P. P. Edwards, *Langmuir*, 1993, **9**, 2301.
11 M. Brust, M. Walker, D. Bethell, D. J. Schiffrin and R. Whyman, *J. Chem. Soc., Chem. Commun.*, 1994, 801.
12 K. V. Sarathy, G. Raina, R. T. Yadav, G. U. Kulkarni and C. N. R. Rao, *J. Phys. Chem. B*, 1997, **101**, 9876.
13 K. V. Sarathy, G. U. Kulkarni and C. N. R. Rao, *Chem. Commun.*, 1997, 537.
14 P. J. Thomas, G. U. Kulkarni and C. N. R. Rao, *J. Phys. Chem. B*, 2000, **104**, 8138.
15 T. Teranishi and M. Miyake, *Chem. Mater.*, 1998, **10**, 594.
16 M. Schulz-Dobricks, K. V. Sarathy and M. Jansen, *J. Am. Chem. Soc.*, 2005, **127**, 12816.
17 (*a*) I. Pastoriza-Santos and L. M. Liz-Marzán, *Langmuir*, 2002, **18**, 2888; (*b*) I. Pastoriza-Santos and L. M. Liz-Marzán, *Pure Appl. Chem.*, 2000, **72**, 83.
18 N. R. Jana and X. Peng, *J. Am. Chem. Soc.*, 2003, **125**, 14280.
19 R. Jin, Y. Cao and C. A. Mirkin, *Science*, 2001, **294**, 1901.
20 R. Jin, Y. Cao, E. Hao, G. S. Métraux, G. C. Schatz and C. A. Mirkin, *Nature*, 2003, **425**, 487.
21 G. S. Metraux and C. A. Mirkin, *Adv. Mater.*, 2005, **17**, 412.
22 E. Hao, R. C. Bailey and G. C. Schatz, *Nano Lett.*, 2004, **4**, 327.
23 C. H. Kuo and M. H. Huang, *Langmuir*, 2005, **21**, 2012.
24 Y. Xiao, B. Shylahovsky, I. Popov, V. Pavlov and I. Wilner, *Langmuir*, 2005, **21**, 5659.
25 M. Zhou, S. Chen and S. Zhao, *J. Phys. Chem. B*, 2006, **110**, 4510.
26 B. J. Wiley, Y. Xiong, Z.-Y. Li, Y. Yin and Younan Xia, *Nano Lett.*, 2006, **6**, 765.
27 T. Tsukatani and H. Fujihara, *Langmuir*, 2005, **21**, 12093.
28 G. S. Fonseca, A. P. Umpierre, P. F. P. Fichtner, S. R. Teixeira and J. Dupont, *Chem.–Eur. J.*, 2003, **9**, 3263.
29 H. Itoh, K. Naka and Y. Chujo, *J. Am. Chem. Soc.*, 2004, **126**, 3026.
30 O. Margeat, C. Amiens, B. Chaudret, P. Lecante and R. E. Banfield, *Chem. Mater.*, 2005, **17**, 107.
31 J. D. Hoefelmeyer, K. Niesz, G. A. Somarjai and T. D. Tilley, *Nano Lett.*, 2005, **5**, 435.
32 C. A. Stowell and B. A. Korgel, *Nano Lett.*, 2005, **5**, 1203.
33 S. Ghosh, M. Ghosh and C. N. R. Rao, *J. Cluster Sci.*, 2007, **18**, 97.
34 O. Margeat, C. Amiens, B. Chaudret, P. Lecante and R. E. Banfield, *Chem. Mater.*, 2005, **17**, 107.
35 H. Song, F. Kim, S. Connor, G. A. Samorjai and P. Yang, *J. Phys. Chem. B*, 2005, **109**, 188.
36 C. Roychowdhury, F. Matsumoto, P. F. Mutolo, H. A. Abruna and F. J. DiSalvo, *Chem. Mater.*, 2005, **17**, 5871.
37 G. De and C. N. R. Rao, *J. Mater. Chem.*, 2005, **15**, 891.
38 M. Chen, J. Kim, J. P. Liu, H. Fan and S. Sun, *J. Am. Chem. Soc.*, 2006, **128**, 7132.

39 J. Rockenberger, E. C. Scher and A. P. Alivisatos, *J. Am. Chem. Soc.*, 1999, **121**, 11595.

40 M. Ghosh, E. V. Sampathkumaran and C. N. R. Rao, *Chem. Mater.*, 2005, **17**, 2348.

41 M. Ghosh, K. Biswas, A. Sundaresan and C. N. R. Rao, *J. Mater. Chem.*, 2006, **16**, 106.

42 M. Ghosh and C. N. R. Rao, *Chem. Phys. Lett.*, 2004, **393**, 493.

43 M. Ghosh, R. Seshadri and C. N. R. Rao, *J. Nanosci. Nanotechnol.*, 2004, **4**, 136.

44 S. Thimmaiah, M. Rajamathi, N. Singh, P. Bera, F. C. Meldrum, N. Chandrasekhar and R. Seshadri, *J. Mater. Chem.*, 2001, **11**, 3215.

45 K. Biswas and C. N. R. Rao, *J. Phys. Chem. B*, 2006, **110**, 842.

46 S. Link and M. A. El-Sayed, *Int. Rev. Phys. Chem.*, 2000, **19**, 409.

47 K. Biswas, S. V. Bhat and C. N. R. Rao, *J. Phys. Chem. C*, 2007, **111**, 5689.

48 J. Park, K. An, Y. Hwang, J. G. Park, H. J. Noh, J. Y. Kim, J. H. Park, N. M. Hwang and T. Hyeon, *Nat. Mater.*, 2004, **3**, 891.

49 W. S. Seo, J. H. Shim, S. J. Oh, E. K. Lee, N. H. Hur and J. T. Park, *J. Am. Chem. Soc.*, 2005, **127**, 6188.

50 S-H Choi, E-G Kim, J. Park, K. An, N. Lee, A. C. Kim and T. Hyeon, *J. Phys. Chem. B*, 2005, **109**, 14792.

51 Z. Hu, D. J. E. Ramirez, B. E. H. Cervera, G. Oskam and P. C. Searson, *J. Phys. Chem. B*, 2005, **109**, 11209.

52 J. Joo, S. G. Kwon, J. H. Yu and T. Hyeon, *Adv. Mater.*, 2005, **17**, 1873.

53 L. S. Panchakarla, A. Govindaraj and C. N. R. Rao, *J. Cluster Sci.*, 2007, (in print).

54 S. O'Brien, L. Brus and C. B. Murray, *J. Am. Chem. Soc.*, 2001, **123**, 12085.

55 H. Wu, Y. Yang and Y. C. Cao, *J. Am. Chem. Soc.*, 2006, **128**, 16522.

56 N. Pinna, G. Garweitner, M. Antonietti and M. Niederberger, *J. Am. Chem. Soc.*, 2005, **127**, 5608.

57 Q. Liu, W. Lu, A. Ma, J. Tang, J. Lin and J. Fang, *J. Am. Chem. Soc.*, 2005, **127**, 5267.

58 J. Tang, F. Redl, Y. Zhu, T. Siegrist, L. E. Brus and M. L. Steigerwald, *Nano Lett.*, 2005, **5**, 543.

59 G. Li, L. Li, J. B. Goated and B. F. Woodfield, *J. Am. Chem. Soc.*, 2005, **127**, 8659.

60 N. Moumen and M.-P. Pileni, *J. Phys. Chem.*, 1996, **100**, 1867.

61 N. Moumen and M.-P. Pileni, *Chem. Mater.*, 1996, **8**, 1128.

62 C. B. Murray, D. J. Norris and M. G. Bawendi, *J. Am. Chem. Soc.*, 1993, **115**, 8706–8715.

63 J. E. Bowen-Katari, V. L. Colvin and A. P. Alivisatos, *J. Phys. Chem.*, 1994, **98**, 4109.

64 U. K. Gautam, M. Rajamathi, F. Meldrum, P. Morgan and R. Seshadri, *Chem. Commun.*, 2001, 629.

65 U. K. Gautam, R. Seshadri and C. N. R. Rao, *Chem. Phys. Lett.*, 2003, **375**, 560.

66 U. K. Gautam and R. Seshadri, *Mater. Res. Bull.*, 2004, **39**, 669.

67 X. Chen and R. Fan, *Chem. Mater.*, 2001, **13**, 802.

68 L. Qu and X. Peng, *J. Am. Chem. Soc.*, 2002, **124**, 2049.

69 Z. A. Peng and X. Peng, *J. Am. Chem. Soc.*, 2001, **123**, 183.

70 Z. A. Peng and X. Peng, *J. Am. Chem. Soc.*, 2002, **124**, 3343.

71 L. Qu, Z. A. Peng and X. Peng, *Nano Lett.*, 2001, **1**, 333.

72 J. Joo, H. B. Na, T. Yu, J. H. Yu, Y. W. Kim, F. Wu, J. Z. Zhang and T. Hyeon, *J. Am. Chem. Soc.*, 2003, **125**, 11100.

73 Y. A. Yang, H. Wu, K. R. Williams and Y. J. Cao, *Angew. Chem., Int. Ed.*, 2005, **44**, 6712.

74 J. Jasieniak, C. Bullen, J. V. Embden and P. Mulvaney, *J. Phys. Chem. B*, 2005, **109**, 20665.

75 P. D. Cozzoli, L. Manna, M. L. Curri, S. Kudera, C. Giannini, M. Striccoli and A. Agostiano, *Chem. Mater.*, 2005, **17**, 1296.

76 J. E. Murphy, M. C. Beard, A. G. Norman, S. P. Ahrenkiel, J. C. Johnson, P. Yu, O. I. Mićić, R. J. Ellingson and A. J. Nozik, *J. Am. Chem. Soc.*, 2006, **128**, 3241.

77 L. Cademartiri, J. Bertolotti, R. Sopienza, D. S. Wiersma, G. Freymann and G. A. Ozin, *J. Phys. Chem. B*, 2006, **110**, 671.

78 L. A. Swafford, L. A. Weigand, M. J. Bowers II, J. R. McBride, J. L. Rapaport, T. L. Watt, S. K. Dixit, L. C. Feldman and S. J. Rosenthal, *J. Am. Chem. Soc.*, 2006, **128**, 12299.

79 S.-H. Choi, E.-G. Kim and T. Hyeon, *J. Am. Chem. Soc.*, 2006, **128**, 2520.

80 A. Ghezelbash and B. A. Korgel, *Langmuir*, 2005, **21**, 9451.

81 See for example: B. Ludolph, M. A. Malik, P. O'Brien and N. Revaprasadu, *Chem. Commun.*, 1998, 1849.

82 N. Pradhan and S. Efrima, *J. Am. Chem. Soc.*, 2003, **125**, 2050.

83 K. Biswas and C. N. R. Rao, *Chem.–Eur. J.*, 2007, **13**, 6123.

84 M. Green, P. Rahman and D. S. Boyle, *Chem. Commun.*, 2007, 574.

85 Y. Xie, Y. Qian, W. Wang, S. Zhang and Y. Zhang, *Science*, 1996, **272**, 1926.

86 K. Sardar and C. N. R. Rao, *Adv. Mater.*, 2004, **16**, 425.

87 (*a*) U. K. Gautam, K. Sardar, F. L. Deepak and C. N. R. Rao, *Pramana*, 2005, **65**, 549; (*b*) K. Sardar and C. N. R. Rao, *Solid State Sci.*, 2005, **7**, 217; (*c*) K. Sardar, F. L. Deepak, A. Govindaraj, M. M. Seikh and C. N. R. Rao, *Small*, 2005, **1**, 91.

88 S. V. Bhat, K. Biswas and C. N. R. Rao, *Solid State Commun.*, 2007, **14**, 325.

89 K. Biswas, K. Sardar and C. N. R. Rao, *Appl. Phys. Lett.*, 2006, **89**, 132503.

90 O. I. Mićić, S. P. Ahrenkiel, D. Bertram and A. J. Nozik, *Appl. Phys. Lett.*, 1999, **75**, 478.

91 J. L. Coffer, M. A. Johnson, L. Zhang, R. L. Wells and J. F. Janik, *Chem. Mater.*, 1997, **9**, 2671.

92 A. C. Frank, F. Stowasser, H. Sussek, H. Pritzkow, C. R. Miskys, O. Ambacher, M. Giersig and R. A. Fischer, *J. Am. Chem. Soc.*, 1998, **120**, 3512.

93 A. Manz, A. Brikner, M. Kolbe and R. A. Fischer, *Adv. Mater.*, 2000, **12**, 569.

94 K. Sardar, M. Dan, B. Schwenzer and C. N. R. Rao, *J. Mater. Chem.*, 2005, **15**, 2175.

95 A. Gomathi and C. N. R. Rao, *Mater. Res. Bull.*, 2006, **41**, 941.

96 (*a*) L. Grocholl, J. Wang and E. G. Gillan, *Chem. Mater.*, 2001, **13**, 4290; (*b*) L. Grocholl, J. Wang and E. G. Gillan, *Nano Lett.*, 2002, **2**, 899.

97 J. Choi and E. G. Gillan, *J. Mater. Chem.*, 2006, **16**, 3774.

98 J. Xiao, Y. Xie, R. Tang and W. Luo, *Inorg. Chem.*, 2003, **42**, 107.

99 M. Yu, X. Hao, D. Cui, Q. Wang, X. Xu and M. Jiang, *Nanotechnology*, 2003, **14**, 29.

100 M. A. Olshavsky, A. B. Goldstein and A. P. Alivisatos, *J. Am. Chem. Soc.*, 1990, **112**, 9438.

101 R. L. Wells, S. R. Aubuchon, S. S. Kher and M. S. Lube, *Chem. Mater.*, 1998, **7**, 793.

102 O. I. Mićić, C. J. Curtis, K. M. Jones, J. R. Sprague and A. J. Nozik, *J. Phys. Chem.*, 1994, **98**, 4966.

103 O. I. Mićić, J. R. Sprague, C. J. Curtis, K. M. Jones, J. L. Machol, A. J. Nozik, H. Giessen, B. Fluegel, G. Mohs and N. Peyghambarian, *J. Phys. Chem.*, 1995, **99**, 7754.

104 A. A. Guzelian, J. E. B. Katari, A. V. Kadavanich, U. Banin, K. Hamad, E. Juban, A. P. Alivisatos, R. H. Wolters, C. C. Arnold and J. R. Heath, *J. Phys. Chem.*, 1996, **100**, 7212.

105 S. C. Perera, P. S. Fodor, G. M. Tsoi, L. E. Wenger and S. L. Brock, *Chem. Mater.*, 2003, **15**, 4034.

106 S. C. Perera, G. Tsoi, L. E. Wenger and S. L. Brock, *J. Am. Chem. Soc.*, 2003, **125**, 13960.

107 S. Wei, J. Lu, W. Yu and Y. Qian, *J. App. Phys.*, 2004, **95**, 3683.

108 S. Xu, S. Kumar and T. Nann, *J. Am. Chem. Soc.*, 2006, **128**, 1054.

109 M. A. Malik, P. O'Brien and M. Helliwell, *J. Mater. Chem.*, 2005, **15**, 1463.

110 K. L. Stamm, J. C. Garno, G. Liu and S. L. Brock, *J. Am. Chem. Soc.*, 2003, **125**, 4038.

111 P. Arumugam, S. S. Shinozaki, R. Wang, G. Maob and S. L. Brock, *Chem. Commun.*, 2006, 1121.

112 J. Lu, Y. Xie, X. Jiang, W. He and G. Du, *J. Mater. Chem.*, 2001, **11**, 3281.

113 W. Schartl, *Adv. Mater.*, 2000, **12**, 1899.

114 See for example: X. Peng, M. C. Schlamp, A. V. Kadavanich and A. P. Alivisatos, *J. Am. Chem. Soc.*, 1997, **119**, 7019.

115 M. A. Hines and P. Guyot-Sionnest, *J. Phys. Chem.*, 1996, **100**, 468.

116 M. A. Malik, P. O'Brien and N. Revaprasadu, *Chem. Mater.*, 2002, **14**, 2004.

117 D. Pan, Q. Wang, S. Jiang, X. Ji and L. An, *Adv. Mater.*, 2005, **17**, 176.

118 R. Xie, U. Kolb, J. Li, T. Basche and A. Mews, *J. Am. Chem. Soc.*, 2005, **127**, 7480.

119 Y. Liu, M. Kim, Y. Wang, Y. A. Wang and X. Peng, *Langmuir*, 2006, **22**, 6341.

120 Y.-W. Cao and U. Banin, *J. Am. Chem. Soc.*, 2000, **122**, 9692.

121 R. H. Morriss and L. F. Collins, *J. Chem. Phys.*, 1964, **41**, 3357.

122 L. Lu, H. Wang, Y. Zhou, S. Xi, H. Zhang, J. Hub and B. Zhaob, *Chem. Commun.*, 2002, 144.
123 L. Rivas, S. Sanchez-Cortes, J. V. Garcia-Ramos and G. Morcillo, *Langmuir*, 2000, **16**, 9722.
124 K. Mallik, M. Mandal, N. Pradhan and T. Pal, *Nano Lett.*, 2001, **1**, 319.
125 Y. W. Cao, R. Jin and C. A. Mirkin, *J. Am. Chem. Soc.*, 2001, **123**, 7961.
126 J.-W. Hu, Y. Zhang, J.-F. Li, Z. Liu, B. Ren, S.-G. Sun, Z.-Q. Tian and T. Lian, *Chem. Phys. Lett.*, 2005, **408**, 354.
127 J. Rivas, R. D. Sanchez, A. Fondado, A. J. Garcia-Bastida, J. Garcia-Otero, J. Mira, D. Baldomir, A. Gonzhlez, I. Lado, M. A. L. Quintela and S. B. Oseroff, *J. Appl. Phys.*, 1994, **76**, 6564.
128 C. T. Seip and C. J. O'Connor, *Nanostruct. Mater.*, 1999, **12**, 183.
129 J. Lin, W. Zhow, A. Kumbhar, J. Wiemann, J. Fang, E. E. Carpenter and C. J. O'Connor, *J. Solid State Chem.*, 2001, **159**, 26.
130 J. Zhang, M. Post, T. Veres, Z. J. Jakubek, J. Guan, D. Wang, F. Normandin, Y. Deslandes and B. Simard, *J. Phys. Chem. B*, 2006, **110**, 7122.
131 L. Wang, J. Luo, Q. Fan, M. Suzuki, I. S. Suzuki, M. H. Engelhard, Y. Lin, N. Kim, J. Q. Wang and C.-J. Zhong, *J. Phys. Chem. B*, 2005, **109**, 21593.
132 W. Stober, A. Fink and E. Bohn, *J. Colloid Interface Sci.*, 1968, **26**, 62.
133 See for example: M. A. Correa-Duarte, M. Giersig, N. A. Kotov and L. M. Liz-Marzán, *Langmuir*, 1998, **14**, 6430.
134 See for example: T. Ung, L. M. Liz-Marzán and P. Mulvaney, *J. Phys. Chem. B*, 1999, **103**, 6770.
135 L. M. Liz-Marzán, M. Giersig and P. Mulvaney, *Langmuir*, 1996, **12**, 4329.
136 K. S. Mayya, D. I. Gittins and F. Caruso, *Chem. Mater.*, 2001, **13**, 3833.
137 I. Pastoriza-Santos, D. S. Koktysh, A. A. Mamedov, M. Giersig, N. A. Kotov and L. M. Liz-Marzán, *Langmuir*, 2000, **16**, 2731.
138 V. Eswaranand and T. Pradeep, *J. Mater. Chem.*, 2002, **12**, 2421.
139 R. T. Tom, A. S. Nair, N. Singh, M. Aslam, C. L. Nagendra, R. Philip, K. Vijayamohanan and T. Pradeep, *Langmuir*, 2003, **19**, 3439.
140 S. Ghosh, K. Biswas and C. N. R. Rao, *J. Mater. Chem.*, 2007, **17**, 2412.
141 V. Salgueiriño-Maceira and M. A. C. Duarte, *J. Mater. Chem.*, 2006, **16**, 3539.
142 O. Masala and R. Seshadri, *J. Am. Chem. Soc.*, 2005, **127**, 9354.
143 C. N. R. Rao, F. L. Deepak, G. Gundiah and A. Govindaraj, *Prog. Solid State Chem.*, 2003, **31**, 5.
144 Y. Xia, P. Yang, Y. Sun, W. Wu, B. Mayers, B. Gates, Y. Yin, F. Kim and H. Han, *Adv. Mater.*, 2003, **15**, 353.
145 C. R. Martin, *Science*, 1994, **266**, 1961.
146 D Almawlawi, C. Z. Liu and M. Moskovits, *J. Mater. Res.*, 1994, **9**, 1014.
147 A. Govindaraj, B. C. Satishkumar, M. Nath and C. N. R. Rao, *Chem. Mater.*, 2000, **12**, 202.
148 M. Zheng, L. Zhang, X. Zhang, J. Zhang and G. Li, *Chem. Phys. Lett.*, 2001, **334**, 298.
149 J. A. Sioss and C. D. Keating, *Nano Lett.*, 2005, **5**, 1779.
150 B. D. Busbee, S. O. Obare and C. J. Murphy, *Adv. Mater.*, 2003, **15**, 414.
151 S. Gou and C. J. Murphy, *Chem. Mater.*, 2005, **17**, 3668.
152 H.-Y. Wu, H.-C. Chu, T.-J. Kuo, C.-L. Kuo and M. L. H. Huang, *Chem. Mater.*, 2005, **17**, 6447.
153 A. Gulati, H. Liao and J. H. Hafner, *J. Phys. Chem. B*, 2006, **110**, 22323.
154 A. Gole and C. J. Murphy, *Chem. Mater.*, 2005, **17**, 1325–1330.
155 B. Basnar, Y. Weizmann, Z. Cheglakov and I. Willner, *Adv. Mater.*, 2006, **18**, 713–718.
156 C.-K. Tsung, X. Kou, Q. Shi, J. Zhang, M. H. Yeung, J. Wang and G. D. Stucky, *J. Am. Chem. Soc.*, 2006, **128**, 5352.
157 A. J. Mieszawska, G. W. Slawinski and F. P. Zamborini, *J. Am. Chem. Soc.*, 2006, **128**, 5622.
158 C. Ni, P. A. Hassan and E. W. Kaler, *Langmuir*, 2005, **21**, 3334.
159 Y. Sun, B. Gates, B. Mayers and Y. Xia, *Nano lett.*, 2002, **2**, 165.
160 Y. Chen, B. J. Wiley and Y. Xia, *Langmuir*, 2007, **23**, 4120.
161 L. Gou, M. Chipara and J. M. Zaleski, *Chem. Mater.*, 2007, **19**, 1755.
162 D. Ung, G. Viau, C. Ricolleau, F. Warmont, P. Gredin and F. Fievet, *Adv. Mater.*, 2005, **17**, 338.
163 W. Z. Wong, B. Poudel, Y. Ma and Z. F. Ren, *J. Phys. Chem. B*, 2006, **110**, 25702.
164 Y. Xiong, H. Cai, B. J. Wiley, J. Wang, M. J. Kim and Y. Xia, *J. Am. Chem. Soc.*, 2007, **129**, 3665.
165 S. R. C. Vivekchand, G. Gundiah, A. Govindaraj and C. N. R. Rao, *Adv. Mater.*, 2004, **16**, 1842.
166 S.-M. Liu, M. Kobayashi, S. Sato and K. Kimura, *Chem. Commun.*, 2005, 4690.
167 D. C. Lee, T. Hanrath and B. A. Korgel, *Angew. Chem., Int. Ed.*, 2005, **44**, 3573.
168 A. I. Hochbaum, R. Fan, R. He and P. Yang, *Nano Lett.*, 2005, **5**, 457.
169 T. Shimizu, T. Xie, J. Nishikawa, S. Shingubara, S. Senz and U. Gosele, *Adv. Mater.*, 2007, **19**, 917.
170 S. Kodambaka, J. B. Hannon, R. M. Tromp and F. M. Ross, *Nano Lett.*, 2006, **6**, 1296.
171 X. Lu, D. D. Fanfair, K. P. Johnston and B. A. Korgel, *J. Am. Chem. Soc.*, 2005, **127**, 15718.
172 H-Y Tuan, D. C. Lee, T. Hanrath and B. A. Korgel, *Chem. Mater.*, 2005, **17**, 5705.
173 P. Nguyen, H. T. Ng and M. Meyyappan, *Adv. Mater.*, 2005, **17**, 549.
174 D. Wang, R. Tu, L. Zhang and H. Dai, *Angew. Chem., Int. Ed.*, 2005, **44**, 2925.
175 H. Gerung, T. J. Boyle, L. J. Tribby, S. D. Bunge, C. J. Brinker and S. M. Han, *J. Am. Chem. Soc.*, 2006, **128**, 5244.
176 U. K. Gautam, M. Nath and C. N. R. Rao, *J. Mater. Chem.*, 2003, **13**, 2845.
177 U. K. Gautam and C. N. R. Rao, *J. Mater. Chem.*, 2004, **14**, 2530.
178 Y. Ma, L. Qi, W. Shen and J. Ma, *Langmuir*, 2005, **21**, 6161.
179 Q. Li and V. W.-W. Yam, *Chem. Commun.*, 2006, 1006.
180 J. M. Song, J. H. Zhu and S. H. Yu, *J. Phys. Chem. B*, 2006, **110**, 23790.
181 H. Zhang, D. Yang, X. Ma, N. Du, J. Wu and D. Gue, *J. Phys. Chem. B*, 2006, **110**, 827.
182 B. Cheng, W. Shi, J. M. R-Tanner, L. Zhang and E. T. Samulski, *Inorg. Chem.*, 2006, **45**, 1208.
183 H. Peng, Y. Fangli, B. Liuyang, L. Jinlin and C. Yunfa, *J. Phys. Chem. C*, 2007, **111**, 194.
184 L. S. Panchakarla, M. A. Shah, A. Govindaraj and C. N. R. Rao, *unpublished results*.
185 Z- Gui, J. Liu, Z. Wang, L. Song, Y. Hu, W. Fan and D. Chen, *J. Phys. Chem. B*, 2005, **109**, 1113.
186 P. X. Gao, Y. Ding, W. Mai, W. L. Hughes, C. Lao and Z. L. Wang, *Science*, 2005, **309**, 1700.
187 Q. Li, V. Kumar, Y. Li, H. Zhang, T. J. Marks and R. P. H. Chang, *Chem. Mater.*, 2005, **17**, 1001.
188 Y. Tak and K. Yong, *J. Phys. Chem. B*, 2005, **109**, 19263.
189 P. X. Gao, C. S. Lao, W. L. Hughes and Z. L. Wang, *Chem. Phys. Lett.*, 2005, **408**, 174.
190 M. Lai and D. J. Riley, *Chem. Mater.*, 2006, **18**, 2233.
191 S. Kar, B. N. Pal, S. Chaudhuri and D. Chakravorty, *J. Phys. Chem. B*, 2006, **110**, 4605.
192 J. H. He, J. H. Hsu, C. W. Wang, H. N. Lin, L. J. Chen and Z. L. Wang, *J. Phys. Chem. B*, 2006, **110**, 50.
193 L.-X. Yang, Y.-J. Zhu, W.-W. Wang, H. Tong and M.-L. Ruan, *J. Phys. Chem. B*, 2006, **110**, 6609.
194 F. L. Deepak, G. Gundiah, Md. M. Shiekh, A. Govindaraj and C. N. R. Rao, *J. Mater. Res.*, 2004, **19**, 2216.
195 P. Chen, S. Xie, N. Ren, Y. Zhang, A. Dong, Y. Chen and Y. Tang, *J. Am. Chem. Soc.*, 2006, **128**, 1470.
196 G. Wang, D.-S. Tsai, Y.-S. Huang, A. Korotcov, W.-C. Yeh and D. Susanti, *J. Mater. Chem.*, 2006, **16**, 780.
197 A. Magrez, E. Vasco, J. W. Seo, C. Dieker, N. Setter and L. Forro, *J. Phys. Chem. B*, 2006, **110**, 58.
198 K. P. Kalyanikutty, F. L. Deepak, C. Edem, A. Govindaraj and C. N. R. Rao, *Mater. Res. Bull.*, 2005, **40**, 831.
199 Y. Hao, G. Meng, C. Ye, X. Zhang and L. Zhang, *J. Phys. Chem. B*, 2005, **109**, 11204.
200 G. Gundiah, A. Govindaraj and C. N. R. Rao, *Chem. Phys. Lett.*, 2002, **351**, 189.
201 J. Zhang, F. Jiang, Y. Yang and J. Li, *J. Phys. Chem. B*, 2005, **109**, 13143.
202 J. Zhan, Y. Bando, J. Hu, F. Xu and D. Goldberg, *Small*, 2005, **1**, 883.
203 J. Joo, S. G. Kwon, T. Yu, M. Cho, J. Lee, J. Yoon and T. Hyeon, *J. Phys. Chem. B*, 2005, **109**, 15297.
204 B. S. Guiton, Q. Gu, A. L. Prieto, M. S. Gudiksen and H. Park, *J. Am. Chem. Soc.*, 2005, **127**, 498.

205 K. P. Kalyanikutty, G. Gundiah, C. Edem, A. Govindaraj and C. N. R. Rao, *Chem. Phys. Lett.*, 2005, **408**, 389.
206 Q. Wan, M. Wei, D. Zhi, J. L. MacManus-Driscoll and M. G. Blamire, *Adv. Mater.*, 2006, **18**, 234.
207 R. Wang, Y. Chen, Y. Fu, H. Zhang and C. Kisielowski, *J. Phys. Chem. B*, 2005, **109**, 12245.
208 Y. M. Zhao, Y.-H. Li, R. Z. Ma, M. J. Roe, D. G. McCartney and Y. Q. Zhu, *Small*, 2006, **2**, 422.
209 Y. Li, B. Tan and Y. Wu, *J. Am. Chem. Soc.*, 2006, **128**, 14258.
210 J. Zhou, Y. Ding, S. Z. Deng, L. Gong, N. S. Xu and Z. L. Wang, *Adv. Mater.*, 2005, **17**, 2107.
211 J.-W. Seo, Y.-W. Jun, S. J. Ko and J. Cheon, *J. Phys. Chem. B*, 2005, **109**, 5389.
212 G. Shen and D. Chen, *J. Am. Chem. Soc.*, 2006, **128**, 11762.
213 G. Xu, Z. Ren, P. Du, W. Weng, G. Shen and G. Han, *Adv. Mater.*, 2005, **17**, 907.
214 S. R. Hall, *Adv. Mater.*, 2006, **18**, 487.
215 A.-M. Cao, J.-S. Hu, H.-P. Liang and L.-J. Wan, *Angew. Chem., Int. Ed.*, 2005, **44**, 4391.
216 S. Kar and S. Chaudhuri, *J. Phys. Chem. B*, 2005, **109**, 3298.
217 J. Hu, Y. Bando and D. Goldberg, *Small*, 2005, **1**, 95.
218 S. Kar and S. Chaudhuri, *J. Phys. Chem. B*, 2006, **110**, 4542.
219 P. Christian and P. O'Brien, *Chem. Commun.*, 2005, 2817.
220 Y. Jeong, Y. Xia and Y. Yin, *Chem. Phys. Lett.*, 2005, **416**, 246.
221 S. G. Thoma, A. Sanchez, P. Provencio, B. L. Abrams and J. P. Wilcoxon, *J. Am. Chem. Soc.*, 2005, **127**, 7611.
222 A. Panda, G. Glaspell and M. S. El-Shall, *J. Am. Chem. Soc.*, 2006, **128**, 2790.
223 S. Kumar, M. Ade and T. Nann, *Chem.–Eur. J.*, 2005, **11**, 2220.
224 Z. Liu, D. Xu, J. Liang, J. Shen, S. Zhang and Y. Qian, *J. Phys. Chem. B*, 2005, **109**, 10699.
225 J.-P. Ge, J. Wang, H-X Zhang, X. Wang, Q. Peng and Y. Li, *Chem.–Eur. J.*, 2005, **11**, 1889.
226 X. Giu, Y. Lou, A. C. S. Samia, A. Devadoss, J. D. Burgess, S. Dayal and C. Burda, *Angew. Chem., Int. Ed.*, 2005, **44**, 5855.
227 R. Chen, M. H. So, C.-M. Che and H. Sun, *J. Mater. Chem.*, 2005, **15**, 4540.
228 F. Gao, Q. Lu and S. Komarneni, *Chem. Commun.*, 2005, 531.
229 M. B. Sigman and B. A. Korgel, *Chem. Mater.*, 2005, **17**, 1655.
230 A. Purkayastha, F. Lupo, S. Kim, T. Borca-Tasciuc and G. Ramanath, *Adv. Mater.*, 2006, **18**, 496.
231 M. Nath, A. Choudhury and C. N. R. Rao, *Chem. Commun.*, 2004, 2698.
232 D. Yu, J. Wu, Q. Gu and H. Park, *J. Am. Chem. Soc.*, 2006, **128**, 8148.
233 Y. H. Yang and Y. T. Chen, *J. Phys. Chem. B*, 2006, **110**, 17370.
234 B. Liu, Y. Bando, C. Tang, F. Xu, J. Hu and D. Goldberg, *J. Phys. Chem. B*, 2005, **109**, 17082.
235 H. Li, A. H. Chin and M. K. Sunkara, *Adv. Mater.*, 2006, **18**, 216.
236 S. Luo, W. Zhou, Z. Zhang, L. Liu, X. Dou, J. Wang, X. Zhao, D. Liu, Y. Gao, L. Song, Y. Xiang, J. Zhou and S. Xie, *Small*, 2005, **1**, 1004.
237 P. V. Radonanvic, C. J. Barrelet, S. Gradecak, F. Qian and C. M. Lieber, *Nano Lett.*, 2005, **5**, 1407.
238 C. J. Novotny and P. K. L. Yu, *Appl. Phys. Lett.*, 2005, **87**, 203111.
239 A. I. Persson, M. T. Björk, S. Jeppesen, J. B. Wagner, L. R. Wallenberg and L. Samuelson, *Nano Lett.*, 2006, **6**, 403.
240 H. Zhang, Q. Zhang, G. Zhao, J. Tang, O. Zhou and L.-C. Qin, *J. Am. Chem. Soc.*, 2005, **127**, 13120.
241 Y. Li, E. Tevaarwerk and R. P. H. Chang, *Chem. Mater.*, 2006, **18**, 2552.
242 Y. S. Hor, Z. L. Xiao, U. Welp, Y. Ito, J. F. Mitchell, R. E. Cook, W. K. Kwok and G. W. Crabtree, *Nano Lett.*, 2005, **5**, 397.
243 Y. Li, M. A. Malik and P. O'Brien, *J. Am. Chem. Soc.*, 2005, **127**, 16020.
244 C. N. R. Rao, A. Govindaraj, F. L. Deepak, N. A. Gunari and M. Nath, *Appl. Phys. Lett.*, 2001, **78**, 1853.
245 K. P. Kalyanikutty, M. Nikhila, U. Maitra and C. N. R. Rao, *Chem. Phys. Lett.*, 2006, **432**, 190.
246 E. J. H. Lee, C. Ribeiro, E. Longo and E. R. Leite, *J. Phys. Chem. B*, 2005, **109**, 20842.
247 R. Li, Z. Luo and F. Papadimitrakopoulos, *J. Am. Chem. Soc.*, 2006, **128**, 6280–6281.
248 K.-S. Cho, D. V. Talapin, W. Gaschler and C. B. Murray, *J. Am. Chem. Soc.*, 2005, **127**, 7140.
249 J. H. Yu, J. Joo, H. M. Park, S-II. Baik, Y. W. Kim, S. C. Kim and T. Hyeon, *J. Am. Chem. Soc.*, 2005, **127**, 5662.
250 A. Gomathi, S. R. C. Vivekchand, A. Govindaraj and C. N. R. Rao, *Adv. Mater.*, 2005, **17**, 2757.
251 Y. Xi, J. Zhou, H. Guo, C. Cai and Z. Lin, *Chem. Phys. Lett.*, 2005, **412**, 60.
252 A. D. LaLonde, M. G. Norton, D. N. McIlroy, D. Zhang, R. Padmanabhan, A. Alkhateeb, H. Man, N. Lane and Z. Holman, *J. Mater. Res.*, 2005, **20**, 549.
253 S. Y. Bae, H. W. Seo, H. C. Choi, D. S. Han and J. Park, *J. Phys. Chem. B*, 2005, **109**, 8496.
254 H.-X. Zhang, J.-P. Ge, J. Wang, Z. Wang, D.-P. Yu and Y.-D. Li, *J. Phys. Chem. B*, 2005, **109**, 11585.
255 M. A. Verheijen, G. Immink, T. de. Smet, M. T. Borgström and E. P. A. M. Bakkers, *J. Am. Chem. Soc.*, 2006, **128**, 1353.
256 M. Afzaal and P. O'Brien, *J. Mater. Chem.*, 2006, **16**, 1113.
257 I. Pastoriza-Santos, J. Pérez-Juste and L. M. Liz-Marzán, *Chem. Mater.*, 2006, **18**, 2465.
258 P. Mohan, J. Motohisa and T. Fukui, *Appl. Phys. Lett.*, 2006, **88**, 133105.
259 C. N. R. Rao and M. Nath, *Dalton Trans.*, 2003, 1.
260 R. Tenne, L. Margulis, M. Genut and G. Hodes, *Nature*, 1992, **360**, 444.
261 Y. Feldman, E. Wasserman, D. J. Srolovitch and R. Tenne, *Science*, 1995, **267**, 222.
262 M. Chhowalla and G. A. J. Amaratunga, *Nature*, 2000, **407**, 164.
263 P. A. Parilla, A. C. Dillon, K. M. Jones, G. Riker, D. L. Schulz, D. S. Ginley and M. J. Heben, *Nature*, 1999, **397**, 114.
264 P. A. Parilla, A. C. Dillon, B. A. Parkinson, K. M. Jones, J. Alleman, G. Riker, D. S. Ginley and M. J. Heben, *J. Phys. Chem. B*, 2004, **108**, 6197.
265 R. Tenne, *Chem.–Eur. J.*, 2002, **8**, 5296.
266 T. Tsirlina, Y. Feldman, M. Homyonfer, J. Sloan, J. L. Hutchison and R. Tenne, *Fullerene Sci. Technol.*, 1998, **6**, 157.
267 M. Nath, A. Govindaraj and C. N. R. Rao, *Adv. Mater.*, 2001, **13**, 283.
268 M. Nath and C. N. R. Rao, *Chem. Commun.*, 2001, 2336.
269 M. Nath and C. N. R. Rao, *J. Am. Chem. Soc.*, 2001, **123**, 4841.
270 M. Nath and C. N. R. Rao, *Angew. Chem., Int. Ed.*, 2002, **41**, 3451.
271 C. Y. Zhi, Y. Bando, C. Tang and D. Goldbeg, *Solid State Commun.*, 2005, **135**, 67.
272 F. L. Deepak, C. P. Vinod, K. Mukhopadhyay, A. Govindaraj and C. N. R. Rao, *Chem. Phys. Lett.*, 2002, **353**, 345.
273 J. Wang, V. K. Kayastha, Y. K. Yap, Z. Fan, J. G. Lu, Z. Pan, I. N. Ivanov, A. A. Puretzky and D. B. Geohegan, *Nano Lett.*, 2005, **5**, 2528.
274 J. Dinesh, M. Eswaramoorthy and C. N. R. Rao, *J. Phys. Chem. C*, 2007, **111**, 510.
275 S.-Y. Zhang, Y. Li, X. Ma and H.-Y. Chen, *J. Phys. Chem. B*, 2006, **110**, 9041.
276 U. K. Gautam, S. R. C. Vivekchand, A. Govindaraj, G. U. Kulkarni, N. R. Selvi and C. N. R. Rao, *J. Am. Chem. Soc.*, 2005, **127**, 3658.
277 C. D. Malliakas and M. G. Kantzidis, *J. Am. Chem. Soc.*, 2006, **128**, 6538.
278 Q. Wu, Z. Hu, C. Liu, X. Wang, Y. Chen and Y. Lu, *J. Phys. Chem. B*, 2005, **109**, 19719.
279 M. S. Sander and H. Gao, *J. Am. Chem. Soc.*, 2005, **127**, 12158.
280 Y. Aoki, J. Huang and T. Kunitake, *J. Mater. Chem.*, 2006, **16**, 292.
281 H. Yu, Z. Zhang, M. Han, X. Hao and F. Zhu, *J. Am. Chem. Soc.*, 2005, **127**, 2378.
282 C. Li, Z. Liu, C. Gu, X. Xu and Y. Yang, *Adv. Mater.*, 2006, **18**, 228.
283 J. M. Macak, H. Tsuchiya and P. Schumuki, *Angew. Chem., Int. Ed.*, 2005, **44**, 2100.
284 C. Ruan, M. Paulose, O. K. Varghese, G. K. Mor and C. A. Grimes, *J. Phys. Chem. B*, 2005, **109**, 15754.
285 G. Armstrong, A. R. Armstrong, J. Canales and P. G. Bruce, *Chem. Commun.*, 2005, 2454.
286 R. Ma, T. Sasaki and Y. Bando, *Chem. Commun.*, 2005, 948.
287 M. A. Khan, H.-T. Jung and O.-B. Yang, *J. Phys. Chem. B*, 2006, **110**, 626.
288 D. Eder, I. A. Kinloch and A. H. Windle, *Chem. Commun.*, 2006, 1448.
289 A. Ghicov, J. M. Macak, H. Tsuchiya, J. Kunze, V. Haeublein, L. Frey and P. Schmuki, *Nano Lett.*, 2006, **6**, 1080.
290 H. Tan, E. Ye and W. Y. Fan, *Adv. Mater.*, 2006, **18**, 619.
291 Q. Ji and T. Shimizu, *Chem. Commun.*, 2005, 4411.

292 J. H. Jung, T. Shimizu and S. Shinkai, *J. Mater. Chem.*, 2005, **15**, 3979.
293 C.-J. Jia, L.-D. Sun, Z.-G. Yan, L.-P. You, F. Luo, X.-D. Han, Y.-C. Pang, Z. Zhang and C. H. Yan, *Angew. Chem., Int. Ed.*, 2005, **44**, 4328.
294 Z. Liu, D. Zhang, S. Han, C. Li, B. Lei, W. Lu, J. Fang and C. Zhou, *J. Am. Chem. Soc.*, 2005, **127**, 6.
295 C. Tang, Y. Bando, B. Liu and D. Goldberg, *Adv. Mater.*, 2005, **17**, 3005.
296 R. H. A. Ras, T. Ruotsalainen, K Laurikainen, M. B. Linder and O. Ikkala, *Chem. Commun.*, 2007, **13**, 1366.
297 R. H. A. Ras, M. Kemell, J. de Wit, M. Ritala, G. ten Brinke, M. Leskela and O. Ikkala, *Adv. Mater.*, 2007, **19**, 102.
298 H. J. Fan, M. Knez, R. Scholz, K. Nielsch, E. Pippel, D. Hesse, M. Zacharias and U. Gosele, *Nat. Mater.*, 2006, **5**, 627.
299 L. Zhao, M. Steinhart, J. Yu and U. Goesele, *J. Mater. Res.*, 2006, **21**, 685.
300 S. V. Pol, V. G. Pol and A. Gedanken, *Adv. Mater.*, 2006, **18**, 2023.
301 C. N. R. Rao, G. U. Kulkarni, V. V. Agrawal, U. K. Gautam, M. Ghosh and U. Tumkurkar, *J. Colloid Interface Sci.*, 2005, **289**, 305.
302 C. N. R. Rao, G. U. Kulkarni, P. J. Thomas, V. V. Agarwal and P. Saravanan, *Curr. Sci.*, 2003, **85**, 1041.
303 Y. Lin, H. Skaff, T. Emrick, A. D. Dinsmore and T. P. Russell, *Science*, 2003, **299**, 226.
304 C. N. R. Rao, G. U. Kulkarni, P. J. Thomas, V. V. Agarwal and P. Saravanan, *J. Phys. Chem. B*, 2003, **107**, 7391.
305 V. V. Agrawal, G. U. Kulkarni and C. N. R. Rao, *J. Phys. Chem. B*, 2005, **109**, 7300.
306 V. V. Agrawal, P. Mahalakshmi, G. U. Kulkarni and C. N. R. Rao, *Langmuir*, 2006, **22**, 1846.
307 U. K. Gautam, M. Ghosh and C. N. R. Rao, *Chem. Phys. Lett.*, 2003, **381**, 1.
308 U. K. Gautam, M. Ghosh and C. N. R. Rao, *Langmuir*, 2004, **20**, 10775.
309 K. P. Kalyanikutty, U. K. Gautam and C. N. R. Rao, *Solid State Sci.*, 2006, **8**, 296.
310 K. P. Kalyanikutty, U. K. Gautam and C. N. R. Rao, *J. Nanosci. Nanotechnol.*, 2007, **7**, 1916.
311 Y. Wu, B. Yang, B. Zong, H. Sun, Z. Shen and Y. Feng, *J. Mater. Chem.*, 2004, **14**, 469.
312 H. T. Nan, J. Li, M. K. Smith, P. Nguyen, A. Cassell, J. Han and M. Meyyappan, *Science*, 2003, **300**, 1249.
313 U. K. Gautam, S. R. C. Vivekchand, A. Govindaraj and C. N. R. Rao, *Chem. Commun.*, 2005, 3995.

COMMUNICATION   www.rsc.org/materials | Journal of Materials Chemistry

Water-solubilized aminoclay–metal nanoparticle composites and their novel properties

K. K. R. Datta, M. Eswaramoorthy and C. N. R. Rao*

Received 24th November 2006, Accepted 22nd December 2006
First published as an Advance Article on the web 9th January 2007
DOI: 10.1039/b617198b

Nanoparticles of metals such as Au, Ag, Pd and Pt embedded in exfoliated sheets of aminoclays of the type $R_8Si_8Mg_6O_{16}$-$(OH)_4$, where $R = CH_2CH_2NH_2$ are entirely water soluble. These sheets of the composite come to the organic–aqueous interface on addition of alkane thiols to the aqueous layer.

Extensive research on metal nanoparticles carried out in the last few years has been partly driven by potential applications in areas such as catalysis[1,2] and sensors.[3,4] Functionalization, solubilization and assembly of metal nanoparticles have been investigated by employing a variety of methods, wherein capping agents like thiols,[5] polymers,[6–8] surfactants,[9] amine–borane complexes,[10] and dendrimers,[11] have been employed. Metal nanoparticles have also been incorporated in inorganic hosts such as zeolites,[12] and clays.[13] Such composites are however not soluble in organic or aqueous media. We have been interested in exploring the synthesis and properties of clay–metal nanoparticle composites which are readily soluble in aqueous media. For this purpose, we have employed Mg-phyllo(organo)silicates containing pendant amino groups. These aminoclays have a structure analogous to 2 : 1 trioctahedral smectites, such as talc, but are covalently linked with the approximate composition $R_8Si_8Mg_6O_{16}(OH)_4$, where $R = CH_2CH_2NH_2$. The structure is made up of octahedrally coordinated brucite layers sandwiched between tetrahedral organosilicate networks.[14]

In the present study the property of the aminoclays wherein protonation of the amino groups in water is accompanied by exfoliation has been exploited.[15] Thus metal nanoparticle composites formed by the exfoliated aminoclay sheets by carrying out the reduction of metal precursors in the presence of the clay have been investigated. Besides being entirely water soluble, the exfoliated sheets of aminoclay–Au nanoparticle composites move to the organic/aqueous interface in the presence of an alkanethiol.

The aminoclay was prepared by the method reported in the literature.[16] Typically, an aminopropyl-functionalized magnesium (organo)phyllosilicate clay was prepared at room temperature by dropwise addition of 3-aminopropyltriethoxysilane (1.3 mL, 5.85 mmol) to an ethanolic solution of magnesium chloride (0.84 g, 3.62 mmol) in ethanol (20 g). The white slurry obtained after 5 min was stirred overnight and the precipitate isolated by centrifugation, washed with ethanol (50 mL) and dried at 40 °C. $HAuCl_4$, $AgNO_3$, H_2PtCl_6 and $PdCl_2$ were used as metal precursors for Au, Ag, Pt and Pd respectively. The composites were prepared in the following manner. The aminoclay was first exfoliated by dispersing 20 mg of clay in 2 mL of millipore water by sonication for

2 minutes. To this transparent clay suspension, 2 mL of 1 mM metal precursor solution was added followed by the addition of 2 ml of 0.1 M of $NaBH_4$ solution. In the case of Ag nanoparticles, the aminoclay was prepared using $Mg(NO_3)_2$ as the Mg source to avoid precipitation of AgCl. The transparent suspension obtained after the reduction of the metal salt by $NaBH_4$ was characterized by UV-visible spectroscopy and transmission electron microscopy (TEM). For TEM analysis, the aqueous clay suspension was first precipitated by the addition of ethanol and redispersed in ethanol by sonication before drop casting on a carbon-coated copper grid.

In the absence of the aminoclay, very large metal particles were obtained which settled within a short time.

In-situ synthesis of the clay stabilized Au nanoparticles was also carried out by dissolving 1.68 g of $MgCl_2·6H_2O$ (8.26 mmol) in 20 mL of 3.8 mM solution of $HAuCl_4$ followed by the addition of 2 mL of 3-aminopropyltrimethoxysilane. The yellow slurry obtained was stirred overnight at room temperature and then kept at 75 °C for 24 h. It slowly turned pink in colour due to the formation of gold nanoparticles by thermal reduction. The pink transparent film obtained was washed with ethanol and then dried again.

The X-ray diffraction (XRD) pattern of the as-synthesized aminoclay shows a low-angle reflection with a d_{001} spacing of 1.4 nm corresponding to the bilayer arrangement of propylamino groups (Fig. 1a). The broad in-plane reflections at higher angles ($d_{020,110} = 0.41$ nm, $d_{130,200} = 0.238$ nm) and the characteristic (060) reflection at 0.156 nm confirm the formation of 2 : 1 trioctahedral Mg–phyllosilicate clay with talc-like structure. The XRD pattern

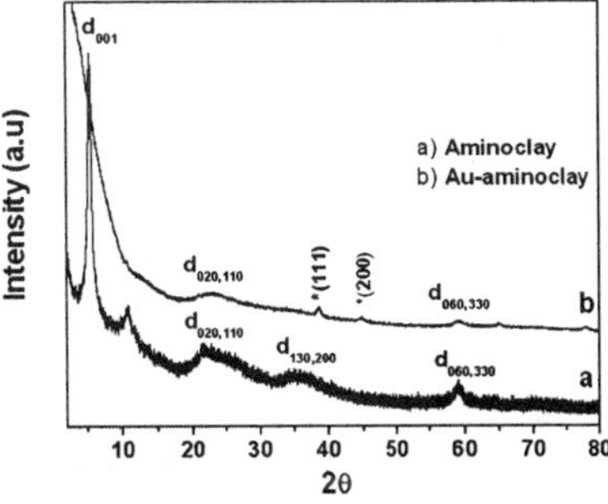

Fig. 1 X-Ray diffraction patterns of the (a) as-synthesized aminoclay and (b) the Au–aminoclay composite. Pattern (a) shows a low-angle reflection with a d_{001} spacing of 1.4 nm corresponding to the bilayer arrangement of propylamino groups. * indicates the peaks corresponding to Au nanoparticles.

Chemistry and Physics of Materials Unit and DST Unit on Nanoscience, Jawaharlal Nehru Centre for Advanced Scientific Research, Bangalore, 560064, India. E-mail: cnrrao@jncasr.ac.in; Fax: +91-080-22082760

Fig. 2 Optical images of aminoclay–metal nanoparticle composites forming clear transparent solutions in water: (a) aminoclay solution, and aminoclay with (b) Au, (c) Ag, (d) Pt and (e) Pd nanoparticles.

of the metal nanoparticle–clay composite did not show the low angle peak demonstrating that exfoliation had occurred, Fig. 1(b).

Fig. 2 shows how the aminoclay–metal nanoparticle composites form clear transparent solutions in water. The solutions are pink and yellow for Au and Ag respectively and dark brown in the cases of both Pt and Pd. The reddish-brown colour observed for Au–clay nanoparticle composites immediately after the addition of NaBH$_4$ changed to pink with time. The solutions exhibit characteristic plasmon bands for the Au– and Ag–clay suspensions at 520 nm and 410 nm respectively as shown in Fig. 3. In the cases of Pt and Pd, the characteristic absorption band for the precursors around 260 to 280 nm was absent thereby confirming the formation of Pt and Pd nanoparticles.[17,18] TEM images of the aminoclay–metal nanoparticle composites deposited on a carbon coated copper grid are shown in Fig. 4. The histograms show the average particle sizes to be around 3.5 and 5 nm respectively in the cases of Au and Ag nanoparticles. We could see the layered arrangements in the cases of Pt and Pd with the interspacing of 1.5 nm commensurate with the bilayer arrangement of aminoclays (see top right inset of Fig. 4b).

The aminoclay–Au nanoparticle composite was also prepared by an *in-situ* procedure, wherein the reduction was carried out thermally without the use of NaBH$_4$. The clay composite so prepared also dissolves in water and shows the characteristic plasmon band at 530 nm.

The effect of addition of hexadecanethiol to an aqueous solution of the aminoclay–Au nanoparticle composite was examined and the result is shown in Fig. 5a. Immediately after the addition, the alkanethiol forms a clear nonaqeous layer on top of the pink

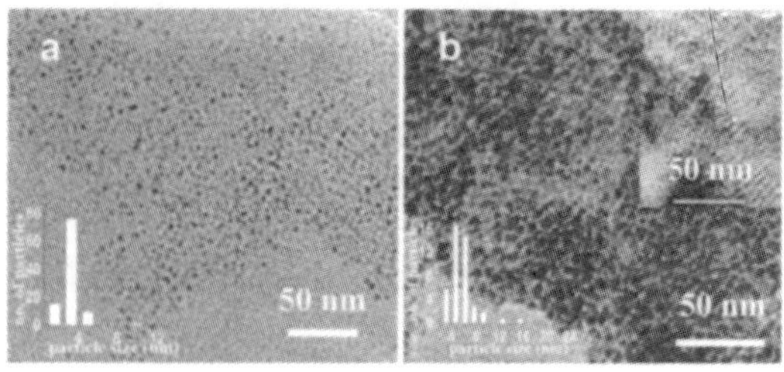

Fig. 4 TEM images of (a) Au–clay nanoparticle composites. Inset: histogram of Au nanoparticles; (b) Ag–clay nanoparticles composite. Insets (bottom left) histogram of Ag nanoparticles; (top right) Pt nanoparticles show layered arrangement.

aqueous layer. After several hours, the thiol interacts with Au nanoparticles embedded in the aminoclay and brings the sheets of composite to the aqueous/organic interface as seen in Fig. 5a(ii). This happens because the binding of metal particles to the aminoclay is very strong and the thiol is unable to dislodge it by forming metal–sulfur bonds. It is interesting that all the aminoclay appears at the interface. The presence of all the exfoliated aminoclay at the interface was confirmed by ensuring the absence of any clay in the aqueous layer. Replacing hexadecanethiol by other organic solvents such as benzene or toluene still retains the composite particles at the interface. We show the process of formation of the clay–Au nanoparticle composite schematically in Fig. 5b. This process occurs with all metal nanoparticles studied, but more effectively in the cases of Ag and Au.

In conclusion, exfoliated sheets of Mg-phyllo(organo)silicates containing pendant amino groups have been used to stabilize Au, Ag, Pd and Pt nanoparticles. The nanoparticle-decorated clay sheets can be easily dispersed in water. These metal nanoparticles

(a)

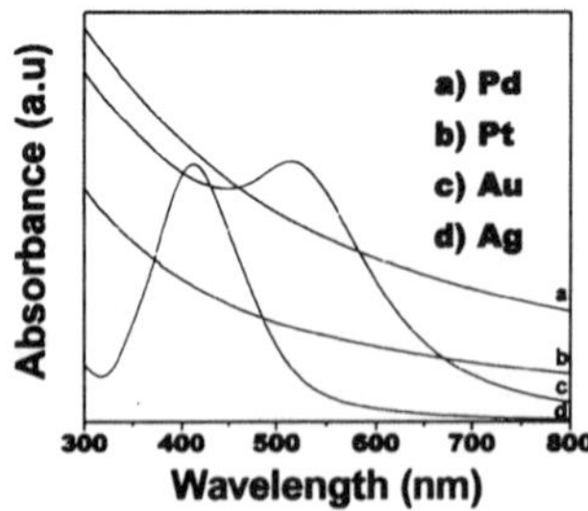

(i)   (ii)

(b)

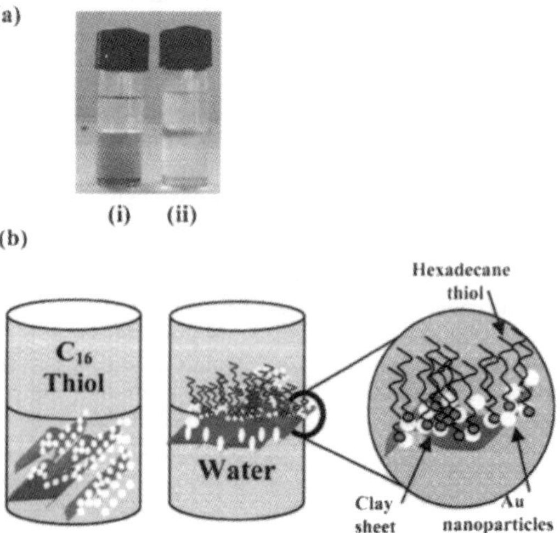

Fig. 5 (a) Optical image of aqueous Au–clay composite on addition of hexadecanethiol (i) at the time of addition (the top transparent layer is the thiol phase and the bottom layer seen as black is the aqueous phase); (ii) after 14 days the composite is at the interface (see dark ring at the interface). (b) Schematic diagram of the process.

Fig. 3 UV-Vis absorption spectra of Pd, Pt, Au, Ag nanoparticles.

can be used to lift the clay sheets to the oil/water interface with the help of an alkanethiol. In many biphasic reactions, such as the biocatalytic transformations of steroids and epoxides, transport limitations can be overcome by carrying out reactions at the aqueous/organic interface.[19] Since clays are good hosts for proteins and enzymes, stabilizing them at the oil/water interface could be useful for carrying out biphasic reactions. Although we have worked here with a family of aminoclays, other suitably functionalized clays may also exhibit similar properties.

Notes and references

1 M. C. Daniel and D. Astruc, *Chem. Rev.*, 2004, **104**, 293.
2 R. Narayanan and M. A. El-Sayed, *J. Phys. Chem. B*, 2005, **109**, 12663.
3 B. K. Jena and C. R. Raj, *Chem.–Eur. J.*, 2006, **12**, 2702.
4 Y. Xiao, V. Pavlov, S. Levine, T. Niazov, G. Markovitch and I. Willner, *Angew. Chem., Int. Ed.*, 2004, **43**, 4519.
5 M. Brust, M. Walker, D. Bethell, D. J. Schiffrin and R. Whyman, *J. Chem. Soc., Chem. Commun.*, 1994, 801.
6 M. K. Corbierre, N. S. Cameron, M. Sutton, S. G. J. Mochrie, L. B. Lurio, A. Ruhm and R. B. Lennox, *J. Am. Chem. Soc.*, 2001, **123**, 10411.
7 S. Porel, S. Singh, S. S. Harsha, D. N. Rao and T. P. Radhakrishnan, *Chem. Mater.*, 2005, **17**, 9.
8 S. W. Kim, S. Kim, J. B. Tracy, A. Jasanoff and M. G. Bawendi, *J. Am. Chem. Soc.*, 2005, **127**, 4556.
9 J. Liu, P. Raveendran, Z. Shervani, Y. Ikushima and Y. Hakuta, *Chem.–Eur. J.*, 2005, **11**, 1854.
10 N. Zheng, J. Fan and G. D. Stucky, *J. Am. Chem. Soc.*, 2006, **128**, 6550.
11 R. W. J. Scott, O. M. Wilson and R. M. Crooks, *J. Phys. Chem. B*, 2005, **109**, 692.
12 N. Zheng and G. D. Stucky, *J. Am. Chem. Soc.*, 2006, **128**, 14278.
13 S. Ayyappan, G. N. Subbanna, R. S. Gopalan and C. N. R. Rao, *Solid State Ionics*, 1996, **84**, 271.
14 E. Muthusamy, D. Walsh and S. Mann, *Adv. Mater.*, 2002, **14**, 969.
15 N. T. Whilton, S. L. Burkett and S. Mann, *J. Mater. Chem.*, 1998, **8**, 1927.
16 A. J. Patil, E. Muthusamy and S. Mann, *Angew. Chem., Int. Ed.*, 2004, **43**, 4928.
17 B. M. Choudary, S. Madhi, N. S. Chowdari, M. L. Kantam and B. Sreedhar, *J. Am. Chem. Soc.*, 2002, **124**, 14127.
18 D. G. Duff, P. P. Edwards, J. Evans, J. T. Gauntlett, D. A. Jefferson, B. F. G. Johnson, A. I. Kirkland and D. J. Smith, *Angew. Chem., Int. Ed. Engl.*, 1989, **28**, 59019.
19 P. Asuri, S. S. Karajanagi, J. S. Dordick and R. S. Kane, *J. Am. Chem. Soc.*, 2006, **128**, 1046.

2404   *J. Phys. Chem. C* **2008**, *112*, 2404–2411

Growth Kinetics of ZnO Nanorods: Capping-Dependent Mechanism and Other Interesting Features

Kanishka Biswas, Barun Das, and C. N. R. Rao*

Chemistry and Physics of Materials Unit, DST nanoscience unit and CSIR Centre of Excellence in Chemistry, Jawaharlal Nehru Centre for Advanced Scientific Research, Jakkur P. O., Bangalore-560064, India, and Solid State and Structural Chemistry Unit, Indian Institute of Science, Bangalore-560012, India

Received: September 18, 2007; In Final Form: November 6, 2007

Although the growth of nanocrystals has been investigated by several workers, investigations of the growth of 1-D nanostructures have been limited. We have investigated the growth kinetics of both uncapped and poly(vinyl pyrollidone) (PVP)-capped ZnO nanorods carefully by a combined use of transmission electron microscopy (TEM) and small-angle X-ray scattering (SAXS) which provide direct information on size and shape and compensate for the deficiency of each other. Values of average length and diameter of the ZnO nanorods obtained by TEM and SAXS are comparable. In the presence of the capping agent, the length of the nanorods grows faster while the diameter becomes narrower. The length distribution shows periodic changes in the width in the case of the uncapped nanorods, a feature absent in the case of the capped nanorods. In the absence of the capping agent, we observe the presence of small nanocrystals next to the nanorods after a lapse of time. The occurrence of small nanocrystals as well as the periodic focusing and defocusing of the width of the length distribution lend support to the diffusion-limited growth model for the growth of uncapped ZnO nanorods. Accordingly, the time dependence of the length of uncapped nanorods follows the L^3 law as required for diffusion-limited Ostwald ripening, while the PVP-capped nanorods show a time dependence which is best described by a combination of diffusion and surface reaction with a $L^3 + L^2$ type behavior. Collapse of all distribution curves obtained at different times of the reaction into a single universal Gaussian in the case of the PVP-capped nanorods also shows that the growth mechanism is more complex than Ostwald ripening.

Introduction

Nanorods and nanowires constitute an important class of nanomaterials[1,2] and chemical routes have proved to be more useful for the synthesis of these one-dimensional materials, just as in the case of nanocrystals.[3,4] The growth mechanism of nanocrystals has attracted the attention of a few workers recently. One of the popular mechanisms employed to explain the growth kinetics of nanocrystals is the diffusion-limited Ostwald ripening process following the Lifshitz–Slyozov–Wagner (LSW) theory.[5,6] An early study of the growth kinetics of ZnO nanocrystals in the absence of capping agents revealed the diffusion-limited growth mechanism to be valid.[7] A more recent study of ZnO nanocrystals in the absence of NaOH has also shown that the variation of particle size follows the LSW model for diffusion-limited coarsening.[8] Alivisatos and co-workers[9] examined the growth of CdSe and InAs nanocrystals by employing UV–visible absorption spectroscopy to determine the size of nanocrystals and by employing the band widths of the photoluminescence spectra to determine their size distribution. These workers observed a focusing and defocusing effect of the size distribution similar to that expected in Ostwald ripening. Qu et al.[10] have shown that the growth of CdSe nanocrystals involves a prolonged formation of relatively small particles (nucleation) followed by focusing and then defocusing of the size distribution because of the growth of the bigger particles and disappearance of the small ones. These workers also found the particle size

distribution to be asymmetric, indicating a diffusion-limited mechanism in the last stage of the growth. In the case of TiO_2 nanocrystals, the average particle radius cubed is reported to increase linearly with time consistent with the LSW model of coarsening.[11] On the basis of an in situ transmission electron microscope (TEM) investigation, El-Sayed and co-workers[12] report a diffusion-controlled growth of small gold nanoclusters.

There are a few reports in the literature where the growth kinetics of nanocrystals are found to deviate from the simple diffusion-limited Ostwald ripening model. Thus, Seshadri et al.[13] propose the growth of gold nanoparticles to be essentially stochastic wherein the nucleation and growth steps are well-separated. Furthermore, these workers observe that the average diameter and the standard deviation of the size distribution exhibit the same time dependence. Theoretical considerations suggest that the growth of the nanocrystals could be either controlled by the diffusion of particles, by the reaction at the surface, or by both factors.[14] Thus, Viswanatha et al.[15] observe the growth of the ZnO nanocrystals in water to follow a growth mechanism intermediate between diffusion-control and surface reaction-control. These workers also report the growth kinetics of the ZnO nanocrystals in the absence of any capping agent to be slower than that predicted by diffusion-controlled Ostwald ripening and to be dependent on the concentration of OH^- in a nonmonotonic manner as well as on the temperature.[16]

Growth kinetics of nanocrystals in the presence of capping agents is determined by several complex factors, and signatures of either the diffusion or the reaction-controlled regimes are

* To whom correspondence should be addressed. E-mail: cnrrao@jncasr.ac.in. Fax: +91 80 22082760.

10.1021/jp077506p CCC: $40.75   © 2008 American Chemical Society
Published on Web 01/29/2008

seen. The effect of capping agents on the growth of nanocrystals have been examined by a few workers.[9,10,17–23] For example, the growth of ZnO nanocrystals in the presence of poly(vinyl pyrollidone) (PVP) is reported to show deviations from Ostwald ripening.[17] To our knowledge, information of the mechanism of the growth of nanorods, especially in the presence of capping agents, is limited. A useful study in this context is that of Peng and Peng[24,25] who examined the growth kinetics of CdSe nanorods by UV–vis spectroscopy and TEM images and found the diffusion-controlled model to be valid when the monomer concentration was sufficiently high. At low monomer concentrations, the aspect ratio of the rod decreases because of intraparticle diffusion on the surface of the nanocrystal. Thoma et al.[26] have observed anisotropic growth of CdSe in the presence of a surfactant. On the basis of TEM studies, Pascholski et al.[27] proposed that small ZnO nanoparticles are converted to rods by the oriented attachment mechanism assisted by Ostwald ripening. On the basis of the estimation of nanorod lengths from XRD peak broadening, Zhu et al.[28] have found the growth kinetics of oleic acid-capped ZnO nanorods to deviate from Ostwald ripening.

In the present study, we have carried out a detailed investigation of the growth of ZnO nanorods prepared solvothermally in the presence and absence of a capping agent by employing TEM and SAXS. Employing two such independent techniques is important because of the limitations of the techniques themselves. While TEM is the most direct probe to observe the size, shape, and the size distribution of nanostructures, it is not possible to carry out in situ measurements. Furthermore, the sampling size in TEM is rather small. SAXS, on the other hand, provides a direct probe to determine the size and shape of nanomaterials, and the sampling size is much larger than what can be used in TEM. The determination of size and shape in SAXS, however, is dependent on the model employed for fitting the data. A combined use of TEM and SAXS provides a satisfactory means to study the growth of one-dimensional (1D) nanorods. We should note that UV–vis, photoluminescence, and such spectroscopic methods are indirect and are strongly affected by the change in the electronic structure of the nanomaterials. Furthermore, the spectroscopic methods are not as suitable for the study of nanorods as for nanocrystals, since the size of the former is determined both by length and by radius. Widths of X-ray diffraction peaks provide only average values of size which are not reliable in the case of nanorods.

The present study of the growth of ZnO nanorods in the presence and absence of a capping agent (PVP) has enabled us not only to determine the growth mechanism of ZnO nanorods but also to determine the effect of the capping agent. The length of the nanorods increases preferentially in the presence of PVP. We find that the growth mechanism of the ZnO nanorods in the absence of PVP follows the diffusion-limited Ostwald ripening mechanism, with evidence of focusing and defocusing of the length distribution. In the presence of PVP, the growth mechanism deviates from the diffusion-limited LSW model requiring an additional contribution from a surface process.

Experimental Section

Synthesis of ZnO Nanorods. In order to carry out the growth study in the absence of any capping agent, ZnO nanorods were prepared by the reaction of zinc acetate dihydrate (Zn(CH$_3$COO)$_2$·2H$_2$O) and sodium hydroxide in ethanol at 100 °C under solvothermal conditions. The reaction was stopped at different times (1, 2, 3, 6, 12, 18, and 24 h), and the products were analyzed by TEM and SAXS. In a typical synthesis, Zn(CH$_3$-COO)$_2$·2H$_2$O (0.05 g. 0.23 mmol) was dissolved in 12 mL ethanol, and NaOH (0.228 g, 5.69 mmol) was added under stirring. The reaction mixture was sealed in a Teflon-lined autoclave of 20 mL capacity (60% filling fraction) and maintained at 100 °C in a hot air oven. The solid products obtained at different times after the reaction were thoroughly washed with ethanol and distilled water. To study the effect of the capping agent on the growth process, we prepared the ZnO nanorods in the presence of PVP (0.25 g, MW ~ 55 000) by maintaining the other reaction parameters the same as in the synthesis of the uncapped ZnO nanorods. The samples were taken out after different reaction times (1, 2, 3, 6, 12, 18, and 24 h) for investigation.

TEM Characterization. The solid products obtained after different reaction periods were dispersed in ethanol by sonication, and the dispersions were taken on holey carbon-coated Cu grids for TEM investigations with a JEOL (JEM3010) microscope operating with an accelerating voltage of 300 kV. The length and diameter distribution were obtained from magnified micrographs by using DigitalMicrograph 3.4 software. Typically, 150–200 well-separated nanorods from three or four micrographs of the same sample were used to arrive at the size distribution.

SAXS Characterization. The average length and diameter of the ZnO nanorods could be readily obtained by SAXS.[29–31] We performed SAXS experiments with a Bruker-AXS NanoSTAR instrument modified and optimized for solution scattering. The instrument is equipped with a X-ray tube (Cu Kα radiation, operated at 45 kV/35 mA), cross-coupled Göbel mirrors, three-pinhole collimation, evacuated beam path, and a two-dimensional (2D) gas-detector (HI-STAR).[31] The modulus of the scattering vector is $q = 4\pi \sin\theta/\lambda$, where 2θ is the scattering angle and λ is the X-ray wavelength. We recorded the SAXS data in the $q = 0.007$ to 2.2 Å^{-1} range, that is, $2\theta = 0.1$ to 3°. Solutions of the ZnO nanorods in ethanol (approximately 0.1 w/v % concentration) obtained after lapse of different reaction times were used for SAXS measurements. The ethanol solutions of nanorods were taken in quartz capillaries (diameter of about 2 mm) for the measurements. A capillary filled with only ethanol was used for background correction. The concentration of the nanorods was sufficiently low to neglect interparticle interference effects. The experimental SAXS data were fitted by Bruker-AXS DIFFRAC^{plus} NANOFIT software by using a solid cylinder model. The form factor of the cylinder used in this software is given by Fournet.[32,33]

Results and Discussion

In order to prepare the ZnO nanorods, we employed a high monomer concentration with a Zn^{2+}/OH$^-$ ratio of 1:25. High monomer concentrations favor the growth of the ZnO nanorods[27,34] because the chemical potential of an elongated structure is generally higher than a dot-shaped nanocrystal because of the lower surface energy of anisotropic structure.[25] As a result, the growth of such an anisotropic structure requires a relatively high chemical potential environment, that is, high monomer concentration in the solution. We could obtain sufficient concentrations of the nanorods with the reactant concentrations and the reaction conditions employed by us. In what follows, we discuss the results of TEM and SAXS studies of the nanorods obtain after different reaction times.

Figure 1a,b shows typical TEM images of the uncapped product after 3 and 12 h of reaction. The inset in Figure 1a indicates the presence of a small nanocrystal on the side wall of an uncapped nanorod in the process of dissolution. The inset

2406 *J. Phys. Chem. C, Vol. 112, No. 7, 2008*

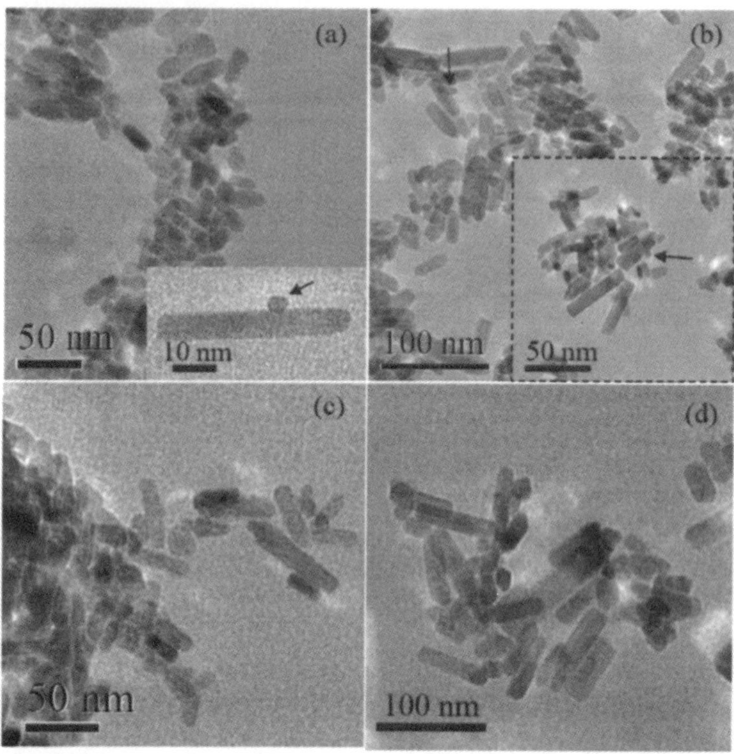

Figure 1. (a,b) TEM images of uncapped ZnO nanorods obtained after 3 and 12 h of reaction. The inset in (a) shows how a small ZnO nanoparticle attached to the side wall of an uncapped ZnO nanorod. The inset in (b) is a TEM image of uncapped ZnO nanorods obtained after 18 h of reaction. The arrow indicates the existence of small nanocrystal along with the nanorods. Several such nanocrystals can be seen in the image. (c,d) TEM images of PVP-capped ZnO nanorods obtained after 3 and 12 h of reaction. Note the absence of nanocrystals sticking to the nanorods.

in Figure 1b shows a TEM image of the uncapped ZnO nanorods after 18 h of reaction. The presence of small nanocrystals with the nanorods is indicated by black arrows in Figure 1b. The existence of small nanocrystals along with the uncapped nanorods after a sufficiently long period of reaction (e.g., 12 or 18 h) suggests that the diffusion-limited Ostwald ripening may be operative, wherein large particles grow at the cost of smaller ones. At the early stage of the growth, the monomer concentration is sufficiently high for the nanorods to grow, and the growth process is fast. After longer times, the monomer concentration gets depleted below a certain critical level, rendering the smaller nanocrystals to dissolve. El-Sayed and co-workers[12] have reported the existence of smaller gold nanocrystals located near gold nanorods and observed shrinkage of the nanocrystals with increasing temperature.

From the TEM studies, we have estimated the distributions of the length as well as the diameter of the ZnO nanorods after different times of the reaction. Figure 2 shows the time evolution of the length distribution of the uncapped ZnO nanorods (in blue). We have fitted each distribution curve by a Gaussian distribution function (solid blue curve). We see from Figure 2, length distribution becomes sharper and broader in a periodic fashion. Such focusing and defocusing of the diameter distribution is a phenomenon noticed earlier in the case of CdSe nanoparticles growth.[9] Since the monomer concentration is high at the beginning of the reaction, focusing of the length distribution (e.g., $t = 3$ h) occurs because of a faster growth rate of the small nanorods. When the monomer concentration gets depleted because of the faster growth of the nanorods, the smaller nanorods start to shrink while the longer ones keep

growing. The size distribution, therefore, becomes broader (e.g., $t = 6$ h). Dissolution of the small nanorods again enriches the monomer concentration in the solution, with the growth of the longer rods continuing through the diffusion of the monomer from solution to the nanorod surface, giving rise to focusing of the length distribution (e.g., $t = 9$, 12 h). Periodic focusing and defocusing of the length distribution is thus dependent on the periodic variation of the monomer concentration in the solution. Focusing and defocusing of the length distribution gives a clean qualitative indication of diffusion-limited growth of uncapped ZnO nanorods.

In Figure 1c,d, we show typical TEM images of PVP-capped ZnO nanorods after 3 and 12 h of reaction. We do not see the presence of small nanocrystals along with the nanorods as found earlier in the case of the uncapped nanorods. We have estimated the length distribution of the PVP-capped ZnO nanorods from the TEM images. In Figure 2, we show the histograms (in black) with Gaussian distribution fittings (solid black curve). In the presence of PVP, we do not observe periodic changes in the width of the distribution curves. Instead, we find the width of the length distribution to become broader continuously with reaction time. This reflects the growth of ZnO nanorods in the presence of PVP to be more complex than simple diffusion-limited growth. Seshadri et al.[13] had observed a similar increase in the width of the distribution with time in the case of gold nanocrystals.

We now examine the results of SAXS measurements at the same reaction time intervals as in the case of TEM studies. Figure 3a,b shows typical plots of intensity versus scattering vector in the logarithmic scale for different times of the reaction

Growth Kinetics of ZnO Nanorods

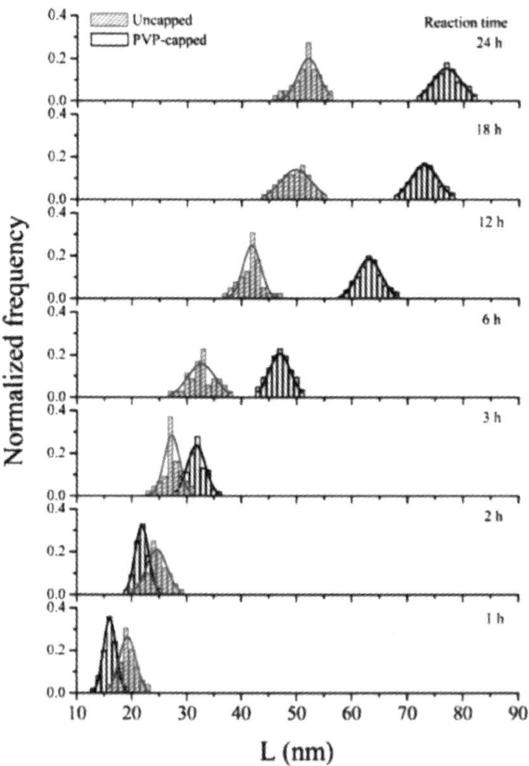

Figure 2. Time evolution of the length distributions of the uncapped (blue) and the PVP-capped (black) ZnO nanorods obtained from TEM images.

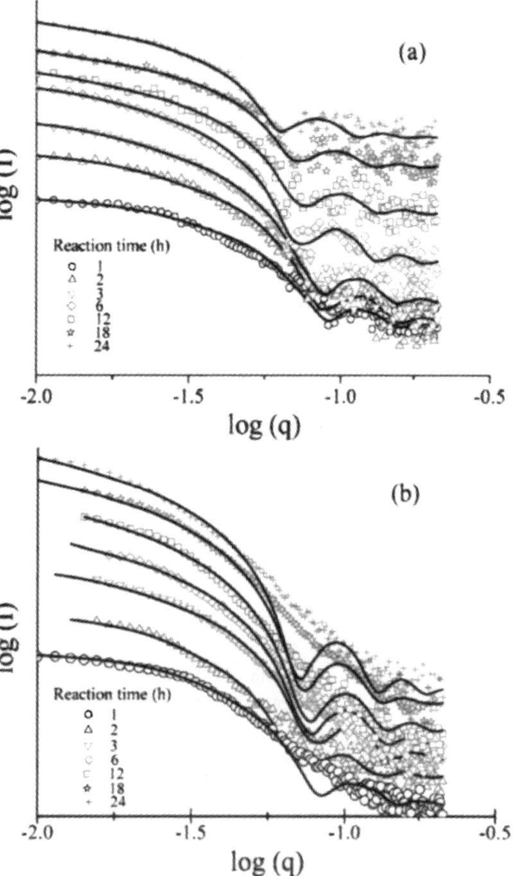

Figure 3. SAXS data of (a) uncapped and (b) PVP-capped ZnO nanorods obtained after different times of reaction. Solid lines are the model fits to the experimental data.

in the case of uncapped and PVP—capped ZnO respectively. Under the reaction conditions employed, the length of the nanorods varies significantly with time, while the diameter varies only slightly. On the basis of this observation, we assume the cylindrical shape for the nanorods to analyze the SAXS data. A cylinder is a close approximation of a hexagonal rod. To date, the hexagonal rod model has not been employed in the literature to analyze SAXS data. In order to estimate the average length and diameter of ZnO nanorods, we have fitted the experimental SAXS data to the cylinder model of Bruker-AXS DIFFRAC[plus] NANOFIT software. The form factor of the cylinder is that due to Fournet:[32,33]

$$P(q,R,L) = \int_0^{\pi/2} \left[\frac{2J_1(qR \sin \alpha)}{qR \sin \alpha} \frac{\sin(qL \cos \alpha/2)}{qL \cos \alpha/2} \right]^2 \sin \alpha \, d\alpha$$

Here, q is the scattering vector, R is the radius of the cylinder, L is the length of the cylinder, α is the orientation dependent parameter (angle between scattering vector and cylinder long axis), and $J_1(x)$ is the first-order Bessel function of the first kind. We have not introduced the length and diameter distribution in the scattering cross-section equation because the form factor amplitude of a cylinder is a more complicated function of length, diameter, and the orientation of the cylinder than the form factor amplitude of a spherical model where the diameter is the only variable. Such a cylindrical model fitting of simulated SAXS data have been reported in the literature.[33,35] The solid lines in Figure 3 are the cylinder model fits of the experimental SAXS data. In the case of the uncapped ZnO nanorods, the theoretical fits are good, but the fits are not as good in the case of PVP-

capped ZnO nanorods. We have constructed our model only considering the cylindrical ZnO nanorods present in the solution. The PVP molecules present on the side walls of the nanorods, may be responsible for not obtaining very good fits of the experimental data.

Figure 4a,b shows the time evolution of the average length calculated from TEM images (filled circles) and SAXS (open circle) of uncapped and PVP-capped ZnO nanorods, respectively. In the absence of PVP, the length of the nanorods increases from 19 to 52 nm after 24 h of reaction. In the presence of PVP, the growth is greater, and the length increases up to 77 nm after 24 h of reaction. In Figure 5a,b, we show the time evolution of the average diameter and aspect ratio calculated from TEM (filled circles) and SAXS (open circles) of uncapped ZnO nanorods, respectively. The diameter of the uncapped nanorods increases slightly starting from 8.3 to 12.3 nm. The aspect ratio increases from 2.3 to 4.3 after 24 h of reaction. Figure 6a,b shows the time evolution of the average diameter and aspect ratio obtained from TEM images (filled circles) and SAXS (open circles) of PVP-capped ZnO nanorods, respectively. In the presence of PVP, we observe an even smaller variation in the diameter (9.1 to 10.4 nm) compared with the capped ones. The aspect ratio, therefore, increases from 1.8 to 7.4 after 24 h of reaction. It is noteworthy that the estimates of lengths and diameters from TEM and SAXS agree closely.

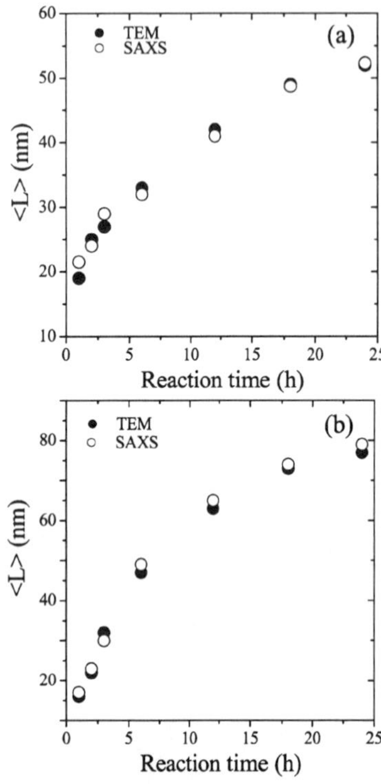

Figure 4. Time evolution of the average length ($\langle L \rangle$) of (a) uncapped and (b) PVP-capped ZnO nanorods obtained from TEM and SAXS.

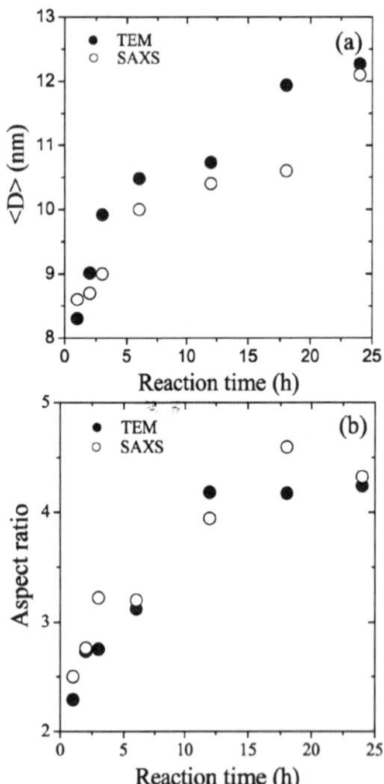

Figure 5. Time evolution of the (a) average diameter ($\langle D \rangle$) and (b) aspect ratio of the uncapped ZnO nanorods obtained from TEM and SAXS.

ZnO with a polar hexagonal wurzite structure can be described as being due to the hexagonal close packing of the

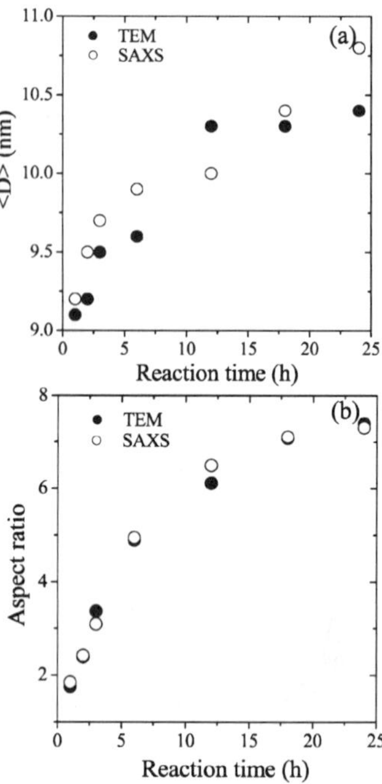

Figure 6. Time evolution of the (a) average diameter ($\langle D \rangle$) and (b) aspect ratio of the PVP-capped ZnO nanorods obtained from TEM and SAXS.

oxygen and zinc atoms, in point group $3m$ and space group $P6_3mc$, the zinc atoms being present in the tetrahedral sites. Thus, the crystal habit of wurzite ZnO exhibits well-defined crystallographic faces; that is, polar-terminated (0001) planes and six side facet are generally bound by the ($10\bar{1}0$) family of planes. The growth rates of different family of planes follow the sequence (0001) > ($10\bar{1}1$) > ($10\bar{1}0$).[36] ZnO anisotropic structures are normally bound by six ($10\bar{1}0$) facets grown along the (0001) direction, that is, along the c axis of the rods. Powder X-ray diffraction patterns of PVP-capped and uncapped ZnO nanorods could be indexed on the hexagonal space group $P6_3mc$ of ZnO. High-resolution electron microscope (HREM) images along with electron diffraction patterns also support the hexagonal crystal structure as well as the preferential c-directional growth of nanorods. Electron diffraction patterns confirm the single crystallinity of nanorods. PVP appears to selectively cap the side facet of the ZnO nanorods, allowing the growth to occur selectively along the c direction compared with the uncapped ZnO nanorods. As a result, we find the length to grow to a greater extent in the presence of PVP compared with the uncapped situation. When PVP is not present in the solution, there is a possibility of diffusion of the monomer to the side walls of the nanorods, rendering the nanorods thicker compared with the capped ones. The inset in Figure 1a gives a typical TEM image showing the presence of a small nanocrystal on the side wall of an uncapped nanorod in the process of dissolution which supports the proposed mechanism. We do not observe the presence of such small nanocrystal on the side of PVP-capped nanorods in the TEM images.

Figure 7a,b shows the time evolution of the standard deviation, σ, of the length calculated from TEM images for uncapped and PVP-capped ZnO nanorods, respectively. In the lower insets of Figure 7, we show the time evolution of the σ of the diameter obtained from TEM images for uncapped and

Growth Kinetics of ZnO Nanorods

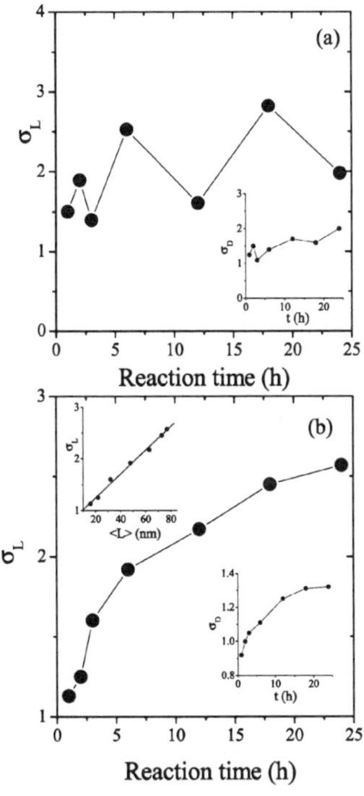

Figure 7. Time evolution of the standard deviation of the length (σ_L) of (a) uncapped and (b) PVP-capped ZnO nanorods. Lower insets in (a) and (b) show the time evolution of the standard deviation of the diameter (σ_D). Upper inset in (b) shows the plot of mean rod length, $\langle L \rangle$, against the σ_L.

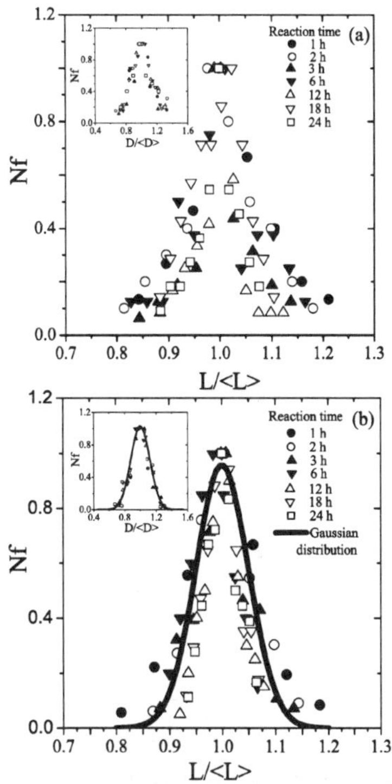

Figure 8. Universal length distribution curves of (a) uncapped and (b) PVP-capped ZnO nanorods obtained by scaling mean lengths at different times. Insets in (a) and (b) show the universal diameter distribution curves of (a) uncapped and (b) PVP-capped ZnO nanorods. (Nf = normalized frequency).

PVP-capped ZnO nanorods, respectively. We observe that σ of the length (σ_L) and the diameter (σ_D) vary periodically with time in the absence of PVP as in Figure 7a. This indicates periodic fluctuation of the monomer concentration in the solution during the growth process of uncapped ZnO nanorods. In the presence of PVP, both σ_L and σ_D increase continuously with time. By plotting the average length as a function of σ_L for different times in the case of PVP-capped ZnO nanorods, we find that the mean and the σ_L bear a linear relationship, with a correlation coefficient (R) of 0.996. This means that both σ_L and average length have the same time dependence (see upper inset in Figure 7b). We have observed the same linear relationship ($R = 0.988$) between σ_D and the average diameter of the PVP-capped ZnO nanorods for different times. This observation suggests that it should be possible to represent the length or the diameter distribution by means of a variable scaled by the mean length or diameter. We do not observe such a relationship between σ and average length or average diameter in the case of uncapped ZnO nanorods.

Figure 8a,b shows plots of normalized frequency (normalized by the maximum counts) against $L/\langle L \rangle$ in the case of uncapped and PVP-capped nanorods, respectively. In the insets in Figure 8a,b, we show the plots of normalized frequency (normalized by the maximum counts) against $D/\langle D \rangle$ in the case of uncapped and PVP-capped nanorods, respectively. In the case of the uncapped nanorods (Figure 8a), the distribution is somewhat asymmetric, and we do not obtain as good a universal curve. The asymmetric nature of universal length as well as diameter distribution curves supports a diffusion-limited Ostwald ripening growth mechanism in the case of the uncapped ZnO nanorods. In the case of PVP-capped nanorods, however, all of the data collapse to a single curve that can be represented by the Gaussian distribution function (solid line) as shown in Figure 8b. Ostwald ripening does not appear to be the sole factor determining growth in the case of the PVP-capped nanorods.

We have attempted to obtain information regarding the growth mechanisms of ZnO nanorods in the presence and the absence of the capping agent based on the time evolution of the average length of nanorods. If the diffusion-limited Ostwald ripening according to LSW theory[5,6] were to be the sole contributor of the growth mechanism, then the rate law would be given by

$$L^3 - L^3_0 = Kt$$

where, L is the average length at time t and L_0 is average initial length of the nanorods. The rate constant K is given by $K = 8\gamma D V_m^2 C_\infty/9RT$, where D is the diffusion constant at temperature T, V_m is the molar volume, γ is the surface energy, and C_α is the equilibrium concentration at flat surface. We have tried to fit the $L(t)$ data obtained from TEM and SAXS to the Ostwald ripening model in the case of the uncapped (solid curve in Figure 9a) and the PVP-capped ZnO nanorods (broken curve in Figure 9b). The fit is reasonably good for the uncapped nanorods (reduced χ^2 of the fit = 0.71 and coefficient of determination, $R^2 = 0.996$) as can be seen from Figure 9a. We have not found it possible to fit the experimental $L(t)$ data of the PVP-capped ZnO nanorods to the diffusion-limited model ($\chi^2 = 58.59$ and $R^2 = 0.903$). The growth process appears to deviate from diffusion-limited Ostwald ripening. We have also tried to fit the $L(t)$ data of the PVP-capped nanorods to the surface-limited reaction model (i.e., $L^2 \propto t$) or by varying the value of the

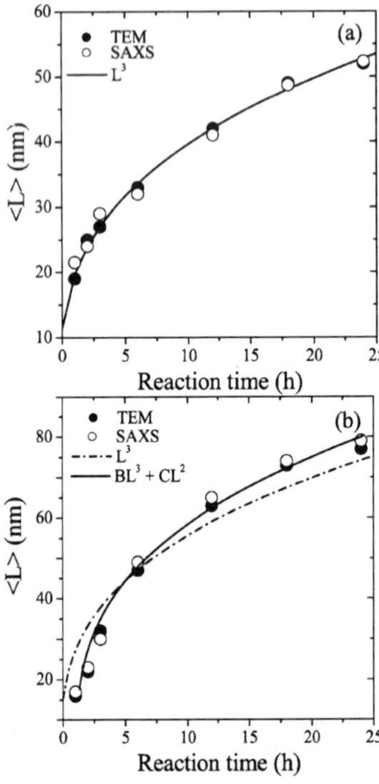

Figure 9. (a) Fits of $L(t)$ data (solid circles from TEM and open circles from SAXS) of uncapped ZnO nanorods to the Ostwald ripening (L^3) model (solid curve). (b) Fits of $L(t)$ data (solid circles from TEM and open circles from SAXS) of PVP-capped ZnO nanorods to the Ostwald ripening (L^3) model (broken curve) and mixed diffusion-surface control growth ($L^3 + L^2$) model (solid curve).

exponent (n in $L^n \propto t$,). We have found the fit of the data either to the surface reaction model ($\chi^2 = 16.97$ and $R^2 = 0.972$) or to a variable n model to be unsatisfactory. In order to fit the experimental $L(t)$ data of the PVP-capped ZnO nanorods, we have, therefore, used a model which contains both the diffusion-limited and surface-limited growth,[15]

$$BL^3 + CL^2 + \text{const} = t$$

where, $B = AT/\exp(-E_a/k_BT)$, $A \propto 1/(D_0\gamma V_m^2 C_\infty)$, $C \propto T/(k_d\gamma V_m^2 C_\infty)$, and k_d is the rate constant of surface reaction. The remarkable goodness of fits ($\chi^2 = 0.95$ and $R^2 = 0.989$) over the entire range of experimental data by the mixed diffusion-reaction control model is shown by a thick solid curve in Figure 9b. Thus, the growth of the PVP-capped ZnO nanorods deviates sufficiently from the diffusion-limited Ostwald ripening model and follows a mechanism involving both diffusion-control and surface reaction control.

The growth of the nanorods mainly occurs by the diffusion of monomers from solution to the nanorod surface or by the reaction at the surface where units of the diffusing particles get assimilated into the growing nanorods. Diffusion and surface reaction are two limiting cases in the growth of nanorods. In the absence of a capping agent, the nanorod growth is essentially controlled by diffusion as suggested by TEM, as well as the goodness of fits of $L(t)$ data to diffusion-limited growth model. Presence of PVP gives rise to a barrier to diffusion. As a result, the contribution of the surface reaction becomes more prominent. The growth of PVP-capped nanorods occurs through a combination of diffusion and surface reaction processes. There is no barrier present to limit the diffusion process in the case of uncapped nanorods.

Conclusions

Several major conclusions can be drawn of the growth mechanism of ZnO nanorods based on the present study. The values of the average length and diameter of the nanorods obtained by TEM and SAXS techniques are close to each other in the cases of both uncapped and PVP-capped ZnO nanorods. The average length of the PVP-capped ZnO nanorods grows at a higher rate compared with the uncapped ones, while the average diameter of the capped nanorods grows at a slower rate. PVP appears to selectively cap the side walls of the ZnO nanorods giving rise to such a preferential increase in the aspect ratio. The observation of small nanocrystals along with the nanorods in TEM images lends support to a growth mechanism based on the diffusion-limited Ostwald ripening process in the case of the uncapped nanorods. The uncapped nanorods exhibit periodic changes in the width of the length distribution curves because of changes in the monomer concentration. The periodic focusing and defocusing as well as the somewhat asymmetric nature of the distribution curves (Figure 8) are also the consequence of the diffusion-limited growth of uncapped ZnO nanorods. Good fits of the $L(t)$ data of uncapped ZnO nanorods with L^3 model confirms the growth mechanism to be mainly diffusion-controlled. In the presence of PVP, however, the nonexistence of small nanocrystals with the nanorods in the TEM images and the continuous broadening of the width of the size distributions with time reveal the growth mechanism to be more complex than a diffusion-limited process. Fits of the $L(t)$ data to $L^3 + L^2$ model suggest that the growth mechanism of the PVP-capped ZnO nanorods involves both diffusion and surface reactions. A similar time dependence of standard deviation σ_L (or σ_D) and the average length (or diameter) as well as the collapse of all distributions onto a single Gaussian also suggest that the basic process contributing to the growth of PVP-capped nanorods to be distinct from Ostwald ripening alone.

References and Notes

(1) Rao, C. N. R.; Govindaraj, A. *Nanotubes and Nanowires*, RSC series on Nanoscience, Royal Society of Chemistry, London, 2006.

(2) Rao, C. N. R., Muller, A., Cheetham, A. K., Eds. *The Chemistry of Nanomaterials*; Wiley-VCH: Weinheim, 2004; Vols. 1,2.

(3) Rao, C. N. R.; Thomas P. J.; Kulkarni, G. U. *Nanocrystals: Synthesis, Properties and Applications*; Springer: New York, 2007.

(4) Burda, C.; Chen, X.; Narayanan, R.; El-Sayed, M. A. *Chem. Rev.* **2005**, *105*, 1025.

(5) Lifshitz, I. M.; Slyozov, V. V. *J. Phys. Chem. Solids* **1961**, *19*, 35.

(6) Wagner, C. *Z. Elektrochem.* **1961**, *65*, 581.

(7) Wong, E. M.; Bonevich, J. E.; Searson, P. C. *J. Phys. Chem. B* **1998**, *102*, 7770.

(8) Hu, Z.; Ramírez, D. J. E.; Cervera, B. E. H.; Oskam, G.; and Searson, P. C. *J. Phys. Chem. B* **2005**, *109*, 11209.

(9) Peng, X.; Wickham, J.; Alivisatos, A. P. *J. Am. Chem. Soc.* **1998**, *120*, 5343.

(10) Qu, L.; Yu, W. W.; Peng, X. *Nano Lett.* **2004**, *4*, 465.

(11) Oskam, G.; Nellore, A.; Penn, R. L.; Searson, P. C. *J. Phys. Chem. B* **2003**, *107*, 1734.

(12) Mohamed, M. B.; Wang, Z. L.; El-Sayed, M. A. *J. Phys. Chem. A* **1999**, *103*, 10255.

(13) Seshadri, R.; Subbanna, G. N.; Vijayakrishnan, V.; Kulkarni, G. U.; Ananthakrishna, G.; Rao, C. N. R. *J. Phys. Chem.* **1995**, *99*, 5639.

(14) Talapin, D. V.; Rogach, A. L.; Haase, M.; Weller, H. *J. Phys. Chem.* **2001**, *105*, 12278.

(15) Viswanatha, R.; Santra, P. K.; Dasgupta, C.; Sarma, D. D. *Phys. Rev. Lett.* **2007**, *98*, 255501.

(16) Viswanatha, R.; Amenitsch, H.; Sarma, D. D. *J. Am. Chem. Soc.* **2007**, *129*, 4470.

(17) Viswanatha, R.; Sarma, D. D. *Chem. Eur. J.* **2006**, *12*, 180.

(18) Pesika, N. S.; Hu, Z.; Stebe, K. J.; Searson, P. C. *J. Phys. Chem.* **2002**, *106*, 6985.

(19) Bullen, C. R.; Mulvaney, P. *Nano Lett.* **2004**, *4*, 2303.

(20) Yu, W. W.; Peng, X. *Angew. Chem., Int. Ed.* **2002**, *41*, 2368.

(21) Qu, L.; Peng, X. *J. Am. Chem. Soc.* **2002**, *124*, 2049.

(22) Dai, Q.; Li, D.; Chen, H.; Kan, S.; Li, H.; Gao, S.; Hou, Y.; Liu, B.; Zou, G. *J. Phys. Chem. B* **2006**, *110*, 16508.

(23) Wong, E. M.; Hoertz, P. G.; Liang, C. J.; Shi, B.; Meyer, G. J.; Searson, P. C. *Langmuir* **2001**, *17*, 8362.

(24) Peng, Z. A.; Peng, X. *J. Am. Chem. Soc.* **2001**, *123*, 1389.

(25) Peng, Z. A.; Peng, X. *J. Am. Chem. Soc.* **2002**, *124*, 3343.

(26) Thoma, S. G.; Sanchez, A.; Provencio, P. P.; Abrams. B. L.; Wilcoxon, J. P. *J. Am. Chem. Soc.* **2005**, *127*, 7611.

(27) Pacholski, C.; Kornowski, A.; Weller, H. *Angew. Chem., Int. Ed.* **2002**, *41*, 1188.

(28) Zhu, Z.; Andelman, T.; Yin, M.; Chen, T.-L.; Ehrlich, S. N.; O'Brien, S. P.; Osgood, R. M. *J. Mater. Res.* **2005**, *20*, 1033.

(29) Pedersen, J. S. *Adv. Colloid Interface Sci.* **1997**, *70*, 171.

(30) Pedersen, J. S. In *Neutrons, X-rays and Light: Scattering Methods Applied to Soft Condensed Matter*; Lindner, P., Zemb, Th., Eds.; North-Holland: New York, 2002, 391.

(31) Pedersen, J. S. *J. Appl. Crystallogr.* **2004**, *37*, 369.

(32) Fournet, G. *Bull. Soc. Franc. Minéral. Crist.* **1951**, *74*, 39.

(33) Pedersen, J. S. *J. Appl. Crystallogr.* **2000**, *33*, 637.

(34) Cao, H. L.; Qian, X. F.; Gong, Q.; Du, W. M.; Ma, X. D.; Zhu, Z. K. *Nanotechnology* **2006**, *17*, 3632.

(35) Kaya, H. *J. Appl. Crystallogr.* **2004**, *37*, 223.

(36) Kar, S.; Dev, A.; Chaudhuri, S. *J. Phys. Chem. B* **2006**, *110*, 17848.

APPLIED PHYSICS LETTERS   VOLUME 84, NUMBER 26   28 JUNE 2004

Dip-pen nanolithography with magnetic Fe$_2$O$_3$ nanocrystals

Gautam Gundiah, Neena Susan John, P. John Thomas, G. U. Kulkarni, and C. N. R. Rao[a]
*Chemistry and Physics of Materials Unit, Jawaharlal Nehru Centre for Advanced Scientific Research,
Jakkur, Bangalore-560064, India*

S. Heun
Sincrotrone Trieste S. c. p. a., I-34012 Trieste, Italy

(Received 18 March 2004; accepted 4 May 2004; published online 17 June 2004)

Dip-pen nanolithography has been employed to obtain magnetic nanopatterns of γ-Fe$_2$O$_3$ nanocrystals on mica and silicon substrates. The chemical and magnetic nature of the patterns have been characterized employing low-energy electron microscopy, x-ray photoemission electron microscopy, and magnetic force microscopy measurements. © *2004 American Institute of Physics.*
[DOI: 10.1063/1.1766399]

Dip-pen nanolithography (DPN), an atomic force microscopy (AFM)-based lithographic technique, developed by Mirkin and co-workers,[1] is emerging as a preferred method to create patterns of nanoscopic dimensions. The method exploits the water meniscus formed between a slowly scanning AFM tip and a substrate to transfer the species on the tip to the substrate by diffusion. Initial studies employed gold–thiol interaction to create robust patterns, with the dimensions being determined by the preset scan area.[2] The use of bifunctional thiols as inks permits tethering of colloids, proteins, and other macromolecules at specific regions on a given surface.[3] Ali *et al.*[4] have reported a DPN-based procedure to deposit small volumes of an Au sol on a substrate, which on evaporation of the solvent leads to circular nanocrystal patterns. Garno *et al.*[5] have suggested methods to obtain linear patterns of Au nanocrystals on Au substrates. DPN patterning of Au and Pd nanocrystals has been accomplished on different substrates starting from hydrosols.[6] Magnetic barium hexaferrite nanostructures have been patterned by coating the AFM cantilever tip with a precursor followed by its deposition on silicon substrates and heating.[7] Based on DPN experiments on organic dyes, Su and Dravid[8] suggest that weak interactions between the substrate and the molecular ink suffice to form DPN patterns. In spite of the progress made with DPN, there is a clear need to develop better inks for different surfaces and fully characterize the nanopatterns by appropriate spectroscopic and microscopic tools.[9]

We have sought to pattern nanoparticles of γ-Fe$_2$O$_3$ by the DPN method on different substrates. It may be interesting to recall that hematite has been traditionally employed as a dye, the prehistoric cave paintings of Lascaux, being a well-known example. Since γ-Fe$_2$O$_3$ is magnetic, the nanopatterns also lend themselves to a magnetic force microscopy (MFM) study. In this letter, we report DPN with γ-Fe$_2$O$_3$ nanoparticles wherein MFM as well as low-energy electron microscopy (LEEM) and x-ray photoemission electron microscopy (XPEEM) (Ref. 10) have been employed to independently characterize the patterns.

Citrate-capped γ-Fe$_2$O$_3$ nanocrystals were prepared by wet-chemical means, by modifying the procedure of Ngo and Pileni.[11] The preparation yields a sol, which can be precipi-

tated and redispersed readily in water. Redispersibility is crucial for a colloidal sol to be used as an ink in DPN experiments. The transmission electron microscopy (TEM) image [see Fig. 1(a)], revealed the nanocrystals to be of an average diameter of 11 nm and a log-normal diameter distribution [see inset in Fig. 1(a)]. The nanocrystals in the powder form were superparamagnetic at room temperature, exhibiting hysteresis at low temperatures [see Fig. 1(b)]. The observed squareness ratio of 0.42 is close to that expected (0.5) for randomly oriented single domain magnetic particles with uniaxial anisotropy.[12]

DPN experiments were carried out under ambient conditions (humidity ~35%–45%) by employing a Digital Instruments Multimode head attached to a Nanoscope-IV controller. Contact mode imaging was carried out in both normal and lateral force modes. Standard Si$_3$Ni$_4$ cantilevers were coated with γ-Fe$_2$O$_3$ nanocrystals by immersing them in a dispersion for 5–10 min followed by drying. Freshly cleaved mica and silicon were used as substrates with and without the native SiO$_x$ layer (the latter by treating with a solution of aqueous HF). Deposition of the nanocrystals was achieved by scanning an area in the contact mode with scan speeds of ~1 μm s^{-1} for a period of 30 min. Subsequently, imaging was carried out using the same cantilever and scanning a larger area with higher scan rates (~10 μm s^{-1}).

The AFM images in Fig. 2 clearly show that patterns of γ-Fe$_2$O$_3$ nanocrystals can be fabricated satisfactorily on different substrates. In each case, high aspect ratio lines, corresponding to the desired scan area are obtained with sharp edges. For example, scanning an area of 9000 nm by 230 nm during deposition produced a pattern of dimensions 9010 nm by 226 nm [Fig. 2(a), central line]. Typical linewidths are in the range of 140–200 nm with lengths extending to 10 μm. It is gratifying that there is no sign of lateral diffusion of the ink in any of the cases. These patterns are therefore distinct from water droplets that are deposited from undipped tips.[13] The water droplets evaporate readily and exhibit noncontinuous structures with diffuse boundaries. The surface plot shown in Fig. 2(b) confirms the observations made in Fig. 2(a). We also see corrugations along the edges corresponding to the particle size. The individual particles are not visible in the patterns, possibly due to the blunt nature of the cantilever coated with the nanocrystals and also due to the mild contact forces employed. The observed height of around 10 nm re-

[a]Author to whom correspondence should be addressed; electronic mail: cnrrao@jncasr.ac.in

5342   Appl. Phys. Lett., Vol. 84, No. 26, 28 June 2004   Gundiah *et al.*

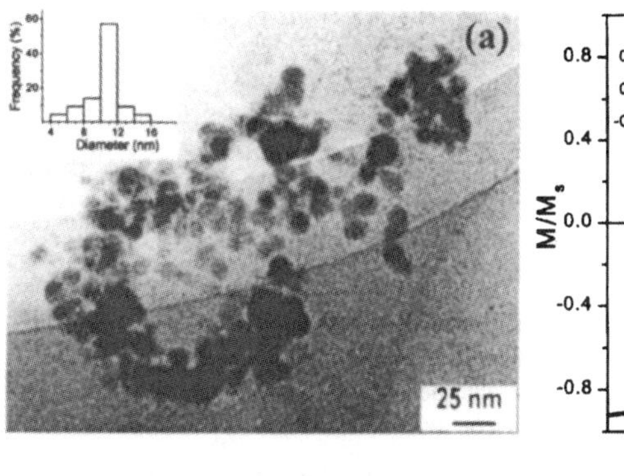

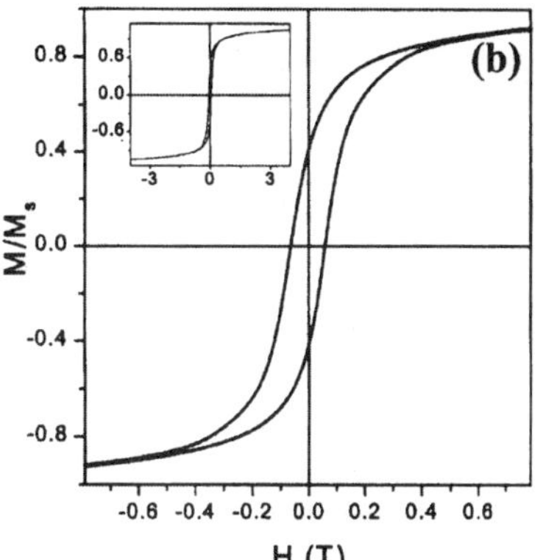

FIG. 1. Citrate-capped γ-Fe$_2$O$_3$ nanocrystals: (a) TEM image along with a histogram showing the distribution in particle size (in the inset) and (b) hysteresis loop at 2 K ($H=4$ T). The inset shows the hysteresis loop in the full range of the magnetic field.

veals that the patterns are made of a single layer of γ-Fe$_2$O$_3$ nanocrystals. It is noteworthy that similar lines are obtained on Si with and without the native oxide layer, although the latter is not ideal for good imaging [see Figs. 2(c) and 2(d)].

The magnetic behavior of the nanopatterns was probed by magnetic cantilever tips, the magnetic nature of the γ-Fe$_2$O$_3$ nanocrystals being exploited to obtain MFM contrast. In this method, a magnetic tip is scanned in the lift mode,[14] at various z values, wherein the magnetic interaction emerges as a difference in the phase shift. Typical tapping mode topography and MFM images are shown in Figs. 3(a) and 3(b). The tapping mode reveals rougher edges arising from the nanocrystals. The change in the phase shift as function of the lift height is shown in Fig. 3(c), to show that the superparamagnetic nanocrystals in the pattern are susceptible to magnetization by the stray field emanating from the magnetic tip. The loss of contrast upon changing the lift height to 10 nm is consistent with the rather weak magnetic nature of the nanocrystals [see Fig. 1(b)].

In order to characterize the DPN patterns by an independent method, we have employed the Nanospectroscopy

Beamline at the synchrotron radiation facility Elettra in Trieste.[15] Electron-beam-based techniques are ideal for such studies[16] in that they have a lateral resolution of 10 nm and atomic depth resolution. In LEEM, a low-energy electron beam is incident on the sample and the backscattered beam is used for imaging in the dark or bright field. The image of Fig. 4(b) was acquired in the mirror electron microscopy (MEM) mode wherein the sample potential is negative with respect to the filament so that the electrons are reflected by the surface potential in front of the sample, causing minimal damage to the nanocrystals. The contrast itself arises from variation in the surface potential and topography. These studies were carried out typically a few weeks after patterning of the nanocrystals on Si surfaces with native oxide layers, demonstrating thereby that the patterns are robust. The patterns were located on the substrates by using suitable micron-sized markers. In Figs. 4(a) and 4(b), we compare the LEEM image in the MEM mode along with the corresponding AFM image obtained right after patterning. The LEEM image corresponds exactly with the AFM image.

We have ascertained the chemical nature of the species by imaging the nanopattern by an independent method, we have recorded XPEEM images. The image in Fig. 4(c) was obtained in x-ray absorption spectroscopy mode with a photon energy of 712 eV using secondary electrons for imaging

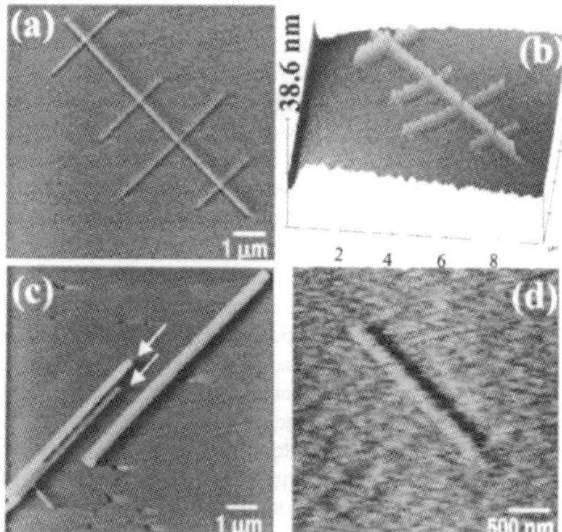

FIG. 2. Contact AFM images of γ-Fe$_2$O$_3$ nanocrystal patterns drawn (a) on mica along with (b) its surface plot, (c) on silicon with the native SiO$_x$ layer (arrows point to closely-drawn patterns) and (d) on etched silicon.

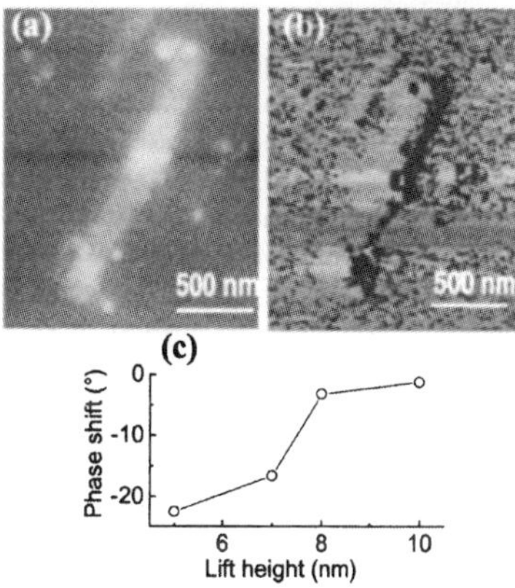

FIG. 3. (a) Tapping mode AFM image of a γ-Fe$_2$O$_3$ nanopattern, (b) the corresponding MFM image at a lift height of 5 nm and (c) plot of the phase shift vs the lift height.

Appl. Phys. Lett., Vol. 84, No. 26, 28 June 2004

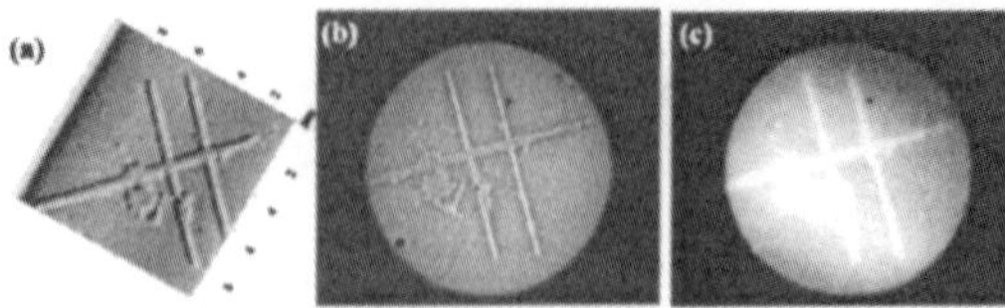

FIG. 4. (a) Tapping mode topographic AFM image of a γ-Fe$_2$O$_3$ nanocrystal pattern along with (b) the corresponding MEM image collected with electron beam of energy 0.8 eV and (c) XPEEM image obtained by collecting the secondary electrons (energy 0.6 eV) following excitation with 712 eV (Fe L_3 edge). The field of view in both cases is 10 μm.

(electron energy 0.6 eV). The photon energy corresponds to the excitation of the Fe L_3 edge in Fe$_2$O$_3$ as verified with a bulk reference sample. The image obtained with photoelectrons corresponds exactly to the AFM image.

In conclusion, we have demonstrated a versatile DPN method for nanopatterning γ-Fe$_2$O$_3$ nanocrystals on different substrates. The patterns are well defined with minimal lateral diffusion of the ink. The patterns are also magnetic and yields well-resolved LEEM and XPEEM images.

The authors thank the Department of Science and Technology, Government of India, for support of this research. One of the authors (N.S.J.) thanks CSIR (India) for financial support.

[1] D. S. Ginger, H. Zhang, and C. A. Mirkin, Angew. Chem., Int. Ed. **43**, 30 (2004) and references therein.

[2] R. D. Piner, J. Zhu, F. Xu, S. Hong, and C. A. Mirkin, Science **283**, 661 (1999).

[3] R. Mckendry, W. T. S. Huck, B. Weeks, M. Fiorini, C. Abell, and T. Rayment, Nano Lett. **2**, 713 (2002); L. M. Demers and C. A. Mirkin, Angew. Chem., Int. Ed. Engl. **40**, 3069 (2001); X. Liu, L. Fu, S. Hong, V. P. Dravid, and C. A. Mirkin, Adv. Mater. (Weinheim, Ger.) **14**, 231 (2002).

[4] M. B. Ali, T. Ondarcuhu, M. Brust, and C. Joachim, Langmuir **18**, 872 (2002).

[5] J. C. Garno, Y. Yang, N. A. Amro, S. Cruchon-Dupeyrat, S. Chen, and G. Y. Liu, Nano Lett. **389**, 3 (2003).

[6] P. J. Thomas, G. U. Kulkarni, and C. N. R. Rao, J. Mater. Chem. **14**, 625 (2004).

[7] L. Fu, X. Liu, Y. Zhang, V. P. Dravid, and C. M. Mirkin, Nano Lett. **3**, 757 (2003).

[8] M. Su and V. P. Dravid, Appl. Phys. Lett. **80**, 4434 (2002).

[9] C. A. Mirkin, MRS Bull. **26**, 535 (2001).

[10] T. Schmidt, S. Heun, J. Slezak, J. Diaz, K. C. Prince, G. Lilienkamp, and E. Bauer, Surf. Rev. Lett. **5**, 1287 (1998).

[11] A. T. Ngo and M. P. Pileni, J. Phys. Chem. B **105**, 53 (2001).

[12] S. Chikuzumi and S. H. Charap, *Physics of Magnetism* (Wiley, New York, 1964).

[13] J. Hu, X.-D. Xiao, D. F. Ogletree, and M. Salmeron, Surf. Sci. **344**, 221 (1995).

[14] Digital Instruments, Santa Barbara, CA.

[15] A. Locatelli, A. Bianco, D. Cocco, S. Cherifi, S. Heun, M. Marsi, A. Pasqualetto, and E. Bauer, J. Phys. IV **104**, 99 (2003).

[16] M. Lazzarino, S. Heun, B. Ressel, K. C. Prince, P. Pingue, and C. Ascoli, Appl. Phys. Lett. **81**, 2842 (2002).

ACCOUNTS
of chemical research

The Liquid–Liquid Interface as a Medium To Generate Nanocrystalline Films of Inorganic Materials

C. N. R. RAO* AND K. P. KALYANIKUTTY

Chemistry and Physics of Materials Unit, DST Nanoscience Unit, and CSIR Centre of Excellence in Chemistry, Jawaharlal Nehru Centre for Advanced Scientific Research, Jakkur P.O., Bangalore 560 064, India

RECEIVED ON AUGUST 31, 2007

CONSPECTUS

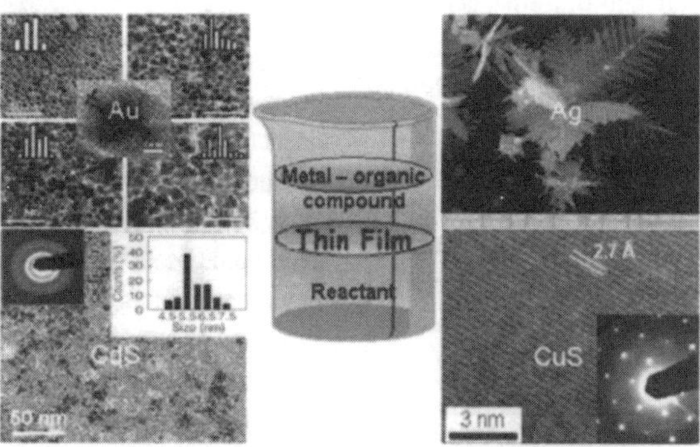

Unlike the air–water interface, the organic–aqueous (liquid–liquid) interface has not been exploited sufficiently for materials synthesis. In this Account, we demonstrate how ultrathin nanocrystalline films of metals such as gold and silver as well as of inorganic materials such as semiconducting metal chalcogenides (e.g., CdS, CuS, CdSe) and oxides are readily generated at the liquid–liquid interface. What is particularly noteworthy is that single-crystalline films of certain metal chalcogenides are also obtained by this method. The as-prepared gold films at the toluene–water interface comprise fairly monodisperse nanocrystals that are closely packed, the nature and properties of the films being influenced by various reaction parameters such as reaction temperature, time, reactant concentrations, mechanical vibrations, and the viscosity of the medium. The surface plasmon band of gold is markedly red-shifted in the films due to electronic coupling between the particles. The shift of the surface plasmon band of the Au film toward higher wavelengths with an accompanying increase in intensity as a function of reaction time marks the growth of the film. Depending on the reaction temperature, the Au films show interesting electrical transport properties. Films of metals such as gold are disintegrated by the addition of alkanethiols, the effectiveness depending on the alkane chain length, clearly evidenced by shifts of the surface plasmon bands. A time evolution study of the polycrystalline Au and CdS films as well as the single-crystalline CuS films is carried out by employing atomic force microscopy. X-ray reflectivity studies reveal the formation of a monolayer of capped clusters having 13 gold atoms each, arranged in a hexagonal manner at the toluene–water interface. The measurements also reveal an extremely small value of the interfacial tension. Besides describing features of such nanocrystalline films and their mode of formation, their rheological properties have been examined. Interfacial rheological studies show that the nanocrystalline film of Ag nanoparticles, the single-crystalline CuS film, and the multilayered CdS film exhibit a viscoelastic behavior strongly reminiscent of soft-glassy systems. Though both CuS and CdS films exhibit a finite yield stress under steady shear, the CdS films are found to rupture at high shear rates. An important advantage of the study of materials formed at the liquid–liquid interface is that it provides a means to investigate the interface itself. In addition, it enables one to obtain substrate-free single-crystalline films of materials.

Published on the Web 03/12/2008   www.pubs.acs.org/acr
10.1021/ar700192d CCC: $40.75   © 2008 American Chemical Society

Vol. 41, No. 4 ■ April 2008 ■ 489-499 ■ ACCOUNTS OF CHEMICAL RESEARCH ■ 489

Introduction

Self-assembly, the process of formation of ordered aggregates, is a powerful strategy for creating novel structures of great academic interest as well as of technological value. Interfaces are an important means to generate two-dimensional self-assemblies of nanocrystals, providing a constrained environment for organized assembly. The air–water interface has been exploited for the preparation of films of metals and semiconductors, which have potential applications in nanodevices. For example, nanocrystal assemblies of gold have been prepared at the air–water interface by employing Langmuir–Blodgett (LB) technique.[1,2] Using a chloroform solution of $Au_{55}(PPh_3)_{12}Cl_6$, Schmid et al.[3] have self-assembled Au_{55} nanocrystals into monolayers in a LB trough. Somorjai and co-workers[4] have employed a LB trough to fabricate monolayers of monodisperse Rh nanocrystals on Si wafers as model 2D catalysts. Catalysts for the oxidation of formic acid have been prepared in the form of Langmuir layers of $Fe_{20}Pt_{80}$ nanoparticles.[5] Bawendi and co-workers[6] have prepared monolayers of monodisperse trioctylphosphine oxide capped CdSe quantum dots by the LB method. By exposing an LB film of lead stearate to H_2S, nanoparticles of PbS have been generated in the form of films.[7] LB films of iron oxide nanoparticles have been obtained by spreading a hexane suspension of oleic acid capped γ-Fe_2O_3 nanoparticles at the air–water interface.[8] Properties of thin films obtained by the LB technique are determined by the nature of the substrate as well as the reaction conditions, and the films are generally polycrystalline.

Unlike the air–water interface, the liquid–liquid (organic–aqueous) interface has not been investigated sufficiently, and it is only recently that there have been concerted efforts to understand the structure of the liquid–liquid interface. The liquid–liquid surface possesses unique thermodynamic properties such as viscosity and density. A liquid–liquid interface is a nonhomogeneous region having a thickness on the order of a few nanometers. The interface is not sharp, since there is always a little solubility of one phase in the other. One of the problems that has been studied in detail at the liquid–liquid interface relates to interfacial charge transfer reactions and dynamics.[9] Distribution of ions and solvent molecules, which determines the structure of a liquid–liquid interface, has recently been investigated by Schlossman and co-workers.[10] Using X-ray reflectivity and molecular dynamics simulations, these workers find that ion sizes and ion–solvent interactions affect the ion distributions near the interface. The relevance of the liquid–liquid interface has been noted in some other areas

such as environmental chemistry, cell biology, and catalysis. The interface between two immiscible liquids offers an important alternative path for the self-assembly and chemical manipulation of nanocrystals.[11] Nanoparticles are highly mobile at the interface and rapidly achieve an equilibrium assembly by reduction in interfacial energy. The three parameters that have been found to influence the energy of the assembly process at the liquid–liquid interface are (i) the nature of the interface, (ii) surface modification of the nanoparticles at the interface, and (iii) the effective radius of the nanoparticles, smaller nanoparticles adsorbing more weakly to the interface than larger ones. Binks and Clint[12] have theoretically treated the wetting of silica particles in terms of the surface energies at the oil–water interface to interpret the interactions between the solid and the liquid phases and to predict the oil–water contact angles for a solid of given hydrophobicity. Only if the contact angle is exactly 90° will the particle be located at the middle of the oil–water interface.

Russel and co-workers[13] have investigated the assembly of phosphine oxide functionalized CdSe nanoparticles of two different diameters by competitive adsorption at the toluene–water interface by employing fluorescence spectroscopy. Benkoski et al.[14] have developed the so-called fossilized liquid assembly for the creation of 2-D assemblies from nanoscale building blocks. By investigating the interactions of a variety of particles such as uncharged, charged, functionalized, and nonfunctionalized, deposited at dodecanediol dimethacrylate/water interface, they have shown that nanoparticles aggregate into a wide variety of complex morphologies. These results provide evidence for the importance of asymmetric dipole interaction in generating the complex morphologies. There are a few assorted reports in the literature where the liquid–liquid interface or a mixture of immiscible liquids has been used for the synthesis or crystallization of nanostructures and other materials. The Brust method,[15] which has been widely employed for the preparation of Au nanocrystals by the reduction of $AuCl_4^-$ by $NaBH_4$ in the presence of an alkanethiol, is carried out in a water–toluene mixture in the presence of a phase-transfer reagent such as tetraoctylammonium bromide. Stucky and co-workers[16] have prepared mesoporous fibers of silica by treating the silica precursor dissolved in an organic phase such as hexane, toluene, or CCl_4 with surfactant molecules dissolved in the aqueous phase. CdS nanoparticles in the form of LB films have been prepared by reacting an aqueous $CdCO_3$ solution with CS_2 in CCl_4.[17] Monodisperse, luminescent nanocrystals of CdS have been prepared by mixing a solution of cadmium–myristic acid and *n*-triphenylphosphine oxide in toluene with an aqueous solution of thiourea, followed by heating under stirring.[18] Langmuir films of silver nanoparticles have been prepared at the water–dichloromethane interface.[19] With Pickering emulsions as a template, dodecanethiol-capped Ag nanoparticles have been self-assembled at the trichloroethylene–water interface.[20] Song et al.[21] have employed the butanol–water interface for the crystallization of Se nanorods. Amorphous Se nanoparticles were first prepared in the aqueous medium and then transported to the butanol–water interface using polyvinylpyrrolidone (PVP), to obtain crystalline nanorods of Se. Cheetham and co-workers[22] have employed a cyclohexanol–water mixture to prepare single crystals of copper adipate. Although the assembly of pregenerated nanoparticles at liquid–liquid interfaces has been examined to some extent, this interface has not been exploited for the synthesis of nanoparticles and their assemblies.

In this Account, we describe a simple but elegant means of generating ultrathin nanocrystalline films of various materials at the organic–aqueous interface. The method primarily involves taking a metal organic compound in the organic layer and a reducing, a sulfiding, or an oxidizing agent in the aqueous layer. The reaction occurs at the interface giving rise to a film at the interface with several interesting features. We describe nanocrystalline films of gold in detail to show how the various reaction parameters affect the films formed at the interface and how alkanethiols bring about the disintegration of the films. We also demonstrate the power of the method in generating ultrathin polycrystalline, as well as single-crystalline, films of some metal chalcogenides at the interface. The formation of single-crystalline films is indeed a noteworthy feature. We believe that this method can be adopted not only for generating nanocrystalline films of various materials but also to study processes occurring at the liquid–liquid interface.

General Experimental Procedure

Use of the liquid–liquid interface for preparing materials in the form of nanocrystalline films is simple and straightforward.[1a,23]The procedure to prepare nanocrystalline films of metals involves taking a metal organic precursor in the organic layer and then injecting an appropriate reducing agent into the aqueous layer to obtain a film comprising metal nanocrystals. In order to prepare nanocrystalline films of metal sulfides, Na_2S is used as the source of sulfur, while Na_2Se or N,N-dimethylselenourea is used as the source of selenium to prepare films of metal selenides.

We shall illustrate the method of preparation of nanocrystalline films with two examples. To prepare nanocrystalline

FIGURE 1. Nanocrystalline films of (a) Au and (b) CdS formed at the toluene–water interface.

gold films, $Au(PPh_3)Cl$ (Ph = phenyl) is found to be a good precursor. In a typical preparation, 10 mL of a 1.5 mM solution of $Au(PPh_3)Cl$ in toluene is allowed to stand in contact with 16 mL of a 3.25 mM solution of NaOH in water in a 100 mL beaker at 300 K. Tetrakishydroxymethylphosphonium chloride (THPC; 300 μL of 50 mM), which acts as the reducing agent,[23a] is slowly injected to the aqueous layer with minimal disturbance to the organic layer. A slight pink coloration of the interface indicates the onset of reduction of the gold salt. As the reaction proceeds, the color of the interface intensifies, finally resulting in a robust film at the interface as shown in Figure 1a. For preparing a nanocrystalline film of CdS, 0.0045 g of Na_2S is dissolved in 30 mL of water (2 mM) in a 100 mL beaker, and 0.0125 g of cadmium cupferronate $[Cd(cup)_2]$ is dissolved in 30 mL of toluene (1 mM) by ultrasonication. A few drops of n-octylamine are added to the $Cd(cup)_2$ solution in order to make it completely soluble. The toluene solution is slowly added to the aqueous Na_2S solution in a 100 mL beaker at 30 °C. The interface attains a yellow color within a few minutes, and a distinct film is formed after 10 h. In Figure 1b, we show a nanocrystalline film of CdS, so formed.

Gold and Other Metals

We shall discuss ultrathin films of gold nanocrystals in some detail to illustrate the features of the method as well as the interesting features of the materials obtained by this method. As-prepared Au films, obtained at 30 °C after 24 h of the reaction at the toluene–water interface comprise fairly monodisperse nanoparticles with a mean diameter of ~7 nm (see the TEM image given in Figure 2a). The nanocrystals are closely packed with a typical interparticle distance of ~1 nm. Furthermore, they are single-crystalline as revealed by the high-resolution electron microscope (HREM) image given as an inset in Figure 2. The image shows the (111) planes of gold, sepa-

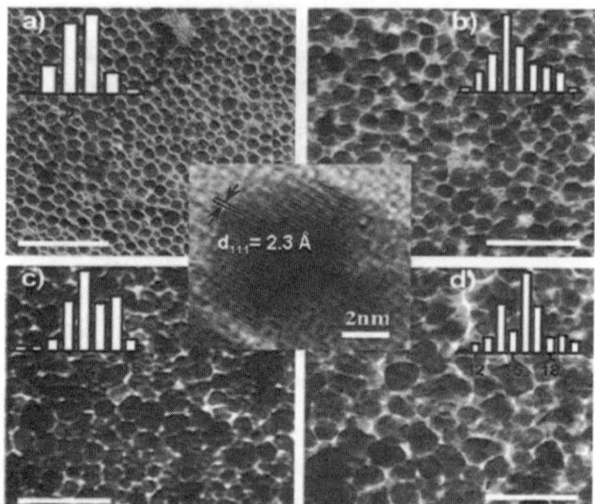

FIGURE 2. TEM images of the ultrathin nanocrystalline Au films obtained at the liquid–liquid interface after 24 h at (a) 30, (b) 45, (c) 60, and (d) 75 °C. Histograms of particle size distribution are shown as insets. The scale bars correspond to 50 nm. A high-resolution image of an individual particle is shown at the center. Reproduced from ref 24. Copyright 2005 American Chemical Society.

rated by a distance of ~2.3 Å. Such nanocrystalline films could also be obtained by using other solvent pairs such as CCl_4–water and butanol–water. Reaction parameters such as the temperature, reaction time, concentrations of the metal precursor and the reducing agent, and the viscosity of the aqueous layer affect the nature and properties of the nanocrystalline films.

The effect of temperature on the size distribution of the Au nanocrystals can be readily seen from the TEM images in Figure 2. The mean diameters of the nanocrystals formed at 30, 45, 60, and 75 °C are 7, 10, 12, and 15 nm, respectively, but the interparticle separation remains nearly the same at ~1 nm. X-ray diffraction measurements show that with increase in temperature, the crystallinity of the film increases (Figure 3). The films obtained at 45 and 60 °C exhibit prominent (111) peaks ($d = 2.33$ Å), while those obtained at 30 °C show weak reflections, probably due to the small particle size. The growth of the (111) peak with temperature indicates an increase in the particle size.

Increasing the concentration of the metal precursor yields nanocrystalline films with a larger number of particles, but the size distribution is essentially unaffected. The thickness of the film also increases with the increase in the metal precursor concentration. The use of high concentrations of the reducing agent results in less uniform films with altered distributions in the nanoparticle diameter. A slight increase in the size of the Au nanoparticles was observed when the viscosity of the aqueous layer was increased by the addition of glycerol.

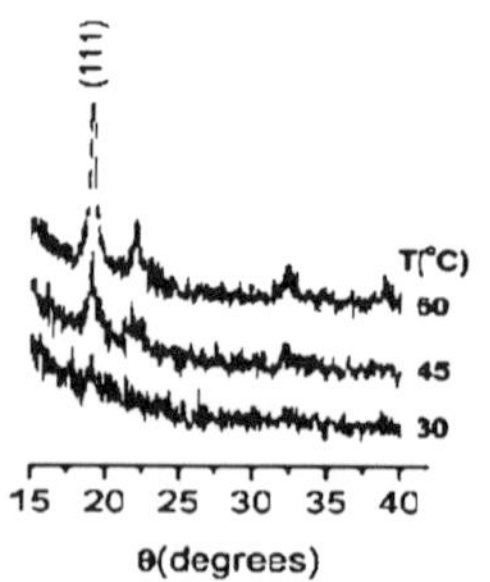

FIGURE 3. X-ray diffraction patterns of nanocrystalline Au films obtained at different temperatures. Reproduced from ref 24. Copyright 2005 American Chemical Society.

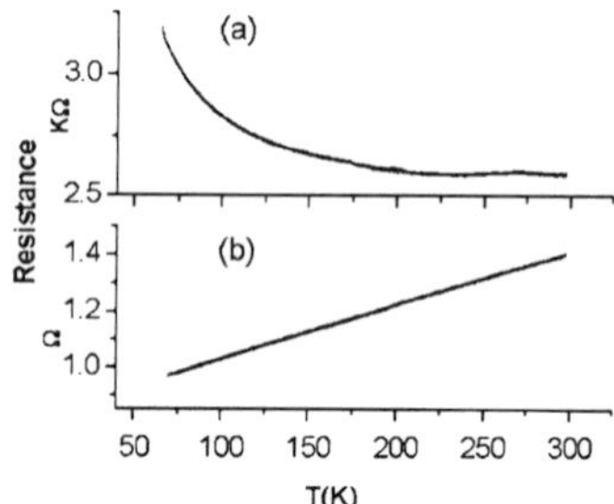

FIGURE 4. Temperature variation of the electrical resistance of the Au films prepared at (a) 45 and (b) 60 °C (current used, 10 mA). Modified from ref 24. Copyright 2005 American Chemical Society.

Reactions at the interface carried out on a vibration-free table yielded nanocrystalline films with reduced roughness, comprising particles of smaller size.

The Au films formed at the interface show interesting electrical transport properties that are dependent on the reaction temperature (see Figure 4).[24] Four-probe electrical resistance measurements on the nanocrystalline films show a metal to insulator transition, metallic behavior being shown by the films formed at high temperatures (>45 °C). The films formed at lower temperatures (≤45 °C) show insulating behavior.

Atomic force microscopy (AFM) shows the thickness of the films to be in the 40–140 nm range. The contact mode AFM image in Figure 5a shows the boundary of a Au film on a mica substrate. The height profile of the film in Figure 5b gives an estimate of the thickness to be ~60 nm. AFM images covering a few micrometers yield a root-mean-square roughness in the range 30–35 nm.

The growth of Au films could be followed as a function of reaction time by UV–visible absorption spectroscopy (Figure 6). The Au plasmon band gets red-shifted due to the increase in the electronic coupling between the particles, accompanied by an increase in the intensity. The absorption band shows

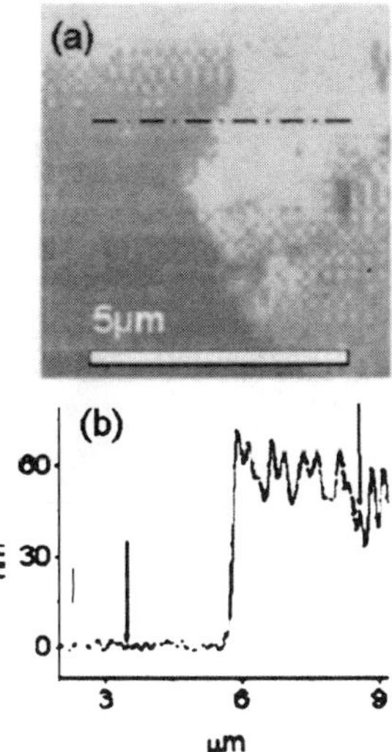

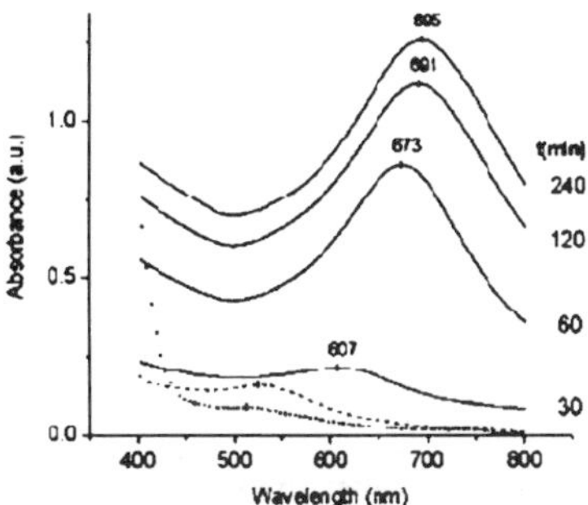

FIGURE 5. (a) Contact-mode AFM image showing the boundary of Au film on a mica substrate and (b) the z-profile. Reproduced from ref 24. Copyright 2005 American Chemical Society.

FIGURE 6. Evolution of the electronic absorption spectra with the growth of the nanocrystalline Au film at the interface at 45 °C (full curves) and spectra of octylamine-capped Au nanocrystals in toluene (broken line) and mercaptoundecanoic acid-capped nanocrystals in water (dotted line). Modified from ref 24. Copyright 2005 American Chemical Society.

negligible change after 120 min. At this stage, the films presumably consist of well-packed nanocrystals.

The effect of surfactants such as tetraoctylammonium bromide (TOAB) and cetyltrimethylammonium bromide (CTAB) on the nanostructures formed when the gold ions present in the organic phase are reduced at the interface by hydrazine (in the aqueous phase) has been investigated.[25] The surfactants give rise to extended fractal networks with a fractal dimension of 1.7 at the interface (Figure 7a). The fractals themselves comprise cauliflower-like spherical units (see top inset in Figure 7a) formed of pentagonal nanorods.

Till now, we have been examining the formation of uniform and robust nanocrystalline films of Au. The films, once

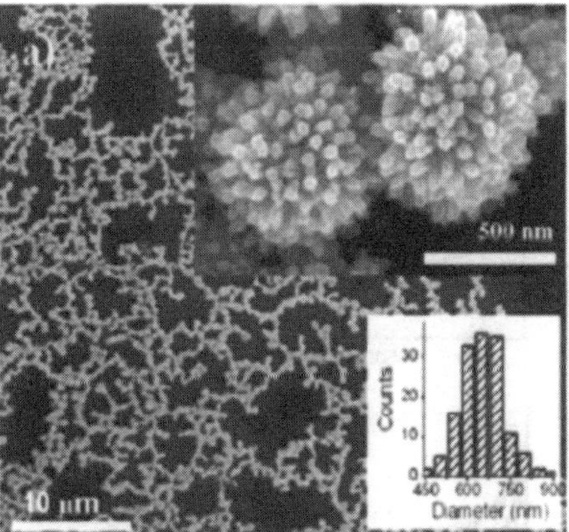

FIGURE 7. Panel a shows an SEM image of the fractal network formed by cauliflower-like gold structures by employing TOAB. The inset on the top right corner shows a high-resolution image of the cauliflower-like structures. The inset at the bottom shows the histogram of the size distribution of cauliflower-like structures. Panel b shows a dendritic nanostructure of Ag. Reproduced with permission from ref 25. Copyright 2008 Elsevier.

formed, can be disintegrated or disordered by the addition of *n*-alkanethiols.[24,26] Addition of alkanethiols is accompanied by a progressive blue shift of the plasmon absorption band of the film, suggesting that the electronic coupling between the nanocrystals gets reduced due to the increased separation between the nanocrystals. Figure 8a shows the variation in the shift of the absorption band brought about by the addition of hexadecanethiol. Interaction with thiols also brings about a change in the morphology of the film as shown in the adjoining AFM images. By varying the chain length of the alkanethiol, one is actually varying the distance between the nanoparticles, thereby giving rise to a blue shift proportional to the length of the alkane chain (Figure 8b). The rate of disordering or disintegration of the Au films is also affected by the chain length of the alkanethiols; the longer the chain

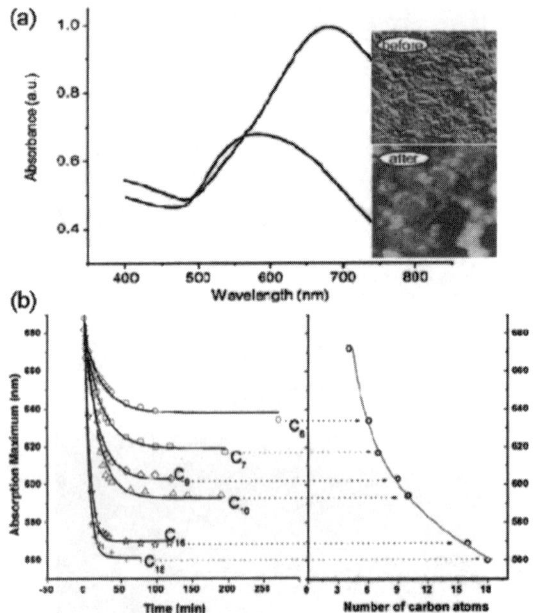

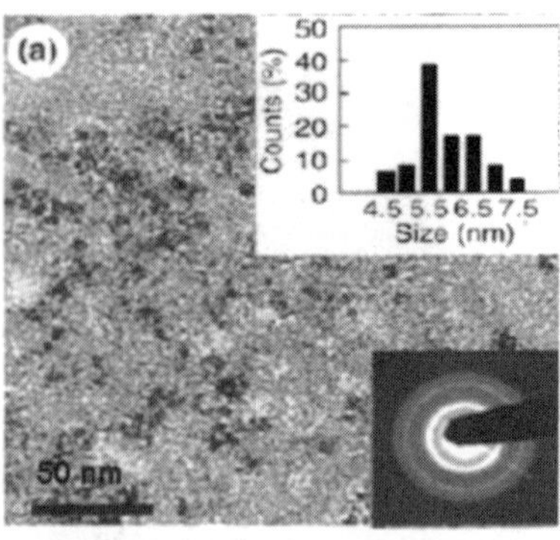

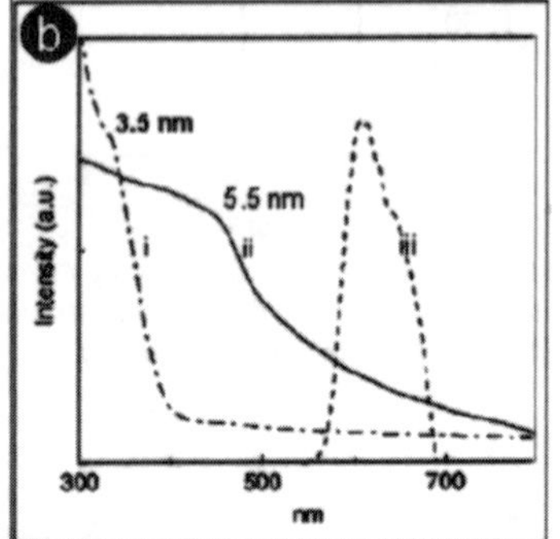

FIGURE 8. Panel a shows the electronic absorption spectra of a Au film before and after treatment with hexadecanethiol solution for 6 h. Tapping mode AFM images are shown alongside (scan area 500 nm × 500 nm). Reproduced from ref 24. Copyright 2005 American Chemical Society. (b) Time variation of the absorption maximum of Au nanocrystalline films on interaction with alkanethiols of different chain lengths (left) and variation in the limiting absorption maximum with chain length of the thiol adsorbed (right). Reproduced from ref 26. Copyright 2008 American Chemical Society.

FIGURE 9. Panel a shows the TEM image of 5.5 nm CdS nanocrystals obtained at room temperature at the interface. Insets give the SAED pattern and the particle size distribution. Panel b shows the UV–visible absorption spectra of (i) 3.4 nm and (ii) 5.5 nm CdS nanocrystals and (iii) the photoluminescence spectrum of 5.5 nm CdS nanocrystals. Modified with permission from ref 28. Copyright 2003 Elsevier.

length, the faster is the rate. Calorimetric measurements also lend evidence for such a process.

It has also been possible to obtain nanocrystalline films of other metals such as Ag, Pd, and Cu at the toluene–water interface by taking $Ag_2(PPh_3)_4$, palladium acetate, and $Cu(PPh_3)Cl$, respectively, in the organic layer.[23] By carrying out the reaction in the presence of TOAB, we have obtained dendritric structures of Ag (Figure 7b).[25] By taking mixtures of the corresponding metal precursors in the organic layer, we have prepared nanocrystalline films of binary Au–Ag and Au–Cu alloys and ternary Au–Ag–Cu alloys.[27]

Metal Chalcogenides

Ultrathin films of metal sulfides such as CdS, CuS, ZnS, and PbS are obtained at the organic–aqueous interface by the reaction of Na_2S in the aqueous layer with the corresponding metal cupferronate in the organic layer.[28–30] We show typical results in the case of CdS. TEM images reveal the films to be composed of nanocrystals of 5.5 nm diameter (Figure

9a). The powder XRD pattern of the film shows that the CdS nanocrystals crystallize in the rock-salt structure rather than in the wurtzite structure. Increasing the reaction temperature and the concentration of the reactants yields bigger nanocrystals. When the viscosity of the aqueous medium is doubled by the addition of glycerol, the size of the nanocrystals is reduced to 3.5 nm. The UV–vis absorption spectrum of the nanocrystals (Figure 9b) shows a broad absorption maximum, which is blue-shifted compared with the bulk CdS due to quantum confinement. The PL spectrum of the 5.5 nm particles shows a peak at 610 nm.

Polycrystalline thin films of CdSe have been prepared at the organic–water interface by reacting cadmium cupferronate in the toluene layer with dimethylselenourea in the aqueous layer.[31] XRD measurements confirm the formation of cubic CdSe at the interface. TEM images reveal the films to be made up of nanocrystals with diameters ranging from 8 to 20 nm (Figure 10a). Time-dependent growth of the CdSe film at 20 °C has been examined by UV–vis absorption spectroscopy. All

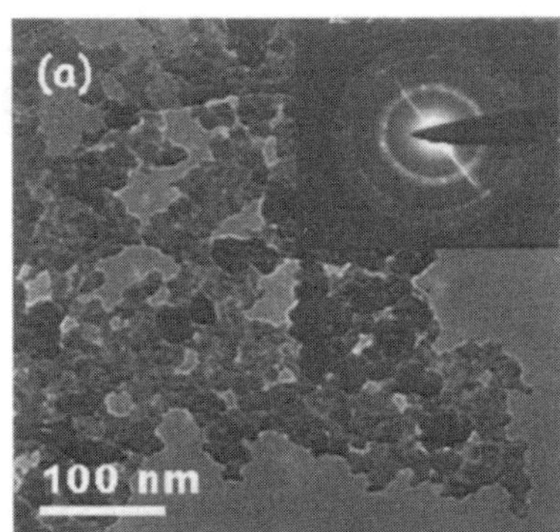

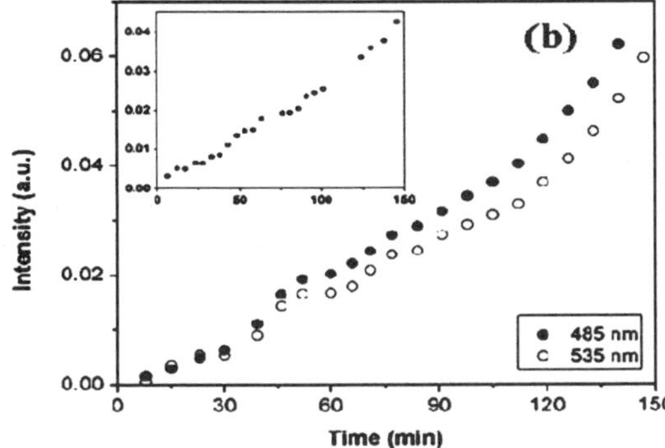

FIGURE 10. (a) TEM image of a CdSe film showing particles obtained by reacting aqueous solution of DMSU at 30 °C and (b) a plot showing the time-dependent growth of the absorption bands at two different wavelengths. Reproduced with permission from ref 31. Copyright 2007 American Scientific Publishers.

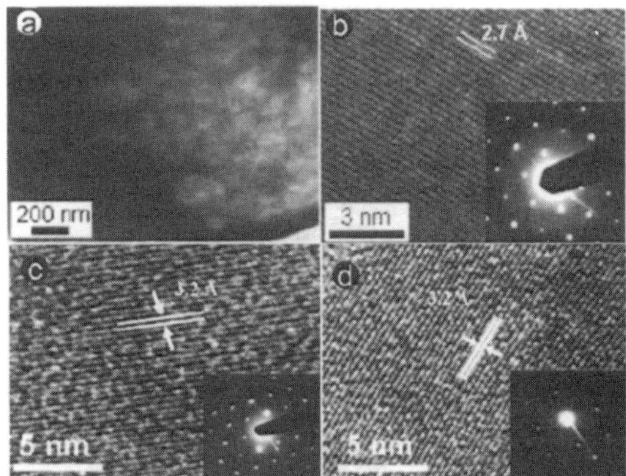

FIGURE 11. (a) TEM image of a CuS film, (b) HREM and SAED pattern of the film shown in panel a (Reproduced from ref 29. Copyright 2004 American Chemical Society), and HREM images and SAED patterns of (c) ZnS film and (d) PbS film (Reproduced with permission from ref 30. Copyright 2006 Elsevier).

the films, including the one obtained after 3 min of the reaction, showed an absorption onset around 700 nm corresponding to the bulk band gap of CdSe and two higher order absorption bands at 485 and 535 nm. With increase in reaction time, the spectra indeed grew in intensity, without any appreciable shift in the positions of the band edge or the absorption bands, suggesting that CdSe nanoparticles possessing a size beyond the quantum confinement effects are formed at the interface within the first few minutes. Accordingly, the intensity versus time plots for the 485 and 535 nm bands, in Figure 10b, show a monotonic increase.

What is truly noteworthy is that we have been able to get single-crystalline films of some of the metal chalcogenides at the interface. In the case of CuS, we obtain continuous films, extending over wide areas (Figure 11a). A HREM image showing the (006) planes of hexagonal CuS and the corresponding SAED pattern given in Figure 11b reveal the single-crystalline and essentially defect-free nature of the CuS films. The thickness of the film was estimated to be ~50 nm from AFM and ellipsometric studies. In Figure 11c, we show a HREM

image of a single-crystalline ZnS film. The (100) planes, separated by a distance of 3.2 Å, as seen in the HREM image correspond to hexagonal ZnS. The Bragg spots in the SAED pattern conform to the 0001 zone axis of hexagonal ZnS. In Figure 11d, we show the HREM image and the SAED pattern of a PbS film generated at the interface to reveal the single-crystalline nature. The HREM image gives a separation of 3.2 Å, corresponding to the separation between the (100) planes of cubic PbS. Single-crystalline CuSe thin films have also been obtained at the interface. Apart from single-crystalline films of metal sulfides, we could obtain metal sulfide bilayers such as CuS–CdS, CuS–PbS, and CdS–PbS at the toluene–water interface by employing the respective metal cupferronates.

Metal Oxides

By reacting $Cu(cup)_2$ in the organic layer with an aqueous NaOH solution, one obtains single-crystalline films of monoclinic CuO at the organic–aqueous interface at 70 °C.[29] It has been possible to obtain crystalline films of ZnO by the reaction of $Zn(cup)_2$ in toluene with an aqueous solution of NaOH at 25 °C.

Mode of Growth of the Films at the Interface

A time evolution study of Au films by AFM shows that just after 10 min of the reaction, a film made up of nanocrystals of size ~5–7 nm is formed. Although the size of the nanoparticles remains nearly the same even after 180 min of the reaction, we have found some evidence for the aggregation of nanoparticles in the films. The organic capped nanocrys-

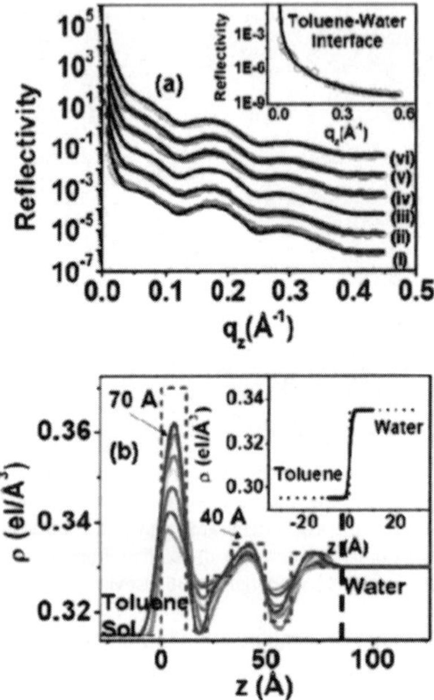

FIGURE 12. Reflectivity curves collected after initiation of the reaction (i, 194; ii, 224; iii, 253; iv, 283; v, 312; vi, 364 min) and the fits (solid line) and (b) extracted electron density profiles for the six reflectivity curves (i–vi) given in panel a as a function of depth. A simple model without (dashed line) and with roughness convolution (solid line) is also shown. The insets in panels a and b show the reflectivity data and the EDPs, respectively, of the bare toluene–water interface and the corresponding fit (solid line). Taken from ref 32.

tals form an ordered arrangement giving rise to large features in the AFM images. The appearance of such large features makes it difficult to carry out a proper AFM study of the Au films. However, the AFM study shows that the thickness of the films increases progressively with time.

The formation of Au nanoparticle films has been investigated by X-ray reflectivity and diffuse scattering employing synchrotron radiation.[32] The oscillations in the reflectivity curves, collected after the initiation of the reaction (see Figure 12a), indicate the presence of a thin film at the toluene–water interface. The extracted final electron density profiles (EDPs; Figure 12b) from the reflectivity curves show three layers of gold clusters at 70 Å, 40 Å, and just above the water surface with central electron density values of 0.37, 0.33, and 0.33 electrons Å^{-3}. The layers have a vertical separation of 30 Å, the electron density between these layers corresponding to that of a typical organic material. A preliminary study shows that small Au nanoparticles covered by organic coatings form hexagonal clusters at the toluene–water interface (Figure 13). Each cluster consists of 13 nanoparticles of

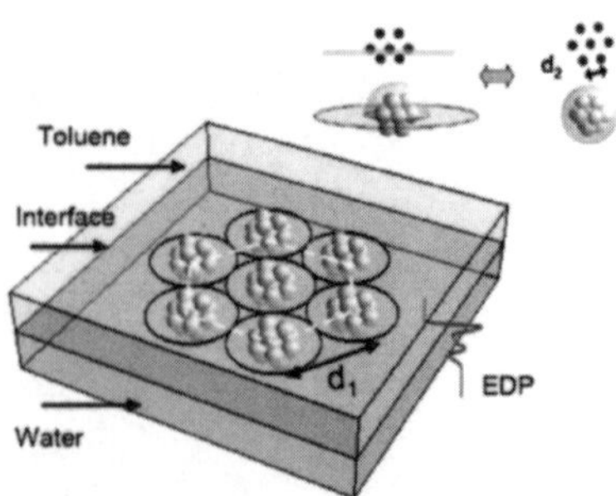

FIGURE 13. Schematic showing the hexagonal arrangement of organic-capped Au nanocrystals at the toluene–water interface. Taken from ref 32.

diameter ~12 Å with an organic capping of ~11 Å. Diffuse scattering measurements give a small value of interfacial tension, indicating an enhancement in the interfacial roughness caused by the presence of an organic layer at the interface.

The growth of the CdS polycrystalline thin films at the organic–aqueous interface has been studied as a function of time by employing AFM. The CdS film obtained after 30 min is made up of nanoparticles of diameter in the range ~5–7 nm. Nanoparticle aggregates of size in the range 80–200 nm are seen on the surface of the film. After 60 min of the reaction, the vertical growth of the aggregates saturates. The nanoparticles constituting the aggregates spread out laterally, thus reducing the overall thickness of the sample. With time, the aggregates gradually self-assemble to form a close-packed layer.

The growth of single-crystalline CuS films has also been examined as a function of reaction time by employing AFM. Nanoparticles of diameter in the range 7–8 nm are found to constitute the films. The amplitude image of the 30 min sample (Figure 14a) shows the presence of nanoparticle aggregates of size mainly in the range 50–100 nm. As the reaction proceeds, the initially formed nanoparticle aggregates come closer to each other and pack themselves at the interface to give flakes or uneven pieces of CuS within 50 min of reaction (denoted by arrows in Figure 14b). In the next 15 min of the reaction, the flakes coalesce to give a continuous and extended film at the interface. The process of formation of flakes and their coalescence repeats, giving rise to a thick film at the interface at the end of 180 min of the reaction (Figure 14c). Here we could hardly observe any individual nanoparticles. This is in contrast to the Au and CdS films, where individual nanoparticles remained.

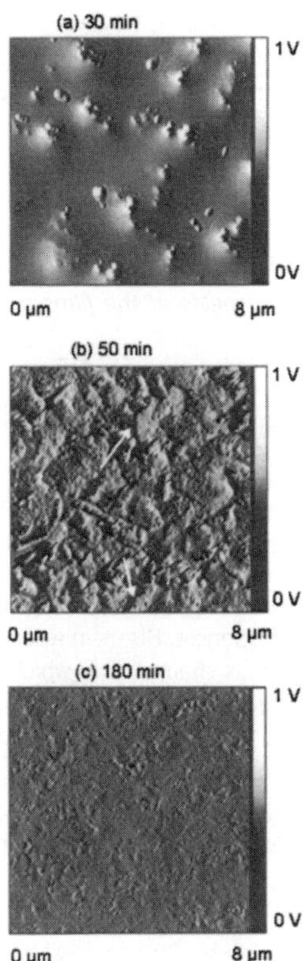

FIGURE 14. Tapping mode AFM images (amplitude channel) of CuS films grown at the toluene–water interface for (a) 30 min and (b) 50 min and (c) contact mode image obtained from the deflection channel of the film grown for 180 min.

Rheological Properties of the Films

Interfacial properties of Ag nanoparticles formed at the toluene–water interface have been investigated by using a biconal bob interfacial rheometer.[33] Strain amplitude measurements carried out on the film reveal a shear-thickening peak in the loss moduli (G'') at large amplitudes, followed by a power law decay of storage (G') and loss moduli with exponents in the ratio 2:1 (see Figure 15a). In the frequency sweep measurements carried out at low frequencies, G' remains nearly independent of the frequency over the range of frequencies probed, whereas G'' shows a power law dependence with a negative slope (Figure 15b). Such a low-frequency response of the 2D film of metal nanoparticles is reminiscent of a soft glassy system with long structural relaxation times. Steady shear measurements carried out on the film revealing a finite yield stress as the shear rate goes to zero, along with a significant deviation from the Cox–Merz rule, confirm that the monolayer of Ag nanoparticles at the interface forms a soft two-dimensional colloidal glass.

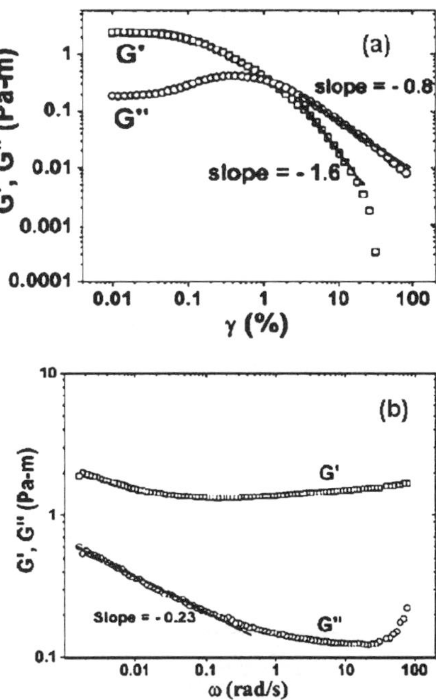

FIGURE 15. Panel a shows the strain amplitude sweep experiment on the 2D film of Ag nanoparticles. The storage modulus, G' (□), is higher than the loss modulus, G'' (○) at low strain amplitudes. Panel b shows the frequency dependence of interfacial storage, G' (□), and loss, G'' (○), moduli of the film. Reproduced from ref 33. Copyright 2007 American Chemical Society.

Interfacial rheological measurements carried out on nanocrystalline films of CdS, as well as on single-crystalline films of CuS, indicate a distinct nonlinear viscoelastic behavior for the films under oscillatory shear.[34] A smooth multilayered CdS film formed at higher concentrations and the single-crystalline CuS film exhibit distinct peaks in the loss modulus above a critical strain amplitude, followed by a power law decay of G' and G'' at higher strain amplitudes, with the decay exponents in the ratio 2:1 (Figure 16). The frequency sweep response of both films, exhibiting a solid-like behavior over the range of angular frequencies probed, is similar to that of the Ag film, a characteristic feature of soft glassy systems. However, the mesoporous CdS film formed at low concentrations, where the pore boundaries form a surface fractal with a fractal dimension of 2.5, exhibits a monotonic decay of storage and loss modulus at large strain amplitudes. Under steady shear, compared with the CuS film, the finite yield stress exhibited by the CdS film is a magnitude lower, and the film is found to rupture under steady shear or at high strain amplitudes.

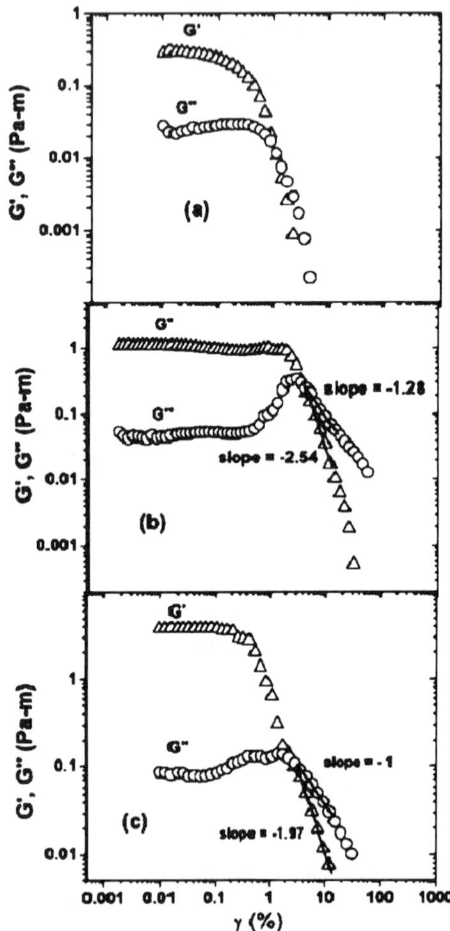

FIGURE 16. Strain amplitude sweep measurements on (a) CdS (low reactant concentrations), (b) CdS (high reactant concentrations), and (c) nanofilms formed at the toluene–water interface at 22 °C. Reproduced from ref 34. Copyright 2008 Elsevier.

Conclusions

This Account should clearly bring out how the liquid–liquid interface can be exploited to prepare nanocrystalline films of various materials with wide ranging properties. What is noteworthy is that all the films are ultrathin with the thickness ranging from 40 to 100 nm. Furthermore, the method enables one to obtain substrate-free films, some of which are single-crystalline. It must be noted that obtaining single-crystalline films is difficult by any other physical or chemical method. The films prepared at the interface can be readily transferred to any substrate. By employing the organic–aqueous interface, one can prepare interesting films of inorganic materials with hydrophilic and hydrophobic coatings on opposite sides. It should also be possible to obtain bilayers of materials at the interface. Preliminary experiments show that CuS–CdS, CuS–PbS, and CdS–PbS bilayers can be prepared at the interface. The study of films of materials at the liquid–liquid interface will also provide a means to understand the nature of the interface itself.

The authors thank their colleagues, Prof. G. U. Kulkarni, Dr. U. K. Gautam, Dr. M. Ghosh, and V. V. Agrawal, who have collaborated on some aspects of the films at the liquid–liquid interface.

BIOGRAPHICAL INFORMATION

C. N. R. Rao is National Research Professor of India, Linus Pauling Research Professor at the Jawaharlal Nehru Centre for Advanced Scientific Research (JNCASR), and Honorary Professor at the Indian Institute of Science, Bangalore. He is a member of many academies including the Royal Society and the U.S. National Academy of Sciences. His main research interests are in solid state and materials chemistry, in which area he has published over 1000 research papers and 30 books. He is the recipient of the Dan David Prize for Materials Research.

K. P. Kalyanikutty received her B.Sc. degree from the University of Calicut and M.S. degree from JNCASR. She is continuing work for her Ph.D. degree in the area of nanomaterials.

FOOTNOTES

*Corresponding author. Fax: 00 91 80 2208 2766. E-mail: cnrrao@jncasr.ac.in.

REFERENCES

1 (a) Petty, M. C. *Langmuir-Blodgett films: An introduction*; Cambridge University Press: Cambridge, U.K., 1996. (b) Gelbart, W. M.; Sear, R. P.; Heath, J. R.; Chaney, S. Array Formation in Nano-colloids: Theory and Experiment in 2D. *Faraday Discuss.* **1999**, *112*, 299–307. (c) Rao, C. N. R.; Agrawal, V. V.; Biswas, K.; Gautam, U. K.; Ghosh, M.; Govindaraj, A.; Kulkarni, G. U.; Kalyanikutty, K. P.; Sardar, K.; Vivekchand, S. R.C. Soft Chemical Approaches to Inorganic Nanostructures. *Pure Appl. Chem.* **2006**, *78*, 1619–1650.

2 Mayya, K. S.; Patil, V.; Sastry, M. Lamellar Multilayer Gold Cluster Films Deposited by the Langmuir–Blodgett Technique. *Langmuir* **1997**, *13*, 2575–2577.

3 Schmid, G.; Baumle, M.; Beyer, N. Ordered Two-Dimensional Monolayers of Au₅₅ Clusters. *Angew. Chem., Int. Ed.* **2000**, *39*, 181–183.

4 Zhang, Y.; Grass, M. E.; Habas, S. E.; Tao, F.; Zhang, T.; Yang, P.; Somorjai, G. A. One-step Polyol Synthesis and Langmuir–Blodgett Monolayer Formation of Size-Tunable Monodisperse Rhodium Nanocrystals with Catalytically Active (111) Surface Structures. *J. Phys. Chem. C* **2007**, *111*, 12243–12253.

5 Chen, W.; Kim, J.; Xu, L.-P.; Sun, S.; Chen, S. Langmuir–Blodgett Thin Films of Fe₂₀Pt₈₀ Nanoparticles for the Electrocatalytic Oxidation of Formic Acid. *J. Phys. Chem. C* **2007**, *111*, 13452–13459.

6 Dabbousi, B. O.; Murray, C. B.; Rubner, M. F.; Bawendi, M. G. Langmuir–Blodgett Manipulation of Size-Selected CdSe Nanocrystallites. *Chem. Mater.* **1994**, *6*, 216–219.

7 Zhu, R.; Min, G.; Wei, Y.; Schmitt, H. J. Scanning Tunneling Microscopy and UV–Visible Spectroscopy Studies of Lead Sulfide Ultrafine Particles Synthesized in Langmuir–Blodgett Films. *J. Phys. Chem.* **1992**, *96*, 8210–8211.

8 Guo, Q.; Teng, X.; Rahman, S.; Yang, H. Patterned Langmuir–Blodgett Films of Monodisperse Nanoparticles of Iron Oxide Using Soft Lithography. *J. Am. Chem. Soc.* **2003**, *125*, 630–631.

9 Benjamin, I. Chemical Reactions and Solvation at Liquid Interfaces: A Microscopic Perspective. *Chem. Rev.* **1996**, *96*, 1449–1475.

10 Luo, G.; Malkova, S.; Yoon, J.; Schultz, D. G.; Lin, B.; Meron, M.; Benjamin, I.; Vanysek, P.; Schlossman, M. L. Ion-Distributions Near a Liquid-Liquid Interface. *Science* **2006**, *311*, 216–218.

11 Binder, W. H. Supramolecular Assembly of Nanoparticles at Liquid-Liquid Interfaces. *Angew. Chem., Int. Ed.* **2005**, *44*, 2–5.

12 Binks, B. P.; Clint, J. H. Solid Wettability from Surface Energy Components: Relevance to Pickering Emulsions. *Langmuir* **2002**, *18*, 1270–1273.

13 Lin, Y.; Skaff, H.; Emrick, T.; Dinsmore, A. D.; Russel, T. P. Nanoparticle Assembly and Transport at Liquid-Liquid Interfaces. *Science* **2003**, *299*, 226–229.

14 Benkoski, J.; Jones, R. L.; Doughlas, J. F.; Karim, A. Photocurable Oil/Water Interfaces as a Universal Platform for 2-D Self-Assembly. *Langmuir* **2007**, *23*, 3530–3537.

15 Brust, M.; Walker, M.; Bethell, D.; Schiffrin, D. J.; Whyman, R. Synthesis of Thiol-Derivatised Gold Nanoparticles in a Two-Phase Liquid-Liquid System. *J. Chem. Soc., Chem. Commun.* **1994**, 801–802.

16 Kleitz, F.; Marlow, F.; Stucky, G. D.; Schuth, F. Mesoporous Silica Fibers: Synthesis, Internal Structure, and Growth Kinetics. *Chem. Mater.* **2001**, *13*, 3587–3595.

17 Sathaye, S. D.; Patil, K. R.; Paranjape, D. V.; Mitra, A.; Awate, S. V.; Mandale, A. B. Preparation of Q-Cadmium Sulfide Ultrathin Films by a New Liquid—Liquid Interface Reaction Technique (LLRT). *Langmuir* **2000**, *16*, 3487–3490.

18 Pan, D.; Jiang, S.; An, L.; Jiang, B. Controllable Synthesis of Highly Luminescent and Monodisperse CdS Nanocrystals by a Two-Phase Approach under Mild Conditions. *Adv. Mater.* **2004**, *16*, 982–985.

19 Schwartz, H.; Harel, Y.; Efrima, S. Surface Behavior and Buckling of Silver Interfacial Colloid Films. *Langmuir* **2001**, *17*, 3884–3892.

20 Dai, L. L.; Sharma, R.; Wu, C. Self-Assembled Structure of Nanoparticles at a Liquid—Liquid Interface. *Langmuir* **2005**, *21*, 2641–2643.

21 Song, J. M.; Zhu, J. H.; Yu, S. H. Crystallization and Shape Evolution of Single-Crystalline Selenium Nanorods at Liquid—Liquid Interface: From Monodisperse Amorphous Se Nanospheres toward Se Nanorods. *J. Phys. Chem. B* **2006**, *110*, 23790–23795.

22 Foster, P. M.; Thomas, P. M.; Cheetham, A. K. Biphasic Solvothermal Synthesis: A New Approach for Hybrid Inorganic—Organic Materials. *Chem. Mater.* **2002**, *14*, 17–20.

23 (a) Rao, C. N. R.; Kulkarni, G. U.; Thomas, P. J.; Agrawal, V. V.; Saravanan, P. Films of Metal Nanocrystals Formed at Aqueous—Organic Interfaces. *J. Phys. Chem. B* **2003**, *107*, 7391–7395. (b) Rao, C. N. R.; Kulkarni, G. U.; Agrawal, V. V.; Gautam, U. K.; Ghosh, M.; Tumkurkar, U. Use of the Liquid-Liquid Interface for Generating Ultrathin Nanocrystalline Films of Metals, Chalcogenides, and Oxides. *J. Colloid Interface Sci.* **2005**, *289*, 305–318.

24 Agrawal, V. V.; Kulkarni, G. U.; Rao, C. N. R. Nature and Properties of Ultrathin Nanocrystalline Gold Films Formed at the Organic—Aqueous Interface. *J. Phys. Chem. B* **2005**, *109*, 7300–7305.

25 Agrawal, V. V.; Kulkarni, G. U.; Rao, C. N. R. Surfactant-Promoted Formation of Fractal and Dendritic Nanostructures of Gold and Silver at the Organic-Aqueous Interface. *J. Colloid Interface Sci.* **2008**, *318*, 501–506.

26 Agrawal, V. V.; Varghese, N.; Kulkarni, G. U.; Rao, C. N. R. Effects of the Change in Interparticle Distance Induced by Alkanethiols on the Optical Spectra and Other Properties of Nanocrystalline Gold Films. *Langmuir*, in press).

27 Agrawal, V. V.; Mahalakshmi, P.; Kulkarni, G. U.; Rao, C. N. R. Nanocrystalline Films of Au—Ag, Au—Cu, and Au—Ag—Cu Alloys Formed at the Organic—Aqueous Interface. *Langmuir* **2006**, *22*, 1846–1851.

28 Gautam, U. K.; Ghosh, M.; Rao, C. N. R. A Strategy for the Synthesis of Nanocrystal Films of Metal Chalcogenides and Oxides by Employing the Liquid-Liquid Interface. *Chem. Phys. Lett.* **2003**, *381*, 1–6.

29 Gautam, U. K.; Ghosh, M.; Rao, C. N. R. Template-Free Chemical Route to Ultrathin Single-Crystalline Films of CuS and CuO Employing the Liquid—Liquid Interface. *Langmuir* **2004**, *20*, 10776–10778.

30 Kalyanikutty, K. P.; Gautam, U. K.; Rao, C. N. R. Ultra-Thin Crystalline Films of ZnS and PbS Formed at the Organic—Aqueous Interface. *Solid State Sci.* **2006**, *8*, 296–302.

31 Kalyanikutty, K. P.; Gautam, U. K.; Rao, C. N. R. Ultra-Thin Crystalline Films of CdSe and CuSe Formed at the Organic—Aqueous Interface. *J. Nanosci. Nanotechnol.* **2007**, *7*, 1916–1922.

32 Sanyal, M. K.; Agrawal, V. V.; Bera, M. K.; Kalyanikutty, K. P.; Daillant, J.; Blot, C.; Kubowicz, S.; Konovalov, O.; Rao, C. N. R. Ordering of Gold Nanoparticles Formed at Liquid-Liquid Interface. *J. Phys. Chem. C*, in press.

33 Krishnaswamy, R.; Majumdar, S.; Ganapathy, R.; Agrawal, V. V.; Sood, A. K.; Rao, C. N. R. Interfacial Rheology of an Ultrathin Nanocrystalline Film Formed at the Liquid/Liquid Interface. *Langmuir* **2007**, *23*, 3084–3087.

34 Krishnaswamy, R.; Kalyanikutty, K. P.; Kanishka, B.; Sood, A. K.; Rao, C. N. R. Nonlinear Viscoelasticity of Ultrathin Nanocrystalline Semiconducting Films of CdS and CuS at Liquid-Liquid Interfaces. *J. Colloid Interface Sci.*, (submitted for publication).

20752

THE JOURNAL OF
PHYSICAL CHEMISTRY B
LETTERS

2006, *110*, 20752−20755
Published on Web 09/28/2006

Use of Fluorous Chemistry in the Solubilization and Phase Transfer of Nanocrystals, Nanorods, and Nanotubes

Rakesh Voggu,[†] Kanishka Biswas,[†,‡] A. Govindaraj,[†,‡] and C. N. R. Rao*,[†,‡]

Chemistry and Physics of Materials Unit, DST Unit on Nanoscience and CSIR Centre of Excellence in Chemistry, Jawaharlal Nehru Centre for Advanced Scientific Research, Jakkur P. O., Bangalore 560064, India, and Solid State and Structural Chemistry Unit, Indian Institute of Science, Bangalore 560012, India

Received: August 19, 2006; In Final Form: September 14, 2006

Fluorous chemistry, involving the use of a fluorous label for the functionalization of a substrate and a fluorous solvent for extraction of the functionalized substrate, is shown to be effective in solubilizing gold and CdSe nanoparticles in a fluorous medium, through phase transfer from an aqueous or a hydrocarbon medium. While these nanoparticles were functionalized with a fluorous thiol, single-walled carbon nanotubes and ZnO nanorods could be solubilized in a fluorous medium by reacting them with a fluorous amine. Fluorous chemistry enables the solubilization of the nanostructures in the most nonpolar liquid medium possible.

Fluorous synthesis is a recent addition to the arsenal of synthetic and separation techniques which integrates solution phase reaction conditions with phase tag separation.[1−4] The fluorous medium constitutes a liquid phase which is orthogonal to organic and aqueous liquid phases. The technique is attractive, since fluorous-tag compounds can be quickly separated from the untagged compounds. Separation and recovery in fluorous chemistry generally utilize fluorous/organic as well as fluorous/aqueous extraction, since the fluorous phase forms a layer distinct from both aqueous and hydrocarbon layers (see Figure 1). One generally attaches the substrate to a fluorous phase label that has sufficient fluorine content and the product extracted into a perfluorohydrocarbon. Such separation is based on fluorine−fluorine interaction between the fluorous reagent and the fluorous separation medium. It is also to be noted that perfluorohydrocarbons provide the most nonpolar medium possible, since they have very low refractive indices (~1.2), even compared to hydrocarbons such as toluene (~1.4). We considered it rewarding to explore how fluorous chemistry can be used effectively for the synthesis and separation/solubilization of nanocrystals, oxide nanorods, and carbon nanotubes. In this communication, we report the success encountered by employing this novel technique. The fluorous technique has enabled us to solubilize various nanostructures in the highly nonpolar perfluorohydrocarbon medium through phase transfer from aqueous as well as hydrocarbon media.

In the literature, metal nanoparticles have been obtained in hydrocarbon media in different ways. Thus, Brust et al.[5] synthesized thiol-derivatized gold nanoparticles in a two-phase liquid−liquid system by transferring the metal salt from the aqueous medium to toluene using tetraoctylammoniumbromide as the phase-transfer reagent. Thiol-derivatized nanoparticles of Au, Pt, and Ag have been prepared as organosols, by the

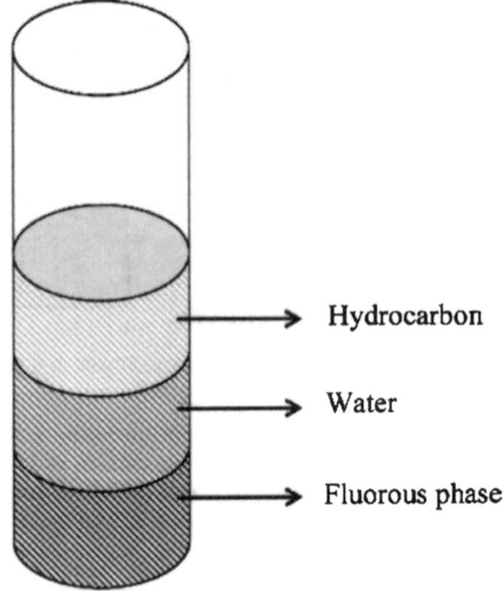

Figure 1. Schematic showing three immiscible layers of a hydrocarbon, water, and a fluorocarbon.

acid-facilitated transfer of the nanoparticles from a hydrosol to a toluene layer containing an alkanethiol.[6] Schulz-Dobrick et al.[7] have dispersed gold nanoparticles prepared by the reduction of tetrachloroaurate by sodium naphthalide in diglyme in various hydrocarbon solvents. Fluorinated alkane thiol-stabilized gold nanoparticles have been prepared in a fluorocarbon solvent,[8] but phase transfer of such nanoparticles from an aqueous or a hydrocarbon medium to a fluorous phase has not been examined.

We shall first present the use of fluorous chemistry with gold nanoparticles. To transfer gold nanoparticles from an aqueous to a perfluorohexane medium, we first prepared a hydrosol containing gold nanoparticles by the reduction of chloroaurate ions (0.55 mL of 25 mM aqueous solution) with partially

* Corresponding author. E-mail: cnrrao@jncasr.ac.in. Fax: (+91) 80 22082760.
† Jawaharlal Nehru Centre for Advanced Scientific Research.
‡ Indian Institute of Science.

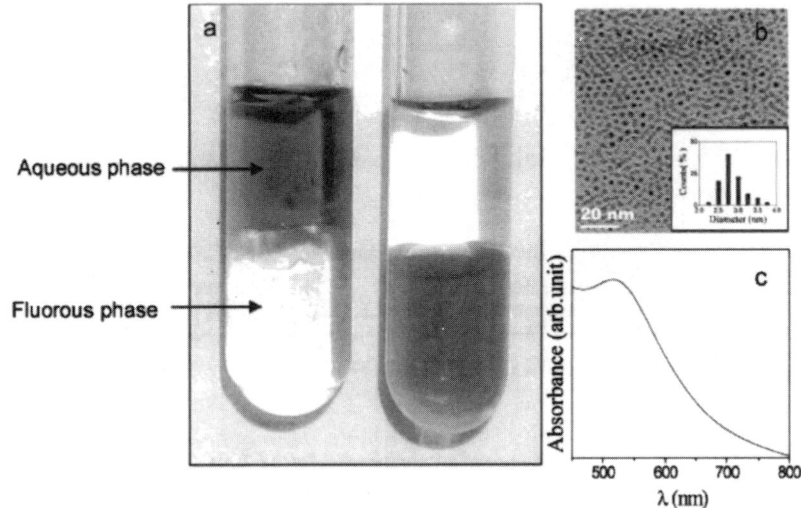

Figure 2. (a) Photograph showing transfer of gold nanoparticles (color) from an aqueous medium to the fluorous medium. (b) TEM image with a size distribution histogram as an inset of ~3 nm gold nanoparticles. (c) UV–vis absorption spectrum of ~3 nm gold nanoparticles in fluorous medium.

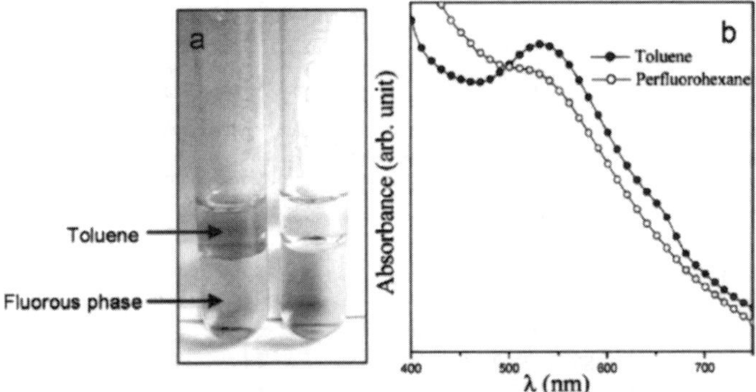

Figure 3. (a) Photograph showing the transfer of gold nanoparticles from toluene to the fluorous medium. (b) UV–vis absorption spectra of the gold nanoparticles in toluene and in perfluorohexane.

hydrolyzed tetrakis(hydroxymethyl)phosphonium choloride (THPC).[9] To the gold sol so obtained, a solution of 10 μL of heptadecafluoro-1-decanthiol (HDFD) in 25 mL of perfluorohexane was added at room temperature, followed by 30 mL of concentrated HCl under vigorous stirring. Within 5 min, gold nanoparticles transferred completely from the aqueous medium to the perfluorohexane medium, as can be seen by the complete transfer of color across the interface, as shown in Figure 2a. In Figure 2b, we show the transmission electron microscope (TEM) images of the gold nanoparticles from perfluorohexane solution, which reveals that the particles have an average diameter close to 3 nm. The UV–vis spectrum of the nanoparticles (Figure 2c) in perfluorohexane shows the characteristic plasmon band at 520 nm.

We have also been able to carry out the phase transfer of gold nanoparticles from toluene to the fluorous phase. For this purpose, we prepared an organosol of gold nanoparticles in toluene by transferring gold nanoparticles from the aqueous medium to the toluene medium by following the literature procedure,[6] but by using the fluorous thiol. We added a solution of 10 μL of HDFD in 20 mL of perfluorohexane to 10 mL of the thiolated gold sol in toluene under vigorous stirring. Within 5 min, the thiolated gold nanoparticles were completely extracted

into the perfluorohexane medium, as can be seen from Figure 3a. The UV–vis absorption spectra of the gold nanoparticles in toluene and perfluorohexane are shown in Figure 3b, with the absorption maximum in the fluorous solvent being blue-shifted compared to that in toluene, due to the lower refractive index of perfluorohexane.

Phase transfer of an organosol of CdSe in a hydrocarbon solvent to the fluorous medium could be carried out as follows. CdSe nanoparticles of 2.5 nm diameter were prepared by reacting cadmium stearate (50 mg, 0.07 mmol) with selenium (6 mg, 0.07 mmol) and tetralin (7.5 μL, 0.05 mmol) in 20 mL of toluene in the presence of HDFD (5.5 μL, 0.02 mmol) for 5 h at 250 °C in a Teflon-lined stainless steel autoclave (40 mL capacity).[10] CdSe nanoparticles of 4.5 nm diameter were prepared by increasing the concentrations of cadmium stearate to 1 g (1.4 mmol), selenium to 120 mg (1.4 mmol), tetralin to 150 μL (1 mmol), and HDFD to 110 μL (0.4 mmol) keeping the amount of toluene and reaction conditions constant. We obtained two different colored CdSe nanoparticles which were precipitated by adding 2-propanol and redispersed in toluene. Upon adding the toluene solution of HDFD-capped CdSe nanoparticles (10 mL) to 10 mL of perfluorohexane under vigorous stirring, we observed a complete transfer of the

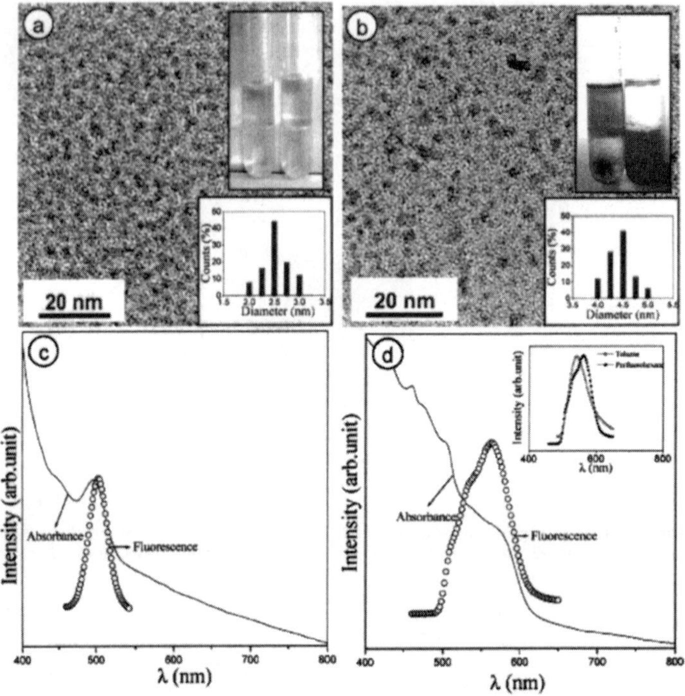

Figure 4. TEM images with size distribution histograms as insets of CdSe nanoparticles of average diameters of (a) 2.5 nm and (b) 4.5 nm solubilized in perfluorohexane. The insets in parts a and b show photographs of CdSe nanoparticles in toluene (upper layer) and the fluorous medium (lower layer). UV–vis absorption spectra and photoluminescence spectra of CdSe nanoparticles of average diameters of (c) 2.5 nm and (d) 4.5 nm in perfluorohexane. The inset in part d shows the PL emission spectra of the 4.5 nm CdSe nanoparticles in toluene and perfluorohexane.

particles from toluene to the perfluorohexane medium. Figure 4a and b shows the TEM images with size distribution histograms (bottom insets) of the 2.5 nm (yellow) and 4.5 nm (red) CdSe nanoparticles solublilized in perfluorohexane. In the insets of Figure 4a and b, we show the photographs of the yellow and red CdSe nanoparticles transferred completely from toluene to the fluorous medium. Figure 4c and d shows the UV–vis absorption and photoluminescence (PL) spectra of the 2.5 and 4.5 nm CdSe nanoparticles solublilized in perfluorohexane. The UV–vis absorption spectrum of the 2.5 nm CdSe nanoparticles clearly shows an absorption band at 495 nm. The band is red-shifted in the case of the 4.5 nm CdSe nanoparticles. The PL band of the 4.5 nm CdSe nanoparticles is also red-shifted compared to that of the 2.5 nm nanoparticles. The inset in Figure 4d compares the PL spectra of the CdSe nanoparticles in toluene and perfluorohexane solutions, showing a shift in the PL maximum due to the different refractive indices of the two liquids.

We could solubilize purified single-walled carbon nanotubes (SWNTs)[11] in the fluorous medium by reacting them with heptadecafluoroundecylamine. For this purpose, we first pre-pared SWNTs by the arc discharge method and purified them by HNO_3 and H_2 treatment.[11] To solubilize the nanotubes, 1.5 mg of SWNTs, 22 μL of the heptadecafluoroundecylamine (SWNTs/amine = 2:1), and 3 mL of perfluorohexanne were sealed in a 7 mL Teflon-lined stainless steel autoclave and heated at 130 °C for 48 h. This process produced a clear solution of SWNTs in perfluorohexane, as can be seen from Figure 5a. The solubilization of SWNTs appears to occur by the interaction of the $-NH_2$ groups of the amine with the $-COOH$ groups present in the surface of the SWNTs, as proposed by Chen et al.[12] The Raman spectrum of the solution gives the characteristic

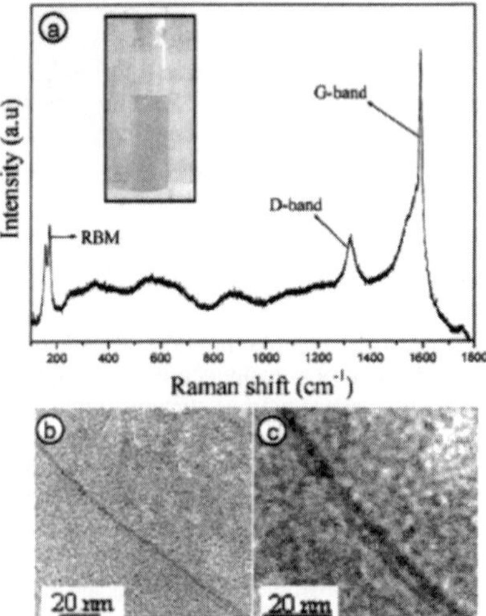

Figure 5. (a) Raman spectrum of SWNTs solubilized in perfluoro-hexane. The inset in part a shows the photograph of the solution of SWNTs in perfluorohexane. TEM images of the solubilized SWNTs are shown in parts b and c.

bands of SWNTs, especially the radial breathing modes (155 and 177 cm^{-1}). The TEM images in Figure 5b and c show the presence of SWNTs in the fluorous medium.

We have extended the method of solubilization of SWNTs to oxide nanorods. We prepared ZnO nanorods by the solvo-thermal decomposition of 250 mg (0.911 mmol) of zinc acetate dihydrate in the presence of 6 mL (0.102 mmol) of ethylene-

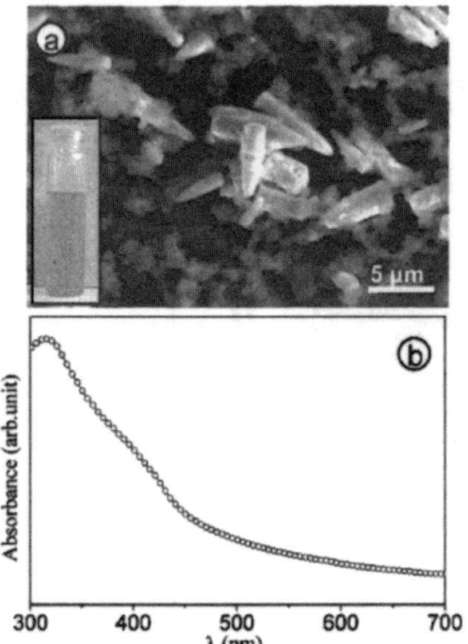

Figure 6. (a) SEM images of ZnO nanorods solubilized in perfluorohexane. The inset in part a shows the photograph of the solution of ZnO nanorods in perfluorohexane. (b) UV–vis spectrum of the solubilized ZnO nanorods.

diamine, 3 mL (0.05 mmol) of ethanol, and 2 mL of Triton X-100 at 335 °C for 5 h in a swagelok autoclave (20 mL capacity). The ZnO nanorods were washed several times with ethanol and precipitated by centrifuging. To solubilize ZnO nanorods in perfluorohexane, 2 mg of ZnO nanorods, 12 μL of heptadecafluoroundecylamine (ZnO nanorods/amine = 1:1), and 3 mL of perfluorohexanne were sealed in a 7 mL Teflon-lined stainless steel autoclave and heated at 130 °C for 48 h. This process produced a clear solution of ZnO nanorods in perfluorohexane. Scanning electron microscope (SEM) images shown in Figure 6a indicate the presence of ZnO nanorods in

perfluorohexane medium. In the inset of this figure, we show the photograph of the solution of ZnO nanorods in perfluorohexane. Figure 6b shows the UV–vis absorption spectrum of ZnO nanorods solubilized in perfluorohexane.

In conclusion, we have successfully demonstrated that, by using a fluorous label and a fluorous solvent, we can affect the phase transfer of gold and CdSe nanoparticles from an aqueous or hydrocarbon medium to the fluorous phase. Single-walled carbon nanotubes and ZnO nanorods can be solubilized in a fluorous solvent after interaction with a fluorous amine. Phase transfer of the nanostructures to a fluorous solvent represents solubilization in a highly nonpolar solvent, accompanied by purification. The high nonpolarity of the fluorocarbon makes it possible to study the optical and other properties of nanostructures in a medium of very low refractive index. Since the fluorocarbon extracts only the species attached to the fluorous label, the process enables one to obtain solely one product in the pure state. We believe that fluorous chemistry may have practical utility in carrying out studies of nanostructures.

References and Notes

(1) Studer, A.; Hadida, S.; Ferritto, S. Y.; Kim, P. Y.; Jeger, P.; Wipf, P.; Curran, D. P. *Science* **1997**, *275*, 823.

(2) Curran, D. P. *Angew. Chem., Int. Ed.* **1998**, *37*, 1175.

(3) Zhang, W. *Chem. Rev.* **2004**, *104*, 2531.

(4) Yoshida, J.; Itami, K. *Chem. Rev.* **2002**, *102*, 3693.

(5) Brust, M.; Walker, M.; Bethell, D.; Schiffrin, D. J.; Whyman, R. *J. Chem. Soc., Chem. Commun.* **1994**, 801.

(6) Sarathy, K. V.; Kulkarni, G. U.; Rao, C. N. R. *Chem. Commun.* **1997**, 537.

(7) Schulz-Dobrick, M.; Sarathy, K. V.; Jansen, M. *J. Am. Chem. Soc.* **2005**, *127*, 12816.

(8) Yonezawa, T.; Onoue, S.; Kimizuka, N. *Langmuir* **2001**, *17*, 2291.

(9) (a) Duff, D. G.; Baiker, A.; Edwards, P. P. *J. Chem. Soc., Chem. Commun.* **1993**, 96. (b) Rao, C. N. R.; Vijayakrishnan, V.; Aiyer, H. N.; Kulkarni, G. U.; Subbanna, G. N. *J. Phys. Chem.* **1993**, *97*, 11157.

(10) Gautam, U. K.; Rajamathi, M.; Meldrum, F.; Morgan, P.; Seshadri, R. *Chem. Commun.* **2001**, 629.

(11) Vivekchand, S. R. C.; Govindaraj, A.; Seikh, M. M.; Rao, C. N. R. *J. Phys. Chem. B* **2004**, *108*, 6935.

(12) Chen, J.; Rao, A. M.; Lyuksyutov, S.; Itkis, M. E.; Hamon, M.; Hu, H.; Cohn, R. W.; Eklund, P. C.; Colbert, D. T.; Smalley, R. C.; Haddon, R. C. *J. Phys. Chem. B* **2001**, *105*, 2525.

PHYSICAL REVIEW B **74**, 161306(R) (2006)

Ferromagnetism as a universal feature of nanoparticles of the otherwise nonmagnetic oxides

A. Sundaresan,* R. Bhargavi, N. Rangarajan, U. Siddesh, and C. N. R. Rao

*Chemistry and Physics of Materials Unit and Department of Science and Technology Unit on Nanoscience,
Jawaharlal Nehru Centre for Advanced Scientific Research, Jakkur P. O., Bangalore 560 064 India*

(Received 18 August 2006; published 20 October 2006)

Room-temperature ferromagnetism has been observed in nanoparticles (7–30 nm diam) of nonmagnetic oxides such as CeO_2, Al_2O_3, ZnO, In_2O_3, and SnO_2. The saturated magnetic moments in CeO_2 and Al_2O_3 nanoparticles are comparable to those observed in transition-metal-doped wideband semiconducting oxides. The other oxide nanoparticles show somewhat lower values of magnetization but with a clear hysteretic behavior. Conversely, the bulk samples obtained by sintering the nanoparticles at high temperatures in air or oxygen became diamagnetic. As there were no magnetic impurities present, we assume that the origin of ferromagnetism may be the exchange interactions between localized electron spin moments resulting from oxygen vacancies at the surfaces of nanoparticles. We suggest that ferromagnetism may be a universal characteristic of nanoparticles of metal oxides.

DOI: 10.1103/PhysRevB.74.161306

PACS number(s): 75.50.Pp, 75.50.Dd, 75.75.+a, 81.07.Wx

Integration of semiconductor with ferromagnetic functionality of electrons has been the focus of recent research in the area of spintronics because of the difficulties associated with the injection of spins into nonmagnetic semiconductors in conventional spintronic devices. Ferromagnetism in semiconductors and insulators is rare, the well-known ferromagnetic semiconductors being the chalcogenides EuX (X=O, S, and Se) ($T_C<70$ K) and $CdCr_2X_4$ (X=S and Se) ($T_C<142$ K) with the rocksalt and spinel structure, respectively.[1,2] Following the theoretical prediction of Dietl *et al.* that Mn-doped ZnO and GaN could exhibit ferromagnetism above room temperature,[3] several studies have focused on films and bulk samples of metal oxides such as TiO_2, ZnO, In_2O_3, SnO_2, and CeO_2 doped with Mn, Co, and other transition metal ions.[4–8]

While the existence of ferromagnetism in transition-metal-doped semiconducting oxides remains controversial,[9] thin films of the band insulator HfO_2 have been reported to exhibit ferromagnetism at room temperature in the absence of any doping.[10] This is puzzling, since pure HfO_2 does not have any magnetic moment and the bulk sample is diamagnetic. Similar ferromagnetism has been reported in other nonmagnetic materials such as CaB_6, CaO, and SiC where the origin of ferromagnetism is believed to be due to intrinsic defects.[11–13] It has been suggested that ferromagnetism in thin films of HfO_2 may be related to anion vacancies.[14] It has been reported very recently that thin films of undoped TiO_2 and In_2O_3 also show ferromagnetism at room temperature,[15] the corresponding bulk forms of these materials being diamagnetic. Thin films of these oxides might have defects or oxygen vacancies that could be responsible for the observed ferromagnetism. *Ab initio* electronic structure calculations using density functional theory in HfO_2 have shown that isolated halfnium vacancies lead to ferromagnetism.[16] Meanwhile, there is a conflicting report attributing the ferromagnetism in HfO_2 to possible iron contamination while using stainless-steel tweezers in handling thin films.[17]

In this Rapid Communication, we report the discovery of ferromagnetism at room temperature in nanoparticles of nonmagnetic oxides such as CeO_2, Al_2O_3, ZnO, In_2O_3, and SnO_2. Our studies show that ferromagnetism is associated only with the nanoparticles while the corresponding bulk samples are diamagnetic. The origin of ferromagnetism in these materials is assumed to be the exchange interactions between localized electron spin moments resulting from the oxygen vacancies at the surfaces of the nanoparticles.

Nanoparticles of CeO_2, Al_2O_3, ZnO, In_2O_3, and SnO_2 were prepared by the methods described in the literature.[18–20] The preparation methods do not involve any magnetic element and therefore we rule out the possibility of contamination of magnetic impurities. For example, the nanoparticles of CeO_2 were prepared by the addition of hexamethylenetetramine to a solution of cerium nitrate $[Ce(NO_3)_3]$ under constant stirring.[18] The nanoparticles of all these oxides were annealed at temperatures between 400 and 500 °C in flowing oxygen to remove organic matter. In order to make bulk samples, these nanoparticles were sintered at high temperatures (1000–1400 °C). Powder x-ray diffraction (XRD) was used to identify the phase and its purity and to determine the grain size. The particle size and morphology were studied by field emission scanning electron microscopy (FESEM) and transmission electron microscopy (TEM). Magnetization measurements were carried out with a vibrating sample magnetometer in a physical property measuring system (PPMS, Quantum Design, San Diego, CA, USA).

XRD patterns of all the samples showed that they were monophasic with broad peaks characteristic of nanoparticles. The lattice parameters and the full width at half maximum of all the reflections were obtained from Rietveld refinement in the pattern matching mode using the program FULLPROF.[21] The lattice parameters of the oxide nanoparticles were generally higher than those of the corresponding bulk forms. For example, the lattice parameter of the CeO_2 nanoparticles (7 nm) is 5.424(3) Å whereas that of the corresponding bulk sample is 5.413(1) Å. This is in agreement with an earlier report that the lattice expands in oxide nanoparticles.[22] The increase of lattice with decreasing particle size might results from the oxygen vacancy associated with nanoparticles. Similar results were obtained for Al_2O_3, ZnO, In_2O_3, and SnO_2 samples. The average particle sizes of CeO_2, Al_2O_3,

SUNDARESAN *et al.*

PHYSICAL REVIEW B **74**, 161306(R) (2006)

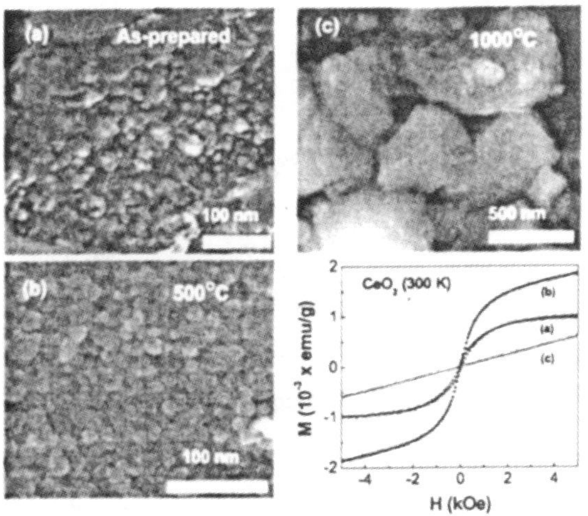

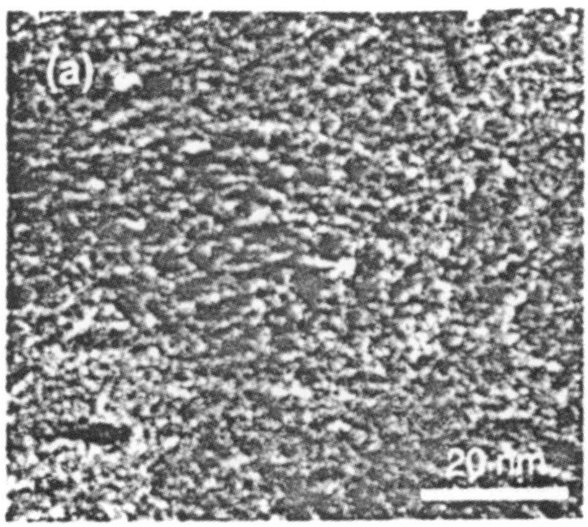

FIG. 1. (Color online) FESEM images of CeO_2 nanoparticles [(a) 7 nm, (b) 15 nm, (c) 500 nm] and their magnetization curves at 300 K. Note the absence of ferromagnetism in the 500 nm nanoparticles, in contrast to the 7 or 15 nm nanoparticles.

ZnO, In_2O_3, and SnO_2 estimated by the Scherrer formula using all diffraction lines were 15, 4, 30, 12, and 20 nm, respectively.

In Fig. 1 we show the room-temperature magnetization-field curves of many CeO_2 samples: (a) as prepared (b) heated at 500 °C for 1 h and (c) heated at 1000 °C for 1 h. We have shown the FESEM images of these three samples in the figure. It can be seen that the as-prepared particles (7 nm) are covered by the organic coating used in the preparation of nanoparticles whereas the 500 °C heated particles (15 nm) are free from such coating. It is obvious from the $M(H)$ curves that the as-prepared and 500 °C heated nanoparticles show ferromagnetic behavior with coercivity ~100 Oe. This is surprising, since bulk CeO_2 is a band insulator with Ce^{4+} in the $4f^0$ electronic configuration. On the other hand, the ferromagnetism is suppressed in the 1000 °C sample with ~500 nm size particles and this sample exhibits a linear $M(H)$ behavior with low magnetic moment, a behavior close to diamagnetism as normally expected of CeO_2.

A TEM image of Al_2O_3 nanoparticles obtained by heating $Al(OH)_3$ at 500 °C is shown in Fig. 2(a). The $M(H)$ curves of these nanoparticles (0.0291 g) recorded at 300 and 390 K are shown in Fig. 2(b). These nanoparticles show ferromagnetism even at 390 K with clear hysteretic behavior. The saturation magnetic moment at 300 K is ~3.5 $\times 10^{-3}$ emu/g, comparable to that reported for Mn-doped ZnO.[5] In order to verify that the room-temperature ferromagnetism is associated only with nanoparticles, the nanoparticles of the sample were pressed into a bar and sintered at 1400 °C for 1 h in air to obtain bulk samples with micrometer-sized particles. The magnetization of the bulk sample thus obtained is shown in Fig. 2(b). It is clear from this figure that the bulk sample is diamagnetic. Similarly, room-temperature ferromagnetism is observed in ZnO nanoparticles heated at 400 °C and diamagnetic behavior in the sample sintered at 1200 °C (Fig. 3).

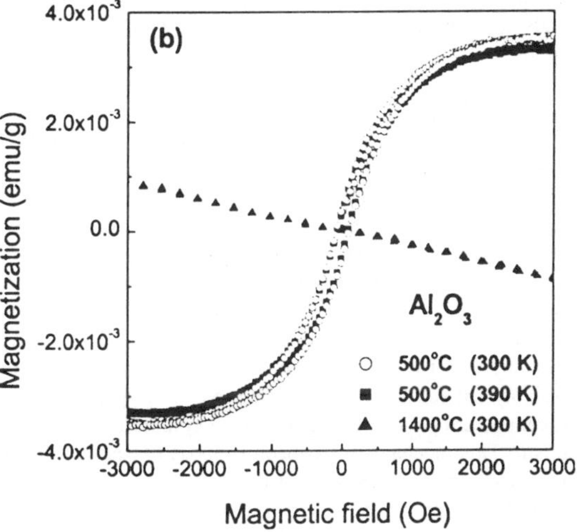

FIG. 2. (Color online) (a) TEM image of Al_2O_3 nanoparticles heated at 500 °C and (b) their magnetization curves showing ferromagnetism even at 390 K. Note that the nanoparticles sintered at 1400 °C exhibit diamagnetic behavior at 300 K.

Unlike CeO_2, Al_2O_3, and ZnO, which are insulators, In_2O_3 and SnO_2 are transparent conductors with a wide band gap (~3.6 eV). Magnetization data of In_2O_3 and SnO_2 nanoparticles are shown in Fig. 4. The magnetization behavior of SnO_2 is slightly different from that of the other oxides, but similar to that observed in thin films of Co-doped SnO_2.[7] It can be seen from this figure that there is a small hysteresis at low fields and that the magnetic moment increases linearly at higher field. The linear behavior may be due to magnetic moments associated with conduction electrons. This is consistent with the observation that the nanoparticles after sintering at 1200 °C show paramagnetic behavior. Though there may be slight differences in the magnetization behavior, nanoparticles of all the oxides studied exhibit room-temperature ferromagnetism. It should be noticed that the nanoparticles of paramagnetic metallic ReO_3 with low mag-

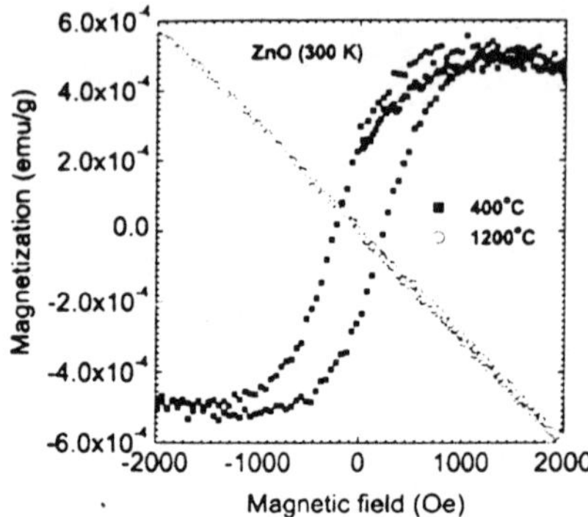

FIG. 3. (Color online) M versus H curves measured at 300 K for nanoparticles of ZnO heated at 400 °C and sintered at 1200 °C.

netic susceptibility are reported to show hysteresis at 5 K.[24] As the magnetic susceptibility of ReO_3 nanoparticles is relatively low, it may show ferromagnetism even at room temperature.

The origin of ferromagnetism in the nanoparticles of these nonmagnetic oxides seems to be similar to that in thin films of HfO_2, TiO_2, and In_2O_3 where the oxygen deficiency results from thin film growth conditions.[10,15] In contrast to thin films, where the contamination of films by handling can vitiate the results, ferromagentism in the oxide nanoparticles is robust and universal. We suggest that the unpaired electron spins responsible for ferromagnetism in the nanoparticles have their origin in the oxygen vacancies, especially on the surfaces of the oxide nanoparticles. The nature of exchange interactions between them is not clear at present. However, one may expect that electrons trapped in oxygen vacancies (F center) are polarized to give room-temperature ferromagnetism. This mechanism has been proposed to explain ferromagnetism in some transparent oxides.[23]

In conclusion, we have shown that nanoparticles of metal oxides such as CeO_2, Al_2O_3, ZnO, In_2O_3, and SnO_2, exhibit room-temperature ferromagnetism whereas the corresponding bulk oxides exhibit diamagnetism. We assume that the origin of ferromagnetism may be the exchange interactions between unpaired electron spins arising from oxygen vacancies at the surfaces of the nanoparticles. We suggest that all metal oxides in nanoparticulate form would exhibit room-temperature ferromagnetism. The ferromagnetism assumed to be associated with oxygen vacancies gives a possible clue

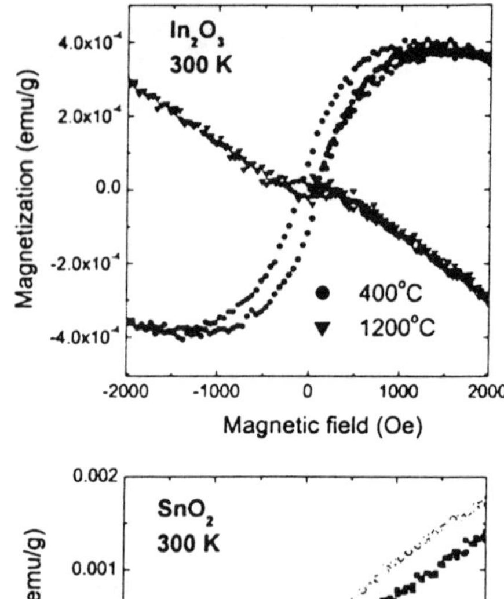

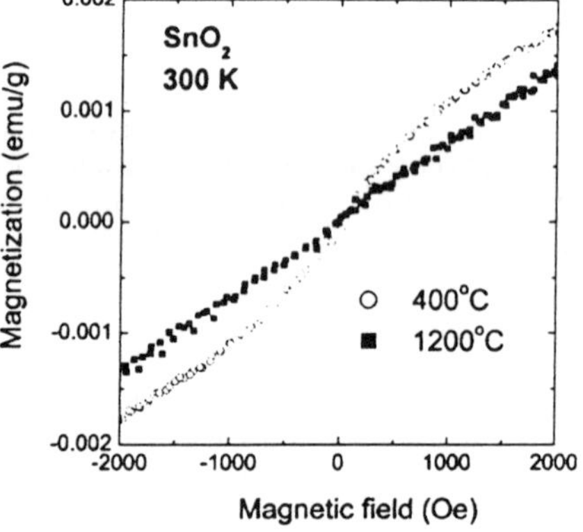

FIG. 4. (Color online) M versus H curves measured at 300 K for nanoparticles of In_2O_3 and SnO_2 heated at 400 °C and sintered at 1200 °C.

to understand some of the contradicting findings in the dilute magnetic semiconducting oxides.

The authors thank R. V. K. Mangalam, Chandra Sekhar Rout, and C. Madhu for their help in the sample preparation and magnetic measurements. R.B., N.R., and U.S. would like to thank JNCASR for providing the opportunity to do research work under the programs Project Oriented Chemical Education (POCE) and Summer Research, respectively. This work was supported by the Department of Science and Technology, India under the nanoscience initiative program.

*Electronic address: sundaresan@jncasr.ac.in

[1] B. T. Matthias, R. M. Bozorth, and J. H. Val Vleck, Phys. Rev. Lett. **7**, 160 (1961).

[2] P. K. Baltzer, H. W. Lehmann, and M. Robbins, Phys. Rev. Lett. **15**, 493 (1965).

[3] T. Dietl, H. Ohno, F. Matsukura, J. Cibert, and D. Ferrand, Science **287**, 1019 (2000).

[4] Y. Matsumoto, M. Murakami, T. Shono, T. Hasegawa, T. Fuku-

mura, M. Kawasaki, P. Ahmet, T. Chikyow, S. Koshihara, and H. Koinuma, Science **291**, 854 (2001).

[5] P. Sharma, A. Gupta, K. V. Rao, F. J. Owens, R. Sharma, R. Ahuja, J. M. O. Guillen, B. Johansson, and G. A. Gehring, Nat. Mater. **2**, 673 (2003).

[6] N. H. Hong, J. Sakai, N. T. Huong, and V. Brizé, Appl. Phys. Lett. **87**, 102505 (2005).

[7] S. B. Ogale, R. J. Choudhary, J. P. Buban, S. E. Lofland, S. R.

SUNDARESAN *et al.*

PHYSICAL REVIEW B **74**, 161306(R) (2006)

Shinde, S. N. Kale, V. N. Kulkarni, J. Higgins, C. Lanci, J. R. Simpson, N. D. Browning, S. Das Sarma, H. D. Drew, R. L. Greene, and T. Venkatesan, Phys. Rev. Lett. **91**, 077205 (2003).

[8] A. Tiwari, V. M. Bhosle, S. Ramachandran, N. Sudhakar, J. Narayan, S. Budak, and A. Gupta, Appl. Phys. Lett. **88**, 142511 (2006).

[9] Ram Seshadri, Curr. Opin. Solid State Mater. Sci. **9**, 1 (2005).

[10] M. Venkatesan, C. B. Fitzgerald, and J. M. D. Coey, Nature (London) **430**, 630 (2004).

[11] R. Monnier and B. Delley, Phys. Rev. Lett. **87**, 157204 (2001).

[12] I. S. Elfimov, S. Yunoki, and G. A. Sawatzky, Phys. Rev. Lett. **89**, 216403 (2002).

[13] A. Zywietz, J. Furthmuller, and F. Bechstedt, Phys. Rev. B **62**, 6854 (2000).

[14] J. M. D. Coey, M. Venkatesan, P. Stamenov, C. B. Fitzgerald, and L. S. Dorneles, Phys. Rev. B **72**, 024450 (2005).

[15] N. H. Hong, J. Sakai, N. Poirot, and V. Brizé, Phys. Rev. B **73**, 132404 (2006).

[16] C. Das Pemmaraju and S. Sanvito, Phys. Rev. Lett. **94**, 217205 (2005).

[17] D. W. Abraham, M. M. Frank, and S. Guha, Appl. Phys. Lett. **87**, 252502 (2005).

[18] F. Li, X. Yu, H. Pan, M. Wang, and X. Xin, Solid State Sci. **2**, 767 (2000).

[19] S. Xia-lan, Q. Peng, Y. Hai-pin, H. Xi, and Q. Guan-zhou, J. Cent. South Univ. Technol. **12**, 536 (2005).

[20] C. S. Rout, A. R. Raju, A. Govindaraj, and C. N. R. Rao, Solid State Commun. **138**, 136 (2006).

[21] J. Roadriguez-Carvajal, Physica B **192**, 55 (1993).

[22] S. Tsunekawa, K. Ishikawa, Z.-Q. Li, Y. Kawazoe, and A. Kasuya, Phys. Rev. Lett. **85**, 3440 (2000).

[23] I. Vinokurov, Z. Zonn, and V. Ioffe, Sov. Phys. Solid State **9**, 2659 (1968).

[24] K. Biswas and C. N. R. Rao, J. Phys. Chem. B **110**, 842 (2006).

PAPER

www.rsc.org/materials | Journal of Materials Chemistry

MnO and NiO nanoparticles: synthesis and magnetic properties

Moumita Ghosh,[ab] **Kanishka Biswas,**[ab] **A. Sundaresan**[a] **and C. N. R. Rao**[*ab]

Received 22nd August 2005, Accepted 22nd September 2005
First published as an Advance Article on the web 13th October 2005
DOI: 10.1039/b511920k

Nanoparticles of MnO with average diameters in the 6–14 nm range have been prepared by the decomposition of manganese cupferronate in the presence of TOPO, under solvothermal conditions. Nanoparticles of NiO with average diameters in the 3–24 nm range have been prepared by the decomposition of nickel cupferronate or acetate under solvothermal conditions. The nanoparticles have been characterized by X-ray diffraction and transmission electron microscopy. Both MnO and NiO nanoparticles exhibit supermagnetism, accompanied by magnetic hysteresis below the blocking temperature (T_B). The T_B increases with the increase in particle size in the case of NiO, and exhibits the reverse trend in the case of MnO.

Introduction

Nanoparticles of transition metal oxides have been investigated by several workers in the last few years. Besides their structural aspects, magnetic properties of the oxide nanoparticles are of particular interest. Thus, it would be of value to know if the nanoparticles of antiferromagnetic oxides generally show evidence for ferromagnetic interaction at low temperatures, a behaviour that has been reported by a few workers.[1,2] A recent study of CoO nanoparticles has shown that small particles of less than 16 nm diameter exhibit magnetic hysteresis below a blocking temperature of $\sim$10 K.[3] We were interested to investigate nanoparticles of the binary metal oxides MnO and NiO, which are antiferromagnetic with Neél temperatures of 122 K and 523 K respectively.[4] There have been a few reports in the literature on MnO nanoparticles. Lee *et al.*[5] prepared MnO nanoparticles by the decomposition of $Mn_2(CO)_8$, but the product always contained Mn_2O_3 as an impurity. These workers, however, found ferromagnetic behaviour in the particles with an average diameter of around 5 nm. Yin and O'Brien[6] synthesized MnO nanocrystals capped with organic ligands by the decomposition of manganese acetate in a mixture of trioctylamine and oleic acid. These workers did not study the magnetic properties of the nanoparticles. Colloidal MnO nanoparticles, prepared by Seo *et al.*[7] by the decomposition of manganese acetylacetonate in oleylamine, showed divergence in the magnetization measured under zero-field-cooled (ZFC) and field-cooled (FC) conditions. Park *et al.*[8,9] prepared MnO nanospheres with diameters in the 5–40 nm range and nanorods of 7–10 nm diameter by the thermal decomposition of Mn-surfactant complexes in trioctylphosphine. These workers also observed divergence in the magnetization recorded under ZFC and FC conditions. We have prepared MnO nanoparticles by the decomposition of manganese cupferronate, $Mn(cup)_2$, in

toluene in the presence of trioctylphosphine oxide (TOPO) under solvothermal conditions and obtained nanoparticles with diameters in the 6–14 nm range. The particles so-prepared had an organic coating and enabled a study of magnetic properties without any surface oxidation.

Small particles of NiO have been known to be superparamagnetic for some time.[10] Magnetic anomalies, such as low temperature hysteresis below a blocking temperature and divergence between the FC and ZFC magnetization data, on NiO nanoparticles have been reported.[2,11,12] The anomalies in the magnetic properties are attributed to uncompensated surface spins causing a change in the magnetic order in the nanoparticles.[1,12] NiO nanoparticles have been prepared by the decomposition of the hydroxide.[10,2,13] and by the oxidation of Ni nanoparticles[14] or by the decomposition of nickel oxalate.[15] We have prepared the nanoparticles of NiO of different diameters in the 3–24 nm range by the decomposition of nickel cupferronate, $Ni(cup)_2$, or nickel acetate in organic solvents under solvothermal conditions. The as-prepared particles have a protective organic coating. We have examined the magnetic properties of MnO and NiO nanoparticles, after characterizing them by X-ray diffraction and transmission electron microscopy.

Experimental

Manganese cupferronate, $Mn(C_6H_5N_2O_2)_2$, or $Mn(cup)_2$ was synthesized as follows.[16] 1 g (5.05 mmol) of $MnCl_2 \cdot 4H_2O$ was dissolved in 100 ml of milli-Q water and 1.56 g (10.05 mmol) of cupferron was dissolved in 50 ml of milli-Q water by sonication. The two solutions were cooled at 0 °C and the cupferron solution added dropwise to the $MnCl_2$ solution under vigorous stirring. After a few minutes, a pale yellow coloured product was obtained. It was filtered, washed thoroughly with milli-Q water and dried at room temperature overnight.

MnO nanoparticles were prepared by the decomposition of $Mn(cup)_2$ at 325 °C in toluene in the presence of trioctylphosphine oxide (TOPO) as the capping agent. In a typical reaction, 0.2 g (0.60 mmol) of $Mn(cup)_2$ and 0.95 g (2.45 mmol) of TOPO were taken in 10 ml of toluene and sealed in a

[a]*Chemistry and Physics of Materials Unit and CSIR Centre of Excellence in Chemistry, Jawaharlal Nehru Centre for Advanced Scientific Research, Jakkur P. O., Bangalore-560064, India. E-mail: cnrrao@jncasr.ac.in; Fax: (+91) 80 22082760*
[b]*Solid State and Structural Chemistry Unit, Indian Institute of Science, Bangalore-560012, India*

stainless steel Swagelok autoclave of 20 ml capacity. The autoclave was kept inside a preheated tube-furnace at 325 °C for 2 h, and was then allowed to cool to room temperature. A brownish black solution was obtained as the product. To this solution, excess ethanol was added and kept overnight. Black coloured TOPO-capped MnO nanoparticles (14 nm diameter) obtained as a solid residue were washed several times with ethanol and acetone.

To vary the particle size of MnO, the concentration of Mn(cup)$_2$, and/or the concentration of TOPO as well as the reaction time were varied, keeping the amount of toluene and the filling fraction of the autoclave constant. The nanoparticles of different diameters were obtained as follows: 6 nm, 0.005 g (0.015 mmol) of Mn(cup)$_2$, 0.045 g (0.012 mmol) of TOPO, 325 °C, 2 h 30 min; 8.5 nm, 0.025 g (0.075 mmol) of Mn(cup)$_2$, 0.18 g (0.046 mmol) of TOPO, 325 °C, 2 h; 10 nm, 0.01 g (0.03 mmol) of Mn(cup)$_2$, 0.09 g (0.024 mmol) of TOPO, 340 °C, 2 h; 14 nm, 0.2 g (0.60 mmol) of Mn(cup)$_2$, 0.95 g (2.45 mmol) of TOPO, 325 °C, 2 h. The nanoparticles could be readily redispersed in toluene by sonication.

Nickel cupferronate, Ni(C$_6$H$_5$N$_2$O$_2$)$_2$, or Ni(cup)$_2$, was prepared as follows. 2 g of Ni(OAc)$_2$·4H$_2$O was dissolved in 150 ml of milli-Q water (0.05 M), and 2.5 g of cupferron dissolved in 100 ml of milli-Q water (0.16 M) by sonication. The two solutions were cooled to 0 °C and the cupferron solution added dropwise to the Ni(OAc)$_2$ solution under vigorous stirring. After a few minutes, the solution became turbid, indicating the formation of complex. It took another 30 min for completion of the reaction. Before filtration, the product was kept at 0 °C for 1 h. The product was filtered and washed with milli-Q water and dried at room temperature.

NiO nanoparticles were synthesized in toluene under solvothermal conditions as follows. In a typical reaction, 0.05 g (0.15 mmol) of Ni(cup)$_2$ were taken in 48 ml of toluene and subjected to sonication for 10 min. The resulting green coloured solution was sealed in a Teflon-lined stainless steel autoclave of 80 ml capacity (allowing 70% filling fraction) and

the autoclave was placed inside a air-oven preheated at 240 °C for 8 h. The autoclave was allowed to cool to room temperature. A brownish coloured solid dispersed in toluene was obtained as the product. Addition of methanol to this toluene dispersion allowed the solid to settle down. The product was washed with methanol by sonication followed by centrifugation. The sample was dried in an air-oven at 50 °C for 1 h. The average diameter of the particle obtained under these preparative conditions was 3 nm. In order to prepare 7 nm NiO nanoparticles, we used 0.4 g (1.2 mmol) of Ni(cup)$_2$ in 48 ml of toluene and carried out the reaction for 24 h, keeping the temperature and filling fraction of the autoclave as constant. In addition to the NiO nanoparticles described above, trioctylphosphine oxide (TOPO)-capped NiO nanoparticles of 3 nm average diameter were prepared by taking 0.5 g (1.5 mmol) of Ni(cup)$_2$ in 1.5 mmolar toluene solution of TOPO.

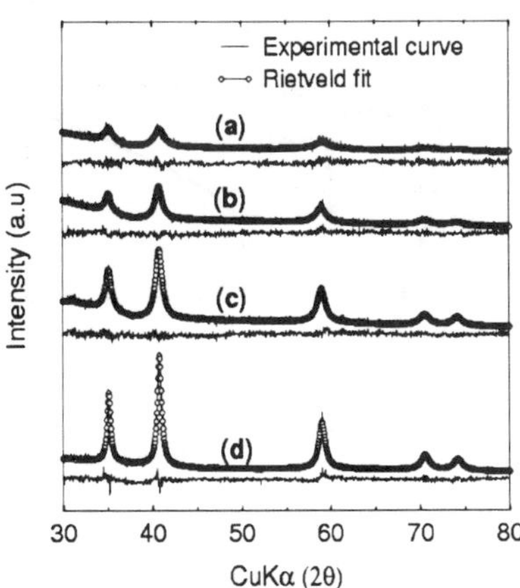

Fig. 1 XRD patterns of the MnO nanoparticles with average diameters of (a) 6, (b) 8.5, (c) 10 and (d) 14 nm along with the Rietveld fits. Difference patterns are shown below the observed patterns.

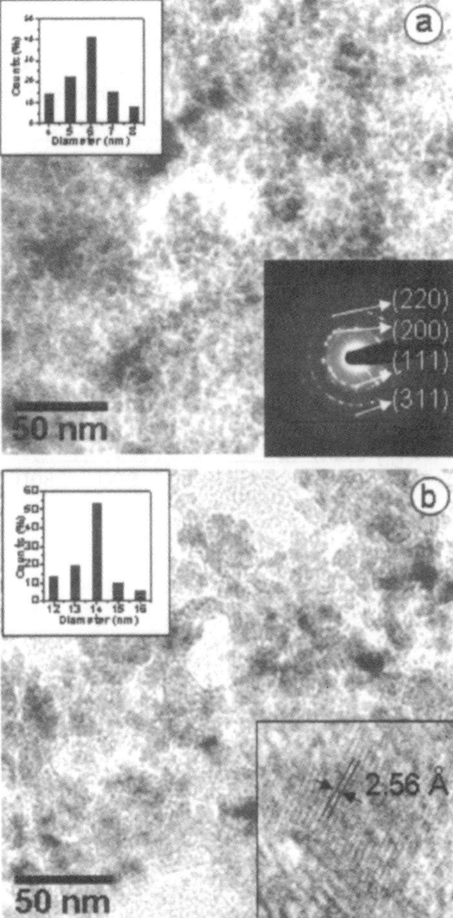

Fig. 2 TEM images of (a) 6 nm and (b) 14 nm particles of MnO. Insets in (a) show the SAED pattern and the size distribution histogram. Insets in (b) show the single particle HREM image and the size distribution.

We have also prepared NiO nanoparticles by the solvothermal decomposition of Ni(OAc)$_2$ using toluene as the solvent in a stainless steel Swagelok autoclave of 20 ml capacity. To prepare 12 nm NiO particles, 0.1 g (0.4 mmol) of Ni(OAc)$_2$·4H$_2$O was taken in 10 ml of toluene and the reaction mixture sealed in a Swagelok autoclave of 20 ml capacity. The autoclave was kept inside a preheated tube furnace at 250 °C. The reaction was carried out for 3 h. A moss green coloured, toluene-insoluble solid residue was obtained as the product. NiO particles of 24 nm diameter were prepared by taking 0.5 g (2.01 mmol) of Ni(OAc)$_2$·4H$_2$O in 10 ml toluene, and carrying out the reaction at 335 °C for 1 h.

The nanoparticles were characterized by powder X-ray diffraction (XRD) using a Phillips X'Pert diffractometer employing the Bragg–Brentano configuration using CuKα radiation with the scan rate of 0.16° 2θ min^{-1}. The data were then rebinned into 0.05° steps. For transmission electron microscopy (TEM), toluene dispersions of the samples were dropped onto the holey carbon-coated Cu grids, and the grids were allowed to dry in the air. The grids were examined using a JEOL (JEM3010) microscope operating with an accelerating voltage of 300 kV. Thermogravimetric analysis (TGA) was carried out using a Mettler Toledo Thermal Analyser. Powder samples of the as-prepared nanoparticles were subjected to magnetic characterization using the VSM in PPMS (Physical Property Measurement System).

Results and discussion

MnO

In Fig. 1, we show the X-ray diffraction patterns of MnO nanoparticles of four different sizes. Fig. 1 also gives the Rietveld profile fits along with the difference patterns obtained by using the Rietveld XND code.[17] The patterns could be indexed with the *Fm3m* space group (JCPDS no 07-0230). By making the use of the line widths from the Rietveld fits, the average particle sizes were estimated by using the Scherrer formula. The particle sizes are indicated in Fig. 1. The cubic lattice parameter of the 8.5 and 14 nm particles are 4.429 and 4.428 Å respectively. The TEM images of the MnO nanoparticles with average diameters of 6 and 14 nm are shown in Fig. 2(a) and (b), respectively. An indexed selected area electron diffraction (SAED) pattern of the 6 nm sample is shown in the inset of Fig. 2(a). The particle size distributions are shown as insets in Fig. 2. Most of the particles in Fig. 2(a) and (b) have diameters of 6 and 14 nm respectively, which compare well with the values calculated from the XRD

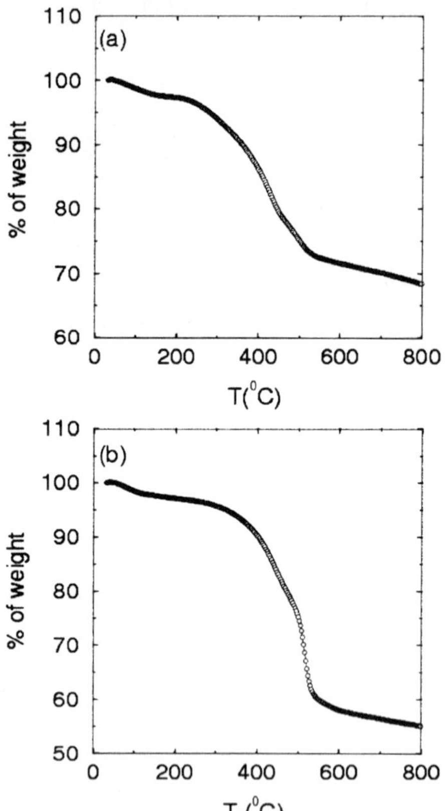

Fig. 3 TGA curves of (a) 14 nm TOPO-capped MnO nanoparticles and (b) 3 nm uncapped NiO nanoparticles. The experiments were carried out in a nitrogen atmosphere at a heating rate of 10 °C min^{-1}.

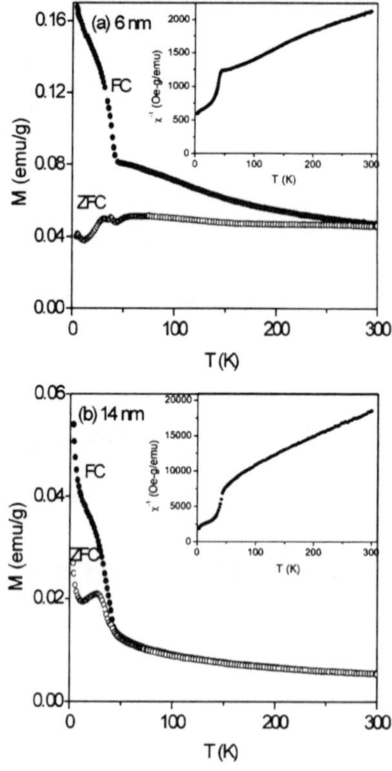

Fig. 4 Magnetization *vs.* temperature curves of (a) 6 nm and (b) 14 nm MnO particles under FC and ZFC conditions using a magnetic field of 100 Oe. The insets show inverse susceptibility *vs.* temperature curves.

patterns. We show an HREM image of a 14 nm MnO nanoparticle as an inset in Fig. 2(b). The lattice spacing of 2.56 Å in the image corresponds to the interplanar distance between (111) planes. The images of the particles are not as good as one would desire because of the organic coating on the particles. The presence of such an organic coating is evidenced from the TGA curve shown in Fig. 3(a), for the 14 nm MnO nanoparticles. The organic coating is removed around 400 °C. The presence of the organic coating helps to avoid oxidation of the MnO nanoparticles.

In Fig. 4(a) and (b), we show the magnetization data of two samples with average diameters of 6 and 14 nm respectively. The FC and ZFC data show divergence at low temperatures as indeed found in the case of other oxide nanoparticles. The magnetization in the FC data shows a marked increase at low temperatures, where a shoulder like feature is also evident. The ZFC data clearly show maxima corresponding to blocking temperatures (T_B) of the nanoparticles. The T_B of the 6 nm particle is higher than that of the 14 nm particles indicating an inverse relationship. A similar dependence has been noticed earlier.[7] The insets in Fig. 4(a) and (b) represent the inverse susceptibility (χ^{-1}) *vs.* temperature data. The magnetic moments estimated from the χ^{-1} data in the high temperature

data are around 4 μ_B, a value close to that expected for MnO. The Curie–Weiss temperatures (Θ_p) obtained from the extrapolation of the high-temperature inverse susceptibility data are −202 and −365 K respectively for the 6 nm and 14 nm samples. The negative sign of Θ_p indicates predominant anti-ferromagnetic interaction. The value of Θ_p decreases with the increase in particle size, probably because the ferromagnetic interaction decreases with the increase in particle size. Accordingly, the value of the magnetization of the 6 nm particles is always higher than that of the 14 nm particles. In Fig. 5(a), we show the variation of magnetization with field at 5 K and 50 K for the 6 nm particles to reveal the occurrence of hysteresis at low temperatures, the H_c value being 1750 Oe at 5 K. At high temperatures ($T > T_B$), the hysteresis disappears.

NiO

In Fig. 6, we show the X-ray diffraction patterns of four different sizes of NiO nanoparticles. The Rietveld fitting of the profiles was carried out using the Rietveld XND code.[17] The diffraction profiles could be indexed on the *Fm3m* space group (JCPDS no 47-1049). The particles sizes were estimated using the Scherrer formula by taking the average of three main linewidths obtained from the Rietveld fitted profiles. The lattice parameters obtained from the Rietveld fits are 4.178 Å, 4.182 Å, 4.192 Å and 4.211 Å respectively for the 24, 12, 7 and 3 nm nanoparticles, reflecting a slight increase in the lattice parameter with the decrease in particle size. The presence of a passive organic coating on the surfaces of the particles is evidenced from the TGA curve shown in Fig. 3(b). The coating is removed around 400 °C, accompanied by the partial reduction of NiO to Ni metal. In Fig. 7(a) and (b), we show the TEM images of the NiO nanoparticles with average diameters of 3 and 7 nm respectively, along with the size distributions.

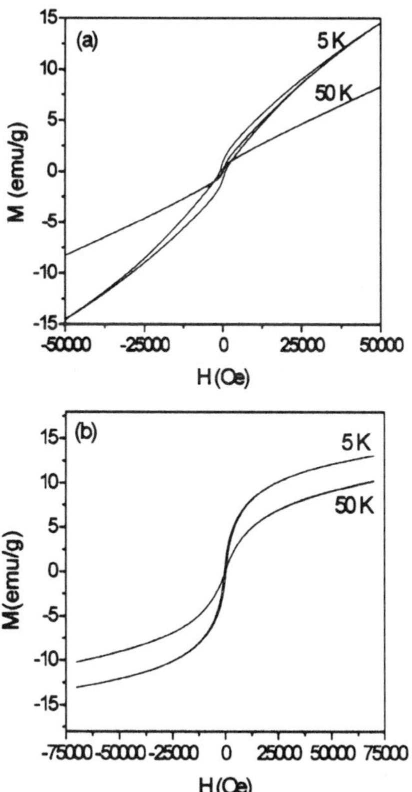

Fig. 5 Field dependence of magnetization of the (a) 6 nm MnO nanoparticles and (b) 3 nm NiO nanoparticles at 5 K and 50 K. Notice the absence of saturation and the disappearance of hysteresis at high temperatures.

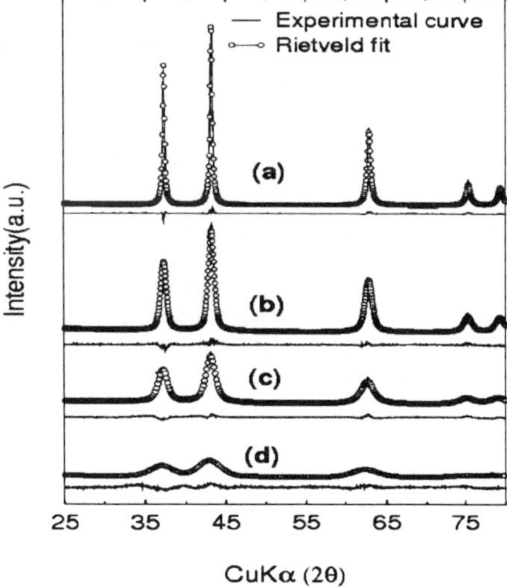

Fig. 6 XRD patterns of the NiO nanoparticles with average diameters of (a) 24, (b) 12, (c) 7 and (d) 3 nm along with the Rietveld fits. The difference patterns are shown below the corresponding observed patterns.

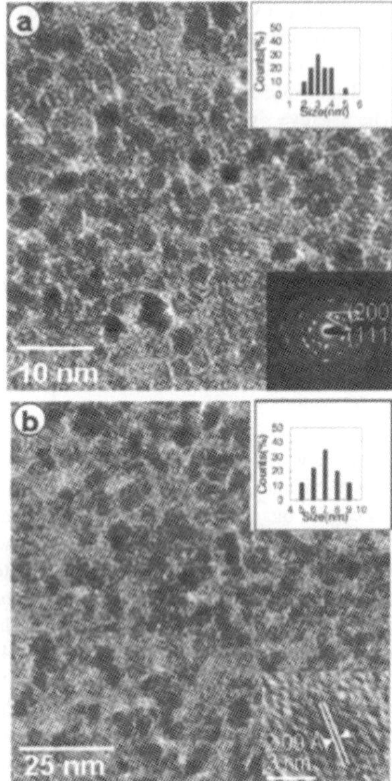

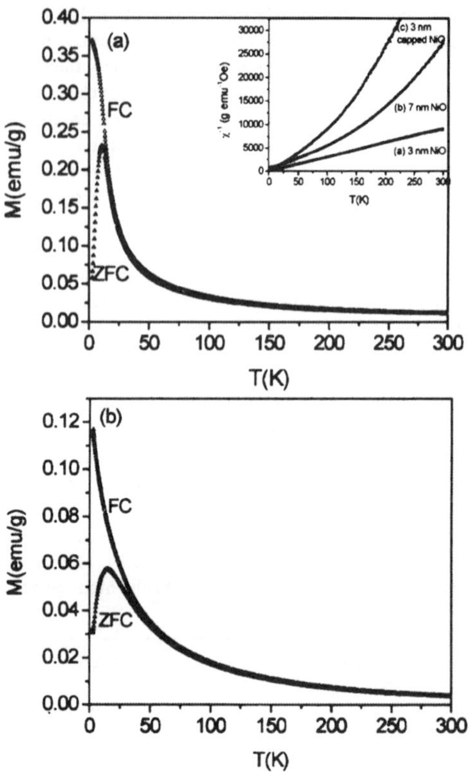

Fig. 7 TEM images of (a) 3 and (b) 7 nm NiO nanoparticles. The insets show the histograms of the particle size distributions (upper panels). The lower panel in (a) shows the SAED pattern. The HREM image of a single NiO nanoparticle of 6 nm is shown in the lower panel in (b).

Fig. 8 The temperature dependence of dc magnetization of (a) 3 nm and (b) 7 nm NiO nanoparticles under ZFC and FC conditions (H = 100 Oe). The inset shows the inverse susceptibility *vs.* temperature curves of the (a) 3 nm and (b) 7 nm particles; (c) shows the data for TOPO-capped 3 nm NiO nanoparticles.

The average particle sizes estimated from the TEM are consistent with the values obtained from the XRD patterns. The indexed SAED pattern shown in Fig. 7(a) suggests the particles to be crystalline. The pattern could be indexed on the basis of the *Fm3m* space group. The HREM image of a 6 nm NiO particle, shown as an inset in Fig. 7(b), shows a lattice spacing of 2.09 Å, corresponding to the interplannar distance between the (100) planes.

We show the temperature dependence of magnetization curves of 3 and 7 nm NiO nanoparticles under ZFC and FC conditions in Fig. 8(a) and (b) respectively. We observe a divergence between the ZFC and FC data, as with the MnO nanoparticles. The ZFC data show maxima corresponding to the blocking temperature (T_B). The T_B increases with the particle size, the values being 10 K and 15 K respectively for the particles with the average diameters of 3 nm and 7 nm. Such a proportionality between T_B and particle size is expected and has been noticed earlier in some cases.[7,14] We show the inverse susceptibility (χ^{-1}) data in the inset of Fig. 8(a). The magnetic moments estimated from the inverse susceptibility data in the high temperature regime are around 3 μ_B, which is close to the value expected for Ni^{2+}. We notice that the Curie–Weiss temperature is close to zero for the 3 nm particles and

negative for the 7 nm particles. The value of the magnetization of the 3 nm particles is also higher than that of the 7 nm particles. The magnetic hysteresis curves for the 3 nm NiO particles are shown in Fig. 5(b). The hysteresis curve at 5 K is associated with a H_c value of 350 Oe. The hysteresis is absent at 50 K. We have measured the magnetic susceptibility of the 3 nm NiO particles capped with TOPO as well. These capped nanoparticles exhibit a lower magnetization than the NiO nanoparticles prepared in the absence of TOPO. The Curie–Weiss temperature of the TOPO-capped nanoparticles is near zero.

Conclusions

MnO and NiO nanoparticles (both in the *Fm3m* space group) of different sizes have been prepared by the decomposition of the metal cupferronate or the acetate in organic solvents under solvothermal conditions. Magnetic properties of the well-characterized nanoparticles of these antiferromagnetic oxides reveal the presence of superparamagnetism (often mentioned as ferromagnetic interactions in the literature), as evidenced by the increasing magnetization with decreasing size as well as the magnetic hysteresis at low temperatures.

Interestingly, the blocking temperature (T_B) increases with the increase in particle size as expected from theory in the case NiO nanoparticles, but shows the opposite relation in the case of MnO nanoparticles. It is not clear why such an inverse relation between T_B and particle size manifests itself. It is noteworthy that supermagnetism at low temperatures is the general feature of small nanoparticles of transition metal oxides, which are otherwise antiferromagnetic with fairly high Néel temperatures.

Acknowledgements

The authors thank Professor E. V. Sampthkumaran for preliminary magnetic measurements.

References

1 R. H. Kodama, *J. Magn. Magn. Mater.*, 1999, **200**, 359.
2 S. A. Makhlouf, F. T. Parker, F. E. Spada and A. E. Berkowitz, *J. Appl. Phys.*, 1997, **81**, 5561.
3 M. Ghosh, E. V. Sampatkumaran and C. N. R. Rao, *Chem. Mater.*, 2005, **17**, 2348.
4 C. N. R. Rao and B. Raveau, *Transition Metal Oxides*, 2nd edn, Wiley-VCH, Germany, 1995.
5 G. H. Lee, S. H. Huh, J.W. Jeong, B. J. Choi, S. H. Kim and H.-C. Ri, *J. Am. Chem. Soc.*, 2002, **124**, 12094.
6 M. Yin and S. O'Brien, *J. Am. Chem. Soc.*, 2003, **125**, 10180.
7 W. S. Seo, H. H. Jo, K. Lee, B. Kim, S. J. Oh and J. T. Park, *Angew. Chem., Int. Ed.*, 2004, **43**, 1115.
8 J. Park, E. Kang, C. J. Bae, J.-G. Park, H.-J. Noh, J.-Y. Kim, J.-H. Park, H. M. Park and T. Hyeon, *J. Phys. Chem. B*, 2004, **108**, 13594.
9 J. Park, K. An, Y. Hwang, J. G. Park, H. J. Noh, J. Y. Kim, J. H. Park, N. M. Hwang and T. Hyeon, *Nature Mater.*, 2004, **3**, 891.
10 J. T. Richardson, D. I. Yiagas, B. Turk, K. Forster and M. V. Twigg, *J. Appl. Phys.*, 1991, **70**, 6977.
11 F. Bødker, M. F. Hansen, C. B. Koch and S. Mørup, *J. Magn. Magn. Mater.*, 2000, **221**, 32.
12 R. H. Kodama, S. A. Makhlouf and A. E. Berkowitz, *Phys. Rev. Lett.*, 1997, **79**, 1393.
13 C. L. Carnes, J. Stipp and K. J. Klabunde, *Langmuir*, 2002, **18**, 1352.
14 B. J. Park, E. Kang, S. Uk. Son, H. M. Park, M. K. Lee, J. Kim, K. W. Kim, H.-J. Noh, J.-H. Park, C. J. Bae, J.-G. Park and T. Hyeon, *Adv. Mater.*, 2005, **17**, 429.
15 W. Xiong, J. Song, L. Gao, J. Jin, H. Zheng and Z. Zhang, *Nanotechnology*, 2005, **16**, 37.
16 K. Tamaki and N. Okabe, *Acta Crystallogr., Sect. C*, 1996, **52**, 1612.
17 J.-F. Bérar and P. Garnier, *NIST Spec. Publ.*, 1992, **846**, 212.

Available online at www.sciencedirect.com

ScienceDirect

Chemical Physics Letters 443 (2007) 118–121

CHEMICAL PHYSICS LETTERS

www.elsevier.com/locate/cplett

Assembling covalently linked nanocrystals and nanotubes through click chemistry

Rakesh Voggu [a], Perumal Suguna [b], Srinivasan Chandrasekaran [b], C.N.R. Rao [a,*]

[a] *Chemistry and Physics of Materials Unit, DST Nanoscience Unit and CSIR Centre of Excellence in Chemistry, Jawaharlal Nehru Centre for Advanced Scientific Research, Jakkur P.O., Bangalore 560 064, India*
[b] *Department of Organic Chemistry, DST Nanoscience Unit, Indian Institute of Science, Bangalore 560 012, India*

Received 19 April 2007; in final form 4 June 2007
Available online 15 June 2007

Abstract

The click reaction involving the Huisgen 1,3-cycloaddition reaction between azide and ethynyl groups has been employed to obtain assemblies of nanostructures. The click reaction between gold nanorods capped with azidoalkane- and alkyne-thiols yields chains and complex assemblies, accompanied by a large red shift of the longitudinal surface plasmon band. A similar reaction between CdSe nanocrystals results in a splitting of the photoluminescence band, possibly due to exciton splitting. Carbon nanotubes decorated by covalently linked gold nanoparticles have been obtained through click chemistry.
© 2007 Elsevier B.V. All rights reserved.

1. Introduction

Click chemistry provides reliable and selective reactions for synthesizing new compounds and to generate combinatorial libraries [1] and the method enables a novel style of organic synthesis. We considered click chemistry as a possible means to assemble or link various types of nanostructures. An important click reaction is the Cu(I)-catalyzed 1,3-dipolar cycloaddition reaction between azido and ethynyl groups to form the 1,2,3-triazole ring, investigated in detail by Huisgen [2,3]. While this reaction has been used to functionalize nanoparticles [4] and single-walled carbon nanotubes (SWNTs) [5], it has not been exploited for assembling nanostructures. We have employed the cycloaddition reaction to generate assemblies of metal nanoparticles, semiconductor nanoparticles as well as covalently attached SWNT–metal nanoparticle composites. The methodology employed in the present study was as follows. An azidoalkane derivative containing a terminal SH or

NH$_2$ group is attached to a nanostructure, A, while an alkyne with a SH or NH$_2$ group is attached to another nanostructure, B. Reaction between A and B gives rise to 1,2,3-triazole, there by causing a covalently attached assembly of the nanostructures. The occurrence of the click reaction is readily ascertained by the absence of the characteristic stretching mode frequencies of the azido and acetylenic units as well as changes in the NMR spectra. When the terminal group is SH, the reaction is given by

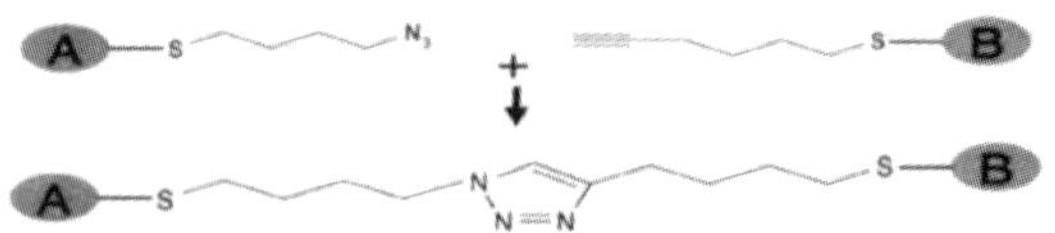

In this Letter, we describe the important initial results obtained by using the Husigen cycloadditon reaction in the case of gold nanorods and CdSe nanocrystals to produce covalently linked assemblies. Carbon nanotubes decorated by covalently linked gold nanoparticles have also been produced by this means.

---

* Corresponding author. Fax: +91 80 23622760.
E-mail address: cnrrao@jncasr.ac.in (C.N.R. Rao).

R. Voggu et al. / Chemical Physics Letters 443 (2007) 118–121

2. Experimental

Gold nanorods were prepared by the photochemical procedure of Kim et al. [6]. Gold nanocrystals with an average diameter of 2.5 nm were obtained by the reduction of chloroaurate ion (0.55 ml of 25 mM aqueous solution) by partially hydrolyzed tetrakis (hydroxymethyl) phosphonium chloride (THPC) [7]. CdSe nanocrystals were prepared by the solvothermal procedure starting with cadmium stearate, selenium and tetralin in toluene [8]. SWNTs prepared by the electric arc method with a Y_2O_3 + Ni catalyst, were purified by acid and hydrogen treatments [9].

4-Azidobutanethiol was prepared starting from 4-bromobutanol adopting following the procedure of Collman et al. [10]. Hex-5-yn-1-thiol was prepared from 5-hexyn-1-ol via the tosylate, followed by reaction with potassium thioacetate to obtain 5-hexyne-1-thioacetate. The latter under acidic conditions gave hex-5-yn-1-thiol. 4-azidobutylamine was prepared following the procedure of Lee et al. [11]. Thiol-capped gold and CdSe nanoparticles were prepared by the reaction of fixed quantities of the nanoparticles in an 1:4 acetonitrile–water mixture (1.5 mL) with 100 µL solution 1 µM of the thiol. The cycloaddition reaction of the capped nanocrystals was carried out with equimolar concentrations (1:1) of A and B. In some of the reactions, the concentration of A was kept constant and the concentration of B was varied. The conditions of the click reaction were as described by Sharpless and coworkers [12].

In order to prepare SWNT–Au nanocrystal composites, SWNTs were functionalized by amidobutane (with a terminal azide group) as follows. Acid-treated SWNTS were reacted with thionyl chloride, followed by reaction with 4-azidobutylamine. These SWNTs were reacted with hexynethiol-capped Au nanocrystals.

Photoluminescence spectra were recorded with a Perkin–Elmer LS-55 spectrometer while the UV–vis absorption spectra were recorded with UV/VIS/NIR Perkin–Elmer spectrometer. Transmission electron microscope (TEM) images were obtained using a JOEL JEM 3010 instrument. Infrared spectra were recorded with a Bruker IFS-66v spectrometer.

3. Results and discussion

In order to assemble gold nanorods (aspect ratio 2.4) by the click reaction, one batch of nanorods capped with 4-azidobutane-1-thiol, A, and another with hex-5-yn-1-thiol, B, were reacted in the acetonitrile–water mixture under standard conditions [12]. The occurrence of the click reaction was established by infrared spectroscopy [13]. The product of the reaction was investigated by electronic absorption spectroscopy and transmission electron microscopy. **Fig.** 1 compares the electronic absorption spectra of the gold nanorods before and after the click reaction between the A and B type nanorods. Isolated gold nanorods show transverse and longitudinal plasmon bands around 520 nm and 630 nm, respectively (see Fig. 1a).

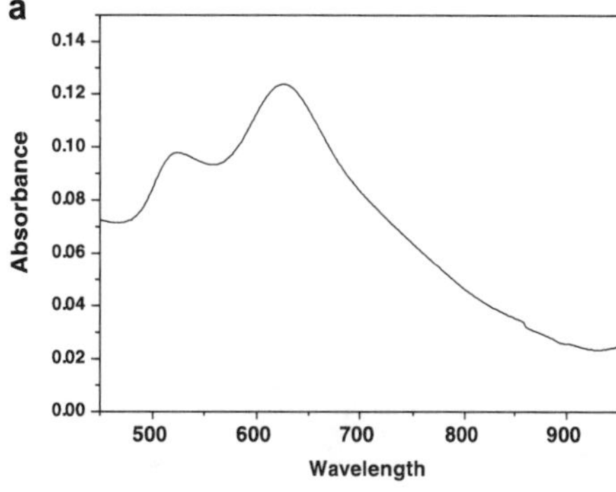

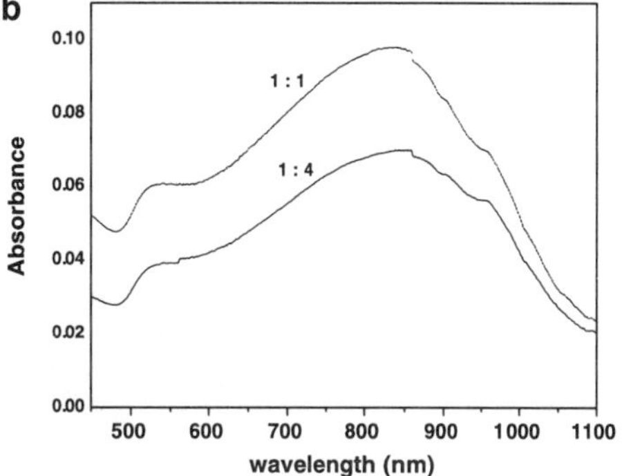

Fig. 1. Electronic absorption spectra of (a) gold nanorods capped by 4-azidobutane-1-thiol before the click reaction and (b) of gold nanorods after the click reaction (at two different (1:1 and 1:4) concentrations of A and B). The spectra of the hexynethiol-capped nanorods were identical to (a).

After the cycloaddition reaction, the spectra show a broad band centered around 800–850 nm due to the longitudinal surface plasmon absorption, the transverse band remaining at 520 nm. The position of longitudinal band changed only little with the relative concentrations of the A and B used in the click reaction. The red shift of the longitudinal plasmon absorption band of the gold nanorods after the click reaction is due to the coupling of the plasmon absorption arising from the assembly of the nanorods [14–16]. The shift observed by us (**Fig.** 1) is comparable to that reported in the case of gold nanorods linked by dithiols or thiol carboxylic acids. In Fig. 2, we show the transmission electron microscope (TEM) images of the gold nanorods before and after the cycloaddition reaction. After the reaction, the image shows chains of nanorods. Thus, the shift in the electronic absorption spectrum to longer wavelengths is accompanied by the assembly of nanorods forming chains. To ensure that the changes in the electronic absorption spectra and the assemblies of nanorods seen in the TEM images were the result of the click reaction, we recorded infrared spectra before and after reaction. The spectra in Fig. 3 clearly show the absence of the azide and acetylenic

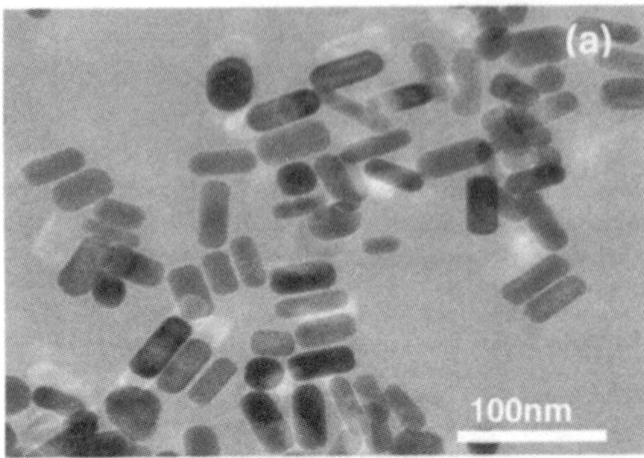

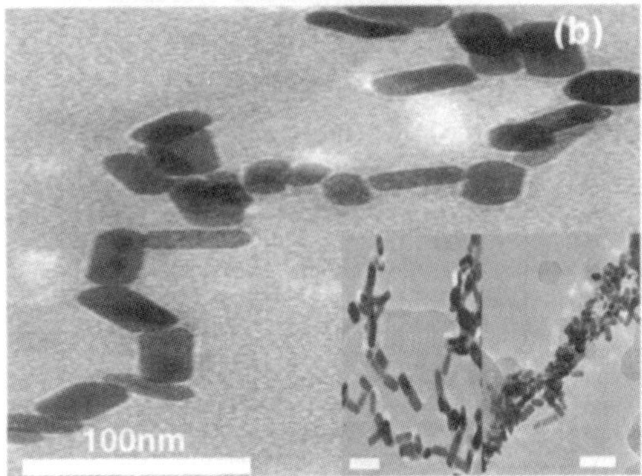

Fig. 2. (a) TEM image of the gold nanorods capped with 4-azidobutane-1-thiol before the click reaction. The TEM image of the hexynethiol-capped nanorods is similar to (a). (b) TEM image of chain like assemblies of gold nanorods obtained after the click reaction with 1:1 concentrations of A and B. The inset in (b) shows more complex assemblies of nanorods obtained by the click reaction with certain concentrations of A and B.

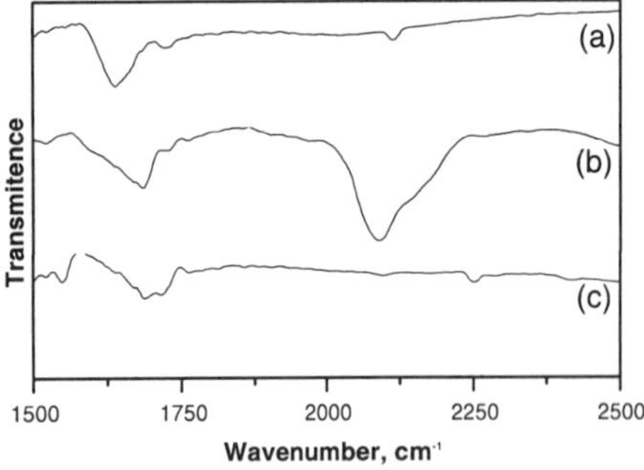

Fig. 3. Infrared spectra in the 1500–2500 cm^{-1} region (a) of the hexynethiol-capped gold nanorods, B, showing the C≡C stretching bond at 2109 cm^{-1}, (b) of the azidothiol-capped gold nanorods, A, showing the asymmetric azide stretching band at 2092 cm^{-1} and (c) of the product of the click reaction (1:1 of A and B) showing the absence of the bands due to the azide and C≡C stretching modes.

stretching frequencies at 2092 cm^{-1} and 2109 cm^{-1}, respectively [13]. The reason for obtaining chains of gold nano-

rods through click chemistry is because the ends of the nanorods comprise (1 1 1) faces of gold which favor thiol binding [17].

In order to link SWNTs with gold nanocrystals, SWNTs functionalized by the amidobutane with a terminal azido group were reacted with Au nanocrystals capped with the hex-5-yne-1-thiol in the acetonitrile–water mixture. The occurrence of the click reaction was ascertained by infrared spectroscopy. This reaction yielded SWNT–Au nanocomposites where in the gold nanoparticles decorate the SWNTs uniformly. This can be seen in the TEM images shown in **Fig. 4**. These images demonstrate the success of click chemistry in generating assemblies comprising covalently linked SWNT–metal nanoparticle composites. Linking gold particles with SWNTs causes perturbation in the electronic spectra of the nanotubes.

The click reaction between spherical CdSe nanocrystals functionalized with the azido and ethynyl thiols gives rise to the cycloaddition products as verified by infrared spectroscopy. In **Fig. 5**, we compare typical photoluminescence

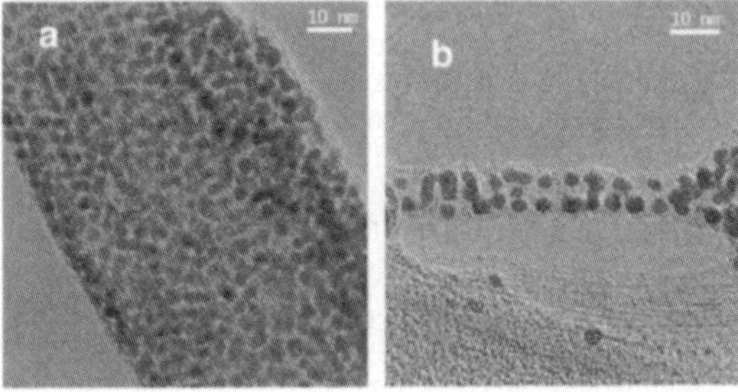

Fig. 4. (a) and (b) TEM images showing covalently attached gold nanocrystals uniformly distributed over SWNT bundles.

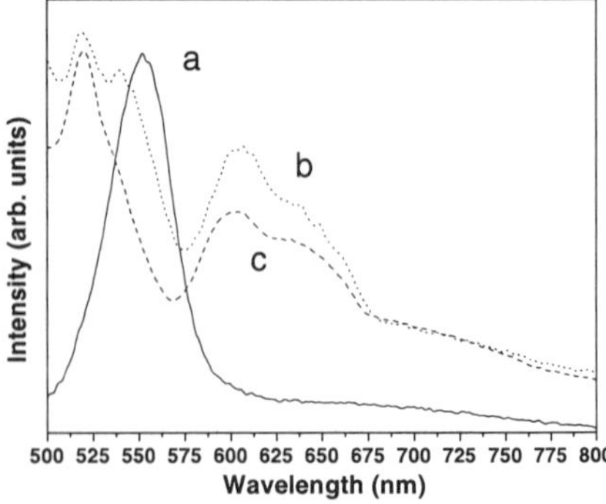

Fig. 5. Photoluminescence spectra of (a) CdSe nanocrystals (3.5 nm diameter) capped by 4-azidobutane-1-thiol before the click reaction and (b) and (c) of the products of the cycloadditon reaction between the CdSe nanoparticles capped with the azidothiol and the hexyne thiol. (b) and (c) correspond to different relative concentrations of the two thiol capped nanoparticles (excitation wavelength 450 nm). The spectrum of the hexynethiol-capped nanoparticles was similar to (a).

R. Voggu et al. / Chemical Physics Letters 443 (2007) 118–121

(PL) spectra of the click reaction product with the spectrum of the CdSe nanocrystals before the reaction. The product of the click reaction shows two broad bands, one at 518 nm and another centered at 620 nm while the isolated thiol-capped nanocrystals give a band at 550 nm. The two bands in **Fig.** 5 are not due to two types of species since we observe one band at a lower wavelength and another at a higher wavelength relative to the band exhibited by the isolated CdSe nanocrystals. Theoretical calculations suggest that the change in the PL spectrum is due to dipolar coupling between neighboring nanocrystals giving rise to exciton splitting somewhat similar to Davydov splitting in CdS nanocrystals [18,19]. These features in the PL spectrum were found in the products obtained with different relative concentrations of the azidothiol- and alkynethiol-capped CdSe nanocrystals. The TEM image of the CdSe nanocrystals subjected to click reaction showed a periodic array.

4. Conclusions

The present results demonstrate the efficacy and use of click chemistry in assembling covalently linked nanostructures. While alkanedithiols and thiolcarboxylic acids are known to assemble gold nanoparticles [15,20], the present method based on click chemistry is more general and can be used with carbon nanotubes, semiconductor nanoparticles and other nanostructures. Further investigations would be necessary to fully characterize the various types of assemblies formed by the use of the click reaction by varying parameters such as the concentrations of the reactants, chain length of the alkane and the nature of the nanostructure. It should be possible to carry out click reactions between other types of nanostructures. Thus, a preliminary study has shown that the click reaction between spherical gold nanocrystals gives a product with a broad plasmon band accompanied by an organized assembly of the nanocrystals, instead of the chains obtained with the gold nanorods (**Fig.** 2). Similarly, the click reaction between suitably capped CdSe and Au nanocrystals causes perturbation of the electronic spectra. In addition to the Husigen cycloaddition reaction, we are exploring other click reactions including the Diels–Alder reaction, to assemble nanomaterials.

Acknowledgements

One of the authors (R.V.) thanks the University Grants Commission for a fellowship. The authors thank Dr. S. Pati for helpful discussion based on theoretical calculations.

References

[1] H.C. Kolb, M.G. Finn, K.B. Sharpless, Angew. Chem. Int. Ed. 40 (2001) 2004.

[2] J.W. Lown, in: A. Padwa (Ed.), 1,3-Dipolar Cycloaddition Chemistry A, Vol. 1, Wiley, New York, 1984.

[3] R. Huisgen, Pure Appl. Chem. 61 (1989) 613.

[4] D.A. Fleming, J.C. Thode, M.E. Williams, Chem. Mater. 18 (2006) 2327.

[5] H. Li, F. Cheng, A.M. Duft, A. Andronov, J. Am. Chem. Soc. 127 (2005) 14518.

[6] F. Kim, J.H. Song, P.D. Yang, J. Am. Chem. Soc. 124 (2002) 14316.

[7] D.G. Duff, A. Baiker, P.P. Edwards, J. Chem. Soc., Chem. Commun. (1993) 96.

[8] U.K. Gautam, M. Rajamathi, F. Meldrum, P. Morgan, R. Seshadri, Chem. Commun. (2001) 629.

[9] S.R.C. Vivekchand, A. Govindaraj, M.M. Seikh, C.N.R. Rao, J. Phys. Chem. B 108 (2004) 6935.

[10] J.P. Collman, N.K. Devraj, C.E.D. Chidsey, Langmuir 20 (2004) 1041.

[11] J.W. Lee, S.I. Jun, K. Kim, Tetrahedron Lett. 42 (2001) 2709.

[12] V.V. Rostovtsev, L.G. Green, V.V. Fokin, K.B. Sharpless, Angew. Chem. Int. Ed. 41 (2002) 2596.

[13] C.N.R. Rao, Chemical Applications of Infrared Spectroscopy, Academic Press, New York, 1963.

[14] P.K. Sudeep, S.T.S. Joseph, K.G. Thomas, J. Am. Chem. Soc. 127 (2005) 6516.

[15] K.G. Thomas, S. Barazzouk, B.I. Ipe, S.T.S. Joseph, P.V. Kamat, J. Phys. Chem. B 108 (2004) 13066.

[16] M. Gluodenis, C.A. Foss, J. Phys. Chem. B 106 (2002) 9484.

[17] K.K. Caswell, J.N. Wilson, U.H.F. Bunz, C.J. Murphy, J. Am. Chem. Soc. 125 (2003) 13914.

[18] A.S. Davydov, Theory of Molecular Excitons, Plenum, New York, 1971.

[19] L. Cao, Y. Miao, Z. Zhang, S. Xie, G. Yang, B. Zou, J. Chem. Phys. 123 (2005) 24702.

[20] S.T.S. Joseph, B.I. Ipe, P. Pramod, K.G. Thomas, J. Phys. Chem. B 110 (2006) 150.

Available online at www.sciencedirect.com

ScienceDirect

Chemical Physics Letters 450 (2008) 340–344

CHEMICAL
PHYSICS
LETTERS

www.elsevier.com/locate/cplett

A calorimetric investigation of the assembly of gold nanorods to form necklaces

Neenu Varghese, S.R.C. Vivekchand, A. Govindaraj, C.N.R. Rao *

*Chemistry and Physics of Materials Unit, DST Unit on Nanoscience and CSIR-Centre of Excellence in Chemistry,
Jawaharlal Nehru Centre for Advanced Scientific Research, Jakkur P.O., Bangalore 560 064, India*

Received 1 October 2007; in final form 7 November 2007
Available online 17 November 2007

Abstract

Interaction of gold nanorods with cysteine as well as 3-mercaptopropionic acid (MPA) has been investigated by isothermal titration calorimetry (ITC), in combination with electronic absorption spectroscopy and electron microscopy. The assembly process with MPA shows two steps, the first due to the binding of MPA to gold nanorods through the sulfur atom, and the second due to assembly of the MPA capped gold nanorods through the formation of cyclic hydrogen bonded dimers between the MPA molecules. In the case of cysteine only a single step is observed in ITC, due to the binding of the molecules to gold nanorods.
© 2007 Elsevier B.V. All rights reserved.

1. Introduction

Self assembly of nanostructures constitutes an important aspect in nanoscience and nanotechnology and several chemical strategies have been employed to create assemblies of nanoparticles [1,2]. Thus, gold nanorods have been assembled using DNA, surfactants and various linker molecules [3–9]. Linking has major consequences on the electronic spectra of the nanorods [5]. Thus when gold nanoparticles are packed closely to form a film, the surface plasmon resonance band of the gold film shifts progressively to longer wavelengths depending on the interparticle separation [10]. Gold nanorods exhibit another unique feature in the electronic spectra, in that they show transverse and longitudinal surface plasmon absorption bands [11]. The transverse band does not vary with the aspect ratio of the nanorods while the longitudinal plasmon band shifts to longer wavelength with increase in aspect ratio. Accordingly, on linking gold nanorods by mercaptopropionic acid (MPA) or cysteine to give rise to chains or necklaces, the longitudinal surface plasmon resonance band shifts signif-

icantly to longer wavelengths [5,6]. Formation of such chains and necklaces using organic linker molecules would involve two processes: (1) interaction of the linker molecules with the gold nanorods and (2) formation of the extended chains or necklaces. When a thiolcarboxylic acid (MPA) and cysteine are used as linkers, the first process involves the formation of the metal–sulfur bond which is an exothermic reaction. The formation of necklaces would also involve an exothermic process. We have investigated the process of assembly of gold nanorods to form necklaces by isothermal titration calorimetry in combination with spectroscopic and transmission electron microscopic studies. For this purpose, we have employed 3-mercapatopropionic acid as the linker molecule wherein the linkage occurs through the formation of six-membered dimeric hydrogen bonded rings between the carboxylic groups [5] and also cysteine where electrostatic interactions lead to the assembly [6] as shown in Scheme 1.

2. Experimental

The gold nanorods were prepared using the photochemical method described in the literature [12]. In a typical preparation, 1.78 mg of tetraoctylammonium bromide

---

* Corresponding author. Fax: +91 80 2208 2760.
E-mail address: cnrrao@jncasr.ac.in (C.N.R. Rao).

For cysteine

For MPA

Scheme 1.

was added to 6 mL 0.08 M aqueous solution of cetyl trimethylammonium bromide, followed by the addition of 120 μL of 0.1 M hydrogen tetrachloroaurate, 130 μL of acetone and 90 μL of cyclohexane. A small amount of silver nitrate solution (90 μL of 0.01 M) was added to the above mixture and the solution was taken in a quartz tube and irradiated with UV light for 14 h. The resulting solution was blue in color and centrifuged at 7000 rpm for 15 min. The precipitate was collected, dispersed in distilled water and centrifuged at 7000 rpm twice. The resulting solution was used for the further experiments. The prepared gold nanorods were characterized using UV–visible spectroscopy and transmission electron microscopy (TEM). All the UV–visible spectral experiments were performed using a Perkin–Elmer Lamda 900 UV/VIS/NIR spectrometer while the TEM images were recorded using JEOL JEM 3010 instrument fitted with a Gatan CCD camera at an accelerating voltage of 300 kV.

Isothermal titration calorimetric (ITC) experiments were performed using a Microcal VP-ITC instrument at 30 °C. The instrument consists of two identical cells, one for sample and the other for reference (distilled water) [13]. In a typical titration, 2 μL of 0.57 mM solution of cysteine or MPA taken in the syringe were injected in equal intervals of 5 min to a solution of 1.59 nM solution of gold nanorods which is taken in the sample cell with 1.47 mL capacity. A total of 130 μL of cysteine or MPA was added to Au nanorod solution and the heat change to the first injection was not considered in the calculations [13]. The concentration of gold nanorods was calculated using the molar extinction coefficient of the surface plasmon resonance band of gold nanorods [14]. The reference cell of the microcalorimeter is filled with distilled water and both the cells were maintained at the same temperature. When cysteine/MPA is injected to the sample cell containing gold nanorods, the difference in heat needed to keep both the sample cell and reference cell at the same temperature is monitored. After the titration the samples were analyzed using UV–visible spectroscopy and transmission electron microscopy. Control experiments were carried out by titrating cysteine or MPA with distilled water under the same conditions and were subtracted with the experimental data to remove the dilution effects of the injectant.

3. Results and discussion

We first carried out the study of the linking of gold nanorods to form necklaces by interactions with cysteine. The raw data obtained during the injections of cysteine

to gold nanorods are shown in Fig. 1a. At each injection, an exothermic reaction occurs. Fig. 1b shows the integrated plot of the data in Fig. 1a, achieved by integrating each peak and normalizing with respect to the concentration of cysteine added. The integrated plot gives the dependence of the heat exchange at each injection in kcal/mole of cysteine on the molar ratio of cysteine to that of gold nanorods. The data in Fig. 1b were obtained after accounting for dilution effects. We observe that the exothermicity of the peaks decreases from one injection to the next due to the binding of cysteine molecules to the gold nanorods forming Au–S bonds. The number of free sites available on the gold nanorods for the binding of cysteine decreases as the reaction progresses, thereby reducing the exothermicity with progressive injections. The actual formation of chains or necklaces of gold nanorods occurs by weak electrostatic interaction of the cysteine molecules which cannot be detected by ITC [15]. We therefore observe only a one step process corresponding to the binding of thiol to Au nanorods.

The process of assembly occurring through the interaction of the gold nanorods with cysteine is directly evidenced in the electronic absorption spectra and

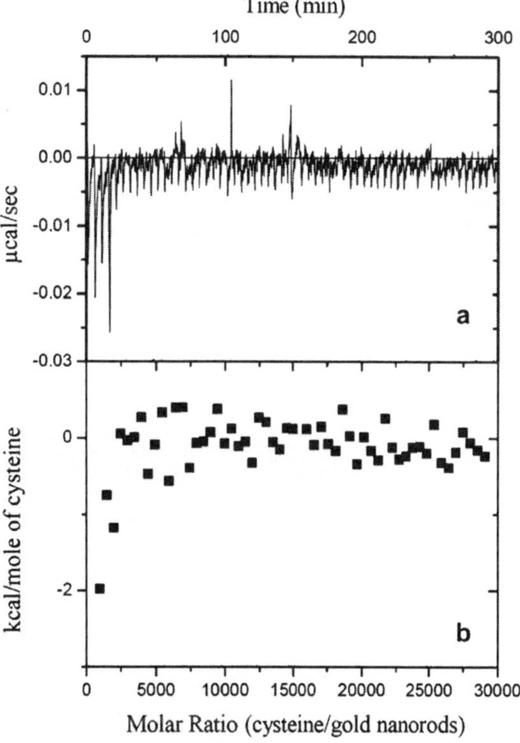

Fig. 1. (a) Raw calorimetric data obtained by the titration of 0.57 mM cysteine with 1.59 nM gold nanorod solution and (b) binding isotherm plot obtained by integrating each peak in raw data and normalizing with cysteine concentration.

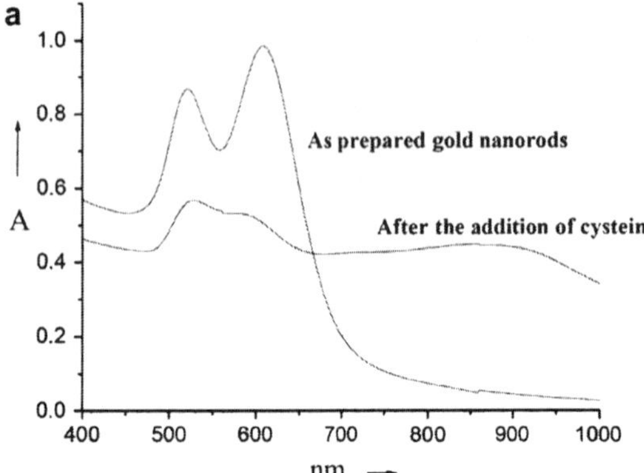

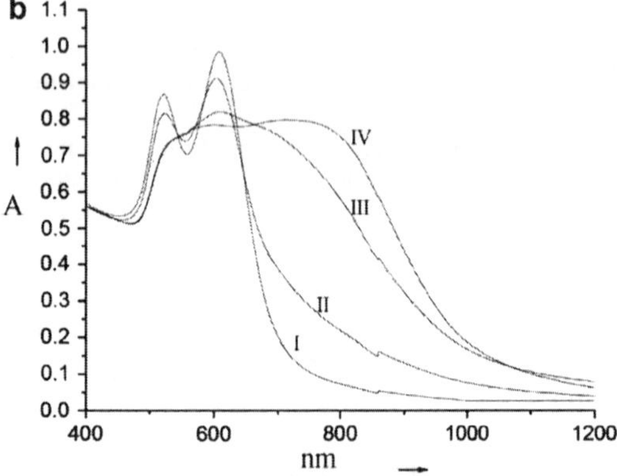

Fig. 2. UV–visible spectra of gold nanorods (a) before and after calorimetric titration with cysteine and (b) taken at different stages of the calorimetric titration with MPA as indicated in Fig. 4b.

transmission electron microscopic images. Fig. 2a shows how the electronic absorption spectra of the as-prepared nanorods changes after the addition of cysteine. The nanorods show the transverse and longitudinal surface plasmon resonance bands at 520 and 605 nm, respectively [11] before the addition of cysteine. After the addition of cysteine, the intensity of the longitudinal surface plasmon band decreases and a new broad band appears in the region 800–900 nm due to formation of chains or necklaces arising from the linking of the gold nanorods by the electrostatic interaction between cysteine molecules. The TEM image in Fig. 3a shows the presence of randomly oriented Au nanorods in the as-prepared gold nanorods. The formation of necklaces of nanorods after the addition of cysteine is evidenced in the image shown in Fig. 3b.

The thermodynamics of interaction of cysteine molecules with the gold nanorods can be understood based on the calorimetric data obtained. Assuming that the energy released is due to the formation of Au–S bond only, since the energy associated with electrostatic interaction is negligible [15], the number of cysteine molecules bonded to each nanorod can be estimated. The binding of thiol to the gold nanorods involves the following reaction:

$$R–S–H + Au \rightarrow R–S–Au + 1/2H_2$$

From the bond energies of RS–H, RS–Au, H–H bond which are 87 kcal/mole, 40 kcal/mole and 104 kcal/mole respectively, the net enthalpy of binding of thiol to Au is −5 kcal/mole [16]. The total heat liberated in the cysteine titration is −966 kcal per mole of Au nanorods. The average number of cysteine molecules attached to Au nanorods thus works out to be around 193.

The results of our calorimetric studies of the interaction of the thiol carboxylic acid, MPA, with gold nanorods are quite different from those with cysteine. In the case of MPA, we observe two steps in the calorimetric data as can be seen in Fig. 4. To understand the binding process of MPA with the gold nanorods, we performed separate experiments wherein the reaction was stopped at different stages (after 12 and 22 injections) and the product analyzed using TEM and UV–visible spectroscopy. In Fig. 2b, we show the absorption spectra of Au nanorods taken at different stages during the binding process indicated in

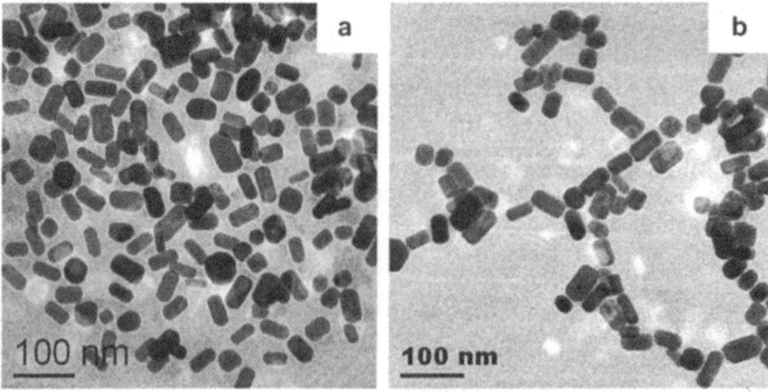

Fig. 3. TEM micrographs of gold nanorods (a) before and (b) after calorimetric titration with cysteine.

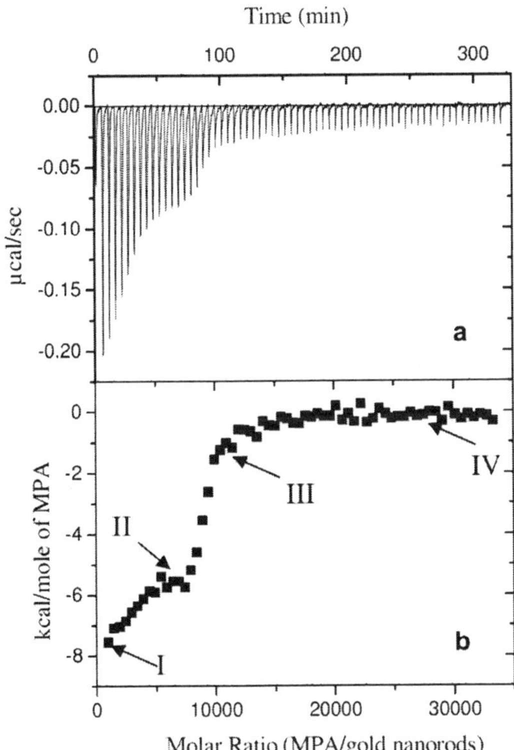

Fig. 4. (a) Raw calorimetric data obtained by the titration of 0.57 mM 3-mercaptopropionic acid (MPA) with 1.59 nM gold nanorod solution and (b) binding isotherm plot obtained by integrating each peak in raw data and normalizing with MPA concentration.

Fig. 4b. The longitudinal plasmon resonance band does not shift after the first few injections of MPA (point II), and shows significant shifts after 22 injections (point III). We assign the first step in the calorimetric data to the binding of MPA molecules to Au nanorods or the formation of Au–S bonds. The second step corresponds to the binding of gold nanorods to form chains or necklaces through the formation of six-membered cyclic hydrogen bonded units between the carboxylic acid groups of MPA. In Fig. 5, we show the TEM image of the as prepared gold nanorods and the necklaces formed by the nanorods after the addition of MPA. Such an occurrence of two steps has been noted in the dynamics of gold nanorod–alkane-dithiol interaction [8].

Using an approach similar to that illustrated for cysteine, we have calculated the average number of MPA molecules attached to each Au nanorod. The heat released in the first step (Fig. 4) is −3650 kcal/mole of Au nanorods. The number of MPA molecules binding to each gold nanorod is therefore around 730. The heat released in the second step is −3015 kcal per mole of gold nanorods. As intermolecular hydrogen bonding between MPA molecules releases approximately −15 kcal/mole, the number of hydrogen bonded units or the number of gold nanorods in the necklace cannot be greater than 200.

4. Conclusions

Isothermal titration calorimetric measurements on the interaction of the thiolcarboxylic acid, MPA, establishes the occurrence of two steps in the assembling process associated with the binding of gold nanorods to the sulfur atom of linker molecules and the formation of the necklace structure. The shifts of the longitudinal plasmon absorption band of the gold nanorods in the electron absorption spectrum as well as transmission electron microscope images carried out at different states of ITC measurements provide direct evidence for the occurrence of the assembly of gold nanorods at the stage expected from ITC measurements. ITC measurements on the binding of cysteine with gold nanorods show only a single step due to the binding of the molecules, since the process of assembly occurs by elec-

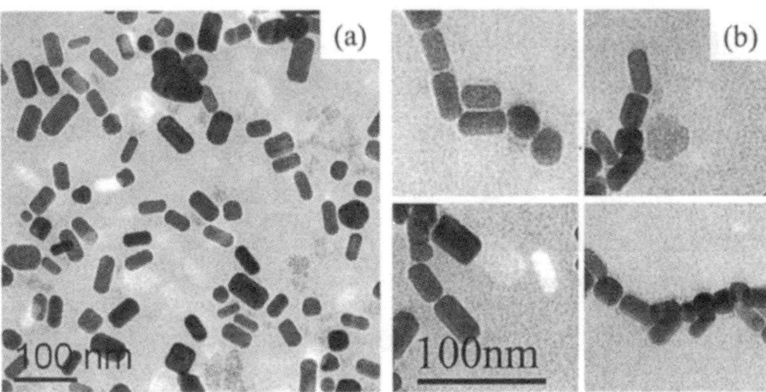

Fig. 5. TEM micrographs of gold nanorods (a) before and (b) after the titration with MPA (scale bar-100 nm).

344   *N. Varghese et al. / Chemical Physics Letters 450 (2008) 340–344*

trostatic interaction. Thus, measurements in the cysteine-Au nanorod systems indirectly support the mechanism for the second step found in the MPA-Au nanorod system.

References

[1] C.N.R. Rao, A. Muller, A.K. Cheetham (Eds.), Recent Advances in the Chemistry of Nanomaterials, Wiley–VCH Verlag GmbH &Co., 2004.
[2] C.N.R. Rao, G.U. Kulkarni, P.J. Thomas, P.P. Edwards, Chem. Soc. Rev. 29 (2000) 27.
[3] E. Dujardin, L.-B. Hsin, C.R.C. Wang, S. Mann, Chem. Commun. (2001) 1264.
[4] B. Nikoobakth, Z.L. Wang, M.A. El-Sayed, J. Phys. Chem. B 104 (2000) 8635.
[5] K.G. Thomas, S. Barazzouk, B.I. Ipe, S.T.S. Joseph, P.V. Kamat, J. Phys. Chem. B 108 (2004) 13066.
[6] S. Zhang et al., Chem. Commun. (2007) 1816.
[7] K.K. Caswell, J.N. Wilson, U.H.F. Bunz, C.J. Murphy, J. Am. Chem. Soc. 125 (2003) 13914.
[8] S.T.S. Joseph, B.I. Ipe, P. Pramod, K.G. Thomas, J. Phys. Chem. B 110 (2006) 150.
[9] R. Voggu, P. Suguna, S. Chandrasekaran, C.N.R. Rao, Chem. Phys. Lett. 443 (2007) 118.
[10] C.N.R. Rao, G.U. Kulkarni, P.J. Thomas, V.V. Agrawal, P. Saravanan, J. Phys. Chem. B 107 (2003) 7391.
[11] S. Link, M.B. Mohamed, M.A. El-Sayed, J. Phys. Chem. B 103 (1999) 3073.
[12] F. Kim, J.H. Song, P. Yang, J. Am. Chem. Soc. 124 (2002) 14316.
[13] J.E. Ladbury, B.Z. Chowdhry, Chem. Biol. 3 (1996) 791.
[14] C.J. Orendorff, C.J. Murphy, J. Phys. Chem. B 110 (2006) 3990.
[15] H. Joshi, P.S. Shirude, V. Bansal, K.N. Ganesh, M. Sastry, J. Phys. Chem. B 108 (2004) 11535.
[16] A. Ulman, Chem. Rev. 96 (1996) 1533.

J. Phys. Chem. B **2004**, *108*, 6935−6937

New Method of Purification of Carbon Nanotubes Based on Hydrogen Treatment

S. R. C. Vivekchand, A. Govindaraj, Md. Motin Seikh, and C. N. R. Rao*

Chemistry and Physics of Materials Unit and CSIR Center of Excellence in Chemistry,
Jawaharlal Nehru Centre for Advanced Scientific Research, Jakkur P.O., Bangalore 560 064, India

Received: March 22, 2004; In Final Form: April 26, 2004

A novel method for the purification of single-walled and multiwalled carbon nanotubes is described. The method involves acid washing followed by hydrogen treatment in the 700−1000 °C range. While acid washing dissolves the metal particles, the hydrogen treatment removes amorphous carbon as well as the carbon coating on the metal nanoparticles. The high quality of the nanotubes obtained after purification has been checked by electron microscopy, X-ray diffraction, and spectroscopic methods.

Multiwalled carbon nanotubes (MWNTs) as well as single-walled carbon nanotubes (SWNTs) are produced by arc evaporation or laser ablation of graphite and by the pyrolysis of hydrocarbons and organometallic precursors.[1−5] The as-synthesized nanotubes usually contain impurities such as amorphous carbon and metal nanoparticles, the latter being prominent when metal catalysts are employed. Purification of carbon nanotubes, therefore, is an important problem in carbon nanotube research. Acid treatment and gas-phase oxidation have been employed to purify SWNTs.[6,7] Micro-filtration and hydrothermal treatment also help in eliminating amorphous carbon and catalyst particles.[8,9] Fullerenes are also removed by solvent extraction. Martinez et al.[10] have used a combination of air oxidation and microwave acid digestion, which is also employed to purify arc-discharge SWNTs. The effect of oxidation conditions on the sample purity on SWNTs prepared by the arc-discharge method has been examined.[11] Due to the importance of purification in nanotube research, we have been exploring an alternative method involving high-temperature hydrogen treatment. The method has been most effective in purifying both SWNTs and MWNTs.

Arc discharge SWNTs (arc SWNTs)[4] were heated in air at 300 °C for 12 h and then stirred in concentrated HNO_3 at 60 °C for 24 h in order to dissolve the metal nanoparticles. The product was washed with distilled water, dried, dispersed in ethanol under sonication, and filtered using Millipore (0.3 μm) filter paper. The filtered product was dried in an oven at 100 °C for 2 h and heated to 1000 °C in a furnace at a rate of 3° per minute, in flowing hydrogen at 100 sccm (standard cubic centimeter per minute) and held at that temperature for 2 h. The resulting sample was again stirred in concentrated HNO_3 at 60 °C for 3 h and heated in a furnace at 1000 °C for 2 h in flowing hydrogen (100 sccm). A similar procedure was employed for the purification of SWNTs obtained by laser ablation (laser SWNTs)[3] as well as HiPCO SWNTs. A CS_2 extraction was carried out on the laser SWNTs followed by washing with 8 N HCl and hydrogen treatment, the last steps being similar to those employed for arc SWNTs. Suspensions of SWNTs for recording visible-NIR spectroscopy were prepared following O'Connell et al.[12] Arc-discharge MWNTs were refluxed in a 2:1 mixture of concentrated HNO_3 and concentrated H_2SO_4 (acid

* To whom correspondence should be addressed. E-mail: cnrrao@ jncasr.ac.in.

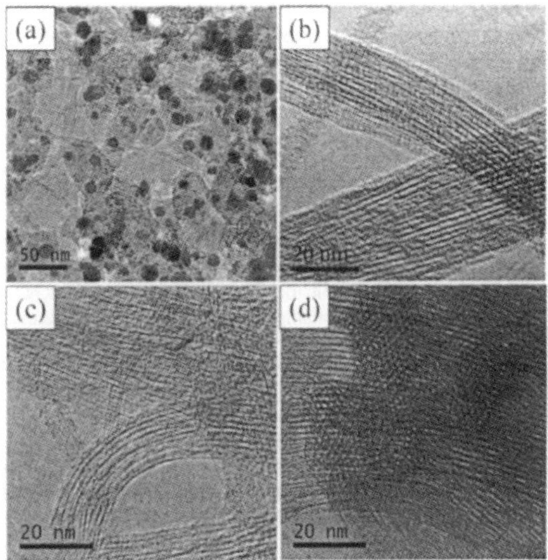

Figure 1. TEM images of (a) as-synthesized SWNTs obtained by the arc-discharge method, (b) after acid treatment, (c) after first hydrogen treatment, and (d) after the second hydrogen treatment.

treated MWNTs) for 20 h, followed by hydrogen treatment at 1100 °C for 2 h. Aligned MWNT bundles obtained by ferrocene pyrolysis[5] were subjected to acid treatment before they were heated in hydrogen at 1100 °C. The nanotubes were characterized at each stage by powder X-ray diffraction, scanning electron microscopy (SEM), transmission electron microscopy (TEM), and Raman spectroscopy.

Figure 1a shows a TEM image of the as-synthesized arc SWNTs containing amorphous carbon and metal nanoparticles apart from the SWNT bundles. The image reveals that the SWNT bundles have a diameter of ~10 nm. Most of the metal nanoparticles get dissolved on acid washing, but the nanotubes are covered with amorphous carbon as seen in the TEM image in Figure 1b. The amorphous carbon is removed by high-temperature hydrogen treatment and the remaining small metal nanoparticles agglomerate into larger particles. We generally find that at this stage of purification the nanotubes have open ends. The absence of amorphous carbon is clearly evident from

6936 *J. Phys. Chem. B, Vol. 108, No. 22, 2004*

the TEM image in Figure 1c. TEM images also reveal that the bundles grow in size and have diameters in the range of 20–50 nm after the hydrogen treatment. The metal nanoparticles agglomeration occurs in the 750–850 °C range and is removed in the second acid treatment carried out for a short duration. The second acid treatment is followed by the high-temperature hydrogen treatment at 1000 °C to obtain pure SWNTs. A TEM image of such purified SWNTs is shown in Figure 1d. We do not see hollow onion-like structures often found in SWNTs purified by other methods,[9] indicating thereby that the carbon covering on the metal particles is etched away by hydrogen.

SWNTs form a triangular lattice and give a distinctive low angle reflection in the XRD pattern.[3] In Figure 2A, we show the low-angle XRD patterns of the as-synthesized and acid-treated SWNTs along with those finally obtained after the second hydrogen treatment. The (1,0) diffraction line is not observed clearly in the case of as-synthesized SWNTs and appears as a small hump in the acid treated SWNTs. We however see an intense peak in the sample after the second hydrogen treatment. The mean diameter of the purified SWNTs is estimated to be 1.52 nm from the (1,0) diffraction line.[13] The visible-NIR spectra of the arc SWNTs at various stages of purification is shown in Figure 2B. Due to their one-dimensional nature, carbon nanotubes exhibit van Hove singularities in the electronic density of states,[14] visible-NIR spectroscopy providing evidence for the one-dimensional nature. The peak centered at 1100 nm is due to the second van Hove singularity transition and the second set of peaks near 700 nm due to the first van Hove singularity transition of metallic nanotubes. The intensities of the two bands increase markedly on hydrogen treatment. The G-band and the radial-breathing modes of SWNTs in the Raman spectra are strong in intensity, whereas the other Raman modes are weak.[15] The Raman spectra of the as-synthesized and purified arc SWNTs are shown in Figure 2C. The as-synthesized sample (curve a) shows a G-band, which is split into bands at 1562 and 1586 cm^{-1}. The D-band appears as a broad peak centered at 1343 cm^{-1} indicating that the sample contains amorphous carbon. The second-order Raman bands appear as weak peaks centered at 941 and 1075 cm^{-1}. The purified sample shows an intense split G-band (1563 and 1585 cm^{-1}) and strong radial-breathing modes. The D-band becomes very weak on hydrogen treatment. The radial breathing modes show the diameters to be in the range of 1.32–1.89 nm,[15] in agreement with the diameter distribution estimated from TEM images and low-angle X-ray diffraction.

The TEM images in Figure 3 show laser SWNTs at various stages of purification. Fullerenes, metal nanoparticles, amorphous carbon, and SWNT bundles are present in the as-synthesized laser SWNTs (Figure 3a). The fullerenes were removed by CS$_2$ extraction and the metal nanoparticles partly dissolved on acid washing. After acid washing, amorphous carbon continues to cover the SWNTs (Figure 3b). This amorphous carbon is effectively removed by the hydrogen treatment at 1000 °C as can be seen in Figure 3c. Similar to the arc SWNTs, we observe an agglomeration of the undissolved metal nanoparticles, which could be removed by the subsequent acid treatment. The nanotubes heated again in hydrogen at 1000 °C were indeed of high purity (Figure 3d). The purity of the nanotubes was also established by the Raman and UV–visible–NIR spectra. SWNTs prepared by the HiPCO could be purified satisfactorily by this method at a relatively lower temperature of 700 °C.

We have examined the efficacy of the high-temperature hydrogen treatment procedure for MWNTs. Acid treated

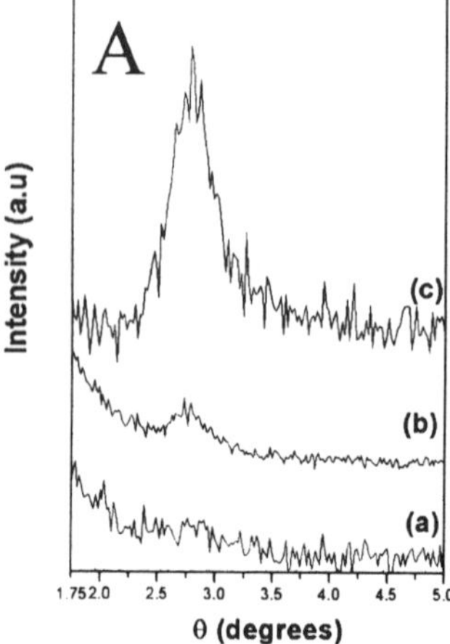

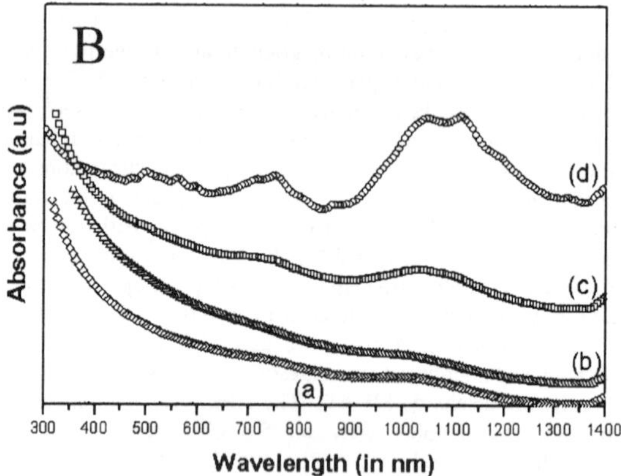

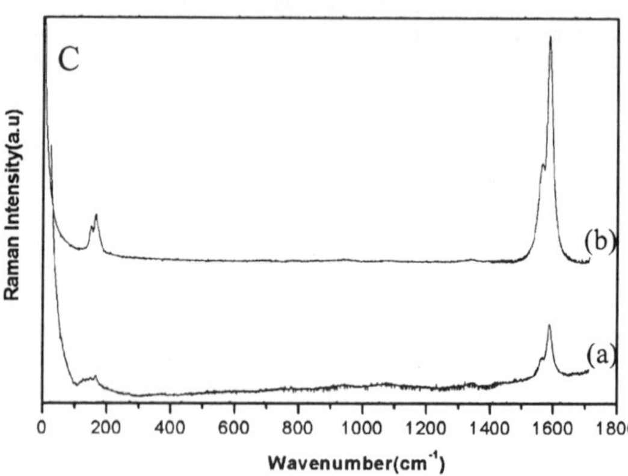

Figure 2. (A) Low-angle X-ray diffraction patterns of (a) as-synthesized arc-discharge SWNTs, (b) after first acid washing, and (c) after second hydrogen treatment. (B) Visible–NIR spectra of (a) as-synthesized arc-SWNTs, (b) after first acid washing, (c) after first hydrogen treatment, and (d) after final purification. (C) Raman spectra of (a) as-synthesized arc-discharge SWNTs and (b) after second hydrogen treatment.

MWNTs contain a large amount of amorphous carbon. A TEM image of the acid treated MWNTs is shown in Figure 4a. The amorphous carbon is removed by the high-temperature hydrogen treatment as can be seen from Figure 4b. The high-resolution electron microscope (HREM) image shown as an inset in Figure

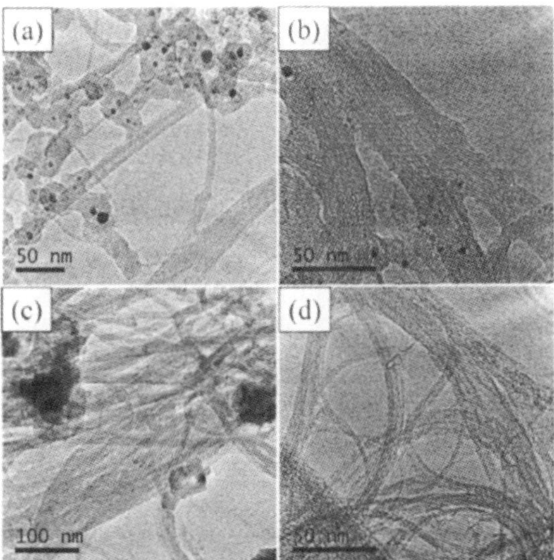

Figure 3. TEM images of SWNTs obtained by laser ablation: (a) as-synthesized, (b) after acid treatment, (c) after first hydrogen treatment, and (d) after final purification.

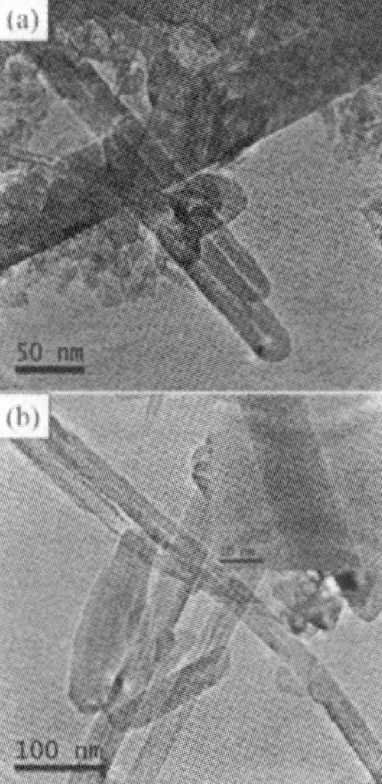

Figure 4. TEM images of arc-discharge MWNTs (a) after acid treatment and (b) after hydrogen treatment at 1100 °C for 2 h. Inset in (b) shows a HREM image of a MWNT after hydrogen treatment.

4b indicates that the crystalline nature of the MWNTs is preserved after the acid and hydrogen treatments. Raman and X-ray diffraction measurements confirmed the high purity of the MWNTs subjected to these treatments. We have found that

aligned carbon nanotubes are also effectively purified by high-temperature hydrogen treatment.

It would be in order to compare the present work with the other procedures reported in the literature. All of the methods make use of acid washing to remove the metal particles where present. In the procedures involving air oxidation, SWNTs are subjected to heat treatment in the 350−500 °C range depending on the method of synthesis. The present method however employs hydrogen treatment around 1000 °C for all SWNTs and MWNTs, except for HiPCO SWNTs which require a lower temperature. Whereas in air oxidation the amorphous carbon is converted into CO_2, it is converted to CH_4 on hydrogen treatment.

In conclusion, we have developed a new and effective method for the purification of SWNTs and MWNTs. The method involves acid washing followed by the high-temperature hydrogen treatment repeated twice. Excellent SWNTs containing little or no amorphous carbon and metal particles are obtained by this means, as verified by microscopy, XRD, and spectroscopic techniques. Equally importantly, thermogravimetric measurements of MWNTs and SWNTs show that the low-temperature weight loss due to amorphous carbon is eliminated after hydrogen treatment. In the case of laser SWNTs, the oxidation temperature is substantially increased after purification.

References and Notes

(1) Iijima, S. *Nature* **1991**, *354*, 56.

(2) Rao, C. N. R.; Satishkumar, B. C.; Govindaraj, A.; Nath, M. *Chem. Phys. Chem.* **2001**, *2*, 78.

(3) Thess, A.; Lee, R.; Nikolaev, P.; Dai, H.; Petit, P.; Robert, J.; Xu, C.; Lee, Y. H.; Kim, S. G.; Rinzler, A. G.; Colbert, D. T.; Scuseria, G. E.; Tománek; Fischer, J. E.; Smalley, R. E. *Science* **1996**, *273*, 483.

(4) Journet, C.; Maser, W. K.; Bernier, P.; Loiseau, A.; Lamy de la Chapelle, M.; Lefrant, S.; Deniard, P.; Lee, R.; Fischer, J. E. *Nature (London)* **1997**, *388*, 756.

(5) Haddon, R. C., Ed.: Special issue on carbon nanotubes. *Acc. Chem. Res.* **2002**, *35*, 997.

(6) Dujardin, E.; Ebbesen, T. W.; Krishnan, A.; Treacy, M. M. *Adv. Mater.* **1998**, *10*, 611.

(7) Chiang, I. W.; Brinson, B. E.; Huang, A. Y.; Willis, P. A.; Bronikowski, M. J.; Margrave, J. L.; Smalley, R. E.; Hauge, R. H. *J. Phys. Chem. B* **2001**, *105*, 8297.

(8) Rinzler, A. G.; Liu, J.; Nikolaev, P.; Huffman, C. B.; Rodríguez-Macías; Boul, P. J.; Lu, A. H.; Heymann, D.; Colbert, D. T.; Lee, R. S.; Fischer, J. E.; Rao, A. M.; Eklund, P. C.; Smalley, R. E. *Appl. Phys. A* **1998**, *67*, 29.

(9) Tohji, K.; Goto, T.; Takahashi, H.; Shinoda, Y.; Shimizu, N.; Jeyadevan, B.; Matsuoka, I.; Saito, Y.; Kasuya, A.; Ito, S.; Nishina, Y. *J. Phys. Chem. B* **1997**, *101*, 1974.

(10) Martínez, M. T.; Callejas, M. A.; Benito, A. M.; Maser, W. K.; Cochet, M.; Andrés, J. M.; Schreiber, J.; Chauvet, O.; Fierro, J. L. G. *Chem. Commun.* **2002**, 1000.

(11) Sen, R.; Rickard, S. M.; Itkis, M. E.; Haddon, R. C. *Chem. Mater.* **2003**, *15*, 4273.

(12) O'Connell, M. J.; Bachilo, S. M.; Huffman, C. B.; Moore, V. C.; Strano, M. S.; Haroz, E. H.; Rialon, K. L.; Boul, P. J.; Noon, W. H.; Ma, J.; Hauge, R. H.; Weisman, R. B.; Smalley, R. E. *Science* **2002**, *297*, 593.

(13) The diameter of the SWNTs was calculated by the following formula $\langle D \rangle = (d_{11}/\cos \theta) - 0.312$ nm.

(14) Ouyang, M.; Huang, J. L.; Lieber, C. M. *Acc. Chem. Res.* **2002**, *35*, 1018.

(15) Dresselhaus, M. S.; Eklund, P. C. *Adv. Phys.* **2000**, *49*, 705.

(16) The diameters of the SWNTs was calculated using the relationship: $D = 248$ cm$^{-1}/\varpi$ (in cm^{-1}).

Nitrogen- and Boron-Doped Double-Walled Carbon Nanotubes

L. S. Panchakarla, A. Govindaraj, and C. N. R. Rao*

Chemistry and Physics of Materials Unit, CSIR Center of Excellence in Chemistry and DST Unit on Nanoscience, Jawaharlal Nehru Centre for Advanced Scientific Research, Jakkur P.O., Bangalore 560 064, India

ABSTRACT Double-walled carbon nanotubes (DWNTs) doped with nitrogen and boron have been prepared by the decomposition of a CH_4 + Ar mixture along with pyridine (or NH_3) and diborane, respectively, over a $Mo_{0.1}Fe_{0.9}Mg_{13}O$ catalyst, prepared by the combustion route. The doped DWNTs have been characterized by transmission electron microscopy (TEM), X-ray photoelectron spectroscopy, electron energy loss spectroscopy, and Raman spectroscopy. The dopant concentration is around 1 atom % for both boron and nitrogen. The radial breathing modes in the Raman spectra have been employed along with TEM to obtain the inner and outer diameters of the DWNTs. The diameter ranges for the undoped, N-doped (pyridine), N-doped (NH_3), and B-doped DWNTs are 0.73–2.20, 0.74–2.30, 0.73–2.32, and 0.74–2.36 nm, respectively, the boron-doped DWNTs giving rise to a high proportion of the large diameter DWNTs. Besides affecting the G-band in the Raman spectra, N- and B-doping affect the proportion of semiconducting nanotubes.

KEYWORDS: carbon nanotubes · doped nanotubes · double-walled nanotubes · Raman spectroscopy · transmission electron microscopy

Double-walled carbon nanotubes (DWNTs), first observed in 1996, constitute a unique family of carbon nanotubes (CNTs).[1,2] DWNTs occupy a position between the single-walled carbon nanotubes (SWNTs) and the multiwalled carbon nanotubes (MWNTs), as they consist of two concentric cylinders of rolled graphene. DWNTs possess useful electrical and mechanical properties with potential applications. Thus, DWNTs and SWNTs have similar threshold voltages in field electron emission, but the DWNTs exhibit longer lifetimes.[3] Unlike SWNTs, which get modified structurally and electronically upon functionalization, chemical functionalization of DWNTs surfaces would lead to novel carbon nanotube materials where the inner tubes are intact. The stability of DWNTs is controlled by the spacing of the inner and outer layers but not by the chirality of the tubes;[4] therefore, one obtains a mixture of DWNTs with varying diameters and chirality indices of the inner and outer tubes. DWNTs have been prepared by several techniques, such as arc discharge[5] and chemical vapor deposition (CVD) using a mixture of ferrocene with a hydrocarbon or alcohol (typical hydrocarbons are methane, alcohol, *n*-hexane, and benzene).[6–10] DWNTs have also been prepared by a sulfur-assisted CVD method using methane as the carbon source.[11,12]

Applications of CNTs based on their electrical properties strongly depend on the diameter and helicity as well as parity.[2,13] Doping of CNTs by boron and nitrogen renders them p-type and n-type, respectively. MWNTs and SWNTs doped with nitrogen[14–17] and boron[18,19] have been reported. Boron-doped carbon nanotubes appear to exhibit enhanced electron field emission due to the presence of the boron atom at the nanotube edges.[20,21] N-doped CNTs show n-type behavior regardless of tube chirality.[22]

We were interested in the synthesis and characterization of boron- and nitrogen-doped DWNTs in view of their potential applications. We have focused on the low-doping regime (~1 atom %), where the fundamental band structure is expected to be unchanged relative to the all-carbon model. To our knowledge, except for a report on the preparation of nitrogen-doped DWNTs by using a mixture of methane, ammonia, and argon over an iron−molybdenum catalyst,[23] there has been no detailed study of these materials. We have prepared nitrogen-doped DWNTs by the thermal decomposition of a CH_4 + NH_3 + Ar mixture as well as a CH_4 + pyridine + Ar mixture over a $Mo_{0.1}Fe_{0.9}Mg_{13}O$ catalyst, prepared by a new procedure. It may be noted that pyridine has been found to be a good nitrogen source to prepare N-doped MWNTs.[16] We have prepared boron-doped DWNTs by the thermal decomposition of a CH_4 + B_2H_6 + Ar mixture over the $Mo_{0.1}Fe_{0.9}Mg_{13}O$ catalyst at 950 °C.

*Address correspondence to cnrrao@jncasr.ac.in.

Received for review September 13, 2007 and accepted November 01, 2007.

Published online November 29, 2007.
10.1021/nn700230n CCC: $37.00

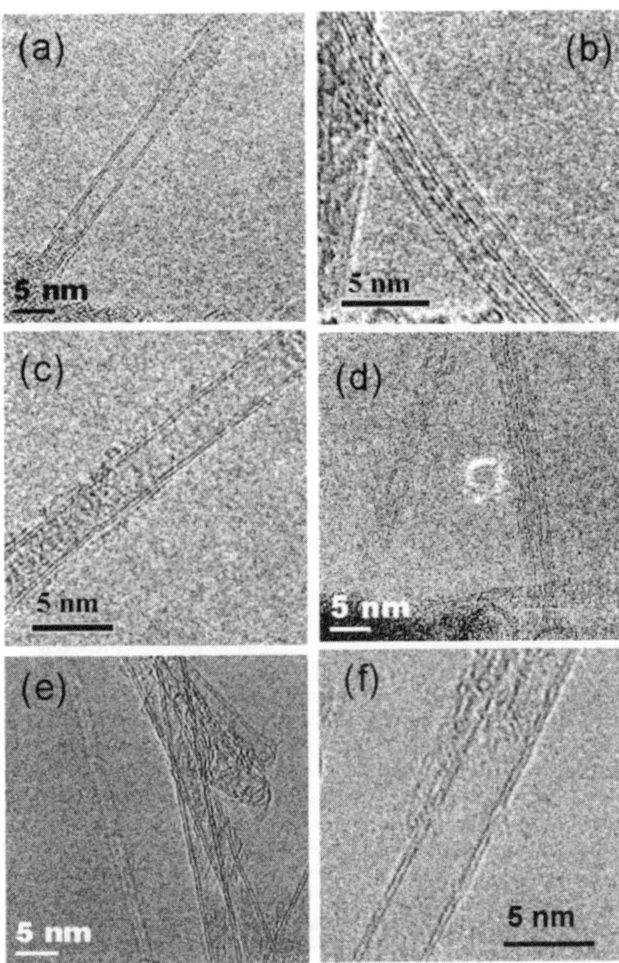

Figure 1. HRTEM images of (a,b) undoped DWNTs, (c) N-doped DWNTs using ammonia, (d) N-doped DWNTs using pyridine, and (e,f) of B-doped DWNTs.

help in producing DWNTs exclusively with only a very small or negligible proportion of SWNTs.

We have used both transmission electron microscopy (TEM) and Raman spectroscopy to characterize the different DWNTs samples. TEM allows direct imaging of the DWNTs and gives indications for the presence of other species along with the DWNTs. In the TEM images, we seldom encountered SWNTs or MWNTs. Besides providing information on the nature and dimensions of DWNTs, Raman spectroscopy helps to characterize the purity and quality of the DWNTs. Electron energy loss spectroscopy (EELS), carried out in a high-resolution electron microscope, and X-ray photoelectron microscopy have been employed to determine the elemental composition of the DWNTs.

Undoped DWNTs obtained by us generally had outer tube diameters of 2.2–2.8 nm and inner tube diameters of 1.4–2.1 nm, as shown by the high-resolution TEM (HRTEM) images in Figure 1a,b. In Figure 1c,d, we show typical HRTEM images of purified nitrogen-doped DWNTs synthesized by using NH_3 and pyridine as the nitrogen source, respectively. In Figure 1e,f we show HRTEM images of the boron-doped DWNTs. The HRTEM images indicate that the purified samples of the DWNTs have well-resolved walls and that most of the amorphous carbon was eliminated from the surface during the purification process. HRTEM images reveal that the outer tube diameters of the N-doped DWNTs prepared by using NH_3 as the nitrogen source are in the 1.7–3.2 nm range and the inner tube diameters are in the 1–2.4 nm range. The interlayer spacing is around 0.38 nm. In the case of N-doped DWNTs prepared by using pyridine as the nitrogen source, the outer tube diameters are generally in the 1.6–2.6 nm range, while the inner tube diameters are in the 0.9–1.8 nm range. The interlayer spacing ranges from 0.34 to 0.41 nm. From the HRTEM studies, we surmise that the diameters of the N-doped DWNTs obtained by using pyri-

The various DWNTs have been characterized with respect to composition and structure. In particular, the effect of B- and N-doping on the dimensions of the nanotubes has been examined by Raman spectroscopy.

RESULTS AND DISCUSSION

While the decomposition of the CH_4 + Ar over the $Mo_{0.1}Fe_{0.9}Mg_{13}O$ catalyst at 950 °C yielded undoped DWNTs, decomposition of the CH_4 + NH_3 + Ar and CH_4 + pyridine + Ar mixtures gave nitrogen-doped DWNTs. Decomposition of the CH_4 + BH_3 + Ar mixture over the catalyst at 950 °C yielded boron-doped DWNTs. The N- and B-doped DWNTs could not be produced at temperatures lower than 950 °C. The combustion method employed for the preparation of the catalyst seems to

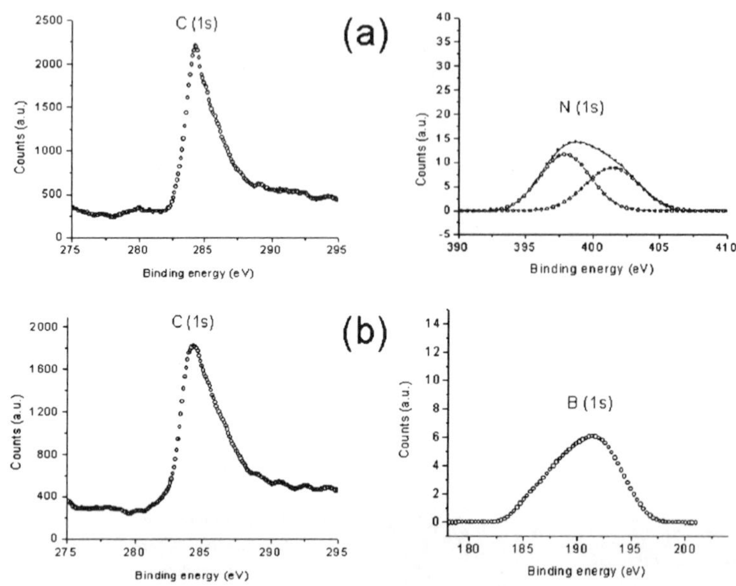

Figure 2. (a) C 1s and N 1s XPS signals of N-doped DWNTs prepared using ammonia. (b) C 1s and B 1s signals of B-doped DWNTs.

dine are smaller than those obtained with NH_3. Thus, the diameters of the N-doped DWNTs appear to depend on the nitrogen source and the reaction conditions employed.

HRTEM images of the boron-doped DWNTs show that they possess larger diameters than the undoped DWNTs as well as the N-doped DWNTs (Figure 1e,f). The outer tube diameters of the B-doped DWNTs range from 2.5 to 4.7 nm, and the inner tube diameters are in the 1.8–3.9 nm range. The interlayer spacing ranges from 0.35 to 0.41 nm. Figure 1f shows the HRTEM image of a large diameter B-doped DWNT with an outer diameter of 4.7 nm and an inner diameter of 3.9 nm.

It has been reported in the literature that boron and nitrogen are incorporated to SWNTs to a smaller extent than in MWNTs.[14–19] We have estimated the compositions of the N- and B-doped DWNTs prepared by us by employing X-ray photoelectron spectroscopy. A core-level X-ray photoelectron spectrumf the N-doped DWNTs obtained by using NH_3 as the N-source is shown in Figure 2a. The C 1s signal is at 284.3 eV, while the N 1s signal is centered at 399.6 eV, indicating of nitrogen substitution in the graphene sheet. It is possible that there is a small amount of amorphous carbon, as suggested by the asymmetry of the C 1s signal, although most of it gets removed on treatment with hydrogen (see Experimental Section). The asymmetric shape of the N 1s peak indicates the existence of at least two components and could be deconvoluted into two peaks at 398 and 401.3 eV. The 398 eV feature is characteristic of pyridinic nitrogen (sp^2 hybridization), while the peak centered at 401.3 eV is due to nitrogen present in graphene sheets.[15] The areas of the two bands bear a ratio of 1:1. On the basis of the total N 1s and C 1s intensities, the nitrogen-to-carbon ratio in the nanotubes samples was calculated by taking the photoionization cross sections of the 1s levels into account. The average composition was thus found to correspond to 1.3 atom % nitrogen. This value is lower than that reported in the literature for DWNTs ($\sim$2.9 atom %)[13] and MWNTs (3–10 atom %).[16] The N 1s spectrum of N-doped DWNTs obtained by using pyridine as the nitrogen source shows mainly the band at 398 eV, the intensity of the 401.3 eV band being very small. Thus, there is an intrinsic difference in the nature of N-substitution between the N-doped DWNTs prepared by using NH_3 and pyridine.

Figure 2b shows the core-level spectra of the B-doped DWNTs. The B 1s feature is at 191.4 eV, and the C 1s signal is at 284.3 eV. The shift of the B 1s signal toward higher binding energy compared to that of pure boron (188 eV) indicates that boron is in the sp^2 carbon network. The slight asymmetry of the B 1s signal would, however, suggest the presence of another possible mode of substitution. The boron content works out to be 1 atom %. Around 3 atom % B-doped MWNTs have been reported.[18] EELS measurements in a high-resolution electron microscope confirmed the presence of nitrogen as well as boron in the respective doped DWNTs. The %B and %N were found to be small ($\sim$1 atom %), consistent with the XPS data.

The resonance Raman spectrum of DWNTs shows three main features: the G band, the D band, and bands due to the radial breathing modes (RBMs). The tangential stretch G-band modes are in the 1550–1600 cm^{-1} range. The disorder-induced D-band is observed between 1200 and 1450 cm^{-1}. The D-band is activated in the first-order scattering process by the presence of in-plane substitutional heteroatoms, vacancies, grain boundaries, or other defects and by finite size effects, all of which lower the crystalline symmetry of the quasi-infinite lattice.[24] RBM frequencies provide information about the nanotube diameter in the case of SWNTS and DWNTs. We have recorded the Raman spectra of the undoped as well as the N- and B-doped DWNTs by using 632.8 nm excitation using a He—Ne laser. The spectra were collected in a backscattering geometry at room

TABLE 1. RBM Frequencies (cm⁻¹) and Diameter (in Parentheses, nm) of Undoped and N- and B-Doped DWNTs

DWNTs[a]	N-DWNTs (Py)[b]	N-DWNTs (NH₃)[c]	B-DWNTs[d]
338 (0.73)	337 (0.74)	334 (0.73)	336 (0.74)
252 (0.98)	241 (1.03)	308 (0.81)	217 (1.14)*
213 (1.16)*	220 (1.13)*	281 (0.88)	201 (1.23)**
199 (1.25)	214 (1.16)*	247 (1.0)	189 (1.31)**
189 (1.31)*	191 (1.30)	213 (1.16)**	177 (1.40)
175 (1.42)	160 (1.55)**	199 (1.25)	163 (1.52)
171 (1.45)	142 (1.75)**	191 (1.30)	151 (1.64)*
158 (1.57)**	134 (1.85)	158 (1.57)*	130 (1.91)**
143 (1.73)**	125 (1.98)	152 (1.63)**	118 (2.10)*
132 (1.88)**	118 (2.10)	147 (1.69)**	105 (2.36)
113 (2.20)	108 (2.30)	134 (1.85)	
		107 (2.32)	

[a]Possible (n,m) values for the intense bands are as follows: 213 [(7,10)], 189 [(3,15); (6,13)], 158 [(14,9); (19,2)], 143 [(17,8)], 132 [(18,9); (21,5)]. [b]Possible (n,m) values for the intense bands are as follows: 220 [(4,12)], 214 [(7,10)], 160 [(18,3); (6,16)], 142 [(20,4); (7,18)]. [c]Possible (n,m) values for the intense bands are as follows: 213 [(7,10)], 158 [(7,10)], 152 [(13,11); (12,12)], 147 [(21,1); (6,18)]. [d]Possible (n,m) values for the intense bands are as follows: 217 [(1,14)], 201 [(3,14); (7,11)], 189 [(3,15)], 151 [(18,5); (15,9)], 130 [(16,12)], 118 [(5,24); (15,16)]. *Highest intensity RBM frequency. **Medium intensity RBM frequency.

temperature. Figure 3a shows the G-bands of the pure as well as doped DWNTs. The G-band of the N-doped DWNTs (Py) appears at a lower frequency (1574 cm⁻¹) compared to that of undoped DWNTs (1575 cm⁻¹), whereas the G-band of the B-doped DWNTs appears at a higher frequency (1579 cm⁻¹). The G-band of the N-doped DWNTs (NH₃) also appears at a lower frequency (1571 cm⁻¹). Thus, the shifts of the G-band are opposite for n- and p-doping of the DWNTs. Such shifts of the G-band have been reported for B- and N-doped SWNTs by Yang et al.[25] and McGuire et al.[26] The small-intensity shoulder around 1540 cm⁻¹ seen in the spectra of undoped DWNTs shows a decrease in intensity in the N-doped DWNTs and is negligible in the case of B-doped DWNTs. This band is related to the metallic nature of the nanotubes,[27] and its near absence in N- and B-doped DWNTs suggests a greater prevalence of semiconducting nanotubes. The inset in Figure 3a shows that the intensity of the D-band is high in the case of the B-doped DWNTs and low in the case of the N-doped DWNTs. The $I(D)/I(G)$ ratios are 0.04, 0.06, and 0.16 for undoped, N-doped (py), and B-doped DWNTs respectively.

We observe several RBM bands in the DWNTs (Figure 3b), resulting from various sizes of the nanotubes, just as in earlier reports.[28] By using the relation $\omega = 248/d$, where ω is the RBM frequency in cm⁻¹ and d is the nanotube diameter in nm, we have obtained the diameters of the DWNTs.[28] The RBM frequencies and the corresponding diameters are tabulated in Table 1 for undoped as well as N- and B-doped DWNTs, along with (n,m) indices for the intense features. From the table, we see that the diameter distribution of the nanotubes is markedly affected by N- and B-doping. The un-

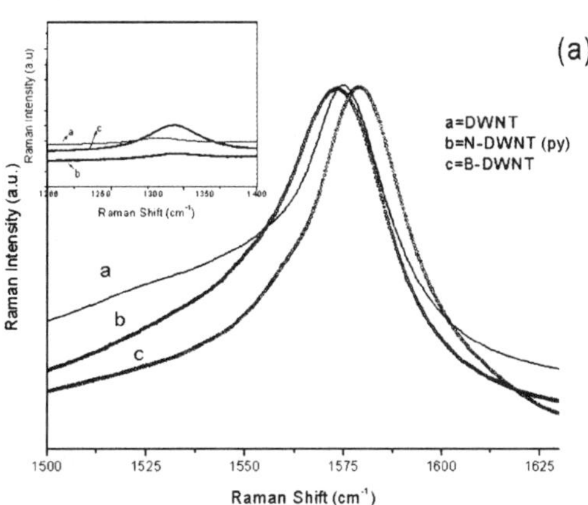

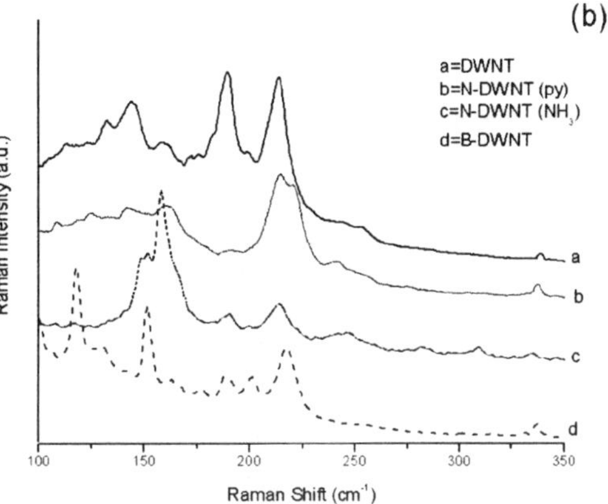

Figure 3. (a) G-bands in the Raman spectra of undoped and doped DWNTs. Inset shows the D-bands of the same. (b) RBM bands of undoped and doped DWNTs.

doped DWNTs show the highest intensity RBM bands centered at 213 and 189 cm⁻¹, corresponding to diameters of 1.16 and 1.31 nm, respectively. The slightly lower intensity or medium-intensity RBM bands are at 158, 143, and 132 cm⁻¹, corresponding respectively to diameters of 1.57, 1.73, and 1.88 nm. The N-doped DWNTs (Py) show the highest intensity RBM bands centered at 220 and 214 cm⁻¹, corresponding to diameters of 1.13 and 1.16 nm, respectively. The slightly lower intensity bands centered at 160 and 142 cm⁻¹ correspond to 1.55 and 1.75 nm diameters, respectively. The diameters of the N-doped DWNTs (Py) are somewhat smaller compared to those of the undoped DWNTs. This is, however, not the case with N-doped DWNTs prepared using NH₃ as the nitrogen source. The DWNTs (NH₃) show the highest intensity RBM band at 158 cm⁻¹, corresponding to a diameter of 1.57 nm. The slightly lower intensity bands centered at 213, 152, and 147 cm⁻¹ correspond to diameters of 1.16, 1.63, and 1.69 nm, respectively. Since the nature of

N-substitution as well as the nature of the defects is different in the N-doped DWNTs prepared by using NH_3 and pyridine, the difference in diameters is not entirely surprising.

The B-doped DWNTs exhibit a high proportion of large-diameter DWNTs compared to the undoped or N-doped DWNTs. The most intense RBM bands of the B-doped DWNTs are at 217, 151, and 118 cm^{-1}, corresponding to diameters of 1.14, 1.64, and 2.1 nm, respectively. The slightly lower intensity bands centered at 201, 189, and 130 cm^{-1} correspond to diameters of 1.23, 1.31, and 1.91 nm, respectively. Due to a cutoff filter, the peaks below 100 cm^{-1} were not detected.

The diameters of the various DWNTs calculated from the RBM modes are comparable with those obtained from the TEM images, but the larger diameter nanotubes seen in the TEM images are not registered in the Raman spectra since the RBM modes below 100 cm^{-1} could not be recorded by us. We can identify DWNT pairs by taking the difference between the inner and outer diameters to be around 0.7 nm. The frequencies (cm^{-1}) of such pairs of the RBM bands in the case of undoped DWNTs are (252,143), (213,132), and (189,113). The metallic (m) and semiconducting (s) natures of these pairs are respectively (s,m), (m,m) or (m,s), and (m,s) or (s,s). In the N-doped DWNTs (py), the pairs are (241,142), (214,125), (220,134), and (160,108), and they are (s,s), (m,s), (s,s), and (m,m) or (m,s), respectively. For N-doped DWNTs (NH_3), the pairs are (281,158), (247,147), (213,134), and (152,107), and these pairs are respectively (s,s), (s,s) or (s,m), (m,s), and (m,m) or (m,s) or (s,m) or (s,s). In the B-doped DWNTs, the pairs are (217,130), (177,118), (189,118), and (151,105), and these pairs are (s,s), (m,s) or (s,s), (m,s), and (m,m) or (m,s) or (s,m) or (s,s), respectively. Taking the semiconducting and metallic nature of all the RBM bands, the ratio of semiconductor to metallic nanotubes in the case of undoped DWNTs works out to be 2:1, while it is 2:1, 2.2:1, and 2:1, respectively, in the case of N-doped DWNTs (py), N-doped DWNTs (NH_3), and B-doped DWNTs. Thus, the RBM modes predict a greater proportion of semiconducting nanotubes in the doped DWNTs as well.

The electronic absorption spectra of undoped as well as doped DWNTs show bands in the 900–1200 nm region due overlapping E_{22}^s (s = semiconductor) features of the outer tubes and E_{11}^s of inner tubes.[29] The absorption bands in the 1600–2400 nm regions are

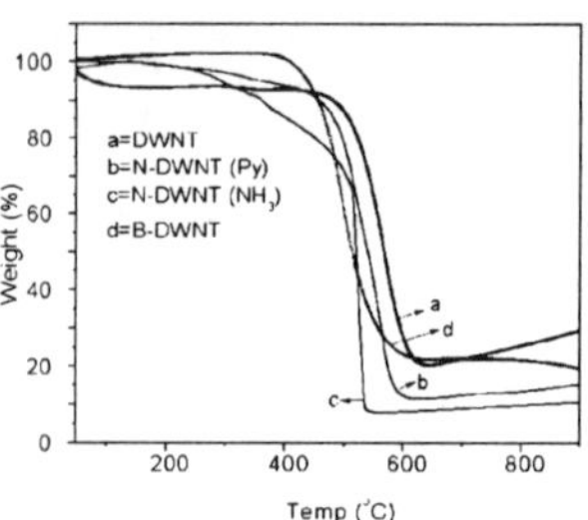

Figure 4. TGA curves of undoped and doped DWNTs.

due to E_{11}^s of the outer tubes. The absorption bands due to E_{11}^m (m = metal) of the outer tubes are found in the 400–600 nm region. The metallic feature seems to prominent in the undoped DWNTs. Accordingly, the 1540 cm^{-1} G^+-band in the Raman spectrum is less prominent in the doped nanotubes.

The smaller diameter carbon nanotubes are known to be less stable than their larger diameter counterparts and tend to oxidize at lower temperatures. Amorphous carbon and carbon nanotubes with defects undergo combustion at lower temperatures. In Figure 4, we show the thermogravimetric analysis (TGA) curves of undoped as well as N- and B-doped DWNTs. The decomposition temperatures of all these doped DWNTs are comparable to but slightly lower that the decomposition temperature of pure DWNTs. Derivative TGA curves also shows the same trend. The slight increase in mass at high temperature may be due to the small metallic impurity.

CONCLUSION

In conclusion, the $Mo_{0.1}Fe_{0.9}Mg_{13}O$ catalyst prepared by the combustion route preferentially yields DWNTs, the proportion of SWNTs being very small or negligible. The use of this catalyst has enabled the synthesis of 1 atom % N- and B-doped DWNTs. The diameters of the nanotubes obtained from the Raman RBM modes and transmission electron microscopy are comparable. The N-doped nanotubes show the G-band in the Raman spectrum at a lower frequency than the undoped ones, while the B-doped nanotubes show an increase in the frequency. The proportion of the metallic nanotubes appears to decrease on N- or B-doping, but the average diameter is substantially larger in the B-doped DWNTs.

EXPERIMENTAL SECTION

Synthesis of DWNTs was carried out in a quartz tube reactor. For each synthesis, 200 mg of the supported Fe–Mo catalyst (Fe–Mo/MgO) was placed in a quartz boat by spreading it uniformly. The quartz boat was inserted into the center of the quartz tube (25 mm i.d. and 1 m long) mounted inside an electrical tube furnace. Subsequently, the furnace was heated to 950 °C in an argon atmosphere at a heating rate of 3 °C/min. A mixture of methane and Ar gas was introduced into the reactor. The

flow rates of methane and Ar were maintained at 50 and 150 sccm (standard cubic centimeters per minute), respectively. After 20 min, the reactor was cooled to room temperature in an Ar atmosphere. The resulting black dense mat contained a homogeneous dispersion of carbon nanotubes around the oxide grains. This crude material was carefully collected from the boat and subjected to purification.

The oxide precursors required to prepare the catalyst for the synthesis of DWNTs were prepared by the combustion route.[30,31] The required amount of $(NH_4)_6Mo_7O_{24} \cdot 4H_2O$ was added to an aqueous solution containing ferric nitrate $(Fe(NO_3)_3 \cdot 9H_2O)$ and magnesium nitrate $(Mg(NO_3)_2 \cdot 6H_2O)$ in a Pyrex dish, keeping the molar ratio of Mo:Fe:MgO at 0.1:0.9:13. To this mixture was added an appropriate amount of urea (three times the stoichiometric ratio), which acts as the fuel in the combustion process. The mixture was dissolved by using a minimum amount of distilled water and kept in an oven at 70 °C for 12 h. The Pyrex dish containing the solution was placed in a furnace preheated at 550 °C. The thick orange-red solution immediately started boiling and underwent dehydration. The resulting thick paste frothed and blazed with a white flame, with the production of a light material which then swelled to the capacity of the Pyrex dish. The total combustion process was over in 10 min. The combustion product was baked at 550 °C for 3 h and ground to a fine powder. Preparation of the catalyst by conventional methods, such as wet impigration or coprecipitation, yields inhomogeneous catalyst particles, whereas the combustion route employed here gives homogeneous catalyst particles.

For obtaining nitrogen-doped DWNTs, the procedure was similar to that used for undoped DWNTs, expect that ammonia or pyridine vapor was taken in mixture with CH_4.[16] For doping nitrogen by using ammonia, the supported Fe—Mo catalyst (200 mg) was placed in a quartz boat at the center of the quartz reactor tube. The quartz tube was heated to 950 °C in an Ar atmosphere. Subsequently, CH_4 (50 sccm), NH_3 (5 sccm), and Ar (150 sccm) were mixed and introduced at the inlet of the reactor tube. After 20 min, the reactor was cooled to room temperature in an Ar atmosphere. For N-doping using pyridine, the supported Fe—Mo catalyst (200 mg) was placed in a quartz boat at the center of the quartz reactor tube. The quartz tube was heated to 950 °C in an Ar atmosphere. Subsequently, 40 sccm of CH_4 was passed through a bubbler containing pyridine, which carries the pyridine vapor to the furnace. These vapors were mixed with 150 sccm of Ar and passed over the MgO-supported catalyst, maintained at 950 °C for 20 min.

For the synthesis of boron-doped DWNTs, diborane (B_2H_6) was used as the boron source, the rest of the procedure being similar to that for undoped DWNTs. B_2H_6 vapor was generated by the addition of BF_3—diethyl etherate to sodium borohydride in tetraglyme.[18] Fifty sccm of CH_4 was mixed and passed along with B_2H_6 vapors. These vapors were further mixed along with 150 sccm of Ar and passed over the MgO-supported catalyst powder, maintained at 950 °C for 20 min.

The percentage of DWNTs in all our preparations was 90% or higher. Such high preferential yields of DWNTs have been reported in the literature.[9,10] In order to dissolve the metal nanoparticles in the DWNTs, the as-prepared nanotubes were treated with concentrated HCl at 60 °C for 24 h. The product was washed with distilled water, dried, dispersed in ethanol under sonication, and filtered using Millipore (0.2 μm) filter paper. The filtered product was dried in an oven at 100 °C for 2 h and heated to 850 °C in a furnace at a rate of 3 °C per minute in flowing hydrogen at 100 sccm and held at that temperature for 6 h to remove the amorphous carbon present on the nanotube walls.[32] The resulting sample was again stirred in concentrated HCl at 60 °C for 3 h and heated in a furnace at 850 °C for 6 h in flowing hydrogen (100 sccm). The same procedure was employed to purify doped DWNTs, except that dilute HCl was used instead of concentrated HCl. It must be noted that hydrogen treatment at high temperatures has been found to be a very good method to eliminate amorphous carbon present on the carbon nanotubes.[32]

We have characterized the undoped and doped DWNTs by various techniques. Field emission scanning electron microscope (FESEM) images were recorded with a FEI NOVA NANOSEM 600. UV–vis absorption spectra of the nanotubes were recorded using a Perkin-Elmer Lambda 900 UV/vis/NIR spectrometer. Thermogravimetric analysis was carried out using a Mettler Toledo TGA 850 instrument. Raman spectra were recorded with a LabRAM HR high-resolution Raman spectrometer (Horiba Jobin Yvon) using a He—Ne laser ($\lambda = 630$ nm). Transmission electron microscope images were obtained with a JEOL JEM 3010 instrument. X-ray photoelectron spectroscopy was recorded using a

VG scientific ESCA Laboratory V spectrometer. EELS were recorded with a transsmision electron microscope (FEI, TECNAI F30) equipped with an energy filter for EELS operating at 300 kV.

Acknowledgment. L.S.P. acknowledges CSIR, New Delhi, for a junior research fellowship.

REFERENCES AND NOTES

1. Dai, H.; Rinzler, A.; Nikolaev, P.; Thess, A.; Colbert, D.; Smalley, R. Single-Wall Nanotubes Produced by Metal-Catalyzed Disproportionation of Carbon Monoxide. *Chem. Phys. Lett.* **1996**, *260*, 471–475.
2. Rao, C. N. R.; Govindaraj, A. *Nanotubes and Nanowires*; RSC Nanoscience & Nanotechnology Series; RSC Publishing: Cambridge, UK, 2005.
3. Kurachi, H.; Uemura, S.; Yotani, J.; Nagasako, T.; Yamada, H.; Ezaki, T.; Maesoba, T.; Loutfy, R.; Moravsky, A.; Nakazawa, T.; Saito, Y. Proceedings of the 21st International Display Research Conference in conjunction with the 8th International Display Workshops, 2001; pp 1237–1240.
4. Saito, R.; Matsuo, R.; Kimura, T.; Dresselhaus, G.; Dresselhaus, M. S. Anomalous Potential Barrier of Double-Wall Carbon Nanotube. *Chem. Phys. Lett.* **2001**, *348*, 187–193.
5. Sugai, T.; Yoshida, H.; Shimada, T.; Okazaki, T.; Shinohara, H. New Synthesis of High-Quality Double-Walled Carbon Nanotubes by High-Temperature Pulsed Arc Discharge. *Nano Lett.* **2003**, *3*, 769–773.
6. Lee, Y. D.; Lee, H. J.; Han, J. H.; Yoo, J. E.; Lee, Y.-H.; Kim, J. K.; Nahm, S.; Ju, B.-K. Synthesis of Double-Walled Carbon Nanotubes by Catalytic Chemical Vapor Deposition and Their Field Emission Properties. *J. Phys. Chem. B* **2006**, *110*, 5310–5314.
7. Liu, B. C.; Lyu, S. C.; Lee, T. J.; Choi, S. K.; Eum, S. J.; Yang, C. W.; Park, C. Y.; Lee, C. J. Synthesis of Single- and Double-Walled Carbon Nanotubes by Catalytic Decomposition of Methane. *Chem. Phys. Lett.* **2003**, *373*, 475–479.
8. Lyu, S. C.; Lee, T. J.; Yang, C. W.; Lee, C. J. Synthesis and Characterization of High-Quality Double-Walled Carbon Nanotubes by Catalytic Decomposition of Alcohol. *Chem. Commun.* **2003**, *1404*, 1405.
9. Lyu, S. C.; Liu, B. C.; Lee, S. H.; Park, C. Y.; Kang, H. K.; Yang, C.-W; Lee, C. J. Large-Scale Synthesis of High-Quality Double-Walled Carbon Nanotubes by Catalytic Decomposition of n-Hexane. *J. Phys. Chem. B* **2004**, *108*, 2192–2194.
10. Lyu, S. C.; Liu, B. C.; Lee, C. J.; Kang, H. K.; Yang, C-W; Park, C. Y. High-Quality Double-Walled Carbon Nanotubes Produced by Catalytic Decomposition of Benzene. *Chem. Mater.* **2003**, *15*, 3951–3954.
11. Wei, J.; Ci, L.; Jiang, B.; Li, Y.; Zhang, X.; Zhu, H.; Xua, C.; Wua, D. Preparation of Highly Pure Double-Walled Carbon Nanotubes. *J. Mater. Chem.* **2003**, *13*, 1340–1344.
12. Ren, W.; Cheng, H.-M. Aligned Double-Walled Carbon Nanotube Long Ropes with a Narrow Diameter Distribution. *J. Phys. Chem. B* **2005**, *109*, 7169–7173.
13. Hassanien, A.; Tokumoto, M.; Kumazawa, Y.; Kataura, H.; Maniwa, Y.; Suzuki, S.; Achida, Y. Atomic Structure and Electronic Properties of Single-Wall Carbon Nanotubes Probed by Scanning Tunneling Microscope at Room Temperature. *Appl. Phys. Lett.* **1998**, *73*, 3839–3841.
14. Sen, R.; Satishkumar, B. C.; Govindaraj, A.; Harikumar, K. R.; Renganathan, M. K.; Rao, C. N. R. Nitrogen-Containing Carbon Nanotubes. *J. Mater. Chem.* **1997**, *7*, 2335–2337.
15. Sen, R.; Satishkumar, B. C.; Govindaraj, A.; Harikumar, K. R.; Raina, G.; Zhang, J.-P.; Cheetham, A. K.; Rao, C. N. R. B-C-N, C-N and B-N Nanotubes Produced by the Pyrolysis of Precursor Molecules over Co Catalysts. *Chem. Phys. Lett.* **1998**, *287*, 671–676.
16. Nath, M.; Satishkumar, B. C.; Govindaraj, A.; Vinod, C. P.; Rao, C. N. R. Production of Bundles of Aligned Carbon and Carbon-Nitrogen Nanotubes by the Pyrolysis of

Precursors on Silica-Supported Iron and Cobalt Catalysts. *Chem. Phys. Lett.* **2000**, *322*, 333–340.

17. Villalpando-Paez, F.; Zamudio, A.; Elias, A. L.; Son, H.; Barros, E. B.; Chou, S. G.; Kim, Y. A.; Muramatsu, H.; Hayashi, T.; Kong, J.; *et al.* Synthesis and Characterization of Long Strands of Nitrogen-Doped Single-Walled Carbon Nanotubes. *Chem. Phys. Lett.* **2006**, *424*, 345–352.

18. Satishkumar, B. C.; Govindaraj, A.; Harikumar, K. R.; Zhang, J.-P; Cheetham, A. K.; Rao, C. N. R. Boron-Carbon Nanotubes from the Pyrolysis of C_2H_2-B_2H_6 Mixtures. *Chem. Phys. Lett.* **1999**, *300*, 473–477.

19. McGuire, K.; Gothard, N.; Gai, P. L.; Dresselhaus, M. S.; Sumanasekera, G.; Rao, A. M. Synthesis and Raman Characterization of Boron-Doped Single-Walled Carbon Nanotubes. *Carbon* **2005**, *43*, 219–227.

20. Charlier, J.-C.; Terrones, M.; Baxendale, M.; Meunier, V.; Zacharia, T.; Rupesinghe, N. L.; Hsu, W. K.; Grobert, N.; Terrones, H.; Amaratunga, G. A. J. Enhanced Electron Field Emission in B-doped Carbon Nanotubes. *Nano Lett.* **2002**, *2*, 1191–1195.

21. Sharma, R. B.; Late, D. J.; Joag, D. S.; Govindaraj, A.; Rao, C. N. R. Field Emission Properties of Boron and Nitrogen Doped Carbon Nanotubes. *Chem. Phys. Lett.* **2006**, *428*, 102–108.

22. Czerw, R.; Terrones, M.; Charlier, J.-C.; Blase, X.; Foley, B.; Kamalakaran, R.; Grobert, N.; Terrones, H.; Tekleab, D.; Ajayan, P. M.; *et al.* Identification of Electron Donor States in N-Doped Carbon Nanotubes. *Nano Lett.* **2001**, *1*, 457–460.

23. Kim, S. Y.; Lee, J.; Na, C. W.; Park, J.; Seo, K.; Kim, B. N-Doped Double-Walled Carbon Nanotubes Synthesized by Chemical Vapor Deposition. *Chem. Phys. Lett.* **2005**, *413*, 300–305.

24. Dresselhaus, M. S.; Eklund, P. C. Phonons in Carbon Nanotubes. *Adv. Phys.* **2000**, *49*, 705–814.

25. Yang, Q. H.; Hou, P. X.; Unno, M.; Yamauchi, S.; Saito, R.; Kyotani, T. Dual Raman Features of Double Coaxial Carbon Nanotubes with N-Doped and B-Doped Multiwalls. *Nano Lett.* **2005**, *5*, 2465–2469.

26. McGuire, K.; Gothard, N.; Gai, P. L.; Dresselhaus, M. S.; Sumanasekera, G.; Rao, A. M. Synthesis and Raman Characterization of Boron-Doped Single-Walled Carbon Nanotubes. *Carbon* **2005**, *43*, 219–227.

27. Das, A.; Sood, A. K.; Govindaraj, A.; Saitta, A. M.; Lazzeri, M.; Mauri, F.; Rao, C. N. R. Doping in Carbon Nanotubes Probed by Raman and Transport Measurements. *Phys. Rev. Lett.* **2007**, *99*, 136803 (1–4).

28. Li, F.; Chou, S. G.; Ren, W.; Gardecki, J. A.; Swan, A. K.; Unlu, M. S.; Goldbeg, B. B.; Cheng, H.-M.; Dresselhaus, M. S. Identification of the Constituents of Double-Walled Carbon Nanotubes using Raman Spectra Taken with Different Laser-Excitation Energies. *J. Mater. Res.* **2003**, *18*, 1251–1258.

29. Kishi, N.; Kikuchi, S.; Ramesh, P.; Sugai, T.; Watanabe, Y.; Shinohara, H. Enhanced Photoluminescence from Very Thin Double-Wall Carbon Nanotubes Synthesized by the Zeolite-CCVD Method. *J. Phys. Chem. B* **2006**, *110*, 24816–24821.

30. Patil, K. C. Advanced Ceramics: Combustion Synthesis and Properties. *Bull. Mater. Sci.* **1993**, *16*, 533–541.

31. Flahaut, E.; Peigney, A.; Bacsa, W. S.; Bacsa, R. R.; Laurent, C. CCVD Synthesis of Carbon Nanotubes from (Mg,Co,Mo)O Catalysts: Influence of the Proportions of Cobalt and Molybdenum. *J. Mater. Chem.* **2004**, *14*, 646–653.

32. Vivekchand, S. R. C.; Govindaraj, A.; Motin Seikh, Md.; Rao, C. N. R. New Method of Purification of Carbon Nanotubes Based on Hydrogen Treatment. *J. Phys. Chem. B* **2004**, *108*, 6935–6937.

Available online at www.sciencedirect.com

SCIENCE @ DIRECT®

Chemical Physics Letters 411 (2005) 468–473

CHEMICAL PHYSICS LETTERS

www.elsevier.com/locate/cplett

Nature and electronic properties of Y-junctions in CNTs and N-doped CNTs obtained by the pyrolysis of organometallic precursors

F.L. Deepak, Neena Susan John, A. Govindaraj, G.U. Kulkarni, C.N.R. Rao *

Chemistry and Physics of Materials Unit and CSIR Centre of Excellence in Chemistry, Jawaharlal Nehru Centre for Advanced Scientific Research, Jakkur P.O., Bangalore 560064, India

Received 4 April 2005; in final form 10 June 2005
Available online 11 July 2005

Abstract

Carbon nanotubes (CNTs) and N-doped CNTs with Y-junctions have been prepared by the pyrolysis of nickelocene–thiophene and nickel phthalocyanine–thiophene mixtures, respectively, the latter being reported for the first time. The junctions are free from the presence of sulfur and contain only carbon or carbon and nitrogen. The electronic properties of the junction nanotubes have been investigated by scanning tunneling microscopy. Tunneling conductance measurements reveal rectifying behavior with regions of coulomb blockade, the effect being much larger in the N-doped junction nanotubes.
© 2005 Elsevier B.V. All rights reserved.

---

1. Introduction

Y-junction carbon nanotubes are considered to be of potential use in the upcoming field of nanoelectronics. In this context, methods that can deliver junction nanotubes of high purity and in good yields are of importance. Satishkumar et al. [1] reported the synthesis of Y-junction nanotubes by the pyrolysis of metallocenes in the presence of thiophene and other sulfur-containing organic compounds. The junctions were later prepared by the pyrolysis of methane over Co supported on MgO by Li et al. [2]. Pyrolysis of organometallics such as metallocenes and metal phthalocyanines in the presence of thiophene, however, appears to be a reliable and efficient route to the junction nanotubes [3,4]. An important aspect of the junction nanotubes relates to the structure and chemical composition of the junction itself, these factors having a bearing on the electronic

properties. In the case of the junction nanotubes prepared by the pyrolysis of organometallics in the presence of organosulphur compounds, it becomes necessary to establish whether sulfur atoms are incorporated at the junctions, considering the high propensity of sulfur to form rings. If sulfur is absent in the junction region, it would imply the presence of five-, seven- or eight-membered rings required to bring about necessary curvature to form a junction [5]. In this context, the disposition of the graphene layers around the junction is a relevant aspect. Since metal nanoparticles are necessary for the formation of the nanotubes, it is of interest to understand their role as well. We have investigated the Y-junction carbon nanotubes prepared by the pyrolysis of nickelocene–thiophene and nickel phthalocyanine–thiophene by transmission electron microscopy (TEM) and electron energy loss spectroscopy (EELS) to throw light on the nature of the Y-junctions.

There have been a few reports on the electronic properties of junction nanotubes [1,6–10]. Rao and coworkers [1] carried out scanning tunneling spectroscopy (STS) measurements on Y-junction carbon nanotubes (CNTs)

---

* Corresponding author. Fax: +91 80 22082760.
E-mail address: cnrrao@jncasr.ac.in (C.N.R. Rao).

F.L. Deepak et al. / Chemical Physics Letters 411 (2005) 468–473

and reported that the junction acts as a diode. A theoretical study of the electron transport properties of doped nanotubes has shown that a doped nanotube with donor atoms on one side and acceptor atoms on the other, can function as a nanodiode [8]. Negative differential resistance behavior predicted for intramolecularly doped carbon nanotube junctions [9] was subsequently observed experimentally in the case of K-doped SWNTs [10]. We have carried out a detailed study of the Y-junctions both in CNTs and in N-doped CNTs by scanning tunneling microscopy (STM). The results show that the Y-junctions, particularly in the N-doped carbon nanotubes, possess rectification behavior. This is the first study of its kind on doped carbon nanotube junctions.

2. Experimental

Y-junction carbon nanotubes were prepared by the pyrolysis of nickelocene–thiophene employing the experimental set-up described earlier [1]. Pyrolysis of nickel phthalocyanine–thiophene mixtures was carried out to obtain N-doped carbon nanotubes with Y-junctions.

The nanotubes were examined with a JEOL JEM-3010 transmission electron microscope (TEM) operating at 300 kV and also with a JEM 4000 EX microscope (Fa. JEOL, Japan) with an accelerating voltage of 400 kV and a LaB_6 cathode. EELS chemical mapping was performed with a energy-filtering microscope (Zeiss 912 Omega, acceleration Voltage, 125 keV and LaB_6 cathode). Electron energy loss spectra were recorded with a Gatan imaging filter system attached to the JEOL microscope and fitted with a CCD camera. X-ray diffraction (XRD) patterns were recorded using a Seifert instrument with Cu Kα radiation.

The carbon nanotube junctions were investigated by tunneling conductance measurements. Highly oriented pyrolytic graphite (HOPG) was used as the substrate. Freshly cleaved HOPG substrates were prepared by peeling the upper layers with an adhesive tape. The nanotubes were then taken as a suspension in an organic solvent (typically CCl_4 and ethanol was used) sonicated for 30 min, after which a drop was deposited onto the substrate. The substrate was left in air but covered, for a period of 12 h for drying. Imaging and spectroscopy were carried out using Au tips prepared by electrochemical etching. The STM and STS studies were carried out at room temperature (using a Omicron Vakumphysik STS) operated in air. Atomically resolved images of HOPG were used for internal calibration. Current–voltage (I–V) data were collected in the spectroscopy mode with the feedback loop turned off (maximum current, 50 nA). Tunneling conductance measurements were carried out by positioning the tip atop a Y-junction (the point of contact between the three arms) as well as on

the individual arms. The I–V data from the clean areas of the HOPG substrate were collected repeatedly as reference to ensure the reliability of the measurements.

3. Results and discussion

3.1. Y-junction CNTs

TEM images of the products of pyrolysis of nickelocene–thiophene mixtures show the presence of highly crystalline Y-junction carbon nanotubes with well-formed arms (Fig. 1a). The images also reveal certain

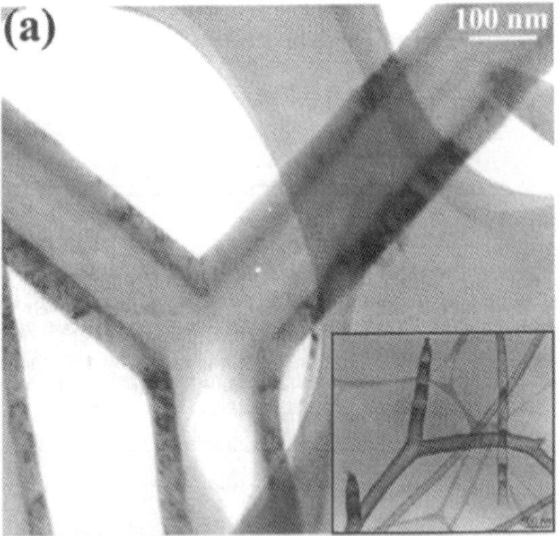

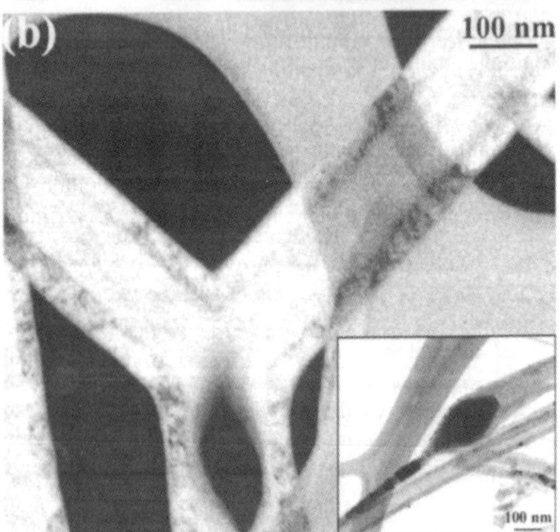

Fig. 1. (a) TEM image of a Y-junction carbon nanotube obtained by the pyrolysis of nickelocene–thiophene mixture. The inset in (a) shows asparagus tips in the Y-junction nanotube. (b) EELS mapping of carbon in the Y-junction carbon nanotube. The inset in (b) shows the TEM image of a Ni nanoparticle inside a carbon nanotube.

470   F.L. Deepak et al. / Chemical Physics Letters 411 (2005) 468–473

unusual nanotube structures such as asparagus-like branches and bamboo structures (see the inset of the Fig. 1a). We show the EELS chemical mapping for carbon in Fig. 1b to demonstrate how the nanotube is made entirely of carbon. We failed to observe any sulfur in the junction region. We have examined the nickel nanoparticles produced in situ by the pyrolysis of the nickelocene–thiophene mixtures. The TEM image of a Ni-nanoparticle is shown in the inset of Fig. 1b. EELS mapping showed the presence of a considerable amount of sulfur, close to 25%, in the nanoparticle. The XRD pattern of the Ni nanoparticles after the reaction with thiophene showed characteristic reflections of rhombohedral Ni_3S_2 with unit cell dimensions of $a = 5.75$ Å and $c = 7.13$ Å (JCPDS file: 44-1418). The sulfur from the thiophene is entirely removed by the nickel particles leaving the thiophene–carbon fragment which could form five-, or seven-membered carbon rings by adding on to different carbon centers in the nascent graphene sheets. Traces of S, in the nanotubes, if at all present, could not be detected at all by EELS or EDX.

The nanotube shown in Fig. 1a was examined by normal-incidence selected area electron diffraction (SAED) with the electron beam perpendicular to the junction. The diffraction pattern of one of the arms close to the junction (Fig. 2a), shows a set of arcs corresponding to four (002) maxima. Similar diffraction patterns were obtained from the other two arms next to the junction, suggesting that the graphene layers close to the junction are well-graphitized and are possibly slanted with some curvature. The electron diffraction pattern of the center shown in Fig. 2b reveals streaks of arcs corresponding to the (002) maxima arising from the three sets of graphitic planes in the three arms close to the junction [11–14]. The intensity distribution is not uniform in the arcs due to the complex nature of the orientation of the graphitic planes at the junction. In Figs. 2c,d, we show the schematic representations of the likely disposition of the graphene sheets around a Y-junction. The structure in Fig. 2c has gradual bends around the junction which can arise due the presence of five- and seven-membered carbon rings in the graphene sheets. It is also possible to have a junction with different fishbone-like orientations of the graphene sheets. From the observations depicted in Figs. 2a,b, it seems likely that the bends at the junctions may not be entirely continuous but instead consist of short straight segments arising from a fishbone-type stacking as in Fig. 2d.

Typical results from the STM measurements on the Y-junction carbon nanotubes are shown in Fig. 3. The image in Fig. 3a shows nanotubes with multiple junctions named J_A, J_B, J_C and J_D, where J_A and J_C are terminal junctions and J_B and J_D are junctions in the middle segments of the nanotubes. A high resolution image of junction J_A, given in the inset, suggests the possible presence of five and eight-membered rings near the

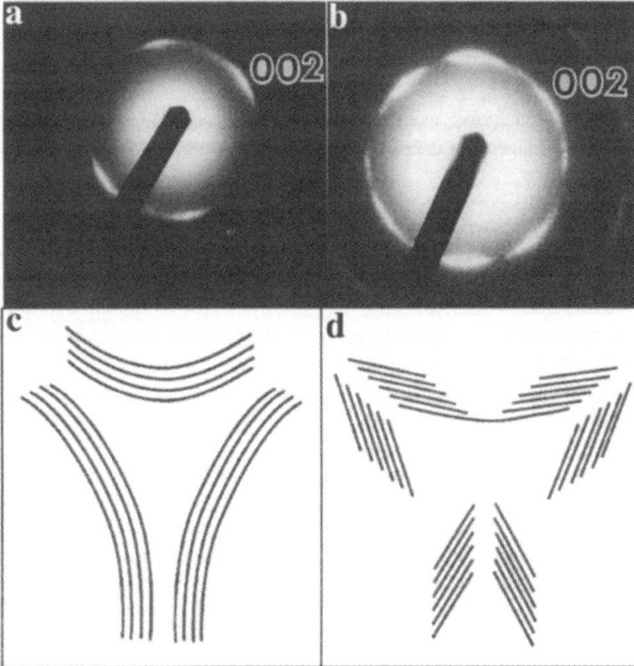

Fig. 2. Selected area electron diffraction (SAED) patterns of the Y-junction carbon nanotubes: (a) of the arms and (b) of the junction region. (c) and (d) Schematic drawings showing different ways of stacking of the graphene sheets in the Y-junction carbon nanotubes.

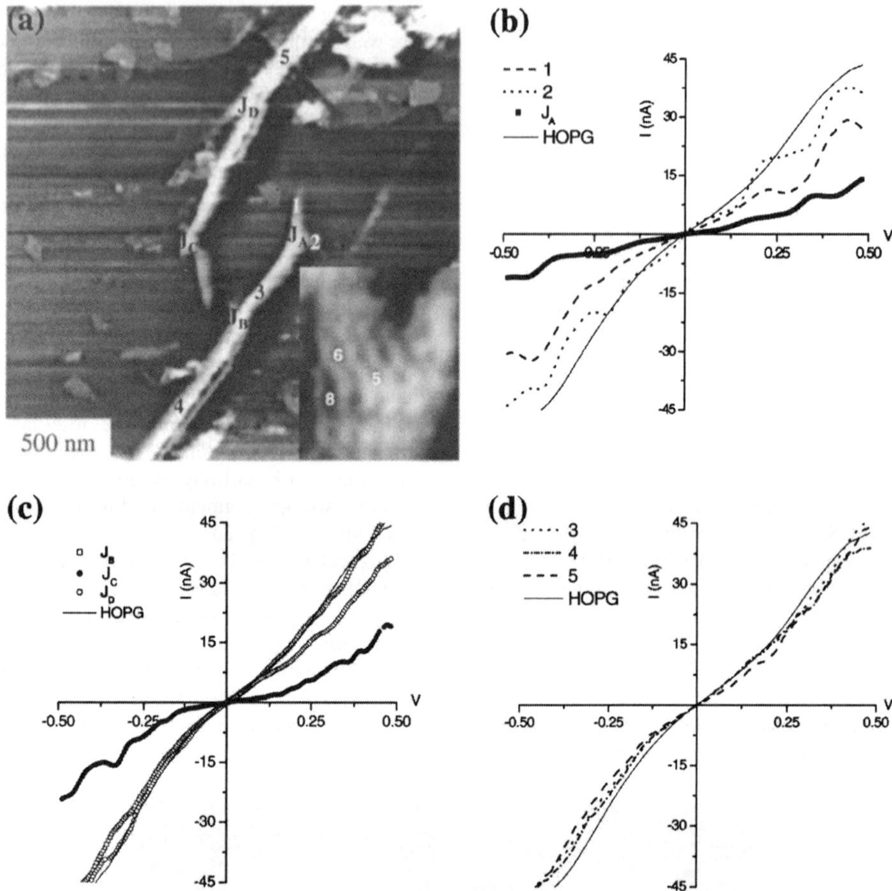

Fig. 3. (a) STM image of undoped carbon nanotubes with multiple junctions. Inset gives a close-up image revealing a five-membered ring (pentagon), and a eight-membered ring (octagon) at the junction region along with many six-membered rings (hexagons) as marked in the figure. (b)–(d) I–V data from the regions marked in (a).

junction region along with the six-membered rings. This is in accordance with the understanding of the structure of junctions in carbon nanotubes, wherein the introduction of five-, seven- or eight-membered rings in an otherwise hexagonal framework is responsible for formation of the junctions. The I–V data collected by positioning the tip over junction J_A exhibits an overall reduction in the current as compared to the HOPG substrate, with steps in the current at ±0.2 and ±0.4 V (Fig. 3b), perhaps due to incremental charging. The low current flat region enclosed by the first step (±0.2 V) corresponds to coulomb blockade at the junction. The curve is also somewhat asymmetric with a rectification ratio (which is the ratio of the forward and reverse currents at a given bias) of 1.3 at ±0.5 V. The data collected from the side arms 1 and 2 showed an overall increase in the current with prominent jumps at similar bias values. The terminal junction J_C exhibits a rectifying behavior similar to J_A (see Fig. 3c) while the interior junctions J_B and J_D

show I–V characteristics comparable to those of the graphite substrate. This is also true of the regions – 3, 4, 5 (Fig. 3d) along the length of the tubes, a behavior characteristic of metallic nanotubes. What is noteworthy is the rectification behavior of the terminal junctions J_A and J_C.

3.2. N-doped Y-junction CNTs

We could obtain good yields of N-doped Y-junction carbon nanotubes by the pyrolysis of nickel–phthalocyanine mixture. These nanotubes have rounded tips and uneven arms (Fig. 4a). Chemical mapping of the tubes revealed a small amount of nitrogen but there was no sulfur in the junction region. Thus, the EEL spectrum (Fig. 4b) shows characteristic edges at 284 and 400 eV corresponding to the K-shell ionization of carbon and nitrogen, respectively. The nitrogen signal is a doublet due to π* and σ* levels, thereby indicating that the nitro-

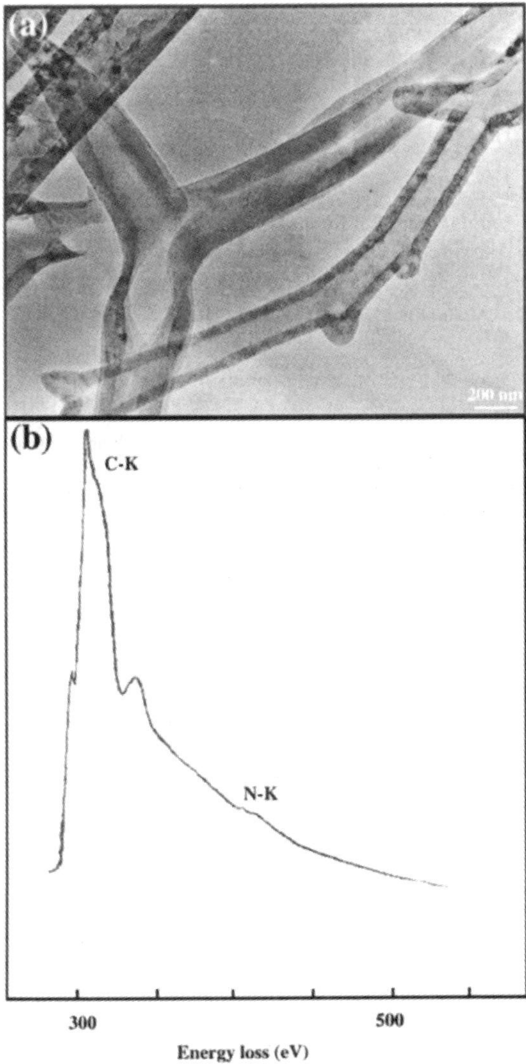

Fig. 4. (a) TEM image of N-doped Y-junction carbon nanotubes prepared by the pyrolysis of a Ni–phthalocyanine–thiophene mixture. (b) EEL spectrum showing the presence of doped nitrogen.

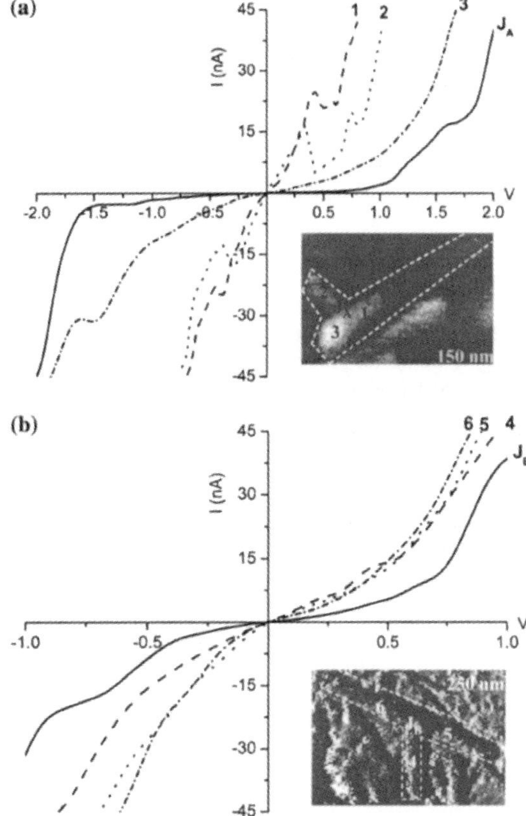

Fig. 5. I–V data of N-doped Y-junction carbon nanotubes shown along with the STM images. The regions where the I–V data were collected are marked with numbers.

gen is present substitutionally in the graphene sheet. Based on the EEL spectrum, the average composition of the nanotubes is estimated to be $C_{45\pm5}N$ [15,16].

We have carried out tunneling conductance measurements on a large number of N-doped Y-junction carbon nanotubes. The junction (J_A) in Fig. 5a exhibits a rectification behavior with distinct change of slope in the I–V data at −1.5 and +1 V. A high value of rectification ratio of 4.75 is obtained for bias voltages of ±1.5 V. Coulomb blockade is easily identifiable in the bias range, −0.4 to +0.6 V. The arms of the tube, however, are devoid of blockade but exhibit varied behavior. The arm-1 shows distinct features at ±0.4 V. In the case of arm-2, similar

features are seen at a slightly lower bias of ±0.3 V. It also exhibits an additional feature in the positive bias region at 0.75 V. The I–V data for arm-3 shows a feature around −1.5 V. While the distinct spectroscopic features corresponding to regions with negative differential resistance would arise from charging or other effects, the occurrence of a wide blockade region at the tube junction is noteworthy. Fig. 5b shows a nanotube with junction J_B, bearing a low-current flat region in the I–V data, from −0.4 to +0.75 V, indicative of a rectification behavior, similar to J_A shown in Fig. 5a. The rectification ratio in this case is 1.21 at ±1.0 V. The I–V curves from the arms (regions 4, 5 and 6) are somewhat asymmetric across zero bias, but do not exhibit any blockade.

4. Conclusions

In conclusion, Y-junction carbon nanotubes prepared by the pyrolysis of nickelocene–thiophene and nickel phthalocyanine–thiophene mixtures do not contain sulfur at the junction but consist entirely of carbon or carbon and nitrogen, suggesting the presence of five-, seven-, or eight-membered rings at the junction. STM investigations show that rectification behavior at the junction is a fairly general feature, although it is more prominent in the N-doped CNTs. In addition, some of the junction nanotubes show coulomb blockade and features of NDR (negative differential resistance) as well.

Acknowledgments

The authors thank Department of Science and Technology (India) for support of this research. N.S.J. thanks CSIR (India) for financial assistance.

References

[1] B.C. Satishkumar, P.J. Thomas, A. Govindaraj, C.N.R. Rao, Appl. Phys. Lett. 77 (2001) 2530.
[2] W.Z. Li, J.G. Wen, Z.F. Ren, Appl. Phys. Lett. 79 (2001) 1879.
[3] F.L. Deepak, A. Govindaraj, C.N.R. Rao, Chem. Phys. Lett. 345 (2001) 5.
[4] C.N.R. Rao, A. Govindaraj, Acc. Chem. Res. 35 (2002) 998.
[5] M. Menon, D. Srivastava, J. Mater. Res. 13 (1998) 2357.
[6] C. Papadopoulos, A. Raitkin, J. Li, A.S. Vedeneev, J.M. Xu, Phys. Rev. Lett. 85 (2000) 3476.
[7] C. Papadopoulos, A.J. Jin, J.M. Xu, Appl. Phys. Lett. 85 (2004) 1769.
[8] K. Esfarjani, A.A. Farajian, Y. Hashi, Y. Kawazoe, Appl. Phys. Lett. 74 (1999) 79.
[9] A.A. Farajian, K. Esfarjani, Y. Kawazoe, Phys. Rev. Lett. 82 (1999) 5084.
[10] C. Zhou, J. Kong, E. Yenilmez, H. Dai, Science 290 (2002) 1552.
[11] S. Amelinckx, A. Lucas, P. Lambin, Rep. Prog. Phys. 62 (1999) 1471.
[12] B. Gan, J. Ahn, Q. Zhang, S.F. Yoon, Rusli, Q.-F. Huang, H. Yang, M.-B. Yu, W.-Z. Li, Diam. Relat. Mater. 9 (2000) 897.
[13] J. Liu, M. Shao, X. Chen, W. Yu, X. Liu, Y. Qian, J. Am. Chem. Soc. 125 (2003) 8088.
[14] J. Liu, L. Xu, W. Zhang, W.J. Lin, X. Chen, Z. Wang, Y. Qian, J. Phys. Chem. B. 108 (2004) 20090.
[15] R. Sen, B.C. Satishkumar, A. Govindaraj, K.R. Harikumar, G. Raina, J.-P. Zhang, A.K. Cheetham, C.N.R. Rao, Chem. Phys. Lett. 287 (1998) 671.
[16] M. Yudasaka, R. Kikuchi, Y. Ohki, S. Yoshimura, Carbon 35 (1997) 195.

Available online at www.sciencedirect.com

ScienceDirect

Chemical Physics Letters 432 (2006) 190–194

CHEMICAL
PHYSICS
LETTERS

www.elsevier.com/locate/cplett

Hydrogel-assisted synthesis of nanotubes and nanorods of CdS, ZnS and CuS, showing some evidence for oriented attachment

K.P. Kalyanikutty [a], M. Nikhila [a], Uday Maitra [b], C.N.R. Rao [a],*

[a] *Chemistry and Physics of Materials Unit, DST Nanoscience Unit and CSIR Centre of Excellence in Chemistry, Jawaharlal Nehru Centre for Advanced Scientific Research, Jakkur P.O., Bangalore 560 064, India*
[b] *Department of Organic Chemistry, Indian Institute of Science, Bangalore 560 012, India*

Received 26 September 2006
Available online 17 October 2006

Abstract

By carrying out the reaction of appropriate metal compounds with Na_2S in the presence of a tripodal cholamide-based hydrogel, nanotubes and nanorods of CdS, ZnS and CuS have been obtained. The nanostructures have been characterized by transmission electron microscopy and spectroscopic techniques. Evidence is presented for the assembly of short nanorods to form one-dimensional chains.
© 2006 Elsevier B.V. All rights reserved.

1. Introduction

Nanotubes and nanowires of semiconducting materials such as CdS, ZnS and CdSe have been prepared by a variety of methods [1,2]. In this context, the surfactant-assisted method is one of the novel innovations [3,4]. In this method, the surfactant molecules spontaneously organize at the critical micellar concentration (CMC) into rod-shaped micelles, thus acting as templates for the formation of nanotubes and nanowires. The surfactant-assisted method can also involve the growth of nanostructures by the oriented attachment of nanoparticles. Oriented attachment of nanoparticles is an effective means of forming anisotropic nanocrystals in solution-based routes. In this mechanism, particles undergo fusion at specific dimensionally similar crystallographic surfaces, leading to the formation of interesting nanostructures [5]. Oriented attachment of nanoparticles has been observed in several materials. Thus, Tang et al. [6] have reported the formation of CdTe nanoparticle chains by spontaneous organization. Formation of nanowires of PbSe by dipole-driven oriented attachment of collections of nanocrystals along identical crystal faces has been reported by Murray and co-workers [7].

By oriented attachment, Yu et al. [8] could obtain ZnS nanorods with cubic blend structure, in which the (1 1 1) planes of the component ZnS nanocrystals are nearly perfectly aligned. Pradhan et al. [9] have reported the formation of colloidal CdSe quantum wires by the oriented attachment of magic-sized clusters of CdSe.

Recently, a hydrogel has been employed as templates to prepare nanotubes of metal oxides and sulfates [10]. The nanotubes so obtained appear to be polycrystalline, and it appears that the growth of nanostructures by the hydrogel route could involve oriented attachment of nanoparticles. Since materials with hexagonal crystal structure with a large difference in the surface energy between the (0 0 0 1) plane and the other planes, are considered to favor anisotropic growth along the [0 0 0 1] crystallographic direction, we have sought to prepare nanotubes and nanorods of CdS, ZnS and CuS by using hydrogel template route. All the three metal sulfides have the hexagonal structure and we therefore considered it feasible to observe oriented growth of nanoparticles in the formation of one-dimensional nanostructures.

2. Experimental

Tripodal cholamide hydrogelator was synthesized according to the procedure reported elsewhere [11]. The

---

* Corresponding author. Fax: +91 80 2208 2766.
E-mail address: cnrrao@jncasr.ac.in (C.N.R. Rao).

average diameter of the hydrogel fibers was in the range 8–10 nm, with lengths extending up to a few hundred nanometers [10]. Nanostructures of CdS, ZnS and CuS were prepared starting with the corresponding metal acetates. Na_2S was used as the sulfur precursor in the synthesis. A sol of the hydrogel was obtained by dissolving 5 mg (0.0075 mmol) in 100 μL of acetic acid and 400 μL of water. In a typical reaction, for the preparation of CdS nanotubes, a gel was formed by adding 11 mg (0.04 mmol) of cadmium acetate to a solution obtained by dissolving 250 mg of KOH in 250 μL water and 25 μL distilled ethanol. To this gel was added a sol of the hydrogel. This was thoroughly mixed under sonication, and warmed slightly to form a sol. A white gel so obtained was allowed to stand for 24 h at 30 °C. An aqueous solution of 3.4 mg (0.043 mmol) of Na_2S dissolved in 100 μL water was added to the sol obtained by warming the white gel slightly. The sol which turned yellow instantly was shaken using a stirrer to ensure complete mixing of the reactants. The yellow gel obtained in this manner was allowed to stand for 24 h at 30 °C.

The preparation of nanostructures of ZnS was carried out with two different concentrations of $Zn(OAc)_2$ and Na_2S, maintaining the gel concentration the same (0.0075 mmol), the procedure being similar to that employed in the case of CdS. In one of the reactions, 8.75 mg (0.04 mmol) of $Zn(OAc)_2$ was reacted with 3.4 mg (0.043 mmol) of Na_2S. This reaction led to the formation of nanorods of ZnS. When the $Zn(OAc)_2$ concentration was reduced to half [4.37 mg (0.02 mmol)], we obtained nanotubes of ZnS. CuS nanostructures were prepared starting with 5 mg (0.0075 mmol) of the gelator, 8 mg (0.04 mmol) of copper acetate and 6.25 mg (0.08 mmol) of Na_2S, the experimental procedure being the same as in the case of CdS. When the blue sol containing $Cu(OH)_2$ was mixed with Na_2S solution, we obtained a black CuS gel. In order to remove the hydrogel template, the products containing CdS, ZnS and CuS nanostructures were washed several times with distilled ethanol.

Energy dispersive analysis of X-rays (EDAX) of the samples was performed with a Oxford microanalysis group 5526 system attached to the SEM employing Links (ISIS) software and a Si(Li) detector. A JEOL JEM 3010 microscope operating at 300 kV was used for transmission electron microscopy (TEM) analysis. TEM samples were prepared by placing a drop of the sample suspensions in C_2H_5OH on a Cu coated holey carbon grid, and allowed to evaporate slowly. UV–visible absorption measurements were carried out at room temperature with a Perkin–Elmer model Lambda 900 UV/Vis/NIR spectrometer. For UV spectroscopic analysis samples were dispersed thoroughly in C_2H_5OH. Photoluminescence spectroscopy (PL) measurements were carried out at room temperature with a Perkin–Elmer model LS50B luminescence spectrometer. The measurements were carried out using CCl_4 suspensions of the samples.

3. Results and discussion

In Fig. 1 we shows the TEM images of nanotubes of CdS, obtained after the removal of the hydrogel template. The images reveal the presence of a number of nanotubes. EDAX analysis of the CdS nanotubes confirmed their purity, giving a Cd:S ratio of 1:1. The TEM image in Fig. 1a demonstrates the hollow nature of the nanotubes. The lengths of the nanotubes extend to a few hundred nanometers while the diameter of the inner tubule is ∼2–3 nm, the outer diameter being in the 20–25 nm range. Electron diffraction patterns showed the nanotubes to be generally polycrystalline. A selected area electron diffraction (SAED) pattern of a single nanotube is given in the bottom inset of Fig. 1a. The diffuse rings correspond to the (100) and (110) Bragg planes of hexagonal CdS. Clearly, the tripodal cholamide gel fibers act as templates, on which the CdS particles get deposited, giving rise to the nanotubes. The low magnification TEM image in Fig. 1b suggests a possible assembly or attachment of the initially formed shorter nanotubes to form linear chains (indicated by arrows in the figure). The hydrogel might be responsible for such

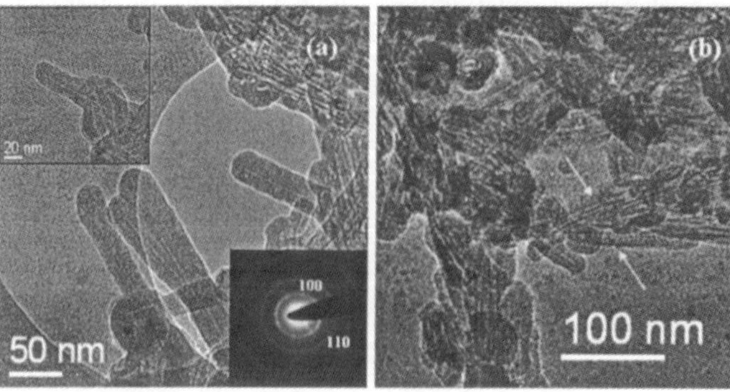

Fig. 1. (a) TEM image of CdS nanotubes obtained after the removal of hydrogel template. Top inset is a high-magnification image of a single nanotube. Bottom inset is the SAED pattern of the nanotubes. (b) TEM image showing a bunch of nanotubes assembled spontaneously, indicated by the arrows.

K.P. Kalyanikutty et al. / Chemical Physics Letters 432 (2006) 190–194

an attachment, the hydrogel playing a dual role of being a template to produce hollow nanotubes as well as favoring the attachment or assembly of the nanotubes.

The UV–visible absorption spectrum of the CdS nanotubes given in Fig. 2a shows a blue-shift in the excitonic absorption band to 460 nm. The blue-shift from the bulk value of 515 nm [12] is due to quantum confinement effects in the CdS nanotubes, the inner diameter of the nanotubes being less than the Bohr-exciton diameter of CdS (6 nm). Xiong et al. [13] have reported an absorption band at 459 nm for CdS nanotubes with an inner diameter $\sim$5 nm prepared by an in situ micelle-template-interface reaction. An absorption maximum around 450 nm has been reported in nanoparticles and hollow spheres of CdS [12,14]. In Fig. 2b, we show the photoluminescence (PL) spectrum of the CdS nanotubes prepared by us, revealing a band centered at 610 nm. This band is due to charge carriers trapped at surface defects of the nanotubes [15,16].

In the case of ZnS, we could obtain nanotubes as well as nanorods depending on the concentration of $Zn(OAc)_2$. The TEM image in Fig. 3a shows nanotubes of ZnS obtained at the lower concentration (0.02 mmol) of $Zn(OAc)_2$. The nanotubes have an inner diameter in the $\sim$4–6 nm range, with lengths going up to a micrometer.

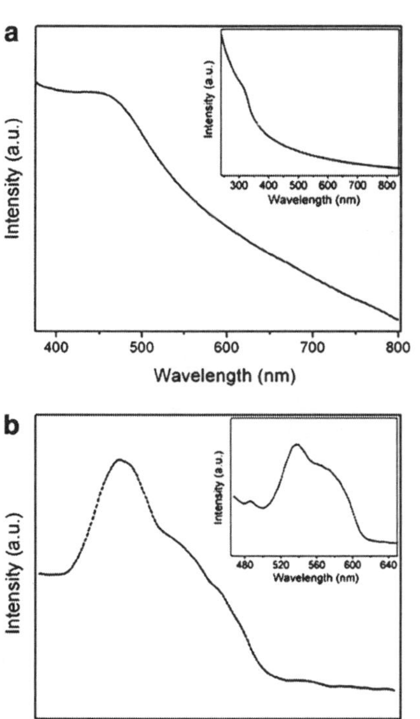

Fig. 2. (a) Electronic absorption spectrum of CdS nanotubes; (b) PL spectrum of the CdS nanotubes. Insets in (a) and (b) are the electronic absorption spectrum and the PL spectrum, respectively of ZnS nanotubes.

The TEM image shows tiny nanocrystals of ZnS making up the walls of the nanotubes. Electron diffraction patterns also reveal the nanotubes to be polycrystalline and of hexagonal structure. The SAED pattern of one of the nanotubes given in the inset of Fig. 3a shows rings arising from the (1 0 0) and (1 1 0) planes of the hexagonal structure. A bunch of ZnS nanotubes is seen in the TEM image in Fig. 3b. When the concentration of $Zn(OAc)_2$ was doubled (0.04 mmol), we obtained nanorods of ZnS, as shown in the TEM image in Fig. 3c. The diameter of the nanorods is in the 10–15 nm range, with lengths extending to a few hundreds of nanometers. The inset in Fig. 3c shows the SAED pattern of the nanorods, with diffuse rings corresponding to the (1 0 0) and (1 1 0) planes of hexagonal ZnS. Fig. 3d shows another TEM image of the ZnS nanorods. The images in Fig. 3c and d suggest the formation of copious quantities of nanorods. EDAX analysis of the ZnS nanostructures gave a Zn:S ratio of 1:1.

The ZnS nanotubes and nanorods were characterized by UV–visible absorption spectroscopy and PL spectroscopy. The inset in Fig. 2a shows the absorption spectrum of the ZnS nanotubes. The band appearing at $\sim$318 nm is blue-shifted relative to that of the bulk ZnS (350 nm) [17]. Nanowires of ZnS of diameter $\sim$5 nm were reported to show an absorption maximum around 326 nm [18]. An absorption band at $\sim$320 nm has been reported in the case of ZnS quantum dots [19]. The PL spectrum of ZnS nanotubes given in the inset of Fig. 2b exhibits two bands, a weak blue emission at $\sim$485 nm and a strong green emission around 538 nm. The 485 nm band is attributed to zinc vacancies in the ZnS lattice. Emission bands at $\sim$470 nm [20] and $\sim$498 nm [21] have been reported in ZnS nanobelts. The 538 emission band is similar to that reported for ZnS nanobelts [22] and is considered to result from vacancy or interstitial states [22,23].

We have obtained nanotubes and nanowires formed by oriented growth in the case of CuS, all the nanostructures having a Cu:S ratio of 1:1 as revealed by EDAX analysis. The TEM image in Fig. 4a shows nanotubes of CuS with a narrow hollow region, obtained after the removal of the hydrogel template. The inner diameter of the nanotubes is $\sim$5 nm with lengths extending to a few hundreds of nanometers. The outer diameter is in the 20–30 nm range. The nanotubes are polycrystalline as found from electron diffraction as well as the TEM images. The SAED pattern of a single nanotube given as an inset displays diffuse polycrystalline rings corresponding to the (1 0 1) and (1 0 7) planes of hexagonal CuS. A few spots are also seen along the rings. The bottom inset in Fig. 4a is a TEM image of a nanotube of CuS with an outer diameter of $\sim$20 nm and a length of $\sim$300 nm, revealing that the walls of the nanotubes are formed from nanoparticles. The hydrogel fibers act as templates leading to the formation of CuS nanotubes comprising nanoparticles, just as in the case of CdS and ZnS. The situation is comparable to that in oriented attachment growth and we do find evidence for oriented attachment of nanocrystals in the TEM images.

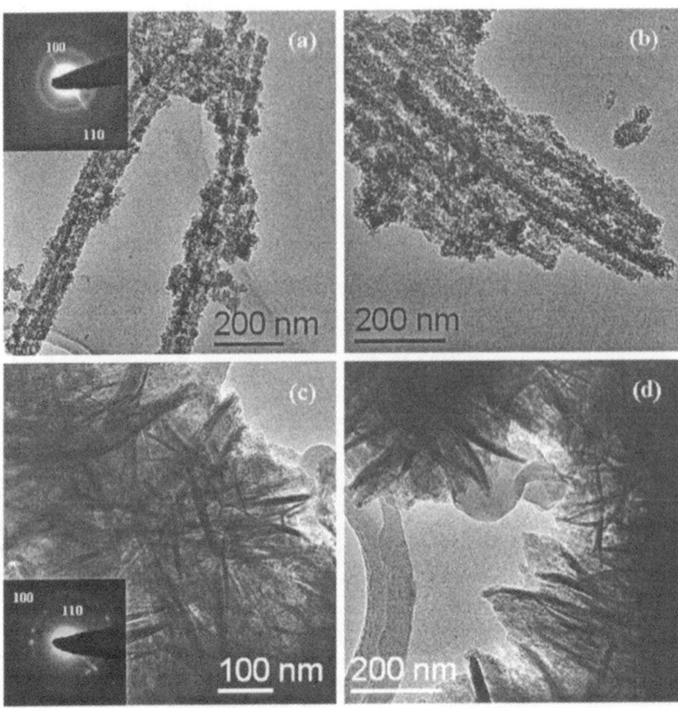

Fig. 3. (a) and (b) TEM images showing nanotubes of ZnS obtained using 0.02 mmol of Zn(OAc)$_2$; (c) and (d), nanorods of ZnS obtained using a higher concentration (0.04 mmol) of Zn(OAc)$_2$. Insets are the corresponding SAED patterns.

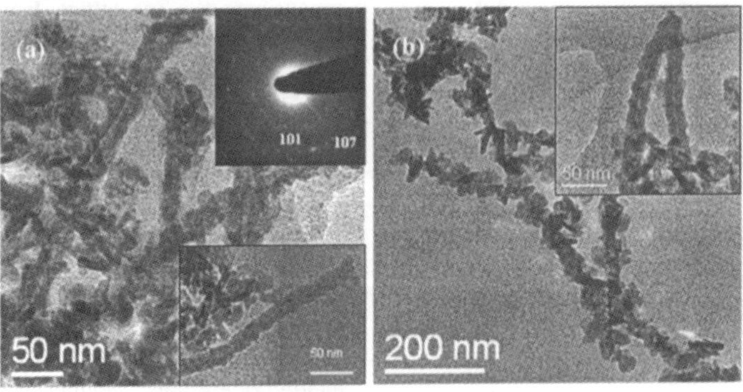

Fig. 4. TEM images showing (a) nanotubes of CuS obtained after the removal of the template (Top inset is the SAED pattern and bottom inset is a TEM image showing a single nanotube). (b) chains of nanorods of CuS formed by self- assembly. A V-shaped nanostructure of CuS formed by the attachment of two nanotubes.

The TEM image in Fig. 4b reveals the formation of chains of short nanorods. The chains are formed by the assembly of CuS nanorods of ~40 nm length, the chains themselves extending over hundreds of nanometers. Zhang et al. [24] have reported the formation of self-assembled rods of Cu$_2$S. Based on the TEM images they noted the rods to be the edges of Cu$_2$S nanodisks that are aggregated with their hexagonal planes assembled together. A similar observation has been reported with hcp-Co nanocrystals by Ali-

visatos et al. [25], wherein magnetic nanodisks are stacked face-to-face giving rise to ribbons. Lee et al. [26] have demonstrated that oriented attachment is an effective mechanism for the formation of chains of SnO$_2$ nanorods. They suggest that either collision of aligned nanocrystals in suspension, or rotation of misaligned nanoparticles in contact toward low-energy configurations might be responsible for the oriented attachments, and that anisotropic growth can be controlled by surfactants adsorbed preferentially on spe-

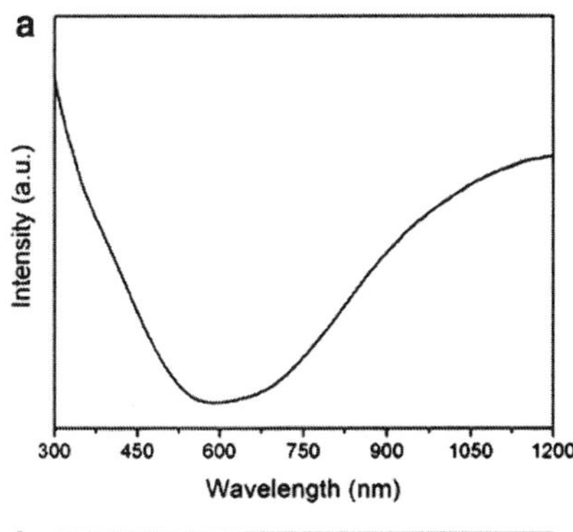

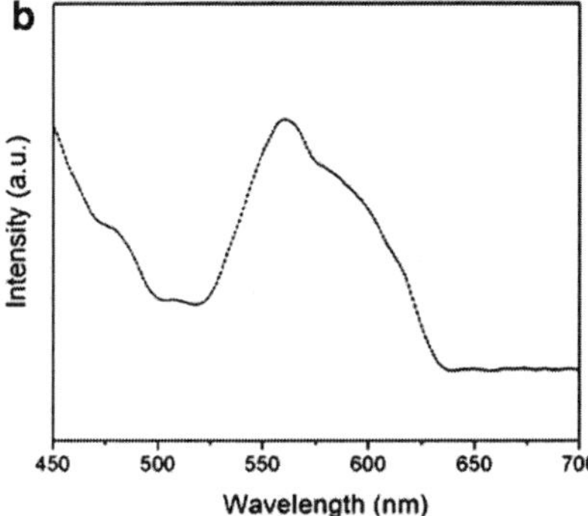

Fig. 5. (a) Electronic absorption spectrum of CuS nanostructures, (b) PL spectrum of the nanostructures.

cific lattice planes of the crystal. The inset in Fig. 4b shows a TEM image of a V-shaped nanostructure of CuS, thought to be formed by the attachment of two nanotubes at their ends.

The electronic absorption spectrum of the CuS nanotubes given in Fig. 5a shows the characteristic broad band of CuS in the near IR region, peaking at ∼1200 nm. The band is attributed to an electron-acceptor state lying within the band gap [17]. A similar broad band has been reported for CuS nanocrystals [27]. The photoluminescence spectrum of CuS nanostructures given in Fig. 5b shows a broad band peaking at 560 nm with a shoulder at 480 nm. Bulk CuS is reported to show a broad band centered at ∼560 nm with a shoulder at ∼587 nm [17]. The absence of any appreciable blue-shift of the emission bands of the CuS nanostructures prepared by us might be due to the formation of chains of nanorods by self-assembly.

4. Conclusions

By making use of a tripodal cholamide-based hydrogel, we have synthesized nanotubes and nanorods of CdS, ZnS and CuS. The nanotubes and nanorods are generally polycrystalline, suggesting the templating role of the hydrogel fibers and possible occurrence of oriented attachment. We have observed self-assembly of nanocrystals giving rise to chains of nanorods, a phenomenon similar to the growth of nanowires by oriented attachment.

Acknowledgements

The authors thank DRDO and the Department of Science and Technology for support of this research. Reji Thomas is thanked for his assistance with hydrogel synthesis.

References

[1] C.N.R. Rao, A. Muller, A.K. Cheetham (Eds.), The Chemistry of Nanomaterials, Wiley-VCH, Weinheim, 2004.
[2] C.N.R. Rao, A. Govindaraj, Nanotubes and Nanowires, Royal Society of Chemistry, London, 2005.
[3] C.N.R. Rao Govindaraj, F.L. Deepak, N.A. Gunari, M. Nath, Appl. Phys. Lett. 78 (2001) 1853.
[4] Y. Xia, P. Yang, Y. Sun, Y. Wu, B. Mayers, B. Gates, Y. Yin, F. Kim, H. Yan, Adv. Mater. 15 (2003) 353.
[5] R.L. Penn, J.F. Banfield, Science 281 (1998) 969.
[6] Z. Tang, N.A. Kotov, M. Giersig, Science 297 (2002) 237.
[7] K.-S. Cho, D.V. Talapin, W. Gaschler, C.B. Murray, J. Am. Chem. Soc. 127 (2005) 7140.
[8] J.H. Yu, J. Joo, H.M. Park, S.I. Baik, Y.W. Kim, S.C. Kim, T.J. Hyeon, J. Am. Chem. Soc. 127 (2005) 5662.
[9] N. Pradhan, H. Xu, X. Peng, Nanoletters 6 (2006) 720.
[10] G. Gundiah, S. Mukhopadhyay, U.G. Tumkurkar, A. Govindaraj, U. Maitra, C.N.R. Rao, J. Mater. Chem. 13 (2003) 2118.
[11] U. Maitra, S. Mukhopadhyay, A. Sarkar, P. Rao, S.S. Indi, Angew. Chem., Int. Ed. 40 (2001) 2281.
[12] U.K. Gautam, M. Ghosh, C.N.R. Rao, Chem. Phys. Lett. 381 (2003) 1.
[13] Y. Xiong, Y. Xie, J. Yang, R. Zhong, C. Wu, G. Du, J. Mater. Chem. 12 (2002) 3712.
[14] J. Huang, Y. Xie, B. Li, Y. Liu, Y. Qian, S. Zhang, Adv. Mater. 12 (2000) 808.
[15] Y. Li, X. Li, C. Yang, Y. Li, J. Mater. Chem. 13 (2003) 2641.
[16] B. Liu, G.Q. Xu, L.M. Gan, C.H. Chew, W.S. Li, Z.X. Shen, J. Appl. Phys. 89 (2001) 1059.
[17] F.L. Deepak, A. Govindaraj, C.N.R. Rao, J. Nanosci. Nanotech. 2 (2002) 417.
[18] X. Jiang, Yi Xie, J. Lu, L. Zhu, W. He, Yitai Qian, Chem. Mater. 13 (2001) 1213.
[19] N. Kumbhojkar, V.V. Nikesh, A. Kshirgar, S. Mahamuni, J. Appl. Phys. 88 (2000) 6260.
[20] P. Hu, Y. Liu, L. Fu, L. Cao, D. Zhu, J. Phys. Chem. B 108 (2004) 936.
[21] W-T. Yao, S-H. Yu, L. Pan, J. Li, Q-S. Wu, L. Zhang, J. Jiang, Small 1 (2005) 320.
[22] L-W. Yin, Y. Bando, J-H. Zhan, M-S. Li, D. Golberg, Adv. Mater. 17 (2005) 1972.
[23] Y. Jiang, X.M. Meng, J. Liu, Z.Y. Xie, C.S. Lee, S.T. Lee, Adv. Mater. 15 (2003) 323.
[24] P. Zhang, L. Gao, J. Mater. Chem. 13 (2003) 2007.
[25] V.F. Puntes, D. Zanchet, C.K. Erdonmez, A.P. Alivisatos, J. Am. Chem. Soc. 124 (2002) 12874.
[26] E.J.H. Lee, C. Riberio, E. Longo, E.R. Leite, J. Phys. Chem. B. 109 (2005) 20842.
[27] U.K. Gautam, B. Mukherjee, Bull. Mater. Sci. 29 (2006) 1.

PAPER   www.rsc.org/materials | Journal of Materials Chemistry

Synthesis, structure and properties of homogeneous BC$_4$N nanotubes

Kalyan Raidongia,[a] Dinesh Jagadeesan,[a] Mousumi Upadhyay-Kahaly,[b] U. V. Waghmare,[b] Swapan. K. Pati,[b] M. Eswaramoorthy[a] and C. N. R. Rao*[a]

Received 14th August 2007, Accepted 14th September 2007
First published as an Advance Article on the web 28th September 2007
DOI: 10.1039/b712472d

BCN nanotube brushes have been obtained by the high temperature reaction of amorphous carbon nanotube (a-CNT) brushes with a mixture of boric acid and urea. The a-CNT brushes themselves were obtained by the pyrolysis of glucose in a polycarbonate membrane. The BCN nanotubes have been characterized by EELS, XPS, electron microscopy, Raman spectroscopy and other techniques. The composition of these nanotubes is found to be BC$_4$N. The nanotubes, which are stable up to 900 °C, are insulating and nonmagnetic. They exhibit a selective uptake of CO$_2$ up to 23.5 wt%. In order to understand the structure and properties, we have carried out first-principles density functional theory based calculations on (6,0), (6,6) and (8,0) nanotubes with the composition BC$_4$N. While (8,0) BC$_4$N nanotubes exhibit a semiconducting gap, the (6,0) BC$_4$N nanotube remains metallic if ordered BN bonds are present in all the six-membered rings. The (6,6) BC$_4$N nanotubes, however, exhibit a small semiconducting gap unlike the carbon nanotubes. The most stable structure is predicted to be the one where BN$_3$ and NB$_3$ units connected by a B–N bond are present in the graphite matrix, the structure with ordered B–N bonds in the six-membered rings of graphite being less stable. In the former structure, (6,0) nanotubes also exhibit a gap. The calculations predict BC$_4$N nanotubes to be overall nonmagnetic, as is indeed observed.

1. Introduction

The fascinating properties of carbon nanotubes[1] gave enormous impetus to researchers to explore analogous materials such as boron nitride (BN), boron carbon nitride (BCN) and boron carbide (BC) nanotubes. Homogeneous BCN nanotubes whose properties can be tuned by varying their composition and the arrangement of B, C and N atoms are expected to have potential applications in electronics, electrical conductors, high temperature lubricants and novel composites.[2] They could be useful in gas sorption applications as well. Stephan *et al.*[3] first reported carbon nanotubes containing B and N prepared through a modified electric arc-discharge method which turn out to be a mixture of graphite, boron and nitrogen. By a similar procedure, Suenaga *et al.*[4] produced BCN nanotubes with well-separated layers of BN and carbon. Redlich *et al.*[5] synthesized BCN nanotubes having a BC$_2$N outer shell and a carbon inner shell by the arc-discharge method. Nanotubes with outer BC$_7$N layers and pure carbon inner layers have been obtained by laser ablation using a composite of BN and carbon as the target in the presence of nickel and cobalt.[6] Rao and co-workers[7] prepared BCN nanotubes of varying compositions of carbon and nitrogen by the pyrolysis of a BH$_3$–trimethylamine adduct. A template based approach has also been reported to prepare BCN nanotubes using graphitic carbon nanotubes and carbon

nitride nanotubes.[8] BC$_4$N powder has been obtained by the nitridation of boric acid and carbonization of saccharose in molten urea.[9] Single walled carbon nanotubes doped with B and N have been prepared by the hot filament method.[10] Multiwalled nanotubes of the composition B$_5$CN$_5$ have been produced by chemical vapor deposition along with nanotubes containing BN layers sheathed with outer carbon layers.[11] While the arc-discharge and laser ablation methods have drawbacks in controlling phase separation and the diameter of the nanotubes, the template based method has limitations in extending the diameter of the BCN nanotubes beyond 20 nm. Furthermore, the surface of pristine carbon nanotubes is generally not reactive. In the present study, we have employed amorphous carbon nanotube (a-CNT) brushes,[12] prepared by the decomposition of glucose, as starting materials to prepare the BCN nanotube brushes. We have introduced BN in to a-CNTs by using the boric acid–urea mixture. Interestingly, the nanotubes obtained by us, with the composition BC$_4$N, are nonmagnetic insulators. To our knowledge, BC$_4$N nanotubes have not been investigated hitherto. We have carried out first-principles density functional calculations to understand the structure and properties of the BC$_4$N nanotubes.

2. Experimental

Synthesis

Amorphous carbon nanotube (a-CNT) brushes were prepared by the following procedure.[12] Polycarbonate membranes with a pore diameter of 220 nm were soaked in 22 mL of a 0.5 M aqueous solution of glucose in a 25 mL Teflon-lined autoclave. The same procedure was repeated with polycarbonate

[a]*Chemistry and Physics of Materials Unit, Jawaharlal Nehru Centre for Advanced Scientific Research, Jakkur P.O., Bangalore, 560064, India. E-mail: cnrrao@jncasr.ac.in; Fax: +91.80.22082760*
[b]*Theoretical Sciences Unit, Jawaharlal Nehru Centre for Advanced Scientific Research, Jakkur P.O., Bangalore, 560064, India*

membranes with a pore diameter of 50 nm. The temperature of the autoclave was maintained at 180 °C for 6 h after which it was allowed to cool to room temperature. The brownish liquid, rich in carbon spheres, was discarded. The membranes that had turned brown were washed with deionized water and ethanol several times and dried at 40 °C for 1 h.

A mixture of boric acid (1 g) and urea (11.8 g) was taken in 40 ml distilled water and heated at 70 °C until the solution became viscous; the a-CNTs were soaked in it for nearly 2 h. They were later separated physically and dried in air at 40 °C overnight. The dried sample was thermally treated at 970 °C for 3 h for 40 nm nanotubes in a N_2 atmosphere, and for 12 h in the case of the larger diameter (170 nm) nanotubes, and then cooled down to room temperature. The product was subsequently heated in an NH_3 atmosphere at 1050 °C in case of 170 nm nanotubes and 900 °C in case of 40 nm nanotubes for three hours to give black-coloured boron–carbon–nitride nanotube brushes. The products were investigated by transmission electron microscopy and other physical techniques.

In order to obtain Au/Pt nanoparticle-covered BCN nanotube brushes, the nanotubes obtained by the template method described earlier were soaked in 2 mL of 5 mM aqueous solutions of hydrogen hexachloroplatinate (IV) or hydrogen tetrachloroaurate (III) for 12 h. The nanotubes were washed with distilled water twice followed by a washing with 10 mM sodium borohydride solution before drying at 40 °C for an hour. The resulting products were examined by electron microscopy.

Characterization

X-Ray diffraction (XRD) patterns were recorded at 25 °C with a Rich-Siefert 3000-TT diffractometer employing Cu Kα radiation. The morphology of the nanotubes was examined by a field emission scanning electron microscope (FESEM, FEI Nova-Nano SEM-600, Netherlands), and scanning electron microscope (SEM) Leica S-440I instrument (U. K.). TEM images were recorded with a JEOL JEM 3010 instrument (Japan) operated at an accelerating voltage of 300 kV. X-Ray photoelectron spectroscopy (XPS) measurements were performed using a ESCALAB MKIV spectrometer employing Al Kα radiation (1486.6 eV). Electron energy loss spectra (EELS) were recorded with a transmission electron microscope (FEI, TECNAI F30) equipped with an energy filter for EELS operating at 300 kV. Raman spectra were recorded with a LabRAM HR with a 633 nm line from HeNe laser. Thermogravimetric analysis was carried out using a Mettler Toledo Star system. Nitrogen adsorption–desorption isotherms were measured using a QUANTACHROME AUTOSORB-1C surface area analyzer at liquid N_2 temperature (77 K). The CO_2 adsorption was carried out at 195 K (1 : 1 mixture of dry ice and acetone). Hydrogen adsorption was carried out at liquid nitrogen temperature (77 K). Magnetization measurements were carried out with a vibrating sample magnetometer in a physical property measuring system (PPMS,Quantum Design, San Diego, CA, USA).

3. Results and discussion

Fig. 1(a) shows a FESEM image of the amorphous carbon nanotubes (a-CNTs) with a well-aligned brush-like

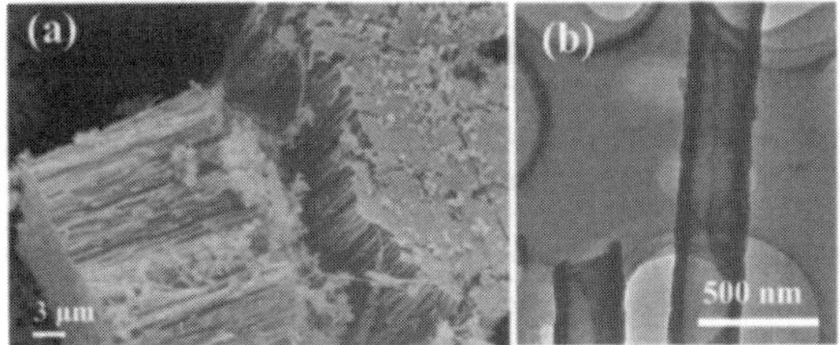

Fig. 1 (a) SEM image of amorphous carbon nanotube brushes. (b) TEM image of individual amorphous carbon nanotubes.

morphology. The TEM image in Fig. 1(b) shows the outer diameter of the a-CNTs to be around 250 nm with a wall thickness of about 50 nm.

After the reaction of a-CNT brushes with the H_3BO_3–urea mixture, we obtain the nanotube structures shown in Fig. 2(a). These structures containing B, C and N replicate the brush-like morphology of the a-CNTs. The diameter of the BCN nanotubes is 170 nm and the lengths are 15 μm. The higher magnification FESEM image in Fig. 2(b) shows the open ends of the BCN nanotubes demonstrating the wall thickness to be around 50 nm. This is further supported by the TEM image of a single nanotube shown in Fig. 2(c). The selected area electron diffraction pattern shows faint rings, with a few spots. The XRD pattern of the BCN nanotube brushes [Fig. 3(a)] shows broad reflections with d spacings of 3.43 Å and 2.13 Å corresponding to (002) and (100) planes respectively, similar to the pattern reported for BC_3N (JCPDS card 35-1292). The broad reflections in the XRD pattern and the diffuse rings in the electron diffraction pattern suggest the turbo static nature of the nanotubes as reported earlier for other preparations of BCN nanotubes.[2] We have prepared a-CNTs using a polycarbonate membrane with a pore size of 50 nm. Using the a-CNTs, we have obtained BCN nanotube brushes. A FESEM image of these nanotube brushes so obtained is shown in the inset in Fig. 2(a). A TEM image of the BCN nanotubes with an outer diameter of 40 nm is shown in Fig. 2(d).

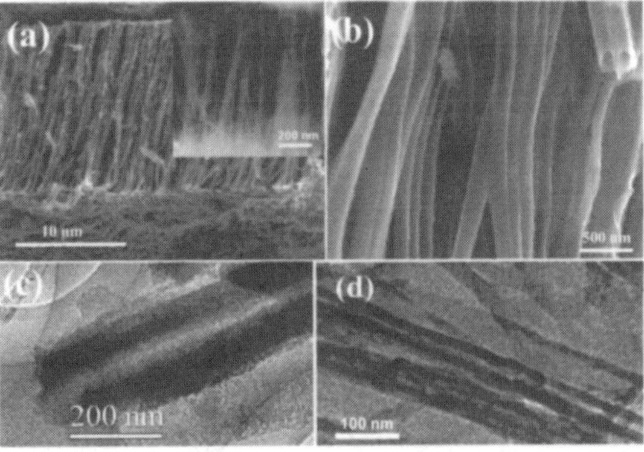

Fig. 2 (a) FESEM images of BCN nanotube brushes with the average diameter of a single tube being around 170 nm. The inset shows a FESEM image of BCN nanotube brushes of 40 nm diameter. (b) Higher magnification FESEM images of BCN nanotube brushes. TEM image of a BCN nanotube (c) 170 nm diameter, (d) 40 nm diameter.

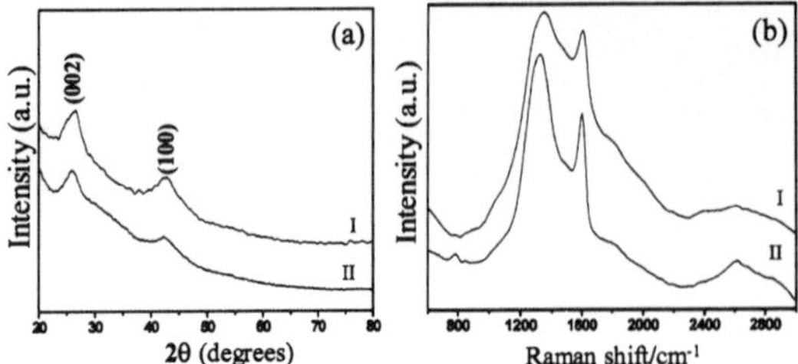

Fig. 3 (a) XRD patterns and (b) Raman spectra of BCN nanotube brushes: (I) 40 nm diameter, (II) 170 nm diameter.

X-Ray photoelectron spectra of the BCN nanotubes in the N, C and B (1s) regions are shown in Fig. 4. We can analyze these data on the lines suggested by Kim *et al.*[11] The N 1s spectrum of the BCN nanotube brushes in Fig. 4(a) shows peaks at 397.7 eV, 400.2 eV and 401.5 eV. The peak at 397.7 eV corresponds to nitrogen bonded to boron (N–B bond), the peak at 400.2 eV corresponds to nitrogen bonded to carbon in a graphite like N–C structure, and the peak at 401.5 eV can be assigned to N bonded to C in a pyridine type structure. The C 1s spectrum in Fig. 4(b) has two peaks with a broad shoulder, the peak at 284 eV is assigned to carbon bonded to a boron atom (C–B bond) and the peak at 286 eV is assigned to carbon bonded to another carbon atom (C–C bond). The long tapering band extending from 286 eV to 289 eV is ascribed to carbon atoms bonded to nitrogen. The B 1s spectrum in Fig. 4(c) has two peaks centered at 191.2 eV and 194 eV, with a shoulder at 189.2 eV. The shoulder at 189.2 eV corresponds to boron bonded to carbon (B–C bond), and the peak at 191.2 eV is due to boron bonded to nitrogen (B–N bond). The peak at 194 eV is assigned to boron bonded to oxygen (B–O bond),

probably arising from the excess B_2O_3. B_2O_3 is, however, X-ray amorphous. By subtracting the contribution of the B–O part from the B 1s signal, we have estimated the composition of the nanotubes taking the capture cross sections into account. Such an analysis gave the approximate composition of the BCN nanotube brushes to be $BC_4N_{1.5}$.

In order to obtain a more reliable elemental analysis, we carried out electron energy loss spectroscopy (EELS) measurements on the K-edge absorption for B, C, and N in a high-resolution electron microscope. The spectrum clearly showed K-shell ionization edges at 188, 284, and 401 eV for B, C and N respectively. Each core edge fine structure consisted of a sharp Π* peak and a well-resolved σ* band, characteristic of sp^2 hybridization.[13] The percentages of B and N were significantly smaller than that of carbon. EELS measurements gave an average chemical composition of the nanotubes to be BC_4N. In Fig. 5 we show the elemental mapping of the nanotubes the red, green and blue colours representing boron, carbon and nitrogen respectively. We see that the colours are randomly distributed across the nanotubes suggesting that the nanotubes are homogeneous with a uniform distribution of B, C and N atoms. The homogeneous nature of the BC_4N nanotubes is also confirmed by the fact that we failed to obtain BN nanostructures after removal of carbon from the BC_4N nanotubes by oxidation.[14]

The Raman spectra of the BC_4N nanotubes were recorded with the 633 nm line from a HeNe laser. The spectra are shown in Fig. 3(b). The observation of two strong peaks at 1324 cm^{-1} and at 1600 cm^{-1} in the Raman spectra are the signatures of the D and G bands of BCN nanotubes.[15,16] The D bands are somewhat broad probably due to the disorder in the BCN layers. The additional peak at around 800 cm^{-1} is similar to that found in BN nanotubes. The band around 2600 cm^{-1} may be due to a combination D + G band or a 2D overtone. Such bands have been observed in BCN nanotubes of other compositions.[15,16]

We have carried out thermogravimetric analysis of BC_4N nanotubes in air. These nanotubes show high thermal stability and we observe no weight loss up to 900 °C (Fig. 6). Amorphous carbon nanotubes get completely oxidized before 750 °C. The high thermal stability of BC_4N nanotubes is noteworthy.

In the literature, there have been theoretical papers describing magnetism of BCN nanotubes. BCN nanotubes

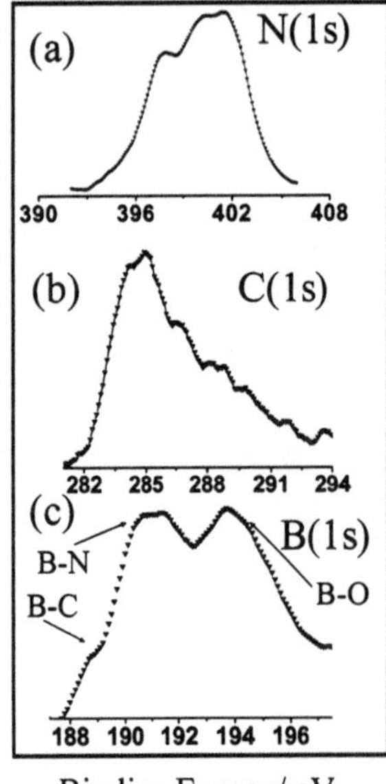

Fig. 4 XPS of the BCN nanotube brushes.

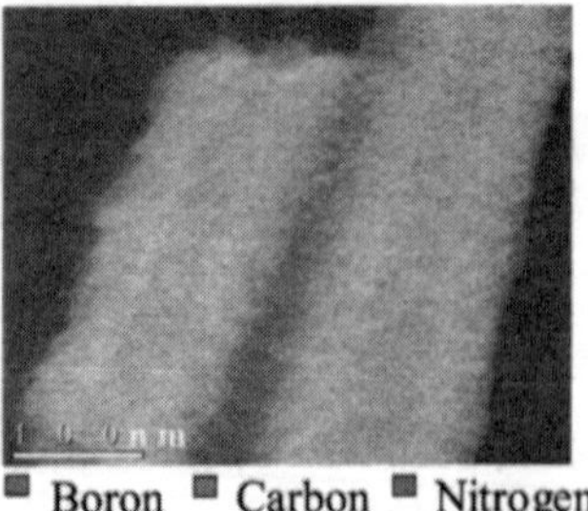

Fig. 5 Elemental mapping of the boron, carbon and nitrogen of BCN nanotubes obtained from EELS.

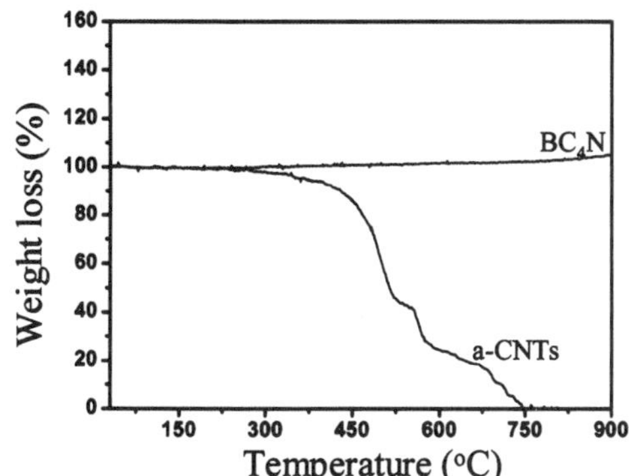

Fig. 6 TGA of BC$_4$N nanotubes and a-CNTs.

having an approximate composition of BC$_2$N have been predicted to be ferromagnetic.[17] Such itinerant ferromagnetism has also been predicted in CBN heterostructured nanotubes.[18] BCN ribbons of the composition BC$_{10}$N have been predicted to be ferrimagnetic.[19] Carbon doping in BN nanotubes is supposed to induce spontaneous magnetism.[20] We have carried out magnetic susceptibility measurements of BC$_4$N nanotube brushes and found them to be nonmagnetic. They show a very small magnetic moment (0.22 μ_B) probably due to defects on the nanotubes. The BC$_4$N nanotubes are highly insulating, the resistivity being in the MΩ region.

Adsorption properties of materials are of importance in gas storage, selective gas recognition and separation. Zeolites and metal–organic frameworks have enjoyed high utility in gas adsorption because of the high surface areas, and well-defined pore shapes. We have measured the surface area of the as synthesized BC$_4$N nanotube brushes using nitrogen adsorption–desorption isotherms at 77 K [Fig. 7(a)]. The surface area measured by the Brunauer–Emmett–Teller (BET) method was 356 m^2 g^{-1} with contributions from micropores (0.7 nm) and mesopores (4 nm) as can be seen from the inset in Fig. 7(a). In Fig. 7(b), the CO$_2$ adsorption on the BC$_4$N nanotube brushes measured at low pressures and low temperature at 195 K are shown. The BC$_4$N nanotube brushes showed remarkably high CO$_2$ uptake of about 23.5 wt% at 195 K. The adsorption does not exhibit saturation even at $P/P_0 = 1$, indicating the presence

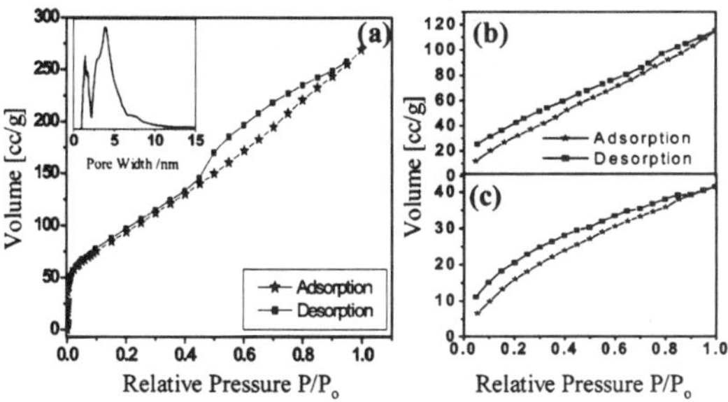

Fig. 7 Adsorption–desorption isotherms of (a) N$_2$ at 77 K (inset shows the pore size distribution), (b) CO$_2$ at 195 K and (c) H$_2$ at 77 K for BC$_4$N nanotube brushes.

of unoccupied pores available for further uptake. The adsorption of hydrogen on the BC$_4$N nanotubes was only 0.4 wt% at 77 K [Fig. 7(c)]. Thus, the BC$_4$N nanotubes prepared by us are selective adsorbents of CO$_2$. New materials having selective adsorption of CO$_2$ are important for environment and industrial applications. Millward and Yaghi[21] have studied CO$_2$ adsorption by various metal–organic frameworks (MOFs), and reported the highest value of adsorption in the case of MOF-177 to be 33.5 mmol g^{-1}, at ambient temperature and high pressures. Among the zeolites, the highest reported value of CO$_2$ adsorption is 7.4 mmol g^{-1} for Zeolite 13X at high pressures and room temperature.[22] Sudik *et al.*[23] measured CO$_2$ adsorption at 195 K and low pressures in certain iron carboxylate MOFs and reported that the highest CO$_2$ uptake was for IRMOP-51 of 74 cm^3 (STP) cm^{-3} (STP = standard temperature and pressure).

In Fig. 8, we show the FESEM images of BC$_4$N nanotubes decorated with Pt and Au nanoparticles. We see a uniform distribution of nearly monodisperse metal nanoparticles on the nanotube walls. The size of the metal nanoparticles is in the range 3–8 nm as found from the TEM images. Such metal nanoparticle decorated BC$_4$N nanotubes could have useful applications.

The BC$_4$N nanotubes can be formed in different ways. They could arise from the substitution of the BN layer in graphitic layers as proposed by earlier reports of BCN nanotubes.[4] Kawaguchi[2] described a BCN structure in terms of an infinite array of six-membered rings of carbon with ordered

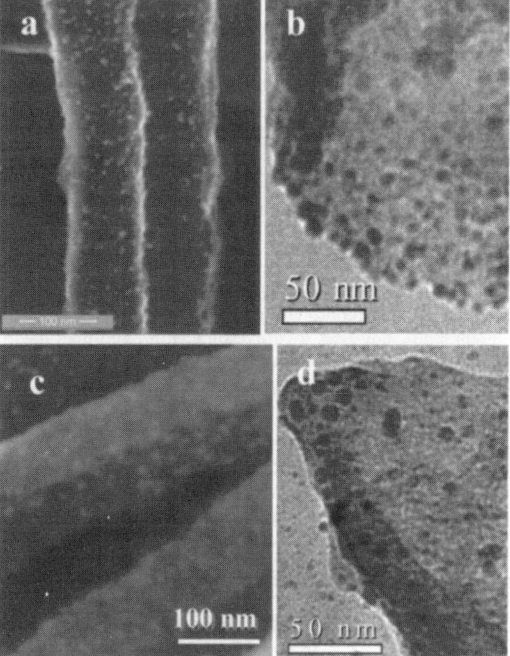

Fig. 8 (a) FESEM image and (b) TEM image of Pt decorated BCN nanotubes. (c) FESEM image and (d) TEM image of Au decorated BCN nanotubes.

substitution of carbon atoms by boron and nitrogen. Zhang *et al.*[6] reported BCN nanotubes with six inner layers of pure carbon and seven outer layers giving BC_7N. The possibility of substitution of BN layers in graphitic layers is discarded, because the selective removal of graphitic layers from BC_4N nanotubes by heating in the presence of oxygen was not possible.[14] We shall now discuss the likely structure and properties of BC_4N nanotubes based on first-principles calculations.

Let us consider benzene with substitution of BN in one of its C–C bonds. Every carbon atom contributes one π-electron to the nearest neighbor bonding in benzene, giving rise to a highly delocalized electron distribution over the entire ring with high resonance stabilization energy and strong C–H chemical shift (ring current). However, in a BN moiety, a complete one electron charge transfer between the B and N atoms gives rise to one π-electron in each of the B and N atoms, thereby causing charge equivalence with the C–C moiety. Thus, in BN substituted benzene (BC_4N), the charge transfer between B and N causes charge localization thereby reducing the overall charge delocalization in the ring, in comparison with benzene. Within a quantum many-body model employing a Hubbard Hamiltonian with realistic parameters (ionization potential and electron affinity) for C, B and N, we find that the lowest molecular gap corresponds to a magnetic excitation (to a spin triplet state), which is 1.8 eV for benzene and 1.7 eV for BC_4N. Charge excitation costs much more energy and the insulating gap in benzene (6.3 eV) reduces to 5.7 eV for BC_4N due to the difference in ionization potentials and electron affinities of the constituent atoms.

Since consideration of BN substitution in a single benzene ring cannot capture the essence of the observed experimental findings in BN substituted carbon nanotubes, we will have to consider BN substitution in each of the benzene rings in both metallic and semiconducting carbon nanotubes. Boron nitride nanotubes (BNNTs) are always insulating unlike CNTs, which can be metallic or semiconducting depending on the chirality. BNNTs doped with a small amount of carbon can be made semiconducting with the appearance of dopant or defect states in the gap. As the concentration of carbon substitution at B and N sites is increased, one expects a crossover from an insulating to a metallic nature of electronic structure. In the present study, however, we are dealing with nanotubes having the well-defined composition, BC_4N. We have carried out first-principles calculations on BC_4N nanotubes with different (m,n) indices.

Our first-principles calculations are based on SIESTA[24,25] implementation of density functional theory; such calculations have been very useful in determination of the electronic structure of single walled carbon nanotubes (SWNTs).[26] We used Troullier–Martins soft pseudopotential[27] with a local spin density approximation (Perdew–Zunger parametrization of the Ceperley–Alder functional) of the density functional theory as implemented in the SIESTA[24,25] method. In these calculations, localized orbitals are used in the basis for representation of the Kohn–Sham[28] wave functions, resulting in sparse representation of the Hamiltonian matrices and efficient diagonalization. We used a grid with an energy cutoff of 200 Ry in representation of the density. Structures were determined through

minimization of energy until the Hellman–Feynman forces on the atoms were smaller in magnitude than 0.03 eV Å^{-1}.

In order to examine various possible structural alternatives for BC_4N, we have calculated properties of metallic CNTs with chirality of (6,6). Our calculations employed periodic boundary conditions with a hexagonal periodic unit cell: the size of this cell along the axis (z direction) of the tube is thrice the periodicity of the tube to accommodate different chemical ordering, and it is kept large enough in the perpendicular directions so that a vacuum of about 20 Å separates the periodic images of the tube ensuring minimal interaction among them. The dimension of the Brillouin zone (BZ) with such a choice of periodic cell is very small in the *ab*-plane and integrations over it were sampled with a 1 × 1 × 20 Monkhorst–Pack k-point mesh.[29] We have used similar calculational parameters in studies of C-doped boron nitride nanotubes and B, N-doped carbon nanotubes.[30]

To facilitate comparison of energies of different structures, we used 72-atom supercell [$(BN)_{12}C_{48}$] in all calculations. We have considered four types of ordering of B and N atoms on the carbon sites of a (6,6) CNT (Fig. 9). In the first one, we have replaced one C–C bond with a B–N bond in each hexagonal ring. Since each of the six C–C bonds in a hexagonal ring is shared between two rings, this gives a correct concentration of atoms. There are many ways to distribute B–N bonds in this way; we have chosen the simple ordered structure S_1 with alternate C–C bonds along the perimeter of CNT replaced with B–N bonds. In the second structure S_2, we have substituted local structural units of BN_3 and NB_3 in a CNT, with no linkage between them [Fig. 9(c)]. In the structure S_3, these local structural units (BN_3 and NB_3) link together with a B–N bond; *i.e.* S_3 exhibits local structural units that are common to both S_1 and S_2 structures. Finally, we also considered a disordered structure D. Since no two boron or nitrogen atoms are expected to be neighbors, we started with BNNTs and substituted ⅙ of the boron and ⅙ of the nitrogen atoms with carbon using a pseudo-random number generator.

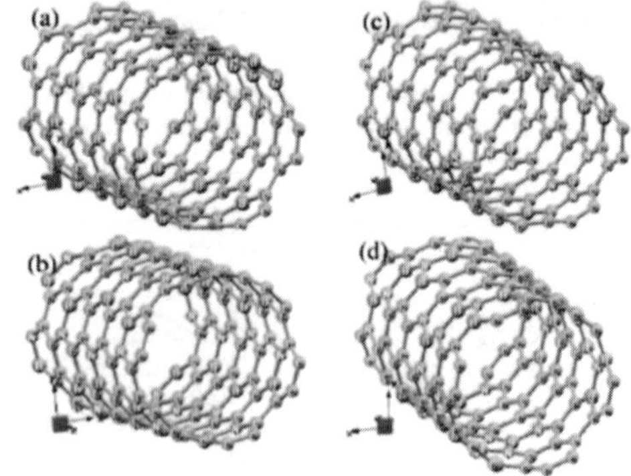

Fig. 9 (6,6) BC_4N nanotubes: (a) a disordered structure D, in which carbon atoms occupy B and N sites in a BNNT randomly with a probability of ⅔, (b) an ordered structure S_1, with one B–N bond present in each carbon ring, (c) an ordered structure S_2, where BN_3 and NB_3 units are distributed on alternate C_6 rings, and (d) an ordered structure S_3, in which BN_3 and NB_3 units are linked with one B–N bond.

Table 1 Relative energies of (6,6) BC$_4$N nanotubes

Nanotube	Energy/eV atom^{-1}
D	0.100
S$_1$	0.139
S$_2$	0.073
S$_3$	0.00

Relative energies (see Table 1) of these BC$_4$N nanotubes clearly show that BN$_3$ and NB$_3$ local units are energetically more favorable (S$_2$ is more stable than S$_1$ by almost 130 meV per B–N bond) than the diatomic BN units. The S$_2$ structure is further stabilized by almost 73 meV per atom through a B–N linkage between BN$_3$ and NB$_3$ local units to yield S$_3$. The disordered configuration D, which consists of different types of local structural units (BN, BN$_3$, NB$_3$, BN$_2$, *etc*), has a stability intermediate to S$_1$ and S$_2$ structures.

We now examine the electronic structure of the BC$_4$N nanotubes through their density of states (see Fig. 10). For each of the four nanotubes, we find a nonmagnetic ground state in the optimized structure. For S$_1$ and S$_2$ BC$_4$N nanotubes, we find features common to CNT and BNNTs: a weak density of states at the Fermi energy as in the (6,6) CNT[26] and two strong peaks within one eV of the fermi energy as in a BNNT.[30] In the disordered BC$_4$N nanotube D, we find a pseudo-gap at the Fermi energy, which is common to many materials lacking long-range order. The S$_3$ BC$_4$N nanotube, however, exhibits a clear gap in its electronic structure of about 1 eV. We find two isolated bands in its band-structure (Fig. 11) that form the HOMO and LUMO electronic states and are interestingly symmetric with respect to the Fermi energy. From visualization of the HOMO and LUMO wave functions at Γ point, we find that the former is centered on the BN$_3$ unit and the latter is centered on the NB$_3$ unit, and they both have an overlap with orbitals of neighboring carbon atoms. The band gap opens up when the two units are linked with a B–N bond yielding a lower energy structure.

Thus, the local structural units of BN$_3$ and NB$_3$ linked with B–N bonds (S$_3$) would appear to be central to the stability of BC$_4$N nanotubes, although chemically speaking S$_1$ with alternate B–N bonds would appear more likely. We predict that a junction between a BNNT and a CNT should have good

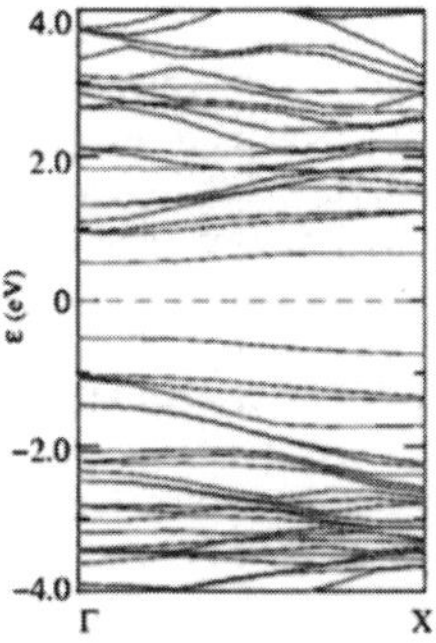

Fig. 11 Electronic structure of a (6,6) BC$_4$N nanotube of S$_3$ type: bands corresponding to up and down spins are represented with black and red lines respectively (both are degenerate).

stability and interesting electronic properties. While the host CNT is metallic, we find that substitutional doping of B and N with large enough concentration (33%), shown here to be feasible experimentally, can make it insulating.

We have also carried out first-principles calculations on (6,0) and (8,0) nanotubes (Fig. 12) on a spin-polarized basis as mentioned above. However, in this case, we have used a cubic unit cell with twice the periodicity in the tube direction and a 1 × 1 × 60 Monkhorst–Pack k-point mesh integration. All other parameters remain the same as has been outlined above. In Fig. 13, we plot the density of states (DOS) for a (6,0) carbon nanotube and projected density of states (pDOS) for the corresponding BC$_4$N nanotube with the same diameter and (*m,n*) indices. As can be seen, the (6,0) carbon nanotube is metallic as expected. Interestingly, the BC$_4$N (6,0) nanotube also remains metallic. In fact, in the BC$_4$N nanotube, the carbon density of states have larger contributions than the same from (6,0) carbon nanotubes at and close to the Fermi energy. The charge transfer induced states from B and N atoms also contribute to the DOS at the Fermi energy. We also find that the charge transfer between B and N affects the neighboring carbon atoms, allowing contributions from the out-of-plane π-orbitals density of carbon, close to the Fermi energy. However, the (6,0) nanotubes with the S$_3$ structure where BN$_3$ and NB$_3$ units are linked show a semiconducting gap.

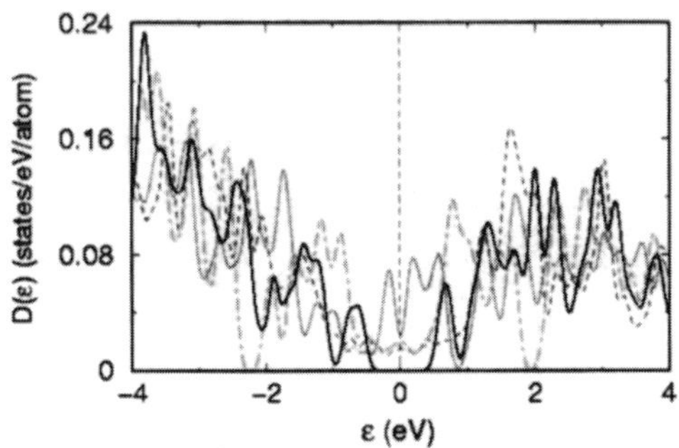

Fig. 10 Density of electronic states of (6,6) BC$_4$N nanotubes in the D (blue dotted lines), S$_1$ (red dotted line), S$_2$ (green dotted line) and S$_3$ (solid black line) structures shown in Fig. 9.

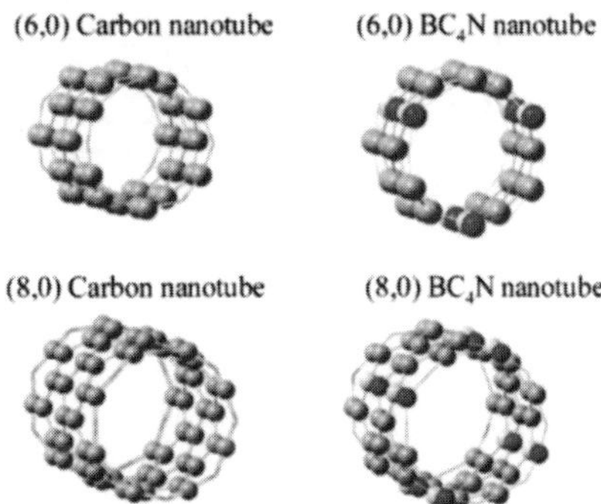

Fig. 12 (6,0) and (8,0) carbon and BC$_4$N nanotubes.

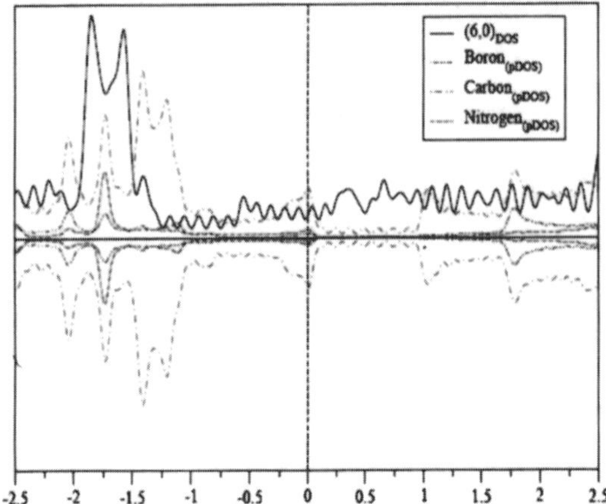

Fig. 13 DOS of the (6,0) carbon nanotube and pDOS of the (6,0) BC$_4$N nanotube.

For the semiconducting (8,0) nanotube, the BN substitution in every hexagonal ring increases the semiconducting gap. For an (8,0) carbon nanotube, the semiconducting gap is 0.33 eV, while for the BC$_4$N (8,0) nanotube, it increases to 0.40 eV. Both at the negative and positive energy regions near the Fermi energy, the carbon density of states is reduced compared to the (8,0) carbon nanotube, while very small but finite density of states from B and N appear at those energies, signifying charge transfer between B and N and with π-orbitals of neighboring C atoms (Fig. 14). These interesting features of metal–semiconductor transition and gap variation arise mainly from the heteroatom substitution leading to charge injection at the local sites with redistribution of charge densities over the entire system.

We have also carried out some initial calculations on BN substituted graphene nanoribbons with passivated edge atoms. We find that the edge atoms have finite magnetic moments, more so if the edge atoms are B and N. In fact, due to the finite magnetic moments at the edge sites, the delocalization of

electrons over the neighboring sites induces some net spin even at the bulk atoms. This shows that if some B, N or C atoms occur at the edges with incomplete coordination, or a few B or N atoms occur as impurities in place of carbon, there is a possibility that the system would show a finite spin polarization, even at finite temperatures. Incorporation of BN in zig-zag graphene nanoribbons, however, makes it insulating.

Conclusions

It has been possible to synthesize nanotubes of the composition BC$_4$N starting with a-CNTs prepared with the porous polycarbonate membranes. The reaction of the a-CNTs with a mixture of urea and boric acid provides an excellent means for the incorporation of boron and nitrogen in the carbon nanotubes. In the first step of the reaction, the decomposition of urea produces NH$_3$, which on reaction with boric acid at high temperatures enables the incorporation of both boron and nitrogen in the carbon nanotubes. The BC$_4$N nanotubes have been characterized by various physical methods and the composition established by EELS carried out in a high-resolution electron microscope. Although the nanotubes may not have an extended ordered structure with graphitic type BCN layers as there is sufficient order to give X-ray diffraction patterns and a well-defined Raman spectra. Surprisingly, BC$_4$N nanotubes have very high thermal stability. They are nonmagnetic insulators with a high propensity for CO$_2$ uptake (up to 23.5 wt%). First principle calculations indicate that the most stable structure may involve a graphitic network containing BN$_3$ and NB$_3$ units connected by a B–N bond. The other structure with a slightly higher energy is the one with ordered B–N bonds in all the six-membered rings. The latter structure is chemically more sensible. Calculations on (6,0), (6,6) and (8,0) BC$_4$N naotubes show that they all are nonmagnetic and both (6,0) and (6,6) BC$_4$N nanotubes open up a small semiconducting gap along with the (8,0) BC$_4$N nanotube.

Acknowledgements

One of the authors (KR) acknowledge CSIR (India) for the fellowship.

References

1 C. N. R. Rao and A. Govindaraj, *Nanotubes and Nanowires*, RSC Publishing, Cambridge, 2005.
2 M. Kawaguchi, *Adv. Mater.*, 1997, **9**, 8.
3 O. Stephan, P. M. Ajayan, C. Colliex, P. Redlich, J. M. Lambert, P. Bernier and P. Lefin, *Science*, 1994, **266**, 1683.
4 K. Suenaga, C. Colliex, N. Demoncy, A. Loiseau, H. Pascard and F. Willaime, *Science*, 1997, **278**, 653.
5 P. Redlich, J. Loeffler, P. M. Ajayan, J. Bill, F. Aldinger and M. Riihle, *Chem. Phys. Lett.*, 1996, **260**, 465.
6 Y. Zhang, H. Gu, K. Suenaga and S. Iijirna, *Chem. Phys. Lett.*, 1997, **279**, 264.
7 R. Sen, B. C. Satishkumar, A. Govindaraj, K. R. Harikumar, G. Raina, J. P. Zhang, A. K. Cheetham and C. N. R. Rao, *Chem. Phys. Lett.*, 1998, **287**, 671.
8 (*a*) W. Q. Han, J. Cumings, X. Huang, K. Bradley and A. Zettl, *Chem. Phys. Lett.*, 2001, **346**, 368; (*b*) M. Terrones, D. Golberg, N. Grobert, T. Seeser, M. Reyes-Reyes, M. Mayne, R. Kamalakaran, P. Dorozhkin, Z. C. Dong, H. Terrones, M. Ruhle and Y. Bando, *Adv. Mater.*, 2003, **15**, 1899.

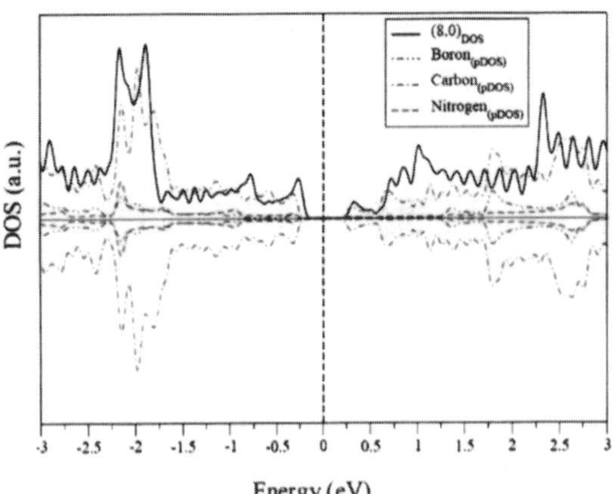

Fig. 14 DOS of the (8,0) carbon nanotube and pDOS of the (8,0) BC$_4$N nanotube.

9 M. Hubacek and T. Sato, *J. Solid State Chem.*, 1995, **114**, 258.
10 W. L. Wang, X. D. Bai, K. H. Liu, Z. Xu, D. Golberg, Y. Bando and E. G. Wang, *J. Am. Chem. Soc.*, 2006, **128**, 6530.
11 S. Y. Kim, J. Park, H. Chul Choi, J. P. Ahn, J. Q. Hou and H. S. Kang, *J. Am. Chem. Soc.*, 2007, **129**, 1705.
12 J. Dinesh, M. Eswaramoorthy and C. N. R. Rao, *J. Phys. Chem. C*, 2007, **111**, 510.
13 X. D. Bai, C. Y. Zhi and E. G. Wang, *J. Nanosci. Nanotechnol.*, 2001, **1**, 35.
14 W. Q. Han, W. Mickelson, J. Cumings and A. Zettl, *Appl. Phys. Lett.*, 2002, **81**, 1100.
15 C. Y. Zhi, X. D. Bai and E. G. Wang, *Appl. Phys. Lett.*, 2002, **80**, 3590.
16 J. Wu, W. Q. Han, W. Walukiewicz, J. W. Ager, III and W. Shan, *Nano Lett.*, 2004, **4**, 647.
17 S. Okada and A. Oshiyama, *Phys. Rev. Lett.*, 2001, **84**, 146803.
18 J. Choi, Y. H. Kim, K. J. Chang and D. Tomanek, *Phys. Rev. Lett.*, 2003, **67**, 146803.
19 J. Nakamura, T. Nitta and A. Natori, *Phys. Rev. Lett.*, 2005, **72**, 205429.
20 R. Q. Wu, L. Liu, G. W. Peng and Y. P. Feng, *Appl. Phys. Lett.*, 2005, **86**, 122510.
21 A. R. Milliward and O. M. Yaghi, *J. Am. Chem. Soc.*, 2005, **127**, 17998.
22 S. Cavenati, C. A. Grande and A. E. Rodrigues, *J. Chem. Eng. Data*, 2004, **49**, 1095.
23 A. C. Sudik, A. R. Millward, N. W. Ockwig, A. P. Cote, J. Kim and M. Yaghi, *J. Am. Chem. Soc.*, 2005, **127**, 7110.
24 P. Ordejon, E. Artacho and J. M. Soler, *Phys. Rev. B*, 1996, **53**, R10441.
25 J. M. Soler, E. Artacho, J. D. Gale, A. Garcia, J. Junquera, P. Ordejon and D. S. Anchez-Portal, *J. Phys.: Condens. Matter*, 2002, **14**, 2745.
26 S. Reich, C. Thomsen and P. Ordejon, *Phys. Rev. B*, 2002, **65**, 155411.
27 N. Troullier and J. L. Martins, *Phys. Rev. B*, 1991, **43**, 1993.
28 W. Kohn and L. J. Sham, *Phys. Rev.*, 1965, **140**, A1133.
29 H. J. Monkhorst and J. D. Pack, *Phys. Rev. B*, 1976, **13**, 5188.
30 M. Upadhyay-Kahaly and U. V. Waghmare, *J. Nanosci. Nanotechnol.*, in press.

COMMUNICATION

www.rsc.org/materials | Journal of Materials Chemistry

Functionalization and solubilization of BN nanotubes by interaction with Lewis bases

Shrinwantu Pal, S. R. C. Vivekchand, A. Govindaraj and C. N. R. Rao*

Received 10th October 2006, Accepted 13th December 2006
First published as an Advance Article on the web 18th December 2006
DOI: 10.1039/b614764j

By interaction with a trialkylamine or trialkylphosphine, BN nanotubes can be dispersed in a hydrocarbon medium with retention of the nanotube structure.

Boron nitride (BN) is an important ceramic material with a wide range of applications. BN crystallizes in a graphite-like structure and can be viewed as replacing the C–C pair in a graphene sheet with the isoelectronic B–N pair. BN does not, however, form five- or seven-membered rings. BN nanotubes have been prepared by several methods.[1–10] During the last year, much interest has been evinced in the functionalization and solubilization of BN nanotubes.[11–15] For example, poly(*m*-phenylenevinylene-co-(2,5-dioctoxy-*p*-phenylenevinylene)) has been used to solublize BN nanotubes in organic solvents, the polymer wrapping around the BN nanotubes leading to its solubilization. Water-soluble BN nanotubes have been prepared by using amine-terminated poly-(ethylene glycol).[12] BN nanotubes have been covalently functionalized using stearoyl chloride and subsequently dissolved in organic solvents.[13] Cycloaddition of dimethyl sulfoxide to BN nanotubes is suggested to weaken the B–N bond and help in the peeling of the nanotubes.[14] BN nanotubes have also been fluorinated.[15] The literature procedures for the functionalization and solubilization of BN nanotubes appear somewhat difficult and we, therefore, felt that there should be a simpler way to accomplish the desired results. We, therefore, considered it desirable to exploit the inherent electron deficiency of boron compounds. If we consider the boron site in BN nanotubes to act as a Lewis acid, it should be possible to form adducts with Lewis bases such as alkylamines and phosphines, which should then enable one to disperse them in suitable solvents. In this communication, we demonstrate that this does indeed happen through a study of the interaction of BN nanotubes with two tri-n-alkylamines and a tri-n-alkylphosphine.

BN nanotubes were prepared by a standard literature procedure.[7] BN nanotubes were prepared by the reaction of B_2O_3 with multi-walled carbon nanotubes (MWNTs) in the presence of ammonia at 1250 °C for 3 h. A grey spongy product was obtained after the reaction indicating the presence of unreacted MWNTs along with BN nanotubes. The product was washed with hot water to remove excess B_2O_3. The excess carbon present in the product was removed by oxidation at 800 °C in low-pressure air (20 mPa). Scanning electron microscope (SEM) and energy dispersive analysis of X-rays (EDAX) were performed with

a Leica S-440I microscope fitted with a Link ISIS spectrometer. Transmission electron microscope (TEM) images were obtained with a JEOL JEM 3010 instrument fitted with a Gatan CCD camera operating at an accelerating voltage of 300 kV. In Fig. 1(a), we show a representative SEM image of the BN nanotubes prepared by us. The nanotubes have diameters in the 80–200 nm range with lengths of tens of microns. BN nanotubes have either bamboo-like or worm-like structures as revealed by the TEM image shown in Fig. 1(b). Raman spectra of the BN nanotubes were recorded using a JobinYvonHoriba HR800 Raman spectrometer using a HeNe laser (632 nm). The spectrum of purified BN nanotubes did not show the D-band (1340 cm^{-1}) or the G-band (1600 cm^{-1}) of carbon nanotubes. The nanotubes exhibit a Raman band centered at 1370 cm^{-1} as shown in Fig. 2(a), this band is assigned to the E_{2g} tangential mode.[9,16]

In a typical functionalization/solubilization experiment, 2 ml of trialkylamine or trialkylphosphine were added to 5 mg of purified BN nanotubes. The mixture was warmed at 70 °C for 12 h and subsequently sonicated for a few minutes. The functionalized BN

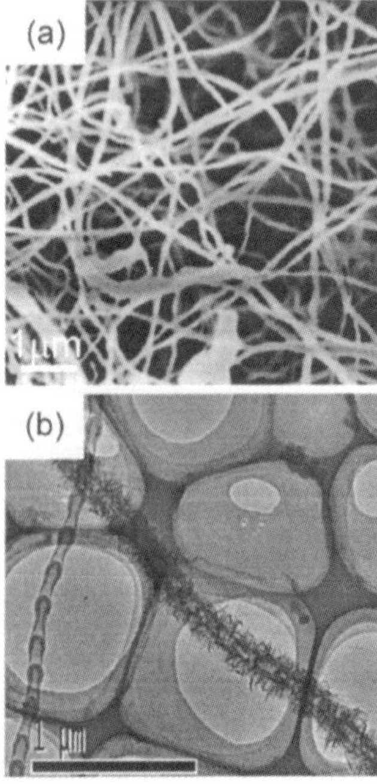

Fig. 1 (a) SEM image of purified BN nanotubes. (b) TEM image of BN nanotubes.

DST unit on Nanoscience, CSIR–Centre of Excellence in Chemistry, Chemistry and Physics of Materials Unit, Jawaharlal Nehru Center for Advanced Scientific Research, Jakkur, Bangalore 560 064, India. E-mail: cnrrao@jncasr.ac.in

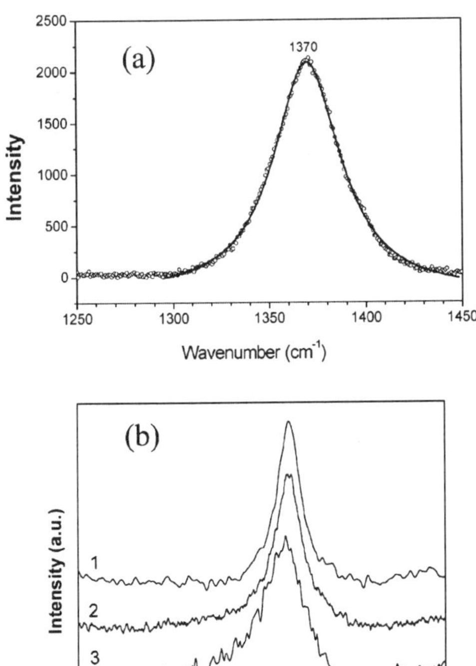

Fig. 2 (a) Raman spectrum of BN nanotubes. (b) Raman spectra of the dispersions of BN nanotubes functionalized with tributylamine (1), trioctylamine (2) and trioctylphosphine (3).

nanotubes were then dispersed in toluene (8 ml) or other hydrocarbons such as benzene, at room temperature. The dispersions were stable over long periods of time. It is not essential to interact the nanotubes with an amine or phosphine at 70 °C. Interaction at laboratory temperature for a long period yielded the same result. In Fig. 3(a), we show a photograph of a toluene dispersion of BN nanotubes treated with trioctylamine. In Fig. 4(a), we show a TEM image of the BN nanotubes obtained by taking a drop of the dispersion on the electronic microscopy grid. We see images of

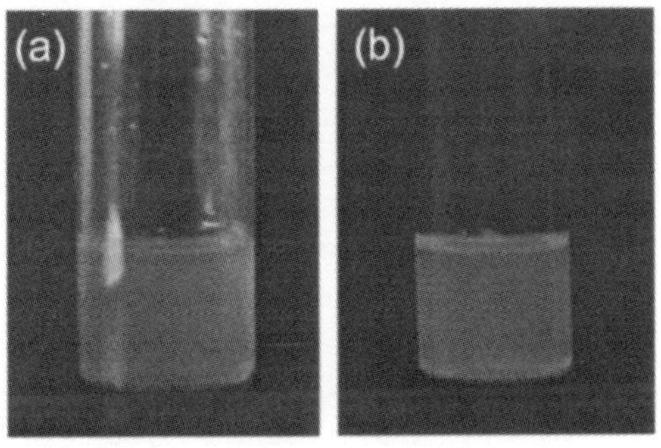

Fig. 3 Photographs of dispersions of BN nanotubes obtained after interaction with (a) trioctylamine and (b) trioctylphosphine.

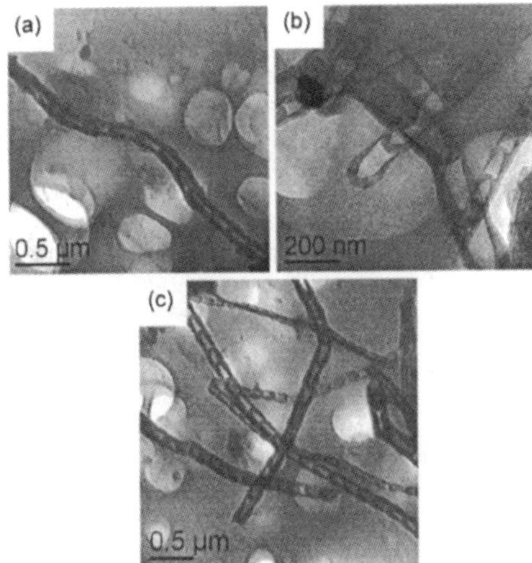

Fig. 4 TEM images of the dispersions of BN nanotubes functionalized with (a) trioctylamine, (b) tributylamine and (c) trioctylphosphine.

nanotubes similar to those in the starting material shown in Fig. 1(b). We could obtain a stable dispersion of BN nanotubes using tributylamine as well. A TEM image of the nanotubes obtained from this dispersion is shown in Fig. 4(b).

Evidence for the interaction between tributylamine and BN nanotubes has been obtained by nuclear magnetic resonance spectroscopy. We have studied the ^{1}H and ^{13}C NMR spectra of tributylamine-functionalized BN nanotubes in comparison with the spectra of tributylamine. We observe a small increase in the ^{1}H chemical shift by ~ 0.02 ppm in the amine–BN adduct. In the case of ^{13}C NMR spectra, we observe a significant increase in the chemical shifts of the γ and δ carbons by ~ 0.4 ppm and a decrease in the chemical shift of the β carbon atom by ~ 0.3 ppm. The chemical shift of the αcarbon is also higher in the amine–BN adduct by ~ 0.1 ppm. The changes in the ^{1}H and ^{13}C spectra of tributylamine found on interaction with BN are comparable to those reported in the literature for similar complexes.[17,18]

We have also carried out solubilization experiments using the as-prepared BN nanotubes containing carbon nanotubes as impurity. Initially, a grey dispersion containing both carbon and BN nanotubes was obtained, but the carbon nanotubes settled to the bottom within a few hours leaving the BN nanotubes in the solution. This indicates that there is no significant interaction between the Lewis base and the carbon nanotubes.

Encouraged by the results obtained with the two trialkylamines, we interacted the BN nanotubes with trioctylphosphine at 70 °C for 12 h or at 30 °C for a longer period. We could disperse the product in toluene as can be seen in the photograph in Fig. 3(b). A TEM image of the nanotubes from the dispersion is shown in Fig. 4(c). Raman spectra of the dispersions of BN nanotubes were recorded after interacting them with an amine or a phosphine. For this purpose, a drop of the dispersion was placed on a glass slide and the spectrum recorded. In Fig. 2(b), we the show Raman spectrum of BN nanotubes treated with trioctylphosphine. We observe a band due to the tangential mode at 1361 cm^{-1}, similar to the starting material. Raman spectra of the toluene dispersions of

BN nanotubes treated with tributylamine and trioctylamine showed a band due to the tangential mode around 1367 cm^{-1}. These results suffice to demonstrate that the interaction of BN nanotubes with Lewis bases helps to solubilize them in nonpolar solvents. We should note that in the absence of interaction with an amine or a phosphine, BN nanotubes could not be dispersed in toluene and the nanotubes settled to the bottom in a short period.

In conclusion we have successfully demonstrated that the Lewis acid nature of boron in BN nanotubes can be exploited to functionalize and solubilize their nanostructures through interaction with Lewis bases. The solubilized nanotubes retain the nanotube features and the hydrocarbon dispersions are stable up to a few days. It is noteworthy that the interaction of the BN nanotubes with the Lewis bases at room temperature suffices to provide good dispersions of the BN nanotubes in hydrocarbon solvents. We also note that any carbon impurity due to carbon nanotubes used as templates in the preparation of the BN nanotubes settles to the bottom after interaction of the BN nanotubes with a Lewis base, thereby providing a means of separation of carbon and BN nanotubes.

Notes and references

1 C. N. R. Rao and A. Govindaraj, *Nanotubes and Nanowires*, RSC, Cambridge, 2005.
2 P. Gleize, M. C. Schouler, P. Gadelle and M. Caillet, *J. Mater. Sci.*, 1994, **29**, 1575.
3 N. G. Chopra, R. J. Luyken, K. Cherrey, V. H. Crespi, M. L. Cohen, S. G. Louse and A. Zettl, *Science*, 1995, **269**, 966.
4 A. Loiseau, F. Williame, N. Demoncy, G. Hug and H. Pascard, *Phys. Rev. Lett.*, 1996, **76**, 4737.
5 O. R. Lourie, C. R. Jones, B. M. Bertlett, P. C. Gibbons, R. S. Ruoff and W. E. Buhro, *Chem. Mater.*, 2000, **12**, 1808.
6 R. Ma, Y. Bando and T. Sato, *Chem. Phys. Lett.*, 2001, **337**, 61.
7 F. L. Deepak, C. P. Vinod, K. Mukhopadhyay, A. Govindaraj and C. N. R. Rao, *Chem. Phys. Lett.*, 2002, **353**, 345.
8 W.-Q. Han, W. Mickelson, J. Cumings and A. Zettl, *Appl. Phys. Lett.*, 2002, **81**, 1110.
9 J. Wu, W.-Q. Han, W. Walukiewicz, J. W. Ager, III, W. Shan, E. E. Haller and A. Zettl, *Nano Lett.*, 2004, **4**, 647.
10 C. W. Chang, W.-Q. Han and A. Zettl, *Appl. Phys. Lett.*, 2005, **86**, 173102.
11 C. Zhi, Y. Bando, C. Tang, R. Xie, T. Sekiguchi and D. Goldberg, *J. Am. Chem. Soc.*, 2005, **127**, 15996.
12 S.-Y. Xie, W. Wang, K. A. S. Fernando, X. Wang, Y. Lin and Y.-P. Sun, *Chem. Commun.*, 2005, 3670.
13 C. Zhi, Y. Bando, C. Tang, S. Honda, K. Sato, H. Kuwahara and D. Goldberg, *Angew. Chem., Int. Ed.*, 2005, **44**, 7932.
14 Q. Huang, Y. Bando, C. Zhi, D. Golberg, K. Kurashima, F. Xu and L. Gao, *Angew. Chem., Int. Ed.*, 2006, **45**, 2044.
15 C. Tang, Y. Bando, Y. Huang, S. Yue, C. Gu, F. Xu and D. Golberg, *J. Am. Chem. Soc.*, 2005, **127**, 6552.
16 S. Saha, D. V. S. Muthu, D. Goldberg, C. Tang, C. Zhi, Y. Bando and A. K. Sood, *Chem. Phys. Lett.*, 2006, **421**, 86.
17 T. D. Coyle and F. G. A. Stone, *J. Am. Chem. Soc.*, 1961, **83**, 4138.
18 M. Ilczyszyn, Z. Latajka and H. Ratajczak, *Org. Magn. Reson.*, 1984, **22**, 419.

COMMUNICATION   www.rsc.org/chemcomm | ChemComm

GaS and GaSe nanowalls and their transformation to Ga_2O_3 and GaN nanowalls

Ujjal K. Gautam,[ab] S. R. C. Vivekchand,[a] A. Govindaraj[ab] and C. N. R. Rao*[ab]

Received (in Cambridge, UK) 12th May 2005, Accepted 9th June 2005
First published as an Advance Article on the web 11th July 2005
DOI: 10.1039/b506676j

Two-dimensional nanowalls of GaS and GaSe are obtained by thermal exfoliation around 900 °C, and transformed to Ga_2O_3 and GaN nanowalls upon reaction with air and ammonia respectively at 800 °C, while maintaining dimensional integrity.

Recent investigations of nanostructures by-and-large pertain to zero-dimensional (0D) nanocrystals, and one-dimensional (1D) nanowires and nanotubes.[1] There are only a few studies on the two-dimensional (2D) nanostructures of materials, which are not only important because of possible technological applications, but also because they may define a stage in the formation of other nanostructures, such as nanotubes. In case of carbon having a layered structure in bulk forms fullerenes and onions which are the 0D nanostructures, while the nanotubes are 1D nanostructures. Carbon nanowalls, which are 2D nanostructures, were reported recently.[2] The growth of carbon nanowalls occurs during the formation of nanotubes by microwave plasma-enhanced chemical vapor deposition.[3] These nanowalls have been used as templates to deposit magnetic nanoparticles.[4] There have also been reports of ZnO nanowalls, and it has been found that ZnO nanorods grow from the nodes of nanowalls.[5] ZnO nanoflowers of wall-like structure have been prepared hydrothermally in the absence of any organic reagent.[6] Besides nanowalls, which generally stand vertically on solid substrates and are interconnected, 2D nanosheets of materials such as Zn, CuS, ZnS, Al_2O_3 and Ga_2O_3 are also known.[7–11] GaS and GaSe, with layered structures similar to graphite, have been considered as ideal for forming nanotubes.[12–14] We would therefore expect these two materials to also form nanowalls. In this communication, we report the successful synthesis of the nanowalls of GaS and GaSe and their reactions with oxygen and ammonia.

On heating GaS powder in a sealed quartz tube at 900 °C at a rate of 1 °C min^{-1}, we obtained solid deposits at the cooler end of the tube where the temperature was $\sim$400 °C.† An investigation of the deposits showed that they contained beautiful wall structures, as revealed by the scanning electron microscope (SEM) images presented in Fig. 1a and Fig. 1b. The walls form smooth curved surfaces and are well connected, creating an extended network, as shown in the high magnification SEM image in Fig. 1b. The identity of the walls was established from energy dispersive X-ray analysis (EDAX) and X-ray diffraction (XRD) patterns. The

[a]Chemistry and Physics of Materials Unit and CSIR Centre of Excellence in Chemistry, Jawaharlal Nehru Centre for Advanced Scientific Research, Jakkur P. O., Bangalore, 540064, India.
E-mail: cnrrao@jncasr.ac.in; Fax: (+91) 80 2208 2760
[b]Solid State and Structural Chemistry Unit, Indian Institute of Science, Bangalore, 560012, India

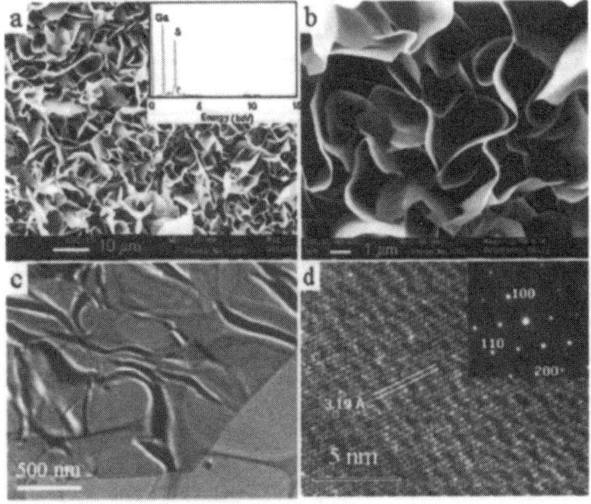

Fig. 1 (a) Low-magnification SEM image of GaS nanowalls covering a large area. Inset shows the EDAX spectrum of the nanowalls. (b) High magnification SEM image of the nanowalls exhibiting a smooth surface—an extended network and intricate curvatures. (c) TEM and (d) HREM images of a nanowall. Inset shows the SAED pattern.

EDAX spectrum (shown as an inset in Fig. 1a), recorded at various locations of the sample, confirmed the Ga : S ratio to be 1 : 1. The XRD pattern of the sample (Fig. 2a) could be indexed on a hexagonal phase (a = 3.59 Å, c = 15.5 Å) in the space group $P6_3/mmc$ (JCPDF no. 30-0576). We show the transmission electron microscope (TEM) image of a nanowall in Fig. 1c. TEM observations reveal that the walls are transparent, especially at the edges, indicating a thickness of around a few nanometers, just as in the case of carbon nanowalls.[4] The nanowalls are single crystalline, as established by the high resolution electron microscope (HREM) images as well as the selected area diffraction (SAED) patterns. Fig. 1d shows a HREM image with an interlayer spacing of 3.19 Å, corresponding to d(100) (3.106 Å). The powder left over at the hot end of the sealed tube did not change in appearance, and was found to be pure GaS by XRD and EDAX analysis. It should be noted that there was no evidence of the formation of Ga_2S_3 in the various zones of the reaction tube.

Heating GaSe powder at 900 °C in a sealed tube at a rate of 1 °C min^{-1} for 6 h gave solid deposits at the cooler (400 °C) end of the tube.† The deposits contained GaSe nanowalls, as shown in Fig. 3a. The GaSe nanowalls cover larger areas than GaS nanowalls, and have a thickness of a few nanometers at the edges. The XRD pattern could be indexed on the hexagonal GaSe phase (a = 3.76 Å, c = 15.91 Å) (JCPDF no. 37-0931). Some of the

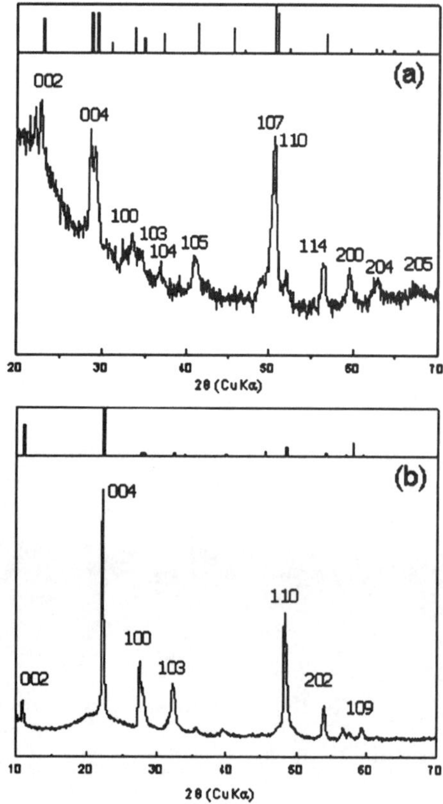

Fig. 2 XRD patterns of the (a) GaS and (b) GaSe nanowalls. The top panels indicate predicted peak positions.

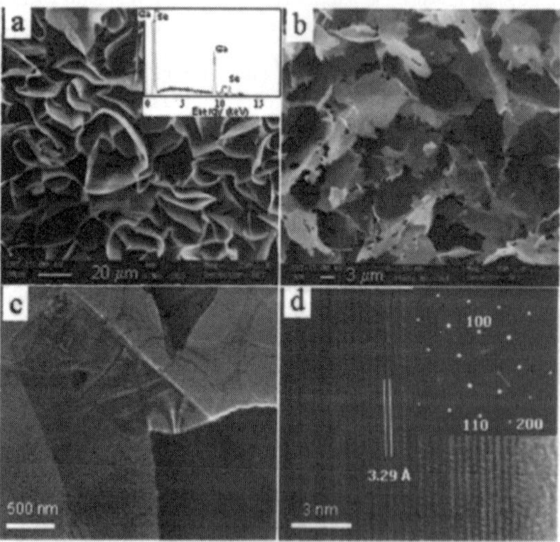

Fig. 3 (a) SEM image of GaSe nanowalls deposited at 400 °C covering a large area. Inset shows the EDAX spectrum of the nanowalls. (b) SEM image of the nanowalls deposited at ~500 °C. (c) TEM and (d) HREM images of a nanowall. Inset shows the SAED pattern.

non-[001] peaks exhibit high relative intensities due to orientational effects, the peak at $2\theta = 48.5°$ being typical. The deposits obtained at the cooler end of the reaction tube, where the temperature was ~500 °C, had nanowalls containing holes (Fig. 3b). However, the composition was stoichiometric throughout, as found by EDAX analysis (inset in Fig. 3a). We show the TEM image of a GaSe nanowall in Fig. 3c. The single crystalline nature of the walls is confirmed by an HREM image (Fig. 3d) and their SAED pattern (inset in Fig. 3d). The lattice spacing observed in the HREM image of 3.29 Å corresponds to the separation between the [100] planes (=3.229 Å) of hexagonal GaSe in the space group $P6_3/mmc$. The SAED pattern also confirms that the GaSe sheets are exfoliated along the c-axis.

Careful observation showed that the cooler end of the reaction tube also contained deposits with the appearance of frozen droplets. Fig. 4a and Fig. 4b show SEM images of these GaS and GaSe droplets respectively, obtained after 1.5 h reaction time. The inset in Fig. 4b shows a flower-like nanostructure formed around a droplet. Extended growth of nanostructures at higher temperatures is known to take place by the vapor–liquid–solid (VLS) mechanism, when a metal catalyst is used, or otherwise by the vapor–solid (VS) mechanism.[1] In the present case, the mechanism of formation of the nanostructures appears to be somewhat different. Thermally-exfoliated sheets of GaS and GaSe fly to the cooler end of the reaction tube due to the temperature-induced pressure gradient. These sheets, initially in a semi-molten state, may form continuous films (underneath the nanostructures), as is evident from the SEM and EDAX observations. Even though the melting points of the bulk materials are high, the exfoliated sheets would be expected to melt at a considerably lower temperature. Smaller nuclei emanate out of the films due to the temperature gradient present between the tube walls and centre. The various nanostructures are thus formed from the nuclei, and accordingly we observe scrolls and tubular structures emerging from the droplets (Fig. 4c, Fig. 4d and Fig. 4e).

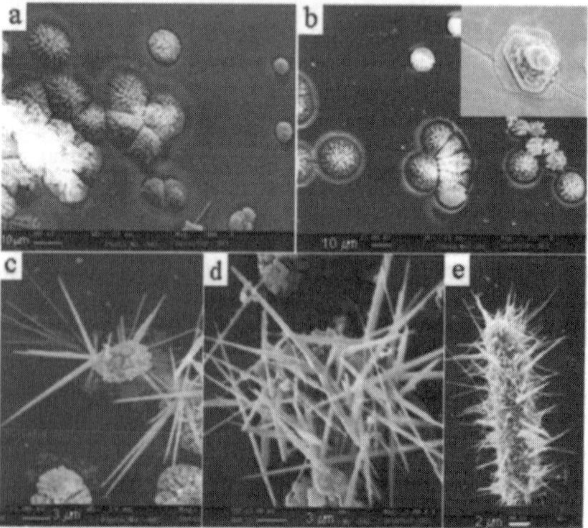

Fig. 4 Droplets of (a) GaS and (b) GaSe obtained at the cooler end of the reaction tube. Inset shows a GaSe nanostructure forming around a droplet. SEM images of (c) GaS and (d)/(e) GaSe scrolls and tubules.

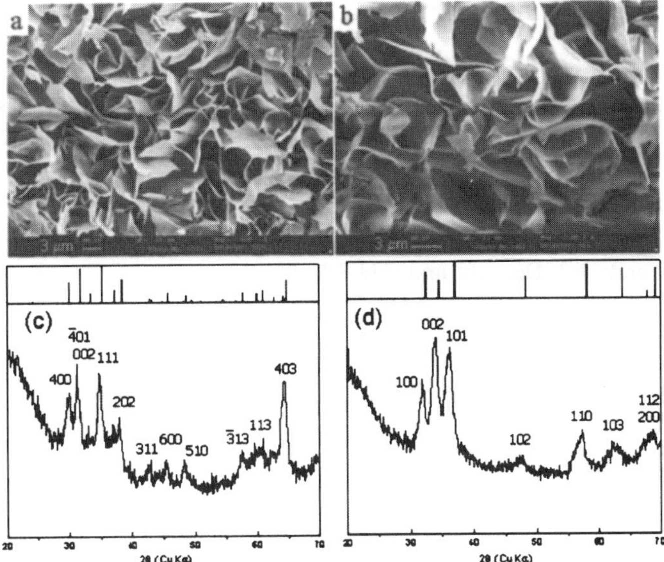

Fig. 5 SEM images of (a) Ga_2O_3 and (b) GaN nanowalls obtained by heating GaS nanowalls in air and ammonia respectively. XRD patterns of the (c) Ga_2O_3 and (d) GaN nanowalls. The top panels indicate predicted peak positions.

We have carried out reactions of GaS and GaSe nanowalls with air and NH_3. We obtained Ga_2O_3 nanowalls by heating GaS and GaSe nanowalls in air (Fig. 5a).† The XRD pattern of the product (Fig. 5c) is readily indexed on monoclinic Ga_2O_3 in the space group $C2/m$ (JCPDF no. 43-1012). The EDAX spectrum confirmed the composition of the sample. On heating the GaS and GaSe nanowalls in NH_3, GaN nanowalls were obtained (Fig. 5b).† The XRD pattern of the product (Fig. 5d) is characteristic of hexagonal GaN (a = 3.20 Å, c =5.19 Å) (JCPDF no. 02-1078). To our knowledge, this is the first report of extended 2D nanostructures of Ga_2O_3 and GaN. It may be noted that Sn nanoflowers are transformed to SnO_2 nanoflowers by thermal oxidation.[15]

In conclusion, GaS and GaSe nanowalls have been obtained by thermal treatment of bulk powders, thereby demonstrating how these materials are quite similar to graphite in that they form 0D, 1D and 2D nanostructures. It is noteworthy that we have obtained Ga_2O_3 and GaN nanowalls from GaS and GaSe nanowalls without the loss of dimensional integrity.

Notes and references

† Nanowalls of GaS were obtained by heating GaS powder in a sealed quartz tube as follows: In a typical reaction, 0.02 g of finely ground GaS was sealed in a 30 cm long quartz tube under vacuum (5×10^{-6} Torr). The tube was placed in a horizontal furnace with a temperature gradient. The furnace was slowly heated to 900 °C at a rate of 1 °C min^{-1} and then maintained at that temperature for 6 h. The temperature of the cooler end of the tube varied over a range of temperatures (300–500 °C), the tip being at 300 °C. Most of the solid deposits studied by us were collected from the region where the temperature was 400 °C. These deposits contained GaS nanowalls. The GaSe nanowalls were obtained in a similar fashion. Nanowalls thus obtained were characterized by scanning electron microscopy, transmission electron microscopy, EDAX analysis and X-ray diffraction. Ga_2O_3 nanowalls were obtained upon heating the

nanowalls at 550 °C in air. GaN nanowalls were obtained by heating the GaS and GaSe nanowalls at 800 °C for 6 h in a furnace under a 100 sccm NH_3 gas flow (99.999% pure). Quartz pieces containing the nanowalls were sputter-coated with gold for SEM imaging using a JEOL scanning electron microscope. TEM images were obtained using a JEOL (JEM3010) transmission electron microscope, operating with an accelerating voltage of 300 kV. For this purpose, the nanowalls were scraped-off the quartz tube, dispersed in CCl_4 and deposited on a holey carbon grid. XRD patterns were recorded on a Siemens 5005 diffractometer employing the reflection Bragg–Brentano geometry with Cu-Kα radiation (λ = 1.5418 Å).

1 *The Chemistry of Nanomaterials*, ed. C. N. R. Rao, A. Mueller and A. K. Cheetham, Wiley-VCH, Weinheim, 2004; C. N. R. Rao, F. L. Deepak, G. Gundiah and A. Govindaraj, *Prog. Solid State Chem.*, 2003, **31**, 5.
2 Y. Wu, B. Yang, B. Zong, H. Sun, Z. Shen and Y. Feng, *J. Mater. Chem.*, 2004, **14**, 469.
3 Y. Wu, P. Qiao, T. Chong and Z. Shen, *Adv. Mater.*, 2002, **14**, 64.
4 B. Yang, Y. Wu, B. Zong and Z. Shen, *Nano Lett.*, 2002, **2**, 751.
5 H. T. Na, J. Li, M. K. Smith, P. Nguyen, A. Cassell, J. Han and M. Meyyappan, *Science*, 2003, **300**, 1249.
6 H. Zhang, D. Yang, X. Ma, Y. Ji, J. Xu and D. Que, *Nanotechnology*, 2004, **15**, 622.
7 Y. C. Zhu and Y. Bando, *Chem. Phys. Lett.*, 2003, **372**, 640.
8 U. K. Gautam, M. Ghosh and C. N. R. Rao, *Langmuir*, 2004, **20**, 10775.
9 X. S. Fang, C. H. Ye, L. D. Zhang, Y. H. Wang and Y. C. Wu, *Adv. Funct. Mater.*, 2005, **15**, 63.
10 X. S. Fang, C. H. Ye, Y. H. Wang, Y. C. Wu and L. D. Zhang, *J. Mater. Chem.*, 2003, **13**, 3040.
11 Z. R. Dai, Z. W. Pan and Z. L. Wang, *J. Phys. Chem. B*, 2002, **106**, 902.
12 M. Cote, M. Cohen and D. Chadi, *Phys. Rev. B: Condens. Matter*, 1998, **58**, 4277.
13 T. Kohler, T. Frauenheim, Z. Hajnal and G. Seifert, *Phys. Rev. B: Condens. Matter*, 2004, **69**, 193403.
14 U. K. Gautam, S. R. C. Vivekchand, A. Govindaraj, G. U. Kulkarni, N. R. Selvi and C. N. R. Rao, *J. Am. Chem. Soc*, 2005, **127**, 3658.
15 A. Chen, X. Peng, K. Koczkur and B. Miller, *Chem. Commun.*, 2004, 1964.

INSTITUTE OF PHYSICS PUBLISHING

Nanotechnology **17** (2006) S287–S290

NANOTECHNOLOGY

doi:10.1088/0957-4484/17/11/S10

Nanorotors using asymmetric inorganic nanorods in an optical trap

Manas Khan[1], A K Sood[1], F L Deepak[2] and C N R Rao[2]

[1] Department of Physics, Indian Institute of Science, Bangalore-560012, India
[2] Chemistry and Physics of Materials Unit, Jawaharlal Nehru Centre for Advanced Scientific
Research, Bangalore-560064, India

E-mail: asood@physics.iisc.ernet.in

Received 27 January 2006, in final form 10 March 2006
Published 19 May 2006
Online at stacks.iop.org/Nano/17/S287

Abstract
We demonstrate how light force, irrespective of the polarization of the light,
can be used to run a simple nanorotor. While the gradient force of a single
beam optical trap is used to hold an asymmetric nanorod, we utilize the
scattering force to generate a torque on the nanorod, making it rotate about
the optic axis. The inherent textural irregularities or morphological
asymmetries of the nanorods give rise to the torque under the radiation
pressure. Even a small surface irregularity with non-zero chirality is
sufficient to produce enough torque for moderate rotational speed. Different
sized rotors can be used to set the speed of rotation over a wide range with
fine tuning possible through the variation of the laser power. We present a
simple dimensional analysis to qualitatively explain the observed trend of the
rotational motion of the nanorods.

(Some figures in this article are in colour only in the electronic version)

1. Introduction

Optical manipulation of micron-sized particles has been a
major success for more than a decade [1]. The technique
of optical tweezers has now become an established tool in
physics and single molecule biophysics. The ability to apply
light force for generating torque on micro-objects making them
rotate in a controlled fashion has great importance in optical
micromachining and biotechnology. Various techniques
have, therefore, been investigated to design optically driven
microrotors. A few schemes in this field have already been
reported [2–14]. In the new age of nanotechnology, it
would obviously be a target to design a laser-assisted rotor
using nanostructures. Here we report a simple scheme,
complementary to the existing literature, describing how
inorganic nanorods can be used to construct laser-assisted
nanorotors.

First we briefly summarize the existing schemes and their
range of applicabilities. In a circularly polarized optical
trap, birefringent microparticles are seen to rotate [2, 3].
Microobjects, when trapped in a spiral optical pattern, have
also been observed to rotate [4–6]. Rotations of specifically
fabricated rather big rotors under an optical trap have been

reported [7–9]. Use of spatial light modulators (SLMs) is
another novel way to rotate multi-particle structures [10] or
to make optical vortex arrays which in turn cause rotation of
spherical microparticles [11]. An asymmetric dumbbell-like
structure of two colloidal particles also rotates in an optical
trap [12]. Nanorods, trapped very close to the cover glass, can
be rotated about their cross-sectional diameter by rotating the
plane of polarization of light [13, 14].

We have trapped ZnO and Al_2O_3 nanorods in a
conventional single beam optical tweezers set-up using linearly
polarized infrared light. The inherent surface irregularities
or asymmetries in shape, bearing non-zero chirality, when
exposed to the radiation pressure experience a net torque,
making the trapped nanorod rotate about the optic axis.
Rotations of different sized particles, ranged from a single
nanorod to a bundle of nanorods, covering a wide spectrum
of rotational speed, have been observed. We observed that
even small extrusions on the surface of a nanorod are sufficient
to generate enough torque to make the nanorod rotate at a
moderate rotational speed. As those small irregularities are
inbuilt to all synthesized nanorods and commonly do not have
any mirror plane of symmetry passing through the optic axis,
any of the nanorods can be used as a nanorotor under the laser

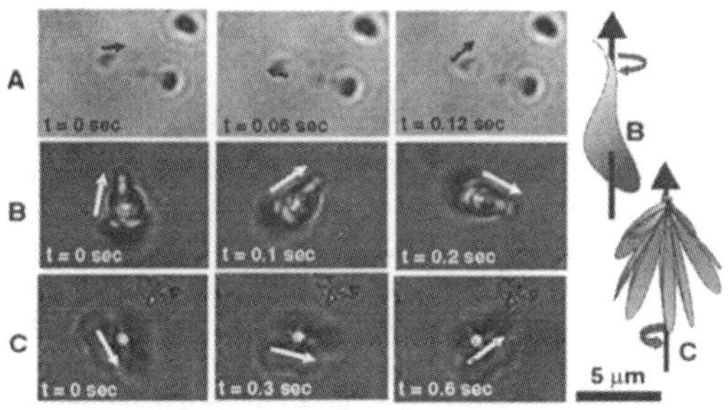

Figure 1. Time sequences of different sized and shaped rotors are shown here. In each time frame the orientation of the rotor is indicated by an arrow. Panels (A), (B) and (C) represent rotations of three Al_2O_3 rotors. The rotor in panel (A) is a typical nanorod whereas the rotor in panel (B) is a bigger asymmetric nanorod, and in panel (C), the rotor is a nanorod bundle. The predicted structures of the rotors in panels (B) and (C) have been depicted in the rightmost column. A size bar has been shown at the bottom right-hand corner. Magnification factors of all the images are same. The images shown in panel (A) are diffraction-limited images and hence they do not convey the real size of the rotor.

trap. A rotor of suitable size and shape is chosen accordingly to set the rotational speed near a desired value and then by varying the trapping beam intensity the speed is finely adjusted. This scheme, therefore, provides a complete and precise control over the rotational speed of the nanorotors.

2. Experimental details

A 1064 nm linearly polarized laser beam from a 2.5 W Nd:YVO$_4$ laser was focused through a 1.4 numerical aperture 63× objective to trap the nanorods with typical diameter 50–60 nm and length $\sim$2 μm. The Al_2O_3 nanorods were obtained by carbon-assisted synthesis [15, 16] starting with a mixture of Al powder and graphite/activated carbon (molar ratio 1:1). For the synthesis of ZnO nanorods, we have employed a procedure involving the solid-state reaction between zinc oxalate and multi-walled carbon nanotubes (MWNTs) [15, 16]. The nanorods were well dispersed in N,N-dimethylformamide by two hours of sonication. A drop of the nanorod suspension was taken on a transmission electron microscope (TEM) grid to observe the morphology of the nanorods. The inset of figure 2 shows a few TEM images of some typical Al_2O_3 and ZnO nanoparticles. The images clearly show that the suspension contains nanorods of different asymmetric shapes or with surface irregularities and some odd bundles of nanorods. A few drops of the dispersion were taken in an annular chamber on a coverglass for observations. We have observed rotations of different types of rotors. The experiments were recorded at 30 frames s^{-1} with a digital CCD camera attached to the microscope. The rotational speeds of the nanoparticles were calculated from the recorded frames and thus the calculation was limited to a maximum value of speed $\sim$10 Hz. The rotational speeds of different sized and shaped single nanorods and nanorod bundles were studied at varying laser power. In many cases, the rotational speed or size of the rotors were beyond the maximum measurement capability through imaging technique. In figure 1 we show rotations of some typical rotors with different shapes and sizes.

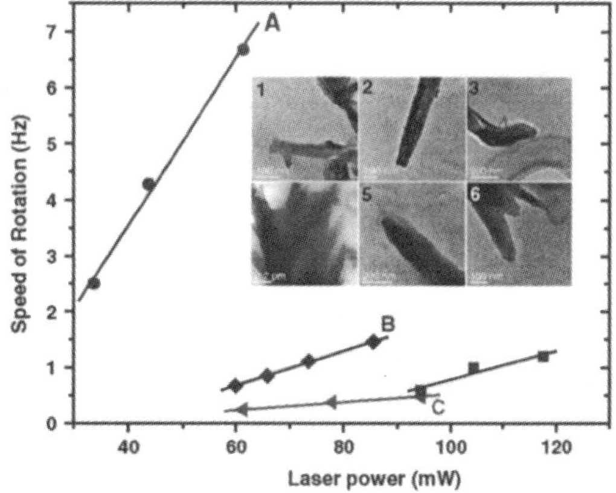

Figure 2. The graph shows the variation of rotational speed with laser power for four different Al_2O_3 rotors. The plots labelled A, B and C correspond to the rotors shown in panels (A), (B) and (C) respectively in figure 1. The inset shows TEM images of some typical nanoparticles. Frames 1 and 2 show Al_2O_3 nanorods with irregularities on the surface, and frame 3 displays an asymmetric Al_2O_3 nanorod. A bundle of Al_2O_3 nanorods is shown in frame 4. Frames 5 and 6 exhibit an asymmetric ZnO nanorod and a bundle of ZnO nanorods respectively.

3. Rotational speed–laser power relationship

The variation of the speed of rotation with laser power for some of the Al_2O_3 rotors is shown in figure 2, corresponding to rotors in panels (A), (B) and (C) of figure 1. In all cases, the speed of rotation increases monotonically with the laser power and there is a non-zero finite value of the laser power at which the rotors start rotating. While the rotational speed of all the rotors follow a linear relationship with the laser power, the slope and the intercept depend on the size and shape of the rotor. The slope is steeper for small and more asymmetric rotors. Bigger rotors have a higher threshold value of laser power for the onset of rotation, along with a flatter variation of rotational speed with laser power. So it can be concluded that small rotors with large asymmetry have more efficiency at all laser power levels whereas the bigger rotors are not so effective at low laser power level as they need more laser power for starting up the rotation.

In addition to this, it is noteworthy that the direction of the rotational motion was not same for all the rotors. Though a rotor always rotated in a particular direction, both right-handed and left-handed rotations were observed with equal probability in randomly selected nanorotors.

4. Theoretical understanding

In figure 3 we depict a very simple model to qualitatively understand the rotational motion of the nanorods when trapped under linearly polarized laser beam. Most of the nanorods have surface irregularities or morphological asymmetries that do not have any symmetry about a mirror plane passing through the optic axis of the trapping beam. In a simple-minded way, here we model a nanorod as a cylinder with some extrusions on its surface. The nanorod is trapped at the point of focus by the gradient force of the tightly focused laser beam. For

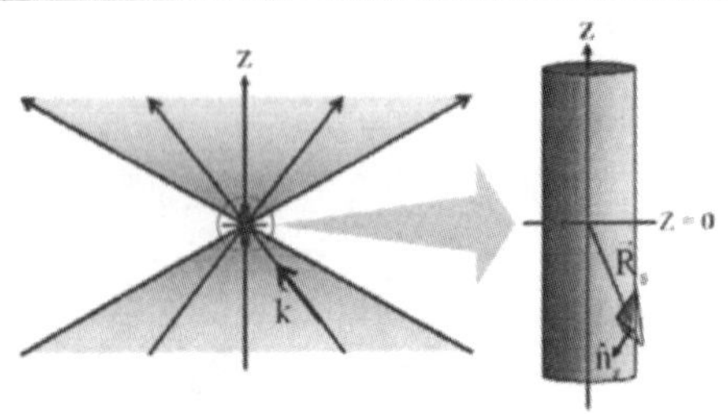

Figure 3. A model of a nanorod in the optical trap is shown here. An enlarged image of the nanorod is depicted at the right. It has a small extrusion with surface area s along $\hat{n}_s = a\hat{r} + b\hat{\phi} + c\hat{z}$ at a position $\vec{R}_s(r_s, \phi_s, z_s)$.

simplicity, in figure 3 we have shown only one extrusion on the surface of the nanorod. As this extrusion is not symmetric about a plane passing through the z-axis, it will provide a non-zero chirality to the nanorod structure. Therefore, the radiation pressure force will generate a non-zero torque on the nanorod. The radial and the azimuthal component of the torque would be nullified by the gradient force and only the axial component would survive to provide a non-zero angular momentum to the nanorod causing its rotation about the optic axis. Now if the surface of the nanorod is full of randomly oriented extrusions, some would produce left-handed torque and the remaining ones would contribute to generate right-handed torque. So the net torque experienced by the nanorod is completely determined by the structure and morphology of the nanorod and hence the direction of rotation as well as the speed of rotation is different for different rotors. Since the absorbance of the inorganic nanorods at 1064 nm is negligible [17, 18], the deformation of the nanorods or any other effects caused by the sample heating are not considered here. The existence of a laser power threshold for the onset of rotation is a bit puzzling. We are unable to understand the origin of the threshold and why it increases with the increasing size of the rotors.

Calculating the electromagnetic field near the focus of a very large aperture beam, and around a particle itself which is neither very small nor very large compared to the wavelength of the radiation, is notoriously very difficult and hence the exact analysis of the radiation pressure force acting on the nanorods becomes nearly intractable. Hence we try to understand the trend of the observed rotational motion by a simple dimensional analysis.

The momentum density of radiation is given by $\vec{g}(r, z) = \frac{1}{2c}\epsilon_0 E^2 \hat{k}$, where $E(r, z)$ is the magnitude of the electric field of light propagating along $\hat{k}$. So the radiation pressure force scales as $E^2/c \equiv P/c$, where P is the laser power and c is the velocity of light. For the nanorod shown in figure 3, the net torque generated in the laser trap can be given by

$$\vec{\tau} = 2Rc \int_s \vec{R}_s \times (-\hat{n}_s)(\vec{g} \cdot \hat{n}_s)\, ds \qquad (1)$$

where R is the reflectance of the nanorod and the integration is done over the total area of the surface irregularities or extrusions. The larger extrusions with greater chirality would contribute more to the net torque. Hence, the magnitude of the

driving torque τ will scale as

$$\tau \sim \chi \left(\frac{P}{c}\right) l. \qquad (2)$$

Here χ is a phenomenological parameter measuring the chirality and l is a size scale factor. Since here the Reynolds number is small ($\sim 10^{-5}$), the Stokes equation can be used to get $\vec{\tau} = D\vec{\Omega}$, where D is the hydrodynamic drag coefficient and Ω is the rotational speed. The drag coefficient for a cylindrical object rotating about its axis with cross-sectional radius r and length L is $D = 4\pi \eta r^2 L$, where η is the viscosity of the medium [19]. Therefore, $D \sim l^3$ and the rotational speed Ω of the rotor will scale as

$$\Omega \sim \chi \left(\frac{P}{c}\right) l^{-2}. \qquad (3)$$

This equation qualitatively explains the observed trend of the rotational motion. For smaller rotors, the rotational speed increases more sharply with increasing laser power compared to the bigger rotors that show a flat variation of rotational speed with laser power. Thus at the same laser power a smaller rotor with more asymmetry (larger χ) would rotate faster.

5. Conclusion

Our observations and the qualitative explanation with dimensional analysis show that inorganic nanorods can be conveniently used as optically driven nanorotors. Different sized rotors can be chosen to cover different rotational speed regimes with the fine tuning made possible by varying the laser power. The inorganic nanorods provide a better choice as nanorotors since even small ones are easily trapped due to their high dielectric constants. The controllability, simplicity and flexibility of the rotational motion of these rotors suggest that this scheme can be applied universally to devise easy-to-use optically driven nanorotors. The availability of inorganic nanorods bearing the desired size and asymmetry would favour the present scheme to design nanorotors with predictable rotational speeds. Recent experiments on trapping and manipulation of carbon nanotubes [20–22] suggest that Y-junction nanotubes [23] may provide another interesting choice to be used as nanorotors under an optical trap. The nanorotors could be used in micropumps to guide fluids through microchannels or as microstirrers. This rotational motion can also be used in experiments related to microrheology and microfluidics.

Acknowledgments

We thank the Department of Science and Technology, India, for support under the DST Nanoscience Initiative.

References

[1] Ashkin A, Dziedzic J M, Bjorkholm J E and Chu S 1986 *Opt. Lett.* **11** 288
[2] Friese M E J, Nieminen T A, Heckenberg N R and Rubinsztein-Dunlop H 1998 *Nature* **394** 348
[3] Cheng Z, Chaikin P M and Mason T G 2002 *Phys. Rev. Lett.* **89** 108303

M Khan *et al*

[4] He H, Friese M E J, Heckenberg N R and
     Rubinsztein-Dunlop H 1995 *Phys. Rev. Lett.* **75** 826
[5] Friese M E J, Enger J, Rubinsztein-Dunlop H and
     Heckenberg N R 1998 *Nature* **394** 348
[6] Paterson L, MacDonald M P, Arlt J, Sibbett W, Bryant P E and
     Dholakia K 2001 *Science* **292** 912
[7] Higurashi E, Ukita H, Tanaka H and Ohguchi O 1994 *Appl.
     Phys. Lett.* **64** 2209
[8] Galajda P and Ormos P 2001 *Appl. Phys. Lett.* **78** 249
[9] Galajda P and Ormos P 2002 *Appl. Phys. Lett.* **80** 4653
[10] Eriksen R L, Rodrigo P J, Daria V R and Glückstad J 2003
     Appl. Opt. **42** 5107
[11] Ladavac K and Grier D G 2004 *Opt. Express* **12** 1144
[12] Luo Z-P, Sun Y-L and An K-N 2000 *Appl. Phys. Lett.* **76** 1779
[13] Bonin K D, Kourmanov B and Walker T G 2002 *Opt. Express*
     10 984
[14] Bishop A I, Nieminen T A, Heckenberg N R and
     Rubinsztein-Dunlop H 2003 *Phys. Rev. A* **68** 033802
[15] Rao C N R, Deepak F L, Gundiah G and Govindaraj A 2003
     Prog. Solid State Chem. **31** 5
[16] Gundiah G, Deepak F L, Govindaraj A and Rao C N R 2003
     Top. Catal. **24** 137
[17] Gu Y, Kuskovsky I L, Yin M, O'Brien S and Neumark G F
     2004 *Appl. Phys. Lett.* **85** 3833
[18] Innocenzi M E, Swimm R T, Bass M, French R H,
     Villaverde A B and Kokta M R 1990 *J. Appl. Phys.* **67** 7543
[19] Cocco S, Monasson R and Marko J F 2002 *Phys. Rev. E*
     66 051914
[20] Plewa J, Tanner E, Mueth D M and Grier D G 2004 *Opt.
     Express* **12** 1978
[21] Tan S, Lopez H A, Cai C W and Zhang Y 2004 *Nano Lett.*
     4 1415
[22] Khan M, Sood A K, Mohanty S K, Gupta P K, Arabale G V,
     Vijaymohanan K and Rao C N R 2006 *Opt. Express* **14** 424
[23] Satishkumar B C, Thomas P J, Govindaraj A and
     Rao C N R 2000 *Appl. Phys. Lett.* **77** 2530

INSTITUTE OF PHYSICS PUBLISHING

NANOTECHNOLOGY

Nanotechnology 17 (2006) S344–S350

doi:10.1088/0957-4484/17/11/S19

Mechanical properties of inorganic nanowire reinforced polymer–matrix composites

S R C Vivekchand[1], U Ramamurty[2] and C N R Rao[1]

[1] Chemistry and Physics of Materials Unit, Jawaharlal Nehru Centre for Advanced Scientific Research, Jakkur PO, Bangalore 560 064, India
[2] Department of Metallurgy, Indian Institute of Science, Bangalore 560 012, India

E-mail: cnrrao@jncasr.ac.in

Received 13 January 2006, in final form 14 February 2006
Published 19 May 2006
Online at stacks.iop.org/Nano/17/S344

Abstract

The mechanical properties of poly(vinyl alcohol) matrix composites incorporating SiC and Al_2O_3 nanowires (NWs) have been investigated. A marked increase in the elastic modulus (up to 90%) has been observed even with the addition of a small quantity (0.8 vol%) of nanowires. This observation cannot be explained by iso-stress analysis, which is appropriate for describing the variation of properties with the reinforcement volume fraction in discontinuously reinforced composites. Crystallization of the polymer induced by the NWs, the high aspect ratio and the surface-to-volume ratio of the NWs as well as the possible in-plane alignment of the NWs during processing are considered to be responsible for the increase in the stiffness. A significant increase in the strength of the composite with the addition of NWs is also observed. This is due to the significant pull-out of the NWs and the corresponding stretching of the matrix due to the complete wetting of the NW surface by the polymer. The increase in tensile strength is found to saturate at higher vol% of NW addition due to the reduced propensity for shear-band induced plastic deformation.

(Some figures in this article are in colour only in the electronic version)

1. Introduction

The synthesis and characterization of one-dimensional nanostructures constitute an area of great interest. Of the various one-dimensional materials, single-walled carbon nanotubes (SWNTs) have been studied extensively. In particular, electrical and mechanical properties of SWNT–polymer composites have been examined in some detail [1, 2]. Incorporation of SWNTs in the polymers generally enhances the stiffness and strength. Although the electrical properties, sensor characteristics and other aspects of inorganic nanowires have been reported [3–6], there has been no systematic study of the mechanical properties of composites with polymers. The mechanical properties of individual SiC nanowires (NWs) have been examined by atomic force microscopy. While the elastic modulus, E, is close to the theoretical limit, the fracture strength is larger than that of the fibre [7]. Wang et al [8]

have studied the mechanical properties of SiC–SiO_2 biaxial nanowires using electric field induced dynamic phenomena, while Yang et al [9] have used carbon-coated SiC nanowires for preparing ceramic composites with high toughness. We have investigated the mechanical properties of composites of SiC and Al_2O_3 nanowires with polyvinyl alcohol (PVA). PVA was chosen for the study in view of practical applications and also because earlier studies of carbon nanotube composites were with PVA [10–12]. For the purpose of the present study, we prepared the PVA–NW composites as a function of the nanowire volume fraction and characterized them by electron microscopy, infrared spectroscopy and differential scanning calorimetry (DSC). Dynamic mechanical analysis (DMA) was conducted to measure the elastic modulus and tensile testing to measure the strength of the polymer composites. To our knowledge, this is the first report on the mechanical properties of inorganic nanowire-reinforced polymer–matrix composites.

2. Experimental details

Nanowires of SiC and Al_2O_3 were synthesized using methods reported earlier in the literature [13, 14]. In a typical preparation of the PVA–SiC NW (0.8 vol%) composite, PVA (1.95 g) and SiC NWs (0.05 g) were added to warm water (50 ml) and the mixture was heated at 70 °C until the polymer dissolved forming a dispersion of the nanowires. The dispersion was dried in Petri dishes at 50 °C over a period of 3 days.[3] As the mechanical properties of PVA are sensitive to the water content, the polymer films were stored in a vacuum desiccator with $CaCl_2$ for at least a week before mechanical testing. Composites with PVA were prepared with 0.2, 0.4 and 0.8 vol% of SiC NWs and 0.4 vol% of Al_2O_3 NWs. Visual as well as optical microscopic examination of the composite strips indicated uniform distribution of the nanowires throughout the matrix.

The composites as well as the nanowires were characterized by several techniques. Scanning electron microscopy (SEM) images and energy dispersive analysis of x-rays (EDAX) were obtained with a Leica S-440I microscope fitted with a Link ISIS spectrometer. Infrared (IR) spectra were recorded on small pieces of the samples embedded in KBr pellets using a Bruker FT-IR spectrometer. DSC was carried out on the samples (~7 mg) with a scanning rate of 20 K min^{-1} between 120 and 260 °C using a Mettler-Toledo DSC.

Mechanical testing was conducted on samples 25 mm in length, 10 mm in width and ~0.1 mm in thickness. Because of the high compliance of the films, it was difficult to mount an extensometer on the samples to measure E. A DMA was therefore used to assess the elastic properties of the composite samples in the tensile mode of loading. A 5 N static tensile load and displacement amplitude of 16 μm at a frequency of 1 Hz were applied. Nine measurements for each sample were made and the average values are reported here.

Tensile stress–strain curves were generated using an electro-mechanical universal testing machine with specially designed flat-ended fixtures that were machined in order the grip the specimens carefully. All the samples were tested for failure under displacement control with a prescribed displacement rate of 1.5 mm min^{-1}. Fractography of the tested samples was carried out using a SIRION field emission SEM.

3. Results

3.1. Characterization

Representative SEM images of the SiC and Al_2O_3 NWs used for producing the composites are shown in figures 1(a) and (b), respectively. The nanowires have diameters in the 90–150 nm range with lengths extending to tens of micrometres. The nanowires were single-crystalline, the growth direction of SiC NWs being $\langle 111 \rangle$ and that of Al_2O_3 NWs forming a 35° acute angle with $\langle 104 \rangle$ [13, 14].

The 1140 cm^{-1} band of PVA in the IR spectrum is known to be sensitive to the crystallinity [15, 16]. Figure 2 shows the IR spectra of PVA and 0.4 vol% NW-reinforced PVA composites. The spectrum of PVA agrees with that reported

---

[3] We have taken the density of PVA, SiC NWs and Al_2O_3 NWs as 1.3, 3.217 and 3.97 g cm^{-3} respectively, corresponding to their bulk values.

Figure 1. SEM images of (a) SiC and (b) Al_2O_3 NWs that were used for producing the PVA–matrix composites.

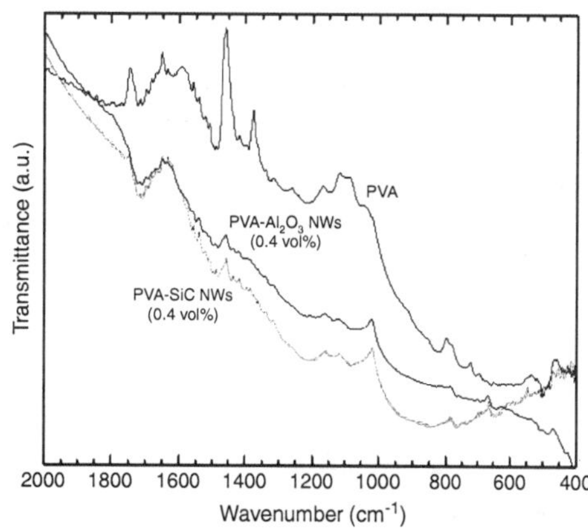

Figure 2. Infrared spectra of PVA and NW composites. Notice the variation in intensity of the 1140 cm^{-1} band relative to that of the 1425 cm^{-1} band.

in the literature [15, 16]. In the NW composites, we observe an increase in the relative intensity of the 1140 cm^{-1} band with respect to the 1425 cm^{-1} band. The observation suggests an increase in the degree of crystallinity of PVA on addition of the NWs.

Quantitative information about the change in polymer crystallinity, χ, due the addition of nanowires was obtained through DSC. Figure 3(a) shows the thermograms of PVA as well as those of the composites, focusing on the broad

Table 1. Summary of the properties of various composites.

Sample (vol%)	Degree of crystallinity (%)	Elastic modulus, E (GPa) (% increase)	Lower and upper bounds for E (GPa)	Tensile strength, σ_U (MPa) (% increase)
Blank PVA	33.9	3.2 ± 0.21	—	72.7 ± 3.4
PVA–SiC NW (0.2)	34.5	3.76 ± 0.21 (17.5)	3.2–4.39	110.6 ± 4.9 (52.1)
PVA–SiC NW (0.4)	34.8	4.58 ± 0.07 (43.1)	3.2–5.59	119.2 ± 3.2 (64.0)
PVA–SiC NW (0.8)	35.0	6.08 ± 0.21 (90)	3.2–7.97	123.1 ± 4.0 (69.3)
PVA–Al$_2$O$_3$ NW (0.4)	38.4	5.06 ± 0.29 (58.1)	3.2–4.94	123.8 ± 6.7 (70.3)

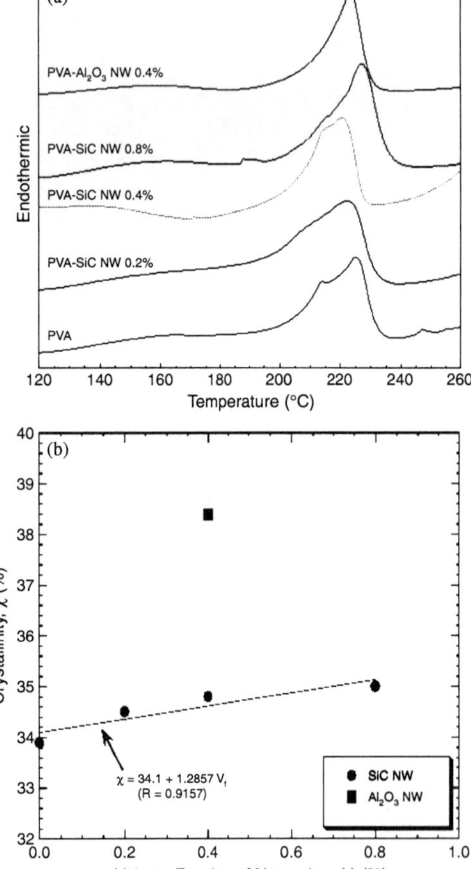

Figure 3. (a) DSC scans of PVA and NW composites showing the crystallization peak in the temperature range of 200–240 °C. (b) Variation of the degree of crystallinity of PVA, χ, as a function of the NW volume fraction, V_f. The change χ with SiC NW V_f is systematic yet small. In contrast, a significant increase in χ was noted with the addition of Al$_2$O$_3$ NWs to PVA.

endothermic peak due to melting centred around 225 °C. The peak becomes narrower as the concentration of nanowires in

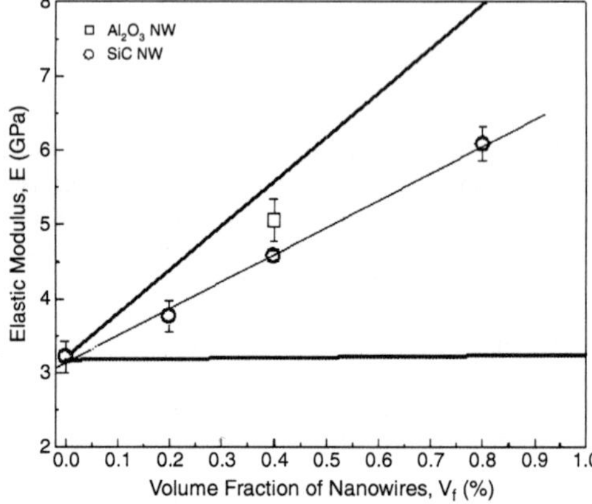

Figure 4. Variation of elastic modulus, E (measured with the DMA technique), as a function of the NW volume fraction, V_f. The upper- and lower-bound predictions (made using iso-strain and iso-stress models, respectively) are also plotted.

the composite increases. From these scans, χ was calculated as equal to $(\Delta H / \Delta H_C)$, where ΔH is the heat required to melt 1 g of dry sample (obtained by integrating the area under the endothermic peak between 190 and 240 °C) and ΔH_C is the standard enthalpy for 100% crystalline PVA ($\Delta H_C = 138.6$ J g^{-1} [17]). The χ values, normalized with respect to the actual PVA content in the samples, are listed in table 1. Figure 3(b) shows the variation of χ with the volume fraction of the NWs, V_f. For the SiC NW composite, χ appears to vary linearly with V_f. However, the variation is small (a maximum of $\sim$1.1% for the 0.8 vol% SiC NW composite) and can be considered negligible. For the Al$_2$O$_3$ NW composite, on the other hand, a 0.4 vol% addition of nanowires leads to a 4.5% increase in χ, which is significant.

3.2. Mechanical properties

Experimental results obtained from the DMA and tensile testing are summarized in table 1. There is a significant increase in E (up to 90%) even with a relatively small addition (0.8 vol%) of NWs. In the case of SiC NWs, the variation of E with V_f is linear (figure 4). The Al$_2$O$_3$ NW composite exhibits

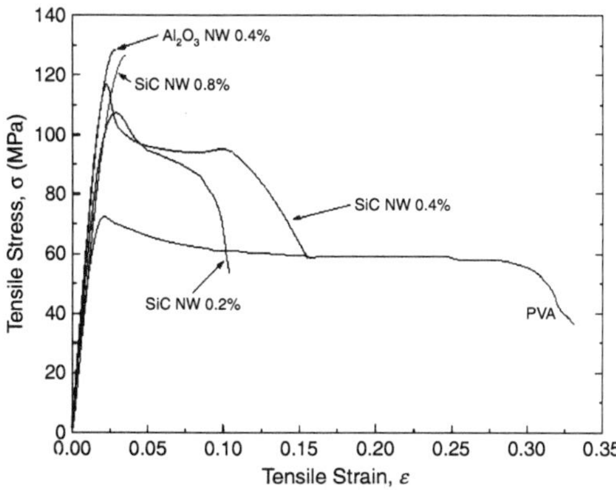

Figure 5. Tensile stress–strain curves of PVA and NW composites, showing an increase in the ultimate tensile strength and reduction in ductility with increasing NW volume fraction.

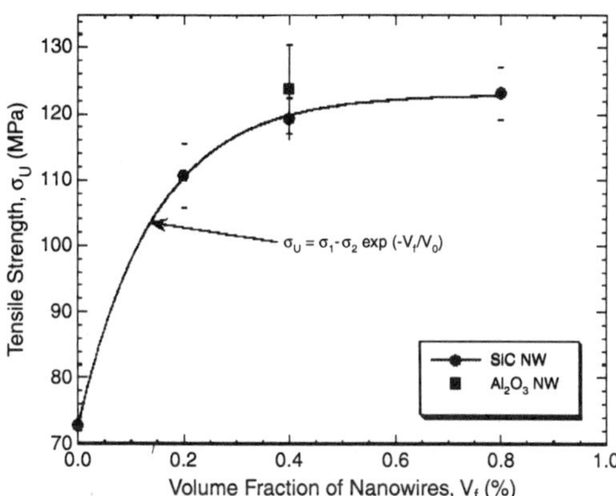

Figure 6. Variation of tensile strength as function of nanowire volume fraction. The solid line represents the least-square fit to the equation given in the figure.

a $\sim$10% higher E than the SiC NW composite with the same amount of reinforcement phase (0.4 vol%).

Figure 5 shows representative tensile stress versus strain (σ versus ε), curves obtained with blank PVA as well as the composites with various V_f. With increasing V_f, a significant increase in the yield strength occurs. However, an estimation of the yield strengths from these plots is difficult due to lack of accurate strain measurement (which requires mounting of either an extensometer or a strain gauge on the sample, both of which are difficult due to the thin and highly compliant samples). Since the peak in the load appears immediately after yielding, the ultimate tensile strength, σ_U, was used as the metric that captures the strengthening changes due to the addition of the NWs. It is seen from figure 5 that, for PVA alone, yielding and maximum load occurs almost simultaneously, with the material sustaining the same stress level for prolonged straining. Essentially, blank PVA behaves like an elastic–perfectly plastic solid. While the composites exhibit higher strengths, a significant strain softening is seen

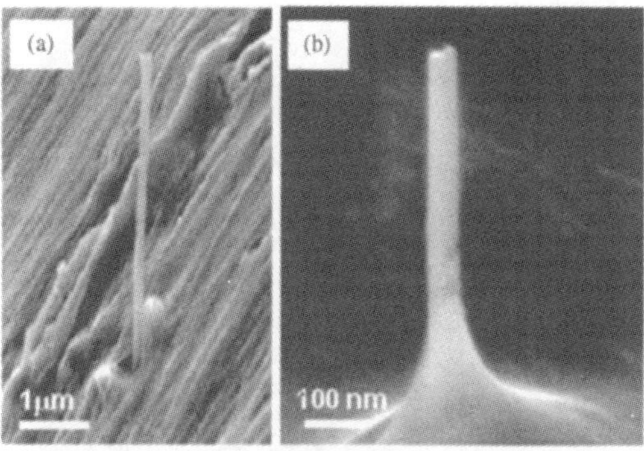

Figure 7. Field emission ESEM images obtained from the fractured PVA–SiC NW (0.8 vol%) composite showing pull-out of the nanowires as well as stretching of the matrix along with the nanowire.

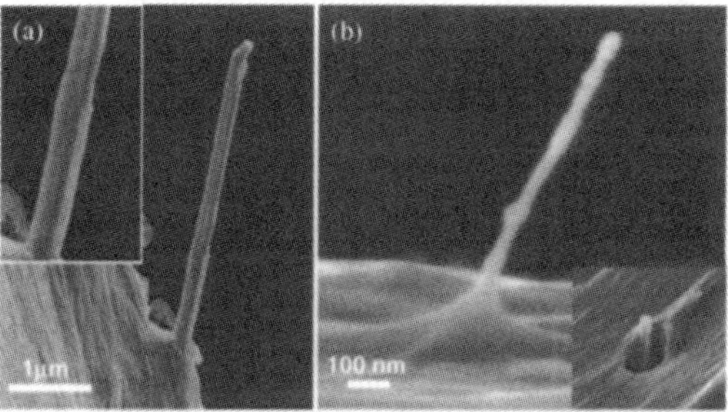

Figure 8. Field emission SEM images obtained from the fractured PVA–Al$_2$O$_3$ NW (0.4 vol%) composite showing pulled-out nanowires. The inset in (b) shows the formation of a hole around the nanowire, indicating retraction of the stretched and debonded matrix material.

immediately after the peak in the load. Localization also sets in rather quickly, *vis-à-vis* blank PVA, and hence the composites exhibit much lower ductility. Thus the 0.4 vol% Al$_2$O$_3$ and the 0.8 vol% SiC NW composites show near-zero ductility, fracturing immediately after yielding.

The variation of σ_U with V_f shown in figure 6 reveals that σ_U reaches a plateau after an initial steep rise, in the case of SiC NW reinforced composite. For the same amount of V_f (0.4 vol%), the Al$_2$O$_3$ NW composite has a similar strength (within experimental scatter) as that of the SiC NW composite. This is in contrast to that observed in E, where a clear 10% difference was noted. An equation of the type,

$$\sigma_U = \sigma_1 - \sigma_2 \exp\left(-\frac{V_f}{V_0}\right) \tag{1}$$

describes the SiC NW data exceptionally well with an R value (that indicates the goodness of fit) of 0.9998. The fit gives 122.96 ± 0.69 MPa, 50.22 ± 0.98 MPa and $0.145 \pm 0.008\%$ for σ_1, σ_2 and V_0, respectively. We discuss the implications of this fit later in section 4.2.

3.3. Fractographic observations

Typical SEM images obtained from the fracture surfaces of tensile-tested samples are shown in figures 7 and 8 for the SiC

Figure 9. Optical micrograph of the deformed PVA–SiC NWs (0.4 vol%) showing shear-band induced deformation.

and Al_2O_3 NW composites, respectively. Fracture surfaces of the matrix (PVA) are generally featureless (even at high resolution), as typified by the backgrounds in figures 7(a) and 8(a). These figures also show extensive pull-out of the NWs from the matrix, with pull-out lengths reaching several micrometres (going up to 10 μm in certain cases). Furthermore, good adhesion between the NWs and the matrix is illustrated by the pulling of the base of the matrix revealed in figures 7(b) and 8(b). In the case of the SiC NW composites, the debonding appears to occur at the NW/matrix interface as reflected by the rather clean appearance of the SiC NWs as shown in figure 7(b). In contrast, the pulled-out Al_2O_3 NWs appear to be coated with the polymer. This can be seen in the inset in figure 8(a) as well as from the corrugated nature of the pulled-out NW (figure 8(b)). In SiC and Al_2O_3 NW composites, holes in the matrix surrounding the broken NWs are seen (see the base of the NW in figure 7(a) and also the inset in figure 8(b)). Qualitatively, these holes are much more pronounced in the case of Al_2O_3 NW composites. It is likely that these holes are due to the recession of the stretched matrix (due to the pull-out of the NWs) that eventually debonds after reaching a critical shear strain.

Visual observations of the tensile-tested samples show a square pattern, corresponding to shear banding of the PVA. Such a shear-band pattern obtained from the 0.4 vol% SiC NW reinforced composite is shown in figure 9. The shear-band morphology is found generally to diminish with increasing V_f. In the case of 0.8 vol% SiC and 0.4 vol% Al_2O_3 NW composites, with near-zero ductility, the shear-band pattern is absent.

4. Discussion

4.1. Stiffness

A significant observation in the present study is the marked increase in the E value, even with relatively small additions of the inorganic NWs. This is to be compared with the report of Zhang *et al* [12] who find that the elastic modulus of the PVA/PVP/SDS polymer increases from 2.5 to 4.0 GPa upon the addition of 5 wt% SWNTs. With the multi-walled carbon nanotubes (MWNTs), a linear increase in E with V_f of the MWNTs has been reported by Coleman *et al* [11]. With a 0.6 vol% addition of MWNTs, the PVA (nearly non-crystalline) modulus was reported to increase from 1.92 to

7.04 GPa. Clearly, the inorganic NWs are significantly better than the SWNTs and are comparable to the MWNTs in their stiffening effect.

The upper and lower bounds of the elastic moduli of the composites, E, can be written in terms of a simple rule-of-mixtures given below (derived by imposing iso-strain and iso-stress conditions, respectively):

$$E = V_f E_f + (1 - V_f) E_M \qquad (2)$$

$$E = \frac{E_f E_M}{V_f E_f + (1 - V_f) E_M}. \qquad (3)$$

Here, E_f and E_M are the moduli of the NW and matrix, respectively. Equation (2) is applicable when the NWs are aligned perpendicular to the loading direction whereas equation (3) is for NWs parallel to the loading direction. The above equations can predict E only for the case where the NWs are continuous throughout the composite. By taking E_f to be 600 GPa for the SiC NWs, corresponding to the bulk single-crystalline modulus of SiC [7] and 3.2 GPa for unreinforced PVA, the upper and lower bounds derived from equations (1) and (2) are plotted in figure 4. We see that the E data of the composites are closer to the upper bound. In the case of the Al_2O_3 NW composite, the measured E is higher (see table 1) than the upper bound predicted using 440 GPa for the elastic modulus of Al_2O_3 [18]. This could be due to the relatively large increase in χ in this case.

Considering that the concentration of the reinforcement phase is small (0.8 vol% at the maximum) and that the NWs are randomly dispersed through the matrix one would expect the E values of the composites to be closer to the lower bound, as the matrix deformation dominates the overall response of the composite. In contrast, a near doubling of E with only 0.8% loading of SiC NWs is observed. There can be several possible reasons for this large increase as detailed below.

- *Large aspect ratio*: Because the reinforcements are nanosized, the large aspect ratio (length/diameter) could make the composite behave like a continuous fibre composite. This scenario is unlikely as the aspect ratios of the NWs used are ∼100–200, which are lower than the value of ∼1000 or so for a transition between short-fibre reinforced composite behaviour and the long-fibre reinforced one [18]. The specimen dimensions are also much larger than the NW dimensions. The anomalous increase from the relatively smaller sized specimens is ruled out.

- *High surface/volume ratio*: Because of their small sizes, nanostructured materials have high surface to volume ratios, which affect not only the functional properties but also the mechanical properties such as E. Miller and Shenoy [19] have investigated the size-dependent elastic properties of nanosized structural elements through continuum modelling that takes the ratio of surface to volume elastic moduli into consideration. Direct atomistic simulations by them on Al and Si nanostructures suggest that the elastic modulus is sensitive to the size only at very small length scales, typically below 10–15 nm. For sizes larger than this, bulk values are obtained asymptotically. The nanowires used by us are 90–150 nm in diameter. Therefore, the surface to volume ratio is unlikely to enhance the elastic modulus of the nanowires. In fact, the E value extracted by fitting equation (1) through the SiC NW data in figure 4, with E_f as an independent variable, gives E_f as 378.5 GPa, which is close to the E value reported for bulk sintered SiC [7].

- *In-plane alignment of the fibres*: Due to the very nature of the technique used for processing the NW composites, in-plane alignment of the NWs is a realistic possibility. From the Krenchel theory of short-fibre reinforcements [20], the orientation and length effects can be incorporated using an efficiency factor to evaluate E,

$$E = K V_f E_f + (1 - V_f) E_M \qquad (4)$$

where K is the Krenchel efficiency factor, which is equal to 3/8 for a random planar arrangement of short fibres. The elastic modulus of the SiC NWs determined by fitting the above equation to the experimental data is ~ 1 TPa, which is significantly higher than that of theoretical modulus of 600 GPa for (111) oriented SiC NWs [6]. It is interesting to note that similar physically inadmissible values have been obtained by Coleman *et al* [10] in the case of MWNT reinforced composites.

- *Inducement of crystallization of the matrix by the NWs*: The NWs act as nucleation sites for the growth of polymer crystals. For the SiC NW composites, the increase in χ is very small and hence NW induced crystallization is ruled out as accounting for the marked changes in E values of the SiC NW composites. For the Al_2O_3 NW reinforced composite, on the other hand, the oxide surface seems to be more favourable for nucleation and growth of the polymer crystals, as a large increase in χ was observed in this case. This, in turn, can be attributed to the E value which is higher than the upper-bound value (predicted using the iso-strain analysis) of E in this composite (see table 1). It therefore appears that surface functionalization can be used to improve the interaction between the polymer and the nanowires and hence the mechanical properties.

From the preceding discussion, it can be summarized that none of the above factors considered *cannot* can alone account for the marked increase in the stiffness of the semi-crystalline PVA. It may be that an optimum combination of these factors results in the high values of E in the composites. It is also possible that there are other possible causes which we have not taken into account.

4.2. Strength

Strength enhancement with the addition of NWs to PVA is also marked. Similar to the stiffness, SWNTs do not appear to cause such a significant strengthening rate (see Zhang *et al* [12] for example) whereas MWNTs do indeed have a similar strengthening effect [11] as that of the inorganic NWs. Small changes in the crystallinity of the polymer can have pronounced effects on the strength. If the interface debonds relatively easily, efficient load transfer cannot take place and hence composites tend to have a low strength. In contrast, high shear strength of the interface means that the reinforcement phase carries considerably more load, leading to superior strength of the composites. The NW surface

induced crystallization means a strong matrix/NW interface (in agreement with the stretching of the matrix by the NWs, shown in figures 7(b) and 8(b)). Effective load transfer across the interface leads to considerable strengthening of the composite. Clearly, MWNTs and inorganic NWs (employed in this study) appear to favour load transfer.

An important observation made in the present study is that σ_U appears to attain a plateau, after an initial steep rise (figure 6), with equation (1) capturing the experimental trends well. In the limit, $V_f = 0$, equation (1) gives $\sigma_U = (\sigma_1 - \sigma_2)$, the strength of the blank PVA. Differentiation of equation (1) with respect to V_f gives the rate of strengthening,

$$\frac{d\sigma_U}{dV_f} = \frac{(\sigma_1 - \sigma_U)}{V_0}. \qquad (5)$$

The highest rate of strengthening is obtained in the limit $V_f \rightarrow 0$ and is equal to σ_2/V_0. From the extracted values (by fitting equation (1) into the σ_U versus V_f data), $\sigma_2/V_0 \sim$ 34.6 GPa. This should, in principle, scale with the strength of the SiC NWs. Wong *et al* [7] reported 53.4 GPa as the strength of SiC NWs (of diameter 23 nm) measured in bending. The value obtained in our work reflects the strength of NW bundles, which tends to be lower than that a single NW [21]. Furthermore, size also plays a role in determining the strength of ceramics: increasing size typically means a lower strength [22]. Given these differences, the agreement between the strength of SiC NWs obtained in this work and that of Wong *et al* [7] seems reasonable. Coleman *et al* [11] report the rate of strengthening in the MWNT reinforced PVA to be 6.8 GPa, which is considerably lower than that found with the NWs. This could be due to the relatively smooth surfaces of MWNTs, which facilitate easy debonding whereas stress transfer is substantial in the case of inorganic NW reinforced composites.

The flattening out of the σ_U versus V_f curve at higher V_f values implies that it may not be possible to increase the strength of the polymer *ad infinitum* by adding more and more NWs. The micromechanical origin of this can be understood with the aid of equation (1) as well as figure **9**. The former suggests that some type of exhaustion process occurs with increasing V_f. Plastic deformation in glassy polymers can occur by crazing, which is common in glassy polymers such as PS, PMMA and PSF wherein polymer chains align perpendicular to the maximum principal stress. It can also occur by shear localization, wherein shear bands originate and propagate along the direction of maximum shear stress [23]. While crazing is considered as a brittle deformation mode, shear banding is considered to be ductile mode of deformation. In the semi-crystalline PVA polymer matrix examined in this study, it is apparent that shear localization is the predominant mode of deformation. The addition of NWs to the polymer appears to suppress the shear localization, with the NWs acting as obstacles for shear-band propagation, eventually exhausting it. In order to design NW reinforced composites that accommodate a higher vol% of NWs and hence higher strengths, it would be necessary to expose the effects of changing the matrix material and tailoring the interface properties.

5. Summary

In conclusion, the present study establishes the occurrence of a significant enhancement of the stiffness and the strength

S R C Vivekchand *et al*

of semi-crystalline PVA due to the incorporation of SiC and Al_2O_3 nanowires. Experimental results show that enhancements of these mechanical properties occur even with a small vol% addition of NWs. Thus, the elastic modulus E of the composites increases linearly with the volume fraction V_f, in accordance with the iso-strain rule-of-mixtures for predicting the composite E. The strength of the composites increases markedly initially and reaches a plateau. The initial hardening rate is commensurate with the strength of the SiC NWs. Shear localization within the matrix and pull-out of the NWs are important micromechanisms of deformation and fracture. With increasing V_f, there is a reduced propensity for shear band mediated plastic deformation (possibly because of the NWs arresting the shear propagation), which leads to the loss in ductility. It is noteworthy that preliminary AFM based nanoindentation measurements show an increase in the hardness of the PVA–SiC NW composites compared to the polymer alone.

Acknowledgments

We are grateful to Messers N Suresh, R Raghavan, M A Azeem, S S Pandit and S Sasidhara for rendering support with various experiments reported in this paper. We thank Dr S Karthikeyan for useful discussions and Dr A Govindaraj for help in sample preparation.

References

[1] Rao C N R and Govindaraj A 2005 *Nanotubes and Nanowires* (London: Nanotechnolgy Royal Society of Chemistry)

[2] Xia Y, Yang P, Sun Y, Wu Y, Mayers B, Gates B, Yin Y, Kim F and Yan H 2003 *Adv. Mater.* **15** 353

[3] Huang Y and Lieber C M 2004 *Pure Appl. Chem.* **76** 2051

[4] Wan Q, Li Q H, Chen Y J, Wang T H, Xe C L, Li J P and Lin Cl L 2004 *Appl. Phys. Lett.* **84** 3654

[5] Hahm J and Lieber C M 2004 *Nano Lett.* **4** 51

[6] Vivekchand S R C, Kam K C, Gundiah G, Govindaraj A, Cheetham A K and Rao C N R 2005 *J. Mater. Chem.* **15** 4922

[7] Wong E W, Sheehan P E and Lieber C M 1997 *Science* **277** 1971

[8] Wang Z L, Dai Z R, Gao R P, Bai Z G and Gole J L 2000 *Appl. Phys. Lett.* **77** 3349

[9] Yang W, Araki H, Tang C, Thaveethavorn S, Kohynama A, Suzuki H and Noda T 2005 *Adv. Mater.* **17** 1519

[10] Shaffer M S P and Whindle A H 1999 *Adv. Mater.* **11** 937

[11] Coleman J N, Cadek M, Blake R, Nicolosi V, Rayan K P, Belton C, Fonseca A, Nagy J B, Gunk'ko Y K and Blau W J 2004 *Adv. Funct. Mater.* **14** 791

[12] Zhang X, Liu T, Shreekumar T V, Kumar S, Moore V C, Hauge R H and Smalley R E 2003 *Nano Lett.* **3** 1285

[13] Gundiah G, Madhav G V, Govindaraj A, Seikh Md M and Rao C N R 2002 *J. Mater. Chem.* **12** 1606

[14] Gundiah G, Deepak F L, Govindaraj A and Rao C N R 2003 *Top. Catal.* **23** 137

[15] Tadokoro H, Nagai J, Seki S and Nitta I 1961 *Bull. Chem. Soc. Japan* **34** 1504

[16] Pappas N A 1977 *Makromol. Chem.* **176** 3433

[17] Pappas N A and Hansen P J 1982 *J. Appl. Polym. Sci.* **27** 4787

[18] Kelly A and MacMillan N H 1986 *Strong Solids* (Oxford: Clarendon)

[19] Miller R E and Shenoy V B 2000 *Nanotechnolgy* **11** 139

[20] Krenchel H 1964 *Fiber Reinforcement* (Copenhagen, Denmark: Akademisk Forlag)

[21] Curtin W A 1999 *Adv. Appl. Mech.* **36** 163

[22] Zweben C and Rosen B W 1970 *J. Mech. Phys. Solids* **18** 189

[23] Kramer E J 1983 *Adv. Polym. Sci.* **52/53** 1

IOP PUBLISHING   NANOTECHNOLOGY

Nanotechnology **18** (2007) 205504 (9pp)   doi:10.1088/0957-4484/18/20/205504

Ammonia sensors based on metal oxide nanostructures

Chandra Sekhar Rout, Manu Hegde, A Govindaraj and C N R Rao[1]

Chemistry and Physics of Materials Unit, DST Unit on Nanoscience and CSIR Centre of Excellence in Chemistry, Jawaharlal Nehru Centre for Advanced Scientific Research, Jakkur PO, Bangalore-5600 64, India

E-mail: cnrrao@jncasr.ac.in

Received 14 February 2007, in final form 29 March 2007
Published 23 April 2007
Online at stacks.iop.org/Nano/18/205504

Abstract

Ammonia sensing characteristics of nanoparticles as well as nanorods of ZnO, In_2O_3 and SnO_2 have been investigated over a wide range of concentrations (1–800 ppm) and temperatures (100–300 °C). The best values of sensitivity are found with ZnO nanoparticles and SnO_2 nanostructures. Considering all the characteristics, the SnO_2 nanostructures appear to be good candidates for sensing ammonia, with sensitivities of 222 and 19 at 300 °C and 100 °C respectively for 800 ppm of NH_3. The recovery and response times are respectively in the ranges 12–68 s and 22–120 s. The effect of humidity on the performance of the sensors is not marked up to 60% at 300 °C. With the oxide sensors reported here no interference for NH_3 is found from H_2, CO, nitrogen oxides, H_2S and SO_2.

1. Introduction

Detection of ammonia in the atmosphere is an extremely important problem with implications to the environment and medical practice as well as the automotive and chemical industries [1]. There have been reports on ammonia sensors, but they do not make use of nanostructures. Thus, thin films of ZnO and ZnO doped with different metals have been found to sense ammonia with sensitivities varying between 4 and 95 for 1–30 ppm of NH_3 over the temperature range 30–300 °C [2, 3]. Surface-ruthenated ZnO films appear to have a sensitivity of ~440 for 1000 ppm of NH_3 at 300 °C [4, 5]. Thin films of sol–gel-derived SnO_2 are reported to exhibit a linear relationship between the logarithm of sensitivity and NH_3 concentration in the range of 0.05–10 volume% at 350 °C [6]. SnO_2 powder modified by Pt or SiO_2 shows sensitivity between 12 and 25 for 200 ppm of NH_3 at 160 °C [7]. In_2O_3 ceramics modified by Ti or loaded with Pt, Au show enhanced selectivity for 5–1000 ppm at 145 °C [8, 9]. Several other materials have also been tested for sensing ammonia, in particular single-walled nanotubes functionalized with poly-aminobenzene sulfonic acid and N-doped carbon nanotubes [10–12]. The nanotubes exhibit a sensitivity of 2 for 5 ppm of NH_3 at 32 °C. Based on the literature, there appears to be a clear need

for reliable and satisfactory ammonia sensors. Our success with the nanostructures of different semiconducting metal oxides for sensing gases such as H_2, oxides of nitrogen and hydrocarbons [13–15], suggested that it would be rewarding to explore nanostructures of certain metal oxides for sensing ammonia. In this paper, we report the gas-sensing properties of the nanostructures of ZnO, In_2O_3 and SnO_2, of which those of SnO_2 are particularly satisfactory, with good sensitivity and relatively short response and recovery times.

2. Experimental details

ZnO nanoparticles were prepared by the sol–gel technique starting with zinc acetate as the precursor as described earlier [16]. ZnO nanorods were synthesized by stirring a fine powder of $Zn(CH_3COO)_2 \cdot 2H_2O$ (Qualigens, 98.5% pure) in 100 ml of methanol at 60 °C for 1 h, followed by dropwise addition of 0.03 M KOH (Ranbaxy) to the solution [13]. The resulting solution was refluxed for 24 h to obtain the product, which was washed with ethanol and dried at 60 °C in air.

In_2O_3 nanoparticles were prepared as follows [17]. 2 g of $InCl_3 \cdot 4H_2O$ (Aldrich, 97% pure) and 5 ml of deionized water were placed in a Teflon-lined autoclave and the autoclave filled up to 80% volume with ethylene diamine (Merck, 99% pure). The autoclave was heated at 200 °C for 24 h and cooled to room

[1] Author to whom any correspondence should be addressed.

Nanotechnology **18** (2007) 205504

temperature. The product was centrifuged and washed with absolute ethanol several times. The product was finally dried in air at 60 °C and then heated at 500 °C in oxygen atmosphere for 2 h.

In_2O_3 nanorods were obtained by using anodic alumina membrane (AAM) templates with a pore size of 20 nm [14]. The method of preparation of these nanorods is as follows. $In(OH)_3$ sol was prepared by adding ethylene diamine dropwise to $InCl_3 \cdot 4H_2O$. The sol was placed in 100 ml deionized water and stirred for 6 h at room temperature. After obtaining a light blue sol, alumina template membranes were immersed in the sol for 5 h under a pressure of 2 atm. The templates were then taken out and dried at 60 °C and heated under argon atmosphere at 600 °C for 4 h. The templates were dissolved in 0.6 M NaOH (Ranbaxy, 98% pure) solution to yield nanorods of In_2O_3.

SnO_2 nanoparticles were prepared by dissolving 1 g of $SnCl_4 \cdot 5H_2O$ (Lobachemie, 98% pure) in 70 ml of H_2O, followed by addition of a few drops of ethylene diamine. The solution was placed in a 100 ml Teflon-lined autoclave and heated at 200 °C for 24 h. After cooling the autoclave, the product was washed with alcohol and water and dried in air at 60 °C overnight. SnO_2 nanorods and flower-like structures were obtained by following methods reported elsewhere [18, 19]. SnO_2 nanorods were prepared starting with a mixture of 1 g of $SnCl_4 \cdot 5H_2O$ in 70 ml of ethanol and water (5:1). After adding 10 M NaOH, the solution was transferred to a 100 ml autoclave and heated at 200 °C for 24 h. After cooling the autoclave, the product was washed with absolute ethanol several times and then it was dried in vacuum to get the final product. Flower-like structures of SnO_2 were obtained by taking 0.5 g of as-prepared SnO_2 nanoparticles and 10 M NaOH in 80 ml ethanol and heating the mixture in an autoclave at 200 °C for 36 h. The product was washed with dilute HCl and water to remove the by-products of sodium, and was dried in vacuum at 60 °C.

The nanostructures of ZnO, In_2O_3 and SnO_2 were characterized by x-ray diffraction (Cu Kα radiation), scanning electron microscopy (Leica S440i), transmission electron microscopy (JEOL JEM 3010), field emission scanning electron microscopy (Nova Nanosem 600) and micro-Raman spectroscopy (Labraman-HR) using an He–Ne laser (632.81 nm) in the back-scattering geometry. To fabricate thick film sensors, an appropriate quantity of diethyleneglycol (Merck, 99% pure) was added to the desired nanostructure of ZnO, In_2O_3 or SnO_2 to obtain a paste. The paste was coated on to an alumina substrate (5 mm × 20 mm, 0.5 mm thick) attached with a comb-type Pt electrode on one side, the other side having a heater. The films were dried at 100 °C and annealed at 300 °C for 1 h. Gas sensing properties were measured using a home-built computer-controlled characterization system consisting of a test chamber, sensor holder, a Keithley multimeter-2700, a Keithley electrometer-6517A, mass flow controllers and a data acquisition system. The test gas was mixed with dry air to achieve the desired concentration and the flow rate was maintained at 200 sccm using mass flow controllers. The current flowing through the samples was measured using a Keithley multimeter 2700. The working temperature of the sensors was adjusted by changing the voltage across the heater side. By monitoring the output

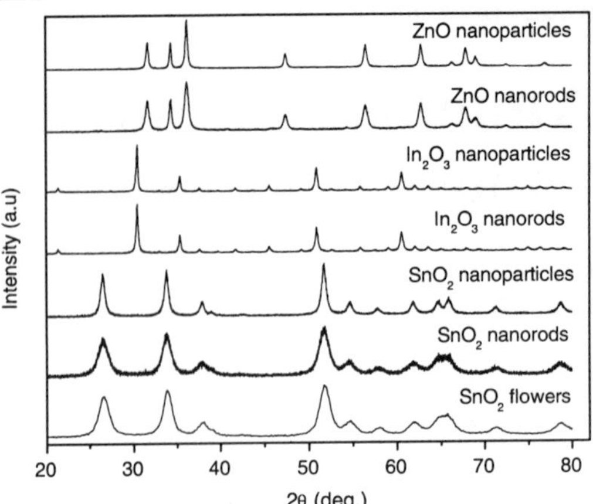

Figure 1. XRD patterns of nanoparticles and nanorods of ZnO, nanoparticles and nanorods of In_2O_3 and nanoparticles, nanorods and flowers of SnO_2.

voltage across the sensor, the resistance of the sensor in dry air or in ammonia can be measured.

The resistance of the oxides decreased on contact with NH_3. The sensitivity (response magnitude), S, was determined as the ratio, $R_{air}/R_{ammonia}$, where R_{air} is the resistance of the thick film sensor in dry air and $R_{ammonia}$ is the resistance in the different concentration of ammonia. The resistance of the sensors based on ZnO nanoparticles and nanorods was in the 0.1–10 MΩ range, whereas the resistance of the sensors based on In_2O_3 and SnO_2 nanostructures was in the 1–100 MΩ range in dry air in the 100–300 °C range. Response time is defined as the time required for the conductance to reach 90% of the equilibrium value after the test gas is injected and recovery time is taken as the time necessary for the sensor to attain a conductance 10% above the original value in air. The sensitivity of thick film sensors was also measured in atmospheres with different relative humidities.

3. Results and discussion

The XRD patterns of ZnO nanoparticles and nanorods, In_2O_3 nanoparticles and nanorods and SnO_2 nanoparticles, nanorods and flowers are shown in figure 1. The x-ray diffraction (XRD) patterns correspond to the wurtzite structure (lattice parameters $a = 3.25$ Å, $c = 5.2$ Å, JCPDS no 36-1451). The XRD pattern of the ZnO nanorods shows sharp 002 reflections, indicating the formation of the rods along the c axis. In figure 2(a), we show a typical TEM image of ZnO nanoparticles with the inset showing the selected area electron diffraction (SAED) pattern. The SAED pattern indicates the particles to be single crystalline. Based on the widths of the reflections in the XRD pattern, the average diameter of the nanoparticle is found to be ~15 nm. Figure 2(b) shows a TEM image of ZnO nanorods, with the SAED pattern as the inset. The SAED pattern indicates the single crystalline nature of the nanorods. The TEM image reveals that the diameter of the nanorods is in the range of 10–20 nm with the length in the 50–250 nm range.

Nanotechnology **18** (2007) 205504

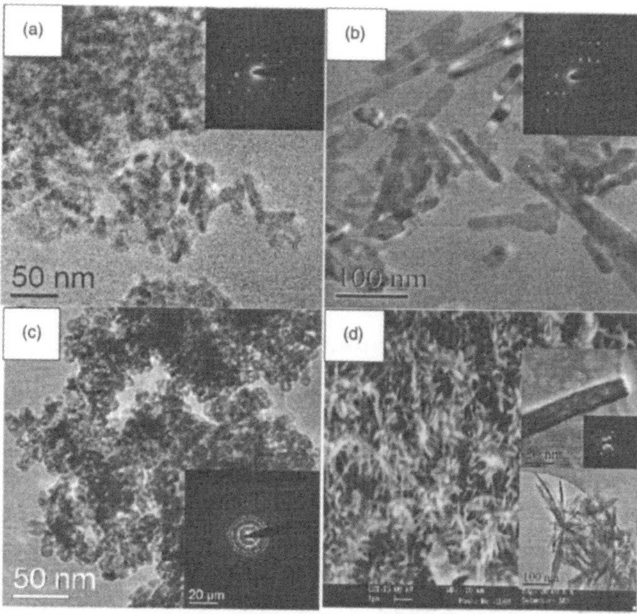

Figure 2. (a) TEM image of ZnO nanoparticles with the inset showing electron diffraction, (b) TEM image of ZnO nanorods with the inset showing electron diffraction pattern, (c) TEM image of In$_2$O$_3$ nanoparticles with the inset showing electron diffraction pattern, (d) SEM image of In$_2$O$_3$ nanorods with the inset showing TEM image and electron diffraction pattern.

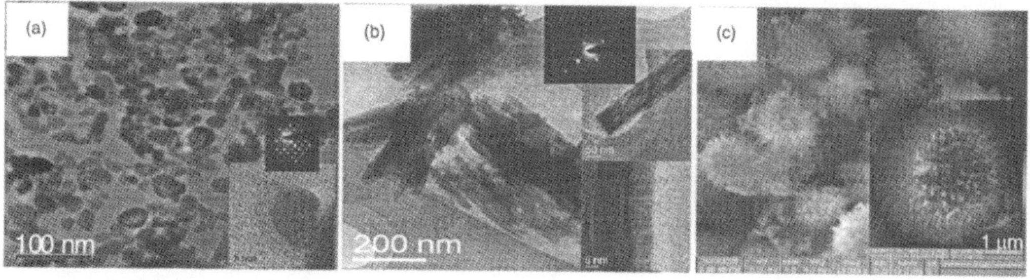

Figure 3. (a) TEM image of SnO$_2$ nanoparticles with the inset showing the electron diffraction pattern, HREM image; (b) TEM image of SnO$_2$ nanorods with the inset showing the electron diffraction pattern and HREM image; (c) FESEM image of SnO$_2$ flowers.

In$_2$O$_3$ nanoparticles and nanorods have the cubic structure (cell parameters $a = 10.11$ Å, JCPDS no 06-0416) as revealed by the XRD patterns (figure 1). The average diameter of the nanoparticles estimated to be ~21 nm from the XRD pattern. Figure 2(c) shows a TEM image of In$_2$O$_3$ nanoparticles, with the inset showing the SAED pattern. Figure 2(d) shows a SEM image of In$_2$O$_3$ nanorods, with the inset showing the TEM image and the electron diffraction pattern. The nanorods are single crystalline as revealed by the SAED pattern and have an average diameter of around 20 nm. XRD patterns of SnO$_2$ nanoparticles, nanorods and flowers could be indexed on the tetragonal rutile structure (cell parameters $a = 4.738$ Å, $c = 3.187$ Å, JCPDS no 41-1445). The average diameter of the nanoparticles estimated from the XRD pattern is ~23 nm. In figure 3(a), we show a typical TEM image of the SnO$_2$ nanoparticles with the inset showing the electron diffraction

pattern and an HREM image, confirming the single crystalline nature of the nanoparticles. In figure 3(b), we show a TEM image of SnO$_2$ nanorods with the inset showing an HREM image and the SAED pattern. The SAED pattern shows the nanorods to be single crystalline in nature. The average diameter of the nanorods is ~25 nm. An FESEM image of flower-like structures of SnO$_2$ is shown in figure 3(c). The inset of figure 3(c) shows a high-resolution picture of a flower. The flowers consist of fine rod-like or fibre-like structures with diameters around 25 nm.

In figure 4 we show the Raman spectra of the various metal oxide nanostructures studied by us. Raman bands are found at 332, 441 and 1076 cm^{-1} for the ZnO nanoparticles and nanorods [20]. Bulk ZnO shows Raman bands at 330 and 439 cm^{-1} [20, 21]. The nanoparticles and nanorods of In$_2$O$_3$ show Raman bands at 305, 364, 495 and 630 cm^{-1}.

Nanotechnology **18** (2007) 205504

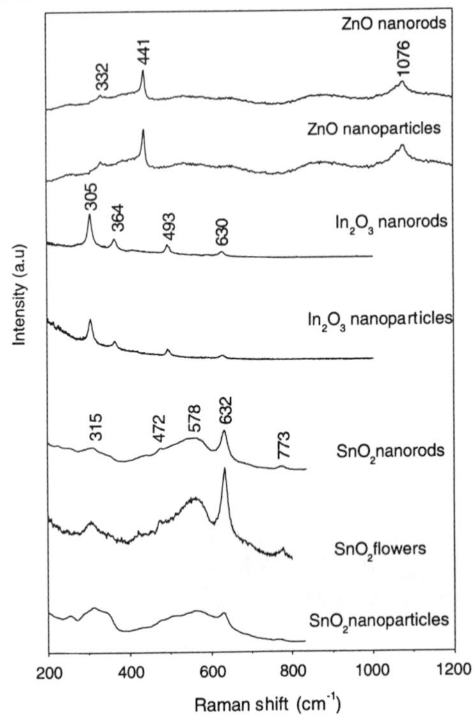

Figure 4. Raman spectra of the various oxide nanostructures.

For bulk In_2O_3, the Raman bands are at 306, 366, 495 and 630 cm^{-1} [22]. Raman bands of SnO_2 nanostructures are observed at 315, 472, 578, 632 and 773 cm^{-1}, in agreement with the literature [23]. Bulk SnO_2 exhibits Raman bands at 472, 632 and 773 cm^{-1} [24], whereas for nanostructures two extra bands are found at 315 and 578 cm^{-1}. The Raman band positions of the nanostructures do not differ significantly from those of the bulk samples. This is not expected since phonon confinement occurs at much smaller sizes.

In figure 5(a), we show the sensing characteristics of In_2O_3 nanorods for 800 ppm of ammonia. The highest sensitivity is 70 at 300 °C and ~32 at 100 °C. The variation of sensitivity of the In_2O_3 nanorods with the concentration of NH_3 (1–800 ppm) at 300 °C is shown in figure 5(b). The nanorods show a sensitivity of 4 for 1 ppm of NH_3 at 300 °C. The inset of figure 5(b) shows the variation of sensitivity of the In_2O_3 nanoparticles with the concentration of NH_3 at 300 °C (1–800 ppm). The response times for In_2O_3 nanoparticles and nanorods are 12 and 18 s respectively for 800 ppm NH_3 at 300 °C; the recovery times are 9 and 15 s respectively for In_2O_3 nanoparticles and nanorods.

Figure 6 shows the sensing characteristics of the ZnO nanoparticles and nanorods. ZnO nanoparticles show a highest sensitivity of ~260 for 800 ppm of NH_3 at 300 °C and at 100 °C the sensitivity becomes ~19. For ZnO nanorods, the observed values of sensitivity are 80 and 18 for 800 ppm of ammonia at 300 °C and 100 °C respectively. The variation of sensitivity with the concentration of NH_3 (1–800 ppm) at 300 °C is shown in figures 7(a) and (b) respectively for ZnO nanoparticles and nanorods. The sensitivity values are 16

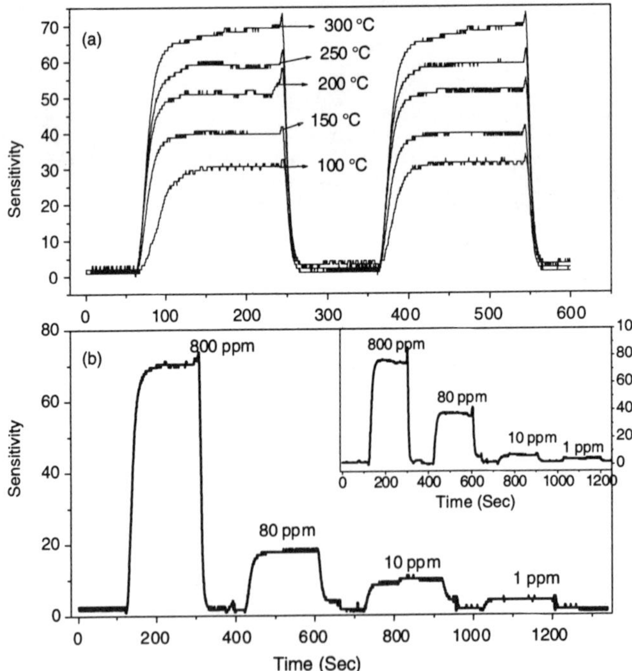

Figure 5. (a) Gas sensing characteristics of In_2O_3 nanorods for 800 ppm of ammonia, (b) variation of sensitivity with concentration of ammonia for In_2O_3 nanorods, the inset showing variation of sensitivity with concentration of ammonia for the nanoparticles at 300 °C.

Nanotechnology **18** (2007) 205504

C S Rout *et al*

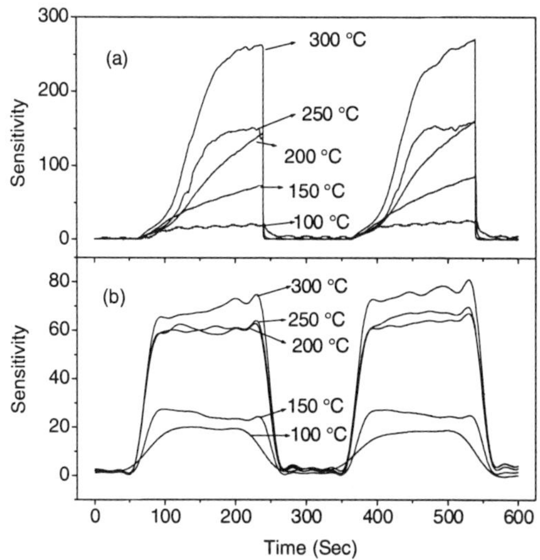

Figure 6. Gas sensing characteristics of (a) ZnO nanoparticles and (b) ZnO nanorods for 800 ppm of ammonia.

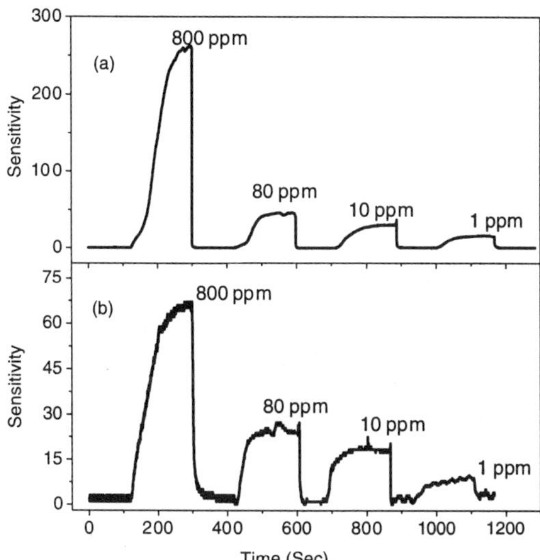

Figure 7. Variation of sensitivity with concentration of ammonia for (a) ZnO nanoparticles and (b) ZnO nanorods at 300 °C.

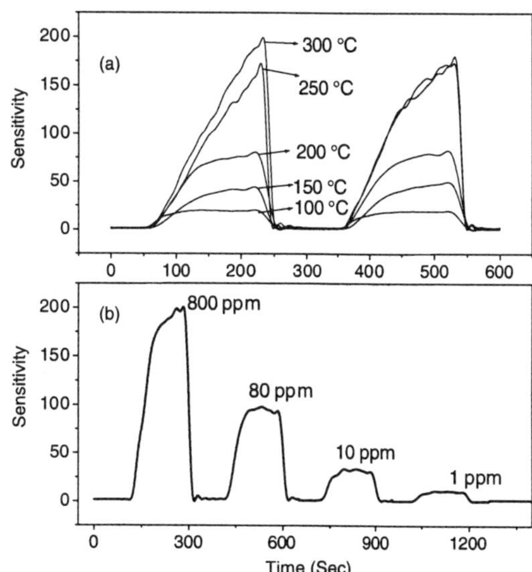

Figure 8. (a) Gas sensing characteristics of SnO_2 nanoparticles for 800 ppm of ammonia; (b) variation of sensitivity with concentration of ammonia for SnO_2 nanoparticles at 300 °C.

and 9 for 1 ppm of NH_3 for the nanoparticles and nanorods respectively. The response and recovery times for the ZnO nanoparticles are 48 and 10 s respectively for 800 ppm NH_3 at 300 °C. For ZnO nanorods, the response and recovery times are 26 and 7 s at 300 °C.

The results of our experiments with the SnO_2 nanoparticles for sensing ammonia are shown in figure 8(a). The sensitivity varies in the range 20–200 for 800 ppm of ammonia in the temperature range 100–300 °C. The variation of sensitivity with the concentration of ammonia for SnO_2 nanoparticles at 300 °C is shown in figure 8(b). The conductance of the SnO_2

nanoparticles does not show saturation behaviour for the first few cycles. This is due to the availability of a greater number of adsorbed oxygen species during initial measurements and after several cycles the sensitivity saturates at a particular value. We did not find any change in sensitivity after 20 cycles. In figure 9(a), we show ammonia-sensing characteristics of SnO_2 nanorods while figure 9(b) shows the variation of sensitivity with concentration. The inset of figure 9(b) shows the variation of sensitivity with the concentration of ammonia for SnO_2 flowers at 300 °C. The sensitivity of SnO_2 nanorods varies between 180 and 20 for 800 ppm of ammonia. For 1 ppm of ammonia a sensitivity of 18 is found at 300 °C. The response and recovery times for the SnO_2 nanoparticles are 22 and 10 s respectively for 800 ppm NH_3 at 300 °C. For the nanorods and flowers, the response times are 36 and 25 s respectively, whereas recovery times are 20 and 12 s.

In figure 10, we compare the temperature variation of sensitivity in the 100–300 °C range for ZnO, In_2O_3 and SnO_2 nanostructures. We see that ZnO nanoparticles show the highest values of sensitivity towards ammonia at 300 °C. SnO_2 nanoparticles, nanorods and flowers also show satisfactory values of sensitivity. The sensitivity is lowest with the In_2O_3 nanoparticles and nanowires. The nanostructures of all the oxides show similar behaviour at low temperatures (<150 °C). The temperature variations of the response and recovery times of the ammonia sensors based on the nanostructures of ZnO, In_2O_3 and SnO_2 are shown in figures 11(a) and (b) respectively. The response times vary in the range of 20–120 s for all the materials. In_2O_3 nanowires and nanoparticles show fast response, but the sensitivity of these nanostructures is rather low compared to SnO_2 and ZnO nanostructures. At 300 °C, the recovery times for the sensors are in the 8–30 s range. At low temperatures, the recovery times are high, but below 70 s at 100 °C.

Nanotechnology **18** (2007) 205504   C S Rout *et al*

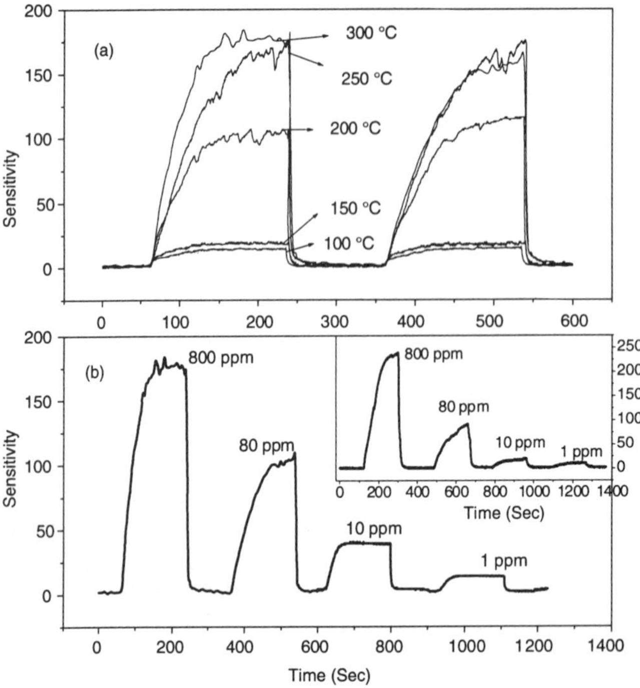

Figure 9. (a) Gas sensing characteristics of SnO$_2$ nanorods for 800 ppm of ammonia; (b) variation of sensitivity with concentration of ammonia for SnO$_2$ nanorods; inset, variation of sensitivity with concentration of ammonia for SnO$_2$ flowers at 300 °C.

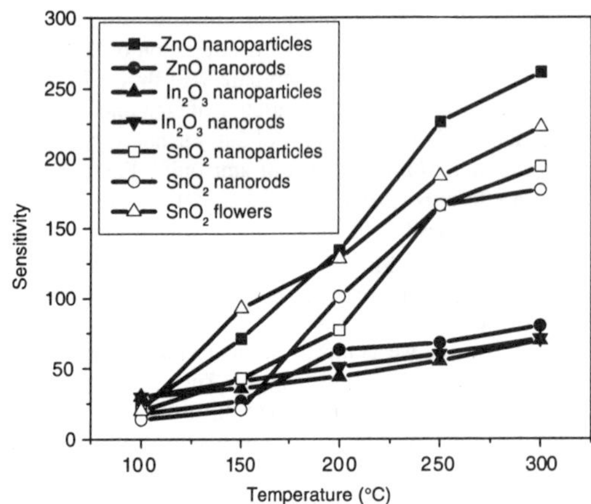

Figure 10. Temperature variation of sensitivity of ZnO, In$_2$O$_3$ and SnO$_2$ nanostructures.

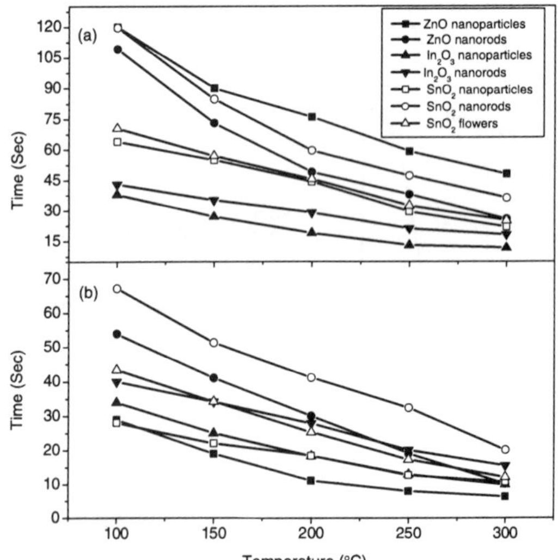

Figure 11. Temperature variation of (a) response and (b) recovery times for ZnO nanoparticles and nanorods, In$_2$O$_3$ nanoparticles and nanorods and SnO$_2$ nanoparticles, nanorods and flowers.

We have studied the effect of humidity on the ammonia sensing characteristics of the ZnO, In$_2$O$_3$ and SnO$_2$ nanostructure sensors in the range of 35–90% relative humidity. We illustrate the effect of humidity on the sensitivity for the In$_2$O$_3$ nanowires at 300 °C for 800 ppm of NH$_3$ in figure 12(a), and for ZnO nanoparticles in figure 12(b). The humidity effect on the sensing characteristics of SnO$_2$ nanoparticles and nanorods is shown in figures 13(a) and (b) respectively. There is a slight decrease in the sensitivity of the sensors with the increase in humidity up to a relative

Nanotechnology **18** (2007) 205504

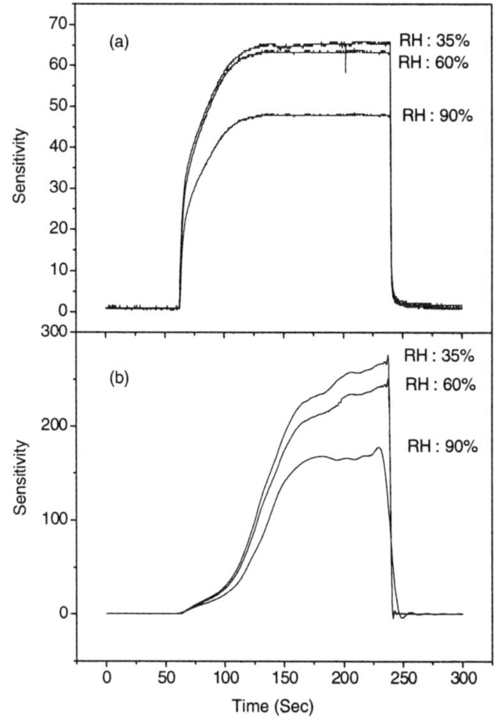

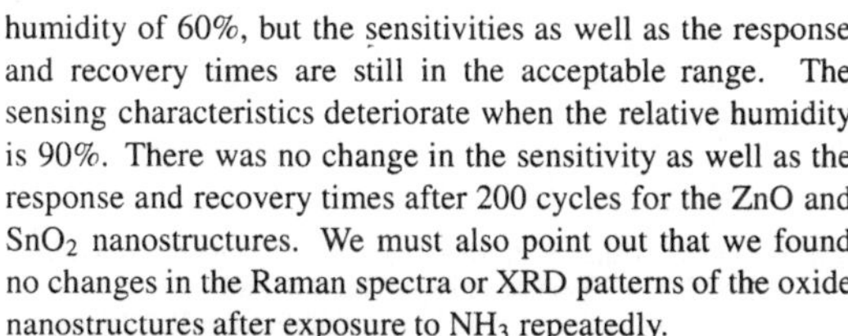

Figure 12. Effect of humidity on the sensitivity of (a) In$_2$O$_3$ nanorods and (b) ZnO nanoparticles at 300 °C for 800 ppm of NH$_3$.

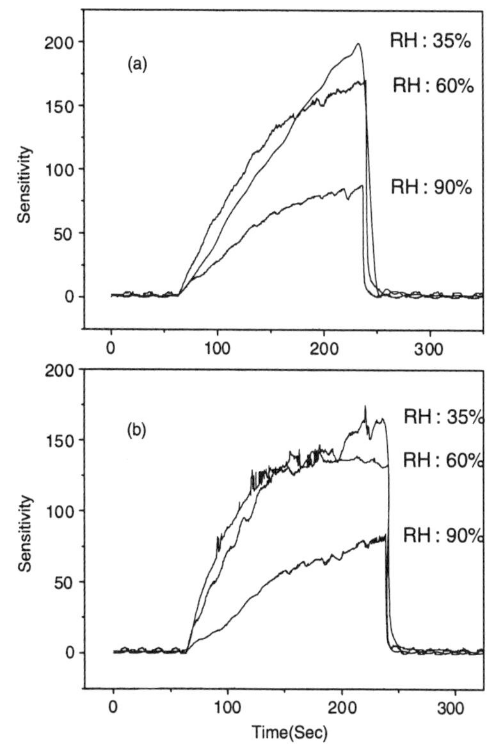

Figure 13. Effect of humidity on the sensitivity of (a) SnO$_2$ nanoparticles and (b) SnO$_2$ nanorods at 300 °C for 800 ppm of NH$_3$.

humidity of 60%, but the sensitivities as well as the response and recovery times are still in the acceptable range. The sensing characteristics deteriorate when the relative humidity is 90%. There was no change in the sensitivity as well as the response and recovery times after 200 cycles for the ZnO and SnO$_2$ nanostructures. We must also point out that we found no changes in the Raman spectra or XRD patterns of the oxide nanostructures after exposure to NH$_3$ repeatedly.

In order to check the selectivity for sensing ammonia, we have studied the sensing characteristics of ZnO, In$_2$O$_3$ and SnO$_2$ nanostructures for NO$_2$, NO, N$_2$O, H$_2$, CO, H$_2$S and SO$_2$. The ZnO nanoparticles and nanorods showed a maximum sensitivity of 2–5 for the nitrogen oxides in the temperature range of 100–300 °C. Maximum sensitivity values of 62, 8 and 17 were observed for 1000 ppm of H$_2$, CO and H$_2$S respectively in the case of ZnO nanoparticles. We did not find In$_2$O$_3$ to have good sensing characteristics for H$_2$, CO and H$_2$S, whereas it sensed nitrogen oxides at the ppm level. The SnO$_2$ nanostructures showed maximum sensitivity values of 43, 11 and 18 for 1000 ppm of H$_2$, CO and H$_2$S, which are considerably low as compared to the sensitivity obtained in the case of ammonia. The nanostructures of ZnO, In$_2$O$_3$ and SnO$_2$ do not sense SO$_2$.

The sensing action by metal oxides depends on several factors such as grain size (available surface area) and surface states as well as the efficiency with which the test gas molecules adsorb on the surface [25–27]. The sensing mechanism of n-type semiconducting metal oxides involves the formation of a charge depletion layer (L_D) on the surface of the oxides due to electron trapping on adsorbed oxygen species O$^-$ and O$_2^-$. The adsorbed oxygens are present on the surface of the metal oxides.

$$O_2\,(g) + e^- \rightarrow O_{2(ads)}^- \tag{1}$$

$$O_{2(ads)}^- + e^- \rightarrow 2O_{(ads)}^-. \tag{2}$$

The adsorbed oxygen species play a crucial role in sensing ammonia. The reaction for sensing ammonia is given by

$$2NH_3 + 3O_{(ads)}^- \rightarrow N_2 + 3H_2O + 3e^-. \tag{3}$$

As expected from equation (3), the resistance of the nanostructured material decreases on contact with ammonia. The thickness of the depletion layers of ZnO, In$_2$O$_3$ and SnO$_2$ is around 5 nm [23, 27, 28]. If the grain size is closer to $2L_D$, electrons in the nanostructures are depleted due to oxygen adsorption from air. Electrons are released when the nanostructures are exposed to ammonia, leading to an increase in the conductance according to equation (3) [23, 26]. In the present study, ZnO nanoparticles show a linear dependence of sensitivity on NH$_3$ concentration and show higher sensitivity compared to the nanorods. Metal oxide nanostructures are known to show non-linear behaviour with respect to the test gas concentration when the particles have an average diameter larger than 20 nm [29].

Nanotechnology **18** (2007) 205504

C S Rout *et al*

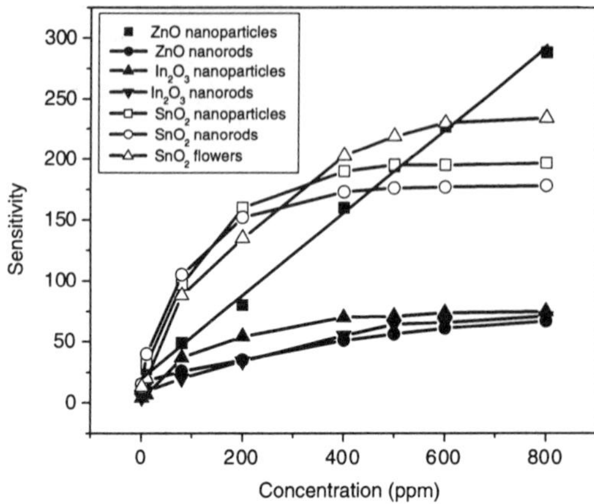

Figure 14. Comparison of the sensitivities of the different oxide nanostructures for sensing NH$_3$.

The nanostructures of SnO$_2$, however, show nearly equal values of sensitivity over a concentration range of 1–800 ppm, and they also have comparable sizes. SnO$_2$ nanostructures also show high sensitivities and a non-linear dependence with respect to the NH$_3$ concentration. This may be due to the higher carrier concentration of SnO$_2$ [25, 30–33]. At higher concentrations of ammonia, the sensitivity of the SnO$_2$ nanostructures seems to saturate. Such a saturation behaviour is related to the relative size of the nanostructures, as reported in the literature for flower-like structures of ZnO [26, 27]. The conductivity of the nanostructures varies according to the relation $\sigma = A[C]^N$, where A is a constant and C is the concentration of the test gas and N varies between 0.5 and 1.0 depending on the grain size (D), which is nearly equal to $2L_D$ [29]. For the particles of very small size ($D \ll 2L_D$), the depletion region extends throughout the whole grain and crystallites are almost fully depleted of mobile charge carriers. As a result the conductivity decreases steeply since the conduction channels between the grains vanish. When $D \gg 2L_D$, the gas sensing mechanism is controlled by the grain boundary barriers. The grain boundary barriers are independent of the grain size and the sensitivity is independent of D. In the case $D \geqslant 2L_D$, the depletion region that surrounds each neck forms a constricted conduction channel within each aggregate. Hence the ammonia sensing characteristics using different nanostructures of ZnO, In$_2$O$_3$ and SnO$_2$ follow the space-charge model.

4. Conclusions

In conclusion, zinc and tin oxide nanostructures exhibit good sensing characteristics for ammonia, as can be seen from the comparative data presented in figure 14. Although ZnO nanoparticles show higher sensitivity, the different nanostructures of SnO$_2$ seem to have a slight edge. The sensitivities and other characteristics are very good at 300 °C, but are quite acceptable even at 200 °C or at a somewhat lower temperature. It is also to be noted that the sensitivity values as well as recovery and response times of the SnO$_2$

nanostructures are generally better than those reported in the literature. Furthermore, the characteristics found by us are in the absence of any noble or transition metal additive. Since humidity does not have a marked effect on the sensitivity up to 60% relative humidity, and the sensing characteristics do not deteriorate on repeated cycling, SnO$_2$ nanostructures emerge as good ammonia sensors.

Acknowledgments

The authors acknowledge Dr Girija Shivakumar from R V College, Bangalore and Mr B R Vinay for their help during synthesis of the nanostructures.

References

[1] Timmer B, Olthuis W and Berg A 2005 *Sensors Actuators* B **107** 666
[2] Nanto H, Minami T and Takata S 1986 *J. Appl. Phys.* **60** 482
[3] Sberveglieri G, Groppelli S, Nelli P, Tintinelli A and Giunta G 1995 *Sensors Actuators* B **24** 588
[4] Aslam M, Chaudhary V A, Mulla I S, Sainkar S R, Mandale A B, Belhekar A A and Vijayamohanan K 1999 *Sensors Actuators* A **75** 162
[5] Wagh M S, Jain G H, Patil D R, Patil S A and Patil L A 2006 *Sensors Actuators* B **115** 128
[6] Teeramongkonrasmee A and Sriyudhsak M 2000 *Sensors Actuators* B **66** 256
[7] Wang Y, Wu X, Su Q, Li Y and Zhou Z 2001 *Solid-State Electron.* **45** 347
[8] Guo P and Pan H 2006 *Sensors Actuators* B **114** 762
[9] Romanovskaya V, Ivanovskaya M and Bogdanov P 1999 *Sensors Actuators* B **56** 31
[10] Bekyarova E, Davis M, Burch T, Itkis M E, Zhao B, Sushine S and Haddon R C 2004 *J. Phys. Chem.* B **108** 19717
[11] Quang N H, Trinh M V, Lee B and Huh J 2006 *Sensors Actuators* B **113** 341
[12] Paez F V, Romero A H, Sandoval E M, Martinez L M, Terrones H and Terrones M 2004 *Chem. Phys. Lett.* **386** 137
[13] Rout C S, Krishna S H, Vivekchand S R C, Govindaraj A and Rao C N R 2006 *Chem. Phys. Lett.* **418** 586
[14] Rout C S, Ganesh K, Govindaraj A and Rao C N R 2006 *Appl. Phys.* A **85** 241
[15] Rout C S, Govindaraj A and Rao C N R 2006 *J. Mater. Chem.* **16** 3936
[16] Rout C S, Raju A R, Govindaraj A and Rao C N R 2006 *Solid State Commun.* **138** 136
[17] Yu D, Wang D and Qian Y 2004 *J. Solid State Chem.* **177** 1230
[18] Chen Y J, Xue X Y, Wang Y G and Wang T H 2005 *Appl. Phys. Lett.* **87** 233503
[19] Chen X and Mao S S 2006 *J. Nanosci. Nanotechnol.* **6** 906
[20] Alim K A, Fonoberov V A, Shamsa M and Babandin A A 2005 *J. Appl. Phys.* **97** 124313
[21] Zhou H, Alves H, Hofmann D M, Kriegseis W, Meyer B K, Kaczmarczyk G and Hoffmann A 2002 *Appl. Phys. Lett.* **80** 210
[22] Korotcenkov G, Brinzari V, Ivanov M, Cerneavschi A, Rodriguez J, Cirea A, Cornet A and Morante J 2005 *Thin Solid Films* **479** 38
[23] Chen Y J, Nie L, Xue X Y, Wang Y G and Wang T H 2006 *Appl. Phys. Lett.* **88** 083105
[24] Chen Z W, Lai J K L and Shek C H 2004 *Phys. Rev.* B **70** 165314
[25] Franke M E, Koplin T J and Simon U 2006 *Small* **2** 36
[26] Feng P and Wang T H 2005 *Appl. Phys. Lett.* **87** 213111
[27] Chen Y, Zhu C L and Xiao G 2006 *Nanotechnology* **17** 4537

Nanotechnology **18** (2007) 205504

[28] Sysoev V V, Bradly K, Button B K, Wepsiec K, Dmitriev S and Kolmakov A 2006 *Nano Lett.* **6** 1584
[29] Ogawa H, Nishikawa M and Abe A 1982 *J. Appl. Phys.* **53** 4448
[30] Yang H S, Norton D P, Pearton S J and Ren F 2005 *Appl. Phys. Lett.* **87** 212106
[31] Allen M W, Alkaisi M M and Durbin S M 2006 *Appl. Phys. Lett.* **89** 103520
[32] Li C, Zhang D, Han S, Liu X, Tang T and Zhou C 2003 *Adv. Mater.* **15** 143
[33] Rothschild A and Komem Y 2004 *J. Appl. Phys.* **95** 6374

PAPER

www.rsc.org/materials | Journal of Materials Chemistry

A study of graphenes prepared by different methods: characterization, properties and solubilization†‡

K. S. Subrahmanyam, S. R. C. Vivekchand, A. Govindaraj and C. N. R. Rao*

Received 26th October 2007, Accepted 6th December 2007
First published as an Advance Article on the web 7th January 2008
DOI: 10.1039/b716536f

Graphene has been prepared by different methods: pyrolysis of camphor under reducing conditions (CG), exfoliation of graphitic oxide (EG), conversion of nanodiamond (DG) and arc evaporation of SiC (SG). The samples were examined by X-ray diffraction (XRD), transmission electron microscopy, atomic force microscopy, Raman spectroscopy and magnetic measurements. Raman spectroscopy shows EG and DG to exhibit smaller in-plane crystallite sizes, but in combination with XRD results EG comes out to be better. The CG, EG and DG samples prepared by us have BET surface areas of 46, 925 and 520 $m^2\ g^{-1}$ respectively and exhibit significant hydrogen uptake up to 3 wt%. EG also exhibits a high CO_2 uptake (34.7 wt%). Electrochemical redox properties of the graphene samples have been examined in addition to their use in electrochemical supercapacitors. Functionalization of EG and DG through amidation has been carried out with the purpose of solubilizing them in non-polar solvents. Water-soluble graphene has been produced by extensive acid treatment of EG or treatment with polyethylene glycol.

Introduction

Graphene has emerged to be an exciting material with potential applications.[1,2] A remarkable feature of graphene is that it is a Dirac solid, with the electron energy being linearly dependent on the wave vector near the crossing points in the Brillouin zone. It also exhibits unusual fractional quantum Hall effect and conductivity behavior. Most physical studies of graphene are carried out on sheets obtained by the 'scotch-tape' technique, wherein a single graphene sheet is removed at a time from highly ordered pyrolytic graphite. This method obviously cannot be employed for large-scale applications or to study many of the chemical and materials properties of graphene. Laboratory synthesis of graphene has been carried out by a few workers. The methods employed include (a) pyrolysis of camphor under reducing conditions,[3] (b) exfoliation of graphitic oxide,[4] and (c) conversion of nanodiamond.[5] An examination of the literature shows that graphene samples prepared by the different methods have not been fully characterized. It is indeed necessary to know the best and most straightforward method(s) to characterize graphene samples and also to have a comparative evaluation of the quality and the characteristics of graphenes obtained by the different preparative methods, to be able to make the right choices. We have, therefore, prepared graphene by employing the three methods mentioned above, designated as CG, EG and DG respectively, and investigated some of their properties

after characterizing them. We have been able to carry out a preliminary study of graphene prepared by the arc evaporation of SiC (SG).[6] For the purpose of characterization, we have employed X-ray diffraction, transmission electron microscopy, atomic force microscopy and Raman spectroscopy. Raman spectroscopy is especially useful in determining the in-plane crystallite size of the graphene samples.[7–9] Our Raman and other studies revealed EG and DG to possess not only a smaller in-plane crystallite size but also desirable properties. We have measured the surface areas and hydrogen uptake of these samples and determined their electrochemical properties including their possible application in electrochemical supercapacitors. Cyclic voltammetric studies reveal that EG exhibits redox behavior similar to that of the basal planes of graphite. Electrochemical supercapacitors fabricated by using an ionic liquid enable a large operating voltage of 3.5 V to be employed and give rise to high values of specific capacitance and energy density. We have carried out investigations on the functionalization and solubilization of EG and DG using methods similar to those reported for carbon nanotubes,[10–12] since such dispersions would be required in many applications and manipulations. It was of interest to us to solubilize graphene in non-polar solvents as well as water.

Experimental

Graphene was prepared by four different methods, namely the reductive pyrolysis of camphor (CG),[3] exfoliation of graphitic oxide (EG),[4] conversion of nanodiamond (DG)[5] and arc evaporation of SiC (SG).[6] In the first method, to prepare CG, camphor was pyrolysed over nickel particles under a reducing atmosphere. The reaction was carried out in a two-stage furnace and camphor was slowly sublimed (170 °C) by heating from the first furnace to the second furnace held at 770 °C where the

Chemistry and Physics of Materials Unit, CSIR-Centre of Excellence in Chemistry and DST Unit on nanoscience, Jawaharlal Nehru Centre for Advanced Scientific Research, Jakkur PO, Bangalore, 560 064, India. E-mail: cnrrao@jncasr.ac.in; Fax: +91 80 2208 2760
† This paper is part of a *Journal of Materials Chemistry* theme issue on carbon nanostructures.
‡ Electronic supplementary information (ESI) available: Thermogravimetric analysis of EG, DG and CG in an oxygen atmosphere. See DOI: 10.1039/b716536f

micron sized nickel particles were placed. The second method to prepare EG involved the thermal exfoliation of graphitic oxide. In this method, graphitic oxide was prepared by reacting graphite (Alfa Aesar, 2–15 µm) with a mixture of conc. nitric acid and sulfuric acid with potassium chlorate at room temperature for 5 days. Thermal exfoliation of graphitic oxide was carried out in a long quartz tube at 1050 °C under an Ar atmosphere. Thermal conversion of nanodiamond (particle size 4–6 nm, Tokyo Diamond Tools, Tokyo, Japan) to graphene was carried out at 1650 °C in a helium atmosphere to obtain DG. In the last method, SG was obtained by arc evaporation of SiC (arc-melted mixture of Si and graphite) in a hydrogen atmosphere (200 Torr) with a DC current of 45 A and 38 V.

The samples were characterized using transmission electron microscopy (TEM), atomic force microscopy (AFM), X-ray diffraction (XRD) and thermogravimetric analysis. Raman spectra were recorded at different locations of the sample using a Jobin Yvon LabRam HR spectrometer with a 514 nm Ar laser. TEM images were obtained with a JEOL JEM 3010 instrument fitted with a Gatan CCD camera operating at an accelerating voltage of 300 kV. AFM measurements were performed using a CP 2 atomic force microscope. Thermogravimetric analysis of the samples was carried out in a flowing oxygen atmosphere with a heating rate of 10 °C min^{-1} using a Mettler-Toledo-TG-850 apparatus. Infrared (IR) spectra were recorded on small pieces of the samples embedded in KBr pellets using a Bruker FT-IR spectrometer. Magnetization measurements were carried out with a vibrating sample magnetometer in a physical properties measuring system (PPMS, Quantum Design, San Diego, CA, USA).

Surface area measurements and low-pressure hydrogen up-take experiments were carried out with a QuantaChrome Autosorb-1 instrument. High pressure hydrogen adsorption experiments were performed using a home-built adsorption set-up as reported by Gundiah et al.[13] The experiments were performed at 300 K and 100 bar using ultra high pure hydrogen (99.99%), with an impurity (e.g. moisture and nitrogen) content of less than 10 ppm. The graphene samples were accurately weighed (in excess of 100 mg) and taken into the sample cell. The sample cell was evacuated to 10^{-5} Torr and heated for 12 h at 125 °C in order to degas the sample before the adsorption study.

Electrochemical measurements were performed using a PG262A potentiostat/galvanostat (Technoscience Ltd, Bangalore, India). Voltammetric properties of graphenes were investigated using a three electrode electrochemical cell containing a graphene paste electrode, platinum foil as the counter electrode and calomel as the reference electrode using 1M KCl solution containing 100 mM potassium ferrocyanide. All graphene paste electrodes were prepared using mineral oil as a binder (25 wt%).[14] Graphene-based supercapacitor cells were fabricated following Conway.[15] The measurements were carried out with a two-electrode configuration, the mass of each electrode being 5 mg and the ionic liquid, N-butyl-N-methylpyrrolidinium bis(trifluoromethanesulfonyl)imide (PYR$_{14}$TFSI) was dried at 80 °C under vacuum for a day prior to the experiment. The fabrication and characterization of cell was done at 60 °C in a mBraun glove box keeping the oxygen and water levels at less that 0.1 ppm. We have performed cyclic voltammetry to characterize the two-electrode supercapacitor cells with the different graphenes. The

specific capacitance is given by the following equation: $C_{CV} = 2(i_+ - i_-)/(m \times$ scan rate), where i_+ and i_- are the maximum currents in the positive scan and the negative scan respectively and m is the mass of the electrode. The energy density is given as $E = CV^2$, where C is the capacitance taking into account both the electrode masses and V is the operational voltage.

To solubilize graphene samples (EG and DG) in non-polar solvents the following procedure was employed. In the first step, conc. nitric acid (2 mL), conc. sulfuric acid (2 mL) and water (16 mL) were added to graphene (50 mg) and subsequently heated in a microwave oven for 10 min under hydrothermal conditions. Further, the sample was heated at 100 °C for 12 h. The product was washed with distilled water and centrifuged repeatedly to remove traces of acid. This yielded graphene that was functionalized with –OH and –COOH groups. The acid treated graphene was refluxed with excess SOCl$_2$ for 12 h and the unreacted SOCl$_2$ was removed under vacuum. The product was treated with dodecylamine (5 mL) under solvothermal conditions at 100 °C.

In order to solubilize graphene in water, EG was further treated with a (1 : 1) mixture of conc. nitric acid and sulfuric acid at 100 °C. This gave some water-soluble graphene in addition to an insoluble fraction. To prepare water soluble graphene, another procedure was also employed. Here, polyethylene glycol (PEG)-functionalized EG was prepared by reacting acid treated graphene with excess of PEG (6 mL) and conc. HCl (2 mL) under solvothermal conditions at 100 °C for 12 h.

Results and discussion

Characterization

In Fig. 1, we show typical TEM images of the different graphene samples, CG, EG, DG and SG, prepared by us. Large crystalline

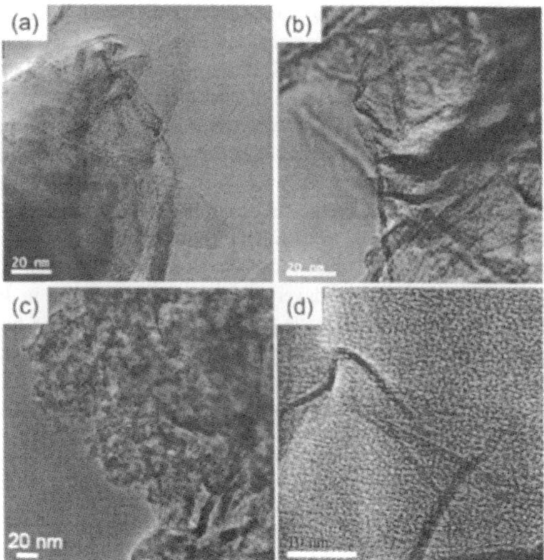

Fig. 1 TEM images of graphene obtained by (a) reductive pyrolysis of camphor (CG), (b) thermal exfoliation of graphitic oxide (EG), (c) thermal conversion of nanodiamond to graphene (DG) and (d) arc evaporation of SiC (SG).

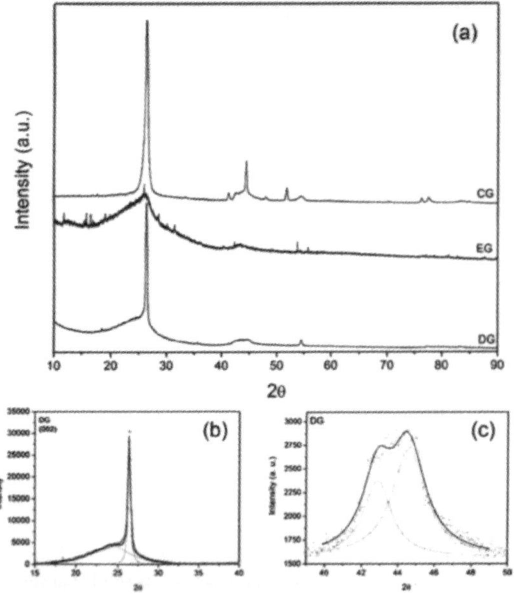

Fig. 2 (a) X-Ray diffraction patterns of the CG, EG and DG. (b) Lorentzian fit for (002) and (c) double Lorentzian fit for (100) and (101) diffraction peaks for DG.

sheets are observed in the case of CG, while disordered graphene sheets are seen in EG as can be seen from Fig. 1 (a) and (b) respectively. Small particles with diameters in the 3–20 nm range along with onion-like nanoparticles are found in the TEM image of DG in Fig. 1(c). The TEM image of SG shows few-layered graphene sheets as seen in Fig. 1(d). The XRD patterns of CG, EG and DG are shown in Fig. 2. The sharp graphitic reflection in the powder X-ray diffraction pattern of CG shows that it is comprised of a large number of layers. EG shows broad peaks in the XRD pattern. The XRD pattern of DG reveals that it contains reflections corresponding to both small and large graphitic particles. We have fitted the (002) and (100) profiles to obtain the number of layers and the in-plane crystallite size of the graphene samples (see Table 1). By fitting the (002) reflection, one can obtain the average number of layers using the Scherrer formula. In Fig. 2(b), we show a typical multiple Lorentzian fit for the (002) reflection in the case of DG. CG primarily consists of ~51 layers. The number of layers in EG is found to be 3 and 16 while in DG it is 6 and 87. We have calculated the in-plane crystallite size of the various graphenes

Table 1 Number of layers and crystallite size from the X-ray diffraction patterns of the graphene samples

Sample	Number of layers from (002) reflection[a]	Crystallite size from (100) reflection/nm
CG	51	6.1
EG	3, 16	4.7
DG	6, 87	5.0

[a] Two values given below are obtained from fits to the two (002) profiles in the XRD patterns

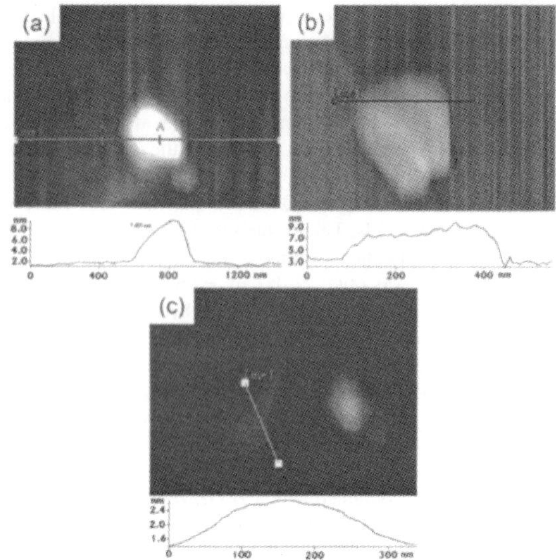

Fig. 3 Typical AFM images and height profiles of graphene samples: (a) CG, (b) EG and (c) DG.

from the (100) reflection. The (100) reflection is overlapped by the (101) reflection and had to be deconvoluted to obtain the position and width. The in-plane approximate crystallite sizes for CG, EG and DG are 6.1, 4.7 and 5.0 nm respectively.

We have carried out AFM studies at different locations of CG, EG and DG. We show typical AFM images of these graphenes in Fig. 3. AFM cross-section height profile analysis indicates that the CG sample consists of more than 20 layers while EG and DG possess 3 to 6 graphene layers.

Raman spectroscopy is a most useful tool to characterize graphene. A single layer graphene shows the well-known G-band around 1560 cm^{-1} and a band around 1620 cm^{-1} (D′). The D′ band is defect induced and not found in graphite. The D-band around 1350 cm^{-1} arising from disorder is very weak in a single layer graphene and increases in intensity with the number of layers. The 2D band (~2600 cm^{-1}) which appears in single layer graphene is also sensitive to the number of layers and shows greater structure (often a doublet) with increasing number of layers. In Fig. 4, we show typical Raman spectra for the graphene samples taken at two different locations and in Table 2, we summarize the Raman band positions with assignments. We see that the D-band is quite intense compared to the G-band, especially in EG and DG. The ratio of the intensity of the G-band to the D-band is related to the in-plane crystallite size, L_a. We have calculated the in-plane crystallite size from the spectra taken at various locations of the graphene samples by employing the ratio, $L_a = 4.4(I_G/I_D)$[8,9] (Table 3). CG shows the largest L_a and DG the smallest with EG falling in between DG and CG. Preliminary studies indicate that the L_a of SG is around 4. One may therefore consider EG and DG to be the samples best suited for further study of various chemical and electrochemical properties. However, if one takes the XRD results (Table 1) into account, EG would be superior to DG. It is noteworthy that the position of the G-band varied in the order

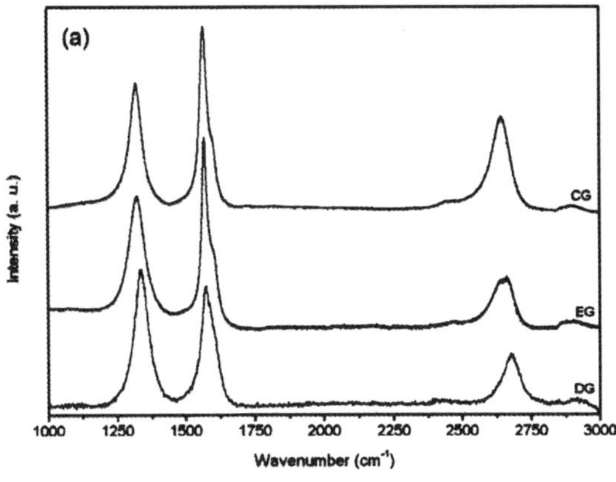

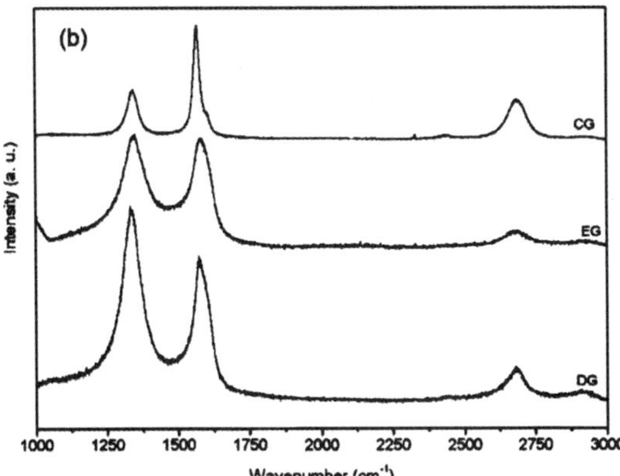

Fig. 4 Raman spectra for CG, EG and DG recorded at different locations (a) and (b) using 514 nm laser excitation.

DG > EG > CG. It is known that the G-band position increases with decreasing number of layers.[7-9] The observed trend in the G-band frequency is consistent with the L_a values calculated from the I_G/I_D intensity ratios (Table 3).

Magnetic susceptibility measurements in the 1–300 K range show that both EG and DG exhibit Curie behavior. The susceptibility values of EG and DG at 300 K are -3.5×10^{-6} and -4.4×10^{-6} emu g^{-1}. These data are comparable to those reported by Andersson *et al.*[5]

The thermogravimetric analysis of CG, EG and DG was carried out. CG undergoes sharp oxidation around 730 °C while DG gets oxidized at 700 °C (see ESI‡). EG exhibits the lowest

Table 2 Raman bands (in cm^{-1}) of graphene samples[a]

Sample		D	G	G'	2D	D + G
CG	(a)	1321	1567	1604	2647	2919
	(b)	1342	1569	1605	2687	2920
EG	(a)	1324	1569	1605	2652	2908
	(b)	1352	1574	1608	2705	2926
DG	(a)	1332	1576	1606	2678	2909
	(b)	1330	1576	1608	2682	2905

[a] (a) and (b) represent data taken from different locations

Table 3 In-plane crystallite sizes, L_a (in nm) of the graphene samples estimated from I_G/I_D ratios in the Raman spectra[a]

Sample	I_G/I_D	L_a
CG	1.5, 2.4, 2.8	7, 10, 12
EG	1.0, 1.4, 1.7	4, 6, 7
DG	0.8, 0.8, 1.1	3, 4, 5

[a] The different I_G/I_D values are from different locations in the sample. A preliminary study of SG gives $L_a = 4$.

oxidation temperature of ~520 °C, exhibiting a sharp mass loss at this temperature followed by a gradual mass loss. The low oxidation temperature of EG is ascribed to the presence of functional groups while the other graphenes consist of pure carbon and are more crystalline in nature as observed in the X-ray diffraction patterns.

Gas adsorption

The Brunauer–Emmett–Teller (BET) nitrogen adsorption isotherms for EG and DG are shown in Fig. 5. Both the samples exhibit Type I + Type II behavior and their surface areas are 925 and 520 m^2 g^{-1} respectively. The pore size distributions of the samples are shown as insets in Fig. 5, and reveal that EG is

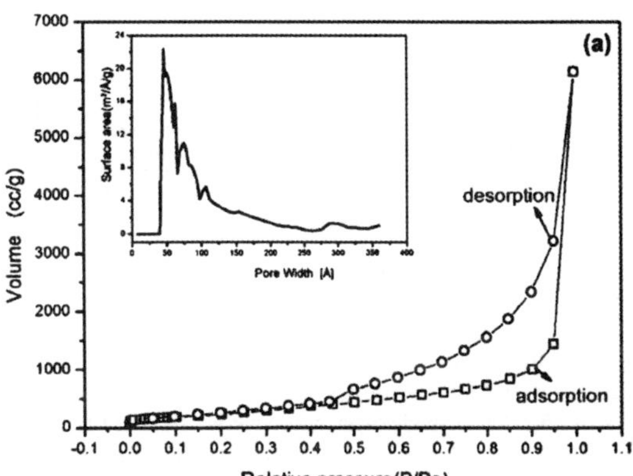

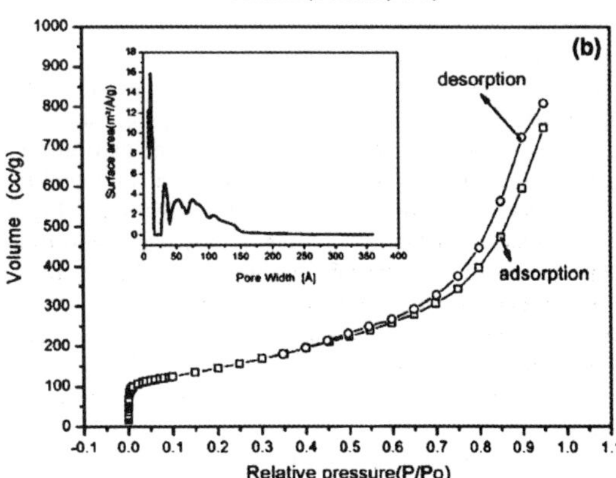

Fig. 5 Nitrogen adsorption isotherms of (a) EG and (b) DG with the insets showing pore size distributions.

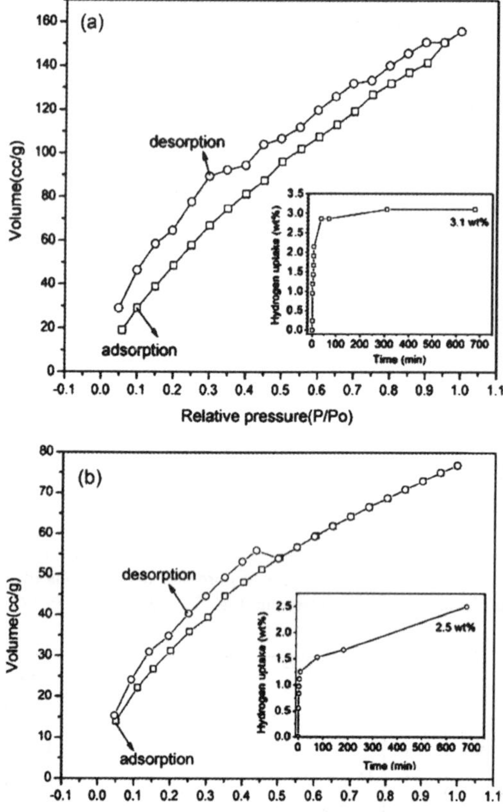

Fig. 6 Hydrogen uptake of (a) EG and (b) DG at 77 K with the insets showing high pressure hydrogen adsorption at ~100 bar.

mesoporous with pores greater than 3 nm while DG has both micropores and mesopores. CG exhibited a rather low surface area of 46 m^2 g^{-1}. Single layer graphene is estimated to have a surface area of 2600 m^2 g^{-1}, thereby indicating that both EG and DG consist of more than one layer as indeed found from Raman and other studies.

We have investigated the hydrogen uptake of EG and DG as they possess large surface areas. The hydrogen adsorption studies were carried out under two different experimental conditions, the first at 1 atm and 77 K and the second experiment at high pressures and 300 K. By the first method, we found that EG and DG can adsorb 1.38 and 0.68 wt% of hydrogen (Fig. 6). These graphene samples showed significantly higher adsorption in the high pressure experiments, yielding 3.1 and 2.5 wt% respectively for EG and DG (see insets in Fig. 6). The adsorption is completely reversible and comparable to that of carbon nanotubes[15] and porous open framework materials.[16] Though the graphene samples examined by us exhibit lower hydrogen uptake compared to the 6.0 wt% target of the Department of Energy (USA), there is significant scope for further improvements, by producing samples with smaller number of layers and significantly higher surface areas. EG is also a good adsorbent for CO_2, the uptake going up to 34.7 wt% at 195 K and 1 atm.

Electrochemistry

We have investigated the electrochemical properties of CG, EG and DG using the redox reactions with potassium ferrocyanide. The peak to peak separation is known to be significantly dependent on the microstructure of the carbon electrode used.[17] A small peak to peak separation (~70 mV) is achieved when the edges of graphite are exposed to the electrolyte, while it is significantly larger (even close to 1 V) when the basal planes are involved. In Fig. 7(a), we show typical cyclic voltammograms of the redox reaction of 100 mM potassium ferrocyanide (in 1 M KCl) carried out using working electrodes of different graphenes. While the peak to peak separation depends on the scan rate, it is found to be the largest in the case of EG. DG and CG exhibit similar peak separations. The behavior of EG is similar to that of the basal plane in graphite. On the other hand, DG and CG exhibit slightly better kinetics.

We fabricated supercapacitor cells with water as well as an ionic liquid, N-butyl-N-methylpyrrolidinium bis(trifluoromethanesulfonyl)imide (PYR$_{14}$TFSI), as electrolytes. EG and DG exhibit specific capacitances of 117 and 35 F g^{-1} respectively in

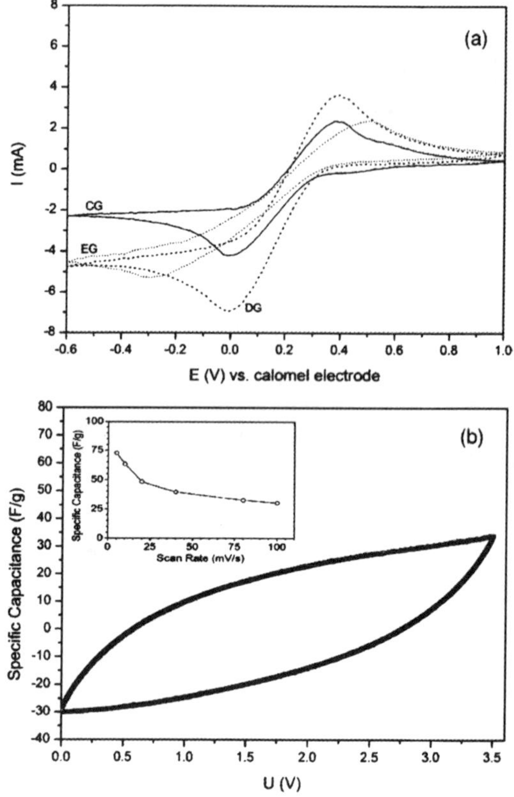

Fig. 7 (a) Cyclic voltammograms of the different graphene electrodes (at a scan rate of 20 mV s^{-1}) for 100mM $K_4Fe(CN)_6$ in 1 M KCl and (b) voltammetry characteristics of a capacitor built from an EG electrode at a scan rate of 100 mV s^{-1} using PYR$_{14}$TFSI. Inset shows the specific capacitance as a function of scan rate.

aq. H_2SO_4 with an operational voltage of 1 V. When we used $PYR_{14}TFSI$, we could achieve a significantly higher operating voltage of 3.5 V compared to that in aq. H_2SO_4 due to the large electrochemical window of the ionic liquid. In Fig. 7(a), we show a typical CV of a graphene–$PYR_{14}TFSI$ supercapacitor obtained at a scan rate of 100 mV s^{-1}. A specific capacitance value of 75 F g^{-1} was obtained with EG. The specific capacitance depends significantly on the scan rate as the ionic liquid is highly viscous (inset in Fig. 7(b)). The maximum value of energy density stored in these capacitors are 31.9 Wh kg^{-1}. This is one of the highest values reported to date and is comparable to those of microporous carbons reported by Balducci *et al.*[18]

Functionalization and solubilization

As-prepared EG shows evidence for the presence of carbonyl and hydroxyl groups on the surface. After further treatment of EG with a nitric and sulfuric acid mixture, we obtained a water solution of EG, along with the insoluble portion of the sample which settled out. The infrared (IR) spectrum of the soluble part obtained after drying shows a prominent band due to carbonyl groups in addition to a broad band due to –OH groups as shown in Fig. 8(a). The Raman spectrum of the water-soluble EG shown in Fig. 8(b) exhibits the characteristic G, D and 2D

bands and show that L_a of the sample is 4 on an average (as calculated from the intensity ratio of the D and G bands). In the inset of Fig 8(b), we show a photograph of the water-soluble graphene. The water-insoluble EG when treated with polyethylene glycol (PEG) yielded water-dispersible graphene. In the inset of Fig. 8(b), we also show a photograph of the water-dispersion of PEG-treated graphene.

By carrying out the amidation reaction similar to that reported for carbon nanotubes, we obtained dispersions of EG in various non-polar solvents. The results are similar to those reported by Haddon and co-workers.[19] In Fig. 9(a), we show photographs of the dispersions of dodecylamide-functionalized EG in dichloromethane, carbon tetrachloride and tetrahydrofuran. In Fig. 9(b), we show photographs of the dispersions prepared with DG. It should be noted, however, that DG requires hasher acid treatment and over longer periods to enable further functionalization to be carried out. In Fig. 9(c), we show the infrared spectra of EG at various stages of the solubilization process. After acid treatment, EG show a carbonyl stretching band at 1710 cm^{-1} due to the carboxyl groups. On functionalization with dodecylamine, the C=O stretching band shifts to 1650 cm^{-1} due to the amide band, in addition to C–H and N–H stretching bands around 2800 and 3300 cm^{-1} respectively.

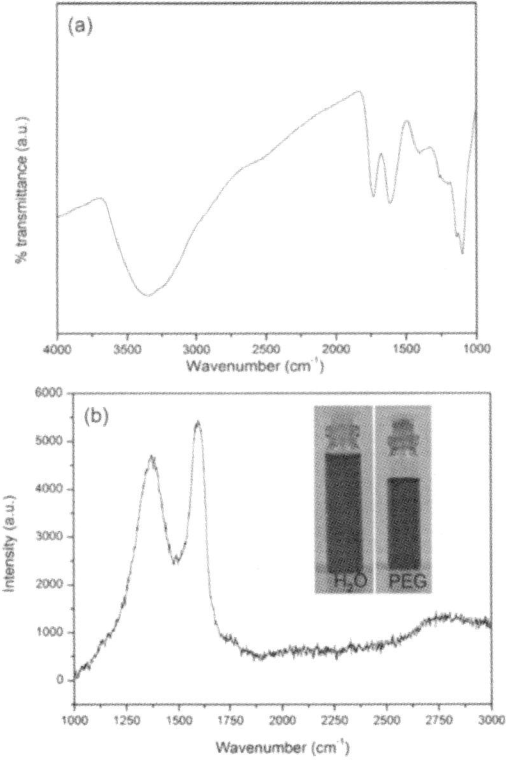

Fig. 8 (a) Infrared and (b) Raman spectra of water-soluble EG. Inset shows photographs of water-soluble (H_2O) and PEG-functionalized EG dispersed in water (PEG).

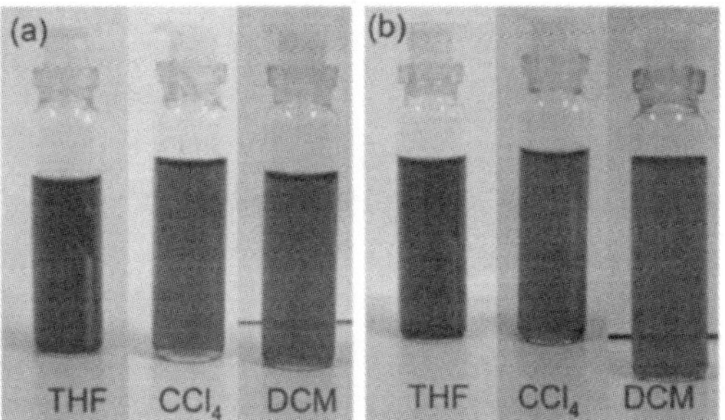

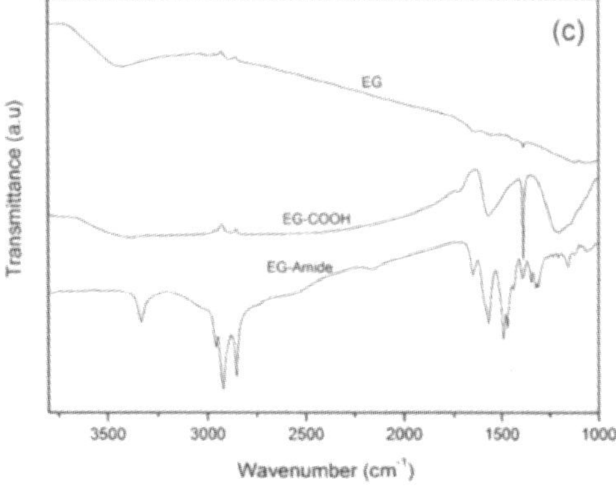

Fig. 9 Photographs of dispersions of (a) the amide-functionalized EG in THF, CCl_4 and dichloromethane, (b) amide-functionalized DG in THF, CCl_4 and dichloromethane. (c) Infrared spectra of pristine EG, acid-treated EG (EG-COOH) and amide-functionalized EG (EG-amide).

Conclusions

In conclusion, a comparative study of graphene samples prepared by different methods has been carried out. In terms of the number of layers, crystallite size as well as surface area, graphene prepared by the exfoliation of graphite (EG) seems to be best, possessing high surface area and a smaller number of layers. Graphene prepared by conversion of nanodiamond (DG) is also satisfactory and has greater thermal stability than EG. Both EG and DG exhibit significant hydrogen uptake. EG shows electrochemical redox behavior similar to that of the basal plane of graphite and can be used for fabrication of electrochemical supercapacitors. Acid-treated EG and DG can both be solubilized in non-polar solvents by reaction with thionyl chloride followed by dodecylamine. Water-soluble EG could be prepared by extensive acid treatment or treatment with polyethylene glycol.

Acknowledgements

The authors acknowledge the help of Mr Saikrishna for surface area and low-pressure hydrogen adsorption measurements and Dr Sundaresan for magnetic measurements.

References

1 A. K. Geim and K. S. Novoselov, *Nat. Mater.*, 2007, **6**, 183.
2 M. I. Katsnelson, *Mater. Today*, 2007, **10**, 20.
3 P. R. Somani, S. P. Somani and M. Umeno, *Chem. Phys. Lett.*, 2006, **430**, 56.
4 H. C. Schniepp, J.-L. Li, M. J. McAllister, H. Sai, M. Herrera-Alonso, D. H. Adamson, R. K. Prud'homme, R. Car, D. A. Saville and I. A. Aksay, *J. Phys. Chem. B*, 2006, **110**, 8535.
5 O. E. Andersson, B. L. V. Prasad, H. Sato, T. Enoki, Y. Hishiyama, Y. Kaburagi, M. Yoshikawa and S. Bandow, *Phys. Rev. B*, 1998, **58**, 16387.
6 Y. Li, S. Xie, W. Zhou, D. Tang, X. P. Zou, Z. Liu and G. Wang, *Carbon*, 2001, **39**, 626.
7 A. Gupta, G. Chen, P. Joshi, S. Tadigadapa and P. C. Eklund, *Nano Lett.*, 2007, **6**, 2667.
8 M. A. Pimenta, G. Dresselhaus, M. S. Dresselhaus, L. A. Cancado, A. Jorio and R. Sato, *Phys. Chem. Chem. Phys.*, 2007, **9**, 1276.
9 A. C. Ferrari, *Solid State Commun.*, 2007, **143**, 47.
10 D. Tasis, N. Tagmatarchis, A. Bianco and M. Prato, *Chem. Rev.*, 2006, **106**, 1105.
11 C. N. R. Rao and A. Govindaraj, *Nanotubes and Nanowires*, RSC Series on Nanoscience, Royal Society of Chemistry, London, 2006.
12 *Nanomaterials Chemistry: Recent Developments*, ed. C. N. R. Rao, A. K. Cheetham and A. Muller, Wiley-VCH, Weinheim, 2007.
13 G. Gundiah, A. Govindaraj, N. Rajalakshmi, K. S. Dhathathreyan and C. N. R. Rao, *J. Mater. Chem.*, 2003, **13**, 209.
14 F. Valentini, A. Amine, S. Orlanducci, M. L. Terranova and G. Palleschi, *Anal. Chem.*, 2003, **75**, 5413.
15 B. E. Conway, *Electrochemical Supercapacitors*, Kluwer Academic, Plenum, New York, 1999.
16 D. J. Collins and H. C. Zhou, *J. Mater. Chem.*, 2007, **17**, 3154.
17 C. E. Banks, T. J. Davies, G. G. Wildgoose and R. G. Compton, *Chem. Commun.*, 2005, 829.
18 A. Balducci, R. Dugas, P. L. Taberna, P. Simon, D. Plee, M. Mastragonstino and S. Passerini, *J. Power Sources*, 2007, **165**, 922.
19 S. Niyogi, E. Bekyarova, M. I. Itkis, J. L. McWilliams, M. A. Hamon and R. C. Haddon, *J. Am. Chem. Soc.*, 2006, **128**, 7720.